Arctic–Subarctic Ocean Fluxes

Defining the Role of the Northern Seas in Climate

Edited by

Robert R. Dickson
Centre for Environment, Fisheries and Aquaculture Science, Lowestoft, UK

Jens Meincke
Institut für Meereskunde, University of Hamburg, Germany

and

Peter Rhines
Oceanography and Atmospheric Sciences Departments, University of Washington, Seattle, WA, USA

Library of Congress Control Number: 2007940853

ISBN 978-1-4020-6773-0 (HB)
ISBN 978-1-4020-6774-7 (e-book)

Published by Springer,
P.O. Box 17, 3300 AA Dordrecht, The Netherlands.

www.springer.com

Cover illustration: Greenland. Photo by Tom Haine, John Hopkins University.

Printed on acid-free paper

Foreword

The Ocean-Atmosphere-Cryosphere system of the Arctic is of unique importance to the World, its climate and its peoples and is changing rapidly; it is no accident that the Arctic Climate Impact Assessment (ACIA) was the first comprehensive regional assessment of climate-impact to be conducted. Reporting in 2005, ACIA concluded that changes in climate and in ozone and UV radiation levels were likely to affect every aspect of life in the Arctic. In effect, the ACIA process was essentially one of prediction: projecting that large climatic changes are likely to occur over the 21st century and documenting what might be their projected impacts.

Although the ACIA Report was based on the most modern synthesis of observations, modelling and analysis by hundreds of Arctic scientists, it notes with clarity that its conclusions are only a first step in what must be a continuing process. Reporting in November 2007, the 2nd International Conference on Arctic Research Planning (ICARP II) has recently made much the same point. To make its projections with higher confidence, --- to take the crucial second step in other words, ---- both reports plainly state the need for a more complete and detailed understanding of the complex processes, interactions, and feedbacks that drive and underlie 'change' at high northern latitudes, including particularly the long-term processes of circulation and exchange in our northern seas where much of the decadal 'memory' for Arctic change must reside.

In this volume, assembled for the first time, we find a detailed description of much of what we believe is essential to take that crucial second step. Here, for example, are described the controls, 'near and remote, short-term as well as long-term' that have been involved in providing the polar basin in recent years with a steady supply of increasingly warmer water through subarctic seas. We find, moreover, a detailed description of the interplay between the storage and release of freshwater from the Central Arctic and its likely impact on the 'workings' of the Ocean's thermohaline 'conveyor'. And throughout the book, we are given a modern account of how well we can simulate the important elements of Arctic-subarctic exchange ---– in some cases very well indeed.

In the future there will, of course, be new stages of understanding and observations, better models and perhaps even a different set of 'driving questions' before society eventually learns to project, adapt and respond to Arctic change. This volume spells out what we now perceive are the driving questions to be addressed if we are to move our skills in prediction to a higher level.

The release of this publication coincides with the [4th] International Polar Year, a comprehensive, international effort in polar science and climate change. As the global scientific community conducts its latest polar study, the reader should be assured that the benchmarks this volume represents are a necessary and a considerable step towards understanding the critical role that the Arctic and subarctic seas play in the global climate system and hence, their importance to humanity everywhere.

Dr Robert W. Corell, Chair,
Arctic Climate Impact Assessment (ACIA),
Chair, International Conference on Arctic Research Planning (ICARP II) and
Director, Global Change Program at the H. John Heinz III
Center for Science, Economics and the Environment.

Contents

Foreword v

Arctic–Subarctic Ocean Fluxes: Defining the Role of the Northern Seas in Climate. A General Introduction 1
Bob Dickson, Jens Meincke, and Peter Rhines

1 **The Inflow of Atlantic Water, Heat, and Salt to the Nordic Seas Across the Greenland–Scotland Ridge** 15
Bogi Hansen, Svein Østerhus, William R. Turrell, Steingrímur Jónsson, Héðinn Valdimarsson, Hjálmar Hátún, and Steffen Malskær Olsen

2 **Volume and Heat Transports to the Arctic Ocean Via the Norwegian and Barents Seas** 45
Øystein Skagseth, Tore Furevik, Randi Ingvaldsen, Harald Loeng, Kjell Arne Mork, Kjell Arild Orvik, and Vladimir Ozhigin

3 **Variation of Measured Heat Flow Through the Fram Strait Between 1997 and 2006** 65
Ursula Schauer, Agnieszka Beszczynska-Möller, Waldemar Walczowski, Eberhard Fahrbach, Jan Piechura, and Edmond Hansen

4 **Is Oceanic Heat Transport Significant in the Climate System?** 87
Peter Rhines, Sirpa Häkkinen, and Simon A. Josey

5 **Long-Term Variability of Atlantic Water Inflow to the Northern Seas: Insights from Model Experiments** 111
Michael Karcher, Ruediger Gerdes, and Frank Kauker

6 **Climatic Importance of Large-Scale and Mesoscale Circulation in the Lofoten Basin Deduced from Lagrangian Observations** 131
Jean-Claude Gascard and Kjell Arne Mork

7 **Freshwater Storage in the Northern Ocean and the Special Role of the Beaufort Gyre** 145
Eddy Carmack, Fiona McLaughlin, Michiyo Yamamoto-Kawai, Motoyo Itoh, Koji Shimada, Richard Krishfield, and Andrey Proshutinsky

8 **Modelling the Sea Ice Export Through Fram Strait** 171
Torben Koenigk, Uwe Mikolajewicz, Helmuth Haak, and Johann Jungclaus

9 **Fresh-Water Fluxes via Pacific and Arctic Outflows Across the Canadian Polar Shelf** 193
Humfrey Melling, Tom A. Agnew, Kelly K. Falkner, David A. Greenberg, Craig M. Lee, Andreas Münchow, Brian Petrie, Simon J. Prinsenberg, Roger M. Samelson, and Rebecca A. Woodgate

10 **The Arctic–Subarctic Exchange Through Hudson Strait** 249
Fiammetta Straneo and François J. Saucier

11 **Freshwater Fluxes East of Greenland** 263
Jürgen Holfort, Edmond Hansen, Svein Østerhus, Stephen Dye, Steingrimur Jonsson, Jens Meincke, John Mortensen, and Michael Meredith

12 **The Changing View on How Freshwater Impacts the Atlantic Meridional Overturning Circulation** 289
Michael Vellinga, Bob Dickson, and Ruth Curry

13 **Constraints on Estimating Mass, Heat and Freshwater Transports in the Arctic Ocean: An Exercise** 315
Bert Rudels, Marika Marnela, and Patrick Eriksson

14 **Variability and Change in the Atmospheric Branch of the Arctic Hydrologic Cycle** 343
Mark C. Serreze, Andrew P. Barrett, and Andrew G. Slater

15 Simulating the Terms in the Arctic Hydrological Budget ... 363
Peili Wu, Helmuth Haak, Richard Wood, Johann H. Jungclaus, and Tore Furevik

16 Is the Global Conveyor Belt Threatened by Arctic Ocean Fresh Water Outflow? ... 385
E. Peter Jones and Leif G. Anderson

17 Simulating the Long-Term Variability of Liquid Freshwater Export from the Arctic Ocean ... 405
Rüdiger Gerdes, Michael Karcher, Cornelia Köberle, and Kerstin Fieg

18 The Overflow Transport East of Iceland ... 427
Svein Østerhus, Toby Sherwin, Detlef Quadfasel, and Bogi Hansen

19 The Overflow Flux West of Iceland: Variability, Origins and Forcing ... 443
Bob Dickson, Stephen Dye, Steingrímur Jónsson, Armin Köhl, Andreas Macrander, Marika Marnela, Jens Meincke, Steffen Olsen, Bert Rudels, Héðinn Valdimarsson, and Gunnar Voet

20 Tracer Evidence of the Origin and Variability of Denmark Strait Overflow Water ... 475
Toste Tanhua, K. Anders Olsson, and Emil Jeansson

21 Transformation and Fate of Overflows in the Northern North Atlantic ... 505
Igor Yashayaev and Bob Dickson

22 Modelling the Overflows Across the Greenland–Scotland Ridge ... 527
Johann H. Jungclaus, Andreas Macrander, and Rolf H. Käse

23 Satellite Evidence of Change in the Northern Gyre ... 551
Sirpa Häkkinen, Hjálmar Hátún, and Peter Rhines

24 The History of the Labrador Sea Water: Production, Spreading, Transformation and Loss ... 569
Igor Yashayaev, N. Penny Holliday, Manfred Bersch, and Hendrik M. van Aken

25 Convective to Gyre-Scale Dynamics: Seaglider Campaigns in the Labrador Sea 2003–2005 613
Charles C. Eriksen and Peter B. Rhines

26 Convection in the Western North Atlantic Sub-Polar Gyre: Do Small-Scale Wind Events Matter? 629
Robert S. Pickart, Kjetil Våge, G.W.K. Moore, Ian A. Renfrew, Mads Hvid Ribergaard, and Huw C. Davies

27 North Atlantic Deep Water Formation in the Labrador Sea, Recirculation Through the Subpolar Gyre, and Discharge to the Subtropics 653
Thomas Haine, Claus Böning, Peter Brandt, Jürgen Fischer, Andreas Funk, Dagmar Kieke, Erik Kvaleberg, Monika Rhein, and Martin Visbeck

28 Accessing the Inaccessible: Buoyancy-Driven Coastal Currents on the Shelves of Greenland and Eastern Canada 703
Sheldon Bacon, Paul G. Myers, Bert Rudels, and David A. Sutherland

List of Contributors 723

Index 731

Arctic–Subarctic Ocean Fluxes: Defining the Role of the Northern Seas in Climate

A General Introduction

Bob Dickson[1], **Jens Meincke**[2], **and Peter Rhines**[3]

1 Background

Almost 100 years ago, Helland-Hansen and Nansen (1909) produced the first complete description of the pattern of oceanic exchanges that connect the North Atlantic with the Arctic Ocean through subarctic seas. At a stroke, they placed the science of the Nordic seas on an astonishingly modern footing; as Blindheim and Østerhus (2005) put it, *'Their work described the sea in such detail and to such precision that investigations during succeeding years could add little to their findings'*. Nonetheless, in the century that followed, oceanographers have gradually persisted in the two tasks that were largely inaccessible to the early pioneers – quantifying the exchanges of heat, salt and mass through subarctic seas and, piecing-together evidence for the longer-term (decade to century) variability of the system.

Evidence of variability was not long in coming. As hydrographic time series lengthened into the middle decades of the 20th century, they began to capture evidence of one of the largest and most widespread regime shifts that has ever affected our waters. For these were the decades of "the warming in the north", when the salinity of North Atlantic Water passing through the Faroe–Shetland Channel reached a century-long high (Dooley et al. 1984), when salinities were so high off Cape Farewell that they were thrown out as erroneous (Harvey 1962), when a precipitous warming of more than 2 °C in the 5-year mean pervaded the West Greenland banks, and when the northward dislocations of biogeographical boundaries for a wide range of species from plankton to commercially important fish, terrestrial mammals and birds were at their most extreme in the 20th century (reviewed in Dickson 2002).

[1] Centre for Environment, Fisheries and Aquaculture Science, Lowestoft, Suffolk NR33 0HT UK

[2] University of Hamburg, Centre of Marine and Climate Research, D-20146 Hamburg, Germany

[3] Department of Oceanography and Atmospheric Sciences, University of Washington, Seattle, WA 98195, USA

R.R. Dickson et al. (eds.), *Arctic–Subarctic Ocean Fluxes*, 1–13

Measuring the ocean fluxes through these waters proved harder, and indeed we are still not quite able to tackle all of them. Recording current meters were not available till the 1960s, and when Worthington (1969) first attempted to capture the violence of overflow through the sill of the Denmark Strait (1967), his moorings were almost all swept away; it was another decade (1975–1976) before year-long records of overflow were successfully recovered from the Greenland–Scotland Ridge by the ICES MONA Project (Monitoring the Overflow in the North Atlantic). The flows through the Canadian Arctic Archipelago (CAA) proved even harder to capture. Understandably so; it is one of the hardest observational tasks in oceanography to measure vigorous flows in a remote complex of narrow passageways with strong seasonal variability in ice-covered seas where the scales of motion are small, where moving ice and icebergs pose a hazard to moored gear and where even the direction of flow is obscured by the proximity of the Earth's magnetic pole. Yet moorings have been maintained in Lancaster Sound since 1998 and in the other five main channels of the CAA since then. Nowadays, making direct flow measurements on the ice-covered subarctic shelves in the presence of heavy fishing activity and grounding bergs remains the last and greatest challenge; though successes have been achieved, it will probably take the development of sub-ice Seagliders to make these shelves routinely accessible to measurement.

The perceived stimulus to making these measurements has also changed with time. Initially, the primary impetus to measuring change, in the European subarctic seas at any rate, was as an aid to understanding the ecosystem, including especially the fluctuations in the great commercial fish stocks. In 1909, Helland-Hansen and Nansen had been concerned with applying what they knew of environmental change to the fluctuating success of the Arcto-Norwegian cod stock. And even in wartime, under the Presidency of Johan Hjort (1938–1948), Martin Knudsen's ICES Sub-committee on Hydrographical and Biological Investigations continued to plan the data collection that would be needed to meet Hjort's aim of fish stock prediction. We retain that legacy today in the small scattering of ultra-long (100-year plus) hydrographic time-series that afford us a glimpse of decade-to-century variation in the hydrography of our northern seas.

Later in the 20th century, it would be fair to say that the primary stimulus for these investigations diversified from this focus on the success of fish stocks to include the ocean's role in climate. Two studies in particular took on the task. Between 1990 and 2002, the WCRP World Ocean Circulation Experiment (WOCE) – the most ambitious oceanographic experiment ever undertaken – circled the globe with the twin aims of establishing the role of the oceans in the Earth's climate and of obtaining a baseline dataset against which future change could be assessed. About 30 nations participated in the observational phase of the programme (from 1990 to 1998) and sophisticated numerical ocean models were developed both to provide a framework for the interpretation of the observations and for the prediction of the future ocean state. The key WOCE scientific goal was thus to develop models useful for predicting climate change and to collect the data necessary to test them.

Overlapping the period of the WOCE Experiment, a second WCRP initiative focused on the more regional study of the high Arctic and its role in global climate. Between 1993 and 2003, the Arctic Climate System Study (ACSYS) attempted to answer two questions in particular: What are the global consequences of natural or human-induced change in the Arctic climate system? Is the Arctic climate system as sensitive to increased greenhouse gas concentrations as climate models suggest? To address these, the twin aims of ACSYS were to understand the interactions between the Arctic Ocean circulation, ice cover, the atmosphere and the hydrological cycle, and to provide a scientific basis for an accurate representation of Arctic processes in global climate models.

2 The Role of the Subarctic Seas in Climate

Despite their global scope, the WOCE and ACSYS initiatives fell short of complete coverage in one important respect. In the Atlantic sector, the measurement programme of WOCE did not extend north of the Greenland–Scotland Ridge, while the ACSYS coverage of the high-latitude ocean was focused north of Fram Strait. The subarctic seas were largely excluded from consideration. Yet we would nowadays strongly assert that the two-way oceanic exchanges that connect the Arctic and Atlantic oceans through subarctic seas are of fundamental importance to climate (and thus to the aims of WOCE and ACSYS). Change may certainly be imposed on the Arctic Ocean from subarctic seas, including a changing poleward ocean heat flux that is central to determining the present state and future fate of the perennial sea-ice. And the signal of Arctic change is expected to have its major climatic impact by reaching south through subarctic seas, either side of Greenland, to modulate the Atlantic thermohaline 'conveyor'.

The global thermohaline circulation (THC), driven by fluxes of heat and freshwater at the ocean surface, is an important mechanism for the global redistribution of heat and salt and is known to be intimately involved in the major changes in Earth climate; thus, a partial shutdown of this worldwide overturning cell appears to have accompanied each abrupt shift of the ocean–atmosphere system towards glaciation (e.g. Broecker and Denton 1989). In turn, the overflow and descent of cold, dense water from the sills of the Denmark Strait and the Faroe–Shetland Channel into the North Atlantic forms a key component of the THC, ventilating and renewing the deep oceans and driving the abyssal limb of this great 'overturning cell'.

Most computer simulations of the ocean system in a climate with increased greenhouse-gas concentrations predict a weakening thermohaline circulation in the North Atlantic as the subpolar seas become fresher and warmer. A representative set of milestones for this prediction might run from the pioneering modelling work of Bryan (1986) and Manabe and Stouffer (1988) through the intermediate complexity of Rahmstorf and Ganopolski (1999), Delworth and Dixon (2000) and Rahmstorf (2003) to the full complexity of earth system modelling by Mikolajewicz et al. (2007). Despite such major advances in simulating the system, we remain undecided on many of the most basic issues that link change in our northern seas to climate, for both observational and modelling reasons. Put simply, uncertainties in our observations are bound to delay the development of our climate models and hinder their critical evaluation.

3 The Development of ASOF

Recognising the importance of Arctic–subarctic exchanges as the source or as the conduit of change in both the high Arctic and in the global ocean, two major collaborative studies in particular set out to meet the deficiencies in our observational coverage throughout the subarctic seas. A first *regional* programme covering all significant ocean fluxes through the Nordic Seas in the EC-VEINS Study (Variability of Exchanges in Northern Seas) of 1997–2000 quickly developed into the full multinational *pan-Arctic* ASOF study (Arctic–Subarctic Ocean Fluxes) from 2000 to the present. The primary scientific objective remained the same – *to measure and model the variability of fluxes between the Arctic Ocean and the Atlantic Ocean with a view to implementing a longer-term system of critical measurements needed to understand the high-latitude ocean's steering role in decadal climate variability*.

Thus, from the outset, it was seen that ASOF had necessarily to be a pan-Arctic exercise if it was to describe the balances of flow entering and leaving the Arctic Ocean through subarctic seas, and that it should involve continuing iteration between the technicalities of observations and the demands of climate models. Since it was already apparent that change was spreading through the subarctic–Arctic system on a timescale of decades (e.g. Morison et al. 2000), it was also clear that our observations of that system had to be of decadal 'stamina' and had if possible to be simultaneous, to the extent permitted by funding. To measure such a system *successively* by moving our focus and our resources from place to place would be to risk confusing spatial changes with temporal ones.

The full pan-Arctic ASOF programme was achieved by instituting task-based planning across the full ASOF domain according to 6 regional task-groups, supported by a 7th system-wide Numerical Experimentation Group (see Fig. 1).

In late June 2006, approximately 10 years after the start of VEINS, the ASOF community met in Thorshaven, Faroe Islands, with two main objectives: *first, to describe progress in quantifying, by both observations and modelling, the two-way exchanges of heat, salt and mass that take place between the Arctic and Atlantic Oceans through subarctic seas*. Within this primary objective were included all aspects of Arctic–subarctic ocean fluxes that seem of importance to the development of our global climate models – the forcing of these oceanic exchanges, their variability at all scales accessible to us, and the interconnected nature of both forcing and variability in space and time. Having assessed progress, our *second*

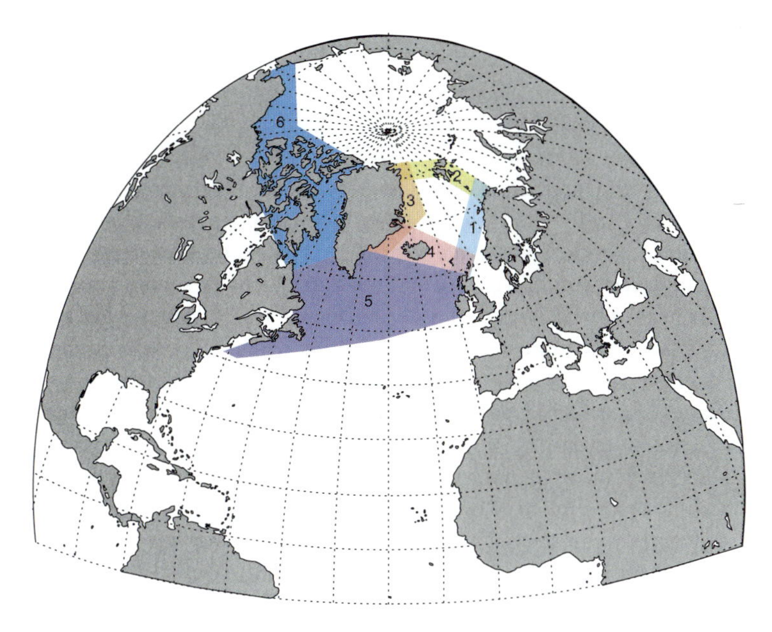

Fig. 1 The full domain of the international ASOF study, showing the distribution of its six regional task-related Working Groups. 1 = warm water inflow to Nordic Seas; 2 = exchanges with the Arctic Ocean; 3 = ice and freshwater outflow; 4 = Greenland–Scotland Ridge exchanges; 5 = Overflows to Deep Western Boundary Current; 6 = Canadian Arctic Archipelago throughflow. A 7th Numerical Experimentation Group covers the whole domain

objective was to redefine the remaining cutting-edge questions regarding the role of the northern seas in climate as the World embarked on its 4th International Polar Year (April 2007–April 2009).

These two objectives are also what have motivated the present volume though here, we have widened our authorship beyond the membership of the ASOF Task Teams in order to provide the fullest possible account of recent achievements in observing and modelling those aspects of our northern seas that seem to have an actual or potential importance to climate. The sub-title of the volume – *Defining the role of the Northern Seas in Climate* – is thus carefully chosen. The volume does not in itself aim to assess the full complexity of that role, and indeed it may well be some time before our observations and our models are capable of doing that, to the point of anticipating future changes in the system. Instead the volume intends to assemble the body of evidence that climate models will need if they are one day to make that assessment, quantifying the ocean exchanges through subarctic seas, describing their importance to climate as we currently understand it, explaining their variability, setting out our current ideas on the forcing of these fluxes and our improved capability in modelling the fluxes themselves and the processes at work. Much of that evidence is assembled here for the first time.

4 Contents, Structure and Rationale

So where are climate models deficient? What aspects of the physics of our northern seas can ocean scientists most usefully contribute to their development? In fact the list is quite long. Climate models are inherently weak in the important subtleties of deep convection, interior diapycnal mixing, boundary currents, shelf circulations (climate models have no continental shelves!), downslope flows that entrain new fluid during their descent, thin cascading overflows, delicate upper ocean stratification by both heat and salt with its strong influence on convective geography, ice dynamics – all of which contribute to a level of uncertainty that may crucially affect our assessment of thermohaline slowdown. And most underline the importance of direct, sustained observations in the regions that lie *between* the dominant polar and subtropical climate programs.

Altogether, in piecing together a modern statement that will define the role of the Northern Seas in climate, we find a need to describe 28 separate facets of the subject, with three main themes (chapters mentioned in this brief introduction are numbered #1–#28).

First we have quantified the fluxes themselves. It is now soundly established, for example that 8.5 million cubic metres per second of warm salty Atlantic Water pass north across the Greenland–Scotland Ridge carrying, on average, some 313 million megawatts of power (relative to 0 °C) and 303 million kilograms of salt per second (# 1; see also Østerhus et al. 2005). A little further north, from a decade of direct measurements off Svinøy (# 2), we find that the inshore (slope) branch of the Norwegian Atlantic Current carries 4.3 Sv and 126 TW poleward, with no sign yet of any trend in transport, while further north still, an Atlantic inflow of 1.8 Sv, increasing by 0.1 Sv per year, carries 48 TW (increasing by 2.5 TW per year) into the Barents Sea. Nine years of intense effort in the Fram Strait complete our accounting of the warm, saline northward flux at its point of entry to the Arctic Ocean (# 3). In fact, while we have resolved the debate about the importance of this poleward ocean heat flux to climate (# 4), the Fram Strait study opens up a new debate about how that flux should be measured. It suggests that the physical concept of oceanic heat transport is only meaningful in terms of its ability to add heat to or take away heat from a defined ocean volume. Thus, oceanic heat transport to the Arctic Ocean, calculated from velocity and temperature measurements at its boundary are only meaningful when the *entire* boundary – and all of the inflows and outflows that cross it – is taken into account, a strong vindication of the ASOF preference for simultaneity of observations; and this will also be true for the temporal variability of that transport. Chapters on our growing ability to simulate the Atlantic water inflow to and through the Northern Seas and its long-term variability (# 5), together with new insights into the strongly mesoscale structure of that Atlantic current west of Norway (# 6) round off this section of the volume.

As with the poleward flux of heat, a similarly broad range of chapters are devoted to describing our improved capability and our changing ideas on the variable and equally important outflows that pass from the Arctic in the opposite sense to modulate the Atlantic Meridional Overturning Circulation – the proximal end of the Ocean's 'Great Conveyor'. No less than 12 chapters describe aspects of the flux of ice and freshwater whose projected increase in recent simulations tends to slow down – and in at least one recent model to shut down – the Atlantic MOC. Here we provide a modern assessment of the freshwater storage in the northern seas (# 7), updating the seminal work by Aagaard and Carmack (1989) before going on to describe the full range of direct measurements of the freshwater flux into and through the six main passageways of the Canadian Arctic Archipelago to the Davis Strait (# 9), the sizeable outflow joining it through Hudson Strait (# 10), and our present (and still incomplete) measures of the equivalent flux passing south to the east of Greenland (# 11 and 28). Companion chapters describe our growing capability in measuring and modelling the terms in the Arctic hydrological budget (# 14 and 15) and in simulating its more important components, the sea-ice and freshwater exports through Fram Strait (# 8 and 17).

In support of the second ASOF goal, we discuss a selection of the cutting edge questions in observing and modelling the Arctic–subarctic system, as we currently perceive them. Real issues remain: for example, although estimates of the total freshwater flux reaching the North Atlantic have recently been published (~300 mSv according to Dickson et al. 2007), there remain real constraints, debated here, on our ability to make such estimates (# 13); equally while there would be general agreement that an increasing freshwater flux to the North Atlantic is likely to be of climatic significance, we remain uncertain as to whether the impact on climate will result from *local* effects on overflow transport (e.g. from the changing density contrast across the Denmark Strait sill; Curry and Mauritzen 2005), from the *regional* effect of capping the water column of the NW Atlantic (leading to a reduction in vertical mixing, water mass transformation, and production of North Atlantic Deep Water), or from *global-scale* changes in the Ocean's thermohaline fields and circulation arising from an acceleration of the Global Water Cycle (Curry et al. 2003). Most fundamental of all, opinion remains divided both on whether thermohaline slowdown is threatened (# 16), is already underway or on whether any variability that we see is natural or anthropogenic (# 12).

Model results are also helping to reshape our thinking on the role of the northern seas in climate; we provide illustrations from two of our most advanced atmosphere–ocean general circulation models. First, the analysis of results from 200 decade-long segments of HadCM3 runs bears the clear implication that a given volume of freshwater, when spread to depth (as, for example, through the descent of the dense-water overflows from the sills of the Greenland–Scotland Ridge into the deep Atlantic) effects a much smaller slowdown of the MOC than when the same freshwater anomaly is spread across the surface – the normal practice and assumption in the 'hosing experiments' (# 12). This observation naturally begs the question as to whether any *future* increase in the freshwater outflow from the Arctic is likely to be incorporated into the overflow system, or (effectively the same thing) *whether any future increase of the freshwater efflux is likely to pass to the west or to the east of Greenland.* And one model study currently makes that prediction. Recent coupled experiments by the Hamburg M-P-I Group using ECHAM 5 and the M-P-I Ocean

Model suggest that although the freshwater flux is expected to increase both east and west of Greenland, the loss of the sea-ice component (which currently dominates the flux through Fram Strait) suggests we should expect a much greater total increase through the CAA by 2070–2099 than through Fram Strait (Haak et al. 2005; # 8 & 12). As a third, intriguing (and perhaps salutary) model result, we revert to HadCM3 which in a large ensemble of experiments has appeared to offer an encouragingly close fit between the density of northern seas and rate of the Atlantic overturning circulation at 45° N (# 12). However, when the density changes are decomposed into those due to changes in temperature and those due to changes in salinity, the three types of experiments ('hosing runs', 'initial perturbation' experiments and greenhouse gas experiments) each behave very differently, suggesting that each class of experiments might involve fundamentally different feedbacks (# 12). If so, how can we be sure that we have yet adequately employed the full range of models that spans the possible and likely behaviour of the real climate system?

Whatever may be the role of the freshwater flux from high latitudes in slowing down the AMOC, it is the overflow and descent of cold dense water from the sills of the Denmark Strait and Faroe–Shetland Channel that ventilate and renew the deep oceans and thus drive the abyssal limb of this overturning cell. Forty years on from Val Worthington's first heroic but unsuccessful attempt to deploy current meters across the violent flow through Denmark Strait, direct measurements in both overflows are now relatively routine. From the longer of the two series (Denmark Strait) a decade of continuous observation shows variability in transport out to interannual timescales, but with no evidence (as yet) of any longer-term trend and no convincing evidence of covariance with the eastern dense overflow through Faroe Bank Channel (# 18 & 19). Observations over many decades have identified a complex of locally and remotely driven large-amplitude variations in the hydrographic character of both overflows and their sources, including a long-sustained trend in salinity of 3–4 decades duration. From the passage of conspicuous thermohaline anomalies (# 21), from the use of novel tracer techniques (# 20) and from a greatly improved modelling capability (# 22), we can now more confidently trace the changing sources and pathways of overflow upstream from the Fram Strait or track them downstream to the abyssal Labrador Sea. It will be downstream, along that track, that the major impact on the global thermohaline circulation will take effect. Through detailed hydrographic analysis of the principal water masses passing through the great storage and transformation basins south of the Greenland–Scotland Ridge, we can now much better describe the combination of local, regional and remote influences that have driven record hydrographic change through the water column of the Northwest Atlantic in recent decades (# 21 & 24). The Irminger Sea is seen to have features of unique global importance for the transfer of ocean climate signals between water masses and to great ocean depths (# 26 & 21). And at the southern boundary of the ASOF domain, the intractable but climatically vital problem of North Atlantic Deep Water formation in the Labrador Sea, its recirculation through the subpolar gyre and its discharge to the subtropics – once described by McCartney (1996) as *'the greatest problem in Oceanography'* – is finally being resolved, through a combination of state-of-the-art observational and modelling techniques (# 27). [It is sobering to reflect, and thus important to acknowledge, that without John Lazier's singular achievement in following the processes of convection and climate change in the Labrador Sea from the first (and only!) three-dimensional hydrographic survey of 1965–1966 to the institution of annual Hudson sections between Hamilton Bank Labrador and Cape Desolation, Greenland, we might have missed the 'greatest change in Oceanography', as it passed through the basin].

The bulk of this brief overview has understandably concerned the task that formed the original primary goal of VEINS and ASOF – the idea of measuring and modelling a complete set of oceanic exchanges between the Atlantic and Arctic Oceans through subarctic seas, simultaneously and with decadal stamina. Our approach to this goal has not been unchanging. In fact, the complexity of our data sets and the need to extract and display its essence have both prompted and required a diversification of technique. Importantly, the strict definition of ocean circulation as integrated volume transport and zonally-averaged overturning streamfunction is now being augmented at many points in this volume by the hydrographer's approach of displaying change on the potential temperature/salinity plane. Transports of heat, freshwater and mass are thus unified on a single diagram, returning us toward articulate description of water masses, their transports across key sections and their transformation and air–sea interaction within boxes bounded by these sections.

Equally important, as our time-series have lengthened, other factors have developed to help sustain these series.

The first is a growing realisation that although the individual flux estimates and their local controls are important, the processes that 'drive' their variability may form part of a full-latitude *system* of change; and that it is the recognition of how that *system* works that will most rapidly advance our ablity to simulate change and predict its onset. One current illustration will make the point, and it concerns some of the largest changes we have ever observed in our waters. Very recently, the temperature and salinity of the waters flowing into the Norwegian Sea along the Scottish shelf and slope have been at their highest values for >100 years. At the 'other end' of the inflow path, the ICES Report on Ocean Climate for 2006 (ICES 2007) will show that temperatures along the Kola Section of the Barents Sea (33° 30′ E) have equally never been greater in >100 years. Shorter records en route and beyond, on the Norwegian arrays off Svinoy (# 2), on the moored array monitoring Fram Strait (# 3), and on Polyakov's NABOS moorings at the Slope of the Laptev Sea (Polyakov 2005, 2007) have all remarked the passage of this warmth; Holliday et al. (2007) have described its continuity along the boundary. It forms part of the rationale for Overpeck's (2005) statement that '*a summer ice-free Arctic Ocean within a century is a real possibility, a state not witnessed for at least a million years.*'

Why? What is driving extreme change through the system? Satellite-based observations seem to provide a plausible explanation: during the whole TOPEX-POSEIDON era (since 1992), as the Labrador Sea Water warmed (# 24), altimeter records reveal a slow rise in sea surface height at the centre of the Atlantic subpolar gyre, suggesting a steady weakening of the gyre circulation (Hakkinen and Rhines 2004; # 23). This weakening, together with a westward retraction of the gyre boundary, appears to have operated as a kind of 'switchgear' mechanism to control the temperature and salinity of inflow to the Nordic seas (Hatun et al. 2005); by that mechanism, when the gyre was strong and spread east (early 1990s), the inflows recruited colder, fresher water direct from the subpolar gyre but when the gyre weakened and shifted west (as in the 2000s), the inflows to Nordic Seas were able to tap-off warmer and saltier water from the subtropical gyre, explaining the recent warmth and saltiness of inflow of Atlantic waters into the Norwegian Sea (# 4). Thus, although the local and the short term have certainly played their part west of Norway – the speed of the Atlantic Current is locally storm-forced so that it tends to change coherently from Ireland to Spitsbergen (# 2) – the ultimate source of the observed changes in the Arctic Ocean lies in a whole system of interactions between polar and sub-polar basins. Near and remote, short-term as well as long-term controls have been involved in providing the Polar Basin with a steady supply of increasingly warmer water through subarctic seas.

We have only just begun to glimpse evidence of this 'system'. But model results too seem to vindicate the view that it is the whole full-latitude system of exchange between the Arctic and Atlantic Ocean – not just spot 'examples' of it – that has to be addressed simultaneously if we are to understand the full subtlety of the role of our Northern Seas in climate. As Jungclaus et al. (2005) conclude from their model experiments using ECHAM5 and the MPI-OM, while *'the strength of the (Atlantic) overturning circulation is related to the convective activity in the deep-water formation regions, most notably the Labrador Sea,.....the variability is sustained by an interplay between the storage and release of freshwater from the central Arctic and circulation changes in the Nordic Seas that are caused by variations in the Atlantic heat and salt transport.'* Likewise, Hakkinen and Proshutinsky (2004) find that *'changes in the Atlantic water inflow can explain almost all of the simulated freshwater anomalies in the main Arctic basin'*.

The final factor that has sustained our time-series has been technical advance. As our understanding of the role of the northern seas in climate has developed in complexity, so the necessary parallel advances have been made in terms of technique. For example, orbiting satellites now contribute an increasingly comprehensive view of ice and circulation. From showing, visually, the areal extent of sea-ice and its remarkable responses to the wind, satellite altimetry now routinely provides maps of the draft/thickness distribution of sea-ice, while retrievals of ocean dynamic topography at the centimetre level (hence measures of the Arctic Ocean circulation) are now possible even in the presence of ice; since 2002, the twin satellites of the Gravity Recovery and Climate Experiment (GRACE) have contributed their own new measure of the grounded-ice mass balance. Within the ocean, there is no better example of technical advance than the evolution of SeaGlider technology to its first uses on survey during ASOF, initially in waters west of Greenland, more recently adding its fine-scale space–time resolution to the classic ship-based hydrography across the Faroe–Shetland Channel and Iceland–Faroe Ridge (# 25).

It is a nice point that the oldest time-series that we have relied on in ASOF, the hydrographic transects of the Faroe–Shetland Channel begun by HN Dickson aboard HMS Jackal on 4 August 1893 and carried-on by the Scots ever since (with Faroese and Norwegian partners), are now supplemented in their coverage by repeat deep SeaGlider sections from the cutting-edge of technical advance (# 25). Its further development to a Deep Glider able to cruise the whole watercolumn of the subpolar gyre is called for (# 19) as a necessary aid to capturing the baroclinic adjustments that cause interannual changes in the transport of overflow from Nordic Seas.

The above rapid tour through the chapters of this volume will justify, or at least explain why ASOF and why this volume have the scope that they do. Why simultaneity and stamina in observation seem key. What the driving questions of the programme now are. And why defining and re-defining the role of the northern seas in climate has become, in itself, a continuing goal of the programme.

Acknowledgements What this introduction does not explain is the structural role played by the ASOF Science Officer, Roberta Boscolo, throughout the life of the ASOF programme from the discussion meetings in 2000 that shaped its goals, through the annual series of outputs and meetings that adjusted its focus, to the actual editing of this volume. Our 7 years and now our 101 authors and co-authors could not have proceeded to this conclusion without her, as we three who are designated editors are happy to acknowledge.

The concept of ASOF was brought to life by the Arctic Ocean Sciences Board under the leadership of Tom Pyle and Lou Brown of the US National Science Foundation and owes much to their enthusiasm. We also gratefully acknowledge the support received from a wide range of national and international funding agencies and individual laboratories whose funding has permitted both the ASOF Conference in the Faroes and the production of this volume. They are identified and thanked by the full list of their logos on the cover.

Finally, it goes without saying that the observations on which this volume is based, won from some of the most difficult waters on Earth, could not have been achieved without the help of the research vessel fleet and their crews over many years. We are most happy to acknowledge their contribution.

References

Aagaard K and E Carmack, 1989. The role of sea-ice and other freshwater in the Arctic Circulation. J Geophys Res, 94, 14485–14498, doi: 10.1029/89JC01375.

Blindheim J and S Østerhus, 2005. The Nordic Seas, main oceanographic features. In: Drange H et al. (Eds) The Nordic Seas: An Integrated Perspective. AGU Monograph 158 American Geophysical Union, Washington, DC, pp. 11–37.

Broecker WS and GH Denton, 1989. The role of ocean-atmosphere reorganizations in glacial cycles. Geochim et Cosmochim Acta, 53, 2465–2501.

Bryan F, 1986. High latitude salinity effects and inter-hemispheric thermohaline circulations. Nature, 323, 301–304.

Curry R, B Dickson and I Yashayaev, 2003. A change in the freshwater balance of the Atlantic Ocean over the past four decades. Nature, 426, 826–829.

Curry R and C Mauritzen, 2005. Dilution of the Northern North Atlantic Ocean in recent decades. Science, 308 (5729): 1772–1774.

Delworth TL and KW Dixon, 2000. Implications of the recent trend in the Arctic/N Atlantic Oscillation for the North Atlantic thermohaline circulation. J Climate, 13, 3721–3727.

Dickson RR, 2002. Variability at all scales and its effect on the ecosystem—an overview. Proc. ICES Hist. Symp. Helsinki, August 2000. ICES Mar Sci Symp Ser, 215, 219–232.

Dickson R, B Rudels, S Dye, M Karcher, J Meincke and I Yashayaev, 2007. Current estimates of freshwater flux through Arctic & subarctic seas. Prog Oceanogr, 73, 210–230.

Dooley H D, JHA Martin and DJ Ellett, 1984. Abnormal hydrographic conditions in the Northeast Atlantic during the 1970s. Rapports et Procès-Verbaux des Réunions du Conseil International pour l'Exploration de la Mer, 185, 179–187.

Haak H, J Jungclaus, T Koenigk, D Svein and U Mikolajewicz, 2005. Arctic Ocean freshwater budget variability. ASOF Newsletter (3), 6–8. [http://asof.npolar.no].

Hakkinen S and A Proshutinsky, 2004. Freshwater content variability in the Arctic Ocean. Journal of Geophysical Research, 109, C03051. doi:10.1029/2003JC001940.

Hakkinen S and PB Rhines, 2004. Decline of subpolar North Atlantic circulation during the 1990s. Science, 304, 555–559.

Harvey JG, 1962. Hydrographic conditions in Greenland waters during August 1960. Annales Biologiques du Conseil International pour l'Exploration de la Mer, 17: 14–17.

Hatun H, AB Sandø, H Drange, B Hansen and H Valdimarsson, 2005. Influence of the Atlantic Subpolar Gyre on the Thermohaline Circulation. Science, 309, 1841–1844.

Helland-Hansen B and F Nansen, 1909. The Norwegian Sea. Its physical oceanography based upon the Norwegian researches 1900–1904. Report on Norwegian Fishery and Marine Investigations, volume II, part I, Chapter 2, 360 pp., tables, figures (total 390 pp.)

Holliday NP, SL Hughes, A Lavin, KA Mork, G Nolan, W Walcowski and A Breszczynska-Moller, 2007. The end of a trend? The progression of unusually warm and saline water from the eastern North Atlantic into the Arctic Ocean. CLIVAR Exchanges, 12 (1) pp 19–20 + figs

ICES, 2007. ICES Report on Ocean Climate 2006. ICES Cooperative Research Report No 289, 59pp.

Jungclaus JH, H Haak, M Latif and U Mikolajewicz, 2005. Arctic-North Atlantic interactions and multidecadal variability of the meridional overturning circulation. J. Climate, 18, 4013–4031.

Manabe S and RJ Stouffer, 1988. Two stable equilibria of a coupled ocean-atmosphere model. J Climate, 1, 841–866.

McCartney MS, 1996. Sverdrup Lecture to AGU Fall Meeting San Francisco, December 1996

Mikolajewicz U, M Groger, E Maier-Reimer, G Schurgers, M Vizcaino and AME Winguth, 2007. Long-term effects of anthropogenic CO_2 emissions simulated with a complex earth system model. Climate Dyn, doi 10.1007/s00382–006–0204-y

Morison J, K Aagaard and M Steele, 2000. Recent environmental changes in the Arctic: a review. Arctic, 53, 359–371.

Østerhus S, WR Turrell, S Jonsson and B Hansen, 2005. Measured volume, heat and salt fluxes from the Atlantic to the Arctic Mediterranean. Geophys Res Lett, 32, L07603,doi:10.1029/2004GL022188.

Overpeck J et al. 2005. Arctic system on trajectory to new seasonally ice-free state. EOS, 86, 34, pp 309, 312, 313.

Polyakov IV, A Beszczynska, E Carmack, I Dmitrenko, E Fahrbach, I Frolov, R Gerdes, E Hansen, J Holfort, V Ivanov, M Johnson, M Karcher, F Kauker, J Morrison, K Orvik, U Schauer, H Simmons, Ø Skagseth, V Sokolov, M Steele, L Timkhov, D Walsh and J Walsh, 2005. One more step toward a warmer Arctic. Geophys Res Lett, 32, L17605, doi:10.1029/2005GL023740.

Polyakov I, L Timokhov, I Dmitrenko, V Ivanov, H Simmons, F McLaughlin, R Dickson, E Fahrbach, J-C Gascard, P Holliday, L Fortier, E Hansen, C Mauritzen, J Piechura, U Schauer and M Steele 2007. The International Polar Year under the banner of Arctic Ocean warming. EOS, 88(40), 398–399.

Rahmstorf S, 2003. Thermohaline Circulation: the current climate. Nature, 421, 699.

Rahmstorf S and A Ganopolski, 1999. Long term global warming scenarios, computed with an efficient climate model. Climate Change, 43, 353–367.

Worthington LV, 1969. An attempt to measure the volume transport of Norwegian Sea overflow water through the Denmark Strait. Deep-Sea Res, 16 (Suppl), 421–432.

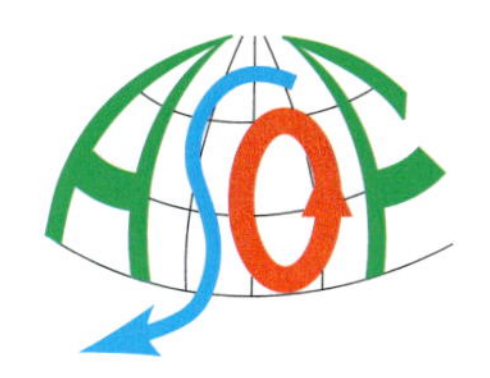

Fisheries and Oceans Canada Pêches et Océans Canada

Science Sciences

Chapter 1
The Inflow of Atlantic Water, Heat, and Salt to the Nordic Seas Across the Greenland–Scotland Ridge

Bogi Hansen[1], Svein Østerhus[2], William R. Turrell[3], Steingrímur Jónsson[4,5], Héðinn Valdimarsson[4], Hjálmar Hátún[1], and Steffen Malskær Olsen[6]

1.1 Introduction

The flow of warm, saline water from the Atlantic Ocean (the *Atlantic inflow* or just *inflow*) across the Greenland–Scotland Ridge into the Nordic Seas and the Arctic Ocean (collectively termed the Arctic Mediterranean) is of major importance, both for the regional climate and for the global thermohaline circulation. Through its heat transport, it keeps large areas north of the Ridge much warmer, than they would otherwise have been, and free of ice (Seager et al. 2002). At the same time, the Atlantic inflow carries salt northwards, which helps maintaining high densities in the upper layers; a precondition for thermohaline ventilation.

The Atlantic inflow is carried by three separate branches, which here are termed: the *Iceland branch* (the North Icelandic Irminger Current), the *Faroe branch* (the Faroe Current), and the *Shetland branch* (Fig. 1.1). These are all characterized by being warmer and more saline than the waters that they meet after crossing the Ridge, although both temperature and salinity decrease as we go from the Shetland branch, through the Faroe branch, to the Iceland branch. All these branches therefore carry, not only water, but also heat and salt across the Ridge.

Systematic investigations on the Atlantic inflow started already at the start of the 20th century with the Shetland branch, which long was treated as by far the dominant inflow branch. These investigations were mainly carried out by Scottish researchers and included measurements of temperature and salinity on two standard sections in the Faroe–Shetland Channel (Turrell 1995). Later, similar investigations were

[1]Faroese Fisheries Laboratory, Tórshavn, Faroe Islands

[2]Bjerknes Centre for Climate Research, University of Bergen, Bergen, Norway

[3]Marine Laboratory, Fisheries Research Services, Aberdeen, UK

[4]Marine Research Institute, Reykjavík, Iceland

[5]University of Akureyri, Akureyri, Iceland

[6]Danish Meteorological Institute, Copenhagen, Denmark

R.R. Dickson et al. (eds.), *Arctic–Subarctic Ocean Fluxes*, 15–43

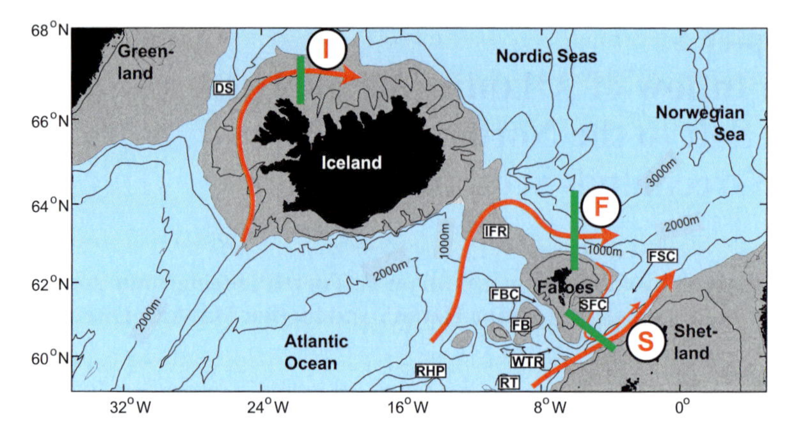

Fig. 1.1 Bottom topography between Greenland and Shetland. Shaded areas are shallower than 500 m. Thick red arrows indicate the three inflow branches: the Iceland branch (I), the Faroe branch (F), and the Shetland branch (S). A thinner red arrow indicates the "Southern Faroe Current (SFC)" and its re-circulation in the Faroe–Shetland Channel (FSC). Thick green lines show the locations of standard sections along which hydrographic and current data have been obtained. Indicated locations are: the Denmark Strait (DS), the Iceland–Faroe Ridge (IFR), the Faroe Bank Channel (FBC), the Faroe Bank (FB), the Faroe–Shetland Channel (FSC), the Wyville–Thomson Ridge (WTR), the Rockall Trough (RT), and the Rockall–Hatton Plateau (RHP)

initiated on the Iceland branch and on the Faroe branch. Sporadic attempts were made to measure currents from research vessels early in the 20th century, but systematic long-term measurements with moored current meters were only initiated in 1985 when Icelandic researchers started monitoring the currents in the Iceland branch (Kristmannsson 1998). For the other two branches, systematic current measurements were initiated with the Nordic WOCE project in the mid-1990s. Building on this, a system has been established, which monitors all the branches of the Atlantic inflow with regular CTD cruises and quasi-permanent current meter moorings. The system is maintained by research vessels from the marine research institutes in Iceland, the Faroes, and Scotland and has received support from the European research programmes through the projects VEINS (Variability of Exchanges In the Northern Seas) and MAIA (Monitoring the Atlantic Inflow toward the Arctic).

This system was further maintained and refined in the MOEN (Meridional Overturning Exchange with the Nordic Seas) project, which was supported by the European FP5, and was a component of ASOF. In the framework of this project, measurements of temperature, salinity, and currents were continued through the ASOF period. ASOF-MOEN also included a numerical modelling component, which studied the exchanges across the Greenland–Scotland Ridge, using an ocean model driven by atmospheric fluxes from reanalysis fields.

The aim of this chapter is to synthesize the information on the Atlantic inflow across the Greenland–Scotland Ridge, based mainly on the results gained by the ASOF-MOEN project and its predecessors, but including other relevant sources, as well. No attempt will be made to repeat the more detailed reviews that have included the Atlantic inflow (Johannesen 1986; Hopkins 1991; Hansen and Østerhus

2000) and neither will we attempt to make a systematic distinction between ASOF and non-ASOF produced results.

1.2 The General Setting

1.2.1 Topographic Constraints

The Greenland–Scotland Ridge separates the Arctic Mediterranean from the Atlantic Ocean and acts as a constraint on all the exchanges across it, the Atlantic inflow as well as the East Greenland Current, and the overflows. On a section (Wilkenskjeld and Quadfasel 2005) following the crest of the Ridge (Fig. 1.2), the warm and saline Atlantic water is seen to be most prominent in the south-eastern parts, where it dominates the section, above the cold and less saline overflow water flowing over the Ridge in many places. In the surface, the Atlantic water extends west of Iceland (Fig. 1.2). The Ridge reaches above the sea surface in Iceland and the Faroes, which split it into three gaps, and this determines the branching structure (Fig. 1.3).

The gap between Greenland and Iceland, the Denmark Strait, is wide and reaches a depth of 640 m. The Atlantic inflow through this gap has to share the cross-sectional area with both the East Greenland Current and the Denmark Strait overflow, and is confined to the easternmost part of the strait.

Between Iceland and the Faroes, the Atlantic water has to flow across the Iceland–Faroe Ridge, which has typical sill depths from 300–480 m along its crest. Atlantic water crosses this ridge over its whole width, in many places passing above the cold overflow water that intermittently crosses the Ridge in the opposite direction.

The Atlantic water that passes between the Faroes and Shetland, can do so along several different routes. The warmest and most saline component flows over the slope as the "Slope Current" (Swallow et al. 1977; Ellett et al. 1979), or "Shelf Edge Current" (New et al. 2001), which has its origin to the south of the Rockall Trough. In addition to this, water of more oceanic origin can pass through the

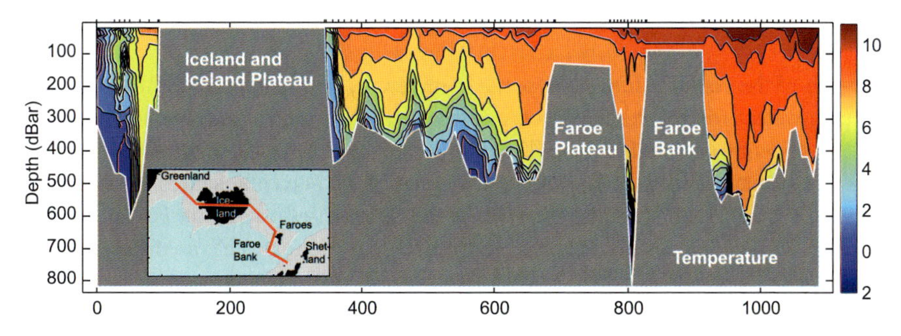

Fig. 1.2 A section following the crest of the Greenland–Scotland Ridge (red line on inset map) showing the temperature in degree Celsius during a cruise in summer 2001

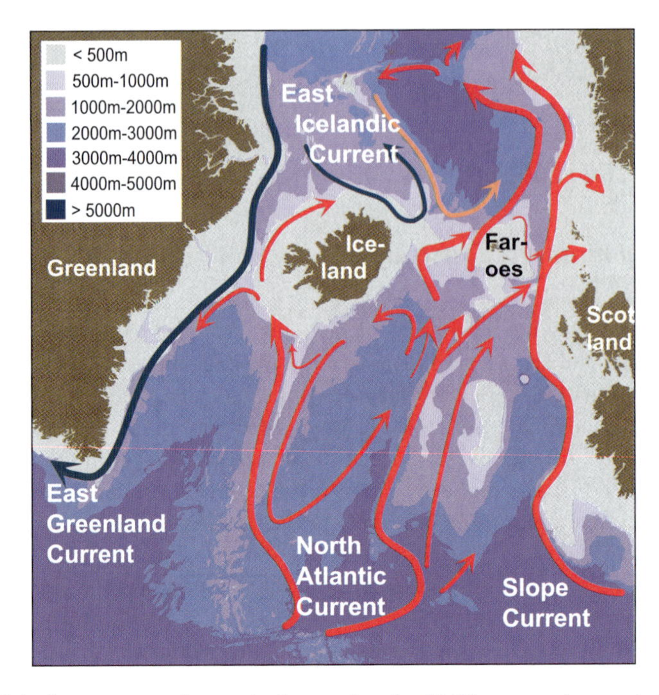

Fig. 1.3 Main flow patterns of warm (red arrows) and cold (blue arrows) currents in the upper layers of the Northeastern North Atlantic. Background colours indicate bottom depth

Rockall Trough, over the Rockall–Hatton Plateau, and even through the Faroe Bank Channel to reach the Faroe–Shetland Channel, although the persistence of some of these pathways is unknown. As these waters pass south of the Faroes, they meet a counter-flow of Atlantic water over the south-eastern Faroe slope. This flow, termed the "Southern Faroe Current" by (Hátún 2004), derives from the Faroe branch. Most of it recirculates in the Faroe–Shetland Channel and joins the other Atlantic water masses in the Shetland branch (Hansen and Østerhus 2000).

1.2.2 The Origin of the Atlantic Inflow Water

In much of the classical literature (see, e.g. review by Hansen and Østerhus 2000), the Atlantic water crossing the Ridge was seen to derive either from an oceanic or from a more continental source (Fig. 1.3). The oceanic source fed the Iceland branch, the Faroe branch, and part of the Shetland branch, whereas the continental source fed the Slope Current and thereby the Shetland branch of the Atlantic inflow. In the Faroe–Shetland Channel, especially, waters from these two sources were treated as different water masses: the "North Atlantic Water (NAW)", carried by the Continental Slope Current, and the "Modified North Atlantic Water (MNAW)", deriving from the oceanic source.

An extreme version of this view, was the proposal by Reid (1979), who suggested a direct import of Mediterranean Water to the Nordic Seas. This suggestion never gained much support and recent observational (McCartney and Mauritzen 2001) and modelling (New et al. 2001) studies have rejected it convincingly.

Distinguishing between an oceanic and a continental source does, however, ignore the continuous exchange between the waters of the Continental Slope Current and the adjacent off-shore waters and time-series show a high degree of coherence between the different Atlantic inflow branches (Section 1.4.2), whether over the continental slope or farther offshore. An alternative view, therefore, does not distinguish between oceanic and continental origin, but rather considers all the Atlantic inflow branches to be fed from two source water masses: the warm and saline ENAW and the colder and less saline WNAW.

The ENAW (Eastern North Atlantic Water) (Harvey 1982; Pollard et al. 1996) gains it properties in the region south of the Rockall Trough, called the "Inter-gyre region" (Ellett et al. 1986; Read 2001; Holliday 2003). This name might indicate a mixed contribution from the two gyres but, certainly, the ENAW has much less input from the Subpolar Gyre than the other source water mass, the WNAW (Western North Atlantic Water), which is carried towards the inflow areas by the North Atlantic Current. The North Atlantic Current is generally considered to originate in the Subtropical Gyre, but it is bounded by the Subpolar Gyre on its northern flank and water from that gyre is admixed into the flow. When it reaches the eastern North Atlantic, it has received sufficient amounts of Sub-Arctic Intermediate Water (SAIW), so that the WNAW is colder and fresher than the ENAW.

1.2.3 The Downstream Fate of the Atlantic Inflow Water

After passing the Greenland–Scotland Ridge, the different branches of Atlantic water progress into the Nordic Seas and from there, parts of the water continue into the Arctic Ocean. The details of the paths and associated water mass changes on route have been reviewed by various authors (Johannesen 1986; Hopkins 1991; Mauritzen 1996; Hansen and Østerhus 2000; Blindheim and Østerhus 2005). The main point to note is that the three different branches affect different regions in the Arctic Mediterranean. The Iceland branch has direct effects only on the southern parts of the Iceland Sea (Swift and Aagaard 1981; Jónsson 1992). The Faroe branch apparently feeds the recirculating water in the southern Norwegian Sea (Fig. 1.3) and thus probably delivers much of its heat and salt to these areas. A part of the Faroe branch also joins with the Shetland branch, which must be considered the main contributor to the North Sea and probably also the Barents Sea.

1.3 Monitoring System

Our knowledge of the Atlantic inflow has been accumulated from a long history of observations, mainly on the hydrography. Here, we focus on the observational system that has been established to monitor the three Atlantic inflow branches and was used in the ASOF-MOEN project.

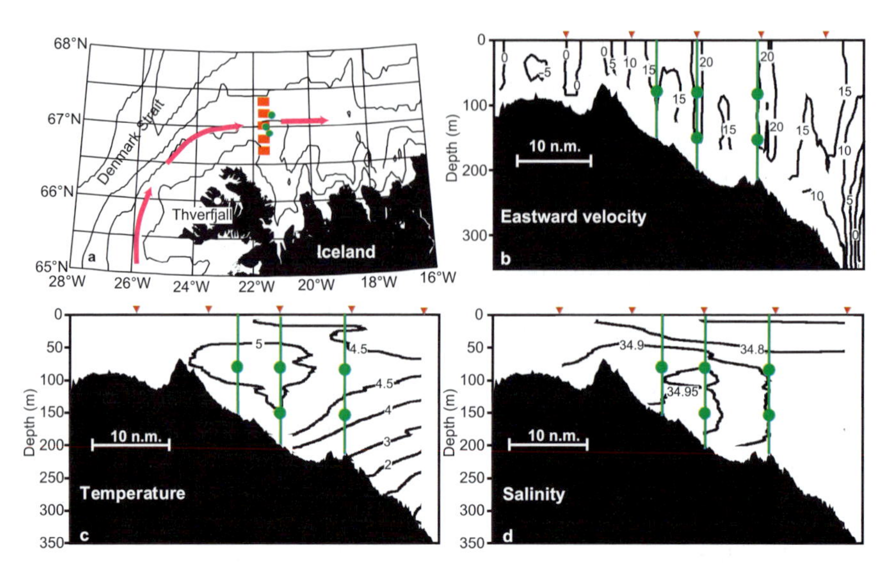

Fig. 1.4 Monitoring system and properties of the Iceland branch. (a) CTD standard stations are indicated by red rectangles. Current meter mooring sites are indicated by green circles. Magenta arrows indicate Atlantic water pathways towards and through the section. (b) Average eastward velocity (cm s^{-1}) based on a total of 20 sections of vessel mounted ADCP data from November 2001–2004, and August 2005 with four sections taken each time. CTD standard stations (red triangles) and current meter moorings (green lines with green circles indicating Aanderaa current meters) are shown. (c, d) Average distributions of temperature in degree Celsius (c) and salinity (d) on the section, based on CTD observations at standard stations (red triangles) in the period 1999–2001

The systematic observations of the Iceland branch have been focused on the Hornbanki section (green line labelled "I" on Fig. 1.1). On this section (Fig. 1.4), CTD profiles have been obtained by the Marine Research Institute in Iceland on several standard stations up to four times a year since 1994 and, during the same period, the inflow of Atlantic water has been monitored by moored current meters. From September 1999, the measurements were extended to three moorings carrying a total of five current meters (Fig. 1.4).

The Faroe branch has been monitored on a section extending northwards from the Faroes along the 6°05′ W meridian (green line labelled "F" on Fig. 1.1). On this section (Fig. 1.5), CTD profiles have been acquired by the Faroese Fisheries Laboratory on several standard stations, at least four times a year since 1988. From the mid-1990s, ADCPs have been moored on the section almost continuously. The number and locations of ADCP moorings have varied somewhat, but since summer 1997, there have always been at least three and sometimes five ADCPs on the section, except for annual servicing gaps.

The observations of the Shetland branch were carried out on a section crossing the channel south of the Faroes (green line labelled "S" on Fig. 1.1). At least four, and before summer 2000, five ADCP moorings have been maintained along the section since November 1994 (Fig. 1.6). These observations have been complemented with ADCP data acquired from oil platforms. Both the Faroese Fisheries Laboratory and the Marine Laboratory in Aberdeen do regular CTD cruises along

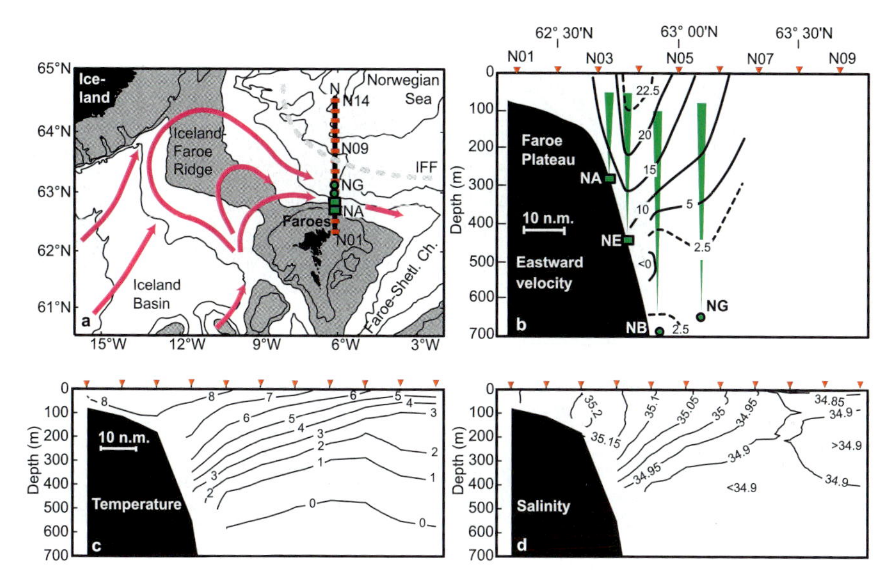

Fig. 1.5 Monitoring system and properties of the Faroe branch. (a) CTD standard stations are indicated by red rectangles, labeled N01–N14. ADCP mooring sites are indicated by green circles (traditional moorings) or rectangles (trawl-proof frames) labeled NA to NG. Shaded areas are shallower than 500 m. The dotted yellow curve indicates the general location of the Iceland–Faroes Front (IFF) and magenta arrows indicate Atlantic water pathways towards and through the section. (b) Average eastward velocity (cm s^{-1}) 1997–2001. The innermost CTD standard stations (red triangles) are indicated as well as the ADCP mooring sites (green circles or rectangles with green cones indicating sound beams). (c, d) Average distributions of temperature in degree Celsius (c) and salinity (d) on the inner part of the section, based on CTD observations at standard stations (red triangles) in the period 1987–2001

this section and altogether four to eight CTD sections have been obtained annually since the mid-1990s.

The region between Iceland and Shetland is heavily fished and traditional current meter moorings have a short survival time in this area. This was the reason for using upward-looking ADCPs instead of more traditional instrumentation. At deep sites, the ADCPs are moored in the top of traditional moorings with the ADCP sufficiently deep to escape trawls. On the slope north of the Faroes, two of the ADCPs are deployed directly on the bottom within frames that protect the ADCPs and other instrumentation from fishing gear (Fig. 1.7).

1.4 Observed Properties

1.4.1 Typical Structure and Properties of the Inflow Branches

The Iceland branch is highly variable but it is of great importance to the regional marine climate and hence the ecosystem in North Icelandic waters (Jónsson and Valdimarsson 2005). There is usually a core of Atlantic water identified by high

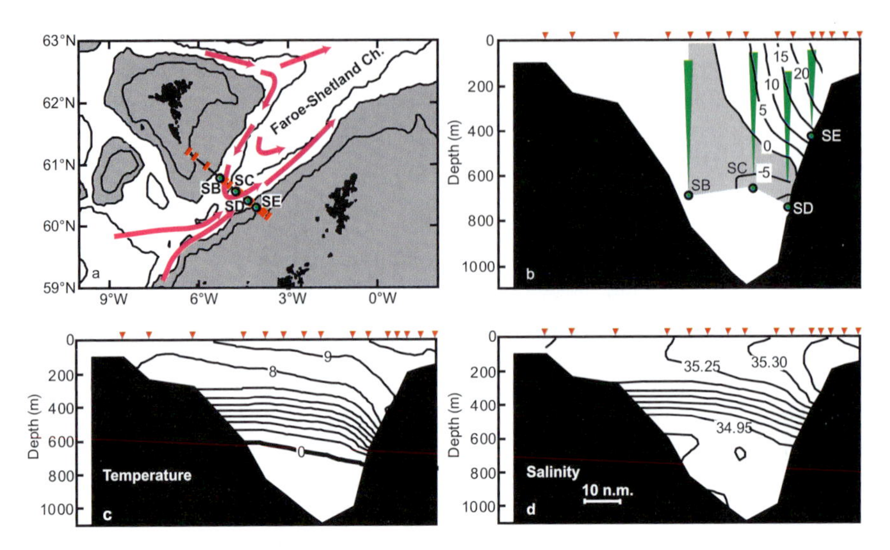

Fig. 1.6 Monitoring system and properties of the Shetland branch. (a) CTD standard stations are indicated by red rectangles. ADCP mooring sites are indicated by green circles labeled SB, SC, SD, and SE. Shaded areas are shallower than 500 m and magenta arrows indicate Atlantic water pathways towards and through the section. (b) Average along-channel velocity (cm s^{-1}) as measured by the ADCP moorings in the period 1994–2005. Shaded area indicates reverse (SW-going) flow. CTD standard stations (red triangles) and ADCP mooring sites (green circles with green cones indicating sound beams) are indicated. (c, d) Average distributions of temperature in degree Celsius (c) and salinity (d) on the section, based on CTD observations at standard stations (red triangles) in the period 1994–2005

Fig. 1.7 Trawl-proof frame containing ADCP, double acoustic releases, ARGOS beacon, and buoyancy, on top of concrete anchor, is being made ready for deployment onboard R/V Magnus Heinason

salinity and temperature, but its location and extent are variable. In Fig. 1.4c, d, the core can be identified over the area covered by the current meters. In this branch, the Atlantic water does not seem to reach deeper than 200 m (Jónsson and Briem 2003). North of the Atlantic water core, the region may variably be dominated by Arctic water masses from the Iceland Sea or Polar water masses from the East Greenland Current.

The Faroe branch carries the Atlantic water that has crossed the Iceland–Faroe Ridge. Northeast of the Ridge, this water meets the much colder and less saline waters of the East Icelandic Current and gets confined into a fairly narrow current, which flows eastwards over the northern slope of the Faroe Plateau. Relatively high temperature and salinity (Fig. 1.5c, d) characterize the Atlantic water, which usually is concentrated on a wedge-shaped area that is bounded by the Iceland–Faroe Front, which hits the Faroe slope at depths 400–500 m, similar to the sill depth of the Iceland–Faroe Ridge. From below, the Atlantic layer is bounded by two water masses (Hansen and Østerhus 2000): the Norwegian Sea Arctic Intermediate Water (NSAIW), which occupies the top of the deep water in the Norwegian Sea, and the Modified East Icelandic Water (MEIW), which derives from the East Icelandic Current and usually is characterized by a salinity minimum.

The Shetland branch carries Atlantic water that has entered the Faroe Shetland Channel from the west in addition to water recirculated from the Faroe branch. A section crossing the channel (Fig. 1.6c, d) has Atlantic water, characterized by high temperature and salinity across the whole channel in the upper layers. Temperature and salinity do, however, increase from the Faroe to the Shetland side of the channel with the highest temperatures and salinities in the core of the Slope Current. Below, the Atlantic layer is bounded by the deep NSAIW and by varying amounts of MEIW from the East Icelandic Current.

1.4.2 Long-Term Variations of Temperature and Salinity

The long-term hydrographic observations allow the generation of long time-series of the properties of the Atlantic inflow. The longest time-series are from the Faroe–Shetland Channel (Fig. 1.8) and show variations on many timescales. Both temperature and salinity peaked around the middle of the 20th century. This feature has been shown to be a characteristic of the northern latitudes (Bengtson et al. 2004). A number of anomalies have been noted, especially in the salinity (Belkin et al. 1998), with the "Great" or "Mid-seventies" anomaly (Dickson et al. 1988) being the most pronounced.

The last decades of the 20th and the beginning of the 21st century show increasing trends in both temperature and salinity, which may perhaps be linked to global change, but which also exhibit large variations on a decadal timescale. Hátún et al. (2005) have shown that these variations to a large extent can be explained by variations in the intensity and extent of the Subpolar Gyre circulation (Häkkinen and

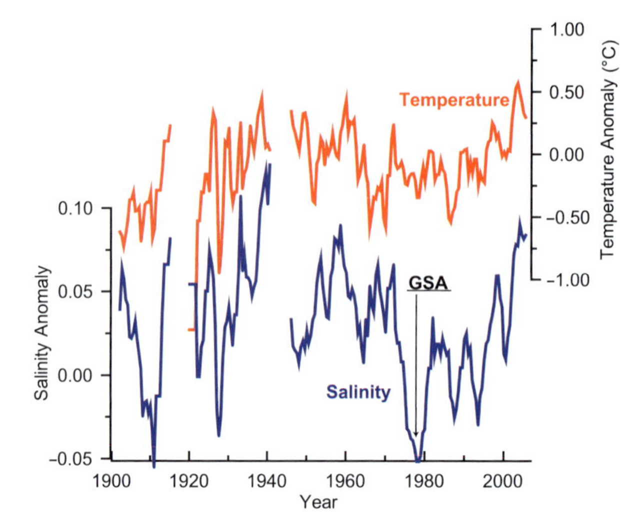

Fig. 1.8 Anomalies of temperature (red) and salinity (blue) over the Scottish shelf. Derived from the temperature and salinity of the water displaying maximum salinity within the Slope Current flowing polewards along the Scottish continental shelf. Anomalies presented are 2-year running means after the average (1961–1990) seasonal cycle has been removed. The "Great Salinity Anomaly (GSA)" is indicated

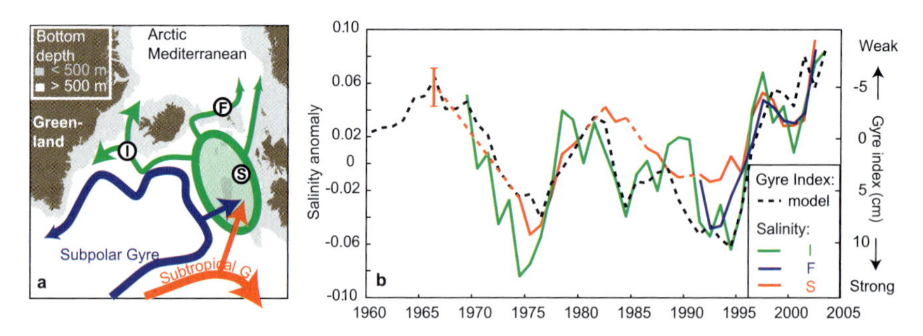

Fig. 1.9 (a) According to Hátún et al. (2005), all of the inflow branches (green arrows) are fed partly from the Subtropical and partly from the Subpolar Gyre in relative amounts that depend on the intensity and extent of the Subpolar Gyre circulation, expressed by the gyre index (Häkkinen and Rhines 2004). (b) The gyre index (dotted black line) from a model is plotted together with observed salinity in the three inflow branches

Rhines 2004). This intensity, the gyre index, is found to correlate well with inflow properties of all the branches (Fig. 1.9) and, using a numerical model, Hátún et al. (2005) could explain this in terms of the source water masses (Section 1.2.2). When the gyre index is high, the Subpolar Gyre extends far towards the east and relatively large amounts of WNAW are transported towards the inflow areas, whereas the warmer and saltier ENAW tends to dominate more, when the gyre index is low. The properties of the Atlantic inflow, thus, seem to be governed by the intensity of the Subpolar

Gyre circulation, which again is believed to depend mainly on the buoyancy flux and convection in the Labrador–Irminger Seas (Häkkinen and Rhines 2004).

1.5 Observed Fluxes

1.5.1 Methods for Flux Estimation

Volume flux through a section is, in principle, simple to calculate as the integral of the normal velocity component over the section. All of the Atlantic inflow branches do, however, flow together with other water masses. For none of the branches, is it possible to define a section that covers all the Atlantic water on the section and no other waters.

To calculate the volume flux of Atlantic water through a section crossing one of the inflow branches, it is therefore necessary to know, not only the normal velocity, but also the fraction of Atlantic water on the section. The methodology is illustrated in Fig. 1.10 and the volume flux of Atlantic water is computed as:

$$V_A(t) = \sum_k \sum_j A_{k,j} \cdot u_{k,j}(t) \cdot \beta_{k,j}(t) \qquad (1.1)$$

where the sum is over all the boxes that the section is subdivided into and $\beta_{k,j}(t)$ is the fraction of Atlantic water in box (k,j) at time t. First of all this requires, of course, a definition of Atlantic water. In the literature on the Nordic Seas and Arctic Ocean, the concept of Atlantic water is often defined by its salinity, e.g. as water more saline than 35. Here, we define the flux of Atlantic water as the flux of water crossing the Greenland–Scotland Ridge into the Nordic Seas.

With comprehensive velocity measurements that provide $u_{k,j}(t)$, the problem is reduced to the determination of $\beta_{k,j}(t)$. This is not a trivial problem, but, in principle it can be solved if not too many different source water masses are involved, and if the characteristics (T, S) of these as well as the waters on the section are known. Differences in data availability and conditions have led to different procedures for the different branches. For the two easternmost branches, the Atlantic water fraction in each sub-area is determined from temperature and salinity measurements by using a three-point mixing model (Hansen et al. 2003; Hughes et al. 2006). Fluxes of volume of the Atlantic water component through each sub-area are then computed and summed. For the Iceland branch, the Atlantic water fraction is determined from temperature observations of the pure Atlantic water and polar water upstream and the temperature observed at the current meters (Jónsson and Valdimarsson 2005).

In addition to volume (mass) flux carried by the various branches, heat and salt fluxes are highly relevant. These fluxes are not, however, meaningful, unless the temperature and salinity of the water returning to the Atlantic are known. Instead of producing heat- and salt fluxes, we therefore compute average values of temperature and salinity of the different inflow branches, where the average is weighted

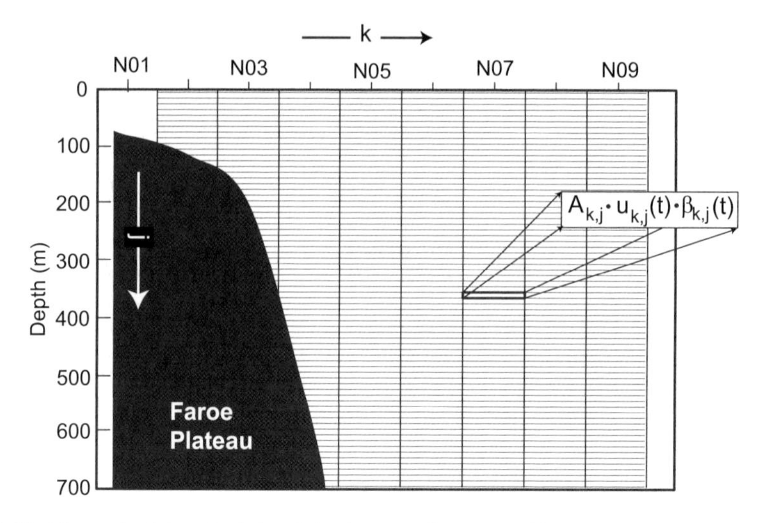

Fig. 1.10 The method for calculating Atlantic water flux exemplified for the section crossing the Faroe branch. The section is subdivided into boxes, which are labelled by indices k and j. One of the boxes is shown in a magnified scale indicating the parameters that must be assigned to each box: The area: $A_{k,j}$, the eastward velocity: $u_{k,j}(t)$, and the fraction of Atlantic water: $\beta_{k,j}(t)$. Each vertical column of boxes is centered around a standard CTD station (labelled N01–N09)

with respect to volume flux. Since the Atlantic water always is found together with water of other origins, this has to be done with care, as discussed by Hansen et al. (2003). If similar values can be produced for all the other exchange flows, meaningful heat- and salt-budgets can be produced.

1.5.2 Flux Estimates for the Individual Inflow Branches

Flux estimates for the individual inflow branches have been given in a number of publications. For the Iceland branch, Jónsson and Valdimarsson (2005) have determined the Atlantic water fraction within the inflow area of the Hornbanki section as a function of time and computed fluxes for the 1994–2000 period. The average volume flux of Atlantic water was found to be 0.8 Sv. No seasonal variation was found in current velocities (Jónsson and Briem 2003), but the Atlantic water fraction varied seasonally, which gave rise to a seasonal amplitude of 0.2 Sv for the volume flux of Atlantic water with a maximum in September. Monthly averaged volume flux ranged between 0 and 1.3 Sv.

For the Faroe branch, Hansen et al. (2003) have analysed the observations from the June 1997 to June 2001 period. On average, the Faroe branch transported a volume flux of 3.5 ± 0.5 Sv of Atlantic water. Monthly averaged volume flux ranged between 2.2 and 5.8 Sv, but with only a small seasonal variation. Daily averages ranged between 0.3 and 7.8 Sv, with not a single flow reversal during the 4-year period.

For the Shetland branch, Hughes et al. (2006) have analysed the fluxes through the Channel. On average, the Atlantic water flux was estimated at 3.9 Sv for the period September 1994–May 2005. The flux was found to have a seasonal variation with an amplitude of 0.8 Sv, which is 21% of the average, and maximum in November. Monthly averaged Atlantic water flux ranged between 0.8 and 7.5 Sv.

1.5.3 The Total Atlantic Inflow 1999–2001

For the 3-year period from 1 January 1999 to 31 December 2001, Østerhus et al. (2005) computed volume fluxes (Fig. 1.11) and average temperature and salinity values for each of the branches and combined them to produce overall values for the total Atlantic inflow (Table 1.1). The average values for the volume fluxes of the various branches differ slightly from previously published values (Østerhus et al. 2001; Hansen et al. 2003; Turrell et al. 2003; Jónsson and Valdimarsson 2005) but the deviations are small and may be due to the different averaging periods.

Østerhus et al. (2005) estimated an uncertainty of about 1 Sv for the average total volume flux of Atlantic water. Within this uncertainty, their estimate of the total volume flux (8.5 Sv) is consistent with the preliminary estimate reported by Hansen and Østerhus (2000) and also remarkably close to the classical value published by Worthington (1970).

For the 1999–2001 period with concurrent measurements, the Iceland branch was found to carry 10% of the Atlantic inflow volume flux, with the other two branches carrying 45% each. Monthly averaged volume fluxes for each branch and for the total inflow during this period are shown in Fig. 1.11. Although they are of similar intensity on the average, Fig. 1.11 indicates larger variations in the Shetland

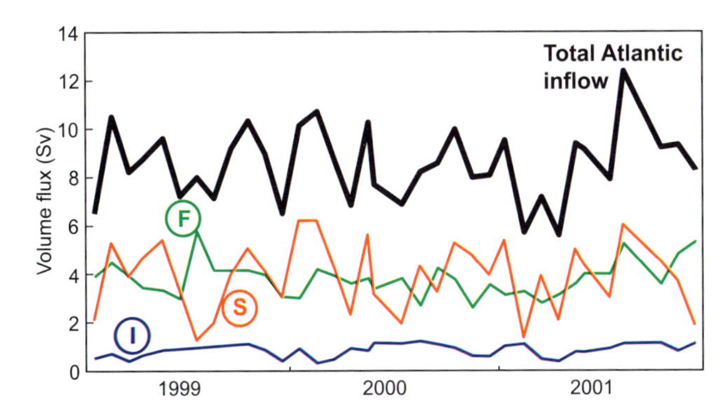

Fig. 1.11 Monthly averaged volume flux of Atlantic water in each of the three branches (coloured lines labelled as the Iceland branch (I), the Faroe branch (F), and the Shetland branch (S)) and in the total Atlantic inflow (black line) for the 1999–2001 period

Table 1.1 Observed characteristics of each of the three Atlantic inflow branches and of the total inflow for the period January 1999–December 2001

	Average			Seasonal var. of vol. flux		
	Vol. flux	Temp.	Sal.	Ampl.	Max.	Signif.
Inflow branch	Sv	°C		Sv	Month	
Iceland branch	0.8	6.0	≤35.00	0.2	Sept.	<0.01
Faroe branch	3.8	8.2	35.23	0.3	Oct.	n.s.
Shetland branch	3.8	9.5	35.32	0.2	Mar.	n.s.
Total Atl. Infl.	**8.5**	**8.5**	**35.25**	**0.4**	**Oct.**	**n.s.**

branch than in the Faroe branch. This might be due to differences in precision of the estimates. Certainly, the Shetland branch is more difficult to monitor accurately due to the recirculation in the Faroe–Shetland Channel and the intensity of meso-scale activity (Sherwin et al. 2006). Results from the ASOF-MOEN numerical modelling activities do, however, show a similar difference between the two branches (Section 1.6).

For the 1999–2001 period, Østerhus et al. (2005) found evidence for a seasonal signal in the Iceland branch with maximum volume flux in September, but the other two branches, as well as the total inflow, showed no statistically significant seasonal variation of the volume flux (Table 1.1). They concluded that a possible seasonal variation of the total Atlantic inflow did not exceed the observational uncertainty, estimated at 1 Sv, in amplitude during this period.

This might seem to conflict with reports of considerably larger seasonal variations in the Norwegian Atlantic Current on the Svinøy section, downstream from the Ridge (Orvik et al. 2001). They only had long-term direct current measurements from the inner branch of this flow, however, and the outer branch has been reported to vary in counter-phase to the inner branch (Mork and Blindheim 2000). The relatively weak seasonal variation of the inflow over the Ridge is therefore consistent with the conclusion of Jakobsen et al. (2003) that the winter intensification of the flow at selected locations like the Svinøy section is primarily linked to spin-up of the local basin gyres.

1.6 Numerical Modelling of the Atlantic Inflow

A number of ocean modelling studies have addressed the Atlantic inflow based on different ocean general circulation models of varying resolution and experimental design (e.g. Karcher et al. 2003; Nilsen et al. 2003; Zhang et al. 2004; Drange et al. 2005). No attempt will be made here to review these studies. Instead, this section presents results from the modelling effort within the ASOF-MOEN project, carried out at the Danish Meteorological Institute. The results are based on an ensemble hindcast simulation for the period 1948–2005 using a global coupled ocean/sea-ice ocean model of relatively coarse resolution (MPI-OM, Marsland et al. 2003), constrained by atmospheric reanalysis data (NCEP/NCAR, Kistler et al. 2001) and observed Arctic river discharges (http://grdc.bafg.de).

The model experiment and results are described in Olsen and Schmith (2007) with a focus on the climatology of the exchanges between the Nordic Seas and the North Atlantic as defined at a set of key sections characterizing the system. The ensemble approach applied in thc model experiment is designed to eliminate the role of internal modes of variability and initial ocean conditions on the simulated ocean climate variability and, thus, to isolate the forced response by known atmospheric changes (e.g. the NAO).

Despite the global domain and coarse average resolution, the displacement of the North Pole onto Greenland in the model grid by making use of the curvilinear coordinates results in relatively high resolution in the Nordic Seas. Therefore all three branches of Atlantic surface inflow to the Nordic Seas (Fig. 1.1) can be identified from the simulated upper ocean velocity and tracer fields (Olsen and Schmith 2007): the Iceland branch north of Iceland, the Faroe branch between the Faroes and Iceland, in the model found close to the Icelandic shelf break turning east upon passage of the Ridge, and finally, the Shetland branch, modeled as a broad inflow extending off the Scottish Slope.

At the defined sections, water mass properties are used to distinguish between transports in individual branches of flow. Model mean exchanges for the period 1948–2005 are shown to compare favourably with existing observational estimates for several flow branches in the area, including the exchanges across the Greenland–Scotland Ridge (Olsen and Schmith 2007; see also Chapter 19 for a comparison between model results and observations). For the Atlantic inflow, this is also illustrated in Fig. 1.12 and Table 1.2 for the recent period 1999–2001 with concurrent, high quality observations of all inflow branches (Fig. 1.11).

It is seen that in total, model inflow is about 0.5 Sv higher than observed, which is linked to excess model transport in the Shetland branch compared to observations. The discrepancy is somewhat lower when comparing long-term mean model results with observations (Table 1.2). According to the model results, the total

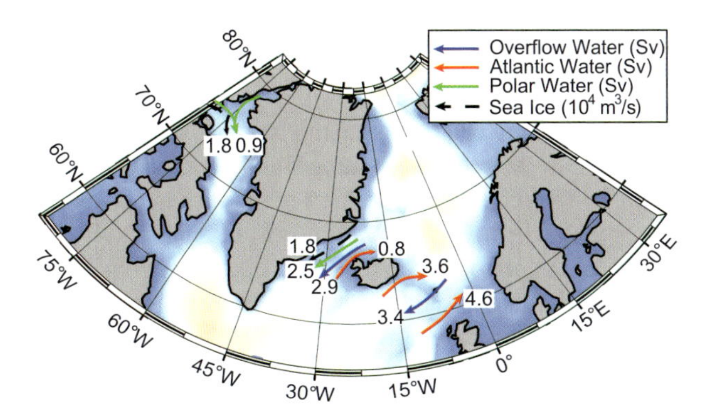

Fig. 1.12 Ensemble- and time-averaged exchanges between the Arctic Mediterranean and North Atlantic for the period 1999–2001 according to the ASOF-MOEN model

Table 1.2 Average values and seasonality (amplitude and time of maximum) of volume fluxes (in Sv) of individual branches and the total Atlantic inflow derived from the ASOF-MOEN model, compared to observed values (based on Østerhus et al. 2005). Observations are only for the 1999–2001 period. Model results are shown partly for this period, partly for the whole (1948–2005) period (All). The correlation coefficients (Corr.) between fluxes from the model and from observations are based on monthly averages and "n.s." indicates "not significant"

Parameter	Average flux			Corr.	Seasonality			
Period	1999–2001		All	1999–2001	1999–2001 (obs)		All (mod)	
Method	Mod.	Obs.	Mod.	Mod.–Obs.	Ampl.	Max.	Ampl.	Max.
Iceland branch	0.8	0.8	0.7	0.74	0.2	Sept.	0.3	Sept.
Faroe branch	3.6	3.8	3.8	−0.38 ns	0.3	Oct.	0.6	Mar.
Shetland branch	4.6	3.8	4.2	0.28 ns	0.2	Mar.	1.2	Dec.
Total Atl. Infl.	9.0	8.5	8.7	0.22 ns	0.4	Oct.	1.2	Dec.

Atlantic inflow did not vary much throughout the 1948–2005 period with only a slight trend, but there was a strengthening of the Shetland branch and a weakening of the Faroe branch (Fig. 1.13b).

The indications in the ASOF-MOEN model results of a nearly constant total Atlantic inflow tend to agree with the findings of Nilsen et al. (2003) from a different model, though in that study, neither of the branches of inflow show robust tendencies, in contrast to the present results. Such stable inflow is, however, at odds with the modeled increase reported by Zhang et al. (2004). Also, the negative correlation between the Shetland branch and the Faroe branch, found by Nilsen et al. (2003), is not supported by the ASOF-MOEN model results.

When comparing model and observations on shorter timescales (Fig. 1.13a), the correspondence is not as good. For the Iceland branch, monthly averaged fluxes in the model and the observations were fairly well correlated, but for the other branches, the correlation coefficients were not significant (Table 1.2). The same conclusion is reached when comparing seasonality in the model and the observations (Table 1.2). Except for the Iceland branch, the model gives higher seasonal amplitudes than the observations and the phases also differ. To some extent, this may be due to different analysis periods. For the 1999–2001 period (Fig. 1.13d), the model does indicate a smaller seasonal amplitude than for the full period (Fig. 1.13c), especially for the Faroe branch. Even for this period, the model still indicates a larger seasonal amplitude than the observations but, when the uncertainties are taken into account, there is no real discrepancy between model and observations.

Summarizing, the ASOF-MOEN model results and the ASOF-MOEN observations show a high degree of correspondence as regards long-term average volume fluxes in the individual branches and the total Atlantic inflow. They also agree on a relatively small seasonal amplitude (<15% of the average flux). They both show fairly similar values for the magnitude of flux variability in the individual branches and total inflow (Fig. 1.13a). It is especially noteworthy, that both observations and model indicate larger monthly variability in the Shetland branch than in the Faroe branch (Fig. 1.13a). When correlating simultaneous monthly averages and seasonal

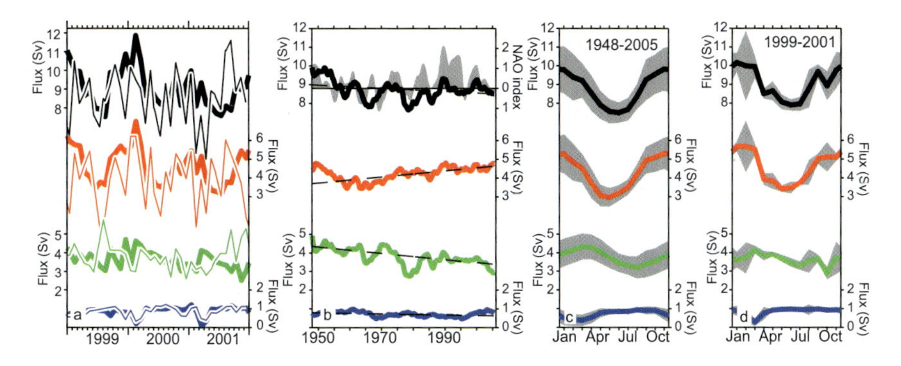

Fig. 1.13 (a) Modeled (thick) and observed (thin) monthly averaged volume flux of Atlantic water in each of the inflow branches and the total Atlantic inflow for the 1999–2001 period. (b) De-seasoned and low-pass filtered (cut-off frequency of 1/24 months^{-1}) modeled volume fluxes and the interpolated, low-pass filtered winter NAO index (gray bars, Jones et al. 1997). The linear trend of each branch is indicated by a thin dashed line. (c, d) Modeled seasonality around the time mean volume flux of each of the inflow branches and the total with the inter-annual spread (2σ) in grey for the periods: 1948–2005 (c), and 1991–2001 (d). The colour coding in all the panels is: Iceland branch (blue), Faroe branch (green), Shetland branch (red), and total Atlantic inflow (black)

variation, however, only the Iceland branch shows significant correlation between the model and the observations. This discrepancy may be due to model deficiencies, or observational inaccuracies, or both, but more work is needed to clarify this.

1.7 Effects of the Atlantic Inflow on the Arctic Mediterranean

It is not the aim of this chapter to give a complete account of the effects of the Atlantic inflow on the Arctic Mediterranean. This topic will be dealt with in other chapters of this book in much more detail, but the Atlantic inflow has tremendous impacts on the area that it enters and no description of it can approach completeness without an overview of the main effects. In the following sections, brief overviews are given for the effects of the Atlantic inflow on the mass (volume), heat, and salt budgets.

1.7.1 Mass Budget

If the estimate by Østerhus et al. (2005) for the 1999–2001 period is used as a basis, the volume flux of the total Atlantic inflow is 8.5 Sv, on the average. In addition to this, 0.8 Sv are reported to enter the Arctic Mediterranean through the Bering Strait (Coachman and Aagaard 1988; Roach et al. 1995) and 0.2 Sv as freshwater (Aagaard and Carmack 1989). Thus, the Atlantic inflow accounts for about 90% of all the water entering the Arctic Mediterranean (Fig. 1.14).

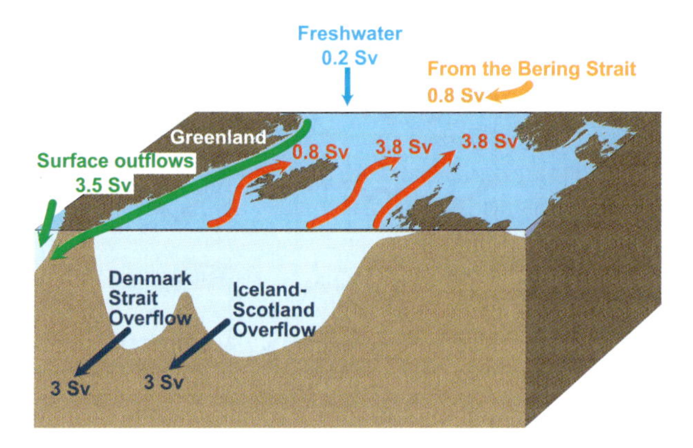

Fig. 1.14 Mass (volume) budget of the Arctic Mediterranean. The value for the volume flux of the surface outflow has been chosen to acquire balance

All of this water has to return to the Atlantic and it does so through several current branches that can be grouped into two main flow systems: the "surface outflow" and the "overflow". The surface outflow includes the East Greenland Current and the flow through the Canadian Archipelago, whereas the overflow includes the deep flow of cold dense water across the Greenland–Scotland Ridge through the Denmark Strait and across the Ridge in different areas east of Iceland. Due to the difficulties of measuring fluxes in shallow ice-covered areas, reliable flux estimates for the total surface outflow have been hard to acquire, but there seems to be a general consensus (Hansen and Østerhus 2000) that the total overflow is around 6 Sv, equally split between the Denmark Strait and the eastern overflow branches (Fig. 1.14).

This is important for understanding the Arctic Mediterranean, but it also has important consequences for the Atlantic inflow, as such. The Bering Strait inflow and the freshwater input are both relatively buoyant and it is not considered likely that they contribute to the overflow (Rudels 1989). This implies that all the overflow water derives from the Atlantic inflow but it also implies that a large fraction of the Atlantic inflow returns as overflow, rather than surface outflow. From Fig. 1.15, this fraction is 71%. This value is, of course, sensitive to uncertainties in the flux estimates, but it is unlikely to be less than 50%. Most of the Atlantic inflow therefore returns as overflow, which has implications for the driving force (Section 1.8).

1.7.2 Heat Budget

The transport of heat to an area by an ocean current can only be determined if the temperatures of all the outflows, as well as the inflows, are known. It is therefore meaningless to consider the heat transport of the Atlantic inflow *per se*. The outflows do, however, have typical temperatures around 0 °C and, with an uncertainty of about 10%, we can therefore estimate the heat import of the Atlantic inflow to the Arctic Mediterranean by using that value for the outflow temperature.

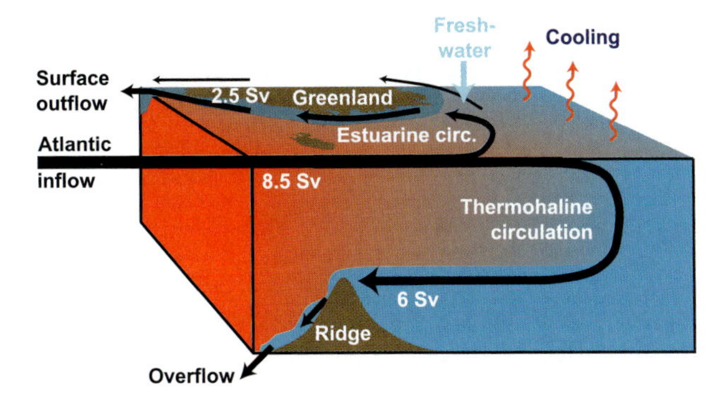

Fig. 1.15 Two circulation systems return the inflowing Atlantic water to the Atlantic Ocean after cooling and freshening. The volume flux of the surface outflow was determined as the difference between the measured Atlantic inflow (this chapter) and the measured overflow (e.g. Hansen and Østerhus 2000). Note that the Bering Strait inflow and its path to the Atlantic is not included in the figure

From this and the average temperatures of the various branches (Table 1.1), the total Atlantic inflow is found to import 310 TW (1 TW = 10^{12} W) of heat. The Shetland branch is the warmest inflow branch and probably contributes most to this heat import, but the different inflow branches may not necessarily contribute equally to the different outflow branches. Hence, the temperature decrease and heat loss of each branch is not well defined without a much more detailed description.

1.7.3 Salt Budget

As for the heat budget, a detailed account of the salt budget requires knowledge of the outflows as well as the inflows. It is, however, possible to make a rough estimate of the Atlantic inflow contribution to the salt budget of the Arctic Mediterranean by a simple calculation. Assuming that 8.5 Sv of Atlantic inflow with salinity 35.25 (Table 1.1) mixes with 0.8 Sv Bering Strait water with salinity 32.5 (Coachman and Aagaard 1988) and with 0.2 Sv of freshwater (Aagaard and Carmack 1989) from runoff and precipitation (P–E), the total outflows must have a volume flux of 9.5 Sv and an average salinity of 34.28.

This implies that the salinity of the Atlantic inflow, on the average, is reduced by ~1, before the water returns to the Atlantic, which may be used to illustrate the (oft-neglected) effect of Atlantic inflow variations on the freshwater balance of the Arctic Mediterranean. Typical variations of the inflow salinity are on the order of 0.1 (Fig. 1.8). A salinity increase of this magnitude would therefore require a 10% increase in freshwater flux in order to maintain a constant average salinity of the Arctic Mediterranean. Similarly, the model results indicate (Fig. 1.13b) that, on decadal timescales, the volume flux of the total Atlantic inflow has varied by about 20% of the average. With constant salinity, this would require a 20% variation in the freshwater flux to maintain balance. These numbers illustrate that both salinity and volume flux variations of the Atlantic inflow need to be taken into account when considering the freshwater (salt) budget of the Arctic Mediterranean.

1.8 Driving Force

All flows in nature require driving forces to accelerate them and maintain them against the retarding effect of friction. This is especially the case for the Atlantic inflow, which exhibits high velocities in the Ridge area (Figs. 1.4–1.6), compared to upstream. These forces may well be affected by future climate change, in which case the inflow may be expected to change. It is therefore important to consider, what forces can drive the Atlantic inflow. The discussion in this section addresses that question, but only as regards the flow across the Ridge, not the circulations in the upstream or downstream basins.

All of the inflow branches are upper layer, surface-intensified, flows, which are fairly uni-directional with depth (Figs. 1.4–1.6). The equations of motion, therefore, include only two external forces that can drive the flow: A surface stress, generated by wind, and a pressure gradient, generated by a sloping sea-surface. By definition, a driving force has to do positive work on the flow and, hence, only the along-flow components of the wind stress or sea-level slope can drive the flow. In the following sections, these two forces are discussed separately and their relative contributions to driving the flow are discussed, although the non-linearity of the system precludes a complete distinction between them.

Both observations (Fig. 1.11) and models (Fig. 1.13) indicate that the total Atlantic inflow is fairly stable with a variable component, superimposed on a constant flow, which seems to contribute considerably more than the variable component, even on timescales as short as a month. The forcing mechanism of the variable component may be studied by correlating flow variations to possible driving forces, but the forcing mechanism of the constant component is more difficult to identify. It is therefore essential to note that the two components may not necessarily have the same forcing.

1.8.1 Wind Forcing

Most upper layer flows in the World Ocean are generally considered to be driven by wind stress and it is natural to assume the same for the Atlantic inflow. This assumption is supported by the fact that the average wind direction in the main inflow region between Iceland and Scotland has a positive component along the inflow path.

The NAO index is commonly used as an indicator of the wind in this region and it may be correlated to the volume flux of the Atlantic inflow. Long time series of the inflow are only available from models and the simulated volume fluxes from the ASOF-MOEN modelling effort can be compared to the NAO index since the ensemble experiment was explicitly designed to disentangle a robust imprint of the variable forcing by the atmospheric reanalysis (Olsen and Schmith 2007). By visual comparison, the total Atlantic inflow shows some similarity to the NAO index (Fig. 1.13b) and the zero-lag correlation coefficient is positive between NAO and the total inflow as well as the Shetland and the Faroe branch, whereas the correlation is negative between NAO and the Iceland branch (Fig. 1.16a). The correlation coefficients are small, however, and not significant statistically, when the autocorrelations of the time-series are taken into account.

To yield further insight into the possible role of the NAO, the correlation analysis is performed for 30-year running segments throughout the hindcast (Fig. 1.16b). This analysis is motivated by the documented shift in the spatial pattern of the NAO in the 1970s, which influenced the marine climate of the Nordic Seas (e.g. Visbeck et al. 2003; Furevik and Nilsen 2005). The results illustrate a near constant imprint of the NAO on the Shetland branch since 1948 with values around 0.3–0.5 though slightly increasing in the latter part. In contrast, a clear shift is seen in the Iceland branch and the Faroe branch from nearly uncorrelated with the index in the early part of the hindcast to being significantly correlated in the recent decades, though with opposite sign; the Faroe branch reaching a positive correlation of 0.72 from 1975 to 2005.

Rather than NAO, it would be preferable to correlate the wind itself with the Atlantic inflow, but what wind parameter? over what region? and with what timelag? (Orvik and Skagseth 2003) addressed that problem by correlating their volume flux measurements off the Norwegian coast with the zonally averaged

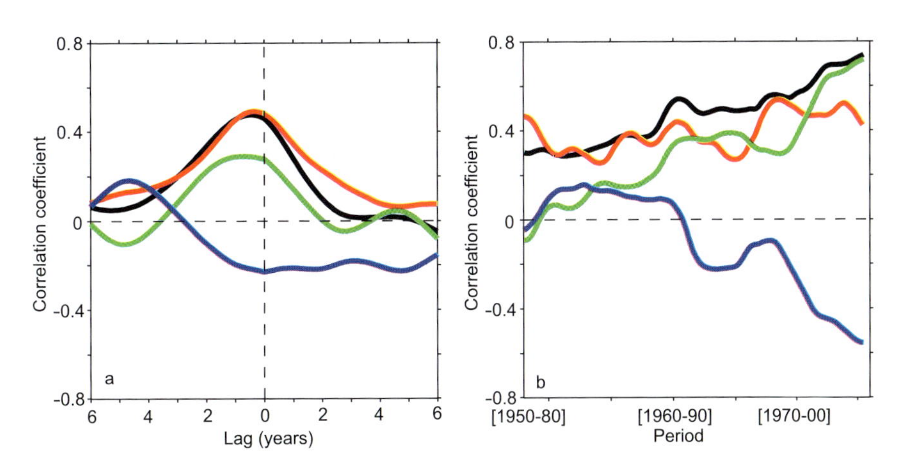

Fig. 1.16 (a) Lagged correlation between the winter NAO index and each individual inflow branch as well as the total Atlantic inflow from the ASOF-MOEN model. (b) Zero-lag correlations of the same parameters for 30-year running segments. Prior to the analysis, the time-series have been de-seasoned and low-pass filtered and the linear trend removed (see Fig. 1.13b). The colour coding in both panels is: Iceland branch (blue), Faroe branch (green), Shetland branch (red), and total Atlantic inflow (black)

North Atlantic wind stress curl at various latitudes and with various lags. They found a maximum correlation coefficient of 0.88 for 55° N and 15 months lag between wind and volume flux. Such a procedure of correlating a variable against several other variables and picking out the maximal correlation does, however, reduce the (already small) number of degrees of freedom and hence the statistical significance of their result. (Sandø and Furevik submitted) were able to partly reproduce these results in an isopycnic coordinate ocean model for the period (1995–2002) considered by (Orvik and Skagseth 2003) but the correlation vanished for the pentad prior to this period.

The directly observed volume fluxes across the Ridge, reported here (Fig. 1.11), are rather short for a comparison to the wind, but for the Iceland branch (Astthorsson et al. submitted) have related the volume flux of Atlantic Water to the wind at Thverfjall in northwest Iceland (Fig. 1.17), indicating that northerly winds reduce the flow of Atlantic water whereas southerly winds increase the flow. This is in accordance with the strong correlation between the spring temperature at Siglunes and the pressure gradient across the Denmark Strait found by Blindheim and Malmberg (2005). This was also suggested by Ólafsson (1999) who reported a significant relationship between hydrographic conditions in spring at the Siglunes transect north of Iceland and the frequency of local northerly/southerly wind directions while he found no correlation with the NAO index.

Thus, there is considerable evidence that variations in the wind stress induce variations in the Atlantic inflow, both as regards the total and individual branches, most clearly seen in the Iceland branch. As noted, however, the variable component of the Atlantic inflow is small compared to the average, whereas the wind stress varies considerably, as illustrated by the seasonal variation. Thus, for the total Atlantic volume flux, the ratio of the seasonal amplitude to its average value is 5% according to the observations

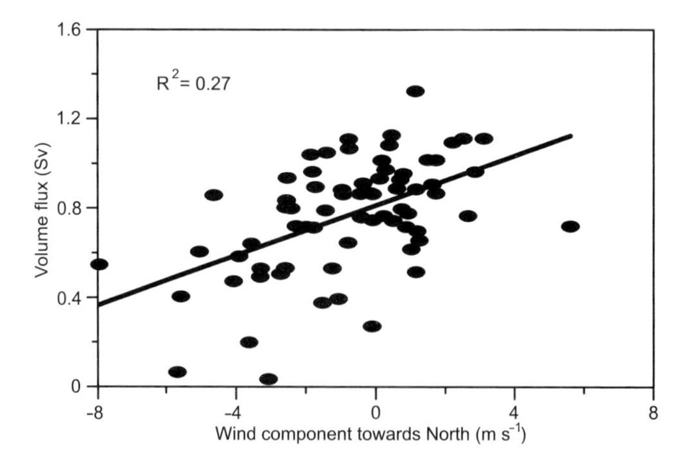

Fig. 1.17 Relationship ($p < 0.01$) between monthly flux of Atlantic water through Denmark Strait and the monthly north–south component of the wind at Thverfjall (Fig. 1.4a), northwest Iceland, for the period 1994–2001. The squared correlation coefficient is indicated

and 14% according to the model. For the wind stress curl averaged over the Nordic Seas, in contrast, this ratio is close to 100% (Jakobsen et al. 2003).

The relative stability of the Atlantic inflow remains also on much smaller timescales than the seasonal. (Hansen et al. 2003) calculated daily Atlantic water volume flux values in the Faroe branch from summer 1997 to summer 2001 and found not a single flow reversal (westward flux) among the 1,348 daily flux estimates. This can be contrasted to the inflow to the Barents Sea, which is much more variable and generally considered to be generated by the wind (Ingvaldsen et al. 2002).

It therefore seems doubtful that wind stress can be the main driving force for the dominant stable component of the Atlantic inflow. This is especially the case for the two main inflow branches but, even for the Iceland branch, wind seems mainly to increase or reduce the volume flux from a basic flow, which is there with no wind (Fig. 1.17) in analogy to the inflow through the Bering Strait (Coachman and Aagaard 1988).

1.8.2 Sea-Level Forcing

In the equations of motion, a water parcel close to the surface is acted on by a force that is proportional to the slope of the surface. Any process that generates a persistent sea-level slope across the Ridge can therefore drive an inflow and two processes within the Arctic Mediterranean can do this (Fig. 1.15). One is the estuarine mechanism (Stigebrandt 2000), which generates the surface outflow. The other is thermohaline ventilation, which generates the overflow. In an alternative terminology, these two processes have been termed positive and negative thermohaline circulation, respectively (e.g. Hopkins 2001).

The outflows, generated by these processes, must be balanced by inflows and the balance has to be maintained on fairly short timescales. Imagine an outflow of ~10 Sv without any inflow. The average sea-level of the Arctic Mediterranean would then sink by ~5 cm a day. This would rapidly establish a sea-level slope across the Ridge. To estimate, how large a sea-level drop is required to drive the observed Atlantic inflow, assume zero initial speed and inviscid flow. This leads to the Bernoulli equation:

$$V^2 = g \cdot \Delta h \tag{1.2}$$

which links the inflow speed V to the sea-level drop Δh across the Ridge. From the observations (Figs. 1.4–1.6), the typical inflow speeds do not exceed 30 cm s^{-1}, which implies a sea-level drop of less than 5 mm. This value is found by ignoring friction, but still, it is so small compared to typical sea-level variations, that this mechanism might seem irrelevant. It remains an inescapable fact, however, that, as long as there is a continuous outflow, this mechanism will turn into effect, if no other mechanism forces an inflow and that it can drive the observed Atlantic inflow with a sea-level drop that is below our observational accuracy.

On the other hand, it is clear, that this mechanism only turns into effect, if no other force maintains an inflow that balances the outflow. The question therefore is, whether there is any evidence for or against this mechanism as an important driver for the Atlantic inflow. An obvious argument against it, is the large variability of the sea-level in the inflow region. From altimetry, the standard deviation of the sea-level is an order of magnitude larger than the 5 mm that are required to drive the inflow.

To investigate this in more detail, a point was chosen downstream of the Ridge. Its location was selected so that it should feel both of the main inflow branches and it was located over the continental slope to keep it relatively unaffected by meandering and eddying. Sea-level height at this point (point A in Fig. 1.18) was then correlated to sea-level height over a wide region (Fig. 1.18a). As could be expected, low correlations were found for the central basins and the Faroe–Shetland Channel, where internal circulation and eddies may dominate, but equation (1.2) is only required to apply when following streamlines and all the upstream inflow region due west of the Ridge was highly correlated to point A. A linear regression analysis, similarly, gave regression coefficients close to 1 (Fig. 1.18b) in this region. This analysis indicates that the typical sea-level drop, as the Atlantic inflow crosses the Ridge, is not as variable as might be expected from a first glance at the altimetry, and it supports the application of equation (1.2).

The next question is, whether sea-level forcing can reproduce established key features of the Atlantic inflow. The discussion above verifies that a sea-level drop of 5 mm across the Ridge should be sufficient to drive the observed total volume flux, but how stable is it? By itself, the pressure gradient generated by a sea-level drop of 5 mm would not seem to be very stable, because an excess inflow of1 Sv would eliminate the sea-level drop in a day. The stability of sea-level forcing, therefore, rests on the stability of the outflows that generate the sea-level drop across the Ridge. As regards the surface outflows, there is little observational evidence on this and, since they are near-surface, variations in wind stress are likely to affect them considerably. Most of the outflow is, however, in the form of

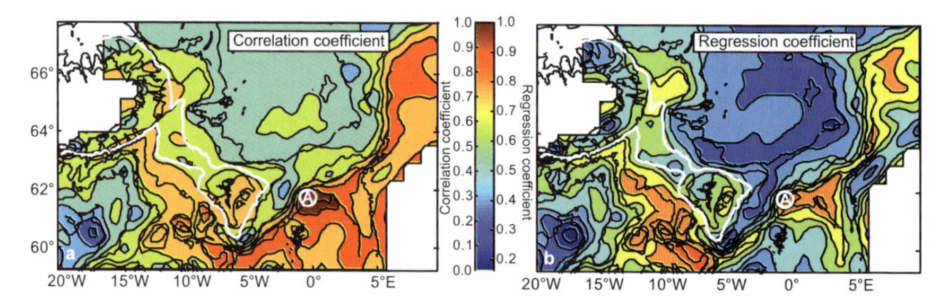

Fig. 1.18 (a) Correlation coefficient between sea-level height at the point A and at other points in the area. (b) Linear regression coefficient (slope) between sea-level height at the point A (y-coordinate) and at other points in the area (x-coordinate). The 500 m depth contour around Iceland and Faroes is enhanced in white to illustrate the Ridge. Based on weekly fields from "The Mapped Sea Level Anomaly (MSLA)" data, produced by the CLS Space Oceanography Division (www.jason.oceanobs.com)

overflow (Fig. 1.14) and, although the overflow has variations, it has been demonstrated to be very persistent (Østerhus et al. 2001; Dickson and Brown 1994).

The overflow stability may be seen in terms of forcing. By generating a barotropic pressure gradient, wind stress can modulate the overflow (Biastoch et al. 2003), but, in addition, there is a pressure gradient at the depth of the overflow, which is generated by the accumulation of dense water in the Arctic Mediterranean (Hansen et al. 2001). This baroclinic pressure gradient is quite clearly responsible for accelerating the main overflow branches to the high speeds (1 m s^{-1}) that are observed.

This point was amply demonstrated by (Biastoch et al. 2003), who, in an idealized model experiment, changed the wind forcing from zero to four times the average observed. With increasing wind forcing, they found a shift in the overflow from east of Iceland to the Denmark Strait, but an essentially constant sum of both parts. They concluded that the total overflow can be changed only by altering the density contrast across the Ridge.

The link between overflow and Atlantic inflow may be illustrated by a simple model (Fig. 1.19). The baroclinic pressure gradient driving the overflow is maintained by the large reservoir of dense water in the Arctic Mediterranean. Even without any renewal, the amount of dense water north of the Ridge is sufficient to maintain an overflow for decades (Hansen and Østerhus 2000), which explains the overflow stability. But, a continuous overflow of 6 Sv will tend to depress the average sea-level of the Arctic Mediterranean by several centimetres each day, which is much more than required to drive the Atlantic inflow across the Ridge. The sea-level forcing will therefore rapidly adjust the volume flux of the Atlantic inflow towards balance.

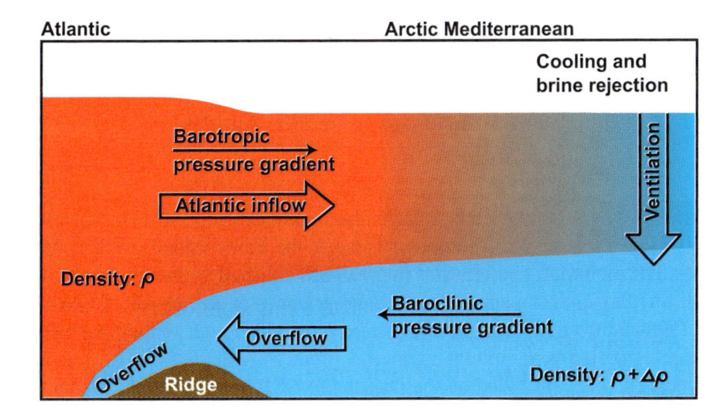

Fig. 1.19 A simple model of the overflow forcing of the Atlantic inflow. Cooling and brine rejection in the Arctic Mediterranean convert the incoming Atlantic water (red) into denser overflow water (blue), which accumulates at depth. The density contrast and sloping isopycnals generate a baroclinic pressure gradient that accelerates overflow water towards and across the Ridge. This removal of water from the Arctic Mediterranean induces a sea-level drop across the Ridge, which reduces the total pressure gradient acting on the overflow slightly (around 10% under present-day conditions) and drives an Atlantic inflow equal to the overflow for stationary conditions. Note that the vertical scales for sea-level slope and isopycnal slope are quite different

The system illustrated in Fig. 1.19 cannot explain all the Atlantic inflow. The surface outflow, associated with the estuarine circulation of the Arctic Mediterranean, and wind stress also contribute, but the fact that most of the Atlantic water returns as overflow (Fig. 1.15) is a clear indication that this is the dominant forcing mechanism and it can explain the relative stability of the Atlantic inflow.

1.9 Conclusions and Outlook

During the last decade, observations and modelling efforts have converged into a consistent description of the properties and intensity of the Atlantic inflow across the Greenland–Scotland Ridge. We know the average temperatures and salinities of the individual branches and have learned to link their decadal variations to the intensity of the Subpolar Gyre. In a series of projects, starting with Nordic WOCE, through VEINS and MAIA, to ASOF-MOEN, we have, for the first time, been able to measure the volume fluxes of all the branches with a relatively high accuracy and the measured average fluxes compare well with those calculated by the ASOF-MOEN model. When considering more rapid variations, the model indicates somewhat larger seasonal flux amplitudes than the observations, but not outside the combined observational and modelling uncertainties. These results highlight the pronounced stability of the Atlantic inflow across the Ridge and indicate that direct wind stress forcing is not likely to be the main driving force for the inflow, although it probably accounts for much of the inflow variability on timescales below a decade.

In the ASOF-MOEN project and its predecessors, an observational system has been established, which allows us to monitor the properties and intensities of all the inflow branches. This system would benefit from additional instrumentation but it can form the backbone of a dedicated monitoring system for the Atlantic inflow. In the coming decades, climate change is going to affect the ocean more and more and such a system will be essential if we are to be able rapidly to identify and quantify any changes to the Atlantic inflow.

Acknowledgements Through the years, a huge observational effort has been spent by the staff on the marine and fisheries research institutes of Iceland, the Faroe Islands, and Scotland and their research vessels. The monitoring system was originally established within the Nordic WOCE project with support from the Environmental Research Programme of the Nordic Council of Ministers (NMR) 1993–1998 and from national Nordic research councils. Later, support has been gained from European research funding programmes in several projects: VEINS, MAIA, and ASOF-MOEN. Continued support has been provided by the Danish DANCEA programme. The authors also wish to thank Beatriz Balino for her dedicated and efficient management of the ASOF-MOEN project.

References

Aagaard K, Carmack EC (1989) The role of sea ice and other fresh water in the Arctic circulation. J. Geophys. Res. 94 (C10): 14,485–14,498.

Astthorsson OS, Gislason A, Jonsson S (Submitted) Climate variability and the Icelandic marine ecosystem. Submitted to Deep-Sea Res. II.

Belkin IM, Levitus S, Antonov J, Malmberg SA (1998) "Great Salinity Anomalies" in the North Atlantic. Prog. Oceanogr. 41: 1–68.

Bengtson L, Semenov VA, Johannesen OM (2004) The early twentieth-century warming in the Arctic – a possible mechanism. J. Climate 17: 4045–4057.

Biastoch A, Käse RH, Stammer DB (2003) The sensitivity of the Greenland-Scotland Ridge overflow to forcing changes. J. Phys. Oceanogr. 33: 2307–2319.

Blindheim J, Malmberg SA (2005) The Mean Sea Level Pressure Gradient Across the Denmark Strait as an Indicator of Conditions in the North Icelandic Irminger Current. In: Drange H et al. (eds.) The Nordic Seas: An Integrated Perspective. AGU Monograph 158, American Geophysical Union, Washington, DC, pp 65–71.

Blindheim J, Østerhus S (2005) The Nordic Seas, Main Oceanographic Features. In: Drange H et al. (eds.) The Nordic Seas: An Integrated Perspective. AGU Monograph 158, American Geophysical Union, Washington DC, pp 11–37.

Coachman LK, Aagaard K (1988) Transports through Bering Strait: Annual and interannual variability. J. Geophys. Res. 93: 15,535–15,539.

Dickson RR, Brown J (1994) The production of North Atlantic deep water, sources, rates, and pathway. J. Geophys. Res. 99: 12,319–12,341.

Dickson RR, Meincke J, Malmberg SA, Lee AJ (1988) The "Great Salinity Anomaly" in the Northern North Atlantic 1968–1982. Prog. Oceanogr. 20: 103–151.

Drange H, Gerdes R, Gao Y, Karcher M, Kauker F, Bentsen M (2005) Ocean General Circulation Modelling of the Nordic Seas. In: Drange H et al. (eds.) The Nordic Seas: An Integrated Perspective. AGU Monograph 158, American Geophysical Union, Washington DC, pp 199–220.

Ellett DJ, Dooley HD, Hill HW (1979) Is there a North-east Atlantic slope current?. ICES CM 1979/C:35: 11pp.

Ellett DJ, Edwards A, Bowers R (1986) The hydrography of the Rockall Channel – an overview. Proc. Royal Soc. 88B: 61–81.

Furevik T, Nilsen JEØ (2005) Large-Scale Atmospheric Circulation Variability and its Impacts on the Nordic Seas Ocean Climate – A Review. In: Drange H et al. (eds.) The Nordic Seas: An Integrated Perspective. AGU Monograph 158, American Geophysical Union, Washington DC, pp 105–136.

Hansen B, Turrell WR, Østerhus S (2001) Decreasing overflow from the Nordic seas into the Atlantic Ocean through the Faroe Bank channel since 1950. Nature 411: 927–930.

Hansen B, Østerhus S (2000) North Atlantic – Nordic Seas exchanges. Prog. Oceanogr. 45: 109–208.

Hansen B, Østerhus S, Hátún H, Kristiansen R, Larsen KMH (2003) The Iceland-Faroe inflow of Atlantic water to the Nordic Seas. Prog. Oceanogr. 59: 443–474.

Harvey J (1982) Θ-S Relationships and water masses in the eastern North Atlantic. Deep-Sea Res. 29 (8A): 1021–1033.

Hátún H (2004) The Faroe Current, Dr. Scient. Thesis, Reports in Meteorology and Oceanography, University of Bergen, Bergen.

Hátún H, Sandø AB, Drange H, Hansen B, Valdimarsson H (2005) Influence of the Atlantic Subpolar Gyre on the thermohaline circulation. Science 309: 1841–1844.

Häkkinen S, Rhines P (2004) Decline of Subpolar North Atlantic circulation during the 1990s. Science 304: 555–559.

Holliday NP (2003) Air-sea interaction and circulation changes in the Northeast Atlantic. J. Geophys. Res. 108 (C8): 3259.

Hopkins TS (1991) The GIN Sea-A synthesis of its physical oceanography and literature review 1972–1985. Earth-Sci Rev. 30: 175–318.

Hopkins TS (2001) Thermohaline feedback loops and Natural Capital. Sci. Mar. 65 (Suppl. 2): 231–256.

Hughes SL, Turrell WR, Hansen B, Østerhus S (2006) Fluxes of Atlantic water (volume, heat and salt) in the Faroe-Shetland Channel calculated from a decade of acoustic doppler current profiler data (1994–2005). Fisheries Research Services Collaborative Report No 01/06, Aberdeen.

Ingvaldsen R, Loeng H, Asplin L (2002) Variability in the Atlantic inflow to the Barents Sea based on a one-year time series from moored current meters. Cont. Shelf Res. 22: 505–519.

Jakobsen PK, Ribergaard MH, Quadfasel D, Schmith T, Hughes CW (2003) Near-surface circulation in the northern North Atlantic as inferred from Lagrangian drifters: Variability from the mesoscale to interannual. J. Geophys. Res. 108(C8): 3251, doi:10.1029/2002JC001554.
Johannesen OM (1986) Brief Overview of the Physical Oceanography. In: Hurdle BG (ed.) The Nordic Seas. . Springer, New York, pp 103–127.
Jones PD, Jonsson T, Wheeler D (1997) Extension to the North Atlantic Oscillation using early instrumental pressure observations from Gibraltar and South-West Iceland. Int. J. Climatol. 17: 1433–1450.
Jónsson S (1992) Sources of fresh water in the Iceland Sea and the mechanisms governing its interannual variability. ICES Mar. Sci. Symp. 196: 62–67.
Jónsson S, Briem J (2003) Flow of Atlantic water west of Iceland and onto the north Icelandic shelf. ICES Mar. Sci. Symp. 219: 326–328.
Jónsson S, Valdimarsson H (2005) The flow of Atlantic water to the North Icelandic shelf and its relation to the drift of cod larvae. ICES J. Mar. Sci. 62: 1350–1359.
Karcher MJ, Gerdes R, Kauker F, Köberle C (2003) Arctic warming: Evolution and spreading of the 1990s warm event in the Nordic Seas and the Arctic Ocean. J. Geophys. Res. 108(C2): 3034, doi:10.1029/2001JC001265.
Kistler R, Kalnay E, Collins W, Saha S, White G, Woollen J, Chelliah M, Ebisuzaki W, Kanamitsu M, Kousky V, van den Dool H, Jenne R, Fiorino M (2001) The NCEP-NCAR 50-year reanalysis: Monthly means CD-ROM and documentation. Bull. Am. Meteorol. Soc. 82: 247–268.
Kristmannsson SS (1998) Flow of Atlantic Water into the northern Icelandic shelf area 1985–1989. ICES Coop. Res. Rep. 225: 124–135.
Marsland SJ, Haak H, Jungclaus JH, Latif M, Röske F (2003) The Max-Planck-Institute global ocean/sea ice model with orthogonal curvilinear coordinates. Ocean Modeling 5: 91–127
Mauritzen C (1996) Production of dense overflow waters feeding the North Atlantic across the Greenland-Scotland Ridge. Deep-Sea Res. 43 (6): 769–835.
McCartney MS, Mauritzen C (2001) On the origin of the warm inflow to the Nordic Seas. Prog. Oceanogr. 51: 125–214.
Mork KA, Blindheim J (2000) Variations in the Atlantic inflow to the Nordic Sea 1955–1999. Deep-Sea Res. I 47: 1035–1057.
New A, Barnard S, Herrmann P, Molines JM (2001) On the origin and pathway of the saline inflow to the Nordic Seas: Insights from models. Prog. Oceanogr. 48: 255–287.
Nilsen JEØ, Gao Y, Drange H, Furevik T, Bentsen M (2003) Simulated North Atlantic-Nordic Seas water mass exchanges in an isopycnic coordinate OGCM. Geophys. Res. Lett. 30 (10), doi:10.1029/2002GL016597.
Ólafsson J (1999) Connections between oceanic conditions off N-Iceland, Lake Myvatn temperature, regional wind direction variability and the North Atlantic Oscillation. Rit Fiskideildar 16: 41–57.
Olsen SM, Schmith T (2007) North Atlantic–Arctic Mediterranean exchanges in an ensemble hindcast experiment. J. Geophys. Res. 112: C04010, doi:10.1029/2006JC003838.
Orvik KA, Skagseth Ø, Mork M (2001) Atlantic inflow to the Nordic Seas: Current structure and volume fluxes from moored current meters, VM-ADCP and SeaSoar-CTD observations 1995–1999. Deep-Sea Res. 48: 937–957.
Orvik KA, Skagseth Ø (2003) The impact of the wind stress curl in the North Atlantic on the Atlantic inflow to the Norwegian Sea toward the Arctic. Geophys. Res. Lett. 30 (17), doi:10.0129/2003GL017932
Østerhus S, Turrell WR, Hansen B, Lundberg P, Buch E (2001) Observed transport estimates between the North Atlantic and the Arctic Mediterranean in the Iceland-Scotland region. Polar Res. 20(1): 169–175.
Østerhus S, Turrell WR, Jónsson S, Hansen B (2005) Measured volume, heat, and salt fluxes from the Atlantic to the Arctic Mediterranean. Geophys. Res. Lett. 32: L07603.
Pollard RT, Griffiths MJ, Cunningham SA, Read JF, Pérez FF, Rios AF (1996) Vivaldi 1991 - A study of the formation, circulation and ventilation of Eastern North Atlantic Central Water. Prog. Oceanogr. 37: 167–192.

Read JF (2001) CONVEX-91: Water masses and circulation of the northeast Atlantic subpolar gyre. Prog. Oceanogr. 48: 461–510.
Reid JL (1979) On the contribution of the Mediterranean Sea outflow to the Norwegian-Greenland Sea. Deep-Sea Res. 26: 1199–1223.
Roach AT, Aagaard K, Pease CH, Salo SA, Weingartner T, Pavlov V, Kulakov M (1995) Direct measurements of transport and water properties through the Bering Strait. J. Geophys. Res. 100 (C9): 18443–18457.
Rudels B (1989) The formation of Polar Surface Water, the ice export and the exchanges through the Fram Strait. Prog. Oceanogr. 22: 205–248.
Sandø AB, Furevik T (Submitted) The relation between the wind stress curl in the North Atlantic and the Atlantic inflow to the Nordic Seas. Submitted to J. Geophys. Res.
Seager R, Battisti DS, Yin J, Gordon N, Naik N, Clement AC, Cane MA (2002) Is the Gulf Stream responsible for Europe's mild winters? Q. J. R. Meteorol. Soc. 128: 1–24.
Sherwin TJ, Williams MO, Turrell WR, Hughes SL, Miller PI (2006) A description and analysis of mesoscale variability in the Faroe-Shetland Channel. J. Geophys. Res. 111: C03003.
Stigebrandt A (2000) Oceanic Freshwater Fluxes in the Climate System. In: Lewis EL et al. (eds.) The Freshwater Budget of the Arctic Ocean. NATO Science series, Kluwer, Dordrecht, The Netherlands, pp 1–20.
Swallow JC, Gould WJ, Saunders PM (1977) Evidence for a poleward eastern boundary current in the North Atlantic Ocean. ICES CM 1977/C:32: 11pp.
Swift JH, Aagaard K (1981) Seasonal transitions and water mass formation in the Iceland and Greenland Seas. Deep-Sea Res. 28A: 1107–1129.
Turrell B (1995). A century of hydrographic pbservations in the Faroe-Shetland Channel. Ocean Challenge 6 (1): 58–63.
Turrell WR, Hansen B, Hughes S, Østerhus S (2003) Hydrographic variability during the decade of the 1990s in the Northeast Atlantic and southern Norwegian Sea. ICES Mar. Sci. Symp. 219: 111–120.
Visbeck M, Chassignet E, Curry R, Delworth T, Dickson B, Krahmann G (2003) The Ocean's Response to North Atlantic Oscillation Variability. In: Hurrell J et al. (eds.) The North Atlantic Oscillation: Climatic Significance and Environmental Impact. American Geophysical Union Geophysical Monograph, pp 113–145, doi:10.1029/134GM06.
Wilkenskjeld S, Quadfasel D (2005) Response of the Greenland-Scotland overflow to changing deep water supply from the Arctic Mediterranean. Geophys. Res. Lett. 32: L21607.
Worthington LV (1970) The Norwegian Sea as a mediterranean basin. Deep-Sea Res. 17: 77–84.
Zhang J, Steele M, Rothrock DA, Lindsay RW (2004) Increasing exchanges at Greenland-Scotland Ridge and their links with the North Atlantic Oscillation and Arctic sea ice. Geophys. Res. Lett. 31: L09307, doi:10.1029/2003GL019304.

Chapter 2
Volume and Heat Transports to the Arctic Ocean Via the Norwegian and Barents Seas

Øystein Skagseth[1,3], Tore Furevik[2,3], Randi Ingvaldsen[1,3], Harald Loeng[1,3], Kjell Arne Mork[1,3], Kjell Arild Orvik[2], and Vladimir Ozhigin[4]

2.1 Introduction

The first comprehensive description of physical conditions in the Norwegian – and the Barents Seas was provided by Helland-Hansen and Nansen (1909), who described both the two areas individually and the relationships between them. They indicated a 2-year delay in the temperature signal from Sognesjøen (west coast of Norway at about 61° N) to the Russian Kola section, and suggested that this time lag could be used to predict temperature conditions in the Barents Sea on the basis of upstream observations. Helland-Hansen and Nansen also pointed out that variations in physical conditions had great influence on the biological conditions of various fish species, and that ocean temperature variations "are the primary cause of the great and hitherto unaccountable fluctuations in the fisheries". The importance of climate impact on marine organisms at high latitudes has recently been well documented in the Arctic Climate Impact Assessment report (ACIA 2005).

The Norwegian Sea, the Greenland Sea and the Iceland Sea comprise the Nordic Seas, which are separated from the rest of the North Atlantic by the Greenland–Scotland Ridge (Fig. 2.1). The Norwegian Sea consists of two deep basins, the Norwegian Basin and the Lofoten Basin, and is separated from the Greenland Sea to the north by the Mohn Ridge. To the west, the basin slope forms the transition to the somewhat shallower Iceland Sea. The upper ocean of the Nordic Seas consists of warm and saline Atlantic water to the east, and cold and fresh Polar water from the Arctic to the west. The Barents Sea, with an average depth of 230 m, is one of the shallow shelf seas that constitute the Arctic continental shelf. Its boundaries are defined by Norway and Russia in the south, Novaya Zemlya in the east, and the continental shelf breaks

[1]Institute of Marine Research, Norway

[2]Geophysical Institute, University of Bergen, Norway

[3]Bjerknes Centre for Climate Research, Norway

[4]Knipovich Polar Research Institute of Marine Fisheries and Oceanography (PINRO), Murmansk, Russia

R.R. Dickson et al. (eds.), *Arctic–Subarctic Ocean Fluxes*, 45–64

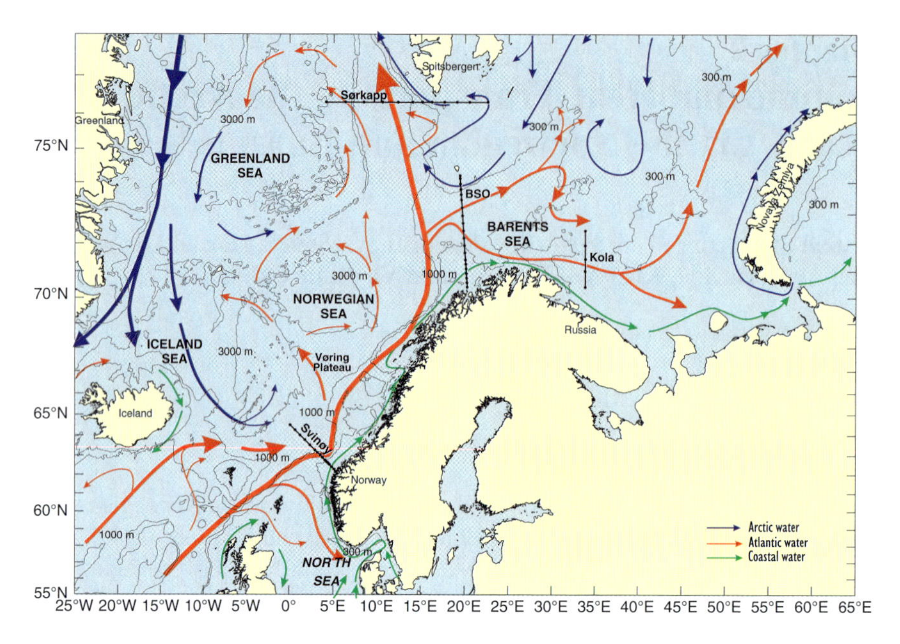

Fig. 2.1 Schematic map of the study area showing the major upper ocean currents and the repeated hydrographic sections used in this chapter; the Svinøy Section, the Barents Sea Opening, the Kola section and the Sørkapp section

towards the Norwegian and Greenland Seas and the Svalbard Acrhipelago in the west and northwest and the Arctic Ocean in the north (Fig. 2.1).

The Norwegian and Barents seas are transition zones for warm and saline waters on their way from the Atlantic to the Arctic Ocean. The major current, the Norwegian Atlantic Current (NwAC), is a poleward extension of the Gulf Stream and the North Atlantic Current, that acts as a conduit for warm and saline Atlantic Water from the North Atlantic to the Barents Sea and Arctic Ocean (Polyakov et al. 2005). As Fig. 2.1 shows, the North Atlantic Current splits into two branches in the eastern North Atlantic before entering the Norwegian Sea over the Iceland–Faeroe Ridge close to the eastern coast of Iceland, and through the Faeroe–Shetland Channel close to Shetland (Fratantoni 2001; Orvik and Niiler 2002). The water then continues in two branches through the entire Norwegian Sea toward the Arctic Ocean (Poulain et al. 1996; Orvik and Niiler 2002). The western branch is a jet associated with the Arctic Front. It tends to feed the interior of the Norwegian Sea via several recirculation branches. The eastern branch, known as the Norwegian Atlantic Slope Current (NwASC), is an approximately 3,500 km long, nearly barotropic shelf edge current flowing along the Norwegian shelf break, that tends to flow into the Barents Sea and Arctic Ocean. The NwASC is thus the major link between the North Atlantic, and the Barents Sea and Arctic Ocean.

In the Barents Sea, the relatively warm Coastal and Atlantic waters that enter between Bear Island and northern Norway, hereafter called the Barents Sea Opening, dominate the southern regions. As they transit the Barents Sea, the Atlantic water masses are modified through mixing, atmospheric cooling, net precipitation, ice freezing and melting, before exiting primarily to the north of Novaya Zemlya (Loeng et al. 1993). This transformation is important for the ventilation of the Arctic Ocean (Schauer et al. 2002; Rudels et al. 2004). The Norwegian Coastal Current mixes with river water to form low-salinity shelf waters (Rudels et al. 2004). Atlantic water has a typical temperature range between 4.5 °C and 6.5 °C but varies seasonally and interannually (Midttun and Loeng 1987). Arctic waters ($T < 0\,°C$, $34.3 < S < 34.7$) dominate the northern Barents Sea, entering between Franz Josef Land and Novaya Zemlya and to a lesser degree between Franz Josef Land and Spitzbergen.

Variations in the properties and volume transport of Atlantic water have a major impact on the oceanographic conditions of the Barents Sea over a broad range of timescales (Loeng et al. 1992), and both in the Barents and the Norwegian Seas large-scale atmospheric circulation changes influence the currents and hydrographic conditions. Since the 1960s, changes in the large-scale wind pattern, principally the North Atlantic Oscillation (NAO), have resulted in a gradual change of the water mass distribution in the Nordic Seas. In particular, this is manifested by the development of a layer of Arctic intermediate waters, deriving from the Greenland and Iceland Seas and spreading over the entire Norwegian Sea (Blindheim et al. 2000). In the Norwegian Basin it has resulted in an eastward shift of the Arctic front and, accordingly, an upper layer cooling in wide areas due to increased Arctic influence. Blindheim et al. (2000) also found that the westward extent of Atlantic water in the Norwegian Sea was less during the high phase of the North Atlantic Oscillation than during the low phase, with the difference between its broadest recorded extent in 1968 and its narrowest extent in 1993 exceeding 300 km. This implies that a stronger cyclonic atmospheric circulation pattern would move the surface waters to the east. This would decrease the area of Atlantic water and thus reduce ocean-to-air heat losses, and could contribute to a warmer Atlantic inflow to the Barents Sea in positive NAO years. In the Barents Sea, higher temperatures are found during positive phases of the NAO index (Dickson et al. 2000). The fluctuations in the strength of the inflow, as measured at the western entrance between northern Norway and Bear Island, depend mainly on the atmospheric circulation (Ingvaldsen et al. 2004a, b).

The present paper offers an overview of the transport of Atlantic water and its properties along the Norwegian Coast and into the Barents Sea. Section 2.2 presents the mean state of currents and hydrography in the Norwegian Sea and in the Barents Sea Opening, followed by an overview of variability at various scales in Section 2.3. Suggested forcing mechanisms for the variability are discussed in Section 2.4 before the paper is summarized and concluded in Section 2.5.

2.2 The Mean State

2.2.1 The Mean Hydrography and Current Structure in the Svinøy Section

The Svinøy section runs northwestward from the Norwegian coast at 62° N and cuts through the entire Atlantic inflow to the Norwegian Sea just to the north of the Iceland–Scotland Ridge. It is thus a key location for comprehensive monitoring of the Atlantic inflow to be used as an upstream reference for the Barents Sea and the Arctic Ocean. Monitoring of the Svinøy section started in the mid-1950s with repeated hydrographic sections, and current measurements commenced in 1995.

We define the Atlantic inflow in the Svinøy section to be water with salinity above 35.0 (Fig. 2.2). This corresponds to a temperature of about 5 °C, and is the definition used by Helland-Hansen and Nansen (1909). By using high-resolution SeaSoar-CTD methodology, the hydrographic field reveals a nearly slab-like extension of warm saline Atlantic water (Orvik et al. 2001). The slab extends about 250 km northwestwards from the shelf break where the interface outcrops the surface and forms a sharp front (the Arctic front) between the Atlantic and Arctic waters. Toward the coast it leans on the shelf slope above the 600 m isobath. This is in contrast to the historical view of the Atlantic water as a wide wedge-shape westward extension. In summer a surface layer of fresh coastal water can be observed in the section. In summers with stronger northerly winds the coastal water tends to extend further westward in the Norwegian Sea than in summers with no or weaker northerly winds (Nilsen and Falck 2006). During the winter this layer disappears as it mixes with the Atlantic water. Arctic intermediate water, situated between the Atlantic and deep-water masses, can also be observed in Fig. 2.2 as a water mass with salinity below 34.9.

The slab-like average hydrographic feature mirrors the baroclinic flow as a frontal jet in accordance with the western branch of the NwAC. By using Vessel Mounted-ADCP transects the western branch of the NwAC has been identified as an unstable and meandering jet in the Arctic Front. In average, the jet is about 400 m deep and 30–50 km wide, located above the 2,000 m isobath. Observations show a maximum speed of 60 cm s^{-1} in the core at a depth of about 100 m.

Over the shelf-slope our moored array has captured the eastern branch of the NwAC as an approximately 30–50 km wide nearly barotropic current, trapped along the topography between 200 and 800 m depth. The annual mean appears as a stable flow, 40 km wide and with a mean velocity of about 30 cm s^{-1} (Fig. 2.2). Accordingly, the volume flux of the slope current can be estimated based on one single current meter in the core of the flow (Orvik and Skagseth 2003a, b), resulting in an average of 4.3 Sv (1 Sv = 10^6 m^3s^{-1}) for the period 1995–2006.

2.2.2 *Mean Structure of the Hydrography and Currents in the Barents Sea Opening*

In the Barents Sea Opening repeated hydrographic sampling has been performed since the mid-1960s, and current measurements since 1997. Atlantic water, defined as water with salinity above 35.0 and temperature above 3 °C,

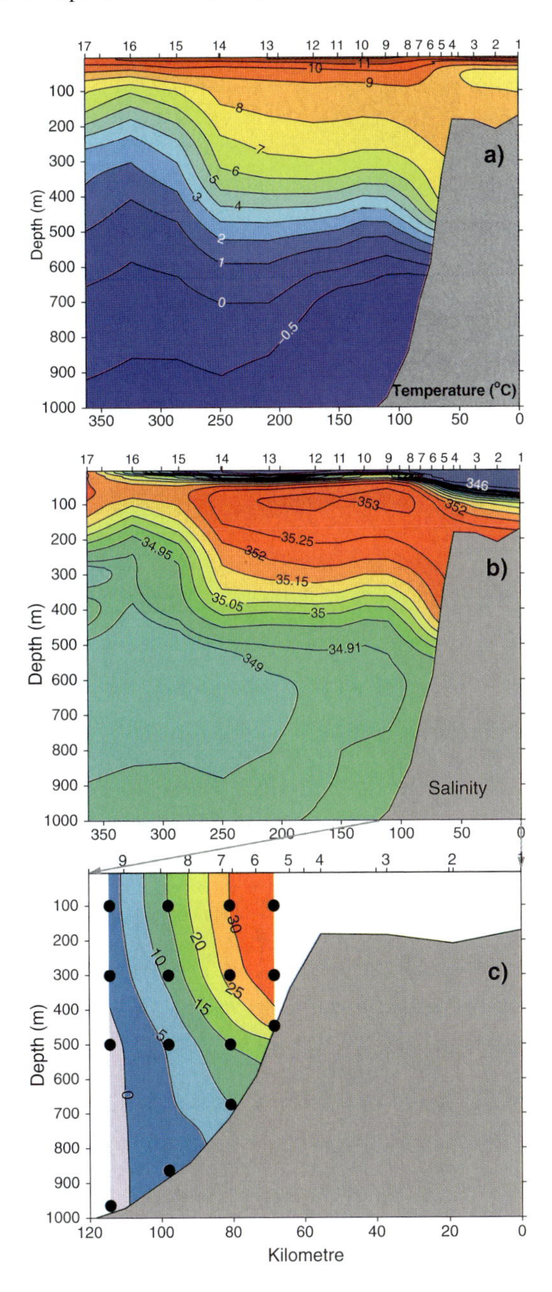

Fig. 2.2 Temperature (upper) and salinity (middle) in July 1998, and mean velocity (lower) in the Svinøy section

occupies most of the section (Fig. 2.3). Above the Atlantic water there is a surface layer of warmer and fresher water. During the winter the surface layer breaks down and Atlantic water extends to the surface. Between the Atlantic

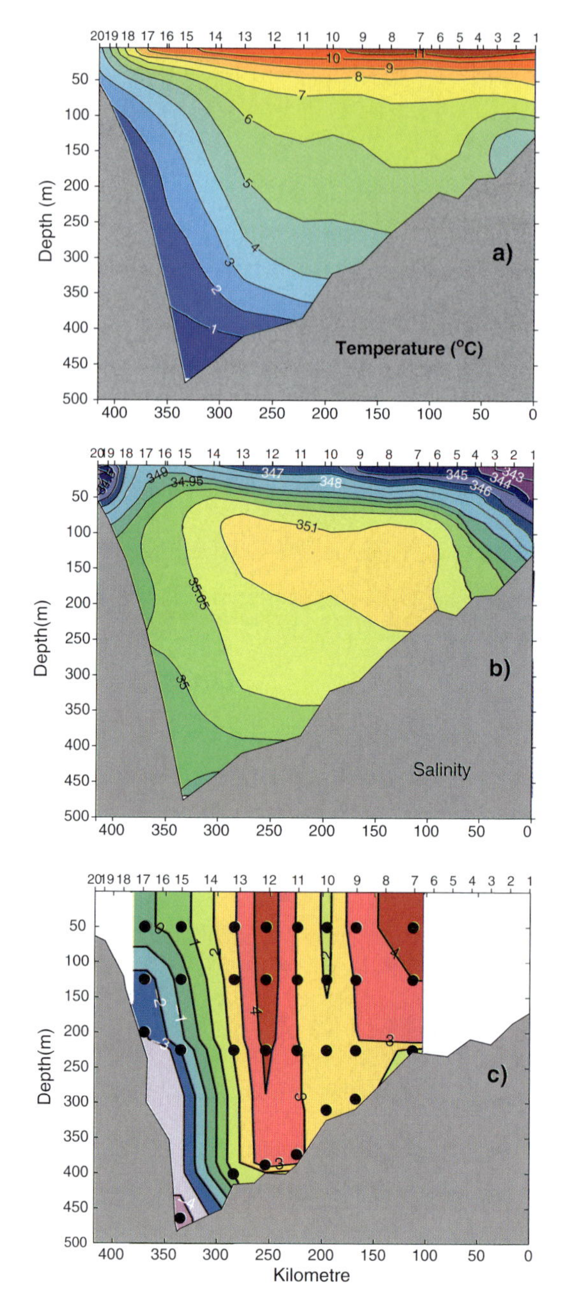

Fig. 2.3 Temperature (upper) and salinity (middle) in August 1998, and mean velocity (lower) in the Barents Sea Opening

inflow and the Norwegian Coast, the Norwegian Coastal Current flows northeastwards into the Barents Sea. The temperature of the Coastal water is about the same as in the Atlantic water, but the salinities are lower (S < 34.7). During winter the Norwegian Coastal Current is deep and narrow, while during summer it is wide and shallow and spreads northwards as a wedge overlying the more saline Atlantic water (Sætre and Ljøen 1971). The northward extent of the upper layer is subject to large inter-annual variations, but in years with favourable wind directions it may reach the middle of the section (Olsen et al. 2003). The shelf slope south of Bear Island is occupied by a mixture of Arctic and modified Atlantic water masses.

Compared to the mean velocities in the Svinøy section of about 30 cm s^{-1} (Fig. 2.2) the mean currents in the Barents Sea Opening are weak (Fig. 2.3). The Atlantic water entering the Barents Sea have a more unstable core and the inflow can form a wide branch, centred close to 72° 30′ N, a relatively narrow inflow in the south accompanied by a wide outflow in the north, or an inflow comprising several branches with return flows or weaker inflows between them (Ingvaldsen et al. 2004a). Daily mean velocities in the Atlantic water core may reach 20 cm s^{-1}, but the long-term means are generally much weaker. In the Norwegian Coastal Current, that is, shoreward of the present mooring array, Blindheim (1989) found mean velocities of about 15 cm s^{-1} based on a 1-month measuring period. The dense current that leaves the Barents Sea in the deepest part of the Barents Sea Opening has mean velocities of 5 cm s^{-1}.

2.3 Variations

2.3.1 Long-Term Hydrographic Changes in the NwAC and the Propagation of Anomalies

The longest instrumental record of the Barents Sea climate is from the Kola section (Bochkov 1982; Tereshchenko 1997, 1999). Focusing on the multidecadal scales, the series shows substantial variations; cold at the beginning of the 20th century, a warm period in the 1930–1940s, followed by a cold period in the 1960–1970s and finally, a still ongoing warming (Fig. 2.4). In order to illustrate the spatial scale of this variation a comparison with the Atlantic Multidecadal Oscillation (AMO) index of Sutton and Hodson (2005), representing the large-scale sea surface temperature variation in the Atlantic, is shown. This record, extending back to the 1870s, shows a remarkable similarity with that of the Kola section, demonstrating that the climatic variation found in the Kola section is a local manifestation of a larger-scale climate fluctuation covering at least the entire North Atlantic Ocean.

The properties of the Atlantic water that enters the Norwegian Sea change as we move northwards along the Norwegian continental slope toward the Barents Sea and the Arctic Ocean. Both temperature and salinity are reduced due to mixing with the fresher Norwegian Coastal Current, the colder, less saline Arctic water from the west, and by net precipitation and heat loss to the atmosphere.

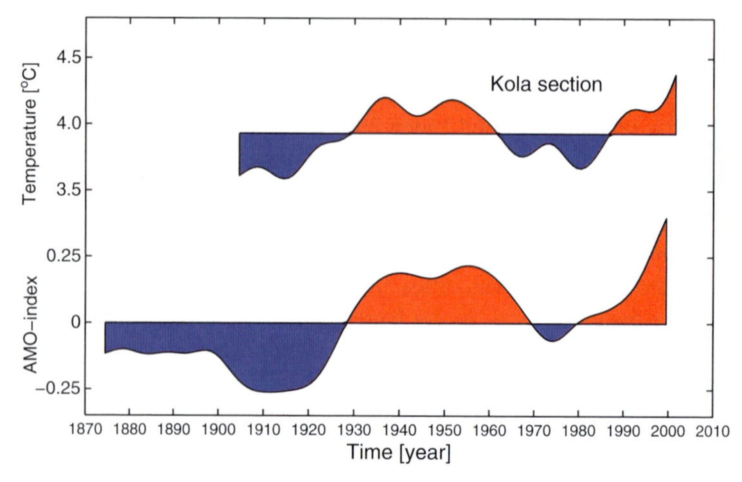

Fig. 2.4 Time series of the Kola section mean temperature (upper graph) and the Atlantic Multidecadal Oscillation (AMO) index (lower graph). The series were filtered using a two-way 14-year Hamming window. The AMO index is based on the sea surface temperature in the region 0–60° N and 7.5–75° W. The Kola section data were obtained from PINRO

In order to identify these changes, temperature and salinity variations in the core of the Atlantic water at three key sections are shown; in the Svinøy section, which represents the starting point of the northward transit, and in the Sørkapp section and the Barents Sea Opening, which represent the two major exits from the Norwegian Sea (Fig. 2.5). The temperature variation in the Barents Sea Opening has been shown to be representative of climate variability in the western Barents Sea (Ingvaldsen et al. 2003). The northward cooling of the Atlantic water is clearly seen in the temperature series, with long-term means of 7.9, 5.3 and 4 °C, respectively for the Svinøy section, Barents Sea Opening and Sørkapp section. Long-term mean salinities are 35.23 for the Svinøy section and about 35.07 for both the Barents Sea Opening and Sørkapp section. Since the late 1970s the temperature has increased in all three sections and all-time high values have been recorded in the past few years, except for a relative cooling in the Svinøy section in 2005. However, in 1960 the temperature for the Svinøy section was at similar level as during the last years. During the period with current measurements in the Svinøy section and the Barents Sea Opening there have been increases of temperature and salinity of 1.0 °C and 0.1, respectively in all three sections. The temperature time series from the current meters in the Svinøy section and Barents Sea Opening display similar trends to those from the hydrographic sections.

The largest change in salinity was during the late 1970s, when the Great Salinity Anomaly (GSA) passed through the Norwegian Sea (Dickson et al. 1988). However, there are several low temperature and salinity anomalies in the time series: in the late 1970s, late 1980s and mid-1990s. Several authors have explained

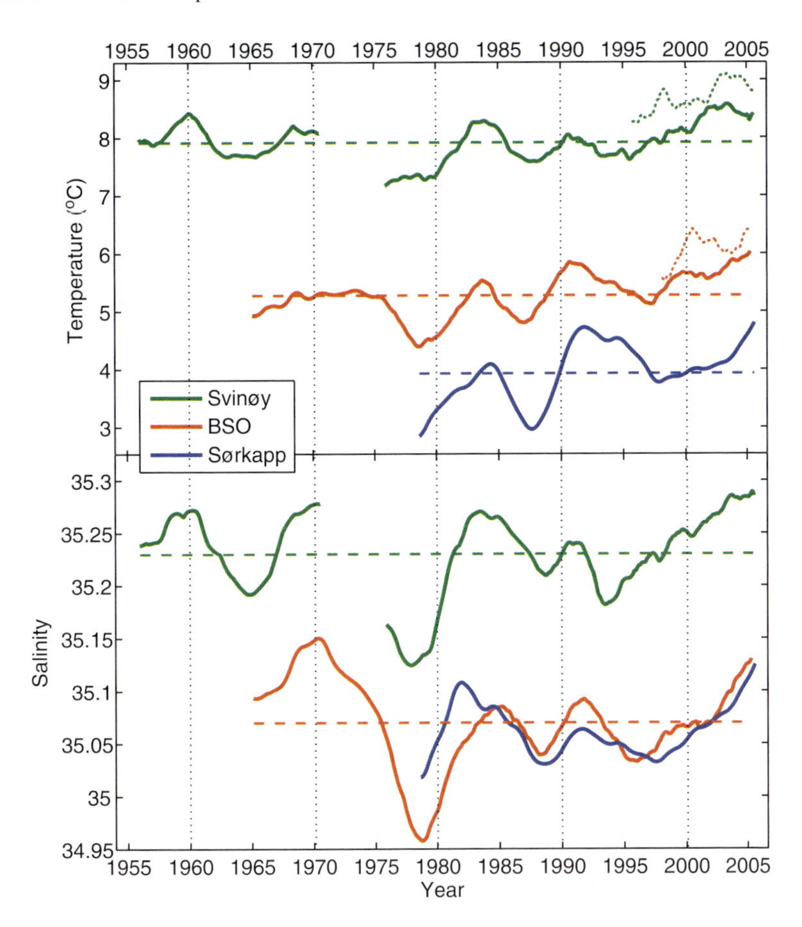

Fig. 2.5 Time series of temperature and salinity in the core of Atlantic water for the Svinøy section, Barents Sea Opening and Sørkapp section. The data from both the Svinøy section and Barents Sea Opening are de-seasoned. The Sørkapp section includes data obtained in August/ September only. Before 1977, the data in the Svinøy section is from winter only and the data from the Barents Sea Opening is from August only. Three years moving averages are applied on all data for the three sections. The temperature data from the current meters in the Svinøy section and the Barents Sea Opening are also included (dotted lines). For these data a 1-year running average was applied, due to the relatively short time series

these anomalies by a strong outflow of Polar water from the Arctic Basin that propagated anti-clockwise around the North Atlantic before reaching the Norwegian Sea several years later (Dickson et al. 1988; Belkin et al. 1998; Belkin 2004). In addition to this advective view of the propagation of salinity anomalies, Sundby and Drinkwater (2006) proposed that salinity anomalies, through the greater gyre of the northern North Atlantic, are caused by changes in volume fluxes along salinity gradients. The high temperature and salinity values observed during the past few years have also been monitored in the Faeroe–Shetland Channel. These extremes are associated with a weakening of the Sub-polar Gyre circulation (Häkkinen and Rhines 2004), resulting in a larger northward flow of subtropical Atlantic water from the northeastern Atlantic to the Nordic Seas (Hátun et al. 2005). The large salinity anomalies observed the last years are not exceptional in the Barents Sea Opening, as the highest salinity value was observed in 1970.

In most cases both the temperature and salinity anomalies fluctuate in phase at the different locations, but with a certain time lag. However, the magnitude of the propagated anomalies might be damped or amplified northward and in some cases the anomalies are also generated within the Nordic Seas (Furevik 2001). While the warm anomaly in the first half of the 1980s weakened to the north, the warm anomaly in the early 1990s became stronger as it propagated northwards, due to anomalous high air temperature in the Nordic Seas associated with an extreme positive NAO index around 1990 (Furevik 2001).

2.3.2 *Variations in Flux Estimates in the Svinøy Section and the Barents Sea*

The NwASC in the Svinøy section and the Atlantic flow in the Barents Sea Opening show fluctuations over a wide range of time scales, from weeks to months, seasons and years (Fig. 2.6). The 12-month moving average values range from 3.7 to 5.3 Sv with a mean of 4.3 Sv for the NwASC, and from 0.8 Sv to 2.9 Sv with a mean of 1.8 Sv for the Barents Sea Opening. Thus the Barents Sea Opening has only 45% of the mean flow of the NwASC but substantially greater inter-annual variability. The volume fluxes in the two sections show some co-variability. Both fell to a minimum in the winter of 2000–2001, and both showed a major increase from mid-2004 to the end of the time series in spring/early summer 2006. However, in 2002–2003, the fluxes diverged. Both increased toward the winter of 2002, but while the NwASC reached a relatively weak local maximum and started decreasing, the flux in the Barents Sea Opening kept increasing toward a strong local maximum in winter 2002–2003. A possible link to the atmospheric forcing is discussed in the following section.

There is a pronounced seasonal signal in both time series, although its strength varies in time. The NwASC seem to have a stronger seasonal signal before 2001 than after, while the opposite is the case in the Barents Sea Opening. These concurrent shifts in the seasonal cycle coincide with large-scale changes in the atmospheric circulation. Before 2001 the NAO winter index was mostly in a positive phase, while since 2001 it has been low and irregular.

When we consider the long-term changes in the volume flux, the NwASC shows no significant trend in the course of the measurement period 1995–2006 (Table 2.1). Orvik and Skagseth (2003b, 2005) found a downward trend of 12% in the velocity field for the period prior to 2005, but due to the strong increase in 2005–2006 (Fig. 2.6), this trend broke down when 2006 was included. In the

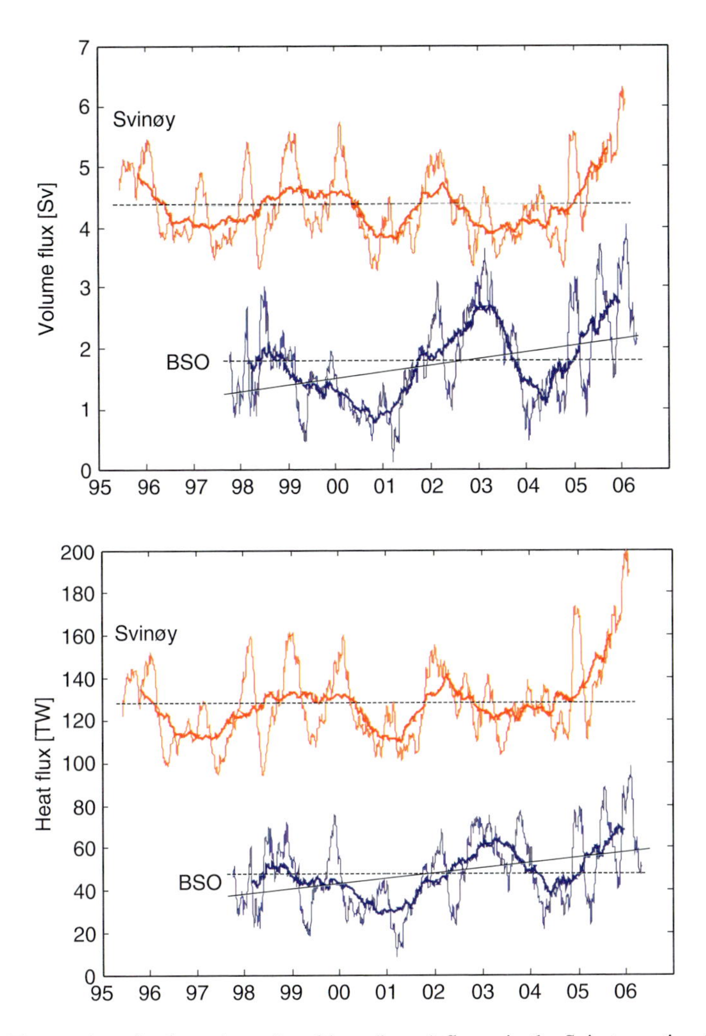

Fig. 2.6 Time series of volume (upper) and heat (lower) fluxes in the Svinøy section (red line) and the Barents Sea Opening (blue line). Lines showing the mean and, when significant, the trend, are included

Table 2.1 Mean fluxes and calculated annual trend when significant at 95% level

	NwASC (1995–2006)		BSO (1997–2006)	
	Volume flux (Sv)	Heat flux (TW)	Volume flux (Sv)	Heat flux (TW)
Mean	4.3	126	1.8	48
Trend ($year^{-1}$)	Not significant	Not significant	0.1	2.5

Barents Sea Opening on the other hand, there is an upward trend of 0.1 Sv per year. Over the 9-year measurement period this trend suggests an increase in volume flux of 45% of the mean value. The strong trend is partly due to a strong increase in 2005–2006, but there was also a significant upward trend before 2005.

The 12-month running average heat flux ranges from 110 TW (1 TW = 10^{12} W) to 160 TW with a mean of 126 TW for the NwASC, and from 29 TW to 70 TW with a mean of 48 TW for the Barents Sea Opening. The variability in the heat flux closely resembles the variability in the volume flux, indicating that the heat flux variations are dominated by velocity fluctuations rather than temperature fluctuations (Fig. 2.6). In particular, this holds for the seasonal scale, where the heat flux is higher in spite of the fact that temperatures are lower during the winter than in summer. An example is during the maximum flux in the Barents Sea Opening in winter 2002–2003. The heat maximum was clearly caused by a velocity maximum (Fig. 2.6), but as the temperature was decreasing at the time (Fig. 2.5), the heat flux maximum was attenuated.

On inter-annual time scales the temperature variations become increasingly important (Orvik and Skagseth 2005). In the NwASC there are no significant trends, either for heat or for volume flux over the 11-year measurement period. Orvik and Skagseth (2005) found that a weak reduction in the velocity field was compensated for by a 1 °C increase in temperature (Fig. 2.5). In the Barents Sea Opening the annual upward trend in heat flux is 2.5 TW, suggesting an increase of 23 TW (48% of the mean value) in the course of 9 years. The trend in volume flux was significant also before 2005, although somewhat weaker. The trend in the heat flux is due to a positive trend in the volume flux combined with a 1 °C increase in temperature (Fig. 2.5).

2.4 Forcing Mechanisms

Identifying the forcing mechanisms for the NwAC into the Arctic is a major task. It will probably depend on the time scales, as we can expect different forcings to be important for mean flow than for fluctuations on for example daily or even monthly timescales. Here we will leave the question of whether the mean flow is wind- or thermohaline-driven, as these forcings are intrinsically linked. Instead we discuss observations and physical mechanisms of relevance to variations in the fluxes, with a major focus on the period of simultaneous mooring records along the Norwegian Coast.

Considering variations in the rate of Atlantic water flow through the Norwegian Sea and into the Barents Sea, there are two main questions. First, what is the driving force for the NwASC, i.e. the topographically trapped current along the Norwegian Continental Slope, and secondly, what determines which fraction of this water that will enter the Barents Sea?

The driving mechanism of the NwASC has been studied in detail on the basis of data from the Svinøy section. A major part of the variation in the Atlantic water flow in the NwASC can be linked to the passage of the atmospheric lows that typically originate southwest of Iceland and propagate northeastwards towards Scandinavia. The accompanying along-slope (coast) component of the wind is found to be a key driver of variations in the flow (Skagseth and Orvik 2002; Skagseth et al. 2004). The mechanism is through surface Ekman transports toward the coast, balanced by a deeper return flow that is transferred into an along-slope current (Adams and Buchwald 1969; Gill and Schumann 1974). On the basis of satellite altimeter sea-level anomaly (SLA) data, Skagseth et al. (2004) found coherent variations in the NwASC from west of Ireland to the entrance to the Barents Sea, forced by wind associated with variations in sea-level pressure (SLP) resembling the NAO pattern. A negligible phase lag clearly indicated barotropic transfer mechanisms.

The volume flux of Atlantic water entering the Barents Sea is highly dependent on the regional wind pattern in the Barents Sea Opening (Ingvaldsen et al. 2004a, b). They found that the variations in the inflow are due to surface Ekman transports toward the coast setting up sea-level gradients that in turn were balanced by flow into the Barents Sea, and argued that these effects were enhanced by divergent Ekman fluxes in the Barents Sea Opening. As for the NwASC, the inflow was strong when the SLP resembled a strong NAO pattern. Additionally, simple theory involving topographic steering implies that during periods of anomalous eastward/westward extent of the NwAC as observed by Blindheim et al. (2000) and Mork and Blindheim (2000), less/more water is recirculated in the Norwegian Sea and more/less water enters the Barents Sea. This has been shown in model runs with idealized (Furevik 1998) and real (Zhang et al. 1998) topography.

With these general considerations of the relevant forcing in mind, it is of interest to compare the records of the Atlantic water flow in the Svinøy Section and in the Barents Sea Opening, and their relationship with atmospheric forcing. The correlation between the monthly filtered mooring records from the Svinøy section and the Barents Sea Opening (starting in 1995 and 1997, respectively) has a maximum for velocity of r = 0.41 with zero lag, and a maximum for temperature of r = 0.65 with

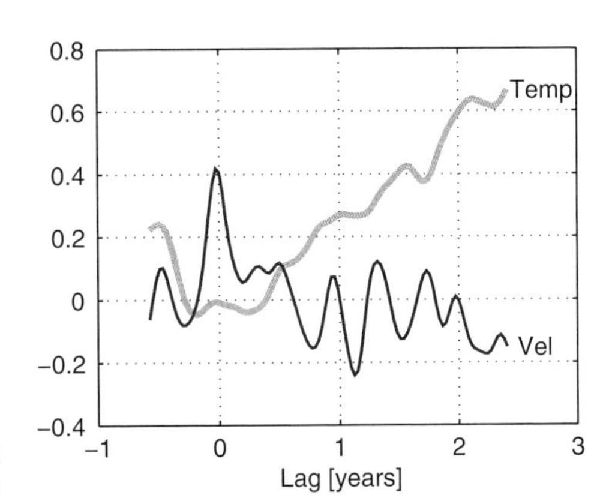

Fig. 2.7 Correlation functions based on the index series of velocity and temperature

a lag of 2 years (Fig. 2.7). The moderate correlation for the currents indicates that the local effective atmospheric forcing of the current at the two sites is different for at least part of the time.

In order to identify the atmospheric variations corresponding to different co-variations of the currents in Svinøy section and the Barents Sea Opening we considered SLP composite fields. The most frequently observed situations occurred when the anomalous currents were simultaneously either strong or weak. The case of anomalous

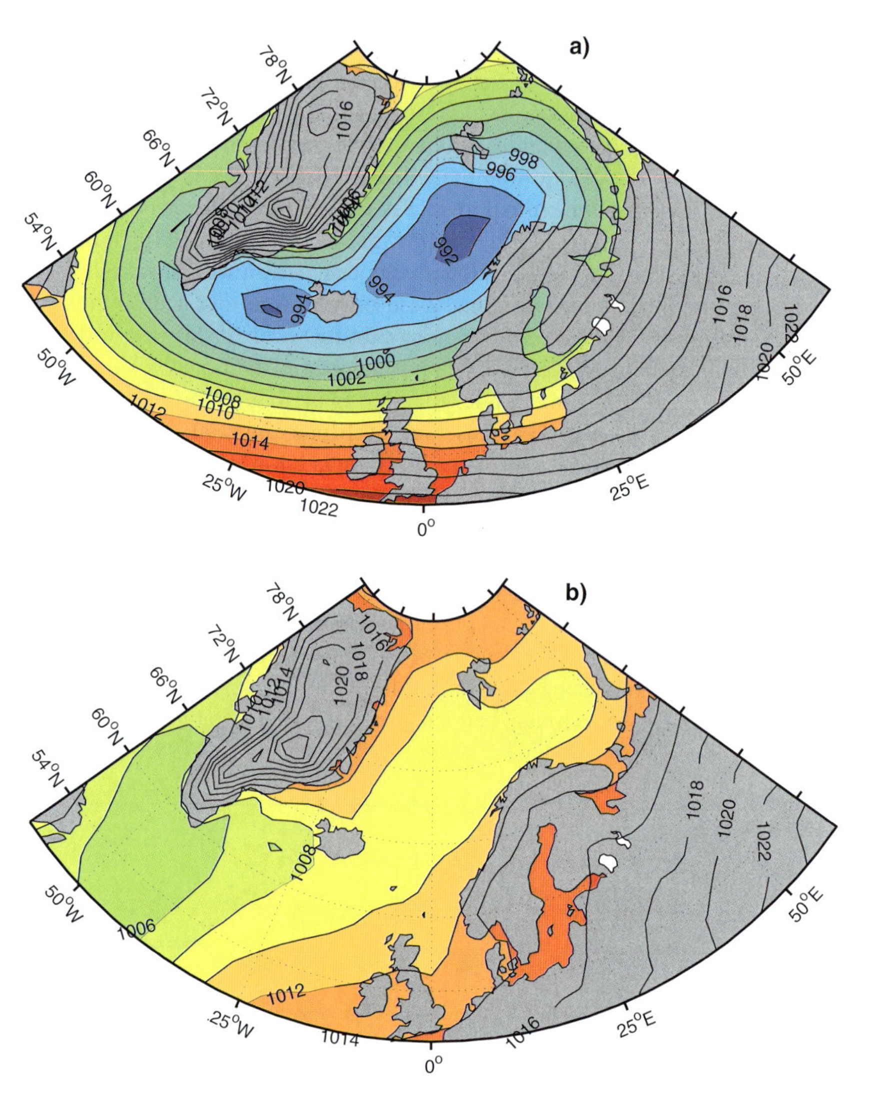

Fig. 2.8 Composite plots of sea-level pressure for periods during which the NwASC in the Svinøy section and the Atlantic inflow to the Barents Sea are either anomalously high (upper panel) or anomalously low (lower panel)

strong currents at both sites was characterized by an atmospheric low extending from southwest of Iceland, seawards along the coast of Norway, and partially into the Barents Sea (Fig. 2.8). This is the most common pathway of Icelandic lows as they propagate into the Norwegian Sea. The case of anomalously negative currents at both sites was characterized by highs over Scandinavia, and generally very weak SLP gradients and hence also weak winds (Fig. 2.8).

The associated changes in sea level were identified by similar sea-level anomaly (SLA) composites (Fig. 2.9). These composite plots reveal marked differences for the two cases. In the case of anomalous strong currents at both sites a strong SLA gradient along the Norwegian continental slope extended into the Barents Sea. As SLA gradients represent geostrophic flow anomalies, this indicates anomalous strong surface currents. Since a significant part of the current is barotropic in the continental slope region (Skagseth and Orvik 2002) this will probably also reflect the deep currents. On the other hand, in the case of anomalous weak surface currents, the SLA gradient along the Norwegian Continental slope and into the Barents Sea was much smaller (possibly in the opposite direction), indicating weaker and a tendency for anomalous negative surface currents.

Currents with opposite phases in the Svinøy section and in the Barents Sea Opening occur less frequently, and the results should be therefore interpreted with some care. The case in which the current was anomalously high in the Svinøy section and anomalously low in the Barents Sea Opening was characterized by strong southwesterly winds towards southern Norway and a local atmospheric low in the Barents Sea (not shown). The effective forcing would be along-slope winds in Svinøy section and northerly winds in the Barents Sea Opening. A high-pressure "blocking" event over Scandinavia characterized the opposite case, with anomalous low currents in the Svinøy section and strong currents in the Barents Sea Opening. In this case the atmospheric low was forced into a more westerly route through the Norwegian Sea, before turning eastward into the Barents Sea, providing winds favourable for Atlantic inflow. The maximum flow in the Barents Sea Opening and weak flow in the Svinøy section in winter 2002–2003 (Fig. 2.6) can be related to the greater influence of such an atmospheric pattern.

The above discussion concerns relatively short time scales (<1 year), but these must be considered as fluctuations in longer timescale variations of various origin. The long-term hydrography series (Fig. 2.5) show that anomalies are a prominent part of the variability. Since these are usually generated to the south of the Greenland–Scotland Ridge they can be regarded as remote forcing in the flux budgets. Prominent examples of upstream forcing are the "Great Salinity Anomaly" (Dickson et al. 1988) and the recently reported circulation changes in the Sub-polar Gyre (Häkkinen and Rhines 2004) with associated water characteristics changes (Hátun et al. 2005). Orvik and Skagseth (2003a) also found a significant relationship on an inter-annual time scale between the wind stress curl in the northern North Atlantic and the volume flux of the NwASC 15 months later between 1995 and 2003.

Finally, a positive internal feedback mechanism has been proposed for the Barents Sea (Ikeda 1990; Aadlandsvik and Loeng 1991). The mechanism is as follows: increased Atlantic inflow to Barents Sea leads to warmer water, more sea ice melting and more open waters in the region, while increased oceanic heat loss and

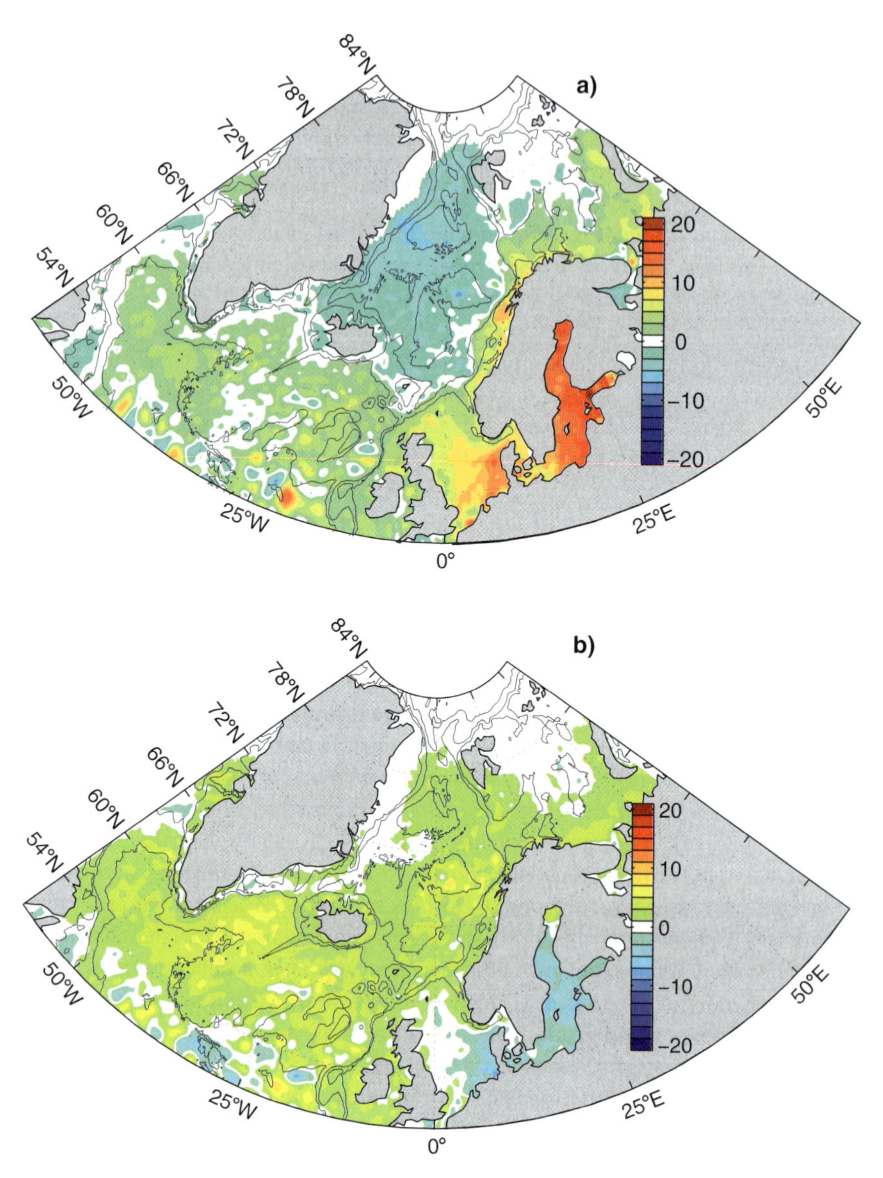

Fig. 2.9 Composite plots similar to those illustrated in Fig. 2.8, but for sea-level anomalies for periods when the NwASC in the Svinøy section and the Atlantic inflow to the Barents Sea were either anomalously high (upper panel) or anomalously low (lower panel)

evaporation create a local atmospheric low, and the associated anomalous cyclonic winds in turn amplifies the Atlantic inflow, thus closing the loop. Based on simulations with a coupled atmosphere-ocean general circulation model, Bengtsson et al. (2004) proposed that a similar feedback mechanism could explain the "early warming" in the 1930s–1940s.

2.5 Summary

The main aim of this paper has been to present a holistic view of the Atlantic water flow along the Norwegian Coast and into the Barents Sea. It has focused on the period starting in the mid-1990s, with simultaneous arrays of moored current meters in the Svinøy section and the Barents Sea Opening. These detailed measurements have provided the bases for improved estimates of means and variations in fluxes, and their forcing mechanisms.

Mean volume and heat fluxes associated with the Atlantic water are 4.3 Sv and 126 TW, respectively for the Svinøy section, showing no significant trends, and 1.8 Sv and 48 TW for the Barents Sea Opening, where positive trends have been found in both measures. The transport series show a prominent, but irregular, seasonal cycle at both sites, mainly determined by variations in the volume flux. The inter-annual changes are both substantial, but are relatively larger in the Barents Sea Opening.

In terms of prediction the data confirm the approximately 2-year lag in anomalies from the Svinøy section to the Barents Sea Opening. This strongly suggests that the recent relative cooling of the Svinøy section will be seen in the Barents Sea Opening in the next few years. However, as the heat loss becomes relatively more important in determining the climate in the eastern part of the Barents Sea, this region is probably less predictable, since atmospheric forcing is basically unpredictable beyond timescales of 1 week.

Hydrographic data along the Norwegian Coast show that the periods of direct current measurements, after 1995 for the Svinøy section and 1997 for the Barents Sea Opening, are the prolongations of a period that started in the late 1970s, since when Atlantic water has become warmer and saltier. This means that, given the assumption of constant volume fluxes, the estimated heat fluxes are higher than the long-term mean.

The close resemblance, throughout the record, between temperature variations in the Kola section and the AMO-index back to the early 20th century illustrates the importance of large-scale long-term variations in the Barents Sea system. Although the magnitudes of these variations are relatively small in comparison with inter-annual variations, other studies have shown them to be of major importance for ecosystem changes (ACIA 2005).

Forcing mechanisms, relating primarily to the wind, of the NwASC and the Atlantic water flow into the Barents Sea, were reviewed. The different forcing effects of the NwASC and the Atlantic inflow to the Barents Sea to similar atmospheric systems are noted. The results strongly suggest that the relative distribution of the NwAC entering the Barents Sea and passing through the Fram Strait is very sensitive to storm tracks. Thus, in a climate change perspective, changes in the predominant storm tracks may trigger major changes, including feedback mechanisms, for the Barents Sea climate and the heat budget of the Arctic Ocean.

Acknowledgements This paper is based on work funded by Research Council of Norway projects NoClim, Proclim and ECOBE and European Union projects VEINS, MAIA, ASOF_N and DAMOCLES.

References

Aadlandsvik, B. and H. Loeng (1991) A study of the climatic system in the Barents Sea. Polar Research, 45–49.

Adams, J.K. and V.T. Buchwald (1969) The generation of cntinental shelf waves. Journal of Fluid Mechanics, 35, 815–826.

ACIA (2005) Arctic Climate Impact Assessment. Cambridge University Press, New York, 1042 pp.

Belkin, I.M., Levitus, S., Antonov, J., and S. Malmberg (1998) "Great Salinity Anomalies" in the North Atlantic. Progress in Oceanography, 41, 1–68.

Belkin, I.M. (2004) Propagation of the "Great Salinity Anomaly" of the 1990s around the northern North Atlantic. Geophysical Research Letters, 31, L08306, doi:10.1029/2003GL019334.

Bengtsson, L., Semenov, V.A., and O.M. Johannessen (2004) The early twentieth-century warming in the Arctic – a possible mechanism. Journal of Climate, 17, 4045–4057.

Blindheim, J. (1989) Cascading of Barents Sea bottom water into the Norwegian Sea, Rapp.P.-v. Reun. Cons. int. Explor. Mer, 188, 49–58.

Blindheim, J., V. Borovkov, B. Hansen, S.A. Malmberg, W.R. Turrell, and S. Østerhus (2000) Upper layer cooling and freshening in the Norwegian Sea in relation to atmospheric forcing, Deep Sea Research, Part I, 47, 655–680.

Bochkov, Y.A. (1982) Water temperature in the 0–200 m layer in the Kola-Meridian section in the Barents Sea, 1900–1981. Sb. Nauchn. Trud. PINRO 46, 113–122 (in Russian).

Dickson, R.R., Meincke, J., Malmberg, S.-A., and A.J. Lee (1988) The Great salinity anomaly in the northern North Atlantic 1968–1982. Progress in Oceanography, 20, 103–151.

Dickson, R.R., Osborn, T.J., Hurrell, J.W., Meincke, J., Blindheim, J., Aadlandsvik, B., Vinje, T., Alekseev, G., and W. Maslowski (2000) The Arctic Ocean Response to the North Atlantic Oscillation. Jounal of Climate, 13, 2671–2696.

Fratantoni, D.M. (2001) North Atlantic surface circulation during the 1990s observed with satellite-tracked drifters. Journal of Geophysical Research, 102, 22,067–22,093.

Furevik, T. (1998) On the Atlantic Water flow in the Nordic Seas: Bifurcation and variability, Dr. scient thesis, University of Bergen, Norway.

Furevik, T. (2001) Annual and interannual variability of Atlantic water temperatures in the Norwegian and Barents seas: 1980–1996, Deep Sea Research, Part I, 48, 383–404.

Gill, A.E. and H. Schumann (1974) The generation of long shelf waves by wind. Journal of Physical Oceanography, 4, 83–90.

Häkkinen, S. and P.B. Rhines (2004) Decline of Subpolar North Atlantic circulation during the 1990s. Science, 304, 5670, 555–559, doi:10.1126/Science 1094917.

Hátun, H., Sandø, A.B., Drange, H., Hansen, B., and H. Valdimarsson (2005) Influence of the Atlantic Subpolar Gyre on the thermohaline circulation, Science, 309, 1841–1844.

Helland-Hansen, B. and F. Nansen (1909) The Norwegian Sea. Its physical oceanography based upon the Norwegian Researches 1900–1904. Report on Norwegian Fishery and Marine Investigations, 2(2), 1–360.

Ikeda, M. (1990) Decadal oscillations of the air-ice-ocean system in the Northern Hemisphere. Atmosphere-Ocean 28, 106–139, 3/1990.

Ingvaldsen, R., Loeng, H., Ådlandsvik, B., and G. Ottersen (2003) Climate variability in the Barents Sea during the 20th century with focus on the 1990s. ICES Marine Science Symposium, 219, 160–168.

Ingvaldsen, R.B., Asplin, L., and H. Loeng (2004a). Velocity field of the western entrance to the Barents Sea. Journal of Geophysical Research, 109, C03021, doi:10.1029/2003JC001811.

Ingvaldsen, R.B., Asplin, L., and H. Loeng (2004b) The seasonal cycle in the Atlantic transport to the Barents Sea during the years 1997–2001. Contintental Shelf Research, 24, 1015–1032.

Loeng, H., Blindheim, J., Ådlandsvik, B., and G. Ottersen (1992) Climatic variability in the Norwegian and Barents Seas. ICES Marine Science Symposia 195, 52–61.

Loeng, H., Sagen, H., Ådlandsvik, B., and V. Ozhigin (1993) Current measurements between Novaya Zemlya and Frans Josef Land September 1991 – September 1992, Institute for Marine Research, Rep. No. 2, ISSN 0804–2128.

Midttun, L. and H. Loeng (1987) Climatic variations in the Barents Sea. In: H Loeng (ed) The effect of oceanographic conditions on the distribution and population dynamics of commercial fish stocks in the Barents Sea, Third Soviet-Norwegian symposium. Murmansk, May 1986, pp. 13–28.

Mork, K.A. and J. Blindheim (2000) Variations in the Atlantic inflow to the Nordic Sea, 1955–1996. Deep Sea Research, Part I, 47, 1035–1057.

Nilsen, J.E. and E. Falck (2006) Variations of mixed layer properties in the Norwegian Sea for the period 1948–1999. Progressin Oceangraphy, 70, 58–90.

Olsen, A., Johannessen, T., and F. Rey (2003) On the nature of the factors that control spring bloom development at the entrance to the Barents Sea, and their interannual variability. Sarsia 88, 379–393.

Orvik, K.A. and P. Niiler (2002) Major pathways of Atlantic water in the northern North Atlantic and Nordic Seas toward Arctic. Geophysical Research Letters, 29(19), 1896, doi:10.1029/2002 GL015002.

Orvik, K.A. and Ø. Skagseth (2005) Heat flux variations in the eastern Norwegian Atlantic Current toward the Arctic from moored instruments, 1995–2005, Geophysical Research Letters, 32, L14610, doi:10.1029/2005GL023487.

Orvik, K.A. and Ø. Skagseth (2003a) The impact of the wind stress curl in the North Atlantic on the Atlantic inflow to the Norwegian Sea toward the Arctic, Geophysical Research Letters, 30(17), 1884, doi:10.1029/2003GL017932.

Orvik, K.A. and Ø. Skagseth (2003b) Monitoring the Norwegian Atlantic slope current using a single moored current meter. Continental Shelf Research, 23, 159–176.

Orvik, K.A., Skagseth, Ø., and M. Mork (2001) Atlantic inflow to the Nordic Seas: Current structure and volume fluxes from moored current meters, VM-ADCP and SeaSoar-CTD observations, 1995–1999. Deep Sea Research, Part I, 48, 937–957.

Polyakov, I.V., Beszczynska, A., Carmack, E., Dmitrenko, I., Fahrbach, E., Frolov, I., Gerders, R., Hansen, E., Holfort, J., Ivanov, V., Johnson, M., Karcher, M., Kauker, F., Morrison, J., Orvik, K., Schauer, U., Simmons, H., Skagseth, Ø., Sokolov, V., Steele, M., Timkhov, L., Walsh, D., J. Walsh (2005) One more step toward a warmer Arctic. Geophysical Research Letters, 32, L17605, doi:10.1029/2005GL023740.

Poulain, P.M., Warn-Varnas, A., and P.P. Niiler (1996) Near-surface circulation of the Nordic seas as measured by Lagrangian drifters, Journal Geophysical Research, 101, 18,237–18,258.

Rudels et al. (2004) Atlantic sources of the Arctic Ocean surface and halocline waters. Polar Research, 23(2), 181–208.

Schauer U, Loeng, H., Rudels, B., Ozhigin, V. and W. Dieck (2002) Atlantic Water flow through the Barents and Kara Seas. Deep-Sea Research, Vol. 49, 2281–2298.

Skagseth, Ø., Orvik, K.A., and T. Furevik (2004) Coherent variability of the Norwegian Atlantic Slope Current derived from TOPEX/ERS altimeter. Geophysical Research Letters, 31, L14304, doi:10.1029/2004GL020057.

Skagseth, Ø. and K.A. Orvik (2002) Identifying fluctuations in the Norwegian Atlantic Slope Current by means of empirical orthogonal functions. Continental Shelf Research, 22, 547–563.

Sundby, S. and K. Drinkwater (2006) On the mechanism behind salinity anomaly signals of the northern North Atlantic. Progress in Oceanography (in press).

Sutton, R.T. and D.L.R. Hodson (2005) Atlantic forcing of the North American and European Summer Climate. Science, 309, 5731, 115–118.

Sætre, R. and R. Ljøen (1971) The Norwegian Coastal Current. In: ANON (ed) Proceeding of the first international Conferanse on Port and Ocean Engening under Arctic Conditions, Vol. II. Norwegian Institute of Technology, Trondheim, pp. 514–535.

Tereshchenko, V.V. (1997) Seasonal and year-to-year variation in temperature and salinity of the main currents along the Kola section in the Barents Sea. Murmansk: PINRO Publ. 71 pp. (in Russian).

Tereshchenko, V.V. (1999) Hydrometeorological conditions in the Barents Sea in 1985–1998. Murmansk: PINRO Publ. 176 pp. (in Russian).

Zhang, J, Rothrock, D.A., and M. Steele (1998) Warming of the Artic Oceam by strengthened Atlantic Inflow: Model results. Geophysical Research Letters, 25(10), 1745–1748, doi: 10.1029/98GL01299.

Chapter 3
Variation of Measured Heat Flow Through the Fram Strait Between 1997 and 2006

Ursula Schauer[1], Agnieszka Beszczynska-Möller[2], Waldemar Walczowski[3], Eberhard Fahrbach[4], Jan Piechura[5], and Edmond Hansen[6]

3.1 Introduction

The northernmost extension of the Atlantic-wide overturning circulation consists of the flow of Atlantic Water through the Arctic Ocean. Two passages form the gateways for warm and saline Atlantic Water to the Arctic: the shallow Barents Sea and the Fram Strait which is the only deep connection between the Arctic and the World Ocean. The flows through both passages rejoin in the northern Kara Sea and continue in a boundary current along the Arctic Basin rim and ridges (Aagaard 1989; Rudels et al. 1994). In the Arctic, dramatic water mass conversions take place and the warm and saline Atlantic Water is modified by cooling, freezing and melting as well as by admixture of river run-off to become shallow Polar Water, ice and saline deep water. The return flow of these waters to the south through the Fram Strait and the Canadian Archipelago closes the Atlantic Water loop through the Arctic.

In the past century the Arctic Ocean evidenced close relation to global climate variation. Global surface air, upper North Atlantic Waters and Arctic intermediate waters showed coherently high temperatures in the middle of the last century and also in the past decades (Polyakov et al. 2003; Polyakov et al. 2004; Delworth and Knutson 2000). A likely candidate for this tight oceanic link is the flow through the Fram Strait. Through the Barents/Kara Sea, only the upper layer (200 m) of Atlantic Water can pass – thereby loosing much of its heat to the atmosphere – while the Fram Strait (sill depth 2,600 m) is deep enough to enable the through-flow of Atlantic Water at intermediate levels.

[1]Alfred Wegener Institute for Polar and Marine Research, Bremerhaven, Germany

[2]Alfred Wegener Institute for Polar and Marine Research, Bremerhaven, Germany

[3]Institute of Oceanology, Polish Academy of Sciences, Sopot, Poland

[4]Alfred Wegener Institute for Polar and Marine Research, Bremerhaven, Germany

[5]Institute of Oceanology, Polish Academy of Sciences, Sopot, Poland

[6]Norwegian Polar Institute, Tromsoe, Norway

R.R. Dickson et al. (eds.), *Arctic–Subarctic Ocean Fluxes*, 65–85

Two currents carry the warm water from the North Atlantic to the Fram Strait: a western branch which is a baroclinic jet in the Polar Front between the Atlantic Water and the central waters of the Nordic Seas, and an eastern branch, called Norwegian Atlantic Slope Current, which is an almost barotropic current along the Norwegian shelf break (Skagseth et al. 2008). They converge in the Fram Strait to form the West Spitsbergen Current (Walczowski and Piechura 2007) but the difference in both their origin (Hansen and Østerhus 2000) as well as their speed and pathway in the Nordic Seas affect their respective impact on the Arctic Ocean.

The complex topography in the Fram Strait itself leads again to a splitting of the West Spitsbergen Current and to a distribution of the Atlantic Water in at least three branches (Quadfasel et al. 1987a). One branch follows the shelf edge and enters the Arctic Ocean north of Svalbard. This branch crosses the Yermak Plateau which limits its depth to approximately 600 m. A second branch flows northward along the northwestern slope of the Yermak Plateau and the third branch recirculates immediately in the Fram Strait between 78° N and 80° N (Perkin and Lewis 1984; Gascard et al. 1995). Evidently, transports and properties of the different branches determine the input of oceanic heat to the Arctic Ocean. While part of the Atlantic Water flows to the central Arctic and is likely to be responsible for observed changes in heat content there, another part returns in a short loop within the northern Fram Strait. Here it can induce ice melt and thus determines the fractions of fresh water entering the Nordic Seas as ice and as water.

On the western side of the Fram Strait, modified Atlantic Water that originates from the West Spitsbergen Current as well as from the Barents Sea (Rudels et al. 1994) leaves the Arctic Ocean augmented by much of the Arctic fresh water surplus both as ice and in liquid form and occasionally some Pacific Water (Falck et al. 2005). This accumulates to a net southward volume transport through the Fram Strait of approximately 2 Sv (Fahrbach et al. 2001).

In the past few decades, the Atlantic Water flowing into the Arctic was not only warmer than earlier (Quadfasel et al. 1991; Schauer et al. 2004) but the influence of Atlantic Water in the Arctic Ocean also became more widespread: in the 1990s the front separating saline Atlantic-derived upper-ocean water from less saline Pacific-derived waters shifted from the Lomonosov Ridge to the Alpha–Mendelyev Ridge (McLaughlin et al. 1996). These changes, together with a reduced ice cover were attributed to a stronger cyclonic atmospheric circulation over the North Atlantic and the Arctic (Dickson et al. 2000). Morison et al. (2006) described the return of the Atlantic Water distribution and properties to near pre-1990s climatology after the cyclonic atmospheric circulation had relaxed. In the same time sea-ice extent continued to decrease and in the late 1990s another warm pulse of Atlantic Water entered the Arctic Ocean that was seen to propagate around the Eurasian Basin (Polyakov et al. 2005).

While the advection of warm Atlantic Water through the Fram Strait has been known since Nansen (1902) its role in the overall heat budget of the Arctic, as well as the role of its anomalies, are not yet understood. Morison (1991) pointed out that the ocean heat transport to the Arctic is an order of magnitude smaller than that of the atmosphere but that it might still be an important contribution to a delicate balance. Thus, one of the tantalizing current questions is whether there was an oceanic contribution to the decrease of the Arctic sea ice of the past decades. In large parts of the Arctic Ocean the Atlantic layer is shielded from sea ice and atmosphere by a fresh surface layer. In and northeast of the Fram Strait, however, the warm water is close to the surface (Aagaard et al. 1987). It may undergo several freezing/melting cycles during its travel in the boundary current along the Eurasian shelf edge (Rudels et al. 1996) before it meets the fresh surface layer that in the Eurasian Arctic is mainly fed by Siberian river-runoff. A shift in the circulation of the run-off on the shelves may increase the area where heat can be released directly from the Atlantic layer to the ice or atmosphere (Martinson and Steele 2001).

An assessment of the warm Atlantic Water impact to the Arctic Ocean – being either a transient feature or a contribution to the surface heat budget – can be made by relating its inflow to its outflow. First estimates of the oceanic heat budget of the Arctic Ocean suffered from a lack of exact volume flux data (Mosby 1962) and also from the erroneous assumption that the volume flux through the Fram Strait is balanced (Aagaard and Greisman 1975). Later attempts did not always follow the concept of heat flux computation in a stringent way so that an evaluation of earlier Fram Strait heat flux computations is difficult.

A prerequisite for the computation of oceanic heat transport is the knowledge of the volume fluxes. Past estimates of transport through the Fram Strait derived from observations were either based on inverse modeling or on velocity measurements at few locations requiring considerable extrapolations. A method-induced bias seems to result in lower volume fluxes from the inverse method (e.g. Schlichtholz and Houssais 1999) than from direct current measurements (e.g. Hanzlick 1983).

In order to examine the exchange of water through the Fram Strait, to quantify the heat transported with the Atlantic Water to the Arctic, and to better understand the mechanisms involved in its variation an intensive mooring programme was established in 1997. An array consisting of 14–16 moorings, covering the Fram Strait from the eastern to the western shelf edge, allows to resolve the complex flow structure. Since 2000, yearly hydrographic surveys took part between 70° N and 79° N. Here we report the results from these observations that form a unique time series of long-term year-round high resolution flux measurements through a key gateway to the Arctic.

3.2 Data

The results reported here are based on a set of regularly repeated observations carried out in the West Spitsbergen Current between 70° N and 79° N in the past decade (Fig. 3.1). Until 2005 the observations were done in the framework of the European Union projects VEINS (Variability of Exchanges in Northern Seas, 1997–2000) and ASOF-N (Arctic–Subarctic Ocean Fluxes, 2002–2005). Since 2006, the work is carried out as a part of EU-DAMOCLES (Developing Arctic Modelling and Observing Capabilities for Long-term Environment Studies).

An array of moorings measuring currents, temperature and salinity has been maintained along 78° 50′ N to 79° 00′ N since 1997. The instruments were RCM7, RCM8 or DCM11 from Aanderaa Instruments, ADCPs from RDI and 3D-ACM from Falmouth Scientific Inc; all registered velocity and temperature at 2-h intervals.

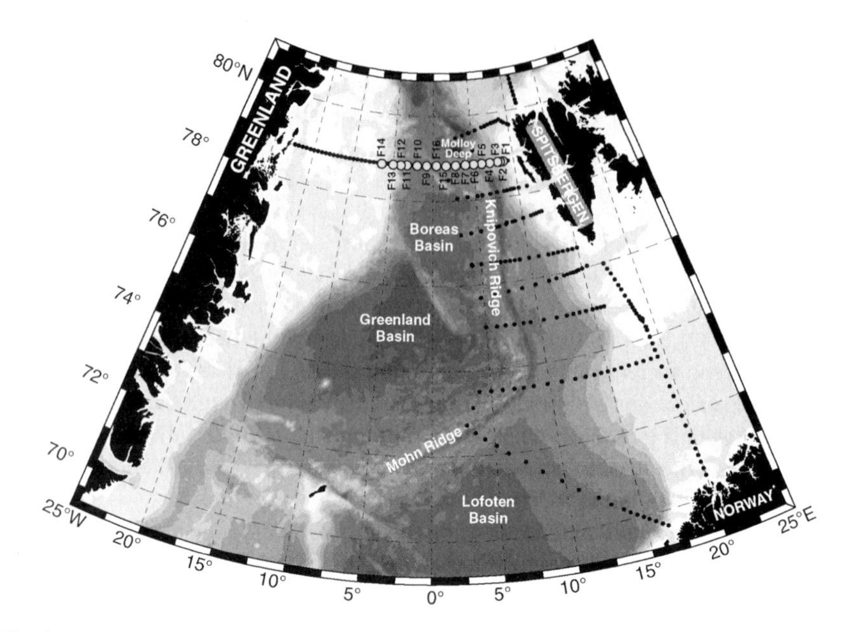

Fig. 3.1 Location of the measurements in the northern Nordic Seas and the Fram Strait between 1997 and 2006. Filled gray circles mark mooring positions during the period September 2002–August 2006; for the respective positions between September 1997 and September 2002, see Fig. 3.2. Black dots show CTD stations taken in August/September. The section overlapping with the moorings along 78° 50′ N was surveyed in the summers 1997–2006. The other sections were taken from 2000 to 2006

The instruments covered the water column from 10 m above the seabed to approximately 50 m below the surface (Fig. 3.2). The measurements extended from 6° 51′ W, the eastern Greenland shelf break, along 79° N to 0° E and continued along 78° 50′ N to 8° 40′ E, the western shelf break off Spitsbergen; since 2002 all moorings were deployed along 78° 50′ N.

The number of moorings and the number of levels equipped with instruments varied over the years (Fig. 3.2). We started with 14 moorings with a relatively narrow horizontal spacing of the moorings over the continental slopes where strong horizontal gradients were expected, and a wider spacing in the interior. It turned out that in this way the return current in the central part of the strait was under-sampled resulting in significant aliasing. Therefore, from 2002 onwards, the number of moorings was increased to 16. In addition, instruments were included at the 750 m level to better identify the lower boundary of the warm Atlantic Water. For more details, see ASOF_N deliverable 6.3. For a description of the data processing we refer to Fahrbach et al. (2001) and Schauer et al. (2004).

The year-round measurements from moored instruments were combined with hydrographic sections, taken along the mooring section during the deployment cruises since 1997 and in addition between 70° N and 79° N since 2000. On all cruises, a Seabird 9/11 CTD system was used. To obtain the horizontal distributions the data were interpolated using the kriging procedure (Walczowski and Piechura 2006). The grids were smoothed with a linear convolution low-pass filter.

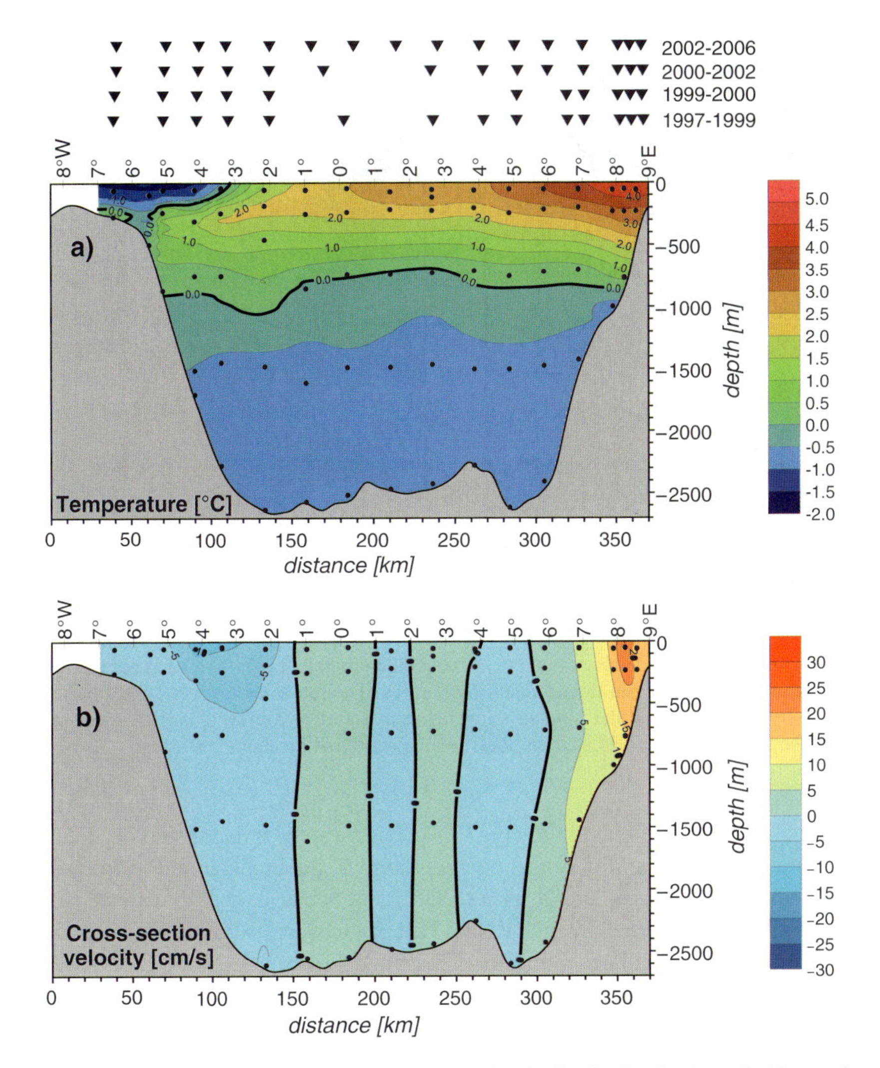

Fig. 3.2 Mean temperature (a) and cross-section velocity (b) distribution for the period September 2002–August 2003 measured from the mooring array. Dots denote the positions of instruments. Triangles on top mark the mooring positions in the different periods

3.3 Flow and Temperature Evolution

3.3.1 Atlantic Water Branches Along the Continental Slope and the Polar Front

Between 72° N and the southern tip of Svalbard, the western branch transporting Atlantic Water along the Arctic front can be derived from the mean baroclinic field while the slope current – due to its barotropic nature – can be identified only from

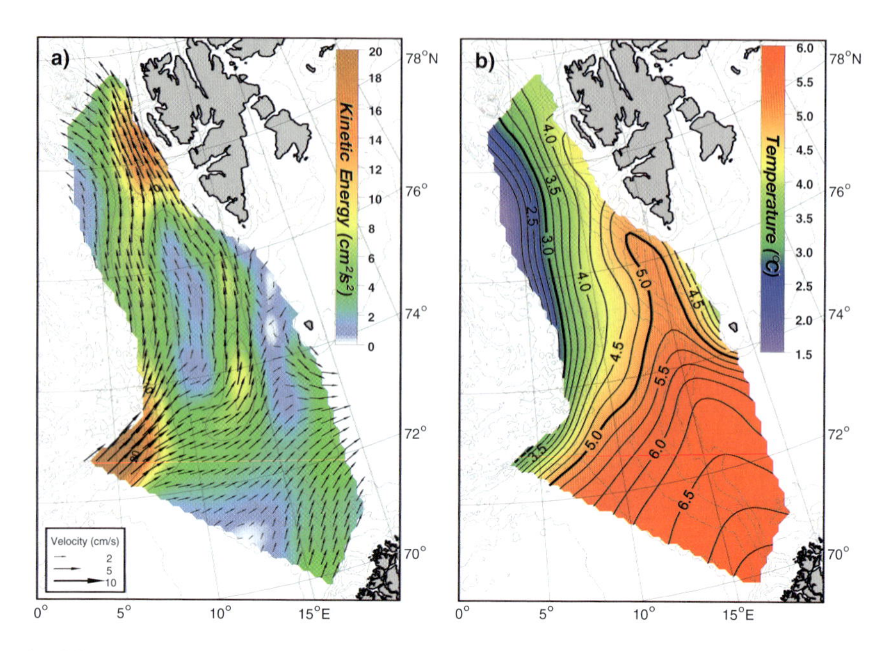

Fig. 3.3 (a) Mean kinetic energy (cm^2/s^2) of baroclinic currents (color scale) and baroclinic currents (arrows) at 100 dbar in the summers 2000–2006. The reference level is 1,000 dbar or the bottom. (b) Distribution of the summer temperature at 100 dbar averaged over the years 2000–2006. The 3 °C and 5 °C isolines are in bold. (Walczowski and Piechura 2007, Fig. 3.2)

the zonal temperature maximum (Fig. 3.3). The western branch is colder than the slope branch due to the difference in temperature of the branches at the entrance to the Nordic Seas (Hansen and Østerhus 2000) and because of cross-frontal mixing with cold water from the Greenland Sea. Part of the warm Atlantic Water from the western branch recirculates along the Greenland Fracture Zone (Quadfasel et al. 1987b) and does not reach the Fram Strait. Between 74° N and 78° N the remaining part of the western branch and the slope branch converge. Their contributions are reflected at the mooring line at 78° 50′ N by the distinct maxima of volume flux density (Fig. 3.4). Due to its position it is mostly the western branch that feeds the immediate recirculation of Atlantic Water while it is mostly the slope branch that crosses the Yermak Plateau to the east.

3.3.2 Flow and Temperature Structure at 78° 50′ N

The mooring data along 78° 50′ N and the hydrographic data clearly show the highly barotropic northward flow of the warm West Spitsbergen Current and the more baroclinic cold East Greenland Current in the western Fram Strait (Fig. 3.5). In the central Fram Strait, the flow is essentially westward, forming one of the recirculation pathways for Atlantic Water. Accordingly, throughout the year the

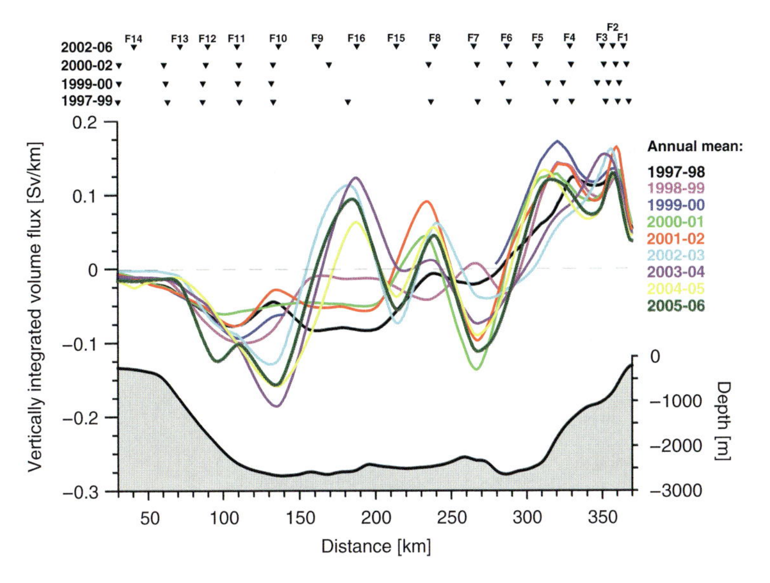

Fig. 3.4 Zonal distribution of annual mean cross-section volume transport per width from 1997 to 2006. The bottom panel shows the bottom depth. Triangles on top mark the mooring positions in the different periods

temperature of the upper layers is highest in the West Spitsbergen Current and decreases towards the west up to the front between the returning West Spitsbergen Current water and the cold Polar Water at about 3° W (Fig. 3.2) – a structure that is known since long from hydrographic summer sections (e.g. Rudels 1987).

The highest velocities were invariably found above the upper slope (water depth <1,500 m) in the West Spitsbergen Current with 9-year mean speed above 20 cm/s. The West Spitsbergen Current also shows the maximum speeds with values above 55 cm/s in the upper 250 m. The East Greenland Current has its core over the base of the continental slope at about 2,500 m where it carries warm modified Atlantic Water southward rather than cold Polar Water (Fig. 3.2). The latter leaves the Arctic Ocean west of this core at somewhat weaker southward velocities.

There is also meridional flow in the central part that is however weaker and more variable than the West Spitsbergen Current and strongly influenced by the complex topography. Immediately west of the West Spitsbergen slope the flow turns southward steered by the southeastern extension of the Molloy Deep. The northward extension of the Knipovich Ridge is likely to be responsible for the northward component at mooring F8. After increasing the lateral resolution in the central Fram Strait in 2002 by adding two moorings also the topographic influence (e.g. that of the Hovgaard Ridge) on the currents further to the west was captured up to the rise of the East Greenland continental slope. While topographic steering is most evident in the near-bottom level (yellow arrows in Fig. 3.5) it also influences the upper-layer meridional component and thus determines the partitioning of Atlantic Water flowing towards the central Arctic Ocean vs the immediate return flow.

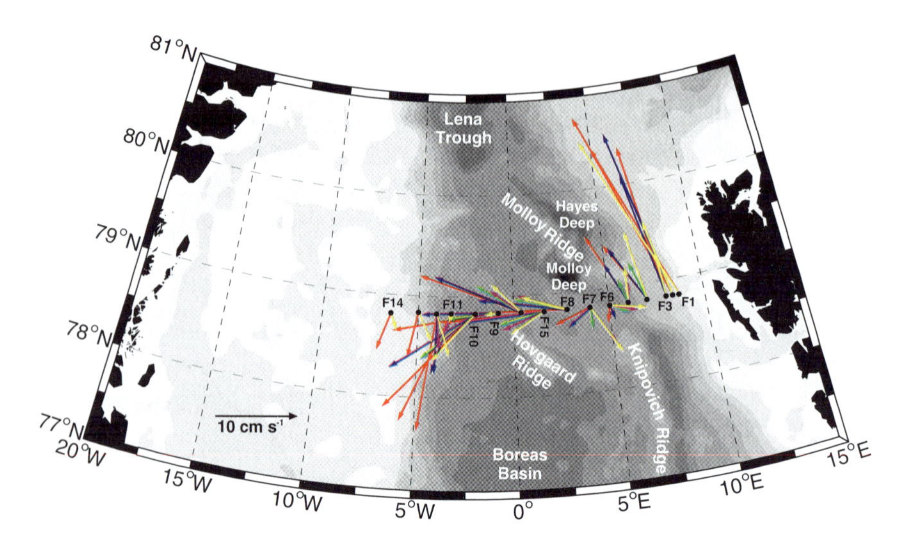

Fig. 3.5 Mean currents obtained from the Fram Strait mooring array. The average was taken over the period 2002–2006 when the mooring positions were kept constant. The color code denotes the nominal depth of the measurement, red: 50 m, blue: 250 m, green: 750 m, magenta: 1,500 m, yellow: near-bottom. With the exception of F15 and F16, the mooring numbering runs from F1 in the east to F14 in the west

Heat can be transported by the mean flow and also by mesoscale features. At irregular intervals southward flow was observed for several weeks at about 8° E (mooring F6) and at the same time northward flow occurred at 7° E (mooring F7), i.e. north of the Knipovich Ridge (not shown, Schauer et al. 2004). South of the mooring line, the current along the Arctic front at times sheds anti-cyclonic baroclinic eddies (Fig. 3.6); these propagate to the north, guided by the topography, and thus explain the intermittent anti-cyclonic features observed in the mooring data.

The strongest flows in the West Spitsbergen Current and in the central Fram Strait occur in winter (not shown, Jónsson et al. 1992; Schauer et al. 2004) which is in accordance with the seasonal spin-up of the cyclonic gyre systems of the Nordic Seas through the wind (Jakobsen et al. 2003). In contrast, the southward volume flow does not show a clear seasonal signal, confirming findings by Jónsson et al. (1992). The upper layer temperatures down to 250 m have a maximum across the entire section in autumn (Fig. 3.7).

3.3.3 Volume Fluxes

The mean net volume flux across the section from 9 years of mooring data is 2 Sv southward, with the standard deviation of 5.9 Sv in the first period with 14 moorings and 2.7 Sv since 2002 when the number of moorings was increased to 16. This residuum is composed of a total of 12 Sv northward flow and 14 Sv southward flow.

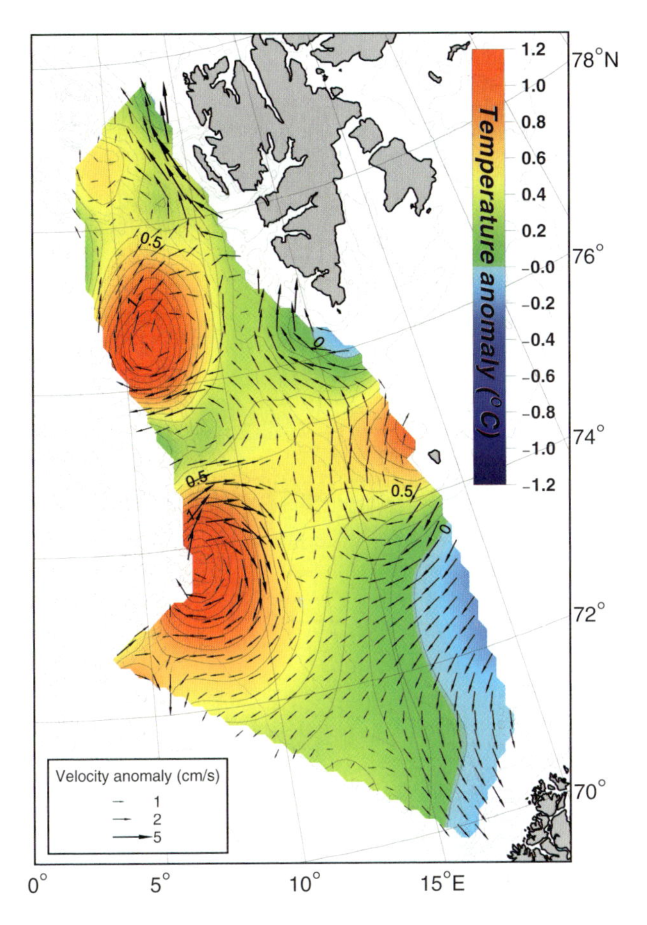

Fig. 3.6 Horizontal distribution of the anomalies of temperature and baroclinic currents in summer 2005 at 100 dbar. The anomalies are with respect to the mean summer values between 2000 and 2006. The baroclinic current is referred to 1,000 dbar or to the bottom

The net southward flow is the compensation for the inflow of Atlantic Water to the Barents Sea opening (Rudels et al. 1994).

While the high velocities of the combined Atlantic Water branches on the West Spitsbergen slope lead to huge volume fluxes in a relatively small area (Fig. 3.4) considerable transports also occur in the current bands in the central part of the strait. Here the mean velocities are low but mostly unidirectional from the surface to the bottom at more than 2,500 m. The weak east–west temperature change in the upper layer (Fig. 3.2) suggests that the banded structure is the projection of meanders of the westward recirculation.

Approximately one third of the northward transport comprises deep water colder than 1 °C that is composed of Greenland and Norwegian Sea Deep Water (Rudels et al. 2008). Part of that water returns within a short loop while the westernmost part of the deep southward flow stems from the interior Arctic Basins.

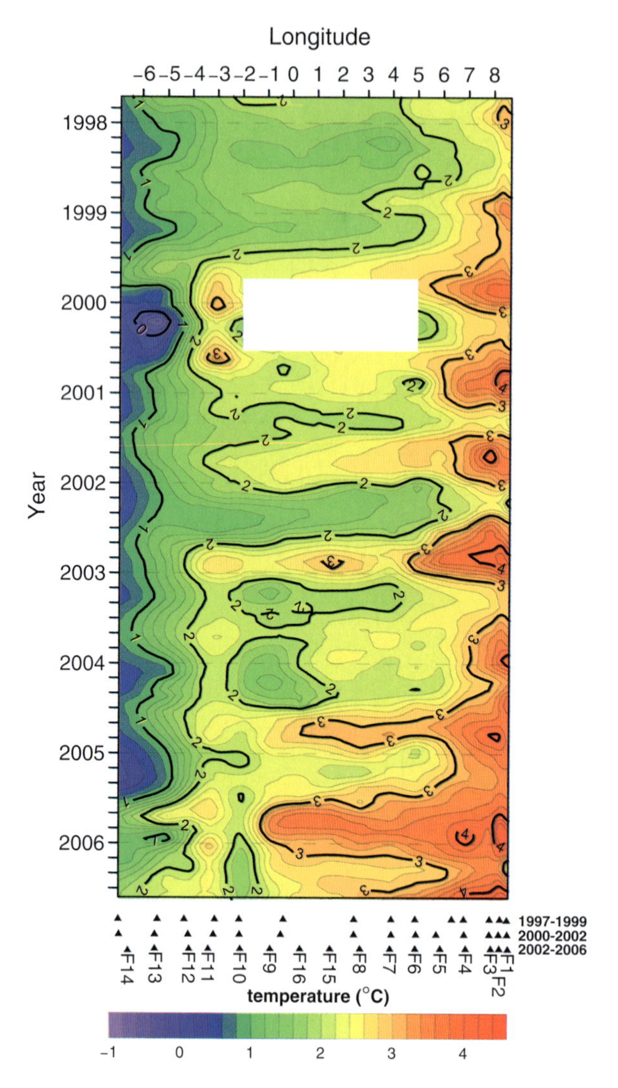

Fig. 3.7 Hovmöller diagram of the monthly mean zonal temperature distribution at 78° 50′ N at 250 m nominal instruments depth from 1997 until 2006

3.3.4 Warming of the Atlantic Water

Both the summer hydrographic data and the year-round mooring data reveal an increase of temperature of the northward flowing Atlantic Water (here water warmer than 1 °C) in the northern Nordic Seas and in the Fram Strait during the decade 1997–2006 (Fig. 3.8). The increase was about 0.5 K between 1998 and 2000 and again about 0.5 K from 2003 to 2006. The significance of this integrated signal is supported by a very coherent course of the time series of individual instruments (not shown). The warming was associated with an increase in salinity and the record maximum values of both properties were observed in summer 2006. With the exception of the first 2 years, the temperature increase was overlaid by a seasonal variation with an amplitude of approximately 0.5 K, but while the summer maxima rose by more than 1 K over 9 years, the winter minima rose much less.

One origin of the warming and the salinity increase are the changes of the sub-polar North-Atlantic with upper ocean temperature and salinity maxima in the Subpolar Gyre and the Faroe–Shetland Channel in 1997/98 and 2003 (Hátún et al. 2005; ICES 2006). On the other hand, changes of the atmospheric cooling of the Atlantic Water during its transfer through the Nordic Seas can mask this signal before it reaches the Fram Strait (Karcher et al. 2008). However, the two temperature maxima occurring both in the Sub-Polar Gyre and in the Fram Strait with a time lag of roughly 2–3 years confirm the fast signal propagation in the boundary current in the Nordic Seas described in (Polyakov et al. 2005).

The hydrographic summer observations at 76° 30′ N (taken here between 2000 and 2006) reveal that warmer water was advected in both the slope current and the frontal current (Fig. 3.9). The average increase of summer temperature at 200 m between 2003 and 2006 was more than 1 K over large parts of the section. This is more than twice as much as the increase in the yearly running mean temperature obtained from the mooring data at 78° 50′ N and also much larger than the increase of the maximum summer temperatures between 2003 and 2006 there. This underlines the difficulty to derive interannual variability from snapshots at a single depth in a region with high seasonality. However, despite being masked by mesoscale features (Fig. 3.6), there is some indication that in 2004 and 2005 the western branch was more warming than the eastern one.

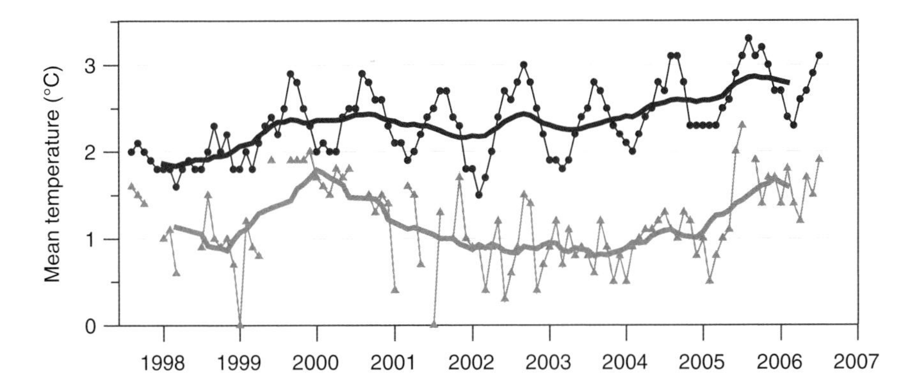

Fig. 3.8 Time series of the cross-section averaged temperature of Atlantic Water derived from mooring data. The black line denotes the temperature of northward flowing water that was warmer than 1 °C. The grey line denotes the temperature of southward flowing Atlantic Water (for explanation see text). Symbols at the thin lines denote monthly mean values, bold lines are 12-month running means

3.4 Heat Transport Through the Fram Strait

3.4.1 Conceptual Remarks About Estimating Oceanic Heat Transport into the Arctic Ocean

Since the volume flux through the Fram Strait, just like the flux through all other passages to the Arctic Ocean, is not balanced the heat flux can not be calculated straightforward. The complexity of the flow through the Fram Strait adds to the difficulties finding a reasonable scheme for computing the heat flux. The principle for the calculation of advective heat transport is described in the oceanographic literature since more than 30 years (Montgomery 1974). Nevertheless, the last decade shows a wealth of publications from which a misconception of this principle is evident (among many others: Schauer et al. 2004; Maslowski et al. 2004; Karcher et al. 2003; Lee et al. 2004). This makes it worthwhile bringing to mind the basic concepts once more.

The physical idea behind oceanic advective heat transport is related to temperature flux convergence. Practically this may be referred to a defined ocean volume (or mass) holding a certain amount of heat. Currents across the boundary of that ocean segment can change the heat content by replacing a certain amount of water of a particular temperature by the same amount of water with (usually) another temperature. The difference of the heat content of the replaced volumes is the heat gain or loss of the considered ocean segment. Such an exchange can be achieved by ocean currents of any scale, by basin-wide gyres or overturning cells as well as by small eddies.

At stationary conditions the heat gain/loss through currents has to be balanced by sinks/sources, S, like, e.g. heat exchange with the atmosphere. This heat balance of the ocean segment is resumed in the equation

$$S = \oint ds \int_0^H c_p \cdot \rho \cdot v_\perp \cdot T dz \qquad (3.1)$$

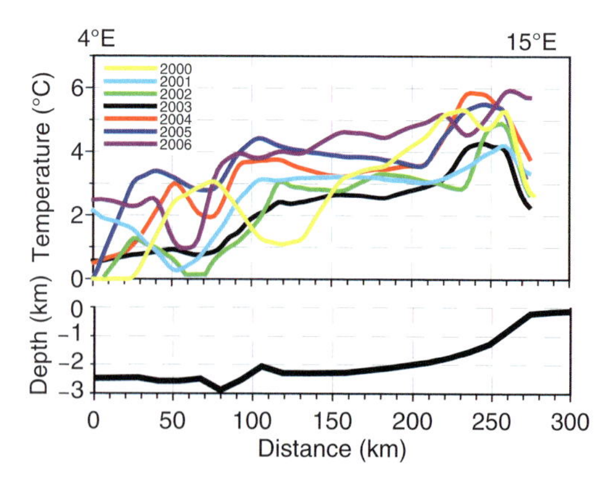

Fig. 3.9 Temperature at 200 dbar along 76° 30′ N between 4° E and 15° E from the summers 2000 to 2006 (Walczowski and Piechura 2007, Fig. 3.5a). The bottom panel shows the bottom depth

with c_p specific heat, ρ density, v_{v1} the velocity component perpendicular to the open ocean boundary confining the segment and T the temperature of the flow. The integral is taken over the full depth, z, from top to the bottom, H, around the entire ocean boundary of which ds is a boundary length element. This concept holds as well for variable conditions in which case also a change of the heat content of the ocean segment, H, with time, t, is possible.

$$\frac{\partial H}{\partial t} + S(t) = \oint ds \int_0^H c_p \cdot \rho \cdot v_\perp(t) T(t) dz \tag{3.2}$$

This concept sounds (and probably is) trivial. It implies that heat transports can be calculated in a system with mass conserved only (Montgomery 1974; Hall and Bryden 1982). However, heat transport computations by evaluating observations and even model results are sometimes far from straightforward. This is partly due to the complexity of ocean currents that often does not allow to determine velocity and temperature along the complete boundary at a high enough resolution. A second problem often arises from the formulation of the advective heat flux term itself. It is extremely tempting to disintegrate the integral over a closed boundary in Equations (3.1) and (3.2) and to calculate "temperature fluxes" over partial cross-sections (Lee et al. 2004). This holds as long as these temperature fluxes are regarded as interim terms required to compute the entire integral. However, it is sometimes argued that temperature fluxes can also be used themselves, e.g. for comparing different cross-section parts (Karcher et al. 2003) or to rate temporal changes through a particular partial cross-section (Schauer et al. 2004). It has also been suggested that certain reference temperatures such as the volume average temperature (Lee et al. 2004) are well suited to derive heat transports from temperature fluxes. However, these as well as any other temperature fluxes are entirely arbitrary and attempts to use them instead of heat fluxes produce wrong results (Schauer and Beszczynska-Möller, in preparation).

Meridional heat transport computed from hydrographic data, e.g. in the North Atlantic south of Greenland has large error bars (Ganachaud and Wunsch 2000) but is reasonable since the Atlantic north of any coast-to-coast zonal section is closed apart from a small influx from the Pacific. This inflow (about 0.8 Sv) (Woodgate et al. 2006) might be neglected in comparison to the meridional flow of O(10–10^2 Sv) through the North Atlantic, and the Bering Strait inflow temperature is similar to that of the deep North Atlantic flow.

With respect to Arctic–Subarctic Ocean fluxes, however, determination of oceanic heat transport principally needs to take into account all openings, Bering Sea, Canadian Archipelago, Fram Strait and Barents Sea Opening, in order to accomplish the requirements of Equations (3.1) and (3.2). Without any further constraints arising from the Arctic Ocean internal circulation heat transport through single straits can not be computed because none of the straits confining the Arctic Ocean has a balanced volume (mass) flux. Consequently one has to define carefully what is meant by "heat transport through the Fram Strait" in order not to deal with an ill-defined term.

The problem does not vanish when "only" temporal changes are compared (Montgomery 1974). Heat transport to the Arctic Ocean can change because of varying temperature difference between inflow and outflow and because of varying flow strengths. Here as well, isolated consideration of the changing properties of individual (in)flow branches leads to arbitrary results (Schauer and Beszczynska-Möller, in preparation).

The only way to elude the necessity of addressing all Arctic Ocean openings simultaneously for heat transport computations evolves if we can use constraints provided through the Arctic Ocean internal circulation. For example, for the inflow of warm Pacific Water through the shallow Bering Strait it has been shown that practically all of this water is cooled to freezing temperature before it exits the Arctic Ocean so that the heat flux can be derived from the inflow only (Woodgate et al. 2006). This is certainly not true for the Atlantic inflow through the Fram Strait. Therefore, only if we can identify compensating in- and outflow branches, i.e. if we can regard them as a stream tube, we can derive the heat flux provided through this pair.

3.4.2 An Approach to Compute the Heat Transported by the West Spitsbergen Current to the Arctic Ocean

With regard to the water carried northward in the West Spitsbergen Current we probably can safely assume that the bulk of this water also leaves the Arctic Ocean through the Fram Strait. Water from the West Spitsbergen Current propagating along the shelf edge into the Nansen Basin might flow on the shelf east of Spitsbergen and return to the northern and then western Barents Sea. This probably is only a small fraction of the water within the upper 150 m since much of the water entering the shelf through a canyon returns in a cyclonic loop to the shelf edge (Gawarkiewicz and Plueddemann 1995). A small fraction might however circulate anti-cyclonically around Svalbard. The flow through the 50 m deep Bering Strait is of the order 1 Sv to the north and there are no reports about Fram Strait water travelling southward to the Pacific (Woodgate et al. 2006). The Canadian Archipelago (sill depth 160 m) is the main gateway for the exit of Pacific Water (Steele et al. 2004) and for a fraction of Barents Sea water (Rudels et al. 2004). Any fraction from the Fram Strait is probably small.

The travel times along the various pathways of West Spitsbergen Current water in the Arctic Ocean, around all basins or only in the northern Fram Strait, last between months and decades. Warm water anomalies that have entered the Arctic Ocean with the West Spitsbergen Current in the nineties have reached the eastern Eurasian Basin 4 years later (Karcher et al. 2003; Polyakov et al. 2005) and we do not know yet which part of the associated additional heat is released to the surface and which part will leave the Arctic Ocean after several years or decades. However, assuming that their remnants finally end up in the Fram Strait we can consider the loops as closed volumes.

This should enable us to use the observations of velocity and temperature in the Fram Strait and compute the heat flux provided to the Arctic by the West Spitsbergen Current by adding the temperature fluxes of northward and southward flow.

Time series of temperature flux can be constructed from the interpolated fields of temperature and cross-section component of the velocity (Schauer et al. 2004). Since the southward volume flow is larger than the northward flow the critical point is how to identify which of the southward flow is returning West Spitsbergen Current water and which water stems from other openings like the western Barents Sea or the Bering Strait.

We assume that owing to continuity, water from any loop of the returning West Spitsbergen Current will flow southward immediately west of the West Spitsbergen Current. There is no indication that the Barents Sea branch crosses any of the West Spitsbergen Current-derived loops. Rudels et al. (1994) and Schauer et al. (2002a) showed that the Barents Sea Water displaces the Fram Strait branch off the slope at the confluence of the two branches in the northern Kara Sea and that further downstream the Fram Strait branch flows at the basin side of the two. If this pattern continues along the entire Arctic Ocean rim, all West Spitsbergen Current-derived southward flow in the Fram Strait would take place immediately west of the northward flow and the Barents Sea water would flow west of that.

While we assume based on continuity reasons (no crossing flow branches) that return flow in the central part east of the westernmost northward branch originates from the West Spitsbergen Current we have to distinguish for the East Greenland Current which part is constituted from West Spitsbergen Current water and which part from other sources. We assume that the warmest water stems from the West Spitsbergen Current.

To avoid volume flux uncertainties that arise from the still poorly resolved deep-water fluxes we limit our computations to the northward flow of upper and intermediate waters and we use a limiting temperature, $T_{DI} = 1\,°C$, for distinction between the two. With the exception of the front around its outcrop the depth of the 1 °C isotherm is below 500 m for northward flow in the West Spitsbergen Current (Fig. 3.2). The argument behind this choice is that water below that depth is very unlikely to reach the surface in the central Arctic Ocean and therefore must return at the same temperature through the deep Fram Strait (of course it can be mixed with other deep water, e.g. generated in the Barents Sea, which would be at similar temperatures). However, it has thus no chance to contribute to the surface heat flux.

The flux of upper layer water warmer than 1 °C is integrated over the entire cross section. The net volume flux can be positive, zero or negative. With zero net volume flux the heat flux of West Spitsbergen Current to the Arctic Ocean is immediately obtained by temperature flux integration over the respective cross section. In the case that the volume flux of water warmer than 1 °C was net northward, obviously West Spitsbergen Current water has been cooled to temperatures below 1 °C before returning. In this case we increased the integration area over water flowing southward to include also colder water. The distinction temperature for returning West Spitsbergen Current water, T_{DO}, was incrementally decreased until the resulting net flux was zero (within ±1 Sv).

A net southward volume flux would mean that there is water warmer than 1 °C flowing southward that does not originate from the West Spitsbergen Current. This is very unlikely: Water that entered through the Bering Strait is cooled to near freezing if it reaches the Fram Strait at all after crossing the entire Arctic. Barents Sea water looses much of its heat in the Barents and Kara seas so that it is densified and sinks to intermediate depths when entering the Eurasian Basin. According to observations taken between the 1960s and mid-1990s (Schauer et al. 2002b) all Atlantic Water that leaves the northern Kara Sea is colder than 1 °C. This might have changed in years thereafter. However, if the water would enter the central Arctic warmer than at 0 °C it would be lighter and closer to the surface. In this case it is exposed to Arctic surface influences more than water from the West Spitsbergen Current because it travels along the shelf edge and is more likely to upwell than the latter is. Furthermore it has the longest pathway. Therefore, in the case of net southward volume flux of water warmer than 1 °C we have to assume that it is caused by a large error of our velocity interpolation and that we can not determine a heat flux in that period from our data.

3.5 Resulting Heat Transports

The result of this approach for computing heat flux to the Arctic by West Spitsbergen Current water is given in Fig. 3.10. The maximum error limits associated with the interpolation between data points are considered to be of equal size as those in (Schauer et al. 2004), ±6 TW, since despite the wrong concept used there the uncertainties arising from the limited spatial resolution remain the same.

For T_{DI} = 1 °C, the distinction temperature for the outflow required to obtain zero net volume flux, T_{DO}, varied between –0.7 °C and 0.7 °C except of 1 month when it was −1.6 °C. Similar as the flux averaged temperature that increased from about 2 °C to almost 3 °C (Fig. 3.8) the annual mean volume flux of the Atlantic Water was rising in the last decade from less than 5 Sv to more than 7 Sv in 2004 and 2005 (Fig. 3.10). Due to the way the Atlantic Water is defined here, the volume flux increase is mostly a consequence of the warming. The temperature increased over the upper 800 m and thus the 1 °C isotherm in the West Spitsbergen Current was found 200 m deeper in 2004 than in 1997.

The annually averaged heat transport increased in the first 2 years from 26 to 36 TW which impressively demonstrates the influence of a wrong method as it was used by Schauer et al. (2004) where the increase was stated to be from 16 to 41 TW. After a dip in 2001, the heat flux increased to its decadal maximum of 50 TW in 2004. While the temperatures of the West Spitsbergen Current water continued to rise to a record high in 2006 the associated heat flux decreased again to 40 TW because much warmer water returned in that year to the Greenland Sea than before (Fig. 3.8). The reason for this can be twofold: Warmer water could finally return from one of the longer loops through the central Arctic Ocean that had entered in previous warming periods like in the early 1990s (Quadfasel et al. 1991). The second possibility is that the anomalously warm Atlantic Water advected in 2005 and 2006 recirculates immediately in the Fram Strait which is suggested from the extraordinarily high temperatures in the central Fram Strait (Fig. 3.7). Then the question must be posed what drives the strengthening of the recirculation vs north- and/or eastward

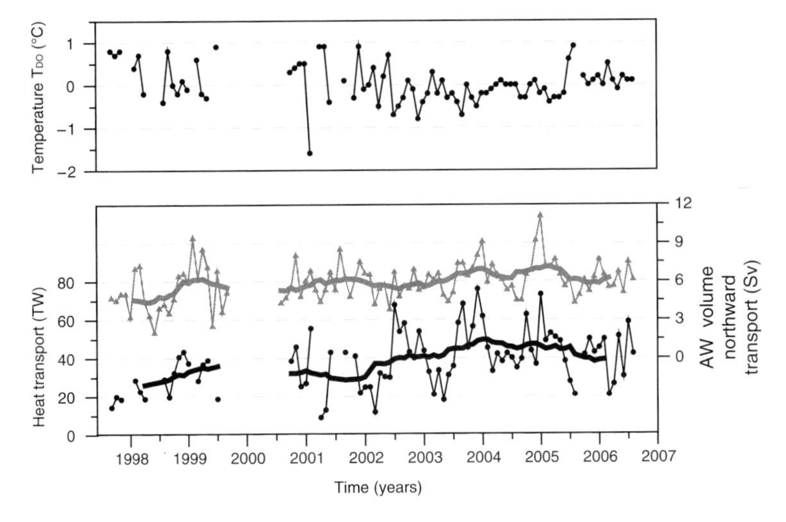

Fig. 3.10 Time series of the volume flux (grey lines) and heat flux (black lines) to the Arctic through Atlantic Water (warmer than 1 °C) in the West Spitsbergen Current. The upper panel gives the outflow distinction temperatures T_{DO} (see text for explanation). Symbols at the thin lines denote monthly mean values, bold lines are 12-month running means. Note that the southward volume flow of Atlantic Water is the same as the northward flow within ±1 Sv. For the uncertainties of the heat flux see discussion in Section 3.6

flow – whether it is a consequence of the change in large-scale atmospheric pattern that returned to a less cyclonic state in recent years or if this is due to a decrease in the pressure gradient across the Fram Strait due to the rising steric height in the Nordic Seas as a consequence of the warming (Jakobsen et al. 2003).

3.6 Critical Discussion of the Limits of the Approach

Besides the volume and heat flux errors inherent in the spatial interpolation, the proposed approach implies several uncertainties.

The choice of the distinction temperature for northward flow, T_{DI}, is somewhat arbitrary. Ideally, T_{DI} should be chosen in a way that the resulting heat flux is not sensitive to small changes. If T_{DI} is too high parts of the West Spitsbergen Current are excluded and the heat flux is underestimated (Schauer and Beszczynska-Möller, in preparation). If T_{DI} is too low many situations arise with non-zero net flow which demonstrate problems with the spatial resolution of the flow. These problems are larger in the first half of the observation period when the mooring number and instrumentation coverage was lower than in the second half.

The most critical point is, however, the disregard of mixing. Diffusion between the Fram Strait and Barents Sea branches during the passage through the Eurasian Basins takes place as double-diffusive layering (Rudels et al. 1999) as well as through mesoscale eddies (Schauer et al. 2002a, b). Also vertical displacement of warm Atlantic Water by entrainment into sinking dense shelf water plumes is a mechanism not explicitly taken into account by the stream tube approach.

Both processes imply that the values as given in Fig. 3.10 are overestimating the heat flux. Entrainment into sinking plumes means that warm West Spitsbergen Current water is returning to the Nordic Seas as deep water which is not considered here. For continuity, the drainage must be replaced by cold deep water upwelling in the central Arctic Ocean. Mixing with Barents Sea water obviously also means that some warm water of the West Spitsbergen Current returns to the Nordic Seas outside of the stream tube.

A gross estimate of the loss of West Spitsbergen Current water (mean temperature 2.5 °C, mean volume transport 6 Sv, Figs. 3.8 and 3.10) to deep waters which have an average temperature of –0.6 °C and –0.7 °C for the northward and southward flow, respectively, yields 0.2 Sv. Assuming thus a contribution of 0.2 Sv of compensating –0.5 °C cold central Arctic deep water included in the return water corresponds to 5% overestimation of the heat flux, i.e. about 2 TW which is within the interpolation induced error limits.

A similar assessment for mixing with the Barents Sea water can hardly be made. According to (Schauer et al. 2002b), 50% of the approximately 2 Sv Barents Sea Water leaving the northern Kara Sea is colder than 0 °C and 50% is warmer. While the cold fraction sinks at the Nansen basin slope deeper than 500 m, the warmer fraction remains in the same depth level as the West Spitsbergen Current water. Assuming the average temperature of the warmer fraction to be 0.5 °C, admixture of this fraction to the West Spitsbergen Current water would explain 10% of the estimated heat flux. If this Barents Sea Water fraction is however cooled to, e.g. –0.5 °C before it is mixed it would effect an overestimation of the heat flux by about one third. Mixture of all Barents Sea water (2 Sv) at –0.5 °C to the West Spitsbergen Current water would imply further reduction by one third and would involve that the heat flux of West Spitsbergen Current water to the Arctic is approximately 10 TW.

These examples show that, for principle reasons, in case of strong mixing the significance of the heat flux variability can hardly be addressed with this approach as long as the variability of the Barents Sea properties at their entrance to and during their passage through the Arctic Ocean are unknown. Would they be known, the stream tube concept for the West Spitsbergen Current could be extended to include the Barents Sea throughflow. Calculation of the heat transports with constant Barents Sea outflow temperature and fluxes in the St. Anna Trough would, however, *a priori* decide upon the variability for which we are searching. In any case, neglecting mixing with Barents Sea Water leads to an overestimation of the heat flux to the Arctic.

3.7 Some Consequences for Observational Strategies

The above considerations point to difficulties inherent to the assessment of the oceanic heat delivered through advection to the Arctic Ocean. They also lead to considerable consequences for observational strategies. First, to compute heat transport variability simultaneous observations are needed at least across those openings that are connected by currents. Second, these observations definitely need to be made at high spatial resolution of the velocity and temperature structure across these openings. In the Fram Strait the lateral variability that is of the scale of tens of kilometres due to the small internal Rossby radius and the complex topography translates directly into the need of a high number of moorings since this is so far the only way for time series of appropriate horizontal resolution. From measurements that spatially integrate properties like temperature or velocity no heat transports can be derived. Furthermore, in order to assess what fraction of the heat is released in the Arctic Ocean vs what fraction is simply passing by time series have to be long enough to cover the maximum travel time of a parcel which in the case of parcels travelling along the entire Arctic continental slope are decades.

Last but not least it should be mentioned that the same considerations, closed volumes or stream tubes, high resolution and long time series, hold also for the assessment of "fresh water fluxes".

Acknowledgements This work was supported by the European Union MAST III Programme VEINS (Variability of Exchanges in the Northern Seas) contract number MAS3-CT96–0070, the Fifth Framework Programme project ASOF-N (Arctic-Subarctic Ocean Flux Array for European Climate: North), contract number EVK2-CT-200200139, and Sixth Framework Programme project DAMOCLES (Developing Arctic Modelling and Observing Capabilities for Long-term Environment Studies), contract number 018509GOCE. Detlef Quadfasel helped to improve the manuscript.

References

Aagaard K, Foldvik A, and Hillmann SR (1987) The West Spitsbergen Current: Disposition and water mass transformation. Journal of Geophysical Research 92(C4): 3778–3784

Aagaard K and Greisman P (1975) Toward new mass and heat budgets for the Arctic Ocean. Journal of Geophysical Research 80(27): 3821–3827

Aagaard K (1989) A synthesis of the Arctic Ocean circulation. Rapports et Procès-verbeaux des Réunions, Conseil International pour l'Exploration de la Mer 188(1): 11–22

ASOF_N deliverable 6.3 http://www.awi.de/de/forschung/fachbereiche/klimawissenschaften/messende_ozeanographie/projekte/asof_n/reports_and_deliverables/list_of_ deliverables/

Delworth TL and Knutson TR (2000) Simulation of early 20th century global warming. Science 287(5461): 2246–2250

Dickson RR, Osborn TJ, Hurrell JW, Meincke J, Blindheim J, Adlandsvik B, Vinje T, Alekseev G, and Maslowski W (2000) The Arctic Ocean response to the North Atlantic oscillation. Journal of Climate 13(15): 2671–2696

Fahrbach E, Meincke J, Østerhus S, Rohardt G, Schauer U, Tverberg V, and Verduin J (2001) Direct measurements of volume transports through Fram Strait. Polar Research 20(2): 217–224

Falck E, Kattner G, and Budéus G (2005) Disappearance of Pacific Water in the northwestern Fram Strait. Geophysical Research Letters 32(14, L14619): 10.1029/2005GL023400

Ganachaud A and Wunsch C (2000) Improved estimates of global ocean circulation, heat transport and mixing from hydrographic data. Nature 408(6811): 453–457

Gascard JC, Richez C, and Rouault C (1995), New Insights on Large-Scale oceanography in Fram Strait: The West Spitsbergen Current, *Arctic Oceanography: Marginal Ice Zones and Continental Shelves*, edited by Smith W and Grebmeier J. American Geophysical Union, Washington DC, pp. 131–182

Gawarkiewicz G and Plueddemann AJ (1995) Topographic control of thermohaline frontal structure in the Barents Sea Polar Front on the south flank of Spitsbergen Bank. Journal of Geophysical Research 100(C3): 4509–4524
Hall MM and Bryden HL (1982) Direct estimates and mechanisms of ocean heat transport. Deep-Sea Research I 29: 339–359
Hansen B and Østerhus S (2000) North Atlantic-Nordic seas exchanges. Progress in Oceanography 45(2): 109–208
Hanzlick DJ (1983) The West Spitsbergen Current, transport, forcing and variability. Ph.D. thesis, University of Washington, Seattle, USA, 127 pp
Hátún H, Sando AB, Drange H, Hansen B, and Valdimarsson H (2005) Influence of the Atlantic Subpolar Gyre on the Thermohaline Circulation. Science 309(5742): 1841–1844
ICES (2006) ICES Report on Ocean Climate 2005. ICES Cooperative Research Report, 280, 47 pp
Jakobsen PK, Ribergaard MH, Quadfasel D, Schmith T, and Hughes CW (2003) Near-surface circulation in the northern North Atlantic as inferred from Lagrangian drifters: Variability from the mesoscale to interannual. Journal of Geophysical Research 108(C8), doi:10.1029/2002JC001554
Jónsson S, Foldvik A, and Aagaard K (1992) The structure and atmospheric forcing of the mesoscale velocity-field in Fram Strait. Journal of Geophysical Research 97(C8): 12585–12600
Karcher MJ, Gerdes R, Kauker F, and Köberle C (2003) Arctic warming – evolution and spreading of the 1990s warm event in the Nordic Seas and the Arctic Ocean. Journal of Geophysical Research 108(C2): 3034
Lee T, Fukumori I, and Tang B (2004) Temperature advection: internal versus external processes. Journal of Physical Oceanography 34(8): 1936–1944
Martinson DG and Steele M (2001) Future of the Arctic sea ice cover: implications of an Antarctic analog. Geophysical Research Letters 28(2): 307–310
Maslowski W, Marble D, Walczowski W, Schauer U, Clement JL, and Semtner AJ (2004) On climatological mass, heat, and salt transports through the Barents Sea and Fram strait from a pan-Arctic coupled ice-ocean model simulation. Journal of Geophysical Research 109(3): C03032 1–16
McLaughlin FA, Carmack EC, Macdonald RW, and Bishop J (1996) Physical and geochemical properties across the Atlantic/Pacific water mass front in the southern Canadian Basin. Journal of Geophysical Research 101: 1183–1197
Montgomery RB (1974) Comments on "Seasonal variability of the Florida Current," by Niiler and Richardson. Journal of Marine Research 32: 533–535
Morison J (1991) Seasonal variations in the West Spitsbergen Current estimated from bottom pressure measurements. Journal of Geophysical Research 96(C10): 18,381–18,395
Morison J, Steele M, Kikuchi T, Falkner K, and Smethie W (2006) Relaxation of central Arctic Ocean hydrography to pre-1990s climatology. Geophysical Research Letters 33(17)
Mosby H (1962) Water, salt and heat balance of the North Polar Sea and the Norwegian Sea. Geophysica Norvegia 24(11): 289–313
Nansen F (1902), Oceanography of the North Polar Basin. The Norwegian North Polar Expedition 1893–1896. Scientific Results (9), 427 pp
Perkin RG and Lewis EL (1984) Mixing in the West Spitsbergen current. Journal of Physical Oceanography 14(8): 1315–1325
Polyakov IV, Walsh D, Dmitrenko I, Colony R, and Timokhov LA (2003) Arctic Ocean variability derived from historical observations. Geophysical Research Letters 30(6), 4 pp
Polyakov IV, Alekseev GV, Timokhov LA, Bhatt US, Colony RL, Simmons HL, Walsh D, Walsh JE, and Zakharov VF (2004) Variability of the intermediate Atlantic water of the Arctic Ocean over the last 100 years. Journal of Climate 17(23): 4485–4497
Polyakov IV, Beszczynska A, Carmack EC, Dmitrenko IA, Fahrbach E, Frolov IE, Gerdes R, Hansen E, Holfort J, Ivanov VV, Johnson MA, Karcher M, Kauker F, Morison J, Orvik KA, Schauer U, Simmons HL, Skagseth R, Sokolov VT, Steele M, Timokhov LA, Walsh D, and Walsh JE (2005) One more step toward a warmer Arctic. Geophysical Research Letters 32(17), doi:10.1029/2005GL023740

Quadfasel D, Gascard JC, and Koltermann KP (1987a) Large-scale oceanography in Fram Strait during the 1984 marginal ice-zone experiment. Journal of Geophysical Research 92: 6719–6728

Quadfasel D and Meincke J (1987b) Note on the thermal structure of the Greenland Sea gyres. Deep-Sea Research 34(11): 1883–1888

Quadfasel D, Sy A, Wells D, and Tunik A (1991) Warming in the Arctic. Nature 350(6317): 385–385

Rudels B (1987) On the mass balance of the Polar Ocean, with special emphasis on the Fram Strait. Norsk Polarinstitutt Skrifter 188: 53 pp

Rudels B, Jones, EP, Anderson, LG, and Kattner G (1994) On the intermediate depth waters of the Arctic Ocean. The Polar Oceans and their role in shaping the global environment. Geophysical Monograph Series 85: 33–46

Rudels B, Anderson LG, and Jones EP (1996) Formation and evolution of the surface mixed layer and the halocline of the Arctic Ocean. Journal of Geophysical Research 101: 8807–8821

Rudels B, Björk G, Muench RD, and Schauer U (1999) Double-diffusive layering in the Eurasian Basin of the Arctic Ocean. Journal of Marine Systems 21(1–4): 3–27

Rudels B, Jones PE, Schauer U, and Eriksson P (2004) Atlantic sources of the Arctic Ocean surface and halocline waters. Polar Research 23(2): 181–208

Schauer U, Rudels B, Jones EP, Anderson LG, Muench RD, Björk G, Swift JH, Ivanov V, and Larsson AM (2002a) Confluence and redistribution of Atlantic water in the Nansen, Amundsen and Makarov basins. Annales Geophysicae 20(2): 257–273

Schauer U, Loeng H, Rudels B, Ozhigin V, and Dieck W (2002b) Atlantic Water flow through the Barents and Kara Seas. Deep-Sea Research I 49(12): 2281–2298

Schauer U, Fahrbach E, Østerhus S, and Rohardt G (2004) Arctic warming through the Fram Strait: Oceanic heat transport from 3 years of measurements. Journal of Geophysical Research 109(C06026), doi:10.1029/2003JC001823

Schlichtholz P and Houssais M (1999) An inverse modeling study in Fram Strait. Part II: water mass distribution and transports. Deep Sea Research II 46(6–7): 1137–1168

Steele M, Morison J, Ermold W, Rigor I, Ortmeyer M, and Shimada K (2004) Circulation of summer Pacific halocline water in the Arctic Ocean. Journal of Geophysical Research 109: No. C02027

Walczowski W and Piechura J (2006) New evidence of warming propagating toward the Arctic Ocean. Geophysical Research Letters 33(L12601), doi:10.1029/2006GL025872

Walczowski W and Piechura J (2007) Pathways of the Greenland Sea warming. Geophysical Research Letters 34, doi:101029/2007GL029974

Woodgate RA, Aagaard K, and Weingartner TJ (2006) Interannual changes in the Bering Strait fluxes of volume, heat and freshwater between 1991 and 2004. Geophysical Research Letters 33(L15609), doi:10.1029/2006GL026931

Chapter 4
Is Oceanic Heat Transport Significant in the Climate System?

Peter Rhines[1], Sirpa Häkkinen[2], and Simon A. Josey[3]

4.1 Introduction

It has long been believed that the transport of heat by the ocean circulation is of importance to atmospheric climate. Circulation of the Atlantic Ocean warms and moistens western Europe, the argument goes, and, because of the pivotal role of the Atlantic/Arctic region, also affects global climate (e.g., Stommel 1979). Indeed, major oceanographic field programs have been launched by many nations, based on this premise. In the US, NOAA issues quarterly assessments of subtropical North Atlantic meridional heat transport (http://www.aoml.noaa.gov/phod/soto/mht/reports/index.php). Estimates from ocean observations show the annual-mean, northward heat transport by the global circulation to decrease by about 1.5 pW (10^{15}W) between latitudes 25° N and 50° N, with nearly 1 pW of that within the narrow Atlantic sector alone (see Bryden and Imawaki 2001, who estimate the uncertainty of individual section heat transports at 0.3 pW). This effect of the oceanic meridional overturning forces an enormous upward flux of heat and moisture in subtropical latitudes, providing a significant fraction of the zonally integrated atmospheric northward energy flux (which peaks at between 3 and 5.2 pW, as discussed further below). Occurring dominantly in wintertime, oceanic warmth and moisture energize the Pacific and Atlantic storm tracks. Combined action of atmosphere and ocean carries this energy northward, with great impact on all facets of high-latitude climate.

Northern Atlantic climate hovers in the midst of debates over the dynamical origins and impacts of the global oceanic meridional overturning circulation (MOC), and its contribution to the coupled atmosphere–ocean system. Still, the

[1] University of Washington, Box 357940, Seattle, Washington 98195,
e-mail: Rhines@ocean.washington.edu

[2] NASA Goddard Space Flight Center, Code 971, Greenbelt, MD 20771,
e-mail: Sirpa@fram.gsfc.nasa.gov

[3] National Oceanography Centre, Southampton, England,
e-mail: Simon.A.Josey@noc.soton.ac.uk

R.R. Dickson et al. (eds.), *Arctic–Subarctic Ocean Fluxes*, 87–109

complexity of the shared process of heat and fresh water transport and exchange by ocean and atmosphere continues to belie oversimplified 'conveyor belt' images. Does 'Bjerknes compensation' occur, under which decadal variability the ocean and atmosphere components of meridional heat transport compensates, one for the other, leaving an unchanging top-of-atmosphere radiation field (e.g., van der Swaluw et al. 2007; Dong and Sutton 2005)?

Taken as a whole, the current debate over the most fundamental principles of the MOC demonstrates how much physical oceanography has yet to learn about its sector of the climate system and how important that sector is. Some of the implied questions currently under debate are:

Is the MOC pushed by buoyancy forcing or pulled by mixing induced by winds and tides (e.g., Toggweiler and Samuels 1995; Wunsch et al. 2004)?

Does deep overturning, the shallow overturning or the lateral wind-driven gyres dominate the meridional transport of heat (e.g., Boccaletti et al. 2005; Talley 2003)?

Where are the pathways of upwelling in the global scheme of the MOC (e.g., Sarmiento et al. 2004; Hallberg et al. 2006)?

How are the cycles of heat and fresh water transport coupled, and how is their dynamical impact on the MOC measured (e.g., Stommel and Csanady 1980)?

Do the zonally integrated overturning streamfunction and its thermal analogues adequately measure the MOC, or can a more penetrating definition be made by analyzing transport across sections and transformation within boxes, on the potential temperature/salinity plane (e.g. Lumpkin and Speer 2000; Fox and Haines 2003; Marsh et al. 2006; Bailey et al. 2005)?

What is the relative importance of the Southern Ocean and the northern Atlantic sinking regions (e.g., Toggweiler and Samuels 1995)?

Some things are not in doubt:

- The existence of 'maritime climates' downwind of the major oceans
- The oceanic moisture source for the entire atmosphere
- The contribution of latent heat associated with this moisture to the heating of the atmosphere
- The presence of storm tracks over the northern Pacific and Atlantic, which channel atmospheric meridional transports of heat and freshwater in these sectors
- The presence, movement and impact on atmospheric climate of sea-ice, in response to atmosphere and ocean circulation and temperatures (see Rhines 2006, for a non-technical discussion)

At the most basic level we are reminded that the ocean is the *dominant* global reservoir of mean thermal energy, water, carbon, anthropogenic thermal energy and is a *significant* reservoir of anthropogenic carbon, primary biological production and respiration. The imprint of physical circulation on the global distribution of ecosystems is widely apparent, and Schmittner (2005) argues that major disruption of the Atlantic MOC will greatly impact ecosystems and global productivity.

Northern Atlantic climate itself involves several nested questions:

- What is the impact of oceanic heat storage on warming the wintertime atmosphere?

- Does northward heat transport by the ocean circulation greatly increase this warming?
- What are the secondary effects through the cryosphere, of the ocean circulation?
- What is the impact of ocean circulation on the climatological mean, the seasonal cycle, and the decade-to-century variability?
- What is the level of dynamical feedback between ocean and atmosphere in the wintertime Atlantic storm track?
- What is the effect of oceanic heat transport on the development of individual cyclonic systems?
- While such general questions can be mind-numbing they come alive when made specific:
- What keeps the Barents and Labrador Seas ice free?
- What caused the 1920s–1930s warming that engulfed the northern Atlantic Ocean and atmosphere and affected ecosystems widely?
- Is explosive cyclogenesis responding to the Gulf Stream front?
- Will the global MOC weaken significantly in the next few decades, as the majority of IPCC climate models predict?
- What are the dominant fresh-water pathways and their impact on deep water formation in the subpolar Atlantic?
- Will the widely predicted (again, by the mean of many IPCC climate models) predominantly zonal bands of precipitation change under global warming and cause a greatly wetter western Europe, great freshening of the northern Atlantic and Arctic, and stronger drought in the subtropical regions of descent in the atmospheric MOC?

Observations required to address these many questions have historically been sparse. Direct and indirect measurement of oceanic heat flux and storage requires time- and space series that have only gradually approached adequate resolution and sustained duration. Fortunately, promising new technologies bearing on thermodynamics are now available. There are multiple ways to constrain ocean–atmosphere heat transport and exchange with the atmosphere, through air–sea flux measurements and bulk-formulas based on wind speed and temperature difference, observations of atmospheric lateral flux, of top-of-the atmosphere radiant flux, of water-column lateral transport, and of water-column heat storage and of regions of water-mass formation and sinking. Such direct and indirect methods are summarised by Bryden and Imawaki (op. cit.). Of particular note is the recent capability provided by ARGO float hydrography and satellite altimetry, which together measure the steric and dynamic height of the oceanic water column (Willis et al. 2003; Hadfield et al. 2007). Inference of air–sea heat exchange by ingenious dynamic use of veering of the thermal wind velocity with depth is also promising (the ‘cooling spiral’ of Stommel 1979). By contrast key sources, sinks and transport pathways of fresh water for the ocean circulation and key sites of global ocean upwelling are far less well observed.

Sophisticated analyses of observations and their assimilation into climate models are aimed not only at refining the numbers, but they also can tell us whether our most basic picture of the workings of the MOC are correct. Two 'back of the envelope' calculations suggest the importance of the Atlantic MOC in the poleward transport of heat and freshwater.

First, heat transport: 16 Sv (Sverdrups or megatonnes second^{-1}) of mass transport (call it F_m) in the Atlantic MOC with a temperature difference of 15 K between upper, northward and deep, southward flowing branches yields a meridional heat transport, $\rho C_p \Delta\theta F_m$, of amplitude 1.0 pW ($1.0 \times 10^{15}$ W), which is comparable with results from both direct and indirect methods (e.g., Bryden and Imawaki 2001; ρ is density, C_p the specific heat capacity at constant pressure, $\Delta\theta$ the potential temperature difference).

Second, fresh-water transport: the same schematic 16 Sv of Atlantic oceanic mass transport is more saline in its northward, upper ocean flow and less saline in its deep Equatorward flow. This difference implies the low latitude evaporation and high-latitude precipitation and runoff which balance a compensating poleward atmospheric fresh-water flux.. The global, east–west integrated value of the water vapor transport, northward in atmosphere, returned southward in the oceans, is fairly convincingly estimated to have a peak value of 0.8 Sv. at about 40° N latitude, representing roughly 2 pW of latent heat transport (e.g., Trenberth and Caron 2001). Wijffels (2001) describes oceanographic determination of the southward return flow of 0.8 Sv of fresh water (riding on top of the net Arctic throughflow communicated through Bering Strait). The Atlantic fraction of this flux is estimated to be roughly 0.4 Sv of fresh water transport difference between 10° N and 50° N. A simple 'box-model' MOC would have a net fresh water transport ½ $(\Delta S/S)F_m$, where ΔS is the salinity difference between upper and deep branches of the flow. Observed ΔS of order 1 psu relative to a mean of 35 psu would support a transport of only 0.23 Sv fresh water, seemingly smaller than observed. Yet the Atlantic also exports moisture westward to the Pacific in the Trade Winds. LeDuc et al. (2007) cite 0.13–0.37 Sv of fresh-water jumping over Central America, which helps to explain this discrepancy. The horizontal-gyre component of the Atlantic circulation above the thermocline also contributes to the equatorward 0.8 Sv of oceanic fresh water transport. This simple reasoning is an example of transport played out on the potential temperature plane, first exploited by Stommel and Csanady (1980).

Some of the relatively new observational resources applicable to these questions are:

- Satellite radiation observations, for example, the ERBE and CERES sensors begun in 1984, and infrared sea-surface temperature measurements, recently extended to long-wave bands which see through cloud cover (AMSR-E sensor).
- Satellite altimetry by NASA Topex/POSEIDON/JASON instruments and European Space Agency instruments since 1992 providing global coverage of sea-surface height, which has a strong contribution from ocean water-column heat storage.
- Satellite scatterometer surface wind-fields, applicable to air–sea momentum and heat fluxes.

- Steady improvement in atmospheric circulation reanalyses providing essential detail for evaluating atmospheric and, as a residual oceanic, heat and moisture fluxes.
- Enhanced ocean observation programs (e.g., WOCE, RAPID and ASOF) targeting key ocean sections with repeated high-resolution hydrography.
- The ARGO float program, now approaching its goal of 3,000 drifting, hydrographic profiling floats in the world ocean.
- Robotic gliders directed to survey key hydrographic sections, boundary currents and convection zones.
- The historic data base of XBT and hydrographic temperatures, mined to reconstruct ocean heat storage time-series (e.g., Levitus et al. 2005).
- Ocean surface flux moorings.
- Deep-sea moorings in key boundary currents providing semi-quantitative mass transports.

Chemical tracer programs, especially CFCs, tritium, radiochemical effluents, carbon and standard nutrients and oxygen, providing quantitative estimates of some of the most difficult elements of the global circulation, particularly the global upwelling sites, diapycnal mixing rates, long-distance boundary current transports, and formation of water masses in 'stable' gyre centers. Tracers also measure air–sea interaction rates in their own way, and can constrain heat- and freshwater exchange across the sea surface.

4.2 The Contribution of the Atlantic Ocean Circulation to Wintertime Climate

The importance of ocean circulation to atmospheric climate has been challenged by Seager et al. (2002), hereinafter 'SO2', in their paper, 'Is the Gulf Stream responsible for Europe's mild winters?' While centering attention on the mild climate of Europe, their work, if correct would have greater consequences. They argue that:

(i) Only a small portion of the total northward heat transport north of 40° N, is accomplished by the ocean in comparison with the atmospheric heat transport.
(ii) Oceanic heat storage is local, with the summer's heating of the mixed layer being the dominant source of wintertime oceanic heat release to the atmosphere, with little contribution from oceanic heat transport.
(iii) Fresh-water transport coupled with heat transport can be neglected.

Here we show that while (i) is true it is misleading, (ii) is based on an analysis which is in error due to comparison of ocean heat transport and surface heat loss on different timescales. (iii) they have missed the most important climate interaction of all.

One could hardly argue against the persuasive reasoning provided by SO2 that the maritime climate maintains the temperature contrast between North America and Europe. However, their conclusions regarding the impact of the oceanic heat transport could be taken to mean that oceanic heat transport has no significant consequence for the climate in Europe and elsewhere, beyond a minor warming of 0–3 °C.

We address points (i)–(iii) in order.

(i) Satellite radiation measurements combined with atmospheric observations assimilated into models give estimates of total, atmospheric and oceanic meridional heat flux. One recent analysis elevates the atmospheric contribution somewhat, the atmospheric transport peaking between 4 and 5 PW (4 to 5 $\times 10^{15}$ W), while the ocean transport peaks at about 2 PW (Trenberth and Caron, op. cit.). However, as Bryden and Imawaki op. cit. emphasize (using transport estimates of Keith 1995), the meridional heat flux is comprised of *three* nearly equal (in amplitude) contributions from latent- and sensible heat flux (the latter known as dry static energy flux) in the atmosphere and sensible heat flux by the ocean. Latent heat is fresh water (2.4 pW per Sverdrup), and its transport is an intrinsically coupled ocean/atmosphere mode. Keith's transports, or the more recent transports, quantitatively similar, by Trenberth et al. op. cit., plotted against latitude show the dominance of subtropical ocean evaporation (typically 1.5 m year^{-1}) in driving the global system; this activity lies poleward of the transition from tropical Hadley circulation to the latitude of the midlatitude, eddy-driven jet stream. Each of the three modes of meridional energy transport has peak amplitude of roughly 2 pW, with the latent-heat mode carrying 0.8 Sv of fresh-water northward, mirrored by equatorward ocean transport. The moisture/latent heat pump of the Atlantic storm track is a crucial part of maritime climate. It is ignored in the thermodynamic discussion of SO2. The Trenberth and Caron, op. cit. discussion uses ERBE top-of-atmosphere radiation data and atmospheric observations/assimilation. Although few error estimates are presented, the occurrence of large heat flux divergence over land is suggestive of significant error.

Wunsch (2005) argues that in fact ERBE radiation observations add significant uncertainty, and provides an error analysis. By taking the ocean observations of heat transport and calculating the atmospheric heat transport as a residual, his analysis revises downward the atmospheric heat transport in the northern hemisphere. The maximum atmospheric transport now averages 4.1 pW (ranging between 3 and 5.2 pW at one standard deviation). Wunsch's analysis also gives a greater ocean transport at high northern latitudes than do Trenberth and Caron.

(ii) If the ocean (here, the Atlantic Ocean) participated in climate only through local, seasonal heat storage and release in the shallow mixed layer, then calculations of ocean circulation would be unnecessary for climate models. Indeed, SO2 state in their abstract that "..the majority of heat released during winter from the ocean to the atmosphere is accounted for by the seasonal release of heat previously absorbed and not by ocean heat flux convergence." This conclusion follows from their comparison of the *annual mean* oceanic heat transport convergence with the *wintertime* release of heat at the sea surface, the latter

> being much larger. Both are inferred using climatological mean surface heat flux fields for the Northern Atlantic, developed from the COADS ship observation dataset by da Silva et al. (1994). Values they estimate (averages north of 35° N) are 37 W m^{-2} heat convergence by the ocean circulation, vs 135 W m^{-2} wintertime heat release from surface observations. This is based on the estimate of 0.8 PW northward oceanic heat transport at 35° N.

The air/sea flux affecting oceanic water-column heat balance includes downward short-wave radiation (corrected for albedo related reflection), net long-wave radiation, sensible heat flux and latent heat flux. The air/sea flux affecting the atmosphere differs from this by the downward short-wave solar radiation, which heats the ocean but does not cool the atmosphere. Thus the maps of air–sea heat flux that matter for the atmosphere show much larger numbers than those we are familiar with, for the ocean. However we want to compare the air/sea heat flux with that of a mixed-layer-only, climatologically steady world, in which no lateral heat transport is allowed, and in this mixed-layer ocean the annual average flux vanishes. Thus to consider the non-seasonal, non-local heat storage and forcing of the atmosphere, the full air–sea heat flux including solar radiation is the relevant field.

Let us assume all these numbers in the paragraphs above are accurate. A model of the annual cycle would include year-round northward heat flux by the Atlantic circulation, *together with its release to the atmosphere in a few winter months.* During summer warming none of this deep heating escapes to the atmosphere. We thus should be comparing the time-averaged heat-flux convergence by the ocean circulation, *multiplied by the ratio 12/(number of months of wintertime heat loss),* with the upward heat flux at the sea surface observed during those winter months, or else simply annualize all the fluxes. The details depend upon the vertical distribution of the north–south heat advection (referenced to the late winter mixed-layer temperature). Using an estimate that half of the transport lies deeper than 100 m (above which depth most of the local, seasonal heating is trapped), suppose we release that heat in 3 winter months and release the other half from the upper 100 m during 6 months of the year. The surface heat flux during winter becomes augmented by a factor $12/6 \times ½ + 12/3 \times ½ = 3$. Multiplying 37 W m^{-2} from the SO2 estimate by 3 gives 111 W m^{-2}, enough to account for much of the observed winter upward heat flux at the sea surface (135 W m^{-2}). This argument shows that oceanic heat advection is plausibly important in warming the atmosphere in winter. Note with a linear model of heat storage in a mixed-layer-only ocean, SO2's procedure would be correct, for the laterally advected heat would be 'available' to the atmosphere in all seasons. The point is that much of it is in fact sheltered below the seasonal mixed layer during the warm months.

The same, or even more dramatic, result follows if we take air–sea heat flux climatology, with monthly surface heat flux averaged in the Atlantic north of 25° N, and integrate the flux with respect to time, Fig. 4.1. Start in spring, when the net surface flux changes sign and begins to warm the upper ocean; then integrate forward. In regions with annual average heat flux that is zero or upward, the integral

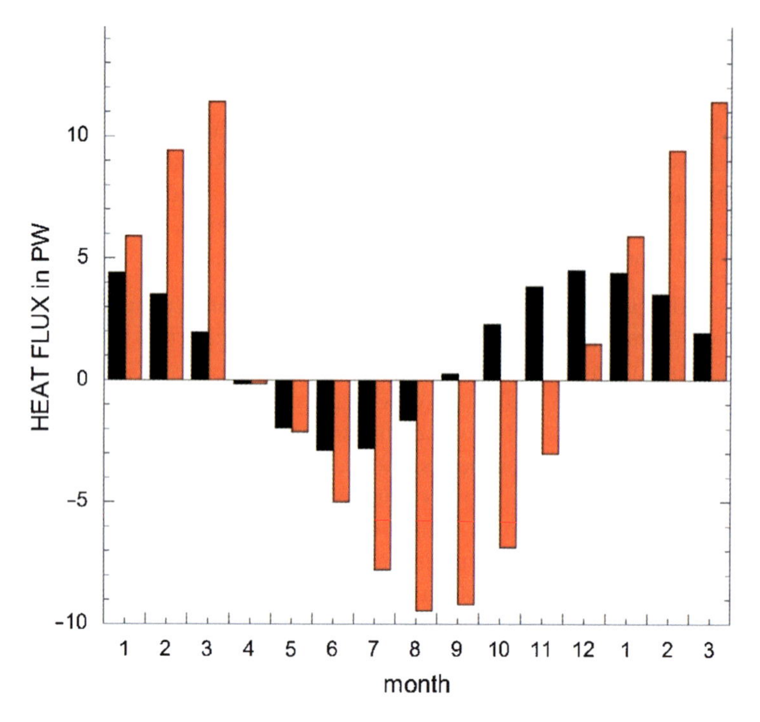

Fig. 4.1 Using the da Silva et al. (1994) air–sea heat flux estimates (black bars) we integrate forward in time (red bars), averaging over the Atlantic north of 25° N. When the integral returns to zero, the local, seasonal heating has been removed by autumnal cooling. On average, by early December the local heat source is exhausted and for the remainder of the winter oceanic warming of the atmosphere relies on heat imported by the ocean circulation. Positive values indicate heat loss from the ocean to the atmosphere. The NOC1.1a heat-flux climatology gives a very similar picture. The ordinate labels refer to the black bars and should range from -1.0 pW to 1.0 pW

will eventually come back to zero, indicating that the locally stored summer's heating has been removed by cooling from above. After this date, continuing upward heat flux must have been imported by the ocean circulation. Averaging north of 25° N latitude (and keeping north of the zero-mean air–sea heat flux line) in the Atlantic we see that by early to mid-December, the locally stored heat is exhausted, and excavation of imported heat dominates the rest of the winter. The geographical distribution of the year days is shown in Figs. 4.2a for NOC/SOC climatology

Fig. 4.2 (continued) subpolar Pacific. Deep red regions (year days >400) the heat balance is local, without significant lateral advection by the ocean circulation. Contour interval: 20 days. (b) Annual mean air–sea heat flux felt by the oceans (short-wave radiation, long-wave radiation, sensible- and latent-heat fluxes), from NOC1.1a data. The total upward heat flux felt by the atmosphere is this map without the net downward short-wave radiation, hence with much larger upward flux. This figure, however, represents the non-local heating of the atmosphere owing to the ocean circulation. Maximum values exceed 150 Wm^{-2} in the Sargasso Sea where roughly 0.5 pW of upward heat flux occurs in winter. Contour interval: 20 Wm^{-2}, zero contour bold black. (c) North Atlantic surface heat flux annual cycle ($W\ m^{-2}$) against year-day at two longitudes: 60° W (cyan) and 65° W (yellow), plotted from the Equator to 60° N. The curves with strongest upward (negative) wintertime heat flux in winter are in the Gulf Stream extension, ~40° N. The integrals of these curves produce the year-day when local seasonal heat storage is exhausted (Fig. 4.2a)

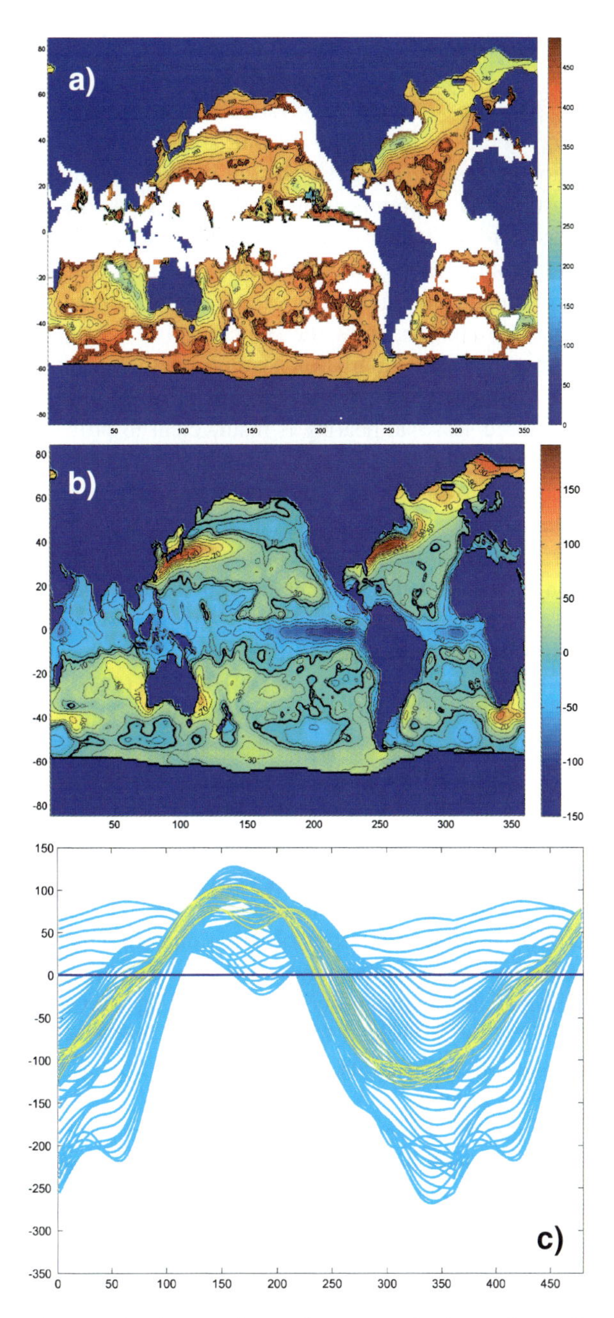

Fig. 4.2 (a) Year-day when local seasonal ocean heat storage has been exhausted by winter cooling. The northern and western Atlantic, Barents, Nordic and Labrador Seas fall in the range, day 225–350. (In the southern hemisphere 180 is subtracted from the year day so that seasonal color pattern is the same as in the northern hemisphere.) Thereafter in much of the late fall and through the winter, warming of the atmosphere by the ocean depends on imported heat flux by the lateral ocean circulation. White regions of annual-mean downward heat flux are never exhausted by wintertime cooling, and heat is exported from them by the ocean circulation. Based on NOC1.1a data. Regions of strong effect of ocean circulation on the atmospheric heat budget appear both east and west of Australia, in the Kuroshio and broadly in the subtropical Pacific, a small region of the

(Grist and Josey 2003, now termed NOC1.1a flux climatology). daSilva/Levitus (1994) climatology yields very similar results (not shown). The Gulf Stream/Sargasso Sea region shows the strongest effect of heat advection by the circulation, with early exhaustion (by September) of the locally stored heat. Yet in a band extending northeastward to the Nordic and Barents Seas, heat flux convergence by the ocean circulation supplies as much heat as does local seasonal heat storage. Plots of the ratio of mean convergence of oceanic lateral heat transport, divided by mean downward solar radiation, show the same northwest Atlantic region, where the contribution from the ocean circulation is significant. Generally speaking, there are large areas of the world ocean within the upward-mean-heat-flux regions, in which the locally stored seasonal heating is insufficient to provide more than half of the upward fall/winter heat flux. The net annual heat transport at the sea surface from NOC1.1a data is shown in Fig. 4.2b, where we include the solar short-wave radiation. The total oceanic warming of the atmosphere omits this term and hence is much larger. It includes contributions from both non-local advection of heat and local re-remission of some of the previously gained solar energy in the form of longwave, latent and sensible heat loss. The construction of Fig. 4.2a is perhaps made clearer by looking at the annual cycle of oceanic heat balance at two longitudes, Fig. 4.2c. Here the deep negative values correspond to the Gulf Stream extension region, where wintertime heat loss exceeds 300 W m^{-2}.

These ideas are all subject to accuracy of the consensus oceanic heat transports, and analysis of air–sea heat flux feedbacks due to the ocean circulation-induced SST. Improvement will occur when water-column heat storage observations become numerous enough. Indeed, wherever winter mixed layers exceed 50 –100 m in depth, we infer that ocean circulation is important, because seasonal surface heating cannot mix down deeper than this, even with the aid of the winds. This is a strong argument for sustained time-series observations of temperature and salinity as can be provided by floats, gliders and moorings. Several parallel arguments given in the SO2 paper, and a similar one given by Wang and Carton (2002) suffer from the same logical error pointed out here, for example when geographical distribution of winter air–sea heat flux is compared with annual-mean heat convergence by the ocean circulation.

SO2 remark also that the wintertime poleward heat transport in their calculation is much reduced in mid-latitude, and attribute this to southward transport in the shallow wind-driven Ekman layer. Other estimates of Ekman heat transport do not support such a large effect, and it is more likely that what they are seeing is the huge (0.5 pW) upward heat flux in the Gulf Stream/Sargasso Sea region which dominates Figs. 4.2 in subtropical latitudes. This upward heat flux reduces the wintertime poleward ocean heat transport, and is a part of the essence of our argument.

There are subjective elements in the model simulations of SO2. With suppressed ocean circulation their models show surface winter temperature changes of 6–12 °C over much of northern Eurasia, reaching 21 °C in Scandinavia. The average temperature change north of 35° N is 6 °C in their GISS-model. We would call these changes ‘large’; yet SO2 argue that they have “little impact”. The great differences apparent between their two simulations (one with a non-dynamical ice model, the other without any ice model) remind us of the complexity and uncertainty of coarsely resolved climate model results, when so many critical high-latitude and upper ocean physical processes are under-represented. And, more to the point, surgical removal of oceanic heat transport has other implications (iii, below).

While this has been a discussion of mean and seasonal wintertime heat balance, some aspects apply also to decadal and secular variability. The warming of northern Asia associated with greenhouse forcing, yet partially associated with strong positive phase of the North Atlantic Oscillation in the early 1990s, is shown by Thompson and Wallace (2001) to involve zonal heat advection: we take this to be a sign of Atlantic oceanic heating penetrating farther eastward over Asia.

(iii) The SO2 model experiments use oceanic mixed layer models which are only governed by heat exchange with the atmosphere and by (diagnosed) heat transport related to the oceanic MOC. This type of model ignores fresh-water flux and fresh-water transport which are known to play an important role in inhibiting heat release from the ocean and determining sinking regions of the meridional overturning circulation (MOC) at subpolar and polar latitudes: Too much fresh water at the surface stabilizes the water column and sea ice can form, changing fundamentally the seasonal cycle of heat exchange between the ocean and atmosphere. On spatial scales beyond the convective regions, the fresh water cycle and heat transport are coupled globally in the atmospheric latent heat flux, and through the thermohaline circulation as was first discussed by Stommel and Csanady (1980). This coupling is played out on the θ–S plane, which is the fundamental 'phase plane' of physical oceanography (e.g., Bailey et al. 2005). It is summarised in maps of integrated buoyancy, integrated from the surface downward (essentially upside-down dynamic height), which we can call 'convection resistance', C_R:

$$C_R(x,y,z_1) = g\int_{z_1}^{0}(\sigma_0(x,y,z) - \sigma_0(x,y,0))dz$$

where σ_0 is surface-referenced potential density, g is gravitational acceleration, and z is vertical coordinate. This quantity shows the amount of buoyancy that must be removed by air–sea interaction in order to convectively mix the water column to a depth z_1. Maps and sections of C_R (Bailey et al. 2005), Fig. 4.3, can be split into its respective salinity and temperature components, assuming an approximately linear equation of state. These maps illustrate how much influence over water-mass formation is provided by thin upper-ocean layers with low salinity. In the Labrador Sea, for example, Hátún et al. (2007) argue that fresh water advected off the west Greenland boundary currents and continental shelf control the geographic distribution of deep convection in winter. Similarly, Häkkinen et al. (2007a) map, for the Greenland–Norwegian seas, the contributions to upper ocean density from salinity and temperature, show-

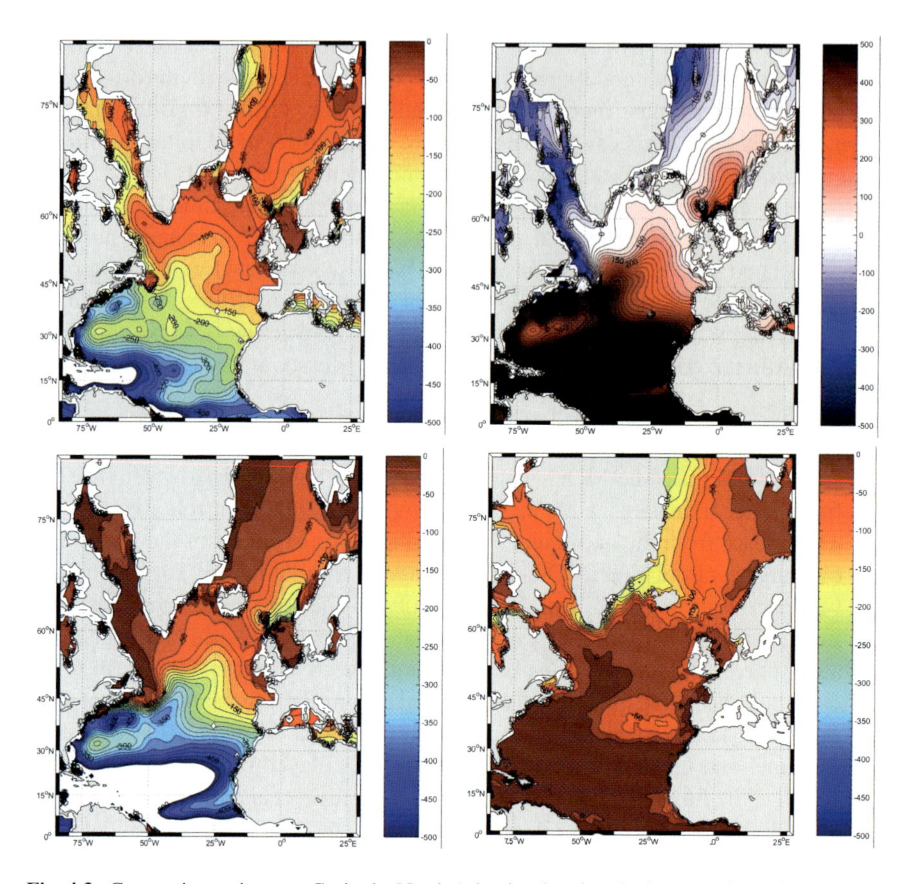

Fig. 4.3 Convection resistance, C_R, in the North Atlantic, showing the integrated density anomaly, relative to the sea surface in winter, integrated from the surface to 500 m depth. *Upper left*: total C_R in units of ppt m; *Lower left*: contribution of thermal stratification to C_R; *lower right*: salinity contribution to C_R; *upper right*: the *difference* between thermal and haline contributions to C_R. In blue regions of the upper right panel salinity stratification dominates, while in red regions temperature stratification dominates buoyant stability of the water column; for shallower depths, z_1 (not shown) upper ocean low-salinity layers more extensively dominate the northwest Atlantic

ing the strong imprint of surface fresh water advection near Greenland, and temperature advection near Norway. These distributions of upper ocean buoyancy control where deep convection occurs in winter, and hence where water-mass formation occurs; yet they are not likely to be modeled well by current climate models.

The global hydrologic cycle has a familiar pattern of high precipitation and runoff at high northern latitudes, evaporation in subtropical oceans, and narrow bands of evaporation and precipitation associated with the ITCZ. A net flux of fresh water from high northern latitudes to the low latitude evaporation sites is needed, even after river pathways are accounted for. The thermohaline MOC provides the return circuit for atmospheric vapor transport. In the North Pacific, low salinity stabilizes the surface layer of the subpolar gyre and there are no truly deep sinking regions. A shallow salinity minimum guided and subducted by the wind-driven Ekman transport, reaches toward the tropics. Yet much of the excess precipitation seems to escape through the Arctic (with the Bering Strait throughflow carrying low-salinity Alaskan coastal current water as well as water from midocean, Woodgate et al. 2006). The robust MOC in the Atlantic illustrates how the θ–S diagram couples the heat- and fresh-water transports, and involves both subpolar and Arctic water-mass transformations. The pioneering study of Stommel and Csanady (1980) gave simple two-degree of freedom illustrations of the nature of these coupled transports. They estimated the northward mass transport of salty waters and the compensating mass transport of less salty deep water using the observational estimates of heat and fresh water transport and water mass properties for the latitudes 40–45° N.

We wish to develop the 'back-of-the-envelope' calculation in the introduction, and reiterate the conclusions of Stommel and Csanady to show that heat transport and fresh water transport are intimately coupled. Removal of only one of them renders the problem meaningless. We consider a two-layer box model of the polar and subpolar oceans bounded by the Bering Strait and 45° N, using information of the 'known' mass fluxes at the surface, river runoff and at the Bering Strait. From the conservation of salt and fresh water we can diagnose the overturning to satisfy the equilibrium conditions and at the same time diagnose the heat transport when the upper-lower layer temperature difference is given. The following computation is done using the definitions of Wijffels et al. (1992) for fresh water and salt transports. For simplicity we assume densities to be 1,000 kg m^{-3} for the ocean and river and P–E fluxes. The inflow (Vbe) of the Bering Strait is 0.8 Sv with salinity (Sbe) 32.5 ppt. P–E flux over the area from the Bering Strait to 45° N is about 0.1 Sv and the runoff (R) from land in the same region amounts to about 0.19 Sv. At 45° N we want to solve the average flow (Vo) and the baroclinic transport V (all velocities are defined positive southward). The upper layer Atlantic salinity (Sa) is 35.3 ppt and the bottom layer salinity (Sb) is 34.9 ppt. The conservation equations for salt and fresh water are:

$$\text{Salt: } Vbe\ Sbe = (-V + Vo/2)\ Sa + (V + Vo/2)\ Sb$$

$$\text{Fresh water: } Vbe\ (1-Sbe) + R + P-E = (-V + Vo/2)\ (1-Sa) + (V + Vo/2)\ (1-Sb)$$

Substituting the above values in the conservation equation gives, for Vo and V, 1.09 Sv and 30.65 Sv, respectively. This simple scheme illustrates the thermohaline nature of the fresh water redistribution, where the northward mass transport in the upper layer is 29.04 Sv and the southward transport in the bottom layer is 30.11 Sv. If the temperature difference between the upper and lower layer is 8 °C, the northward heat transport would be about 0.9 PW at 45° N, which is close to the current estimates of ocean heat transport at 40° N representative of the present climate (e.g., Bryden and Imawaki 2001). Thus based solely on conservation of salt and fresh water, with hydrographic data we can diagnose the overturning and the associated heat transport to satisfy the equilibrium conditions when the various fresh water fluxes of the present climate are given. This traditional overturning picture of meridional heat transport does conflict with recent arguments suggesting that the deep branches of the MOC are unimportant for transporting heat (Boccaletti et al., 2005); their argument continues to depend on a particular choice of reference temperature, which has little effect on the arguments given here.

We have arrived at the crux of the problem not considered in the numerical experiments of SO2. Removal of the oceanic heat transport due to the thermohaline circulation means also that the redistribution of the fresh water is blocked which in the real world would lead to accumulation of fresh water at the high latitudes. The lack of the thermohaline circulation intensifies freshening because no salt is transported northwards. Fresh water accumulation will eventually build an extensive sea ice cover north of 40° N and influence the seasonal uptake of heat in the ocean. This is consistent with the paleo-records showing that periods of extensive ice cover over the high latitude ocean, and over the European and North American continents, were associated with weak production of North Atlantic deep water (Boyle and Keigwin 1982, 1987) and thus a weak thermohaline circulation. So in fact during the height of the last glaciation, the maritime effect was reduced to a minimum, and the temperature gradient across the Atlantic vanished.

In summary, accounting for the fresh water accumulation at the high latitudes alters significantly the picture suggested by climate models that would neglect the oceanic MOC: It is the existence of the oceanic heat transport that allows the maritime effect to operate in the northern North Atlantic and to create a milder European climate than in the North America; without the heat transport, ice would likely extend over much greater areas of ocean and land. Since the northward heat transport and southward fresh water transport in the Atlantic are strongly tied together, removing oceanic heat transport influences the climate and atmospheric circulation in ways that are not possible to simulate with a simple mixed layer model coupled to an atmospheric model. This also suggests that use of this type of model with a fixed oceanic heat transport (today's climate) is not suitable to describe climatic states where the thermohaline circulation is expected to change significantly from the present, as might happen for instance in doubled CO2 scenarios where the fresh-water input at high latitudes can increase by 40% or more (Manabe and Stouffer 1994). The signature of oceanic heat transport is deep convective mixing in winter, which accesses energy well below the ~50–100 m penetration of local summertime warming. Improved global mapping of winter mixed-layer depth using ARGO, XBT lines and other water column observations should go far toward identifying these regions.

Removal of one piece of a complex machine (here, the oceanic heat transport) can have unforeseen consequences. We have pointed out some, and there may be others, such as effects on cloudiness, atmospheric standing waves and storm tracks. The conclusion of SO2 that the particular climate feature of interest, the warming of western Europe, is 'fundamentally caused by the atmospheric circulation interacting with the oceanic mixed layer', and thus 'does not require a dynamical ocean' is flawed in the three aspects described above.

4.3 Atmosphere: Ocean Fluxes of Heat and Freshwater

4.3.1 Currently Available Estimates

Estimates of the ocean–atmosphere fluxes of heat and freshwater are available from a number of sources. Gridded monthly mean surface heat flux datasets were first produced from voluntary observing ship and buoy meteorological observations using a bulk formula approach (e.g., Bunker 1976; da Silva et al. 1994; Josey et al. 1999). More recently, atmospheric model reanalyses have provided an alternative, widely used source of flux estimates, the two principal datasets being the NCEP/NCAR (Kistler et al. 2001) and ECMWF reanalyses (Uppala et al. 2005). Attempts are now also being made to produce flux datasets by applying the bulk formula approach to combinations of reanalysis and satellite based meteorological fields (Yu and Weller 2007). Indirect estimates of the net air–sea heat flux have also been obtained using residual techniques that employ top-of-the atmosphere radiative flux measurements from satellites and estimates of the atmospheric flux divergence from reanalyses (e.g., Trenberth and Caron 2001).

Precipitation estimates are also available from the reanalyses and, for 1979 onwards, from satellite observations. The Global Precipitation Climatology Project Version-2 (GPCPV2) Monthly Precipitation Analysis dataset (Adler et al. 2003) incorporates precipitation measurements from satellite and rain gauges which are merged in an analysis that retains the best features of each dataset. The resulting dataset is independent of the reanalyses and is the leading satellite/rain gauge based set of precipitation fields currently available. Note, however, that there remain large differences between the various precipitation datasets and thus the freshwater flux field is more poorly determined than the net heat flux.

Significant differences exist between the various datasets in many regions of the ocean and these reflect the difficulty in obtaining accurate estimates of the fluxes. Specific problems include:

(i) Poor sampling in regions away from the major shipping lanes (e.g., the high latitude North Atlantic (see Josey et al. 1999, Fig. 4.2; see also Gulev et al. 2007).
(ii) Uncertainty over the values of the transfer coefficients which appear in the bulk formula for the sensible and latent heat fluxes (although significant progress has been made with the development of the COARE algorithm, Fairall et al. 2003).
(iii) Differences in the spatial and temporal averaging methods used to produce the gridded flux fields.
(iv) Poor representation of clouds in the atmospheric models used for the reanalyses which can have a major impact on shortwave and longwave flux estimates (e.g., Cronin et al. 2006).

A detailed review of flux estimation techniques and associated sources of error is provided in the report of the WMO/SCOR Working Group on Air–Sea Fluxes (WGASF 2000).

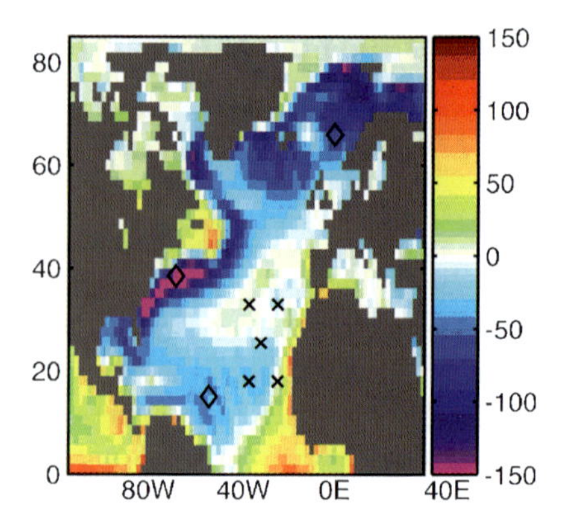

Fig. 4.4 Annual mean heat flux field from NCEP reanalysis for the period 1949–2001, in units of Wm^{-2}. Also shown are the presently maintained flux reference sites (black diamonds) and the NTAS and CLIMODE moorings and OWS M, and the earlier Subduction Experiment flux buoy array (crosses)

The main features of the net heat flux field in the North Atlantic are common to each of the various datasets currently available and are illustrated in Fig. 4.4, which shows the annual mean field from the NCEP/NCAR reanalysis (Kistler et al. 2001) for the period 1949–2001. In particular, there is strong net heat loss over the Gulf Steam region, of order 150 Wm^{-2} in the annual mean, with winter month averages (not shown) up to 400 Wm^{-2}, and a transition to heat gain at more southerly latitudes. The Nordic and Labrador Seas also experience strong cooling, but it should be noted that these regions are poorly sampled and thus the values there may be biased low as the reanalysis surface flux estimates are reliant to a certain extent on the assimilation of surface observations. Furthermore, they have difficulty (because of the relatively coarse spatial resolution in the atmospheric models employed) in representing small spatial scale features such as the central Labrador Sea and Greenland tip-jet that may be key to fully understanding the location and processes by which deep ocean convection occurs (e.g., Hátún et al. 2007; Lilly et al. 1999, 2003; Pickart et al. 2003). Use of subsurface observations of water-column heat storage are promising for the future, as ARGO, repeat hydrography lines and glider-based hydrography become more plentiful.

A measure of the uncertainty in the net heat flux field is provided by Fig. 4.5 which shows the variation with latitude in the North Atlantic of the zonal mean net heat flux for five recent climatological datasets: NCEP/NCAR, ECMWF, Trenberth residual (Trenberth and Caron 2001), NOC1.1a (formerly termed the adjusted SOC climatology, Grist and Josey 2003) and adjusted UWM/COADS (da Silva et al. 1994). There is some dispersion between the datasets with typical differences at the 20–30 Wm^{-2} level in the zonal annual mean; these differences are likely to be further amplified when monthly means for specific locations are considered. In the absence of high quality independent flux measurements it has not been possible to firmly establish the reasons for these differences and thereby narrow the gap between the different estimates. However, there is now the prospect for significant progress on this front as a result of the increasing number of moorings within the surface flux reference site array (also shown on Fig. 4.4). Of particular interest is the CLIMODE mooring deployed in November 2005 at 38.5° N, 65° W, which samples the strong

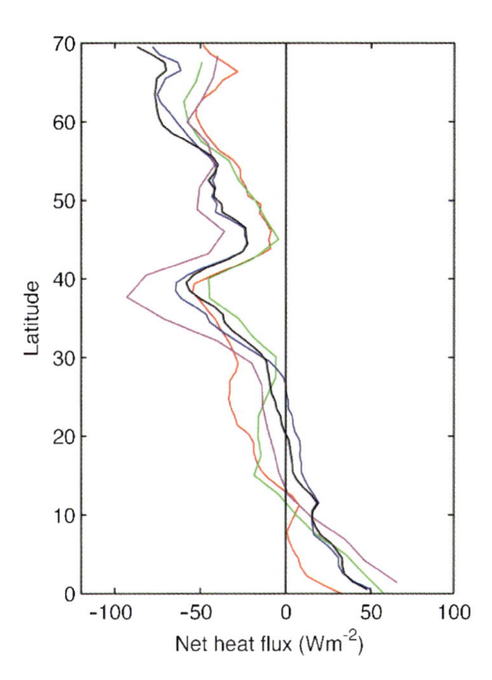

Fig. 4.5 Zonally averaged annual mean net heat flux in the North Atlantic for five different surface flux climatologies: NCEP/NCAR (red), ECMWF (green), Trenberth residual (magenta), NOC1.1a (black) and UWM/COADS adjusted (blue)

heat loss region towards the western boundary of the Atlantic which is a major source of uncertainty (Josey et al. 1999). Detailed analysis of the flux time series from this and other reference sites in the next few years is expected to firmly establish the causes of uncertainty (biased flux algorithms, differences in analysis procedures, sampling issues) and ultimately lead to more accurate flux estimates.

4.3.2 Evaluation Methods

Given the differences between the gridded flux datasets discussed above, and the advent of hybrid products obtained through various combinations of reanalysis, satellite and ship fields (Yu and Weller 2007; Large and Yeager 2004) together with flux fields from ocean synthesis (e.g., Stammer et al. 2004), a common method of evaluation is needed to provide a means by which their accuracy can be compared and potential biases identified. To this end, a set of guidelines for evaluation of flux products has recently been developed (Josey and Smith 2006). Previous studies have been limited by the availability of high quality reference observations which comprise both

(i) Local measurements of the fluxes from research buoys/vessels.
(ii) Large-scale constraints, principally estimates of heat and freshwater transports across hydrographic sections, from which regionally averaged fluxes can be inferred.

However, there has been a significant increase in the number of reference observations in recent years as noted above which will enable significant progress towards a more accurate picture of ocean–atmosphere interaction. Specific examples of flux evaluations using the limited amount of data available to date are now discussed.

(a) Comparisons with Local Flux Reference Data

Renfrew et al. (2002) found that in the Labrador Sea, NCEP overestimates the sensible and latent heat fluxes by 51% and 27%, respectively. They ascribed these biases to an inappropriate choice for the roughness length formula in the NCEP reanalysis under large air–sea temperature difference and high wind speed conditions. Thus, they were able to extend conclusions drawn from an analysis in a specific region to provide an indication of biases that are likely to arise in other regions experiencing similar conditions (e.g., the Gulf Stream and Kuroshio in winter).

(b) Evaluation Using Ocean Heat Transports

Hydrographic estimates of the heat and freshwater transport typically along zonal sections may be used to identify biases in net air–sea heat flux and net evaporation datasets by comparison with the climatologically implied property transport. Grist and Josey (2003) carried out such an evaluation of the heat transport for various datasets and Fig. 4.6 is an updated version of their Fig. 4.9a. Agreement within the error bars is typically obtained in the North Atlantic but the implied heat transport for the ECMWF dataset becomes unrealistically high in the South Atlantic. Further insight is obtained by considering regional differences using section pairs, as discussed by Grist and Josey (2003), which reveals that there is an underestimate of the ocean heat gain in the Tropical Atlantic in the ECMWF reanalysis. With a bit of oceanic chauvinism we point out that classic hydrographic ocean observations first analyzed by Hall and Bryden (1982) at 24° N in the Atlantic so accurately portrayed the MHT of the ocean that the atmospheric scientists were forced to re-evaluate the atmospheric MHT upward by almost 50%.

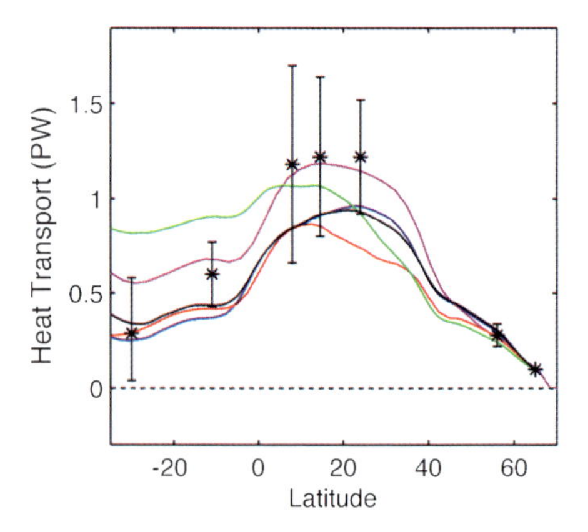

Fig. 4.6 Climatologically implied ocean heat transport (in pW, 10^{15} W) in the North Atlantic for five different surface flux climatologies: NCEP/NCAR (red), ECMWF (green), Trenberth residual (magenta), NOC1.1a (black) and UWM/COADS adjusted (blue). Hydrographic estimates are indicated by star symbols with error bars

4.3.3 Specific Mid-High Latitude Regions

(a) North Atlantic Subpolar Gyre. The subpolar gyre has shown significant decadal variability in both gyre strength and the salinity of the major water masses. Long-term freshening of the gyre from the 1960s through to the mid-1990s as part of a wider pattern of change in the freshwater balance of the Atlantic is now well documented (e.g., Curry et al. 2003). This freshening has recently been linked to increases in net precipitation over the ocean, river input, ice attrition and glacial melt over a broader domain including the Arctic Ocean (Peterson et al. 2006). The change in precipitation is driven partly by the multidecadal upward trend in the North Atlantic Oscillation (NAO) although detailed analysis of the eastern gyre region has shown that the second mode of sea level pressure, the east Atlantic Pattern also plays a significant role (Josey and Marsh 2005). A few regions of intense deep convection have been studied with dedicated observations over many years (e.g., Lilly et al. op. cit.). Gyre fluctuations have been inferred from satellite altimetry and hydrography (Häkkinen et al. 2004; Hátún et al. 2005).

The freshening trend appears to have partly reversed over the last decade from about 1995 onwards as the water entering the Nordic Seas from the gyre has become more saline (Hátún et al. 2005). This increase in salinity may reflect an increase in the amount of subtropical gyre water being advected north as a result of a weakening of the subpolar gyre (Häkkinen et al. 2004, 2007a, b).

(b) Nordic Seas. Decadal variability in the surface forcing of the Nordic Seas is particularly difficult to quantify given the lack of observations in this region. The major influence is likely to be the NAO as observations of deep convection in the Greenland Sea show a strong anticorrelation with the NAO (Dickson et al. 2000). Convective activity was particularly strong in the 1970s during which time the NAO was predominantly negative leading to enhanced heat loss in the Greenland Sea. The relative roles of heat loss and wind stress in controlling deep convection and subsequent variability of the deep outflows to the Atlantic remains to be fully established. Grist et al. (2007) find from a coupled model analysis variations in the Denmark Strait transport of up to 30% in response to Greenland Sea heat flux variability. However, other studies have suggested that variations in the wind field are the dominant factor controlling the overflow (e.g., Biastoch and Käse 2003).

4.4 Conclusion

Evaluation of the meridional transports of heat and fresh-water in the ocean, using several independent means, suggests that fundamentals of Earth's climate are indeed responsive to the ocean circulation, in the mean and seasonally, and likely (though not discussed here) at decadal time-scales. The degree of active feedback between ocean and atmosphere in each case is still controversial, yet will be refined by rapidly improving models and observations. The upward heat flux at the sea surface, in places reaching wintertime averages of hundreds of W m^{-2}, and exceeding the net solar radiation at the surface (annual mean of order 100 W m^{-2}, and far weaker in winter), is a significant contribution to atmospheric climate. Both decadal variability and persistent global warming have the potential to alter these heating and moistening patterns greatly. Indications of an increasing hydrologic cycle are already documented (e.g., Liu and Curry 2006; Curry et al. 2003).

The imprint of oceanic upward heat flux, and its enhancement by the ocean circulation, on the atmosphere is so apparent in diagnosed diabatic heating maps for the atmosphere (e.g., Held et al. 2002), in the existence of ice-free ocean at high latitude, like the Norwegian, Barents and Labrador Seas, and in the general warmth and moisture content of the storm track winds that it is inconceivable that climate models could neglect it. Meridional energy-transport is a strongly interacting collaboration of warm, moist storm track winds and warm underlying ocean. The fresh water cycle, coupled with the heat flux, involves massive subtropical evaporation from the oceans, followed by poleward transport in the storm track circulations, demonstrably concentrated over the oceanic sectors of the northern hemisphere, and finally by precipitation at high latitude. Indeed, the 3 km ice-mountain of Greenland is a living record of this Atlantic storm-track transport. Were the dynamical ocean to be replaced by a thin mixed layer, some aspects of seasonal heat and local evaporative forcing would remain. It is difficult to believe, however, that the intense warming and moistening of the northward moving air masses would continue, nor would the distributions of deep convection and water mass formation, nor the geography of sea-ice cover, unless those model mixed layers were artificially forced to mimic the true surface conditions of the ocean. Papers like SO2 stimulate us to observe more accurately the vertical structure of energy and fresh-water transport, and, just as important, to move toward descriptions of the ocean circulation through transports across sections and transformation within 'boxes', played out on the θ–S plane.

Acknowledgments PBR is supported by National Science Foundation's Office of Polar Programs and by NOAA's Arctic Research Office, both in connection with the ASOF program. SH is supported by NASA, and SJ by the UK Natural Environment Research Council at the National Oceanography Centre.

References

Adler, R.F., G.J. Huffman, A. Chang, R. Ferraro, P-P. Xie, J. Janowiak, B. Rudolf, U. Schneider, S. Curtis, D. Bolvin, A. Gruber, J. Susskind, P. Arkin and E. Nelkin (2003) The Version-2 Global Precipitation Climatology Project (GPCP) Monthly Precipitation Analysis (1979–Present). J. *Hydromet.*, **4**: 1147–1167.

Bailey, D.A., P.B. Rhines and S. Häkkinen (2005) Pathways and formation of North Atlantic Deep Water in a coupled model of the Arctic-North Atlantic Oceans. *Clim. Dyn.*, **24**, doi:10.1007/s00382-005-0050-3

Biastoch, A. and R.H. Käse (2003) The sensitivity of the Greenland-Scotland Ridge overflow to forcing changes. J. *Phys. Oceanogr.*, **33**: 2307–2319.

Boccaletti, G., R. Ferrari, A. Adcroft, D. Ferreira and J. Marshall (2005) The vertical structure of ocean heat transport. *Geophys. Res. Lett.*, **32**, doi:10.1029/2005GL022474.

Boyle, E. and L.D. Keigwin (1982) Deep circulation of the North Atlantic over the last 20,000 years: geochemical evidence. *Science*, **218**: 784–787.
Boyle, E. and L.D. Keigwin (1987) North Atlantic thermohaline circulation during the past 20,000 years linked to high latitude surface temperature. *Nature*, **334**: 333–335.
Bryden, H. and S. Imawaki (2001) Ocean heat transport. *In* Ocean Circulation and Climate, Siedler, Church and Gould (Eds.). Academic Press, London; 455–474.
Bunker, A.F. (1976) Computations of surface energy flux and annual air-sea interaction cycles of the North Atlantic Ocean. *Mon. Wea. Rev.*, 104: 1122–1140.
Cronin, M.F., N.A. Bond, Fairall, C.W. and R.A. Weller (2006) Surface cloud forcing in the East Pacific stratus deck/cold tongue/ITCZ complex. J. *Clim.*, 19(3): 392–409.
Curry, R., B. Dickson and I. Yashayaev (2003) A change in the freshwater balance of the Atlantic Ocean over the past four decades. *Nature*, 426: 826–829.
da Silva, A.M., C.C. Young and S. Levitus (1994) Atlas of Surface Marine Data Vol. 1: Algorithms and Procedures. *NOAA Atlas series*, pp. 74.
Dickson, R.R., T.J. Osborn, J.W. Hurrell, J. Meincke, J. Blindheim, B. Adlandsvik, T. Vinje, G. Alekseev and W. Maslowski (2000) The Arctic Ocean response to the North Atlantic Oscillation. J. *Clim.*, **13**: 2671–2696.
Dong, B. and R.T. Sutton (2005) Mechanism of interdecdal thermohaline circulation variability in a coupled ocean-atmosphere GCM. J. *Clim.*, **18**: 1117–1135.
Fairall, C.W., E.F. Bradley, J.E. Hare, A.A. Grachev and J.B. Edson (2003) Bulk parameterization of air-sea fluxes: Updates and verification for the COARE algorithm. J. *Clim.*, **16**(4): 571–591.
Fox, A.D. and K. Haines (2003) Interpretation of water mass formation deduced from data assimilation. *J.Clim.*, **33**: 485–498.
Grist, J.P. and S.A. Josey (2003) Inverse analysis of the SOC air-sea flux climatology using ocean heat transport constraints, J. *Clim.*, **16**: 3274–3295.
Grist, J.P., S.A. Josey and B. Sinha (2007) Impact on the ocean of extreme Greenland Sea heat loss in the HadCM3 coupled ocean-atmosphere model. J. *Geophys. Res.*, 112, C04014, doi:10.1029/2006JC003629.
Gulev, S., T. Jung and E. Ruprecht (2007) Estimation of the impact of sampling errors in the VOS observations on air–sea fluxes. Part I: uncertainties in climate means. J. *Clim.*, **20**: 279–301.
Hadfield, N.C. Wells, S.A.Josey and J. J-M Hirschi (2007) Accurac of North Atlantic temperature and heat storage fields from ARGO. J. *Geophys Res.*, **112**, C01009, doi:10.1029/2006JC003825.
Hall, M.M. and H.L. Bryden (1982) Direct estimates and mechanisms of ocean heat transport. *Deep-Sea Res.*, **29**: 339–359.
Häkkinen, S. and P.B. Rhines (2004) Decline of the North Atlantic subpolar circulation in the 1990s. *Science*, **304**: 555–559.
Häkkinen, S., F. Dupont, M. Karcher, F. Kauker, D. Worthen J. and Zhang (2007a) Model simulation of Greenland Sea upper-ocean variability *J. Geophys. Res.*, **112**, No. C6, C06S9010.1029/2006JC003687
Häkkinen, S. and P.B. Rhines (2007b) Shifting circulations in the northern Atlantic Ocean. *Science*, submitted.
Hallberg, R. and Gnanadesekin, A. (2006) The role of eddies in determining the structure and response of the wind-driven Southern Hemisphere overturning: results from the modeling eddies in the Southern Ocean (MESO) project. J. *Phys. Oceanogr.*, **36**: 2232–2252.
Hátún, H., A.B. Sandø, H. Drange, B. Hansen and H. Valdimarsson (2005) Influence of the Atlantic subpolar gyre on the thermohaline circulation. *Science*, **309**: 1841–1844.
Hátún, H., C.E. Eriksen, P.B. Rhines and J. Lilly (2007) Buoyant eddies entering the Labrador Sea observed with gliders and altimetry. *J. Phys. Oceanogr.*, Dec.
Held, I., M.F. Ting and H. Wang (2002) Northern stationary waves: theory and modeling. J. *Clim.*, **15**: 2125–2144.
Keith, D.W. (1995) Meridional energy transport: uncertainty in zonal means. *Tellus*, **47A**: 30–44.
Josey, S.A., E.C. Kent and P.K. Taylor (1999) New insights into the ocean heat budget closure problem from analysis of the SOC air-sea flux climatology. J. *Clim.*, **12**(9): 2856–2880.

Josey, S.A. and R. Marsh (2005) Surface freshwater flux variability and recent freshening of the North Atlantic in the Eastern Subpolar Gyre. J. *Geophys. Res.*, **110**, C05008, doi:10.1029/2004JC002521.

Josey, S.A. and S.R. Smith (2006) Guidelines for Evaluation of Air-Sea Heat, Freshwater and Momentum Flux Datasets, CLIVAR Global Synthesis and Observations Panel (GSOP) White Paper, July 2006, pp. 12. Available at http://www.noc.soton.ac.uk/JRD/MET/gsopfg.pdf.

Kistler, R., E. Kalnay, W. Collins, S. Saha, G. White, J. Woollen, M. Chelliah, W. Ebisuzaki, M. Kanamitsu, V. Kousky, H. van den Dool, R. Jenne and M. Fiorino (2001) The NCEP–NCAR 50-year reanalysis: monthly means CD-ROM and documentation. *Bull. Am. Met. Soc.*, 82: 247–267.

Large, W.G. and S.G. Yeager (2004) Diurnal to decadal global forcing for ocean and sea-ice models: the data sets and flux climatologies. NCAR Technical Note NCAR/TN-460 + STR, 111 pp.

Leduc, G., L. Vidal, K. Tachikawa1, F. Rostek, C. Sonzogni, L. Beaufort and E. Bard (2007) Moisture transport across Central America as a positive feedback on abrupt climatic changes. *Nature*, **445**, 908–911, doi:10.1038/nature05578

Lilly, J.M., P.B. Rhines, M. Visbeck, R. Davis, J.R.N. Lazier, F. Schott, and D. Farmer (1999) Observing deep convection in the Labrador Sea during winter, 1994–1995. *J. Phys. Oceanogr.*, 29: 2065–2098.

Lilly, J.M., P.B. Rhines, F. Schott, K. Lavender, J. Lazier, U. Send and E. d'Asaro (2003) Observations of the Labrador Sea eddy field. *Prog. Oceanogr.*, 59: 75–176.

Liu, J. and J. Curry (2006) Variability of the tropical and subtropical ocean surface latent heat flux during 1989–2000. *Geophys. Res. Lett.*, **33**, doi:10.1029/2005GL023809.

Levitus, S., J. Antonov and T. Boyer (2005) Warming of the world ocean, 1955–2003. *Geophys. Res. Lett.*, **32**, doi:10.1029/2004GL021592

Lumpkin, R. and K. Speer (2003) Large-scale vertical and horizontal circulation of the Atlantic Ocean. J. *Phys. Oceanogr.*, **33**: 1902–1920.

Manabe, S. and R. Stouffer (1994) Multiple century response of a coupled ocean-atmosphere model to a n increase of atmospheric carbon dioxide. J. *Clim.*, **7**: 5–23.

Marsh, R., S.A. Josey, A.J.G. Nurser, B.A. de Cuevas, and A.C. Coward (2006) Water mass transformation in the North Atlantic over 1985–2002 simulated in an eddy-permitting model. *Ocean Sci.*, 1: 127–144.

Peterson, B.J., J. McClelland, R. Curry, R.M. Holmes, J.E. Walsh, K. Aagaard (2006) Trajectory shifts in the Arctic and Subarctic freshwater cycle. *Science*, **313**: 1061–1066.

Pickart, R.S., M.A. Spall, M.H. Ribergaard, G.W.K. Moore and R.F. Milliff (2003) Deep convection in the Irminger Sea forced by the Greenland tip jet *Nature*, **424**: 152–156.

Rhines, P.B. (2006) Sub-Arctic oceans and global climate. *Weather*, **61**: 109–118.

Renfrew, I.A., G.W.K. Moore, P.S. Guest and K. Bumke (2002) A comparison of surface-layer and surface heat flux observations over the Labrador Sea with ECMWF and NCEP reanalyses. J. *Phys. Oceanogr.*, **32**: 383–400.

Sarmiento, J.L., N. Gruber, M. Brzezinski and J.P. Dunne (2004) High-latitude controls of thermocline nutrients and low latitude biological productivity. *Nature*, **427**: 56–60.

Schmittner, K. (2005) Decline of the marine ecosystem caused by a reduction in the Atlantic overturning circulation. *Nature*, **434**: 628–632.

Seager, R., D.S. Battisti, J. Yin, N. Gordon, N. Naik, A.C. Clement, and M.A. Cane (2002) Is the Gulf Stream responsible for Europe's mild winters? *Q. J. Roy. Met. Soc.*, **128**: 2563–2586.

Stammer, D., K. Ueyoshi, A. Kohl, W.G. Large, S.A. Josey and C. Wunsch (2004) Estimating air-sea fluxes of heat, freshwater and momentum through global ocean data assimilation. J. *Geophys. Res.*, **109**, C05023, doi:10.1029/2003JC002082.

Stommel, H.M. (1979) Oceanic warming of western Europe. *Proc. Nat. Acad. Sci.*, **76**: 2518–2521.

Stommel, H.M. and G.T. Csanady (1980) A relation between the T-S curve and global heat and atmospheric water transports. J. *Geophys. Res.*, **85**: 495–501.

van der Swaluw, E., S.S. Drijfhout and W. Hazeleger (2007) Bjerknes compensation at high Northern latitudes: the ocean forcing the atmosphere. J. *Clim.* submitted.

Talley, L. (2003) Shallow, deep and intermediate components of the global heat budget. J. *Phys. Oceanogr.*, **33**: 530–560.

Thompson, D. and J.M. Wallace (2001) Annular modes in the extratropical circulation: Part I: Month-to-month variability. J. *Clim.*, **13**: 1000–1016.

Toggweiler, R. and B.Samuels (1995) Effect of Drake Passage on the global thermohaline circulation. *Deep-Sea Res.*, **42**: 477–500.

Toggweiler, R. and B. Samuels (1998) On the ocean's large-scale circulation near the limit of no vertical mixing. J. *Phys. Oceanogr.*, **28**: 1832–1852.

Trenberth, K. and J.M. Caron (2001) Estimates of meridional atmosphere and ocean heat transports. J. *Clim.*, **14**: 3433–3443.

Trenberth, K.E., J.M. Caron and D.P. Stepaniak (2001) The atmospheric energy budget and implications for surface fluxes and ocean heat transports. *Clim. Dyn.*, **17**: 259–276.

Uppala, S.M., P.W. Kållberg, A.J. Simmons, U. Andrae, V. da Costa Bechtold, M. Fiorino, J.K. Gibson, J. Haseler, A. Hernandez, G.A. Kelly, X. Li, K. Onogi, S. Saarinen, N. Sokka, R.P. Allan, E. Andersson, K. Arpe, M.A. Balmaseda, A.C.M. Beljaars, L. van de Berg, J. Bidlot, N. Bormann, S. Caires, F. Chevallier, A. Dethof, M. Dragosavac, M. Fisher, M. Fuentes, S. Hagemann, E. Hólm, B.J. Hoskins, L. Isaksen, P.A.E.M. Janssen, R. Jenne, A.P. McNally, J.-F. Mahfouf, J.-J. Morcrette, N.A. Rayner, R.W. Saunders, P. Simon, A. Sterl, K.E. Trenberth, A. Untch, D. Vasiljevic, P. Viterbo and J. Woollen (2005) The ERA-40 re-analysis. *Q. J. Roy. Met. Soc.*, **131**: 2961–3012. doi:10.1256/qj.04.176.

Wang, J. and J.A. Carton (2002) Seasonal heat budget of the Pacific and Atlantic Oceans. *J. Phys. Oceanogr.*, **32**: 3474–3489.

WGASF (2000) Intercomparison And Validation Of Ocean-Atmosphere Energy Flux Fields, Final Report Of The Joint WCRP/SCOR Working Group On Air-Sea Fluxes SCOR Working Group 110, P.K. Taylor (Ed). 312.

Wijffels, S.E. (2001) Ocean transport of fresh water. *In* Ocean Circulation and Climate, Siedler, Church and Gould (Eds.). Academic Press, London: 475–488.

Wijffels, S.E., R.W. Schmitt, H.L. Bryden and A. Stigebrandt (1992) Transport of freshwater by the oceans. *J. Phys. Oceanogr.*, **22**: 155–162.

Willis, J., D. Roemmich and B. Cornuelle (2003) Combining altimetric height with broadscale profile data to estimate steric height, heat storage, subsurface temperature, and sea-surface temperature variability. J. *Geophys. Res.*, **108**, doi:10.1029/2002JC001755.

Woodgate, R., K. Aagaard and T. Weingartner (2006) Interannual changes in the Bering Strait fluxes of volume, heat and freshwater between 1991 and 2004. *Geophys. Res. Lett.*, **33**, L15609, doi:10.1029/2006GL026931.

Wunsch, C. (2005) The total meridional heat flux and its oceanic and atmospheric partition. J. *Clim.*, **18**: 4374–4380.

Wunsch, C. and R. Ferrari (2004) Vertical mixing, energy and the general circulation of the oceans. *Ann. Revs. Fluid Mech.*, **36**: 281–314.

Yu, L. and R.A. Weller (2007) Objectively analyzed air–sea heat fluxes for the global ice-free oceans (1981–2005). *Bull. Am. Meteor. Soc.*, **88**: 527–539.

Chapter 5
Long-Term Variability of Atlantic Water Inflow to the Northern Seas: Insights from Model Experiments

Michael Karcher[1], Ruediger Gerdes[2], and Frank Kauker[3]

5.1 Introduction

The inflow of water from the Atlantic Ocean has long been known to be an essential source of heat and salt for the Nordic Seas and the Arctic Ocean (Hansen and Østerhus 2000). However, only in the period of the projects VEINS (Variability of Exchanges in the Northern Seas) and ASOF (Arctic /Subarctic Ocean Flux Study) a coordinated effort to quantify the oceanic lateral fluxes in the Arctic/Subarctic domain has been attempted. In VEINS and ASOF, the simultaneous observation of the fluxes linking the Nordic Seas with adjacent oceans was combined with numerical modelling, providing an opportunity to synthesise the observations in a larger-scale context.

In older literature on the oceanography of the Nordic Seas and the Arctic Ocean temporal variability had largely been ignored. An exception is the oceanic response to seasonally and interannually variable wind fields (e.g. Proshutinsky and Johnson 1997 and references therein). Only very few investigations have addressed variability of temperature and salinity fields due to variable fluxes at the upper boundaries or the lateral gateways.

This changed dramatically in the recent decade after Quadfasel et al. (1991) had measured a significant increase of temperatures in the eastward Atlantic Water (AW) boundary current on the northern Barents Sea slope, relative to historical values. Then, it was not clear whether the observed warming was part of a long-term trend or an expression of low-frequency oscillations in the inflow water properties.

Subsequent analysis of temperature and salinity observations in the West Spitsbergen Current (WSC) had shown that signals approaching from the south were fluctuating interannually (Grotefendt et al. 1998; Saloranta et al. 2001; Furevik 2001). However, a possible link of the fluctuating signals in the WSC with the patchy

[1]Alfred Wegener Institute for Polar and Marine Research, Bremerhaven, Germany
O.A.Sys – Ocean Atmosphere Systems GbR, Hamburg, Germany

[2]Alfred Wegener Institute for Polar and Marine Research, Bremerhaven, Germany

[3]Alfred Wegener Institute for Polar and Marine Research, Bremerhaven, Germany
O.A.Sys – Ocean Atmosphere Systems GbR, Hamburg, Germany

R.R. Dickson et al. (eds.), *Arctic–Subarctic Ocean Fluxes*, 111–130

temperature anomalies in the Arctic basins remained elusive. It was also unclear how far the WSC anomalies would travel into the Arctic, and what were the causes or the consequences. Specifically the sources and pathways of the exceptionally strong warming observed in the early 1990s were important to understand. Was it a unique situation or just another warm event in a series of warm and cold anomalies?

The questions of sources, pathways, and time scales of AW anomalies were also addressed with numerical model experiments. After initial efforts to reproduce the observed variability, the potential predictability associated with propagating signals was assessed. More recently, emphasis has shifted to the discussion of consequences of changing Atlantic Water inflow for the Nordic Seas and the Arctic Ocean.

In a numerical experiment, Zhang et al. (1998) found an increase of AW flow from the Nordic Seas into the Arctic which commenced in the late 1980s. They identified a warmer and more intense inflow, mainly through the Barents Sea, as a source for the observed warming in the Eurasian and Makarov basins in the 1990s (Quadfasel et al. 1991; Carmack et al. 1995). Other model calculations (e.g. Häkkinen and Geiger 2000; Karcher et al. 2003a) confirmed the increase of AW volume inflow in the early 1990s, and highlighted its association with the high state of the North Atlantic Oscillation (NAO). The reason for the spatially inhomogeneous signature of the warming in the Eurasian Basin remained open.

The following overview of model studies is closely linked with the chapters by Schauer et al. on the observations of Atlantic Water inflow and by Gerdes et al. on the simulation of freshwater exchanges between the Arctic Ocean and the Nordic Seas. Most of the results presented here stem from a version of the NAOSIM (North Atlantic/Arctic Ocean Sea Ice Models) hierarchy of coupled ice-ocean models. For applications of NAOSIM in the Arctic/Subarctic domain, see for example Köberle and Gerdes (2003), Karcher et al. (2003a) and Kauker et al. (2005). Most of the modelling results presented here are from an experiment with forcing and setup as described by Kauker et al. (2003). The simulation covers the period 1948–2005. Drange et al. (2005) present a comparison of the general circulation in the Nordic Seas simulated by the NERSC MICOM and this NAOSIM experiment.

In the next section, we discuss the propagation of temperature anomalies with the inflowing Atlantic water, combining numerical model results and observations to better understand the development of Atlantic Water temperature in the last decades. We then discuss the influence of variability in the subpolar North Atlantic on the conditions in the Nordic Seas by highlighting the inflow and circulation of salinity anomalies. The most recent development of AW inflow will be taken up in the following section. We close the chapter with a summary and what we regard as the cutting edge questions concerning the long-term variability of the Atlantic water inflow into the Northern Seas.

5.2 Propagation and Sources of Temperature Anomalies

Gerdes and Schauer (1997) listed observations of AW core temperature at the Siberian slope north of the Barents Sea from the 1930s to 1993. They found the pattern to be inhomogeneous and inconclusive with respect to long-term variability or trends. In a different approach, comparing modern (1980–1995) and historic datasets, Grotefendt et al. (1998) identified a warm anomaly in the inflowing water through Fram Strait in the modern data to be most likely a consequence of decadal variability in the inflow of Atlantic Water.

Despite the patchiness of the observations and the difficulties in the interpretation, such data can be used to validate model results which hindcast the development of Arctic Ocean temperatures in recent decades. In case of good agreement, the model can be used to develop hypotheses and find mechanisms responsible for the simulated (and observed) variability.

In a study with the coupled ice-ocean model NAOSIM driven with atmospheric data from 1979 to 1999, the observed pattern of anomalies in the AW layer of the Eurasian Basin was reproduced (Karcher et al. 2003a). In the model, the pattern resulted from a sequence of warm and cold pulses of inflowing water propagating with a flow of variable intensity and spatial structure. The idea of a temporally steady and spatially continuous boundary current turned out to be invalid (Fig. 5.1).

According to the simulation, from about 1989 to the mid-1990s the northward flowing Atlantic water in the Nordic Seas was not only anomalously warm, but was also associated with significantly larger volume transport. In addition, the heat loss of the Atlantic water to the atmosphere was reduced due to anomalously high air temperatures during that period. These sources for anomalosly large heat inflow to the Arctic were associated with the unprecedented high positive index state of the NAO in the early 1990s, which caused anomalously strong northward transport of the oceanic North Atlantic Current (NAC) and the Norwegian Atlantic Current (NwAC) and the northward atmospheric transport over the northeastern North Atlantic.

Several models agree in the strong relation of net northward volume transport between Faroe and Scotland and the NAO during recent decades (Karcher et al. 2003a; Nilsen et al. 2003; Zhang et al. 2004). However, Nilsen et al. (2003) find the northward volume flux across the Iceland–Faroe Ridge and the Faroe–Scotland Ridge to be out of phase, the SLP pattern associated with the Faroe–Scotland Ridge inflow is NAO-like while the SLP regression pattern for the Iceland–Faroe Ridge inflow does not resemble the NAO pattern.

The further fate of the 1990s warm event could be followed in the NAOSIM simulation. In consistence with observations (e.g. Swift et al. 1996) the warm anomaly enters the Makarov and Canadian Basins. An extended experiment of the same model driven with NCEP reanalysis data from 1948 to 2001 (Gerdes et al. 2003) showed several distinct temperature maxima in the WSC. Each temperature maximum was associated with a warming of the Atlantic layer in the western Nansen Basin. Only a warm phase in the 1960s and the strong warming event of the 1990s, however, impacted the entire Eurasian Basin and only during the latter warm phase, the Makarov and Canadian Basins were affected. The study thus confirmed the uniqueness of the intensity and spatial extent of the 1990s warm event on a temporal scale of more than five decades.

Both studies (Karcher et al. 2003a; Gerdes et al. 2003) supported obervational evidence of the importance of water mass transformation in the Barents Sea for the mid-depth Atlantic Water layer in the Arctic Ocean. In the Barents Sea the inflowing Atlantic Water masses are cooled before they enter the deep basins of the Arctic Ocean.

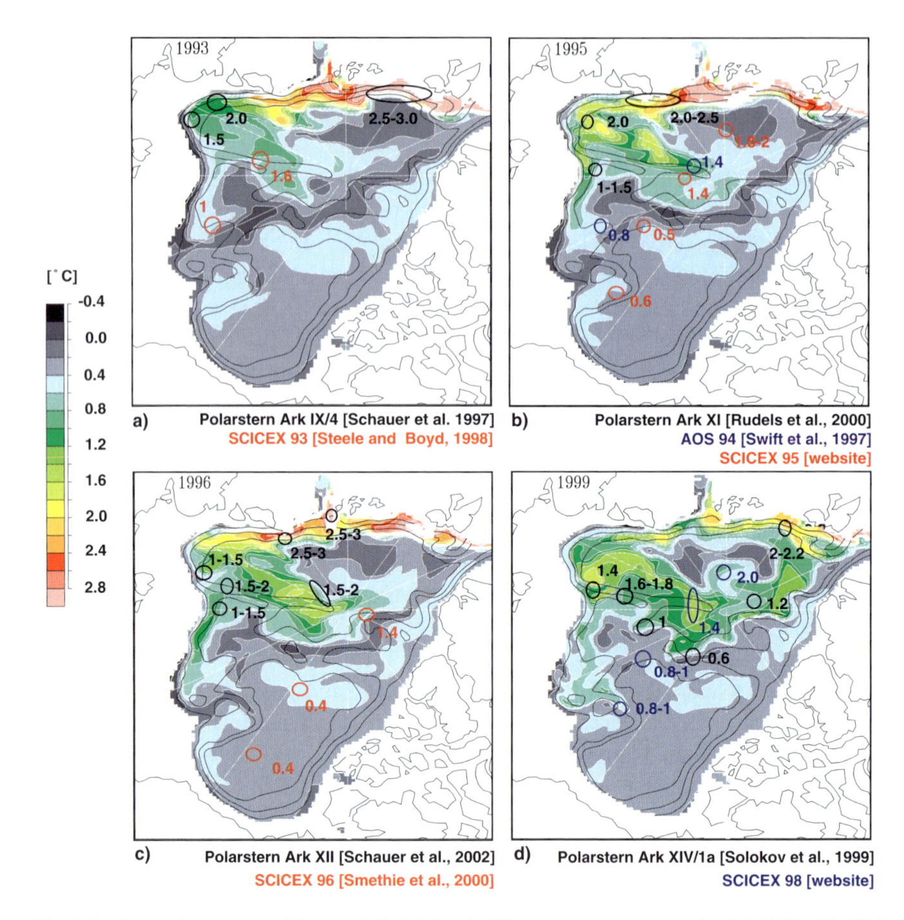

Fig. 5.1 September means of the modelled Atlantic Water core temperature for (a) 1993, (b) 1995, (c) 1996, and (d) 1999. The AW core is defined as the maximum temperature in the water column. Areas shallower than 500 m and profiles with the maximum temperature at depths shallower than 150 m have been omitted. Numbers in the maps are observed core temperatures at the indicated locations taken from published observations of the same year (see citations below each panel). Numbers and text in blue refer to observations made 1 year earlier. For the observations, only single locations from the published data are shown. The reference "website" refers to http://www.ldeo.columbia.edu/SCICEX/Media/3tdif.jpg. (Copyright 2003 American Geophysical Union. Modified from Karcher et al. (2003a). Reproduced by permission of American Geophysical Union)

This leads to a damping of AW temperature anomalies advected into the Barents Sea. Due to the variability of local processes, however, the intensity of this cooling process can vary considerably. In the 1960s a large import of sea ice from the central Arctic into the Barents Sea led to a capping with meltwater and reduced heat loss of the Atlantic Water (Karcher et al. 2003b). Less cold AW left the Barents Sea via the St. Anna Trough and contributed to an anomalously warm Eurasian Basin (Fig. 5.2). The latter was also found in an analysis of historic hydrographic data (Swift et al. 2005). Consequently, monitoring of Atlantic Water properties at the

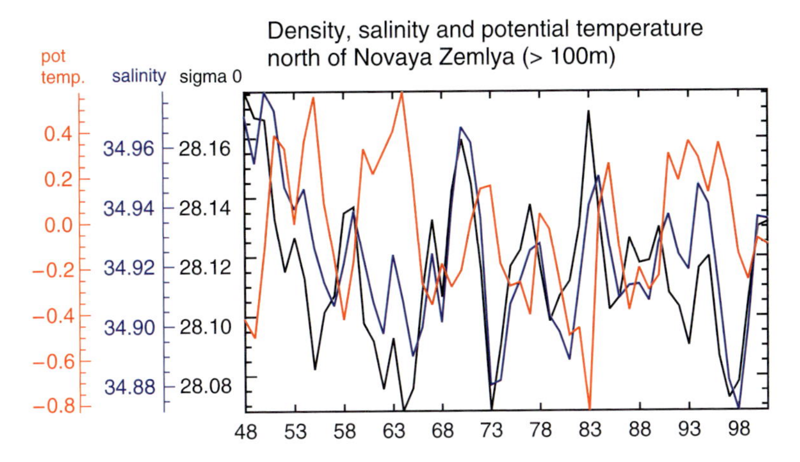

Fig. 5.2 Annual mean potential temperature (red), salinity (blue) and potential density (black) of deep Barents Sea Water passing the section Franz Josef Land to Nowaja Semlja before it enters St. Anna Trough (from Karcher et al. 2003b)

Fram Strait and the Barents Sea Opening has to be enhanced with downstream observation when the fluxes relevant for the interior deep basins of the Arctic are to be documented.

The above investigations have shown that it is possible to follow temperature anomalies of the AW propagating from Fram Strait into the Eurasian Basin. The results were consistent with observations, from which alone such propagation could only have been hypothesized. In a study which combined temperature observations from moorings along the AW path at Svinøy, the WSC in Fram Strait and at the Laptev Sea slope in the Eurasian Basin, Polyakov et al. (2005) estimated travel times for two recent step-like increases of AW temperature. Polyakov et al. used hindcast data from the NAOSIM simulation employed in Gerdes et al. (2003), extended to the year 2004. The model data compared well with the observations in timing and amplitude of the signals. See Fig. 5.4 for a temporally further extended timeseries. The estimated observed travel time was 1.5 years from Svinøy to Fram Strait and 6.5 years from Svinøy to the Laptev Sea slope. The respective times for the model were 1.5 and 7 years (Polyakov et al. 2005). These results were based on model timeseries covering the late 1990s to 2004.

Based on the same model data as in Polyakov et al. (2005) but further extended in time to cover the period 1948–2005 we perform a lagged regression of the simulated large-scale temperature pattern at 280 m depth on the temperature in the WSC at the same level (Figs. 5.3 and 5.4a). Instantaneous regression (lag 0) reveals simultaneous anomalies in the WSC and the western Barents Sea, delineating the Fram Strait and Barents Sea branches of AW after separation at the continental slope near the Norwegian coast (Fig. 5.3). The forward regression for 1 and 2 years lag (lag + 1, lag + 2) exhibits high correlation in both branches when the signal has reached the slope at St. Anna Trough, where both branches realign. This does not

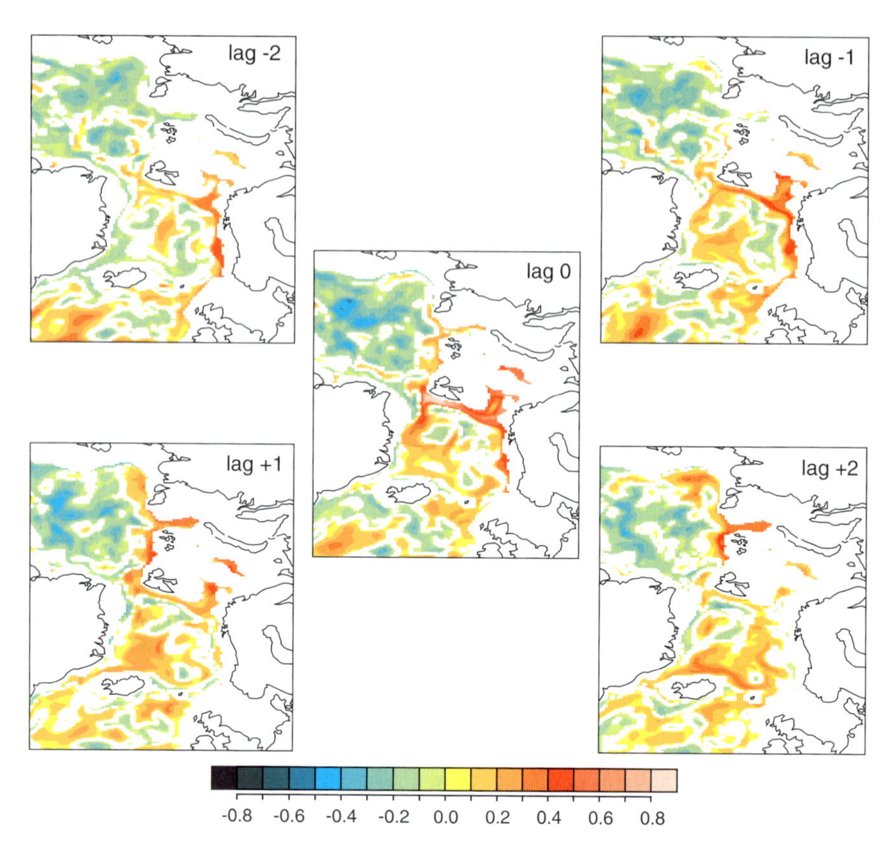

Fig. 5.3 Regression maps of simulated 280 m temperature fields on the temperature in the core of the WSC (black line in Fig. 5.4a). Shown is the normalized regression slope, i.e. the correlation. Lag 0 stands for instantaneous correlation, lag + 1 and + 2 (−1 and −2) for the WSC time series leading (lagging) by 1 and 2 years, respectively. Data used in this analysis are detrended and filtered with 7-month running mean. Correlations not significant at the 99% level are left white (F-test with 17 degrees of freedom of the denominator)

mean that travel times for temperature anomalies along both paths are similar. The Barents Sea branch needs less time than the Fram Strait branch as can be seen from tracer experiments (Karcher et al. 2005). Despite the strong damping of the anomalies due to the surface fluxes in the Barents Sea (e.g. Gerdes et al. 2003), the temperature variability can still be recovered in St. Anna Trough where the Barents Sea branch descends to fill the Atlantic Water Layer of the Eurasian Basin.

How far upstream relative to Fram Strait the anomalies can be traced to assess the predictive potential of temperature measurements in the NwAC? The regression backward in time (lag −1, lag −2) allows tracing the signals back to Svinøy, where correlations up to 0.4 with the WSC temperature time series can be found (Fig. 5.3). When tracing the signal further upstream to the Iceland–Faroe–Scotland gap, correlations fall to 0.3. The maximum correlation between the WSC time series and the inflow core of AW in the Faroe–Scotland section at 280 m depth (Fig. 5.4a) can be found for a lag of 15 months (Fig. 5.4b). The difference to the lag of 18 month for the slightly shorter distance from Fram Strait to Svinøy, as mentioned in the analysis of Polyakov et al. (2005), is due to the different length of the employed timeseries. The use of the long timeseries, covering here five decades, also leades

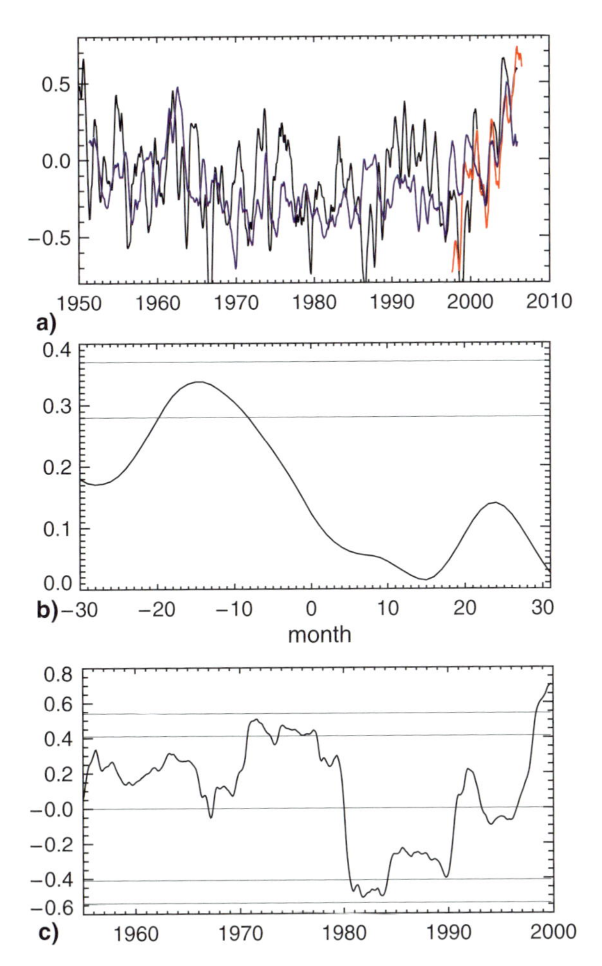

Fig. 5.4 (a) Temperature anomalies of the inflowing Atlantic Water at 280 m depth relative to the 1997–2005 mean. Temperatures in the core of the WSC in Fram Strait: simulated (black), observed from moorings in the eastern core of the WSC (red; Beszczynska, personal communication, 2007). These timeseries are based on the same model and observational data as discussed in Polyakov et al. (2005). The blue line shows temperatures from the eastern part of the Faroe– Scotland section plotted with a time lag of 15 months to the WSC time series. Maximum correlation of the simulated WSC and Faroe–Scotland time series occurs at 15 months as shown in (b). (c) Running correlation of the two simulated timeseries with a 10-year window. The time axis refers to the center of the window. Broken lines show the 95% (dotted) and 99% (dashed) confidence intervals estimated with a Monte Carlo test by fitting an AR(1) random time series. Data are 7-month running means

to other differences. In contrast to Polyakov et al. (2005) we find only a weak correlation (0.34 at most) when using the entire simulated period, indicating almost no predictability. However, a running correlation for the WSC and the Faroe–Scotland temperatures at 280 m depth with a 10-year window shows that a statistically highly significant correlation does exist for the last 10–15 years, the period which was covered by the data used in Polyakov et al. (2005).

5.3 Intrusion of Signals from South of the Sills

The source for the Atlantic water flowing into the Nordic Seas is the North Atlantic Current, which carries subtropical water, modified in the subpolar gyre (SPG) by surface fluxes and mixing processes.

The relation of changes in the subpolar North Atlantic to developments in the Nordic Seas is not easy to detect. Kauker et al. (2005) had shown that advection of temperature and salinity anomalies from the subpolar gyre into the Nordic Seas is quite possible and subpolar temperature and salinity signals can have a substantial impact on the conditions in the Nordic Seas. However, they also pointed out as unlikely that detection of signal propagation in the NAC could lead to a prediction of oceanic conditions in the Nordic Seas and the Arctic Ocean with several years lead time. The reason are influences from local atmospheric forcing and the complexity of North Atlantic–Nordic Seas advection pathways. The situation is different, however, when it comes to large-scale changes of the hydrography just south of the sills.

Based on model results from the DYNAMO project, New et al. (2001) describe the inflow between Faroe and Scotland as consisting of North Atlantic Current (NAC) waters and the very saline Eastern North Atlantic Water transported with the Shelf Edge Current. According to Hátún et al. (2005), the position of the front between the two water masses largely determines the salinity of the inflow into the Nordic Seas. At Rockall Trough south of the sills, observations show a pronounced increase in salinity after a fresh anomaly which occurred in the early 1990s (ICES 2006). This recent increase is linked to changes in the SPG as detected in SSH data from the 1990s and early 2000s (Häkkinen and Rhines 2004). Häkkinen and Rhines attributed the weakening of the SPG in the 1990s to changes in Labrador Sea convection. In a simulation with the Nansen Center version of MICOM, Hátún et al. (2005) showed that the same relation between SPG strength and northeastern Atlantic salinities held as far back as 1960. Brauch and Gerdes (2005) described the reaction of the SPG to a sudden change from the positive to the negative phase of the NAO as it happened from 1995 to 1996. They pointed at the role of the horizontal gyres in the heat transport changes and the role of the thermohaline surface fluxes in forcing the changes in SPG strength. Observations on repeat sections from Greenland to Scotland indicate a fast response of frontal positions in the northeastern North Atlantic to the sudden transition from the positive to the negative NAO phase in the mid-1990s (Bersch et al. 2007).

In model simulations with NAOSIM the same response mechanisms are active. In the early 1990s the frontal shift leads to a minimum salinity south of the sills (Fig. 5.5). The minimum salinities which fill the Iceland basin and Rockall Trough

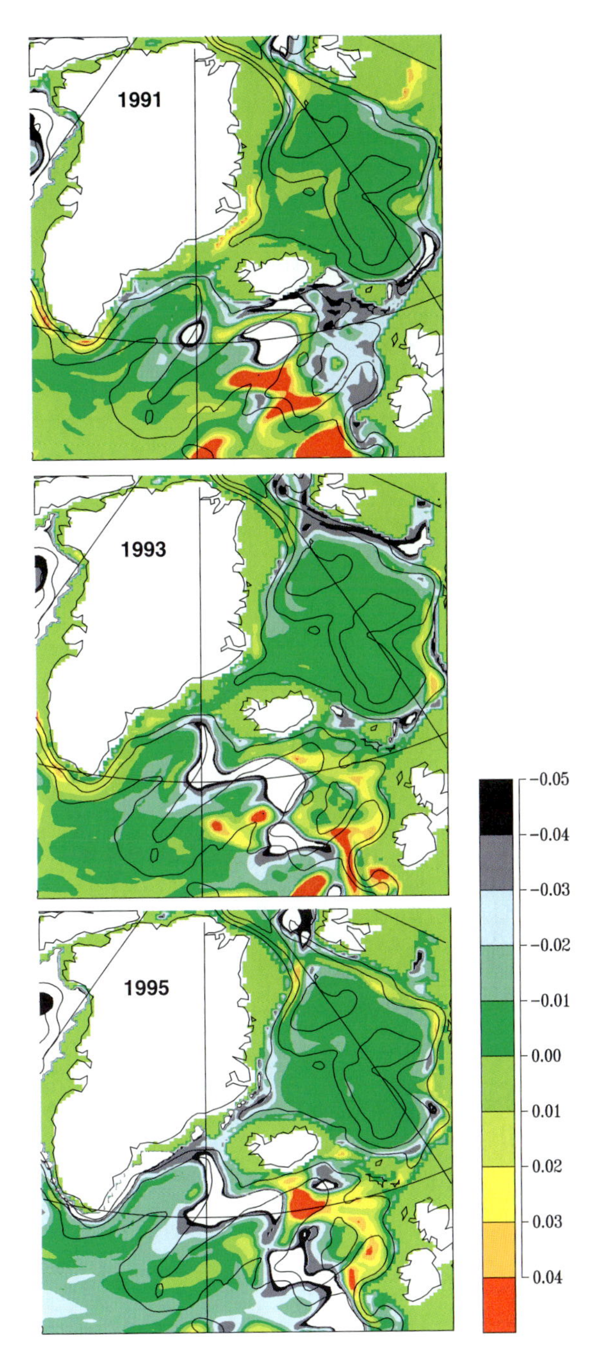

Fig. 5.5 Anomaly of annual mean salinity from the 1948–2005 mean averaged over 312–580 m depth. The panels show the large-scale fresh anomaly occupying the Iceland Basin and the Rockall Trough (1991), its propagation to the WSC (1993) and partial recirculation towards the Denmark Strait sill where it contributes to a record salinity minimum (1995). Please note that this water is bound to descend to greater depth after passing the sill. The larger anomaly south of the sill at the chosen depth level, though larger, does not contribute to the overflow

in the early 1990s lead to a fresh intrusion into the Nordic Seas. The advection of a salinity minimum with the Iceland–Faroe branch in that period is confirmed by observations north of the Faroe islands (Hansen et al. 2003). In the model simulation the salinity minimum is advected further with the northward flowing NwAC at mid-depth. Subsequently it recirculates with the return AW in Fram Strait southward to the Denmark Strait overflow sill (Figs. 5.5 and 5.6a). Surprisingly the low salinity anomaly is able to survive at intermediate depths over this long distance. After the recirculation the low salinity anomaly merges with a large freshwater

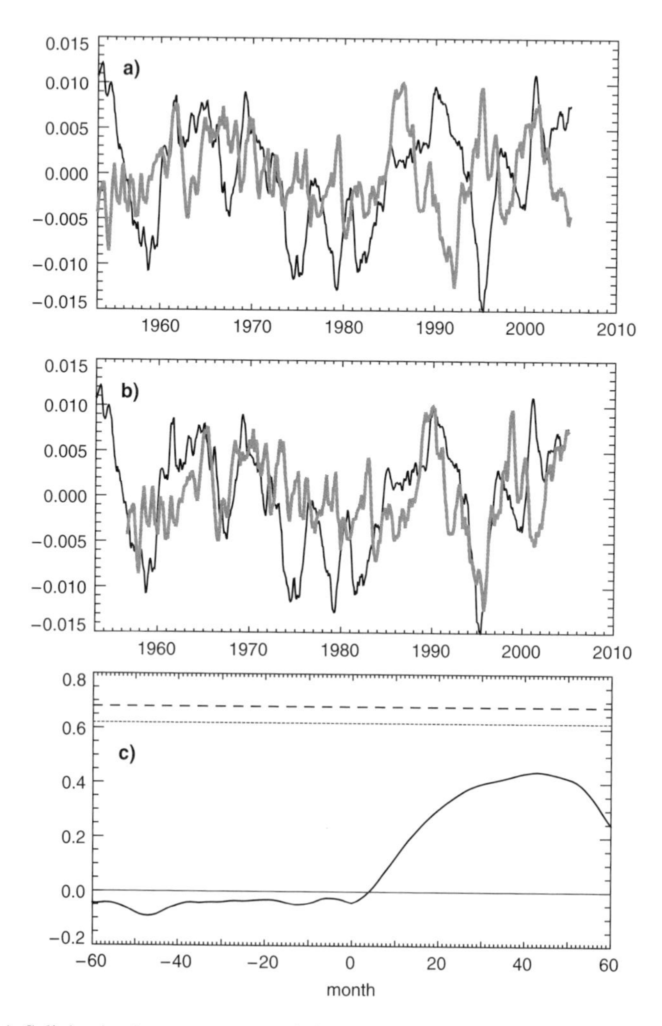

Fig. 5.6 (a) Salinity in the eastern part of the Faroe–Scotland section (grey) and above the Denmark Strait sill (black), both averaged over the 312–580 m depth interval in a NAOSIM simulation. (b) The same two timeseries with Denmark Strait shifted 42 months. (c) Lagged correlation of the two timeseries. Broken lines show the 95% (dotted) and 99% (dashed) confidence intervals. Data are 7-month running means. Maximum correlation occurs when Faroe–Scotland is lagging Denmark Strait by 42 months

release from the Arctic Ocean that was partially mixed down from the upper levels of the East Greenland Current (EGC) (Karcher et al. 2005; Gerdes et al. 2008). Consequently, the salinities at the overflow sill in Denmark Strait were at a record low in the mid-1990s. As observed (Dickson et al. 2002), the simulated salinity abovc the Denmark Strait sill shows decadal anomalies with minima in the late 1970s and the mid-1990s.

For the simulation, we can calculate the lagged correlation of the salinities at the inflow between Faroe and Scotland and the outflow at the Denmark Strait sill. Maximum correlation exists for a lag of 42 months. The comparison of the shifted timeseries (Fig. 5.6b) exhibits that on long timescales some large events from the inflow signal can be detected at the overflow sill. This holds for example for the minimum salinities in the mid-1960s and the mid-1990s, and the maxima in the early 1970s and the late 1980s. Because of the high level of high frequency variability and the shortness of the model experiment compared to the decadal time scale of the anomalies, the statistical relationship is not significant (Fig. 5.6c). It will be necessary to move to longer simulation times to arrive at more robust statements (see also Gerdes et al. this issue).

5.4 The Arctic Ocean as a Buffer

For the period after the mid-1960s, the model simulations reveal an upward trend in the net inflow of volume and heat through the Faroe–Scotland gap into the Nordic Sea (Figs. 5.7 and 5.8). For Fram Strait such no long-term trend in net volume transport is apparent. Outstanding maximum northward heat transports and Atlantic Water temperatures occurred in the early 1990s and the early 2000s. To hold up comparison with observations in the analysis of heat transports through Fram Strait, we use the monthly mean temperature of the Arctic Ocean proper, including the shelves, as a reference temperature. In this we follow Lee et al. (2004) who provide a more thorough discussion of the reference temperature choice. Reference temperatures in heat transport calculations can only be avoided when ocean volumes with closed mass balance are considered. It should be noted that the basic results remain unchanged when we consider all lateral fluxes in and out of the Arctic Ocean accordingly.

The observational record of heat transport estimates through the passages of the Nordic Seas (e.g. Schauer et al. 2004 and several arcticles this issue) is too short to allow an assessment of long-term developments. For the last decade, however, the increase of net heat flux in the year 1999 and again in the period 2002–2005 (Schauer et al., this issue) goes along with the step-like temperature increases in the WSC mentioned above (Polyakov et al. 2005). The most recent years see the emergence of another warm inflow event into the Arctic Ocean in Fram Strait mooring observations as well as in NAOSIM simulations (Fig. 5.4).

The Atlantic water needs O(10 years) to circulate through the Arctic Ocean to Fram Strait (Schlosser et al. 1995; Smethie et al. 2000; Karcher and Oberhuber 2002). The Atlantic Water layer (AWL) in the central Arctic is shielded from heat

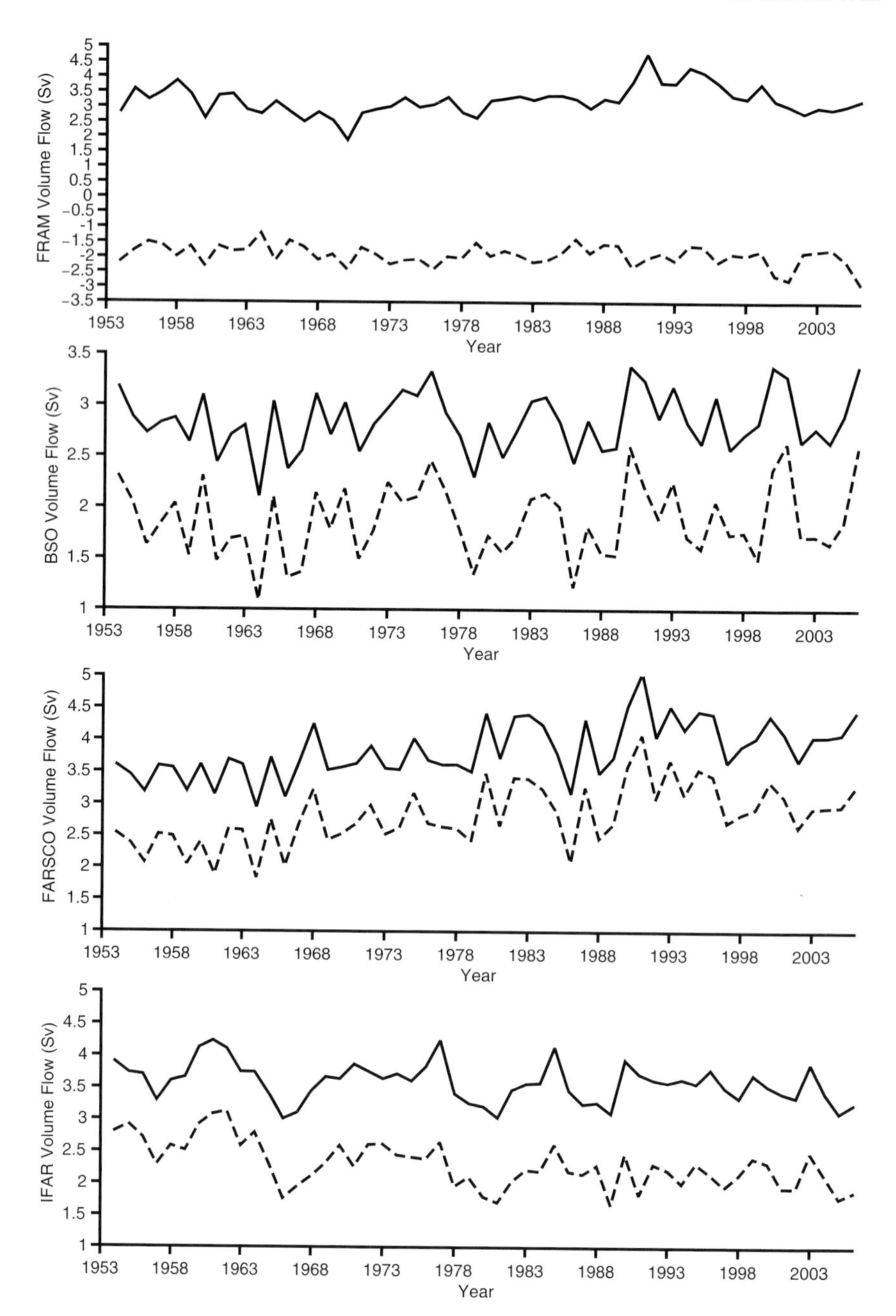

Fig. 5.7 Vertically integrated volume transport [Sv] across (top to bottom) the Fram Strait (FRAM), Barents Sea Opening (BSO), the Faroe–Scotland ridge (FARSCO), and the Iceland–Faroe ridge (IFAR). Solid lines denote north- or eastward flow, dashed lines represent net transport. Simulation period is 1948 to 2005 but the first 5 years have been suppressed in the plot because the initial adjustment period in the model cannot be properly interpreted

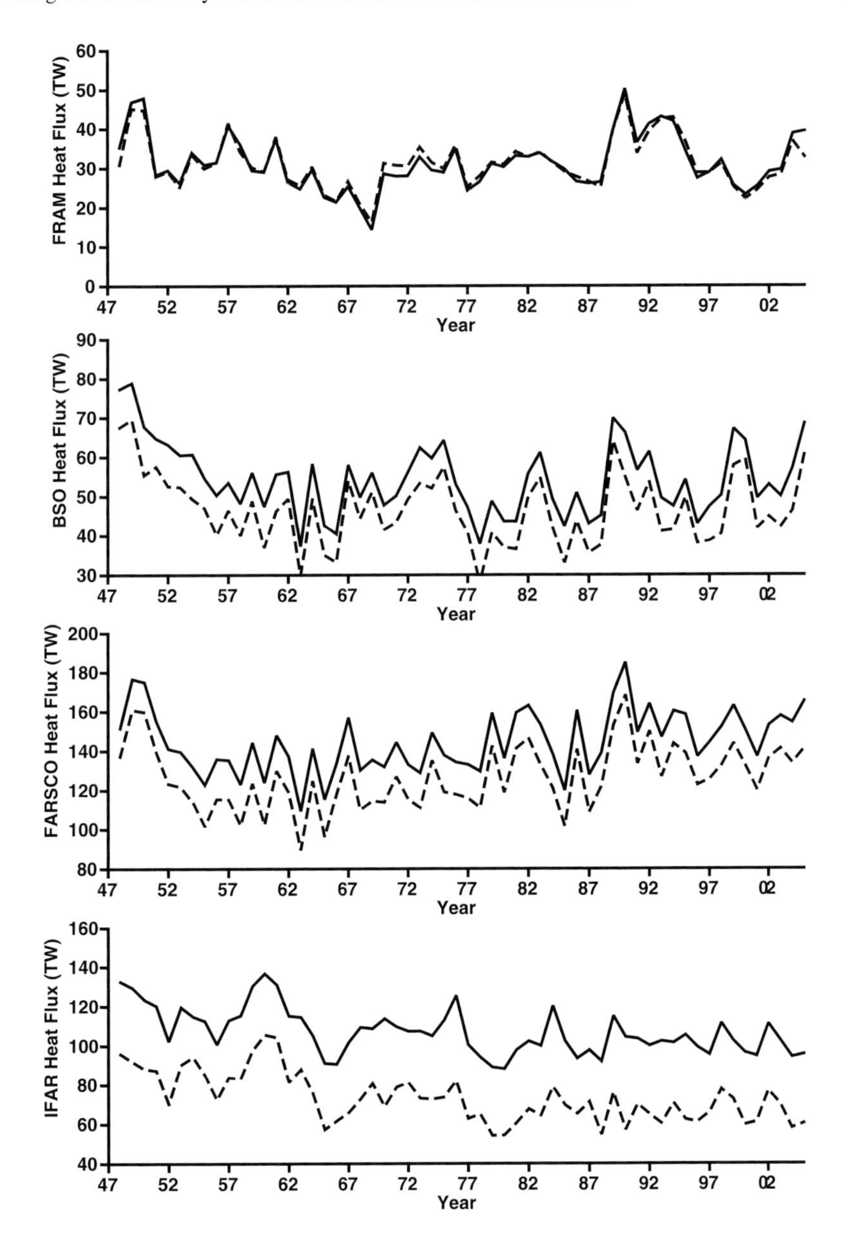

Fig. 5.8 Vertically integrated heat transport [TW] across (top to bottom) Fram Strait (FRAM), Barents Sea Opening (BSO), the Faroe–Scotland ridge (FARSCO), and the Iceland–Faroe ridge (IFAR). Solid lines denote north- or eastward flow; dashed lines represent net heat transport. Simulation period is 1948 to 2005 but the first 5 years have been suppressed in the plot (see Fig. 5.7). The reference temperature is the time-varying spatial mean temperature over the Arctic Ocean including the shelves

loss to the mixed layer by a strong halocline (Aagaard and Carmack 1989). Thus, its heat is largely conserved and only redistributed in the Arctic Ocean. The bulk of the heat which had entered from the south can be expected to leave the Arctic Ocean via Fram Strait after some 10–20 years delay. For the period 1948 to the mid-1980s the model results show a sequence of almost balanced warm and cold heat inflow anomalies of a few years duration each (Fig. 5.8). The situation differs, however, in the recent time. The prolonged inflow of anomalously warm water from 1989 to 1995 and again from 1999 to today has lead to to a temperature increase of large parts of the interior Arctic. Part of this temperature increase has not been accompanied by a corresponding salinity increase such that the density of the AWL has decreased over this period. The present Arctic Ocean therefore hosts a large volume of anomalously light and warm water at AWL depth, which will be exported to the Nordic Seas in the future (Fig. 5.9). Most likely, this export

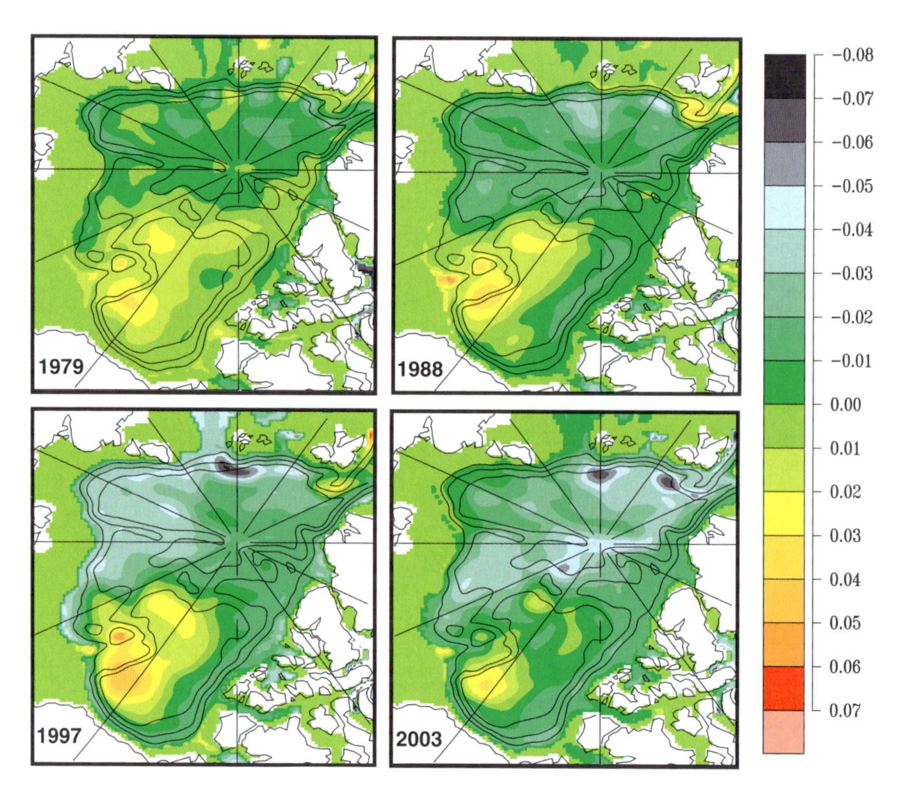

Fig. 5.9 Potential density σ_0 anomaly relative to the 1948–2005 mean averaged over the periods: 1978–1980, 1987–1989, 1996–1998 and 2002–2004 (center years are given on the panels). Depth averaging is performed for 250 to 600 m. The large-scale negative density anomalies associated with the two periods of warm inflows in the 1990s and 2000s appear in green to blue colour along the Lomonosov Ridge and in the Makarov basin, and along the Barents Sea slope, respectively. The positive anomaly in the Beaufort Sea is not propagating and associated with decadal changes in the depth of Beaufort Gyre. This signal partially hides the low density anomaly which also passes through the Canadian Basin

will lead to anomalously light overflow water at the Greenland–Scotland sills. Densities observed upstream of the overflow sills already decreased by 0.05 kg/m^3 since the 1970s due to fresh water accumulating in the Nordic Seas (Curry and Mauritzen 2005). The outflow of the anomalously warm, light water now populating large parts of the interior Arctic basins will introduce a density anomaly of the Arctic Intermediate Water in the Nordic Seas of similar magnitude. This means that overflow densities will likely remain below their pre-1990s values for decades unless local dense water production in the Nordic Seas and changes in recirculating water south of Fram Strait compensate the warming effect. This might have an effect on the overturning circulation, since the dense water export rate across the sills is roughly proportional to the density gradient across the sills (Whitehead 1998; Curry and Mauritzen 2005). How large the impact on the large-scale overturning circulation in the Atlantic may be remains to be investigated. On even longer time scales the feedback processes involving changes in freshwater storage and release in the upper water column as described by Jungclaus et al. (2005) have to be considered, too.

5.5 Recent Development

The most recent series of warm inflow events after 1999 occurs in a phase of low NAO index (Hurrell and Deser 2006). NAOSIM results for the recent years show no increased northward volume transport across the sills, as was the case in the early 1990s. The volume flux through the Barents Sea Opening is increased by 30% in 2004/5 while Fram Strait shows a slight increase of volume transports. Since 1989, the total heat flux across the Iceland–Scotland ridge has been on a higher level than during the 1970s and 1980s. In the 2000s, the heat transport is still high but does not quite reach the maximum values of the early 1990s. In Fram Strait and the Barents Sea Opening, on the other hand, the net lateral heat transports into the Arctic in 2005 have reached the high level of the early 1990s again.

The cause of this recent increase in heat transport into the Arctic Ocean is not fully understood. No significant rise in net lateral heat inflow into the Nordic Seas is apparent recently, although there is a large positive temperature anomaly in the inflowing water to the Nordic Seas at mid depths over the southern sills in 2004/5. A calculation of the complete heat budget for this period awaits to be done. Here, we may point to the surface heat flux as a possible source (Fig. 5.10). It exhibits positive anomalies not only for the northeast North Atlantic south of the sills between 2002 and 2005, but also for large parts of the NwAC in the Nordic Sea in 2004 and 2005. The reduced heat loss to the atmosphere could result from anomalously warm SAT in recent years (see, e.g. http://data/giss.nasa.gov/gistemp) and would leave more heat in the ocean to be advected to the north.

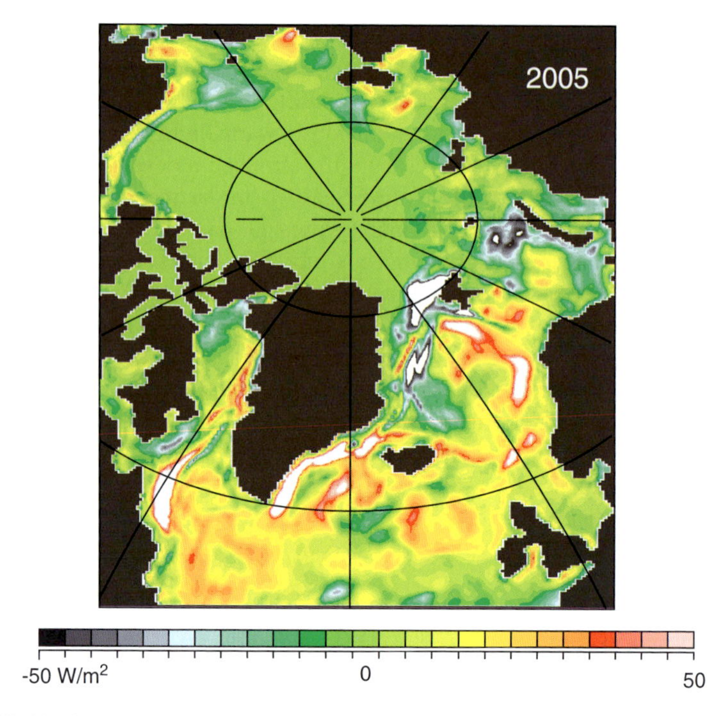

Fig. 5.10 Yearly mean surface heatflux anomaly (W/m^2) for 2005 from a NAOSIM experiment driven with NCEP reanalysis atmospheric data (SAT, windstress, scalar wind, cloudiness and precipitation). Positive anomalies denote less heat loss of the ocean to or more heat gain from the atmosphere compared to the long-term mean

5.6 Summary and Outlook

The transport of salt and heat with the Atlantic Water far north into the Nordic Seas and the Arctic Ocean is an outstanding feature of the North Atlantic circulation. The necessity of a broad view on the variability of this feature, which is at the same time long-term and large-scale, makes numerical models a momentous tool. The last decade of the VEINS and ASOF projects has seen a step forward in the application of such modelling work, which has been specifically useful when combined with observations. We have shown results of investigations from this period including very recent examples, which build upon research in the VEINS/ASOF projects.

The 1990s stood out in comparison to the previous four decades with high temperatures and large volume transport of Atlantic Water from the sills through the Nordic Seas and in the Arctic Ocean. Since the late 1990s, the heat input has been rising again. Inflow temperatures in Fram Strait have risen to record high values. In contrast to earlier periods, the warming of the most recent years is not associated with a strong positive NAO phase. Complete heat balances, which would help to understand the cause of this recent development, have to be performed. However, there are indications that anomalous surface air temperatures over the region are responsible.

Model results covering more than six decades have demonstrated that anomalies are carried with the circulation system over large distances, regardless whether they enter at the southern sills or are produced locally in the Nordic Seas. The results are consistent with the sparse observations in the central basins and time series in the boundary currents. Current regional ocean–sea ice models are able to simulate the amplitudes, pathways and propagation speed of temperature and salinity anomalies with good accuracy.

The possible tracing of signals from Svinøy to Fram Strait and even into the Arctic Ocean suggests a prognostic potential for temperature anomalies if continuous observations are performed upstream in the propagation pathways. However, the correlation between Fram Strait signals and those far upstream at the Iceland–Faroe sills and Rockall Trough is very weak. Local forcing effects do not allow a robust relationship to be detected. Nevertheless, individual events could still be traced and the last 10–15 years showed a common trend.

For salinity at mid-depth, where the strong local surface fluxes in the Nordic Seas have reduced impact, short timescale variability is still apparent and superposed on incoming decadal signals from the sills. Several large events of such timescale nevertheless have been shown to travel from the sills to the recirculation in Fram Strait and back south to the Denmark Strait overflow. There they directly feed into the lower limb of the MOC. These salinity fluctuations apparently originate from frontal shifts linked to changes of SPG intensity (Häkkinen and Rhines 2004; Hátún et al. 2005; Bersch et al. 2007).

Our results thus point to a possible link between variability in the two components of the thermohaline circulation in the northern North Atlantic, the SPG and the MOC.

Another potential influence of AW transport in the Nordic Seas and the Arctic Ocean on the MOC has been found in the warming of the intermediate water of the Arctic Ocean. The intense inflow of warm water in the early 1990s and from 1999 to today has not been fully density compensated. This led to a widespread low-density anomaly passing slowly through the Eurasian and Canadian basins. When this anomaly exits the Arctic Ocean, it will decrease the density of the overflow water. We estimate the length of this period of reduced overflow density as one to two decades. It remains to be investigated how large the actual impact on the overflow and the overturning could be.

The large research effort of the last decade has brought doubtless progress in terms of data sampling as well as improved understanding of processes in which the northward moving water of Atlantic origin is involved. An important future issue is a better evaluation of the predictive potential for the propagation of T, S anomalies on one hand and of volume flux anomalies on the other hand. We have shown indications for some predictive potential looking upstream into the NAC and for the properties of Arctic intermediate water back along the boundary currents in the Arctic basins.

The buffering capacity of the interior Arctic ocean for fluctuations of heat and salt transport influences the overflows into the subpolar North Atlantic. Thus, we need to learn more on the processes which govern the time- and space-scales of hydrographic changes at mid-depth. This includes the processes that govern the separation of flow into branches in the Nordic Seas (WSC and Barents Sea inflow) and the Arctic interior (AW boundary current recirculation in Eurasian basin or passage into the Canadian Basin). Are situations possible in which the AW circulation intensity in the Arctic Ocean changes dramatically (slowdown or reversal of flow in single basins) and what are the consequences for the feeding of the overflows?

Acknowledgements The authors would like to express their gratitude toward the agencies which funded part of this work: the European Union under EC MAST III programme (grant MAS3-CT96-0070 VEINS) and the FP6 programme (grant ASOF-N), and the German Ministry for Education and Research (BMBF) through grant 01 LA 9823/7. Further support has been received from the SFB 512 "Cyclones and the North Atlantic Climate System" of the DFG.

This material is based upon work partially supported by the National Science Foundation under agreements OPP-0002239 and OPP-0327664 with International Arctic Research Center, University of Alaska Fairbanks (Arctic Ocean Model Intercomparison Project).

References

Aagaard K, Carmack EC (1989) The role of sea ice and other fresh water in the arctic circulation. J. Geophys. Res. 94: 485–14,498

Bersch M, Yashayaev I, Koltermann KP (2007) Recent changes of the thermohaline circulation in the subpolar North Atlantic. submitted to Deep Sea Research

Brauch JP, Gerdes R (2005) Response of the northern North Atlantic and Arctic oceans to a sudden change of the North Atlantic Oscillation. J. Geophys. Res. 110, doi:10.1029/2004JC002436

Carmack EC, MacDonald RW, Perkin RG, McLaughlin FA, Pearson RJ (1995) Evidence for warming of Atlantic Water in the southern Canadian Basin of the Arctic Ocean: Results from the Larson-93 expedition. Geophys. Res. Lett. 22: 1061–1064

Curry R, Mauritzen C (2005) Dilution of the northern North Atlantic Ocean in recent decades. Science 308: 1772–1774

Dickson RR, Yashayaev I, Meincke J, Turrell W, Dye S, Holfort J (2002) Rapid freshening of the Deep North Atlantic over the past four decades. Nature 416: 832–837

Drange H, Gerdes R, Gao Y, Karcher M, Kauker F, Bentsen M (2005) Ocean General Circulation Modelling of the Nordic Seas. In: Drange H, Dokken T, Furevik T, Gerdes R and Berger W (eds.) The Nordic Seas: An Integrated Perspective. AGU Monograph 158, American Geophysical Union, Washington DC, pp199–220

Furevik T (2001) Annual and interannual variability of Atlantic Water temperatures in the Norwegian and Barents Seas: 1980–1996. Deep Sea Res., Part I 48: 383–404, doi:10.1016/S0967-0637(00)00050-9

Gerdes R, Schauer U (1997) Large-scale circulation and water mass distribution in the Arctic Ocean from model results and observations. J. Geophys. Res. 102: 8467–8483

Gerdes R, Karcher MJ, Kauker F, Schauer U (2003) Causes and development of repeated Arctic Ocean warming events. Geophys. Res. Lett. 30(19), doi:10.1029/2003GL018080

Gerdes R, Karcher M, Köberle C, Fieg K (2008) Simulating the long term variability of liquid freshwater export from the Arctic Ocean. Arctic-Subarctic Ocean Fluxes: Defining the role of the Northern Seas in Climate, Editors: B. Dickson, J. Meincke and P. Rhines, Springer.

Grotefendt K, Logemann K, Quadfasel Q, Ronski S (1998) Is the Arctic Ocean warming? J. Geophys. Res. 103: 27,679–27,687
Häkkinen S, Geiger CA (2000) Simulated low-frequency modes of circulation in the Arctic Ocean. J. Geophys. Res. 105: 6549–6564
Häkkinen S, Rhines PB (2004) Decline of Subpolar North Atlantic circulation during the 1990s. Science 304: 555–559; doi: 10.1126/science.1094917
Hansen B, Østerhus S (2000) North Atlantic-Nordic Seas exchanges. Prog. Oceanogr. 45: 109–208
Hansen B, Østerhus S, Hátún H, Kristiansen R, Larsen KMH (2003) The Iceland–Faroe inflow of Atlantic water to the Nordic Seas. Prog. Oceanogr. 59: 443–474
Hátún H, Sandø AB, Drange H, Hansen B, Valdimarsson H (2005) Influence of the Atlantic Subpolar Gyre on the Thermohaline Circulation. Science 309: 1841–1844, doi: 10.1126/science.1114777
Hurrell JW, Deser C (2006) North Atlantic climate variability. submitted to J. Marine Systems.
ICES (2006) ICES Report on Ocean Climate 2005. ICES Cooperative Research Report No. 280
Jungclaus J, Haak H, Latif M, Mikolajewiczc U (2005) Arctic–North Atlantic interactions and multidecadal variability of the Meridional overturning circulation. J. Clim. 18: 4013–4031
Karcher MJ, Oberhuber JM (2002) Pathways and modification of the upper and intermediate water of the Arctic Oceans. J. Geophys. Res. 107, doi: 10.1029/2000JC000530
Karcher MJ, Gerdes R, Kauker F, Köberle C (2003a) Arctic warming: Evolution and spreading of the 1990s warm event in the Nordic Seas and the Arctic Ocean. J. Geophys. Res. 108, doi:10.1029/2001JC001265
Karcher MJ, Gerdes R, Kauker F, Köberle C, Schauer U (2003b) Transformation of Atlantic Water in the Barents Sea between 1948 and 2002. Seventh Conference on Polar Meteorology and Oceanography and Joint Symposium on High-Latitude Climate Variations, Extended Abstract (CD-ROM) 12–16 May 2003, Hyannis, USA
Karcher M, Gerdes R, Kauker F, Köberle C, Yashayev I (2005) Arctic Ocean change heralds North Atlantic freshening. Geophys. Res. Lett. 32, doi:10.1029/2005GL023861
Kauker F, Gerdes R, Karcher M, Köberle C, Lieser JL (2003) Variability of Arctic and North Atlantic sea ice: a combined analysis of model results and observations from 1978 to 2001. J. Geophys. Res. 108, doi: 10.1029/2002JC001573
Kauker F, Gerdes R, Karcher M, Köberle C (2005) Impact of North Atlantic Current changes on the Nordic Seas and the Arctic Ocean. J. Geophys. Res. 110, doi:10.1029/2004JC002624
Köberle C, Gerdes R (2003) Mechanisms determining the variability of Arctic sea ice conditions and export. J. Clim. 16: 2843–2858
Lee T, Fukumori I, Tang B (2004) Temperature advection: internal versus external processes. J. Phys. Oceanogr. 34: 1936–1944
New AL, Barnard S, Herrmann P, Molines J-M (2001) On the origin and pathway of the saline inflow to the Nordic Seas: insights from models. Prog. Oceanogr. 48: 255–287
Nilsen JEØ, Gao Y, Drange H, Furevik T, Bentsen M (2003) Simulated North Atlantic-Nordic Seas water mass exchanges in an isopycnic coordinate OGCM. Geophys. Res. Lett. 30, doi: 10.1029/2002GL016597
Polyakov IV, Beszczynska A, Carmack EC, Dmitrenko IA, Fahrbach E, Frolov IE, Gerdes R, Hansen E, Holfort J, Ivanov VV, Johnson MA, Karcher M, Kauker F, Morison J, Orvik KA, Schauer U, Simmons HL, Skagseth Ø, Sokolov VT, Steele M, Timokhov LA, Walsh D, Walsh, JE (2005) One more step toward a warmer Arctic. Geophys. Res. Lett. 32, doi: 10.1029/2005GL023740
Proshutinsky AY, Johnson MA (1997) Two circulation regimes of the wind-driven Arctic Ocean. J. Geophys. Res. 102: 12,493–12,514
Quadfasel D, Sy A, Wells D, Tunik A (1991) Warming in the Arctic. Nature 350: 385
Rudels B, Muench RD, Gunn J, Schauer U, Friedrich HJ (2000) Evolution of the Arctic Ocean boundary current north of the Siberian Shelves. J. Mar. Syst. 25: 77–99
Saloranta TM, Haugan PM (2001) International variability in the hydrography of Atlantic water northwest of Svalbard. J. Geophys. Res. 106, doi:10.1029/2000JC000478
Schauer U, Muench RD, Rudels B, Timokhov L (1997) The impact of eastern Arctic shelf waters on the Nansen Basin. J. Geophys. Res. 102: 3371–3382

Schauer U, Rudels B, Jones EP, Anderson LG, Muench RD, Björk G, Swift JH, Ivanov V, Larsson A-M (2002) Confluence and redestribution of Atlantic Water in the Nansen, Amundsen and Makarov basins. Annales Geophysicae 20: 257–273

Schauer U, Fahrbach, E., Østerhus S, Rohardt, G (2004) Arctic warming through the Fram Strait – Oceanic heat transport from three years of measurements. J. Geophys. Res. 109, doi:10.1029/2003JC001823

Schlosser P, Swift JH, Lewis D, Pfirman SL (1995) The role of the large- scale Arctic Ocean circulation in the transport of contaminants. Deep Sea Res., Part II 42: 1341–1367

Smethie WM Jr, Schlosser P, Boenisch G, Hopkins TS (2000) Renewal and circulation of intermediate waters in the Canadian Basin, observed on the SCICEX 96 cruise. J. Geophys. Res. 105: 1105–1121

Sokolov V, Pivovarov S, Schneider W (1999) Oceanography. In ARCTIC'98: The Expedition ARK-XIV/1a of RV'Polarstern' in 1998, Chap. 7, Rep. on Polar Res. 308, Alfred Wegener Inst. für Polar- und Meeresforschung, Bremerhaven, Germany.

Steele M, Boyd T (1998) Retreat of the cold halocline layer in the Arctic Ocean. J. Geophys. Res. 103: 10,419–10,435

Swift JHE, Jones P, Aagaard K, Carmack EC, Hingston M, Macdonald RW, McLaughlin FA, Perkin RG (1996) Waters of the Makarov and Canada basins. Deep Sea Res., Part II 44: 3371–3382

Swift JH, Aagaard K, Timokhov L, Nikiforov EG (2005) Long-term variability of Arctic Ocean waters: Evidence from a reanalysis of the EWG data set. J. Geophys. Res. 110, doi:10.1029/2004JC02312

Whitehead JA (1998) Topographic control of oceanic flows in deep passages and straits. Rev. Geophys. 36, No. 3: 423–440

Zhang J, Rothrock DA, Steele M (1998) Warming of the Arctic Ocean by a strengthened Atlantic Inflow: Model results. Geophys. Res. Lett. 25: 1745–1748

Zhang JL, Steele M, Rothrock DA, Lindsay RW (2004) Increasing exchanges at Greenland-Scotland Ridge and their links with the North Atlantic Oscillation and Arctic Sea Ice. Geophys. Res. Lett. 31, doi: 10.1029/2003GL019304

Chapter 6
Climatic Importance of Large-Scale and Mesoscale Circulation in the Lofoten Basin Deduced from Lagrangian Observations

Jean-Claude Gascard[1] and Kjell Arne Mork[2]

6.1 Introduction

The Nordic Seas (Norwegian, Iceland and Greenland Seas) is one of the regions that have been best covered and continuously monitored with hydrographic observations. The view of the large-scale ocean circulation in the Nordic Seas has traditionally been based on hydrography due to the relatively few direct current measurements. It has been known for a century that the ocean circulation in the Nordic Seas is influenced by the basin topography (Helland-Hansen and Nansen 1909). However, the large number of surface drifters that have been released during the last 10–15 years have increased our knowledge of the surface circulation in the Nordic Seas (Orvik and Niiler 2002). The main features of the upper circulation in the Nordic Seas are a northward flow of warm water on the eastern side and a cold current flowing southward on the western side (Helland-Hansen and Nansen 1909). The flow of warm waters into the Nordic Seas represents the final poleward transport of the global thermohaline circulation system before being transformed by cooling processes into intermediate and deep waters that flow back into the North Atlantic. As the Fig. 6.1 shows, two main branches of warm, saline Atlantic water of approximately equal magnitude enter the Norwegian Sea.

The Norwegian Atlantic Current (NwAC) is revealed as a two branch current system through the entire Norwegian Sea (Poulain et al. 1996; Orvik and Niiler 2002). The eastern branch follows the shelf edge as a barotropic slope current while the western branch is a polar jet current associated with the Arctic Front (Orvik et al. 2001). While the inshore branch passes north against the Norwegian Continental Slope and is covered by current meter arrays in the Faroe–Shetland Channel, off Svinøy, across the Barents Sea Opening and in eastern Fram Strait, the offshore branch, passing north through the Norwegian Sea as a free jet, is unmeasured. Both will be involved in the spread of warmth to the Barents Sea and Arctic Ocean, and the issue of determining what might control this warm, saline

[1] LOCEAN, Université Pierre et Marie Curie, Paris, France

[2] Institute of Marine Research and Bjerknes Centre for Climate Research, Bergen, Norway

R.R. Dickson et al. (eds.), *Arctic–Subarctic Ocean Fluxes*, 131–143

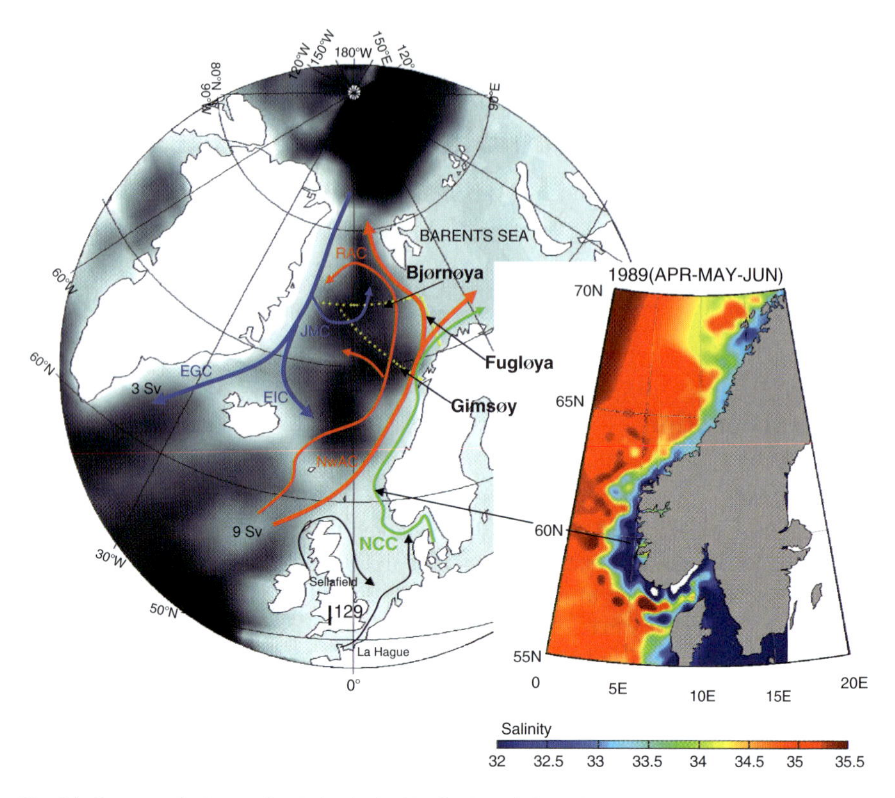

Fig. 6.1 Large-scale Ocean circulation in the Nordic Seas (left) and observed surface salinity along the Norwegian coast (right). The Norwegian Atlantic Current (NwAC), East Icelandic Current (EIC), Norwegian coastal Current (NCC) and the three setions (Gimsøy, Bjørnøya and Fugløya) are indicated

flux and its variability is of such central importance to understanding the imposition of change on the Arctic Ocean from subarctic seas that its solution must be of first priority. Yet although Orvik and Skagseth (2003) have now recovered lengthy (10-year) time-series of transport for the along-Slope branch and have developed some sense of its local and remote forcing, the offshore jet and its forcing remain less known. A mix of modern and classical methods has become available to tackle these issues such as floats, gliders, bottom pressure gauges, PIES, remote sensing, conventional hydrography and tracers, profiling CTDs, shipborne and moored ADCPs, etc. Here we describe the application of lagrangian techniques for understanding the circulation in this vital area.

6.2 Hydrographic Structures

The general hydrography of the Nordic Seas has been described and reviewed in Blindheim and Østerhus (2005). The transition zone between the domains of the NwAC and the Arctic waters to the West is known as the Arctic Front, located both

South and North of Jan Mayen. The front along the Mohn Ridge, northeastward from Jan Mayen, is topographically controlled and shows only small fluctuations in position. The position of the Arctic Front South of Jan Mayen is to a large extent controlled by variations in the volume of Arctic waters carried by the East Icelandic Current (EIC) and thus experiences large shifts (Blindheim et al. 2000).

The Norwegian Sea consists of the Norwegian and Lofoten Basins that have rather different hydrographic conditions. The Norwegian Basin in the South is occupied by Atlantic Water in the East and Arctic waters deriving from the EIC in the West. While Atlantic Water reaches westward ~250 km from the shelf edge in the southern Norwegian Basin it covers the whole ~500 km width of the Lofoten Basin. In the Norwegian Basin, Atlantic water typically reaches to 500 m depth, depending on the sill depth of the Faeroe–Shetland Channel. In contrast the whole Lofoten Basin is occupied by Atlantic Water in the upper ~800 m depth (e.g. Blindheim and Rey 2004; Blindheim and Østerhus 2005). This makes the Lofoten Basin the major reservoir of Atlantic Water. Orvik (2004) explained the difference in the Atlantic Water thickness between the Basins by a deep counter current influencing the northward volume transport of Atlantic Water in Lofoten Basin. The Lofoten Basin is also characterized by a large eddy activity and a long residence time (Poulain et al. 1996).

The hydrographic conditions of the Lofoten Basin and the Greenland Sea are shown in two sections, taken in June 2000 (Fig. 6.2). One section, 'Gimsøy-NW', runs from the Norwegian coast, crossing the Lofoten Basin, and into the Greenland Sea while the other section, 'Bjørnøya-W' (equals to 'Bear Island-W'),

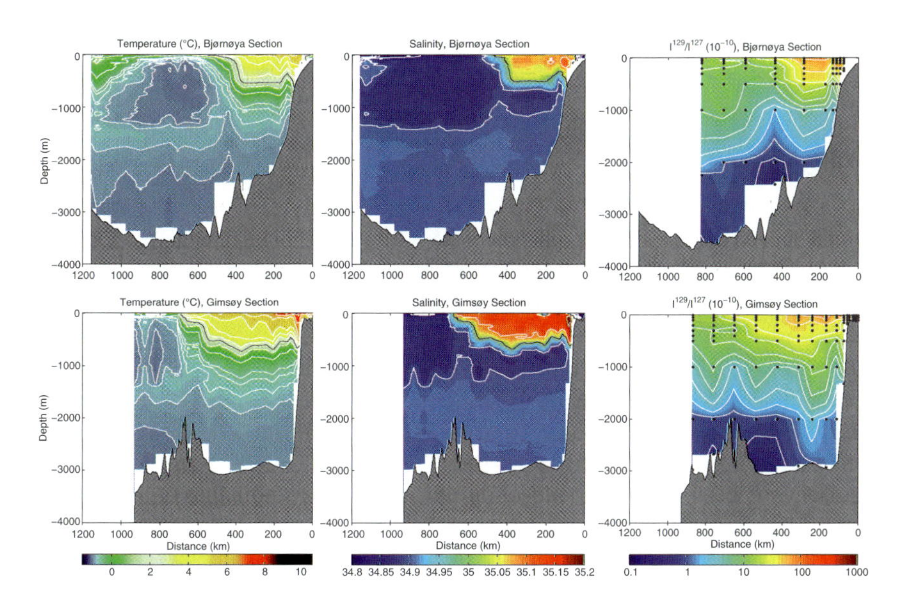

Fig. 6.2 Hydrological sections of temperature, salinity and Iodine ratio (I^{129}/I^{127}) taken from the Lofoten Islands (Gimsøy, lower figures) and from Bear Island (Bjørnøya, upper figures) to the Greenland Sea in June 2000. See also Fig. 6.1 for locations of the sections

runs westward at 74.5° N from Bear Island into the Greenland Sea (Fig. 6.1). Atlantic Water, with salinities above 35 and temperatures above 2 °C, reaches down to approximately 700 m depth. The sharp front that separates the Atlantic and Arctic water masses in the Lofoten Basin and Greenland Sea, respectively, is the Arctic Front located over the Mohns Ridge. Arctic intermediate water (AIW) is seen as a tongue between the Atlantic and deep layers with salinities less than 34.90 (Blindheim 1990).

In addition to the NwAC, the Norwegian Coastal Current (NCC) is a well defined current structure covering most of the shelf regions to the south, west and north of Norway (Fig. 6.1). The NCC is mainly characterized by fresh water originating from the Baltic Sea. The NCC fresh water is also tagged by Iodine 129 an anthropogenic tracer originating from nuclear waste re-treatment plant in France (La Hague) and UK (Sellafield) as shown in Fig. 6.1.

The hydrographic conditions of the Barents Sea Opening are shown on the Fugløya section (Fig. 6.3 upper panels) extending from the northern coast of Norway up to Bear Island 400 km northwards. The NCC fresh water (blue) is clearly visible on the salinity section near the coast of Norway. The NCC anthropogenic tracer Iodine 129 enriched water, clearly identified on the Gimsøy section (lower panel Fig. 6.3), is now well spread all over the Fugløya section (Fig. 6.3 upper panel) which indicates a very efficient mixing process occurring in the Lofoten Basin. This is also confirmed by a strong dilution of

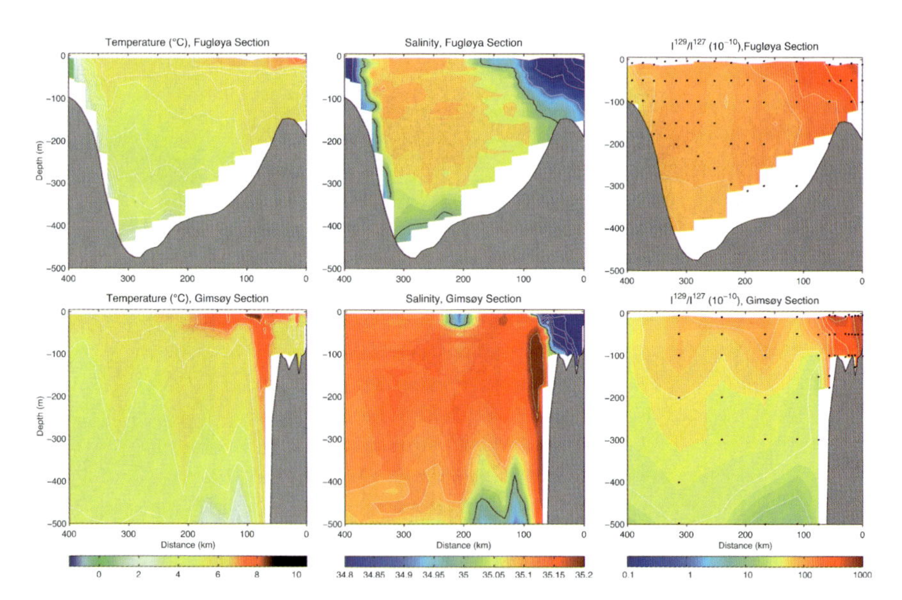

Fig. 6.3 Fugløya (upper figures) and Gimsøy (lower figures) sections of temperature, salinity and Iodine ratio (I^{129}/I^{127}) in June 2000. Variability of the Iodine ratio depends mainly on the Iodine 129 distribution

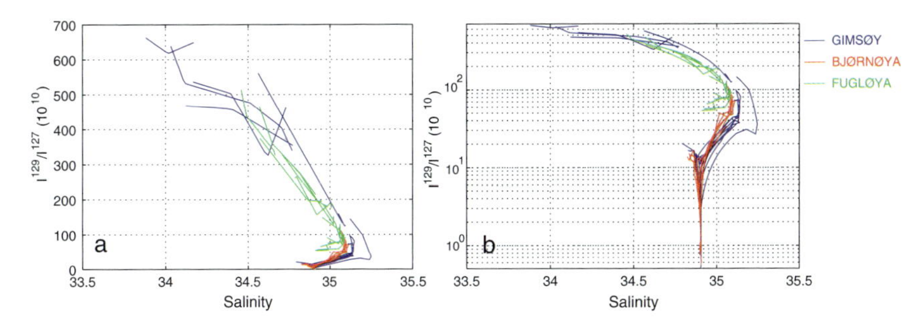

Fig. 6.4 The anthropogenic tracer ratio I^{129}/I^{127} as function of salinity in the Lofoten Basin along the Gimsøy, Bjørnøya and Fugløya sections. (a) linear scale, (b) logarithmic scale

the NCC fresh water (blue) into the NwAC salty water (red) producing a fresher Atlantic current (orange) entering in the Barents Sea. Figure 6.4 shows the remarkable distribution of anthropogenic tracer ratio as function of salinity in the Lofoten Basin along the three sections (Gimsøy, Bjørnøya and Fugløya). The variability of the Iodine ratio depends mainly on the Iodine 129 distribution. A striking split appears in Iodine 129 distribution between Bjørnøya section representing the Fram Strait branch of the NwAC and the Fugløya section representing the Barents Sea branch of the NwAC. Much larger concentration of Iodine is observed in the Atlantic water along the Fugløya section compared to the Bjørnøya section. The Gimsøy section is the sum of the two (before surface fresh NCC waters mix with subsurface salty NwAC waters).

6.3 Lagrangian Observations

The oceans have traditionally been monitored with measurements from ships and moored instruments. Observations from ships are weather- and ice-dependent which means that a preponderance of observations has been made during summer. Collecting oceanographic data of high quality is also both time and effort consuming. The need for systematic and near-real-time monitoring of ocean climate has resulted in an increased attempt to take advantage of new technology. This has led to the development of autonomous floats that can be deployed in areas where there is little cruise activity (and therefore few ship-measurements) and provides measurements throughout the year. In addition, Lagrangian techniques are particularly effective since trajectories provide much detailed spatial information that is almost impossible to get in any other way, including the sensitivity of fluid motion to topography since there is every reason to believe bottom relief plays a major role in shaping the circulation.

6.3.1 ARGO Floats

Within the international Argo programme the Institute of Marine Research, Bergen, has deployed eleven Argo floats in the Norwegian Sea drifting with ocean currents at 1,500 m depth. The first floats were deployed in 2002 while the last two floats were deployed in March/April 2006. Of the eleven floats, eight were deployed in the Norwegian Basin while the other three floats were deployed in the Lofoten Basin. However, several floats drifted from the Norwegian to the Lofoten Basin and vice versa. In addition, the last years University of Hamburg deployed more than 20 Argo floats in the Nordic Seas drifting at 1,000 m depth. At present there are about 20 active Argo floats in the Nordic Seas drifting at 1,000–1,500 m depth.

An Argo float drifts passively with the ocean currents at a chosen reference depth, usually at 1,500 m depth in the Norwegian Sea. The float is battery driven with a life time of about 4 years and is programmed to ascend to the surface every 10th day. During the ascent it measures pressure, temperature and salinity (i.e. a vertical profile of temperature and salinity) with the potential to add oxygen and fluorescence (chlorophyll) sensors as well. When the float surfaces, the data, together with its position, are sent to land via satellite. The float positions can be used to estimate the ocean currents at the reference depth. After the data are transmitted, the float descends to its reference depth, repeating this cycle every 10 days. The data transmission rates are such as to guarantee error free data reception and location, and in all weather conditions the Argo float must, in the Nordic Seas, spend about 6 h at the surface. The float positions are accurate to ~100 m depending on the number of satellites within range and the geometry of their distribution.

Trajectories of four Argo floats drifting at a reference depth of 1,500 m in the Lofoten Basin are shown on Fig. 6.5. The 1,500 m reference depth corresponds to the Norwegian Sea Deep Water, below the Arctic intermediate water, with potential temperature less than −0.5 °C and salinity near 34.91. Two of these floats were deployed in the Lofoten Basin while the other two were deployed in the Norwegian Basin but drifted into the Lofoten Basin. For all four floats a deep cyclonic circulation is revealed in the Lofoten Basin. One of the floats (id: 6900218) circulated cyclonically two and half times around the Basin before ending at the Mohn Ridge nearly 3 years after deployment. The other floats circulated between one and two times around the Basin before ending in the Lofoten or the Norwegian Basin. All floats followed nearly constant isobaths over long periods. For instance float 6900219 followed the 3,000 m isobath for about 2 years switching abruptly to the 3,500 m isobath when reaching the Norwegian Basin. Typically drift speeds of the floats are from a few cm/s up to 10 cm/s but in some cases reached 15 cm/s. The float 6900218 took about 1 year to complete one cycle around the Basin and its mean drift speed was 6.7 cm/s. The mean drift speed was calculated as the average of all drift speeds between two neighbouring locations. The mean speeds for all floats were estimated between 4 and 7 cm/s with lowest values for the two floats that also drifted in the Norwegian Basin. All floats show the strong influence of topography, but they also exhibit different behaviour in different areas. In the Eastern part of the Lofoten Basin the floats have a more irregular pattern of motion than in the other areas due to mesoscale turbulence as we will see later on.

Argo floats thus reveal a large-scale deep cyclonic circulation in the Lofoten Basin and a strong topographic influence. Near the Mohn Ridge, between the Greenland and the Lofoten basins, the direction in the deeper layer is also in the opposite sense to the surface current (Orvik and Niiler 2002). Using wind stress and density fields, Nøst and Isachsen (2003) modelled the stationary bottom geostrophic circulation in the Nordic Seas. Their results revealed a cyclonic circulation in the Greenland and Norwegian Seas with typical speeds of 5–10 cm/s which is in agreement with the Argo floats as far as the Lofoten Basin is concerned.

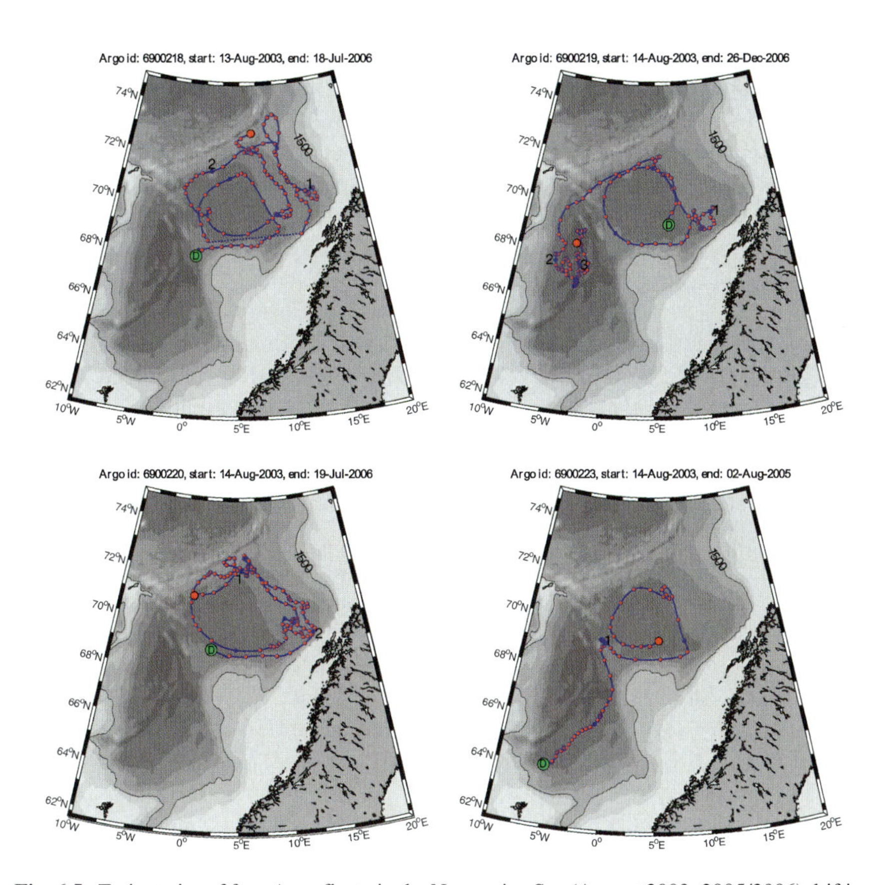

Fig. 6.5 Trajectories of four Argo floats in the Norwegian Sea (August 2003–2005/2006) drifting at 1,500 m depth. Dots indicate surfacing of the float and the interval between each surface position is 10 days. Dashed line is missing positions. Location of deployment is marked by "D" in a green dot while red dot indicates last position. There are blue dots every 6 months after deployment and the numbers (1–3) indicate number of years after deployment. The averaged drift speeds are estimated to 6.7, 4.3, 6.4 and 4.5 cm/s for Argo floats 6900218, 6900219, 6900220 and 6900223, respectively. Bathymetry shades change at every 500 m. The trajectories are smoothed before plotting (20 days moving averages)

6.3.2 RAFOS Floats

In five float deployments between April 2003 and November 2004 (April 2003, October 2003, April 2004, June 2004 and November 2004) a total of 42 RAFOS floats were deployed for periods of 6 months approximately as part of the ASOF-N programme. Figure 6.6 indicates the location of float deployments (circles) and the end-points (crosses) where the floats popped up to the surface 6 months later for transmitting data to satellites. In addition to ASOF, we also show observations obtained from May 2001 to October 2001 during the EU-MAIA project (Monitoring the Atlantic Inflow towards the Arctic) using the same lagrangian techniques. It is quite remarkable that the floats split equally between those heading northward before merging with the West Spitsbergen Current and those turning eastward before entering in the Barents Sea.

Detailed float trajectories are shown on Figs. 6.7 and 6.8. Most of the deployments occurred west of the Lofoten Islands in the NwAC and above the continental slope where bottom depths vary from 1,000 m down to 2,500 m. Most of the ASOF floats (34) were ballasted to sink and drift at a constant depth of about 300 m according to the following prescribed initial conditions: P = 300 dbar, T = 5.667 °C, S = 35.133 psu, in situ density = 29.0805, while during the MAIA experiment in

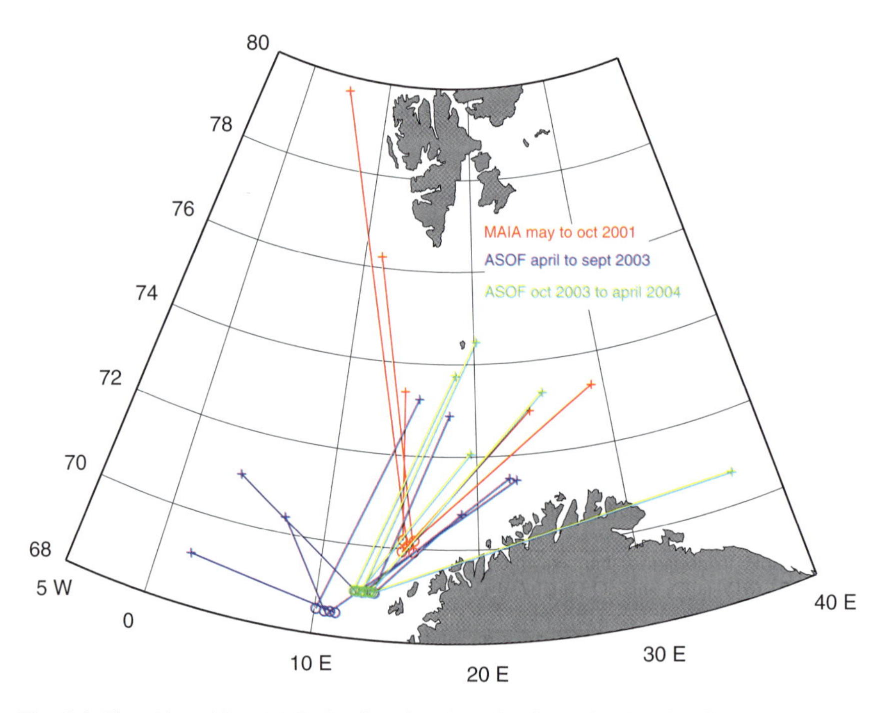

Fig. 6.6 First (o) and last (+) Rafos float locations for 6 months duration from May–October 2001, April–September 2003 and October 2003–April 2004

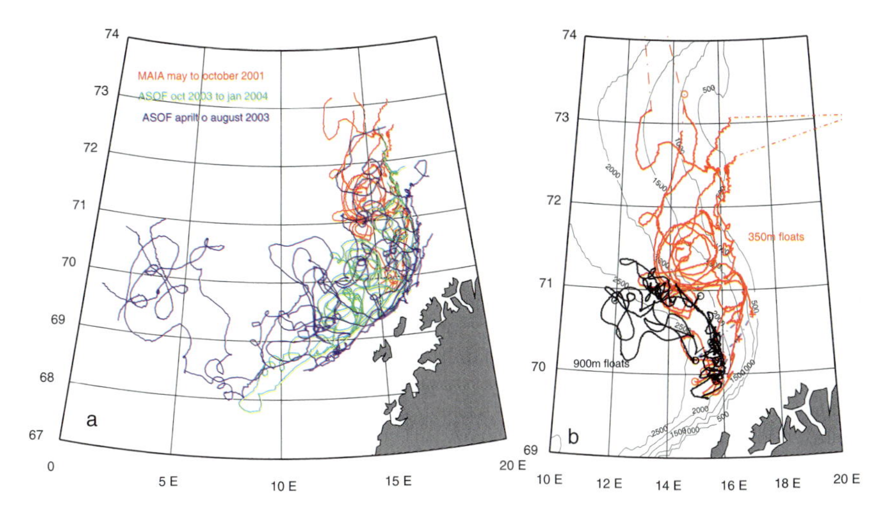

Fig. 6.7 (a) Rafos Floats trajectories at 300 m depth from May to October 2001, April to August 2003 and October 2003 to January 2004. (b) Rafos Floats trajectories at 350 m depth (red) and 900 m depth (black) from May to October 2001. On Fig. 6.7b, one can easily distinguish between the general drift pattern of the shallow floats (350 m depth) compare with the deep floats (900 m depth) drift pattern more constrained by topography

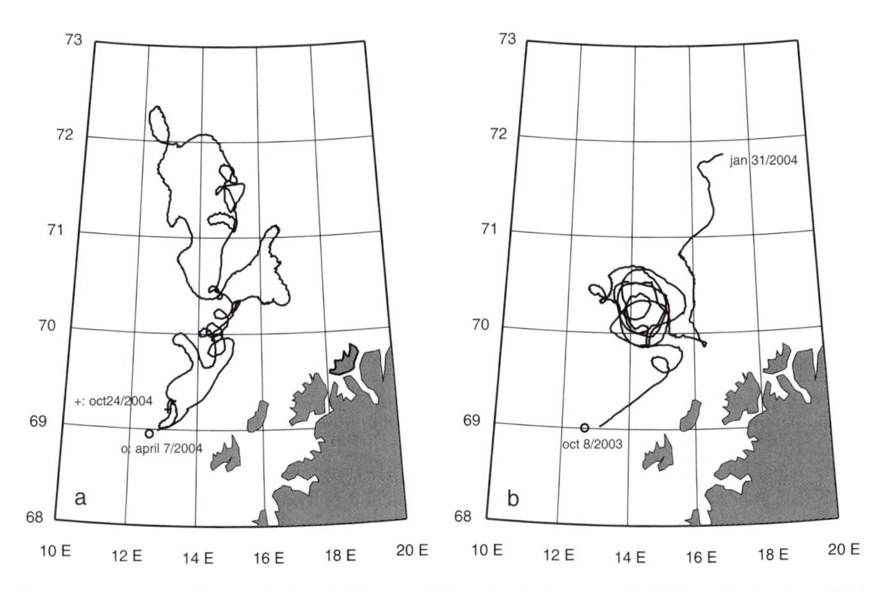

Fig. 6.8 (a) Rafos float (RF 518) drifting at 300 m depth from April 2004 until October 2004. Strikingly this float drifted over 1,000 km but corresponding to a net drift close to zero after 6 months total drift period. (b) Rafos float (RF12) drifting at 300 m depth from October 2003 to January 2004 indicating the presence of a quasi stationary anticyclonic mesoscale eddy (50 km diameter and 1 week period) for about 3 months

2001, float depths were slightly deeper at ~500 m depth. A few ASOF-N floats (8) were ballasted for 1,000 m depth and the duration of the drift for these deep floats was extended to about 1 year. Deployments usually occurred during the spring and the fall of each year (2003 and 2004). After 6 months (1 year) drifting at an average depth of about 300 m (1,000 m), floats were released to pop up at the surface and start to transmit 6 months (1 year) worth of data to the satellites (Argos link). At depth, floats were recording in situ temperature and pressure every hour and every 4 h they recorded the time of arrival (TOA), of acoustic signals transmitted by sound sources deployed in April 2003 and September 2003, the first year of the experiment.

From the float trajectories (Figs. 6.7 and 6.8), mesoscale turbulence, characterized by large-scale eddies, 50–100 km diameter and 1–2 weeks rotating period, is clearly identified as the dominant and ubiquitous mechanism influencing the general circulation in this part of the Lofoten Basin. It appears that most of these large-scale anticyclonic eddies were capped off by a layer of relatively fresh water originating from the NCC as shown on Fig. 6.9. The mean transport associated with a mesoscale turbulent vein, 100 km wide, 700 m deep moving at an average speed of 6 cm/s would correspond to about 4 Sverdrups (1 Sverdrup = 10^6 m^3 s^{-1}).

Figure 6.9 represents a large-scale Lofoten eddy identified in July 2001 by (a) trajectories of five RAFOS isobaric floats drifting at 350 m depth approximately

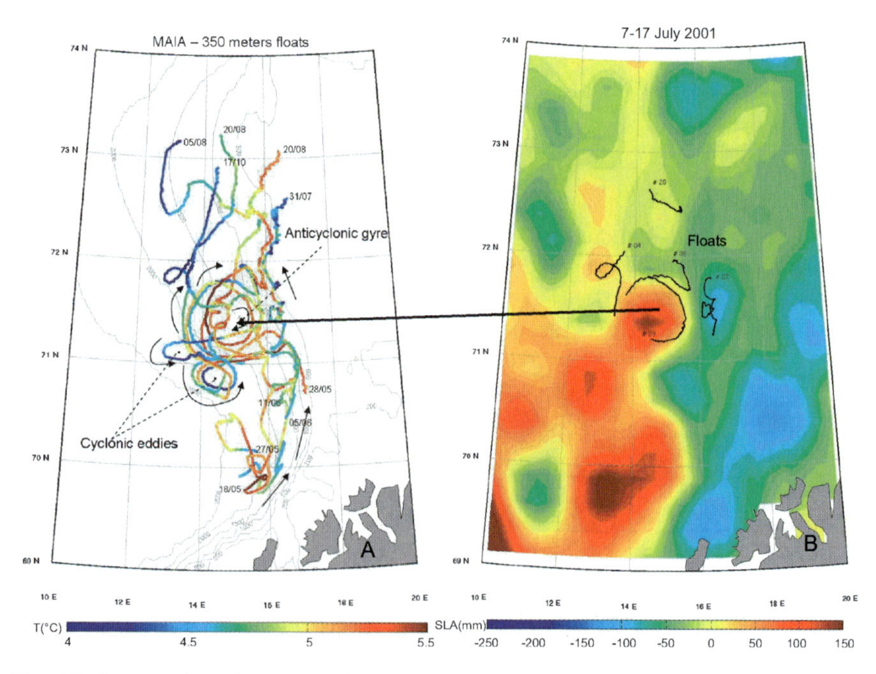

Fig. 6.9 Large-scale Lofoten eddy identified by 5 RAFOS floats (a) and Sea Level Anomaly from satellite (b)

and (b) Sea Level Anomaly (SLA) observed by satellite altimetry. The RAFOS floats are influenced by a large and persistent anticyclonic eddy (100 km diameter, 2 weeks period, located between 13° E and 16° E longitude and 71° N and 72° N latitude) and by unstable mesoscale cyclonic eddies migrating around the main anticyclonic eddy feature. The anticyclone is capped off by a thick layer of relatively buoyant fresh water originating from the NCC, also clearly visible on the SLA maps. Similar anticyclonic eddies were observed during ASOF in November–December 2003 located between 13° E and 16° E longitude and 70° N and 71° N latitude (Fig. 6.8b).

This sea level anomaly was created by the large buoyancy input of the fresh water layer inducing a strong deviation of the sea surface topography. This fresh water originates from the Baltic Sea in addition to the run-off from Norwegian Fjords distributed all along the Norwegian coast. This fresh water is tagged with anthropogenic tracers such as Iodine 129 originating from La Hague in France and Sellafield in UK. Gascard et al. (2004) published a detailed analysis of Iodine 129 concentrations all along the coast of Norway. The high mixing rate between NwAC and NCC water masses passing the Lofoten islands and entering the Barents Sea is clearly identified along the Fugløya section from the distribution of temperature, salinity and Iodine properties. The large-scale eddies generated by the interaction between the NwAC and NCC northwest of Norway are the most likely cause, as illustrated on Figs. 6.7, 6.8b and 6.9. This area is characterized by intense mixing between the Norwegian coastal water masses and Norwegian Atlantic water masses as illustrated by the Fugløya section (Tromsø to Bear Island; Fig. 6.3) showing an intense mixing across the whole Fugløya section and through the entire water column from top to bottom. This is also a sound explanation for the fact that higher concentrations of Iodine 129 spreading away from the NCC affect most of the Atlantic water masses entering in the Barents Sea. This intense mixing was reported by Gascard et al. (2004) but not the process responsible for it, i.e. a very active and sustained interaction between the fresh Iodine-enriched NCC and salty NwAC triggered by intense mesoscale activity entraining NCC coastal fresh water offshore past the Lofoten Islands. This mesoscale interaction between NCC and NwAC, developing a large-scale stationary eddy offshore, might also be involved in controlling the overwintering of the copepod *Calanus Finmarchicus* at great depths under the Atlantic water layer (Halvorsen et al. 2003) – an important issue for a region that is among the most productive regions in the world ocean.

6.4 Conclusions

Though the analysis is incomplete, three main results of climatic importance have emerged from this set of Argo and RAFOS quasi lagrangian observations:

1. The first concerns the deep recirculation in the Lofoten Basin (900–1,500 m), and its control by bottom topography. This cyclonic recirculation has the important

effect of storing a large quantity of Atlantic Water and increasing the residence time of Atlantic water masses circulating in the area. Even some of the shallow (RAFOS) floats injected in the core of the Norwegian Atlantic Current close to the Lofoten Islands, revealed a very strong recirculation component. The deep Argo and RAFOS floats confirmed the topographic influence on the deep cyclonic circulation that characterizes the Lofoten Basin. The kinetic energy associated with this deep circulation is weak but the mass transport is important due to large-scale horizontal spreading and deepening of the Atlantic layer across the whole Lofoten basin.

2. The second result concerns the mesoscale eddies dominance in the inshore branch of the Norwegian Atlantic Current. In consequence, of this, the Norwegian Atlantic Current does not resemble the narrow, swift boundary jet that the literature often describes, but rather a turbulent, broad (100 km) and slow current (~6 cm/s mean velocity) progressing to the north, in the Lofoten Basin or passing east into the Barents Sea. Half of the shallow floats (300 m depth) launched west of the Lofoten Islands entered through the Barents Sea Opening, the other half continuing North towards Fram Strait.
3. The third result of general importance to our understanding of Arctic–subarctic ocean fluxes concerns the strong interaction between the relatively fresh water of the Norwegian Coastal Current and the Norwegian Atlantic Current offshore, particularly West and North of the Lofoten Islands. The mesoscale interactive processes are well described by the RAFOS float trajectories and satellite altimetry North of the Lofoten Islands and West of the Tromsøflaket. This is a region prone to a high mesoscale turbulence activity and intense mixing between Norwegian Coastal Current and Norwegian Atlantic Current water masses as also revealed by anthropogenic tracer distribution and temperature-salinity properties.

None of these results would have been acquired without the extensive use of a new Lagrangian High Technology represented by neutrally buoyant isobaric floats (Argo and RAFOS) drifting at depth in the Ocean in addition to more conventional techniques.

Acknowledgements This paper was funded by the European Union under FP5 (ASOF). We are grateful to the crew on board R/V Johan Hjort and G.O. Sars/Sarsen, who made this study possible. We are also grateful to Catherine Rouault and Sandra Sequeira from LOCEAN and Harald Loeng from IMR who contributed to the preparation of the experiments and the data processing.

References

Blindheim, J (1990) Arctic intermediate water in the Norwegian Sea. Deep-Sea Research I, 37, 1475–1489

Blindheim J, Rey F (2004) Water-mass formation and distribution in the Nordic Seas during the 1990s. ICES Journal of Marine Science, 61 (5): 846–863, doi: 10.1016/j.icesjms.2004.05.003

Blindheim J, Østerhus S (2005) The Nordic Seas, Main Oceanographic Features. In *Climate Variability in the Nordic Seas*, H. Drange, T.M. Dokken, T. Furevik, R. Gerdes, and W. Berger, Eds., Geophysical Monograph Series, 158, AGU, 10.1029/158GM03

Blindheim J, Borovkov V, Hansen B, Malmberg SAa, Turrell WR, Østerhus S (2000) Upper layer cooling and freshening in the Norwegian Sea in relation to atmospheric forcing. Deep-Sea Research, 47: 655:680

Gascard JC, Raisbeck G, Sequeira S, Yiou F, Mork KA (2004) The Norwegian Atlantic Current in the Lofoten basin inferred from hydrological and tracer data (^{129}I) and its interaction with the Norwegian Coastal current. Geophysical Research Letters, vol 31, LO1308, doi:10.1029/2003 GL018303

Halvorsen E, Tande KS, Edvardsen A, Slagstad D, Pedersen OP (2003) Habitat selection of over-wintering *Calanus finmarchicus* in the NE Norwegian Sea and shelf waters off Northern Norway in 2000–02. Fisheries Oceanography, 12 (4–5): 339–351, doi:10.1046/j.1365-2419.2003.00255.x

Helland-Hansen B, Nansen F (1909) The Norwegian Sea: Its Physical Oceanography based on Norwegian Researches 1900–1904. In Report on Norwegian fishery and marine investigations, vol. 2., Bergen, Norway, 390 pp + 25 plates

Nøst OA, Isachsen PE (2003) The large-scale time-mean circulation in the Nordic Seas and Arctic Ocean estimated from simplified dynamics. Journal of Marine Research, 61, 175–210

Orvik KA (2004) The deepening of the Atlantic water in the Lofoten Basin of the Norwegian Sea, demonstrated by using an active reduced gravity model. Geophysical Research Letters, 31, L01306, doi:10.1029/2003GL018687

Orvik KA, Niiler PP (2002) Major pathways of Atlantic water in the northern North Atlantic and Nordic Seas toward Arctic. Geophysical Research Letters, 29, 1896, doi:10.1029/2002 GL015002

Orvik, KA, Skagseth Ø (2003) The impact of the wind stress curl in the North Atlantic on the Atlantic inflow to the Norwegian Sea toward the Arctic. Geophysical Research Letters, 30(17), 1884, doi:10.1029/2003GL017932

Orvik KA, Skagseth Ø, Mork M (2001) Atlantic inflow to the Nordic Seas: Current structure and volume fluxes from moored currentmeters, VM-ADCP and SeaSoar-CTD observations, 1995–1999. Deep-Sea Research, 48, 937–957

Poulain PM, Warn-Varnas A, Niiler PP (1996) Near-surface circulation of the Nordic Seas as measured by Lagrangian drifters. Journal of Geophysical Research, 101, 18237–18258

Chapter 7
Freshwater Storage in the Northern Ocean and the Special Role of the Beaufort Gyre

Eddy Carmack[1], Fiona McLaughlin[1], Michiyo Yamamoto-Kawai[1], Motoyo Itoh[2], Koji Shimada[2], Richard Krishfield[3], and Andrey Proshutinsky[3]

7.1 Introduction

As part of the global hydrological cycle, freshwater in the form of water vapour inexorably moves from warm regions of evaporation to cold regions of precipitation and freshwater in the form of sea ice and dilute seawater inexorably moves from cold regions of freezing and net precipitation to warm regions of melting and net evaporation. The global plumbing that supports the ocean's freshwater loop is complicated, and involves land–sea exchanges, geographical and dynamical constraints on flow pathways as well as forcing variability over time (cf. Lagerloef and Schmitt 2006). The Arctic Ocean is a central player in the global hydrological cycle in that it receives, transforms, stores, and exports freshwater, and each of these processes and their rates both affect and are affected by climate variability. And within the Arctic Ocean, the Canada Basin (see Fig. 7.1) is of special interest for three reasons: (1) it processes freshwater from the Pacific, from North American and Eurasian rivers and from ice distillation; (2) it is the largest freshwater storage reservoir in the northern oceans; and (3) it has exhibited changes in halocline structure and freshwater storage in recent years.

In this chapter we examine the distribution of freshwater anomalies (relative to a defined reference salinity) in northern oceans by reviewing criteria that have been used to construct freshwater budgets and then by comparing freshwater disposition in the subarctic Pacific, subarctic Atlantic and Arctic oceans. This comparison provides a useful basis for the interpretation of Arctic Ocean flux measurements and affirms that the Canada Basin is a significant freshwater reservoir (Section 7.2). We next examine various hydrographic data sources within the Canada Basin (a geographical feature) to define the role of the Beaufort Gyre (a wind-forced dynamical feature) in freshwater storage and release (Section 7.3). Due to this latter feature, the upper

[1] Fisheries and Oceans Canada, Institute of Ocean Science, 9860 W. Saanich Road, Sidney, B.C., V8L 4B2

[2] Japan Agency for Marine-Earth Science and Technology, Yokohama, Japan

[3] Woods Hole Oceanographic Institution, Woods Hole, MA, USA

R.R. Dickson et al. (eds.), *Arctic–Subarctic Ocean Fluxes*, 145–169

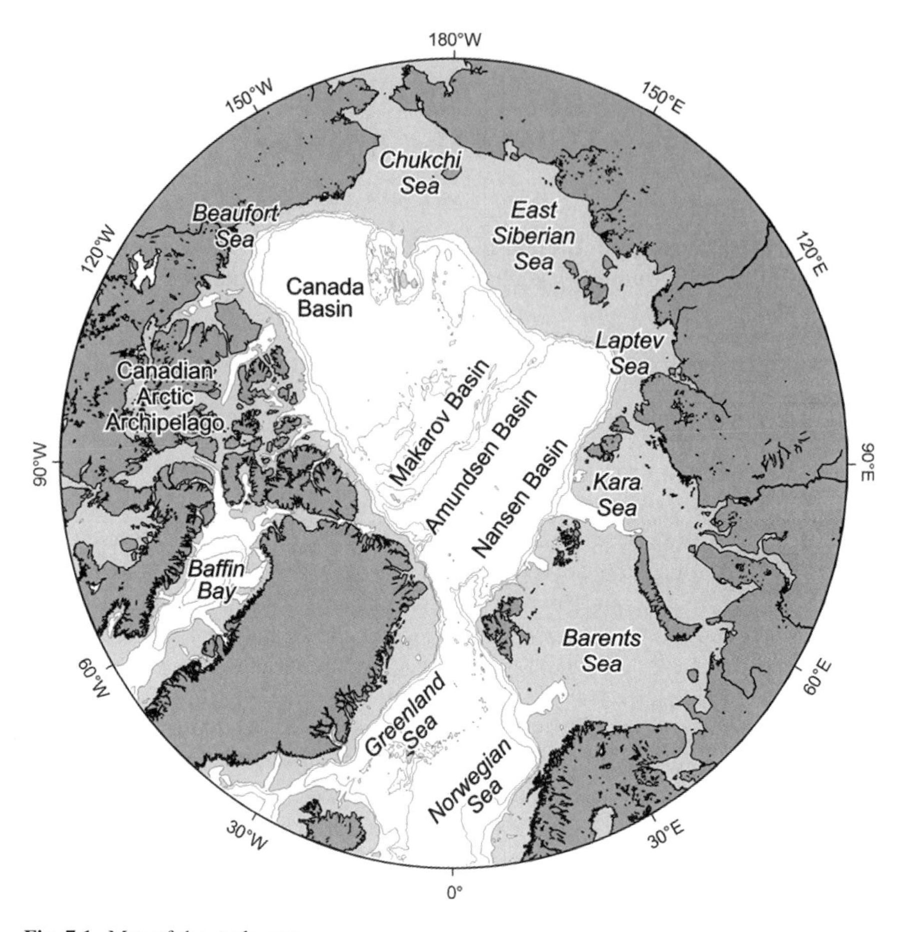

Fig. 7.1 Map of the study area

layer circulation in the Beaufort Gyre is anticyclonic whereas circulation elsewhere in the Arctic Ocean is cyclonic. Then we examine the Canada Basin's role as a reservoir with respect to sources of its freshwater components (e.g. meteoritic (runoff and precipitation), sea-ice melt and Pacific throughflow), and also to its water mass structure, within which freshwater components are stored (Section 7.4). This distinction among source components and among water mass affiliations is a prerequisite to interpreting downstream freshwater fluxes and to predicting the response of the Arctic system to climate variability. Finally, we combine geochemical data and recent freshwater budget estimates to calculate the relative contributions of freshwater components from the Canada Basin to other Arctic basins (Section 7.5). A summary and outlook is given in Section 7.6.

7.2 Freshwater Anomaly Definition

The first step in formulating freshwater budgets lies in defining a useful measure of freshwater content. The standard method defines a "freshwater anomaly", based on the selection of a defined reference salinity. This approach has often been used to construct budgets of confined seas and has the advantage that it relates directly to stratification which, in turn, constrains the dynamics of the system. The challenge, however, is to select an appropriate reference salinity. One method supposes a stirred box system and the freshwater anomaly is calculated with respect to a mean salinity within the box (Fig. 7.2a); for example, Aagaard and Carmack (1989) chose a reference salinity S = 34.8 for the Arctic Ocean. Alternatively the salinity of the saline end-member entering the confined sea is selected; for example, Dickson et al. (2007) chose S = 35.2 to represent the salinity of inflowing Atlantic water. Another strategy supposes estuarine circulation in the confined sea and here the reference salinity is chosen to be that of the lower layer which forms the base of the halocline (Fig. 7.2b). This approach was employed by Tully and Barber (1960) who used S = 33.8 to estimate the quantity of freshwater stored in the upper layers of the north Pacific, north of the subarctic front.

The weakness of the reference salinity approach becomes evident however, when ocean basins are not confined but are connected via sills and passageways. Thus an appropriate choice for one basin may be meaningless for the adjoining basin. Here it may be necessary to define a "practical" reference salinity for the upstream basin based on the salinity it can export above sill depth (Fig. 7.2c). Fortunately, budget calculations within a given confined basin (e.g. the Arctic Mediterranean) are not overly sensitive to small differences in the choice of reference salinity. Alternative approaches that are independent of a reference salinity have been advanced by Wijffels et al. (1992), who constructed a global budget for total freshwater, and by Walin (1977), who formulated conservation equations for an

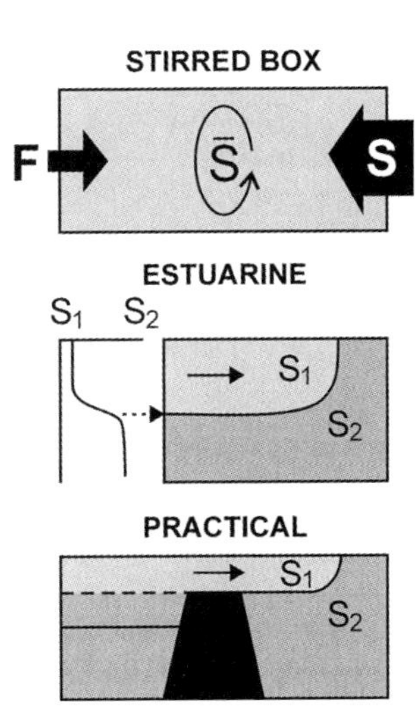

Fig. 7.2 Schematic showing various approaches used to define reference salinity

estuarine system using salinity and time as independent variables (natural coordinates). All of these approaches have merits and limitations and a choice must be based on intent and application.

An initial perspective on freshwater storage in the northern oceans can be obtained by mapping the distribution of salinity in arctic and subarctic seas at selected sill depths from climatological data (Conkright et al. 2002; World Ocean Data (WOD) 2001). The map of salinity at 20 m (Fig. 7.3a), taken to represent the salinity of the near-surface mixed-layer, shows that the Pacific at this depth is much fresher than the Atlantic ($\Delta S \sim 2$) and this low salinity water enters the Arctic Ocean through Bering Strait. Saline water from the North Atlantic crosses the Iceland–Scotland Ridge, flows northward through the Norwegian Sea and branches into the Barents and Greenland seas. Together these two sources define the large-scale estuarine forcing of the Arctic Ocean (cf. Stigebrandt 1984). The freshest near-surface water is found in the Canadian Basin (which includes both the Makarov and

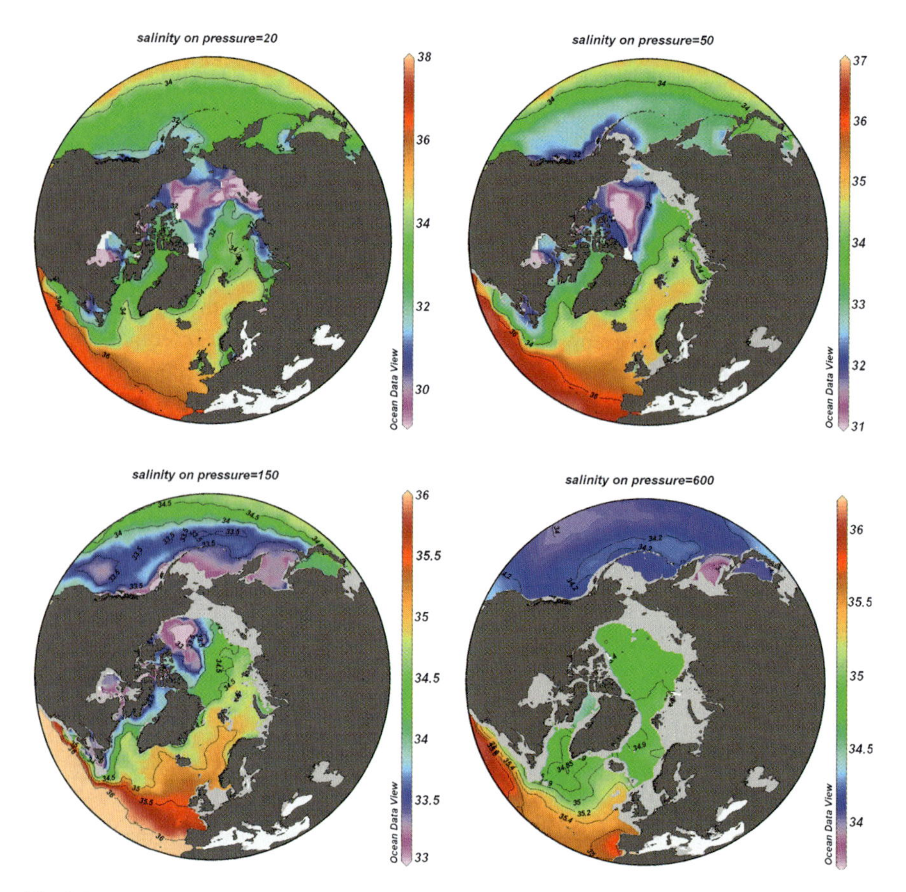

Fig. 7.3 Maps created from WOD 2001 showing the horizontal distribution of (a) salinity at 20 m, (b) salinity at 50 m, (c) salinity at 150 m and (d) salinity at 600 m

Canada basins) and is separated from a more saline upper layer found in the Eurasian Basin (Nansen and Amundsen basins) by a primary front that defines the baroclinic structure of the Transpolar Drift. The very low salinity waters lying above the East Siberian and Laptev seas, associated with Russian river inputs, are proximal to the western Canada Basin. Also evident are low salinity sources in the Kara Sea and southern Hudson Bay. The map of salinity at 50 m, the deepest depth connecting upper ocean waters of the Pacific, Arctic and Atlantic, defines the depth of free exchange of water masses among the three oceans (Fig. 7.3b). These two near-surface distributions of salinity also imply the key role of coastal-trapped and shelf-break currents in the transport of low salinity waters (cf. Griffiths 1986; Cenedese and Linden 2002; Williams et al. 2006; Bacon et al. 2007). (Because the resolution used in these mappings is insufficient to strictly distinguish between coastal-trapped and shelf break currents, we will use the term 'near-coastal flows in reference to them.') Such near-coastal flows are particularly fresh along the northeast Pacific, through the Canadian Archipelago and eastern coast of Canada and Greenland. Previous studies have demonstrated the regional importance of near-coastal flows forced by local freshwater discharge: for example, see Royer (1982) for the Northeast Pacific, Woodgate and Aagaard (2005) for the Bering Sea, McLaughlin et al. (2006) for the Canadian Arctic Archipelago, Chapman and Beardsley (1989) for the west coast of Greenland and Labrador Sea, and Bacon et al. (2002) for southern Greenland. Although no study has yet demonstrated the connectedness of these flows from a full, Northern Hemisphere perspective, we speculate that, jointly, such flows form a *contiguous* band of baroclinic flow around northern North America and constitute a substantial component of the freshwater transport (cf. Bacon et al., this volume, for the western subarctic Atlantic). The term contiguous reflects the fact that the forcing of individual components by freshwater inputs and wind is phased seasonally from one local current system to its downstream neighbour according to local supply of fresh water (cf. Carmack and McLaughlin 2001).

At 150 m (Fig. 7.3c), the approximate sill depth of passageways connecting the Arctic Ocean with the North Atlantic via the Canadian Archipelago, low salinity water is found almost exclusively within the Canada Basin. The fact that low salinity water is still evident at 150 m indicates how thick and therefore robust the reservoir of low salinity water in the Canada Basin is. The export pathway of these deeper, low salinity waters appears to be primarily through Nares Strait and the Canadian Archipelago. Salinity at 600 m (Fig. 7.3d), approximately the deepest depth connecting the Arctic Ocean with the global ocean via flow through Demark Strait, the Iceland–Scotland Ridge and Davis Strait, shows the relative uniformity of deep waters (S ~ 34.9) within the Arctic Ocean, Nordic Sea and Irminger Sea. The large-scale field of dynamic topography 20/600 dbar (Fig. 7.4, also see Steele and Ernold 2007) illustrates the 'downhill' journey, from Pacific to Arctic to the convective regions of the Nordic, Labrador and Irminger seas, that is largely responsible for sustaining arctic and subarctic fluxes.

One motivation for investigating freshwater distributions in high-latitude northern oceans lies in the dominant contribution of salinity to stratification. Stratification is

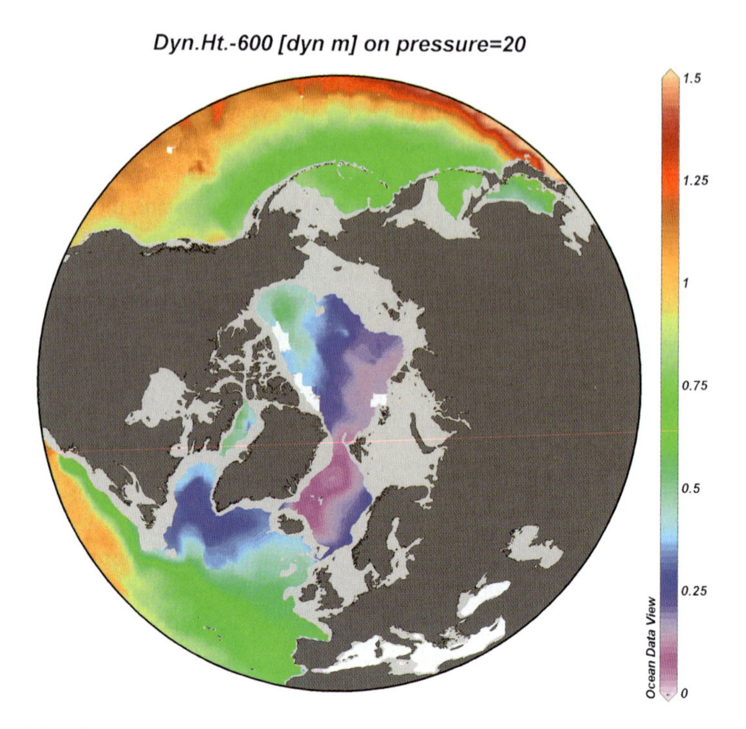

Fig. 7.4 Map showing dynamic topography (20/600 db)

typically expressed by the buoyancy frequency, $N^2 = g(d\rho/dz)$ where g is gravity and ρ is density. The density gradient can be further expressed as $d\rho/dz = \alpha(dT/dz) + \beta(dS/dz)$, where α is the thermal expansion coefficient, β is the haline contraction coefficient, dT/dz is the vertical gradient of temperature and dS/dz is the vertical gradient of salinity. Then $N^2 = N_T^2 + N_S^2$; where $N_T^2 = g\alpha(dT/dz)$ and $N_S^2 = g\beta(dS/dz)$. Because the magnitude of α decreases with decreasing temperature, the upper layers of the warm and saline subtropical seas are permanently stratified mainly by temperature; likewise the upper layers of cold, relatively fresh subarctic and arctic seas are permanently stratified mainly by salinity (cf. Carmack 2007). Because N_S^2 is, in fact, negative in subtropical seas, positive values of N_S^2 indicate freshwater storage. Thus the boundary defining the southern limit of salinity control on stratification and also the southern limit of freshwater storage in the upper ocean can be roughly identified by mapping the mean value of N_S^2 averaged between appropriate depth levels. Figure 7.5a, a map of N_S^2 averaged between 50 and 300 m, shows that the southern limit of salinity control roughly traces the boundary between the subtropical and subarctic gyres of both the Pacific and Atlantic oceans, and in the Atlantic it extends further northward into the Nordic Seas. High values of this stratification parameter are found in the Canada Basin and in western Baffin Bay. Moderately high values are found regionally in the North Pacific, in areas associated with Arctic outflow in the Canadian Archipelago, and along the east coast of Greenland. Moderately high values are also associated with river outflow in Hudson Bay and the Gulf of St. Lawrence.

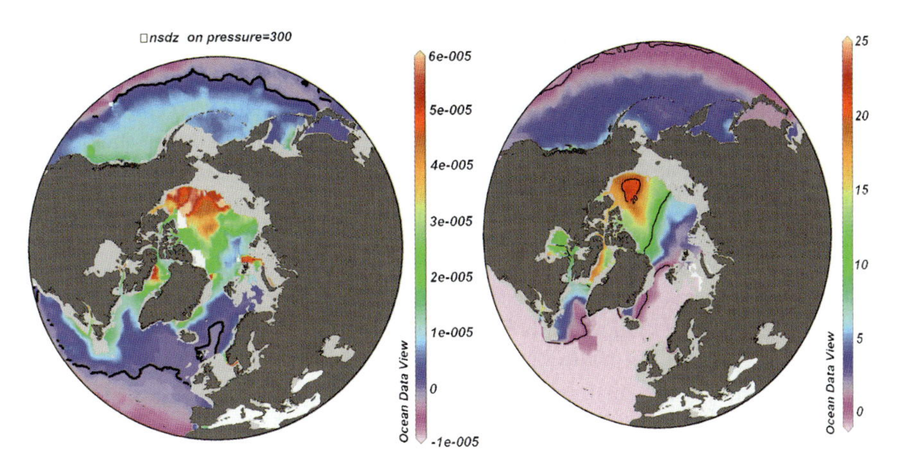

Fig. 7.5 Maps showing (a) the horizontal distribution of the contribution of salinity to mean stability $N_S^2 = g\beta(dS/dz)$ between 50 and 300 m, and (b) the freshwater equivalent height in northern oceans using the practical approach of selecting a reference salinity of 34.8 and using the following upstream sill depths: for Bering Strait (50 m); the Canadian Arctic Archipelago (150 m); Hudson Bay (100 m); and Denmark Strait (600 m). Although selective withdrawal over sills will occur, the use of upstream sill depth is useful as a first approximation. Black lines indicate freshwater equivalent heights of 0, 10 and 20 m

Based on the above discussion and incorporating the so-called practical approach of constraining the depth of integration depth according to 'upstream' sill depths for Bering Strait (50 m), the Canadian Arctic Archipelago (150 m), Hudson Bay (100 m) and Denmark Strait (600 m), the integrated freshwater content (equivalent height) relative to a reference salinity of 34.8 in the northern ocean is then calculated (Fig. 7.5b). From this figure it is evident that the North Pacific is a substantial upstream reservoir whose mean freshwater, expressed in equivalent height, is ~3–5 m (cf. Tully and Barber 1960; Aagaard et al. 2006). In contrast, the North Atlantic contains little freshwater apart from Arctic Ocean exit pathways. Within the Arctic Ocean the major reservoir of freshwater is the Canada Basin where 15–20 m is stored within the halocline and, moving toward the Atlantic, the equivalent height decreases from the Makarov (~10 m) to the Amundsen (~5 m) to the Nansen Basin (0–2 m). Given the magnitude of freshwater stored in the Canada Basin, changes in storage volume over time can significantly impact downstream fluxes and, at the same time, mask short-term imbalances in inflow and outflow rates.

7.3 Time Variability of Freshwater Storage in the Canada Basin

The volume of freshwater stored in the Arctic Ocean is roughly equal to that stored in all lakes and rivers of the world and is 10–15 times greater than the annual export of freshwater (including ice and water) from the Arctic Ocean (Aagaard and Carmack 1989). The bulk of this storage is located in the Canada Basin in association with the anticyclonically driven and topographically steered Beaufort Gyre. Here, fresh water is accumulated by wind-forced Ekman convergence of low salinity waters from various proximal sources including Pacific inflow, river discharge and sea-ice melt (Proshutinsky et al. 2002). Hence the atmospheric forcing and mechanisms of air/ice/ocean coupling that affect its storage and release over time are of major importance. Proshutinsky et al. (2002) argued that the freshwater storage in the Beaufort Gyre varied according to the strength of anticyclonic wind-forcing in that freshwater would accumulate under strong anticyclonic forcing and would be released under weak forcing. Indeed, the release of only 5% of this freshwater could cause a change in the salinity in the North Atlantic similar to that of the Great Salinity Anomaly of the 1970s (Dickson et al. 1988). Based on changes in water mass distributions, the relative fresh water outflow between the Fram Strait and Canadian Arctic Archipelago gateways is now believed to vary on interannual timescales (McLaughlin et al. 2002; Steele et al. 2004; Falck et al. 2005). A detailed analysis of freshwater storage variability in the Arctic Ocean over the last 100 years has recently been completed by Polyakov et al. (2007). In this section we will briefly examine existing data for evidence of temporal variability, from multi-decadal to decadal to interannual, with focus on the Canada Basin and recognizing the limitations of the sparse historical data set.

We begin by looking for evidence of any large-scale changes in the distribution of freshwater within the Arctic Ocean over the past half century. For example, Swift et al. (2005) noted that the persistent (over several decades) wide-spread presence of Pacific water in the central Arctic Ocean halocline was followed by its abrupt disappearance from a large area in 1985 (also see McLaughlin et al. 1996). Decadal variability has been emphasized by Proshutinsky et al. (2005), Richter-Menge et al. (2006) and Polyakov et al. (2007). Accordingly, we examine freshwater content computed from gridded historical data from the 1950s to the 1980s (EWG) and

Table 7.1 List of expeditions

Year	Month	Expedition
1993	Aug.–Sept.	Scientific Ice Expedition (SCICEX)
1994	July–Aug.	Arctic Ocean Section
1995	May–Sept.	SCICEX
1996	Oct.–Sept.	SCICEX
1997	Oct.–Sept.	SCICEX
1997–1998	Oct.–Sept.	Surface Heat Budget of the Arctic Ocean/Joint Ocean Ice Study (SHEBA/JOIS)
1998	Aug.–Sept.	SCICEX
1999	Apr.–May	SCICEX
2000	Oct.	SCICEX
2002	Aug.–Oct.	Joint Western Arctic Climate Study (JWACS)
2003	Aug.–Sept.	JWACS/Beaufort Gyre Exploration Project (BGEP)
2004	Aug.–Oct.	JWACS/BGEP
2005	Aug.–Oct.	JWACS/BGEP
2006	Aug.–Oct.	JWACS/BGEP

from ship and submarine observations in the 1990s and 2000s (see Table 7.1) for evidence of decadal change in the Canada Basin. We note that data from the 1990s are sparse and there are no observations from 76–80° N along 140° W. The observational data are extracted every 0.5° from gridded fields produced by fitting a polynomial surface to the observations and smoothed slightly in latitude with a 1° running mean triangular filter. Sections along 140 and 150° W (Fig. 7.6) reveal no discernable decadal trend in the cumulative freshwater content from the 1950–1980s.

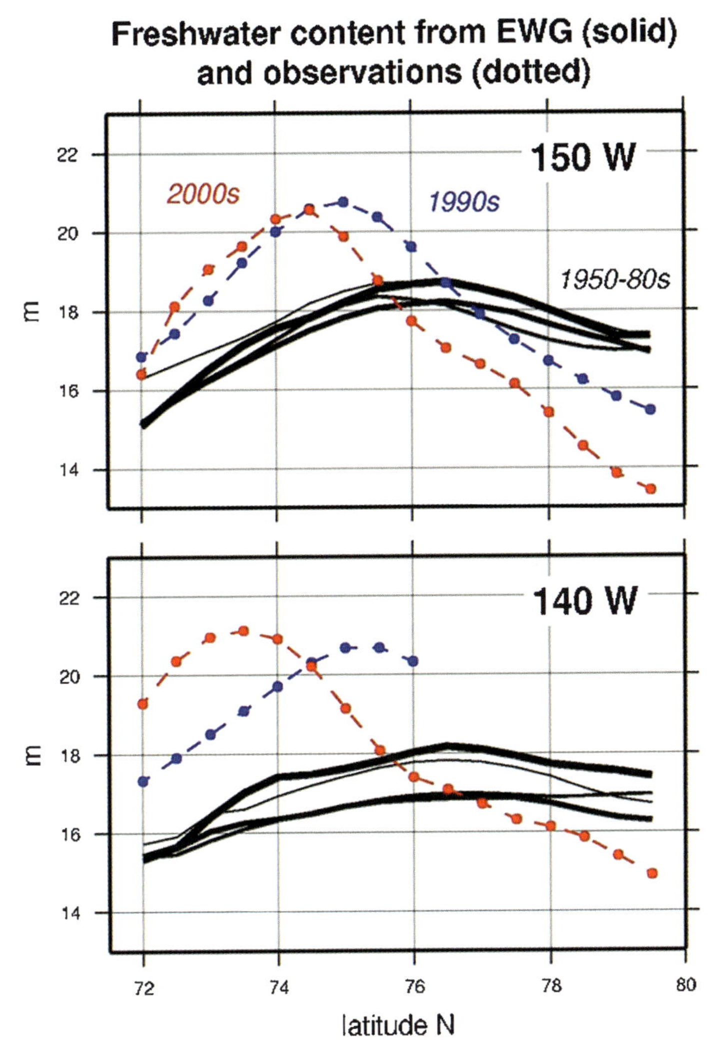

Fig. 7.6 Freshwater content computed from gridded historical data (EWG) from the 1950s to the 1980s, from ship and submarine expeditions in the 1990s and from the Canada /Japan/US Joint Western Arctic Climate Study in the 2000s. The black lines that represent the EWG data are thinnest in the 1950s and thickest in the 1980s

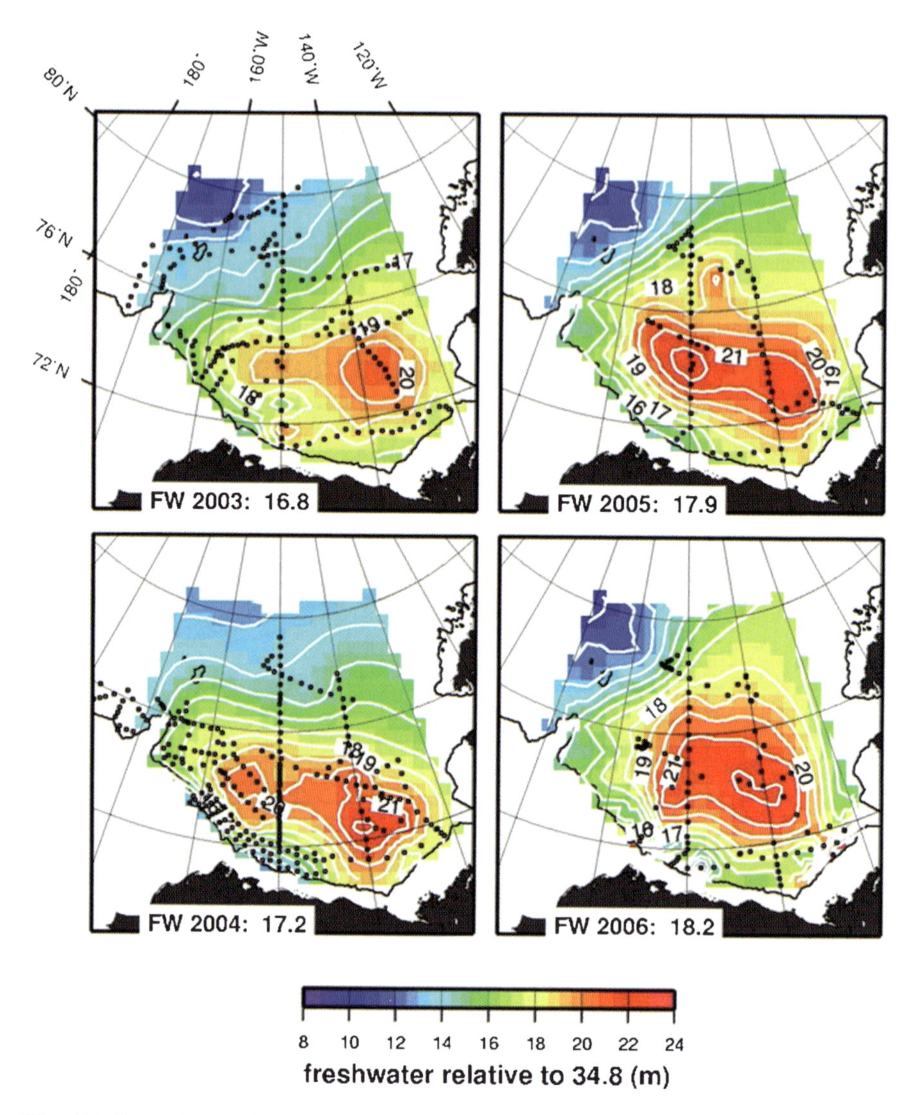

Fig. 7.7 Comparison of freshwater content in the Beaufort Gyre in (a) 2003, (b) 2004, (c) 2005 and (d) 2006. Numbers at the bottom of each figure indicate the total freshwater content in the gridded region ($\times 1{,}000\,km^3$) for each year

In the 1990s, however, there was a shift in freshwater distribution, with an increase in freshwater content in the southern portions of the basin and a compensating decrease in the northern portion. In the early 2000s the freshwater content maximum shifted toward the east, away from the Northwind Ridge (near 150° W) and southward towards Banks Island (near 140° W). We acknowledge that these temporal changes are speculative, as they may include spatial and seasonal variability due to data sparseness.

Finally we examine recent data (2003–2006) from the southern Canada Basin (south of 80° N) for evidence of interannual variability (Fig. 7.7). Maps of freshwater equivalent height show variability in the Beaufort Gyre, identified by freshwater equivalent heights of approximately 20 m. In 2002 (not shown), the gyre was located between 72° and 76° N and 140° and 150° W. In 2003 the gyre shifted slightly eastward by approximately 2–3°, however a secondary maximum in freshwater content remained along 150° W near 74° N. In 2004 the shape of the gyre is more elongated and two relative maxima are evident, the larger being near 73° N and 140° W (>22 m) and the smaller near 74° N and 155° W. In 2005 the core remains elongated, the region covered by 20 m is larger and the freshwater content maximum is located along 150° W. In 2006, the maximum again shifted eastward and the core of the gyre spread northward. Overall, the total freshwater content in the region appears to have increased slightly over this 5-year period. These recent data suggest interannual variability in the spatial distribution of freshwater and indicate that the Beaufort Gyre may be tightly coupled to interannual changes in wind forcing and air–sea-ice coupling (cf. Shimada et al. 2006).

7.4 Freshwater Components and Distributions in the Canada Basin

Thus far we have examined freshwater content by integrating the salinity anomaly relative to a reference salinity of 34.8, and have reported variability on a number of spatial and temporal scales. It is also important to understand where and how each freshwater *component* stored in the present ocean is derived to predict the effects of future change. Although the main approach used here identifies source constituents (i.e. meteoritic, sea-ice melt, Pacific), it is initially important to recognize freshwater storage on the basis of its water mass distributions (e.g. with the mixed layer, Pacific summer water, Pacific winter water, lower halocline). Stratification in the Canada Basin is especially complicated and the halocline is comprised of a series of layers (modes and clines) from the surface down to about 300 m (Fig. 7.8). The temperature and salinity structure is characterized by a seasonal mixed layer found in the upper ~40 m, wherein the effects of sea-ice melt and river plume spreading in summer and sea-ice formation in winter are manifest. Below, from ~40 to ~200 m, lie both summer and winter influxes of Pacific-origin water. Pacific-origin winter water is further characterized by high nutrient levels and a distinct N/P relationship. Below ~200 m the transition to Atlantic-origin waters occurs, first with the Lower Halocline layer which in turn overlies the Atlantic water.

Following from the seminal work of Östlund (1982) on $\delta^{18}O$ partitioning, a number of authors have applied this and other geochemical tracers to examine the constituents of Arctic Ocean freshwater by source (cf. Macdonald et al. (1995) used $\delta^{18}O$; Guay and Falkner (1997) used barium; Jones et al. (1998) used nitrate/phosphate

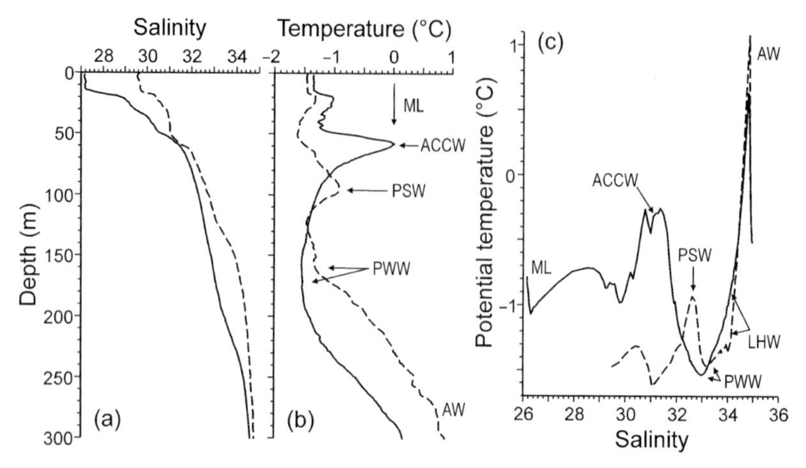

Fig. 7.8 Two representative stations from the Canada Basin illustrating regional water mass structure: (a) profile of salinity, (b) profile of temperature and (c) the corresponding θ/S correlation diagram. The winter mixed-layer (ML) extends to 40–50 m depth. A number of different Pacific-origin water masses are present: summer inflows include Alaska coastal current water (ACCW) that enters near Barrow Canyon, Pacific summer water inflows (PSW) and Pacific winter water (PWW). Atlantic-origin waters include lower halocline water (LHW); and the Atlantic layer (AW) which is comprised of Fram Strait branch waters indicated by the temperature maximum and Barents Sea branch waters (not shown)

relationships; Ekwurzel et al. (2001) used $\delta^{18}O$ and PO_4*; Yamamoto-Kawai et al. (2005) used $\delta^{18}O$ and alkalinity). Additional refinements have been used to identify source origin, for example, Anderson et al. (1994) used alkalinity and silicate to distinguish between Pacific and river waters in the upper waters of the Eurasian Basin and Guay and Falkner (1994) used barium to distinguish between Eurasian and North American river discharge. Yamamoto-Kawai et al. (2005) showed that alkalinity and $\delta^{18}O$ are interchangeable tracers of freshwater and brine in the Arctic Ocean, with the exception in regions where waters are influenced by Mackenzie River discharge.

The main objective of such analysis is to partition a seawater sample into its primary constituents – meteoritic water, sea-ice melt and a meaningfully defined saline end-member – and this is typically done using $\delta^{18}O$/S correlations. With regards to the saline end-member a number of different choices have been made. For example Macdonald et al. (1995) and Macdonald et al. (2002) used the polar mixed layer ($S = 32.2$) and middle halocline water ($S = 33.1$), respectively, although these layers are already diluted with sea-ice melt/brine and meteoric water. Melling and Moore (1995) and Yamamoto-Kawai et al. (2005) used Atlantic water as the saline end-member since Pacific water appears roughly as a mixture of Atlantic water and meteoric water on a $\delta^{18}O$/S correlation diagram. However, this approach cannot distinguish Pacific water from meteoric water. Bauch et al. (1995) and Ekwurzel et al. (2001) used silicate and PO_4*, respectively, to estimate the contribution of Pacific water because values are much higher in Pacific water than in

Atlantic water. As nutrients undergo significant seasonal variability while crossing the Bering and Chukchi seas this method may underestimate the contribution of Pacific water. To avoid these effects, Jones et al. (1998; 2003) used the nitrate–phosphate (N/P) relationship as a tracer of Pacific water. They noted that the N/P correlation diagram consists of three straight-line segments. Two near-parallel lines, that follow the Redfield ratio, represent Atlantic and Pacific sources and the offset between them arises from denitrification in Pacific inflow during transit across the Bering and Chukchi shelves. These two lines are connected by a third line that represents mixing between the superimposed Pacific and Atlantic water masses. Water at any point on this mixing line can thus be divided into its Atlantic and Pacific fractions. Yamamoto-Kawai et al. (2008) modified the Jones method and used dissolved inorganic nitrogen instead of nitrate so as to include the ammonium associated with regeneration and high production on the Chukchi Shelf. In the Canada Basin the mixing point line corresponds to salinities from S ~33.0 to ~34.8.

The most recent freshwater inventory of the Canada Basin, calculated by Yamamoto-Kawai et al. (2008) from data collected in 2003–2004, are used to investigate constituent distributions. They combined an analysis of N/P correlations, thus identifying the saline end-members, with an analysis of $\delta^{18}O$/S correlations and a three component mixing model for meteoritic water, Sea-ice meltwater and saline end-member water, using the following approach. Pacific water is the saline end-member for $S \leq 33$ waters. In $S > 33$ waters, the saline end-member is a mixture of Pacific and Atlantic waters, and the mixing ratio is calculated using the Jones et al. (1998) approach (see Yamamoto-Kawai et al. 2008 for equation). Next, S and $\delta^{18}O$ values for the saline end-member are calculated using the fraction of Pacific water and values for inflowing Pacific-origin ($S = 32.5$, $\delta^{18}O = -0.80$) and Atlantic-origin water ($S = 34.87$, $\delta^{18}O = 0.24$, see Yamamoto-Kawai et al. 2007 for selection criteria). The three component mixing model is then applied to estimate freshwater fractions of meteoritic water, sea-ice melt and saline end-member components, and the saline end-member is further divided into its Pacific and Atlantic parts.

The resulting mean vertical distributions of freshwater components (calculated as the mean depth of a given fraction, $<\mathbf{Z}(f)>$ for Pacific and Atlantic waters, and the mean fraction at a given depth $<\mathbf{f}(Z)>$ for meteoritic and sea-ice melt components) computed from all stations deeper than 1,000 m are shown in Fig. 7.9. To further show the association of freshwater components with distinct water masses, these fractions are then plotted on a θ/S correlation diagram from each station in the Canada Basin (Fig. 7.10). The seasonal mixed layer (S < ~31, cf. Carmack et al. 1989) is mainly comprised of Pacific water (>80%), freshened by the addition of meteoric water (10–20%). Sea-ice melt slightly freshens (<10%) the upper 30 m of the seasonal mixed layer whereas the addition of brine (i.e. negative sea-ice melt) makes the lower 10–20 m more saline. The transition from freshening by melting to increasing the salinity by brine injection occurs above the base of the seasonal mixed layer at 30 m and S ~ 30. Summer and winter influxes of Pacific-origin water dominate (>80%) the water column to ~175 m and S ~ 33.3 and are greater than 70% to ~190 m and S ~ 33.8, the later being due to the injection of hypersaline polynya water (cf. Weingartner et al. 1998) and diapycnal mixing of Pacific water (Woodgate

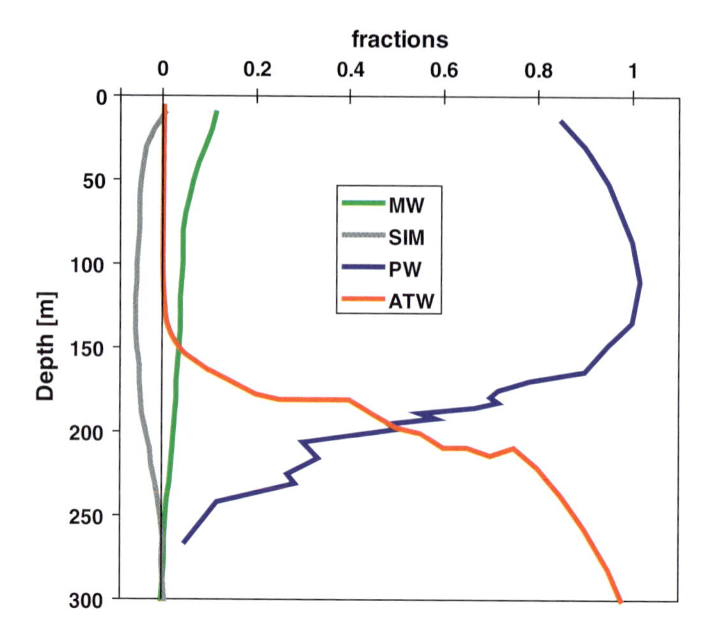

Fig. 7.9 The mean vertical profile of source water fractions in the Canada Basin 2003–2004: meteoric (MW), sea-ice melt water (SIM), Pacific water (PW) and Atlantic water (ATW). Only stations >1,000 m depth are used. Mean profiles are calculated as the mean fraction at a given depth <**Z**(f)> for MW and SIM, and the mean depth of a given fraction, <**f**(Z)> for PW and ATW

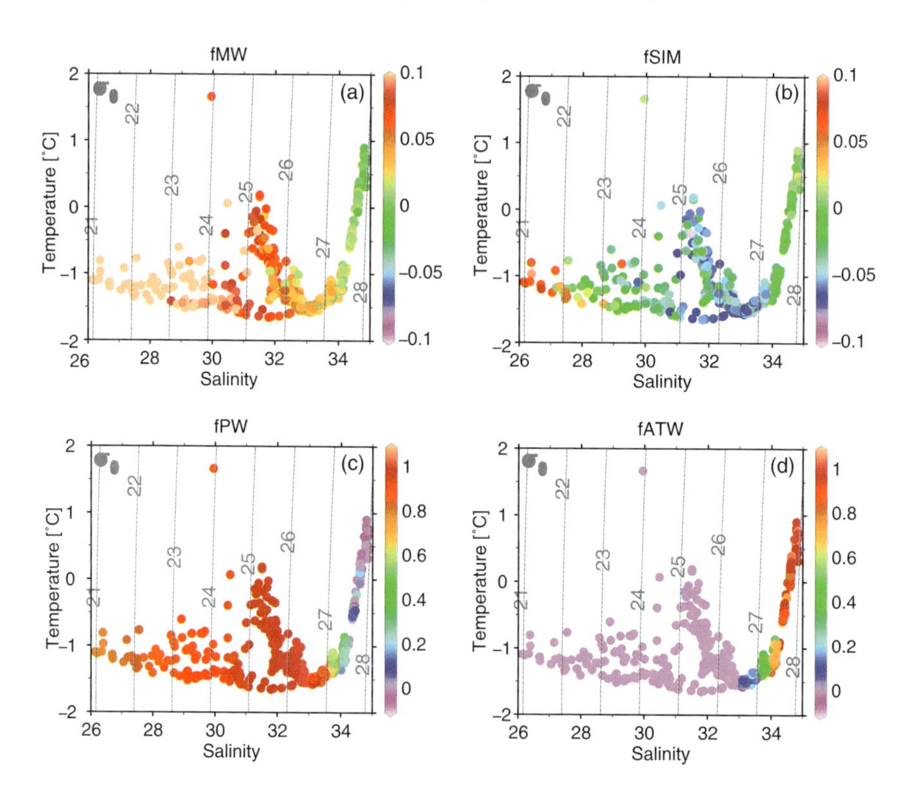

et al. 2005) into Lower Halocline Water. The transition to more than 80% Atlantic water occurs sharply and at ~220 m and S = 34.2. It is interesting to note that very small amounts of meteoric water and brine are present in both summer and winter Pacific waters and in Atlantic waters to S ~ 34.2. As there is no freshwater component below 300 m it is the depth of integration used in the following calculations.

Integrating the fraction of Pacific water at each station, the horizontal distribution of the equivalent thickness of Pacific water is calculated and when mapped is found to be >200 m in the south and <150 m in the north (Fig. 7.11a). The geographic difference in the depth of Pacific water corresponds to the apparent influx of Atlantic water around the northern perimeter of the Northwind Ridge by topographically steered boundary currents (Fig. 7.11b; also see McLaughlin et al. 2002; McLaughlin et al. 2004; Häkkinen and Proshutinsky 2004; Shimada et al. 2004) and spatial variability in Ekman pumping associated with the large-scale wind field and air/ice/sea/coupling (cf. Shimada et al. 2006).

To examine the horizontal distribution of freshwater by component, the freshwater equivalent fractions at every station are calculated using S = 34.87 as the reference salinity, integrated and mapped (see Yamamoto-Kawai et al. 2008 for selection of reference salinity and error analysis). It should be noted that use of S = 34.87 instead of S = 34.8 as a reference salinity results in a difference in integrated content of 1–2%. The equivalent thickness of total freshwater is ~20 m in the southeastern Canada Basin and decreases to ~14 m in the northwest (Fig. 7.12a). The equivalent thickness of meteoritic water is highest (>15 m) near the Mackenzie River and

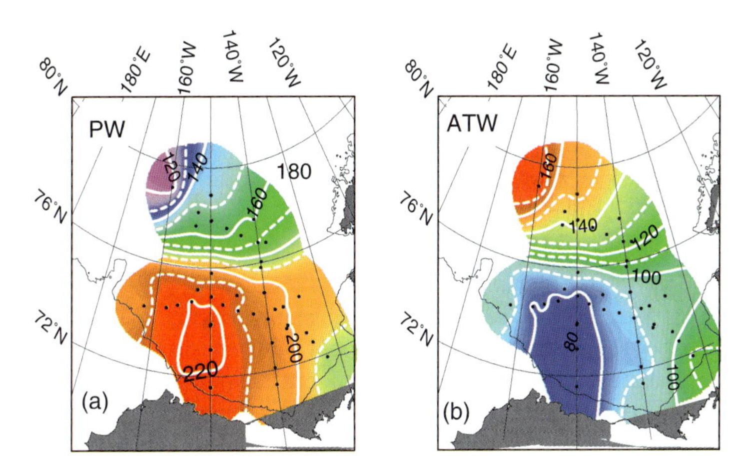

Fig. 7.11 Horizontal distribution of equivalent thickness of (a) Pacific water and (b) Atlantic water integrated from 0 to 300 m. Only stations >1,000 m depth are used

Fig. 7.10 Distribution of freshwater components plotted on a T/S correlation diagram; only stations >1,000 m depth are used: (a) fraction of meteoritic water; (b) fraction of sea-ice melt; (c) fraction of Pacific water; and (d) fraction of Atlantic water. Gray lines indicate isopycnal contours

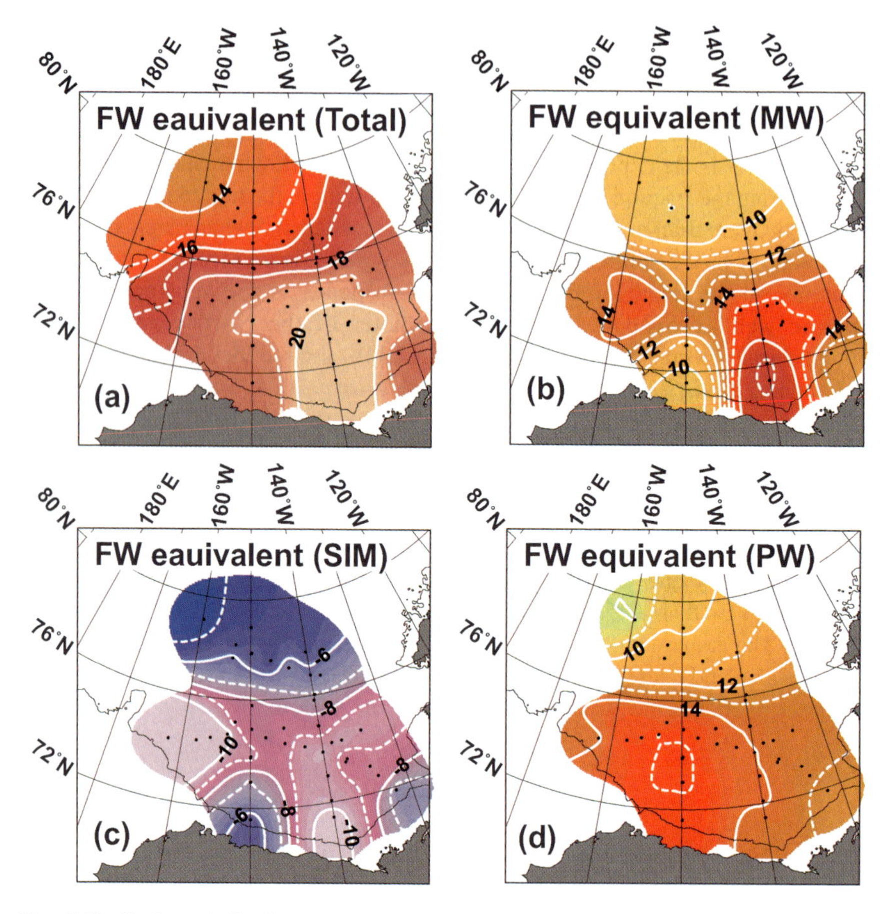

Fig. 7.12 Horizontal distribution of freshwater equivalent thickness of (a) total freshwater equivalent; (b) meteoritic water, (c) sea-ice melt; and (d) freshwater equivalent of Pacific water. Only stations >1,000 m depth are used

decreases to ~10 m in both the northwest and southwest (Fig. 7.12b). The equivalent thickness of sea-ice melt is negative throughout the basin and this indicates that freshwater is removed by sea-ice formation and net export from the Canada Basin (Fig. 7.12c). The equivalent thickness of freshwater removed as sea-ice is ~6 m in the north and ~9 m in the south with higher values (~10 m) near the northern Chukchi and Beaufort shelves and lower values (~6 m) near the coast at Point Barrow. However, as brine can be laterally transported into the basin by shelf-basin exchange mechanisms, the distribution of *net* sea-ice formation does not necessarily represent *in situ* ice formation but instead reflects the history of the water mass. The equivalent thickness of Pacific-origin freshwater is ~14 m in the south and ~10 m in the north (Fig. 7.12d). The mean inventories of meteoric water, the freshwater equivalent of sea-ice melt and the freshwater equivalent of Pacific water for the study area are found to be 13 m, −8 m and 13 m, respectively. The mean for the entire Canada Basin is likely lower than this because the freshwater content in the northern part of Canada Basin is lower than in the southern part, and therefore values of 10.5 m, −6.5 m and 12 m – the mean freshwater inventories for the central Canada Basin (75–80° N) – are used in the calculations below (see Yamamoto-Kawai et al. 2008).

In summary, freshwater in the Canada Basin is comprised primarily of meteoric water and the freshwater equivalent of Pacific water, and they contribute almost equally to the total freshwater found in the upper 300 m. The net effect of sea-ice formation and melting is to remove ~30% of the freshwater contributed by both meteoric and Pacific water.

7.5 Freshwater Storage, Flux and Residence Time

7.5.1 Canada Basin

The mean freshwater inventories for the central Canada Basin (75–80° N) now can be used to calculate the storage, flux, and residence time of freshwater components (Fig. 7.13). Multiplying the mean freshwater inventories by the surface area of the

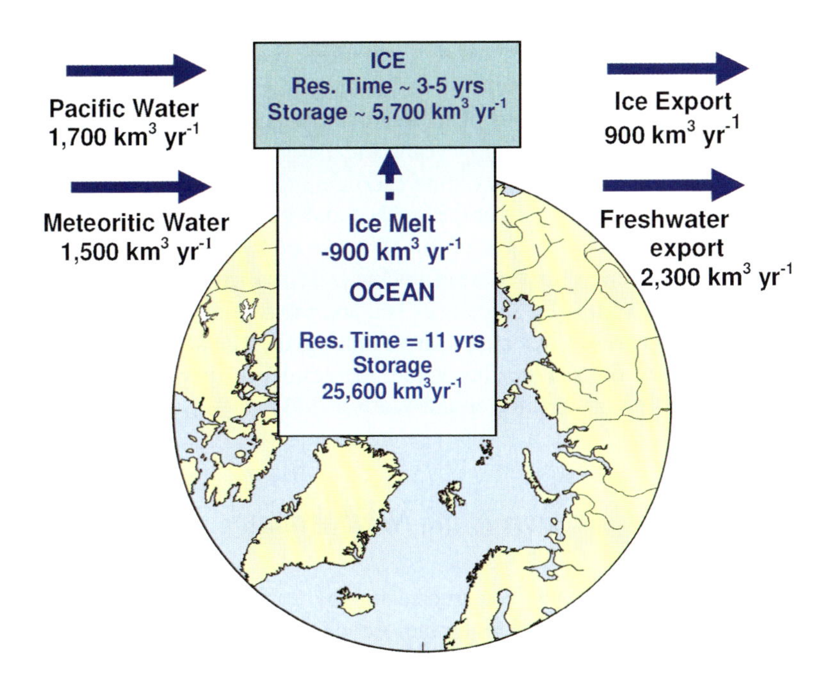

Fig. 7.13 Schematic of the freshwater budget for the Canada Basin

Canada Basin deeper than 1,000 m (1.6×10^6 km^2), the volumes of meteoritic and freshwater equivalent of Pacific water are computed to be 16,800 km^3 and 19,200 km^3, respectively, whereas 10,400 km^3 of freshwater have been removed by sea-ice export. The total (net) freshwater storage in the top 300 m of the Canada Basin is thus 25,600 km^3 and this corresponds to approximately one third of the total freshwater stored in the Arctic Ocean (Zhang and Zhang 2001). The volumes of meteoric water and freshwater equivalent of Pacific water actually stored in the ocean depend on the fractions of each of these freshwater sources that are removed as ice. Given that the water in the winter mixed-layer is approximately a 1:9 mixture of meteoric water and Pacific water, and assuming that sea-ice is formed equally from both sources, then the volume of meteoric water and freshwater equivalent of Pacific water actually stored is 15,800 and 9,800 km^3, respectively.

The addition of 19,200 km^3 of freshwater equivalent of Pacific water corresponds to approximately 11 years of Pacific inflow through Bering Strait, and this residence time is calculated as follows. The mean transport of water through Bering Strait is ~0.8 Sv with a mean salinity of S = 32.5 (Woodgate et al. 2005), supplying ~1,700 km^3 year^{-1} of freshwater (relative to S = 34.87) into the Arctic Ocean. Although the Alaskan Coastal Current (cf. Woodgate and Aagaard 2005) also carries ~700 km^3 year^{-1} of additional freshwater into the Canada Basin, this freshwater component will be included in the meteoric water component in our three component analysis. Assuming that all inflowing Pacific water enters the Canada Basin interior, the mean residence time in the upper 300 m is thus about 11 years (19,200 km^3/1,700 km^3 year^{-1}). Applying this residence time to the volumes of other stored components, the fluxes of meteoric water and freshwater equivalent of sea-ice melt are 1,500 km^3 year^{-1} (of this, ~700 km^3 year^{-1} enters through Bering Strait) and –900 km^3 year^{-1}, respectively. Admittedly the residence time of the near-surface layer, where the fractions of meteoric water and sea-ice melt are higher (see Fig. 7.10), is likely shorter than 11 years, and thus our estimates above represent lower limits of flux. This residence time estimate of 11 years is consistent with tritium–helium ages of <4 years at the surface and ~18 years at 300 m (Smethie et al. 2000) and with tritium ages of 10–16 years of the freshwater component (Östlund 1982). Assuming that a mean thickness of sea-ice in the Canada Basin deeper than 1,000 m (1.6×10^6 km^2) is 2–3 m, and the salinity of sea-ice is S = 4, then the volume of freshwater stored in sea-ice in the Canada Basin is 2,800–4,300 km^3. Applying the sea-ice flux value, the residence time of sea-ice in the Canada Basin is 3–5 years (cf. Rigor and Wallace 2004).

7.5.2 Arctic Basin Export to the North Atlantic

We can now combine budgets constructed for the Canada Basin with published estimates of the boundary conditions for the Arctic Ocean (e.g. basin-wide estimates of river discharge, net precipitation and ice and liquid water export to the North Atlantic) to develop a rough budget for freshwater component fluxes in the whole arctic basin. Estimates of river discharge, net oceanic precipitation and exports of ice and liquid freshwater through Fram Strait and the Canadian Arctic Archipelago are taken from Serreze et al. (2006), fluxes through Bering Strait are taken from Woodgate and Aagaard (2005) and fluxes of freshwater components into and out of the Canada Basin are taken from the above budget (also see Yamamoto-Kawai et al. 2008); this formulation, specific to the Canada Basin, is shown schematically in Fig. 7.13.

Now, to arrive at a budget for the entire Arctic Ocean, the region is divided into two principal domains: the Canada Basin and all other basins (Fig. 7.14). This distinction is physically relevant because it separates the halocline of the Arctic Ocean into its anticyclonic (Beaufort Gyre) and cyclonic (Trans Polar Drift) components. The freshwater equivalent of Pacific water that enters via Bering Strait is 1,700 km^3 year^{-1}, the total volume of runoff that enters the Arctic Ocean is 3,300 km^3 year^{-1} and the net oceanic precipitation is 2,000 km^3 year^{-1}. The flux of meteoritic water into the Canada Basin is ~1,500 km^3 year^{-1} and ~700 km^3 year^{-1} of this enters through Bering Strait. Thus ~800 km^3 year^{-1} of the total meteoritic influx (5,300 km^3 year^{-1}) should enter the Canada Basin and

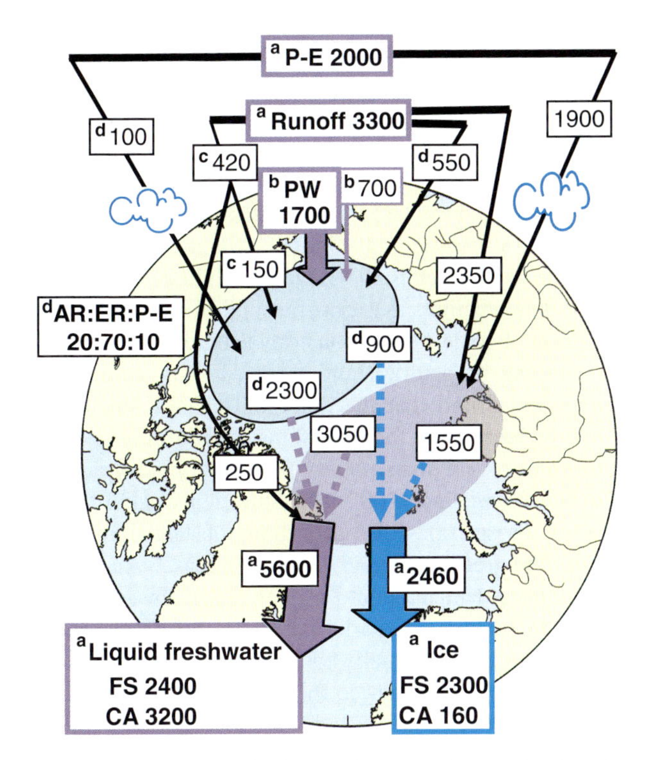

Fig. 7.14 Schematic of the freshwater budget of the Arctic Ocean, showing the partitioning of freshwater export. Superscripts indicate the references used flux values (km^3 year^{-1}): (a) is Serreze et al. 2006; (b) is Woodgate and Aagaard 2005; (c) is Lammers et al. 2001; and (d) is Yamamoto-Kawai et al. 2008

4,500 km^3 year^{-1} must therefore enter into the other basins. Using the component ratios of meteoric water obtained by Yamamoto-Kawai et al. (2008: American Rivers: Eurasian Rivers: Net Precipitation = 20:70:10), and assuming accuracies of ±50 km^3 year^{-1} in rates estimated here and Yamamoto-Kawai et al. (2008), then 150 km^3 year^{-1} of the meteoric water in the Canada Basin is from the discharge of North American rivers and 550 km^3 year^{-1} from Eurasian Rivers. This suggest that 250 km^3 year^{-1} of total discharge from North American rivers (420 km^3 year^{-1}; Lammers et al. 2001) might flow out from the Arctic Ocean without entering the deep basins. The remnant 2,350 km^3 year^{-1} of runoff must enter the other basins. As 100 km^3 year^{-1} of the net oceanic precipitation (2,000 km^3 year^{-1}) enters the Canada Basin then 1,900 km^3 year^{-1} must enter the other basins. Waters exiting the Canada Basin are partitioned into ice (900 km^3 year^{-1}) and liquid water (2,300 km^3 year^{-1}) components, and the ratio of ice export to liquid water export is 0.4. Using the outflow boundary conditions of ice (2,460 km^3 year^{-1}) and liquid (5,600 km^3 year^{-1}) exiting the Arctic Ocean given by Serreze et al. (2006), the fluxes of waters exiting the other basins can thus be calculated according to their ice (1,550 km^3 year^{-1}) and liquid water (3,050 km^3 year^{-1}) components and the ratio of ice export to liquid water for the other basins is 0.5. This then allows closure of the budget (Fig. 14).

7.6 Summary

In this chapter we examined the large-scale distribution of freshwater in northern oceans (subarctic Pacific, Arctic and subarctic Atlantic) and found that the main storage reservoir for freshwater is the Beaufort Gyre of the Canada Basin, and therefore that small perturbations in export from this reservoir could well dominate interannual and decadal scale fluctuations downstream. We then focused on the Canada Basin, looking at gridded and observational data for variability in freshwater content and found that the dominant change since the 1990s was a southward shift in the location of the core of the Beaufort Gyre. However, the data available for this analysis are sparse. Only the repeat hydrography carried out in the southern Canada Basin since 2002 is of sufficient spatial resolution to make reliable comparisons. These 2002–2006 data show substantial interannual variability and it appears that the freshwater content has increased marginally during this time. We next used geochemical data to investigate freshwater components in the Canada Basin and found that meteoric water and the freshwater equivalent of Pacific water contribute almost equally to the total freshwater found in the upper 300 m, and that the net effect of sea-ice formation and melting removes ~30% of the annual supply. Finally we calculated volumes and residence times of the various source components that comprise the freshwater inventory of the Canada Basin, and then combined these findings with the overall freshwater budget of the Arctic Ocean compiled by Serreze et al. (2006) to arrive at a basin-wide description of source water fluxes. As storage within the basin is large and variable there is not reason to suppose a priori that inflows and outflows must balance on annual to decadal time scales.

What are the existing knowledge gaps? Our estimates here are based on available data sets and it must be admitted that the uncertainty in many of these numbers is high and could be improved. For example, the historical context could be improved if the raw EWG data were made available, as this would allow direct comparisons between the past and present surveys and provide more insight about seasonal and interannual variability. Data from recent programs in the Canada Basin suggest circulation and storage within the Beaufort Gyre is highly variable and these findings demonstrate the value and importance of maintaining long-term observational programs. Ideally, the geographic reach of the observational program should be increased so that the entire Canadian Basin is surveyed and the northern reaches of the Beaufort Gyre determined. The basins immediately north of the Canadian Archipelago remain unexplored. Such extended surveys might also reveal if the shift in the Atlantic/Pacific water mass boundary away from the Lomonosov Ridge in the mid-1980s initiated a new permanent circulation mode or is only a transitory event on a yet-to-be determined time scale. As the majority of recent measurements have been collected during summer there is little known about the magnitude of seasonal variability, and moorings in the Canada Basin, both anchored and drifting, would provide such information. In terms of budgets, sea-ice estimates are rough calculations and require more detailed measurement and mapping to reflect recent and future changes. Data from a few moorings has shown that Bering Strait inflow has significant seasonal variability in all components (Woodgate et al. 2005) and therefore assigning a canonical transport value is a challenge. An array of moorings across Bering Strait would improve estimates of freshwater transport from the Pacific and include information about the Alaska Coastal and Anadyr currents.

The importance of identifying source components and how they are processed within the arctic basin becomes especially clear if one attempts to project what future freshwater exports would be under scenarios of global warming. The distinction is made all the more important by the fact that the ocean, with its longer residence times and recirculation rates, lags the atmosphere in response to climate forcing (cf. Peterson et al. 2006). How will climate change affect the processing and storage of freshwater components within the Arctic Basin, and thus quantitatively impact on export rates? Ice distillation (separation of ice and brines) accounts for ~30% of the freshwater export and therefore changes in the annual formation (thickness and extent) will impact the Arctic's freshwater budget. Changes in sea-ice cover may also trigger abrupt changes in air/ice/sea coupling (e.g. Carmack and Chapman 2003; Shimada et al. 2006). Changes in the supply of river water are also expected (Peterson et al. 2006; Déry and Wood 2005). Feedbacks among the interior Arctic and its marginal oceans will play a role (Dukhovskoy et al. 2004; Häkkinen and Proshutinsky 2004; Polyakov et al. 2007). Clearly, it is necessary to monitor not just the quantities of freshwater exiting the Arctic Ocean, but also their composition, their mode of formation and history, their mechanism of storage and release and the diverse physical constraints that limit their southward spreading as surface waters into the North Atlantic. This last issue cannot be over emphasised because regional stratification within future subarctic seas will not increase (decrease) simply because river discharge increases (decreases) but instead is critically dependant upon the extent and dynamics of the reservoir to which it is confined.

Acknowledgements We are deeply indebted to the Captains and crews of the CCGS Louis S. St-Laurent for their undaunted efforts in completing our ambitious Canada Basin expeditions. We are also deeply indebted to Sarah Zimmerman and Bon van Hardenberg, who served as Chief Scientists, and the many dedicated technicians who carefully collected and analyzed samples.. Organisational support was provided by the National Centre for Arctic Aquatic Research Expertise, Fisheries & Oceans Canada. Partial funding for Eddy Carmack, Fiona McLaughlin and Michiyo Yamamoto-Kawai was provided by Fisheries and Oceans Canada, for Koji Shimada and Motoyo Itoh by the Japan Agency for Marine-Earth Science and Technology, and for Andrey Proshutinsky and Rick Krishfield by the National Science Foundation.

References

Aagaard K, Carmack EC (1989) The role of freshwater sea ice and other fresh waters in the Arctic circulation. Journal of Geophysical Research, 94, 14485–14498.

Aagaard K, Weingartner TJ, Danielson SL, Woodgate RA, Johnson GC, Whitledge TE (2006) Some controls on flow and salinity in Bering Strait. Geophysical Research Letters, submitted.

Anderson LG, Björk G, Holby O, Jones EP, Kattner G, Koltermann KP, Liljeblad B, Lindegren R, Rudels B, Swift J (1994) Water masses and circulation in the Eurasian Basin: Results frm the Oden-91 expedition. Journal of Geophysical Research, 99, 3273–3283.

Bacon S, Reverin G, Rigor IG, Snaith HM (2002) A freshwater jet on the east Greenland shelf, Journal of Geophysical Research, 107, doi: 10.1029/2001JC00935.

Bauch D, Schlosser P, Fairbanks R (1995) Freshwater balance and sources of deep and bottom water in the Arctic Ocean inferred from the distribution of $H_2{}^{18}O$, Progress in Oceanography, 35, 53–80.

Carmack EC (2000) The Arctic Ocean's freshwater budget: sources, storage and export. In: The Freshwater Budget of the Arctic Ocean. EL Lewis, EP Jones, P Lemke, TD Prowse and P Wadhams (eds.), Kluwer, Dordrecht, The Netherlands, pp. 91–126.

Carmack EC (2007) The alpha/beta ocean distinction: a perspective on freshwater fluxes, convection, nutrients and productivity in high-latitude seas, Deep-Sea Research II, 54, 2578–2598.

Carmack EC, Chapman DC (2003) Wind-driven shelf/Basin exchange on an Arctic Shelf: The joint roles of ice cover extent and shelf-break bathymetry. Geophysical Research Letters, 30, 1778, doi: 10 1029/2003GL017526.

Carmack EC,McLaughlin FA (2001) Arctic Ocean change and consequences to biodiversity: a perspective on linkage and scale, Memoirs of National Institute of Polar Research, Special Issue, 54, 365–375.

Carmack EC, Macdonald RW, Papadakis JE (1989) Water mass structure and boundaries in the Mackenzie Shelf Estuary, Journal of Geophysical Research, 94, 18043–18055.

Cenedese C, Linden PF (2002) Stability of a buoyancy driven coastal current at the shelf-break. Journal of Fluid Mechanics, 452, 97–121.

Chapman DC,Beardsley RC (1989) On the origin of shelf water in the Middle Atlantic Bight. Journal of Physical Oceanography, 18, 384–391.

Conkright ME, Antonov JI, Baranova O, Boyer TP, Garcia HE, Gelfeld R, Johnson D, Locarnini RA, Murphy PP, O'Brien TD, Smolyar I, Stephens C (2002) World Ocean Data Base 2001 vol.

1. In: Introduction. Sydney Levitus (ed.) NOAA Atlas NESDIS 42, US Government Printing Office, Washington, DC, 167 pp.
Dickson RR, Meincke J, Malmberg S-A,Lee AJ (1988) The "Great Salinity Anomaly" in the northern North Atlantic 1968–1982, Progress in Oceanography, 20, 103–151.
Dickson RR, Dye S, Karcher M, Meincke J, Rudels B, Yashayaev I (2007) Current Estimates of freshwater flux through Arctic and Subarctic seas. Progress in Oceanography, 73, 210–230, doi:10.1016/j.pocean.2006.12.003.
Dukhovskoy DS, Johnson MA, Proshutinsky A (2004) Arctic decadal variability: An auto-oscillatory system of heat and fresh water exchange. Geophysical Research Letters, 31, doi:10.1029/GL019023.
Déry SJ, Wood EF (2005) Decreasing river discharge in Northern Canada, Geophysical Research Letters, 32, doi:10.1029/GL022845.
Ekwurzel B, Schlosser P, Mortlock R, Fairbanks R, Swift J (2001) River runoff, sea-ice meltwater, and Pacific water distribution and mean residence times in the Arctic Ocean. Journal of Geophysical Research, 106, 9075–9092.
Environmental Working Group, Joint U.S.-Russian Atlas of the Arctic Ocean for the Winter Period [CD-ROM] (1998) National. Snow and Ice Data Center. Boulder, CO.
Falkner KK, O'Brien M, Carmack E, McLaughlin F, Melling H, Muenchow A, Jones EP (2006) Implications of nutrient variability in passages of the Canadian Archipelago and Baffin Bay for freshwater throughflow and local productivity. Journal of Geophysical Research, submitted.
Falck E, Kattner G, Budéus G (2005) Disappearance of Pacific Water in the northwestern Fram Strait. Geophysical Research Letters, 32, doi:10.1029/2005GL023400.
Griffiths RW (1986) Gravity currents in rotating systems. Annual Reviews in Fluid Mechanics 18, 59–89.
Guay CK, Falkner KK (1997) Barium as a tracer of Arctic halocline and river waters. Deep Sea Research, II, 44, 1543–1570.
Häkkinen S, Proshutinsky A (2004) Freshwater content variability in the Arctic Ocean. Journal of Geophysical Research, 109, doi:10.1029/2003JC001940.
Jones EP, Anderson LG, Swift JH (1998) Distribution of Atlantic and Pacific waters in the upper Arctic Ocean: implications for circulation. Geophysical Research Letters, 25, 765–768.
Jones EP, Swift JH, Anderson LG, Lipizer M, Civitarese G, Falkner KK, Kattner G, McLaughlin FA (2003) Tracing Pacific water in the North Atlantic, Journal of Geophysical Research, 108, doi:10.1029/2001JC001141.
Lagerloef G, Schmitt R (2006) Role of ocean salinity in climate and near-future satellite measurements, Eos 87.
Lammers RB, Shiklomanov AI, Vorosmarty CJ, Fekete BM, and Peterson BJ (2001) Assessment of contemporary Arctic river runoff based on observational discharge records, Journal of Geophysical Research, 106, 3321–3334.
Macdonald RW, Paton DW, Carmack EC, Omstedt A (1995) The freshwater budget and under-ice spreading of Mackenzie River water in the Canadian Beaufort Sea based on salinity and $^{18}O/^{16}O$ measurements in water and ice. Journal of Geophysical Research, 100, 895–919.
Macdonald RW, Mclaughlin FA, Carmack EC (2002) Freshwater and its sources during the SHEBA drift in the Canada Basin of the Arctic Ocean. Deep-Sea Research, II, 49, 1769–1785.
McLaughlin FA, Carmack EC, Macdonald RW, Bishop JKB (1996) Physical and geochemical properties across the Atlantic/Pacific front in the southern Canadian Basin. Journal of Geophysical Research, 101, 1183–1197.
McLaughlin FA, Carmack EC, Macdonald RW, Weaver AJ, Smith J (2002) The Canada Basin 1989–1995: Upstream events and far-field effects of the Barents Sea. Journal of Geophysical Research, 107, doi:10.1029/2001JC000904.
McLaughlin FA, Carmack EC, Macdonald RW, Melling H, Swift JH, Wheeler PA, Sherr BF, Sherr EB (2004) The joint roles of Pacific and Atlantic-origin waters in the Canada Basin, 1997–1998. Deep Sea Research, I, 51, 107–128.
McLaughlin FA, Carmack EC, Ingram RG, Williams WJ, Michel C (2006) Oceanography of the Northwest Passage, Chapter 31. In: The Sea Vol 14: The Global Coastal Ocean, Interdisciplinary

Regional Studies and Syntheses. AR Robinson, KH Brink (eds.), Harvard University Press, 1213–1244.
Melling H, Moore RM (1995) Modifications of halocline source waters during freezing on the Beaufort Sea shelf: evidence from oxygen isotopes and dissolved nutrients. Continental Shelf Research, 15, 89–113.
Östlund HG (1982) The residence time of the freshwater component in the Arctic Ocean. Journal of Geophysical Research, 89, 6373–6381.
Peterson BJ, McClelland J, Curry R, Holmes RN, Walsh JE, Aagaard K. (2006) Trajectory shifts in the Arctic and Subarctic freshwater cycle. Science, 313, 1061–1066.
Polyakov IV, Alexeev V, Belchansky GI, Dmitrenko IA, Ivanov V, Kirillov S, Korablev A, Steele M, Timokhov LA, Yashayaev I (2007) Arctic Ocean freshwater changes over the past 100 years and their causes. Journal of Climate, in press.
Proshutinsky A, Bourke RH, McLaughlin FA (2002) The role of the Beaufort Gyre in Arctic climate variability: Seasonal to decadal climate scales. Geophysical Research Letters, 29, 2100, doi:10.1029/2002gl015847.
Proshutinsky A, Yang J, Krishfield R, Gerdes R, Karcher M, Kauker F, Koeberle C, Hakkinen S, Hibler W, Holland D, Maqueda M, Holloway G, Hunke E, Maslowski W, Steele M, Zhang J (2005) Arctic Ocean Study: Synthesis of Model Results and Observations. Eos Transactions of the AGU, 86, 368, doi:10.1029/2005EO400003.
Richter-Menge J, Overland J, Proshutinsky A, Romanovsky V, Gascard JC, Karcher M, Maslanik J, Perovich D, Shiklomanov A, Walker D (2006) Arctic. In: State of the Climate in 2005. K.A. Shen (ed.), Special Supplement to the Bulletin of the American Meteorological Society, 87, S46–S52.
Rigor I, Wallace JM (2004) Variations in the age of Arctic sea-ice and summer sea-ice extent. Geophysical Research Letters, 31, doi:10.1029/2004GL019492.
Royer TC (1982) Coastal fresh water discharge in the Northeast Pacific. Journal of Geophysical Research, 87, 2017–2021.
Serreze MC, Barrett AP, Slater AG, Woodgate RA, Aagaard K, Lammers RB, Steele M, Moritz R, Meredith M, Lee C (2006) The large-scale freshwater cycle of the Arctic. Journal of Geophysical Research, 111(C11), doi:10.1029/2005JC003424.
Shimada K, McLaughlin FA, Carmack EC, Proshutinsky A, Nishino S, Itoh M (2004) Penetration of the 1990s warm temperature anomaly of Atlantic water in the Canada Basin. Geophysical Research Letters, 31, doi:10.1029/2004GL020860.
Shimada K, Kamoshida T, Nishino S, Itoh M, McLaughlin FA, Carmack EC, Zimmerman S, Proshutinsky A (2006) Pacific Ocean Inflow: influence on catastrophic reduction of sea ice cover in the Arctic Ocean. Geophysical Research Letters, 33, L08605, doi:10.1029/2005GL025624.
Smethie WM, Schlosser P, Bönisch G, Hopkins TS (2000) Renewal and circulation of intermediate water in the Canadian Basin observed on the SICEX 96 cruise. Journal of Geophysical Research, 105, 1105–1121.
Steele M, Ernold W (2007) Steric sea level change in the Northern Seas. Journal of Climate, 20, 403–417.
Steele M, Morison J, Ernold W, Rigor I, Ortmeyer M, Shimada K (2004) Circulation of summer Pacific halocline water in the Arctic Ocean. Journal of Geophysical Research, 109, doi:10.1029/JC002009.
Stigebrandt A (1984) The North Pacific: A global scale estuary. Journal of Physical Oceanography, 14, 464–470.
Swift JH, Aagaard K, Timokhov L, Nikiforov EG (2005) Long-term variability of Arctic Ocean waters: evidence from a reanalysis of the EWG data set. Journal of Geophysical Research, 110, doi:10.1029/2004JC002312.
Tully JP, Barber FG (1960) An estuarine analogy of the subarctic Pacific Ocean. Journal of the Fisheries Research Board of Canada 17, 91–112.
Walin G (1977) A theoretical framework for the description of estuaries. Tellus, 29, 128–136.
Weingartner TJ, Cavalieri DJ, Aagaard K, Sasaki Y (1998) Circulation, dense water formation and outflow on the northeast Chukchi shelf. Journal of Geophysical Research, 103, 7647–7661.

Wijffels SE, Schmitt RW, Bryden HL, Stigebrandt A (1992) Transport of freshwater by the oceans, Journal of Physical Oceanography, 22, 155–162.
Williams WJ, Weingartner TJ, Hermann AJ (2006) Idealized 3-dimensional modeling of seasonal variation in the Alaska Coastal Current. Journal of Geophysical Research, in revision.
Woodgate R, Aagaard K (2005) Revising the Bering Strait freshwater flux into the Arctic Ocean. Geophysical Research Letters, 32, doi:10.1029/2004GL021747.
Woodgate R, Aagaard K, Weingartner TJ (2005) Monthly temperature, salinity, and transport variability of the Bering Strait through flow. Geophysical Research Letters, 32, doi:10.1029/2004GL21880.
Yamamoto-Kawai M, Tanaka N, Pivovarov S (2005) Freshwater and brine behaviours in the Arctic Ocean deduced from historical data of $\delta^{18}O$ and alkalinity (1929–2002 A.D.). Journal of Geophysical Research, 100, doi:10.1029/2004JC002793.
Yamamoto-Kawai M, McLaughlin FA, Carmack EC, Nishino S, Shimada K (2008) Freshwater budget of the Canada Bain of the Arctic Ocean from geochemical tracer data. Journal of Geophysical Research, in press.
Zhang X, Zhang J (2001) Heat and freshwater budgets and pathways in the Arctic Mediterranean in a coupled ocean/ sea-ice model. Journal of Oceanography, 57, 207–234.

Chapter 8
Modelling the Sea Ice Export Through Fram Strait

Torben Koenigk, Uwe Mikolajewicz, Helmuth Haak, and Johann Jungclaus

8.1 Introduction

The Arctic plays an important role in the climate system. The sea ice controls most of the heat, momentum and matter transfers in the ice-covered Arctic regions. Furthermore, melting and freezing of sea ice have a considerable impact on the ocean stratification. Only a small fraction of the salt is included in the sea ice during freezing processes while the majority is released to the underlying ocean layer. The density of the seawater is increased, which may lead to a destabilization of the ocean stratification. In contrast, melting of sea ice represents a freshwater input into the ocean. The density is reduced and the ocean stratification stabilized. It is of great importance for the ocean where sea ice is freezing and melting. The formation area is not necessarily the same as the melting area. The transport of ice along with the associated freshwater and negative latent heat plays a critical role in the climate system.

The largest sea ice export out of the Arctic Ocean takes place through Fram Strait. It represents a very important flux of freshwater into the North Atlantic Ocean. After passing Fram Strait, the sea ice/freshwater propagates along the east coast of Greenland to the south and into the Labrador Sea. Dickson et al. (1988) and Belkin et al. (1998) suggested that the Great Salinity Anomaly (GSA) observed in the Labrador Sea in the early 1970s was caused by previous large positive ice export anomalies through Fram Strait. Häkkinen (1999) simulated this process by prescribing idealized freshwater pulses in the East Greenland Current in an ocean model. The observed salinity anomalies and the decrease in the oceanic convection were reproduced. Haak et al. (2003) concluded, from simulations with the ocean model MPI-OM that the GSA's in the 1980s and 1990s were caused by anomalous large ice export events through Fram Strait as well.

In a recent paper, Koenigk et al. (2006) showed with a global coupled atmosphere–ocean model that large ice export events through Fram Strait have a significant impact on the atmosphere. The reduced convection in the Labrador Sea after positive ice export anomalies leads to colder ocean surface temperatures, an increased ice cover and consequently a reduced ocean heat release to the atmosphere. Air temperature in the

Max-Planck-Institut für Meteorologie, Bundesstraβe 53, 20146 Hamburg, Germany

R.R. Dickson et al. (eds.), *Arctic–Subarctic Ocean Fluxes*, 171–191

Labrador Sea is therefore significantly reduced and large-scale atmospheric circulation is influenced 1 and 2 years after high ice exports through Fram Strait. Based on these results, Koenigk et al. (2006) suggested a high predictive skill for atmospheric and oceanic climate in the Labrador Sea.

Variations of the ice export through Fram Strait have a considerable effect on ice cover in the Greenland Sea (Walsh and Chapman 1990) and can lead to large and long lasting anomalies. Observational analysis of Deser et al. (2000) suggested a northward shift in the storm track as a consequence of low ice concentration in the Greenland Sea. They argued that SLP in the Greenland Sea is decreased due to enhanced heat fluxes from ocean to atmosphere in areas of reduced sea ice. In contrast, model results of Magnusdottir et al. (2004) and Deser et al. (2004) showed a negative NAO pattern as response to reduced sea ice cover in the Greenland and Barents Sea.

Observations in the Arctic are rather sparse and exist only for the last decades, which were characterized by an unusual state of the general atmospheric circulation and large trends in Arctic climate parameters. Hence, in this study, a 500-year control integration of the global coupled atmosphere–ocean–sea ice model ECHAM5.0/MPI-OM is used to analyze the ice export through Fram Strait and its interannual to decadal variability. The length of the integration provides the possibility to perform statistical analyses on different time scales. The results are compared both with observational data and other model studies.

8.2 Model Description

The model used in this study is the Max-Planck-Institute for Meteorology's global coupled atmosphere–ocean–sea ice model ECHAM5.0/MPI-OM. It consists of the fifth cycle of the atmosphere model ECHAM (ECmwf HAMburg) and the ocean model MPI-OM (Max-Planck-Institute Ocean Model). The atmosphere model ECHAM5.0 (Roeckner et al. 2003) is run at T42 resolution, which corresponds to a horizontal resolution of about $2.8° \times 2.8°$. It has 19 vertical levels up to 10 hPa. The ocean model MPI-OM (Marsland et al. 2003) includes a Hibler-type dynamic-thermodynamic sea ice model. The grid has a resolution of about 2.8° but with an increasing refinement of the meridional grid spaces between 30° N to 30° S up to 0.5° from 10° N to 10° S. The North Pole is shifted towards Greenland (30° W, 80° N) to avoid the grid singularity at the geographical North Pole. Thus, the model resolution in Fram Strait and the deep convection areas of Greenland and Labrador Sea is relatively high. The model's South Pole is located at 30° W, 80° S.

The atmosphere and the sea ice–ocean model are coupled by the OASIS coupler (Valcke et al. 2003). The coupler transfers fluxes of momentum, heat, and freshwater from the atmosphere to the ocean and sea surface temperature and sea ice properties from the ocean to the atmosphere. The climate model includes a river runoff scheme (Hagemann and Düemenil 1998, 2003). Glacier calving is included in a way, that the amount of snow falling on Greenland and Antarctica is instantaneously transferred into the nearest ocean point. In the coupled model no flux adjustment is used.

A 500-year control integration from this model has been used in this study. Analyses of multidecadal scale changes in the North Atlantic thermohaline circulation by Latif et al. (2004) and analyses of the impacts of the Fram Strait ice export on climate by Koenigk et al. (2006) are based on the same control integration.

8.3 Results

8.3.1 Mean Sea Ice Export Through Fram Strait

Above, we discussed the importance of the Fram Strait sea ice export for the climate system. Table 8.1 shows observation-based estimates and parameterizations of the ice export from different studies. Aagaard and Carmack (1989) used ice volume flux measurements from moored upward looking sonars (ULS) by Vinje and Finnekåsa (1986) at 81° N and results from Untersteiner (1988) to estimate the Fram Strait ice export. They found a mean export of about 100,000 m^3/s. Vinje et al. (1998) used ULS to obtain the ice thickness at 79° N for the time period 1990–1996. Together with the velocity, derived from the cross-strait sea level pressure (SLP) gradient, they calculated the ice export through Fram Strait. The mean of the 7-year period was 83,000 m^3/s. Annual mean values vary substantially between about 60,000 m^3/s and 150,000 m^3/s. The export is largest in March with slightly below 120,000 m^3/s and smallest in August with about 40,000 m^3/s. They determined an error for monthly measurements of 8–17% for ice area flux and about 0.1 m for ice thickness. This amounts to an error of approximately 12% for the highest and 20% for the smallest monthly fluxes. Vinje (2001) parameterized the ice export through Fram Strait for the period 1950–2000 by using the close relationship between the SLP gradient across Fram Strait and the ice export. The mean export of the 50-year period was 92,000 m^3/s with a standard deviation of about 21,000 m^3/s. Schmith and Hansen (2003) used sea ice observations from the southwest coast of Greenland to reconstruct the ice export through Fram Strait. The extent of the summer sea ice depends on the ice export through Fram Strait in the previous winter. The authors found an average ice export of 100,000 m^3/s and both a strong interannual variability and a marked multi-decadal variability

Table 8.1 Observation-based estimates of the mean sea ice export through Fram Strait in m^3/s

Author	Time	Ice export (m^3/s)
Aagaard and Carmack (1989)	1953–1984	100,000
Vinje et al. (1998)	1990–1996	83,000
Vinje (2001)	1950–2000	92,000
Schmith and Hansen (2003)	1820–2000	100,000
Kwok et al. (2004)	1991–1998	70,000

with low ice exports around 1920/1930 and high exports in the 1950s and 1960s. Kwok et al. (2004) used ULS data from 1991 to 1998 and found a mean ice export through Fram Strait of 70,000 m^3/s with a standard deviation of approximately 15,000 m^3/s. The authors indicated that 50% of the ice export takes place during December and March while the summer export is weak. It has to be noted that both the mean export and the standard deviation is substantially smaller in Kwok et al. (2004) than in Vinje et al. (1998) although the same ULS data and almost the same time period has been used. Kwok et al. (2004) derived the ice motion from satellite passive microwave data while Vinje et al. (1998) used the SLP gradient across Fram Strait to estimate the ice velocity.

Table 8.2 summarizes model simulations of the Fram Strait ice export. Häkkinen (1993) used an ocean–sea ice model for the Arctic and the northern North Atlantic, forced with monthly means of NCEP/NCAR-reanalysis data (Kalnay et al. 1996). The ice export is relatively small with 63,000 m^3/s because ice thickness is slightly underestimated in the model. Simulations with sea ice models (Hilmer et al. 1998; Arfeuille et al. 2000) and ocean–sea ice models (Koeberle and Gerdes 2003; Haak et al. 2003), forced by 40- or 50-year reanalysis data, all indicate a high interannual to decadal variability. All model simulations show pronounced ice export events in 1967/68 and in 1994/95. The mean exports are similar (83,000–104,000 m^3/s), except for the model of Arfeuille et al. (2000) that simulated an average ice export of 160,000 m^3/s. Nevertheless, their ice export anomalies compare well with the other model simulations.

In this study, a global coupled atmosphere–ocean–sea ice model is used. Hence, only statistics of the time series (Fig. 8.1a) can be compared to observations and other studies. The mean export amounts to 97,000 m^3/s, which is in the upper range of observation-based estimates and model simulations. The ice export is highly variable on interannual time scales with a standard deviation of 21,000 m^3/s for annual mean exports. The monthly mean ice export through Fram Strait, averaged over the 500-year control integration (Fig. 8.1b), shows a pronounced seasonal cycle. The maximum occurs in March with an average of 147,000 m^3/s and the minimum in August with 35,000 m^3/s. This agrees with observation-based estimates by Vinje et al. (1998) and parameterizations by Vinje (2001). The standard deviation has been calculated for each month.

Table 8.2 Model simulations of the mean sea ice export through Fram Strait in m^3/s

Author	Time	Ice export (m^3/s)
Häkkinen (1993)	1955–1975	63,000
Hilmer et al. (1998)	1958–1997	91,000
Arfeuille et al. (2000)	1958–1998	160,000
Koeberle and Gerdes (2003)	1948–1998	83,000
Haak et al. (2003)	1948–2001	104,000
Koenigk et al. (2006)	500-year ctrl-run	97,000

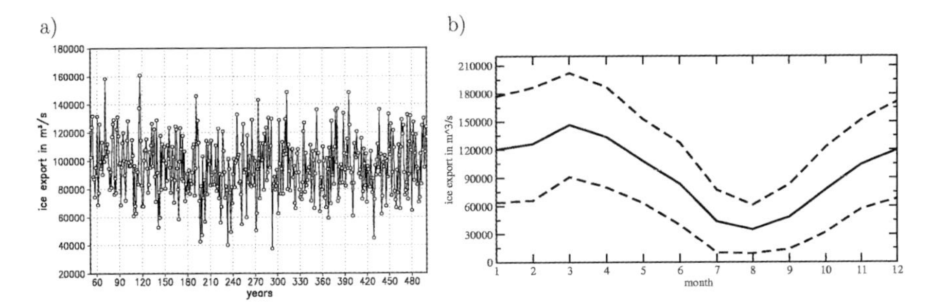

Fig. 8.1 (a) Annual mean ice export through Fram Strait in m^3/s. (b) Monthly mean ice export (solid) and ice export ± 1 standard deviation (dashed) in m^3/s, averaged over the 500-year control integration for each month

Table 8.3 Correlation between seasonal mean ice exports through Fram Strait. Seasons, written in the horizontal, lead seasons, written in the vertical. The last row indicates the correlation between the annual mean ice export (averaged from September to August) and the single seasons

	DJF	MAM	JJA	SON
DJF	1	0.34	0.10	0.04
MAM	−0.04	1	0.15	0.14
JJA	0.12	0.01	1	0.31
SON	0.25	0.00	0.11	1
Year	0.75	0.64	0.50	0.57

It is largest in late winter/early spring with 50,000–60,000 m^3/s, but even in late summer it amounts to more than half of the winter values and contributes considerably to the interannual variability. Hence, the ice export in late summer/early autumn should not be neglected.

The correlation among single seasons is presented in Table 8.3 Sequenced seasons are weakly positively correlated with coefficients between 0.15 for spring (MAM) – summer (JJA) and 0.34 for winter (DJF) – spring (MAM). The correlation between non-sequenced seasons is very low. All seasons are significantly and highly positively correlated with the annual mean ice export. The correlation is largest in winter but still reaches 0.5 and 0.57 in summer and autumn, respectively.

The same correlation analysis is performed for ice exports of single months. Sequenced months are significantly positively correlated, whereas highest correlations occur between July–August ($r = 0.41$), August–September ($r = 0.42$) and September–October ($r = 0.41$) ice exports. In this time period, during summer and early autumn, wind variability is much weaker than in winter and the ice export depends largely on the amount of ice remaining from the previous winter.

8.3.2 Variability of Sea Ice Export

As shown above, the interannual to decadal variability of the ice export through Fram Strait is very high. Figure 8.2 shows the energy spectrum of annual mean ice exports through Fram Strait in the 500-year integration of our model. Three peaks in the ice export at time scales of about 3–4 years, 9 years and 15 years can be observed.

These three peaks, although shifted towards slightly shorter time scales, can be found in the reconstructed ice export time series of Schmith and Hansen (2003). Several other studies (e.g. Venegas and Mysak 2000; Hilmer and Lemke 2000; Polyakov and Johnson 2000) also found peaks at roughly 10 years in Fram Strait ice export and Arctic atmospheric circulation regimes. Venegas and Mysak (2000) and Goosse et al. (2002) reported significant variability in the Arctic ice volume at a timescale of 15–20 years, which might fit to the 15-year peak in the ice export in this study.

The interannual variability of the Fram Strait ice export is highly related to the local wind forcing. Figure 8.3a presents a correlation analysis between annual mean ice exports through Fram Strait and SLP anomalies in our model simulations. In the area of the Kara Sea, correlation exceeds –0.6. A smaller positive correlation exists over the Canadian Archipelago. This pattern is related to anomalous winds from the coasts of Laptev, East Siberian, Chukchi and Beaufort Seas across the Arctic towards Fram Strait and enhanced northerly wind stress in Fram Strait. Consequently, ice is anomalously transported towards Fram Strait in the entire Arctic Basin (Fig. 8.3b). The correlation between annual mean ice export and SLP gradient across Fram Strait is 0.86. The SLP gradient explains therefore about three quarters of the annual mean ice export variability. This is in agreement

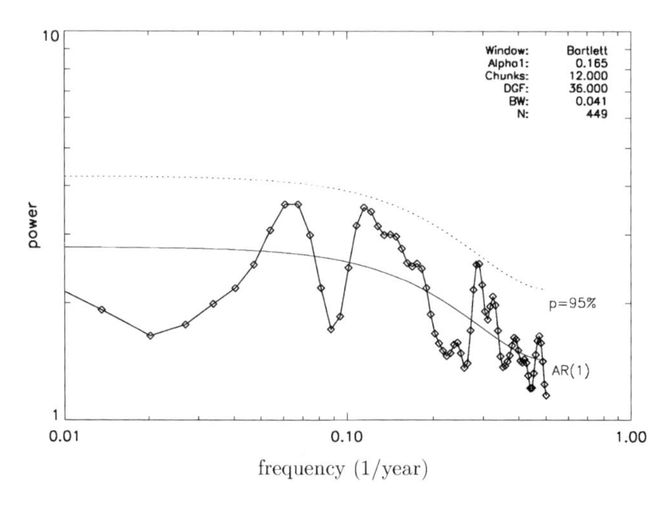

Fig. 8.2 Spectral analysis of annual mean Fram Strait ice export (Based on Koenigk et al. 2006)

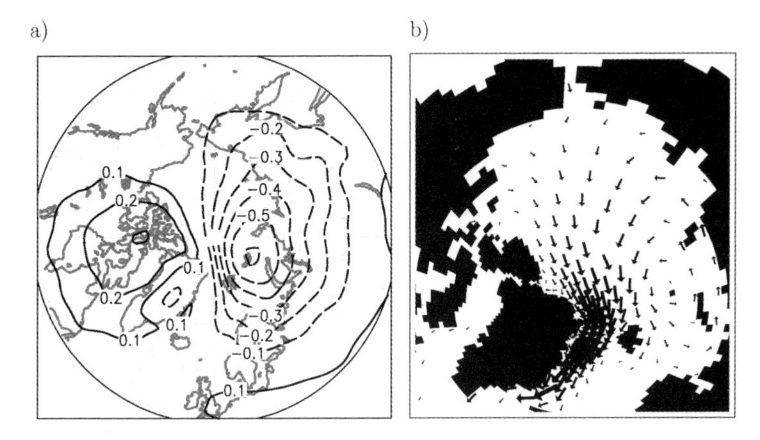

Fig. 8.3 (a) Correlation pattern between annual mean ice export through Fram Strait and sea level pressure. (b) Regression pattern between Fram Strait ice export and sea ice transport in the Arctic. The areas of the arrows show the amount of ice transport. The smallest arrow shown represents an ice transport of 0.1×10^{-2} m^3/s per standard deviation ice export, the largest 6.3×10^{-2} m^3/s. Each 16th arrow is shown (Based on Koenigk et al. 2006)

with observation-based estimates of Kwok and Rothrock (1999) who found an explained variance of 80%.

8.3.2.1 North Atlantic Oscillation

The North Atlantic Oscillation (NAO, Hurrel and van Loon 1997; Bojariu and Gimeno 2003) governs important parts of climate variability in high northern latitudes. However, its impact on the ice export through Fram Strait is still under debate. Kwok and Rothrock (1999) found a correlation coefficient of 0.86 for the period 1978–1998. Simulations with a sea ice model (Hilmer and Jung 2000) indicated similar results for this time period but no significant correlation between ice export and NAO before 1978. Analysis of a 300-year control run of the atmosphere–ocean–sea ice model ECHAM4/OPYC3 by Jung and Hilmer (2001) showed no significant correlation either. The changing character of the relation between ice export and NAO can be explained by an eastward shift in the extension of the Icelandic Low into the Arctic since the late 1970s. This shift leads to an increased pressure gradient across Fram Strait in the positive NAO case. Before 1978, the NAO did not affect the SLP gradient across Fram Strait at all. Whether the shift in the Icelandic Low is due to anthropogenic climate changes or natural variability cannot yet be determined. Ostermeier and Wallace (2003) analyzed the trends in the NAO over the 20th century. They found a negative trend from 1920 to 1970 and a strong positive trend since. In our model simulations, the NAO has neither influenced the ice volume export through Fram Strait nor the pressure gradient across Fram Strait. Nevertheless, the anomalous SLP pattern and the associated wind anomalies lead to strong anomalous ice transports from the Barents Sea across the North Pole to the Beaufort Sea. Ice transport to the south is enhanced in the Baffin Bay and in the Labrador Sea (not shown).

To elucidate the temporal relationship between NAO-index and sea ice export through Fram Strait, running 30-year intervals of the 500-year control run are analyzed (not shown). In the entire 500 years, not a single 30-year period with a high correlation between NAO and ice export or SLP gradient can be found. As no anthropogenic forcing is used in the model, this supports the presumption of Jung and Hilmer (2001) that the recent state of the NAO may be a response to anthropogenic forcing.

8.3.2.2 Stratospheric Polar Vortex

Several studies have recently discussed the effect of the stratospheric circulation on tropospheric climate. Christiansen (2001) as well as Graversen and Christiansen (2003) showed that zonal wind anomalies from the stratosphere propagate downward to the troposphere in about 10–15 days. Thompson et al. (2002) and Baldwin et al. (2003) proposed an increased skill from the stratospheric circulation to predict northern hemisphere tropospheric conditions on this time scale. Norton (2002) performed sensitivity experiments with an atmospheric general circulation model and altered the mean state and variability of the stratosphere. The wintertime SLP responded with a lag of 10–25 days with a pattern that is similar to the AO-pattern.

In this study, the impact of the stratospheric polar vortex on the ice transport in the Arctic and especially the ice export through Fram Strait is analyzed with ECHAM5/MPI-OM. The polar vortex index has been defined, according to Castanheira and Graf (2003), as the zonal mean zonal wind speed in 50 hPa height and 65° N.

In contrast to the NAO, the annual mean stratospheric polar vortex index is significantly positively correlated ($r = 0.34$) with the annual mean ice export through Fram Strait in the control integration although the explained variance is rather small.

Figure 8.4a displays the difference of annual mean SLP between strong and weak stratospheric polar vortex regimes (exceeding the mean ± 1 standard deviation). The largest differences occur over the Barents Sea with more than −2 hPa. Smaller positive values appear over the North Atlantic, Western Europe and the Bering Strait. This SLP pattern compares well with results of the sensitivity experiments of Norton (2002). The maximum pressure anomaly in ECHAM5/MPI-OM is slightly shifted towards the Barents Sea. It should be noted that annual mean values are used in this study while Norton focused on winter means.

The SLP anomalies lead to an increased SLP gradient across Fram Strait and stronger northerly winds during a strong polar vortex regime. The ice export through Fram Strait is consequently enhanced and vice versa during weak vortex regimes (Fig. 8.4b).

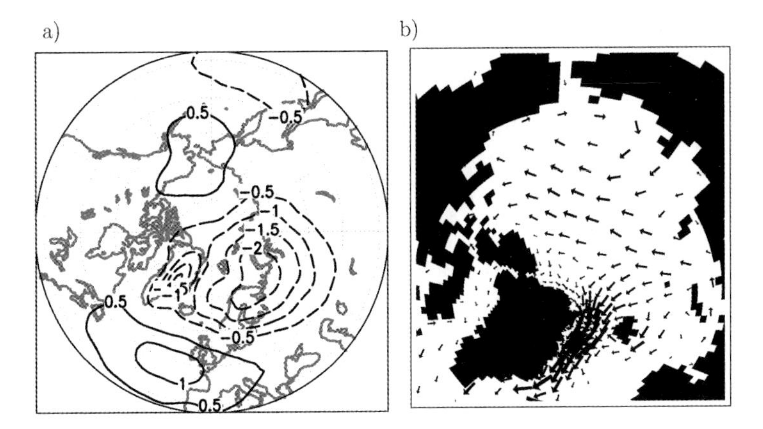

Fig. 8.4 Annual differences between strong and weak stratospheric polar vortex regimes: (a) SLP in hPa; (b) sea ice volume transport. The areas of the arrows show the amount of ice transport. The smallest shown arrow presents an ice export of 0.1×10^{-2} m^3/s per standard deviation ice export, the largest 5.2×10^{-2} m^3/s. Each 16th arrow is shown

8.3.2.3 Atmospheric Planetary Waves

Large-scale atmospheric planetary waves in the northern hemisphere are mainly caused by the topography and land–sea distribution (O'Hanlon 2002). Cavalieri and Häkkinen (2001) investigated the relationship between atmospheric planetary waves and Arctic climate variability. They performed a zonal Fourier analysis over a monthly averaged 50-year SLP-record from 1946 to 1995 for different latitude bands. They showed that the phase of the first wave for the latitude band from 70° to 80° N in January is well correlated with the ice export through Fram Strait. The Siberian High and the Icelandic Low determine the first wave. A ridge of the Siberian High that extends into the East Siberian and Chukchi Seas and a trough of the Icelandic Low into the Arctic form maximum and minimum of the first wave. A shift in the positions of the pressure systems to the east is associated with reduced pressure in the Barents and Kara Seas. Hence, the pressure gradient across Fram Strait is increased. In contrast to the NAO, the high correlation between ice export and the first wave in January held for the entire 50-year period. Cavalieri (2002) attributes this consistency to the sensitivity of the first wave phase to the presence of secondary low pressure systems in the Barents Sea that serve to drive Arctic sea ice southward through Fram Strait. Figure 8.5 shows the relation between sea ice export and first wave in the ocean–sea ice model MPI-OM (Haak 2004) forced by NCEP/NCAR-reanalyses. The correlation of both time series exceeds 0.6 considering the period 1948–2002. The interannual variability of the wave-1 phase seems to be related to the cyclonic and anti-cyclonic regimes proposed by Proshutinsky and Johnson (1997).

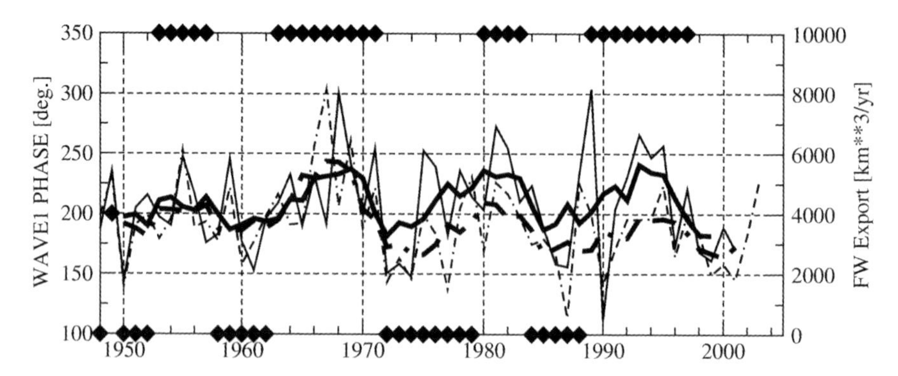

Fig. 8.5 Fram Strait solid freshwater export (solid) and phase of zonal SLP-wave-1 in 70–80° N (dashed) for January (taken from Haak 2004). Thick lines are smoothed by a 5-year running average. Black squares indicate the cyclonic (top) and anti-cyclonic (bottom) Arctic circulation regimes described by Proshutinsky and Johnson (1997). Units are (km^3/year) and (° lon.), respectively

In this study, the relationship suggested by Cavalieri and Häkkinen (2001) has been analyzed in the 500-year control integration. In accordance with their results, the ice export through Fram Strait is highly correlated with the phase of the first SLP wave for the latitude band from 70° to 80° N. This relation holds for the entire year but is highest in winter ($r = 0.6$ in February) while the correlation is quite weak in June and July ($r = 0.2$ resp. 0.15, Table 8.4). Annual mean values are correlated with 0.59.

In spite of the high correlation between phase and ice export, usage of the phase as index has some disadvantages. In months with small amplitude, the first wave explains only a minor part of the SLP variability and the phase is of minor importance. Furthermore, it is difficult to qualify large shifts in the phase as positive or negative because phase anomalies of nearly –180° or 180° both describe the same SLP pattern. A shift in the phase exceeding 90° is not further increasing the SLP gradient across Fram Strait and the ice export is not enhanced anymore. To avoid these difficulties, a new index containing both the phase and the amplitude of the wave is introduced here:

$$WI1 = A \cdot \sin(\Phi')$$

We call this index wave index 1 (WI1). A is the amplitude of the first wave and $\Phi'$the phase anomaly. Use of $\sin(\Phi')$ instead of the phase anomaly has two effects: WI1 is decreased if the phase anomaly exceeds 90° and the function is continuously differentiable at the location $\Phi' = 180°$. Φ' is weighted with the amplitude to reduce the noise of years with small explained variances of the first SLP wave. Table 8.4 displays the correlation between monthly mean WI1 and monthly mean ice exports through Fram Strait. The correlation is generally high between September and May with a maximum of 0.7 in February. It exceeds the correlation between wave phase and Fram Strait ice export in all months. During summer,

Table 8.4 Correlation between monthly means of the phase of wave number 1 and Fram Strait ice export (FP), WI1 and Fram Strait ice export (FI) and WI1 and SLP gradient across Fram Strait (FG)

	Jan.	Feb.	Mar.	Apr.	May	June	July	Aug.	Sept.	Oct.	Nov.	Dec.
FP	0.45	0.60	0.45	0.43	0.40	0.20	0.15	0.29	0.35	0.38	0.37	0.40
FI	0.66	0.70	0.61	0.66	0.59	0.34	0.26	0.35	0.49	0.54	0.56	0.59
FG	0.73	0.73	0.69	0.67	0.59	0.31	0.31	0.53	0.66	0.67	0.60	0.61

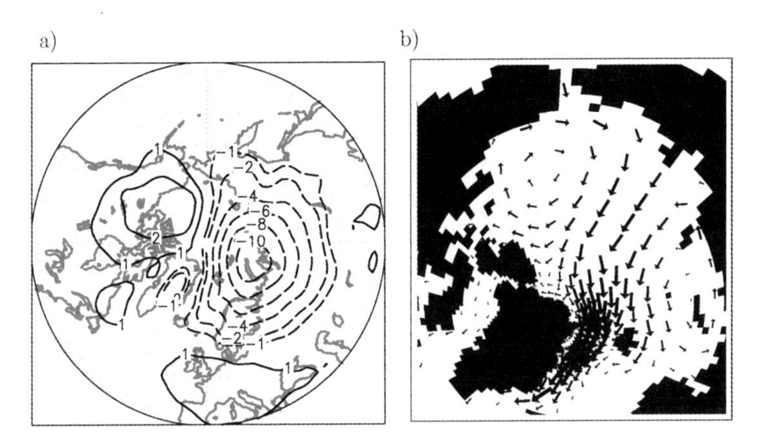

Fig. 8.6 Differences between high and low annual WI1 for (a) winter SLP in hPa, (b) annual sea ice thickness transport. The areas of the arrows show the amount of ice transport. The smallest arrow shown represents an ice export of 0.2×10^{-2} m³/s per standard deviation ice export, the largest 12×10^{-2} m³/s. Each 16th arrow is shown

correlation is significant but does not exceed 0.35. For comparison, the relation between WI1 and SLP gradient across Fram Strait is shown. This correlation is reduced in summer as well.

Figure 8.6a shows the SLP difference pattern between large positive and negative WI1 (exceeding the mean ± 1 standard deviation) for winter means. The pattern is characterized by negative anomalies of up to −10 hPa at the Siberian coast centered in the Kara Sea and much smaller positive anomalies of about 2 hPa in the western Arctic and in Western Europe. Obviously, the WI1 is mainly governed by SLP variations in the Kara Sea. This results in a steepened SLP gradient across Fram Strait during a positive WI1 and vice versa. Dorn et al. (2000) depicted from simulations with a regional coupled model that warm and cold Arctic winters are connected with two distinct circulation states of the Arctic atmosphere. Cold Januaries are characterized by the extension of the Icelandic Low into Barents and Kara Sea while warm Januaries are linked to a more pronounced Siberian High. These two states fit well to the WI1 pattern.

The associated large SLP anomalies affect the ice transport in the entire Arctic. Figure 8.6b displays the annual mean ice transport differences between positive and

Table 8.5 Correlation between DJF ice transport divergence in the Arctic regions and WI1

	Ba	Ka	La	Sib	Chu	Bea	CA	AB
WI1	−0.45	0.20	0.67	0.61	0.60	0.24	−0.75	0.57

Ba = Barents Sea, Ka = Kara Sea, La = Laptev Sea, Sib = East Siberian Sea, Chu = Chukchi Sea, Bea = Beaufort Sea, CA = Central Arctic, AB = Arctic Basin

negative WI1 (mean ± 1 standard deviation). Sea ice is anomously advected from the Siberian coast over the Central Arctic towards Fram Strait and Barents Sea. In the Beaufort Gyre a weak anticyclonic circulation occurs. Ice transport differences are largest in Fram Strait and East Greenland Current where they reach 0.1 m^2/s. The probability distribution of annual mean ice exports through Fram Strait for highly anomalous WI1 (not shown) provides evidence that almost all extreme export events in the 500-year control run are related to WI1. The distributions for the cases of large positive and negative WI1 show a distinct shift towards corresponding positive and negative ice export anomalies.

The composite patterns for SLP and ice transport resemble the correlation and regression patterns between ice export through Fram Strait and SLP and ice transport, respectively (Fig. 8.3). The ice transport anomalies due to the WI1-variability are associated with variations in the ice transport divergence (Table 8.5). During positive WI1, ice transports diverge anomalously from the Laptev Sea to the Chukchi Sea. Ice transports converge in the Barents Sea and particularly in the Central Arctic. The entire Arctic shows a loss of ice volume due to the large ice export through Fram Strait.

It has been demonstrated above that the state of the WI1 is mainly characterized by the SLP in the Kara Sea. The persistence and the source of these SLP anomalies are analyzed below.

The correlations among consecutive months of WI1 are very weak. The highest correlation coefficient of 0.15 is obtained between WI1 of January and February. Furthermore, daily winter (DJF) SLP values in the Kara Sea are compared for months with high positive and negative WI1. Two features are particularly striking: The SLP for positive WI1 in the Kara Sea is generally lower than for negative WI1 and the variability is larger. A sequence of short, relatively large negative anomalies occurs during positive WI1. Contrary, high SLP can persist for a time of 1–2 weeks in the Kara Sea during a negative WI1. During a positive WI1, more cyclones are active in the Barents and Kara Seas while in the negative case, longer periods with stable anticyclonic regimes occur.

Storm tracks are calculated from daily winter SLP data to determine the cyclonic activity. In this study, the standard deviation of the 2–6 days band-pass filtered daily SLP data is defined as storm track (Blackmon 1976). A composite analysis of these storm tracks for winter means of the WI1 (Fig. 8.7) shows distinct differences between the phases of WI1. During positive WI1, the storm track over the North Atlantic extends far into Barents and Kara Sea, whereas it is much more zonal during negative WI1 and the standard deviation of the band-pass filtered daily SLP in the Barents and Kara Seas is small. The standard deviation in the formation area of the North Atlantic low-pressure systems over northeastern North America is slightly increased in the positive WI1 case. Thus, an intensification of the North Atlantic storm track and especially a deflection to the north are the main reasons for a positive WI1. The cyclones propagate mainly from the North Atlantic into the Barents and Kara Sea and are not formed locally. This is in agreement with results of Serreze and Barry (1988). They analyzed the winter synoptic activity in the Arctic Basin and found the largest activity in the European sector of the Arctic. Most of the cyclones migrated from the North Atlantic into the Arctic. As the Fram Strait is located on the western side of the storm track, an enhanced pressure gradient across it and anomalously northerly winds are the consequence. This implies that single cyclones are of great importance for the ice export through Fram Strait, which fits well to observations of cyclones in the Fram Strait by Brümmer et al. (2001, 2003).

During a negative WI1 the Siberian High extends further into the Kara and Barents Seas and the storm track is more zonal. It remains unclear whether the Siberian High can extend further to the northwest due to a weaker, more zonal storm track or if a strong Siberian High blocks the cyclones. One possible mechanism affecting the Siberian High may be related to snow anomalies over Siberia in early fall. Cohen et al. (2000) showed that they influence SLP in the northern hemisphere and affect the AO in the following winter. Gong et al. (2003) affirmed these results with model experiments but found a much smaller amplitude in SLP anomalies. However, no significant correlation could be found between autumn snow cover over Siberia and the wintertime SLP or the storm track in the control integration of this model. The impact on the atmospheric circulation is very weak even after extreme snow cover anomalies.

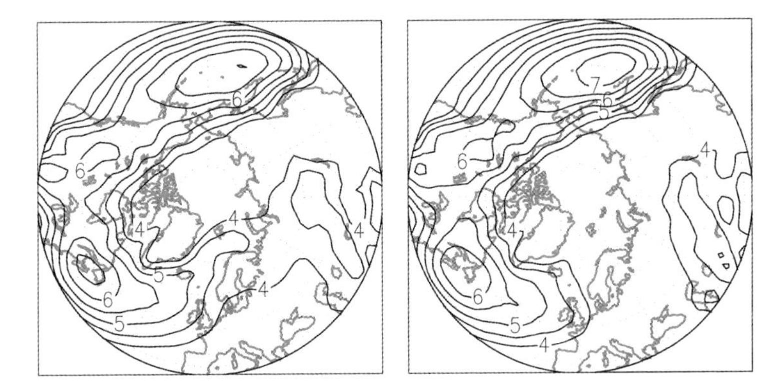

Fig. 8.7 Composite analysis of the standard deviation of the bandpass filtered (2–6-day periods) winter (DJF) daily SLP in hPa for the cases of large positive (left) and negative (right) annual WI1

8.3.3 *Role of Sea Ice Thickness for the Export*

In the previous sections, the strong influence of the atmospheric circulation on the interannual variability of Fram Strait ice export has been demonstrated. However, studies by Koeberle and Gerdes (2003) and Arfeuille et al. (2000) pointed out that sea ice thickness anomalies have a considerable impact on the export through Fram Strait as well. In both studies, the ice export anomalies have been divided into a part, related to ice thickness anomalies and a part related to ice velocity anomalies. The results indicate an almost equal importance of ice thickness anomalies for the entire export anomalies. Our model results show an increased ice thickness in Fram Strait if the cross-strait SLP-gradient is large. Assuming the same SLP-gradient, sea ice velocity is slightly smaller in Fram Strait with thick ice than with anomalously thin ice. Both, ice velocity and thickness are to large extent driven by the wind. Hence, the results of Koeberle and Gerdes (2003) and Arfeuille et al. (2000) need not to contradict to the high correlation of SLP-gradient and ice export.

To further analyze the relation between ice thickness anomalies in the Arctic and Fram Strait sea ice export, we performed a lag regression analysis (Fig. 8.8). Five

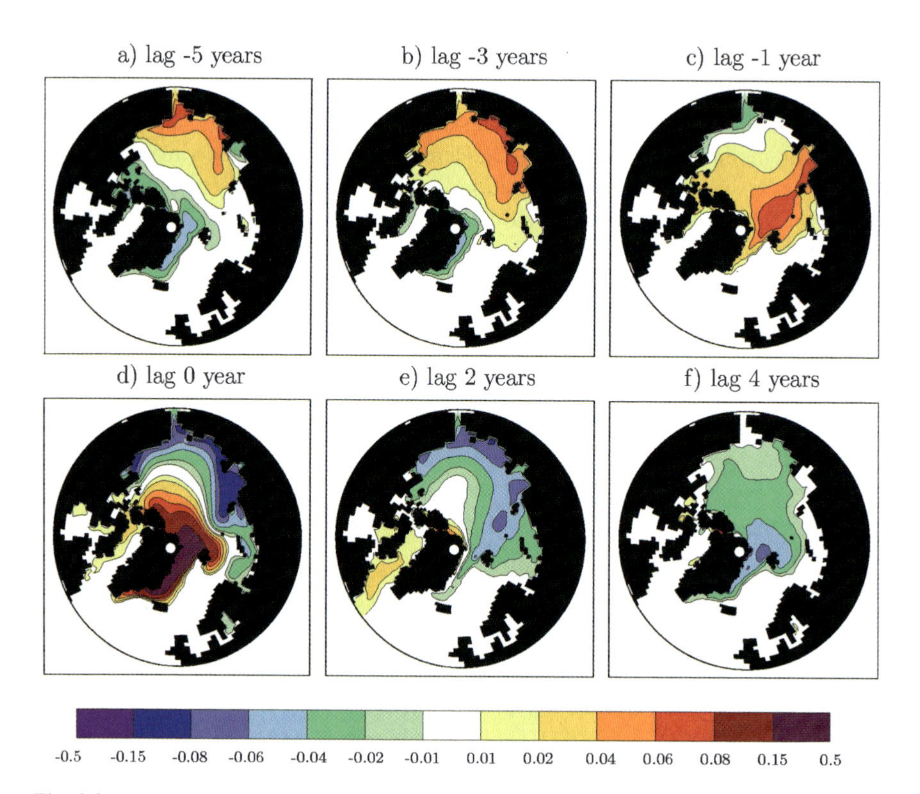

Fig. 8.8 Regression coefficient between annual mean ice exports through Fram Strait and ice thickness anomalies in cm per standard deviation ice export. (a) ice export lags 5 years, (b) ice export lags 3 years, (c) ice export lags 1 year, (d) lag 0, (e) ice export leads 2 years, (f) ice export leads 4 years (Based on Koenigk et al. 2006)

years before high ice exports, positive ice thickness anomalies are formed at the coasts of Chukchi and East Siberian Sea (Fig. 8.8a). In agreement with results by Tremblay and Mysak (1998) and Haak et al. (2003), these anomalies are caused by a convergent ice transport due to an anomalous wind field and are associated with a negative ice export through Fram Strait. In the next 2 years, the positive ice thickness anomaly slowly propagates clockwise along the Siberian coast (Fig. 8.8b) and crosses the Arctic to reach Fram Strait leading the ice export by 1 year (Fig. 8.8c). High ice exports themselves are associated with large anomalous ice transports all across the Arctic towards Fram Strait (Fig. 8.3b) caused by the anomalous atmospheric forcing described above (Fig. 8.3a). A negative ice thickness anomaly occurs at the Siberian coast as a consequence of the divergence in ice transports. It propagates across the Arctic to Fram Strait in the next years, which leads to a decreased ice export (Fig. 8.8e and f) 4 years later. One further year later, the ice export is still reduced and ice thickness at the Siberian coast is again increased. The entire cycle takes about 9 years and matches the peak in the power spectrum of the ice export at the same time scale (Fig. 8.2). A detailed description of this process is given in Koenigk et al. (2006).

This mode has the potential for predictability of the ice export through Fram Strait. Apparently, large ice exports are characterized by previous ice volume anomalies at the Siberian coast and vice versa. Statistical analyses show the largest predictability for the ice export through Fram Strait if ice thickness is increased 2 years before in the Laptev Sea. Figure 8.9 displays the probability distribution of the annual mean ice export 2 years after 69 years with positive and 71 years with negative ice volume anomalies (exceeding the mean ± 1 standard deviation) in the Laptev Sea. After positive anomalies, a considerable shift in the mean ice export towards positive values can be seen and vice versa. The skewness of the distribution is negative after thick ice and positive after previously thin ice in the Laptev Sea. The probability for negative ice export events through Fram Strait is highly

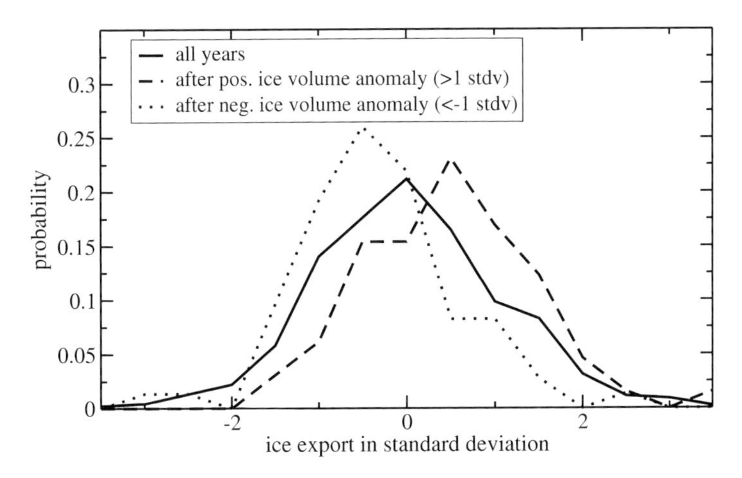

Fig. 8.9 Probability distribution of annual mean Fram Strait ice export 2 years after positive (dashed) and negative ice volume anomalies (dotted) (exceeding one standard deviation) in the Laptev Sea. The solid line gives the mean ice export distribution for all years

decreased while probability for extreme positive events increases only slightly after thick ice in the Laptev Sea.

Ice thickness in the Laptev Sea before extreme ice exports events through Fram Strait, which exceed the mean +2 standard deviations, have been analyzed: In four out of five cases, ice volume has increased by more than one standard deviation 2 years before. All four extreme negative exports have been led by largely reduced ice thickness in the Laptev Sea. The formation of ice thickness anomalies at the coasts of Laptev and East Siberian Sea can be regarded as preconditioning for extreme ice export events through Fram Strait.

8.3.4 Sensitivity Experiment

The Siberian coast is an important source region for the formation of ice volume anomalies. To analyze the propagation of such signals across the Arctic and their interactions with the atmosphere, ice volume anomalies of 2,000 km^3 were prescribed at the Siberian coast in model experiments. Twenty runs were performed, initialized from 1 May of 20 different years with basically normal Fram Strait ice exports. This assures that the initial conditions are not relevant for the ensemble mean. The ice volume anomaly was produced by increasing the ice thickness by 1 m in an area along the Siberian coast and 0.5 m in a transition region (two grid points) to the Central Arctic relative to the initial conditions.

Results of the ensemble mean of these experiments are discussed below. A lag of 1 year is defined as the mean from August in the year in which the experiment starts to July of the following year. Figure 8.10 shows the development of the ice thickness anomaly in the first 3 years after initialization of the experiment. The main part of the anomaly propagates in the transpolar drift stream across the Arctic towards Fram Strait. Already after 1 year, parts of the anomaly reach Fram Strait. After 2 years, the ice anomaly detaches from the Siberian coast and another year later it passes Fram Strait. At the same time, a negative ice thickness anomaly develops at the Siberian coast. This fits well with the regression analysis between Fram Strait ice export and ice thickness (Fig. 8.8).

The ice export through Fram Strait is enhanced in the first 5 years after experiment start with a maximum in the third year. In this period, the anomalous ice export through Fram Strait amounts to two thirds of the imposed sea ice volume anomaly. As the ice export over the Barents Shelf into the North Atlantic is enhanced as well, one can conclude that the Arctic reaches its balance mainly by dynamical reduction of sea ice and subsequent melting in the northern North Atlantic.

Figure 8.11 shows the impact of the imposed ice anomaly after 3 and 4 years. As described above, sea ice export is especially strong after 3 years. This freshwater signal propagates in the East Greenland Current to the south and into the Labrador Sea. Salinity is strongly reduced, which leads to a reduced oceanic convection and more sea ice in the Labrador Sea. Consequently, the oceanic heat release decreases and the air temperature is significantly colder than usual. The SLP responds, especially

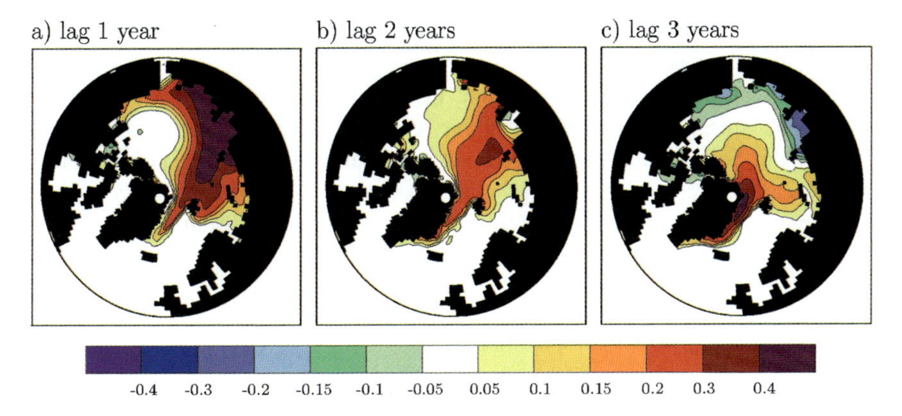

Fig. 8.10 Annual mean sea ice thickness anomalies (in meters) 1–3 years after addition of the ice thickness anomaly at the Siberian coast. Mean of the 20 ensemble runs

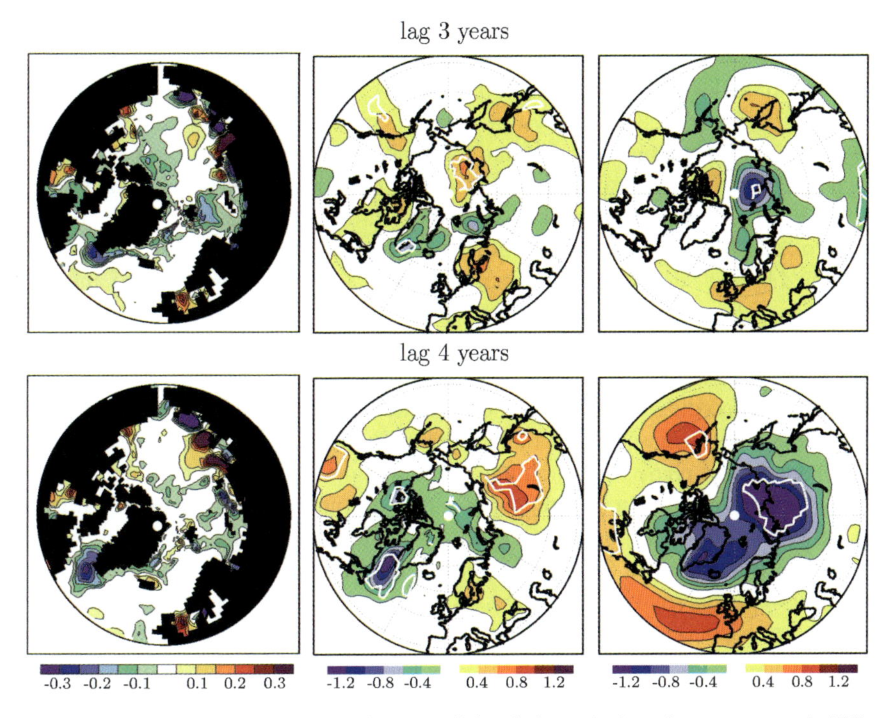

Fig. 8.11 Annual mean anomalies of 10 m salinity (left, psu), 2 m air temperature (middle, Kelvin) and SLP (right, hPa) 3–4 years after addition of the ice thickness anomaly at the Siberian coast. Mean of the 20 ensemble runs. The white lines indicate significance at 95% (for SLP and air temperature, salinity is significant in all colored areas)

in the fourth year, with positive anomalies over the North Atlantic and North Pacific and reduced values further north. This pattern resembles the NAO or AO-pattern. Obviously, the formation process of GSAs can be caused by ice thickness anomalies at the Siberian coast.

In another experiment (detailed description in Koenigk et al. 2006), the effect of extreme export events through Fram Strait has been prescribed. A 3,000 km^3 ice volume anomaly has been implemented in the East Greenland Current south of Fram Strait. Strong and long lasting salinity and temperature anomalies occurred in the Labrador Sea in the following years. Atmospheric circulation responded with a NAO-like pattern 2 years after the mimicked ice export. These results suggest a high skill of predictability after large sea ice exports through Fram Strait.

8.4 Summary and Conclusions

The sea ice export through Fram Strait and its variability have been studied by analyzing a 500-year control integration of a global coupled atmosphere–ocean–sea ice model and by sensitivity studies.

The Fram Strait constitutes the main passage for sea ice out of the Arctic. A comparison of the simulated Fram Strait ice export with observation-based estimates and other model studies has been performed. The estimates of the mean export vary between 70,000 and 100,000 m^3/s, while the spread of the model simulations is slightly larger. Our simulation presents a mean export of 97,000 m^3/s with a standard deviation of 21,000 m^3/s, which also fits well to the observation-based estimates.

Analyses of the variability of the ice export through Fram Strait confirm results of Kwok and Rothrock (1999) that almost 80% of the ice export variability can be explained by the SLP gradient across Fram Strait. In contrast to the NAO, the first planetary-scale zonal SLP wave, meridionally averaged over 70–80° N, is closely related to the ice export through Fram Strait. The phase of the first wave is determined by the position of the extensions of Icelandic Low and Siberian High into the Arctic. According to Cavalieri and Häkkinen (2001), a shift of the phase to the east leads to an increased SLP gradient across Fram Strait in winter. A new index (WI1) combining phase and amplitude of the first planetary wave between 70° and 80° N has been defined in this study. WI1 and ice export are significantly correlated year-round with highest correlation in winter. Moreover, the first zonal wave turned out to be very important for climate variability in the entire Arctic. It is therefore essential to further analyze the processes determining its variability.

The stratospheric polar vortex has been identified as another source of Fram Strait ice export variability. During periods with strong polar vortex, both the SLP gradient across Fram Strait and the ice export are enhanced and vice versa during weak vortex regimes.

In spite of the close relationship between atmospheric forcing and sea ice export through Fram Strait, the atmospheric variability cannot fully explain the 9-year peak in the ice export. This study has presented a sea ice mode on a decadal time scale. It is characterized by the propagation of ice thickness anomalies within the Arctic Basin and leads to decadal ice export variability through Fram Strait. The mechanism of the mode is as follows: onshore winds form an ice thickness anomaly at the coasts of the Siberian and Chukchi Seas. This anomaly propagates with the mean ice drift along the Siberian coast to the west and crosses the Arctic in the transpolar drift. It reaches Fram Strait 4–5 years after the formation and increases the ice export. The mode develops particularly well if atmospheric forcing strengthens the propagation of the ice anomaly. Simultaneously to the increased ice export, negative ice thickness anomalies occur at the Siberian coast due to offshore winds during high ice exports. They take the same way to Fram Strait in another few years.

In a sensitivity experiment, an ice volume anomaly at the Siberian coast has been prescribed. About two thirds of the prescribed ice volume anomaly are anomalously exported through Fram Strait in the following years. The ice export anomalies provoke the process of GSA-formation in the Labrador Sea, which in turn also affects atmospheric climate conditions. The anomalies in atmospheric circulation force divergent sea ice transports at the Siberian coast, causing anomalously low ice thickness and setting the stage for a negative ice export anomaly through Fram Strait a few years later.

Knowledge of the decadal sea ice mode provides a good framework for predictability. A considerable increase of high ice exports through Fram Strait occurs after previous ice thickness anomalies at the coast of the Laptev Sea. However, the predictability of Labrador Sea climate using the ice export through Fram Strait as predictor seems to be even more promising because the associated processes are less affected by the highly variable atmosphere.

Acknowledgements This work was supported by the Deutsche Forschungsgemeinschaft through the Sonderforschungsbereich 512. The computations have been performed of the Deutsches Klima Rechenzentrum (DKRZ).

References

Aagaard K, Carmack E (1989) The role of sea ice and other fresh water in the Arctic circulation. J Geophys Res 94: 14485–14498

Arfeuille G, Mysak L, Tremblay LB (2000) Simulation of the interannual variability of the wind-driven Arctic sea-ice cover during 1958–1998. Geophys Res Lett 30, doi:10.1029/2002GL016271

Baldwin M, Stephenson D, Thompson D, Dunkerton T, Charlton A, O'Neill A (2003) Stratospheric memory and skill of extended-range weather forecasts. Science 301: 636–640

Belkin I, Levitus S, Antonov J, Malmberg SA (1998) "Great Salinity Anomalies" in the North Atlantic. Progr Oceanogr 41: 1–68

Blackmon M (1976) A climatological spectral study of the 500 mb geoptential height of the Northern Hemisphere. J Atmos Sci 33: 1607–1623

Bojariu R, Gimeno L (2003) Predictability and numerical modeling of the North Atlantic Oscillation. Earth Sci Rev 63 (1–2): 145–168

Brümmer B, Müller G, Affeld B, Gerdes R, Karcher M, Kauker F (2001) Cyclones over Fram Strait: impact on sea ice and variability. Polar Res 20 (2): 147–152

Brümmer B, Müller G, Hober H (2003) A Fram Strait cyclone: Properties and impact on ice drift as measured by aircraft and buoys. J Geophys Res 108 (D7): 6/1–6/13

Castanheira J, Graf HF (2003) North Pacific-North Atlantic relationships under stratospheric control. J Geophys Res 108 (D1): 11/1–11/10

Cavalieri D, Häkkinen S (2001) Arctic climate and atmospheric planetary waves. Geophys Res Lett 28 (5): 791–794
Cavalieri D (2002) A link between Fram Strait sea ice export and atmospheric planetary waves. Geophys Res Lett 29 (12): Art. No. 1614
Christiansen B (2001) Downward propagation of zonal wind anomalies from the stratosphere to the troposphere: Model and reanalysis. J Geophys Res 105: 27307–27322
Cohen J, Saito K, Entkhabi D (2000) The role of the Siberian high in the Northern Hemisphere climate variability. Geophys Res Lett 28 (2): 299–302
Deser C, Walsh J, Timlin M (2000) Arctic sea ice variability in the context of recent atmospheric circulation trends. J Clim 13: 607–633
Deser C, Magnusdottir G, Saravanan R, Phillips A (2004) The effects of North Atlantic SST and sea ice anomalies on the winter circulation in CCM3. Part 2: Direct and indirect components of the response. J Clim 17: 2160–2176
Dickson R, Meincke J, Malmberg SA, Lee A (1988) The "Great Salinity Anomaly" in the northern North Atlantic, 1968–1982. Progr Oceanogr 20: 103–151
Dorn W, Dethloff K, Rinke A, Botzet M (2000) Distinct circulation states of the Arctic atmosphere induced by natural climate variability. J Geophys Res 105 (D24): 29659–29668
Gong G, Entekhabi D, Cohen J (2003) Modeled Northern Hemisphere winter climate response to realistic Siberian snow anomalies. J Clim 16: 3917–3931
Goosse H, Selten F, Haarsma R, Opsteegh J (2002) A mechanism of decadal variability of the sea-ice volume in the Northern Hemisphere. Clim Dyn 19: 61–83
Graversen R, Christiansen B (2003) Downward propagation from the stratosphere to the troposphere: A comparison of two hemispheres. J Geophys Res D24: Art. No. 4780
Haak H, Jungclaus J, Mikolajewicz U, Latif M (2003) Formations and propagation of great salinity anomalies. Geophys Res Lett 30 (9), doi:10.129/2003GL017065
Haak H (2004) Simulation of Low-frequency Climate Variability in the North Atlantic Ocean and the Arctic. Max-Planck-Institute for Meteorology, Reports on Earth System Science, No. 1
Hagemann S, Dümenil L (1998) A parameterisation of the lateral waterflow for the global scale. Clim Dyn 14 (1): 17–31
Hagemann S, Dümenil-Gates L (2003) Improving a subgrid runoff parameterisation scheme for climate models by the use of high resolution data derived from satellite observations. Clim Dyn 21 (3–4): 349–359
Häkkinen S (1993) An Arctic source for the Great Salinity Anomaly: A simulation of the Arctic-ice-ocean system for 1955–1975. J Geophys Res 98 (C9): 16397–16410
Häkkinen S (1999) A simulation of thermohaline effects of a Great Salinity Anomaly. J Clim 6: 1781–1795
Hilmer M, Harder M, Lemke P (1998) Sea ice transport: A highly variable link between Arctic and North Atlantic. Geophys Res Lett 25 (17): 3359–3362
Hilmer M, Jung T (2000) Evidence for a recent change in the link between the North Atlantic Oscillation and Arctic sea ice export. Geophys Res Lett 27 (7): 989–992
Hilmer M, Lemke P (2000) On the decrease of Arctic sea ice volume. Geophys Res Lett 27 (22): 3751–3754
Hurrel J, van Loon H (1997) Decadal variations in climate associated with the North Atlantic Oscillation. Clim Change 36: 301–326
Jung T, Hilmer M (2001) The link between North Atlantic Oscillation and Arctic sea ice export. J Clim 14 (19): 3932–3943
Kalnay E, Kanamitsu M, Kistler R, Collins W, Deaven D, Gandin L, Iredell M, Saha S, White G, Woollen J, Zhu Y, Chelliah M, Ebisuzaki W, Higgins W, Janowiak J, Mo K, Ropelewski C, Wang A, Leetmaa J, Reynolds R, Jenne R, Joseph D (1996) The NCEP/NCAR 40-Year Reanalysis Project. Bull Am Meteor Soc 77 (3): 437–471
Koeberle C, Gerdes R (2003) Mechanisms Determining the Variability of Arctic Sea Ice Conditions and Export. J Clim 16: 2843–2858
Koenigk T, Mikolajewicz U, Haak H, Jungclaus J (2006) Variability of Fram Strait sea ice export: Causes, impacts and feedbacks in a coupled climate model. Clim Dyn 26 (1):17–34, doi: 10.1007/s00382–005–0060–1

Kwok R, Rothrock, DA (1999) Variability of Fram Strait ice flux and North Atlantic Oscillation. J Geophys Res 104 (C3): 5177–5189
Kwok R, Cunningham G, Pang S (2004) Fram Strait sea ice outflow. J Geophys Res 109 (C01009), doi:10.1029/2003JC001785
Latif M, Roeckner E, Botzet M, Esch M, Haak H, Hagemann S, Jungclaus J, Legutke S, Marsland S, Mikolajewicz U, Mitchell J (2004) Reconstructing, monitoring and predicting multidecadal scale changes in the North Atlantic thermohaline circulation with sea surface temperatures. J Clim 17 (5): 857–876
Magnusdottir G, Deser C, Saravanan R (2004) The Effects of North Atlantic SST and Sea Ice Anomalies on the Winter Circulation in CCM3. Part 1: Main features and storm track characteristics of the response. J Clim 17 (5): 857–876
Marsland S, Haak H, Jungclaus J, Latif M, Roeske F (2003) The Max-Planck-Institute global ocean/sea ice model with orthogonal curvilinear coordinates. Ocean Modelling 5: 91–127
Norton W (2002) Sensitivity of Northern Hemsiphere surface climate to simulations of the stratospheric polar vortex. Atmospheric, Oceanic and Planetary Physics, University of Oxford, Oxford, pp. 1–5
O'Hanlon L (2002) Making waves. Nature 415: 360–362
Ostermeier G, Wallace J (2003) Trends in the North Atlantic Oscillation-Northern Annular Mode during the twentieth century. J Clim 16: 336–341
Polyakov I, Johnson M (2000) Arctic decadal and interdecadal variability. Geophys Res Lett 27 (24): 4097–4100
Proshutinsky A, Johnson M (1997) Two circulation regimes of the wind-driven Arctic Ocean. J Geophys Res 102 (C6): 12493–12514
Roeckner E, Bäuml G, Bonaventura L, Brokopf R, Esch M, Giorgetta M, Hagemann S, Kirchner I, Kornblueh L, Manzini E, Rhodin A, Schlese U, Schulzweida U, Tompkins A (2003) The atmosphere general circulation model ECHAM5, part 1: Model description. Max-Planck-Institute for Meteorology, Report No. 349, 127 pp
Schmith T, Hansen C (2003) Fram Strait ice export during the 19th and 20th centuries: Evidence for multidecadal variability. J Clim 16 (16): 2782–2791
Serreze M, Barry R (1988) Synoptic activity in the Arctic Basin, 1979–85. J Clim 1: 1276–1295
Thompson D, Baldwin M, Wallace J (2002) Stratospheric connection to the Northern Hemisphere wintertime weather: Implication for prediction. J Clim 15 (12): 1421–1428
Tremblay L, Mysak L (1998) On the origin and evolution of sea-ice anomalies in the Beaufort-Chukchi Sea. Clim Dyn 14: 451–460
Untersteiner N (1988) On the ice and heat balance in Fram Strait. J Geophys Res 93: 527–531
Valcke S, Coubel A, Declat D, Terray L (2003) OASIS Ocean Atmosphere Sea Ice Soil user's guide. CERFACS, Tech Rep TR/CGMC/03/69, Toulouse, France, 85 pp
Venegas S, Mysak L (2000) Is there a dominant timescale of natural climate variability in the Arctic? J Clim 13: 3412–3434
Vinje T (2001) Fram Strait ice fluxes and atmospheric circulation: 1950–2000. J Clim 14: 3508–3517
Vinje T, Finnekåsa O (1986) The ice transport through the Fram Strait. Norsk Polarinst. Skrifter 186: 1–39
Vinje T, Nordlund N, Kvambeck A (1998) Monitoring the ice thickness in Fram Strait. J Geophys Res 103: 10437–10449
Walsh J, Chapman W (1990) Arctic contribution to upper ocean variability in the North Atlantic. J Clim 3 (12): 1462–1473

Chapter 9
Fresh-Water Fluxes via Pacific and Arctic Outflows Across the Canadian Polar Shelf

Humfrey Melling[1], Tom A. Agnew[2], Kelly K. Falkner[3], David A. Greenberg[4], Craig M. Lee[5], Andreas Münchow[6], Brian Petrie[7], Simon J. Prinsenberg[8], Roger M. Samelson[9], and Rebecca A. Woodgate[10]

9.1 Introduction

Observations have revealed persistent flows of relatively low salinity from the Pacific to the Arctic and from the Arctic to the Atlantic (Melling 2000). It is customary to associate fluxes of fresh-water with these flows of brine, as follows: the fresh-water flux is the volume of fresh water that must be combined with a volume of reference-salinity water to yield the volume of seawater of the salinity observed. As with sensible heat flux, the choice of reference is arbitrary, but the value 34.8 is often used in discussions of the Arctic. This value is an estimate of the mean salinity of the Arctic Ocean by Aagaard and Carmack (1989) for a time period and averaging domain that were not specified. Because the salinity of seawater flowing across the shallow Bering, Chukchi and Canadian Polar shelves is typically lower than 34.8, these flows transport fresh-water from the Pacific to the Atlantic Ocean.

[1] Fisheries and Oceans Canada, Institute of Ocean Sciences, Box 6000 Sidney BC Canada V8S 3J2

[2] Environment Canada, Meteorological Service of Canada, 4905 Dufferin St. Downsview ON Canada M3H 5T4

[3] College of Ocean and Atmospheric Science, Oregon State University, Corvallis, OR 97331-5503, USA

[4] Fisheries and Oceans Canada, Bedford Institute of Oceanography, Box 1006 Dartmouth NS Canada B2Y 4A2

[5] Applied Physics Laboratory, University of Washington, 1013 NE 40th St. Seattle WA 98105, USA

[6] College of Marine Studies, University of Delaware, 44112 Robinson Hall, Newark DE 19716, USA

[7] Fisheries and Oceans Canada, Bedford Institute of Oceanography, Box 1006 Dartmouth NS. Canada B2Y 4A2

[8] Fisheries and Oceans Canada, Bedford Institute of Oceanography, Box 1006 Dartmouth NS Canada B2Y 4A2

[9] College of Ocean and Atmospheric Science, Oregon State University, Corvallis OR 97331-5503, USA

[10] Applied Physics Laboratory, University of Washington, 1013 NE 40th St. Seattle WA 98105, USA

R.R. Dickson et al. (eds.), *Arctic–Subarctic Ocean Fluxes*, 193–247

The transfer of waters from the Pacific to the Atlantic has been attributed to the higher sea level of the Pacific (Stigebrandt 1984; Wijffels et al. 1992), which is in turn the steric manifestation of lower salinity in the North Pacific relative to the North Atlantic. Steele and Ermold (2007) have examined the steric anomaly field derived from hydrographic data in the North Pacific Arctic and North Atlantic Oceans, 1950–2000. Their calculations, referenced to 1,000-db, suggest that Pacific Sea level at 55° N (zonal mean) is 0.55 m higher than Arctic sea level at 75° N (Beaufort gyre) and 1 m higher than Atlantic sea level at 85° N (Greenland Sea).

The magnitude and variability of volume and fresh-water fluxes through Bering Strait and the Canadian Archipelago are not well known (Melling 2000). The earliest geostrophic calculations of volume fluxes, based on bottle casts in the 1960s, were frequently cited until the late 1990s. At this time, volume flux had not been measured with established accuracy in any channel. The few estimates of fresh-water flux were inadequate, being products of long-term averages (volume flux and fresh-water anomaly) rather than averages of products. The wide variation (32–34: Aagaard and Carmack 1989; Prinsenberg and Bennett 1987; Sadler 1976) in the assumed salinity of through-flow for these estimates is indicative of their large uncertainty, equivalent to a factor of 3 in fresh-water flux. Geostrophic calculations were not referenced to measured currents until the 1980s (Prinsenberg and Bennett 1987; Fissel et al. 1988).

There were good reasons for the inadequate state of knowledge less than a decade ago (Melling 2000). One was political, the bisection of Bering Strait by a national jurisdictional boundary. Another was a geographic peculiarity, namely the proximity of the magnetic pole (80° N 105° W) to the Canadian Archipelago, which renders the geomagnetic field unreliable as a direction reference. Others were logistical – remoteness, harsh climate, persistent pack ice – or technical challenges to observation – hazard from moving sea ice and icebergs. Some arose from the nature of the flows themselves, such as small scales of motion, re-circulation and dramatic annual and inter-annual variability. Constraints on numerical simulation included computing capacity and deficient bathymetric and hydrographic information.

Lack of observations, attributable in large part to deficient technology, had been the principal impediment to scientific progress for many years. However by the late 1990s, improved technological capability provided the incentive for a renewed initiative to measure Pacific–Arctic through-flow. Doppler sonar, which offered the potential to measure near-surface current from a safe depth provided that zooplankton scatterers were sufficiently abundant, had been proven effective for year-round use in Arctic waters (Melling et al. 1995). Developments in microprocessors had opened up possibilities for smart instruments and new low-power electronics promised much longer operating intervals for sub-sea instruments. New all-weather microwave sensors offered higher resolution for ice reconnaissance, and developments in software permitted the automated tracking of pack drift and deformation.

Some technological challenges remain: how to measure current and salinity in the zone of extreme hazard from drifting ice, the upper 30 m of the ocean where much of the fresh-water flux occurs; how to recover moored instruments from remote areas of the Canadian Archipelago that are rarely free of ice; how to build affordable arrays that resolve the baroclinic scale of motion (5 km) across wide channels; how to measure the direction of current in the vicinity of the geomagnetic pole in the Canadian Arctic. Moreover, numerical simulation of circulation in geography of such complexity is in its infancy.

The ultimate objective of ASOF is understanding the Arctic branch of the global hydrologic cycle. Moreover, useful predictions of changing climate are dependent on realistic parameterization of the relevant oceanic processes for computer simulation. For this we need greatly improved understanding of the forcing and controls on oceanic fresh-water flux from Pacific to Arctic to Atlantic. Specific topics where we need improved theoretical knowledge are:

- Sea-level differences between Pacific, Arctic and Atlantic basins
- Through-flow forcing by inter-basin differences in sea level
- Through-flow forcing by wind and atmospheric pressure
- Dynamics of rotating flow through channels of realistic geometry
- Boundary stress at the seafloor and the ice canopy in tidal channels
- Buoyant boundary flow through a network of 'wide' interconnected channels
- Lagrangian aspects of mixing in channels

A unique aspect of Arctic channel flows is their seasonally varying canopy of pack ice. When the pack is comprised of small floes at moderate concentration, its main impact is on the stress exerted by wind on the ocean surface. However, when large thick floes are present at high concentration, they can jam within the channel (Sodhi 1977; Pritchard et al. 1979). As ice drift continues downstream of the blockage, an arch becomes evident marking the boundary between open water and fast ice. In addition to its obvious effect of stopping ice flux through straits, a fast-ice canopy reduces the oceanic flux by imposing additional drag at the upper boundary of the flow. Ice cover introduces several additional theoretical challenges:

- Dynamics of pack-ice flow through channels of realistic geometry
- Stable ice-arch formation in channels of realistic geometry
- Dynamical interactions between the flows of water and of pack ice in channels

In the context of climate change, it is interesting to compare the mobility of pack ice that populates the three principal exits routes of ice from the Arctic Ocean. Ice within the channels of the Canadian Archipelago is fast for 8–10 months of the year, when it completely blocks ice export from the Arctic. Within Fram Strait, the pack ice is never fast and there is export of fresh-water to the Atlantic as ice year-round. Ice in Nares Strait flickers between these extremes, sometimes providing an export route year-round and in some years blocking ice drift from December through July.

If the Canadian Arctic channels were simply plumbing, carrying water without modification from ocean to ocean, Davis Strait would offer the appeal of metering the total through-flow on a single section, although its great width (2.5 times the total of other gateways) and cross-section (five times the total) would present challenge. However, the vastness of the Bering and Chukchi Seas and of the Canadian polar shelf precludes their simplification to conduits that convey water without modification from the Arctic to the Atlantic. At least three check points are needed to develop a useful understanding the Arctic's role in the global hydrologic cycle: at entry to

Arctic and exit from Pacific, at outflow from the Arctic Ocean over the Canadian polar shelf and at exit from the Arctic and entry to the Atlantic.

For practical reasons, observations have been focused at constrictions along the pathways joining the North Pacific to the Atlantic through the Arctic. These are circled and labeled on the map in Fig. 9.1. All inflow from the Pacific Ocean passes through Bering Strait, a short wide (85 km, with two islands obstructing about 9 km) channel separating the Bering and Chukchi Seas; the greatest depth in the strait is 60 m, but there is a sill of 47-m depth about 200 km to the southwest. In contrast, the channels of the Canadian Archipelago are much longer than they are wide. The Archipelago occupies slightly less than half of the Canadian polar continental shelf, which at 2.9×10^6 km^2 represents almost a quarter of the Arctic Ocean area (13.2×10^6 km^2). Its many channels have been deepened by glacial action to form network of basins as deep as 600 m, separated by sills. Deep (365–440 m) sills at the western margin of the continental shelf are the first impediment to inflow from the Canada Basin, but the shallowest sills are in the central and southern parts. For flux measurement, there is an optimal set of relatively narrow, shallow straits

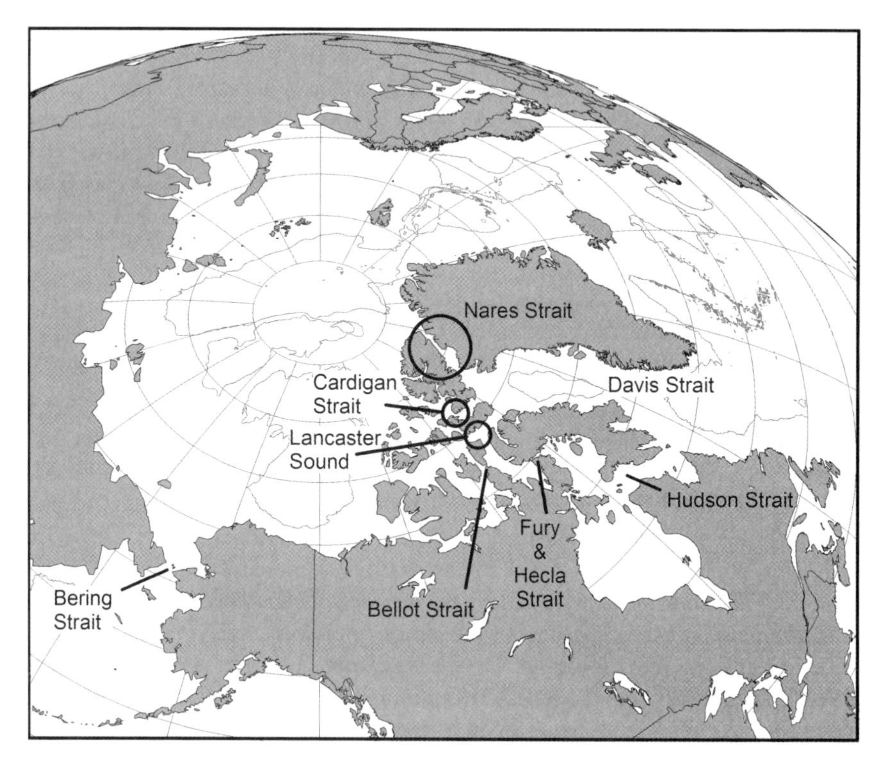

Fig. 9.1 The Arctic Ocean with focus on the North American shelves. The gateways for Pacific Arctic through-flow are indicated. To reduce congestion, the Lancaster Sound tag has been used as a single identifier of Barrow Strait to the west and Wellington Channel to the north-west; the Cardigan Strait tag also represents nearby Hell Gate. The 1,000-m isobath is plotted to delineate the continental shelves, ridges and ocean basins

through which all flow must pass: Bellot Strait, Barrow Strait (east of Peel Sound), Wellington Channel, Cardigan Strait, Hell Gate and Kennedy Channel. Among these, Bellot Strait is probably of little importance because it has such a small cross-section at the sill (less than 24 m deep, 1.9 km wide).

Although the net flux of volume is towards the Atlantic, water is exchanged in both directions between Baffin Bay and the Arctic Ocean. It is modified by mixing, freezing and melting during the months spent over the shelf and may ultimately be re-circulated back to its source. In contrast to Bering Strait, where re-circulation is usually dependent upon temporal reversals in flow direction, that within the Canadian Arctic is implicit in the spatial pattern of the circulation and the strength of tidally forced mixing and entrainment. The important net fluxes of volume and fresh-water must, therefore, be calculated as the differences between the much larger fluxes in opposing directions through adjacent parts of the cross-section.

Understanding of the fresh-water flux through the North American Arctic is presently inadequate to permit prediction of its sensitivity to climate change. The science is at the stage of basic research, during which monitoring of through-flow to detect variation and change must be a stand-in for simulation and forecast. However, the infrastructure needed to measure fluxes at all gateways for through-flow is not sustainable in the long run. A more tractable observing system will likely involve the integration of data from a few points of prolonged observation and realistic simulations of through-flow by numerical ocean circulation models; these must be driven in the greater part by observations that are readily available. We anticipate an opportunity to relax observational diligence when a capability in numerical simulation of Pacific Arctic through-flow has been demonstrated.

This chapter starts in a geographic progression from west to east around the North American continent, exploring recent advances in the empirical knowledge of volume and fresh-water through-flows via Bering Strait into the Arctic and via the gateways of the Canadian Archipelago that open into Baffin Bay. In order from southwest to northeast these are Lancaster Sound, Cardigan Strait, Hell Gate and Nares Strait. The subsequent sections review progress in relation to three issues common to all gateways – numerical simulation of Canadian Arctic through-flow, sea-ice budget for the Canadian polar shelf, mesoscale orographic influence on wind forcing in Arctic sea straits and trace chemicals in seawater as indicators of the sources, mixing and transit times for Pacific–Arctic through-flow. The final geographically oriented section examines Davis Strait, where Arctic fresh water is delivered to the convective gyre of the Labrador Sea. A closing section takes stock of our progress in the Arctic sub-Arctic Ocean fluxes study and identifies the issues that impede our understanding of fresh-water flows and dynamics in the North American Arctic.

9.2 Pacific Arctic Inflow via Bering Strait

Bering Strait is the only gateway between the Pacific and the Arctic Oceans. On an annual average, the flow through the strait is northwards; it is likely a consequence of decreasing sea level from south to north, Pacific to Arctic. The steric anomaly computed from hydrographic data support this interpretation, but the difference in geopotential across the Bering and Chukchi shelves has yet to be measured. Regional winds, which are southward on average, oppose flow into the Arctic (Coachman and Aagaard 1966, 1981; Woodgate et al. 2005b). Melling (2000) provides an overview of early studies.

Since 1990, measurements of temperature, salinity, current have been made in Bering Strait almost continuously at one site and sometimes at two or three sites simultaneously (Fig. 9.2; Roach et al. 1995; Woodgate et al. 2005a). Instruments have been positioned near the seabed to avoid damage from ice keels that can extend to 20-m depth. Before 2000, hydrographic sections were measured only sporadically and only in summer (Coachman et al. 1975). Since 2000, sections have been measured every year, but again only in summer. Snapshots of flow structure at high spatial resolution have been measured several times by ship-mounted ADCP. Such detailed views, though transient, are essential for justifying (or otherwise) the validity of flux estimates based on long-term data acquired at only one or two points across the section.

The observations since 1990 have revealed an average annual flux of volume through Bering Strait of about 0.8 Sv towards the Arctic (Coachman and Aagaard 1981; Roach et al. 1995; Woodgate et al. 2005a). Higher estimates from earlier times (e.g. 1.2 Sv in the 1950s: Mosby 1962) likely reflect the greater uncertainty of measurement using the technology and methods then available. The best estimate

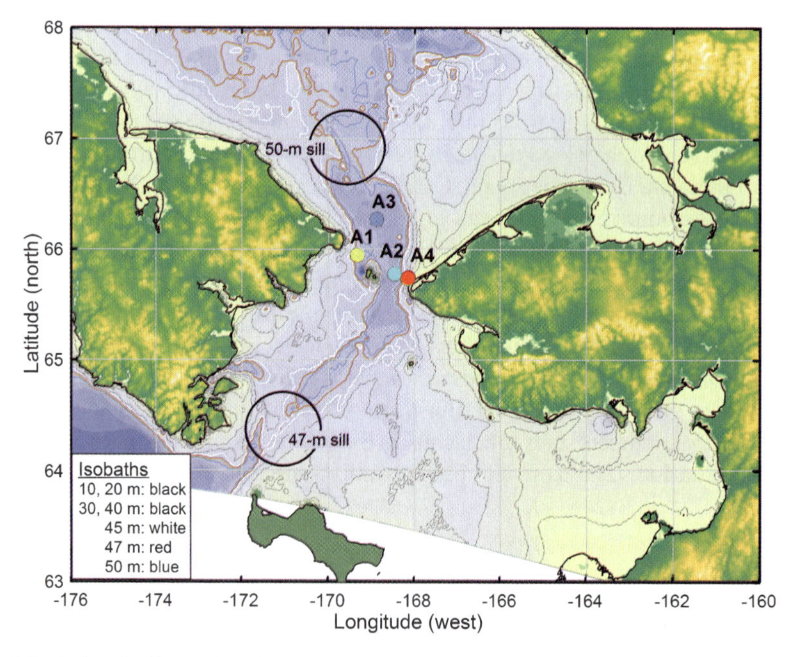

Fig. 9.2 Bering Strait showing the locations of moorings for determining through-flow (coloured discs). The sills limiting through-flow from the Pacific Ocean are circled

of the fresh-water flux through Bering Strait, circa 1990, was 1,670 km^3/year (53 mSv) relative to 34.8-salinity, calculated by Aagaard and Carmack (1989) for their review of Arctic Ocean fresh-water; these authors used Mosby's value for the average volume flux and an assumed annual average salinity of 32.5 (based mainly on hydrographic measurements in summer during the 1960s and 1970s).

Simultaneous observations at several sites during the last 5 years have provided some new information on the structure and variability of the Bering Strait through-flow (Woodgate and Aagaard 2005). In particular, an ADCP moored near the Alaskan coast has revealed the important contribution, previously unacknowledged, of the low salinity Alaskan Coastal Current (Paquette and Bourke 1974; Ahlnäs and Garrison 1984) to the fresh-water flux through Bering Strait. Woodgate and Aagaard (2005) now estimate that this stream contributes 220–450 km^3/year (7–14 mSv) to the fresh-water flux. Moreover, a previously ignored decrease in salinity towards the surface in mid strait is responsible for a second fresh-water flux increment of 350 km^3/year (11 mSv). These new contributions increase the flux of fresh-water via Bering Strait by about 50%, to 2,500 ± 300 km^3/year (80 ± 10 mSv), equivalent to three quarters of the fresh-water inflow to the Arctic Ocean via rivers. The contribution from ice flux through Bering Strait remains unknown.

Annual average values conceal strong annual cycles in fluxes through Bering Strait. Monthly mean volume flux is typically highest in summer (1.3 Sv in June), when the prevailing north wind of this region is weakest (Roach et al. 1995; Woodgate et al. 2005a). The flux decreases in winter under the influence of stronger north winds and reaches a minimum of about 0.4 Sv in January. A concurrent increase in salinity contributes to a much reduced northward flux of fresh water in winter; the minimum is 100 km^3/month (38 mSv) in January (Serreze et al. 2006). A lower near-bottom salinity, the presence of the Alaskan Coastal Current (April–December) and stronger salt stratification throughout the Strait (salinity decreases by 0.5–1 from seabed to surface: Woodgate et al. 2005a) act in concert with the stronger northward current to increase the fresh-water flux in summer; the maximum is 300–400 km^3/month (115–150 mSv) in June.

A model operating at 9-km resolution has been successful in simulating a seasonal cycle although it is weaker and lagged by 2 months relative to observations: the modelled fresh-water flux reaches a maximum at 220 km^3/month in July or August and a minimum at 80 km^3/month in March or April (Clement et al. 2005). The discrepancy between the model and observations has been attributed to the model's lower northward volume flux (only 0.65 Sv) and its poor resolution of the Alaskan Coastal Current. On the other hand, the observational basis for flux estimates within the Alaskan Coastal Current is also meagre. Prolonged measurements of the flow and stratification of the Alaskan Coastal Current (now viable using new technology) are needed to reduce uncertainty in this component of the Bering Strait through-flow.

Although the seasonal cycles in fresh-water flux through Bering Strait is strongly linked to that in volume, the former is also independently forced by seasonal sources of fresh water to the south, the Yukon River for example. The maximum monthly outflow of the Yukon River is only 40 km^3 and the total outflow of all rivers into the Bering Sea is 300 km³ each year (http://nwis.waterdata.usgs.gov/usa/nwis/discharge; Lammers et al. 2001). Therefore, other sources of fresh water must contribute to the 220–450 km³/year that the Alaska Coastal Current carries through Bering Strait. The fresh-water influx from the Gulf of Alaska to the Bering Sea (500 km³/year: Weingartner et al. 2005) is large enough to be the unrecognized contributor, but it is difficult to reconcile the 2-month lag for transit from the Aleutians to Bering Strait with observed seasonal variation in the Bering Strait through-flow.

Over periods of years, the variation of fresh-water flux is influenced by variations in both volume flux and in seawater salinity (Fig. 9.3: Woodgate et al. 2006). The highest annual mean volume flux occurred in 1994 (1 Sv), whereas the annual mean salinity at the seabed was highest in 1991 (32.8). Since 1998, when a better observational array was established, the fresh-water flux has ranged from 2,000 km³/year

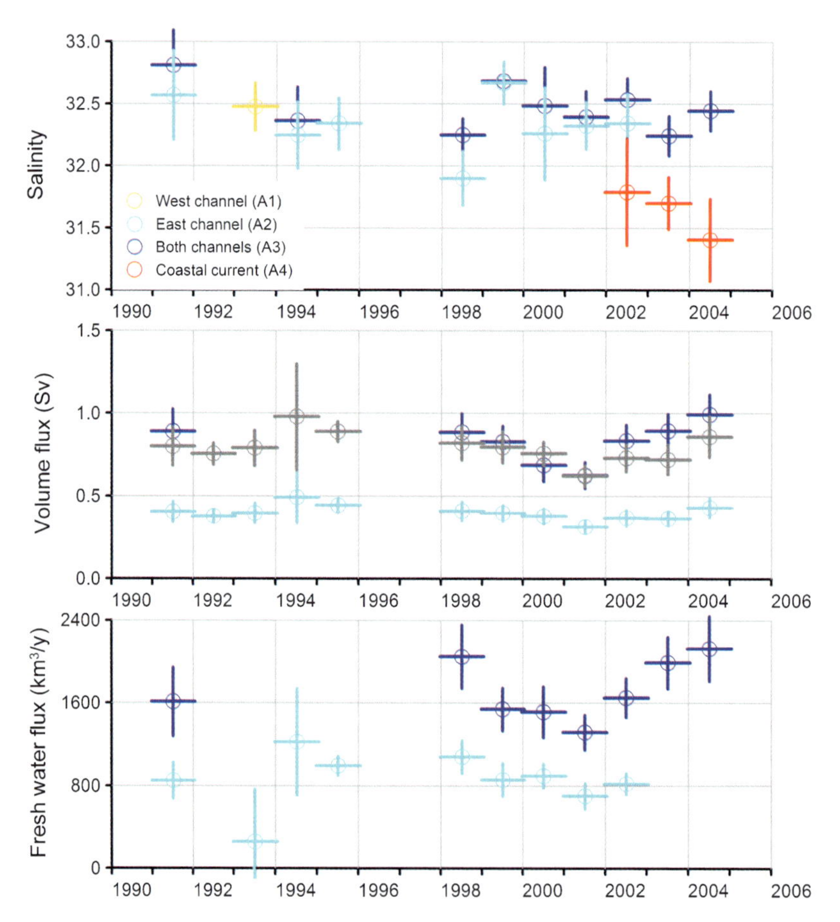

Fig. 9.3 Annual mean values of near-bottom salinity, volume flux and fresh-water flux derived from Bering Strait moorings as indicated by colouration. For flux estimates, blue (from A3) represents the entire strait, cyan (from A2) only the eastern channel and grey the entire strait, estimated from A2 only. Dashed lines indicate uncertainty in the means (Adapted from Woodgate et al. 2006)

Table 9.1 Fluxes of volume and fresh water through Bering Strait. Contributions of the Alaskan Coastal Current to the fresh-water flux have been included except in the estimate for inter-annual variation, for which there are insufficient data

	Volume (Sv)	Fresh Water (mSv)
Annual minimum (January)	0.4	38
Annual maximum (June)	1.3	115–150
Long-term mean	0.8	80
Inter-annual variation	±0.2	±10

(63 mSv) in 1998, to 1,400 km^3/year (44 mSv) in 2001 and back to 2,000 km^3/year in 2004. The 43% increase between 2001 and 2004 equals almost one quarter of the total annual inflow to the Arctic from rivers. Weakened north winds and consequent increased volume flux (0.7–1.0 Sv) explains 80% of the increase in fresh-water flux at this time (Woodgate et al. 2006). Clearly atmospheric variability in the Bering–Chukchi region has important influence on fluctuations in the Arctic fresh-water budget.

Current best estimates of fluxes through Bering Strait are summarized in Table 9.1. In this table, and in the preceding paragraph, the magnitude of inter-annual variation in fresh-water flux (1,400–2,000 km^3/year) does not include the fresh-water transported within the Alaskan Coastal Current and within the low-salinity surface layer, because such observations were initiated only recently. Investigators now suggest that these components are likely more than one third of the total.

New autonomous instruments (notably IceCAT, an upper-layer sensor in a trawl-resistant housing that transfers data to a recorder at safe depth) may provide the means for year-round measurement of the important fresh-water flux near the ocean surface where risk from storm waves and ice-ridge keels is high. Information from sensors on Earth satellites can also be valuable. For example thermal sensors have been used to delineate the northward flow of warm (and river-freshened) seawater in the coastal current, and radar altimeters have provided estimates of the atmospherically variability in the flux through Bering Strait via direct measurements of sea level on assumption of geostrophy (Cherniawsky et al. 2005).

International politics have been an impediment to flux measurement in Bering Strait, which is split between the Exclusive Economic Zones of the United States and Russia. Since 2004, a joint US–Russian scientific programme RUSALCA (Russian–American Long-term Census of the Arctic), lead in the USA by NOAA, has facilitated the installation of instruments on moorings to measure fluxes in the western channel of the Bering Strait.

9.3 Flux and Variability in Lancaster Sound

Lancaster Sound is the southernmost of the three principal constrictions to flow across the Canadian polar shelf between the Arctic Ocean and Baffin Bay (Fig. 9.1). It ranks second to Bering Strait in the duration of ocean flux measurements. Current meters were moored in Barrow Strait, to the west of Lancaster Sound, for 4 years during the early 1980s, providing data for the calculation of volume and fresh-water fluxes, subject to limitations of the technology of the time (Prinsenberg and Bennet 1987; Fissel et al. 1988). An array of new generation instruments was established in 1998 about 100 km further east (western Lancaster Sound) with intent to measure the combined outflows of seawater from Barrow Strait to the west and Wellington Channel to the north-west. Lancaster Sound is 68 km wide at this location and has a maximum depth of 285 m (Fig. 9.4). The array continues to evolve with the development and proving of new technology for this challenging application.

The location in Lancaster Sound is ice covered for as long as 10 months every year and typically lies beneath fast ice for half this time. It is well positioned logistically because it can be conveniently serviced in August via the icebreakers of the Canadian Coast Guard that routinely operate near Resolute Bay. Moorings have been recovered and redeployed annually and a modest hydrographic survey has been completed via CTD, with water sampling for analysis of geochemical tracers.

Arctic surface water occupies the upper part of the instrumented section. In summer, the coldest water (−1.7 °C, 32.8–33.0 salinity) is at 50–100 m depth, a remnant of winter (Prinsenberg and Hamilton 2005). Above this layer lies less dense surface water formed by addition of ice melt-water and runoff and by warming through insolation. The lightest water is organized into buoyancy boundary currents that flow in opposite directions along the northern and southern shores. The temperature and salinity increase with depth below the remnant winter waters. Some of this deeper water has arrived from the north and west (Melling et al. 1984; de Lange Boom et al. 1987) but the warmest and most saline is derived from the West Greenland Current in Baffin Bay to the east.

Because the keels of ice ridges threaten near-surface instruments, moorings have not extended above 30-m depth. For this reason, the array incorporates an ICYCLER in addition to the familiar instruments for measuring current, temperature and salinity. The ICYCLER periodically deploys a buoyant temperature-conductivity module

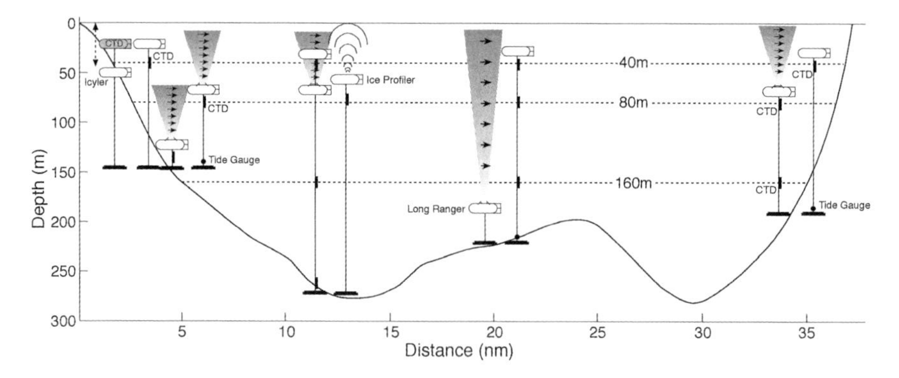

Fig. 9.4 The expanded array of moorings in Lancaster Sound used for through-flow measurements during 2005–2006. The instruments were concentrated near the southern shore (left of figure) in order to detect the buoyancy boundary current which carries much of the Arctic through-flow

upwards to the ice, measuring fresh-water and heat in the hazardous part of the water column. A comparison has revealed that a fresh-water inventory calculated using ICYCLER data is 20% larger during June to October than that inferred by extrapolation from data recorded at a fixed depth of 30 m. Since flow speed also increases towards the surface, the impact of accurate surface data on computed fresh-water flux is quite dramatic. During the cold part of the year in Lancaster Sound, when the surface mixed layer is deeper than 30 m, a sensor at 30-m depth provides a better measure of the near-surface fresh-water inventory. Since 2004, the array has also included ice-profiling sonar (IPS). Pack-ice draft data from this instrument in combination with ice tracking by the ADCP provide the component of fresh-water flux moved by pack-ice (e.g. Melling and Riedel 1996).

Reliance on the magnetic compass for a reference direction is standard practice in oceanography. However in western Lancaster Sound only 800 km from the north magnetic pole, the horizontal component of the Earth's field is less than 2,500 nT, the inclination of field lines is almost vertical (87.6°) and the magnetic declination is significantly perturbed by ionospheric effects over a range of time scales. To use a geomagnetic reference under such conditions, instrument orientation must be measured using a precise three-axis flux gate compass and the instantaneous geomagnetic vector must be monitored. Fortuitously for installations in Lancaster Sound, there is a geomagnetic observatory in nearby Resolute Bay. Details are provided by Prinsenberg and Hamilton (2005).

Based on a hydrographic section measured in August 1998, geostrophic calculations revealed an eastward current that extended across two thirds of the sound with highest speed at the surface near the southern shore (Prinsenberg and Hamilton 2005). There was weak westward flow at depth on the northern side. Subsequent study has shown that flow through the northern third of the section is quite variable and contributes little to net flux on a long-term average. In recognition of this apparent broad structure to the flow, the array of moored instruments provides observations of current, temperature and salinity at only 2–4 positions across the section.

In computing flux, it has been assumed that data from each location and depth represent average conditions across a specified sub-area of the cross-section, so that flux is the sum of area-weighted data. The selection of sub-sectional areas was guided by data from an expanded array of four sites in place during 2001–2004. This array provided the usual observations at sites in the coastal boundary currents near the southern and northern shores, and additional observations of near-surface (0–60 m) current at the quarter and half-way points from the southern shore (Fig. 9.4). Figure 9.5 displays the average of currents measured at 10, 30 and 50 m as weekly averages for three sites at 15-km spacing in the southern half of the section. At times, most often during November through May, upper ocean flow was similar at all three sites. However during the summer the shear across the channel was large; the speed at the southernmost mooring was almost twice the average value for the 3 sites. A seasonally varying weighting of data from the southernmost mooring has therefore been used in calculating fluxes at times when only two sites were established.

Estimated fluxes through Lancaster Sound are listed in Table 9.2 and plotted in Fig. 9.6; the reference value for fresh-water is 34.8. Volume flux has a 6-year mean

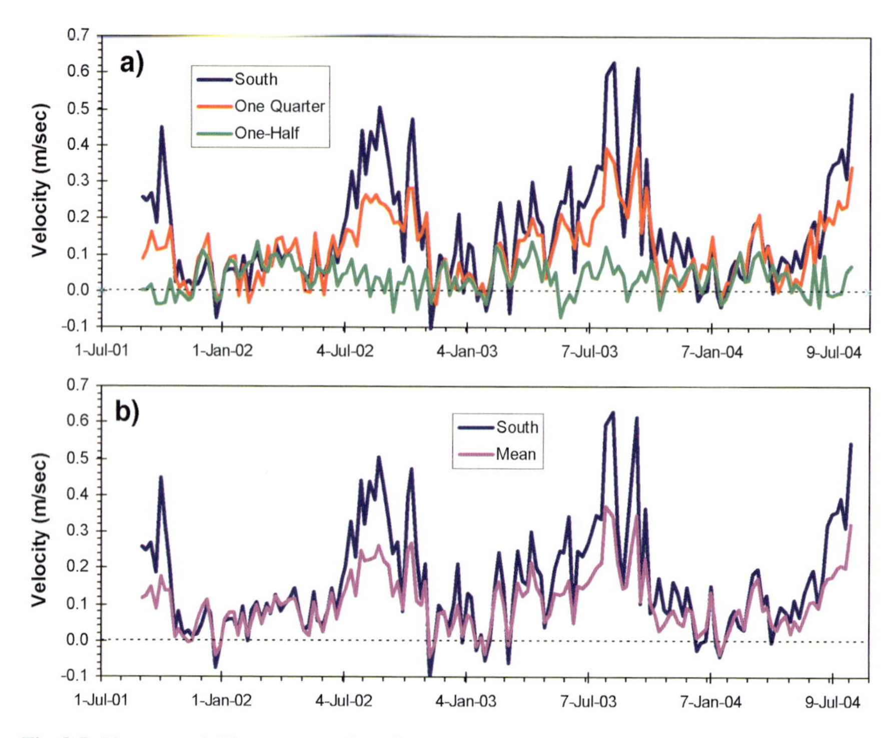

Fig. 9.5 Upper panel: The average value of currents measured at 10, 30 and 50 m for three locations in the southern half of Lancaster Sound, labelled by fractional distance from the southern shore. Lower panel: Average of the three curves in the upper panel compared with the time series from the southernmost site. All values are week-long averages

Table 9.2 Fluxes of volume and fresh-water through Lancaster Sound as seasonal and annual averages for August 1998 to August 2004. The reference salinity for fresh-water flux is 34.8. Arctic exports have positive value

		Fall	Winter	Spring	Summer	Year
1998–1999	Volume (Sv)	–0.01	0.37	0.48	0.70	0.39
	Fresh-water (mSv)	**10**	**26**	**31**	**44**	**26**
1999–2000	Volume (Sv)	0.25	0.91	1.09	1.32	0.89
	Fresh-water (mSv)	**20**	**56**	**65**	**81**	**56**
2000–2001	Volume (Sv)	0.97	0.82	0.81	1.19	0.95
	Fresh-water (mSv)	**59**	**51**	**51**	**72**	**58**
2001–2002	Volume (Sv)	0.11	0.35	0.87	0.93	0.56
	Fresh-water (mSv)	**14**	**22**	**55**	**70**	**40**
2002–2003	Volume (Sv)	0.60	0.54	1.18	1.13	0.86
	Fresh-water (mSv)	**45**	**34**	**77**	**92**	**62**
2003–2004	Volume (Sv)	0.31	0.45	0.63	1.24	0.57
	Fresh-water (mSv)	**35**	**34**	**43**	**93**	**45**

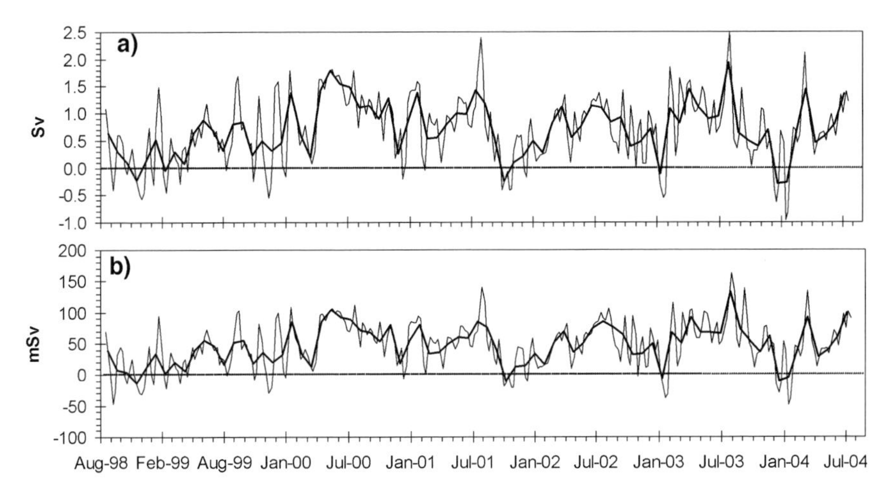

Fig. 9.6 Time series of weekly and monthly averaged fluxes through western Lancaster Sound, August 1998–2004

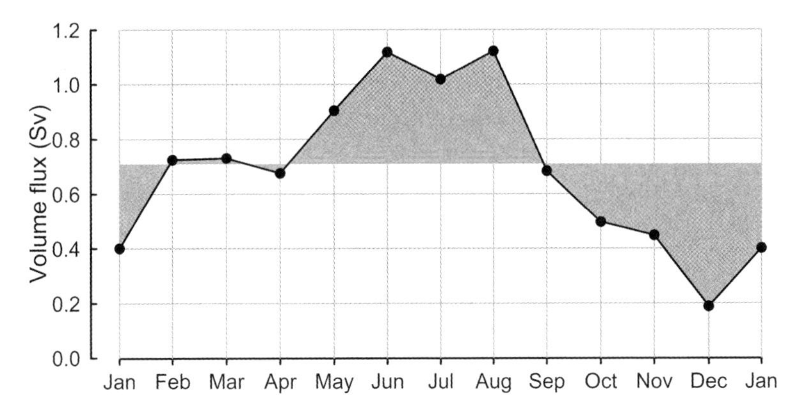

Fig. 9.7 Annual cycle in volume flux through western Lancaster Sound, as means for each month computed for the 6-year record of measurements, August 1998–2004

of 0.7 Sv, with yearly averages spanning a range of 0.4–1.0 Sv. There is a strong annual cycle (Fig. 9.7), ranging between low values in autumn and winter (0.2 Sv) and high values in summer (1.1 Sv). The fresh-water flux is typically about 1/15 of the volume flux – 6-year mean of 48 mSv (1,510 km^3/year) – and has a similar seasonal cycle. The range of variation in annual means is 36 mSv (1,140 km^3/year).

Atmospheric variability is one possible driver of flow variability in Lancaster Sound. Figure 9.8 displays an obvious co-variation of 12-month running averages of the NAO Index and of fresh-water flux through Lancaster Sound; the former has been delayed by 8 months. One possible linking mechanism is the oceanic response to the AO, mediated primarily via Ekman pumping and via lateral displacement of the Beaufort gyre. The associated cycle in the ocean circulation pattern has been

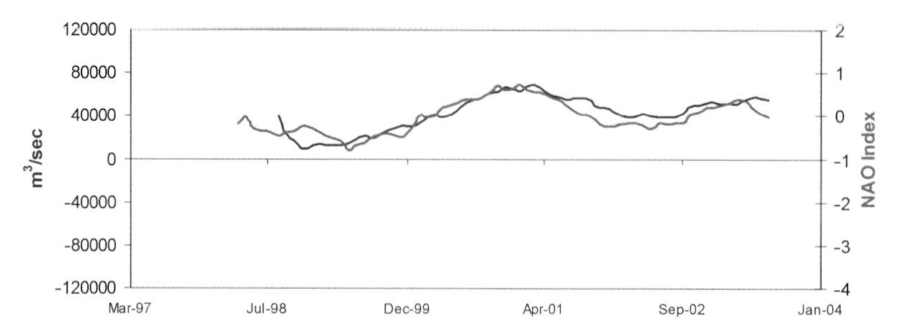

Fig. 9.8 12-month running averages of the fresh-water flux through Lancaster Sound and of the NAO index, with the latter delayed by 8 months

labelled the Arctic Ocean Oscillation (AOO) by Häkkinen and Proshutinsky (2004). Under this interpretation, the 8-month lag of the flow surge in Lancaster Sound could represent the spin-up time of the AOO. The possible role of the AOO in forcing Canadian Arctic through-flow is discussed further in the section on chemical tracers.

One goal of present study is the demonstration of a minimal array of moored instruments that could monitor fluxes through Lancaster Sound over the long term with help from numerical ocean models. The relative magnitudes of flows at three locations in the southern half of the section have already been discussed. The lower panel of Fig. 9.5 shows that the upper ocean flow 22 km from the southern shore was close to the average of values from all the three sites during a 3-year period of trial. This demonstration is the basis of a proposed flux-monitoring installation at this location: 300-kHz ADCP at 75 m to measure upper-ocean current and ice drift, bottom-mounted 75-kHz ADCP (with pressure sensor) to measure deep current, IPS at 50 m to measure ice draft, temperature-conductivity recorders at several depths below 50 m and an ICYCLER to determine profiles of temperature and salinity in the upper 50 m and a pressure gauge.

Ultimately, if Canadian Arctic through-flow is found to be predominately barotropic, then precise, geodetically referenced sea-level stations around the Canadian Archipelago could provide the information needed by numerical models to determine the oceanic fluxes.

9.4 Structure of Flow in Hell Gate/Cardigan Strait

In 1998, Fisheries and Oceans Canada began a study of current in Cardigan Strait with two goals that are fundamental to the successful measurement of fluxes through the Archipelago: (1) a reliable and cost-effective method of measuring current direction near the geomagnetic pole, and (2) a better knowledge of the spatial structure of Arctic channel flows. The latter information is essential to the design of sparse arrays of moored instruments for accurate measurement of oceanic through-flow.

Cardigan Strait had advantages as an experimental site. Because of its simple geometry and narrow width (8 km: Fig. 9.9), the through-flow could perhaps be resolved at the internal Rossby scale using a small number of moorings. Mixing by strong tidal currents (2 m/s) could be expected to weaken the density stratification and thereby to reduce the importance of the difficult-to-measure baroclinic component of flow. Moreover, strong tides provided a key to measuring current direction in the Canadian Arctic because tidal ellipses in a narrow strait are necessarily flat and aligned with the strait's axis. Nearby Hell Gate was an experimental control with half the width and contrasting 'dog-leg' geometry.

The study in Cardigan Strait was planned in phases of 2-year duration. The objective of the first phase, 1998–2000, was evaluation of a new torsionally rigid mooring for ADCPs; that of the second was investigation of co-variability between flows in Cardigan Strait and in Hell Gate; that of the third was a look at the cross-sectional structure of flow within Cardigan Strait. In response to presently ambiguous results, the third phase has been continued beyond 2005.

A unique mooring (Fig. 9.10) was designed to meet the special challenges of this environment. It was torsionally rigid to keep the ADCP on a fixed geographic heading throughout the deployment; a universal joint in the backbone allowed the mooring to stand upright regardless of seabed roughness and slope. The ADCP itself was mounted in gimbals to remain zenith-pointing during lay-over of the mooring in

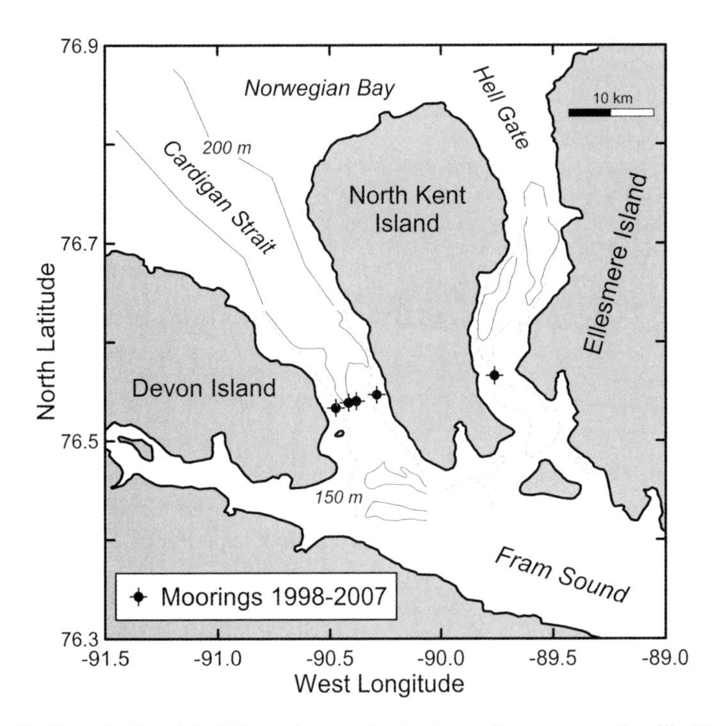

Fig. 9.9 Cardigan Strait and Hell Gate, showing the locations of moorings at the sills. The plotted 150 and 200-m isobaths are based on sparse soundings

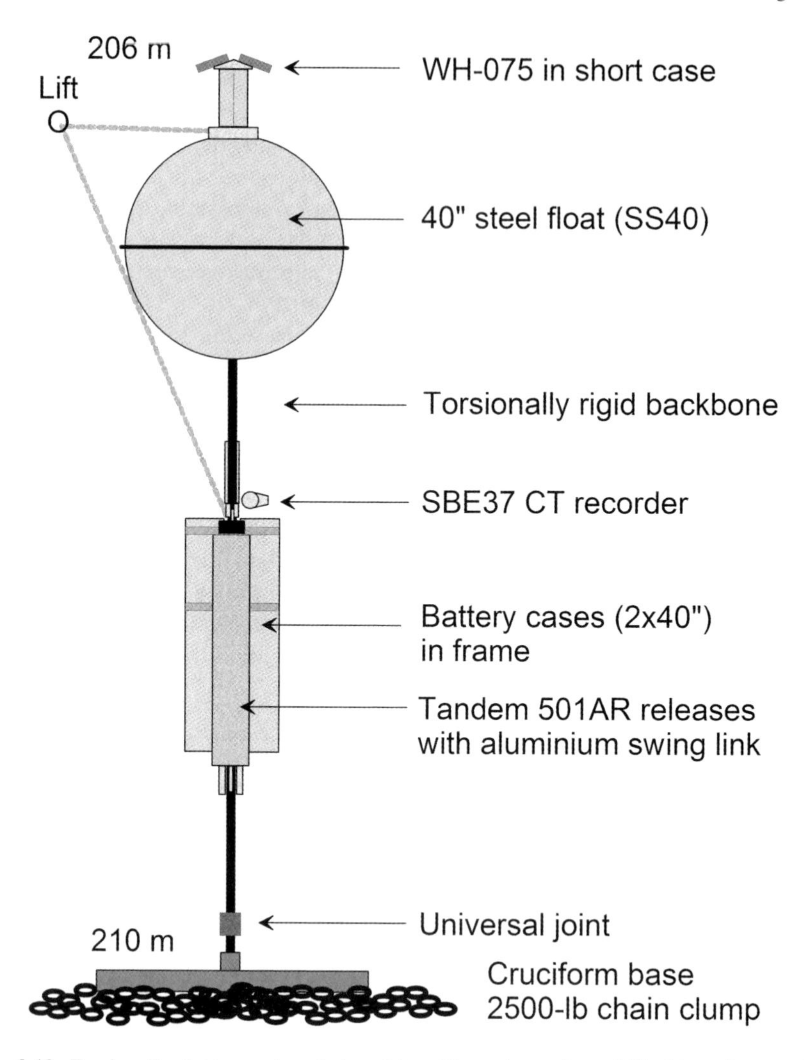

Fig. 9.10 Torsionally rigid mooring designed to address the various difficulties of measuring current in the Canadian Archipelago

strong current. The mooring rose only 3 m from the seafloor so as to minimize its sensitivity to the drag from current and its vulnerability to icebergs. The mooring was designed for free fall from the surface, enabling expeditious deployment in fast current and drifting ice; heavy chain arranged in loops as part of the deadweight anchor cushioned the shock of landing at 3 m/s.

Phase 1 provided proof of the value of the new mooring, which was over the side and deployed in 30 s, survived impact at the seabed and held the ADCP within ± 1° of upright in 2 m/s current and at constant heading for 2 years. The latter result justifies our reliance on a tidal-stream analysis of the recorded data to infer the ADCP's orientation. Current were measurable using backscattered sound to a range of 100 m from early July to late January, but the effective range shrank to about 70 m for 3 months (April through June) when echoes were weak. The strong diurnal variation of echoes implies a biological explanation for the weak back-scatter in late winter. During the second phase, our trial with a 75-kHz ADCP was successful in providing current profiles to the surface (185-m range) in all seasons.

Annual mean currents at three locations across Cardigan Strait are shown in Fig. 9.11. Measurements were made on the western slope during August 1998–2000, on the central axis during August 2000–2005 and on the eastern slope during August 2002–2004. The observations reveal uniform current in the middle depth range and sheared flow near the seafloor and the surface. Benthic drag or hydraulics at the sill may influence the lower shear layer and baroclinicity or wind action the upper. The small year-to-year variation between 1998 and 1999 (western slope), between 2000 and 2001 (channel axis) and between 2002 and 2003 (both eastern slope and channel axis) initially prompted an interpretation that differences between sites were indications of a stable spatial structure for the flow. For example, the left and centre panels of the figure (data not synoptic) suggest a halving of speed in only 2.4 km; such a steep gradient raised doubt about fluxes calculated using data from a single location in this 8-km wide channel. However, on presumption that data from 1998–2002 provided a valid representation of a constant spatial structure in the flow, we estimated volume fluxes of 0.2 Sv and 0.1 Sv through Cardigan Strait and Hell Gate (2000–2002 data not shown), respectively.

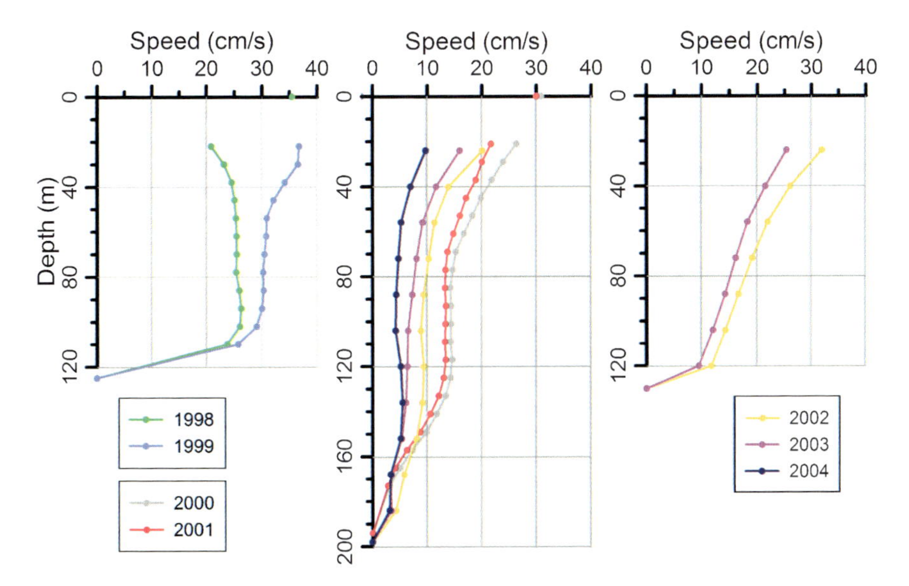

Fig. 9.11 Mean annual profiles of along-channel flow at three locations in Cardigan Strait, 1998–2005. An ADCP operated on the western slope during August 1998–2000 (left panel), on the central axis during August 2000–2005 (middle panel) and on the eastern slope during August 2002–2004 (right panel). Labels denote the starting August for the 12-month average. Positive values indicate flow towards Baffin Bay

Prolonged observation has provided new perspectives. Although the annual mean current at mid channel was much the same during 2000 and 2001 and during 2002 and 2003, the five annual values span a three-fold range. Clearly our early assumptions regarding a static cross-sectional variation and temporal constancy are invalid. Moreover, during 2002–2004, the mean (southward) current along the eastern slope of the Strait was about 50% stronger than the mean on the channel axis (Fig. 9.11); the flow along the western slope may also be stronger than on the axis. This pattern of cross-channel variation is consistent neither with a wall-bounded buoyancy current following the western slope nor with a frictionally controlled flow wherein current would be fastest at mid channel. We conclude that observations at more than two locations are required to calculate flux even in a channel as narrow as Cardigan Strait.

Results concerning seasonal variation in current are ambiguous; some data reveal an obvious annual cycle and some do not. One of the more definitive records, acquired on the eastern slope of Cardigan Strait during August 2002–2004, is plotted in Fig. 9.12. There is a strong Arctic outflow from January through September in both years (strongest in June), but during the autumn and early winter the average flow is weaker and the direction of flow reverses at times. This cycle is roughly in phase with that reported from Lancaster Sound as an average over 6 years of measurement. If this result survives more thorough analysis it will lend credence to a common forcing mechanism for both gateways, perhaps a seasonally varying pressure gradient from the Canada Basin to Baffin Bay that is weakest in the late autumn.

The difficulty of calculating volume flux through Cardigan Strait and Hell Gate has just been described. The challenge of calculating fresh-water flux as the covariance of flow velocity and salinity anomaly integrated across the channel section is even greater. The difficulty of delineating the cross-sectional variation of current is clear; measurement of the time-varying cross-section of salinity is even more problematic. Strong hydrodynamic drag (current up to 3 m/s in Hell Gate) effectively

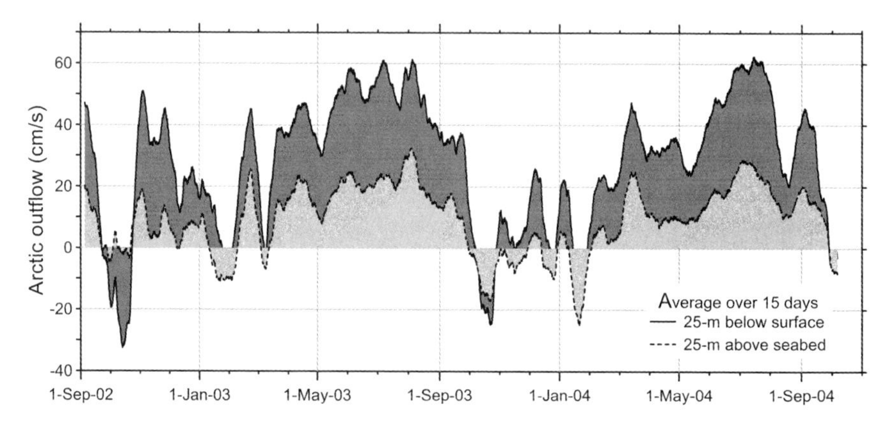

Fig. 9.12 Current at two levels on the eastern slope of Cardigan Strait, August 2002 to October 2004. Note the strong Arctic outflow during February through October. The record has been filtered to attenuate tides

precludes the use of conventional taut-line moorings to suspend temperature-conductivity recorders at fixed depths. Moreover, hydrographic fields are strongly forced by tidal flow over the sloping topography of the straits. At fixed depth near the seafloor, where temperature and salinity are presently being measured, the range in the value of these parameters over a tidal cycle is comparable to the range in values that might be measured via an instantaneous CTD cast from surface to seabed. This is perhaps indicative of fresh-water flux contributions from covariance at tidal frequency. An implied necessity to resolve variation of the salinity section at such high frequency cannot be met with present technology.

9.5 A Snapshot of Flux via Nares Strait

Because new long-term measurements of current and salinity from which fluxes in Nares Strait might be calculated have only recently been retrieved from the sea, the work of Sadler (1976) remains for now the standard reference. A 10-month record of current was acquired by an ADCP deployed near the Canadian shore in Smith Sound during the North Water Project in 1997–1998, but calculation of fluxes from single-point data in this wide sound is not defensible (Melling et al. 2001).

We do have an excellent set of observations acquired using ship-based ADCP during August 2003 which provide a detailed description of the cross-sectional structure of the through-flow. As discussed earlier, such information is essential to the use of data from widely spaced moorings in the calculation of fluxes and in the estimation of sampling error. It has also provided demonstrably accurate values, albeit short-term, of volume and fresh-water fluxes as benchmarks against which to assess values based on less well resolved measurements by moored instruments. The data from the high resolution surveys in Nares Strait and their significance for ocean-flux measurement using moored instruments are the subjects of this section.

LeBlond (1980) proposed that the generally cyclonic circulation of icebergs across the mouth of Lancaster Sound was a manifestation of buoyancy concentrated in narrow boundary currents of low salinity. Direct observations of these currents revealed an approximately geostrophic balance of flow and cross-channel pressure gradient (Prinsenberg and Bennett 1987; Sanderson 1987) on a 10-km scale comparable to the local internal Rossby radius of deformation.

In August 2003, a team on USCG Healy completed simultaneous surveys of current and salinity in Nares Strait. Flow data were acquired at high resolution using vessel-mounted acoustic Doppler current profiler (ADCP) and conventional hydrographic casts provided temperature and salinity (and therefore density) at 5-km spacing on selected sections. A notable feature of the salinity and density sections was the spreading of isopycnals at about 130-m depth within 10-km of Ellesmere Island (Münchow et al. 2006): isopycnals above this depth sloped upward toward the coast whereas those below it sloped downward. Such hydrographic structure is indicative of a sub-surface baroclinic jet hugging the western side of the channel. A weaker counter-flow of similar width was measured near the Greenland coast.

Healy's ADCP used a 75-kHz, hull-mounted, phased array. Echoes received at 2-s intervals were processed to yield a vertical profile of velocity relative to the ship. The ship's motion was derived from an independent bottom-tracking pulse (or via high precision GPS tracking) as described by Münchow et al. (2006). Because sonar beams were directed obliquely downwards, velocity could not be measured in the lowest 15% of the water column where there is interference from the seabed. Also, there are no data for the uppermost 25 m because the hull-mounted transducer was 8 m below the surface, because signals from the first 10 m of range were obscured by ring-down of the transmitter and because the sonar pulse averaged flow over 15 m increments in range.

Measured current was the sum of components at tidal and lower frequencies which vary with the position of the ship and with the time of measurement. Sub-tidal current was masked in each instantaneous measurement by tidal current which was generally much larger. However, because the tide is predictable in space and time (Padman and Erofeeva 2004) its contribution can be removed from each observation via collective analysis of the observations from all places and times. We fitted oscillations at tidal frequencies to velocities measured separately at different times for each depth of interest; this approach allowed realistic vertical variations in tidal current with friction and density stratification.

The continuous measurements of current from the slowly moving ship easily resolved flow features on the scale of the internal Rossby radius. The along-channel flow at sub-tidal frequency was observed to be spatially coherent with a Rossby number of 0.13, indicating near-geostrophic balance. Approximately one third of the total volume flux was associated with cross-channel slope of the sea surface (barotropic mode) and two thirds with across-channel slope of isopycnal surfaces (baroclinic mode).

One section at 80.5° N (Fig. 9.13) was measured repeatedly over several tidal cycles. The sub-tidal flow was southward with much of the flux in the western half of the channel above 200-m depth. The principal feature was a sub-surface jet that reached a maximum speed of 0.3 m/s about 12 km from the Ellesmere coast. The calculated net flux of seawater averaged over several days of observation was 0.8 ± 0.3 Sv towards Baffin Bay. The southward net flux of fresh water was 25 ± 12 mSv (790 km^3/year). These values are dependent upon assumption of current speed within the upper 30 m of the water column, which could not be measured. The fresh-water flux is particularly sensitive to the assumption because the low salinity of surface water strongly weights the current in this layer upon integration. The quoted confidence limit for fresh-water flux is the difference between a lower bound that neglects flux above the shallowest depth of measurement and an upper bound for which current was assumed uniform in the top layer and equal to the average flow in the shallowest two levels of measurement (18–48 m).

A second section with good observational coverage was completed in Robeson Channel at the northern end of Nares Strait. This survey encompassed the locations where current was measured for 6 weeks in the spring of 1971. Data from three sites at this time provided the often cited 0.6 ± 0.1 Sv through-flow value for Nares Strait (Sadler 1976); the attribution of more than 50% of the calculated flux to the record

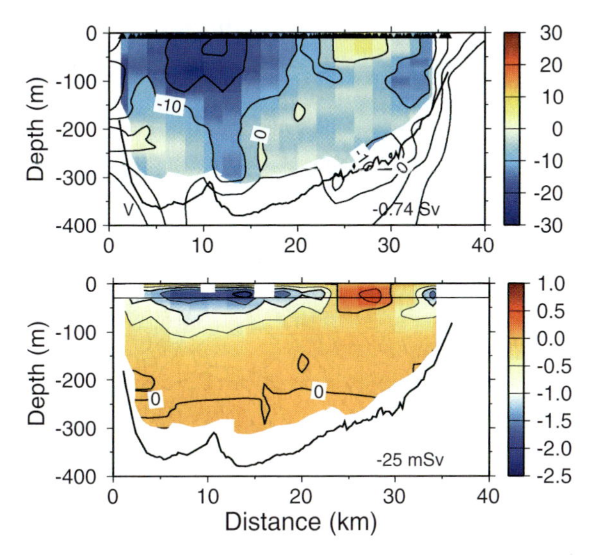

Fig. 9.13 Distribution of volume flux (upper panel) and fresh-water flux (lower panel) through a cross-section near 80.5° N in Kennedy Channel. Data were collected over a short period in August 2003. Flow out of the Arctic has negative value. The horizontal line near the top of the lower frame marks the depth above which current was estimated, not measured

from one instrument at 100-m depth near Ellesmere Island has long been disquieting. Figure 9.14 shows the locations along the track of USCGC Healy where current profiles were measured during 7–11 August 2003; the coordinate axes associated with along and cross-channel flow are also shown. The observations were de-tided using tidal predictions (Padman and Erofeeva 2004), then averaged at each level within bins spanning 1 km across the channel and 50 km along it. Figure 9.15 is a cross-section of the along-channel current, which shows the dominant feature to be a southward subsurface jet peaking at 0.4 m/s only 2 km from Ellesmere Island. The depth of maximum speed was 150 m where the jet was 10-km wide. Current through the eastern part of the section was weaker, 0.05 m/s, and northwards. The calculated flux of volume through this section was also about 0.7 Sv in early August 2003, with the principal part within baroclinic subsurface jet on the Ellesmere side.

Prior to and during the survey in 2003, winds were persistent from the south-west (towards the Arctic), promoting down-welling on the Greenland side. Because the subsurface jet below 50-m depth ran counter to the wind, atmospheric conditions may have weakened the down-channel flow from values prevalent under more typical north-east wind. Three-year time series from Doppler sonar recently recovered from Nares Strait reveal a strong modulation of current at periods typical of synoptic meteorological forcing (Fig. 9.16). How the data from the surveys of August 2003 fit into this strong pattern of variability has yet to be determined. Nonetheless, the volume flux through Nares Strait at this time was comparable to the long-term average inflow through Bering Strait (0.8 Sv); the fresh-water flux was about half the estimated Bering Strait inflow (Woodgate and Aagaard 2005).

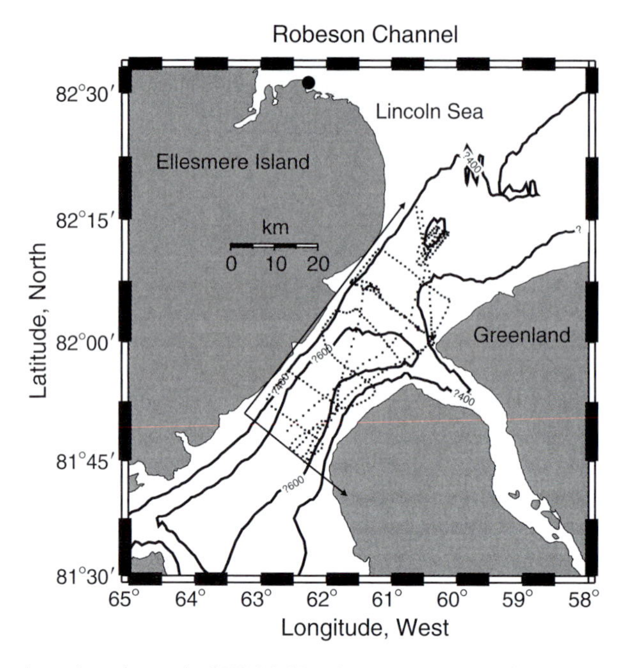

Fig. 9.14 Locations along the track of USCGC Healy where current profiles were measured during 7–11 August 2003

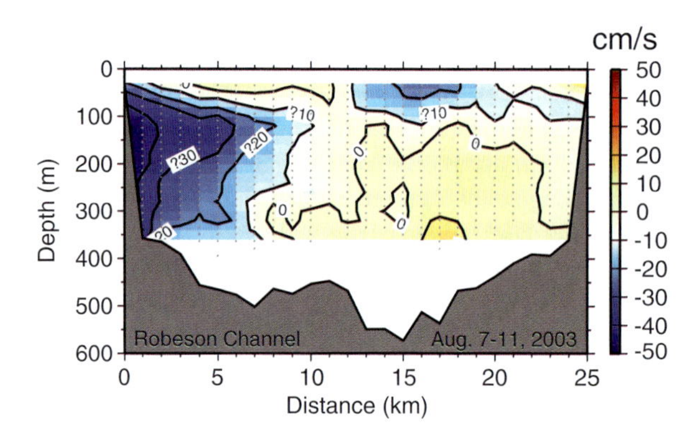

Fig. 9.15 Non-tidal current through Robeson Channel, 7–11 August 2003. Flow out of the Arctic has negative value. Note the wall-bounded southward jet on the Canadian side. The blank area near the top of the frame is too shallow for measurement by the hull-mounted sonar; that below 350 m is beyond the effective operating range of the sonar

The close correspondence in value between our volume flux and that of Sadler (1976) is fortuitous. However, the detailed picture from the 2003 survey does indicate that the dominant flux contribution in 1971 was from an instrument optimally positioned to measure the core of the sub-surface jet. We conclude that the suspicion attached

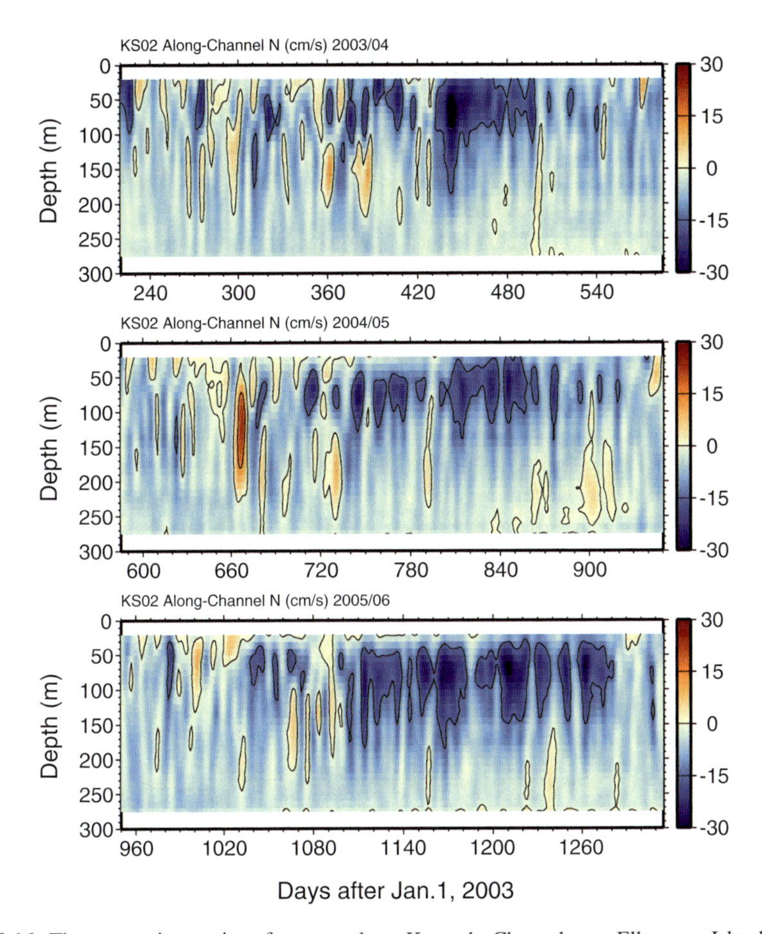

Fig. 9.16 Three-year time series of current along Kennedy Channel near Ellesmere Island from August 2003–2006; tides have been removed. Strong variability at periods typical of synoptic weather is obvious

to the uncharacteristically strong current measured by Sadler's instrument at 100-m depth was probably unwarranted. By inference this jet is apparently a persistent feature of Nares Strait through-flow.

9.6 Insights from Simulation of Canadian Arctic Circulation

Numerical models of fresh-water and ice movement through the Canadian Archipelago face formidable challenges. Principal among these are: (1) the scarcity of data to represent the three-dimensional structure of temperature and salinity and its seasonal variation; (2) the difficulty of resolving necessary detail in the many small but important passages while maintaining a correct dynamical interaction between the modelled domain and bordering seas; (3) the weakness of sea-ice models in representing ice drift through channels, including the appearance and break-up of fast-ice and its influence on oceanic through-flow; and (4) realistic wind forcing of oceanic circulation. Until these challenges are met, our preoccupation is the realistic simulation of present conditions. Predictions of flow under future changed climate are fraught with uncertainty.

Nonetheless, there has been notable progress in the numerical simulation of fresh-water and ice movements through the Canadian Archipelago in recent years. Model-based flux estimates for seawater volume and fresh water are converging and models of pack-ice dynamics in island-studded waters have improved.

Advances have emerged from modern coastal-ocean models that have been implemented at high spatial resolution within the Canadian Archipelago. One model, Fundy, is linear and harmonic and a second Quoddy is non-linear and prognostic. The models have been built around the finite element method to best represent the geographic complexity of the area. In the present (2006) implementation, the horizontal triangular mesh has 76,000 nodes and 44,000 elements and a resolution ranging from 1.1 km in narrow straits to 53 km in Baffin Bay (Fig. 9.17). The vertical coordinate is resolved via a hybrid mesh with fixed levels over the upper 150 m, where the vertical stratification is strongest, and terrain-following computational surfaces at greater depths. The gridded density field has been developed iteratively,

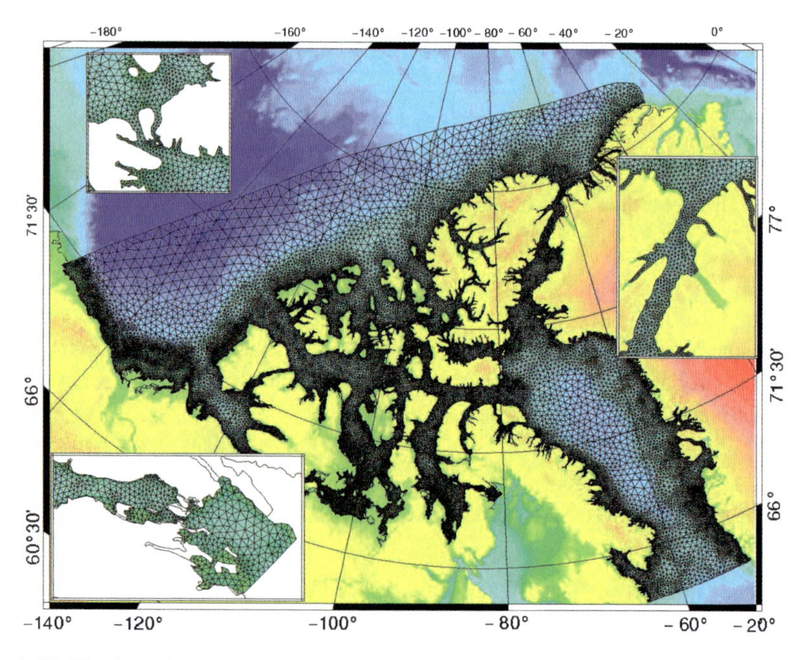

Fig. 9.17 The irregular triangular grid used for numerical simulation of circulation over the Canadian polar shelf. There are 76,000 elements (triangles) and 44,000 nodes (computation points). The mean separation of nodes is 7.8 km; the minimum and maximum are 1.1 and 80 km. Kliem and Greenberg (2003) used 20,000 elements and 12,000 nodes with 2.3–83 km separation

with the horizontal correlation scale inversely dependent on the density of hydrographic observations and directly proportional to the speed and orientation of calculated tidal flow. Fields of potential temperature and salinity for two seasons, summer and late winter, have been constructed from observational archives that span four decades. At present, sea ice appears only via a retarding effect on through-flow appropriate to the season; in other respects it is passive.

The tides are important to circulation within the Canadian Archipelago. They drive mixing and dissipation and control the boundary stresses (drag) in confined waterways. The properties of the tide vary with ice cover particularly near amphidromes where small changes in the amplitudes of incident and reflected waves can have a large impact on phase (Prinsenberg 1988; Prinsenberg and Bennett 1989). Dunphy et al. (2005) have computed a tidal mixing parameter based on modelled tides in the Archipelago. A map of this parameter reveals the regions of most intense tidal influence on mixing (Fig. 9.18), which match in some instances the locations of

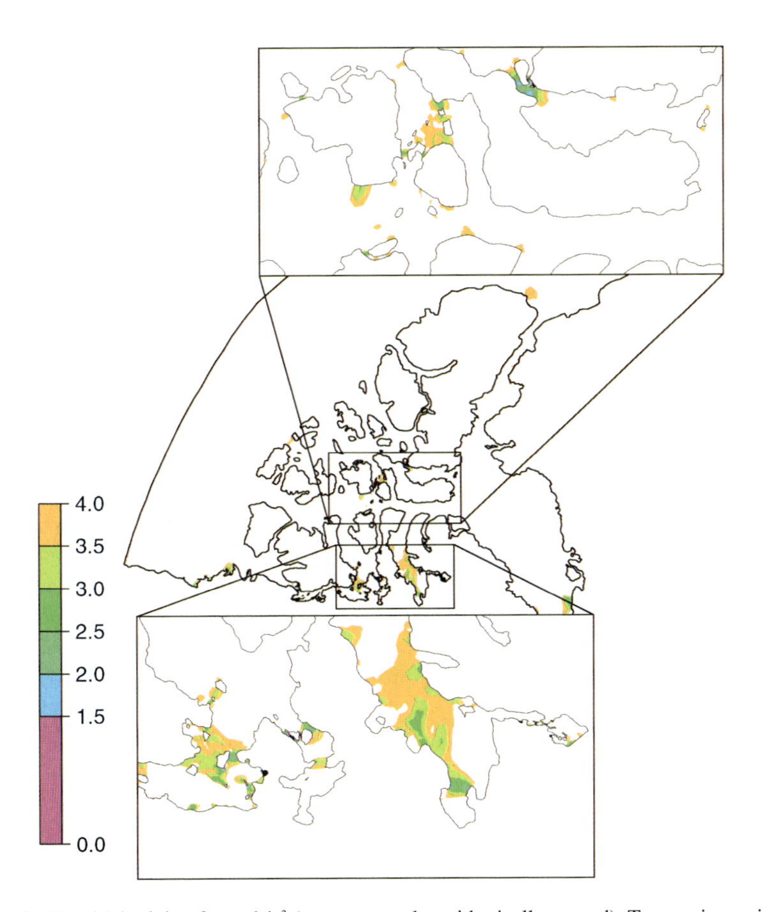

Fig. 9.18 The tidal mixing factor h/u^3 (contours are logarithmically spaced). Two regions with the smallest values (strongest mixing) are expanded for detail. The upper inset is centred on Hell Gate, Cardigan Strait, Queens Channel and Penny Strait. The lower is the region around the Boothia Peninsula (After Dunphy et al. 2005)

wintertime polynyas within the fast ice of the Archipelago. The dramatic inhomogeneity in tidal influence implies differences in water-mass evolution via mixing, heat loss and freeze-thaw cycling during through-flow depending on the path taken.

There are three runs involved in the simulation of the equilibrium through-flow. In the first, Fundy provides initial fields of sea-surface elevation and velocity from gridded fields of temperature and salinity. In the second, Quoddy is run diagnostically to incorporate tides and non-linear effects. The third run is prognostic. The simulations run for 10 days during each of the two observation-rich seasons, March–April and August–September. For sea-surface elevation along the inflow (viz. Arctic Ocean) boundary, average June–August values over 52 years have been used. These were computed using an updated version of the large-scale ocean model described by Holloway and Sou (2002). The value varies along the boundary and has an average value of about 0.1 m.

Not surprisingly, the diagnostic model reveals that the partition of through-flow among available pathways depends on the elevation difference between the Arctic Ocean and Baffin Bay and on baroclinic pressure gradients (viz. the distribution of temperature and salinity). For a representative 10-cm difference in sea level, the models yield a mean total through-flow of 0.9 Sv in summer (Kleim and Greenberg 2004). This value is smaller than numbers derived from observations (Melling 2000) but larger than the Steele et al. (1996) value derived from a simple ice–ocean model driven by observations of ice drift and concentration. Of the modelled total flux, 46% passes via Nares Strait, 20% via Cardigan Strait/Hell Gate and 34% via Lancaster Sound (Table 9.3). The model indicates that outflow via Lancaster Sound is supplied mostly from the Sverdrup Basin, with little contribution from Viscount Melville Sound and channels to the south (Fig. 9.19). This interesting outcome is consistent with observations reported by Fissel et al. (1988). The relationship between flux and sea-level difference is linear in the models (wherein hydrographic fields are fixed): a 5-cm increase in the sea level of the Arctic relative to Baffin Bay doubles the flux of volume.

The net volume flux reflects a balance between the barotropic pressure gradient, which drives water from the Arctic Ocean toward Baffin Bay, and the baroclinic pressure gradient which forces flow in the other direction. This is clear from Table 9.3 where the diagnostic result and that of a barotropic calculation using the same sea-surface elevation along the Arctic boundary are compared: the volume flux associated with barotropic forcing alone is five times larger. Clearly the baroclinic mode is an important aspect of circulation in the Canadian Archipelago. However, this result should not be viewed as an accurate measure of the relative contributions of the barotropic and baroclinic modes to flux. The ratio is suspect because it was derived using the diagnostic mode wherein the density field was specified (and of necessity grossly smoothed) and unresponsive to the circulation. Only a fully prognostic model can provide a realistic value for the ratio.

The diagnostic calculation has also provided values for the fresh-water flux. Values provided in Table 9.3 are subject to the cautions raised in the preceding paragraph. Typically, the model totals for all three routes of through-flow are approximately equal to values derived from observations in western Lancaster Sound alone (see Table 9.2), namely about 50 mSv (1,580 km^3/year) re 34.8.

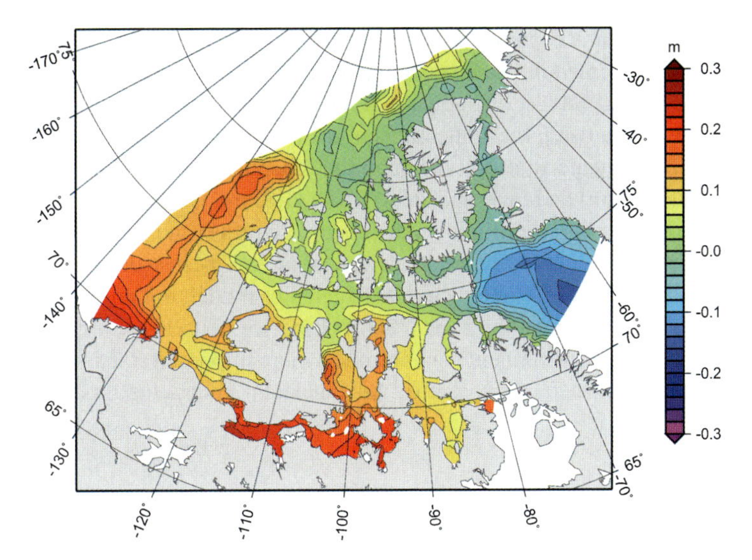

Fig. 9.19 Surface elevation as a proxy for transport, from a 3D non-linear diagnostic calculation. (Figure. 9.9 courtesy of Nicolai Kliem, DMI: http://ocean.dmi.dk/staff/nk/ArcticArchipelago/)

Table 9.3 Model-simulated fluxes of volume and fresh water through the Canadian Archipelago (after Kleim and Greenberg 2003). The reference salinity for fresh-water flux is 34.8. Arctic exports have positive value

	Diagnostic		Barotropic
	Volume (Sv)	Fresh-water (mSv)	Volume (Sv)
Barrow Strait	0.3	20	1.8
Jones Sound	0.2	10	0.6
Nares Strait	0.4	20	2.4
Total	0.9	50	4.8

There have been few modelling studies with spatial mesh sufficiently fine to represent baroclinicity adequately within the narrow channels of the Canadian Archipelago. The coupled ice–ocean model of the US Navy Postgraduate School, which has 1/12th degree resolution (about 9 km), has been used for pan-Arctic simulations of the period 1979–2002 (Williams et al. 2004). The results indicate that the Canadian Arctic through-flow is the greater contributor (relative to Denmark Strait) of oceanic fresh water to the North Atlantic. According to the simulation, the fresh-water flux through the Archipelago has increased over the period studied, a trend that has perhaps contributed to decreasing salinity in the Labrador Sea.

The goal of future work is a prognostic model with time evolving fields of temperature and salinity. This is not a trivial undertaking, particularly on a terrain-following mesh; methods with acceptable truncation error have been sought for many years. Such capability is essential for realistic simulation of baroclinic effects, including fresh-water and heat fluxes. An increase in resolution is also desirable, best accomplished for this area using the finite element method. The present best resolution is 1.1 km, barely adequate to represent important channels such a Hell Gate (4 km), Cardigan Strait (8 km) and Fury and Hecla Strait (1.8 km). Ocean circulation models need to be forced using wind fields that adequately reflect the important influence of topography and boundary-layer stratification on the mesoscale. Lastly, there is need for a realistic and fully interactive ice dynamics model; not only is pack ice an important element of ocean dynamics, but moving ice is itself a component of the fresh-water flux. The ice element may become a more important fraction fresh-water flux in a warmer climate, when ice of the Canadian Archipelago may be mobile longer each year (Melling 2002).

9.7 Ice Flux Across the Canadian Polar Shelf

Moving pack ice transports a fresh-water flux disproportionate to its thickness, by virtue of its low salinity (less than one tenth that of seawater) and of its position at the ice–atmosphere interface where it moves readily in response to wind. Both ice thickness and drift velocity are needed to calculate the sea-ice fresh-water flux. At present, ongoing observations of ice thickness are not available for any of the gateways discussed in this chapter. Here we concentrate on using satellite-based sensors to measure the movement of pack ice through the Canadian Archipelago. With supplementary guesses of pack-ice thickness, approximate values for the accompanying fresh-water flux can be provided.

The geography of the Canadian Archipelago is too complex for effective use of satellite-tracked drifters to measure the through-flow of pack ice. Methods based on the tracking of features in sequential images from satellite-borne sensors are better suited to the task. Microwave sensors provide the least interrupted time series of ice flux at key locations because they are relatively unaffected by cloud and wintertime darkness. However, the tracking of ice movement may be error-prone at times when ice features have poor contrast or when the pack is deforming appreciably as it moves; the latter is a common circumstance during rapid drift through narrow channels.

The displacement of sea ice over the interval between two images is derived by the method of maximum cross correlation (Agnew et al. 1997; Kwok et al. 1998). The technique works with sub-regions or patches on the two images that are 5–50 pixels on a side, depending on resolution. The underlying premise is that difference between consecutive images is the result of displacement only, the same for all features. Any additional rotation and straining of the ice field or creation of new ice features (e.g. leads) degrade the correlation.

Two long-term studies of ice movement through the Canadian Arctic have been completed. One used scenes acquired by synthetic-aperture radar at 0.2-km resolution (Radarsat: Kwok 2005; Kwok 2006) and the other utilized images from a passive microwave scanner, which resolves ice features at approximately 6-km resolution (89 GHz AMSR-E: Agnew et al. 2006). Both approaches yield estimates of ice displacement and ice concentration at intervals of 1–3 days, constrained by the interval between repeated orbital sub-tracks.

The utility of AMSR-E is marginal in some parts of the Archipelago where channels are only a few pixels wide. Moreover, the 89-GHz channel is of little value during the thaw season (July–August) when the wet surface of the ice and high atmospheric moisture degrade image contrast; data acquired during the shoulder-months of June and September may also be poor at times. Microwave radar produces images of better contrast than microwave scanners during the thaw season, but the identification of floes and ice features from Radarsat can still be challenging during summer.

The flux estimates derived from microwave-emission images only incorporate ice motion that occurred during the cold months (October–May or September–June). Since this period overlaps significantly with fast-ice conditions within the Canadian Archipelago, the months of most active ice movement may have been missed. The flux estimates derived from Radarsat nominally span the entire year. However, it is noted that feature-tracking algorithms return a null result (low correlation) when the quality of images is poor or ice-field deformation is large; this fact may contribute a low bias to average displacement during the summer, when image contrast is poor and low ice concentration permits rapid movement and deformation of the pack.

Radarsat transmits microwaves and detects the energy back-scattered from the rough surface or upper few centimetres of the ice; it is not sensitive to ice thickness. AMSR-E detects natural microwave emission at several frequencies and polarizations, which can be manipulated to yield information on ice type and concentration. In general, satellite-based data on ice movement must be augmented by ice-thickness values from other sources if the flux of ice volume and fresh-water are to be estimated.

Kwok et al. (1999) calculated an area budget for Arctic multi-year ice during 1996–1997 using observations made from space by microwave scatterometer (NSCAT). They estimated an annual outflow from Nares Strait of 34×10^3 km^2 by mapping multi-year ice in northern Baffin Bay, presumed to have arrived here via Smith Sound. Subsequently, Kwok (2005) has used Radarsat images over a 6-year period (1996–2002) to measure directly the drift of ice through a 30-km wide gate at the northern end of Robeson Channel (Fig. 9.20). During these years, the average annual flux of ice from the Lincoln Sea into Nares Strait was 33×10^3 km^2, with an inter-annual span of ±50%. There was a strong annual cycle in ice drift, with the bulk of the transport during August through January; ice is typically fast in Nares Strait between mid winter and late July. The average volume flux of an assumed 4-m thickness of ice would have been 130 km^3/year (4 mSv).

For the years 1997–1998 to 2001–2002, Kwok (2006) has estimated ice-area transport across the main entrances to Canadian Archipelago from the west (Fig. 9.20): Amundsen Gulf, M'Clure Strait, Ballantyne Strait plus Wilkins Strait plus Prince Gustaf Sea (cf. Queen Elizabeth Islands south) and Peary Channel plus Sverdrup Channel (Queen Elizabeth Islands north). His results are summarized in Table 9.4. On average during the 5-year study, Amundsen Gulf was a source of ice for the

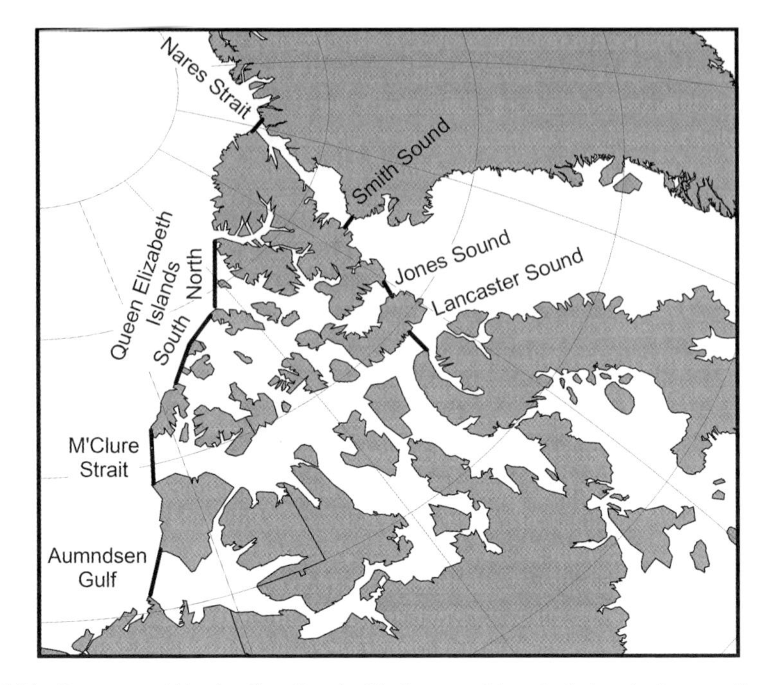

Fig. 9.20 Gateways within the Canadian Archipelago used in calculating the ice-area flux from sequential satellite images

Table 9.4 Annual average areal flux of ice between the Arctic Ocean and the Canadian polar shelf during the last decade. The unit is 1,000 km². Exports from the Arctic Ocean to the shelf have positive value

	Amundsen Gulf	M'Clure Strait	QEI south	QEI north	Nares Strait
1996–2002 (Sept.–Aug.)[a]	–	–	–	–	33 ± 13
1997–2002 (Sept.–Aug.)[b]	−85 ± 26	−20 ± 24	6 ± 5	2 ± 6	–
2002–2006 (Sept.–June)[c]	−14 ± 19	−5 ± 14	30 ± 8	6 ± 4	–

[a] Kwok (2005)

[b] Kwok et al. (2006)

[c] Agnew et al. (2006)

Arctic Ocean. Since the Gulf was ice-free during the summer, as typical, most of the export would be first-year ice leaving during autumn and winter. On assumption of 1-m average thickness (perhaps high because the gate traverses the Bathurst polynya), the average export would have been 85 km³/year. There was also an average export of ice from M'Clure Strait to the Beaufort Sea, although in smaller quantity and with occasional reversals (there was net import from the Beaufort in 2000). The average export would have been 80 km³/year, on assumption of 4-m average thickness

(McLaren et al. 1984). Only the entry points to the Sverdrup Basin accepted a net influx of ice to the Canadian polar shelf, but the amount was small (8×10^3 km^2/year or 7 km^3/year if ice was 3.4 m thick). This net influx is consistent with the analysis of Melling (2002), although its value is only about 20% of that implied by Melling's analysis.

The analysis has been extended to the cold months of 2002–2003 to 2005–2006 using AMSR-E (Agnew et al. 2006). The pattern of flux, with export from Amundsen Gulf and M'Clure Strait and import into the Sverdrup Basin, was continued during this period. However, the average out-fluxes from Amundsen Gulf and M'Clure Strait during this 4-year period (14 and 5×10^3 km^2/year) were smaller than during the preceding 5-year period (85 and 20×10^3 km^2/year) and the influxes to the Sverdrup Basin (30 and 6×10^3 km^2/year) were larger (6 and 2×10^3 km^2/year). There is obviously inter-decadal variability, as inferred by Melling (2002), which may respond to cycles in atmospheric circulation; it may also be that ingress of pack ice to the Sverdrup Basin was easier after the extensive loss of old ice within the Archipelago in 1998.

On the other side of the Canadian Archipelago, ice generally moves from the Canadian polar shelf into Baffin Bay. Agnew et al. (2006) have also used images acquired via AMSR-E to estimate ice flux into Baffin Bay during the colder months of 2002–2003 to 2005–2006: annual average fluxes were 48, 10 and 9×10^3 km^2/year via Lancaster, Jones and Smith Sound, respectively. The associated fluxes of volume were 49, 10 and 9 km^3/year per metre of ice thickness.

Agnew and Vandeweghe (2005) have calculated the ice flux during 2002–2004 through a gate across central Baffin Bay; the average over the 2-year interval was 690×10^3 km^2/year southward. Clearly the efflux of ice from the Canadian polar shelf during the last decade has been larger than the influx, implying that much of the ice exported to the Labrador Sea has been formed there and not in the Arctic Ocean itself. Moreover, the southward flux of ice through Baffin Bay actually exceeded that through Fram Strait over the same period in terms of area (590×10^3 km^2/year: Agnew and Vandeweghe 2005). However because the Fram Strait flux is primarily old ice and that the Baffin flux is primarily seasonal, the export of ice volume through Baffin Bay is probably the lesser.

Table 9.5 summarizes the ice flux values discussed here.

Table 9.5 Annual average areal flux of ice between the Canadian polar shelf and Baffin Bay during the last decade. The unit is 1,000 km^2. Arctic exports have positive value

	Lancaster Sound	Jones Sound	Smith Sound	Baffin Bay	Davis Strait
1996–2002 (Oct.–Apr.)[a]	–	–	34 ± 9	–	–
2002–2004 (Oct.–May)[b]	–	–	–	690 ± 80	610 ± 70
2002–2006 (Sept.–June)[c]	48 ± 6	10 ± 3	9 ± 2	–	–

[a] Kwok et al. (1999)

[b] Agnew and Vandeweghe (2005)

[c] Agnew et al. (2006)

9.8 Terrain-Channelled Wind and Oceanic Fluxes

The probable prime mover of the Pacific-Arctic through-flow is a decrease in sea level between the Pacific and the Atlantic. However, evidence for supplementary forcing of flows via internal gradients of pressure in the ocean and by winds has already been discussed. A strong channelling of airflow through Arctic straits, with consequent amplification of wind forcing on the ocean is a recent discovery. Its effect is discussed here in relation to Nares Strait, where it is possibly most influential. However, it is likely a factor at all gateways of interest to the Arctic fresh-water budget.

The flow of seawater through the Canadian Archipelago is variable but persistent. However, the flow of ice through the narrow waterways is strongly constrained by material stresses within the pack. In most channels, high ice concentration and low ice temperature during the cold season are sufficient to halt ice drift. One consequence is the cessation of fresh-water flux via moving ice. A second is the isolation of oceanic flows from stresses exerted by wind. A third is increased drag on the flow of water imposed by friction at the ice-water interface.

Pack ice in Nares Strait usually consolidates in winter behind an ice bridge at its southern end in Smith Sound (Agnew 1998). Consolidation can occur any time between November and April, and may occur in stages, with bridges forming consecutively in Robeson Channel (northern end), Kennedy Channel (middle section) and Smith Sound, perhaps to collapse a few weeks later or perhaps to remain as late as August. Such variability suggests that the fast-ice regime of Nares Strait is of marginal stability in the present climate, flitting between the permanent mobility typical of Fram Strait and the reliably static winter ice of the western Archipelago.

It is plausible that topographically amplified winds in Nares Strait contribute to the intermittent instability of fast ice in the channel. However because there are no systematic long-term observations of wind in the area, present insights have been derived via numerical simulation (Samelson et al. 2006) using the Polar MM5 mesoscale atmospheric model (Bromwich et al. 2001). This is a version of the Pennsylvania State/NCAR MM5 (non-hydrostatic, primitive-equation, terrain-following, full moist physics) which has been optimized for the polar environment (Cassano et al. 2001; Guo et al. 2003). The configuration is triply nested, from 54- to 18-km to 6-km grids. It has been run daily at Oregon State University since August 2003 in a 36-h forecast mode, with initial and time-dependent boundary conditions taken from the operational AVN model of the US National Center for Environmental Prediction.

Strong radiational cooling at the surface in polar regions commonly creates a stable planetary boundary layer in winter (Bradley et al. 1992; Kahl et al. 1992), wherein wind may be strongly channelled through areas of low terrain. The mesoscale model commonly generates an intense boundary-layer jet at elevation below that of the confining terrain. Moreover, the along-channel wind speed is well correlated with the difference in sea-level pressure along Nares Strait (Samelson et al. 2006). The along-channel balance indicates that the atmospheric jet is an ageostrophic response to orography. The drop in sea-level pressure along the 550-km long strait can exceed 25 mb, with simulated winds reaching 40 m/s at 300-m elevation.

Mesoscale processes are clearly influential in accelerating airflow through Nares Strait: high terrain on both sides, the unusual length of the channel and its narrow width isolate air flow from the synoptic-scale geostrophic constraint; the strong ageostrophic response to pressure gradient is only weakly damped by momentum transfer through the stable boundary layer and is locally amplified by effects of varying channel width. Moreover regional synoptic climatology is a contributing factor because Nares Strait is a short-cut between two different synoptic regimes, the Polar high and the Icelandic low. Figure 9.21, depicting the regional variation in sea-level pressure from the MM5, clearly reveals both synoptic-scale and mesoscale factors: the large difference in pressure between the Lincoln Sea and Baffin Bay and the two zones of steep pressure gradient and strong along-channel wind, in Kennedy Channel and in northern Baffin Bay. The probable along-channel force balance involves the pressure gradient, inertia and friction while the cross-channel balance is geostrophic (on the mesoscale). Boundary stress likely fades to insignificance above a few hundred meters, leaving an inviscid balance in the upper part of the jet.

The dynamical explanation for the wind maxima at two locations, where Kennedy Channel widens into Kane Basin and again where Smith Sound widens into Baffin Bay may be super-critical flow. This phenomenon is known to create similar expansion fans in summer in the lee of capes on the US west coast (Winant et al. 1988; Samelson and Lentz 1994). Pressure gradients develop as the inversion-capped

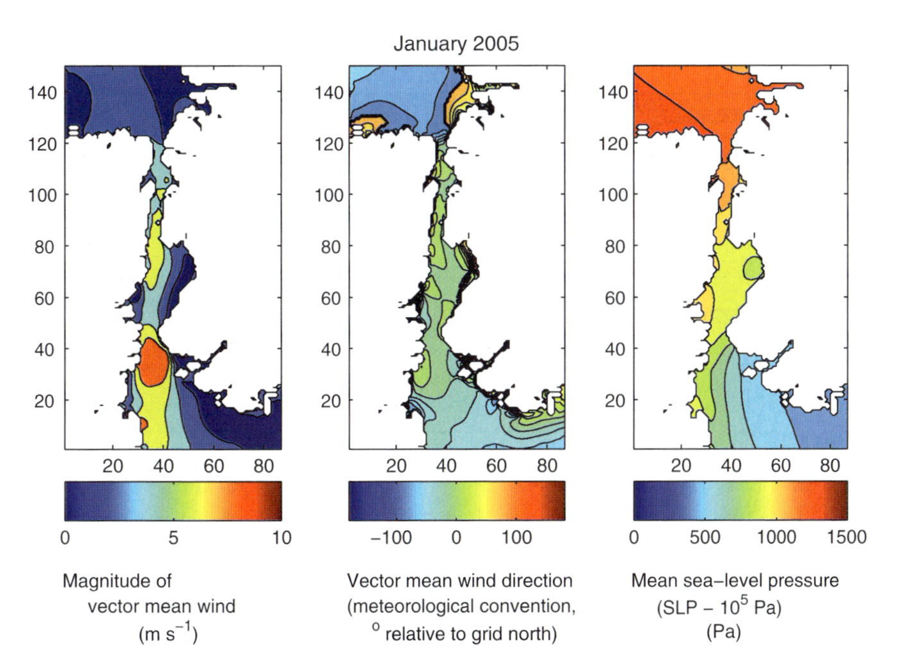

Fig. 9.21 Average fields of vector wind and sea-level pressure for January 2005. These data from simulations using the MM5 clearly reveal meteorological features on both synoptic and meso scales

marine boundary layer thins where the channel widens; these gradients in turn force ageostrophic acceleration.

Empirical orthogonal functions computed from monthly averages of simulated airflow and surface stress over a 2-year period (Fig. 9.22) show that the time-dependent flow has a spatial structure very similar to that of the mean flow, shown for January 2005 in Fig. 9.21. The annual cycle was energetic during this particular period: the average airflow alternated between strongly southward during October through January and northward in July and August.

Variance in the synoptic band of frequency was suppressed by the monthly averaging applied in the preparation of Fig. 9.22. Nonetheless, this band is very energetic in Nares Strait. Figure 9.23 displays the along-channel surface wind for a 1-year period. Values have been derived from the along-channel difference in sea-level pressure (Carey Islands minus Alert) using the regression line calculated by Samelson et al. (2006), but comparable fluctuations are apparent in simulated winds.

Simulations of mesoscale atmospheric flow within the Canadian Archipelago have been focussed to date on Nares Strait. However, it is likely that each of the six constrictions to through-flow in the oceanic domain – Nares Strait, Hell Gate, Cardigan Strait, Lancaster Sound, Bering Strait and Davis Strait – have some impact on the speed and direction of winds. The intensity of mesoscale influence likely differs within the group, since the straits encompass a wide range of dimensions in terms of height of terrain (200–2,000 m), width of strait (8–350 km), length of strait

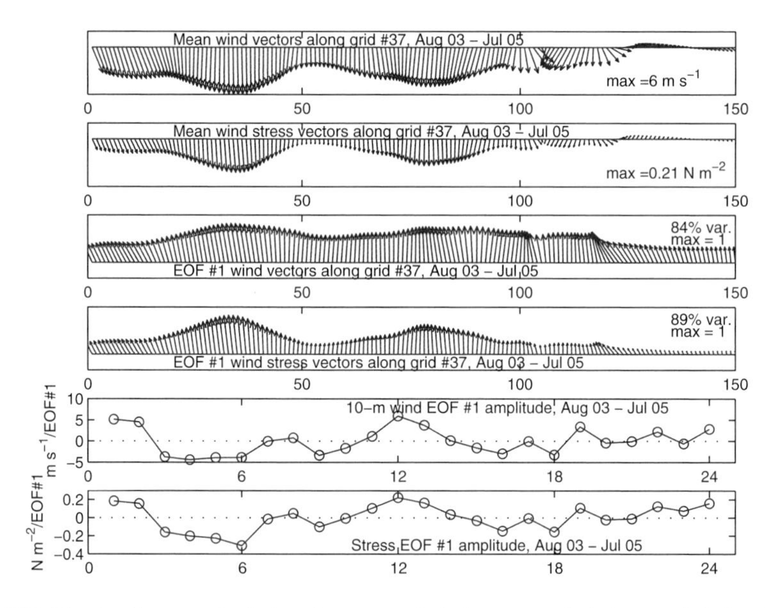

Fig. 9.22 The top four panels display the mean wind, the mean wind stress and their primary empirical orthogonal functions. The horizontal coordinate (grid node) increases along a line running up Nares Strait from Baffin Bay to the Lincoln Sea; node 50 and 80 are in Smith sound and southern Kennedy Channel, respectively. The bottom two panels display the eigenvalues plotted against month, for 2 years beginning in August 2003

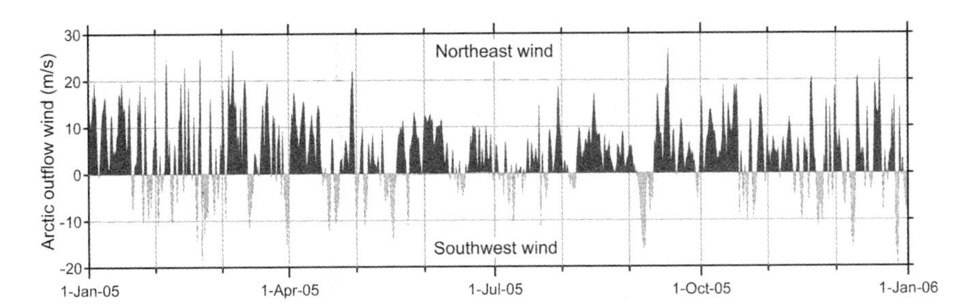

Fig. 9.23 Year-long series of along-channel surface wind in Nares Strait, calculated using the linear dependence of wind on the along-channel difference in sea-level pressure established by Samelson et al. (2006). Pressure was measured at Alert and on the Carey Islands

(0–550 km) and latitude. The latter may influence boundary-layer stability through its direct and indirect effects on insolation, surface albedo and surface emissivity. Based on our presently incomplete understanding of these effects within Nares Strait, we rank the straits in the following sequence of decreasing sensitivity to wind amplification on the mesoscale: Nares Strait, Cardigan Strait/Hell Gate, Lancaster Sound, Davis Strait, Bering Strait.

9.9 Geochemical Identification of Sources for Canadian Arctic Outflow

Knowledge of the magnitude and causes of fresh-water flux through the North American Arctic is the primary objective of the present study. However, knowledge of the sources of the fresh water is essential to understanding the roles of the fresh-water flux in the global hydrologic cycle and climate system. The trace geochemical signatures of seawater can provide clues about the sources, transit times and history of the through-flow.

Although the primary indicator of fresh water in the ocean is salinity, a number of trace chemical constituents can provide insight into fresh-water origin and transport within the Arctic. Recent studies that illustrate the application of chemical tracers to Arctic fresh-water issues have been published by Cooper et al. (1997), Jones et al. (1998), Smith et al. (1999), Schlosser et al. (2000), Ekwurzel et al. (2001), Amon et al. (2003), Jones et al. (2003), Taylor et al. (2003), Alkire et al. (2006), Falkner et al. (2006), Yamamoto-Kawai et al. (2005), Yamamoto-Kawai et al. (2006) and Jones and Anderson (2007). Exploited dissolved trace chemicals include nutrients, molecular oxygen, alkalinity, chlorofluorocarbons, natural and artificial radionuclides, barium and other trace metals, organic matter and heavy isotopes ^{18}O and ^{2}H in water molecules.

The interpretation of the first exploratory sampling of tracers in Arctic waters was constrained by poor geographic coverage. Data from several expeditions, perhaps spanning several years, were typically aggregated or averaged to draw maps of tracer distributions. Interpretation was necessarily based on the assumption of steady ocean circulation. Increased effort in data collection over the last decade has permitted a more rewarding focus on temporal variability. Here we discuss new knowledge emerging from tracer hydrography in the western hemisphere of the Arctic, with particular attention to temporal variability in the relative contributions from various sources of fresh water. In future years, a significantly improved understanding should emerge from the time series of strictly comparable data that are now being produced.

The interpretation of oceanographic tracers in the North American Arctic presents special challenges. For example, the wide range in surface conditions from year-round ice to seasonal ice zones, from fast ice to ice-free seas may render inappropriate simple assumptions applied elsewhere regarding the impacts of biology and ventilation on tracer concentrations. Interpretation of tracer distribution can be ambiguous. Baffin Bay for example receives Arctic waters via two paths, from the north via the Canadian Archipelago and from the south via the West Greenland Current. Moreover, fresh water with large and variable $\delta^{18}O$ anomalies from melting ice sheets in Greenland and northern Canada (which also contribute glacial flour) increases the complexity of geochemical interpretation.

Dissolved nutrients and oxygen have the longest history among all chemical tracers used in ocean science, in the Arctic as in temperate waters. A relatively high concentration of silicic acid ([Si] $\geq$ 15 mmol m^{-3}) has long been known to distinguish waters that enter the Arctic from the Pacific via Bering Strait; this influx can be traced as a relative maximum in dissolved silica concentration (coincident with a maximum in dissolved phosphorus [P] and coupled with a minimum in dissolved oxygen [O_2]) in the halocline (Kinney et al. 1970; Codispoti and Lowman 1973; Jones and Anderson 1990). Recent interpretation that additionally utilizes $\delta^{18}O$ has revealed that the dissolved nutrient and oxygen in the Arctic halocline result primarily from the Bering Strait inflow in winter (Cooper et al. 1997, 2006). In the sunlit half of the year, biological cycles of growth and decay change the concentrations of dissolved nutrients and oxygen. Biological impact on the Pacific inflow is further amplified in summer when the inflow is less saline (therefore closer to the surface) and free of light-obstructing ice cover. Tracing the movement of Bering Sea water that enters the Arctic during spring, summer and autumn demands ingenuity in geochemical interpretation.

Within the Canadian Archipelago, the earliest reliable cross-sections of dissolved silica were observed in the summer of 1977. The concentration was highest in Lancaster Sound, intermediate in Jones Sound and lowest in Smith Sound. This gradation was taken to indicate that water (and fresh water) from the Pacific was unlikely to reach the Lincoln Sea and contribute to the flow through Nares Strait (Jones and Coote 1980).

This conclusion was revised when the co-variation of dissolved nitrate and phosphate was developed as a discriminant of Pacific from Atlantic-derived waters within the Arctic (Jones et al. 1998; Alkire et al. 2006; Yamamoto-Kawai et al. 2006). De-nitrification of inflowing seawater occurs over the shallow shelves of the

Chukchi and northern Bering Seas. This process renders the Pacific inflow deficient in fixed inorganic nitrogen relative to Atlantic water. Within the Arctic Ocean, biological action tends to move the nitrate and phosphate concentrations within each contributing water mass (Pacific and Atlantic) along lines of constant "Redfield-like" slope on a nitrate-versus-phosphate diagram. Because mixtures of Pacific and Atlantic waters have concentrations of these two constituents that fall between the source-water reference lines; the position between the reference lines on the diagram indicates the fractions of Pacific and Atlantic water in the mixture.

Complications arise with the contribution of water from other sources, such as rivers and melting ice. To a first approximation, however, studies of $\delta^{18}O$ reveal that these interfering contributions are generally less than 10% within the Arctic Ocean (Östlund and Hut 1984) and that rivers provide nutrients in proportions resembling those characteristic of Atlantic water (Jones et al. 1998). The nitrate–phosphate (N–P) method for discriminating Pacific from Atlantic waters has recently been refined to include the contribution of ammonium and nitrite to the fixed inorganic nitrogen. The quality of the analysis has improved because ammonium is an appreciably component of the nitrogen dissolved in Pacific-derived seawater (Yamamoto-Kawai et al. 2006).

Jones et al. (2003) have applied the N-P method to track the Pacific influence in Arctic through-flow within the Canadian Arctic and the Labrador Sea and through Fram Strait to Denmark Strait. From sections measured in August 1997, they concluded that Pacific inflow completely dominated the seawater end member in Barrow Strait and provided at least three quarters of this end member in the topmost 100 m of Jones and Smith Sounds; Pacific water was similarly prevalent that year within 100 km of Baffin Island in Davis Strait. It was detected in diluted (50%) form with somewhat variable extent over the Labrador shelf in 1993, 1995 and 1998 and as far south as the Grand Banks in 1995.

The magnitude of Pacific influence in waters south of Davis Strait may be an over-estimate because de-nitrification likely occurs also in the relatively shallow waters of Hudson Bay. N*[1] is a nutrient-based parameter that has negative value in de-nitrified water. In the North Atlantic, values of N* are near zero or positive. Values of N* are negative for Pacific waters passing through Bering Strait and about −12 μM/kg for water in Barrow Strait (Falkner et al. 2006). Unpublished data from Hudson Bay in the summer of 1982 (Bedford Institute of Oceanography, DFO Canada) reveal water with N* even more negative (−23 to −12 μM/kg). Because the deepest waters of Hudson Bay (S ~ 33.5) are replaced on a time scale of about a decade (Roff and Legendre 1986), Atlantic water supplied to Hudson Bay via Hudson Strait could be de-nitrified to a Pacific-like signature before re-emergence into the Labrador Sea. Such occurrence would complicate the N–P interpretation wherever there is influence from Hudson Bay.

---

[1] N* (μM/kg) = $[NO_3] - 16 \cdot [PO_4] + 2.90$. N* was first defined by Gruber and Sarmiento (1997) and modified by Deutsch et al. (2001).

A recent analysis that combines measurements of tracer concentration and current velocity on several sections across Nares Strait is a valuable innovation (Falkner et al. 2006). Current measurements by ship-mounted ADCP in August 2003 delineated a southward flowing jet at 100–200 m depth along the western side of Nares Strait; simultaneous hydrographic sections revealed an enrichment of silica and phosphorus in this jet that was indicative of origin as wintertime Arctic inflow through Bering Strait. N–P analysis has shown that the only seawater end-member in the upper 100 m at the northern end of Nares Strait was Pacific water, but that Pacific water shared equal status with Atlantic water in the marine component at the southern end (Smith Sound). Clearly, there had been appreciable mixing between the south-flowing Pacific water and north-flowing Atlantic water (from the West Greenland Current) during transit. Pacific influence in the mixture at this section in 2003 was appreciably more dilute than in 1987. Subsequent comparison of nutrient measurements in August of various years has revealed inter-annual variability (Fig. 9.24), comparable for silica to that proposed as seasonal in the interpretation of a 10-month series from the North Water in 1997–1998 (Tremblay et al. 2002).

Figure 9.24 displays nutrient concentration for various years in Canadian Arctic straits with inflows from both the Arctic Ocean and Baffin Bay. The

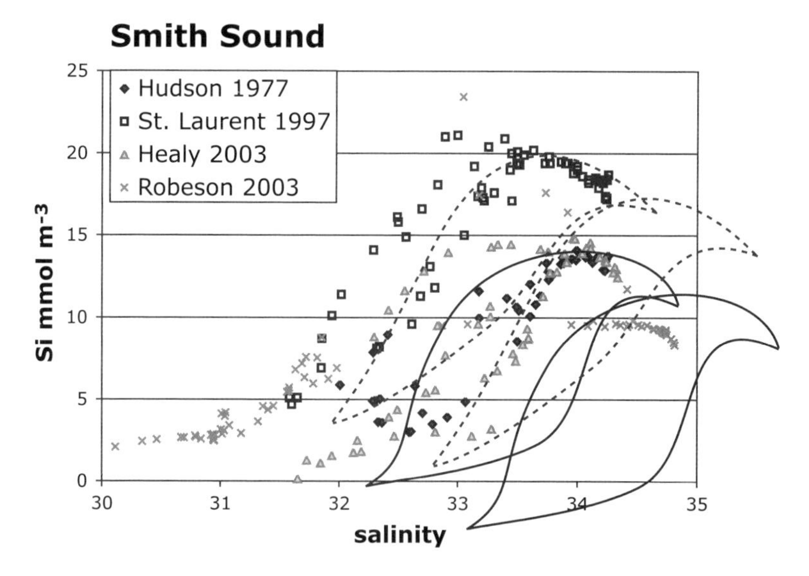

Fig. 9.24 Silicate versus salinity for seawater samples acquired in Smith Sound during August of several years. Added curves envelope data from 1977 and 2003 (solid lines) and from 1997 (dashed line). Within both envelopes, the concentration is highest on the western side of the straits. Note the high silica concentration (strong Pacific influence) in Robeson Channel in 2003, values comparable to those measured 600 km to the south in Smith Sound in 1997 (Falkner et al. 2006)

envelopes that enclose these data are shifted toward higher nutrient concentration (viz. greater Arctic influence) in August 1997 than in 1977 and 2003 (Falkner et al. 2006). The simplest interpretation is that the flux of nutrient rich Pacific water (plus meteoric and ice-melt waters mixed with it) was higher in all the straits in August 1997 than in 1977 and 2003. The higher flux occurred just after the prolonged positive anomaly in the Arctic Oscillation Index (AO) during 1989–1995.

What is the mechanism via which the AO, which is an expression atmospheric pressure distribution over the northern hemisphere in winter, might influence oceanic circulation and fluxes in summer? Proshutinsky and Johnson (1997) have used a barotropic ocean model to demonstrate that the Arctic Ocean responds to the AO in a basin-wide oscillation with cyclonic and anti-cyclonic anomalies: higher peripheral sea level results from the set-up of low salinity water against the ocean boundary under high AO forcing. A subsequent study based on a more realistic ice-ocean model was forced by NCEP-reanalysis winds for 1951–2002 (Häkkinen and Proshutinsky 2004). Various measures of the Arctic Ocean Oscillation, including sea surface height, all co-varied with the Arctic (Atmosphere) Oscillation. During the years of unusually high AO, 1989–1996, the model indicated a sustained loss of fresh water from the Arctic Ocean, which had by 1997 created the most negative fresh-water anomaly of the entire 50-year simulation. Although Häkkinen and Proshutinsky (2004) do not comment on whether the exported fresh water passed to the east or to the west of Greenland, the timing meshes with the inference of Falkner et al. (2006) based on geochemical analysis of Canadian Arctic through-flow in 1997. Interestingly, the inflow of Atlantic water was an essential element in the wind-driven barotropic response to the AO; it was the factor most strongly correlated with fresh-water anomalies within the basin.

Additional trace compounds can be used to distinguish the meteoric (river inflow plus precipitation) and ice-melt components of Canadian Arctic through-flow. For example, Jones and Anderson (this volume) discuss the use of seawater alkalinity for this purpose. Östlund and Hut (1984) pioneered the mass-balance analysis of the seawater isotopic composition in the Arctic to distinguish run-off from ice melt-water as freshening agents. Within the Canadian Archipelago and east of Greenland, a more complicated analysis may be required. As discussed by Strain and Tan (1993), the separation of salt and water by the freeze-thaw process can, in combination with mixing under conditions prevalent in Baffin Bay, generate a seasonal cycle in the $\delta^{18}O$ value for the zero-salinity end-member. The $\delta^{18}O$ values can vary from that typical of summertime precipitation ($\delta^{18}O \approx -10$) to that typical of glacial melt-water ($\delta^{18}O \leq \approx 25$). In the big picture, direct contributions of fresh-water via precipitation and ablation of ice sheets are small relative to those via Arctic rivers ($\delta^{18}O \approx -20$) and Pacific inflow ($\delta^{18}O \approx -1$). However, they may be important in the principal Arctic fresh-water outflows because of proximity to the ice sheets of Greenland and of the Canadian Arctic Cordillera. An ideal analysis would be expanded to incorporate additional tracers and contextual information so that artefacts can be identified.

9.10 Gateway to the Atlantic, Davis Strait

All streams of Arctic water that cross the Canadian polar shelf enter Baffin Bay, with the exception of about 0.1 Sv that is diverted along the western side of Baffin Island via Fury and Hecla Strait (Barber 1965; Sadler 1982). These streams join the cyclonic circulation of the West Greenland Current (itself fed by Arctic outflow via Fram Strait) to form the Baffin Current, which follows the continental slope of Baffin Island and enters the Labrador Sea through Davis Strait.

The properties of Arctic seawater and ice are modified by freezing, thawing, terrestrial and glacial run-off and mixing during their transit across the Canadian polar shelf (more than 1 million square kilometers of ocean area) and through Baffin Bay (an additional 2/3 million square kilometers) to Davis Strait. Although the residence times for through-flowing water and ice are not known, they are likely significantly longer than the most rapid transit (by ice from the Lincoln Sea to Davis Strait), which requires about a year. Ultimately, it is this modified Arctic water mass that affects deep water formation in the Labrador Sea. Davis Strait is a suitable location to measure the sum of all Arctic outflows via routes west of Greenland at a single section just prior to their entry into the convective gyres of the north-west Atlantic.

The operation of a moored array to measure volume and fresh-water fluxes through Davis Strait is not a trivial undertaking. The narrowest part of the strait is 330 km, with 200 km of this span deeper than 500 m; the maximum depth is close to 1,000 m at the narrowest point, but shoals to 700 m at the sill. There is a topographic spur that extends along the axis of the Strait and likely influences flow near the sill. Relative to the internal Rossby scale (here about 25 km), the Strait is dynamically wide, admitting small eddies and recirculation that must be resolved to obtain accurate estimates of fluxes. The upper few hundred metres, particularly on the Canadian side, are swept by a broad stream of icebergs moving south with the current; this is a big risk to instrumented sub-sea moorings within the Arctic outflow. There is a strong counter-flow (the West Greenland Current) on the eastern side of the Strait, with a front, eddies and re-circulation features in the region where the two currents interact over the broad flat sill.

The water masses and circulation within Davis Strait during the ice-free season have been mapped using hydrographic surveys and satellite-based temperature scanners. The 500-m isobath on Fig. 9.25 reveals the broad extent of the continental shelf (150 km on the Greenland side) and the coloured underlay depicts the mean sea-surface temperature in September. Water warmer than 4 °C (fresher than 33) in the Labrador Sea and over the Greenland shelf is carried northward by the West Greenland Current; it is a mixture of Atlantic water and outflow from Fram Strait. Most of this stream turns west and then south following isobaths into the northern Labrador Sea; some continues northward along the Greenland shelf. Vectors in Fig. 9.25 represent depth-averages of measured current; they confirm the inference from sea-surface temperature of a northward flow on the Greenland side and a southward flow of Arctic water on the Canadian side.

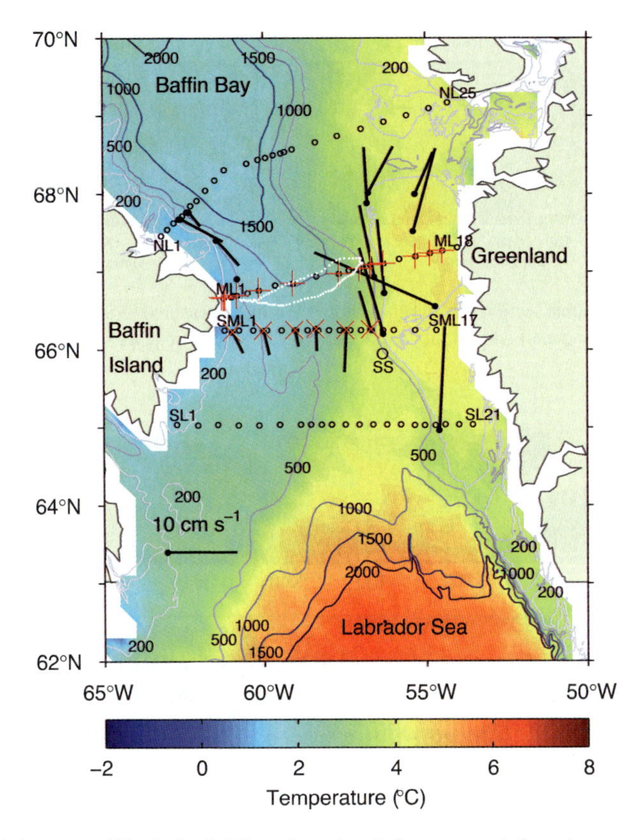

Fig. 9.25 Bathymetry of Davis Strait. The coloured underlay represents long-term mean sea-surface temperature. Red 'x' mark the positions of the 1987–1990 moorings (Ross 1992); Red '+' mark the locations of moorings placed in September 2004 in the new initiative to measure oceanic fluxes. Open circles mark recent hydrographic surveys. The white dotted line is the 2006 Seaglider track. Vectors depict depth-averaged current from instruments on moorings in the 1980s

There are three principal water masses in Davis Strait (Tang et al. 2004): Arctic Water, West Greenland Intermediate Water originating in the Atlantic and Baffin Bay Deep Water (below 800 m). Figure 9.26 displays their distribution across a hydrographic section measured at 25-km resolution in September 2004. Here West Greenland Intermediate Water is warmer than 2 °C, more saline than 34.5 and extends from the Greenland slope into mid-strait below 50-m depth; a smaller core of this water over the Baffin slope is likely a recently separated filament that is returning southward. Arctic water colder than 0 °C and fresher than 33.5 fills the upper 250 m of the western half of Davis Strait; here the salinity anomaly (referenced to 34.8) is quite large in a thin layer within 50 m of the surface. Both this thin layer and the sharp front that separates Arctic from Atlantic-derived water present significant challenges to the measurement of the flow and salinity structure needed to calculate fresh-water flux.

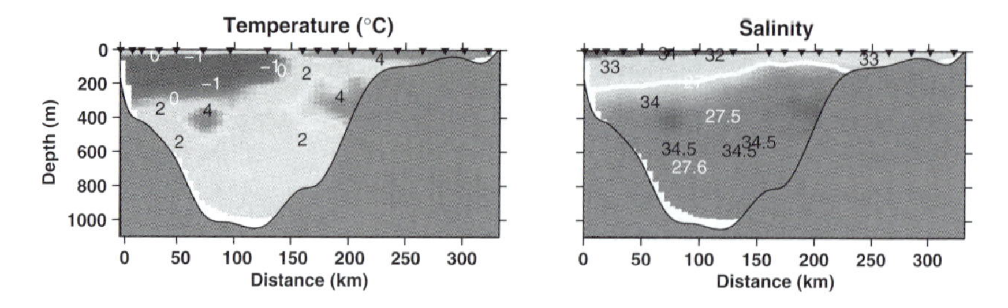

Fig. 9.26 Hydrographic section across Davis Strait (ML line) measured in September 2004, showing temperature (°C, left panel) and salinity (right panel). The station spacing was about 25 km

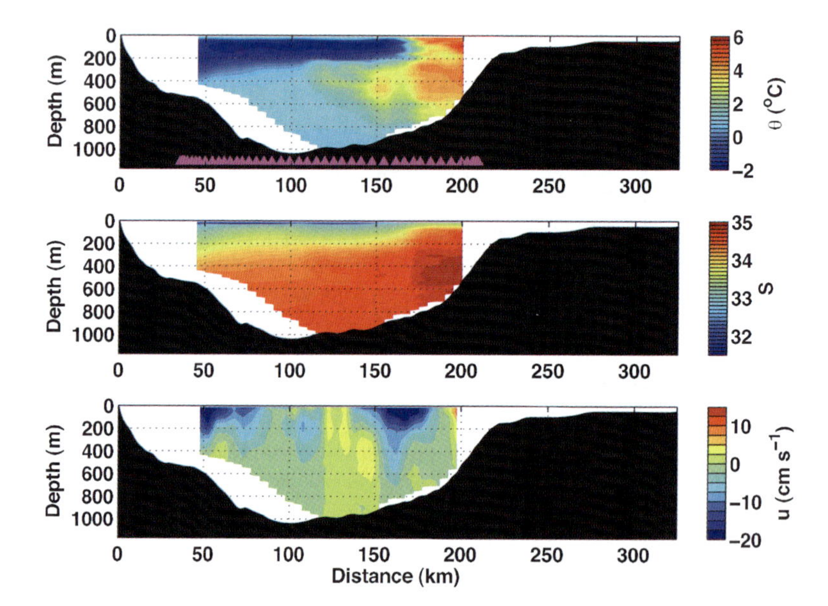

Fig. 9.27 Hydrographic structure within the deep trough of Davis Strait measured by Sea Glider at approximately 5-km resolution in September 2006

Hydrographic sections measured recently at much higher resolution (5 km) by Seaglider (see Chapter 25) illustrate the challenge posed by meso-scale structure within Davis Strait (Fig. 9.27). Even at this fine station spacing, there is plentiful detail in temperature, salinity and geostrophic shear at the limit of resolution; average values of current for the upper 1,000 m (estimated from glider navigation) uncover analogous variation in flow. The high-resolution section also reveals the large fresh-water anomaly of the thin surface layer. The rapid movement of this surface layer represents a substantial fraction of the fresh-water flux in Arctic waters (Melling 2000). For example, use of the salinity at 100-m depth (the shallowest measurement from the present array) to represent the salinity above this level over a 150-km span with a 0.1 m/s current results in an under-estimate of fresh-water flux by 23–32 mSv, depending on the month.

During 1987–1990, Fisheries and Oceans Canada maintained an array of conventional current meters (current, temperature, salinity at 150, 300 and 500 m) on five moorings along the 66.25° N (Ross 1992). The array spanned the deep central trough at roughly 50-km spacing. Because instruments were not placed shallower than 150 m, where iceberg risk is high, the array did not cross the shelves or sample the low-salinity Arctic outflow. In addition, Tang et al. (2004) and Cuny et al. (2005) report low correlations between time series from instruments on different moorings, indicating that the array failed to resolve flows at the scale of variability within the Strait. Because of these shortcomings, the data were ill-suited to flux estimation. Nonetheless, Cuny et al. (2005) calculated fluxes on assumption that: (1) temperature and the salinity were constant above 150 m when the sea was ice-covered; (2) seasonally appropriate recent or archived data provided valid vertical gradients above 150 m during ice-free months; (3) upper ocean profiles could be estimated by shifting climatological data to match daily values observed at 150 m; (4) measured daily current speed at 150 m provided known motion at a reference level for calculated geostrophic current; and (5) values varied linearly between moorings. Fluxes over adjacent continental shelves were ignored. Tang et al. (2004) have used the same data under slightly different assumptions; principally they substituted climatological values for salinity gradient in the upper ocean year-round.

The upper part of Table 9.6 summarizes volume and fresh-water flux estimates out of the Arctic derived from these older data. Values based on the long-term, but under-resolved, direct observations average about 3.1 Sv and 125 mSv (3,940 km^3/year), respectively (Loder et al. 1998; Tang et al. 2004; Cuny et al. 2005). Fluxes passing along the Greenland shelf have not been included; Cuny et al. (2005) estimate these as −0.8 Sv and −38 mSv (1,200 km^3/year), so that their corrected net fluxes for the entire Strait are 2.3 Sv and 87 mSv (2,750 km^3/year). The hydrographic surveys in September provide better horizontal resolution and include the shelves but because they are snapshots, the derived flux estimates are more variable year to year, ranging over 1.5–5.7 Sv and 126–286 mSv (3,980–9,020 km^3/year). It is plausible (and consistent with some observations within the Archipelago) that the fluxes might actually be larger in September than in annual average.

The lower part of Table 9.6 summarizes preliminary flux estimates derived from the ongoing USA–Canada Fresh-water Initiative that is acquiring data from a new and larger moored array and from hydrographic surveys by ship and Seaglider. The present summary is preliminary, derived from an independent consideration of each source of data. Geostrophic calculations referenced to zero at the seabed and averaged over four sections yield volume fluxes of 1.8 ± 1.5 Sv and 2.3 ± 0.9 Sv for September 2004 and 2005. A simplistic estimate based on data from moored instruments during 2004–2005 (including some measurements over the shelves) is an annual mean value of 2 Sv, with large uncertainty. Preliminary estimates of fresh-water flux from the ship CTD survey in September 2004 and the September/October 2005 transects by

Table 9.6 Estimates of the net fluxes of volume and fresh-water through Davis Strait. Those for 2004–2006 in the lower part of the table are preliminary. Arctic exports have positive value

Method	Data source	Location	Includes shelves?	Year	Timing	Volume flux (Sv)	Fresh-water flux (mSv)
Geostrophy[a]	CTD section	66.25° N	–	1987	September	5.7	195
Geostrophy[a]	CTD section	66.25° N	–	1988	September	1.5	126
Geostrophy[a]	CTD section	66.25° N	–	1989	September	5.7	286
Currents and geostrophy[a]	Current meters	66.25° N	–	1987–1990	3-Year mean	2.6	92
Currents and geostrophy[a]	Current meters	66.25° N	No	1987–1990	3-Year mean	3.4	130
Currents and geostrophy[b]	Current meters	66.25° N	No	1987–1990	3-Year mean	2.6	99
Currents and geostrophy[c]	Current meters	66.25° N	No	1987–1990	3-Year mean	3.3	120
Currents and geostrophy[d]	Current meters	66.25° N	No	1987–1990	3-Year mean	3.1	
1/12° simulation[e]	Ocean model			1979–2001	21-Year mean		76
Geostrophy[f]	CTD section	Northern line	–	2004	September	2.5	130
Geostrophy[f]	CTD section	Mooring line	–	2004	September	3.1	110
Geostrophy[f]	CTD section	66.25° N	–	2004	September	2.0	98
Geostrophy[f]	CTD section	Southern line	–	2004	September	–0.3	34
Currents[f]	ADCP	Mooring line	–	2004–05	1-Year mean	2.0	–
Geostrophy[f]	CTD section	Northern line	–	2005	September	2.8	–
Geostrophy[f]	CTD section	Mooring line	–	2005	September	2.8	–
Geostrophy[f]	CTD section	66.25° N	–	2005	September	2.5	–
Geostrophy[f]	CTD section	Southern line	–	2005	September	0.9	–
Geostrophy[f]	SeaGlider CTD	Mooring line	No	2006	September	–	72
Geostrophy[f]	SeaGlider CTD	Mooring line	No	2006	September	–	102
Geostrophy[f]	SeaGlider CTD	Mooring line	No	2006	September	–	115

[a]Cuny et al. (2005)

[b]Tang et al. (2004)

[c]Loder et al. (1998)

[d]Ross (1992)

[e]Maslowski et al. (2003)

[f]APL-UW, unpublished data

Seaglider are 93 ± 40 mSv (2,930 km^3/year), including shelves and 96 ± 20 mSv (3,030 km^3/year), excluding shelves. These values are smaller than those estimated from data in the late 1980s, but the magnitude of error is unknown and likely large.

The fluxes listed are net values. The West Greenland Current has a northward flow of approximately 2 Sv in Davis Strait, and relative to 34.8 reference salinity it carries fresh-water northward at roughly 60 mSv (1,890 km^3/year) (Cuny et al. 2005). Therefore, based on Cuny's numbers, the fluxes southward within the Baffin Current are 4.6 Sv and 150 mSv (4,730 km^3/year).

Narrow buoyancy-driven flows may carry appreciable fresh-water during summer within 10 km of the Greenland and Baffin coasts. For example, a coastal current fed by ice-sheet run-off along southeast Greenland apparently transports volume and fresh-water at 1 Sv and 60 mSv (1,890 km^3/year) during the thaw season (Bacon et al. 2002). Components of the present observational array may detect such currents but will not likely resolve their extent and rate of transport.

The USA–Canada Fresh-water Initiative is addressing the principal challenges to accurate measurement of volume and fresh-water fluxes through Davis Strait. Among these are: (1) a small baroclinic deformation scale that permits decorrelation of flow variations on a scale of order 10 km; (2) a pronounced concentration of fresh-water flux in a thin (25 m) fast-moving surface layer where current and salinity are difficult to measure; (3) the risk to moorings from moving ice keels and icebergs at depths as great as 200 m; and (4) the fresh-water flux carried by pack ice. The initiative has brought new technology to bear on these challenges.

Instruments on six sub-surface moorings measure ice draft (upward looking sonar), ice velocity and profiles of upper ocean current (ADCP) from a relatively safe depth of 105 m, current at specific depths in the lower part of the water column (conventional current meters) and seawater temperature and conductivity from sensors at discrete depths (Fig. 9.28). There are also three bottom-mounted ADCPs paired with temperature-conductivity sensors to measure the full velocity profile in shelf waters, two on the Greenland side and one on the Baffin. There are temperature-conductivity sensors at five additional shallow sites. At some shelf sites (1 in 2004/2005, 2 in 2005/2006, 4 in 2006/2007) there is an additional temperature-conductivity sensor at roughly 25-m depth in a package (IceCAT) developed at APL-UW; because this sensor measures within the low salinity layer near the ice, at significant risk of damage, it relays its data to a recording module at the seabed. If the sensor is snagged by ice, a weak link in the mooring line fails, permitting loss of the sensor while protecting the data module for later recovery.

Seagliders complement the moored instruments by providing fields of temperature and salinity at appropriate spatial resolution, right up to the surface, year-round and without ongoing ship support. The highly resolved hydrography in combination with time series of velocity, salinity and temperature provides a detailed picture of spatial and temporal variation. This information is essential for the accurate estimation of fluxes and of their empirical uncertainty. The measurements are already practical in the absence of pack ice and effort is now focussed on developing acoustic navigation and communication to provide the same capability when ice prevents communication via satellite.

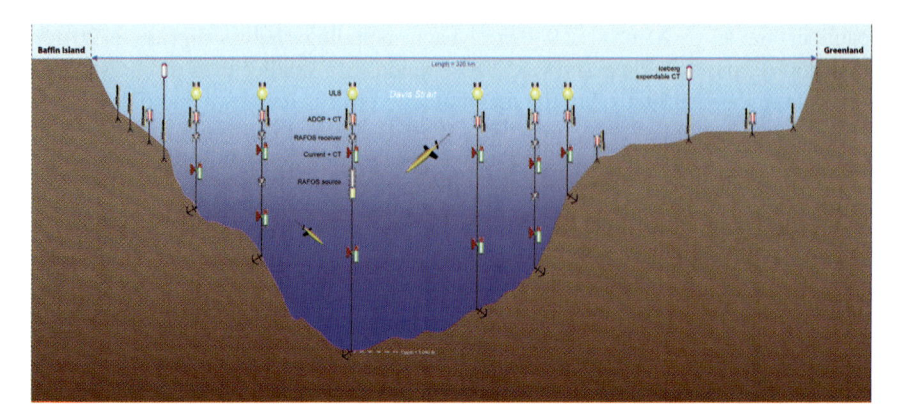

Fig. 9.28 Schematic representation of the array of instruments placed in September 2004 to measure fresh-water flux through Davis Strait

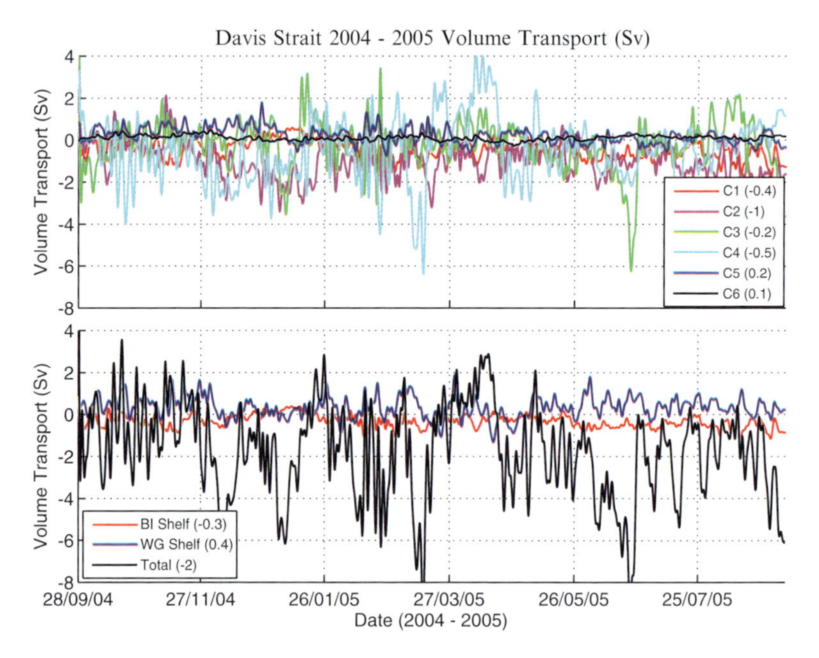

Fig. 9.29 Time series of area-weighted contributions to volume flux through Davis Strait, based on data from individual moorings during September 2004–2005

Although this is a respectable array with instruments at 14 sites and with current measured at 9, the average site spacing of 40 km is still greater than the decorrelation scale of ocean variability (Tang et al. 2004). This is apparent when considering how data from individual moorings contribute to the 2-Sv volume flux for 2004–2005 (Fig. 9.29). The plotted time series are the area-weighted contributions to the volume flux, each based on data from a single mooring. The time series are obviously poorly correlated with the result that there are long-lived and spurious fluctuations in the estimated flux. The accurate point measurements clearly require the complementary data from Seagliders to resolve variability within the Strait and thereby provide the hydrographic detail needed for intelligent interpolation between time series at fixed and widely spaced locations.

9.11 Summary and Outlook

The tabulation (Table 9.7) of volume and fresh-water fluxes through the gateways for Pacific Arctic through-flow is the outcome of our work in its most concise form.

All of the ASOF initiatives in the North American Arctic are clearly works in progress. Our research is advancing along learning curves in measurement, in interpretation of observations and in modeling. Our confidence to integrate with other ASOF sub-programmes and to explore the impact of global change is growing. Nonetheless, manifest environmental, logistical and technical complexity makes Pacific Arctic through-flow a big topic for research.

The following sections summarize progress towards desired outcomes.

9.11.1 *Quantitative Knowledge of Flux Magnitude and Variability*

We continue to benefit from promising new observational tools – ADCP, ice-profiling sonar, ICYCLER, IceCat, Sea Glider, methods for direction reference – and developing

Table 9.7 Summary of fluxes through the gateways for Pacific Arctic through-flow, estimated as described in this chapter and subject to many cautions – buyer beware. The value is positive for Arctic out-flow

	Seawater Volume (Sv)	Oceanic fresh-water (mSv)	Ice area (1,000s km^2)	Fresh-water as ice[a] (mSv)
Bering Strait	–0.8	–80	–	–
Amundsen Gulf	–	–	–53	–1.7 [1 m]
M'Clure Strait	–	–	–13	–0.8 [2 m]
Sverdrup Basin	–	–	20	2.5 [4 m]
Lancaster Sound	0.7	48	48	1.5 [1 m]
Cardigan Strait and Hell Gate	0.3	–	10	0.3 [1 m]
Nares Strait	0.8[b]	25	33	4.2 [4 m]
Baffin Bay	–	–	690	22 [1 m]
Davis Strait[c]	2.0	100	610	19 [1 m]

[a] Ice thickness has been estimated

[b] Snap-shot in time

[c] Not including flux over the Greenland shelf

numerical models. We have derived new values for fluxes, but values are less forthcoming for fresh-water than for volume. The bias and uncertainty of flux estimates are poorly known. Time series are far shorter than a decade in most instances and are non-existent for fresh-water at some key gateways. In consequence, the temporal overlap of time series within the North American Arctic is not yet sufficient to balance the budgets of Arctic seawater or fresh-water.

9.11.2 Forcing and Controls on Pacific Arctic Through-Flow

Researchers favour the steric anomaly of the North Pacific as the prime mover of Pacific Arctic through-flow (Steele and Ermold 2007), but renewed effort to define the magnitude and temporal variation of the absolute geopotential anomaly would be beneficial. Wind may augment or oppose steric forcing. There has been significant recent advance in the understanding of wind amplification in sea straits via mesoscale atmospheric effects and of its consequences for Pacific Arctic through-flow.

Models and observations agree that baroclinicity is an important attribute of Pacific Arctic through-flow. In baroclinic flows the width of low-density boundary currents is comparable to the internal Rossby scale (here about 10 km: Leblond 1980). With this constraint, wider channels cannot necessarily carry larger fluxes. We note that the flux through Lancaster Sound, nearly 70 km wide, is apparently only three times that through Cardigan Strait which has one ninth the width.

Numerical simulation has demonstrated that flow through Nares Strait is strongly influenced by atmospheric forcing that has been amplified via local orography and mesoscale atmospheric dynamics. Measurements of wind and temperature are needed in the planetary boundary layer to evaluate the simulations and to promote the understanding of oceanic and pack-ice responses. Because these processes may be important to through-flow in other areas, there is need for modelling at high resolution over a widened geographic domain.

Oceanic flows through the Pacific–Arctic gateways are thought to be controlled by friction and perhaps by rotational hydraulic effects. Numerical simulation of circulation within the Canadian Archipelago using a simple parameterization of drag has illustrated the importance of tidal current as a source of turbulence kinetic energy and therefore of resistance to flow at sub-tidal frequency. Since details are poorly developed, a future focus on these mechanisms in the context of Pacific Arctic through-flow is recommended.

Sea ice, as pack ice or as fast ice, covers the North American Arctic for much of the year. The impact of ice on channel flow is highly non-linear. It can range from an enhancement of wind forcing in the presence of rough mobile pack ice to a complete isolation from wind forcing by fast ice. In the latter instance, the immobile ice sheet exerts additional drag on oceanic flow. We recommend an initiative to understand the intermittent flow and blockage of sea ice in straits, with the ultimate objective of a reliable predictive capability.

9.11.3 Theory and Simulation of Through-Flow

There are many relevant theoretical topics to be addressed, ranging from the practical representation of rotational stratified flow in tidal channels to ocean hydrography and the circulation of fresh-water and sea ice on a hemispheric scale. In particular, we need a clearer notion of the hydrologic asymmetry between the Atlantic and Pacific that creates the steric anomaly that may drive the Pacific Arctic through-flow. An improved theoretical understanding will contribute to the numerical models that must ultimately provide our capability to hind-cast and predict Pacific Arctic through-flow, and to generate spatially complete and temporally continuous perspectives that are inaccessible via direct measurement.

9.11.4 Response to Global Change

The Pacific Arctic through-flow apparently responds to atmospheric and hydrologic forcing on a hemispheric scale. Our understanding of this forcing, of the varying storage of fresh-water within the Arctic Ocean and its ice cover and of the controls on out-flows to the Labrador Sea is not sufficient at present to support plausible hypotheses regarding the impact of changing climate on Pacific Arctic through-flow over the next century.

The practical task of ocean-flux measurement could benefit from continued focus on several issues.

There is need for a proven and agreed methodology for ocean-flux estimation. Table 9.8 summarizes the arrays presently installed in various gateways to measure Pacific Arctic through-flow. With two exceptions, the arrays fail to resolve the flow at the baroclinic Rossby scale (10 km) and all fail to measure salinity in the upper 30 m, where a large fraction of the fresh-water flux occurs (Melling 2000). Two arrays do use a prototype instrument to sample the upper layer, but at too few locations. Table 9.8 joins discussion earlier in the chapter to illustrate that we have yet to justify our methodology for flux measurement. Arrays with

Table 9.8 Summary of the arrays now installed to measure fresh-water flux

	Width of gateway (km)	Number of moorings (current)	Mooring separation (km)	Maximum depth (m)	Number of levels of salinity[a]	Top level of salinity (m)
Bering Strait	76	1–2	38–76	50	1	40
Lancaster Sound	68	2–4	17–34	280	1–3+	30 (5)[b]
Cardigan Strait, Hell Gate	12	2	6	180	1	100
Nares Strait	38	8	5	380	5	30
Davis Strait	360	9	40	1,000	1–3+	50 (25)[b]

[a] '+' indicates more levels measured by prototype near-surface instrument

[b] Shallow depth measured by prototype instrument at one site only

improved resolution of flow structure across sections and improved delineation of shallow salinity structure can assist in this task. The data that we acquire with improved arrays may ultimately provide the justification for the simplifying assumptions that are now being made, a priori.

One expected outcome of a proven methodology is an error model for ocean-flux estimation. Expanded arrays will provide the redundancy required to advance our understanding of sampling error and observational bias. At present we lack the ancillary data to understand why computed fluxes on adjacent sections differ and cannot check the consistency of our results via independent means.

An integrated approach to measuring and modelling the fresh-water fluxes moved as sea ice and as low salinity seawater is strongly advised. Fresh-water cycles between the seawater and ice phases with the annual freeze-thaw cycle as it moves across the North American Arctic. At times the ice and ocean may transport fresh-water in opposite directions. Ice measurements lag those in the ocean except in the aspect of geographic coverage; otherwise, ice velocity is only coarsely resolved in time (3 days) and ice thickness is rarely measured.

The Arctic Sub-Arctic Ocean Fluxes study has recommended a decade of synoptic observation. We have been late starting in the west and the only time series to achieve the 10-year target is that in Bering Strait (Fig. 9.30). The period of synoptic observation at all gateways is 3 years (2003–2006). A prolongation of existing time series is necessary to meet the original ASOF target.

At present we work hard to determine fresh-water flux, perhaps resolved as weekly or monthly averages. However, the ultimate impact of fresh-water in the receiving basins is critically dependent on the form in which it is delivered; the effect of a large seawater flux at salinity near 34.8 is very different from that of a small flux at near zero salinity, such as melting sea ice. In many cases we actually have the data in hand to report histograms of fresh-water flux according to salinity (Melling 2000) – the separate reporting of ice and seawater contribution is a first step. We recommend that a breakdown of the fresh-water flux by salinity become standard practice in reporting.

We are beginning to exploit the potential of trace chemical and isotope anomalies to reveal the sources of fresh-water, decadal variability and the time scales of transit.

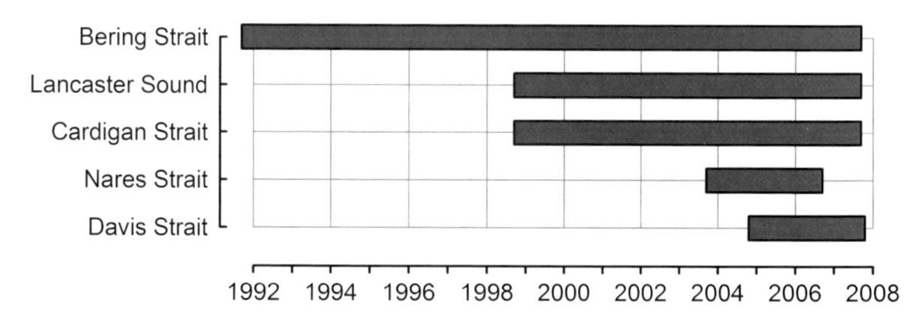

Fig. 9.30 Lifetimes of moored arrays within the gateways for Pacific-Arctic through-flow. The bars span those years during which current was measured for much of the time, but perhaps not in sufficient detail to permit the calculation of fluxes

The reward of this work will increase with the number of repeated geochemical surveys. With caution we can use some of the earlier (1985–1995) data, but sampling and analyses to modern standards have been completed only three times and these during the last decade. At a 3-year repetition interval, change in some components of the ice-ocean system is already aliased. Continued regional surveys, at annual intervals in certain areas, and efforts to resolve the strong seasonal cycles in fresh-water components, are needed to move understanding forward at this time.

There is continuing need for new observational technology. Preliminary work with sea-level signals (via pressure recorders and satellite altimetry) shows promise and should be pursued, in conjunction with programmes of in situ observation. The challenge of measuring fresh-water flux within the top 30 m remains with us – new technological approaches are always welcome. Above all, there is a strong incentive for new instruments and methods that provide needed data at reduced cost.

Ocean circulation models of the Canadian Archipelago are afflicted by shortage in three domains, bathymetry, hydrography and surface meteorology. The first, required to build a realistic geometry for the Canadian polar shelf, is plentiful in some areas, but patchy or non-existent in others. In some areas the need could be addressed by facilitating the migration of existing survey data into an accessible digital archive; in other areas new surveys are required that meet the reasonable needs of numerical simulation – needs that are much more modest than those of navigation. A modest objective for hydrographic information is the acquisition of temperature-salinity data sufficient to prepare a synoptic picture for the entire Canadian polar shelf for each season of the year. From meteorology, we need ocean-relevant observations of surface wind and temperature, to evaluate mesoscale atmospheric models and to promote understanding of the seasonal cycle of sea-ice growth, consolidation, break-up and decay in the largest fast-ice domain in the world.

References

Aagaard K, EC Carmack (1989) The role of sea ice and other fresh-water in the Arctic circulation. Journal of Geophysical Research 94:14485–14498

Agnew TA (1998) Drainage of multi-year sea ice from the Lincoln Sea. CMOS Bulletin 26(4):101–103

Agnew T, AH Le, T Hirose (1997) Estimation of large scale sea ice motion from SSM/I 855 GHz imagery. Annals of Glaciology 25:305–311

Agnew TA, J Vandeweghe (2005) Report on estimating sea-ice transport into the North Atlantic using the Advanced Microwave Scanning Radiometer (AMSR-E). Unpublished report, 21 pp

Agnew TA, J Vandeweghe, A Lambe (2006) Estimating the sea-ice-area flux across the Canadian Arctic Archipelago using the Advanced Microwave Scanning Radiometer (AMSR-E). Unpublished report, 20 pp

Ahlnäs K, GR Garrison (1984) Satellite and oceanographic observations of the warm coastal current in the Chukchi Sea. Arctic 37:244–254

Alkire MB, KK Falkner, I Rigor, M Steele, J Morison (2006) The return of Pacific waters to the upper layers of the central Arctic Ocean. Deep-Sea Research, submitted

Amon RMW, G Budéus, B Meon (2003) Dissolved organic carbon distribution and origin in the Nordic Seas: Exchanges with the Arctic Ocean and the North Atlantic. Journal of Geophysical Research 108, doi:101029/2002JC001594

Bacon S, SG Reverdin, IG Rigor, HM Snaith (2002) A freshwater jet on the east Greenland shelf. Journal of Geophysical Research 107, doi:101029/2001JC000935

Barber FG (1965) Current observations in Fury and Hecla Strait. Journal of the Fisheries Research Board of Canada 22:225–229

Bradley RS, FT Keimig, HF Diaz (1992) Climatology of surface-based inversions in the North American Arctic. Journal of Geophysical Research 97:15699–15712

Bromwich DH, JJ Cassano, T Klein, G Heinemann, KM Hines, K Steffen, JE Box (2001) Mesoscale modeling of katabatic winds over Greenland with the Polar MM5. Monthly Weather Review 129:2290–2309

Cassano JJ, JE Box, DH Bromwich, L Li, K Steffen (2001) Evaluation of Polar MM5 simulations of Greenland's atmospheric circulation. Journal of Geophysical Research 106:33867–33890

Cherniawsky JY, WR Crawford, OP Nikitin, EC Carmack (2005) Bering Strait transports from satellite altimetry. Journal of Marine Research 63:887–900

Clement JL, W Maslowski, LW Cooper, JM Grebmeier, W Walczowski (2005) Ocean circulation and exchanges through the northern Bering Sea: 1979–2001 model results. Deep-Sea Research II 52:3509–3540, doi:101016/jdsr2200509010

Coachman LK, K Aagaard (1966) On the water exchange through Bering Strait. Limnology and Oceanography 11:44–59

Coachman LK, K Aagaard (1981) Re-evaluation of water transports in the vicinity of Bering Strait. In: The Eastern Bering Sea Shelf: Oceanography and Resources vol 1. DW Hood and JA Calder (eds). National Oceanic and Atmospheric Administration, Washington, DC:95–110

Coachman LK, K Aagaard, RB Tripp (1975) Bering Strait: The Regional Physical Oceanography. University of Washington Press, Seattle, WA

Codispoti LA, D Lowman (1973) A reactive silicate budget for the Arctic Ocean. Limnology and Oceanography 18:448–456

Cooper LW, LA Codispoti, V Kelly, GG Sheffield, JM Grebmeier (2006) The potential for using Little Diomede Island as a platform for observing environmental conditions in Bering Strait, Arctic 59:129–141

Cooper LW, TE Whitledge, JM Grebmeier, T Weingartner (1997) The nutrient salinity and stable oxygen isotope composition of Bering and Chukchi Seas waters in and near Bering Strait. Journal of Geophysical Research 102:12563–12573

Cuny J, PB Rhines, R Kwok (2005) Davis Strait volume freshwater and heat fluxes. Deep-Sea Research I 52:519–542

de Lange Boom BR, H Melling, RA Lake (1987) Late Winter Hydrography of the Northwest Passage: 1982 1983 and 1984. Canadian Technical Report of Hydrography and Ocean Sciences No 79, 165 pp. Unpublished report available from Institute of Ocean Sciences Box 6000 Sidney Canada V8L 4B2

Deutsch C, N Gruber, RM Key, JL Sarmiento, A Ganachaud (2001) De-nitrification and N_2 fixation in the Pacific Ocean. Global Biogeochemical Cycles 15:483–506, doi:101029/2000GB001291

Dunphy M, F Dupont, CG Hannah, D Greenberg (2005) Validation of Modeling System for Tides in the Canadian Arctic Archipelago. Canadian Technical Report of Hydrography and Ocean Sciences 243: vi + 70 pp. Unpublished report available from Bedford Institute of Oceanography Box 1006 Dartmouth Canada B2Y 4A2

Ekwurzel B, P Schlosser, RA Mortlock, RG Fairbanks (2001) River runoff sea ice meltwater and Pacific water distribution and mean residence times in the Arctic Ocean. Journal of Geophysical Research 106:9075–9092

Falkner KK, MC O'Brien, H Melling, E C Carmack, F A McLaughlin, A Münchow, E P Jones (2006) Interannual variability of dissolved nutrients in the Canadian Archipelago and Baffin Bay with implications for fresh-water flux. Journal of Geophysical Research Bio-Geosciences, submitted

Fissel DB, JR Birch, H Melling, RA Lake (1988) Non-tidal Flows in the Northwest Passage. Canadian Technical Report of Hydrography and Ocean Sciences No 98:143 pp. Unpublished report available from Institute of Ocean Sciences Box 6000 Sidney Canada V8L 4B2

Gruber N, JL Sarmiento (1997) Global patterns of marine nitrogen fixation and de-nitrification. Global Biogeochemical Cycles 11:235–266

Guo Z, DH Bromwich, JJ Cassano (2003) Evaluation of Polar MM5 simulations of Antarctic atmospheric circulation. Monthly Weather Review 131:384–411

Häkkinen S, A Proshutinsky (2004) Fresh-water content variability in the Arctic Ocean. Journal of Geophysical Research 109 doi:101029/2003JC001940

Holloway G, T Sou (2002) Has Arctic sea ice rapidly thinned? Journal of Climate 15:1691–1701

Jones EP, LG Anderson (1990) On the origin of the properties of the Arctic Ocean halocline north of Ellesmere island: results from the Canadian Ice Island. Continental Shelf Research 10:485–498

Jones EP, LG Anderson, JH Swift (1998) Distribution of Atlantic and Pacific waters in the upper Arctic Ocean: Implications for circulation. Geophysical Research Letters 25:765–768

Jones EP, AR Coote (1980) Nutrient distributions in the Canadian Archipelago: Indicators of summer water mass and flow characteristics. Canadian Journal of Fisheries and Aquatic Science 37:589–599

Jones EP, JH Swift, LG Anderson, M Lipizer, G Civitarese, KK Falkner, G Kattner, FA McLaughlin (2003) Tracing Pacific water in the North Atlantic Ocean. Journal of Geophysical Research 108, doi:101029/2001JC001141

Kahl JD, MC Serreze, RC Schnell (1992) Tropospheric low-level temperature inversions in the Canadian Arctic. Atmosphere-Ocean 30:511–529

Kinney P, ME Arhelger, DC Burrell (1970) Chemical characteristics of water masses in the Amerasian Basin of the Arctic Ocean. Journal of Geophysical Research 75:4097–4104

Kliem N, DA Greenberg (2003) Diagnostic simulations of the summer circulation in the Canadian Arctic Archipelago. Atmosphere-Ocean 41:273–289

Kwok R (2005) Variability of Nares Strait ice flux. Geophysical Research Letters 32 L24502, doi:101029/2005GL024768

Kwok R (2006) Exchange of sea ice between the Arctic Ocean and the Canadian Arctic Archipelago. Geophysical Research Letters 33 L16501, doi:101029/2006GL027094

Kwok R, GF Cunningham, S Yueh (1999) Area balance of Arctic Ocean Perennial Ice Zone: October 1996 – April 1997. Journal of Geophysical Research 104 25747, doi:101029/1999JC900234

Kwok R, A Schweiger, D A Rothrock, S Pang, C Kottmeier (1998) Sea ice motion from satellite passive microwave imagery assessed with ERS SAR and buoy motions. Journal of Geophysical Research 103:8191–8214

Lammers RB, AI Shiklomanov, CJ Vorosmarty, BM Fekete, BJ Peterson (2001) Assessment of contemporary Arctic river runoff based on observational discharge records. Journal of Geophysical Research 106:3321–3334

LeBlond PH (1980) On the surface circulation in some channels of the Canadian Arctic Archipelago. Arctic 33:189–197

Loder JW, B Petrie, G Gawarkiewicz (1998) The coastal ocean off north-eastern North America: A large-scale view. In: The Sea 11:105–133, Chapter 5.

Maslowski W, JL Clement, W Walczowski (2003) Modeled Arctic sub-Arctic ocean fluxes during 1979–2001. Abstract #9554 EGS-AGU-EUG Joint Assembly. Nice, France: 6–11 April 2003

McLaren AS, P Wadhams, R Weintraub (1984) The sea ice topography of M'Clure Strait in winter and summer of 1960 from submarine profiles. Arctic 37:110–120

Melling H (2000) Exchanges of fresh-water through the shallow straits of the North American Arctic. In: The Fresh-water Budget of the Arctic Ocean. Proceedings of a NATO Advanced Research Workshop. Tallinn Estonia. 27 April–1 May 1998. EL Lewis et al. (eds). Kluwer, Dordrecht, The Netherlands, pp. 479–502

Melling H (2002) Sea ice of the northern Canadian Archipelago. Journal of Geophysical Research 107:3181, doi:101029/2001JC001102

Melling H, Y Gratton, RG Ingram (2001) Ocean circulation within the North Water polynya of Baffin Bay. Atmosphere-Ocean 39:301–325

Melling H, PH Johnston, DA Riedel (1995) Measurement of the topography of sea ice by moored sub-sea sonar. Journal of Atmospheric and Oceanic Technology 12:589–602
Melling H, RA Lake, DR Topham, DB Fissel (1984) Oceanic thermal structure in the western Canadian Arctic. Continental Shelf Research 3:233–258
Melling H, DA Riedel (1996) Development of seasonal pack ice in the Beaufort Sea during the winter of 1991–92: A view from below. Journal of Geophysical Research 101:11975–11992
Mosby H (1962) Water salt and heat balance of the North Polar Sea and of the Norwegian Sea. Geophysica Norvegica 24:289–313
Münchow A, H Melling, KK Falkner (2006) An observational estimate of volume and freshwater flux leaving the Arctic Ocean through Nares Strait. Journal of Physical Oceanography 36:2025–2041
Östlund HG, G Hut (1984) Arctic Ocean water mass balance from isotope data. Journal of Geophysical Research 89:6373–6381
Padman L, S Erofeeva (2004) A barotropic inverse tidal model for the Arctic Ocean. Geophysical Research Letters 31 L02303, doi:101029/2003GL019003
Paquette RG, RH Bourke (1974) Observations on the coastal current of Arctic Alaska. Journal of Marine Research 32:195–207
Prinsenberg SJ (1988) Damping and phase advance of the tide in western Hudson Bay by the annual ice-cover. Journal of Physical Oceanography 18:1744–1751
Prinsenberg SJ, EB Bennett (1987) Mixing and transports in Barrow Strait the central part of the Northwest Passage. Continental Shelf Research 7:913–935
Prinsenberg SJ, EB Bennett (1989) Vertical variations of tidal currents in shallow land fast ice-covered regions. Journal of Physical Oceanography 19:1268–1278
Prinsenberg SJ, J Hamilton (2005) Monitoring the volume freshwater and heat fluxes passing through Lancaster Sound in the Canadian Arctic Archipelago. Atmosphere-Ocean 43:1–22
Pritchard RS, RW Reimer, MD Coon (1979) Ice flow through straits. In: Proceedings of POAC'79 vol 3:61–74. Norwegian Institute of Technology, Trondheim, Norway
Proshutinsky AY, M A Johnson (1997) Two circulation regimes of the wind-driven Arctic Ocean. Journal of Geophysical Research 102:12493–412514
Roach AT, K Aagaard, CH Pease, SA Salo, T Weingartner, V Pavlov, M Kulakov (1995) Direct measurements of transport and water properties through the Bering Strait. Journal of Geophysical Research 100:18443–18457
Roff JC, L Legendre (1986) Physico-chemical and biological oceanography of Hudson Bay. In: Canadian Inland Seas. IP Martini (ed). Elsevier, New York, pp. 265–291
Ross C (1992) Moored current meter measurements across Davis Strait. NAFO Research Document 92/70
Sadler HE (1976) Water heat and salt transports through Nares Strait Ellesmere Island. Journal of the Fisheries Research Board of Canada 33:2286–2295
Sadler HE (1982) Water flow into Foxe Basin through Fury and Hecla Strait. Le Naturaliste Canadian 109:701–707
Samelson RM, SJ Lentz (1994) The horizontal momentum balance in the marine atmospheric boundary layer during CODE-2. Journal of the Atmospheric Sciences 51:3745–3757
Samelson RM, T Agnew, H Melling, A Münchow (2006) Evidence for atmospheric control of sea-ice motion through Nares Strait. Geophysical Research Letters 33 L02506, doi:101029/2005GL025016
Sanderson BG (1987) Statistical properties of iceberg motion at the western entrance of Lancaster Sound. In: Proceedings of Oceans'87 MTS/IEEE, vol 19:17–23
Schlosser P, B Ekwurzel, S Khatiwala, B Newton, W Maslowski, S Pfirman (2000) Tracer studies of the Arctic fresh-water budget. In: The Fresh-water Budget of the Arctic Ocean. EL Lewis (ed). Kluwer, Dordrecht, The Netherlands, pp. 453–478
Serreze MC, AP Barrett, AG Slater, RA Woodgate, K Aagaard, R Lammers, M Steele, R Moritz, M Meredith, CM Lee (2006) The large-scale fresh-water cycle of the Arctic Journal of Geophysical Research, submitted

Smith JN, KM Ellis, T Boyd (1999) Circulation features in the Central Arctic Ocean revealed by nuclear fuel reprocessing tracers from SCICEX 95 and 96. Journal of Geophysical Research 104:29633–29677

Sodhi DS (1977) Ice arching and the drift of pack ice through restricted channels. CRREL Report No 77–18. US Army Cold Regions Research and Engineering Laboratory Hanover NH. Available NTIS, 11 pp

Steele M, W Ermold (2007) Steric sea level change in the northern seas. Journal of Climate 20:403–417

Steele M, D Thomas, D Rothrock, S Martin (1996) A simple model of the Arctic Ocean fresh-water balance 1979–1985. Journal of Geophysical Research 101:20833–20848

Stigebrandt A (1984) The North Pacific: a global-scale estuary. Journal of Physical Oceanography 14:464–470

Strain PM, FC Tan (1993) Seasonal evolution of oxygen isotope-salinity relationships in high-latitude surface waters. Journal of Geophysical Research 98:14589–514598

Tang CL, CK Ross, T Yao, B Petrie, BM DeTracy, E Dunlop (2004) The circulation water masses and sea ice of Baffin Bay. Progress in Oceanography 63:183–228

Taylor JR, KK Falkner, U Schauer, M Meredith (2003) Quantitative considerations of dissolved barium as a tracer in the Arctic Ocean. Journal of Geophysical Research 108, doi:101029/2002JC001635

Tremblay J-É, Y Gratton, EC Carmack, CD Payne, NM Price (2002) Impact of the large-scale Arctic circulation and the North Water polynya on nutrient inventories in Baffin Bay. Journal of Geophysical Research 107, doi:101029/2000JC000595

Weingartner TJ, SL Danielson, TC Royer (2005) Fresh-water variability and predictability in the Alaska Coastal Current. Deep-Sea Research II 52 169–191, doi:1101016/jdsr1012200410091030

Wijffels SE, RW Schmitt, HL Bryden, A Stigebrandt (1992) Transport of freshwater by the oceans. Journal of Physical Oceanography 22:155–162

Williams CE, W Maslowski, JC Clement, AJ Semtner (2004) Fresh-water Fluxes from the Arctic into the North Atlantic Ocean: 1979–2002 model results. American Geophysical Union Fall Meeting 2004 2004AGUFMC54A-05W

Winant CD, CE Dorman, CA Friehe, RC Beardsley (1988) The marine layer off northern California: An example of supercritical channel flow. Journal of Atmospheric Sciences 45:3588–3605

Woodgate RA, K Aagaard (2005) Revising the Bering Strait fresh-water flux into the Arctic Ocean. Geophysical Research Letters 32 L02602, doi:101029/2004GL021747

Woodgate RA, K Aagaard, TJ Weingartner (2005a) Monthly temperature salinity and transport variability of the Bering Strait through-flow. Geophysical Research Letters 32 L04601, doi:101029/2004GL021880

Woodgate RA, K Aagaard, TJ Weingartner (2005b) A year in the physical oceanography of the Chukchi Sea: Moored measurements from autumn 1990–1991. Deep-Sea Research II 52:3116–3149 101016/jdsr2200510016

Woodgate RA, K Aagaard, TJ Weingartner (2006) Inter-annual changes in the Bering Strait fluxes of volume heat and fresh-water between 1991 and 2004. Geophysical Research Letters 33 L15609, doi:101029/2006GL026931

Yamamoto-Kawai M, N Tanaka, S Pivovarov (2005) Fresh-water and brine behaviors in the Arctic Ocean deduced from historical data of $\delta^{18}O$ and alkalinity (1929–2002). Journal of Geophysical Research 110, doi:101029/2004JC002793

Yamamoto-Kawai M, FA McLaughlin, EC Carmack, S Nishino, K Shimada (2006) Fresh-water budget of the Canada Basin Arctic Ocean from geochemical tracer data. Journal of Geophysical Research, submitted

Chapter 10
The Arctic–Subarctic Exchange Through Hudson Strait

Fiammetta Straneo[1] and François J. Saucier[2]

10.1 Introduction: The Hudson Bay System: An Extensive Arctic Basin

One major export of (fresh) water from the Arctic region to the North Atlantic is due to surface-intensified currents that flow along the topographic margins. These enter the North Atlantic through three major straits – Fram, Davis and Hudson – which therefore provide ideal gateways for monitoring the exchange. Of these straits, the first two link the North Atlantic Ocean with the Arctic Ocean while the third, Hudson Strait, connects it to an extensive Arctic region, the Hudson Bay System (HBS), which, in its northwest corner, is also connected to the Arctic Ocean (via Fury and Hecla Strait – Fig. 10.1). The lack of a direct connection with the Arctic Ocean is, likely, the reason why Hudson Strait's contribution to the Arctic/North Atlantic exchange has, until recently, been overlooked. In this chapter, we present estimates for the net, as well as the inflow and outflow transports, of volume, heat and freshwater through Hudson Strait. These are based both on a review of the inputs into the basin and on the first year-long measurements of the outflow from Hudson Strait to the Labrador Sea. This analysis shows not only that the HBS provides a substantial net input of Arctic (fresh) water to the North Atlantic but, also, that a significant fraction of the export through Davis Strait is recirculated in the HBS before it effectively flows into the Labrador Sea.

The outflow from Hudson Strait emerges as a highly stratified flow, even after transiting the turbulent region at the mouth of the Strait (LeBlond et al. 1981), along the Labrador coast. Here it merges with the 'direct' Davis Strait outflow and the offshore continuation of the West Greenland Current into the Labrador Current (Mertz et al. 1993) – a freshwater laden current which flows close to the Labrador Sea's deepest convection region (Clarke and Gascard 1983; Pickart et al. 2002). This current is recognized as an important source for the freshwater that rapidly re-stratifies the convection region in the spring (Lazier et al. 2002; Straneo 2006).

[1] Woods Hole Oceanographic Institution, Woods Hole, MA, USA

[2] Université du Québec à Rimouski, Canada

R.R. Dickson et al. (eds.), *Arctic–Subarctic Ocean Fluxes*, 249–261

Freshwater anomalies in the flow, which make their way to the interior, can thus have a significant impact on the extent of convection (Lazier 1980; Straneo 2006) and hence on the large-scale ocean circulation. Further downstream, the outflow from Hudson Strait is thought to have a profound influence on the highly productive regions of the Labrador and Newfoundland shelves (Sutcliffe et al. 1983).

The Hudson Bay System (HBS) is a large inland sea that includes Hudson Bay, James Bay, Foxe Basin, Ungava Bay and Hudson Strait (Fig. 10.1). Though its meridional extension spans roughly 20° of latitude across the Arctic Circle (roughly 52–70° N), the entire basin is characterized by typical Arctic (oceanic and atmospheric) conditions and, as such, is the southernmost extension of the Arctic region as defined, for example, by the Arctic Monitoring and Assessment Programme (AMAP; http:/ www.amap.no). One typical Arctic feature of the basin is its complete seasonal ice-cover, which makes it the largest inland body of water in the world (1 million square kilometers, one fifth of the size of the Arctic Ocean) to seasonally freezes over and

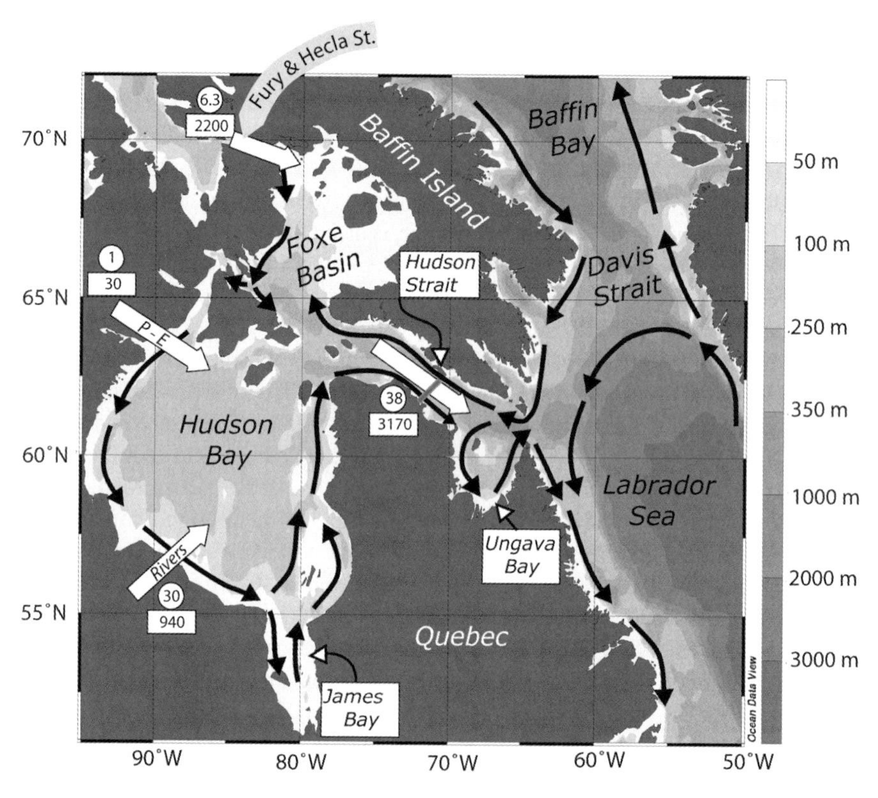

Fig. 10.1 Topography and schematic circulation of the Hudson Bay System, Labrador Sea and Baffin Bay region. White arrows overlaid show the net volume (rectangles, in km^3/year) and freshwater (circles, in mSv referenced to a salinity of 34.8) transports due to the input of rivers, precipitation minus evaporation and Fury and Hecla Strait into the HBS. The resulting estimated net transports out of the system, through Hudson Strait, are also shown

then be virtually ice-free in the summer (Prinsenberg 1988). This southern extension of Arctic conditions, well beyond the Arctic Circle, has a large impact on the climate of the surrounding land masses and oceanic basins as indicated, for example, by the southern displacement of the tree line due to the HBS (United Nations Environment Program 2006, http://maps.grida.no/go/graphic/treeline_in_the_arctic).

The large size of the HBS, alone, makes it a likely significant contributor to the Arctic/North Atlantic exchange. Its contribution, furthermore, is greatly enhanced by its considerably larger watershed. Indeed, the HBS' drainage basin occupies an area roughly four times larger than HBS itself, which extends from Alberta to Quebec and from Baffin Island to south of the Canada/USA border (roughly 4 million square kilometers; Déry et al. 2005). This large catchment area drains approximately 900 km^3/year of freshwater into the HBS (McClelland et al. 2006; Déry et al. 2005) which, for comparison, is approximately one fifth of the river discharge into the Arctic Ocean (Déry et al. 2005).

The exchange between the HBS and the North Atlantic or more specifically the Labrador Sea, occurs through Hudson Strait, a narrow (~100 km) and long (~400 km) channel, with depths ranging from 900 m (east) to 200 m (west), and HBS' primary opening. Water flows through the strait in two opposite directions. Along its northern shores, the Baffin Island Current flows *into* the HBS from the Labrador Sea. Along its southern shores, buoyant (fresh) waters flow *towards* the Labrador Sea (Fig. 10.1; LeBlond et al. 1981; Drinkwater 1988). Both flows participate in the exchange between the HBS and the Labrador Sea, and the *net* transports out of Hudson Strait (into the North Atlantic) must be calculated as the difference between the *outflow* (the flow towards the Labrador Sea, on the southern side) and the *inflow* (the flow along Baffin Island).

If observations of the fluxes on either side of the strait were available, along with their variability, the estimate of the net transports through Hudson Strait would be straightforward. In practice, data in the strait are scarce and the only simultaneous measurements of the flow on *both* sides of the strait are due to a moored array deployed for 2 months, in the summer of 1984, which measured the volume flux alone (Drinkwater 1988). Also, the strong seasonality of the high-latitudes and the lack of direct current measurements do not allow one to estimate the mean annual transports from the limited available summer hydgrographic surveys.

In order to estimate the net transports through Hudson Strait, an alternate approach is to assume the HBS in steady state. Given that the system has only two open-boundaries (Fury and Hecla and Hudson Straits), it follows that if one knows the air–sea fluxes (heat, evaporation, precipitation), the river input and the exchange through one of the straits, then one can estimate what flows through the other. Here, we use this approach to infer the net transports out of Hudson Strait. These estimates are then compared with the observed transports in the outflow, on the southern side of the strait, obtained by deploying a moored array from August 2004 to August 2005 (Straneo and Saucier 2007). The difference between the net and the outflow gives us a measure of the inflow into the HBS.

This analysis reveals that the net exchange of volume is typically one order of magnitude smaller than the inflow/outflow transports, and that care must be taken in using them to characterize an exchange. For freshwater, this analysis shows that much more freshwater is outflowing than is input by rivers, air–sea fluxes, or the direct exchange with the Arctic Ocean (through Fury and Hecla Strait). The implication is that a significant portion of the Davis Strait outflow recirculates into the HBS instead of flowing directly towards the Labrador shelf. This is a new result that suggests re-drawing the Arctic export pathways west of Greenland since this recirculation can, potentially, add a significant lag to the emergence of anomalies from the Arctic Ocean. A discussion of these results is presented in the last part of this chapter.

10.2 Volume, Heat and Freshwater Budgets for the Hudson Bay System

Estimates for the input of volume, heat and freshwater into the HBS from all external sources (rivers, air–sea fluxes) and from Fury and Hecla Strait, based on published data, are presented below.

10.2.1 River Discharge

The most recent assessment of river discharge into the HBS can be found in Déry et al. (2005). They used data from 42 rivers draining into Hudson, James, Ungava Bays and Hudson Strait from 1964 to 2000, compiled in Environment Canada's Hydrometric Database (HYDAT), to assess the mean discharge rates and their interannual variability. The observed mean annual discharge is 714 km^3/year. Since these rivers occupy roughly 80% of the overall drainage area of this region, and assuming the same rate of discharge per unit area, this implies that the net annual discharge is 892 km^3/year. This number is reasonably close to a previous estimate by Shiklomanov and Shiklomanov (2003) who report a mean of 938 km^3/year from 1966 to 1999. In either case, these contributions are likely underestimated since they do not include the contribution from the islands (including Baffin Island) that surround the HBS (where no data are available). In this review we thus assume that a reasonable estimate of the river input is 940 km^3/year. This is equivalent to a volume transport of 0.03 Sv (1 Sv = 10^6 m^3/s) and to a freshwater flux of 30 mSv (milli Sverdrup).

We did not find, in the literature, any reference to the rivers' contribution to the heat budget of the region. While this is unlikely to be large, it may still play a role. If we assume, for example, that the river water has a mean inflow temperature of 1.5 °C, then it would contribute 0.15 TW of heat to the HBS. As shown below, this input is of the same order of magnitude (indeed it offsets it) of the negative heat transport that we estimate through Fury and Hecla Strait.

10.2.2 Air–Sea Fluxes

Air–sea fluxes contribute to the volume and freshwater budget of the HBS through the difference between evaporation and precipitation. Overall direct measurements of precipitation over the entire system are poor, and evaporation estimates depend on the algorithm used – making the difference between the two quite uncertain. Earlier estimates suggested that the Hudson/James Bay region is characterized by a net evaporative loss over precipitation except for James Bay, where the two balance (Prinsenberg 1977). A more quantitative assessment, by the same author, suggests that the mean annual freshwater loss over Hudson Bay and Foxe Basin is of the order of 6 mSv (Prinsenberg 1980).

More recent estimates, on the other hand, suggest that precipitation exceeds evaporation in the region (Saucier et al. 2004). These are based on the operational analyses (NOWCAST) and 12-h forecast cycles issued from the Canadian Meteorological Centre using the Global Environmental Multiscale Model (GEM) from 1997 to 1999. These estimate the net precipitation to be of the order of 10 to 50 km^3/year. Clearly, there is some discrepancy between these estimates and, still, a large uncertainty associated with these fluxes. At the same time, the estimated volume (and freshwater) contribution from the air–sea exchange appears to be an order of magnitude smaller than the river input. Below, we make use of the estimate based on reanalyses data, as the most recent, in the volume and freshwater budgets for the region.

Similar discrepancies are found in the literature for estimates of the annual heat loss over the HBS. Prinsenberg (1983) claims that Hudson Bay gains heat from the atmosphere (in an annual mean sense) and estimates the annual gain to be 1.8 W/m^2 – equivalent to a net input of 1.8 TW. The same re-analyses products described above (see Saucier et al. 2004) suggest, on the other hand, that the region including HS and FB undergoes a net heat loss on the order of 10 W/m^2. Much of this heat loss, however, occurs to the east of the mooring section where warmer waters are recirculating at the mouth the strait. Given their uncertainty, we feel that one cannot rely on these numbers to infer, for example, the inflow of heat into the HBS. Instead, as discussed below, we will do the opposite (only for heat) and use an estimated inflow mean temperature, combined with the estimated transport of the inflow to infer the net heat flux into the HBS.

10.2.3 Transports via Fury and Hecla Strait

Fury and Hecla is a narrow, shallow strait that connects the Gulf of Boothia, in the Arctic Ocean, to Foxe Basin, the northern extension of the HBS. Its width varies from approximately 15 to 30 km, and it is approximately 120 km long and 170 m deep. Observations in the strait are limited to current meter data collected in April–May 1976 (Sadler 1982) and in the summer of 1960 (Barber 1965) yielding mean residual transports, towards Foxe Basin, of 0.04 Sv and 0.1 Sv respectively.

These transports are associated with typical Arctic Ocean temperatures of −1.7 °C and salinities of 32.0–32.1, in late winter, and of 0.5–0.75 °C and 31.0–32.0, in summer (Ingram and Prinsenberg 1998). If we take the late winter conditions to be representative of the period from December to May, and the summer conditions of the remaining 6 months, the estimated volume and freshwater (relative to 34.8) transports are of 0.07 Sv (2,200 km^3/year) and 6.3 mSv, respectively. The heat transport, relative to 0 °C, is −0.15 TW into the HBS.

10.2.4 Summary

The estimates of volume and freshwater transports out of HBS are summarized in Table 10.1 and shown schematically in Fig. 10.1. The heat flux contribution is omitted given its large uncertainty. For volume and freshwater the sum of the contributions, listed in Table 10.1, must be balanced by the net transports through Hudson Strait. It is interesting to note that the volume and freshwater balances are maintained by two different input terms. For volume, the net flux through Hudson Strait mostly offsets the inflow via Fury and Hecla Strait. For freshwater, the balance is between the river input and the export through Hudson Strait.

10.3 Transports Through Hudson Strait

The estimated contributions listed in Table 10.1, to the volume and freshwater budgets for the HBS imply that there must be a net volume transport out of Hudson Strait (towards the Labrador Sea) of approximately 3,200 km^3/year (or equivalently 0.1 Sv) and a freshwater transport of 38 mSv (relative to a salinity of 34.8). The net volume transport out compares well with the measurements of Drinkwater (1988) who found a mean residual northwestward transport of 0.82 Sv (along Baffin Island) and of 0.93 Sv towards the Labrador Sea (along the Quebec shore), even if these measurements were based on a 2-month survey only.

Next, we compare these estimated net transports with those observed in the outflow alone by Straneo and Saucier (2007). The outflow transports were obtained from a three mooring array, deployed across the southern portion of the strait, roughly mid-strait (Fig. 10.1) from August 2004 to August 2005 as part of a collaboration

Table 10.1 Mean annual inputs of volume and freshwater (relative to a salinity of 34.8) due to air–sea interaction, to rivers and to Fury and Hecla Strait

Input to HBS	Volume (km^3/year) (Sv)	Freshwater (mSv rel 34.8)
River	940 (0.03)	30
Air–sea	30 (0.001)	1.0
Fury and Hecla St.	2,200 (0.07)	6.3
Total	3,170 (0.1)	38

between the Woods Hole Oceanographic Institution (WHOI) and Fisheries and Oceans Canada (the Canadian MERICA program for the monitoring of the HBS). The section occupied by the array was only slightly more inland of the one occupied by Drinkwater in 1984. This location was chosen over the mouth of the strait for a number of practical and scientific reasons. The mouth of the strait is characterized by strong, turbulent flow and persistent eddies, associated with the interaction of tides with numerous channels and islands (LeBlond et al. 1981). Such strong circulation makes mooring deployment difficult and hazardous, and makes defining the mean flow more problematic. Similarly, the region extending west of the of the mouth of the strait to the section is characterized by a strong recirculation (Ingram and Prinsenberg 1998) which would not only affect the measurements but also threaten the moorings since it carries numerous icebergs (Drinkwater 1986). Finally, it should be noted that the section chosen is upstream of the river input from Ungava Bay. Because these are relatively small compared to the total input to HBS, and given the uncertainty on the input, we have not attempted to factor this into our calculations.

The moorings were positioned across the fresh outflow current that is characteristic on this side of the strait (Fig. 10.2). They consisted of a combination of profiling salinity and temperature recorders, acoustic Doppler current profilers and fixed depth instruments. The most difficult problem, typical when attempting to estimate freshwater transports, was to reconstruct properties in the upper part of the water column, over a layer of about 40 m, where no measurements were made. Instead of simply using a mixed layer approach for these upper 40 m, the authors made use of the observed dynamic characteristics of the flow to infer the density (and hence the salinity) distribution. This method and, in general, the transport calculations are described in detail in Straneo and Saucier (2007).

The section shown in Fig. 10.2 is representative of summer conditions across the strait. It shows the fresh, and strongly stratified, outflow wedged across the sloping topography on the southern side of the strait and a much more weakly stratified,

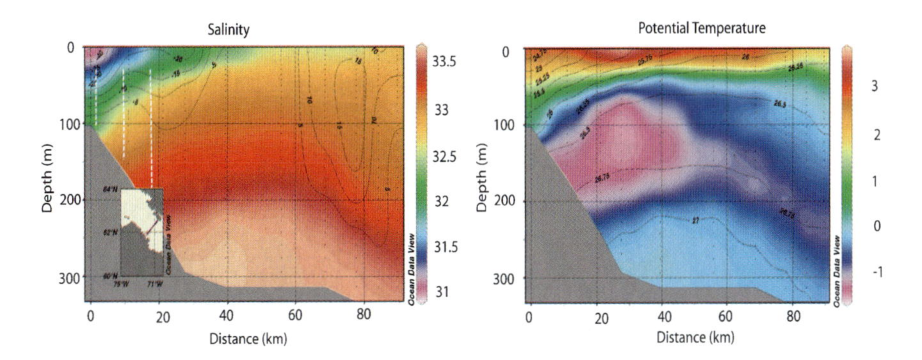

Fig. 10.2 Hydrographic section across Hudson Strait collected in August 2005. Left panel: Salinity and geostrophic velocity contours overlaid (black lines), distance is from the coast of Quebec. Also shown are the mooring locations of Straneo and Saucier (2007). Right panel: Potential temperature and potential density contours overlaid

barotropic and saltier inflow along the coast of Baffin Island. The geostrophic velocity overlaid was calculated assuming zero flow at the bottom and should, therefore, be regarded as the baroclinic portion of the flow only. The outflow was characterized by a marked seasonal variability in properties with the freshest waters transiting from June to March, with salinities as low as 28.8 observed at the uppermost instrument of the most onshore mooring (Fig. 10.3). The along-strait flow was found to increase (at least in the surface layers) during the passage of the freshest waters. The temperature of the water flowing past the moorings was close to freezing for much of the year, except during a short period between July and November when it reached about 2 °C (Fig. 10.3).

The flow is dominated by the tidal cycle due to mostly barotropic, semi-diurnal tides with speeds in excess of 1 m/s and tidal ranges of the order of 8 m. Once the tides are removed, the mean flow is aligned with the strait and has the characteristics of a buoyant, gravity current over a sloping bottom with a depth of 150 m and

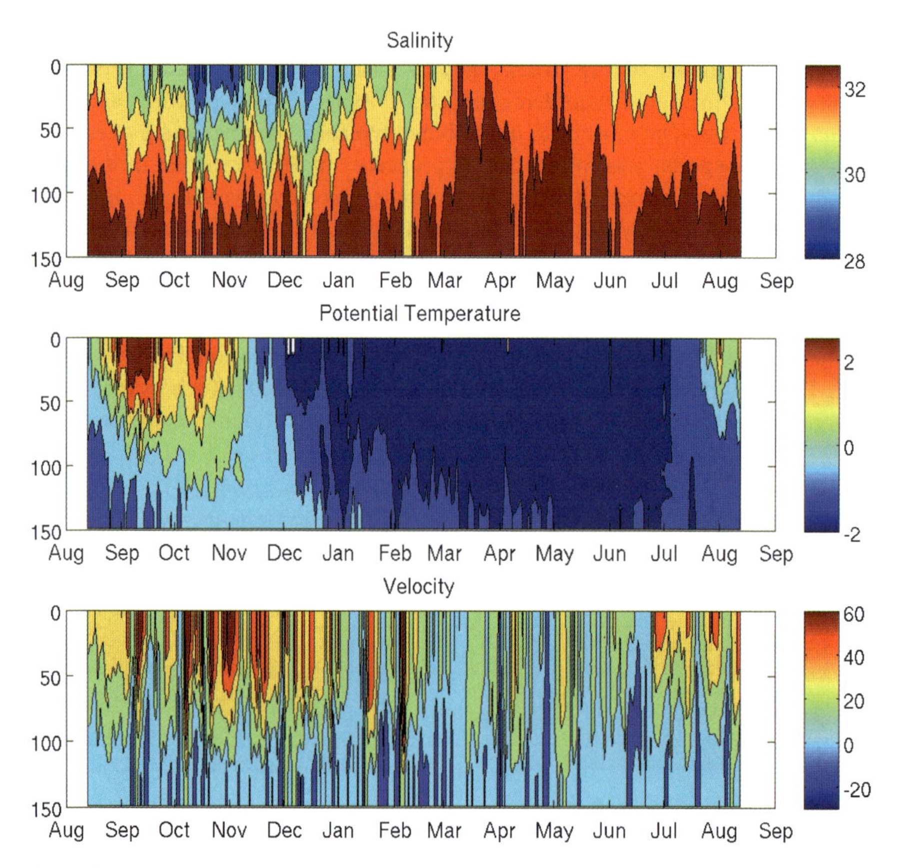

Fig. 10.3 Reconstructed Salinity, Potential Temperature and Along-strait Velocity (cm/s) at the central mooring of the array of shown in Fig. 10.2, details of the data analysis can be found in Straneo and Saucier (2007)

a horizontal width of 30 km. Superimposed on the seasonal variability, temperature, salinity, and especially velocity are characterized by strong variations over several days. These appear to be mostly barotropic and not determined by changes in the density field (Fig. 10.3; Straneo and Saucier 2007). The outflow exhibits very high lateral coherence and the time series of salinity, temperature and currents from the other two moorings are qualitatively very similar to those shown in Fig. 10.3. The salinity decreases onshore, as shown in the hydrographic section, while the maximum flow in the surface layers occurs more offshore (Straneo and Saucier 2007).

The mean annual transports of volume, heat and freshwater in the outflow calculated from these data are of 0.94 Sv (~30,000 km^3/year), –2.2 TW and 79 mSv, respectively. Details of how they were determined can be found in Straneo and Saucier (2007). We note that the observed freshwater transport is due to the liquid portion of the outflow since no sea-ice thickness data were available. Model simulations suggest that when included the sea-ice portion may contribute a net additional 6 mSv (Saucier et al. 2004).

These transports can be used, in combination with the net transports derived from the volume and freshwater budget of the HBS, to infer the inflow transports into HBS, for which we have no measurements. To balance volume, the inflow volume transport must be on the order of 0.84 Sv (26,800 km^3/year) which agrees with Drinkwater's 1984 measurements. To balance freshwater, we need a freshwater transport, along the coast of Baffin Island, of 41 mSv and likely more if we had taken into account the freshwater transport due to sea-ice in the outflow. This means that as much freshwater (relative to a salinity of 34.8) is carried into the HBS through Hudson Strait by the inflow as is input by the rivers throughout. Given that both the freshwater outflow and the river discharge are obtained from direct observations, this estimate is likely fairly reliable (excluding the missing sea-ice contribution and the interannual variability). For the inflow volume transport (0.84 Sv) to contribute 41 mSv of freshwater, the mean salinity of the inflowing water must be of the order of 33.1. This is well within the range of what is observed in the summer hydrographic section shown in Fig. 10.2. It is also in agreement with the mean salinity of the waters flowing out of Davis Strait ($32.5 < S < 33.5$; Cuny et al. 2005).

Given the uncertainty on the annual mean net heat flux estimates for HBS, we use a different approach to estimate the heat transports through Hudson Strait. First we ask what the mean temperature of the inflow waters must be in order to balance what comes out – assuming that the HBS' net annual heat loss is small. Given the inflowing transport of 0.84 Sv, such condition can be maintained with a mean inflow temperature of –0.65 °C. Next, from a review of the historical data found in the World Ocean Atlas 1998 (http://www.nodc.noaa.gov/OCL/woa1998.html),we find that the temperature at the mouth of HS is around 1.2 °C between 500 and 900 m, cooling to –0.7 °C at 150 m. On the other hand, from the few historical profiles we have, waters are colder (0.8 °C in August, –1.7 °C during winter) throughout the water column in the narrow strait 100 m deep (Annapolia and Gabriel Straits) just north of the mouth of HS. Across the wider section at the mouth of HS, these profiles show the temperature at depths over 500 m is also colder, from –0.4 °C in the north to –0.9 °C in the south, warming as we move up from 500 to 110 m to reach –0.4 in the north to 0.1 °C in the south. We note that while the core of the fresh outflow

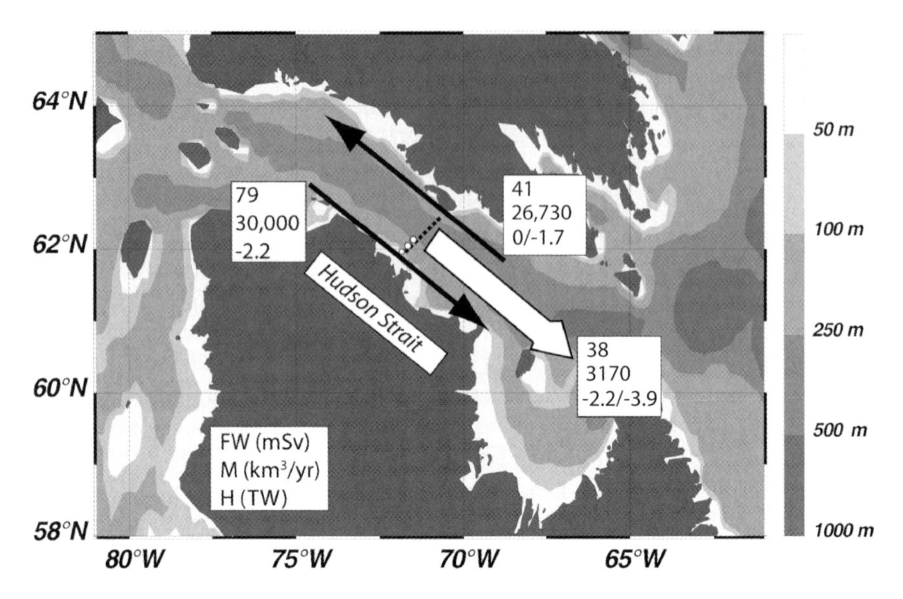

Fig. 10.4 Freshwater, volume and heat transports for the inflow and outflow through Hudson Strait (black arrows) and for the net (white arrow). Units are shown in the key. Also shown are the mooring locations (white circles) and the location of the section shown in Fig. 10.2 (dashed line)

from Davis Strait is cold (−1.5 °C; Cuny et al. 2005), its heat content is likely to modified as it mixes with warmer waters of the Labrador Sea at the entrance of Hudson Strait. Finally, if we look at the hydrographic section in Fig. 10.3, this shows that the mean inflow temperature is on the order of 0–0.5 °C, which would, in turn, yield a heat flux of 0–1.7 TW. If we assume that this is a reasonable estimate for the heat flux into the system it would result in a net heat transport out of Hudson Strait in the range of −2.2 to −3.9 TW. These numbers, in turn, imply that the HBS (west of the section) is a region of net heat loss with a mean annual heat flux out of −2 to −3.9 W/m^2. Clearly, these numbers are highly uncertain and should only be viewed as a preliminary attempt.

The transports of volume, freshwater and heat for the inflow, outflow and net flow through Hudson Strait are represented schematically in Fig. 10.4. Not surprisingly, the net transports are a poor indicator of the flow through the strait – especially for volume where they are an order of magnitude less than the actual circulation. For freshwater, the amount that flows out is equally due to the inputs into the HBS as to the inflow on the northern side.

10.4 Summary and Discussion

In this chapter, we have provided estimates for the net transports of heat, volume and freshwater through Hudson Strait as well as the respective contributions of the inflow and outflow. These are summarized in Fig. 10.4. While the estimates for the heat transports (except for measured outflow) remain strongly uncertain, we believe the remaining numbers to be reasonable.

The transports shown in Fig. 10.4 support the statement made in the introductory paragraph: that the HBS is an important contributor to the Arctic/North Atlantic exchange. They also show, however, that care must be taken in how one assesses the size of a contribution. If we look at the net contribution alone, the volume transport through Hudson Strait is small compared with the net 2–3 Sv that flow through Davis Strait (Cuny et al. 2005). This reflects the fact that the HBS is essentially an enclosed basin in contrast to the Arctic Ocean which has a net exchange both via Bering and Fram Straits. Yet, in terms of freshwater, the net freshwater flux through Hudson Strait is non-negligible and of the order of one third of the Davis Strait contribution (130 mSv according to Cuny et al. 2005).

As is obvious from Fig. 10.4, however, the net transports alone are a limited indicator of the pathways of Arctic water. The volume outflow from Hudson Strait is one order of magnitude larger than the net and about a third of that from Davis Strait (3–4 Sv according to Cuny et al. 2005 and Loder et al. 1998). The freshwater outflow is about two thirds of that flowing out of Davis Strait (120–150 mSv according to the same studies). Care must clearly be taken in interpreting these numbers since, in reality, what these results suggest is that approximately 1/3 (according to both freshwater and volume estimates) of the Davis Strait outflow recirculates into Hudson Strait instead of directly joining the Labrador Current further downstream. This recirculation must be taken into consideration when we list the contributors to the volume or freshwater flux of the Labrador Current – clearly the two outflows cannot be simply summed.

Why should we care if some of the Davis Strait outflow is recirculated through Hudson Strait? First, this implies re-drawing the Arctic export routes and adding a time lag for at least a portion of the Davis Strait outflow between the time it exits Davis Strait and when it merges into the Labrador Current. Second, the passage through the mouth of Hudson Strait will change the characteristics of the transiting water masses given the intense, tidally driven, mixing that occurs there. Third, and perhaps most important, the recirculation that is being discussed here has made its way to about the middle of Hudson Strait and is, possibly, on the way to Foxe Basin and Hudson Bay. If this is the case it will then participate in the water mass transformation processes of the HBS and re-emerge, transformed, several years later. At a time when we are seeking to understand how the variability observed in the Arctic region will propagate to the North Atlantic, and potentially impact global climate, the transit through Hudson Strait may modify the signal in a non-linear and, hence, non-trivial way.

The transports derived in this chapter represent our best estimates to this day. This analysis, however, highlights the chronic lack of data in a region and the imperative need to make additional measurements. The most important gap is due to the lack of measurements that cover a full seasonal cycle on the northern side of the strait. The second biggest gap is the lack of simultaneous flux measurements on both sides of Hudson Strait. Given the considerable interannual variability observed in the river outflow (Déry et al. 2005), for example, the sea-ice cover (Parkinson and Cavalieri 2002), or the atmospheric forcing (Saucier et al. 2004) – it is clear that observing both sides of the strait simultaneously is important. Hudson Strait is a rough working environment due to the strong tides, the large sea-ice ridges and, in general, its inaccesibility for much of the year. But, as the 2004–2005 measurements show, we now have the adequate technology (through moored, profiling instrumentation) to measure these transports.

Finally, like much of the Arctic Ocean, the HBS is undergoing rapid change. The river discharge into the HBS is decreasing (Déry et al. 2005) thus offsetting approximately 50% of the increased river discharge into the Arctic Ocean (McClelland et al. 2006). The sea-ice cover has been steadily decreasing (Laine 2004) and, in general, models show that this region is likely to undergo accelerated change towards ice-free conditions (Gagnon and Gough 2005a, b). The Intergovernmental Panel on Climate Change (IPCC 2001) identified the HBS as one particulary prone to climate change with important consequences for the indigeneous populations which depend on the stability of the region's ecosystem. Thus, not only is it important to assess the mean transports through this Arctic gateway but, also, we need to monitor its variability.

Acknowledgements The authors would like to thank the Department of Fisheries and Oceans, the captains and crews of the NGCC Pierre Radisson and NGCC DesGroseillers, technicians Jim Ryder, Sylvain Cantin, Roger Pigeon, Rémi Desmarais and Simon Senneville. Straneo would like to acknowledge support from the Woods Hole Oceanographic Institution's Ocean and Climate Change Institute and the Comer Foundation in particular. Also, Straneo would like to thank R. Pickart and B. Beardsley for use of their instruments, and T. McKee for help with the figures. Support to FJS from NSERC Research Grant and the Canadian Program on Energy Research and Development.

References

Barber FG (1965) Current observations in Fury and Hecla Strait. J. Fish. Res. Board Can. 22:225–229

Clarke RA, Gascard JC (1983) The formation of Labrador Sea Water. Part I: Large-scale processes. J. Phys. Ocean. 13:1764–1788

Cuny J, Rhines PB, Kwok R (2005) Davis Strait volume, freshwater and heat fluxes. Deep Sea-Res. I 52:519–542

Déry SJ, Stieglitz M, McKenna EC, Wood EF (2005) Characteristics and trends of river discharge into Hudson, James, and Ungava bays, 1964–2000. J. Climate 18:2540–2557

Drinkwater KF (1986) Physical oceanography of Hudson Strait and Ungava Bay. In: Martini IP (ed), Canadian Inland Seas. Elsevier, Amsterdam, pp. 237–264

Drinkwater KF (1988) On the mean and tidal currents in Hudson Strait. Atmos.-Ocean 26:252–266

Gagnon AS, Gough WA (2005a) Climate change scenarios for the Hudson Bay region an intermodel comparison. Climatic Change 69:269–297

Gagnon AS, Gough WA (2005b) Trends and variability in the dates of freeze-up and break-up over Hudson Bay and James Bay. Arctic 58:370–382

Ingram RG, Prinsenberg S (1998) Coastal oceanography of Hudson Bay and surrounding Eastern Canadian Arctic Waters. In: Robinson AR, Brink KN (eds), The Sea, vol 11. Wiley, pp. 835–861

IPCC (2001) Climate Change 2001 synthesis report: contribution of Working Groups I, II, and III to the third assessment report of the Intergovernmental Panel on Climate Change. Cambridge University Press, Cambridge

Laine V (2004) Arctic sea ice regional albedo variability and trends, 1982–1998. JGR, 109: C06027, doi:10.1029/2003JC001818
Lazier JRN (1980) Oceanographic conditions at Ocean Weather Ship Bravo, 1964–1974. Atmos-Ocean. 18:227–238
Lazier JRN, Hendry R, Clarke A, Yashayaev I, Rhines PB (2002) Convection and Restratification in the Labrador Sea, 1990–2000. Deep-Sea Res. I 49:1819–1835
Leblond PH, Osborn TR, Hodgins DO, Goodman R, Metge M (1981) Surface circulation in the western Labrador Sea. Deep Sea-Res. 28A:683–693
Loder JW, Petrie B, Gawarkiewicz G (1998) The coastal ocean off North-eastern North America: a large-scale view. In: Robinson AR, Brink KH (eds), The Sea, vol. 11. Wiley, New York, pp. 105–133
McClelland JW, Déry S, Peterson BJ, Holmes RM, Wood E (2006) A pan-arctic evaluation of changes in river discharge during the latter half of the 20th century. Geophys. Res. Lett. 33, doi:10.1029/2006GL025753
Mertz G, Narayanan S, Helbig S (1993) The freshwater transport of the Labrador Current. Atmos-Ocean 31:281–295
Parkinson CL, Cavalieri DJ (2002) A 21 year record of arctic sea-ice extents and their regional, seasonal and monthly variability and trends. Ann. Glaciol. 34:441–446
Pickart RS, Torres DJ, Clarke RA (2002) Hydrography of the Labrador Sea During Convection. J. Phys. Ocean. 32:428–457
Prinsenberg SJ (1977) Fresh water budget of Hudson Bay. Manuscript Report Series No. 5, Research and Development Division, Ocean and Aquatic Sci., Central Region, Environment Canada, Burlington, p. 71
Prinsenberg SJ (1980) Man-made changes in the freshwater input rates of Hudson and James Bays. Can. J. Fish. Aquat. Sci. 37:1101–1110
Prinsenberg SJ (1983) Effects of the hydroelectric developments on the oceanographic surface parameters of Hudson Bay. Atmos-Ocean 21:418–430
Prinsenberg SJ (1988) Ice-cover and ice-ridge contributions to the freshwater contents of Hudson Bay and Foxe Basin. Arctic 41:6–11
Sadler HE (1982) Water flow into Foxe Basin through Fury and Hecla Strait. Naturaliste Can. 109:701–707
Saucier FJ, Senneville S, Prinsenberg S, Roy F, Smith G, Gachon P, Caya D, Laprise R (2004) Modelling the sea ice-ocean seasonal cycle in Hudson Bay, Foxe Basin and Hudson Strait, Canada. Climate Dyn. 23:303–326
Shiklomanov IA, Shiklomanov AI (2003) Climatic change and the dynamics of river runoff into the Arctic Ocean. Water Resour. 30:593–601 (Translated from Russian)
Straneo F (2006) Heat and freshwater transport through the central Labrador Sea, J. Phys. Ocean. 36:606–628
Straneo F, Saucier FJ (2007) The outflow from Hudson Strait into the Labrador Sea. Deep Sea Res., in revision
Sutcliffe WH Jr, Loucks RH, Drinkwater KP, Coote AR (1983) Nutrient flux onto the Labrador Shelf from Hudson Strait and its biological consequences. Can. J. Fish. Aquat. Sci. 40:1692–1701

Chapter 11
Freshwater Fluxes East of Greenland

Jürgen Holfort[1,7], Edmond Hansen[1], Svein Østerhus[2], Stephen Dye[3], Steingrimur Jonsson[4], Jens Meincke[5], John Mortensen[5], and Michael Meredith[6]

11.1 Introduction

The northern North Atlantic features areas of strong surface cooling which sets up a southward flow of sinking cold, dense waters in the deep ocean. A northward flow of warm surface waters replaces the sinking southbound waters; a loop often termed the meridional overturning circulation (MOC). The associated northward heat transport is an important moderator of the high latitude climate along the flow. It is a major concern that excessive amounts of freshwater added to the northern North Atlantic could alter the dense water formation and associated ocean density contrasts driving this part of the MOC (Häkkinen 1999; Haak et al. 2003). The region has been undergoing a remarkable freshening since the mid-1960s (Curry and Mauritzen 2005; Curry et al. 2003; Dickson et al. 2002). Sources of freshwater input are runoff from Greenland, net precipitation, and export of freshwater from the Arctic in the form of sea ice and melt water through Fram Strait and the Canadian Archipelago. In the late 1960s a major freshening event contributed with as much as half of the extra freshwater required to account for the observed 1965–1995 freshening (Curry and Mauritzen 2005). The event was labeled the Great Salinity Anomaly (GSA) (Dickson et al. 1988), and appeared as extraordinarily fresh water circulating in the Subpolar gyre during the 1970s. The freshwater release has been attributed to an anomalous export of sea ice through Fram Strait during the late 1960s (Häkkinen 1993; Karcher et al. 2005). Pulses of excess fresh-water and sea ice appear to have been emitted from the Arctic also after the GSA;

[1]Norwegian Polar Institute, Tromsø, Norway

[2]University of Bergen and Bjerknes Center for Climate Research, Bergen, Norway

[3]Centre for Environment Fisheries and Aquaculture Sciences, Lowestoft, UK

[4]University of Akureyri and Marine Research Institute, Akureyri, Iceland

[5]University of Hamburg, Hamburg, Germany

[6]British Antarctic Survey, Cambridge, UK

[7]now at BSH, Germany

R.R. Dickson et al. (eds.), *Arctic–Subarctic Ocean Fluxes*, 263–287

both the 1980s and 1990s featured appearances of low salinity water in the Subpolar gyre (Belkin 2004). Karcher et al. (2005) attributed the salinity anomaly of the 1990s to a large release of liquid freshwater from the Arctic through Fram Strait.

Aagaard and Carmack (1989) provided a complete accounting of the freshwater budget of the Arctic Ocean. Dickson et al. (2007) updated this landmark report by collating new estimates of freshwater flux through Arctic and subarctic seas, while Serreze et al. (2006) introduced ERA-40 reanalysis and land surface and ice–ocean models to synthesize the understanding of the Arctic's large-scale freshwater cycle. In this chapter we address some of the components in the Arctic freshwater budget: the export of freshwater through Fram Strait and the temporal and meridional evolution of the freshwater transport south along East Greenland. This transport occurs in the form of sea ice or freshwater, i.e. water fresher than some reference salinity. It does not follow a continuous path, several factors modulate the content and transport of freshwater at a given section. A large portion of the freshwater that leaves the Arctic in the form of sea ice is converted into a liquid mode due to melting while flowing south (Vinje et al. 2002). During winter there is also local ice production, with a resulting salinification of the water below. Some of the ice is diverted into the Nordic Seas and does not reach, at least not directly, the North Atlantic. Additional freshwater may enter/leave the EGC as precipitation/evaporation or as runoff from Greenland. Once the freshwater has attained a liquid form it may mix vertically or laterally. Even if it ultimately reaches the North Atlantic, the mixing may at that stage occur near the bottom as part of the Denmark Strait Overflow Water (DSOW). Lateral mixing may export freshwater out of the EGC and into the Nordic Seas. Whereas the corresponding ice export can be observed by remote sensing, the liquid freshwater leakage from the EGC is difficult to observe.

Estimates of sea ice fluxes through Fram Strait have been made based on ice thicknesses measured by upward looking sonars (ULS) (Vinje et al. 1998) and drift velocities from in-situ measurements or remote sensing. The liquid freshwater transport is somewhat more difficult to observe directly. The bulk of the transport occurs in the very upper layers, where ridged sea ice and drifting icebergs represent a hazard to the instrumentation. Earlier estimates in Fram Strait were mainly based on budget calculations or estimated mean volume transports with some mean salinity. With the advent of the EU-VEINS project in 1997 the first efforts towards direct observations of the liquid freshwater transport through Fram Strait at 79° N were initiated. In 2000 similar observations were initiated at 74° and 63° N. The observations were continued into the ASOF period, partly funded through the EU-ASOF projects. The first freshwater flux estimates east of Greenland were published by Holfort and Meincke (2005), who presented estimates for the 74° N observation site. Similar estimates from the 79° N section are in the process of being published (Holfort and Hansen, in preparation)

In addition to a brief account on the various estimates of freshwater transport east of Greenland, this chapter should provide an overview of results from the freshwater observations at the three ASOF arrays at 63°, 74° and 79° N. Several open questions from the pre ASOF years may now be addressed. How are the freshwater observations along the freshwater path east of Greenland connected? What are the similarities, what are the differences? Do we see signals propagating in the Polar Water (PW) of the EGC? Are we able to determine changes in the freshwater transport along the EGC, with the final goal of estimating the freshwater leakage to the Nordic Seas? What is the temporal variability and are there long-term trends in the signal? With this chapter we take the opportunity to address these questions and thereby summarize the status of the ASOF freshwater observational work east of Greenland.

11.2 "Freshwater" in the Ocean?

What does the term "freshwater" mean in this context? If we, in a "traditional" or strict sense, use the term freshwater as a synonym for pure H_2O, then most of the seawater is indeed freshwater with a minor salt constituent. In this case the freshwater transport would be approximately equal to the entire volume transport. The term "freshwater" is here rather used as a relative quantity, or a deviation from a certain mean salinity of an ocean water mass. This mean salinity is often used as the *reference salinity* for freshwater content and flux calculations.

A water column of some salinity can be viewed as if it was composed of two fractions: One with water equal or above a given reference salinity, and a second fraction of zero salinity water that is required to dilute the water above the reference salinity to the actual salinity of that water mass. The amount of zero salinity water that is required to achieve this dilution is termed "freshwater" and may be expressed in terms of a freshwater thickness layer FW_T measured in meters. Mathematically this would be expressed as an integration over the depth of the water column; $FW_T = \int((S_{ref}-S_z)/S_{ref})dz$, where S_{ref} is the reference salinity and S_z the salinity of the water at depth z. Integrating the product of the freshwater thickness and the cross section velocity component over a section yields the freshwater transport through this section.

Only freshwater entering as runoff and precipitation is independent of the reference salinity. When calculating oceanic freshwater content and flux, one must select a reference salinity that is of general applicability to the waters being addressed. In the Arctic Aagaard and Carmack (1989) used 34.8, the estimated mean salinity of the Arctic Ocean and the most commonly adopted value in the literature. In their review report, Dickson et al. (2007) used the same value when reviewing the historic point estimates, but they used a value of 35.2 when calculating the freshwater balance of the Arctic Mediterranean. The latter value is the salinity of the inflowing Atlantic water (Hansen and Østerhus 2000). In Fram Strait the mean salinity is 34.9. For the purpose of calculating the export of freshwater through the strait at 79° N, this value has been used as reference salinity. This work is now in the process of being published (Holfort and Hansen, in preparation). The same value of 34.9 was also used for estimating the freshwater flux at 74° N (Holfort and Meincke 2005).

Sea ice is included in the oceanic freshwater budget, but is considered separately from the liquid component. In this contribution we focus on the liquid part, sea ice is only discussed when it interacts with the liquid component. This division between sea ice and liquid freshwater is also discussed in Section 11.7 from tracer estimates.

11.3 Freshwater Input and Fluxes East of Greenland

A modern review of current estimates of freshwater fluxes through Arctic and Subarctic seas already exists; the compilation by Dickson et al. (2007). Still, for the completeness of this chapter we quote some numbers from the historic literature on freshwater fluxes east of Greenland. The ASOF sections at 63°, 74 and 79° N are the main focus in this contribution and are discussed in more detail later. But, again for the sake of completeness, these sites are listed here as well. We follow the geographic division of Dickson et al. (2007), and refer to this publication for a more complete account.

11.3.1 Fram Strait

Recent estimates of sea ice flux range from 2,218 km^3/year (70 mSv) (Kwok et al. 2004), via 2,400 km^3/year (76 mSv) (Widell et al. 2003) to 2,850–2,900 km^3/year (90–92 mSv) (Vinje et al. 1998; Vinje 2001). Aagaard and Carmack (1989) quoted a sea ice flux of 2,790 km^3/year (88 mSv). Dickson et al. (2007) point out that the estimates are in reasonable agreement with the average value of the 1990s of 96 mSv from the NAOSIM model (Karcher et al. 2005).

The liquid freshwater flux through Fram Strait at 79° N has been in focus during the ASOF years, and is discussed in more detail below. From direct observations using moored instruments, Holfort and Hansen (in preparation) arrived at an annual average of ~1,000 km^3/year (32 mSv) in the EGC, relative to 34.9. This is in good agreement with the NAOSIM model, which also reproduces the annual cycle and interannual variability of the freshwater transport well (Hansen et al. 2006). The transport on the shelf is still an open question, although NAOSIM suggests that the transport there is about 700 km^3/year (22 mSv). Based on two hydrographic and $\delta^{18}O$ transects, Meredith et al. (2001) derived meteoric water fluxes of 2,000–3,680 km^3/year (63–117 mSv). This is further discussed in Section 11.7 on tracer estimates. Aagaard and Carmack (1989) estimated the liquid freshwater import to the Greenland, Iceland and Norwegian Seas (GIN) as 1,160 km^3/year (37 mSv), relative to 34.93 as reference salinity. The latter being their estimate of the mean salinity of the Nordic Seas. Relative to 34.8, Aagaard and Carmack (1989) estimate a flux of 820 km^3/year (26 mSv). Jónsson (2003) arrived on an estimate of ~3,940 km^3/year (125 mSv) for the total freshwater flux through Fram Strait.

11.3.2 Greenland Ice Sheet

The Greenland ice sheet is indeed the largest single freshwater storage of this region, and even of the northern hemisphere. Modeling (e.g. Fichefet et al. 2003) and remote sensing (e.g. Thomas et al. 2006) studies show that there is an progessive increase in the ice loss from this freshwater storage. Based on a mix of modeling techniques reported in the literature, Dickson et al. (2007) conclude that a value of ~570 km^3/year (18 mSv) is appropriate for the present annual freshwater flux from Greenland.

11.3.3 East Greenland Current at 74° N

Based on direct observations by moored instruments in 2001 and 2002, Holfort and Meincke (2005) estimated a mean liquid freshwater transport in the EGC of 869 km^3/year (28 mSv). Assuming that the total freshwater content of the water column stays constant over the year, and that the ice drift is closely correlated with the upper layer currents, they estimated a total liquid plus solid freshwater transport of 1,250–1,750 km^3/year (40–56 mSv).

11.3.4 Jan Mayen Current

Jónsson (2003) gave an estimate of ~315 km^3/year (10 mSv) for freshwater being directed from the EGC into the Jan Mayen current, based on historical literature.

11.3.5 East Icelandic Current

Based on hydrography and direct current observations, Jónsson and Briem (2003) estimated the diversion of freshwater from the EGC into the East Icelandic Current as ~158 km^3/year (5 mSv).

11.3.6 Denmark Strait

Based on literature and cited personal communication, Aagaard and Carmack (1989) argue that half of the sea ice exported through Fram Strait melts before 73° N. Extrapolating this to Denmark Strait, they conclude that the freshwater flux through Denmark Strait in the form of sea ice is 560 km^3/year (18 mSv). NAOSIM modeling results (Karcher et al. 2005) suggest a flux of roughly double this value; the 1950–2000 mean is ~1,200 km^3/year (38 mSv).

For the liquid freshwater transport through Denmark Strait, Aagaard and Carmack (1989) estimate a value of 1,520 km^3/year (48 mSv). This estimate is relative to the salinity 34.93. Karcher et al. (2005) gave timeseries of liquid freshwater flux through Denmark Strait produced by the NAOSIM model, but relative to 34.8. There is an increasing trend and strong interdecadal variability over the 1950–2000 integration period, with values typically between 2,500 and 3,000 km^3/year (79–95 mSv) during the second half of the period. During the anomalous freshwater release in the 1990s the flux peaked to about 3,600 km^3/year (114 mSv).

11.3.7 South East Greenland Shelf

Dickson et al. (2007) report a preliminary and partial estimate from moorings at 63° N; ~2,020 km^3/year (64 mSv). This is further discussed in the following sections. Bacon et al. (2002) provide a snapshot estimate for an observed freshwater jet on the shelf at Cape Farewell on the southern tip of Greenland. The jet, driven by melt water runoff from Greenland, carried 1,800 km^3/year (57 mSv) relative to a "transatlantic mean salinity" of 34.956.

11.4 The ASOF Freshwater Mooring Arrays Along East Greenland

Although data from the shelf and shelf slope region east of Greenland (Fig. 11.1) in general are sparse, hydrographic observations within the EGC and on the east Greenland shelf have been performed on several recent cruises. However, most observations were obtained in summer, and even then mostly in ice-free waters. Wintertime hydrographic data from the region are very rare, and even more so for the regions with a wintertime ice cover.

Maintaining long-term observations of freshwater by moorings are hampered by drifting sea ice, for two main reasons. First, deploying and recovering moorings within the ice field requires ships with sufficient ice class and machinery. Second, the bulk of the freshwater is located in the near surface layers. Ice bergs and deep pressure ridges represent a hazard to the instrumentation in layers above, say, 50 m. In the case of ice bergs, even 100–150 m is not a safe distance from the surface.

Nevertheless, long-term monitoring of ice thickness and ice drift in Fram Strait was initiated by the Norwegian Polar Institute (NPI) in 1989 (Vinje et al. 1998). Monitoring of liquid freshwater was added in 1997 under the EU-VEINS project, as a part of the long-term mooring line across Fram Strait (see, e.g. Schauer et al. this volume). The continuation of these observations was secured from various national sources and the 2003–2006 EU-ASOF projects. The liquid freshwater

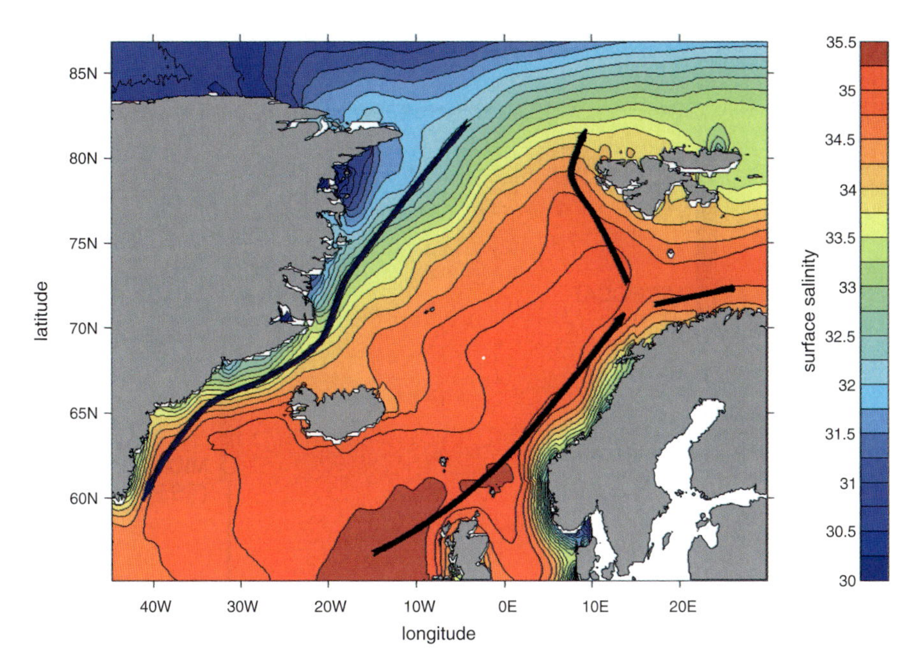

Fig. 11.1 Map of the seas east of Greenland, with surface salinities from WOA, and a sketch of the surface circulation

transport is monitored using moorings equipped with temperature (T) and salinity (S) sensors as well as current meters. Upward looking sonars (ULS) facilitate the estimation of ice thickness, which, along with the ice drift estimated by ADCPs (Acoustic Doppler Current Profilers), enable us to calculate the solid freshwater fluxes.

The freshwater mooring array close to the shelf edge at 79° N in Fram Strait provides the longest time series and best resolved record of freshwater within the EGC (Holfort and Hansen, in preparation). But it was only after 2003 that this mooring array covers the wide shelf at this latitude. In the presence of at least a seasonal ice cover, measurements of temperature and salinity near the surface became available only after the instruments and the flotation were protected from being destroyed by ice. This was achieved using long (~40 m) polyethylene tubes, with instruments and flotation inside (Fig. 11.2). The first such moorings, devised by University of Hamburg and dubbed "tube moorings" or just "tubes", were deployed in 2000 at 74° N (Holfort and Meincke 2005) and at 63° N. These are the two other latitudes where multi-year mooring data are available, maintained by the University of Hamburg and CEFAS at Lowestoft.

At 79° N there is continuous mooring data addressing the liquid freshwater flux since 1997, from nominally four moorings covering the EGC. From 2003 three moorings were added to this array on the continental shelf further towards Greenland, as an effort to capture some of the freshwater transport that might take place here. In addition to hydrographic sections in summer, which in 1997 and

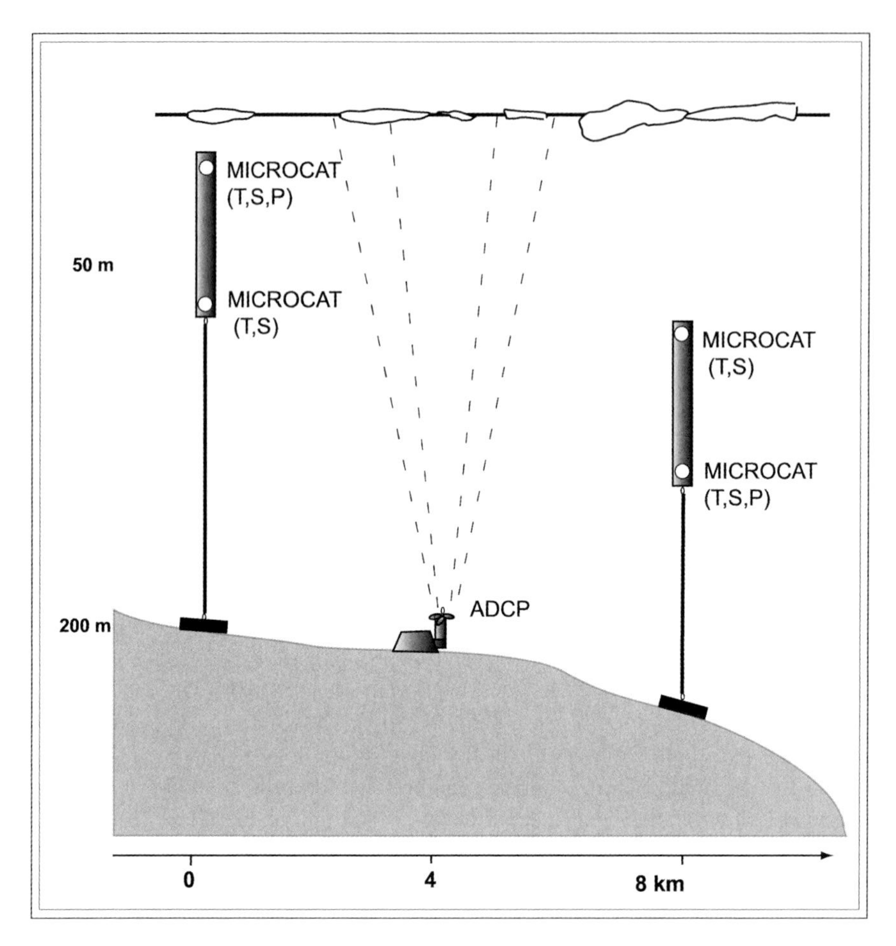

Fig. 11.2 Schematic representation of tube moorings and ADCP velocity measurements as used at 74° N (Holfort and Meincke 2005)

1998 included tracer observations (Meredith et al. 2001) also wintertime hydrographic data are available.

At 74° N mooring data exists since 2000. A maximum of two tube moorings have been out simultaneously, with an ADCP moored between the tubes measuring the velocity profile. Hydrographic sections were done only in summer, but then also covering the ice covered part of the section.

Also at 63° N mooring data exists since 2000. A maximum of two tube moorings were out at the same time to measure temperature and salinity. Single point current meters below the tubes measured velocities, but never recorded a full seasonal cycle. Hydrographic sections have been done over several summers, but not into the ice covered part of the section.

Most of the freshwater is found within the Polar Water (PW). PW is a cold and low saline water mass originating from the Arctic Ocean, its low salinity reflects its large freshwater content. At the three ASOF sections it is seen at the surface and down to about 200–300 m (Fig. 11.3), covering the East Greenland shelf and portions of the shelf break. The sea ice, also mainly originating from the Arctic Ocean, is closely interacting with PW through melting and new ice formation. The largest

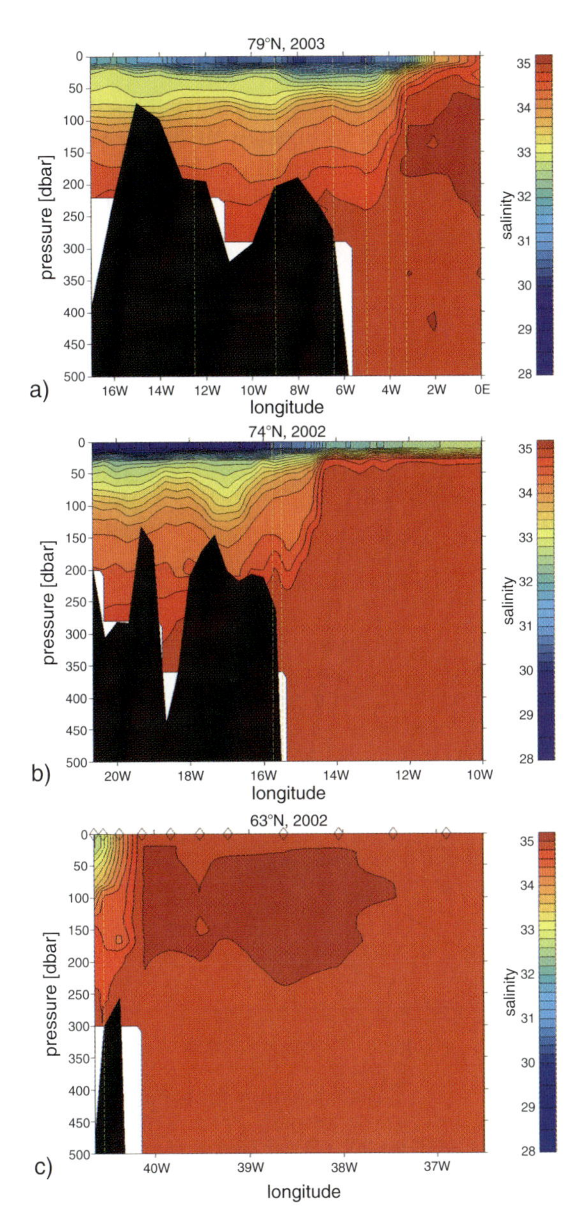

Fig. 11.3 Sections of salinity across 79° N (a, summer 2003), 74° N (b, summer 2002) and 63° N (c, summer 2002) with approximate mooring positions

signal in the PW salinity is indeed the annual cycle, which reflects the transition between solid and liquid freshwater. The best correlations between salinity and local ice cover is observed at 74° N, where large parts of the local salinity changes can be attributed to local ice formation and melting (Holfort and Meincke 2005). At 79° N this correlation is lower, ice advection and recirculating Atlantic water are additional factors influencing both the salinity and the ice cover there. At 63° N the mooring is situated very near the frontal zone between PW and Atlantic water, hence frontal shifts are dominating the variability.

Figure 11.4 shows the freshwater thickness at the three latitudes from summers when there were cruises to all three sections. At all the three latitudes the freshwater thickness reaches its maximum near the coast, where it amounts to about 8–10 m at all three sections. In the EGC, just outside the shelf break indicated in Fig. 11.4 by

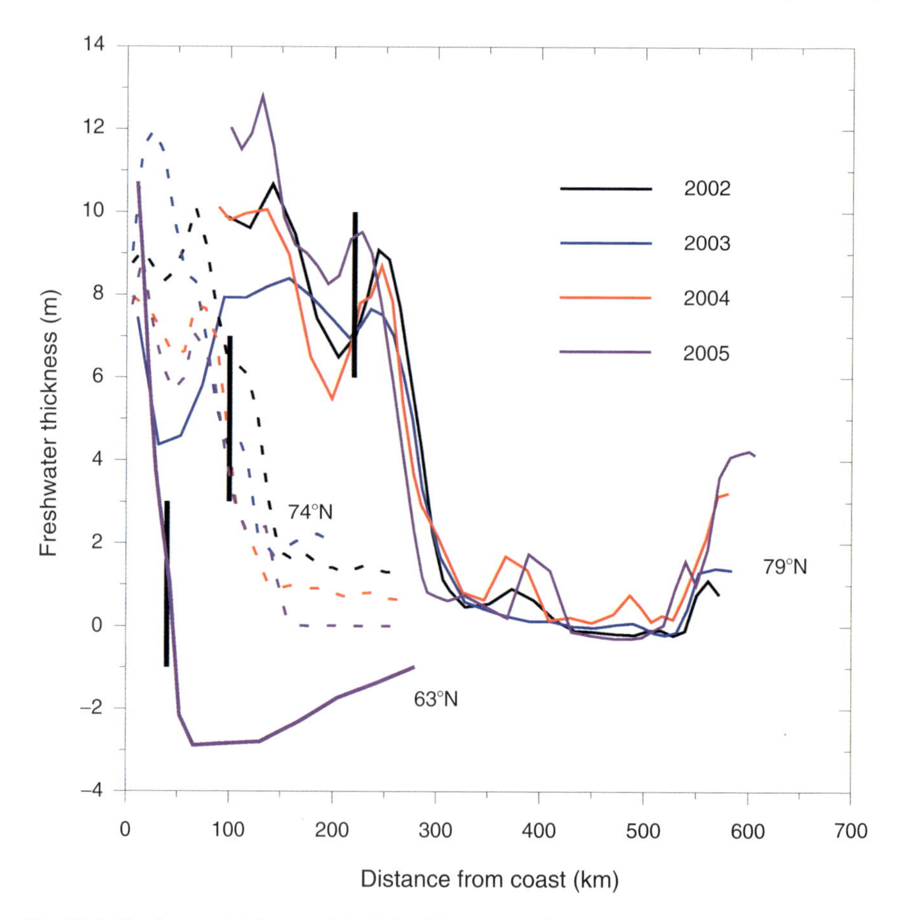

Fig. 11.4 Freshwater thickness of the Polar Water layer along the sections at 79° N, 74° N and 63° N between 2002 and 2005. Based on a mean between two neighboring stations from the surface down to the σ_θ = 27.7 kg m^{-3} isopycnal and relative to a reference salinity of 34.90. Vertical black lines indicates the position of the east Greenland shelf edge at 63°, 74° and 79° N, respectively

a black vertical bar, the freshwater thickness is decreasing as we go south. Associated with this decrease is also a change in the TS characteristics. The processes responsible for this change are salinification of PW from interaction with Atlantic water, cooling and salinification from ice formation in winter and escape of some freshwater into waters with densities above $\sigma = 27.7\,\mathrm{kg\ m^{-3}}$. The latter was used as a lower boundary for the freshwater thickness calculation.

The relative maximum in freshwater thickness found near the shelf break, as well as the minimum west of it at 74° N and 79° N, is due to topography. Although the salinity is generally decreasing towards the west, the water column and thereby the freshwater thickness decreases due to decreasing depth. The freshwater content in the deeper regions also depends on the chosen lower bound of the water column in consideration, but appropriate choices normally encompass the whole water column on the shelf itself. At 63° N the freshwater is restricted to the inner shelf, at 74° N it reaches the shelf break and at 79° N it extends even farther out. In addition the shelf width decreases towards south, so that the freshwater content decreases going southwards.

11.5 Mean Liquid Freshwater Transport from the ASOF Moorings

At 63° N the best mooring coverage available at the same time is two tube moorings. A tube mooring carries a maximum of three salinity/temperature sensors and two pressure sensors within the tube, and one current meter below the tube. At other times only one mooring is available for the calculations. With only one tube mooring covering the transport we must assume that the data from the mooring is representative for a certain region around that mooring, and that it is within this region that most of the transport takes place. With two moorings available one can linearly interpolate between the moorings and extrapolate further assuming constant values, or select a representative width for each mooring and add the transports found from each mooring. A similar procedure must be adopted in the vertical. As only one current meter is available on each mooring, we must assume that the current is barotropic. With a barotropic current only the depth integrated salinity is of importance for the freshwater transport. The error from various interpolation methods of the salinity in the vertical is small compared to other error sources, for example assumptions about the width of the current. Assumptions about representative depth intervals, or more advanced interpolation methods where also the summertime high resolution CTD profiles are included, would therefore not reduce the error bars.

For the mooring-based freshwater transport estimates we therefore assume a barotropic current, and use the mean salinity in the 10–110 m interval from the tube measurements. The current width/depth used in the calculations are 36 km/250 m for the inner tube and 24 km/350 m for the outer tube whenever two tubes are available. With only one tube available the corresponding current width/depth assumption is 60 km/300 m.

The transports in the overlapping months (July–October) (Table 11.1a and b) for all years vary considerably, between ~150 and 2,800 km^3/year (8–89 mSv) southwards. As the outer mooring is situated near the frontal zone separating PW from the surrounding water, there are times where this mooring is outside the PW. This can also explain the large variability between months and even between years. The 2003 values, especially July–September, are very small. During this time the front was probably situated west of the mooring, and only small parts or nothing of the PW transport was measured at the mooring position. Our best estimate therefore is the 2001 calculation with two tubes, where at least one tube is within the PW most of the time. Compared to the yearly mean, this summer value of ~2,200 km^3/year is on the high side, since during winter some of the liquid freshwater will be in the form of ice and the corresponding liquid freshwater transport is smaller than in summer.

At 74° N there are tube moorings at two positions on the shelf measuring upper ocean temperature and salinity, but there is only one position/mooring with current measurements. Nevertheless, since the current measurements are done using an upward looking ADCP, covering the whole water column, this current measurement is better suited for freshwater transport calculations than the two point measurements from 63° N. The observation period to be dealt with here covers 2 years.

Table 11.1a Southward freshwater transport (in km^3/year) at 63° N with a reference salinity of 34.9. The values represent monthly means from July to October, the mean of this 4 months and the mean of the full time span with data, which differs from year to year. "Stddev" provides the corresponding standard deviations

Dataset	July	August	September	October	Mean JASO	Full series
2001, 2 tubes	−2,505	−1,963	−2,173	−2,216	−2,214	−2,196
stddev	933	1,058	925	2,012	223	1,284
2001, outer	30	−55	−728	−500	−313	−531
stddev	471	469	747	939	361	1,009
2002, outer	−1,892	−647	−890	−956	−1,096	−1,303
stddev	2,619	858	1,184	780	547	1,861
2003, outer	−22	−42	−37	−512	−153	−674
stddev	43	44	104	507	239	979

Table 11.1b Same as Table 11.1a, but with a reference salinity of 35.2

Dataset	July	August	September	October	Mean JASO	Full series
2001, 2 tubes	−3,000	−2,408	−2,767	−2,794	−2,742	−2,723
stddev	1,087	1,266	1,105	2,324	246	1,500
2001, outer	71	−62	−1,083	−696	−443	−736
stddev	718	673	1,030	1,283	543	1,354
2002, outer	−2,183	−784	−1,167	−1,411	−1,386	−1,628
stddev	3,104	1,026	1,490	1,029	591	2,196
2003, outer	207	161	3	−885	−129	−998
stddev	117	72	262	714	512	1,376

Additionally both tube moorings and the ADCP in between them were well within the PW and near the current core most of the time. For the first year (2000–2001) Holfort and Meincke (2005) estimated a mean transport of 869 km^3/year (28 mSv), scaling the current width to get 1 Sv of total volume transport. For the second year of data similar calculations yield a mean freshwater transport of 700 km^3/year (22 mSv), again assuming a 1 Sv volume transport. Including the shelf, and assuming a total volume transport of 1.9 Sv, they arrived on a total freshwater transport (EGC + shelf) of 1,400 km^3/year (44 mSv) (see Fig. 11.5).

Some summer months are ice free at 74° N, when the liquid freshwater transport is observed to be approximately twice as large as the annual mean. With no solid phase freshwater present, this equals the total freshwater transport. Based on the good correlation between salinity and regional ice cover concentration (Holfort and Meincke 2005), we assume that during the year the total freshwater content is constant with an interchange between liquid and solid phase. Further assuming that the sea ice is carried with the same speed as the surface currents, we conclude that the annual mean freshwater transport in the form of sea ice is approximately the same as the annual mean liquid freshwater transport.

At 74° N the shelf is wider and the longitudinal extent of PW carrying the freshwater is larger then at 63° N. Further north at 79° N the shelf widens even more. Large uncertainties about the amount of freshwater on this shelf still exist. Is there a coastal jet carrying larger amounts of FW southward, like observed by Bacon et al. (2002) at 60° N? At 79° N a recirculating current/eddy on the shelf leads to a northward transport of low salinity water near the coast; what is the shelf circulation

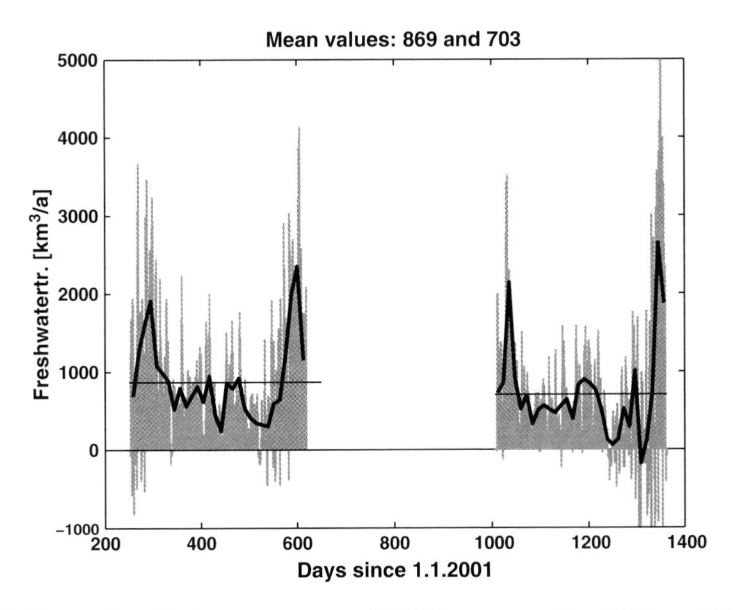

Fig. 11.5 Time series of freshwater transport at 74° N for a cross section of 34 km, which represents only a part of the total transport

at 74° N like? Is the shelf current pattern persistent over the year, does it have only local effects or does it contribute more significantly to the mean freshwater transport? These questions cannot yet be answered due to lack of data.

The data coverage is best at 79° N. Nominally four moorings cover the shelf break, where the main core of the EGC is situated. Although salinity measurements are missing in the uppermost water column (above ~50–60 m), the thickness of the PW layer is so large that the measurements at 50–60 m depth catch the PW salinity changes quite well. ADCPs on most of the moorings capture the upper ocean velocity structure. The time series starts in 1997, and although mooring losses have occurred in several years, a good estimate of the freshwater transport is available since then.

The liquid freshwater transport estimates from this array (Holfort and Hansen, in preparation) result in a long-term mean value of ~1,000 km³/year (32 mSv) for the EGC. The shelf is wide at this latitude, no time series from moorings are available for shelf transport calculations. Based on geostrophic calculations from CTD sections including the shelf (Fig. 11.6), we conclude that the summertime total liquid (EGC and shelf) freshwater transport is ~1,500–3,000 km³/year (48–95 mSv). As seen in Fig. 11.6 there is a good agreement between geostrophic calculations and the slope mooring estimates. NAOSIM modeling results give a long-term mean freshwater transport over the shelf of ~700 km³/year (22 mSv) (Hansen et al. 2006; see also

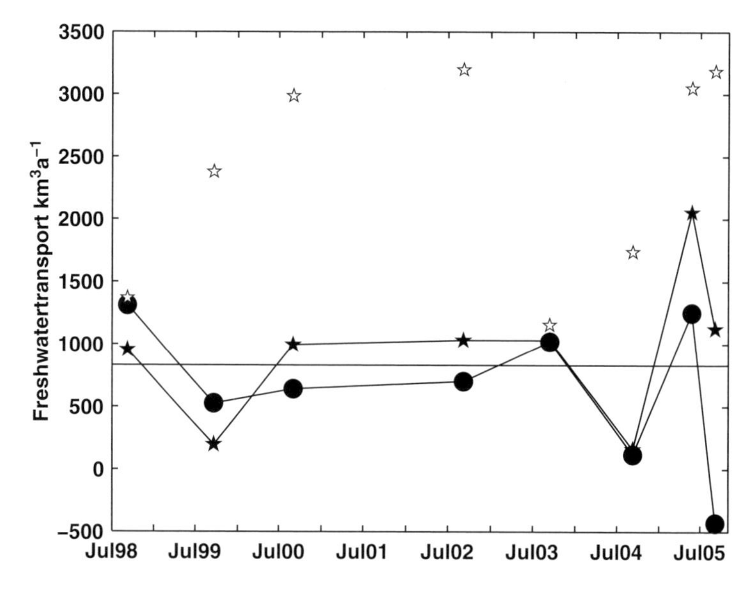

Fig. 11.6 Freshwater transport at 79° N. Stars are geostrophic calculations from CTD sections using a level of no motion at the deepest common depth. Filled stars are the transports across the area of the mooring array (in the EGC, over the slope), while hollow stars are the transports across the whole CTD section including varying portions of the east Greenland shelf. Filled circles are mean moorings transport around the time of the CTD sections

Gerdes et al. in this book). During summer the PW carries more liquid freshwater than during winter. Based on the information at hand, we conclude that the total (EGC plus shelf) liquid annual mean freshwater transport through Fram Strait is ~2,000 km^3/year (63 mSv).

11.6 Temporal Variability and Signal Propagation Along East Greenland

At 74° and 79° N, where 2 year or longer freshwater transport estimates are available, the largest temporal signal is the seasonal cycle. This reflects the seasonal cycle of salinity. At 74° N (Fig. 11.7) the transport minimum is found around May/June when the upper water salinity is at its maximum, the transport maximum is found in September when the upper salinity is at its minimum. In the seasonal cycle of the total freshwater flux, the liquid signal will be in opposite phase of the solid phase signal: The ice concentration is at its minimum, in some years even zero, in September. The freshwater transport in form of sea ice is therefore at its minimum when the liquid freshwater transport is at its maximum. Correspondingly, with more sea ice present in winter the solid freshwater transport is largest when the liquid transport is at its minimum. The second largest factor influencing freshwater transport is the total volume transport. At 74° N this is the main source of freshwater transport variability in late winter/early spring, when the variability in salinity is small. At 79° N, where the measurements include also deeper parts of the ocean, we observe that it is only the mass transport in the upper few hundred meters that is important for the freshwater transport. As the salinity in the deeper part of the water column is near the reference salinity, the changes in volume flux have only a minor impact on the freshwater transport. The variability of both these parameters explain most of the variability of the freshwater transport.

In the longer time series at 79° N, after subtracting the mean seasonal cycle, Holfort and Hansen (in preparation) report variability of about 500 km^3/year (~50% of the long-term mean) for periods of several months, but no longer term trend. This corresponds to Vinje (2001) observations for the ice export through Fram Strait.

As long-term continuous (1997–2006) freshwater transport estimates are only available at 79° N we cannot directly address the propagation of this signal to the south using the other two mooring arrays. But as the transport variability is correlated with salinity changes, we can use the available longer term time series of salinity at the three latitudes to try to identify a signal propagating within the EGC southwards. A propagating advective signal along the EGC would first appear in the north and appear at 74° N about 3 months later, using a propagating speed of 10 cm/s. It would show up at 63° N another 6 months later. As the seasonal cycle is the major signal in both temperature and salinity, monthly means were calculated and subtracted. In the resulting salinity and temperature anomalies (Fig. 11.8) we can not observe any propagation of anomalies. In general any anomalies occur at

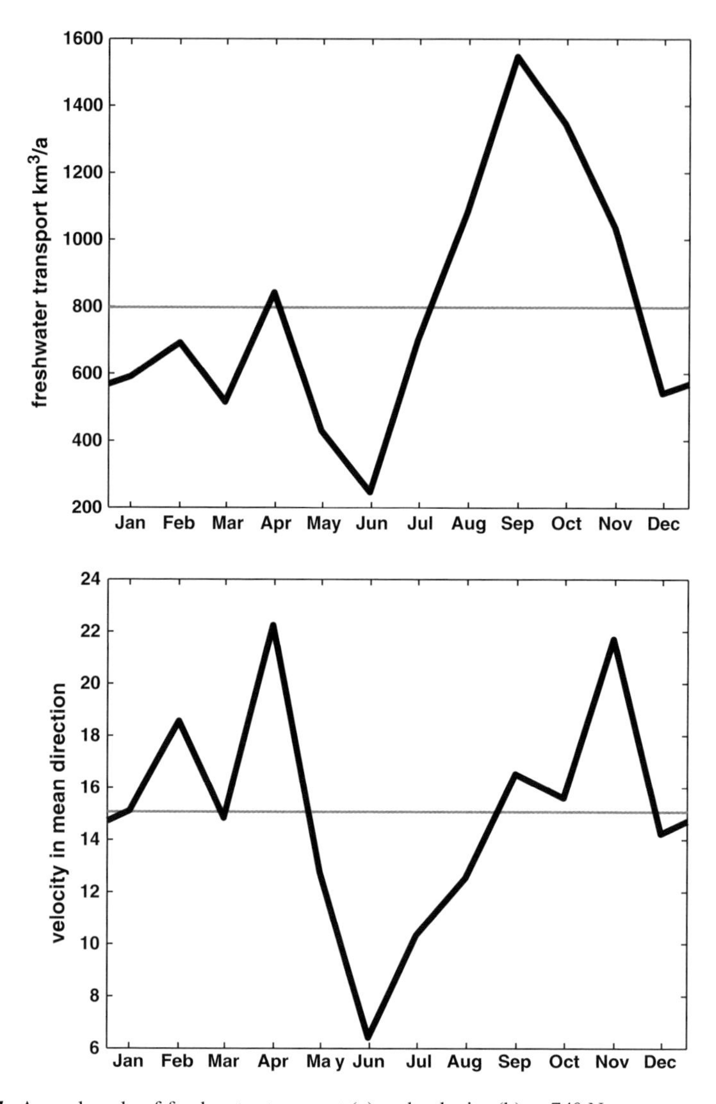

Fig. 11.7 Annual cycle of freshwater transport (a) and velocity (b) at 74° N

more or less simultaneously. This indicates that the variability is due to large-scale atmospheric forcing (heating, wind, etc.). There is one small signal in the temperature which could be due to advection, propagating from 79° N to 74° N near the end of 2002. It does not propagate further, or the associated signal is too small to be distinguishable within the large fluctuations at 63° N.

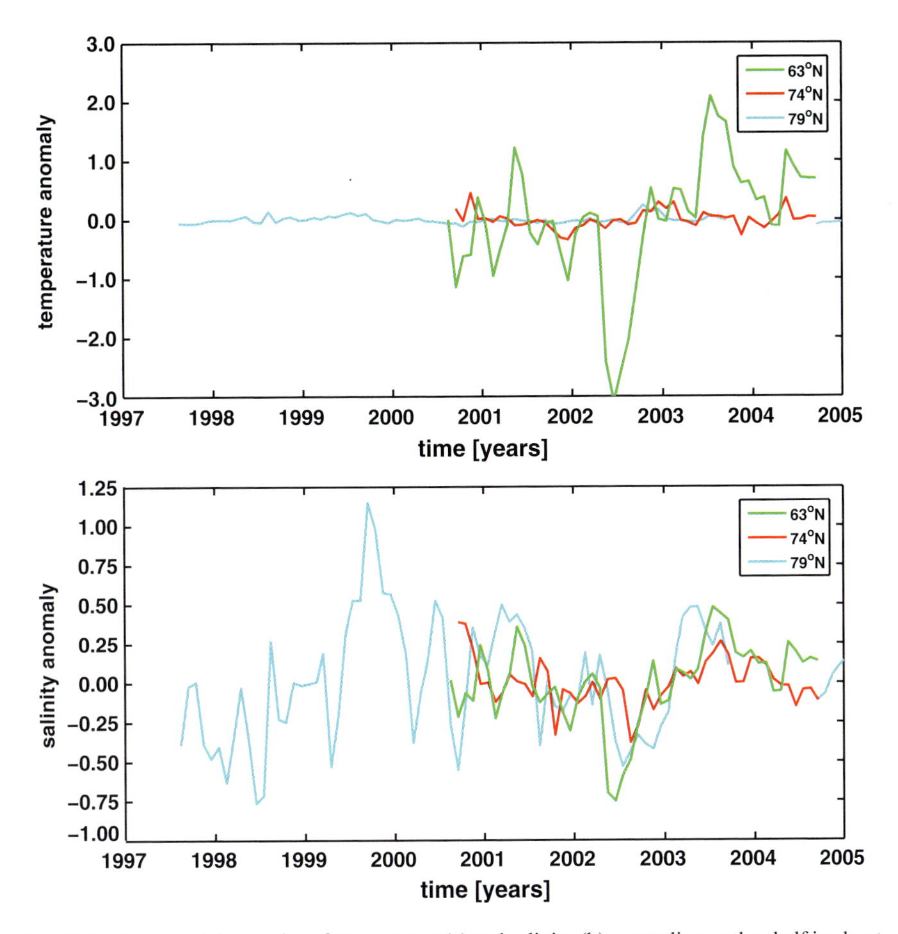

Fig. 11.8 Smoothed time series of temperature (a) and salinity (b) anomalies on the shelf in about 60 m depth at 63°, 74° and 79° N

11.7 Tracer Estimates

Measurements of salinity are vital in determining the freshwater content and transport in the ocean, but by themselves cannot provide information on the sources of the freshwater being measured. For this, other tracers must also be incorporated. A particularly powerful tracer for this purpose is $\delta^{18}O$, the standardized ratio of stable oxygen isotopes in seawater ($H_2{}^{18}O$ to $H_2{}^{16}O$) (Craig and Gordon 1965). This tracer was first used in a systematic, quantitative way for Arctic freshwater studies by Ostlund and Hut (1984), who exploited the fact that meteoric water inputs (river runoff and precipitation) are isotopically lighter than sea ice melt (around 21‰ versus approximately 2‰), despite having very similar salinities (low or zero).

Ostlund and Hut (1984) used a three-end member mass balance to quantify the percentages of meteoric water versus sea ice melt, based on combined sampling for salinity and $\delta^{18}O$:

$$Fa + Fsim + Fmet = 1$$

$$Fa*Sa + Fsim*Ssim + Fmet*Ssim = Smeasured$$

$$Fa*\delta a + Fsim*\delta sim + Fmet*\delta sim = \delta measured$$

where Fa,Fsim,Fmet are the fractions of Atlantic water, sea ice melt and meteoric water; Sa,Ssim,Smet are the respective salinities of the undiluted forms of these waters; δa, δsim, δmet are the corresponding $\delta^{18}O$ values of the undiluted waters and Smeasured, δmeasured are the measured salinity and $\delta^{18}O$ values of the samples.

This technique was subsequently employed to great effect by a succession of authors (e.g. Bauch et al. 1995; Schlosser et al. 1994; Schlosser et al. 2002), but a restriction of such works was that they were typically constrained to identifying solely percentages of freshwater from different sources, rather than their fluxes. The difficulty was that derivation of reliable fluxes requires concurrent velocity data, and such information was lacking.

Meredith et al. (2001) was the first to overcome this problem, using combined salinity and $\delta^{18}O$ measurements with concurrent velocity data from current meter moorings at 79° N in Fram Strait. Two tracer sections were occupied (in August/September 1997 and 1998), during which period an array of moorings was in place across the EGC and broader Fram Strait. From the tracer sections, it was found that up to ~16% of meteoric water was present in the EGC in both years, but that the waters there had become relatively saline by the net integrated formation of up to ~11 m of sea ice. A large area of isotopically light water was observed occupying the East Greenland Shelf, indicative of significant meteoric water there.

By integrating the covariance of the derived freshwater fractions and the velocity, Meredith et al. (2001) derived meteoric water fluxes of 3,680 km^3/year from the 1997 section, and 2,000 km^3/year from the 1998 section, with typical errors of a few hundred km^3/year. It is important to note, however, that these values are specific to the times of the sections, and not necessarily indicative of a long-term mean: whilst the tracer fields are likely to evolve comparatively slowly, the velocity field will contain significant variability at seasonal, interannual and other periods.

Fluxes of sea ice are not able to be derived directly using the isotope method, due to the separation of the sea ice and the water column from which it was formed. However, it was notable that the tracer budget revealed a consistent ratio of 2:1 for the prevalence of meteoric water to sea ice melt (the sea ice melt percentages being negative since there had been a net sea ice formation from the waters sampled). This ratio was strikingly similar to that derived from previous isotope sections across Fram Strait, from several years earlier (Bauch et al. 1995), indicating a robust result.

Interpreting these results requires care. Most studies of freshwater exiting the Arctic derive values for liquid freshwater flux and sea ice flux, and the former is not directly comparable with a meteoric water flux. The key to understanding this difference is to appreciate that derivation of a meteoric water flux quantifies the total liquid freshwater input to the ocean from river runoff and precipitation, and ignores any additional salinification or freshening due to sea ice formation or melt. Conversely, a liquid freshwater flux calculated using salinity alone includes the freshening or salinifying effects of sea ice processes. In the present case (of a 2:1 ratio for meteoric water to sea ice melt at Fram Strait), this corresponds to a ratio of around 1:1 for liquid freshwater flux to sea ice flux. Thus, if the long-term sea ice flux through Fram Strait is of order 2,300 km^3/year (see Serreze et al. 2006, for a review), the liquid freshwater flux will be broadly comparable to this, albeit with errors of a few hundred km^3/year.

Other tracers can be added to the suite being measured in order to generate further information on freshwater sources. For example, measurements of barium yield information on the source of river runoff (North American versus Eurasian). Taylor et al. (2003) exploited this for one of the Fram Strait sections used by Meredith et al. (2001), and found that Eurasian river runoff dominated the meteoric water of the EGC in 1998. In addition, measurements of dissolved nutrients were used to quantify the contribution of Pacific water (from Bering Strait) to the water mass composition at Fram Strait. In 1998, the contribution of Pacific Water was seen to be very much larger than previous work (from a section in 1987) had indicated, suggesting significant variability of this input.

It has been demonstrated recently that alkalinity has great potential in tracing meteoric water inputs to the Arctic, as a direct analogue to $\delta^{18}O$ (Yamamoto-Kawai et al. 2005). This tracer was used by Jones et al. (2008), along with measurements of dissolved nutrients and salinity, to trace the varying freshwater composition of the EGC as it flows south from Fram Strait to Denmark Strait and beyond. One of the major signals found was an increase in oceanic sea ice melt with decreasing latitude, consistent with the melting of sea ice into the water column as it moves south. This is in agreement with recent results obtained from $\delta^{18}O$ (Dodd 2006). Further aspects the evolution of freshwater content in the EGC are discussed separately (Jones et al. 2008; see also chapter by E.P. Jones, this volume).

Whilst significant progress has been made in recent years in understanding the nature of tracers and what information they can provide concerning fluxes of Arctic freshwater, significant challenges remain. Tracer sections are almost invariably conducted during the summer months, and the consequent lack of information concerning seasonality in the tracer fields makes it very difficult to produce reliable annual mean estimates for fluxes. Furthermore, given that the timing of the sections varies within the summer months, it is not easy to assess inter annual variability separately from possible aliased seasonal variability. Producing estimates for the fluxes of separate freshwater constituents on the East Greenland shelf is also difficult at present, due to the problematic nature of obtaining concurrent velocity and tracer data there.

The ultimate goal is to achieve long time series of fluxes of the separate freshwater components, both in the EGC and shelf regions east of Greenland, and across all the regions of freshwater export from the Arctic. To achieve this, we need *in situ* moored systems capable of simultaneously measuring velocity and salinity, and capturing water samples for subsequent tracer analysis from different depths (ideally right to the surface). Although technically challenging in icy waters, such a capability would be central to a sustained monitoring system for Arctic freshwater fluxes.

11.8 FW Transport from the EGC into the Nordic Seas

There are two main escape routes for freshwater from the EGC into the Greenland and Iceland Seas respectively. In the Greenland Sea the Jan Mayen Polar Current flows north of the Jan Mayen Fracture Zone towards Jan Mayen and within the Iceland Sea the East Icelandic Current (EIC) flows towards the east, north of Iceland. The ASOF estimates of the freshwater transport on and along the east Greenland shelf have too large uncertainties; we can not deduce from them the amount of freshwater lost along these two routes.

The freshwater within the EIC reflects the amount of Polar Water present and has profound effects on the physics and biology in the area. It stratifies the water column, making deep convection less likely to occur and thus affecting the thermohaline circulation. It was shown by Jónsson (1992) that the fresh water present in the EIC originates mainly from the EGC. He also showed in accordance with Swift and Aagaard (1981) that the Atlantic water that enters the North Icelandic shelf through Denmark Strait is closely confined to the shelf and leaves the Iceland Sea to the southeast, without entering or otherwise contributing significantly to the EIC. This also means that the fresh water runoff from Iceland is also confined to the shelf and does not contribute to the fresh water content of the EIC.

As a part of the EU-VEINS project, two current meter moorings were deployed on a standard CTD section (Langanes section) in the EIC northeast of Iceland (Fig. 11.9). The moorings were put out in June 1997 and were recovered in June 1998. Using the geostrophic velocity referenced to the current meters as explained in Jónsson (2006) and the salinity measurements, the fresh water flux above 170 m was calculated for 5 different CTD coverings of the section during the deployment period (Fig. 11.10). The average of the flux was 5.5 mSv or about 4.4% of the freshwater flux through Fram Strait as estimated by Aagaard and Carmack (1989). Most of the flux occurs over the slope where the current was strongest and the geostrophic shear was largest. The fresh water transport was smallest in March 1998, 1.2 mSv, and this was due to a combination of both less fresh water and lower speed over the slope. The maximum fresh water transport of 8.3 mSv was observed in August 1997.

In an attempt to put the numbers for the transport of fresh water presented here into a longer term perspective, the fresh water thickness at a station on the Langanes section in May/June is shown in Fig. 11.11. It is seen that the period 1996–1998

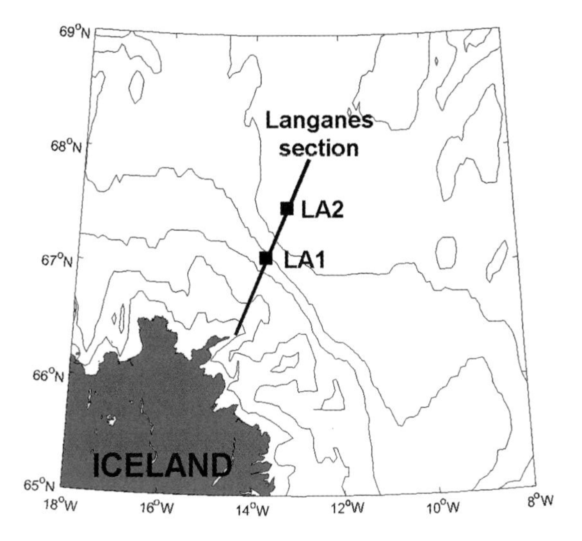

Fig. 11.9 A map showing the current meter positions LA1 and LA2 as well as the Langanes CTD section. The depth contours are 100, 200, 500, 1,000 and 2,000 m

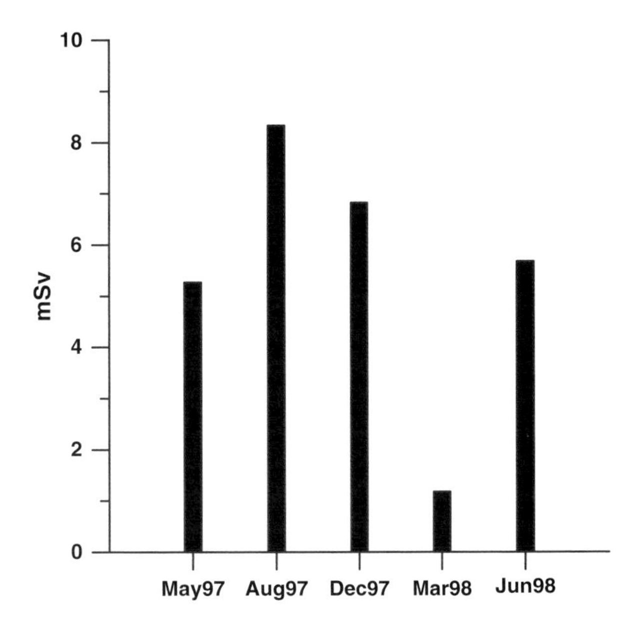

Fig. 11.10 The total transport of fresh water north of Island above 170 m for the times when CTD stations were occupied

was characterized by relatively high values. Therefore it is likely that the estimates of the fresh water fluxes presented here are higher than the long-term average. The highest values were observed in the mid-1970s when polar water was dominating the area (Malmberg 1984).

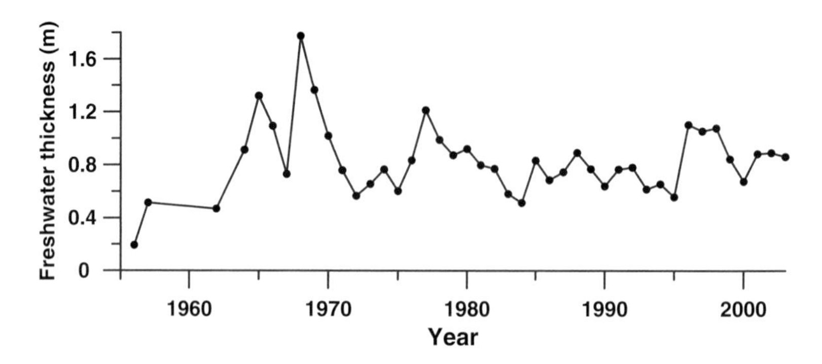

Fig. 11.11 The fresh water thickness north of Island above 150 m in May/June 1956–2003 relative to a salinity of 34.93 at a station on the Langanes section

The forcing of the variability of the flow of freshwater from the EGC into the Iceland Sea was studied by Jónsson (1992). He found that the main force was the wind stress curl over the Iceland Sea. This was interpreted in such a way that increased cyclonic winds keep the fresh water tighter to the Greenland shelf while a decrease in the cyclonic wind field reduces the gradients and allows for increased advection of fresh water into the Iceland Sea. A similar relationship was suggested for the Greenland Sea by Meincke et al. (1992) and by Malmberg and Jónsson (1997).

It can be concluded that a relatively small proportion of the fresh water flux through the Fram Strait is deflected into the Nordic Sea and most of it continues uninterrupted through Denmark Strait into the North Atlantic. The amount incorporated into the DSOW and therefore lost to the deep ocean, is also only small. Assuming a transport of 3 Sv of DSOW with a salinity of 34.8 the freshwater transport relative to S = 34.9 is only ~9 mSv, being less then 10% of the total flux through Fram Strait.

11.9 Summary and Conclusions

This chapter has adressed the oceanic freshwater fluxes east of Greenland. In addition to a brief review of previous estimates, recent estimates from direct observations performed during the ASOF years are presented. According to these direct observations ~1,000 km^3/year of liquid freshwater escapes the Arctic Ocean in the EGC through Fram Strait at 79° N. Based on geostrophic calculations from available CTD sections and numerical modeling, we estimate that there is an additional transport over the shelf of the same magnitude. The total flux of liquid freshwater through Fram Strait is therefore ~2,000 km^3/year. Further south at 74° N an estimated ~800 km^3/year of liquid freshwater passes by the section over the EGC, averaged over the 2 available years. Including the shelf, the estimates point to a total liquid freshwater transport across 74° N of 1,400 km^3/year. Although the instrumentation at the 63° N ASOF section is sparse, rough estimates of the liquid freshwater transports here point to mean values ranging between 150, 1,100 and 2,200 km^3/year for the summer months in 2001, 2002 and 2003, respectively.

The instrumentation and corresponding data have proved to be too sparse to estimate the divergence in freshwater transport between the different latitudes, and hence address the leakage of freshwater into the Nordic Seas as time series. Based on the mean numbers and estimates of diversions of liquid freshwater in the Jan Mayen and East Icelandic Currents, we conclude that only a relatively small proportion of the fresh water flux through Fram Strait is deflected into the Nordic Seas. Most of it continues uninterrupted through Denmark Strait into the North Atlantic.

From the 79° and 74° N section the availability of data allow us to construct time series of liquid freshwater transport. The annual cycle at 74° N show a minimum in June and a maximum in September. At 79° N this work is still in preparation. From the data at hand, some published (Holfort and Meincke 2005) and some in preparation (Holfort and Hansen, in preparation), we see no trend or systematic development in the liquid freshwater transport over the years of observation.

Comparing the temperatures and salinities observed by the 79°, 74° and 63° N moorings, we detect no signal propagation along the east Greenland shelf. The variability is dominated by the seasonal cycle, and the anomalies we may observe after subtracting the seasonal cycle from the signal tend to occur at all three latitudes simultaneously. We attribute this to effects of large-scale atmospheric forcing.

References

Aagaard K, Carmack EC (1989) The role of sea ice and other fresh water in the Arctic circulation. Journal of Geophysical Research 94, 14,485–14,497

Bacon S, Reverdin G, Rigor IG, Snaith HM (2002) A freshwater jet on the east Greenland shelf. Journal of Geophysical Research 107, doi:10.1029/2001JC000935

Bauch D, Schlosser P, Fairbanks RG (1995) Freshwater balance and the sources of deep and bottom waters in the Arctic Ocean inferred from the distribution of H-2O-18. Progress in Oceanography 35, 53–80

Belkin I (2004) Propagation of the "Great Salinity Anomaly" of the 1990s around the northern North Atlantic. Geophysical Research Letters 31, L08306, doi:10.1029/2003GL019334

Craig H, Gordon L (1965) Deuterium and oxygen-18 variations in the ocean and the marine atmosphere, in Stable Isotopes in Oceanographic Studies and Paleotemperatures, E. Tongiorgio (ed), pp. 9–130, Spoleto

Curry R, Mauritzen C (2005) Dilution of the Northern North Atlantic Ocean in recent decades. Science 308 (5729), 1772., doi: 10.1126/science.1109477

Curry R, Dickson B, Yashayaev I (2003) A change in the freshwater balance of the Atlantic Ocean over the past four decades. Nature 426, 826–829

Dickson RR, Meincke J, Malmberg S-A, Lee AJ (1988) The 'Great Salinity Anomaly' in the northern North Atlantic 1968–1982. Progress in Oceanography 20, 103–151

Dickson R, Rudels B, Dye S, Karcher M, Meincke J, Yashayaev I (2007) Current estimates of freshwater flux through Arctic and subarctic seas. Progress in Oceanography, doi:10.1016/j.pocean.2006.12.003

Dickson RR, Yashayaev I, Meincke J, Turrell B, Dye S, Holfort J (2002) Rapid freshening of the deep North Atlantic Ocean over the past four decades. Nature 416, 832–837

Dodd P (2006) Ph.D. thesis, in preparation, University of East Anglia, Norwich

Fichefet T, Poncin C, Goosse H, Huybrechts P, Janssens I (2003) Implications of changes in freshwater flux from Greenland ice sheet for the climate of the 21st century. Geophysical Research Letters 30, 3498–3510, doi:10.1029/2003GLO178826

Hansen B, Østerhus S (2000) North Atlantic-Nordic Seas exchanges. Progress in Oceanography 45, 109–208

Hansen E, Holfort J, Karcher M (2006) Comparing observed and modeled freshwater fluxes: Preliminary results. ASOF-N WP4 Project Report, 2006

Häkkinen S (1993) An arctic source for the Great Salinity Anomaly: A simulation of the arctic ice-ocean system for 1955–1975. Journal of Geophysical Research 98, 16397–16410

Häkkinen S (1999) A simulation of thermohaline effects of a great salinity anomaly. Journal of Climate 12, 1781–1795

Haak H, Jungclaus J, Mikolajewicz U, Latif M (2003) Formation and propagation of great salinity anomalies. Geophysical Research Letters 30(9), 1473, doi:10.1029/2003GL017065

Holfort J, Meincke J (2005) Freshwater transport on the east Greenland Shelf at 74°N. Meteorologische Zeitschrift 14, No.6, 703–710

Holfort J, Hansen E (2007) Freshwater transport through Fram Strait. In preparation

Jones EP, Anderson LG, Jutterström S, Swift JH (2008) Sources and distribution of freshwater in the East Greenland current, *Prog. Ocean.*, accepted

Jónsson S (2006) Volume flux and fresh water transport associated with the East Icelandic Current. Progress in Oceanography, accepted

Jónsson S (2003) The transport of water and freshwater within the East Icelandic Current. Poster on the European Geophysical Society meeting, Nice

Jónsson S, Briem J (2003) The transport of water and freshwater within the East Icelandic Current. ICES CM 2003/T:05

Jónsson S (1992) Sources of fresh water in the Iceland Sea and the mechanisms governing its interannual variability. ICES Marine Science Symposia, 195, 62–67

Karcher M, Gerdes R, Kauker F, Köberle C, Yashayaev I (2005) Arctic Ocean change heralds North Atlantic freshening. Geophysical Research Letters 32, L21606, doi:10.1029/2005GL023861

Kwok R, Cunningham GF, Pang SS (2004) Fram Strait sea ice outflow. Journal of Geophysical Research 109. C01009, doi:10.1029/2003JC001785

Malmberg SA (1984) Hydrographic conditions in the east Icelandic current and sea ice in north Icelandic waters, 1970–1980. Rapp. P.-v Réun. Cons. Int. Explor. Mer. 185, 170–178

Malmberg SA, Jónsson S (1997) Timing of deep convection in the Greenland and Iceland Seas. ICES Journal of Marine Science 54, 300–309

Meincke J, Jónsson S, Swift JH (1992) Variability of convective conditions in the Greenland Sea. ICES Marine Science Symposia, 195, 32–39

Meredith, MP, Heywood KJ, Dennis PF, Goldson LE, White RMP, Fahrbach E, Schauer U, Østerhus S (2001) Freshwater fluxes through the western Fram Strait. Geophysical Research Letters 28 (8), 1615–1618

Ostlund HG, Hut G (1984) Arctic Ocean water mass balance from isotope data. Journal of Geophysical Research 89, 6373–6381

Schlosser P, Bauch D, Fairbanks R, Bonisch G (1994) Arctic river runoff: Mean residence time on the shelves and in the halocline. Deep-Sea Research I 41, 1053–1068

Schlosser P, Newton R, Ekwurzel B, Khatiwala S, Mortlock R, Fairbanks R (2002) Decrease of river runoff in the upper waters of the Eurasian Basin, Arctic Ocean, between 1991 and 1996: Evidence from δ18O data. Geophysical Research Letters 29 (9), 10.1029/2001GL013135

Serreze MC, Barrett AP, Slater AG, Woodgate RA, Aagaard K, Lammers RB, Steele M, Moritz R, Meredith MP, Lee CM (2006) The large-scale freshwater cycle of the Arctic. Journal of Geophysical Research 11, C11010, doi:10.1029/2005JC003424

Swift JH, Aagaard K (1981) Seasonal transitions and water mass formation in the Iceland and Greenland seas. Deep-Sea Research 28A, 1107–1129

Taylor JR, Falkner KK, Schauer U, Meredith MP (2003) Quantitative considerations of dissolved barium as a tracer in the Arctic Ocean. Journal of Geophysical Research 108 (C12), 3374, doi:10.1029/2002JC001635

Thomas R, Frederick E, Krabill W, Manizade S, Martin C (2006) Progressive increase in ice loss from Greenland. Geophysical Research Letters 33, L10503, doi:10.1029/2006GL026075

Vinje T, Nordlund N, Kvambekk A (1998) Monitoring ice thickness in Fram strait. Journal of Geophysical Research C 103, pp. 10, 437–10

Vinje T (2001) Fram Strait ice fluxes and atmospheric circulation: 1950–2000. Journal of Climate 14, 3508–3517

Vinje T, Løyning TB, Polyakov I (2002) Effects of melting and freezing in the Greenland Sea. Geophysical Research Letters 29(23), 2129, doi:10.1029/2002GL015326

Widell K, Østerhus S, Gammelsrød T (2003) Sea ice velocity in the Fram Strait monitored by moored instruments. Geophysical Research Letters 30, No. 19, doi:10.1029/2003GL018119

Yamamoto-Kawai M, Tanaka N, Pivovarov S (2005) Freshwater and brine behaviors in the Arctic Ocean deduced from historical data of d18O and alkalinity (1929–2002 A.D.). Journal of Geophysical Research 110 (C10003), doi:10.1029/2004JC002793

Chapter 12
The Changing View on How Freshwater Impacts the Atlantic Meridional Overturning Circulation

Michael Vellinga[1], Bob Dickson[2], and Ruth Curry[3]

12.1 Introduction

These days, it would be generally accepted that through its northward transport of warm tropical waters, the Atlantic Meridional Overturning Circulation (AMOC) contributes effectively to the anomalous warmth of northern Europe (Large and Nurser 2001; see also Rhines and Hakkinen 2003; Rhines et al., this volume). The oceanic fluxes of mass, heat and salt that pass north across the Greenland–Scotland Ridge from the Atlantic to the Arctic Mediterranean have now been soundly established by direct measurement under the EC VEINS and ASOF/MOEN programmes, as have the corresponding fluxes to the Arctic Ocean (Ingvaldsen et al. 2004a, b; Schauer et al. 2004). We now know that the 8.5 million cubic metres per second of warm salty Atlantic Water that passes north across this Ridge carries with it, on average, some 313 million megawatts of power and 303 million kilograms of salt per second (Østerhus et al. 2005). As it returns south across the Ridge in the form of the two dense overflows from Nordic Seas, its salinity has decreased from about 35.25 to 34.88 and its temperature has dropped from 8.5 °C to 2.0 °C or less. Not surprisingly, surrendering this amount of heat is of more than local climatic importance. To quantify its contribution to climate the AMOC was deliberately* shut down in the HadCM3 Atmosphere-Ocean General Circulation Model by artificially releasing a large pulse of freshwater in the northern North Atlantic (Wood et al. 2003; Vellinga 2004; Wood et al. 2006). The cooling of mean air temperature over the northern Norwegian Sea and Barents Sea in the first 10 years after shutdown exceeds −15 °C, and some lesser degree of cooling is evident over the entire Hemisphere. In addition, significant changes in rainfall are evident (especially at low latitudes, Vellinga and Wood 2002), as well as changes in sea level height

[1]Met Office Hadley Centre, Fitzroy Road, Exeter EX1 3PB UK, e-mail: michael.vellinga@metoffice.gov.uk

[2]Centre for Environment, Fisheries and Aquaculture Science, Pakefield Road, Lowestoft NR33 0HT UK

[3]Woods Hole Oceanographic Institution, Woods Hole, MA 02543, USA

R.R. Dickson et al. (eds.), *Arctic–Subarctic Ocean Fluxes*, 289–313

(Levermann et al. 2005; Vellinga and Wood 2007). [*note that this is a 'what if' experiment. The response of the AMOC to more plausible scenarios of gradual anthropogenic greenhouse gas increase is discussed in Section 12.3.2 of this chapter.]

The obvious follow-up questions are much harder to answer: what is the physical basis for a slowdown in the AMOC? and is the AMOC actually slowing?

Most computer simulations of the ocean system in a climate with increasing greenhouse-gas concentrations predict that the AMOC will weaken as the subpolar seas become fresher and warmer in the 21st century and beyond (e.g. Manabe and Stouffer 1994; Rahmstorf and Ganopolski 1999; Delworth and Dixon 2000; Rahmstorf 2003), but opinions are divided both on whether thermohaline slowdown is already underway or on whether any variability that we see is natural or anthropogenic. From the current literature for example, we have the results from HadCM3 (Wu et al. 2004) that the recent freshening of the deep N Atlantic occurs in conjunction with an *increase* in the AMOC, diagnostically associated with an increased north–south density gradient in the upper-ocean; from studies with the GFDL model Delworth and Dixon (2006) proposed the idea that anthropogenic aerosols may actually have delayed a greenhouse-gas-induced weakening of the AMOC; from the Kiel Group (Latif et al. 2006), the suggestion that the expected anthropogenic weakening of the thermohaline circulation will be small, remaining within the range of natural variability during the next several decades; and from the Southampton Group (Bryden et al. 2005), the claim that the AMOC has already slowed by 30% between 1957 and 2004. None of these opinions – and there are others! – is controversial in the sense that they are all based on established and accepted techniques. But the more extreme are certainly controversial in their interpretation of events. Our observational series are simply too short or gappy or patchy to deal unambiguously with the complex of changes in space, time and depth that the Atlantic is exhibiting, and even the closely observed line that Bryden et al. rely on is not immune. Modelling the same Atlantic transect (26° N), Wunsch and Heimbach (2006) find a strengthening of the outflow of North Atlantic Deep Water since 1992 (i.e., including the layers and years where Bryden et al. 2005 had observed their major decrease), and from the month-to-month variability that they encounter are forced to conclude that *single section determinations of heat and volume flux are subject to serious aliasing errors*. Such uncertainties in our observations are bound to hinder a critical evaluation of our models. Thus in their recent assessment of the risk of AMOC shutdown, Wood et al. (2006) can go no further than conclude that shutdown remains a *high impact, low probability event* and that *assessing the likelihood of such an event is hampered by a high level of modeling uncertainty*.

The present chapter concerns itself with the two types of advance that seem necessary to reducing these present uncertainties. We start with a review of the history of progress in modeling the role of the Northern Seas in climate through their influence on the AMOC. The aim of this review is to assess the basis in both numerical experimentation and observational constraints for present ideas. Some of the earlier advances are discussed in Section 12.2, more recent improvements of our understanding are discussed in Section 12.3. In Section 12.3 we also present examples of recent model experiments that raise intriguing questions about simulating future change of the AMOC. Those questions lead us to Section 12.4, in which we conclude this Chapter with an attempt to identify the next steps – both in observations and modeling – that we believe are necessary to reduce the present uncertainties regarding future change of the AMOC.

12.2 Advances in Modelling to the Mid-1990s

The ability of the ocean to integrate high-frequency atmospheric surface flux variability into a red energy spectrum (e.g. Hasselmann 1976) points to the importance of the ocean in generating low-frequency climate variability. However, as already mentioned, our incomplete data coverage in space and time make it difficult to obtain a complete understanding of the underlying mechanisms from ocean observations alone. Numerical models are the obvious tool to help increase our qualitative understanding of observed phenomena, though ideally, observations and models should go hand in hand. Although models have improved greatly over recent years they have their own deficiencies, due to underlying simplifications and assumptions. Here, we present a (by necessity incomplete) overview of some of the progress that has been made since the 1990s in our understanding of the variability and stability of the North Atlantic meridional overturning circulation ('AMOC').

The AMOC was considered part of a global system of ocean currents (e.g. Gordon 1986), driven by surface buoyancy fluxes that are balanced by upward diffusion of heat and freshwater. It involved a few localized areas of deep convection together with the overflows and entrainment that ventilate the deep basins of the North Atlantic. In the modern ocean, the AMOC transported mass, heat, and salt northward inter-hemispherically, being responsible for around 1 PW of heat transport across 24° N.

Stommel (1961) had conceptualized the notion of salt advection feedbacks as an important factor in modulating the strength of AMOC and its stability, which he characterized as non-linear with multiple equilibrium states. Welander (1982) had described the idea of "flip-flop" convective feedbacks, whereby decreased surface density reduced vertical convection leading to accumulation of fresh water, which decreased surface density still more. Stommel's findings of the AMOC as a system with the capability of having multiple equilibria were confirmed in studies with ocean-only GCMs (Bryan 1986; Marotzke and Willebrand 1991) and with an early version of the GFDL coupled climate model (Manabe and Stouffer 1988), suggesting that multiple equilibria can exist even in presence of 3D ocean dynamics and coupled ocean–atmosphere feedbacks, respectively. Rahmstorf (1995) demonstrated in an ocean GCM that this multiplicity caused hysteresis behaviour of the AMOC to anomalous surface freshwater forcing.

Deep sea sediment cores provided evidence for millennial-scale reorganizations of deep ocean circulation, with greatly reduced NADW production during the Last Glacial Maximum (Curry and Lohmann 1982). Broecker (1997) proposed that turning the "ocean conveyor" on and off could explain certain rapid global climate shifts (Dansgaard–Oeschger cycles, Last Glacial Maximum, Younger Dryas). The classic modeling studies of Manabe and Stouffer (1993, 1994) showed that the AMOC could essentially shut down as a consequence of strong greenhouse gas forcing in the GFDL climate model.

The notion that the AMOC might exhibit significant decadal variability, with implications for the state of the North Atlantic (e.g. SST) was just emerging, for example in a model study by Delworth et al. (1993). Decadal variability of deep convective activity and watermass characteristics appeared to be organized around the structure of the NAO forcing, anti-phased between GIN Seas and Labrador Sea (Dickson et al. 1996).

12.3 Recent Advances in Understanding the Variability of the AMOC

Understanding the causes of simulated variability of the AMOC enables us to quantify possible implications of observed changes in the ocean. By carrying out model experiments with and without changes in anthropogenic forcing of climate (e.g. greenhouse gases, aerosols and ozone) we can interpret observed changes in the oceans: i.e. are they anthropogenic or due to internal variability, or a combination of the two? If modeled and observed changes agree then this provides an important model validation, demonstrating that all model processes add up to give the right (or at least plausible) feedbacks. This should enhance our confidence in the usefulness of models to project future changes to the ocean. Validation is complicated by the chaotic nature of climate: a single model simulation is unlikely to reflect observed changes, even if the model were perfect, so we need ensembles of simulations. Running ensembles allows a better estimate (and characterization) of model internal variability, against which the characteristics of a particular observation can be compared. Also, by averaging over several model realizations the presence of internal variability can be smoothed out, thus making it easier for any forced response to emerge from the noise. In terms of signal-to-noise ratio for forced response, some regions (e.g. high-latitude oceans) are probably better than others for this (Banks and Wood 2002; Vellinga and Wood 2004), and models can be helpful in identifying such regions.

12.3.1 Internal Variability

The North Atlantic Oscillation is the leading mode of interannual sea-level pressure variability in the North Atlantic domain (Hurrell 1995), and thus plays an important role in modifying air–sea interaction in this area (Cayan 1992). For this reason many studies of ocean variability focus on the ocean's response to NAO-variability, but it is important to remember that the NAO can not explain all observed inter-annual variability of SST and surface fluxes over the Atlantic domain (e.g. Krahmann et al. 2001; Bojariu and Reverdin 2002). Mechanisms by which the North Atlantic responds to changes in surface flux caused by the NAO, have been explored in many studies. At inter-annual to decadal time scales (Häkkinen 1999; Eden and Willebrand 2001) fluctuations in the NAO cause AMOC anomalies of a few Sv, attributed primarily to surface wind stress and heat flux variability, with both a fast barotropic and a delayed baroclinic response.

We note that many of the above studies employ regional ocean models rather than a global coupled model such as was used by Delworth et al. (1993). The advantage is that a regional ocean model can be run at higher resolution than a global model, and re-run with different kinds of surface forcing, so that the relative importance of the different fluxes (heat, freshwater, momentum, etc.) can be identified. Furthermore, the direct feedback of the ocean on the atmosphere is excluded, making it easier to understand the ocean response (although part of the ocean feedback may implicitly be incorporated in the surface flux forcing that is generally taken from atmosphere reanalyses). A disadvantage of regional models is that their forcing needs to be prescribed at lateral boundaries. For example, Eden and Willebrand's model domain is bounded by 70° N, where water mass properties are fixed to climatology across all depths, thus eliminating variability in the overflows from Nordic Seas and in the Arctic Ocean inflows. Also, re-analyses fluxes are not necessarily balanced over the domain (Häkkinen 1999), or in balance with the ocean transports. This causes ocean drifts that need damping by surface relaxation, which may affect the model's variability.

At longer, multi-decadal time scales, the ocean is also susceptible to NAO forcing involving the gyre and overturning circulations (as examples: Timmermann et al. 1998; Eden and Jung 2001; Cheng et al. 2004; Dong and Sutton 2005; Häkkinen 1999; Latif et al. 2006). Surface heat flux forcing by the atmosphere emerges as an important process to excite decadal variability in the AMOC (either through NAO-like forcing over the subpolar gyre in the GFDL_R15 model; Delworth and Greatbatch 2000), or through atmospheric heat flux variability unrelated to the NAO (e.g. over the Greenland/Norwegian Sea in HadCM3 (Dong and Sutton 2005). The fundamental agreement as to mechanism, if not regions and time scales, suggests that overall the processes responsible for this type of decadal variability are robust across a range of climate models. Other details (which are the most effective forcing patterns and time scales for ocean response, etc.) are model dependent, and appear to be linked to where in a particular model deep-water is formed preferentially (Cheng et al. 2004; Dong and Sutton 2005). Surface heat flux changes typically emerge as dominant over freshwater or momentum surface flux changes in driving interannual-to-interdecadal variability in the North Atlantic (e.g. Eden and Jung 2001; Delworth and Greatbatch 2000). Salinity changes resulting from anomalous transports associated with the heat flux anomalies, are, however, often instrumental in variability of the AMOC (Delworth et al. 1993; Timmermann et al. 1998; Dong and Sutton 2005).

Sometimes aspects of simulated variability fail to stand observational tests. For example, the Parallel Climate Model ('PCM') (Dai et al. 2005) has a sharp spectral peak of AMOC variability at ~24 years, forced by NAO variability in the model at this frequency. However, in the (admittedly limited) instrumental NAO record such a persistent spectral peak in this frequency band is not evident (Hurrell and van Loon 1997; Higuchi et al. 1999; Gamiz-Fortis et al. 2002) implying that in this particular model the air–sea interaction is perhaps over-emphasized. Generally, models succeed in reproducing the NAO as the dominant pattern of internal variability over the North Atlantic domain, as well as certain observed aspects of impact on the rest of climate (such as SST and precipitation). However, in inter-comparison studies coupled models are often reported to fail in reproducing the magnitude of the observed upward trend of the NAO between the 1960s and 1990s when greenhouse gas concentrations are fixed, or increasing at 1% per year (e.g. Osborn 2004; Kuzmina et al. 2005; Stephenson et al. 2006).

Model intercomparison studies typically only have access to limited amounts of model output (e.g. 80 years are requested for CMIP integrations, which are the data used by Kuzmina et al. 2005; Stephenson et al. 2006; although Osborn 2004 uses 240 years for his study). From a nearly 2,500-year-long integration of HadCM3 at 1xCO_2 we can estimate the low-frequency, internal winter NAO variability in this model rather better. We compare the model NAO time series to that derived from station data from Iceland and the Azores (Jones et al.1997; Fig. 12.1). For clarity we show 10-year average data only. Neither model data nor observations have been normalized so that the actual magnitude of the trend in model and observations can be compared. The observed low-frequency NAO trend (8.6 hPa/30 years for the period 1955–1995) is indeed large compared to the model trends (median of upward model trends is 3.3 hPa/30 years). However, the observed 30-year trend does fall within the 95th percentile of the model data. The magnitude of the observed NAO trend is therefore consistent with internal variability at the 95% level. While this is seemingly at odds with the results of some of the studies referred to previously that included shorter segments of the same HadCM3 control run, the conclusion must be that one needs rather long segments of model integrations to draw any conclusions about the observed upward trend in the NAO, since it may well lie in the tail of a model's distribution; at least it does so in the case of HadCM3.

Multi-decadal to centennial scale variability in the AMOC has been linked to shifts in the Atlantic ITCZ and the ocean advection of low-latitude salinity anomalies caused by such shifts (Vellinga and Wu 2004). The slow time scale is set by the time it takes for salinity anomalies to propagate from low to high latitudes. Indirect support for the existence of this kind of low-frequency AMOC variability comes from the similarity between observed SST records and anomalies driven by the AMOC in coupled simulations (Delworth and Mann 2000; Latif et al. 2004; Knight et al. 2005). The capability of the low-latitude Atlantic for generating salinity anomalies that eventually affect the AMOC has been described in several other studies, either as a response to global warming (Latif et al. 2000; Thorpe et al. 2001) or to low-frequency modulations of ENSO variability (Mignot and

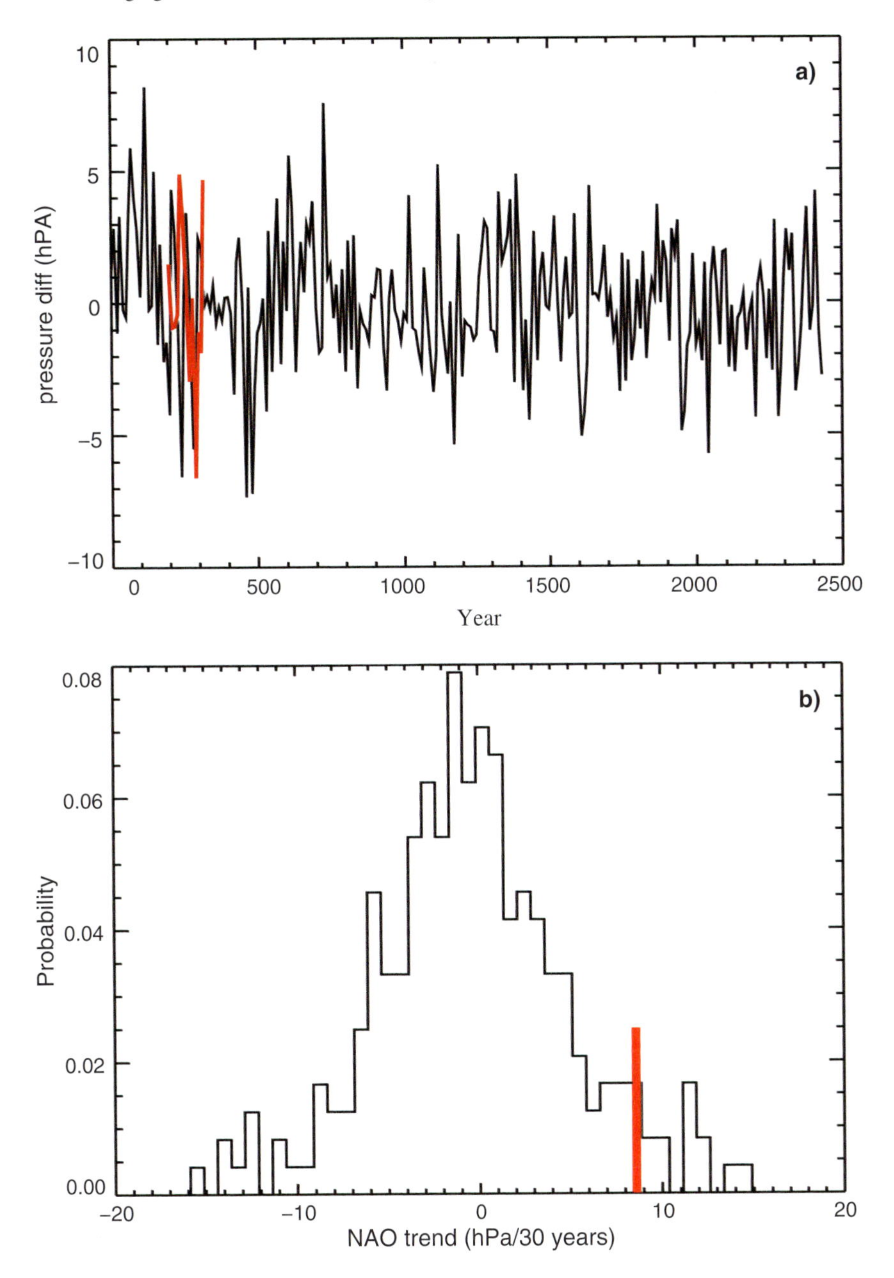

Fig. 12.1 (a) Time series of decadally averaged, un-normalised winter (DJF) values of the pressure difference between Iceland and the Azores from the HadCM3 control run (thin). The red line overlying this series represents the observed data for the period 1865–1995, from Jones et al. 1997. (b) PDF of 30 year trends for model data shown in (a); the vertical bar indicates the 30-year trend in the observed data for the period 1955–1995

Frankignoul 2005). In contrast, Jungclaus et al. (2005) link near-centennial (70–80 years) variability in the AMOC to a delayed response in Arctic freshwater storage/release, where the long time scale is presumably set by the time it takes the Arctic basin to freshen. As in Delworth et al. (1993) their mechanism depends on interaction between the meridional overturning and the gyre circulation and transports. But here the emphasis is more on Greenland/Norwegian Sea and Arctic gyre circulation. The issue of whether and to what extent low or high latitude regions are crucial to centennial AMOC fluctuations is not yet resolved. The inherently long time-scales involved make it difficult to use ocean observations to assess this, and one might have to rely on multi-model inter-comparisons to investigate any model robustness. Hunt and Elliot (2006) describe a 10,000-year simulation with a coarse (5.6° × 3.2°) resolution climate model. Such a long integration could be useful for studying low-frequency variability. They offer a tantalizing view of internal variability of the AMOC, with spectral peaks at decadal and centennial time scales, but not an analysis that would allow comparison with other studies.

Understanding the multi-decadal time-scale of internal AMOC variability is useful in exploring possible mechanisms for observed changes (e.g. Wu et al. 2004; Hu and Meehl 2005). Furthermore, the red spectrum of the AMOC and its heat transport yield the potential for decadal climate prediction, although the skill appears to be largest over the ocean and limited over land (Collins and Sinha 2003; Collins et al. 2006). If low-frequency internal variability of the AMOC has a sufficiently large amplitude this could affect the onset of the projected weakening under anthropogenic climate change (Latif et al. 2004).

12.3.2 Stability of the AMOC Under Anthropogenic Climate Change

None of the comprehensive climate general circulation models, when forced by more or less plausible (Cubasch et al. 2001; Schmittner et al. 2005) or idealised (Gregory et al. 2005) greenhouse gas scenarios project a full shutdown of the AMOC by 2100. In a limited number of studies coarse-resolution climate GCMs have been run well beyond the year 2100. When CO_2 concentrations have reached high values (typically four times pre-industrial levels) a gradual spin-down of the AMOC was simulated (Manabe and Stouffer 1994; Mikolajewicz et al. 2007), sometimes followed by a recovery after several millennia (Stouffer and Manabe 2003). There remains a large spread in the projected weakening for the 21st century among models, which is indicative of the uncertainty in model formulation. In most models of a multi-model study, the AMOC weakening under increasing CO_2 concentrations is dominated by the effects of heating (Gregory et al. 2005). Global warming tends to reduce ocean heat loss at high latitudes, which adds an anomalous buoyancy flux to the ocean. Anthropogenic changes in freshwater fluxes add to AMOC weakening. The amount to which the latter contributes varies between models (Gregory et al. 2005), reflecting uncertainty about how global warming will affect the hydrological cycle (Cubasch et al. 2001; Allen and Ingram 2002), both in magnitude and spatial structure. Efforts to quantify the effects of this uncertainty on climate projections is a relatively recent development and we will return to this topic in Section 12.4.2.

To address the uncertainty associated with changes in the surface freshwater forcing and any implications for the AMOC it is necessary to understand what positive and negative feedbacks act on the AMOC. Many modeling groups have carried out sensitivity experiments to understand these feedbacks. In this type of experiment freshwater is added to the ocean artificially: either as a prolonged surface flux anomaly, often referred to as 'hosing' (e.g. Schiller et al. 1997; Ottera et al. 2004; Cheng and Rhines 2004; Dahl et al. 2005), or instantaneously (e.g. Vellinga et al. 2002). By reducing density in the deep-water formation regions such freshwater perturbations are an efficient way to weaken the AMOC. Idealized experiments like these allow one to establish what model feedbacks are triggered by the AMOC weakening. Like hydrological sensitivity, such feedbacks tend to be model-dependent, and typically involve an atmospheric response in different parts of the world. There are perhaps indications that the dominant feedbacks in a specific model are linked to its preferred mode of internal low-frequency AMOC variability (Schiller et al. 1997; Timmermann et al. 1998 in the case of the ECHAM3/LSG model; Vellinga et al. 2002; Vellinga and Wu 2004 in the case of HadCM3). Standardized 'hosing' experiments have been carried out across a multi-model ensemble to try to map out where models agree or disagree in their response (Stouffer et al. 2006). For 100 years of hosing at a rate of 0.1 Sv between 50–70° N, none of the models show a permanent AMOC shutdown, but some models do so for 100 years of 1 Sv hosing. Work to understand the basis of the disagreements is ongoing. Evidence from an ocean-only model study (Rahmstorf 1996) and an experiment with a flux-adjusted coupled model (Manabe and Stouffer 1997) suggests that freshwater perturbations at low-latitudes are less effective in affecting the AMOC than those at high-latitudes, because of their dilution. This is apparently confirmed for the transient response in a study by Goelzer et al. 2006 who found that the AMOC responds more quickly to freshwater fluxes that are applied near the northern convection sites than to those applied over the tropical Atlantic. At long time scales, low-latitude anomalies do reach the northern Atlantic, and the difference in sensitivity diminishes for equilibrium response to sustained freshening. These studies apparently confirm each other, but it should be realized that Rahmstorf (1996) and Goelzer et al. (2006) use ocean models that are coupled to idealized atmospheric models, so do not necessarily share the atmospheric response to hosing that is seen in GCMs. Indeed, changes in surface freshwater flux in response to hosing are smaller in models with more simplified dynamics than in GCMs (Stouffer et al. 2006).

Several hosing experiments have been carried out with HadCM3 in which various amounts of freshwater were applied over various parts of the North Atlantic as sustained surface fluxes lasting for at least 100 years (Vellinga 2004). This allows us to estimate the AMOC sensitivity as a function of the magnitude of the freshwater forcing, Fig. 12.2. The regions to which the flux was applied are shown in Fig. 12.3. In one experiment (F04), salt was added (negative hosing rate) to the Southern

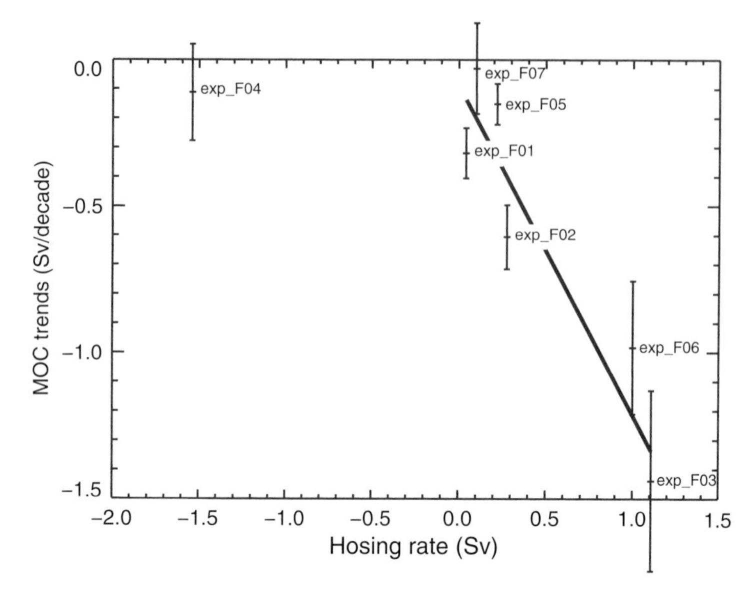

Fig. 12.2 Sensitivity of the AMOC at 48° N (expressed as weakening rate in Sv/decade) against the magnitude of freshwater forcing in HadCM3 hosing runs

Ocean to see if increasing density in the south is as effective as reducing it in the north in causing AMOC weakening.

As shown, the AMOC sensitivity has a near-linear dependency on the freshening rate, and the slope of the regression line is -1.1 ± 0.2 (Sv decade^{-1}/Sv). This simple regression suggests that 1 Sv of hosing applied for about 16 decades should reduce the AMOC in HadCM3 from 18 Sv to 0 Sv. From this limited number of experiments, it is difficult to say if there is a geographical dependency, although experiments in which the hosing is applied over the convection areas of the Greenland–Norwegian Seas appear to have sensitivities that are slightly stronger than expected from the regression (F01, F02 and F03). Experiment F04 (Southern Ocean salting) shows no appreciable AMOC response.

Freshwater perturbations used in 'hosing' experiments are typically applied at the surface, over a very large area. It is possible to conjecture that in the real world, more moderate amounts of high latitude freshwater anomalies (e.g. from glacial melt) might find their way to depth through entrainment into the dense-water overflow system. Could the ocean's sensitivity be different to this type of freshening as opposed to surface freshening? As far as we are aware, no direct numerical experiments have addressed this issue. However, the sensitivity of the AMOC response to the vertical distribution of fresh anomalies in the North Atlantic can be estimated from a suite of experiments with HadCM3. In 15 experiments, various freshwater perturbations were applied to different parts of the North Atlantic (Vellinga 2004).

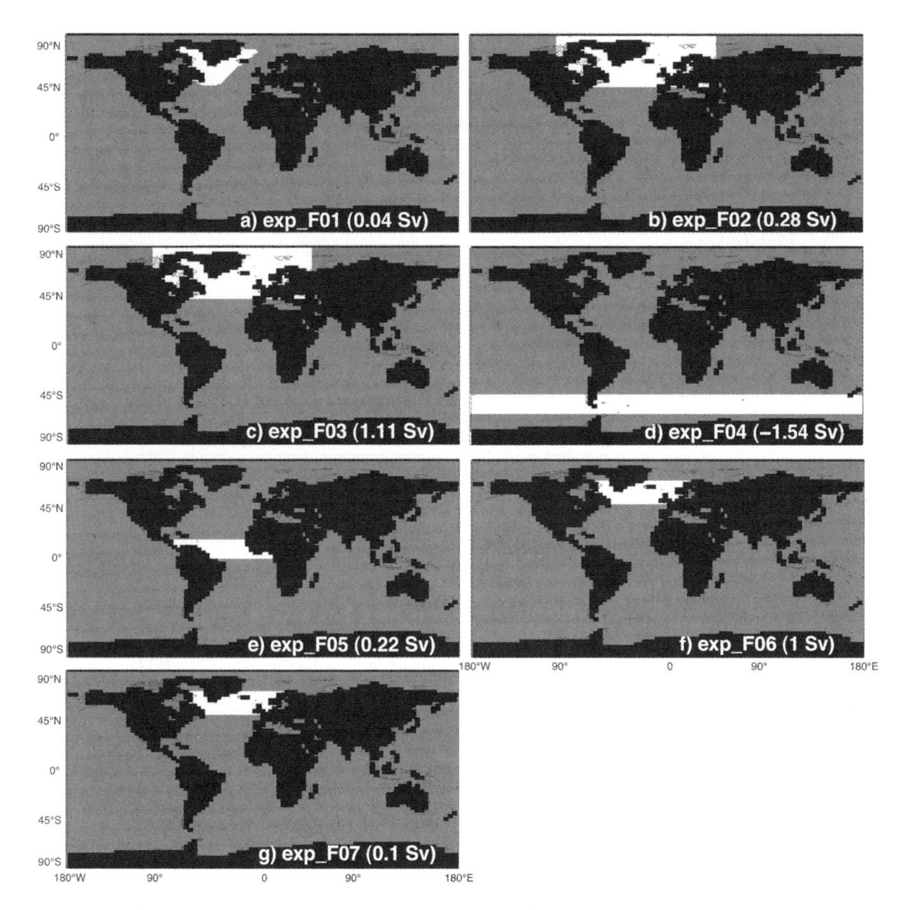

Fig. 12.3 Shown in white are areas to which freshwater forcing is applied in the HadCM3 hosing runs of Fig. 12.2. Figure inserts show the experiment name and the magnitude of the flux

The perturbations were applied as an instantaneous pulse (as in Vellinga et al. 2002) or as a continuous anomalous surface flux (as those applied in Fig. 12.2), and differed in strength and location. The perturbations are mainly applied to the upper 1,000 m of the water column, although ocean dynamics will mix some of the anomalies to greater depths. By pooling all experiments (amounting to over 200 decades of data) we sample a range of model states, through which we can quantify the AMOC dependence on the vertical distribution of salinity anomalies.

Using individual decadal mean data from all experiments, salinity is averaged over an area south of the overflows, between 45–0° W, 50–60° N, and for each depth this mean salinity is plotted against the AMOC strength at 50° N. A quadratic curve is then fitted to the data using least-squares regression (examples for two depths are shown in Fig. 12.4). The empirical quadratic relation between AMOC strength and salinity at each depth is then used to quantify the AMOC weakening associated with a freshening of 0.5 psu relative to normal conditions (cf. the two

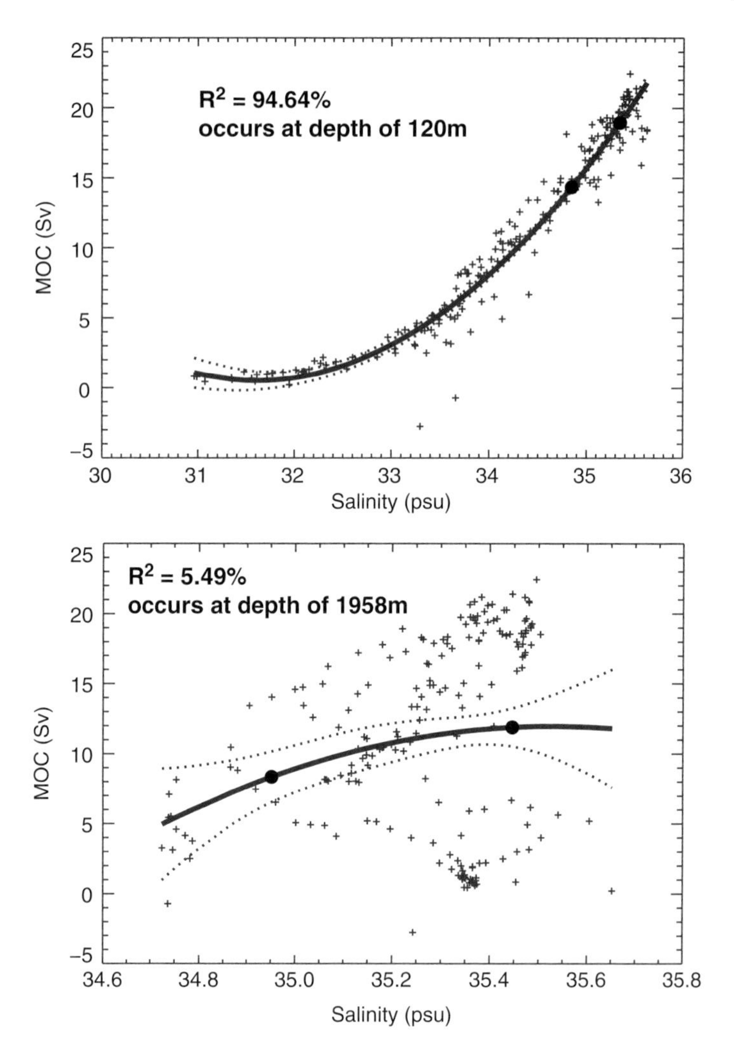

Fig. 12.4 (a) AMOC strength vs. salinity averaged over the region (45–0° W, 50–60° N) for two particular ocean depths (120 and 1,958 m). Here, AMOC strength is defined as the total meridional volume transport in the Atlantic across 50° N between the surface and 666 m (i.e. near to where maximum transport normally occurs in the model). Dotted curves indicate the two-sided 90% confidence intervals of the regression mean. Solid circles show the points on the curve for the model's normal salinity (higher value), and after it is freshened by 0.5 psu

black circles in Fig. 12.4). Dependence of this 'AMOC sensitivity' on where in the water column freshening occurs is shown in Fig. 12.5a. Sensitivity increases with depth from the surface down to about 600 m, then decreases to become near-zero at intermediate depths around 1,500 m. Towards abyssal depths the AMOC sensitivity grows again, but there the quadratic fit is hardly useful, as quantified by the R^2 curve or as seen in the scatterplot of Fig. 12.4b; by the nature of the perturbations,

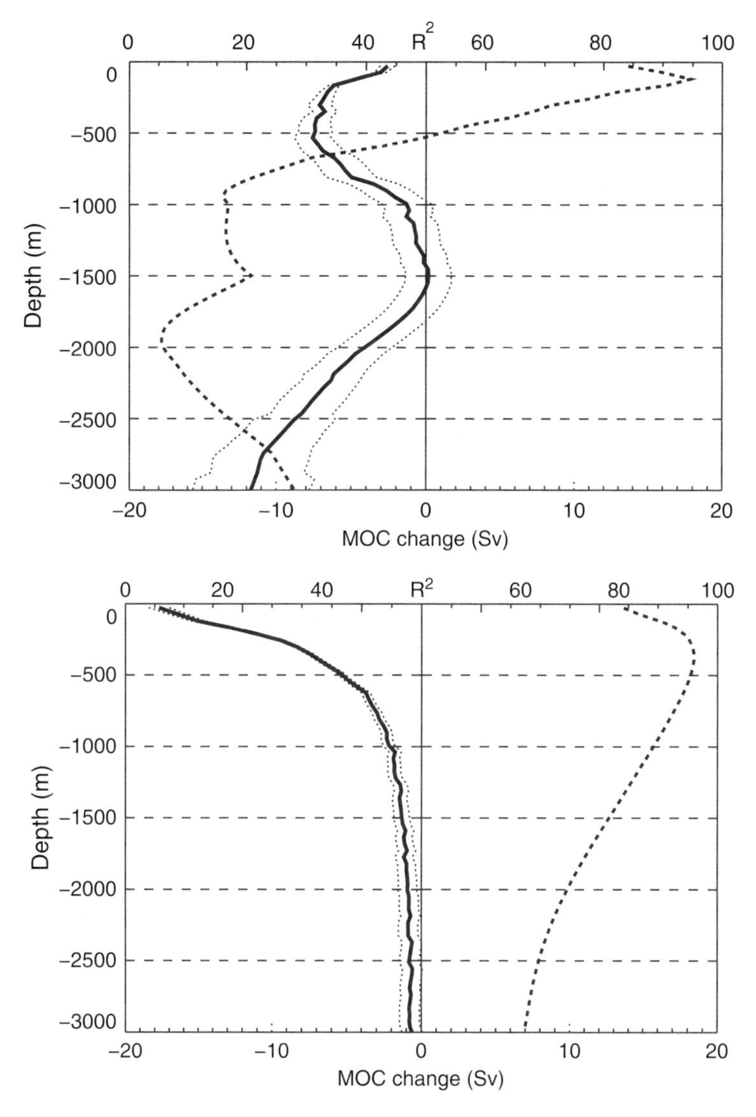

Fig. 12.5 (a) AMOC weakening (solid line, lower horizontal axis) in response to a freshening of 0.5 psu, applied at a single *spot* depth (vertical axis). Dotted curves denote the range based on the uncertainty estimate of the regression at each depth (cf. Fig. 12.4). The dashed curve (upper horizontal axis) shows R^2, the fraction of variance that is explained by each quadratic fit at each depth). (b) As in (a), but for 0.1 Sv*year (about 3^*10^{12} m^3) of fresh water distributed uniformly *between* the surface and the indicated depth

model states with freshening at greater depths are probably less well-sampled. The results suggest that freshening is more effective in weakening the AMOC if it occurs at shallower depths, and less effective at the depth occupied by the overflow water south of the Ridges (1,000 m and deeper).

One can also ask if a given anomalous freshwater loading is more or less effective in affecting the AMOC if it is spread out over a larger vertical section of the water column. The above analysis was repeated, but now for salinity anomalies averaged *between* the surface and different depths. The AMOC sensitivity was then determined for a given freshwater anomaly of 0.1 Sv*year, by converting that into a salinity anomaly based on the ocean volume occupied by that part of the water column (effectively diluting it with depth). As shown in Fig. 12.5b, the greatest sensitivity occurs if the fresh anomaly is confined to a shallow layer near the top of the water column. If the anomaly is distributed over depth and the salinity anomaly is smaller, AMOC weakening is reduced. The goodness of the quadratic fit between AMOC and salinity turns out to be particularly strong at depths between 400 and 500 m.

One of the motivations to do sensitivity experiments in the form of 'water hosing' is to quantify the effects on the AMOC of any future increases in freshwater flux that may be missed by models due to model imperfections (Stouffer et al. 2006). These might include, for example, the aforementioned uncertainty in projected precipitation change, or in the melt of the Greenland ice sheet which is not usually simulated directly in GCM climate change experiments (although, recently, several groups have begun to include in their climate simulations some of the processes that affect the Greenland ice sheet mass balance: Ridley et al. 2005; Swingedouw et al. 2006).

It seems appropriate to verify how comparable is the model response to freshwater hosing (typically carried out under pre-industrial greenhouse gas concentrations) to that under anthropogenic climate change, where both surface heat and freshwater fluxes are changing. For this we use several experiments carried out with HadCM3. These include the same freshwater experiments used in the hosing sensitivity study (Fig. 12.3) and in the study of sensitivity to the vertical distribution of freshening (Figs. 12.4 and 12.5). In addition we use data from idealized CO_2 and SRES forcing scenario experiments for the 21st century. Decadally averaged data from all these experiments show a close relation between the ocean density of the combined Nordic Seas/Arctic Ocean (averaged over the top 3,000 m), and the AMOC (Fig. 12.6a), similar to what has been found in other studies (Hughes and Weaver 1994; Rahmstorf 1996; Thorpe et al. 2001). The relation is approximately linear for density changes of magnitude less than 0.5 kg m^{-3}. For greater density changes the effect on the AMOC saturates. Crucially, all experiments (hosing, initial perturbations, greenhouse gas) follow the same empirical relation.

If, however, the density changes in this region are decomposed into those stemming from changes in temperature ($\Delta\rho_T$), and those due to changes in salinity ($\Delta\rho_S$) the different experiments start to fan out, as described in Fig. 12.6b. For instance, in greenhouse gas experiments (red circles) warm temperature anomalies dominate density changes. In hosing runs (black squares) fresh anomalies dominate density changes, though we also note from this figure that the most-extreme freshening effects on density are those with accompanying warm anomalies. In initial perturbation experiments (black triangles) temperature and salinity changes work in opposite ways, but salinity effects dominate. In Fig. 12.6, we also show data from

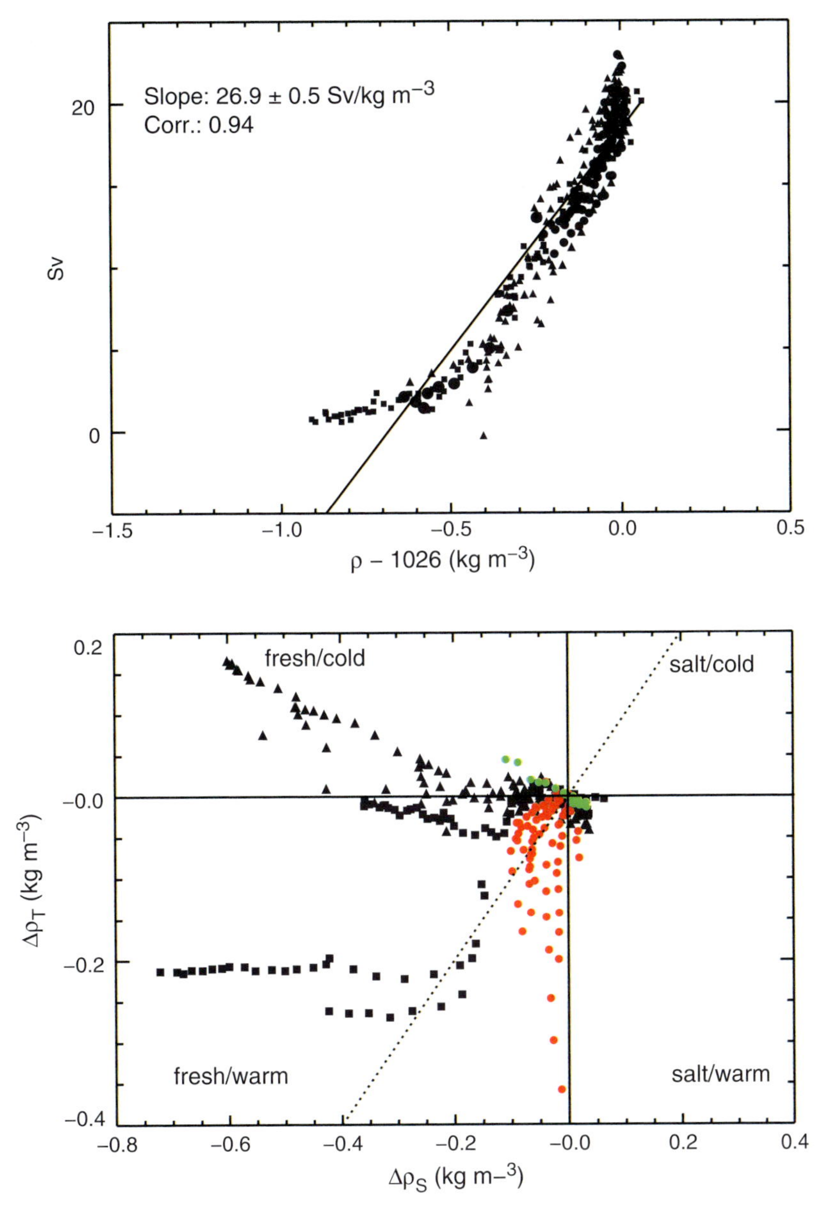

Fig. 12.6 (a) AMOC strength at 45° N (vertical) against density, averaged over the top 3,000 m in the Nordic Seas/Arctic Ocean. Circles indicate greenhouse gas experiments, squares hosing runs and triangles 'initial perturbation' experiments. (b) Density anomalies (relative to long-term mean of control run) caused by temperature ($\Delta\rho_T$, vertical) and salinity ($\Delta\rho_S$, horizontal). Legend as in (a), but greenhouse gas runs are in red, control run is in green. The dotted line denotes values $\Delta\rho_T = \Delta\rho_S$

30 decades of the control run (green circles) to allow comparison with anomalies associated with internal variability (e.g. stemming from centennial oscillations of the AMOC (Vellinga and Wu 2004)). Hosing, greenhouse gas and initial perturbation experiments all cluster in their own regions of the $\Delta\rho_S - \Delta\rho_T$ plane. The three types of experiments sample distinct model states. This result suggests that each class of experiments might involve fundamentally different feedbacks. To what extent this is the case requires further analysis. At this stage we can only suggest that care be taken in transferring conclusions about feedbacks in one class of experiments to those of another.

12.4 Cutting-Edge Questions and Implications for Future Work

As regards future model improvements, there exists a large choice of plausible numerical schemes, parameterizations, parameter values, etc. that could be used in climate models. This kind of uncertainty is inherent to modelling, and can only be quantified using observational constraints (e.g. Knutti et al. 2002; Bony et al. 2006). Our suggestions for 'future work' are therefore by no means exclusive, but we base them on the two results just described (Figs. 12.5b and 12.6) since they are novel, summarise the results of a wide range of model experiments, and seem to pose clear questions for the observer- and modelling-communities that are of more-than-local significance.

12.4.1 For the Observational Community

Despite major advances in observing and simulating the system, we remain undecided on many of the most basic issues that link change in our northern seas to climate. For example, while there is agreement that an increasing freshwater flux through Fram Strait to the North Atlantic is likely to be of climatic significance, we remain uncertain as to whether the impact on climate will result from *local* effects on overflow transport (e.g. from the changing density contrast across the Denmark Strait sill; Curry and Mauritzen 2005), from the *regional* effect of capping the water column of the NW Atlantic (leading to a reduction in vertical mixing, water mass transformation, and production of North Atlantic Deep Water), or from *global-scale* changes in the Ocean's thermohaline fields and circulation arising from an acceleration of the Global Water Cycle (Curry et al. 2003). Equally, we have yet to reconcile the subtleties of cause and effect revealed in our simulations of Arctic–Atlantic exchanges; for example, the finding by Oka and Hasumi (2006) that the deep-convective seesaw between the Labrador and Greenland Seas (Dickson et al. 1996) is controlled by changes in the freshwater transport through Denmark Strait, with the finding of Wu and Wood (2007, submitted) that the freshening recently observed in subpolar seas may ultimately be triggered by Labrador Sea deep convection. Despite this, there would probably be general acceptance of the conclusion of Jungclaus et al. (2005; from model experiments using ECHAM5 and the MPI-OM), that while

> the strength of the (Atlantic) overturning circulation is related to the convective activity in the deep-water formation regions, most notably the Labrador Sea, … the variability is sustained by an interplay between the storage and release of freshwater from the central Arctic and circulation changes in the Nordic Seas that are caused by variations in the Atlantic heat and salt transport.

The significance of Fig. 12.5b is that it leads us into a complex of fairly specific questions relevant to the latitudinal exchange of freshwater with the Arctic through subarctic seas, and the way it might interface with the watercolumn of the NW Atlantic.

It suggests, fundamentally, that the impact on the AMOC will depend on the extent to which the freshwater efflux from the Arctic will be spread to depth on its arrival in the NW Atlantic. We already know from half a Century of repeat hydrography that the system of dense-water overflows from the Nordic seas *has* been the vehicle for the freshening of the deep and abyssal layers of the Labrador Basin, below the limits of convection (2,300 m or so) since the mid-1960s (Dickson et al. 2002). And this observation lends point to the more-specific questions posed by Fig. 12.5b: whether any future increase in the freshwater outflow from the Arctic is likely to be incorporated into the overflow system, or (effectively the same thing) *whether any future increase of the freshwater efflux is likely to pass to the west or to the east of Greenland.*

We know of only one model study that currently makes that prediction. Recent coupled experiments by Helmuth Haak and the MPI Group using ECHAM 5 and the MPI-OM (1.5 deg; l 40) suggest that although the freshwater flux is expected to increase both east and west of Greenland, the loss of the sea-ice component (which currently dominates the flux through Fram Strait) suggests we should expect a much greater total increase through the CAA by 2070–2099 (+48%) than through Fram Strait (+3% only; see Table 12.1). Such a stark shift in the balance of outflow should be evident even in intermittent observations, and the validation of this prediction should be one general task of a future observing system.

Both east and west of Greenland, the historical hydrographic record and some novel observing techniques are beginning to identify the more-localised processes

Table 12.1 Simulated Arctic Ocean freshwater flux (km^3 year^{-1}) through Fram Strait and the Canadian Arctic Archipelago in 2070–2099 compared with 1860–1999. Results of coupled experiments using ECHAM 5 and the MPI-OM (1.5°; l 40) (Adapted from Haak et al. 2005. See also Koenigk et al., Chapter 8, this volume)

	1860–1999			2070–2099		
	Solid	Liquid	Total	Solid	Liquid	Total
Fram Strait	2543	1483	4026	317 (−87%)	3840 (+159%)	4157 (+3%)
CAA	495	1975	2470	187 (−62%)	3461 (+75%)	3648 (+48%)

that control the interface between the freshwater outflows and the Atlantic circulation. East of Greenland, for example, predictive analysis based on the historical record has provided insight into the likelihood of future direct effects on the strength of overflow through Denmark Strait. Recognising that it is the density contrast across the Denmark Strait sill that drives the overflow and noting that both overflows have undergone a remarkably rapid and remarkably steady freshening over the past four decades (Dickson et al. 2002), Curry and Mauritzen (2005) use Whitehead's (1998) hydraulic equation to ask how much more fresh water would have to be added to the western parts of the Nordic seas to produce significant slowdown. They find that it's not going to happen anytime soon:-

> At the observed rate, it would take about a Century to accumulate enough freshwater (e.g. 9000 km^3) to substantially affect the ocean exchanges across the Greenland-Scotland Ridge, and nearly two Centuries of continuous dilution to stop them. In this context, abrupt changes in ocean circulation do not appear imminent.

The fact that the freshening trend of both overflows at the sill has slowed to a stop over the last 10 years (see Yashayaev and Dickson 2007) has merely reinforced this conclusion.

West of Greenland, results remain much more equivocal regarding the local-to-regional impact of an increased flux of freshwater through the CAA. Though the relatively coarse global models of Goosse et al. (1997) and Wadley and Bigg (2002) find decreases of 10% and 35% (respectively) in the strength of the overturning circulation between closing and opening the CAA, Myers (2005) has subsequently used a high resolution regional model to suggest that very little (6–8%) of the freshwater exported from the Canadian Arctic gets taken up in the Labrador Sea Water of his model. In general terms then, it remains an open question as to whether a future increase in the freshwater outflow through Davis Strait would spread across the surface or skirt around the boundary of the Labrador Basin; a more complete observing system south of Davis Strait will be necessary to developing that understanding.

In summary then, the watercolumn of the Labrador Sea is of global climatic importance, acting as the receiving volume for time-varying inputs of fresh- and other watermasses from Northern Seas which are then stored, recirculated, transformed and discharged to modulate the abyssal limb of the Atlantic Meridional Overturning Circulation (AMOC). The extreme amplitude of anomalous conditions throughout the watercolumn of the Labrador Sea over the past four decades and the importance of their claimed effects for the thermohaline circulation and for climate justify a sustained ocean-observing effort to understand and test the behaviour of this system in climate models. Here we have placed emphasis on monitoring the changing balance of freshwater fluxes east and west of Greenland, and on investigating how each of these main freshwater outflows interfaces with the watercolumn of the NW Atlantic. In practice of course, each of the watermasses recruiting to the Labrador Basin will carry with them the imprint of time-varying climatic forcing in their source regions and of modifications en route, and their properties (volume, temperature, salinity, density, tracer-loading) will also be subject to alteration by the processes of horizontal and vertical exchange within the Labrador Basin itself. The key issue for climate may lie not so much in describing and attributing the diverse sources of change in this vertical stack of watermasses but in understanding whether and to what extent they interact and the effect of such interactions on deep ocean hydrography and circulation.

12.4.2 For the Modeling Community

The top-end, climate general circulation models include what are believed to be the most important (physical) processes in the coupled ocean–atmosphere–sea ice system. These models allow us to make a 'best estimate' of what future climate will be like for a given choice of future anthropogenic changes in greenhouse gas and aerosol concentrations. It is natural to assume that models improve if more sophisticated schemes are used, or if their resolution is increased. To what extent that translates into more reliable projections of climate change is another matter, but there is no doubt that improved model formulation has led to the ability of global climate models to simulate some of the large changes observed in the oceans during the 20th century (e.g. Barnett et al. 2001; Gregory et al. 2004; Wu et al. 2004).

Clearly models need sufficient resolution to resolve geometry (such as the overflow sills from the Nordic Seas (e.g. Böning et al. 1996; Roberts and Wood 1997), important ocean bathymetry (e.g. Banks 2000) and boundary currents and other narrow currents (Oka and Hasumi 2006). The need in climate studies for eddy-resolving ocean resolution has not been established, but little work has been done in this field. Regional eddy-resolving ocean models are becoming more widely used (e.g. Smith et al. 2000), often to be employed in short-range ocean forecasting (Johannessen et al. 2006), rather than lengthy climate runs. Comparing the behaviour of a global eddy-permitting ($1/3° \times 1/3°$) and a non-eddying ($5/4° \times 5/4°$) version of the same coupled model to rising CO_2 concentrations, Roberts et al. (2004) show that the response of the AMOC and its heat transport to global warming depend on this particular increase in model resolution. Only one study with a global, eddy-resolving ocean model has been reported to date, integrated for 13 years in stand-alone mode (Maltrud and McClean 2005), with promising results in terms of eddy statistics in the model compared to altimeter observations. Variable or perhaps adaptive grids (i.e. finer resolution where and when it is needed, Pain et al. 2005) might provide computationally manageable solutions for high-resolution climate modelling, but are still under development.

Since, as already mentioned, the future development of climate models is liable to involve a large choice of plausible numerical schemes and an equally wide range of observational constraints, the concept of working towards a single best model is not particularly meaningful. It is more helpful to think of a range of models, that spans the possible and likely behaviour of the real climate system (Allen and Ingram 2002). Several groups have already started, through 'perturbed-physics' experiments, to quantify how the uncertainty in model formulation creates uncertainty in climate projections (e.g. Murphy et al. 2004, Schneider von Daimling et al. 2006). But two questions remain.

First, how can we be sure that we have adequately employed 'the full range of models that spans the possible and likely behaviour of the real climate system'? Figure 12.6, just described, provides a clear example. Although, from a large ensemble of model experiments, Fig. 12.6a offered an encouragingly close fit between the density of northern seas and rate of the Atlantic overturning circulation at 45° N, in fact (Fig. 12.6b) the factors controlling density were found to be quite distinct in the three constituent types of experiment ('hosing runs', 'initial perturbation' experiments and greenhouse gas experiments). As a first step, it would be very useful to verify if the distinct trajectories in the $\Delta\rho_S - \Delta\rho_T$ plane are found in other models for similar experiments. If so, then the next step would be for the modelling community to validate the processes that control how a model state evolves along the respective trajectories, by seeking observational analogues for these trajectories (e.g. over a seasonal cycle, or during the Great Salinity Anomaly).This will clearly not be easy in the case of the full spatial domain used to calculate the data in Fig. 12.6, but it may be possible to use spatially degraded coverage, taking data from key regions only.

Second, how can we weigh the contributions of individual models in a multi-model ensemble, such as those contributing to reports by the IPCC? Perturbed-physics multi-model ensembles are likely to become increasingly important in quantifying the impact of model uncertainty on climate projections. Such ensembles are only meaningful if a suitable, observationally based model weighting is applied. Schmittner et al. (2005) provide an example for this, but the absence of repeated, observed realisations of the predictand in the real world prevents us from determining model skill, in the same way as is done for numerical weather prediction. It is a non-trivial task to ascertain what the relevant observations are that constrain prediction of quantities at climate time scales, such as Arctic summer sea ice cover by the 2050s, or AMOC heat transport at 30° N by 2100. One answer may be observational 'weighting by proxy': by identifying model skill in simulating fields for which there are observations, and that are proven to also provide skill measures for the unobserved quantities that we wish to predict.

Acknowledgements Michael Vellinga was supported by the Joint Defra and MoD Programme, (Defra) GA01101 (MoD) CBC/2B/0417_Annex C5. Bob Dickson was supported by the Department for Environment, Food and Rural Affairs under the Defra Science and Research Project OFSOD-iAOOS, contract SD0440. Thanks to Jonathan Gregory for providing model data and to Jochem Marotzke for useful comments on this Chapter.

References

Allen MR, WJ Ingram (2002) Constraints on future changes in climate and the hydrologic cycle. Nature 419: 224–232

Banks HT (2000) Ocean heat transport in the South Atlantic in a coupled climate model. J. Geoph. Res. 105(C1): 1071–1091

Banks HT, RA Wood (2002) Where to look for anthropogenic climate change in the ocean? J. Climate 15: 879–891

Barnett TPD, W Pierce, R Schnur (2001) Detection of anthropogenic climate change in the world's oceans. Science 292: 270–274

Bojariu R, G Reverdin (2002) Large-scale variability modes of freshwater flux and precipitation over the Atlantic. Clim. Dyn. 18: 369–381

Böning CW, FO Bryan, WR Holland, R Döscher (1996) Deep water formation and meridional overturning in a high-resolution model of the North Atlantic. J. Phys. Oceanogr. 26: 515–523

Bony S, R Colman, VM Kattsov, RP Allan, CS Bretherton, JL Dufresne, A Hall, S Hallegate, MM Holland, WJ Ingram, DA Randall, BJ Soden, G Tselioudis, MJ Webb (2006) How well do we understand and evaluate climate change feedback processes? J. Climate 19: 3445–3482

Broecker WS (1997) Thermohaline circulation, the Achilles heel of our climate system: Will man-made CO_2 upset the current balance? Science 278: 1582–1588

Bryan F (1986) High-latitude salinity effects and interhemispheric thermohaline circulations. Nature 323: 301–304

Bryden HL, HR Longworth, SA Cunningham (2005) Slowing of the Atlantic meridional Overturning Circulation at 25° N. Nature 438: 655–657

Cayan DR (1992) Latent and sensible heat flux anomalies over the northern oceans: The connection to monthly atmospheric circulation. J. Climate 5: 354–369

Cheng W, PB Rhines (2004) Response of the overturning circulation to high-latitude fresh-water perturbations in the North Atlantic. Clim. Dyn. 22(4): 359–372

Cheng W, R Bleck, C Rooth (2004) Multi-decadal thermohaline variability in an ocean-atmosphere general circulation model. Clim. Dyn. 22: 573–590.

Collins M, B Sinha (2003) Predictability of decadal variations in the thermohaline circulation and climate. Geophys. Res. Lett. 30(6): 1413, doi:10.1029/2002GL016776

Collins M, A Botzet, A F Carril, H Drange, A Jouzea, M Latif, S Masina, OH Otteraa, H Pohlmann, A Sorteberg, R Sutton, L Terray (2006) Interannual to decadal climate predictability in the north Atlantic: A multimodel-ensemble study. J. Climate 19: 1195–1203

Cubasch U, GA Meehl, GJ Boer, RJ Stouffer, M Dix, A Noda, CA Senior, SCB Raper, and KS Yap (2001) Projections of future climate change. In JT Houghton, Y Ding, DJ Griggs, M Noguer, P van der Linden, X Dai, K Maskell, CI Johnson (eds.) Climate Change 2001: The Scientific Basis. Contribution of Working Group I to the Third Assessment Report of the Intergovernmental Panel on Climate Change. Cambridge University Press, Cambridge, pp 525–582

Curry B, GP Lohmann (1982) Carbon isotopic changes in benthic foraminifera from the western South Atlantic: Reconstruction of glacial abyssal circulation patterns, Quat. Res. 18: 218–235

Curry R, C Mauritzen (2005) Dilution of the Northern North Atlantic Ocean in Recent Decades. Science 308 (5729): 1772–1774

Curry R, RR Dickson, I Yashayaev (2003) A change in the freshwater balance of the Atlantic Ocean over the past four decades. Nature 426: 826–829

Dahl K, A Broccoli, R Stouffer (2005) Assessing the role of North Atlantic freshwater forcing in millennial scale climate variability: A tropical Atlantic perspective. Clim. Dyn. 24(4): 325–346

Dai A, A Hu, GA Meehl, WM Washington, WG Strand (2005) Atlantic thermohaline circulation in a coupled general circulation model: unforced variations versus forced changes. J. Climate 18: 3270–3293

Delworth TL, KW Dixon (2000) Implications of the recent trend in the Arctic/N Atlantic Oscillation for the North Atlantic thermohaline circulation. J Climate 13: 3721–3727

Delworth TL, KW Dixon (2006) Have anthropogenic aerosols delayed a greenhouse gas-induced weakening of the North Atlantic thermohaline circulation? Geophys. Res. Lett. 33, LO2606, doi:10.1029/2005Glo24980

—Delworth TL, RJ Greatbatch (2000) Multidecadal thermohaline circulation variability driven by atmospheric flux forcing. J. Climate 13: 1481–1495

Delworth TL, ME Mann (2000) Observed and simulated multidecadal variability in the North Atlantic. Climate Dyn. 16 (9): 661–676

Delworth, TL, S Manabe, RJ Stouffer (1993) Interdecadal variations of the thermohaline circulation in a coupled ocean-atmosphere model. J. Climate 6: 1993–2011

Dickson RR, J Lazier, J Meincke, P Rhines, J Swift (1996) Long-term coordinated changes in the convective activity of the North Atlantic. Prog. Oceanogr. 38: 241–295

Dickson RR, I Yashayaev, J Meincke, W Turrell, S Dye, J. Holfort (2002) Rapid freshening of the deep North Atlantic over the past four decades. Nature 416: 832–837

Dong B, R Sutton (2005) Mechanism of interdecadal thermohaline circulation variability in a coupled ocean-atmosphere GCM. J. Climate 18: 1117–1135

Eden C, T Jung (2001) North Atlantic interdecadal variability: oceanic response to the North Atlantic oscillation (1865–1997). J. Climate 14: 676–691

Eden C, J Willebrand (2001) Mechanism of interannual to decadal variability of the North Atlantic circulation. J. Climate 14: 2266–2280

Gamiz-Fortis SR, D Pozo-Vazquez, MJ Esteban-Parra, Y Castro-Diez (2002) Spectral characteristics and predictability of the NAO assessed through Singular Spectral Analysis. J. Geoph. Res. 107(D23): 4685–4699

Goelzer H, J Mignot, A Levermann, S Rahmstorf (2006) Tropical versus high latitude freshwater influence on the Atlantic circulation. Clim. Dyn. 27(7–8): 715–725

Goosse H, T Fichefet, J-M Campin (1997) The effects of the water flow through the Canadian Archipelago in a global ice-ocean model. Geophysical Research Letters 24: 1507–1510, doi:10.1029/97GL01352

Gordon AL (1986) Inter-ocean exchange of thermocline water. J. Geophys. Res. 91: 5037–5046

Gregory JM, HT Banks, PA Stott, JA Lowe, MD Palmer (2004) Simulated and observed decadal variability in ocean heat content. Geophys. Res. Lett. 31: L15312, doi:10.1029/2004GL020258

Gregory JM, KW Dixon, RJ Stouffer, AJ Weaver, E Driesschaert, M Eby, T Fichefet, H Hasumi, A Hu, JH Jungclaus, IV Kamenkovich, A Levermann, M Montoya, S Murakami, S Nawrath, A Oka, AP Sokolov, and RB Thorpe (2005) A model intercomparison of changes in the Atlantic thermohaline circulation in response to increasing atmospheric CO_2 concentration. Geophys. Res. Lett. 32, L12703, doi:10.1029/2005GL023209

Haak H, J Jungclaus, T Koenigk, D Svein, U Mikolajewicz (2005) Arctic Ocean freshwater budget variability. ASOF Newsletter (3): 6–8. http://asof.npolar.no

Häkkinen S (1999) Variability of the simulated meridional heat transport in the North Atlantic for the period 1951–(1993) J. Geoph. Res. 105(C5): 10,911–11,007

Hasselmann K (1976) Stochastic climate models. Part I: Theory. Tellus 28: 473–485

Higuchi K, JP Huang, A Shabbar (1999) A wavelet characterization of the North Atlantic oscillation variation and its relationship to the North Atlantic sea surface temperature. Int. J. Climatol. 19(10), 1119–1129

Hu A, GA Meehl (2005) Reasons for a fresher northern North Atlantic in the late 20th century. Geophys. Res. Lett. 32, doi:10.1029/2005GL022900

Hughes TMC, AJ Weaver (1994) Multiple equilibria of an asymmetric two-basin model. J. Phys. Oceanogr. 24: 619–637

Hunt BG, TI Elliott (2006) Climatic trends. Clim. Dyn. 26: 567–585

Hurrell JW (1995) Decadal trends in the North Atlantic Oscillation: regional temperatures and precipitation. Science 269: 676–679

Hurrell JW, H van Loon (1997) Decadal variations in climate associated with the North Atlantic Oscillation. Clim. Change 36: 301–326

Ingvaldsen RB, L Asplin, H Loeng (2004a) Velocity field of the western entrance to the Barents Sea. J. Geophys. Res. 109, C03021, doi:101029/2003JC001811

Ingvaldsen RB L Asplin, H Loeng (2004b) The seasonal cycle in the Atlantic transport to the Barents Sea during the years 1997–2001. Continent. Shelf Res. 24: 1015–1032

Johannessen, JA, PY Le Traon, I Robinson, K Nittis, MJ Bell, N Pinardi, P Bahurel (2006) Marine environment and security for the European area – Toward operational oceanography. Bull. Am. Met. Soc. 87(8): 1081

Jones PD, T Jonsson, D Wheeler (1997) Extension to the North Atlantic Oscillation using early instrumental pressure observations from Gibraltar and south-west Iceland. Int. J. Climatol. 17: 1433–1450

Jungclaus J, M Haak, H Latif, U. Mikolajewicz (2005) Arctic-North Atlantic interactions and multidecadal variability of the meridional overturning circulation. J. Climate 18: 4013–4031

Knight, JR, RJ Allan, CK Folland, M Vellinga, and ME Mann (2005) A signature of persistent natural thermohaline circulation cycles in observed climate. Geophys. Res. Lett. 32, doi:10.1029/2005GL024233

Knutti R, TF Stocker, F Joos, GK Plattner (2002) Constraints on radiative forcing and future climate change from observations and climate model ensembles. Nature 416: 719–723

Krahmann G, M Visbeck, G Reverdin (2001) Formation and propagation of temperature anomalies along the North Atlantic Current. J. Phys. Oceanogr. 31(5): 1287–1303

Kuzmina SI, L Bengtsson, OM Johannessen, H Drange, LP Bobylev, MW Miles (2005) The North Atlantic Oscillation and greenhouse-gas forcing. Geoph. Res. Let. 32(4), doi:10.1029/2005GL04703

Large WG, AJG Nurser (2001) Ocean surface water mass transformations, pp 317–335. In G Siedler, J Church and J Gould (Eds) Ocean Circulation and Climate. Academic Press International Geophysics Series, 77, 715 pp.

Latif M, E Roeckner, U Mikolajewicz, R Voss (2000) Tropical stabilisation of the thermohaline circulation in a greenhouse warming simulation. J. Climate 13: 1809–1813

Latif M, E Roeckner, M Botzet, M Esch, H Haak, S Hagemann, J Jungclaus, S Legutke, S Marsland, U Mikolajewicz, J. Mitchell (2004) Reconstructing, monitoring and predicting multidecadal-scale changes in the North Atlantic thermohaline circulation with sea surface temperature. J. Climate 17: 1605–1614

Latif M, C Böning, J Willebrand, A Biastoch, J Dengg, N Keenlyside, Schweckendiek (2006) Is the themohaline circulation changing? J. Climate 19: 4632–4637

Levermann AA, M Griesel, M Hofmann, M Montoya, S Rahmstorf (2005) Dynamic sea level changes following changes in the thermohaline circulation. Clim. Dyn. 24: 347–354

Maltrud ME, JL McClean (2005) An eddy resolving global ocean simulation. Ocean Model 8: 31–54

Mikolajewicz U, M Groger, E Maier-Reimer, G Schurgers, M Vizcaino and AME Winguth (2007) Long-term effects of anthropogenic CO_2 emissions simulated with a complex earth system model. Clim Dyn., doi 10.1007/s00382-006-0204-y

Manabe S, RJ Stouffer (1988) Two stable equilibria of a coupled ocean-atmosphere model. J. Climate 1: 841–866

Manabe S, RJ Stouffer (1993) Century-scale effects of increased atmospheric $C0_2$ on the ocean–atmosphere system. Nature 364: 215–218, doi:10.1038/364215a0

Manabe S, RJ Stouffer (1994) Multiple century response of a coupled ocean-atmosphere model to an increase of atmospheric carbon dioxide. J. Climate 7: 5–23

Manabe S, RJ Stouffer (1997) Coupled ocean-atmosphere model response to freshwater input: comparison with Younger Dryas event. Paleoceanography 12: 321–336

Marotzke J, J Willebrand (1991) Multiple equilibria of the global thermohaline circulation. J. Phys. Oceanogr. 21: 1372–1385

Mignot J, C Frankignoul (2005) On the variability of the Atlantic meridional overturning circulation, the NAO and the ENSO in the Bergen Climate Model. J. Climate 18: 2361–2375

Murphy JM, DMH Sexton, DN Barnett, GS Jones, MJ Webb, M Collins, DA Stainforth (2004) Quantification of modelling uncertainties in a large ensemble of climate change simulations. Nature 430: 768–772

Myers PG (2005) Impact of freshwater from the Canadian Arctic Archipelago on Labrador Sea Water formation. Geophys. Res. Lett. 32, L06605, doi:10.1029/2004GL022082

Oka A, H Hasumi (2006) Effects of model resolution on salt transport through northern high-latitude passages and Atlantic meridional overturning circulation. Ocean Model 13: 126–147

Osborn TJ (2004) Simulating the winter North Atlantic Oscillation: the roles of internal variability and greenhouse gas forcing. Clim. Dyn. 22: 605–623

Østerhus S, WR Turrell, S Jonsson and B Hansen (2005) Measured volume, heat and salt fluxes from the Atlantic to the Arctic Mediterranean. Geophys. Res. Lett. 32, L07603, doi:10.1029/2004GL022188

Ottera OH, H Drange, M Bentsen, NG Kvamsto, DB Jiang (2004) Transient response of the Atlantic meridional overturning circulation to enhanced freshwater input to the Nordic Seas-Arctic Ocean in the Bergen Climate Model. Tellus (A) 56(4): 342–361

Pain CC, MD Piggott, AJH Goddard, F Fang, GJ Gorman, DP Marshall, MD Eaton, PW Power, CRE de Oliveira (2005) Three-dimensional unstructured mesh ocean modelling. Ocean Model 10(1–2): 5–33

Rahmstorf S (1995) Bifurcations of the Atlantic thermohaline circulation in response to changes in the hydrological cycle. Nature 378: 145–149

Rahmstorf S (1996) On the freshwater forcing and transport of the Atlantic thermohaline circulation. Clim. Dyn. 12: 799–811

Rahmstorf S (2003) Thermohaline Circulation: The current climate. Nature 421: 699

Rahmstorf S, A Ganopolski (1999) Long term global warming scenarios,computed with an efficient climate model. Clim. Change 43: 353–367

Rhines P, S Hakkinen (2003) Is the Oceanic heat transport in North Atlantic irrelevant to the climate in Europe? ASOF Newsletter #2: 13–17

Ridley J, Huybrechts P, Gregory JM, Lowe JA (2005) Elimination of the Greenland ice sheet in a high CO2 climate. J. Climate 18: 3409–3427

Roberts MJ, RA Wood (1997) Topography sensitivity studies with a Bryan-Cox type ocean model. J. Phys. Oceanogr. 27: 823–836

Roberts, MJ, H Banks, N Gedney, J Gregory, R Hill, S Mullerworth, A Pardaens, G Rickard, R Thorpe, R Wood (2004) Impact of an eddy-permitting ocean resolution on control and climate change simulations with a global coupled GCM. J. Climate 17: 3–20

Schauer U, E Fahrbach, S Østerhus, G Rohardt (2004) Arctic warming through the Fram strait: Oceanic heat transports from 3 years of measurements. J. Geophys Res. 109, C06026, doi:10.1029/2003JC001823

Schiller A, U Mikolajewicz, R Voss (1997) The stability of the thermohaline circulation in a coupled ocean-atmosphere general circulation model. Clim. Dyn. 13: 325–347

Schmittner A, M Latif, B Schneider (2005) Model projections of the North Atlantic thermohaline circulation for the 21st century assessed by observations. Geophys. Res. Lett. 32, doi:10.1029/2005GL024368

Smith RD, ME Maltrud, FO Bryan, MW Hecht (2000) Numerical simulations of the North Atlantic Ocean at 1/10 degree. J. Phys. Oceanogr. 30: 1532–1561

Schneider von Deimling T, H Held, A Ganapolski, S Rahmstorf (2006) Climate sensitivity estimated from ensemble simulations of glacial climate. Clim. Dyn. 27: 149–163, doi:10.1007/s00382-006-0126-8

Stephenson DB, V Pavan, M Collins, MM Junge, R Quadrelli (2006) North Atlantic oscillation response to transient greenhouse gas forcing and the impact on european winter climate: a cmip2 multi-model assessment. Clim. Dyn. 27: 401–420

Stommel HM (1961) Thermohaline convection with two stable regimes of flow. Tellus 13: 224–230

Stouffer RJ, J Yin, JM Gregory, KW Dixon, MJ Spelman, W Hurlin, AJ Weaver, M Eby, GM Flato, H Hasumi, A Hu, J Jungclaus, IV Kamenkovich, A Levermann, M Montoya, S Murakami, S Nawrath, A Oka, WR Peltier, DY Robitaille, A Sokolov, G Vettoretti, N Weber (2006) Investigating the causes of the response of the thermohaline circulation to past and future climate changes. J. Climate 19: 1365–1387

Swingedouw D, P Braconnot, O Marti (2006) Sensitivity of the Atlantic meridional overturning circulation to the melting from northern glaciers in climate change experiments. Geophys. Res. Lett. 33, L07711, doi:10.1029/2006GL025765

Thorpe, RB, JM Gregory, TC Johns, RA Wood, and JFB Mitchell (2001) Mechanisms determining the Atlantic thermohaline circulation response to greenhouse gas forcing in a non-flux-adjusted coupled climate model. J. Climate 14: 3102–3116

Timmermann, A, M Latif, RVA Grötzner (1998) Northern Hemisphere interdecadal variability: a coupled air-sea mode. J. Climate 11: 1906–1931

Vellinga M (2004) Robustness of climate response in HadCM3 to various perturbations of the Atlantic meridional overturning circulation. Hadley Centre Technical Note CRTN 48, Met Office Hadley Centre, FitzRoy Road, Exeter EX1 3PB, United Kingdom (available via: URL http://www.metoffice.gov.uk/research/hadleycentre/pubs/HCTN/HCTN_48.pdf)

Vellinga M, RA Wood (2004) Timely detection of anthropogenic change in the Atlantic meridional overturning circulation. Geophys. Res. Lett. 31, doi:10.1029/2004GL020306

Vellinga M, RA Wood (2007) Impacts of thermohaline circulation shutdown in the twenty-first century. Clim. Change, doi:10.1007/s10584-006-9146-y

Vellinga M, P Wu (2004) Low-latitude fresh water influence on centennial variability of the thermohaline circulation. J. Climate 17: 4498–4511

Vellinga M, RA Wood, JM Gregory (2002) Processes governing the recovery of a perturbed thermohaline circulation in HadCM3. J. Climate 15: 764–780

Wadley MR, GR Bigg (2002) Impact of flow through the Canadian Archipelago and Bering Strait on the North Atlantic and Arctic circulation: An ocean modelling study. Q. J. Roy. Met. Soc. 128: 2187–2203

Welander P (1982) A simple heat salt oscillator. Dyn. Atmos. Oceans 6: 233–242

Whitehead JA (1998) Topographic control of oceanic flows in deep passages and straits. Rev. Geophys. 36: 423–440

Wood RA, M Vellinga, R Thorpe (2003) Global warming and thermohaline circulation stability, Phil. Trans. R. Soc. Lond. (A) 361: 1961–1975

Wood RA, M Collins, J Gregory, G Harris, M Vellinga (2006) Towards a risk assessment for shutdown of the Atlantic Thermohaline Circulation. In HJ Schellnhuber et al. (eds.)'Avoiding Dangerous Climate Change. Cambridge University Press, Cambridge, 392 pp

Wu P, RA Wood (2007) Convection-induced long term freshening of the Subpolar North Atlantic Ocean. Climate Dyn. submitted

Wu P, RA Wood, P Stott (2004) Does the recent freshening trend in the North Atlantic indicate a weakening thermohaline circulation? Geophys. Res. Lett. 31, Lo2301, doi: 0.1029/2003GLO18584

Wunsch C, P Heimbach (2006) Estimated decadal changes in the North Atlantic Meridional overturning circulation and heat flux 1993–2004. J. Phys. Oceanogr. 36: 2012–2024

Chapter 13
Constraints on Estimating Mass, Heat and Freshwater Transports in the Arctic Ocean: An Exercise

Bert Rudels, Marika Marnela, and Patrick Eriksson

13.1 Introduction

The ASOF programme, with its study of the transports between the Arctic Ocean and the North Atlantic via the subarctic seas – the Nordic Seas, Baffin Bay and the Labrador Sea –, also provides an opportunity to examine the mass (volume), freshwater and heat budgets of the Arctic Ocean. The exchanges between the two passages between the Arctic Ocean and the Nordic Seas, Fram Strait and the Barents Sea opening between Norway and Bear Island, have been measured continuously since 1997, first in the VEINS programme (Variability of Exchanges in the Northern Seas) and then in ASOF and the observations are presently continued within the DAMOCLES (Developing Arctic Modelling and Observing Capabilities for Long-term Environmental Studies) programme. The transports through two of the three main channels in the Canadian Arctic Archipelago, the Lancaster Sound and the Jones Sound, have been directly measured for a couple of years now (Prinsenberg and Hamilton 2005), and the instruments from the first year-long measurements in Nares Strait have been brought in. The fluxes through Bering Strait have also been studied intensely the last 10–15 years (e.g. Woodgate and Aagaard 2005). The work within ASOF has shown that the transports through Fram Strait and through the Canadian Arctic Archipelago are those most difficult to determine. The Archipelago because of the severe climate, the remoteness of the area and the nearby location of the magnetic North Pole, Fram Strait because of its depth, the transports in both directions, and the presence of baroclinic and barotropic eddies leading to high spatial and temporal variability.

The estimates of the mean transport through Bering Strait obtained since the mid-1980s have ranged around 0.8 Sv (1×10^6 m^3 s^{-1}), but large seasonal variations have been reported, 1.2 Sv in summer and 0.4 Sv in winter (Coachman and Aagaard 1988; Woodgate and Aagaard 2005). The mean transport of Atlantic water to the Arctic Ocean through the Barents Sea opening has been estimated to 1.5 Sv from

Finnish Institute of Marine Research, Erik Palménin aukio 1, P.O. Box 2, FI-00561 Helsinki, Finland

R.R. Dickson et al. (eds.), *Arctic–Subarctic Ocean Fluxes*, 315–341

the observations in VEINS and ASOF, but with large short periodic variations (Ingvaldsen et al. 2004a, b). A longer time variation with a period of 3–4 years also appears to be present, causing the transport to change from below 1 Sv to slightly above 2 Sv (ASOF-N Final Report 2006). In addition to the inflow of Atlantic water there is also the contribution from the Norwegian Coastal Current, which amounts to 0.7 Sv with salinity 34.4 (Aagaard and Carmark 1989 based on Blindheim 1989). The Arctic Ocean also receives a freshwater input from runoff and net precipitation amounting to 0.15–0.2 Sv (Serreze et al. 2006). Assuming these estimates to be close to reality, the total transport through Bering Strait and the Barents Sea and the freshwater input, adding up to 3.2 Sv, can be used, together with requirements of mass and freshwater balance, to evaluate the transport estimates derived from the observations in Fram Strait. The passages and the transports are indicated in Fig. 13.1.

We begin by examining some of the estimates obtained in Fram Strait during different phases of the VEINS and ASOF programs and what these transports imply for the Arctic Ocean mass and freshwater budgets. In fact, this exercise was provoked

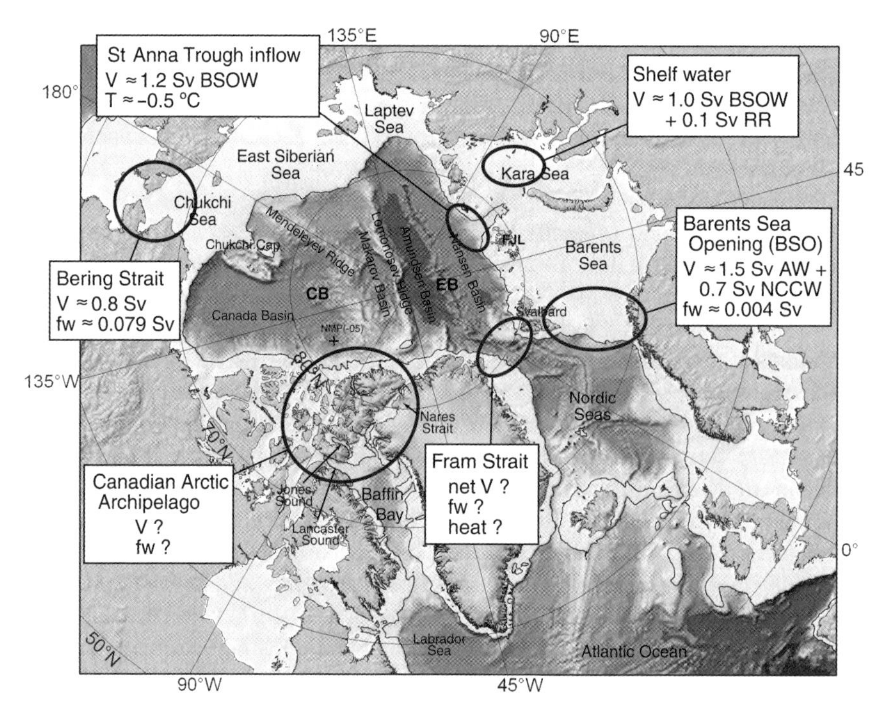

Fig. 13.1 The four main passages between the Arctic Ocean and the world ocean. The Bering Strait inflows are adopted from Woodgate and Aagaard (2005) and the inflows through the Barents Sea Opening (BSO) are taken from Ingvaldsen et al. (2004a) Atlantic Water (AW) and Blindheim (1989) Norwegian Coastal Current Water (NCCW). For the separation of the BSO inflow into a deep inflow via St Anna Trough and a less saline shelf water see discussion in Section 13.6. The freshwater is computed relative to 34.92. The river runoff is taken from Dickson et al. (2007). CB (Canadian Basin), EB (Eurasian Basin), FJL (Franz Josef Land), MNP (Magnetic North Pole). The Lambert equal area projection has been provided by M. Jakobsson

by the report that observations from the current meter array showed a net northward transport persisting for more than 1 year (ASOF-N 2nd Annual Report 2005). Is such result compatible with the transports found through the other passages?

Concentrating on the net transport through Fram Strait, presently ignoring the total northward and southward fluxes, the long-term mean net transport is southward and estimated from the mooring array to be 0.6 Sv (ASOF-N 2nd Annual report 2005). (This value was later adjusted to 1.7 Sv (ASOF-N Final report 2006)). Using 0.6 Sv mass conservation demands a mean outflow of 3.2 – 0.6 = 2.6 Sv through the Canadian Arctic Archipelago. The commonly cited estimates for the outflow through the Canadian Arctic Archipelago range between 1 and 2 Sv converging toward 1.7 Sv (e.g. Melling 2000; Prinsenberg and Hamilton 2005). Can the straits in the Archipelago sustain a mean outflow of 2.5 Sv? Suppose that this is not the case. There is then an imbalance and water is accumulating in the Arctic Ocean at a rate of 1 Sv. The area of the Arctic Ocean is, including the shelves, 10×10^{12} m^2 and imbalances of this order would raise (lower) the sea surface by 25 cm in 1 month. One month is then probably the longest period such an imbalance can prevail.

These speculations can be extended further. A net northward flow (inflow) of 0.4 Sv was estimated from the Fram Strait array in 2002–2003 and this situation prevailed for more than 1 year (ASOF-N 2nd Annual Report 2005). This amounts to a total inflow of 3.6 Sv, which, to maintain mass balance and assuming an ice export of ~0.1 Sv, requires an outflow of ~3.5 Sv through the Canadian Arctic Archipelago. Only water from the upper 250 m can pass through the straits in the Archipelago, and even if there is a net inflow through Fram Strait the East Greenland Current will still carry low salinity upper water out of the Arctic Ocean at a rate of ~1 Sv. This implies a total outflow of ~4.5 Sv of Polar surface water, more than twice the available input of low salinity water from Bering Strait, from river runoff, from the Norwegian Coastal Current, and from the interaction between sea ice and the Fram Strait Atlantic inflow (see below for details). The outflow would reduce a 100 m thick upper low salinity layer in the deep basins by 10 m in 1 year. The net inflow through Fram Strait was observed during a period, when the Barents Sea inflow was close to its maximum (ASOF-N Final Report 2006). A net inflow can therefore not be explained by smaller transport through the Barents Sea.

If the upper layer thickness is to be maintained, sea ice must be melted and mixed into the entering Atlantic water to re-supply the exported low salinity water. To produce the 2.5 Sv of additional upper water with salinity 33.2, assuming this to be a realistic mean value of the salinity of the outflows in the East Greenland Current and through the Archipelago, requires an ice melt rate of 0.12 Sv, taking the Atlantic water salinity to be 35. This is of the same order as the present ice export and implies that the ice volume over the deep basins ($3 \times 5 \times 10^{12}$ m^3), using a mean ice thickness of 3 m, would be reduced by 20–25% in 1 year. The ice melt would also require that 40 TW of the heat entering the Arctic Ocean goes to ice melt. This is about equal to the heat released by cooling 2.4 Sv of Atlantic water (3 °C) to the freezing point. Furthermore, melting sea ice by sensible heat stored in the water column may not be possible without also supplying a substantial amount of heat to the atmosphere (Rudels et al. 1999a). The required heat input would then be even larger.

It should be kept in mind that these numbers and scenarios describe possible responses of the Arctic Ocean to large perturbations and do not represent the present situation in the Arctic Ocean, which is one where about 0.1 Sv liquid freshwater is transformed into ice (equal to the ice export). The excessive ice melt is needed, if the stratification in the Arctic Ocean basins shall be maintained during a major inflow event. A more likely effect is a thinning of the upper layer.

The examples described above show that some questions may still be asked and some insight might still be gained by studying basic mass, heat and salt balances. To be specific; we shall examine the contributions from Fram Strait to the mass (volume), heat and freshwater budgets of the Arctic Ocean using geostrophically determined transports through hydrographic sections obtained in Fram Strait between 1980 and 2005.

The reasons for using geostrophy instead of the results from the current meter array are: (1) Hydrographic observations are easier to work with and to interpret. (2) The time series of the hydrographic observations is considerably longer than the period of direct current measurements. (3) The spatial resolution on the hydrographic sections is finer than for the current meter array and allows for a better identification of water masses. On the other hand, the temporal resolution (about once a year) is considerably worse than that of the array. (4) The geostrophic transports are undetermined with respect to the reference velocity. If the transports do not fulfill obvious required budget constraints, it is then possible, and permissible, to deduce where an error might reside and also to suggest plausible corrections of the computed transports. Such corrections are much more difficult to defend with direct current measurements, which, when treated correctly, should give an optimal estimate.

In Section 13.2 we discuss the assumptions made when estimating the geostrophic transports through Fram Strait (Section 13.2.1) and then determine the exchanges of volume (Section 13.2.2). The choice of reference temperature and reference salinity is presented in Section 13.3. The distribution of the transports in different areas of the strait and the exchanges of different water masses are examined in Section 13.4. The mean Θ–S properties of the in- and outflow of the different water masses are computed for each crossing, and their variations with time and in the different part of the strait are discussed in Section 13.5. The heat transport is studied in Section 13.6 and the freshwater transport in Section 13.7. In Section 13.8 the obtained transports through Fram Strait are used, together with the requirement of mass, heat and freshwater balances of the Arctic Ocean, to examine if they lead to realistic outflows of mass (volume) and freshwater through both Fram Strait and the Canadian Arctic Archipelago. The inflows through Bering Strait (Woodgate and Aagaard 2005) and through the Barents Sea opening (Ingvaldsen et al. 2004a, b), as well as the river runoff and the net precipitation (Serreze et al. 2006; Dickson et al. 2007) are then assumed known (see Fig. 13.1). If not both the freshwater balance and the volume balance are acceptable, we will re-examine and adjust the geostrophic transport through Fram Strait to establish more realistic balances. The results of the study are summarised in Section 13.9.

13.2 Transports

13.2.1 The Geostrophic Calculations

The transports through 16 hydrographic sections taken between 1980 and 2005 are determined using the dynamic method. The first section was obtained in 1980 from the Swedish icebreaker Ymer, and the 1983 and 1984 crossings were made by RV Lance on regular Norwegian Polar Institute cruises. The 1988 and 1993 sections were taken by RV Polarstern on AWI expeditions and from 1997 onwards the sections have been obtained within the VEINS and ASOF programs. The sections taken in the 1980s used Neil Brown CTDs and the station spacing was generally larger than on the sections from 1997 onwards. SeaBird CTDs have been used since 1993. The data quality improved significantly between the 1980s and the 1990s. All sections run along the sill at about 79° N except 1983 which was taken along 79° 15 N (over the Molloy Deep). All sections were obtained in late summer, August–September except 1988 (June) and 1993 (March). For further details see Table 13.1.

On the sections the depth at each station is assumed constant halfway to the neighboring stations and the temperatures and salinities (1 or 2 db average) observed at the station are taken to extend halfway to the neighboring stations. Between stations of unequal depth the method of Jacobsen and Jensen (1926) is used to estimate the density anomaly correction below the deepest common level. Direct current measurements have shown that both the West Spitsbergen Current and the East Greenland Current are largely attached to the continental slope and follow the isobaths with shallow water to the right. To mimic this behavior within

Table 13.1 Information on sections and number of stations

Year	Vessel	Institute/programme	Stations 9° E – 6° W	Stations shelf
1980	IB Ymer	Ymer – 80	15	5
1983	RV Lance	NPI	23	–
1984	RV Lance	NPI	17	3
1988	RV Polarstern	AWI	18	–
1993	RV Polarstern	AWI U. Hamburg	17	–
1997	RV Lance	VEINS	16	2
1998	RV Polarstern	VEINS	20	14
1999	RV Polarstern	VEINS	26	8
2000	RV Polarstern	VEINS	16	20
2000	RV Lance	VEINS	22	9
2001	RV Polarstern	AWI	27	12
2002	RV Polarstern	AWI	49	23
2003	RV Polarstern	ASOF	50	–
2004	RV Polarstern	ASOF	42	7
2005	RV Polarstern	ASOF	50	24

the geostrophic framework we set the velocity to zero at the bottom of that station of the pair, which results in a flow at the deeper station, below the deepest common level, that has the shallower station to the right, looking in the direction of the flow.

A variational approach with auxiliary constraints on the deep-water exchanges is finally applied to the deep part of the strait. The Arctic Ocean is known as a source of dense water, warmer and more saline than the deep-water masses formed in the Nordic Seas (the Greenland Sea). We expect the deep-water formation to have a relaxation time scale comparable to the ventilation times of the deep basins, ranging from about 30 years in the Greenland Sea to perhaps 400 years in the Canada Basin. This is long enough to expect a fairly constant, baroclinic exchange of the deep waters during the observation period. The increase in temperature and salinity observed in the deep waters in the strait, however, suggests that the deep transports might be changing during the period. If so, it is ignored.

The circulation in the deeper layers is largely confined to the Arctic Ocean and the Nordic Seas and the Arctic Ocean appears at present to be a more active source of deep water than the Greenland Sea. We postulate that a production of 0.4 Sv of deep water with a mean salinity of 34.9325 takes place in the Arctic Ocean by brine rejection on the shelves and subsequent sinking of dense saline plumes down the slope, entraining warmer intermediate water on their way to their equilibrium density levels (Rudels 1986; Rudels et al. 1994; Jones et al. 1995). The dense water production by open convection in the Greenland Sea is assumed strong enough to generate an inflow of 0.2 Sv of deep water ($\sigma_\theta \geq 28.06$) with salinity 34.910 from the Nordic Seas to the Arctic Ocean. Since we do not expect that any deep water advected into the Arctic Ocean to be mixed upward into the overlying layers, this implies an outflow of 0.6 Sv with salinity 34.925 through Fram Strait from the Arctic Ocean. The volume and salt constraints on the deep exchanges then become $M = -0.4 \times 10^6\ m^3\ s^{-1}$ and $S = -13.973 \times 10^6\ kg^{-1}$. The flow field with the least added kinetic energy below the density surface $\sigma_\theta = 28.06$, fulfilling these constraints, is then determined.

The minimization of the added kinetic energy below the 28.06 isopycnal leads to a weak flow field, and the constraints on the deep water exchange are mainly introduced to ascertain that the more saline Arctic Ocean deep waters, to the west, leave and the Nordic Seas deep waters, mainly located to the east, enter the Arctic Ocean. A stronger outflow could be obtained by increasing the net deep water export, and a more intense deep circulation would be generated by increasing the salt export while keeping the net volume flux. However, the deep exchanges between the Arctic Ocean and the Nordic Seas as well as the deep-water production in the two areas are essentially unknown and the constraints have therefore been kept small. They force the deep outflow to take place in the west and the inflow to the east consistent with the locations of the East Greenland Current and the West Spitsbergen Current, but, because of the small added barotropic velocities, ~0.01 m s^{-1}, they do not unduly influence the transports in the upper layers, the main concern in this work, which are then essentially geostrophic. If reliable estimates of the deep-water productions in the two areas become available a more realistic barotropic flow field can be determined.

13.2.2 Volume Transports

The geostrophic transports are shown in Fig. 13.2, central panel. The total in- and outflows range from 5 Sv to almost 15 Sv with an average inflow of ~6 Sv and an outflow close to 9 Sv. This is smaller than the transports obtained from the direct current measurements, but not alarmingly so. The net outflow, 2.5 Sv, is, however, larger than that reported from the current meter array (e.g. Schauer et al. 2004; ASOF-N 2nd annual report 2005; ASOF-N final report 2006). The total in- and outflows estimated here include everything that is moving north and south and do not discriminate between eddies and more organized exchanges. The slight increase in total transports that is noticed in recent years might then be due to the closer station spacing on the later sections.

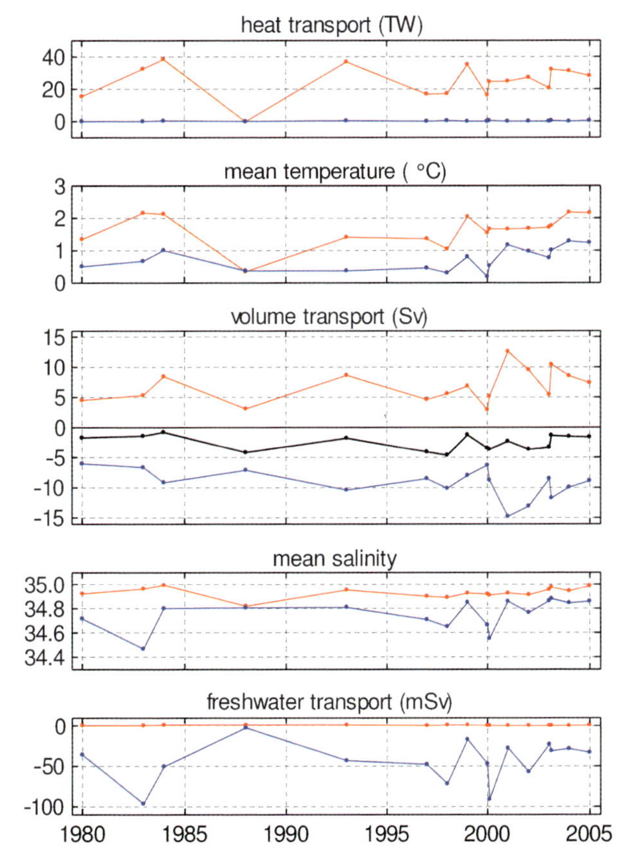

Fig. 13.2 Centre frame: Total in (red), out (blue), and net transports (black) in Sv obtained from the geostrophic computations. Upper frames: Mean inflow temperature (red) and mean outflow (reference) temperature (blue) and heat transport into the Arctic Ocean (red), the heat export (blue) is zero. Lower frames: Mean inflow (reference) salinity (red) and mean outflow salinity (blue) and the liquid freshwater export (blue). The freshwater import (red) is zero

Fram Strait probably contributes more than 60% of the inflow and 80–90% of the outflow volumes, if the deep exchanges are included. Although the Barents Sea inflow supplies intermediate and deep water to the Arctic Ocean, these dense waters are created, by cooling and also by freezing, in the Barents Sea. Similarly a small amount of Pacific water is made dense enough on the Chukchi Sea to enter the Canada Basin deep water. However, Fram Strait is the only passage that allows deep water to enter, and perhaps more important, the only passage that permits an outflow of deep water.

13.3 Reference Temperatures and Reference Salinities

To properly assess the Fram Strait contribution to the heat and freshwater balances of the Arctic Ocean all in- and outflows have to be accounted for, and a mass balance must first be established. Although this is one of the ultimate aims of ASOF, it has, as yet, not been accomplished. Without mass balance the heat and freshwater transports will depend upon the choice of reference temperature and reference salinity. Often these have been set as –0.1 °C and 34.80, taken as representing the mean temperature and the mean salinity of the Arctic Ocean (e.g. Aagaard and Greisman 1975; Aagaard and Carmack 1989; Simonsen and Haugan 1996; Schauer et al. 2004; Serreze et al. 2006). These values were determined in the 1970s, if not earlier, when the observational basis for forming such averages was very slim, and the variability in space and time of the Arctic Ocean water masses that has become evident during the last 10–15 years (e.g. Quadfasel et al. 1991; Polyakov et al. 2005) makes it doubtful that values determined 30 years ago can still be used without qualification.

Acknowledging the fact that we do not have, at present, sufficient observations from the other passages to formulate a mass balance of the Arctic Ocean, and taking into consideration the temporal variations of the Arctic Ocean mean temperature and salinity, we here choose a different approach. In view of the overreaching importance of the exchanges through Fram Strait we deem it sensible to estimate the inflow of heat to the Arctic Ocean and the outflow of freshwater from the Arctic Ocean through each section in Fram Strait relative to the mean outflow temperature and the mean inflow salinity determined on that section. This implies that no heat is transported by the outflowing water and no freshwater is transported by the inflowing water through the sections in Fram Strait. It should be noted that since the outflow is larger than the inflow, these choices give the largest transports of heat and freshwater through Fram Strait, unless reference temperatures, higher than the mean outflow temperature, and reference salinities, higher than the mean inflow salinity, are used. To compare the results obtained here with other estimates using different reference values, the differences in reference values should be multiplied with the net volume transport.

This does not eliminate the necessity to close the mass (volume) budget for the Arctic Ocean to really determine the fate of the heat entering the Arctic Ocean through Fram Strait and to estimate the relative contribution of the momentary export of liquid freshwater through Fram Strait in the Arctic Ocean freshwater budget. To use these varying reference salinities and temperatures might therefore appear a futile exercise. However, they bring the balances down to simple inflow/outflow terms, which makes it possible to discuss the mass imbalance, its origin and what it can reveal about the redistribution of the heat carried by the entering Atlantic water.

Furthermore, by comparing the time series of the heat transport, the reference temperature and the inflow and outflow volumes different factors contributing to the variability of the heat transport can be assessed. In a similar manner the variability of the freshwater export can be related to the variability of the reference salinity and the exchanged volumes (Fig. 13.2). These tasks have not been attempted here. Before we turn our attention to the heat and freshwater fluxes, we shall further discuss the exchange of different water masses through Fram Strait and how the transports are distributed in different parts of the strait.

13.4 Exchanges of Different Water Masses

The obtained estimates do not, so far, say anything about the exchanges of different water masses, nor where in the strait the main transports take place. A detailed water mass definition for the Arctic Mediterranean Sea has been formulated elsewhere (Rudels et al. 2005), but for the transports here we introduce a simplified water mass classification of 6 water masses, Surface water (SW), Atlantic water (AW), dense Atlantic water (dAW), Intermediate water (IW), Deep water I (DWI) and Deep water II (DWII) separated mainly by isopycnals but in the case of dAW and IW by the 0 °C isotherm (Table 13.2 and the Θ–S diagrams in Fig. 13.5).

The net outflow occurs as surface water and in the dense Atlantic water and the intermediate water ranges. It appears reasonable that waters from other passages that leave the Arctic Ocean through Fram Strait create net outflows with properties that at least partially reflect their initial characteristics. The low salinity of the less dense surface outflow (see Fig. 13.5a) reveals that it originates from the part of the Barents Sea inflow, mainly comprising Norwegian Coastal Current water, that stays on the shelves and incorporates most of the Siberian river runoff. Some ice melt might also be present as well as low salinity Pacific water from Bering Strait,

Table 13.2 Simplified water mass classification

Surface water (SW)	$\sigma_\theta < 27.70$
Atlantic water (AW)	$27.70 \leq \sigma_\theta < 27.97$
Dense Atlantic water (dAW)	$27.97 \leq \sigma_\theta, \sigma_{0.5} < 30.444, 0 < \theta$
Intermediate water (IW)	$27.97 \leq \sigma_\theta, \sigma_{0.5} < 30.444, \theta < 0$
Deep water I (DWI)	$30.444 \leq \sigma_{0.5}, \sigma_{1.5} < 35.142$
Deep water II (DWII)	$35.142 \leq \sigma_{1.5}$

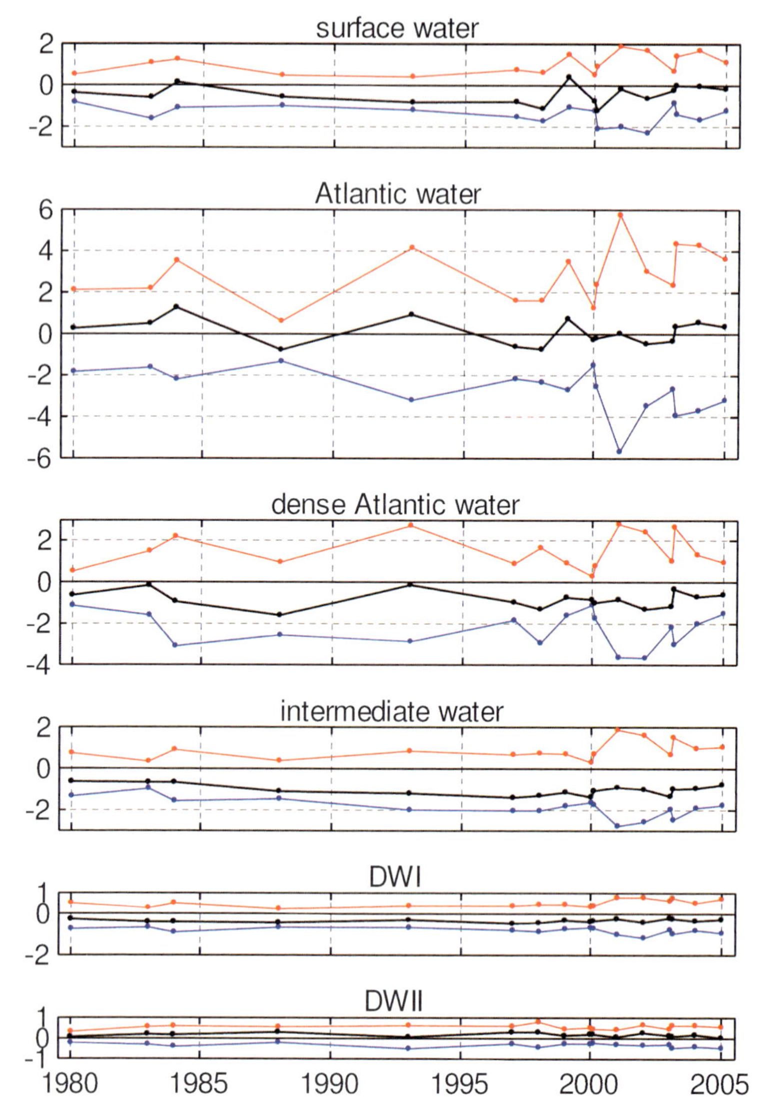

Fig. 13.3 Transports in Sv of different water masses based on geostrophic calculations. Inflow (red), outflow (blue) and net transport (black)

although the Pacific water mainly leaves the Arctic Ocean through the Canadian Arctic Archipelago (Jones et al. 2003). The net outflow in the denser, intermediate water range largely derives from the part of the Barents Sea inflow that enters the deeper Arctic Ocean water column via the St Anna Trough (Fig. 13.3).

The Fram Strait sections are subdivided into five different areas. Four of them, the eastern slope, the eastern deep part, the western deep part and the western slope

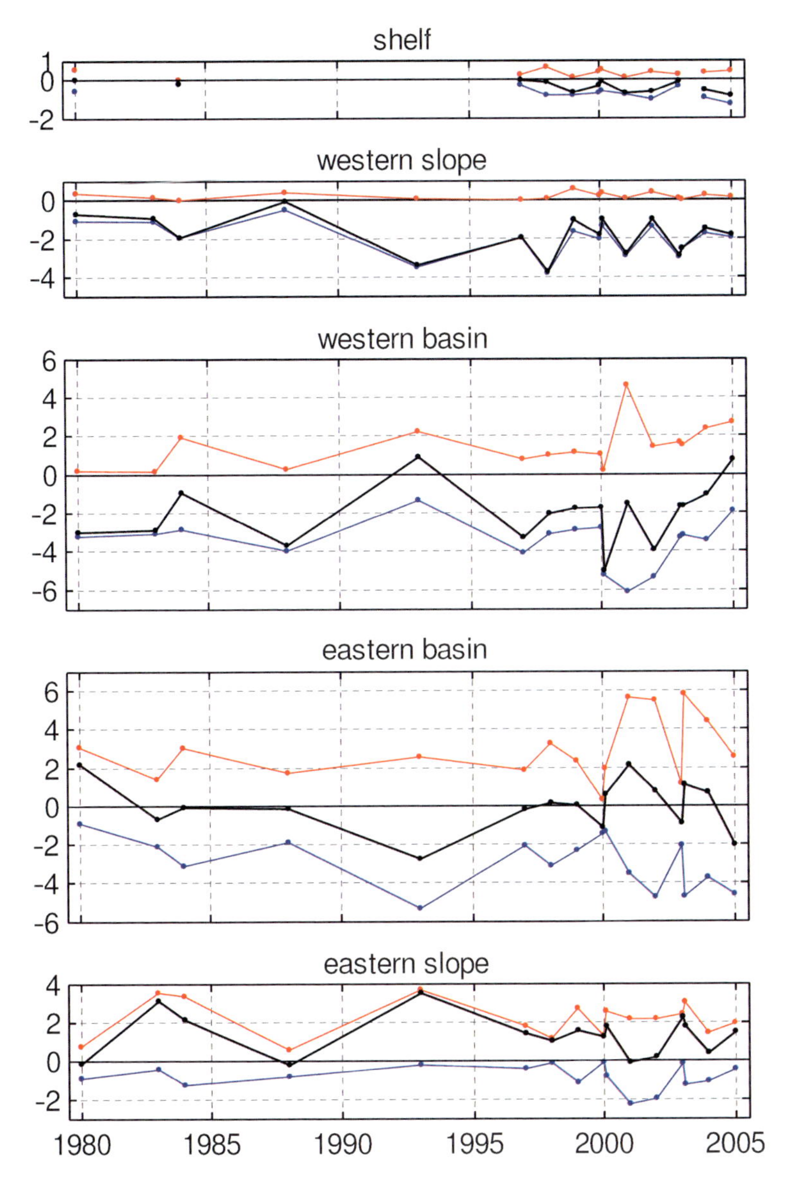

Fig. 13.4 Transports in Sv in different parts of Fram Strait between 6° W and 9° E based on geostrophic calculations. The western and eastern slopes extend down to 2,200 m and the eastern and western basins are separated by the Greenwich meridian. Inflow (red), outflow (blue) and net transport (black). The shelf transports are determined from geostrophic calculations with the velocity set to zero at the bottom

are approximately the same for all sections. East and west are separated by the Greenwich meridian, and the slopes and deep parts by the 2,200 m isobath. The eastern slope area reaches 9° E and the western slope area is taken to extend to 6° W. The fifth part is the shelf area west of 6° W, where the extent of the observations varies from year to year depending upon ice conditions. Here only surface water and occasionally Atlantic water are encountered. The Svalbard shelf east of 9° E was only observed twice. The net northward transport was in both cases less than 0.1 Sv. The transport here can therefore be ignored. The total northward and southward flow and the net transports in each part are shown in Fig. 13.4. The most consistent southward flow occurs over the western slope, and the outflow over the shelf varies between almost zero and occasionally close to 1 Sv. A net inflow is found in the West Spitsbergen Current at the Svalbard slope. The central parts generally indicate outflows, the western part more so. However, when an inflow is observed in the west, the eastern deep part shows a compensating outflow.

13.5 Variations in Water Mass Properties

So far we have considered in- and outflows but not, in detail, examined the characteristics of the water masses involved in the exchanges. Are the exchanges connected with small-scale eddy motions, which practically make the same water mass cross the section in both directions? Is there a systematic recirculation in the strait with most of the inflow taking place in one part, the outflow in another? Are there large differences between the in- and outflow characteristics, suggesting that the water masses have been long enough in the Arctic Ocean for substantial water mass transformations to occur? The total transports shown in Fig. 13.4 suggest that at least in the two central areas the exchanges largely compensate each other, and a northward transport in the west is mirrored by a southward transport in the east and vice versa. The East Greenland Current on the western slope consistently shows an outflow, while an inflow is concentrated to the West Spitsbergen Current in the east.

We only consider the main part of the strait, from 9° E to 6° W, and presently ignore the Greenland shelf, which is occupied mostly by outflowing low salinity water. The Θ–S characteristics of the northward and southward flowing water masses are determined by dividing the heat and salt transports with the volume (mass) transport in each water mass class. The transports of the different water masses in each area are indicated in Θ–S diagrams by bubble plots, where the location of the bubbles gives the Θ–S properties and their size indicates the transport. We have here included additional water masses in the classification. The surface water (SW) is sub-divided into Polar surface water (PSW) and warm Polar surface water (PSWw) by the 0 °C isotherm, and the Atlantic water (AW) is separated by the 2 °C isotherm into the colder Arctic Atlantic water (AAW) present in the Arctic Ocean and warmer Atlantic water (AW) from the south, which partly enters the Arctic Ocean, partly recirculates in Fram Strait. In the deep water ranges water more saline than 34.915 in the DWI class is defined as Canadian Basin Deep Water (CBDW) and in the DWII class as Eurasian Basin Deep Water (EBDW), while the water less saline than 34.915 in both classes is denoted Nordic Seas Deep Water (NDW) (Fig. 13.5).

Because of the widely different potential temperature and salinity ranges of the water masses we present the layers separately. The panels show the Θ–S properties for the upper, the Atlantic and the dense Atlantic, and the intermediate and deep waters respectively in all sub-areas for all years in eight Θ–S diagrams. Two Θ–S diagrams are given for each sub-area, since also the in- and outflows are shown separately. The years are distinguished by colour coding to indicate the temporal variability.

In the upper layers the difference between the areas is very distinct (Fig. 13.5a). The cold, low salinity surface water is located over the Greenland continental slope and is advected southward by the East Greenland Current. The deep western part is also dominated by outflow but the transport of PSW is smaller and the water, although still cold, is more saline and warmer than over the slope. In the eastern deep part the upper waters are warmer still and more saline. The characteristics of the northward flowing water are well clustered, while the southward transports have more varying properties. To the east, over the Svalbard slope, warm, saline and well-clustered inflows are observed, while the southward flow shows slightly more diverse characteristics and are smaller. The transports observed to the east are smaller than those to the west, especially the net transports.

The Atlantic waters over the Greenland slope mainly flow southward (Fig. 13.5b) and the low temperatures and salinities imply that water from the Atlantic layer in the Arctic Ocean here is carried out of the Arctic Ocean by the East Greenland Current. The Svalbard slope, by contrast, is dominated by northward flows and the Atlantic water is warmer and more saline. The transports here appear to be larger than over the Greenland slope.

The dense Atlantic water shows larger Θ–S variations on the Svalbard slope than on the Greenland side, where the Θ–S relations are tight except occasional years, when the recirculating Atlantic Water from the south extends onto the Greenland slope. The transports are northward over the Svalbard slope, southward over the Greenland slope and the net transports are fairly equal. The differences in Θ–S properties indicate that the Atlantic waters have become cooler and less saline, reflecting the mixing, and cooling that the Atlantic water experiences in the Arctic Ocean.

In the central parts the transports are as large as over the Svalbard slope. The range of the Θ–S characteristics between the different years found in the central areas is wider than at the Svalbard slope. The Atlantic water is slightly colder than over the slope and perhaps the western part is colder than the eastern, indicating a weak cooling and freshening from east to west. These differences are, however, smaller than the annual variability, indicating that the temporal variability over most of the strait is larger than the spatial variability across the strait. This suggests that part of the water from the West Spitsbergen Current recirculates westward in the strait on time-scales of months rather than years. In the deep central parts the inflow and outflow are of similar magnitude. The location of the in- and outflows appears to shift in time and often a large inflow in the deep western area is balanced by a strong outflow in the deep eastern area and the opposite. This is consistent with a pattern similar to that seen in the current meter array, where narrow barotropic eddies drift westward along the sill (ASOF-N Final report 2006).

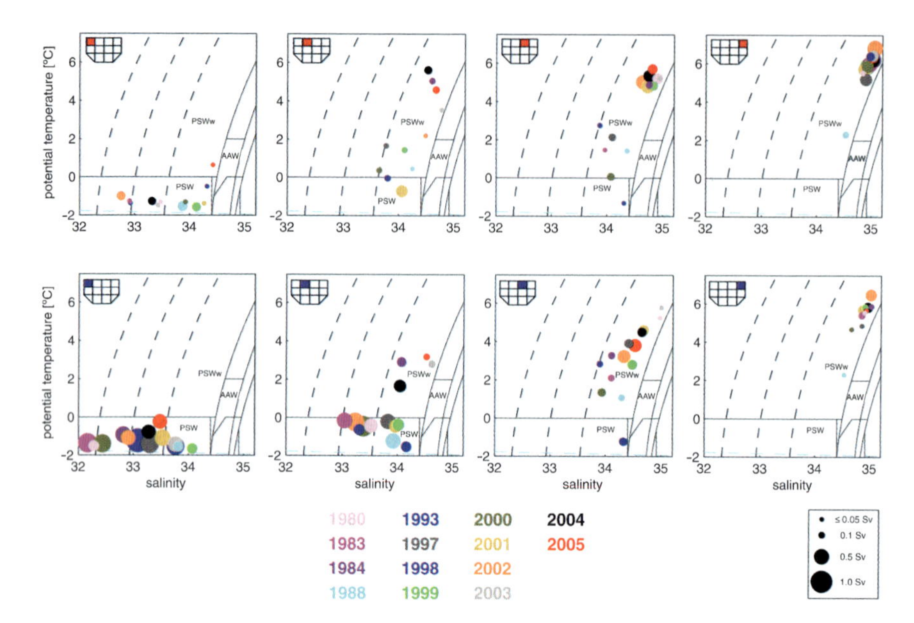

Fig. 13.5a Θ–S characteristics and transports in the surface waters for the different parts of Fram Strait. The upper four diagrams give, from left to right, the inflow over the western (Greenland) slope, the western deep part, the eastern deep part and the eastern (Svalbard) slope. The four lower panels give the outflow for the same areas. The different years are colour coded and the size of the bubbles indicates the transports. All transports ≤0.05 Sv are shown as the same size

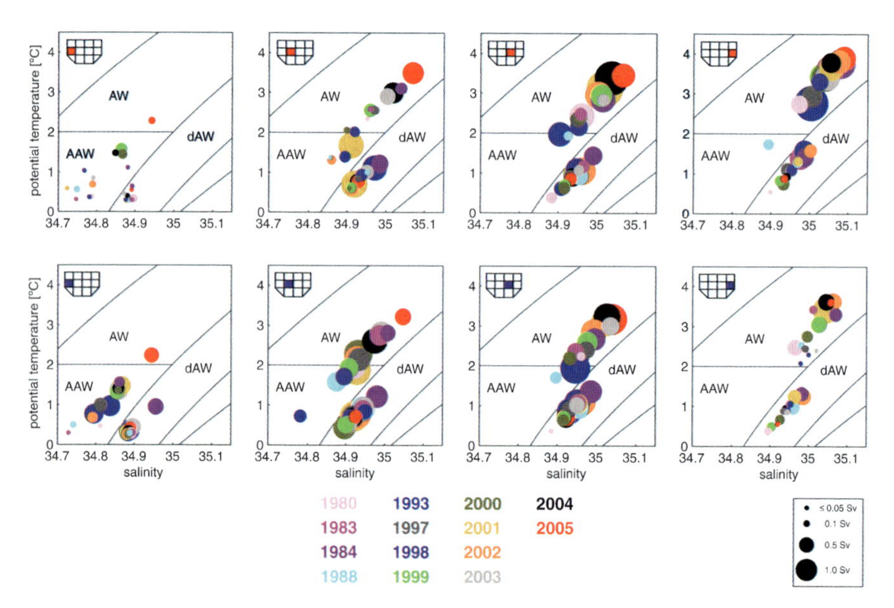

Fig. 13.5b Θ–S characteristics and transports in the Atlantic and dense Atlantic waters for the different parts of Fram Strait. The upper four diagrams give, from left to right, the inflow over the western (Greenland) slope, the western deep part, the eastern deep part and the eastern (Svalbard) slope. The four lower panels give the outflow for the same areas. The different years are colour coded and the size of the bubbles indicates the transports. All transports ≤0.05 Sv are shown as the same size

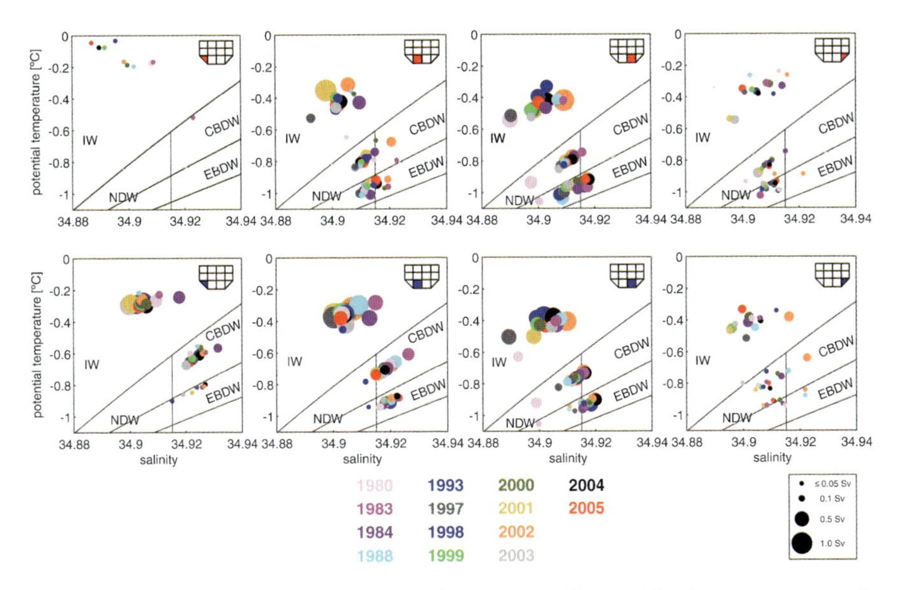

Fig. 13.5c Θ–S characteristics and transports in the intermediate and the deep-water masses for the different parts of Fram Strait. The upper four diagrams give, from left to right, the inflow over the western (Greenland) slope, the western deep part, the eastern deep part and the eastern (Svalbard) slope. The four lower panels give the outflow for the same areas. The different years are colour coded and the size of the bubbles indicates the transports. All transport ≤0.05 Sv are shown as the same size

In the intermediate and deep water ranges a similar but weaker pattern is detected (Fig. 13.5c). In the intermediate range outflow dominates over the western slope, while in the other parts of the strait the in- and outflow are about equal. The southward flows are slightly warmer and more saline, especially over the western slope, which agrees with the upper Polar Deep water (uPDW) of the Arctic Ocean being warmer and more saline than the Arctic Intermediate Water (AIW) of the Nordic seas.

The outflow of CBDW is concentrated to the western part, the western slope and the western deep area. Farther to the east the NDW becomes more prominent. The NDW dominates the inflows but also the outflows at the Svalbard slope, the inflow being stronger. In the deep areas the NSD is more strongly represented in the inflow than in the outflow. The inflow occurs mostly in the eastern but is also fairly strong in the western deep area. The EBDW is present in the outflow in both the eastern and the western deep area. However, it also takes part in the inflow, especially in the deep eastern area, suggesting some recirculation. Because of the small cross sectional areas the deep transports are comparatively small over the slopes, and the strongest deep exchanges occur in the deep areas. This can partly be explained by the larger areas, although the existence of strong, barotropic eddies could also contribute, adding recirculation to the north–south exchanges. However, the deep transports are slightly forced by the volume and mass constraints that have been imposed on the deep exchanges and too much should not be read into smaller features seen in the transports of the deep waters.

13.6 The Heat Transports

The heat transports, except in 1988, vary between above 35 TW and below 15 TW with an average around 25 TW and the mean reference temperature lies around 0.7 °C. The temperature has risen during recent years and the mean outflow temperature for the last 5 years is above 1 °C (Fig. 13.2). The time series is still rather short and contains lots of gaps in the early part of the observation period. We will therefore here not examine the time variation in transport and reference temperatures but concentrate on the mean transports and mean reference temperature. We shall especially discuss the net outflow volume and what that discloses about the distribution of the heat transported into the Arctic Ocean. For this discussion the mean heat transport (25 TW), the mean net volume flux (2.5 Sv) and the mean reference temperature (0.7 °C) are sufficient.

The obtained mean heat transport is clearly less than the >40 TW estimated from the current meter array using −0.1 °C as reference temperature (ASOF-N Final report 2006). If we adjust for the use of different reference temperatures the heat transport obtained here should be reduced by $c \times (0.7-(-0.1)) \times 2.5 \times 10^9 = 8$ TW, c being the heat capacity of sea water (4,000 J kg^{-1} K^{-1}). The difference between the results from the direct current observations and the geostrophic computations thus become larger.

The excess volume leaving the Arctic Ocean through Fram Strait must derive from the inflow over the Barents Sea and/or through Bering Strait. The Barents Sea inflow partly forms, together with the river runoff, the low salinity shelf water that eventually contributes to the low salinity surface water in the Arctic Ocean, partly supplies a denser inflow down the St Anna Trough, which cools the Atlantic water of the Fram Strait branch and forms the bulk of the underlying intermediate water mass, the upper Polar Deep Water (uPDW). The Bering Strait inflow contributes low salinity surface and upper halocline waters, which presently are mainly confined to the Canada Basin (Jones et al. 1998). The entire Pacific inflow, perhaps excluding the Bering Strait Summer Water, and about half of the Barents Sea inflow are eventually cooled to freezing temperatures within the Arctic Ocean. The denser St Anna Trough inflow is cooled at least to below zero in the Barents Sea and we tentatively set this deep inflow to 1.2 Sv with temperature −0.5 °C.

This is slightly larger than the 0.75 Sv of dense water that Schauer et al. (2002) estimated passing between Novaya Zemlya and Franz Josef Land. However, Schauer et al. only give the transports with temperature below 0 °C both for the dense deep water, 0.75 Sv, and the less dense surface water, 0.75 Sv. To have volume balance the rest of the inflow through the Barents Sea opening must either pass between Franz Josef Land and Novaya Zemlya with temperatures above 0 °C, or enter the Arctic Ocean west of Franz Josef Land and the Kara Sea south of

Novaya Zemlya. We have therefore increased the deep inflow estimate given by Schauer et al. (2002) from 0.75 Sv to 1.2 Sv and the less dense part from 0.75 Sv to 1 Sv. The less dense part will eventually be cooled to freezing temperature, and we do not expect any high temperatures to be present in the deep inflow and the postulated −0.5 °C should be a reasonable mean temperature for the denser inflow to the Arctic Ocean over the Barents Sea. The details of these "known" transports are summarized in Fig. 13.1.

The net outflow Y Sv through Fram Strait would then comprise 1.2 Sv of intermediate water with temperature −0.5 °C, since the deep Barents Sea inflow can only exit through Fram Strait, and (Y – 1.2) Sv of surface water at the freezing point (the seasonal heating of the surface water is ignored). To attain the mean outflow temperature – the reference temperature – the temperature of the cold, net outflow Y has to be compensated by a comparably warm return flow of Fram Strait branch Atlantic water. In a heat balance based on the mean outflow temperature in Fram Strait the amount F, $F = c \times (T_{out} - T_f) \times (Y - 1.2) + c \times (T_{out} - (-0.5)) \times 1.2.$, of the inflowing heat has to be used to increase the temperature of the excess volume Y to the mean outflow temperature T_{out}. Again c is the heat capacity of seawater (4,000 J kg^{-1} K^{-1}) and T_f the freezing temperature (−1.8 °C). Taking the mean outflow temperature (0.7 °C) and the mean net outflow volume Y = 2.5 Sv F becomes ~19 TW. If instead all the added water would be upper layer water the heat needed to compensate for the outflow becomes 26 TW and if all added water is upper Polar Deep water 14 TW is required. If choosing a mean temperature of the deeper outflow to 0 °C or −1.0 °C the corresponding heat requirement becomes 17 TW and 22 TW respectively.

A large heat loss of the inflowing Atlantic water occurs in the area just north of Svalbard, the Whalers' Bay. The heat is lost to ice melt and to the atmosphere, and Rudels et al. (1999a) suggested that when ice is melting on warmer water and the air temperature is below the freezing temperature of sea water, the heat loss of the ocean is distributed in such a way that the ice melt rate is a minimum. With a linear equation of state this implies that the fraction, f, of the heat loss that goes to ice melt is given by $f \approx 2\alpha L(c\beta S_A)^{-1}$. S_A is the salinity of the underlying water, L (336,000 J kg^{-1}) is the latent heat of melting and α and β are the coefficients of heat expansion and salt contraction respectively (Rudels et al. 1999a). About one third of the oceanic heat loss then goes to ice melt. The ice melt dilutes the upper part of the inflowing Atlantic water and creates an upper layer with lower salinity, ~34.3, which in the Nansen Basin is cooled to freezing temperature in winter and homogenised down to the Atlantic layer by (mainly) haline convection (Rudels et al. 1996; Rudels et al. 2005). Farther to the east this mixed layer is overrun by less saline and less dense shelf water and becomes the Fram Strait branch lower halocline (Rudels et al. 1996; Rudels et al. 2004).

Untersteiner (1988) estimated the formation of low salinity upper water in Whalers' bay due to ice melt to at least 0.5 Sv. The estimated salinity in the water was less than the ~34.3 normally encountered in the area, and the amount of low salinity water created north of Svalbard is probably larger. We shall assume a formation rate of 0.7 Sv, and using the difference, ~5 K, between the entering Atlantic water temperature T_A ~3 °C and the freezing temperature, the amount of inflowing oceanic heat lost during the initial formation of the lower halocline water can be estimated from $c \times (T_A - T_f) \times 0.7 \times 10^9$ to 14 TW. Here the heat going to ice melt as well as that being lost to the atmosphere is accounted for.

The amount of ice melted, I, can be found in two ways. Either by computing the dilution of the Atlantic water from salt conservation $(0.7 + I) \times 34.3 = 0.7 \times 35$, giving I = 0.015 Sv, or by using the expression from Rudels et al. (1999a) (given above) for the fraction of heat going to ice melt. With $\alpha = 0.6 \times 10^{-4}$ and $\beta = 8 \times 10^{-4}$ f becomes 0.36, and the heat lost to ice melt $0.36 \times 14 = 5$ TW. This also corresponds to a melting rate of 0.015 Sv. By contrast, Untersteiner (1988) deduced a much larger melting rate, 0.06 Sv, in Whalers' Bay, based on ice transport estimates by Vinje and Finnekåsa (1986).

These two heat sinks then use most (all) of the heat advected into the Arctic Ocean. In some years the heat loss is larger, in some years it is smaller than the heat import. This points to a further factor to consider in the Arctic Ocean heat balance, the change in temperature in the Atlantic layer in the Arctic Ocean. The higher temperatures of the Atlantic layer, first noticed in the early 1990s (Quadfasel et al. 1991), suggest an increase in heat storage in the Arctic Ocean. The continued studies in the Arctic Ocean have shown that this warm inflow pulse lasted perhaps close to a decade and gradually spread around the gyres in the different basins. Return flows were encountered in the northern Nansen Basin and in the Amundsen Basin (Rudels et al. 1999b), along the Lomonosov Ridge (Swift et al. 1997). It was observed in the Makarov Basin, first at the Siberian continental slope and at the Mendeleyev Ridge (Carmack et al. 1995), and then around the basin, and presently it is returning along the Lomonosov Ridge from North America towards Siberia (Kikuchi et al. 2005). The pulse also penetrated from the Chukchi Cap into the northern Canada Basin (Smethie et al. 2000). The spreading into the southern Canada Basin appears to occur differently (Shimada et al. 2004), perhaps through interleaving structures (Carmack, 2006) rather than circulating along the continental slope. Similar ideas have been advanced for the spreading of heat from the boundary current into the central Nansen Basin (Carmack et al. 1997; Swift et al. 1997). For the present discussion the spreading mechanisms are of little importance.

The long, warm inflow event was eventually followed by the arrival of colder Atlantic water. A comparison between sections taken in Fram Strait 1984 and 1997 (e.g. Rudels et al. 2000; Rudels 2001) indicate that a cooling and freshening of the Atlantic water has taken place. This is perhaps not so obvious in the time series from Fram Strait (Fig. 13.2) because of the gaps in the time series between 1984 and 1997 and because after 1997 the temperature gradually increases, indicating that the cold pulse has passed. The presence of colder water was noticed at the NABOS moorings north of the Laptev Sea in 2002 (Dmitrenko et al. 2005; Polyakov et al. 2005). Another warm pulse was observed around 2000 in Fram strait (ASOF-N Final report 2006) and a sudden, strong increase in the Atlantic water temperatures was detected at the NABOS moorings in 2004 (Dmitrenko et al. 2005; Polyakov et al. 2005). Still warmer Atlantic water was observed in Fram

Strait in 2004 suggesting the arrival of another warm inflow pulse. This pulse was found to partly recirculate in Fram Strait (ASOF-N Final report 2006).

The temperature increase in the Atlantic layer in the Arctic Ocean is uneven. Some of the warm Atlantic water has already left the Arctic Ocean and the rest is redistributed around the different gyres. Roughly assessing an overall temperature increase of 0.3 °C over a 200 m thick Atlantic layer over 15 years, this corresponds to a storage rate of 2 TW. Polyakov et al. (2004) estimated the change in heat content of the Atlantic layer between 1970s and the late 1990s as 4.3×10^{8} J m^{-2}, which corresponds to 2.7–3.4 TW, reasonably close to the back of the envelope calculation above.

13.7 Freshwater Transports

The freshwater export estimated relative to the inflow salinity has three components, the salinity difference between the in- and outflows, and the volume transport, which can be separated into two parts: one part corresponding to the inflow volume, and a second part representing the net outflow volume. The inflow salinities range between 34.8 and 35 but cluster around 34.92. The outflow salinity tends to co-vary with the inflow salinity and averages around 34.8 (Fig. 13.2).

As with the heat transport we can consider the freshwater export partly as a dilution of the inflow, partly as the addition of water from other sources with different freshwater content. The freshwater outflow, excluding 1988, ranges between 0.02 and 0.1 Sv, is highly variable but the mean appears to be somewhere between 0.03 and 0.05 Sv. Almost all the freshwater export occurs in the surface water, suggesting that the Barents Sea inflow, combined with river runoff and ice melt, contributes most of the net outflow volume with occasionally some Bering Strait inflow water added. The dilution of the upper part of the Fram Strait inflow to 34.3 is mainly due to ice melt and creates 0.7 Sv of halocline water (Section 13.6) but only 0.015 Sv. of freshwater is added by this process.

The outflowing Arctic Atlantic water (AAW) is, as expected, less saline than the inflowing Atlantic water. In fact, the crossover point in a Θ–S diagram, where the Arctic Ocean water column changes from being less saline than the entering Nordic Seas water column to becoming more saline than the Nordic Seas water column occurs close to 0 °C, which, according to our water mass definitions, separates dense Atlantic water (dAW) from the intermediate water.

The inflowing deep waters are less saline than the reference salinity and the deep inflow will add freshwater to the Arctic Ocean (Fig. 13.5c). This freshwater is largely re-exported by the outflowing Arctic Ocean deep and intermediate waters. Only if the reference salinity lies between the deep inflow and outflow salinities will both deep transports result in a freshwater flux into the Arctic Ocean. The salinity anomalies are then small and no large deep freshwater transports take place. The freshwater flux below the Atlantic layer is thus small and can safely be ignored.

In the Arctic Ocean the dilution of the entering Atlantic water occurs by ice melt north of Svalbard and perhaps also, but to a much smaller degree, in the entire Nansen Basin. A freshening of the Atlantic layer core takes place through convection of cold, dense shelf water, reaching the Atlantic layer. This occurs north of Svalbard (Rudels et al. 2005) and also at the Barents Sea slope between Svalbard and Franz Josef Land (Rudels 1986; Schauer et al. 1997). This freshwater input is restricted to the Atlantic layer. For the slope convection to reach deeper the initial salinities on the shelf have to be higher than the salinity of the Atlantic water and no freshwater is exported to the deeper layers.

The major freshening occurs downstream of the St Anna Trough. Here the denser part of the Barents Sea branch inflow joins the boundary current. It forms a colder and less saline water column extending from the surface to about 1,200 m. It is initially confined to the slope and depresses the deep isopycnals and the denser underlying Arctic Ocean deep water (Schauer et al. 1997). The upper part derives from the mixed layer in the eastern Barents Sea and the northern Kara Sea. Like the mixed layer in the Nansen Basin, it is initially formed by sea ice melting on warm Atlantic water (Rudels et al. 2004). The denser part of the inflow eventually mixes with the Fram Strait branch, cools and freshens the Atlantic core and creates the intermediate salinity minimum observed in the Eurasian Basin (Rudels and Friedrich 2000).

Farther to the east the river runoff and the rest of the Barents Sea inflow enter the central basins as low salinity shelf water, capping the boundary current and reducing its interaction with the sea surface and the ice cover. The mixed layer of the Nansen Basin and the boundary current deriving from the Fram Strait branch, as well as the mixed layer of the Barents Sea branch, are then covered by less saline water, the Polar Mixed Layer (PML), and become halocline waters. The two lower halocline waters as well as the Atlantic derived part of the Polar Mixed Layer return towards and exit through Fram Strait, although the Barents Sea branch halocline water moves along the North American slope and partly passes through the Nares Strait, contributing to the deep and bottom waters of Baffin Bay (Rudels et al. 2004).

The Bering Strait inflow provides the second largest freshwater source to the Arctic Ocean, larger than the net precipitation and almost as large as the river runoff (Woodgate and Aagaard 2005; Serreze et al. 2006). It supplies most of the water that passes through the straits in the Canadian Arctic Archipelago into the Baffin Bay. Pacific water also exits the Arctic Ocean through Fram Strait (Jones et al. 2003) but not continuously. Some years Pacific water is absent (Falck et al. 2005). The main contributions to the liquid freshwater transport through Fram Strait then come from river runoff and from the Barents Sea inflow, mainly the Norwegian Coastal Current. Some ice melt is exported in the halocline but this is likely to be a smaller part, ~0.015 Sv, if the same estimates as for the heat transport are used.

A considerable fraction of the Arctic Ocean freshwater export occurs as ice and about 90% of the ice export from the Arctic Ocean is estimated to pass through Fram Strait (Vowinckel and Orvig 1970). The passages in the Canadian Arctic Archipelago are narrow and often blocked by landlocked ice. In the northern

Barents Sea the opening between Svalbard and Franz Josef Land usually freezes early in fall, and the ice cover prevents the multi-year ice from the Arctic Ocean to pass into the Barents Sea. Occasionally it happens, but the sea ice, as well as the low salinity water of the East Spitsbergen Current, will be brought northward by the West Spitsbergen Current to the Arctic Ocean and are not really exported. A problem for the volume balance could therefore arise, if this transport is measured and included in the inflow but not accounted for as an outflow. Its contribution might be as large as 1 Sv, at least in winter (Rudels et al. 2005).

Freezing extracts freshwater from the surface water and the sea ice comprises river runoff from the Siberian shelves as well as water drawn from the Pacific water and the runoff from the North American continent. The ice gradually thickens, as it is advected towards Fram Strait indicating that freshwater is extracted from the PML in the entire Arctic Ocean. It is also likely that the net precipitation on the Arctic Ocean mainly falls on the sea ice and ends up in the solid phase, not in the water column.

How the freshwater export is distributed between the liquid and solid phases in Fram Strait has, so far, not been determined. Commonly the ice export has been assumed the largest, and results from ASOF-N indicate that the ice export in Fram Strait could be three times the liquid freshwater export (ASOF-N Final report 2006). However, tracer studies have suggested that the liquid freshwater export could be as large or larger than the ice export (Meredith et al. 2001). The results from the geostrophic calculations here indicate large variability in the liquid freshwater export, ranging from 0.1 Sv, which is close to the most cited value (0.09 Sv) for the ice export, down to 0.01 Sv. The mean value (~0.04 Sv) is close to that obtained by direct measurements in ASOF-N. However, the transport estimates given so far are for the standard section between 9° E and 6° W. The transport over the Greenland shelf has, because of the different extent of the section during different years, to be estimated separately. The transport over the shelf, which only comprises low salinity upper waters, is occasionally almost as large as the outflow of upper water in the rest of the strait, while in other years it is much weaker (compare Figs. 13.3. and 13.4.). Taking the mean of the freshwater transports over the shelf from the existing shelf sections (not shown) we get 0.025 Sv, which, added to the 0.04 Sv obtained for the strait proper, increases the freshwater flux to 0.065 Sv, or almost 75% of the ice export.

The reference salinity has been determined only for the deep part of the strait, and even if some inflow occurs on the shelf, it mainly involves a recirculation of the same low salinity water masses, which derive from passages other than Fram Strait. The choice of reference salinity, based on the inflow salinity, would therefore be the same, also when the shelf transports are included. The fact the transports over the shelf were excluded, when the reference temperature was determined, should also not seriously affect the discussion about the heat balance given above. The transports over the shelf almost exclusively involve waters from other passages than Fram Strait, which have lost their heat to the atmosphere being cooled to freezing temperature within the Arctic Ocean. They therefore say more about the fate of the heat fluxes through the Bering Strait and the Barents Sea than about the distribution of the heat transport through Fram Strait.

13.8 Updated Fram Strait Exchanges

The obtained freshwater transports can be used, together with external information about the freshwater budget, to re-examine the calculated volume transports through Fram Strait. A freshwater budget for the Arctic Ocean and for the Nordic Seas has recently been compiled by Dickson et al. (2007). Taking the values given in Dickson et al. (2007) for runoff, net precipitation, the Bering Strait inflow, the inflow through the Barents Sea opening, the export through the Canadian Arctic Archipelago, and the ice export, recomputed to the mean reference salinity (34.92) applied here, we obtain the transports presented in Table 13.3. The Fram Strait net outflow has been increased to 2.8 Sv as compared to 2.5 from Fig. 13.2 to accommodate the transport over the shelves. It should also be mentioned that the outflow through the Canadian Arctic Archipelago in this estimate is 0.25 Sv lower than the most often cited value, 1.7 Sv (Prinsenberg and Hamilton 2005).

The freshwater budget is balanced to within 2%, while the volume budget indicates a large net outflow through Fram Strait. The use of geostrophy underestimates the transports, since strong barotropic current will not be adequately accounted for. In Fram Strait the exchanges are known to have large barotropic components, in the central part of the strait as well as in the two main currents, the West Spitsbergen Current and the East Greenland Current. The applied constraints, combined with the requirement of minimum added kinetic energy in the deep exchanges, obviously cannot reproduce the barotropic transports.

However, the East Greenland Current is more stratified than the West Spitbergen Current and likely to be more baroclinic and better represented by the geostrophic computations. We therefore hypothesize that the volume imbalance in Fram Strait is due solely to underestimation of the inflow volume. By adding 1.1 Sv with the mean inflow characteristics to the inflow, we obtain an approximate balance also in volume. Since the inflow salinity is the same as the reference salinity this will not affect the freshwater balance, which continues to hold.

Table 13.3 Volume fluxes, salinity and freshwater fluxes Black numbers from Dickson et al. (2007) red numbers from this work

Contribution	Volume (Sv)	Salinity	Freshwater (mSv)
Runoff	0.1	0	102
Net precipitation	0.065	0	65
Bering Strait	0.8	31.49	79
Barents Sea	2.2	34.84	4
Canadian AA	−1.44	32.7	−92
Fram Strait ice export	−0.09	4	−88
Fram Strait net outflow and liquid freshwater export	−2.8	–	−65
–	–	–	–
Net transport	−1.17	–	−5

However, the heat transport through the strait, and the distribution of the heat within the Arctic Ocean will change. The average difference between the inflow temperature (1.6 °C) and the outflow (reference) temperature is 0.9 K. This gives $0.9 \times 1.1 \times 10^9 \times 4{,}000 \approx 4$ TW and the average transport of heat through Fram Strait into the Arctic Ocean increases from 25 TW to 29 TW. The net outflow that has to be heated to the reference temperature is reduced from 2.5 to 1.7 Sv, and if we keep the estimate of 1.2 Sv of intermediate water added by the deep Barents Sea inflow at –0.5 °C, only 0.5 Sv of surface water at freezing temperature needs to be heated to 0.7 °C. The amount of heat required to warm the net outflow volume then becomes 10 TW. The formation of 0.7 Sv. of halocline water still needs 14 TW and the heat storage rate remains 2 TW. This leaves 3 TW to be lost to the atmosphere, which corresponds to a surface heat transfer of 0.6 W m^{-2}, much less than the 2 W m^{-2} often quoted for the oceanic heat loss to the atmosphere (Maykut and Untersteiner 1971; Maykut 1986). The resulting mass, heat and freshwater transports are summarized in Fig. 13.6.

We may also note that by adding the net outflow of low-density surface water to the halocline water, formed by the entering Atlantic water, the export of low

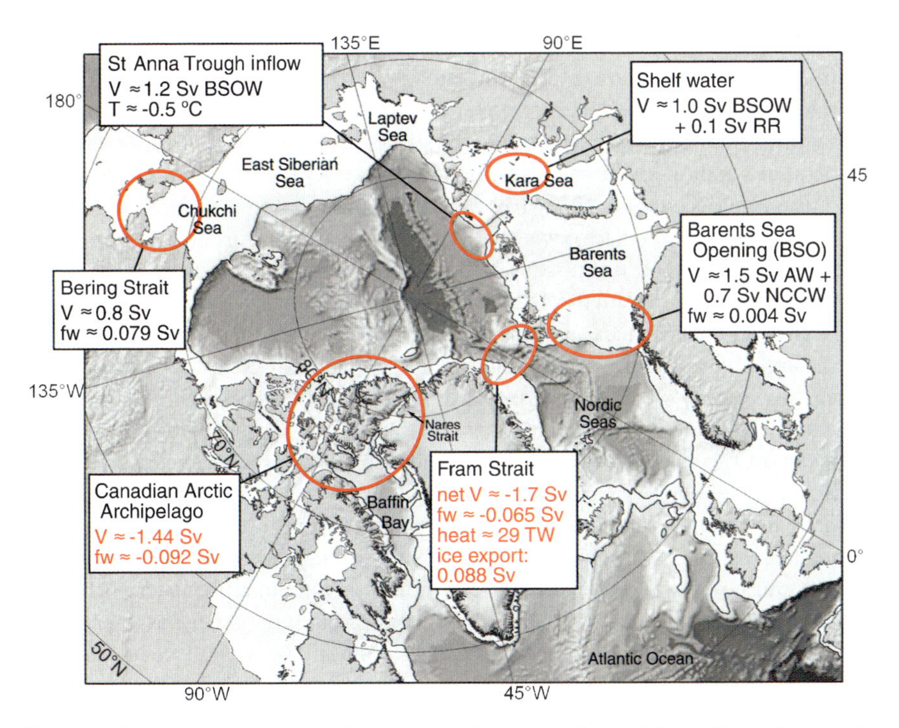

Fig. 13.6 Volume and freshwater balances for the Arctic Ocean. The outflows through the Canadian Arctic Archipelago and the Fram Strait ice export are taken from Dickson et al. (2007), while the net outflow, the heat transport and the export of liquid freshwater through Fram Strait are based on the discussions in the present work

salinity upper layer water becomes 1.2 Sv. This is 0.2 Sv less than the low salinity outflow through the Canadian Arctic Archipelago (Table 13.2). It is somewhat low, but about the same as was calculated in Dickson et al. (2007), and the number is not unreasonable. The estimate of the rate of halocline water formation is a guess and it might be smaller or larger. The surface water fraction provided by the Barents Sea could also be larger than the 1 Sv used here. However, it is not possible to extract more information from these data without becoming excessively speculative and it is time to stop.

13.9 Summary

Transports of volume, heat and freshwater through Fram Strait have been determined from geostrophic velocities computed on sixteen hydrographic sections taken in the strait between 1980 and 2005. To find the unknown reference velocities the deep water exchanges have been determined, which have the least kinetic energy while fulfilling prescribed volume transport and salt transport constraints in the deeper layers. The obtained northward and southward transports are smaller than those estimated from the current meter array, while the net southward transport is larger.

The heat and freshwater fluxes through the strait are calculated relative to the mean outflow temperature and the mean inflow salinity on each section. This choice of reference values removes the northward transport of freshwater and the southward transport of heat through the sections, but it leads to varying reference temperatures and reference salinities. In this study only the mean reference salinity and mean reference temperature over the observation period have been used.

The computed liquid freshwater export, combined with existing estimates of other freshwater sources and sinks in the Arctic Ocean (e.g. Serreze et al. 2006; Dickson et al. 2007), shows that the freshwater transport in Fram Strait almost fulfils the freshwater balance and thus appears realistic. However, there is an imbalance in the volume fluxes, and the net volume export through Fram Strait is found to be too large. As a remedy we hypothesize that the inflow through Fram Strait is underestimated by the geostrophic calculations, and an inflow through Fram Strait, with the mean inflow characteristics, is added to establish volume balance in the Arctic Ocean.

Since the northward and southward transports in Fram Strait do not balance, a unique heat transport through the strait cannot be found. However, the net outflow volume can be examined separately. This simplifies the interpretation of the heat transport, because most of the water that enters through the other passages is less dense surface water that is cooled to freezing point in the Arctic Ocean. The only exception is the large fraction of the Barents Sea inflow, which is dense enough to supply the intermediate layer. This volume has been set to 1.2 Sv at −0.5 °C. These considerations then allow for a discussion of the fate of the heat entering the Arctic Ocean through Fram Strait. As long as all inflows and outflows are not successfully monitored, such approach should provide some insight on the importance of Fram Strait for the Arctic Ocean heat budget.

Much of the barotropic transports that dominate the deep water exchange may be associated with barotropic eddies, implying that the deep water exchange between the Arctic Ocean and the Nordic Seas is smaller than the direct current observations indicate. The geostrophic transports, since they depend upon the density differences between the northward and southward flowing waters, can be seen as mirroring the effects of the water mass transformation processes active in the Arctic Ocean and in the Nordic Seas. This then describes the transport of the water in Θ–S space and thus partly represents the oceanic transport having impact on climate. The fact that additional constraints are needed to obtain a realistic volume balance for the exchanges between the Arctic Ocean and the world ocean shows that the transports through Farm Strait are not just caused by the density changes, but are also forced by large-scale wind fields and sea level slopes. The variational approach applied here, which minimizes the kinetic energy of the exchanges, will remove, or at least diminish, this "external" forcing and thus require additional constraints or information on the freshwater and/or volume transports to become realistic.

Acknowledgements This work has received support from the EU projects: ASOF-N (contract No EVK2-CT-2002-00139) (BR, MM, PE), ASOF-W (Contract No EVK2-CT-2002-00149) (BR, MM, PE), the 6th Framework Programme for Research and Development DAMOCLES (Contract No 018509) (BR, PE) and from the Academy of Finland (No. 210551) (MM).

References

Aagaard K, Greisman P (1975) Towards new mass and heat budgets for the Arctic Ocean. J Geophys Res 80:3821–3827

Aagaard K, Carmack EC (1989) The role of sea ice and other fresh water in the Arctic circulation. J Geophys Res 94:14485–14498

ASOF-N 2nd Annual Report (2005) Bremerhaven, 55 pp

ASOF-N Final Report (2006) Bremerhaven, 112 pp

Blindheim J (1989) Cascading of Barents Sea bottom water into the Norwegian Sea. Rapp P-v Reun Cons Int Explor Mer 188:49–58

Carmack EC, Macdonald RW, Perkin RG, McLaughlin FA (1995) Evidence for warming of Atlantic water in the southern Canadian Basin. Geophys Res Lett 22:1961–1964

Carmack, EC, Aagaard K, Swift JH, Macdonald RW, McLaughlin FA, Jones EP, Perkin RG, Smith JN, Ellis KM, Killius LR (1997) Changes in temperature and tracer distributions within the Arctic Ocean: Results from the 1994 Arctic Ocean section. Deep-Sea Res II 44:1487–1502

Coachman LK, Aagaard K (1988) Transports through Bering Strait: Annual and interannual variability. J Geophys Res 93:15535–15539

Dickson RR, Rudels B, Dye S, Karcher M, Meincke J, Yashayaev I (2007) Current estimates of freshwater flux through Arctic & Subarctic seas. Prog Oceanogr 73:210–230, doi:10.1016/j.pocean.2006.12.003

Dmitrenko, I. Timokhov L, Andreev O, Chadwell R, Churkin O, Dempsey M, Kirillov S, Smoliansky V, Mastrukov S, Nitishinskiy M, Polyakov I, Repina I, Ringuette M, Vetrov A, Walsh D (2005) NABOSD-04 Expedition on the Northern Laptev Sea aboard the Icebreaker Kapitan Dranitsyn (September 2004). IARC Technical report 2, 113 pp

Falck E, Kattner G, Budéus G (2005) Disappearance of Pacific Water in the northwestern Fram Strait. Geophys Res Lett 32:L14619, doi:101029/2005GL023400

Ingvaldsen RB, Asplin L, Loeng H (2004a) Velocity field of the western entrance to the Barents Sea. J Geophys Res 109:C03021, doi:101029/2003JC001811

Ingvaldsen RB, Asplin L, H. Loeng H (2004b) The seasonal cycle in the Atlantic transport to the Barents Sea during the years 1997–2001. Cont Shelf Res 24:1015–1032

Jacobsen JP, Jensen AJC (1926) Examination of hydrographic measurements from research vessels Explorer and Dana during the summer of 1924, Cons Per Int p l'Explor de la Mer, Rapp et Pr-Ver 39:31–84

Jones EP, Rudels B, Anderson LG (1995) Deep Waters of the Arctic Ocean: Origins and Circulation. Deep-Sea Res 42:737–760

Jones EP, Anderson LG, Swift JH (1998) Distribution of Atlantic and Pacific waters in the upper Arctic Ocean: Implications for circulation. Geophys Res Lett 25:765–768

Jones EP, Swift JH, Anderson LG, Lipizer M, Civitarese G, Falkner KK, Kattner G, McLaughlin FA (2003) Tracing Pacific water in the North Atlantic Ocean. J Geophys Res 108:3116, doi:10.1029/2001JC001141

Kikuchi T, Inoue J, Morison J (2005) Temperature differences across the Lomonosov Ridge: Implication for the Atlantic Water circulation in the Arctic Ocean. Geophys Res Lett 32: L20604, doi:10.1029/2005GL023982

Maykut GA (1986) The surface heat balance. In: Untersteiner N (ed) Geophysics of Sea Ice. Plenum, New York, pp 395–464

Maykut GA, Untersteiner N (1971) Some results from a time-dependent thermodynamic model of sea ice. J Geophys Res 76:1550–1575

Melling H (2000) Exchanges of freshwater through the shallow straits of the North American Arctic. In: Lewis EL, Jones EP, Lemke P, Prowse TD, Wadhams P (eds) The Freshwater Budget of the Arctic Ocean. Kluwer Academic, Dordrecht, pp 479–502

Meredith MP, Heywood KJ, Dennis PF, Goldson LE, White R, Fahrbach E, Østerhus S (2001) Freshwater fluxes through the western Fram Strait. Geophys Res Lett 28:615–618

Polyakov IV, Alekseev GV, Timokhov LA, Bhatt US, Colony RL, Simmons HL, Walsh D, Walsh JE, Zakharov VF (2004) Variability of the Intermediate Atlantic Water of the Arctic Ocean over the last 100 years. J Climate 17:4485–4494

Polyakov IV, Beszczynska A, Carmack EC, Dmitrenko IA, Fahrbach E, Frolov IE, Gerdes R, Hansen E, Holfort J, Ivanov VV, Johnson MA, Karcher M, Kauker F, Morison J, Orvik KA, Schauer U, Simmons HL, Skagseth Ø, Sokolokov VT; Steele M, Timokhov LA, Walsh D, Walsh JE (2005) Geophys Res Lett 32:L17605, doi:10.1029/2005GL023740

Prinsenberg SJ, Hamilton J (2005) Monitoring the volume, freshwater and heat fluxes passing through Lancaster Sound in the Canadian Arctic Archipelago. Atmosphere-Ocean 43:1–22

Quadfasel D, Sy A, Wells D, Tunik A (1991) Warming in the Arctic. Nature 350:385

Rudels B (1986) The Θ-S relations in the northern seas: Implications for the deep circulation. Polar Res 4 ns:133–159

Rudels B, Jones EP, Anderson LG, Kattner G (1994) On the intermediate depth waters of the Arctic Ocean. In: Johannessen OM, Muench RD, Overland JE (eds) The Role of the Polar Oceans in Shaping the Global Climate. American Geophysical Union, Washington, DC, pp 33–46

Rudels B, Anderson LG, Jones EP (1996) Formation and evolution of the surface mixed layer and the halocline of the Arctic Ocean. J Geophys Res 101:8807–8821

Rudels B, Friedrich HJ, Hainbucher D, Lohmann G (1999a) On the parameterisation of oceanic sensible heat loss to the atmosphere and to ice in an ice-covered mixed layer in winter. Deep-Sea Res II 46:1385–1425

Rudels B, Björk G, Muench RD, Schauer U (1999b) Double-diffusive layering in the Eurasian Basin of the Arctic Ocean. J Marine Syst 21:3–27

Rudels B, Friedrich HJ (2000) The transformations of Atlantic water in the Arctic Ocean and their significance for the freshwater budget. In: Lewis EL, Jones EP, Lemke P, Prowse TD, Wadhams P (eds) The Freshwater Budget of the Arctic Ocean. Kluwer, Dordrecht, The Netherlands, pp 503–532

Rudels B, Meyer R, Fahrbach E, Ivanov VV, Østerhus S, Quadfasel D, Schauer U, Tverberg V, Woodgate RA (2000) Water mass distribution in Fram Strait and over the Yermak Plateau in summer 1997. Ann Geophys 18:687–705

Rudels B (2001) Ocean Current: Arctic Basin Circulation. In: Steele J, Thorpe S, Turekian K (eds) Encyclopedia of Ocean Sciences. Academic Press, London, pp 177–187

Rudels B, Jones EP, Schauer U, Eriksson P (2004). Atlantic sources of the Arctic Ocean surface and halocline waters. Polar Res 23:181–208

Rudels B, Björk G, Nilsson J, Winsor P, Lake I, Nohr C (2005) The interactions between waters from the Arctic Ocean and the Nordic Seas north of Fram Strait and along the East Greenland Current: results from the Arctic Ocean-02 Oden expedition. J Marine Sys 55:1–30

Schauer U, Muench RD, Rudels B, Timokhov L (1997) Impact of eastern Arctic Shelf water on the Nansen Basin intermediate layers. J Geophys Res 102:3371–3382

Schauer U, Loeng H, Rudels B, Ozhigin VK, Dieck W (2002) Atlantic Water flow through the Barents and Kara Seas. Deep-Sea Res I 49:2281–2298

Schauer U, Fahrbach E, Østerhus S, Rohardt G (2004) Arctic warming through the Fram Strait: Oceanic heat transports from 3 years of measurements. J Geophys Res 109:C06026, doi:10.1029/2003JC001823

Serreze MC, Barrett A, Slater AJ, Woodgate RA, Aagaard K, Lammers R, Steele M, Moritz R, Meredith M, Lee C (2006) The large-scale freshwater cycle of the Arctic. J Geophys Res 111: C11010, doi:10.1029/2005JC003424

Shimada K, McLaughlin FA, Carmack EC, Proshutinsky A, Nishino S, Itoh M (2004) Penetration of the 1990s warm temperature anomaly of the Atlantic Water in the Canada basin. Geophys Res Lett 31:L20301, doi:10.1029/2004GL020860

Simonsen K, Haugan PM (1996) Heat budget of the Arctic Mediterranean and sea surface flux parameterization for the Nordic seas. J Geophys Res 101:6553–6576

Smethie WM, Schlosser P, Bönisch G (2000) Renewal and circulation of intermediate waters in the Canadian Basin observed on the SCICEX 96 cruise. J Geophys Res 105:1105–1121

Swift JH, Jones EP, Carmack EC, Hingston M, Macdonald RW, McLaughlin FA, Perkin RG (1997) Waters of the Makarov and Canada Basins. Deep-Sea Res II 44:1503–1529

Untersteiner N (1988) On the ice and heat balance in Fram Strait. J Geophys Res 92:527–532

Vinje T, Finnekåsa Ø (1986) The ice transport through the Fram Strait. Norsk Polarinstitutt Skrifter 186, 39 pp

Vowinckel E, Orvig S (1970), The climate of the North Polar Basin. In: Orvig S (ed) World Climate Survey vol 14: Climates of the Polar Regions. Elsevier, Amsterdam, 370 pp

Woodgate RA, Aagaard K (2005) Revising the Bering Strait freshwater flux into the Arctic Ocean. Geophys Res Lett 32:L02602, doi:1029/2004GL021747

Chapter 14
Variability and Change in the Atmospheric Branch of the Arctic Hydrologic Cycle

Mark C. Serreze, Andrew P. Barrett, and Andrew G. Slater

14.1 Introduction

Water evaporated in low and middle latitudes is transported poleward via the atmospheric circulation. Through convergence and uplift, some of it condenses, and falls to the surface as precipitation. Further evaporation returns some of this water back to the atmosphere, which may again fall as precipitation. However, annual evaporation rates in high latitudes are in general modest. The end result is that most of the north polar region is characterized by positive net precipitation (precipitation minus evaporation, or P – E) in the annual mean (Fig. 14.1). Ultimately, this freshwater excess must be returned to lower latitudes via the ocean, with river discharge representing an intermediate step. In its broadest sense, the major features of the Arctic's mean annual freshwater budget (Fig. 14.2) reflect this large-scale balance requirement.

The devil is in the details. There is strong seasonality in the pathways of freshwater associated with atmospheric processes. Consider river discharge to the Arctic Ocean. For the long-term annual mean, this is approximately equal to P–E over the terrestrial drainage, and represents the largest single input of freshwater to the Arctic Ocean (38% of the estimated annual input from all sources relative to a salinity of 34.8, see Fig. 14.2). However, discharge is not evenly distributed through the year, but arrives as a strong pulse in late spring and early summer, due to melt of the winter snowpack. Although annual P–E is positive over land areas, it is actually negative over much of this area in summer, i.e., summer is a period of net drying. About 24% of the annual freshwater input to the ocean is from positive P–E over the Ocean itself (Fig. 14.2). This input has a summer maximum and cold season minimum, quite different than the pattern over land.

Oceanic transports add to the complexity (Carmack 2000; Stigebrandt 2000). The major freshwater exports through Fram Strait and the straits of the Canadian Arctic Archipelago, along with salty Atlantic inflow, not only compensate for

National Snow and Ice Data Center, Cooperative Institute for Research in Environmental Sciences, Campus Box 449, University of Colorado, Boulder CO, 80309–0449, USA

R.R. Dickson et al. (eds.), *Arctic–Subarctic Ocean Fluxes*, 343–362

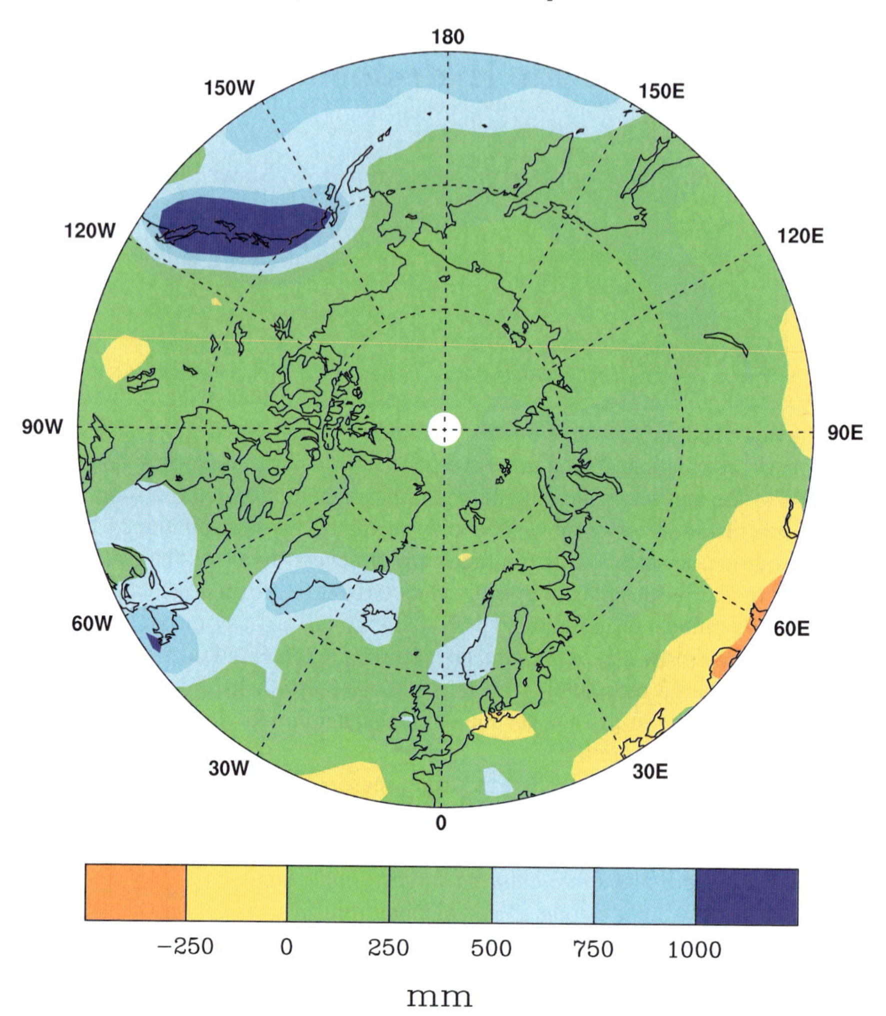

Fig. 14.1 Mean annual P–E for the region north of 50° N from ERA-40, based on aerological calculations over the period 1979–2001

freshening from river runoff, P–E over the Arctic Ocean, and from other minor terms, but also for the large freshwater inflow into the Arctic Ocean via Bering Strait (Fig. 14.2). This represents 30% of annual freshwater input and manifests several processes, including (on the large scale) differences in P–E over the North Pacific and North Atlantic, associated with upper-ocean salinity differences that help maintain a gradient in sea surface height, and discharge from the Yukon river

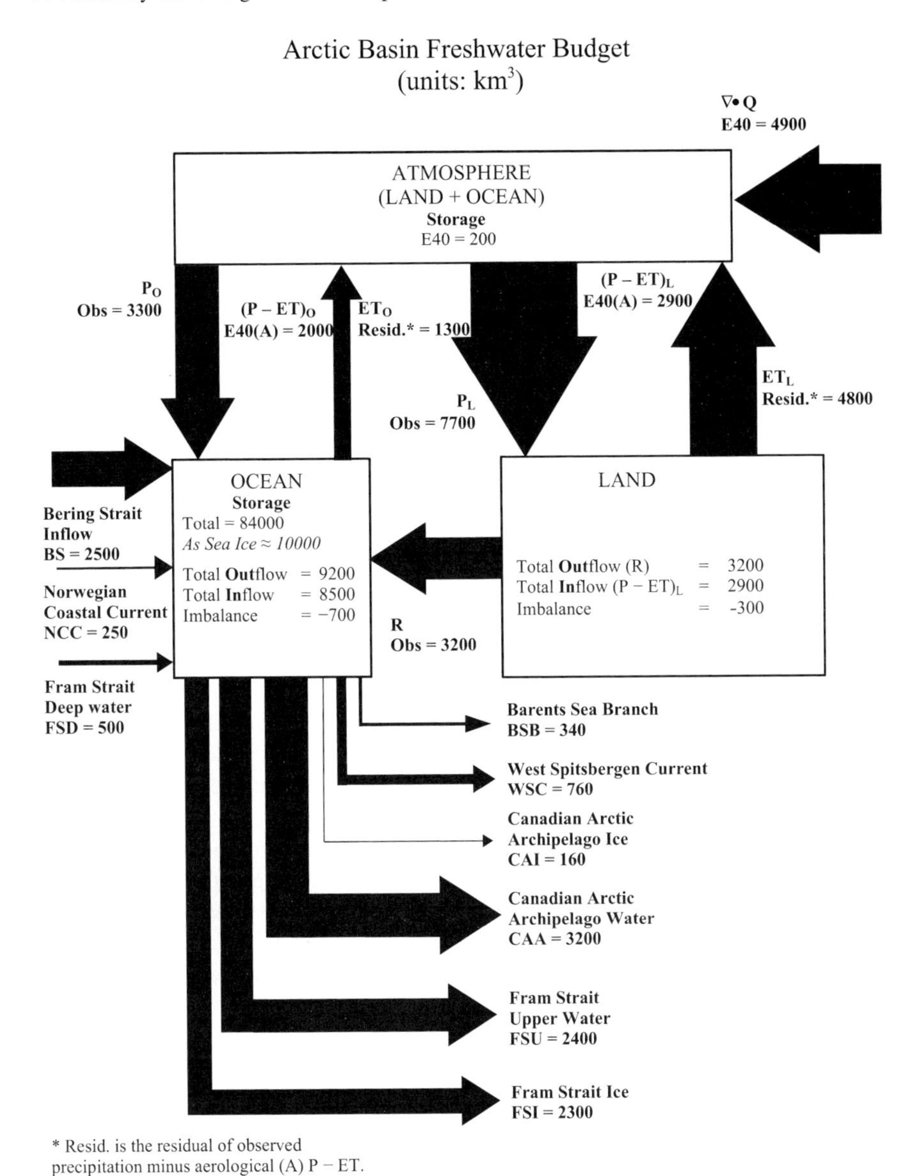

Fig. 14.2 Mean annual freshwater budget for the Arctic, based on a reference salinity of 34.8. (Adapted from Serreze et al. 2006.) The ocean domain (area of 9.6 × 10^6 km^2) is defined by lines across Fram Strait, from Svalbard to northern Scandinavia, across the Bering Strait, and along the northern coast of the Canadian Arctic Archipelago. The land region draining into this ocean domain (area of 15.8 × 10^6 km^2) was defined using a digital river network. The atmospheric box combines the land and ocean domains. The boxes for land and ocean are sized proportional to their areas. All transports are in units of km^3 per year. Stores are in km^3. These are based on the best available estimates drawn from recent literature or computed as part of the Serreze et al. (2006) study. This includes data from the ERA-40 reanalysis

that appears as part of the seasonal Alaskan Coastal Current (Woodgate et al. 2006). Another aspect of atmospheric forcing is that ocean transports are sensitive to the regional wind field. This has been well documented for the Fram Strait ice flux (Vinje 2001) and Bering Strait inflow (Aagaard et al. 1985; Woodgate et al. 2005).

This paper reviews the atmospheric branch of the Arctic hydrologic cycle with a focus on the integrating theme of P–E. We examine seasonal and spatial patterns of P and E, and how these are expressed as net precipitation. Aspects of freshwater storage are addressed, as well as observed variability and projected future states of the freshwater system. Use is made of output from the European Centre for Medium Range Weather Forecasts (ECMWF) ERA-40 reanalysis, results from land surface models (LSMs), time series of observed precipitation, and findings from recent published studies.

14.2 Primary Data Sources

Atmospheric reanalyses such as ERA-40 (Uppala et al. 2005) and from the National Centers for Environmental Prediction/National Center for Atmospheric Research (NCEP/NCAR) (Kalnay et al. 1996; Kistler et al. 2001) are retrospective forms of numerical weather prediction, whereby gridded fields of atmospheric and surface variables are compiled using fixed versions of a forecast/data assimilation system. ERA-40 provides 6-hourly fields on a grid of approximately 125 km from September 1957 through July 2002.

Fields of free-air variables, such as tropospheric pressure heights, winds, and humidity, are compiled by assimilating observations within a short term atmospheric forecast. These "blended" products are generally the most reliable aspects of reanalysis. Fields of surface variables such as precipitation, evaporation and other terms of the surface energy budget, do not involve blending with observations. In an operational setting (i.e., in routine weather forecasting), the forecast/data assimilation system is constantly refined to improve forecast skill. This can lead to non-climatic jumps and trends in archived fields. By using fixed systems, archives from reanalysis are more consistent, but temporal inconsistencies are still present due to changes in observing networks (e.g., rawinsonde and satellite data bases).

Atmospheric reanalyses allow for a full accounting of the atmospheric hydrologic budget. P–E can be obtained in two ways. The first is from the model forecasts of P and E. The second and preferred method (Cullather et al. 2000; Rogers et al. 2001) is the aerological approach. Consider an atmospheric column, extending from the surface to the top of atmosphere. Its water budget can be expressed as:

$$\partial W/\partial t = E - P - \nabla \cdot Q \quad (14.1)$$

where $\partial W/\partial t$ represents the change in precipitable water (W) in the atmosphere (the water depth of the vapor in the column), and $-\nabla \cdot \mathbf{Q}$ is the convergence of the vertically integrated horizontal water vapor flux **Q**. In the aerological approach, P–E is obtained by adjusting the vapor flux convergence by the tendency in precipitable water. For long-term annual means and assuming a steady-state, the tendency term can be dropped, so that net precipitation equals the vapor flux convergence. Small effects of phase transformations in the atmosphere represented by clouds, as well as convergence of water in liquid and solid phases, are ignored. A number of other studies (e.g. Walsh et al. 1994; Göber et al. 2003) have used the aerological approach to assess P–E averaged for large domains (such as the region north of 70° N) using data from rawinsonde profiles. An advantage of using reanalysis is that one can obtain gridded fields of P–E.

High-latitude precipitation fields from ERA-40 are known to be greatly improved over those from NCEP/NCAR, and, at least for most regions, capture observed interannual variability, although the model has generally less precipitation than observations (Serreze et al. 2005; Betts et al. 2003). Evaporation estimates from ERA-40 seem reasonable, at least for land (Slater et al. 2007). However, P–E based on the aerological method and from the forecasts of P and E are not in balance, with lower P–E in the latter. This results primarily from nudging the model humidity toward observations. For annual means over the period 1979–1993, Cullather et al. (2000) cite an imbalance over the polar cap (the region north of 70° N) of 50 mm for ERA-15 and 82 mm for NCEP/NCAR. For ERA-40, Serreze et al. (2006) calculate a smaller imbalance of 15 mm. These issues should be kept in mind when interpreting our results.

Our approach is to view aerological P–E as the best representation of truth, and then assess its components using ERA-40 fields of P and E. This recognizes that surface observations of P and E are insufficient to obtain gridded fields over the entire north polar region. The primary focus is on the period 1979–2001 for which fields are most reliable due to the wealth of satellite data for model assimilation. However, data back to 1958 are used to examine longer-term variations in aerological P–E. To complement ERA-40, estimates of zonally averaged P for the region 55–85° N are examined for 1900–2004, based on land station records contained in the Global Precipitation Climatology Centre (GPCC) database (http://gpcc.dwd.de).

Snowpack water equivalent (SWE) over the terrestrial drainage is a key aspect of the hydrologic system. Surface observations are insufficient to compile gridded fields, and those based on satellite remote sensing are of questionable fidelity. We use estimates of SWE (seasonal storage of P–E) based on averaging output from five different land surface models (LSMs), each driven with ERA-40 inputs for 1979–2001 (precipitation, temperature, low level humidity and winds, and downwelling solar and longwave radiation). The five model average should give a better representation of SWE than from any one model.

14.3 Seasonal Aspects

14.3.1 Fields of P, E and P–E

Figure 14.3 shows mean fields (1979–2001) for winter (December–February) and summer (June–August) of precipitation, evaporation and P–E from ERA-40 for the region north of 50° N. Recall that P–E is computed via the aerological method.

For winter, the highest precipitation totals, exceeding 450 mm, are found in the northern North Atlantic, well south of the Arctic, associated with the primary North Atlantic storm track. High totals in the North Pacific are associated with the East Asian storm track. Precipitation is also locally high along the coasts of southeastern Greenland, Scandinavia and the Pacific Northwest. These areas are characterized by orographic uplift of moist airmasses. By sharp contrast, winter precipitation totals of less than 50 mm characterize the northern Canadian Arctic, the Arctic Ocean and northeastern Eurasia. This manifests distance from oceanic moisture sources (continentality) and generally anticyclonic atmospheric conditions.

As the primary storm tracks weaken through spring and summer, the Atlantic and Pacific precipitation maxima become less prominent. By contrast, over most land areas, precipitation increases through spring to a summer maximum. This is associated with solar heating of the surface that promotes convective precipitation, and, especially over northern Eurasia, a summer peak in extratropical cyclone activity (Serreze and Etringer 2003). Precipitation over the Arctic Ocean also has a summer to early autumn maximum (depending on the region), mostly associated with the migration of lows into the region formed over Eurasia. Cyclogenesis (the formation of lows) occurs throughout northern Eurasia in summer, and is especially common over the northeast part of the continent along the summer Arctic Frontal Zone, which develops in response to differential heating of the atmosphere over the Arctic Ocean and snow-free land (Serreze et al. 2001).

The areas of low winter precipitation over Canada, the Arctic Ocean and Eurasia are also regions with low evaporation. Note the sharp contrast in winter evaporation over these cold snow and ice covered regions and the ice-free ocean, where open water fosters strong vertical vapor gradients. These spatial contrasts weaken in spring. The pattern of summer evaporation stands in stark contrast to winter. Low summer evaporation over the Arctic Ocean follows as the surface temperature over melting sea ice stays near the freezing point, limiting the magnitude of vertical vapor gradients. Evaporation is much higher over land, where there is strong solar heating of the snow-free surface, and exceeds that for most open-ocean areas.

We are now poised to see how spatio-temporal variations in P and E combine as net precipitation. It is apparent that the positive P–E which characterizes nearly all of the north polar region in the annual mean (Fig. 14.1) does not hold on seasonal time scales. Although winter P–E is strongly positive (>200 mm) over parts of the North Atlantic and North Pacific, it is negative over other ocean areas, such as Baffin Bay, and the Norwegian and Barents seas. Although winter precipitation is rather high over some of these areas, this is countered by stronger evaporation.

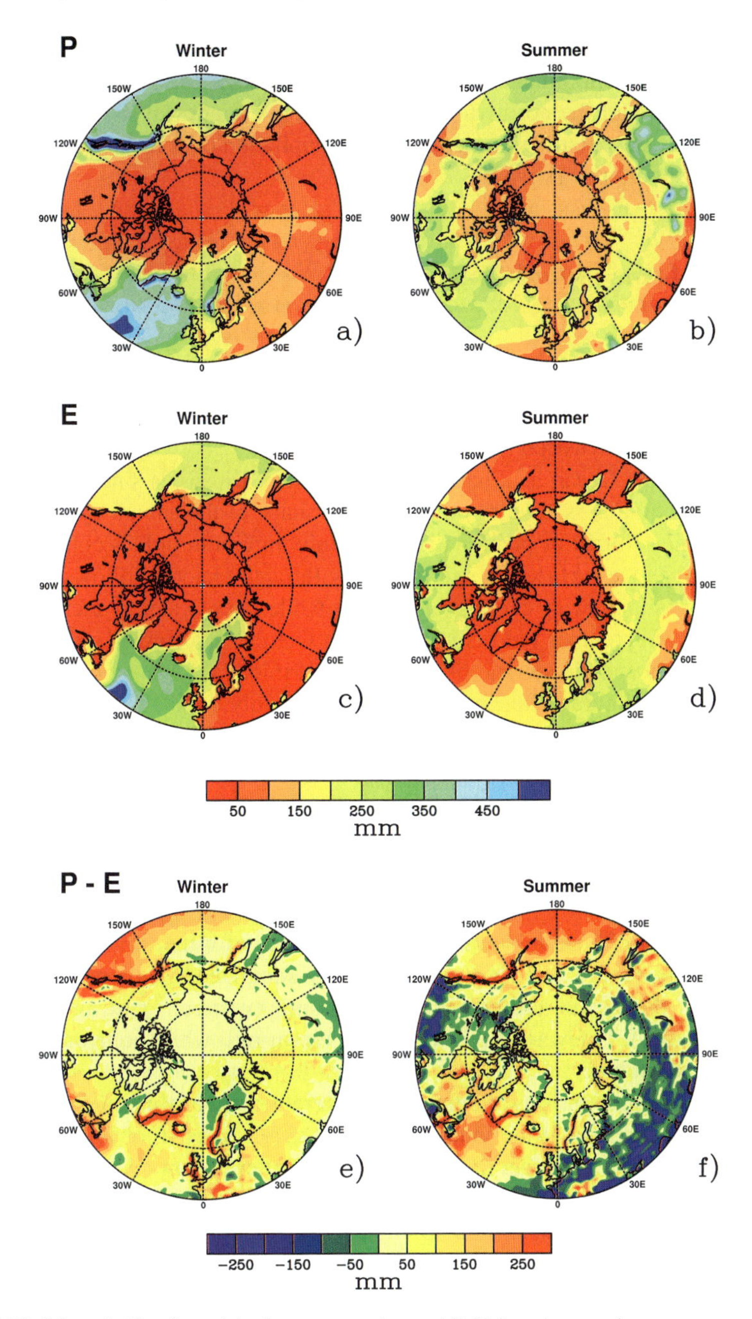

Fig. 14.3 Mean fields of precipitation, evaporation and P–E for winter and summer

By comparison, winter P–E is modestly positive over most land regions as well as the Arctic Ocean. In summer, P–E is strongly positive over the northern North Pacific and parts of the northern North Atlantic, and modestly so over the central Arctic Ocean. Despite the summer peak in precipitation over most land areas, P–E tends to be small or negative.

14.3.2 Mean Annual Cycles

The mean annual cycles of freshwater budget components from ERA-40 (P, E, P–E and atmospheric storage) averaged for the Arctic Ocean and its contributing terrestrial drainage (the same as used in Fig. 14.2, see caption) help to summarize some of the preceding discussion (Fig. 14.4).

For the Arctic Ocean as a whole, P–E peaks in July and is smallest in March. Because evaporation is always rather low, the shape of the annual cycle in P–E is broadly similar to that of precipitation. The July minimum in E is consistent with the melting sea ice surface. Its rise from July through October follows as specific humidity is falling (with cooling of the air) while open water is increasing to a maximum in September (and is still large in October), fostering stronger vapor gradients. Atmospheric storage of water vapor exhibits a fairly symmetric annual cycle, with a minimum in January and maximum in July, following the annual cycle of tropospheric temperatures and the water-holding ability of the atmosphere. The prominent feature of the terrestrial drainage (see also Walsh et al. 1994 and Serreze and Etringer 2003) is the opposing annual cycles of P and P–E.

14.3.2.1 Water Vapor Pathways

It is useful to briefly examine the dominant pathways for the flow of water vapor into the Arctic. Figure 14.5 shows the vertically integrated meridional water vapor flow across the 70° N latitude circle by month (vertical axis) and longitude (horizontal axis). Poleward flows (inflows) are in red, while equatorward flows (outflows) are in blue.

For every month, inflows dominate, i.e., there is a vapor flux convergence into this "polar cap" domain. In the annual mean, this equates to a P–E of 193 mm (average water depth). Flows are larger in summer than in winter. Inflows during summer are prominent in four regions, near the prime meridian, about 90° E, about 165° W, and about 50° W. The peak at around 90° E is slightly east of the Urals trough, while the feature at about 165° W is located just east of the east Asian trough. Prominent inflows at about 50° W and near the prime meridian are separated by a region of equatorward flow in most months. This separation manifests blocking by the Greenland ice sheet. Most of the moisture flow occurs below 700 hPa (roughly 3,000 m). At 70° N, the highest ice sheet elevations of about 2,900 m are found at about 35° W longitude. The outflow centered at about 110° W corresponds to the descending leg of the western North American ridge.

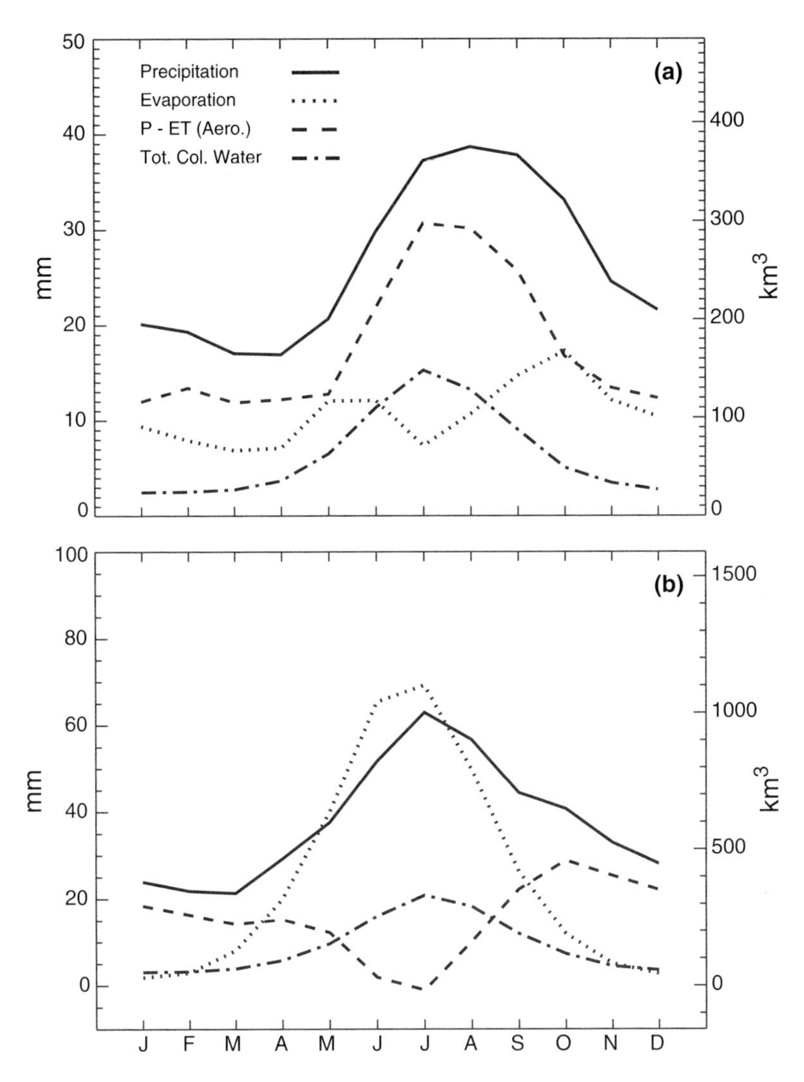

Fig. 14.4 Mean annual cycles of precipitation, evaporation, P–E and atmospheric water vapor storage for: (a) the Arctic Ocean; (b) the terrestrial drainage

14.3.3 Seasonal Storage of P–E

The snowpack, be it over land or atop the sea ice cover, can be viewed as seasonal storage of P–E. Over land, this P–E is released as meltwater in spring and summer, to appear, minus further losses on its journey by evaporation and infiltration, as a pulse of river discharge to the Arctic Ocean. The hydrograph for the Lena River, at the gauging station nearest the mouth of the river, serves as an example (Fig. 14.6).

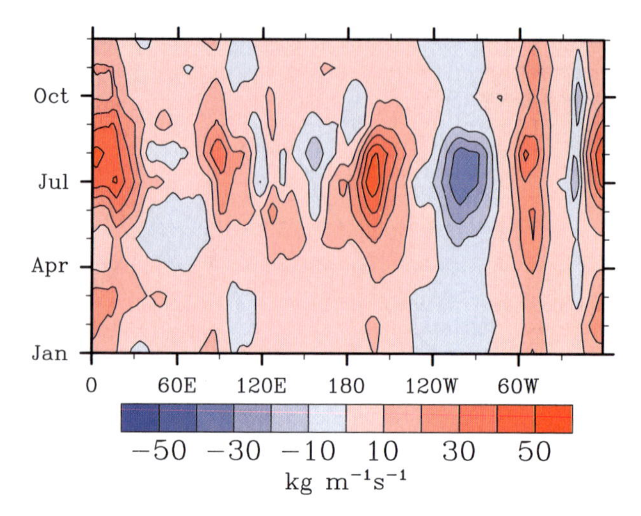

Fig. 14.5 Vertically integrated water vapor flows across 70° N by longitude and season

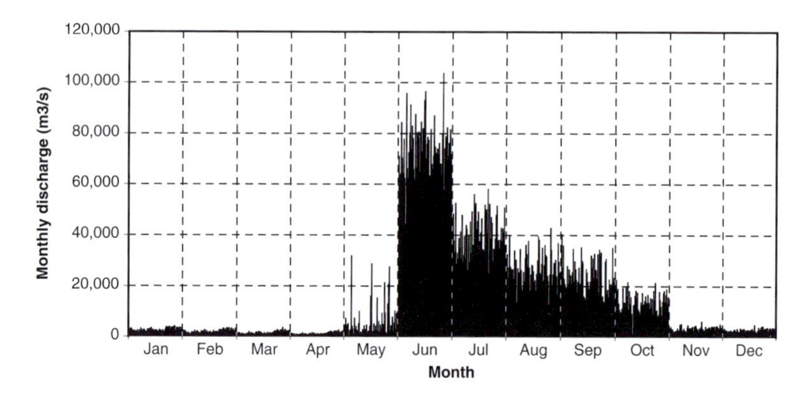

Fig. 14.6 Monthly discharge (m^3 s^{-1}) at the mouth of the Lena river over the period 1936–1999. For each month, the plot shows discharge for all years (From Yang et al. 2002)

This figure shows monthly discharge for all years individually from 1936–1999. There is a strong June peak, associated with the major snowmelt period, followed by a decline through summer and autumn. Hydrographs for other major rivers flowing into the Arctic Ocean (the Ob, Yenisey, Lena in Eurasia, and the Mackenzie in North America) differ in detail, but not in basic form. For example, the June peak is flatter for the Ob, as conditions over this watershed are warmer (promoting more discharge earlier in the season and larger losses from evaporation) and there is less permafrost (promoting more infiltration of meltwater) (Serreze et al. 2003).

Figure 14.7 shows the estimated terrestrial snowpack water equivalent (SWE, in mm) for March and May, based on averaging output from five different LSMs

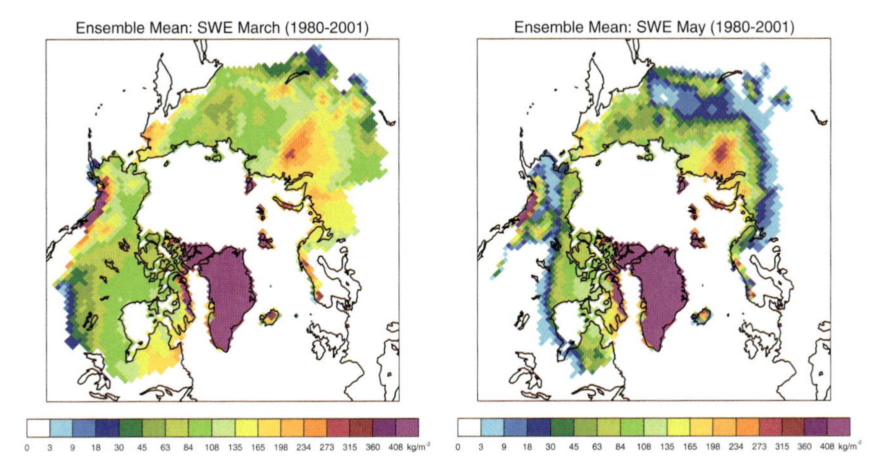

Fig. 14.7 Snow water equivalent over the Arctic terrestrial drainage for (a) March and (b) May, based on the average of five land surface models driven by ERA-40 forcings. The five models are CHASM, Noah, CLM, VIC and ECMWF (See Slater et al. 2007 for model details and simulation setup)

driven by ERA-40 data (see Section 14.2). The terrestrial drainage as defined for these simulations is different than employed in previous discussion as it includes areas draining into Hudson Bay, James Bay and Baffin Bay/Davis Strait. Areas in purple (off the scale) are glaciated.

For the terrestrial drainage as a whole, March (Fig. 14.7a) can be considered as the month of peak SWE. For Eurasia, the largest snowpack storage is in the western sector, broadly corresponding to the watersheds of the Ob and Yenisey, with lower values to the east. This is broadly in accord with the pattern of autumn and winter P–E. Over North America, and ignoring glaciated regions, SWE generally increases eastward across central Canada, again reflecting patterns of precipitation and P–E. Over both continents, SWE is low along parts of the southern fringes of the drainage, where snowfall is less frequent, and often quickly melts once fallen.

By May (Fig. 14.7b), most of the snowpack has melted in the warmer southern parts of the drainage. Over the colder northern regions, much of it has yet to melt, and in some regions it has actually grown in comparison with March. By June (not shown) a snowpack lingers only over the Canadian Arctic Archipelago and extreme northern Eurasia. That discharge peaks in June when the snowpack is already mostly depleted is explained in that it takes about 1 month for meltwater in the upper headwaters of the drainage to make its way to the mouths of the rivers.

For March, the spread in SWE between the five models is fairly small in cold regions (e.g., eastern Siberia and northern Canada), but larger in areas with a relatively deep snowpack, which tend to be warmer. Each model uses the same inputs from ERA-40, but they have different architectures. For example, they differ in how precipitation is partitioned between snowfall and rainfall, and in the treatment of melt processes. It follows that the spread between models is small in the very cold regions, where nearly all precipitation falls as snow and there is little melt, and greater in warmer areas. It also follows that the spread between is larger in May, when conditions are warmer.

14.4 Variability and Trends

14.4.1 Atmosphere/Ocean Links

Research over the past decade has documented a number of links between variability in P, P–E, river discharge and modes of atmospheric and ocean variability. Most emphasis has been placed on the role of the North Atlantic Oscillation (NAO) and its larger-scale counterpart, the Northern Annular Mode (NAM).

The NAO refers to co-variability between the strength of the Icelandic Low and Azores High. When both atmospheric centers of action are strong (weak), the NAO is in its positive (negative) mode. In the framework of the NAM (also known as the Arctic Oscillation, see Thompson and Wallace 1998), the NAO is viewed as the North Atlantic component of a more fundamental Northern Hemisphere mode of circulation variability, characterized by a mass oscillation between the Arctic and middle latitudes. When the NAM is positive, surface pressures are relatively low over the Arctic and relatively high in mid latitudes, associated with strengthening of the high-latitude westerlies. High latitude cyclone activity is enhanced, especially in the northern North Atlantic.

Index time series of the NAO and NAM are highly correlated, and there has been debate as to whether the NAO and NAM are different expressions of the same thing. While the following discussion views them interchangeably, their regional climate expressions do in fact differ somewhat. The NAO and NAM are present throughout the year, but most research has focused on the cold season, when their climate impacts are especially pronounced. Time series of the NAO index, based on station records of sea level pressure near the centers of the Icelandic Low and Azores High, extend back to 1870. Those of the NAM, based on principal component analyses of sea level pressure fields, extend back to about 1900.

The NAO/NAM has exhibited complex temporal behavior. Year-to-year and month to month variability is superimposed on multiyear to decadal variations. Portis et al. (2001) identify four epochs in the winter NAO index. From about 1870 though 1900 the NAO was mostly negative, followed by generally positive values from about 1900 to 1950, a negative period from about 1960 to 1980, and then a strongly positive epoch, peaking in the late 1980s through mid-1990s.

The period encompassing the last two epochs, in particular the mid-1960s through mid-1990s, appears as a strong upward trend. This change was first articulated in a brace of papers by Hurrell (1995, 1996), who further showed that associated changes in wind fields help to explain winter warming over Eurasia and parts of northern North America, as well as partly-compensating cooling over parts of Greenland and northeast North America. Shortly thereafter, Thompson and Wallace (1998) addressed this in the framework of the Arctic Oscillation, or NAM.

Connections between the NAO/NAM, P, P–E, and river discharge are complex and cannot be fully reviewed here. In a broad context, the positive phase fosters not only a generally warmer Arctic, but a wetter one as well. While studies by Dickson et al. (2000), Rogers et al. (2001) and others show that the positive phase fosters prominent positive high-latitude anomalies in precipitation and P–E along the longitudes of the Nordic seas, positive anomalies are also manifested over large parts of the northern high latitudes (Thompson et al. 2000; Rogers et al. 2001; Peterson et al. 2006). This follows in that while cyclone activity becomes pronounced in the northern North Atlantic, particularly in the vicinity of the Icelandic Low (the NAO framework), it also increases over the northern high latitudes in more general sense (the NAM framework)

These high latitude signals were recently addressed as part of the remarkable study of Peterson et al. (2006). They asked if observed freshening of the North Atlantic (e.g., Curry and Mauritzen 2005) could be accounted for by freshwater inputs from P–E and melting Arctic ice. The answer is that it can, and that processes involved can be associated with the NAO in various ways.

Their study used P–E from ERA-40 back to 1958 (based on the forecasts of the two terms), river discharge, and information on glacier and ice sheet mass balance and sea ice thickness to examine the time history of anomalies in freshwater input into the high latitude oceans. They found that the sum of these freshwater sources matched the amount and rate at which freshwater accumulated in the Atlantic Nordic-Subpolar Subtropical (NSSB) basins for much of the period 1965–1995. Of particular note is that as the NAO rose from a negative to positive state, there was a dominance of positive anomalies in P–E over the Arctic Ocean, the region encompassing Hudson Bay and the Canadian Arctic Archipelago, and over the subpolar Atlantic basins. This period also saw increasing river discharge into the Arctic Ocean (for details of changing discharge of Eurasian rivers see Peterson et al. (2002)).

Since the mid-1990s, the NAO/NAM has regressed from its high values to a more neutral state. Peterson et al. (2006) document a decline in freshwater storage over the NSSB over the period 1996–2001. In part, this reflects a change to negative and zero anomalies in P–E over the Arctic Ocean and the Hudson Bay Canadian Archipelago region, respectively. It also appears that as the NAO went from positive to neutral, patterns of winds and ocean currents that had previously helped to transport freshwater from the Arctic to the NSSB changed so that more freshwater was sequestered in the Arctic Ocean. As a results of these process, and the fact that discharge anomalies from rivers draining into the Arctic Ocean have continued to stay positive, freshwater is likely now accumulating in the Arctic Ocean and will be exported southward when (and if) the NAO enters a new positive phase (Peterson et al. 2006).

While fascinating, the NAO/NAM does not explain everything. An obvious example is the continued positive anomalies in river discharge to the ocean. One must also consider impacts of other patterns, such as the Pacific Decadal Oscillation (PDO) and the North Pacific Oscillation (NPO) which link with the El-Nino Southern Oscillation (ENSO) in various ways. The index of the PDO is defined on the basis of sea surface temperatures (SSTs) in the North Pacific. During the positive PDO phase, SSTs in the central North Pacific are relatively cool, those along the west coast of North America are relatively warm, and the Aleutian low tends to be strong. The negative phase has roughly opposing signals. The NPO index relates to the area-weighted mean sea level pressure in the extratropical Pacific. It is a good measure of the strength of the Aleutian Low and is related to the PDO in this respect.

Hartmann and Wendler (2005) document consistency between warming and increased precipitation over parts of Alaska from 1951–2001 and a shift in the PDO from a negative phase from 1951 to 1976 to a primarily positive phase from 1977–2001. The deeper Aleutian Low during the latter period helped to transport warm, moist air into the region. As shown by Rogers et al. (2001), the NPO has signals over Alaska, but also northwestern Canada and areas to the north. About 40% of the variance in P–E that includes northeastern Canada can be linked to combined influences of the PDO and NAO.

Figure 14.8 provides time series of aerological P–E from ERA-40 over the period 1958–2001. Values are expressed as anomalies with respect to 1970–1999 and are broken down for the same Arctic Ocean and terrestrial drainage domains used in compiling Figs. 14.2 and 14.4. Cold season (September–May) and warm season (June to August) anomalies are plotted for the terrestrial drainage to account for the strong seasonality in evaporation. Only annual anomalies are shown for the ocean, where evaporation is low throughout the year.

P–E is variable over both land and ocean (more so over land), and no long-term trends are obvious. Looking at the latter half of the record, terrestrial P–E for the cold season shows a general, albeit modest rise from mostly negative to mostly positive anomalies from the 1970s through the mid-1990s, followed by a decline, in broad accord with the NAO time series over this period. A similar shift is seen for the warm season. These results are compatible with generally rising river discharge over this period. Ocean P–E also shows the same general change from the 1970s through the mid-1990s, but with fairly pronounced negative anomalies for the most recent years.

While like Fig. 14.8, Peterson et al. (2006) show negative P–E anomalies for the ocean at the beginning and end of the ERA-40 record, they report positive anomalies for pentads throughout the period 1970 to 1995. In particular, they do not show negative ocean anomalies from 1975 to 1980. These inconsistencies may result from the choice of domains, their different baseline for computing anomalies (1936–1955, based on regressing P–E against the NAO index), averaging over pentads, and the use of forecasts of P and E instead of aerological calculations. Unfortunately, the ERA-40 record ends in July 2002, which largely precludes assessing the positive river discharge anomalies since 2000 reported by Peterson et al. (2006).

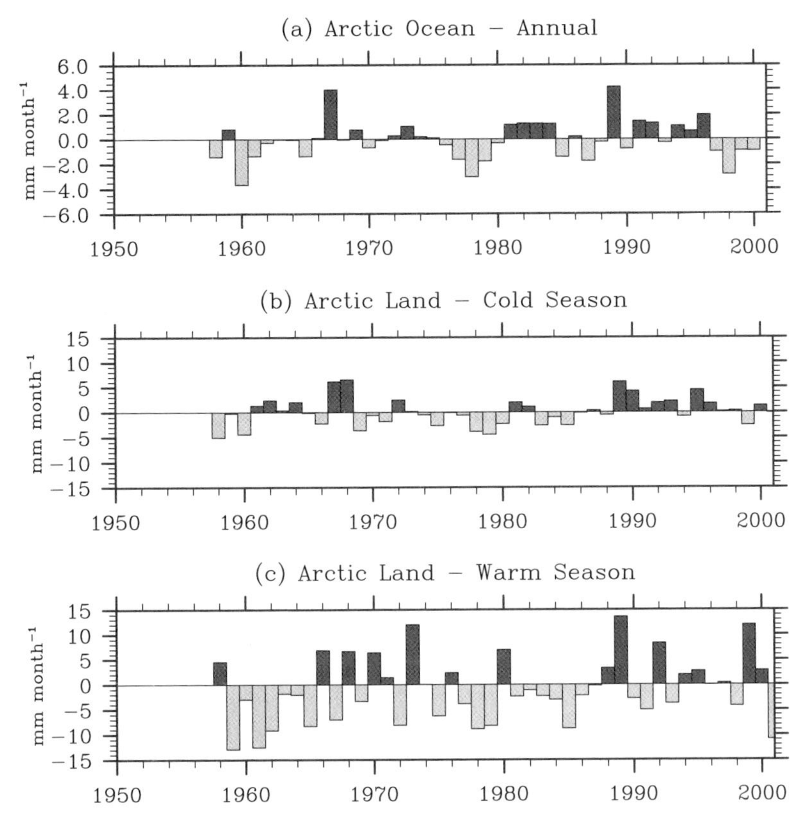

Fig. 14.8 Aerological P–E expressed as anomalies (referenced to 1970–1999) for the Arctic Ocean (annual) and terrestrial drainage, the latter for the cold season (September–May) and warm season (June–August)

14.4.2 A Longer View

The conclusion from model studies, such as those conducted for the Arctic Climate Impact Assessment (ACIA) and more recently from models participating in the Intergovernmental Panel on Climate Change Fourth Assessment Report (Holland et al. 2007), is that precipitation will increase over the north polar region through the 21st century. This is fundamentally explained in that a warmer atmosphere can hold more water vapor. Although warming will also increase evaporation, the increase in precipitation dominates. P–E and river discharge increase as a result.

These projected changes are nevertheless generally modest and there is considerable scatter between model projections. For example, expressed as annual means over the period 2040–2059 minus a 1950–1959 base period, the eight models examined by Holland et al. (2007) show an increase in P–E over the terrestrial drainage of near zero to 900 km^3 (57 mm). Over the Arctic ocean, the range is from 100 to

350 km^3 (10–36 mm). This scatter arises from different model treatments of hydrologic processes, that different models depict different changes in aspects of the atmospheric circulation, and that for any decadal time slice, different models will be in different phases of their own internal variability, which may include NAO-like behavior.

Is the Arctic's hydrologic system beginning to respond to greenhouse forcing? The evidence from Fig. 14.8 and river discharge records is inconclusive. Change detection is of course hampered by the short time series. However, records from land, when zonally averaged, allow for at least a broad assessment of precipitation changes since 1900. The conclusion, which must be viewed with the caveat that data are exceedingly sparse in the early part of the record, is that annual precipitation for the zonal band 50–85° N has increased (Fig. 14.9). The annual pattern is most strongly driven by changes during winter, summer and autumn.

However, the larger changes occurred over the first half of the century. As atmospheric greenhouse gas concentrations have risen most sharply in recent decades, it is reasonable to expect that the precipitation changes would be greater in the later part of the record. On the other hand, Wu et al. (2005) argue that greenhouse-forced increases in precipitation have already occurred and can help to explain the increased Siberian river discharge noted by Peterson et al. (2002) and Peterson et al. (2006).

14.5 Summary and Conclusions

The Arctic hydrologic system is fundamentally shaped by the fact that high-latitude P–E is positive in the annual mean. However, P–E varies strongly on a seasonal basis. River discharge to the ocean represents the release of P–E accumulated in the winter snowpack. By contrast, summer is a period of net drying over most land areas, despite fairly high precipitation. Precipitation over the northern North Atlantic tends to peak in winter, but this can also be strongly countered or exceeded by evaporation.

Variability in freshwater input to the Arctic Ocean, and the ultimate delivery of this freshwater to the North Atlantic, is strongly tied to atmospheric variability, for which the North Atlantic Oscillation and the Northern Annular Mode play key roles. These links are complex, involving not only influences of the atmospheric circulation on P–E, but attendant variations in winds and ocean currents that determine the extent to which freshwater is sequestered in the Arctic Ocean, or transported southward.

The Fram Strait ice sea ice flux, for example, is quite sensitive to the wind field, tending, in general (but not always), to be larger under the positive NAO phase (Kwok and Rothrock 1999). Proshutinsky et al. (2002) suggest that release of only a few percent of the Arctic Ocean's freshwater store that could occur from changes in this outflow could cause a North Atlantic salinity anomaly comparable to the Great Salinity Anomaly of the late 1960s to early 1970s. Dukhovskoy et al. (2004)

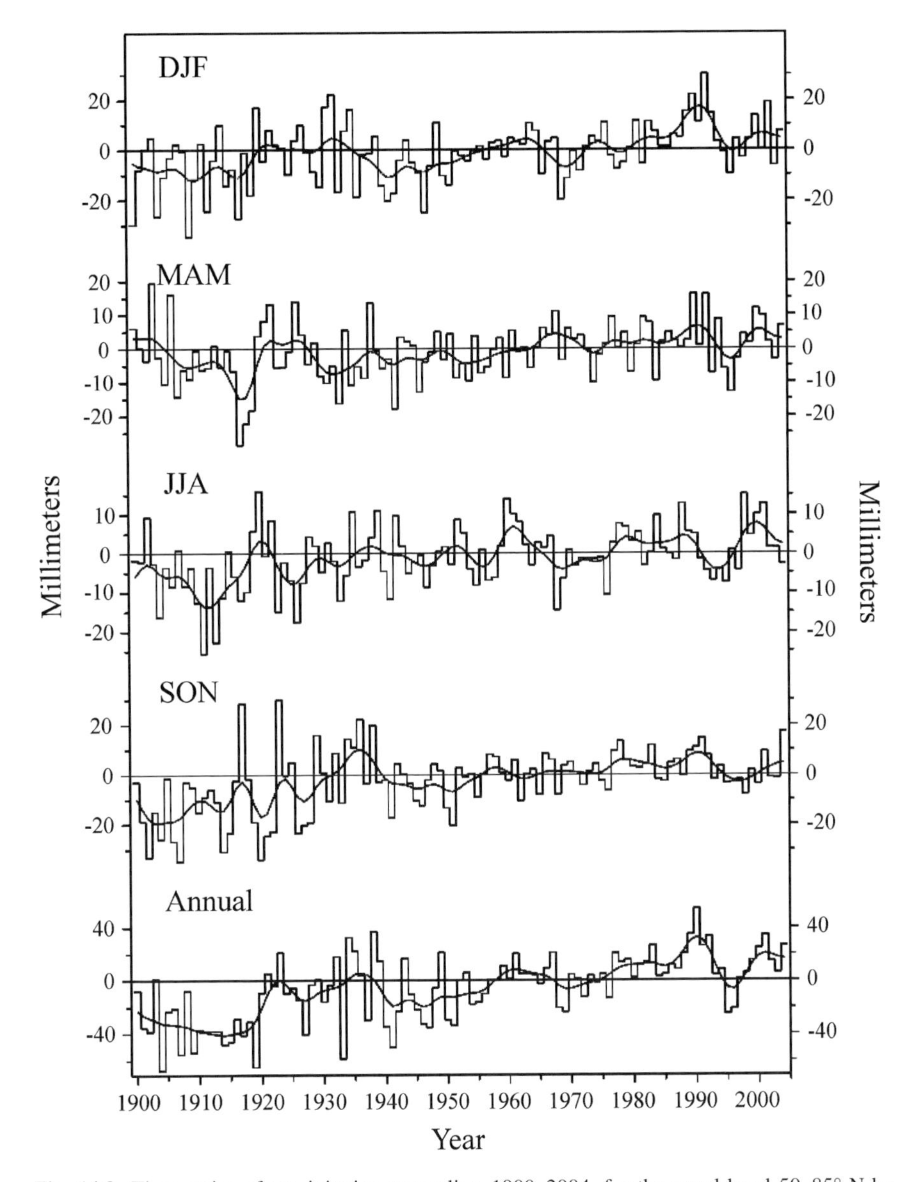

Fig. 14.9 Time series of precipitation anomalies, 1900–2004, for the zonal band 50–85° N by season and for annual totals. Anomalies are computed with respect to 1951–1980 means. The smoothed lines represent results from a nine-point low-pass filter. No adjustments have been applied for gauge biases (Courtesy of J. Eischeid, Climate Diagnostics Center, Boulder, CO)

extend some of these ideas, and suggest that the phase of the NAO/NAM may itself be in part determined by changes in freshwater input to the Greenland–Iceland–Norwegian seas that alter oceanic convection and vertical heat fluxes to the atmosphere.

It is expected that P–E will increase over northern high latitudes as the climate warms in response to greenhouse gas loading. Observational evidence that this change is emerging from the "noise" of natural variability is at present inconclusive. While the basic argument behind these projections is that a warmer atmosphere holds more water vapor, changes in atmospheric circulation are likely to be important. In this sense, an important "wild card" in the system is the future behavior of the NAO/NAM.

Despite recent return of the NAO/NAM to a more neutral phase, there is evidence that external forcing may favor an increased frequency of its positive state. A number of studies (e.g., Baldwin and Dunkerton 1999; Fyfe et al. 1999; Thompson et al. 2000; Gillett et al. 2003; Yumikoto and Kunihiko 2005; Kuzmina et al. 2005) argue that cooling of the stratosphere in response to increasing carbon dioxide and methane concentrations, or even through ozone destruction by chlorofluorocarbons, may "spin up" the polar stratospheric vortex, resulting in lower Arctic surface pressures and a positive shift in the NA&Mtilde;. A different idea is that the NAM could be bumped to a preferred positive state via increases in sea surface temperatures (SSTs) in the tropical Indian Ocean which themselves may be partly driven by greenhouse gas loading (Hoerling et al. 2001, 2004; Hurrell et al. 2004).

A protracted positive NAO state in the future, coupled with an increasingly warm Arctic, implies high P–E over northern high latitudes attended by strong exports of freshwater into the North Atlantic. This would represent an intensified hydrologic cycle similar to that suggested for the past high NAM state, but perhaps more vigorous.

Acknowledgements This study was supported by NSF grants ARC-0229769, ARC-0229651, OPP-0242125, and NASA contracts NNG04GH04G and NNG04GJ39G.

References

Aagaard K, Roach A, Schumacher JD (1985) On the wind-driven variability of the flow through Bering Strait. J Geophys Res 90: 7213–7221

Baldwin MP, Dunkerton TJ (1999) Propagation of the Arctic Oscillation from the stratosphere to the troposphere. J Geophys Res 104(D24): 30,937–30,946

Betts AK, Ball JH, Viterbo P (2003) Evaluation of the ERA-40 surface water budget and surface temperature for the Mackenzie river basin. J Hydrometeorology 4: 1194–1211

Carmack EC (2000) The Arctic Ocean's freshwater budget: sources, storage and export. In: Lewis EL, Jones PE, Lemke P, Prowse TD, Wadhams P (eds) The freshwater budget of the Arctic Ocean. Kluwer, Dordrecht, The Netherlands, pp 91–126

Cullather RI, Bromwich DH, Serreze MC (2000) The atmospheric hydrologic cycle over the Arctic basin from reanalyses. Part I: Comparisons with observations and previous studies. J Climate 13: 923–937

Curry R, Mauritzen C (2005) Dilution of the North Atlantic Ocean in recent decades. Science 17: 1772–1774

Dickson RR, Osborn TJ, Hurrell JW, Meincke J, Blindheim J, Adlandsvik B, Vinje T, Alekseev G, Maslowski W (2000) The Arctic Ocean response to the North Atlantic Oscillation. J Climate 13: 2671–2696

Dukhovskoy DS, Johnson MA, Proshutinsky A (2004) Arctic decadal variability: An auto-oscillatory system of heat and fresh water exchange. Geophys Res Lett 31: L03302, doi:10.1029/2003GL019023

Fyfe JC, Boer GJ, Flato GM (1999) The Arctic and Antarctic oscillations and their projected changes under global warming. Geophys Res Lett 26: 1601–1604

Gillett NP, Graf HF, Osborn TJ (2003) Climate change and the North Atlantic Oscillation. In: Hurrell JW, Kushnir Y, Ottersen G, Visbeck, M (eds) The North Atlantic Oscillation: Climatic significance and environmental impact. Geophysical Monograph 134, American Geophysical Union, pp 193–209

Göber M, Hagenbrock P, Ament F, Hense A (2003) Comparing mass-consistent atmospheric moisture budgets on an irregular grid: An Arctic example. Quart J Roy Meteorol Soc 129: 2383–2400

Hartmann B, Wendler G (2005) The significance of the 1976 Pacific climate shift in the climatology of Alaska. J Climate 18: 4824–4839

Hoerling MP, Hurrell JW, Xu T (2001) Tropical origins for recent North Atlantic climate change. Science 292: 90–92

Hoerling MP, Hurrell JW, Xu T, Bates GT, Phillips A (2004) Twentieth century North Atlantic climate change. Part II: Understanding the effect of Indian Ocean warming. Climate Dyn 23: 391–405

Holland MM, Finnis J, Barrett AP, Serreze MC (2007) Projected changes in Arctic Ocean freshwater budgets. J Geophys Res 112: GO4S55, doi:10.1029/2006JG000354

Hurrell JW (1995) Decadal trends in the North Atlantic Oscillation: regional temperatures and precipitation. Science 269: 676–679

Hurrell JW (1996) Influence of variations in extratropical wintertime teleconnections on Northern Hemisphere temperature. Geophys Res Lett 23: 665–668

Hurrell JW, Hoerling MP, Phillips A, Xu T (2004) Twentieth century North Atlantic climate change. Part I: Assessing determination. Climate Dyn 23: 371–379

Kalnay E, Kanamitsu M, Kistler R, Collins W, Deaven D, Gandin L, Iredell M, Saha S, White G, Woolen J, Zhu Y, Chelliah M, Ebisuzaki W, Higgens W, Janowiak J, Mo KC, Ropelewski C, Wang J, Leetma A, Reynolds R, Jenne R, Joseph D (1996) The NCEP/NCAR 40-year reanalysis project. Bull Am Meteorol Soc 77: 437–471

Kistler R, Kalnay E, Collins W, Saha S, White G, Woolen J, Chelliah M, Ebisuzaki W, Kanamitsu M, Kousky V, van den Dool H, Jenne R, Fiorino M (2001) The NCEP-NCAR 50-year reanalysis: Monthly means CD-ROM and Documentation. Bull Am Meteorol Soc 82: 247–267

Kuzmina S, Bengtsson L, Johannessen OM, Drange H, Bobylev LP, Miles MW (2005) The North Atlantic Oscillation and greenhouse-gas forcing. Geophys Res Lett 32: L04703, doi:10.1029/2004GL021064

Kwok R, Rothrock DA (1999) Variability of Fram Strait ice flux and North Atlantic Oscillation. J Geophys Res 104(C3): 5177–5189

Peterson BJ, Holmes RM, McClelland JW, Vorosmarty CJ, Lammers RB, Shiklomanov AI, Shiklomanov IA, Rahmstorf S (2002) Increasing river discharge to the Arctic Ocean. Science 298: 2171–2173

Peterson BJ, McClelland J, Curry R, Holmes RM, Walsh JE, Aagaard K (2006) Trajectory shifts in the Arctic and subarctic freshwater cycle. Science 313: 1061–1066

Portis DH, Walsh JE, El Hambly M, Lamb P (2001) Seasonality in the North Atlantic Oscillation. J Climate 14: 2069–2078

Proshutinsky A, Bourke RH, McLaughlin FA (2002) The role of the Beaufort Gyre in Arctic climate variability: Seasonal to decadal climate scales. Geophys Res Lett 29: 2100, doi:10.1029/2002GL015847

Rogers AN, Bromwich DH, Sinclair EN, Cullather RI (2001) The atmospheric hydrologic cycle over the Arctic basin from reanalyses. Part II: Interannual variability. J Climate 14: 2414–2429

Serreze MC, Etringer AJ (2003) Precipitation characteristics of the Eurasian Arctic drainage system. Int J Climatol 23: 1267–1291

Serreze MC, Lynch AH, Clark MP (2001) The Arctic frontal zone as seen in the NCEP-NCAR reanalysis. J Climate 14: 1550–1567

Serreze MC, Bromwich DH, Clark MP, Etringer AJ, Zhang T, Lammers R (2003) The large-scale hydroclimatology of the terrestrial Arctic drainage. J Geophys Res 108(D2): 8160, doi:10.1029/2001JD000919

Serreze MC, Barrett A, Lo F (2005) Northern high latitude precipitation as depicted by atmospheric reanalysis and satellite retrievals. Mon Wea Rev 133: 3408–3430

Serreze MC, Barrett AP, Slater AG, Woodgate RA, Aagaard K, Lammers R, Steele M, Moritz R, Meredith M, Lee CM (2006) The large-scale freshwater cycle of the Arctic. J Geophys Res 111: C11010, doi:10.1029/2005JC003424

Slater AG, Bohn TJ, McCreight JL, Serreze MC, Lettenmaier DP (2007) A multi-model simulation of pan-Arctic hydrology. J Geophys Res 112: GO4S45, doi:10.1029/2006JG000303

Stigebrandt A (2000) Oceanic freshwater fluxes in the climate system. In: Lewis EL, Jones PE, Lemke P, Prowse TD, Wadhams P (eds) The freshwater budget of the Arctic Ocean. Kluwer, Dordrecht, The Netherlands, pp 1–20

Thompson DWJ, Wallace JM (1998) The Arctic Oscillation signature in the wintertime geopotential height and temperature fields. Geophys Res Lett 25: 1297–1300

Thompson DWJ, Wallace JM, Hegerl G (2000) Annular modes in the extratropical circulation. Part II: Trends. J Climate 13: 1018–1036

Uppala SM, Kalberg PW, Simmons AJ, Andrae U, da Costa Bechtold V, Fiorino M, Gibson JK, Haseler J, Hernandez A, Kelly GA, Li X, Onogi K, Saarinen S, Sokka N, Allan RP, Andersson E, Arpe K, Balmaseda MA, Beljaars ACM, van de Berg L, Bidlot J, Bormann N, Caires S, Chevallier F, Dethof A, Dragosavac M, Fisher M, Fuentes M, Hagemann S, Holm E, Hoskins BJ, Isaksen L, Janssen PAEM, Jenne R, McNally AP, Mahfouf J-F, Morcrette J-J, Rayner NA, Saunders RM, Simon P, Sterl A, Trenberth KE, Untch A, Vasiljevic D, Viterbo P, Woollen J (2005) The ERA-40 reanalysis. Quart J Roy Meteorol Soc 131: 2961–3012

Vinje T (2001) Fram Strait ice fluxes and atmospheric circulation: 1950–2000. J Climate 14: 3508–3517

Walsh JE, Zhou X, Portis A, Serreze MC (1994) Atmospheric contribution to hydrologic variations in the Arctic. Atmos-Ocean 32: 733–755

Woodgate RA, Aagaard K, Weingartner T (2005) A year in the physical oceanography of the Chukchi Sea: Moored measurements from autumn 1990–1991. Deep-Sea Res II 52: 3116–3149

Woodgate RA, Aagaard K, Weingartner TL (2006) Interannual changes in the Bering Strait fluxes of volume, heat and freshwater between 1991 and 2004. Geophys Res Lett 33: L15609, doi:10.1029/2006GL026931

Wu P, Wood R, Stott P (2005) Human influence on increasing Arctic river discharges. Geophys Res Lett 32: L02707, doi:10.1029/2005GL021570

Yang D, Kane DL, Hinzman LD, Zhang X, Zhang T, Ye H (2002) Siberian Lena river hydrologic regime and recent change. J Geophys Res 107(D23): 4692, doi:10.1029/2002JD002542

Yumikoto S, Kunihiko K (2005) Interdecadal Arctic Oscillation in twentieth century climate simulations viewed as internal variability and response to external forcing. Geophys Res Lett 32: L03707, doi:10.1029/2004GL021870

Chapter 15
Simulating the Terms in the Arctic Hydrological Budget

Peili Wu[1], Helmuth Haak[2], Richard Wood[1], Johann H. Jungclaus[2], and Tore Furevik[3,4]

15.1 The Arctic Hydrological Budget

The hydrological cycle in the Arctic is composed of three branches: atmosphere, land and ocean. Figure 15.1 schematically shows the estimated annual rate of transport in cubic kilometer per year carried by the individual branches. Bold numbers are observational estimates taken from Aagaard and Carmack (1989), Carmack (2000) and Woodgate and Aagaard (2005), while other numbers are taken from freshwater budget analysis of several century long climate model simulations: (a) ECHAM5/MPIOM under pre-industrial conditions and (b) Bergen Climate Model (BCM) present-day control experiment. These numbers should be treated with caution because of the large uncertainties involved in various observational estimates and the realism of the relatively coarse resolution climate models in representing the hydrological process and the complicated geometry of the narrow straits. The choice of a reference salinity used to calculate freshwater content in the ocean can also introduce discrepancies in ocean transport and storage changes, although a common reference of 34.8 psu is adopted here. It is clear that significant differences exist between modelled and observational estimates as well as among different models at the present stage. It is, however, useful to show the relative importance of each individual component in the overall hydrological budget. Nonetheless, both models and observations agree on the leading terms of contribution.

The atmospheric branch provides a freshwater input through direct net precipitation (precipitation minus evaporation: P – E) and the land branch via river discharges (R). The combination of (P – E + R) is the leading term of freshwater sources shown in Fig. 15.1. Direct P–E is relatively small, compared to river discharges, which play a far bigger role as the Arctic is the only ocean with a contributing land area greater than its own surface. River discharges provide twice (or even several

[1] Met Office Hadley Centre, Exeter, UK

[2] Max Planck Institute for Meteorology, Hamburg, Germany

[3] Geophysical Institute, University of Bergen, Norway

[4] Bjerknes Centre for Climate Research, Bergen, Norway

R.R. Dickson et al. (eds.), *Arctic–Subarctic Ocean Fluxes*, 363–384

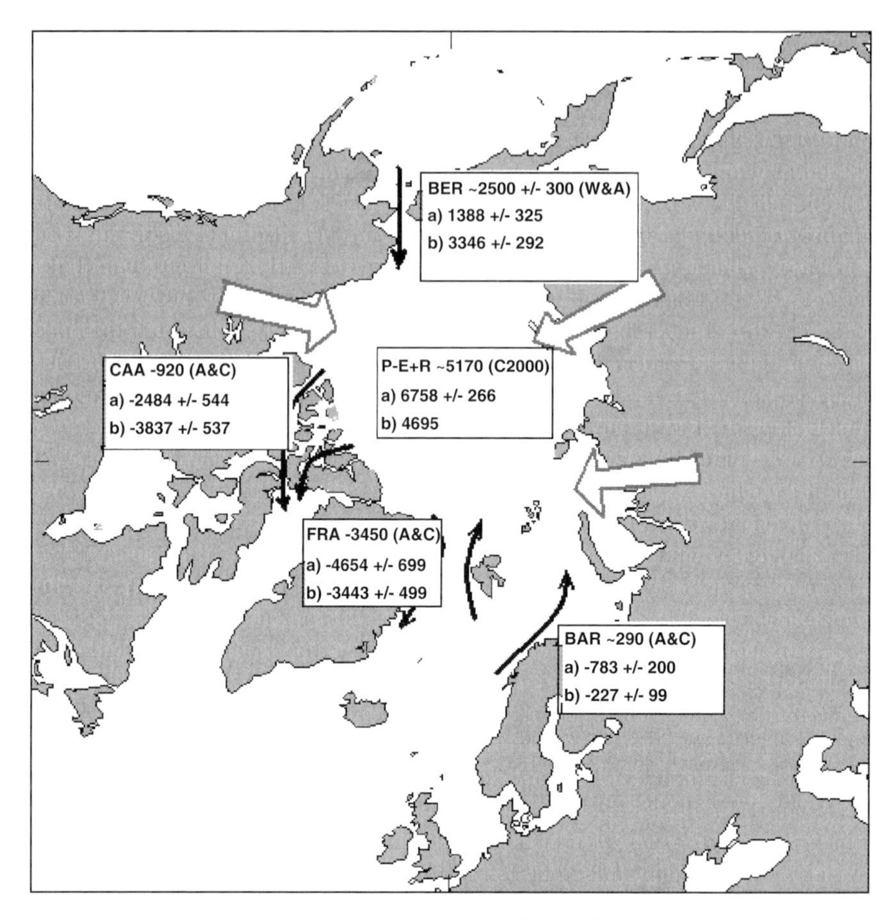

Fig. 15.1 A schematic diagram showing the mean hydrological budget of the Arctic Ocean from various observational estimates (bold) in comparison with simulated budgets from two coupled climate models (a) and (b) using a reference salinity of 34.8 psu. Data sources are A&C: Aagaard and Carmack (1989); C2000: Carmack (2000); W&A: Woodgate and Aagaard (2004); (a) ECHAM5/MPIOM and (b) BCM

times according to the Arctic Climate Impact Assessment report, 2005) as much as the freshwater input coming from direct P–E. Ocean transport is the largest individual carrier among all through both liquid water and ice fluxes: the Bering Strait transport of the relatively fresher Pacific inflow (noted as BER in Fig. 15.1) provides another freshwater source while exports via Fram Strait (FRA), the Canadian Archipelago (CAA) and the salty Atlantic inflow via the Barents shelf (BAR) are freshwater sinks. A more detailed estimate of large-scale Arctic freshwater budget from observations can be found in Serreze et al. (2006) and Dickson et al. (2007). In the following sections, we will discuss each of these three components in more detail as simulated by major coupled climate models. Each section focuses on three aspects of the specific subject: simulation of the climatological mean, future projections under global warming scenarios, and detection and attribution of changes occurring during the recent 5–6 decades.

15.2 Simulated Precipitation–Evaporation

Direct measurements of precipitation in the Arctic have large uncertainties due to the sparse network and adjustment errors to the gauge undercatch of solid precipitation. Measurement errors may reach 50–100% (Serreze et al. 2005). Evaporation from ocean surface or evapo-transpiration from land and plants is equivalent to surface latent heat flux. Direct estimates of evaporation or latent heat flux are sparse. Evapo-transporation is sometimes estimated as residual between precipitation and runoff assuming no soil moisture change. Runoff is obtained from gauged river discharge data, dividing it by the area of the watershed. Net precipitation or P–E as a combined variable can also be estimated from atmospheric moisture budget using the "aerological method" (Cullather et al. 2000). Assuming no escape to space, the difference between water content (precipitable water) changes and lateral moisture convergence for a given air column must equal to the water flux through the lower boundary surface, i.e. P–E. The average P–E in the Arctic can be estimated from water vapour transport through a given latitude circle (e.g. 70° N). The average P–E for the polar cap north of 70° N is estimated to be 188 ± 6 mm/year (equivalent to 2,095 ± 67 km^3/year) from the NCEP and ERA-15 reanalysis data compared to the rawinsonde-derived estimate of 163 mm/year (Serreze and Barry 2005).

Precipitation and evaporation are not only difficult to measure in the physical world, but also difficult to simulate in climate models. This is mainly due to the sporadic nature of the precipitation process, which can only be parameterized to some degree in relatively coarse resolution climate models. In the Arctic, this becomes even more complicated due to interactions between the atmosphere, the ocean, sea ice, snow cover and permafrost. Validation of model simulations is also hampered by the large uncertainties in observational estimates. Models generally overestimate the rate of the hydrological cycle. This is the case for the HadCM3 model (Pardaens et al. 2003). Walsh et al. (2002) have compared simulated Arctic precipitation across two groups of models: the AMIP-II atmosphere only models (AGCM) and the 2001 IPCC coupled models (AOGCM). The AGCMs are constrained through their lower boundary conditions by prescribing sea surface temperature (SST) and sea ice concentrations to observational data. Such constraints do not exist in fully coupled climate models, so large differences are more likely among coupled model simulations. Both the AGCMs and AOGCMs can capture the general large-scale features although there are significant differences in the details. Much of the error in coupled model simulations is due to the simulated strength and position of the storm tracks and the underestimate of sea ice.

According to the IPCC report (2001), an increase in global mean precipitation and evaporation is to be expected during the 21st century. The spatial distribution of such increase varies with latitude. While precipitation increases in the tropics and the high latitudes it decreases in the subtropics. Figure 15.2 is an example using a six-member 80-year ensemble simulation by the Bergen climate model (Furevik et al. 2003; Bethke et al. 2006) with a 1% per year increase in atmospheric CO_2 concentrations, leading to doubling after 70 years. Shown by colour shading in Fig. 15.2 are the mean trends in P–E (mm day^{-1} per decade) and arrows are water vapour transports. The banding structure is clearly visible: positive trends in the tropics and polar regions and negative trends in the subtropics. The simulations also show larger water vapour transports associated with the storm tracks, bringing water from the northeast Pacific towards Alaska and western Canada, and from the North Atlantic towards Siberia. The latitudinal dependency can also be seen from predicted changes in zonal mean freshwater fluxes (P – E + R) into the ocean from the HadCM3 model simulations (Wu et al. 2005). The model is run under a projected scenario forcing following the IPCC SRES B2. The increasing trend in the high latitudes, particularly the northern high latitudes, is clear. Trends in the tropics are dominated by the strong short term variability linked to El Niño-Southern Oscillation (ENSO). Some parts of the anomalous fluxes come from river discharges, which are discussed in the following section.

Given the projected trend, researchers are keen to detect any early signs of such expected changes. Kattsov and Walsh (2000) investigated Arctic precipitation changes during the 20th century. They reported an increasing trend in total Arctic precipitation in both observational data and a model simulation with prescribed SST, sea ice cover and CO_2 concentrations. They have also noted the poor reliability of the observed precipitation data and the observations used to force the atmosphere model. In the meantime, such increasing trend seems inconsistent with global mean land precipitation changes reported by Allen and Ingram (2002) and Lambert et al. (2004), which do not show an obvious trend.

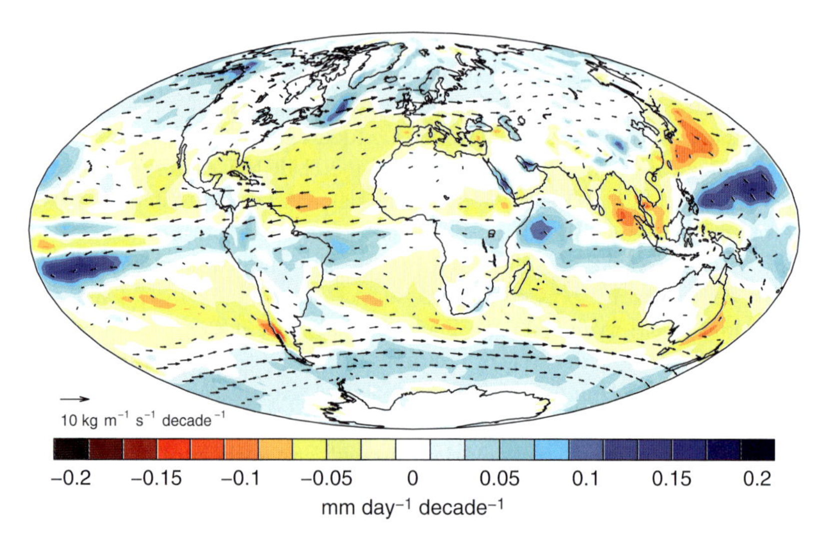

Fig. 15.2 Trends in P–E (mm day^{-1} per decade, colour scale) and water vapour transports (kg m^{-1} s^{-1}, arrows) in a six-member ensemble of greenhouse gas experiments (After Bethke et al. 2006)

Since there are not enough reliable observations in either precipitation or evaporation for the past, indirect measurements such as ocean salinity changes (e.g. Curry et al. 2003) and Arctic river discharges (Peterson et al. 2002) are used to infer changes in the global hydrological cycle. Curry et al. (2003) found signs of freshening polar oceans and salinity increases in the subtropics. The coincidence between the observed ocean salinity changes and the overlaying P–E climatological distribution is suggestive of a link with surface freshwater forcing. Peterson et al. (2002) found an increasing trend of river discharges into the Arctic Ocean from the six largest Eurasian rivers. A more detailed study using both observational data and modelling scenarios by McClelland et al. (2004) has suggested that increasing northward transport of atmospheric moisture as a result of global warming is the most viable explanation for the observed increasing trends in Eurasian river discharges into the Arctic Ocean. Both studies have suggested that the expected changes in surface freshwater fluxes are detectable and may have started to appear in ocean salinities already during the late 20th century.

The HadCM3 "all forcings" ensemble simulation of 20th century climate has been used in various recent studies for detection and attribution of climate change. It has succeeded in realistically simulating many aspects of observed changes in the 20th century, such as global mean surface air temperature (Stott et al. 2000), mean land precipitation (Allen and Ingram 2002), Arctic river discharges (Wu et al. 2005) and water mass property changes in the Labrador Sea (Wu et al. 2004) and the subpolar North Atlantic (Wu et al. 2007). Figure 15.3 shows the simulated linear trends

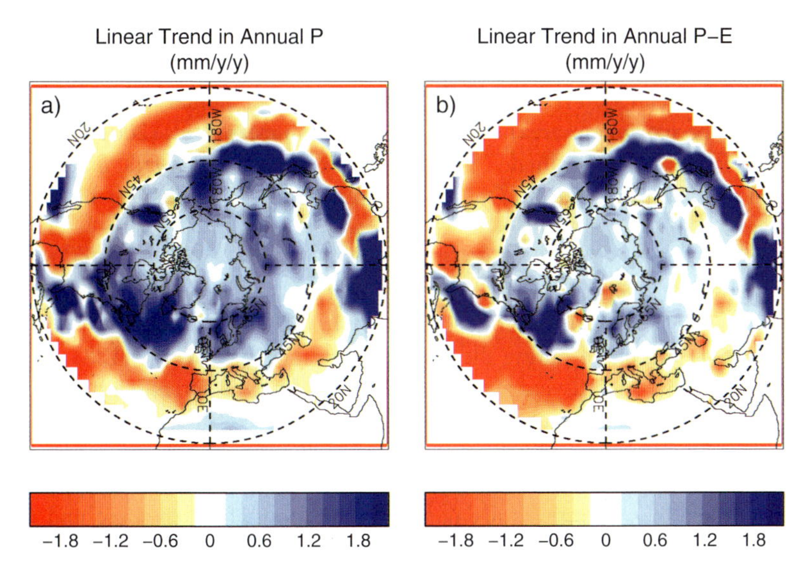

Fig. 15.3 Linear trends in precipitation (a) and precipitation–evaporation (b) between 1949 and 2001 in a four-member ensemble simulation by the HadCM3 coupled climate model with all historical external forcings

for P and P–E over the northern hemisphere during the later half of the 20th century. As Allen and Ingram have reported earlier, there is no clear trend in total global land precipitation. However, spatial variations are large and there are clear latitudinal bands of opposite trends in both P and P–E: negative trends in the subtropics and positive trends over the polar cap suggesting an increasing trend in northward moisture transport by the atmosphere. The large-scale spatial patterns of trends in precipitation (Fig. 15.3a) are very similar to the trends in P–E (Fig. 15.3b). Regional differences do exist in some areas, the Nordic Seas and northern Canada in particular. Both regions show very weak trends in P, but clearly negative trends in P–E. The subtropical downward trends in P are amplified further in P–E, implying decreasing precipitation and increasing evaporation. These opposite patterns in P–E are consistent with the recent observed salinity trends in the North Atlantic Ocean reported by Curry et al. (2003).

15.3 River Discharges

River discharge contributes two thirds of the total freshwater fluxes into the Arctic Ocean. On average, it inputs about 3,300 km^3/year north of Fram Strait and another 420 km^3/year into the Greenland–Iceland–Norwegian (GIN) Seas (Aagaard and Carmack 1989). The ACIA report (2005) gives a larger amount of ~4,300 km^3/year. Given the fact that precipitation and evaporation in the Arctic are very difficult to measure and observational records are thus so sparse and unreliable, river runoff may provide a better alternative to P–E for inferring integral changes in the hydrological cycle over the catchment areas. There are nearly 2,000 individual rivers flowing into the Arctic Ocean system, encompassing over 22 million square kilometer of land area (Serreze and Barry 2005). The mean annual runoff (discharge divided by the catchment area) can be found in Lammers et al. (2001). The total discharge is, however, dominated by a small number of large rivers, among them are the Lena, Yenisey and the Ob' from Eurasia and the Mackenzie from North America.

Changes in Arctic river discharges have come into the spotlight since the high profile paper by Peterson et al. (2002). Combining monitoring data from the six largest Eurasian rivers from 1936 to 1999, they found a linear trend of 2 ± 0.7 km^3/year in annual discharge during the period, accumulating to a 7% (128 km^3/year) increase by 1999. The 10-year running average of the time series seems to follow both the global mean surface air temperature and the North Atlantic Oscillation (NAO) index, suggesting influence of both global warming and the upward trend of the NAO during the late 20th century. Other factors that could have affected the total Eurasian river discharges are dam building, permafrost thawing and forest fires. But these have all been excluded as major drivers by the work of McClelland et al. (2004). They have concluded that increasing northward moisture transport as a consequence of global warming remains the most likely explanation for the upward trend.

Coupled climate models can be a very useful tool in attributing the causes of certain observed changes. Using the Hadley Centre's coupled climate model HadCM3, Wu et al. (2005) have carried out the analyses of three separate ensemble simulations of the 20th century climate under different observed forcing factors. Including all historical factors, natural (estimated changes in solar irradiance and volcanic aerosols) and anthropogenic (greenhouse gases, sulphate aerosols and ozone), the model is capable of reproducing the observed trend in Eurasian river discharges. A similar trend is also simulated for the total discharges of pan-Arctic rivers. With climate model simulations, one can separate the forcing factors to determine which are responsible for the increasing trend. Wu et al. (2005) find that the increasing trend can only be simulated when anthropogenic factors are included. The major results of both Peterson et al. (2002) and Wu et al. (2005) are summarized in Fig. 15.4, where the Eurasian observations are rescaled in order to compare the observed and simulated upward trends from the 1960s.

River discharges from North America (Dèry et al. 2005; Dèry and Wood 2005), however, have shown a downward trend, in contrast with the Eurasian rivers (Peterson et al. 2002). McClelland et al. (2006) have now combined all the available data from both Eurasian and North American rivers and analysed them in a same time frame between 1964 and 2000 using the same method. It is confirmed that there is a small downward trend of 0.4 km^3/year in annual discharges from the

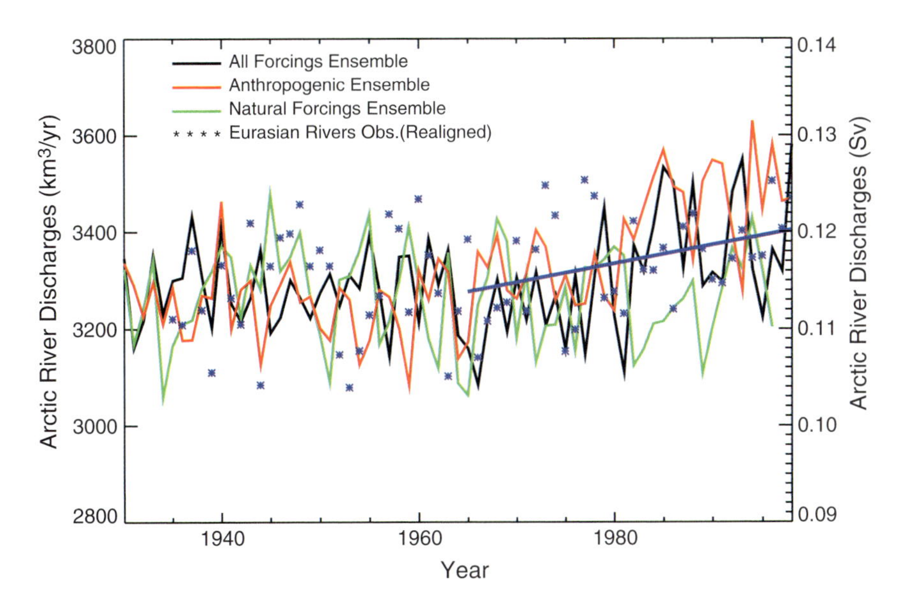

Fig. 15.4 Simulated river discharges into the Arctic Ocean by the HadCM3 coupled climate model (after Wu et al. 2005) in comparison with observations from the six largest Eurasian rivers. (After Peterson et al. 2002). The observations are shown in blue stars with a shift in the mean and the blue solid line is a linear fit to the data from 1965 to 1999. The upward trend from the 1960s seen in observations (shown in blue stars) can only be simulated by the model when anthropogenic factors are included (black and red curves)

North American rivers into the Arctic Ocean. That does not cancel out a much stronger positive trend of 6.3 km^3/year from the Eurasian rivers during the same period, resulting in an increasing trend of 5.6 km^3/year for pan-Arctic river discharges into the ocean. However, their data does not include all the rivers discharging into the Arctic Ocean. By taking into account the estimated 26% Eurasian and 15% North American discharges not included in their analysis, McClelland et al. (2006) have scaled up their estimated trend to be 7.4 km^3/year in total annual discharges into the Arctic Ocean. That is close to the model simulated trend of 8.7 km^3/year by Wu et al. (2005).

To statistically test the significance of the recent upward trend, we have analysed the long HadCM3 control integration with fixed pre-industrial greenhouse gas concentrations. Figure 15.5 shows the decadal mean total Arctic river discharges from 1,610 years of the control simulation in comparison to the model simulated decadal average of the 1990s from the "all forcings" ensemble. It is clear that the recent level of discharges is higher than any decades during the 1,600 years of control simulation. It also suggests that the integral of Arctic river discharges is a good indicator of climate change.

It remains to explain the contrasting trends between North American rivers and the Eurasian rivers and the role of the NAO. HadCM3, as most other coupled climate models, does not simulate the observed upward trend of the NAO during the late 20th century when forced with all historical factors. So it is not clear how the NAO may

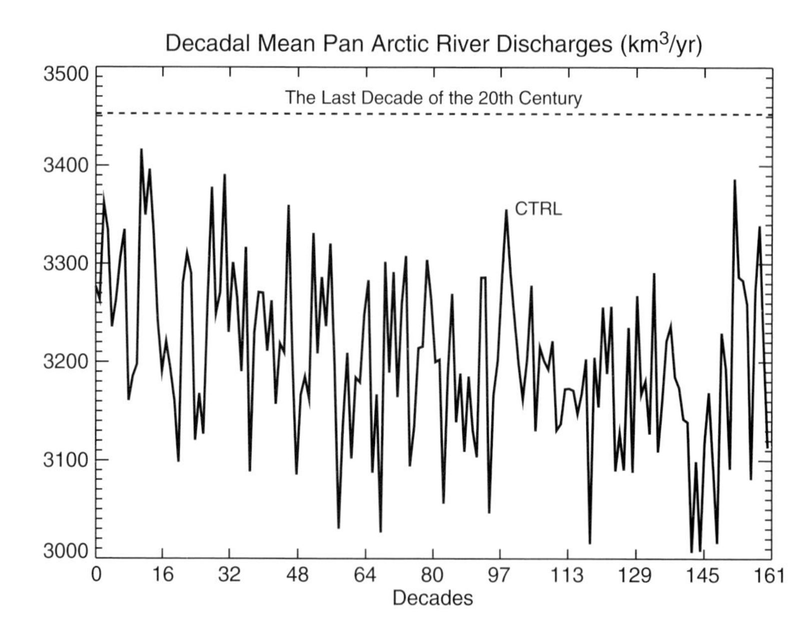

Fig. 15.5 Decadal mean river discharges into the Arctic Ocean simulated in a 1610 year control integration by the HadCM3 model with fixed pre-industrial greenhouse gas concentrations (solid line) in comparison with the average level for the 1990s (dashed line) simulated by the same model under all transient historical forcings

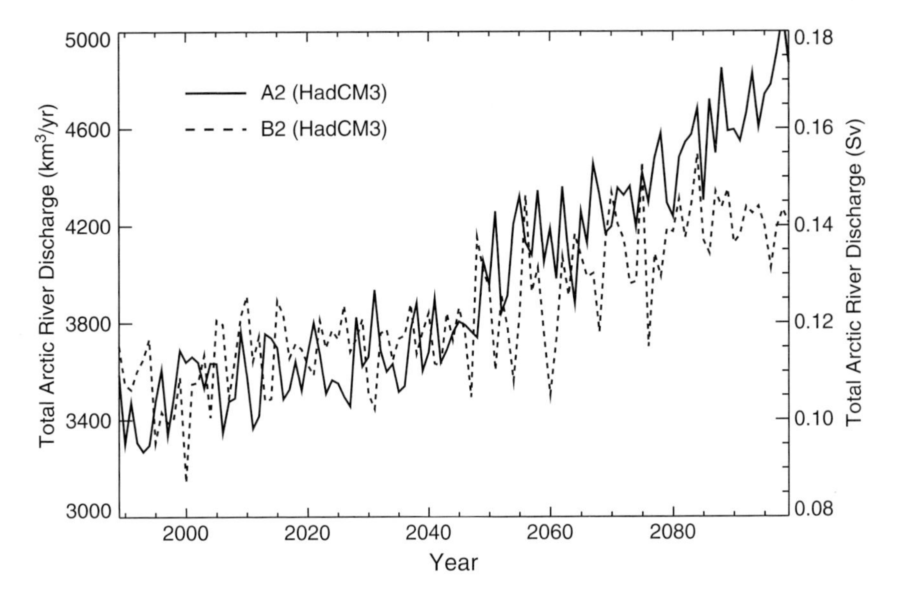

Fig. 15.6 Simulated river discharges into the Arctic Ocean projected by the HadCM3 coupled climate model for the 21st century under the IPCC SRES A2 (solid line) and B2 (dashed line)

have affected the observed and simulated trends. We may speculate that the NAO makes no difference in the total pan-Arctic river discharges. However, an east–west contrast may have been produced by the upward NAO trend in both precipitation and river runoffs. A strong positive NAO favours enhanced precipitation over northwest Europe, but reduced rainfall over northeast America. This may explain the differing trends in observed river discharges from Eurasian and North American continents.

Arnell (2005) has analysed simulations from six climate models including HadCM3, ECHAM4 and GFDL-R30 under two IPCC SRES scenarios: low emissions (B2) and high emissions (A2), and found an increase in total Arctic river discharges between 24% and 31% by 2080s. Figure 15.6 shows the time series of HadCM3 simulated total Arctic river discharges for the 21st century under the B2 and A2 scenarios. Models predict a continuing rise throughout the century, but the rate of increase has a jump around 2050 from a slower pace in the first half century to a much steeper rise towards the end. By 2010, we could see a further increase about 5.4% relative to the 1990s, but by 2090s there could be 30% more river discharges into the Arctic Ocean.

15.4 Arctic Sea Ice

In this section we discuss simulations using coupled climate models of variations in Arctic sea ice during the late 20th century, and projections for the 21st century. Arzel et al. (2006) and Zhang and Walsh (2006) discuss the simulation of sea ice in the 20th and 21st centuries in the most recent generation of coupled climate models used in the IPCC 4th Assessment Report (AR4). The sophistication of the sea ice components of such models has increased in recent years, so that many models now include better representations of sea ice dynamics and rheology, including ridging, and allow multiple ice thickness categories within a grid box (e.g. see Johns et al. 2006). While these sea ice components are not comparable in sophistication to the most complex schemes available, they may be sufficient to capture the large-scale changes that have been observed over recent decades.

The 14 climate models studied by Arzel et al. (2006) and Zhang and Walsh (2006) show a range of simulations of mean sea ice area and extent. Generally there is less spread of the models about the observed values in winter than in summer. Based on satellite (Johannessen et al. 1999; Comiso 2002) and in situ (Rayner et al. 2003) observations, ice extent and area are believed to have declined in recent decades, albeit against a background of significant interannual variability. Most of the models studied, when run with the history of natural and anthropogenic climate forcings, suggested a declining trend over the period 1981–2000, although there is considerable variation among models and a few showed a slight increase in extent. Figure 15.7 shows the HadCM3 simulations, where comparison is made between observations and model simulation for sea ice extent (see Fig. 15.7a), but not for sea ice volume (Fig. 15.7b), for which observations are limited. In most cases the trends seen over this period were outside the 1 standard deviation range of variability in the model control runs, suggesting that the climate forcings were playing a role. A few studies have taken the attribution of the observed changes further using individual models, and conclude that anthropogenic forcings have probably been playing a role in the observed decline (e.g. Gregory et al. 2002; Johannessen et al. 2004). Nonetheless, the important processes of internal variability that are believed to operate in the Arctic (e.g. variations in transport through the Fram Strait) mean that a full explanation of the observed changes is likely to be complex.

Assessment of changes in sea ice volume is harder because of the lack of complete observations. Some studies have suggested a substantial decline in recent decades (Rothrock et al. 1999), while Winsor (2001) suggests no trend during the 1990s. The model simulations studied by Arzel et al. (2006) again showed a wide range of responses over 1981–2000, with a decrease in nearly all models. The mean reduction in annual mean ice volume across all the models was 2.2×10^3 km^3, and if this is considered to be a supply of fresh water to the Arctic ocean it represents around 0.004 Sv over that period. Gregory et al. (2002), using the HadCM3 model, showed that the simulated loss of ice volume in that model was inconsistent with the model's internal variability, and that although the sea ice may respond on shorter timescales to volcanic eruptions, there appeared to be no long-term trend due to natural forcings. Again this suggested a role for anthropogenic forcings in the simulated decline in recent decades.

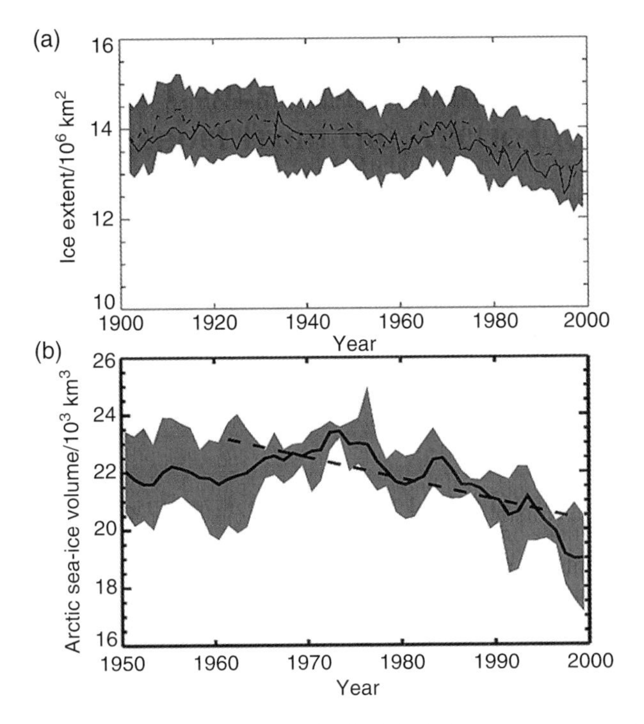

Fig. 15.7 Observed (solid) and HadCM3 simulated (dashed) Arctic sea ice extent (a) and model simulated Arctic sea ice volume (b) for the 20th century (from Gregory et al. 2002). Both observations and modelling indicate a decline in Arctic sea ice from the 1970s. The shaded region in both (a) and (b) delimits the envelope of the four ensemble members. Dashed line in (a) and solid line in (b) are the ensemble averages

Models suggest further reductions in sea ice volume and extent through the 21st century, in response to increasing greenhouse gas concentrations. Most models studied by Zhang and Walsh (2006) and Arzel et al. (2006) suggested increased rate of reduction in ice area. They also suggested that ice area or extent would decline at a faster rate for summer than for winter, resulting in an increased seasonal range. Many models suggested an ice-free Arctic in late summer by the end of the 21st century, under mid-range forcing scenarios. Figure 15.8 shows again the HadCM3 projections, as an example, under a number of IPCC scenarios. While the above results appear qualitatively robust among models, there is again substantial variation among models in the projected magnitude of the changes. The multi-model mean projected change in annual mean sea ice volume over the 21st century, under the IPCC A1B forcing scenario, was 13.1×10^3 km^3 (Arzel et al. 2006), corresponding to an average melting rate of 0.0042 Sv, with some models predicting up to twice this value. This is less than 15% of the projected increase in total river discharges of ~0.025 Sv (see Fig. 15.6).

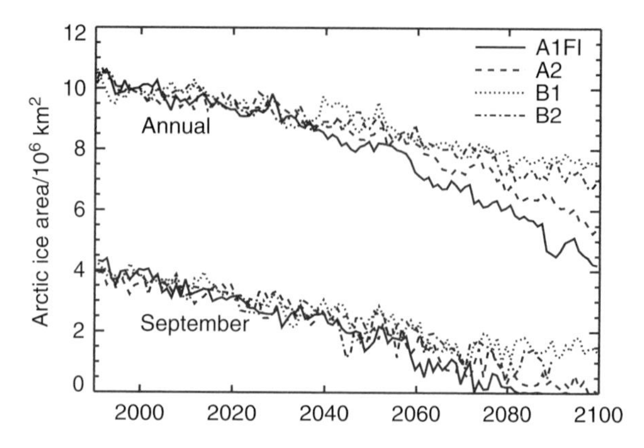

Fig. 15.8 HadCM3 projected Arctic sea ice extent for the 21st century under various IPCC scenarios. The model predicts a continuous decline in sea ice cover towards the end of this century and by 2080 the Arctic will become ice free in the autumn under certain IPCC scenarios

15.5 Greenland Ice Sheet

Ice sheets play a dynamic role in Earth's climate system, influencing regional climate and global sea level and responding to climate change on time scales of millennia. The Greenland ice sheet holds enough freshwater to raise the oceans seven metres if it all melts. An imbalance between new ice formed from falling snow and melting will have an important impact on the Arctic hydrological budget.

Climate models forced with IPCC projected greenhouse gas concentrations predict that the Greenland ice sheet is likely to lose its stable state, although a warmer climate comes with more precipitation. For an annual average global warming of more than 3.1 ± 0.8 K and 4.5 ± 0.9 K regional warming over Greenland, the net surface mass balance of the Greenland ice sheet becomes negative in the AR4-scenarios (Gregory and Huybrechts 2006). Greenhouse gas concentrations will probably have reached levels before the year 2100 that are sufficient to raise the temperature past this warming threshold (Gregory et al. 2004). The most extreme scenario considered in the third assessment report (TAR) of IPCC involves a warming of 8 °C over Greenland, in which case most of the ice sheet will be eliminated within the next 1,000 years (IPCC 2001). Toniazzo et al. (2004) have shown that the loss of Greenland ice sheet is possibly irreversible. By simulating the pre-industrial climate without the Greenland topography, which is equivalent to a cut of greenhouse gas concentrations to the pre-industrial level after the elimination of the Greenland ice sheet, they are not able to generate a long-term snow accumulation over Greenland. On the other hand, Lunt et al. (2004) using a higher-resolution model suggest that reglaciation may also be possible.

Many recent observations indicate increased ablation of the Greenland ice sheet. Repeat-pass airborne laser altimetry measurements indicate that Greenland is losing ice at a rate of $-80 \pm 12\,km^3$/year during the period 1997–2003, mostly from the periphery (Krabill et al. 2004). A more recent study based on satellite interferometry (Rignot and Kanagaratnam 2006) suggests that the tide-water glaciers (those grounded below sea level) are accelerating and the resulting dynamical loss may have increased the mass imbalance to $-224 \pm 41\,km^3$/year in 2005 compared to an estimate of $-91 \pm 31\,km^3$/year in 1996 using the same method. Based on gravity measurements from the GRACE satellite mission, Chen et al. (2006) have estimated a total ice melting rate of $-239 \pm 23\,km^3$/year for the period from April 2002 to November 2005. While other studies have suggested that the interior may be stable, Johannessen et al. (2005) have shown an increase of 6.4 ± 0.2 cm/year in the vast interior areas above 1,500 m, using altimeter height data from European Remote Sensing satellites (ERS-1 and ERS-2) for the period 1992–2003. They have also confirmed the thinning of ice sheet margins below 1,500 m (-2.0 ± 0.9 cm/year), but on average they find 60 cm increase over 11 years for the area studied. Because the measurement does not completely cover the marginal areas, it is not possible to know the integral change. The short observational records of some satellite measurements must be treated with caution as they are unable to account for the considerable natural variability in surface mass balance (Hanna et al. 2002).

Characterising the response of ice sheets, to various degrees of climatic forcing, is usually conducted using an off-line model with idealised temperature and precipitation. Such models are used because of the need for high spatial resolution at the steep ice sheet margins, a resolution not available from the coarse resolution climate model grids. A significant step forward has been to implement a fully coupled high resolution (20 km) Greenland ice sheet model within the Hadley Centre climate model HadCM3 (Ridley et al. 2005). Figure 15.9 shows the state of the Greenland ice sheet at various stages in a stabilisation experiment at $4 \times CO_2$. The Greenland ice sheet would be half its size in 850 years (3.5 m of sea level rise) and reduce to a small ice cap in the eastern mountains within 3,000 years. Local climate feedbacks indicate that the ice sheet will melt slower than expected by offline simulations with the same forcing.

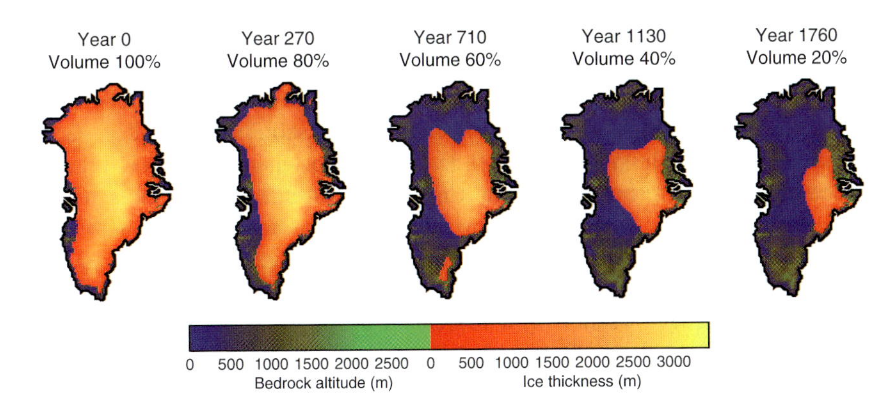

Fig. 15.9 The state of the Greenland ice sheet during various stages of its decline under $4\times CO_2$ simulated by the HadCM3 model coupled with a high-resolution ice sheet model (After Ridley et al. 2005; figure courtesy of Jonathan Gregory)

The reversibility of the decline has been further investigated by reverting to pre-industrial levels of CO_2 during various stages of the ice sheet decline. These experiments, conducted off-line, have indicated that the ice sheet can return (over ~10,000 years) to its current state if it has not lost more than 30% of its mass. It is unlikely to recover, unless temperatures are colder than under pre-industrial concentrations of CO_2, if 70% or more of the ice sheet has been lost. Stable intermediate ice sheets may form, during the climatic cooling, for ice sheets between 30% and 70% of the current mass.

Increasing meltwater flux from the Greenland ice sheet could potentially accelerate the weakening of the THC under global warming conditions. Depending on the scenario, model studies show flux rates between 0.01 Sv (Fichefet et al. 2003) and 0.1 Sv (Ridley et al. 2005) that would dilute the North Atlantic surface waters and potentially suppress deep water formation. However, the simulated THC weakening appears to be highly model-dependent. Fichefet et al. (2003) found a strong and abrupt weakening of the AMOC at the end of the 21st century. In contrast, Ridley et al. (2005) analyzed a climate with four times the pre-industrial CO_2 level and found relatively minor changes in the THC. Jungclaus et al. (2006b) calculated meltwater flux rates from IPCC A1B scenarios and found a flux of about 0.03 Sv by 2100. In a sensitivity experiment this additional meltwater input caused only a slight acceleration of the THC weakening.

15.6 Freshwater Content Changes

Observational estimates indicate an average Arctic Ocean fresh water storage of 100,000 km^3 with respect to an Arctic mean reference salinity of 34.8 psu. Most of this is stored in liquid form. Solid sea ice is thought to contribute roughly in the order of 30,000 km^3. The largest source comes from the combination of surface flux and river discharge, followed by an inflow of relatively low salinity water through the Bering Strait (see Fig. 15.1). The largest individual sink is Fram Strait sea ice export. Recent studies, however, indicate that the CAA outflow might be of the same order (Prinsenberg and Hamilton 2004). From time to time, large pulses of freshwater (in the order of several thousand km^3/year) move from the Arctic to the North Atlantic in the form of the so-called Great Salinity Anomalies (GSA, Dickson et al. 1988; Belkin et al. 1998), leading to substantial changes in regional surface salinity and freshwater storage (Curry and Mauritzen 2005; Peterson et al. 2006). There have been various modelling studies using ocean-only or ocean–sea–ice models forced with atmospheric reanalysis data to investigate driven mechanisms of Arctic freshwater content changes (e.g. Zhang and Zhang 2001; Zhang et al. 2003; Hakkinen and Proshutinsky 2004 and Karcher et al. 2005), but here we mainly focus on modelling efforts with fully coupled climate models.

Table 15.1 shows the Arctic mean freshwater budgets and Table 15.2 the volume transports for the 20th, 21st and 22nd centuries simulated by the ECHAM5/MPIOM model (Haak et al. 2005; Koenigk et al. 2007). Terms are averages for each time window and error bars are standard deviations of the annual means. Percentages

Table 15.1 Simulated Arctic Ocean fresh water budget (km^3 year^{-1}) with S_{ref} = 34.80 psu

	1860–1999			2070–2099			2170–2199		
Period	Solid	Liquid	Total	Solid	Liquid	Total	Solid	Liquid	Total
Bering Strait inflow	−82 ± 137	1412 ± 206	1330 ± 317	−16 ± 48 **−81%**	1836 ± 246 **+30%**	1820 ± 278 **+37%**	−13 ± 27 **−84%**	2104 ± 355 **+49%**	2090 ± 367 **+57%**
Barents Shelf inflow	−161 ± 111	−1199 ± 216	−1361 ± 210	0 ± 2 **−100%**	−1169 ± 264 **−2%**	−1168 ± 264 **−14%**	0 ± 0 **−100%**	−694 ± 406 **−42%**	−694 ± 406 **−49%**
CAA	−495 ± 140	−1975 ± 388	−2470 ± 511	−187 ± 39 **−62%**	−3461 ± 453 **+75%**	−3648 ± 469 **+47%**	−234 ± 35 **−52%**	−4481 ± 611 **+126%**	−4615 ± 634 **+86%**
Fram Strait	−2543 ± 555	−1483 ± 527	−4026 ± 785	−317 ± 92 **−87%**	−3840 ± 567 **+158%**	−4157 ± 612 **+3%**	−195 ± 75 **−92%**	−5591 ± 685 **+277%**	−5787 ± 738 **+43%**
P − E + R	–	–	6718 ± 279	–	–	8268 ± 328 **+23%**	–	–	8891 ± 344 **+32%**
NET	–	–	191	–	–	1115	–	–	−115

Table 15.2 Simulated Arctic Ocean volume transports (mSv)

	1860–1999		2060–2099		2160–2199	
Period	Solid	Liquid	Solid	Liquid	Solid	Liquid
Bering Strait inflow	−3.6 ± 5.1	404 ± 53	−0.6 ± 2.0 **−83%**	379 ± 60 **−6%**	−0.5 ± 1 **−86%**	383 ± 73 **−5%**
Barents Shelf inflow	−6.4 ± 4.3	3723 ± 258	0 ± 0 **−100%**	4623 ± 386 **+24%**	0 ± 0 **−100%**	4733 ± 453 **+27%**
CAA	−17.5 ± 4.5	−928 ± 209	−7.0 ± 1.4 **−57%**	−1012 ± 172 **+9%**	−4.9 ± 1.3 **−72%**	−1195 ± 169 **+29%**
Fram Strait	−97.5 ± 20.7	−3269 ± 317	−13.9 ± 5.8 **−86%**	−4226 ± 450 **+29%**	−7.2 ± 2.7 **−92%**	−4191 ± 507 **+28%**
NET	−125	−70	−21.5	−236	−12.6	−270

give the relative change in respect to the 1860–1999 period. The model was forced with observed CO_2 and aerosol concentrations up to year 2000 and then follows the IPCC A1B scenario to 2100 with stabilisation thereafter. The model produces a realistic mean state and variability for the pre-industrial control integration and for the 20th century (Jungclaus et al. 2006a). The relative importance of individual terms is consistent with observations (see Fig. 15.1). Bering Strait inflow and transport of Atlantic water via the Barents Shelf (positive) vary little with a standard deviation of only 200–300 km^3/year. Exports (negative) through Fram Strait and the CAA show much larger fluctuations with standard deviations of ~800 km^3/year and ~500 km^3/year respectively. CAA transport is about 70% of that from Fram Strait.

The simulated mass flux through Bering Strait is by 50% too small compared to the 800 mSv from observational estimates (Aagaard and Carmack 1989). The mass transport through the Canadian Archipelago is weaker than observational estimates of 1,500–2,000 mSv (Prinsenberg and Hamilton 2004). Fram Strait mass flux seems to be about right compared to observational estimates of 2,000–4,000 mSv (Fahrbach et al. 2003). The simulated Barents Shelf Inflow agrees with the estimates of 3,100 mSv (Blindheim 1989). Note that the later three transports increase in the future projection experiments by roughly 20–30%. Associated is an increase in sea level difference between the Artic and the North Atlantic, indicating an intensified communication between the two ocean basins under global warming conditions.

Figure 15.10a shows the time evolution of the different terms making up the Arctic hydrological budget simulated by ECHAM5/MPIOM. We do not see a significant trend in total Arctic freshwater content from 1860 to 1999. However, there is a clear upward trend during the 21st century mainly due to the intensification of atmospheric hydrological cycle under increased greenhouse gas warming. Surface P–E plus river discharge increase by 20% from ~6,700 to ~8,200 km^3/year, while Bering Strait fresh inflow increases by 30%. Note that the mass flux through Bering Strait does not change (see Table 15.2). In total this leads to a 25% increase in total freshwater input to the Arctic from ~8,000 to ~10,000 km^3/year. Freshwater export through the CAA almost doubles (from ~2,700 to 5,000 km^3/year), while transports via Fram Strait and the Barents Shelf remain fairly constant. By the late 21st century, freshwater export through the CAA reaches the same level of Fram Strait. Such an increase is mostly due to a fresher halocline. However the increased sea level difference and the disappearance of sea ice that is blocking the narrow strait results in approximately a 20% increase in mass flux through CAA. By 2070, the Arctic becomes ice-free in summer with multi-year gaps of total ice disappearance. The reduction of sea ice also suppresses the variability of freshwater exports and the total storage. On interannual to decadal timescales, Haak et al. (2005) found that the combination of the cumulative freshwater transport anomalies through the CAA and Fram Strait dominate the Arctic freshwater content variations (see Fig. 15.10b). Associated with shifts in large-scale surface pressure patterns and winds, a relationship of roughly antiphase between CAA export and Fram Strait export occurs for several large events during the simulated period.

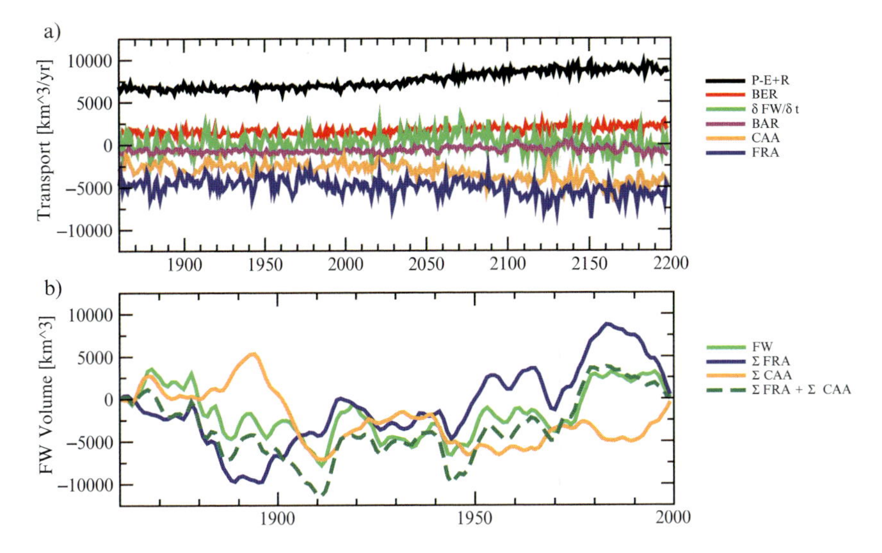

Fig. 15.10 (a) Evolution of the separate terms of the Arctic freshwater budget simulated by the ECHAM5/MPIOM and (b) time integrated freshwater transport anomalies for the 20th century. The model was forced with transient CO2 and aerosols from pre-industrial to the end of the 20th century and then following the IPCC A1B scenario till 2100, after which forcing was kept constant. Freshwater transport is calculated with a reference salinity of 34.8psu (After Haak et al. 2005)

Changes in both surface sources (P − E + R and sea ice melting) and lateral transports can lead to freshwater content changes in the Arctic Ocean. Surface fluxes are likely to be the dominant factors for future changes as global warming and the atmospheric hydrological cycle intensify. For the 20th century changes, lateral transports may have played a more important role (Wu et al. 2007). Proshutinsky et al. (2002) suggested that anomalous freshwater storage within the anticyclonic Beaufort Gyre can be potentially larger than changes in river runoff and sea ice export. Häkkinen and Proshutinsky (2004) and Hátún et al. (2005) emphasise the contribution of the Atlantic water inflow. Jungclaus et al. (2005), Wu and Wood (2007) have all realised the importance of freshwater exchange between the Arctic and the subpolar North Atlantic in affecting basin scale freshwater content changes. Wu and Wood (2007) have shown that anomalous atmospheric conditions such as the winter of 1971/72 may cause a circulation regime change within the Arctic/subpolar North Atlantic Ocean system that has a long lasting effect on water exchanges through the Greenland–Iceland–Scotland (GIS) straits. Freshwater redistribution following such circulation changes can lead to substantial freshwater content changes comparable to the recent freshening trend reported by Dickson et al. (2002), Curry and Mauritzen (2005). The GSA can now be well simulated in climate models (Haak et al. 2003; Koenigk et al. 2006a; Wadley and Bigg 2004, 2006). Haak et al. (2003) suggested from their model simulations that the GSA is linked to anomalous sea ice export through Fram Strait driven by anomalous atmospheric circulation. On the other hand Houghton and Visbek (2002), Wadley and Bigg (2006) have recently questioned the advective nature of the GSA.

15.7 Conclusions

Increasing greenhouse gas concentrations in the atmosphere have a disproportionate impact on polar climates relative to global warming. Enhanced warming due to polar amplification, first pointed out by Manabe and Stouffer (1980), is now a well recognised phenomenon (see, e.g., Holland and Bitz 2003; ACIA 2005). Additional freshwater input due to increased moisture transport from the subtropics and river discharges has made another distinction for the polar regions under an accelerating global hydrological cycle (Wu et al. 2005; Stocker and Raible 2005). Changes in the global freshwater cycle will directly affect the distribution of water resources worldwide (see, Oki and Kanae 2006, for a recent review). Changing patterns and severity of droughts and floods will be parts of its climate impact on regional scales. Extra freshwater input into the Arctic/subarctic oceans has another worrying consequence on the climate system. This is its potential of diluting the northern polar oceans where deep convection occurs and the associated weakening the Atlantic thermohaline circulation (THC, e.g. Vellinga and Wood 2002; Wu et al. 2004; Curry and Mauritzen 2005). Moreover, meltwater input from a disintegrating Greenland ice sheet could further accelerate the THC weakening (Fichefet et al. 2003; Jungclaus et al. 2006b).

There are already signs of systematic changes in the Arctic/subarctic freshwater cycle (Dickson et al. 2002; Curry et al. 2003; Curry and Mauritzen 2005; Peterson et al. 2006). In order to understand and attribute the observed changes to different causes, long climate records and comprehensive computer models are needed to expand our research into further depth and accuracy. Having described the progress in simulating the terms in the Arctic hydrological budget above, it is clear that there are weak areas in both observations and modelling. Because the polar regions are highly sensitive parts of the global hydrological cycle, we need observations to be more reliable, continuous with better coverage for monitoring global changes. We need climate models to resolve more detailed processes and feedbacks in simulating precipitation, evaporation, sea ice and land hydrology. We need better estimates of the magnitude and variability of the Arctic/subarctic hydrological budgets. As modellers, we would like to use increasingly more observational measurements to validate and constrain climate model simulations. In the meantime, we would also like to use our models to help understand the mechanisms of observed variability and change, to attribute them to different possible causes, and to use model projections to guide future observational efforts.

There are competing sources of freshwater adding to the Arctic/subarctic oceans as global warming continues. At the present, there are considerable uncertainties even for the climatological means for the individual contributors from both observational estimates and climate model simulations (see Fig. 15.1). Large differences also exist between model simulated budget terms and observationally based estimates, as well as among different models. Those uncertainties in the means will undoubtedly overshadow any predicted budget and trends. We should aim to achieve an observationally constrained, multi-model ensemble prediction of the Arctic freshwater budget such as the one shown in Table 15.1 in the near future. It will enable us to answer the following questions: What is the likely upper bound of freshwater input into the Arctic/subarctic oceans? How much of that is likely to be realised over the next 50 or 100 years? Which component is likely to play a leading role? Besides, how can we best use the Arctic as an indicator for monitoring the global hydrological cycle? To complete these tasks will require concerted efforts from both the observational and the modelling communities.

Acknowledgements Peili Wu and Richard Wood are funded by the UK Department of Environment, Food and Rural Affairs under the Climate Prediction Programme (PECD/7/12/37). Tore Furevik has been supported by the Norwegian Research Council through the NoClim project. We thank Jeff Ridley and Jonathan Gregory for helpful comments on the review of Greenland ice sheet.

References

Aagaard K, Carmack EC (1989) The role of sea ice and other fresh-water in the Arctic circulation. J Geophys Res, 4 (C10):14485–14498.

ACIA (2005) Arctic Climate Impact Assessment: Scientific report. Cambridge University Press, Cambridge.

Allen MR, Ingram WJ (2002) Constraints on the future changes in climate and the hydrological cycle. Nature, 419:224–232.

Arnell NW (2005) Implications of climate change for freshwater inflows to the Arctic Ocean. J Geophys Res, 110:D07105.

Arzel O, Fichefet T, Goosse H (2006) Sea ice evolution over the 20th and 21st centuries as simulated by current AOGCMs. Ocean Modelling, 12 (3–4):401–415.

Belkin IM, Levitus S, Antonov J (1998) Great salinity anomalies in the North Atlantic, Prog Oceanogr, 41:1–68

Bethke I, Furevik T, Drange H (2006) Towards a more saline North Atlantic and a fresher Arctic under global warming. Geophys Res Lett, 33:L21712.

Blindheim J (1989) Cascading of Barents Sea bottom water into the Norwegian Sea, Rapports et Procès-Verbaux des Réunions/ Conseil Permanent International pour l'Exploration del la Mer, 17:161–189

Carmack EC (2000) The Arctic oceans freshwater budget: Sources, storage and export. In: Lewis EL et al. (eds.), *The Freshwater Budget of the Artic Ocean*, Kluwer, Dordrecht, The Netherlands, pp 91–126.

Chen JL, Wilson CR, Tapley BD (2006) Satellite gravity measurements confirm accelerated melting of Greenland ice sheet. Science, 313:1958–1960.

Comiso JC (2002) A rapidly declining perennial ice cover in the Arctic. Geophys Res Lett, 29:1956.

Cullather RI, Bromwich DH, Serreze MC (2000) The atmospheric hydrologic cycle over the Arctic basin from reanalyses. Part I: Comparison with observations and previous studies. J Clim, 13:923–937.

Curry R, Mauritzen C (2005) Dilution of northern North Atlantic Ocean in recent decades. Science, 308:1772–1774.

Curry R, Dickson B, Yashayaev I (2003) A change in the freshwater balance of the Atlantic Ocean over the past four decades. Nature, 426:826–829.

Dèry SJ, Wood EF (2005) Decreasing river discharge in northern Canada. Geophys Res Lett, 32: L10401.

Dèry SJ, Stieglitz M, McKenna EC, Wood EF (2005) Characteristics and trends of river discharge into Hudson, James, and Ungava Bays, 1964–2000. J Clim, 18:2540–2557.

Dickson B, Yashayaev I, Meincke J, Turrel B, Dye S, Holfort J (2002) Rapid freshening of the deep North Atlantic Ocean over the past four decades. Nature, 416:832–837.

Dickson R, Rudels B, Dye S, Karcher M, Meincke J, Yashayaev I (2007) Current estimates of freshwater flux through Arctic and subarctic seas. Prog Oceanogr, 73:210–230.

Dickson RR, Meincke J, Malmberg SA, Lee AJ (1988) The Great Salinity Anomaly in the northern North Atlantic. Prog Oceanogr, 20:103–151.

Fahrbach E, Schauer U, Rohard G, Meincke J, Osterhus S (2003) How to measure oceanic fluxes from North Atlantic through Fram Strait. ASOF Newsletter, 1: 3–7 (unpublished manuscript).

Fichefet T, Poncin C, Goosse H, Huybrechts P, Janssens I, Le Treut H (2003) Implications of changes in freshwater flux from the Greenland ice sheet for the climate of the 21st century. Geophys Res Lett, 30:1911.

Furevik T, Bentson M, Drange H, Kindem I, Kvamsto N, Sorteberg A (2003) Description and evaluation of the Bergen Climate Model: ARPEGE coupled with MICOM. Climate Dyn, 21(1):27–51.

Gregory JM, Huybrechts P (2006) Ice-sheet contributions to future sea-level change. Phil Trans R Soc Lond, A(364):1709–1731.

Gregory JM, Stott PA, Cresswell DJ, Rayner NA, Gordon C, Sexton DMH (2002) Recent and future changes in Arctic sea ice simulated by the HadCM3 AOGCM. Geophys Res Lett, 29:2175.

Gregory JM, Huybrechts P, Raper SCB (2004) Threatened loss of the Greenland ice-sheet. Nature, 428:616.

Haak H, Jungclaus J, Mikolajewicz U, Latif M (2003) Formation and propagation of great salinity anomalies. Geophys Res Lett, 30:1473.

Haak H, Jungclaus JH, Koenigk T, Sein D, Mikolajewicz U (2005) Arctic Ocean freshwater budget variability. ASOF Newsletter, 3:6–8 (unpublished manuscript).

Hanna E, Huybrechts P, Mote TL (2002) Surface mass balance of the Greenland ice sheet from climate-analysis data and accumulation/runoff models. Ann Glaciol, 35:67–72.

Hakkinen S, Proshutinsky A (2004) Freshwater content variability in the Arctic Ocean. J Geophys Res, 109 (C3):C03051.

Hátún H, Sandø AB, Drange H, Hansen B, Valdimarsson H (2005) Influence of the Atlantic subpolar gyre on the thermohaline circulation. Science, 16:1841–1844.

Holland MM, Bitz CM (2003) Polar amplification of climate change in the Coupled Model Intercomparison Project. Climate Dyn, 21:221–232.

Houghton RW, Visbek M (2002) Qasi-decadal salinity fluctuations in the Labrador Sea. J Phys Oceanogr, 32:687–701.

IPCC (2001) Climate Change. The Scientific Basis. Houghton JT, Ding Y, Griggs DJ, Noguer M, van der Linden PJ, Dai X (eds.), Cambridge University Press, Cambridge, pp 525–582.

Johannessen OM, Shalina EV, Miles MW (1999) Satellite evidence for an Arctic sea ice cover in transformation. Science, 286:1937–1939.

Johannessen OM et al. (2004) Arctic climate change: Observed and modelled temperature and sea ice variability. Tellus, 56A:328–341.

Johannessen OM, Khvorostovsky K, Miles MW, Bobylev LP (2005) Recent ice-sheet growth in the interior of Greenland. Science, 310:1013–1016.

Johns TC et al. (2006) The new Hadley Centre climate model (HadGEM1): Evaluation of coupled simulations. J Clim, 19:1327–1353.

Jungclaus JH, Haak H, Latif M, Mikolajewicz U (2005) Arctic-North Atlantic interactions and multidecadal variability of the meridional overturning circulation. J Clim, 18:4013–4031.

Jungclaus JH, Botzet M, Haak H, Keenlyside N, Luo JJ, Latif M, Marotzke J, Mikolajewicz U, Roeckner E (2006a) Ocean circulation and tropical variability in the coupled model ECHAM5/MPI-OM. J Clim, 19:3952–3972.

Jungclaus JH, Haak H, Esch M, Roeckner E, Marotzke J (2006b) Will Greenland melting halt the thermohaline circulation? Geophys Res Lett, 33:L17708.

Karcher M, Gerdes R, Kauker F, Koberle C, Yashayaev I (2005) Arctic Ocean change heralds North Atlantic freshening. Geophys Res Lett, 32, L21606.

Kattsov VM, Walsh JE (2000) Twentieth-century trends of arctic precipitation from observational data and a climate model simulation. J Clim, 13:1362–1370.

Krabill W et al. (2004) Greenland ice sheet: increased coastal thinning. Geophys Res Lett, 31, L24402.

Koenigk TU, Mikolajewicz U, Haak H, Jungclaus J (2006a) Variability of Fram Strait sea ice export: Causes, impacts and feedbacks in a coupled climate model. Climate Dyn, 26(1):17–34.

Koenigk TU, Mikolajewicz U, Haak H, Jungclaus J (2007) Arctic freshwater export and its impact on climate in the 20th and 21st Century. J Geophys Res, 112, G04S41, doi:10.1029/2006JG000274.

Lambert FH, Stott PA, Allen MR, Palmer MA (2004) Detection and attribution of changes in 20th century land precipitation. Geophys Res Lett, 31, L10203.

Lammers AR, Shiklomanov AI, Vorosmarty CJ, Fekete BM, Peterson BJ (2001) Assessment of contemporary Arctic river runoff based on observational discharge records. J Geophys Res, 106(D4):3321–3334.

Lunt DJ, de Noblet-Ducoudre N, Charbit S (2004) Effects of a melted Greenland ice sheet on climate, vegetation, and the cryosphere. Climate Dyn, 23 (7–8): 679–694.

Manabe S, Stouffer RJ (1980) Sensitivity of a global climate model to an increase of CO2 concentration in the atmosphere. J Geophys Res, 85:5529–5554.

McClelland JW, Holmes RM, Peterson BJ (2004) Increasing river discharge in the Eurasian Arctic: Consideration of dams, permafrost thaw, and fires as potential agents of change. J Geophys Res, 109:D18102.

McClelland JW, Dèry SJ, Peterson BJ, Holmes RM, Wood EF (2006) A pan-arctic evaluation of changes in river discharge during the latter half of the 20th century. Geophys Res Lett, 33: L06715.

Oki T, Kanae S (2006) Global hydrological cycles and world water resources. Science, 313:1068–1072.

Pardaens AK, Banks HT, Gregory JM, Rowntree PR (2003) Freshwater transports in HadCM3. Climate Dyn, 21:177–195.

Peterson BJ, Holmes RM, McClelland JW, Vorosmarty CJ, Lammers RB, Shiklomanov AI, Shiknomanov IA, Rahmstorf S (2002) Increasing river discharge to the Arctic Ocean. Science, 298:2171–2173.

Peterson BJ, McClelland J, Curry R, Holmes RM,Walsh JE, Aagaard K (2006) Trajectory shifts in the Arctic and subarctic freshwater cycle. Science, 313:1061–1066.

Prinsenberg SJ, Hamilton J (2004) The Oceanic fluxes through Lancaster Sound of the Canadian Archipelago, ASOF Newsletter No 2 (unpublished manuscript).

Proshutinsky A, Bourke R, MacLaughlin F (2002) The role of the Beaufort gyre in Arctic climate variability: seasonal to decadal climate scales. Geophys Res Lett, 29:2100.

Rayner NA, Parker DE, Horton EB, Folland CK, Alexander V, Rowell DP, Kent EC, Kaplan A (2003) Global analyses of sea surface temperature, sea ice, and night marine air temperature since the late nineteenth century. J Geophys Res, 108(D14):4407.

Ridley JK, Huybrechts P, Gregory JM, Lowe JA (2005) Elimination of the Greenland ice sheet in a high CO_2 climate. J Clim, 18:3409–3427.

Rignot E, Kanagaratnam P (2006) Changes in the velocity structure of the Greenland ice sheet. Science, 311:986–990.

Rothrock DA,Yu Y, Maykut GA (1999) Thinning of the Arctic sea ice cover. Geophys Res Lett, 26:3469–3472.

Serreze MC, Barry RG (2005) The Arctic Climate System. Cambridge University Press, Cambridge.

Serreze MC, Barrett AP, Lo F (2005) Northern high-latitude precipitation as depicted by atmospheric reanalyses and satellite retrievals. Mon Wea Rev, 133:3407–3430.

Serreze MC, Barrett AP, Slater AG, Woodgate RA, Aagaard K, Lammers RB, Steele M, Moritz M, Meredith M, Lee CM (2006) The large-scale freshwater cycle of the Arctic. J Geophys Res, 111 (C11), Art No C11010.

Stocker TF, Raible CC (2005) Water cycle shifts gear. Nature, 434:830–832.

Stott PA, Tett S, Jones G, Allen M, Mitchell J, Jenkins GJ (2000) External control of 20th century temperature by natural and anthropogenic focings. Science, 290:2133–2137.

Toniazzo T, Gregory JM, Huybrechts P (2004) Climatic impact of a Greenland deglaciation and its possible irreversibility. J Clim, 17:21–33.

Vellinga M, Wood RA (2002) Global climatic impacts of a collapse of the Atlantic thermohaline circulation. Climatic Change, 54:251–267.

Wadley MR, Bigg GR (2004) "Great Salinity Anomalies" in a coupled climate model. Geophys Res Lett, 31, L18302.

Wadley MR, Bigg GR (2006) Are "Great Salinity Anomalies" advective? J Clim, 19:1080–1088.

Walsh JE, Kattsov VM, Chapman WL, Govorkova V, Pavlova T (2002) Comparison of Arctic climate simulations by uncoupled and coupled global models. J Clim, 19:1429–1446.

Winsor P (2001) Arctic sea ice thickness remained constant during the 1990s. Geophys Res Lett, 28:1039–1041.

Woodgate RA, Aagaard K (2005) Revising the Bering Strait freshwater flux into the Arctic Ocean. Geophys Res Lett, 32:L02602.

Wu P, Wood R (2007) Convection induced long term freshening of the subpolar North Atlantic Ocean. submitted.

Wu P, Wood R, Stott P (2004) Does the recent freshening trend in the North Atlantic indicate a weakening thermohaline circulation? Geophys Res Lett, 31:L02301.

Wu, P, Wood R, Stott P (2005) Human influence on increasing Arctic river discharges. Geophys Res Lett, 32:L02703.

Wu, P, Wood R, Stott P, Jones GS (2007) Deep North Atlantic freshening simulated in a coupled climate model. Prog Oceanogr, 73:370–383.

Zhang X, Walsh JE (2006) Towards a seasonally ice-covered Arctic Ocean: Scenarios from the IPCC AR4 model simulations. J Clim, 19:1730–1747.

Zhang X, Zhang J (2001) Heat and freshwater budgets and pathways in the Arctic Mediterranean in a coupled ocean/sea ice model, J Oceanogr, 57:207–234.

Zhang X, Ikeda M, Walsh JE (2003) Arctic sea ice and freshwater changes driven by the atmospheric leading mode in a coupled sea ice-ocean model. J Clim, 16:2159–2177.

Chapter 16
Is the Global Conveyor Belt Threatened by Arctic Ocean Fresh Water Outflow?

E. Peter Jones[1] and Leif G. Anderson[2]

16.1 Introduction

Understanding climate and climate change is a main motive for determining fresh water budget of the Arctic Ocean, specifically the sources, distributions and pathways of fresh water. Most fresh water within the Arctic Ocean occurs as a result of there being more evaporation than precipitation in the Atlantic Ocean. Much of the excess evaporation from the Atlantic Ocean falls as rain into the Pacific Ocean and into river drainage basins that feed into both the Pacific and Arctic Oceans. The climate change concern is that, in returning to evaporation sites in the Atlantic Ocean, the fresh water passes through regions of deep convection in the Nordic and Labrador seas, the "headwaters" of the Global Conveyor Belt (Fig. 16.1). To quote Aagaard and Carmack (1989), "We find that the present-day Greenland and Iceland seas, and probably also the Labrador Sea, are rather delicately poised with respect to their ability to sustain convection." Under climate change, we can anticipate changes in fresh water fluxes from the Arctic Ocean. A main motive for trying to determine Arctic Ocean fresh water sources and their distributions is to try to assess the vulnerability of the Atlantic thermohaline circulation to such changes. How the fresh water sources are redistributed within the Arctic Ocean together with the place and timing of their exit from the Arctic Ocean are of direct relevance to the development of models giving scenarios of changes, possibly abrupt, in the Atlantic thermohaline circulation (e.g., Rahmstorf 1996). Fortunately, tracers allow us to distinguish among the different sources of fresh water being exported from the Arctic Ocean thereby allowing changes in each to be separately accommodated in climate change model scenarios.

Fresh water in the Arctic Ocean, whose sources are sea ice meltwater, river water, and Pacific water (Pacific water entering the Arctic Ocean is fresher, S ~32,

[1]Bedford Institute of Oceanography, Dartmouth, NS, B2Y 4A2, Canada, e-mail: jonesp@mar.dfo-mpo.gc.ca

[2]Department of Chemistry, Göteborg University, SE-412 96 Göteborg, Sweden, e-mail: leifand@chem.gu.se

R.R. Dickson et al. (eds.), *Arctic–Subarctic Ocean Fluxes*, 385–404

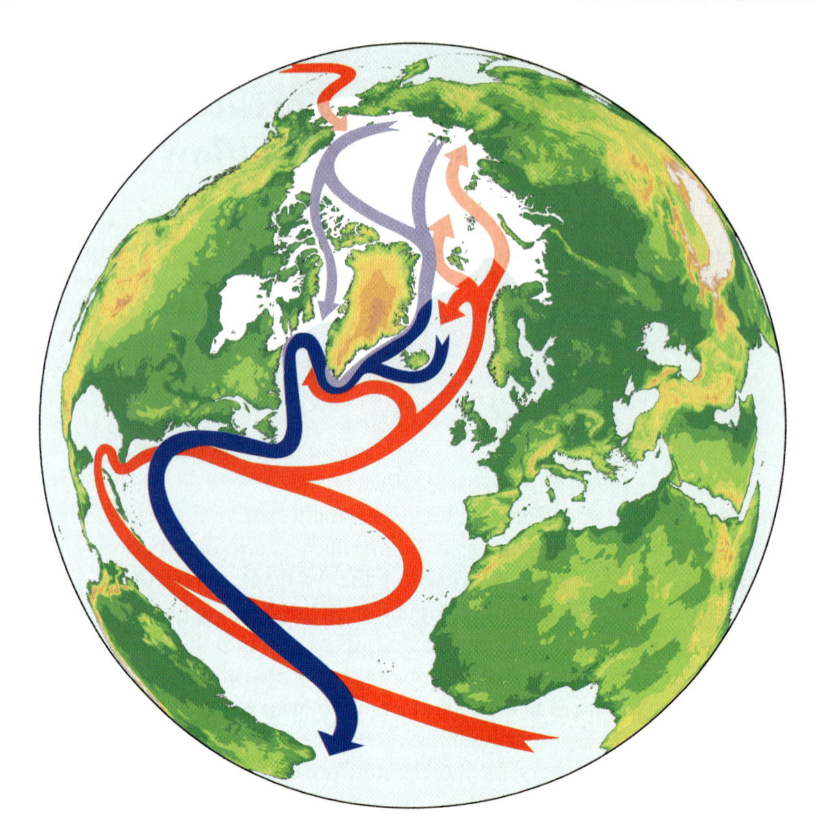

Fig. 16.1 The Global Conveyor Belt emphasizing the Arctic Ocean and the return of fresh water from the Pacific Ocean to the Atlantic Ocean. (After Holloway and Proshutinsky 2007.) Dark red traces the northward flow of warm relatively salty and into the Arctic Ocean. Dark blue traces the deep flow south. Light shades illustrate the flow from the Arctic Ocean and hint at a possible influence on thermohaline circulation

than Atlantic water, S ~34.85), is exported from the Arctic Ocean through Fram Strait and through the Canadian Arctic Archipelago. Each of these fresh water sources will likely respond differently under climate change. While sea ice is also a significant component of the total fresh water export, only the liquid form of fresh water is discussed here.

16.2 Approach

Salinity, alkalinity, and nutrients (nitrate and phosphate) can distinguish among Pacific water, Atlantic water, river water, and sea ice meltwater (Jones et al. 2003, 2006a; Taylor et al. 2003). Three equations using salinity and alkalinity relate the relative fractions of Atlantic water, Pacific water, sea ice meltwater, and river water:

$$f^{AW}+f^{PW}+f^{si}+f^{r}=1 \tag{16.1}$$

$$A_T^{AW}f^{AW}+A_T^{PW}f^{pw}+A_T^{si}f^{si}+A_T^{r}f^{r}=A_T^{m} \tag{16.2}$$

$$S^{AW}f^{AW}+S^{PW}f^{PW}+S^{si}f^{si}=S^{m} \tag{16.3}$$

A_T is alkalinity, f is a water fraction, and the superscripts, *PW, AW, si* and *r*, designate end-members respectively of Pacific water, Atlantic water, sea ice meltwater and river water, and m indicates measured values.

A fourth independent equation is required to determine the fractions f. Nutrient relationships distinguish the relative fraction of Pacific water from the other three components. In the Arctic Ocean, waters of Pacific and Atlantic origin have their own linear phosphate (PO_4) vs. nitrate (NO_3) relationship (Equations 16.4 and 16.5). From limited data, we presume that river water and sea ice meltwater have nitrate-phosphate relationships similar to those of Atlantic source water (Jones et al. 1998, 2003). Thus the fraction of Pacific water, f^{PW}, in a sample with particular nitrate and phosphate concentrations can be determined using the nitrate phosphate relationships:

$$PO_4^{PW}=PW_{slope}\times NO_3^{PW}+PW_{intercept} \tag{16.4}$$

$$PO_4^{AW*}=AW^{*}_{slope}\times NO_3^{AW*}+AW^{*}_{intercept} \tag{16.5}$$

$$f^{PW}=\frac{PO_4^{m}-PO_4^{AW*}}{PO_4^{PW}-PO_4^{AW*}} \tag{16.6}$$

$AW*$ represents Atlantic water together with river water and sea ice meltwater. Once the Pacific water fraction is determined, salinity and alkalinity can distinguish the other three component fractions (Equations 16.1–16.3). The method gives the net sea ice formation or melt, i.e., the fresh water not otherwise accounted for. A "negative" sea ice meltwater value represents sea ice formation.

The total fresh water fraction is the amount of fresh water that must be mixed with Atlantic water to give the measured salinity values and the Pacific fresh water fraction is the fraction of the total fresh water carried by Pacific water, i.e., the amount of fresh water referenced to end-members $S^{PW}=32.0$ and $S^{AW}=34.85$:

$$f_{fresh}^{total}=1-S^{m}/S^{AW} \tag{16.7}$$

$$f_{fresh}^{PW}=f^{PW}(1-S^{PW}/S^{AW}) \tag{16.8}$$

The end-member slopes and intercepts of the nitrate–phosphate relationships (Equations 16.4 and 16.5) as well as the salinity and alkalinity end-members (Equations 16.1–16.3) are determined from regions of the Arctic Ocean where the water type is well determined, i.e., from data inside the Arctic Ocean not far from

Bering Strait at depths where no Atlantic water is present, and from north of St. Anna Trough where no Pacific water is present. Pacific water entering the Arctic Ocean follows two paths (Jones et al. 1998; Shimada et al. 2001; Steele et al. 2004) with slightly different slopes and intercepts of the nitrate–phosphate relationship for each path (Jones et al. 2003). The slightly different relationships give results within the expected precision of this approach.

Uncertainties in calculated fractions can result from uncertainties in end-member values as source waters are defined by a range of values. We believe that the chosen salinity, nutrient, and alkalinity end-member values reasonably well represent Pacific and Atlantic waters, but the different rivers entering the Arctic Ocean can have fairly different alkalinity values. River alkalinity values vary from river to river and season to season (PARTNERS http://ecosystems.mbl.edu/partners/data.html). A recently published representative average river alkalinity value for the Arctic Ocean of 831 ± 100 μmol kg^{-1} (Yamamoto-Kawai and Tanaka 2005) is lower than that appearing in some publications (e.g., Anderson et al. 2004) and lower than the average (~1,100 μmol kg^{-1}) reported in PARTNERS. In an attempt to choose the single alkalinity value best representing river water in the Arctic Ocean, we compared estimates of river water fractions from both the Eurasian and Canadian basins for which both alkalinity and oxygen-18 data are available. We used the approach outlined above but with oxygen-18 end-members (Ekwurzel et al. 2001; Macdonald et al. 1999) in place of alkalinity values. By comparing calculated results using data from three expeditions (Ekwurzel et al. 2001; Macdonald et al. 1999) we found representative alkalinity river water values of 1,000 μmol kg^{-1} give consistent results with oxygen-18 results, the value we chose for this work. With the uncertainties in end-members the computed source water concentrations should be considered somewhat approximate, likely no better than ±0.02. Also ascertaining uncertainties in the calculation of fresh water fractions is not straight-forward as the calculated fractions are interdependent and uncertainties vary according to the magnitude of the calculated fractions (Jones et al. 1998; Ekwurzel et al. 2001; Taylor et al. 2003). We consider the absolute values of higher water fractions, >0.03, to be reasonably valid with uncertainties of ±0.01, while lower fractions, <0.01, may not be reliable.

16.3 Results

Measurements in the Arctic Ocean were made on several expeditions from 1991 through 2005, while those in the Canadian Arctic Archipelago and in the East Greenland Current were made in 1997 and 2002 respectively (Fig. 16.2). All data except those from the East Greenland Current were collected under summer conditions, typically from late July to mid-September. Data from the East Greenland Current were collected in early spring when winter conditions persisted.

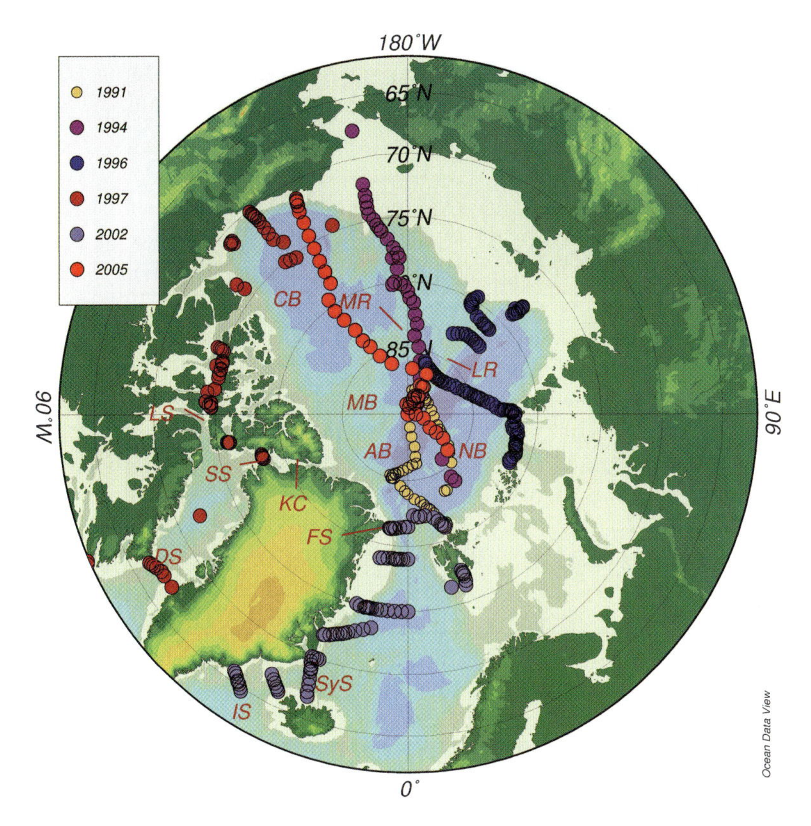

Fig. 16.2 Map showing all stations of this study. Labels indicate Canada Basin (CB), Makarov Basin (MB), Amundsen Basin (AB), Nansen Basin (NB) Mendeleyev Ridge (MR), Lomonosov Ridge (LR), Smith Sound (SS), Davis Strait (DS), Lancaster Sound (LS), Kennedy Channel (KC), Fram Strait (FS), Scoresby Sound (SyS), and Irminger Sea (IS)

We present sections showing vertical distributions of the fresh water components in the Eurasian Basin (1996) and Canadian Basin (2005) (Fig. 16.3a, b) as representative of the Arctic Ocean, in the Canadian Arctic Archipelago (1997) as representative of fresh water exiting through that region (Figs. 16.3c–d), and in the East Greenland Current (2002) as representative of fresh water exiting through Fram Strait (Figs. 16.3e–g). All data are presented in fresh water inventory plots (Fig. 16.4). It should be noted that these data span several years, and, because of variability in water mass distributions (e.g., Anderson et al. 2004; Falck et al. 2005), the results do not give a synoptic view of conditions of fresh water in the Arctic Ocean and exiting from it.

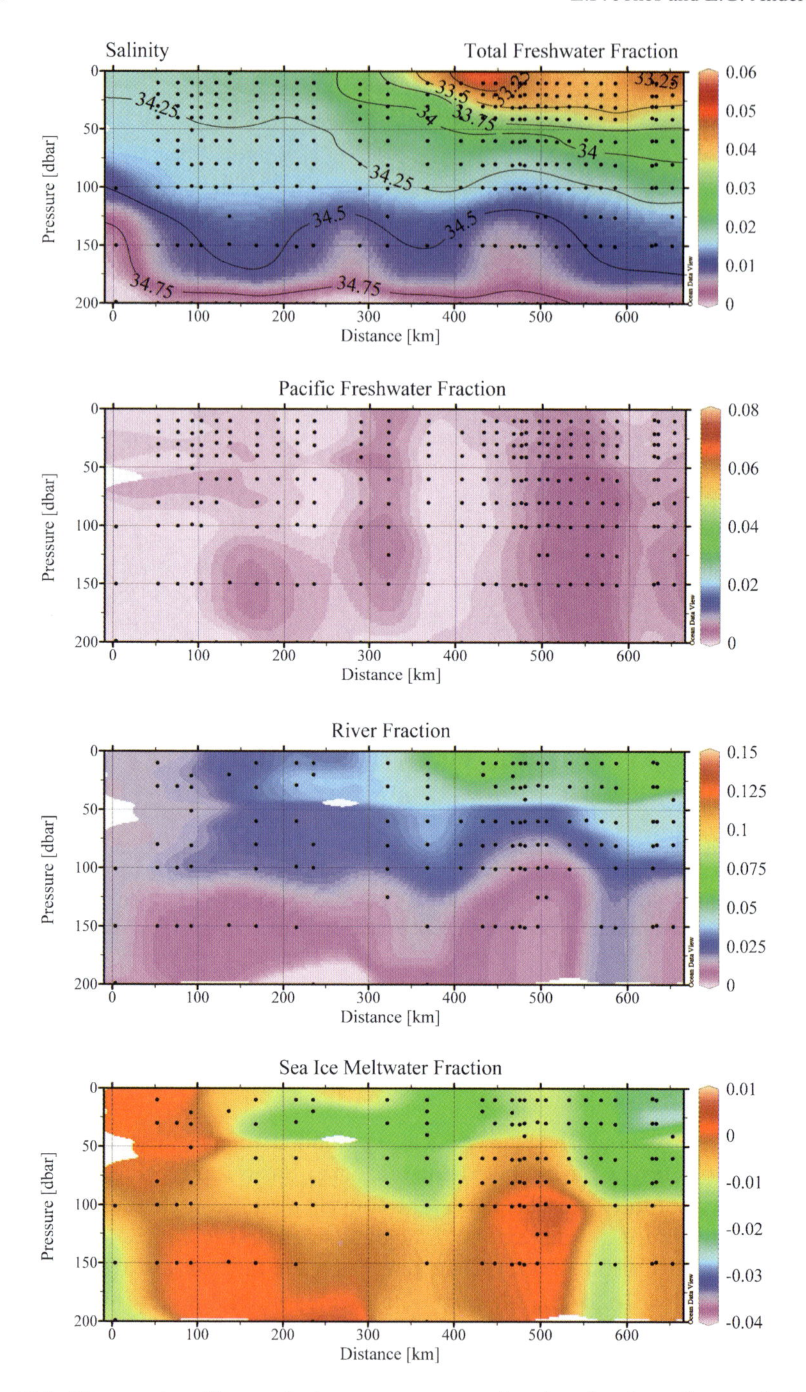

Fig. 16.3 These sections illustrate fresh water components in selected regions. Contours in Total Fresh Water sections represent salinity. See Fig. 16.4 for inventories at stations shown in Fig. 16.2. (a) Eastern Eurasian Basin Section (1996). The section begins on the slope north of the Barents and Kara seas, crosses the Eurasian Basin, and extends into the Makarov Basin

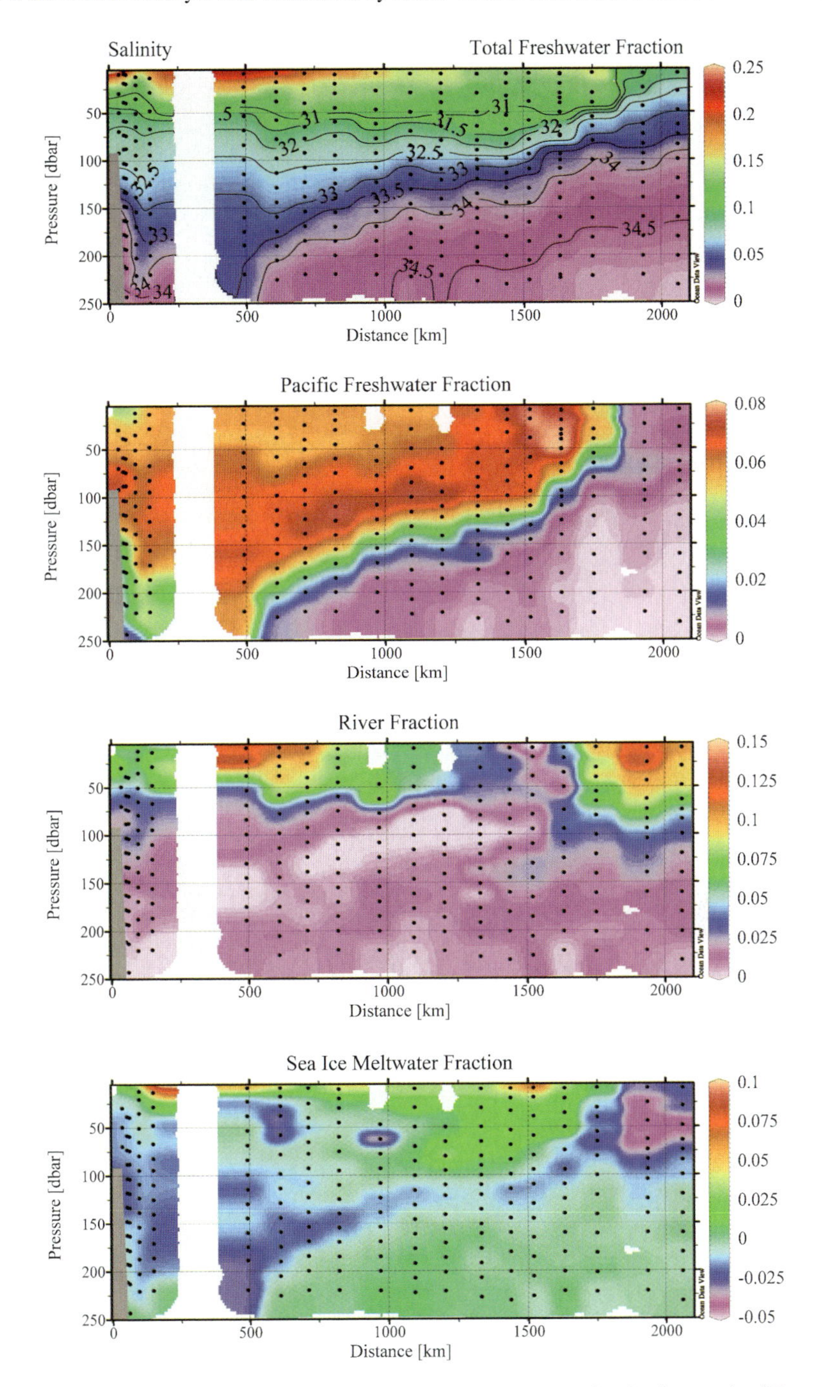

Fig. 16.3 (continued) (b) Canadian Basin Section (2005). The section begins north of Barrow, Alaska, crosses the Canada and Makarov basins, and ends at the Lomonosov Ridge

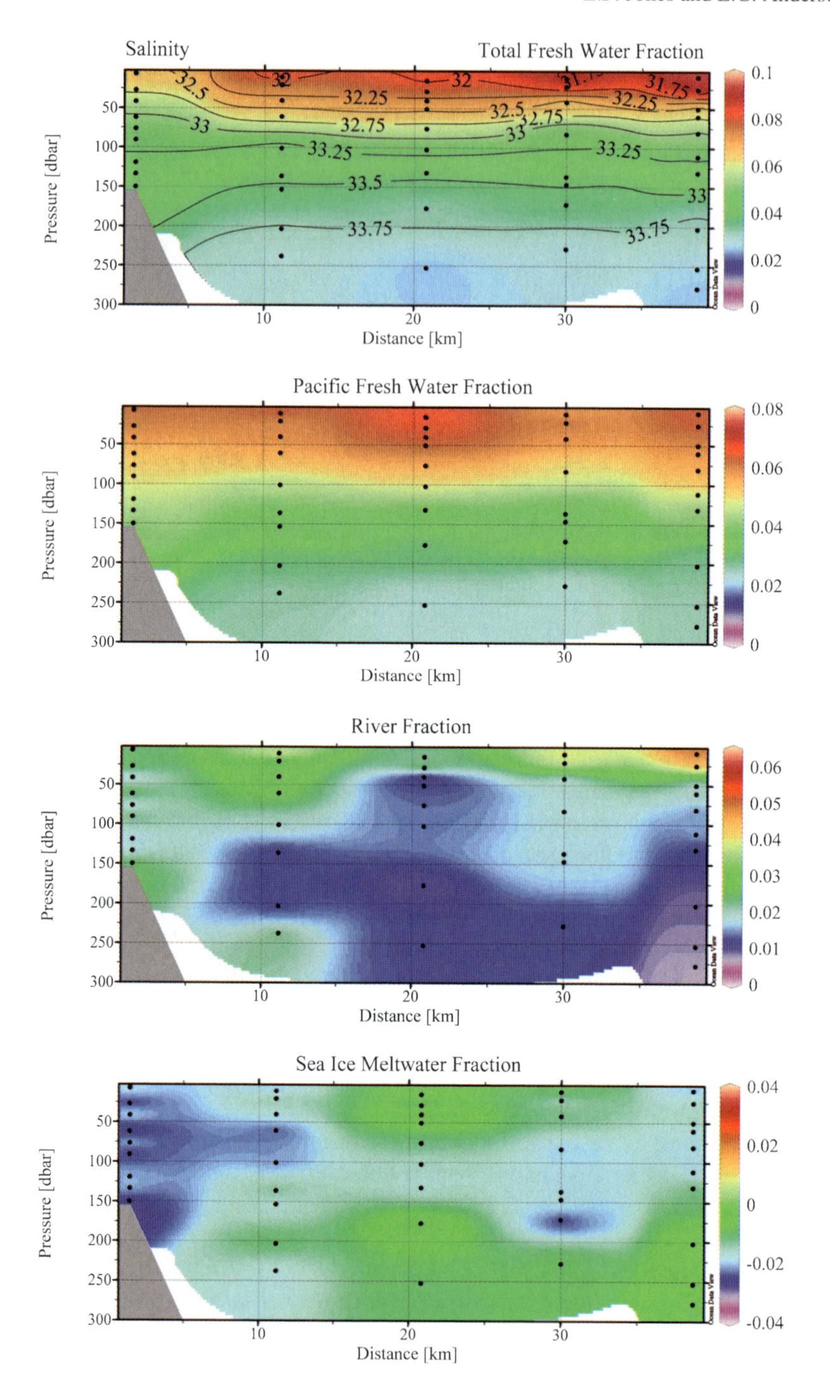

Fig. 16.3 (continued) (c) Smith Sound Section (1997). The section extends across southern Nares Strait

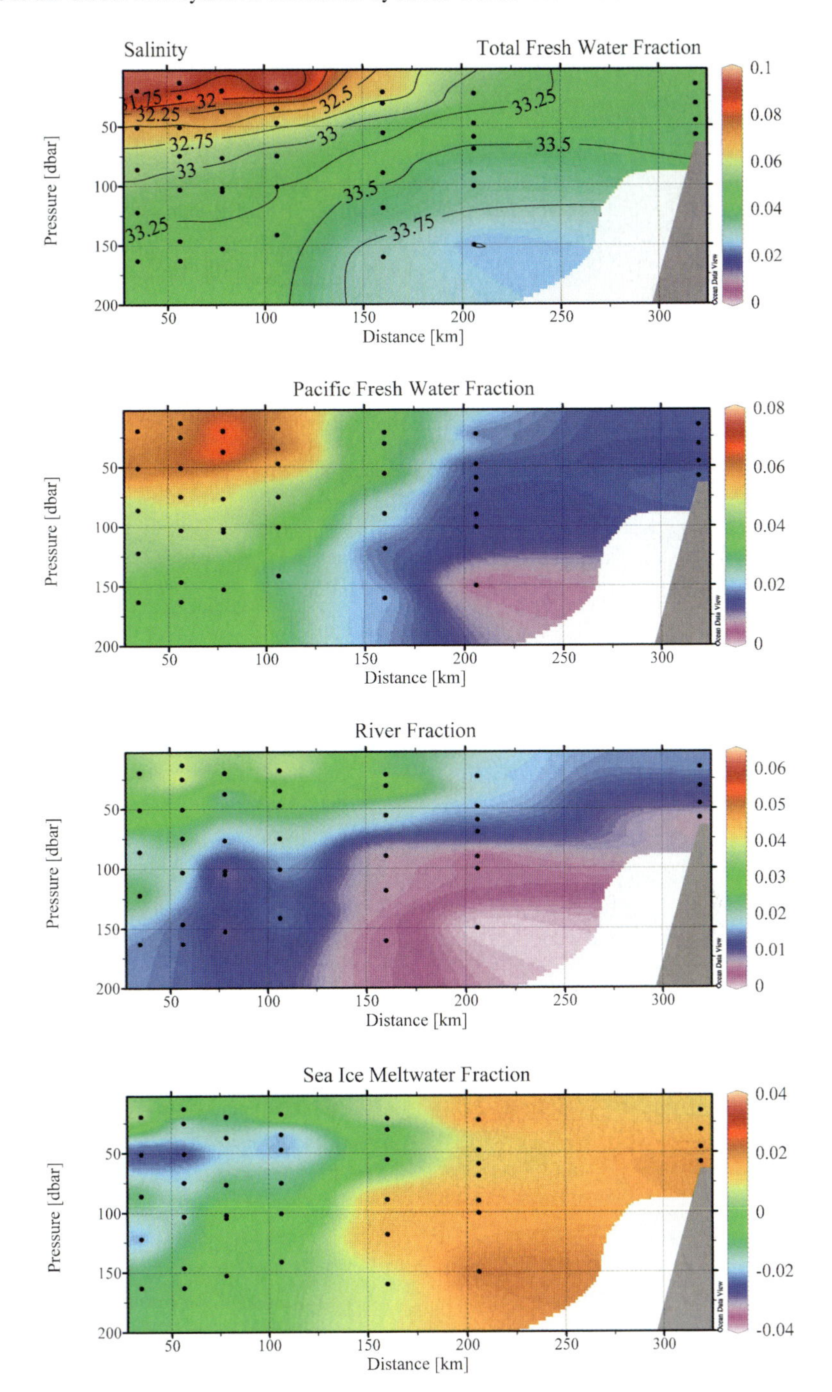

Fig. 16.3 (continued) (d) Davis Strait Section (1997). The section extends across Davis Strait just north of the sill between Baffin Bay and the Labrador Sea

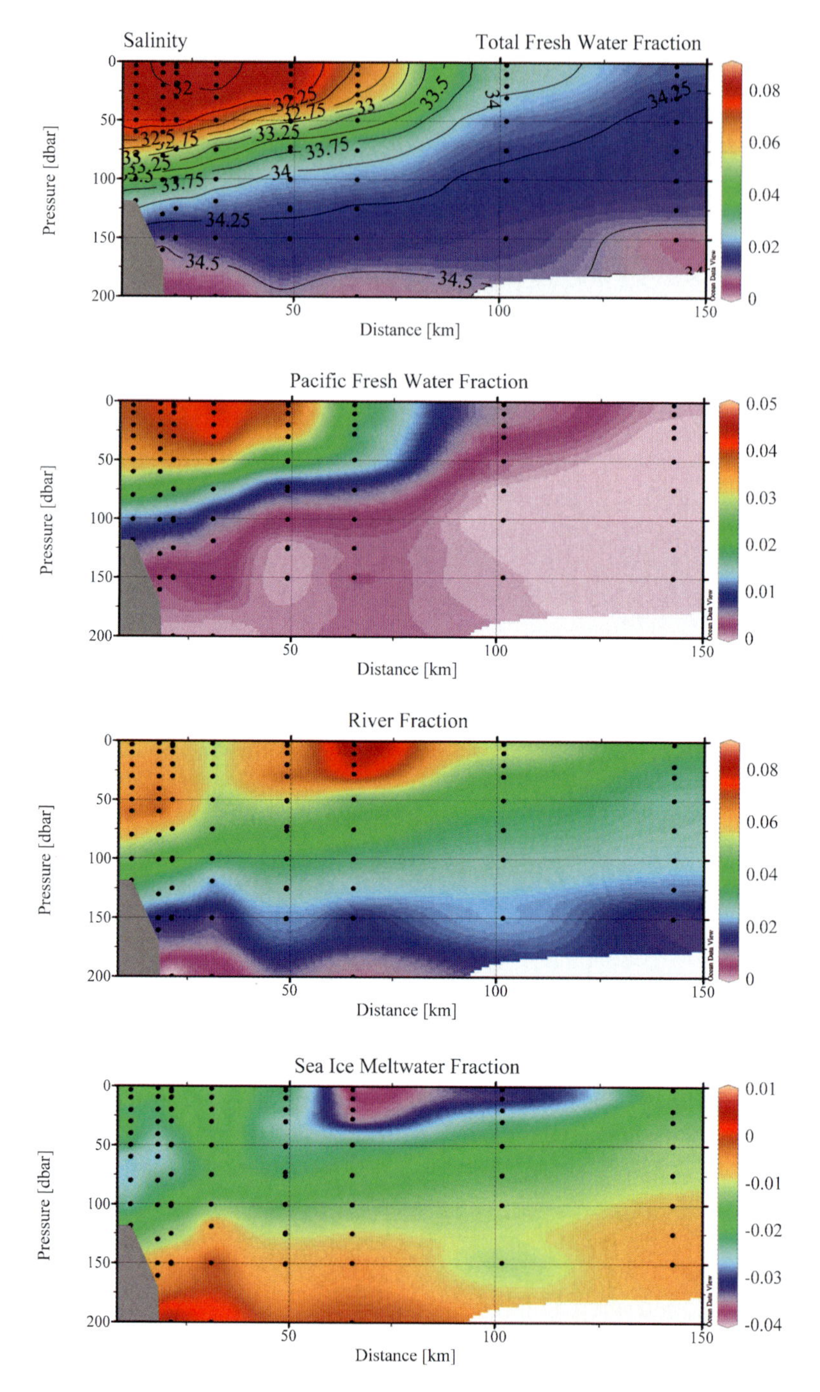

Fig. 16.3 (continued) (e) Fram Strait Section (2002). The section extends into Fram Strait from Greenland at 82° N

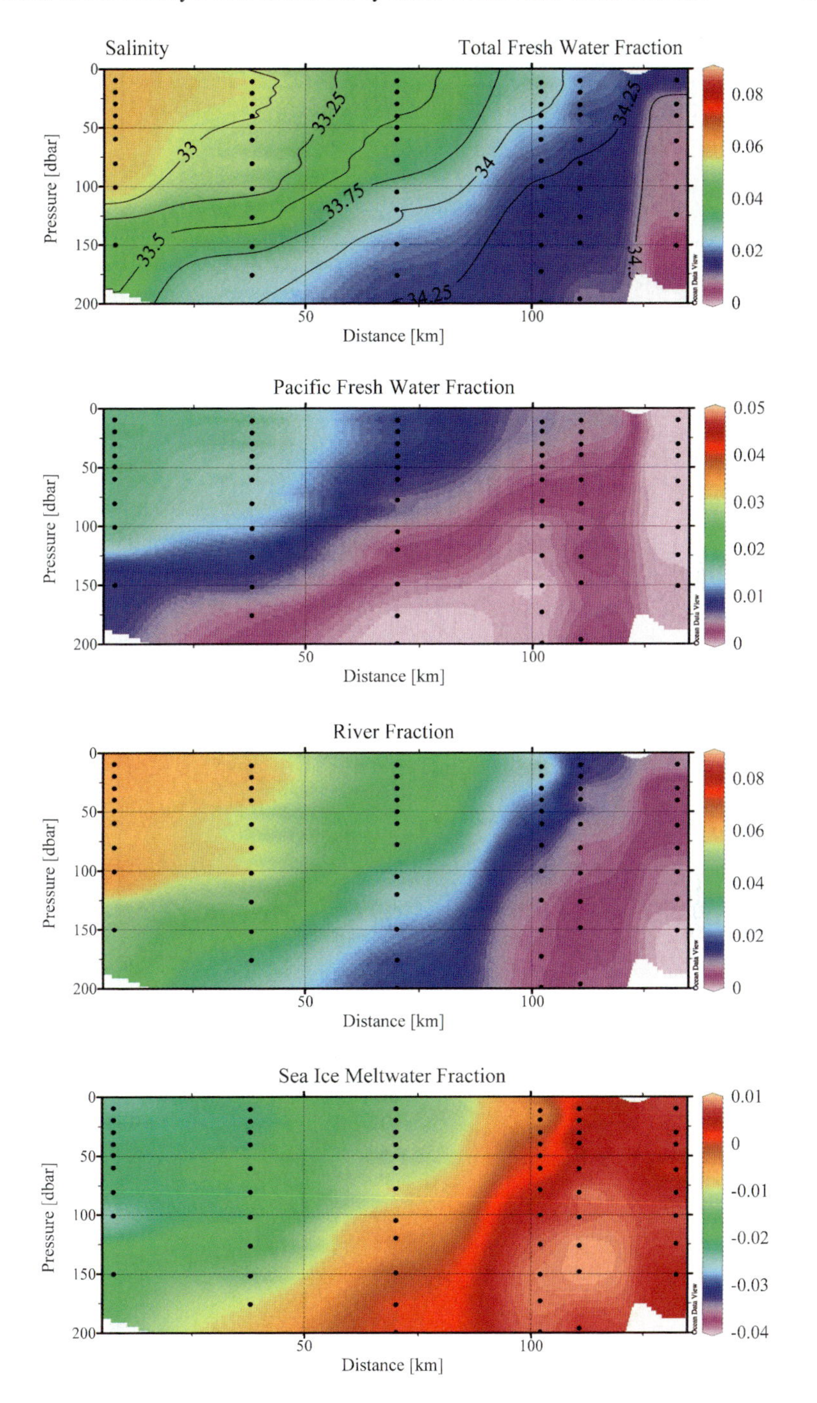

Fig. 16.3 (continued) (f) Scoresby Sound (2002). The section extends from Greenland between 67° N and 65° N

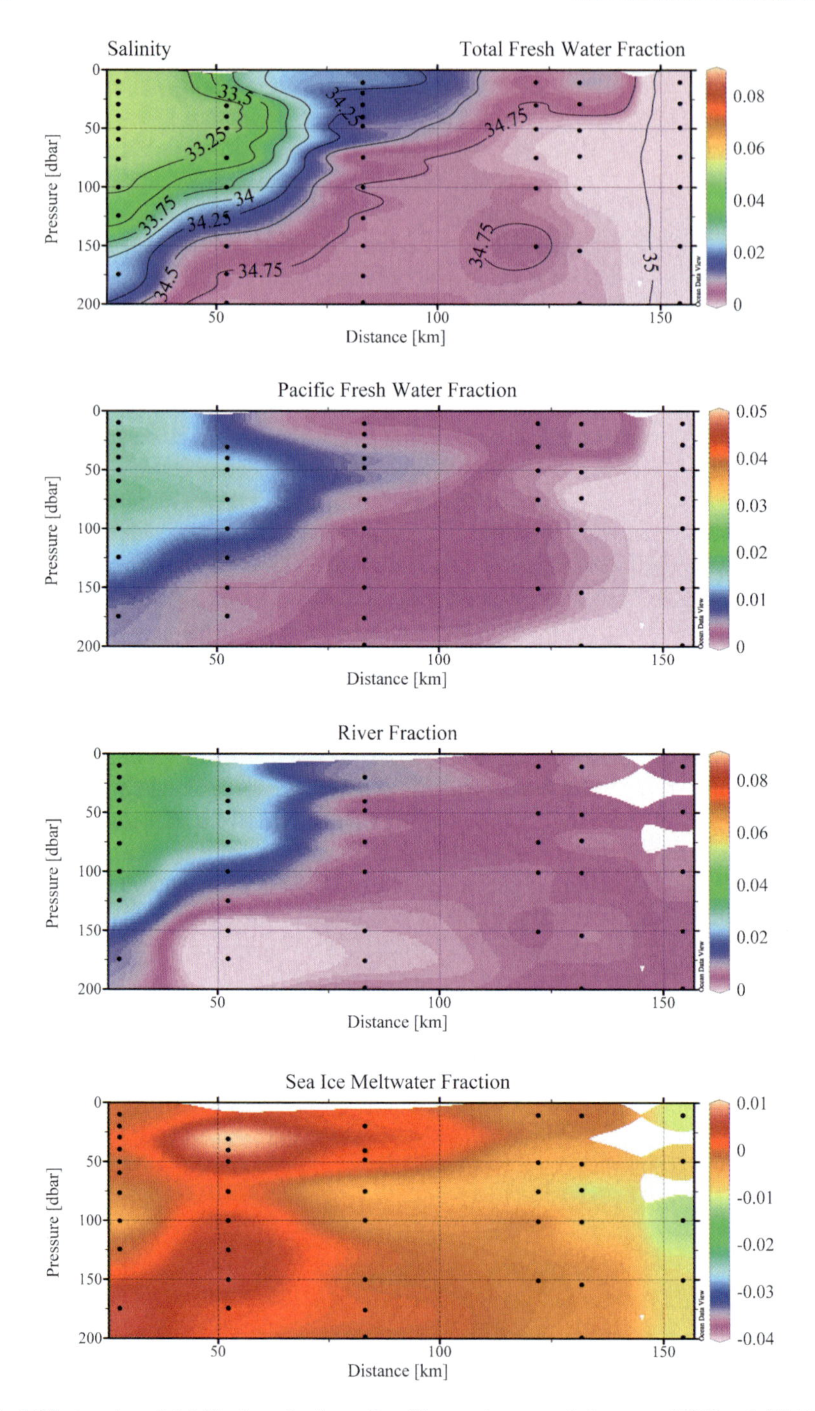

Fig. 16.3 (continued) (g) Northern Irminger Sea. The section extends between 66° N and 65° N

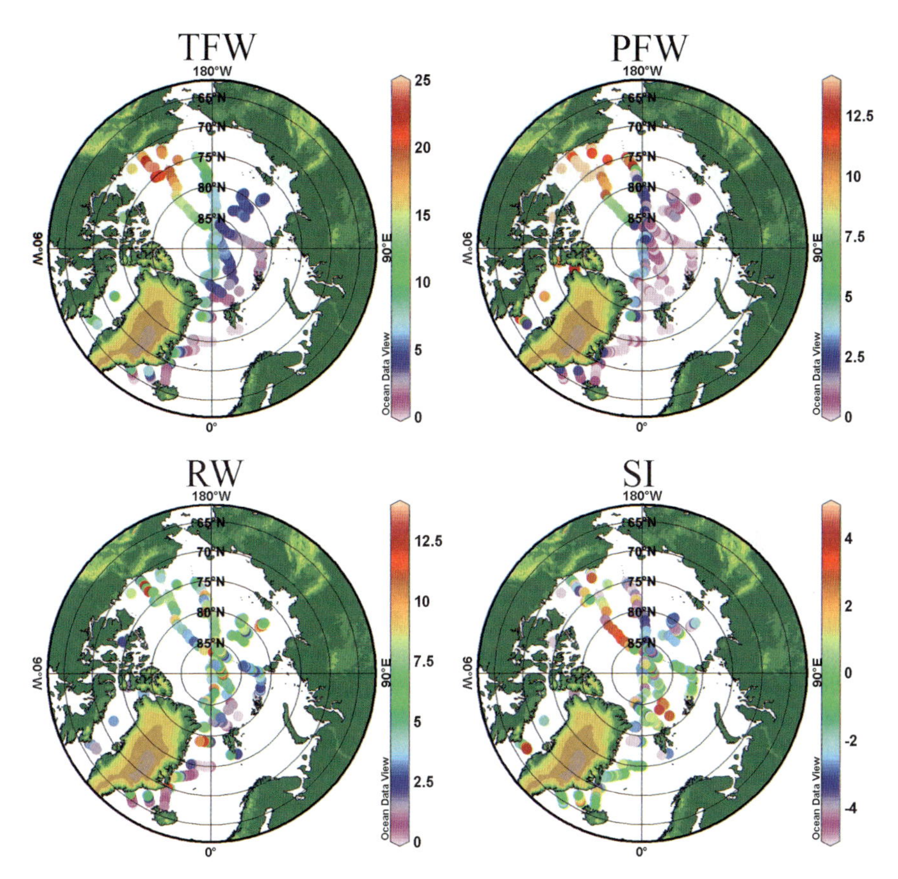

Fig. 16.4 Fresh water inventories: Total Fresh Water (TFW), Pacific Fresh Water (PFW), River Water (RW), and Sea Ice Meltwater (SI), within the Arctic Ocean, the East Greenland Current and the Canadian Arctic Archipelago. The units of the color bars are meters

16.3.1 Arctic Ocean Sections

We chose two sections in the Arctic Ocean to illustrate the fresh water components in the general regions of Atlantic water inflow (Eurasian Basin) and Pacific water inflow (Canadian Basin).

16.3.1.1 Eurasian Basin (1996)

This section begins at the shelf north of the Barents and Kara seas, crosses the Eurasian Basin and enters into the Makarov Basin (Fig. 16.3a). It is the region where Atlantic water dominates. The total fresh water fraction is relatively small throughout the Eurasian Basin, ~0.02, except near the Lomonosov Ridge. Pacific fresh water is essentially non-existent. River water fractions are small, between 0.01 and 0.04, with the higher values near the coast and in the vicinity of the Lomonosov Ridge. And while some sea ice meltwater is found near the coast, sea ice formation is present in most locations, particularly in the vicinity of the Lomonosov Ridge, and is roughly coincident with river water.

16.3.1.2 Canadian Basin (2005)

This section begins in the Canada Basin north of Barrow, Alaska, crosses the Canada and Makarov basins, and ends at the Lomonosov Ridge (Fig. 16.3b). The total fresh water fraction is greatest in the Canada Basin, with highest amounts in the southern part of the section, where near surface salinities are as low as 27. Pacific water concentrations in near surface water were high in the central Canadian Basin, a region where earlier only speculations on their concentrations existed (Jones et al. 1998). Pacific fresh water is most abundant in the southern Canada Basin, where near surface fractions approach 0.07. The high concentrations extend over the Mendeleyev Ridge into the Makarov Basin.

River water distributions differ somewhat from Pacific fresh water. River water extends across the Canada Basin and is most abundant from about the middle of the southern Canada Basin to offshore from the Chukchi Cap. River water is generally confined to the top 50 m, whereas Pacific fresh water generally extends to depths of nearly 300 m in the southern Canada Basin. River water displaces some Pacific fresh water above about 50 m in this region. The highest river water fractions, up to 0.12, are about double the highest Pacific fresh water concentrations. There is very little river water in the general vicinity of the Mendeleyev Ridge. River water concentrations increase in the Makarov Basin, with maxima in the Canada Basin and in the central Makarov Basin and decrease towards the Lomonosov Ridge. The high concentrations of river water in the Canada Basin coincide with fresh water pool in the central Beaufort Gyre. The high concentrations in the Makarov Basin are likely outside the gyre.

Sea ice meltwater with fractions up to 0.1 is found over much of the Canada Basin and generally coincides with Pacific fresh water. Lower concentrations extend to depths near 50 m. Sea ice formation is evident in the Amundsen Basin to depths approaching 100 m except near the North Pole, where there is some sea ice meltwater and Pacific fresh water.

16.3.2 Canadian Arctic Archipelago

Sections across Smith Sound, at the southern end of Nares Strait, and across Davis Strait at the southern end of Baffin Bay were chosen to represent waters flowing through the Canadian Arctic Archipelago.

16.3.2.1 Smith Sound (1997)

The Smith Sound section (78.3° N) is in the southern part of Nares Strait, south of the sill near 81° N in Kennedy Channel (Fig. 16.3c). Pacific fresh water is the dominant source of fresh water in this section. Pacific fresh water and river water distributions more or less overlap and are fairly uniform across the section. The highest concentrations of Pacific fresh water and river water are comparable (0.06) but higher river water concentrations are confined to very near the surface. Sea ice meltwater is mostly negative. Its distribution is also roughly coincident with Pacific fresh water and river water suggesting that it reflects sea ice formation that has occurred in the Arctic Ocean. Distributions of total fresh water and Pacific fresh water in Kennedy Channel in 2001 (Jones and Eert 2006b) are similar to those in Smith Sound in 1997. From this we infer that there is not much change in near surface waters traversing from the Arctic Ocean through Kennedy Channel to Smith Sound.

16.3.2.2 Davis Strait (1997)

The Davis Strait section is at the southern end of Baffin Bay just north of the sill between Baffin Bay and the Labrador Sea (Fig. 16.3d). Fresh water from the Arctic Ocean in the Baffin Island Current on the west side extends nearly half way across the strait. The near surface water is slightly fresher than that in Nares Strait. Pacific fresh water and river water distributions roughly overlap, and concentrations are comparable to those in Nares Strait thus indicating little dilution of the Arctic Ocean water flowing south. As in Nares Strait, sea ice formation is seen at depth coinciding with the Pacific fresh water and river water, again suggesting that this is reflecting sea ice formation in the Arctic Ocean. The slightly fresher surface water may be reflecting a contribution of fresh water passing through the other several Canadian Arctic Archipelago channels into Lancaster Sound as well as possibly some local melting as indicated by the lesser amount of sea ice formation near the surface.

16.3.3 East Greenland Current

Three sections were chosen to illustrate the fresh water constituents in the East Greenland Current from the Arctic Ocean to south of Iceland: the most northerly at 82° N, one near the entrance to Denmark Strait at 70–67° N, and the most southerly in the northern Irminger Sea at 66–65° N. The data were collected under winter conditions with surface freezing apparent almost everywhere. Because of ice conditions not all of the sections reached close enough to the coast to capture the inshore Polar Surface Water flowing from the Arctic Ocean (Rudels et al. 2005). The same concentration scales are maintained for all East Greenland Current section plots.

16.3.3.1 Fram Strait (2002)

This section is just at the boundary of Fram Strait and the Arctic Ocean (Fig. 16.3e). Most of the fresh water exiting the Arctic Ocean lies above 100 m. As in the Canadian Arctic Archipelago, Pacific fresh water and river water distributions roughly overlap, with river water extending deeper and somewhat farther offshore. In contrast to the Canadian Arctic Archipelago, here river water concentrations are nearly twice that of Pacific fresh water. In this section, as in Smith Sound, sea ice meltwater is essentially non-existent. Sea ice formation, likely having occurred within the Arctic Ocean, is clearly present.

16.3.3.2 Scoresby Sound (2002)

This section north of Denmark Strait covers much of the Greenland shelf and likely captures most of the fresh water in the East Greenland Current (Fig. 16.3f). Surface salinities are higher than in Fram Strait and the fresh water extends to greater depths. River water and Pacific fresh water concentrations are lower than in Fram Strait; however both river water and Pacific fresh water are found at depths greater than those in Fram Strait, with the concentrations and extent of river water relative to Pacific fresh water being similar to what is seen in Fram Strait. Sea ice meltwater is non-existent.

16.3.3.3 Northern Irminger Sea (2002)

The Irminger Sea section was not close enough to the coast to capture all of the East Greenland Current fresh water (Fig. 16.3g). The trends of increasing surface salinities and deeper penetration of fresh water toward the south continued in this section and the relative distributions of river water and Pacific fresh water were maintained. Although possibly within the uncertainty of the measurements, there was an indication of sea ice meltwater at the surface. This could be a result of local melting rather than sea ice meltwater exported from the Arctic Ocean.

16.3.4 Fresh Water Inventories

Another way to describe fresh water distributions is by water column inventories of fresh water. The fresh water inventories represent integrated fractions of each fresh water component from the surface to a depth of 200 m (Fig. 16.4).

In the Arctic Ocean all but sea ice meltwater inventories are highest in the Canada Basin north of Alaska. The total fresh water inventory is more than 20 m with river water reaching nearly 20 m in some locations and Pacific fresh water values nearly 10 m over a large region of the Canada Basin. Values for these sources are generally low in most of the Amundsen and Nansen basins. Sea ice meltwater inventories can be as much as 4 m, though inventories are generally negative (representing sea ice formation) over most of the Arctic Ocean.

In the Canadian Arctic Archipelago, Pacific and river fresh water inventories are roughly the same in Lancaster Sound and Nares Strait. In the East Greenland Current, river water inventories are up to 10 m, typically three times greater than in the Canadian Arctic Archipelago. This is likely a reflection of significant Pacific water draining off through the Canadian Arctic Archipelago (Rudels et al. 2005). In the near surface waters of the Arctic Ocean along the North American coast, river water tends to be farther offshore than Pacific water, likely a sign of their dominant Eurasian source. The relatively greater amounts of river water in the East Greenland Current may be reflecting this.

16.4 Summary

The data span a significant time period over which changes in fresh water distributions have been reported (e.g., Schlosser et al. 2002; Anderson et al. 2004). Nevertheless, a general picture of distributions does emerge. Pacific fresh water and river water are the two main contributors to the total fresh water within the Arctic Ocean and in near surface waters exiting from it. The greatest amount of Pacific fresh water is in the Canada Basin. High concentrations in the near surface in Makarov Basin diminish to very little in the Amundsen Basin. River water distributions, at least in 2005, suggest two pathways of Eurasian rivers, one into the central Canada Basin as part of the Beaufort Gyre and another into the central Amundsen Basin likely associated with the Transpolar Drift. This is consistent with two paths of near surface flow suggested in earlier work (Jones et al. 1998; Steele et al. 2004). The lesser amounts of river water near the North American coast in 2005 (near 152° W) and relatively large amounts farther east in 1997 (near 141° W) (Macdonald et al. 1999 and satellite observations referred to in this paper) may suggest that these larger amounts being of North American origin.

There is a distinct difference between relative amounts of Pacific fresh water and river water exiting the Arctic Ocean via the Canadian Arctic Archipelago and those exiting via the East Greenland Current, with Pacific fresh water fractions being much more dominant in the former and river water much more dominant in the latter. Based on the relative Pacific halocline depths north of the Canadian Arctic Archipelago and in Fram Strait, Rudels et al. (2004) suggested that much of Pacific water exits through the Canadian Arctic Archipelago. Also, this may be reflecting two paths for Eurasian rivers, one toward the east into the Beaufort Sea and one headed more directly to the north and west towards Fram Strait. This is consistent with the finding that, at least in 1998, North American river water was not present in Fram Strait (Taylor et al. 2003).

Sea ice meltwater is the least abundant of the fresh water sources in the Arctic Ocean, where it was present in the near surface waters of the Canada Basin in 1994 and 2005 as well as in the Makarov Basin in 2005. It was consistently observed north of Svalbard, where inflowing relatively warm Atlantic water entered the Arctic Ocean, in 1991, 1994, 2002 and 2005. Sea ice formation was evident in most other locations.

Sea ice meltwater is essentially absent from out flowing Arctic Ocean water. Most regions outside the Arctic Ocean show evidence of sea ice formation that probably has occurred both locally and in the Arctic Ocean. Within the East Greenland Current the negative sea ice meltwater values are likely reflecting sea ice formation in the Arctic Ocean and possibly also the freezing conditions in the East Greenland Current when the data were obtained with the implication that no residual sea ice meltwater is flowing from the Arctic Ocean.

16.5 Implications

Is a change in fresh water flux from the Arctic Ocean a threat to the Global Conveyor Belt, i.e., is fresh water from the Arctic Ocean controlling deep convection in the Nordic and Labrador seas? The usual thinking stemming from the analysis of Aagaard and Carmack (1989) is that an increase in the fresh water outflow from the Arctic Ocean could have a major impact on deep convection in the Nordic and Labrador seas. This scenario may need further investigation, however. In 2002, the salinity of near surface waters of the Greenland Sea was much higher than that of the Polar Surface Water in the East Greenland Current, thereby precluding the possibility of Arctic Ocean fresh water reaching the Greenland Sea region of deep convection (Rudels et al. 2005). Further, Rudels et al. (2005) point out that there was no apparent diminishment of the fresh water in the East Greenland Current as it progressed to south of Denmark Strait. This could suggest that liquid fresh water from the Arctic Ocean may have little influence on deep convection in these regions. If so, ice export (~0.1 Sv, Vinje 2001) and precipitation remain the only possible Arctic Ocean contributions to surface freshening.

A similar situation seems to exist for the Labrador Sea, where polar waters from the Canadian Arctic Archipelago also seem to be strongly constrained to near the shelf and away from where deep convection takes place. A recent model simulation (Myers 2005) also suggests that fresh water export from the Arctic Ocean has little impact on Labrador Sea deep convection. Here precipitation, sea ice meltwater from Baffin Bay, and perhaps fresh water from the East Greenland Current entering via the West Greenland Current would more likely be influencing deep convection.

This and other similar work make clear that climate change scenarios dealing with fresh water budgets of the Arctic Ocean should not use a single parameter representation of fresh water since the fresh water has different sources and distributions, all subject to different forcing. The differing geographical sources of fresh water components and their potentially differing response to climate change requires that each be separately considered in climate change scenarios. Of the three components that affect thermohaline circulation, the inflow of Pacific fresh water might seem to be the least open to change, though what fraction of it exits through Fram Strait can change dramatically (Falck et al. 2005; Steele et al. 2004). River water and sea ice meltwater may be the components that change most in climate change scenarios because of changes in precipitation in the large river drainage basins feeding into the Arctic Ocean and because of the changes in sea ice meltwater arising from ice-free Arctic Ocean summers.

Acknowledgements Many people on several expeditions helped to collect relevant data. Their contributions are greatly appreciated. This work presented here was supported in part by the Canadian Panel on Energy Research and Development and the National Centre for Arctic Aquatic Research Excellence (EPJ) and by the Swedish Natural Research Council (LGA). Most figures were produced using Ocean Data View (Schlitzer 2006).

References

Aagaard K, Carmack EC (1989) The role of sea ice and other fresh water in the Arctic circulation. J Geophys Res 94: 14485–14498

Anderson LG, Jutterström S, Kaltin S, Jones EP, Björk G (2004) Variability in river runoff distribution in the Eurasian Basin of the Arctic Ocean. J Geophys Res 109: C01016, doi:10.1029/2003JC001773

Ekwurzel B, Schlosser P, Mortlock RA, Fairbanks RG (2001) River runoff, sea ice meltwater, and Pacific water distribution and mean residence times in the Arctic Ocean. J Geophys Res 106: 9075–9092

Falck E, Kattner G, Budéus G (2005) Disappearance of Pacific Water in the northwestern Fram Strait. Geophys Res Lett 32: L14619, doi:10.1029/2005GL023400

Holloway G, Proshutinsky A (2007) Role of tides in Arctic ocean/ice climate. J Geophys Res 112: C04S06, doi:10.1029/2006JC003643

Jones EP, Anderson LG, Swift JH (1998) Distribution of Atlantic and Pacific waters in the upper Arctic Ocean: Implications for circulation. Geophys Res Lett 25: 765–768

Jones EP, Swift JH, Anderson LG, Lipizer M, Civitarese G, Falkner KK, Kattner G, McLaughlin FA (2003) Tracing Pacific water in the North Atlantic Ocean. J Geophys Res 108 (C4): 3116, doi:10.1029/2001JC001141

Jones EP, Anderson LG, Jutterström S, Swift JH (2006a) Sources and distribution of fresh water in the East Greenland Current. Submitted to Progress in Oceanography

Jones EP, Eert AJ (2006b) Waters of Nares Strait in 2001. Polarforschung 74: 185–189

Macdonald, RW, Carmack EC, McLaughlin FA, Falkner KK, Swift JH (1999) Connections among ice, runoff and atmospheric forcing in the Beaufort Gyre. Geophys Res Lett 26: 2223–2226

Myers PG (2005) Impact of freshwater from the Canadian Arctic Archipelago on Labrador Sea Water formation. Geophys Res Lett 32, L06605, doi:10.1029/2004GL022082

Holloway, G, Proshutinsky, A (2007) Role of tides in Arctic ocean/ice climate. J Geophys Res 112: C04S06, doi:10.1029/2006JC003643

Rahmstorf, S (1996) On the freshwater forcing and transport of the Atlantic thermohaline circulation. Climate Dyn 12: 799–811

Rudels B, Jones EP, Schauer U, Eriksson P (2004) Atlantic sources of the Arctic Ocean surface and halocline waters. Polar Research 23: 181–208

Rudels B, Björk G, Nilsson J, Winsor P, Lake I, Nohr C (2005) The interaction between water from the Arctic Ocean and the Nordic Seas north of Fram Strait and along the East Greenland Current: Results from the Arctic Ocean-02 Oden Expedition. J Mar Sys 55: 1–30

Schlitzer R (2006) Ocean Data View, http://www.awi-bremerhaven.de/GEO/ODV
Shimada K, Carmack EC, Hatakeyama K, Takizawa T (2001) Varieties of shallow temperature maximum waters in the western Canadian Basin of the Arctic Ocean. Geophys Res Lett 28: 3441–3444
Schlosser P, Newton R, Ekwurzel B, Khatiwala S, Mortlock R, Fairbanks R (2002) Decrease of river runoff in the upper waters of the Eurasian Basin, Arctic Ocean, between 1991 and 1996: Evidence from $\delta^{18}O$ data. Geophys Res Lett 29 (9), 1289: doi:10.1029/2001GL013135
Steele M, Morison J, Ermold W, Rigor I, Ortmeyer M, Shimada K (2004) Circulation of summer Pacific halocline water in the Arctic Ocean. J Geophys Res 109: C10003, doi:10.1029/2003JC002009
Taylor J R, Falkner KK, Schauer U, Meredith M (2003) Quantitative consideration of dissolved barium as a tracer in the Arctic Ocean. J Geophys Res 108: C12, 3374, doi:10.1029/2002JC001635
Vinje T (2001) Fram Strait ice fluxes and the atmospheric circulation: 1950–2000. J Clim 14: 3508–3517
Yamamoto-Kawai M, Tanaka N (2005) Freshwater and brine behaviors in the Arctic Ocean deduced from historical data of D18O and alkalinity (1929–2002 A.D.). J Geophys Res 110: C10003, doi:10.1029/2004JC002793

Chapter 17
Simulating the Long-Term Variability of Liquid Freshwater Export from the Arctic Ocean

Rüdiger Gerdes, Michael Karcher, Cornelia Köberle, and Kerstin Fieg

17.1 Introduction

The fresh water export from the Arctic has not been measured yet. The major problem lies in the transport over the shallow East Greenland shelf that is not easily accessible for oceanographic vessels and so far has been off-limits for moored instrumentation. Even if we would be able to start measurements now, we would have no statistics to evaluate trends and natural variability of the transport. For long time series and for predictions of future changes, there is no other means than numerical models of the oceanic circulation and the water mass distribution. For past times, models can perhaps be combined with observations of different variables to yield better reconstructions of long-term variability in fresh water fluxes between the Arctic and the sub-polar North Atlantic.

The liquid fresh water export from the Arctic Ocean through the passages of the Canadian Archipelago, Fram Strait and the Barents Sea is constrained by the fresh water fluxes entering the Arctic Ocean and by changes in the fresh water contents in the Arctic halocline. If one knew the fluxes entering the Arctic Ocean and the changes in the salinity very precisely, the export rates could be determined as a residual. (We use this technique to derive export rates in a coupled climate model in Section 17.5.) Different components of the Arctic Ocean fresh water balance exhibit very different long-term variability. Serreze et al. (2006) provide a recent compilation of estimates of the interannual variability of river discharge, net precipitation, Bering Strait inflow, and Fram Strait ice flux. Fram Strait ice transport shows by far the largest standard deviation of these fresh water fluxes. River run-off into the Arctic Ocean has increased over the last 50 years by approximately 5% (Peterson et al. 2002). Interannual variability as shown by Peterson et al. is of similar or smaller magnitude. Compared to fluctuations in other components of the fresh water balance, this is a small variability. The variability in river discharge is also indicative of the variability of the total atmospheric moisture convergence at

Alfred-Wegener-Institut für Polar- und Meeresforschung, Bremerhaven, Germany,
e-mail: Ruediger.Gerdes@awi.de

R.R. Dickson et al. (eds.), *Arctic–Subarctic Ocean Fluxes*, 405–425

high northern latitudes and thus the net precipitation over the Arctic Ocean. The fresh water flux from the Pacific into the Arctic Ocean fluctuates seasonally, but interannually fluctuations are small around a mean of 2,500 ± 300 km^3/year[1] (relative to a reference salinity of 34.8; Woodgate et al. 2005). This means that over recent decades, the fresh water balance of the Arctic was determined by lateral exchanges with lower latitudes, temporal changes in the fresh water content, and rather constant sources of fresh water.

The size of the Arctic Ocean liquid fresh water reservoir of 74,000 km^3 (Serreze et al. 2006, using a reference salinity of 34.8) and an average export rate of 3,000–6,000 km^3/year gives an average renewal time for the reservoir of 10–20 years. This implies that the Arctic Ocean system is capable of sustaining substantial anomalies in the fresh water export rate over decades (e.g. Proshutinsky et al. 2002). In model simulations, occasional high liquid fresh water export events exceed the long-term mean by at least 1,000 km^3/year (Karcher et al. 2005) and last for several years. Köberle and Gerdes (2007) found that the simulated liquid fresh water export from the Arctic between 1970 and 1995 was 500 km^3/year larger than on average over the second half of the 20th century. This long-term enhanced export rate corresponds to a decline of the Arctic liquid fresh water reservoir by 12,500 km^3 between 1970 and 1995.

This review will commence with an assessment of the uncertainties and their causes in current ocean–sea ice models for the Arctic Ocean. It is important to be aware of the consequences of uncertainties in the forcing fields (like precipitation and run-off), their implementation in different models, and their impact on the simulation of liquid fresh water export rates from the Arctic. The specific effects associated with numerical resolution in the Arctic Ocean and the passages connecting it with the global ocean will be discussed. The following section describes the variability in liquid fresh water export over the last five to six decades as it is simulated in models of the NAOSIM (North Atlantic/Arctic Ocean Sea Ice Models) hierarchy. This includes two outstanding events that are responsible for much of the long-term changes in the Arctic Ocean liquid fresh water reservoir. Possible downstream effects of such fresh water export events and the possible development of liquid fresh water export from the Arctic Ocean during the 21st century are the topics of Sections 17.4 and 17.5, respectively. In the last section, we summarize the current state of the art and try to identify the most important problems affecting the modeling of fresh water exports from the Arctic Ocean.

17.2 Uncertainties in Model Estimates of Arctic Liquid Fresh Water Export

Although numerical models are our primary if not only means to assess the long-term variability in the liquid fresh water export from the Arctic, there are relatively few model results documented in the literature. The reasons for this shortage are

[1] Fresh water fluxes are given in $km^3\ year^{-1}$ or Sv with $1\ Sv = 10^6\ m^3\ s^{-1} = 31,536\ km^3\ year^{-1}$.

fourfold. There is a lack of data to validate this aspect of the models, thus a natural way to communicate model results is blocked. Secondly, one of the major pathways for fresh water from the Arctic to lower latitudes, the Canadian Archipelago, needs extremely high horizontal and vertical resolution to be properly represented. Thirdly, ocean–sea ice as a climate sub-system must be provided with proper boundary conditions. This is a general problem that affects all ocean–sea ice simulations and which can have severe consequences for the stability of the large-scale oceanic circulation in a model. Finally, surface fresh water fluxes and their variability over decades are poorly known over the Arctic.

In the following, we shall briefly address these four items that remain an obstacle for model based statements about Arctic fresh water export rates. Figure 17.1 compares observational and model based fresh water transports in the East Greenland Current at Fram Strait. Hansen et al. (2006) analysed data from the moorings F11–F14 which are located in the core of the EGC over and east of the shelf break at 79° N. Their calculated average fresh water transport from July 1997 to July 2005 is southward with around 1,000 km^3/year. NAOSIM (North Atlantic/Arctic Ocean–Sea Ice Models) freshwater transport results (Karcher et al. 2003, 2005) for the same period and sub-sampled for the area that is covered by the moorings are very similar except that the model does not capture all the high frequency variability that is in the observed data. However, the total southward fresh water transport in the model is almost twice as high as the observational estimate. The model transport is enhanced by contributions from outside the area covered by the mooring array, namely from southward flow of very fresh water over the shallow East Greenland shelf and still relatively fresh water east of the core of the EGC. Hydrographic sections and geostrophic calculations indicate that the fresh water transports east of the 0° E should be very small and that the model overestimates the fresh water transport there. However, there are no observations over the East Greenland shelf to validate the model. Transports over the shallow shelf can be substantial according to the model. This notion is reinforced by hydrographic and $\delta^{18}O$ measurements by Meredith et al. (2001) who found a large volume of meteoric water on the East Greenland shelf.

Holfort and Meincke (2005) report continuous measurements of salinity and velocity at 74° N on the East Greenland shelf. They describe the uncertainties involved in estimating fresh water transports from these measurements. Among other factors, uncertainties are due to the incomplete coverage of the shelf (the two moorings are just 8 km apart with a bottom mounted ADCP between them) and the extrapolation of the measured salinity profile to the surface. For the Arctic Ocean liquid fresh water balance a further uncertainty lies in the unknown amount of sea ice that melted between Fram Strait and 74° N. A strong seasonal cycle in the salinity time series from the uppermost instrument (at 20 m depth) indicates a strong influence of sea ice melt at and upstream of the mooring site. Overall, we have to state that observationally based estimates of liquid fresh water transport in the EGC are not suited to validate model results at present.

Simulated liquid fresh water transports through the second large export pathway, the Canadian Archipelago, differ substantially from model to model (Dickson et al.

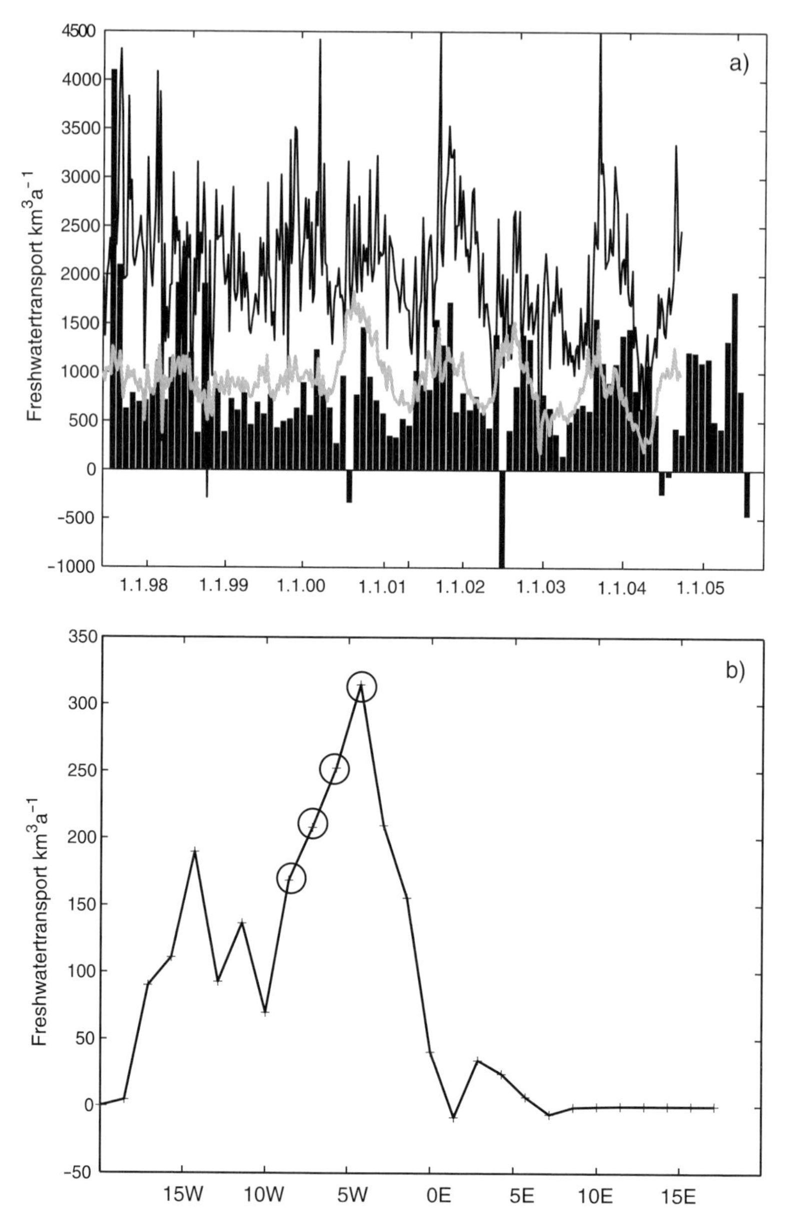

Fig. 17.1 (top) Comparison of observed and modeled (NAOSIM) freshwater fluxes through Fram Strait. Black bars: Monthly average of the observed fresh water flux in the EGC at 79° N. Thick gray line: Modelled fresh water flux for the same section as covered by the moorings. Thin black line: Modelled total fresh water flux, including the Greenland shelf region. (bottom) Average (1997–2005) modelled freshwater flux across Fram Strait in km^3 a^{-1} per grid box. The circles indicate the position of moorings F11–F14 across the East Greenland Current. All fluxes are calculated using a reference salinity of 35.0. (Observed data courtesy of E. Hansen and J. Holfort, personal communication, 2007)

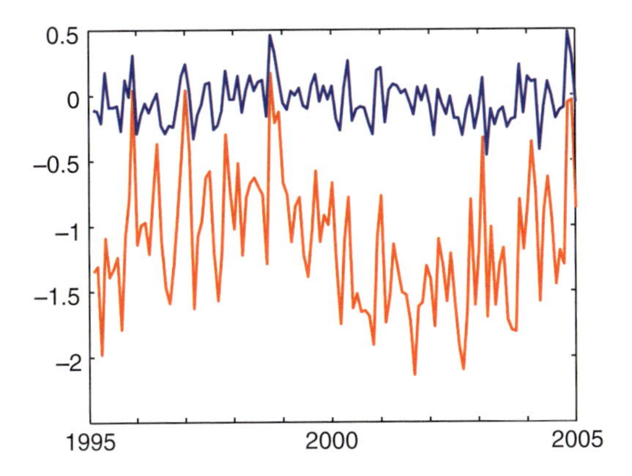

Fig. 17.2 Time series of net volume transport through Davis Strait in the 1/4° resolution version (blue) and in the 1/12° resolution version of NAOSIM (Fieg et al., manuscript in preparation). Northward transports are positive; axis labels are in Sv

2007). Considering the narrow channels that connect the Arctic Ocean with Baffin Bay, model resolution is an obvious candidate for the differences. Figure 17.2 shows volume transport time series of net volume transport through Davis Strait (which is identical to that through the Canadian Archipelago) in two versions of NAOSIM. In the lower resolution version (1/4° resolution in a rotated spherical grid) the volume transport almost vanishes whereas the higher resolution version (1/12°) has a more realistic southward transport of around 1 Sv. Results from the even lower resolution version of NAOSIM with 1° resolution suggest, however, that an exaggerated Channel width may at least help in getting a realistic net outflow of water though the Archipelago (Köberle and Gerdes 2007).

The resolution dependent representation of the Canadian Archipelago topography is a critical issue in modelling the Arctic freshwater balance. This is aggravated by the possible resolution dependence of the fresh water export distribution between the Archipelago and Fram Strait.

The magnitude of the liquid fresh water flux through the Canadian Archipelago is around 0.1 Sv or approximately 3,000 km^3/year in the high-resolution NAOSIM version. This is of the same order of magnitude as the liquid Fram Strait export and about twice as large as the estimate of Aagard and Carmack (1989). Newer observational estimates put the total liquid fresh water transport through the Canadian Archipelago at around 3,000 km^3/year (Prinsenberg and Hamilton 2004, 2005). These estimates rely on 3 years of mooring data in Lancaster Sound and assumptions about the additional flow through other channels, especially Nares Strait. The transport through the Canadian Archipelago is not solely determined by resolution. The passages in the model have to be well resolved. This might be achieved in coarse resolution models by overly wide channels through the Canadian Arctic. In a 1°-resolution model, Prange and Gerdes (2006) find an average southward fresh water transport of almost 1,400 km^3/year while Köberle and Gerdes (2007) simulate a transport of more than 2,200 km^3/year with different surface forcing and a somewhat different land–sea configuration. Another important factor is the representation of flow through Bering Strait. In a simple two-dimensional model, Proshutinsky et al. (2007) show that inflow through Bering Strait sets up a surface elevation pattern with highest amplitudes along the North American coast and indicating strong flows through the Canadian Archipelago and Fram Strait. Simulations with a comprehensive model (Karcher and Oberhuber 2002) that includes an artificial tracer for Pacific Water confirm this direct path from Bering Strait to the Canadian Archipelago. All these relatively coarse resolution models have a prescribed volume influx at Bering Strait while the higher resolution NAOSIM models have a closed boundary where only hydrographic properties are imposed.

Many Arctic Ocean models employ 'virtual salt fluxes' instead of fresh water fluxes to represent precipitation, melt water, and continental run-off. This is imposed by a rigid-lid condition that leads to volume conservation and cannot accommodate volume fluxes across the surface. In this case, a choice of reference salinity is necessary to convert fresh water fluxes into salt fluxes. Usually, a constant value or the local surface salinity is used. A constant value S_{ref} with which the surface fresh water flux becomes $F^S = (-P + E - R)S_{ref}$, allows tracer conservation when the total surface fresh water fluxes (including evaporation) sum up to zero over the model surface. However, locally very large errors are possible. This includes the possible occurrence of negative salinities near strong fresh water sources like the Siberian river mouths during summer. Local surface salinity SSS, $F^S = (-P + E - R)SSS$, avoids these errors but involves a spatially variable weighting of the fresh water fluxes which implies a deviation from the originally specified surface fresh water fluxes. Prange and Gerdes (2006) discuss these choices and their consequences for the Arctic Ocean fresh water balance. Depending on the chosen surface boundary condition, Fram Strait liquid fresh water transports differ by up to 1,000 km^3/year. In the case of prescribed volume fluxes through the surface, the Arctic Ocean is becoming saltier while in case of 'virtual salt fluxes' with the local SSS for conversion from fresh water fluxes, the Arctic Ocean is getting fresher in Prange and Gerdes' calculation.

In equilibrium, the exchanges of volume and salt between the subpolar North Atlantic and the Arctic Ocean are strongly constrained by the mass and salt balances of the Arctic Ocean. On short time scales, inflow and outflow salinities do not change substantially and an increase in the run-off, precipitation minus evaporation, or Bering Strait inflow will result in increasing transports of both the Atlantic inflow and the outflow of Polar Water. Besides other processes, this exchange will eventually lead to a new equilibrium. Important questions are how long the adjustment processes will last (determined by the size of the involved fresh water reservoirs and the magnitude of the flux anomaly) and what changes in fresh water content in the Arctic Ocean will develop during the transition phase. Over decadal or longer time scales, the outflow with the EGC can be described by a simple formula derived from a 1.5-layer model of the Polar Water flow (Köberle and Gerdes 2007). The volume transport is proportional to the square of the upstream thickness of the Polar Water layer. An adjustment of the lateral fluxes thus likely involves changes in the thickness of the

Arctic halocline. In a model with prescribed fresh water input through precipitation, run-off and Bering Strait inflow (usually with prescribed salinity), the lateral fluxes will adjust accordingly to reach equilibrium. A bias in the prescribed fluxes will result in a bias in the lateral fluxes as well as in the Arctic hydrography. Even in a perfect model, the biases in fresh water fluxes prescribed as forcing will introduce biases in the distribution of salinity and the lateral fluxes in a model. The equilibrium response in the volume transports of in- and outflows to a change in run-off and precipitation is amplified by a factor $S_{ref}/\Delta S$ where S_{ref} is the salinity of the inflow or the outflow and ΔS is the salinity difference between inflow and outflow. Because of the large salinity contrast between inflow and outflow in the case of the Arctic, this factor is only O(10) for current conditions. However, the uncertainty in precipitation over the Arctic Ocean as expressed in the different integral numbers of fresh water flux from different data sets is almost 0.1 Sv. For the ocean area north of 65° N with the exception of the Nordic Seas and the Barents Sea south of 79° N and east of 50° E we calculate 5,600 km^3/year in the Large and Yeager (2004) dataset, 2,900 km^3/year in the ERA40 reanalysis data based Röske (2006) atlas, and 5,000 km^3/year in the satellite-based NASA GPCP V1DD data set. An ocean model confronted with a precipitation data set that is perhaps 0.1 Sv off will react either with a bias in the exchanges between the Arctic and adjacent seas of around 1 Sv or a corresponding change in the outflow salinities, i.e. a massive bias in the Arctic Ocean hydrography.

Because of the above difficulties to satisfactorily combine prescribed fresh water fluxes, lateral exchange rates, and hydrograhy in the interior Arctic, many modellers have relied on additional artificial fresh water sources. Perhaps the most frequently used device is the restoring of modelled surface salinity to climatological values. Steele et al. (2001) discuss the effect of surface salinity restoring in different Arctic Ocean models. Köberle and Gerdes (2007) discuss the spatial and temporal distribution of the restoring flux in their model under NCAR/NCEP reanalysis forcing. Biases in the model that were compensated for include a lack of fresh water originating at the Siberian rivers and following the transpolar drift into the interior Arctic Ocean. Run-off in their model is around 1,000–2,000 km^3/year less than more recent estimates (Shiklomanov et al. 2000). More important, however, was the failure of the model to disperse the fresh water away from the coasts. The insufficient communication between shallow shelf seas and the deep interior is a common problem in this class of ocean models. River water is accumulating near the river mouths, leading to unrealistically low salinities. This diminishes the efficiency of the fresh water flux that is transformed into a salt flux by multiplying with the local surface salinity. In other areas of the Arctic, the flux adjustment is typically less than 0.5 m/year in each direction. These values still are comparable to the annual mean precipitation in this area.

The flux adjustment partly compensates for a mismatch between the climatological surface salinities, based on observations mainly between 1950 and 1990, and the forcing period that extends to 2001. For instance, north of the strong fresh water input through the flux adjustment Köberle and Gerdes (2007) find an area where fresh water is extracted. This can be ascribed to the changed pathways of river water in times of the strongly positive North Atlantic Oscillation (NAO) towards the end of the 20th century (Steele and Boyd 1998) that is not well represented in the climatological surface salinities. Similarly, the climatology might not reflect completely the supposed high ice export rates from the Arctic during positive NAO phases, thus featuring relatively low surface salinities in the sea ice formation regions and relatively high salinities in the melting regions of the EGC.

Restoring introduces a negative feedback that acts against surface salinity anomalies. With time-varying atmospheric forcing the restoring term represents a strongly varying component in the Arctic Ocean fresh water balance. To avoid the feedback that damps variability, surface fresh water fluxes are prescribed. A naïve application of fresh water fluxes will lead to large biases in simulated hydrography and lateral exchanges as explained above. A flux-compensation can be introduced as described for instance in Köberle and Gerdes (2007). Basically, the restoring term is evaluated for an experiment run and averaged over a certain period. In a repetition of the run with otherwise identical forcing, this climatology of the restoring term is applied as a fixed salt flux to the surface box of the ocean model. This is an artificial fresh water flux that, however, compensates for biases in the forcing fields and deficiencies of the model. Since the flux is constant in time and there is no connection with the surface salinity, the former feedback is no longer present. This allows much larger variability in all components of the fresh water balance. As an example we show in Fig. 17.3 time series for Arctic fresh water content in a model run with restoring and a model run with flux adjustment.

While this procedure seems a feasible solution to the problem, potentially it has a grave drawback. Prescribing surface temperature through bulk formulae that tie the SST to fixed atmospheric temperatures and surface fresh water fluxes (mixed boundary conditions) is known to cause too high sensitivity in large-scale models of the oceanic circulation (Zhang et al. 1993; Rahmstorf and Willebrand 1995; Lohmann et al. 1996). Regional models are more constrained by lateral boundary conditions where large-scale transports are prescribed. In the example of Fig. 17.3, we see that the fresh water content is systematically higher in the flux-adjusted case but the value at the end of the integration is close to that of the restored case again. This indicates that no substantial shift in the circulation regime has occurred due to the change in the surface boundary conditions. We conclude that this model

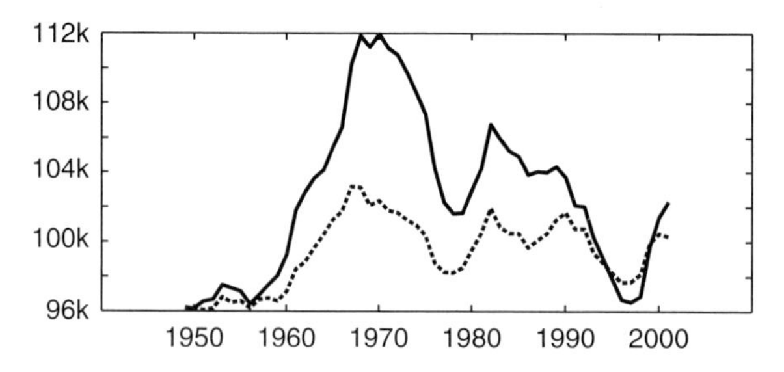

Fig. 17.3 Arctic Ocean liquid fresh water content from the NAOSIM hindcast simulation of Köberle and Gerdes (2007). The solid line shows the result with surface fresh water flux adjustment while the dashed line shows the results under restoring of surface salinity

apparently does not suffer from the tendency to unrealistically high sensitivity of the large-scale oceanic circulation under mixed boundary conditions.

17.3 Variability of Liquid Fresh Water Export Since 1950

As an example of the variability in ocean–sea ice models that are forced with realistic atmospheric forcing for the last decades we show in Fig. 17.4, the lateral fresh water fluxes from the flux-adjusted simulation of Köberle and Gerdes (2007).

Fram Strait export dominates the variability of lateral transports of liquid fresh water. The Fram Strait fresh water export in turn is determined by the fresh southward component because the northward volume transport of Atlantic water is less than one third of the East Greenland Current and the salinities of the inflow are much closer to the reference value than those of the outflow in the EGC. Fram Strait export is responsible for the extremely low total export rates in the mid-1960s and for the large export of the mid-1970s.

In this model result, the export through the Canadian Archipelago is somewhat smaller than Fram Strait export and shows less variability. However, between the mid-1980s and the mid-1990s, this component contributes significantly to the large fresh water exports during that period (Belkin et al. 1998). It is also largely responsible for overall decreasing exports after 1995. Because of the limited resolution of the model, the representation of the passage through the Canadian Archipelago is

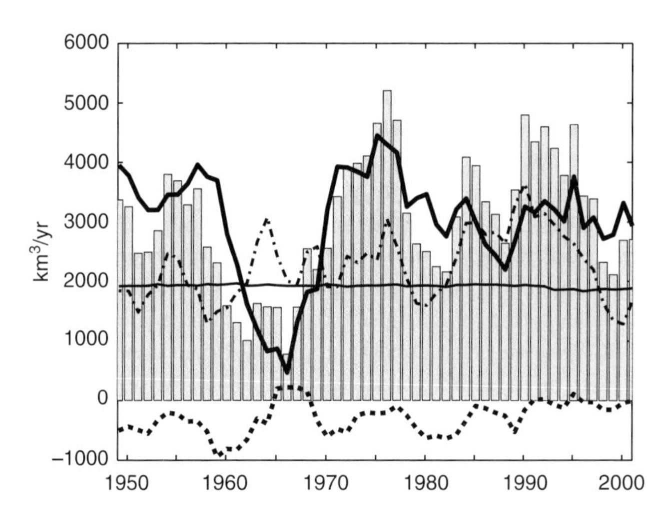

Fig. 17.4 Time series of the lateral liquid fresh water fluxes out of the Arctic Ocean. All fluxes are given in km^3/year. Bars represents the total fresh water export, the solid line is the transport through Fram Strait, the dash-dotted line is the transport through the Canadian Archipelago, the dashed line is the transport through the Barents Sea, and the thin solid line is the transport through the Bering Strait. Transports are calculated using a reference salinity of 35

rather crude. However, as noted above, the simulated mean fresh water transport through the Canadian Archipelago is within the range of recent observational estimates. Surface fluxes exhibit large interannual variability while the export rate is rather smoothly varying with a quasi-decadal time scale. On decadal to multidecadal time scales, the Arctic liquid fresh water reservoir responds mainly to changes in the export rate. The liquid fresh water export rate from the Arctic Ocean was extremely low in the 1960s and showed two periods of high values afterward. Especially the late 1970s and early 1980s were characterized by large export rates.

Häkkinen and Proshutinsky (2004) have analyzed similar hind cast simulations with the Goddard Space Flight Center (GSFC) model. There is no restoring or flux adjustment in this model. They do not show fresh water transport rates but show that liquid fresh water content changes in the Arctic Ocean can largely be explained by the oceanic exchanges with lower latitudes. As in Köberle and Gerdes (2007), their model result features accumulation of liquid fresh water in the Arctic Ocean in the early 1960s, in the early and late 1980s, and a strong decline afterwards. Both papers identify the export through Fram Strait as the most important component of the Arctic fresh water balance responsible for these fluctuations although Köberle and Gerdes point at reduced sea ice formation in the late 1960s and the early 1980s contributing to the increase in fresh water content during those periods. Relatively little ice formation in the late 1990s also contributed to the increase in fresh water content at the end of the integration period.

In both simulations, variability of Bering Strait inflow, continental run-off and precipitation are neglected. Häkkinen and Proshutinsky (2004) give a detailed justification of these omissions. The known anomalies in these forcing functions are clearly much smaller than those resulting in the model for lateral fresh water fluxes and for surface fluxes associated with fluctuations in sea ice formation.

The robust results in these and other simulations with different versions of the NAOSIM system are the increase in Arctic Ocean fresh water content during the first half of the 1960s, a dramatic reduction in fresh water content in the mid-1990s, and an overall downward trend from maximum fresh water content in the mid-1960s to a minimum in the mid-1990s. All these changes seem to be associated mostly with changes in Fram Strait liquid fresh water export.

The reduced fresh water export during the early 1960s allowed the Arctic liquid fresh water reservoir to increase, a prerequisite for the following fresh water export events and the long period of enhanced export rates to the subpolar North Atlantic. According to Köberle and Gerdes (2007), this important event was caused by low volume transports in the EGC. Averaged from mid-1963 to mid-1969 the mean southward volume transport in the EGC was only 2.4 Sv while it increased to 4 Sv for the period mid-1975 to mid-1980. The initial trigger of this volume transport anomaly in the 1960s was an anomalous sea ice export from the Barents Sea into the northward flowing Atlantic waters that determine the salinity of the West Spitzbergen Current (WSC). The fresh Polar Water in the west and the deep reaching saline Atlantic Water in the east characterizes salinity in Fram Strait at a time of normal fresh water export. In the early 1960s, on the other hand, there is little zonal salinity contrast in the upper 200 m (Fig. 17.5). Unfortunately, these salinity anomalies cannot be verified with historical hydrographic data. The published salinity time series from the Sørkapp section across the WSC near the southern tip of Svalbard begin in 1965 (Dickson et al. 1988) and thus would just have missed the event.

A strong reduction in the zonal density gradient in Fram Strait resulted in a strong reduction in the sea surface height difference between Greenland and Svalbard and a corresponding drop in the barotropic transport through the strait. The overall atmospheric situation during the minimum export event is characterized by an anomalous high SLP over most of the Arctic Ocean and the Nordic Seas. The anomalous atmospheric circulation favoured ice transport from the interior Arctic Ocean through the Barents Sea to the Norwegian Sea. Stratification in the Barents Sea was enhanced, heat losses over the Barents Sea reduced. The transport of Atlantic water into the Arctic Ocean through both pathways, Fram Strait and Barents Sea, was reduced.

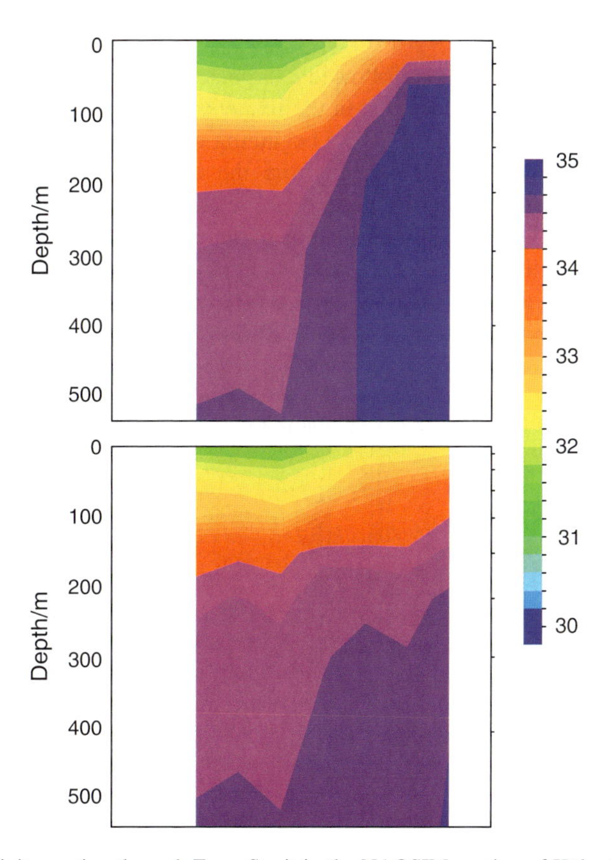

Fig. 17.5 Salinity section through Fram Strait in the NAOSIM version of Köberle and Gerdes (2007). The lower panel shows the reduced upper ocean salinity gradient between Greenland and Svalbard during the low export event of the early 1960s. The upper panel shows the more normal salinity distribution in the 1970s with a pronounced salinity contrast between Atlantic and Polar waters

Analyzing Atlantic layer warming events, Gerdes et al. (2003) had identified an inflow of sea ice into the Barents Sea from the interior Arctic Ocean during the early 1960s (Fig. 17.6). This inflow resulted in a very stable stratification and reduced heat loss from the ocean to the atmosphere. The time series of ice transport through a section from Svalbard to the northern tip of Novaja Semlja shows southwestward ice transport in excess of 1,000 km^3/year for several years in the early 1960s.

A second outstanding liquid fresh water export event happened during the mid-1990s. Karcher et al. (2005) have diagnosed this event in a NCAR/NCEP driven simulation with the 1/4° resolution NAOSIM version. This model was run with restoring of surface salinity (180 days relaxation time) towards climatology. Thus, it likely underestimates the variability of components of the Arctic Ocean fresh water balance. A long-term increasing trend in Fram Strait liquid fresh water transport culminates in an event in the mid-1990s where the transports in several years exceeded the background value by 500–1,000 km^3/year (Fig. 17.7). Most of the freshwater exported during this event continued with the East Greenland Current (EGC) to Denmark Strait.

The liquid export maximum followed a large-scale change of the hydrographic structure in the Arctic as illustrated in the sequence of vertically integrated freshwater content maps in Fig. 17.8. In the beginning of the 1990s, a large freshwater deficit relative to the 1980s extends from the eastern Eurasian Basin to the Mendeleev Ridge. This is consistent with the 'retreat of the cold halocline' (Steele and Boyd 1998), a widespread salinification of the eastern Eurasian Basin observed in the first half of the 1990s. It was attributed to a diversion of Laptev Sea-origin river water eastward along the Siberian shelf sea instead of into the interior of the Arctic Ocean. The changed river water pathway as well as increasing inflow of

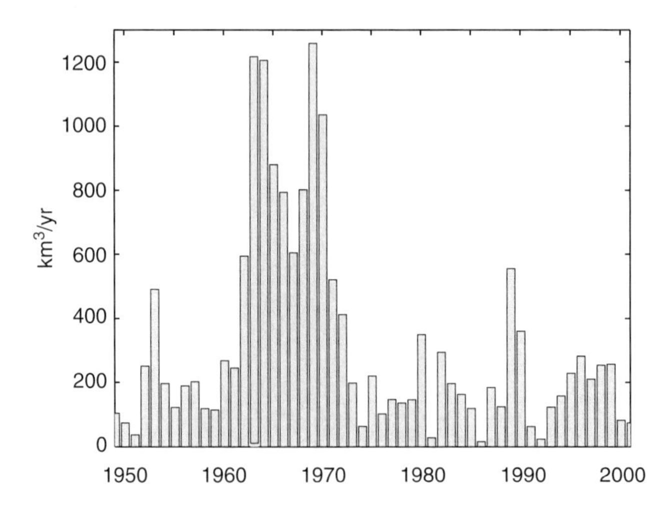

Fig. 17.6 Time series of net sea ice transport through a section between Svalbard and Novaja Semlja. Positive values indicate south-westward transport. The mean transport over the duration of the simulation is 314 km^3/year

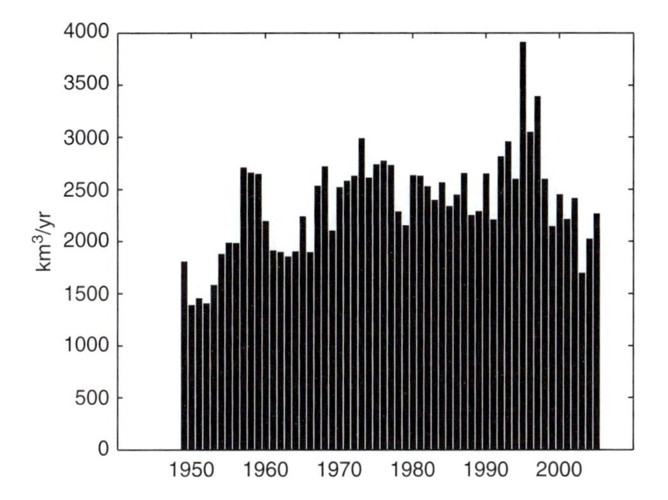

Fig. 17.7 Time series of annual mean liquid fresh water transport through Fram Strait after Karcher et al. (2005). Transports are given in km^3/year. Positive values indicate southward transport anomalies

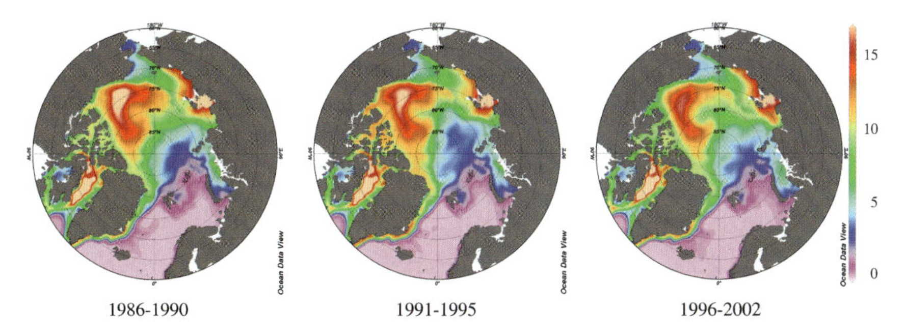

Fig. 17.8 Pentadal averages of simulated (Karcher et al. 2005) liquid fresh content during the second half of the 1980s (left), the first (middle) and the second half (right) of the 1990s. The scale is in meters of pure fresh water that is must be added to a water column of salinity 34.8 to arrive at the actual salinity

Atlantic Water into the Arctic (Karcher et al. 2003) was due to exceptionally large positive index state of the North Atlantic Oscillation until the mid-1990s. In conjunction with the more cyclonic wind stress, this led to an eastward shift of the Atlantic Water boundary in the lower halocline beyond the Lomonosov Ridge towards the Canadian Basin (McLaughlin et al. 2002). The fresh water previously residing in the central Canada Basin and the Beaufort Sea was pushed towards Fram Strait. The thicker layer of Polar Water in Fram Strait geostrophically forced a larger outflow of fresh water into the EGC.

After the export event, the fresh water distribution in the Arctic returned to its more normal state (Fig. 17.8). Downstream, the freshwater export event of the mid-1990s was characterized by a freshening that occurred over a larger depth interval than the GSA signal that was more confined to shallower levels. The signature that has been verified by measured salinity time series in Denmark Strait and is thought to relate to the origin of the low salinity in the liquid fresh water export from the Arctic rather than the export of ice and subsequent melt (Karcher et al. 2005).

To help evaluate model results, it is instructive to compare the liquid fresh water export time series through Fram Strait in the low resolution (Fig. 17.4) and medium resolution (Fig. 17.7) versions of NAOSIM. First inspection reveals that the reduced export of the early 1960s is much less pronounced in the medium resolution model. On the other hand, that model produces a far larger Fram Strait export around 1995 than the low resolution model. Do these differences imply that the model results are arbitrary and that model specifics have a larger influence on the outcome than the atmospheric forcing? Largest differences are due to the negative feedback for salinity anomalies affected by the surface restoring term. Another source of differences are different spin-up histories that produce different initial conditions and that are felt for 10–20 years, corresponding to the renewal time of the Arctic liquid fresh water reservoir.

When we compare results with surface salinity restoring in both model versions and similar spin-up procedures (Fig. 17.9), we find high export rates in the 1970s and in the mid-1990s in both cases. The minimum export in the early 1960s is now somewhat hidden in an adjustment period from very low exports at the end of the spin-up to the higher exports in the 1970s. Overall, the results are quite similar considering the different resolutions. Differences between the models in the relative magnitude of the export in the 1970s and 1990s are partly due to the trend in the fresh water transport through the Canadian Archipelago in the low-resolution version. Here, the flow through the Canadian Archipelago carries an increasing amount of the total

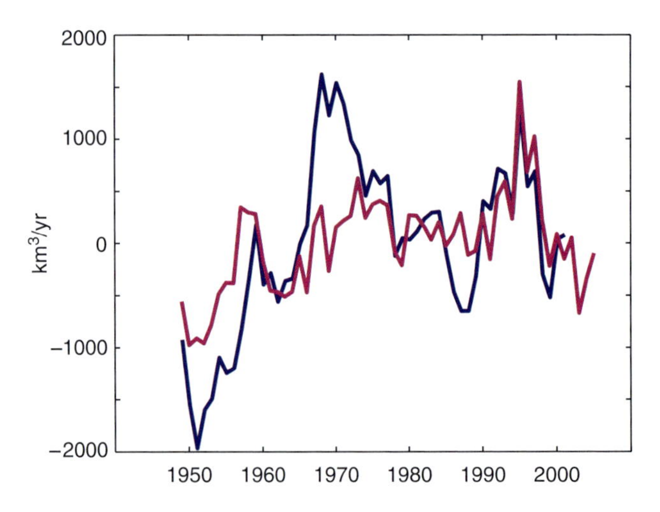

Fig. 17.9 Net liquid fresh water transport anomalies in Fram Strait for the low-resolution version (blue) and the medium-resolution (magenta) version of NAOSIM. Both models use restoring to climatological surface salinities (in contrast to the low-resolution results shown earlier) and are initialised with climatological hydrography

fresh water export from the Arctic Ocean while the medium resolution version has an unrealistically small transport. The export through Fram Strait is thus higher in the low-resolution version during the 1970s but lower during the 1990s.

This comparison again highlights some of the difficulties that still exist in hindcast simulations of the Arctic Ocean: Spin-up, resolution, and treatment of surface fresh water fluxes and run-off. We believe that a flux adjustment as in Köberle and Gerdes (2007) is a viable way to perform hindcasts in regional models. It is obvious that the flux adjustment is something to be documented and interpreted. The possible nonlinear instability of this kind of boundary condition has to be checked. Resolution to resolve the transports through the Canadian Archipelago is achievable now or in the near future. Eddy resolving resolution is certainly in reach for regional models of the Arctic and the sub-polar North Atlantic. Ideally, a model spin-up would be carried out using atmospheric forcing for a long time before the period of interest begins. Given the strong multi-decadal variability, the period of interest is usually as long as consistent and area-wide forcing data, namely the reanalysis data, exist. Kauker et al. (2007) have constructed atmospheric forcing data for the whole 20th century that could be used to spin-up models that are used to investigate the last decades of the century.

17.4 Downstream Effects of Fresh Water Export Events

Increasing fresh water export and large export events from the Arctic Ocean to the subpolar seas are potentially important processes for the deep water formation in the northern North Atlantic. Both major pathways, Fram Strait and the Canadian Archipelago, have been identified as sources for observed large-scale freshenings in the Nordic Seas and the subpolar North Atlantic (Belkin et al. 1998).

There is wide agreement that the GSA of the 1970s was triggered by release of large amounts of sea ice from the Arctic through Fram Strait and to some degree from the Barents Sea. Numerical simulations confirm this picture (Häkkinen 1993; Haak et al. 2003; Köberle and Gerdes 2003). For liquid fresh water export events, the relationship with deep water formation is less clear. Gerdes et al. (2005) show that the deep convection in the Labrador Sea occurred during phases of strongly positive NAO. The only exception in their 50 years hindcast was in the early 1980s when a delay of convection compared to the NAO index (their Fig. 2) was caused by relatively fresh water reaching the convection site of the interior Labrador Sea from the boundary. Gerdes et al. also show (their Fig. 7) salinity variability in the East and West Greenland Currents compared with the interior of the Labrador Sea. The time series of the EGC and WGC are well correlated, indicating propagation of salinity anomalies around the southern tip of Greenland. However, the boundary current time series are uncorrelated with the signals in the interior Labrador Sea. It appears that only certain freshening events in the boundary currents are filtered out and reach the interior Labrador Sea. A similar diagnostic is shown in Fig. 17.10 for the higher resolution version of the NAOSIM model (Fieg et al. manuscript in preparation).

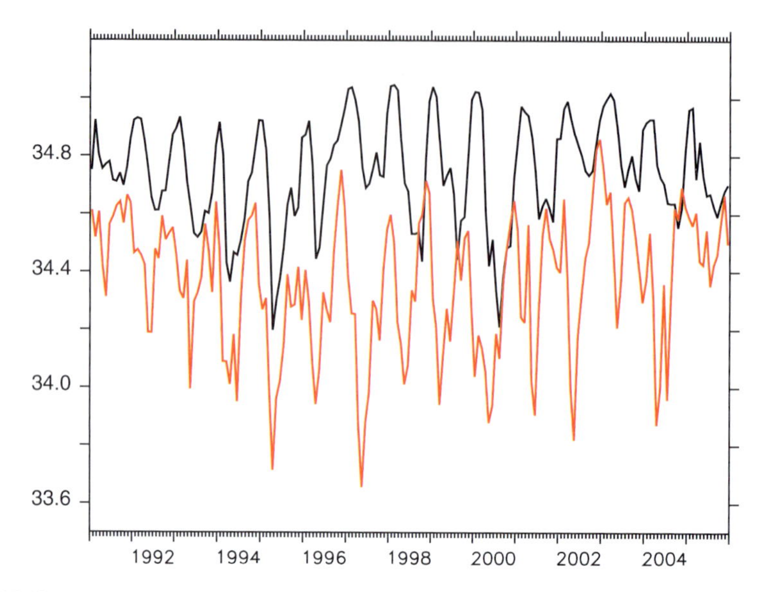

Fig. 17.10 Salinity time series from the 1/12° resolution NAOSIM model (red: West Greenland Current; black: interior Labrador Sea). Note that the model was initialised in 1990 from results of an integration with the corresponding ¼° model

The interannual variability in the interior of the Labrador Sea is much reduced compared to the variability in the boundary current. Thus, we see the decoupling of the interior Labrador Sea from the WGC even in a model that resolves local eddies very well.

Apparently, not every fresh water export event will affect convection and deep water production in the Labrador Sea. In model results, the same is true for the Greenland Sea. The circumstances under which a freshening in the EGC or WGC will affect the adjacent deep basins are not well understood. Sensitivity experiments with a regional eddy-permitting model indicate that fresh water exports that propagate through Davis Strait are not likely to impact Labrador Sea convection directly. Myers (2005) found that enhancing the freshwater export through Davis Strait had little effect on the fresh water content of the Labrador Sea interior and on Labrador Sea Water formation. Similar conclusions were drawn by Komuro and Hasumi (2005) who compare model simulations with and without an open passage connecting the Arctic Ocean and Baffin Bay. Only salinity anomalies moving through Fram Strait directly affect deep water formation while anomalies through the Canadian Archipelago are carried with the rather tight and topographically constrained Labrador Current along the periphery of the Labrador Sea.

17.5 Possible Future Developments

As the hydrologic cycle increases because of global warming, we expect the Arctic fresh water balance and especially the fluxes to lower latitudes to change. Scenario calculations for the development during the 21st century are our best estimate how these changes will evolve – despite all inadequacies still present in these calculations. Results submitted for the fourth assessment report of the IPCC are available from a number of climate research centers. Here, we cannot produce a comprehensive analysis of all these results and must confine ourselves to examine just one example. Figure 17.11 shows results from the A1B scenario calculation with the UK Met Office model HadCM3. The total surface fluxes (including run-off and exchanges with sea ice) and monthly salinity fields are publicly available on the PCMDI (Program for Climate Model Diagnosis and Intercomparison) server (http://www-pcmdi.llnl.gov/ipcc/about_ipcc.php). From this information, we calculated the temporal change of fresh water content in the Arctic and the lateral transports across the boundaries of the Arctic Ocean.

The surface fluxes increase by around 30% over the 21st century. Partly, this is due to sea ice effects. The ice volume is decreasing, as is the difference in ice volume between winter and the preceding summer. This indicates that formation and export of sea ice decrease over the 21st century, which implies that sea ice contributes a positive trend on the surface fresh water fluxes. From the information available to us, we are not able to further distinguish between sea ice exchanges and meteoric fresh water. Precipitation over the Arctic Ocean increases from around 5,500 km^3/year to around 7,500 km^3/year in this scenario calculation.

The lateral exchanges shown in Fig. 17.11 are very variable and determine the higher frequency variability in liquid Arctic Ocean fresh water content. There are a few episodes of very low export rates, comparable to the early 1960s event in the Köberle and Gerdes (2007) hindcast for the second half of the 20th century. These episodes are associated with pronounced increases in Arctic Ocean fresh water

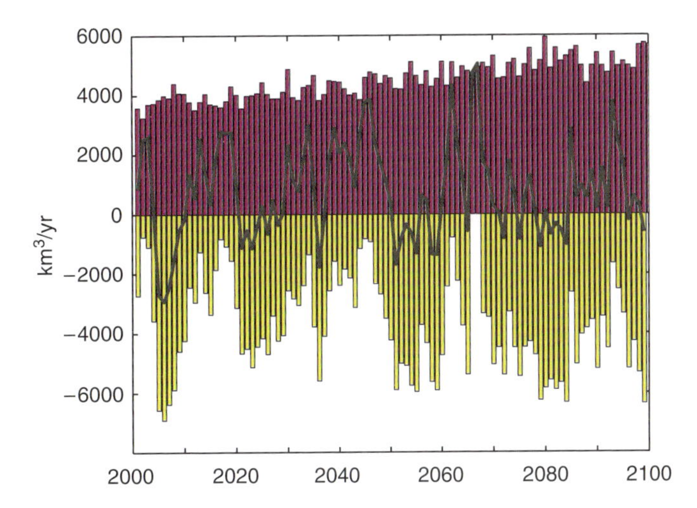

Fig. 17.11 Components of the Arctic Ocean fresh water balance in the A1B scenario calculation for the 21st century with the UK Met Office model HadCM3. Red bars indicate annual surface fresh water fluxes including continental run-off and exchanges with the sea ice. Yellow bars represent the sum of all lateral net liquid fresh water transports to lower latitudes. The solid black line is the rate of change of the Arctic Ocean liquid fresh water content (calculated with a reference salinity of 34.8)

content. Overall, however, we see little change in the liquid fresh water export rate from the Arctic Ocean in the first half of the 21st century. The fresh water content increases over this period. Only in the last decades of the 21st century do the exports gain in strength and counterbalance the increase in surface fluxes. The fresh water balance has gained a new equilibrium with larger fluxes of liquid fresh water through the Arctic Ocean and a thicker Arctic halocline.

For an early detection of global change effects in the northern high latitude seas, sea ice, surface fluxes, and salinity distribution are more suitable than lateral fresh water fluxes because the latter reacts slowest.

17.6 Discussion and Conclusion

Numerical models have been increasingly used with the aim to reconstruct the state of the Arctic ocean–sea ice system and its variability over recent decades. With prescribed atmospheric forcing as well as continental run-off so-called hindcast simulations have been performed. A good overview of many recent simulations can be found in a special issue of the *Journal of Geophysical Research* dedicate to the Arctic Ocean Model Intercomparison Project (AOMIP; Holloway et al. 2007). Results from these calculations can be directly compared with many observational data and estimates. The analysis and the validation of these simulations are still ongoing. Despite its undisputed importance for the high-latitude Atlantic and the large-scale oceanic circulation, the liquid fresh water transport from the Arctic Ocean is not the focus of any detailed study done in the AOMIP framework yet. This is due to problems many ocean–sea ice models have with representing the fresh water balance of the Arctic Ocean and the lack of validation data.

Here, we have presented mostly results from the NAOSIM simulations that, however, are representative for the class of models that are involved in AOMIP. The family of models provides the opportunity to investigate the influence of different model choices, especially of the horizontal and vertical resolution. Salient results of these simulations are the multidecadal variability of the fresh water export and the decreasing trend in the Arctic liquid fresh water from the mid-1960s to the mid-1990s that paralleled the decreasing ice volume (Köberle and Gerdes 2003, 2007).

The time series of fresh water transport through Fram Strait is punctuated by events that potentially have a large downstream impact. A better understanding of the triggers of these events, the frequency of the events and how these conditions will change in the future is necessary. Model simulations indicate that fresh water export events are preceded by redistribution of salt in the Arctic Ocean (Karcher et al. 2005). Should this relationship be confirmed to be robust, it would provide the opportunity to estimate Arctic liquid fresh water export events from interior Arctic hydrographic conditions. This implies predictive potential of conditions in the Arctic for the downstream basins. The temporal variability in the division of fresh water export between the Canadian Archipelago and Fram Strait is an essential process in this respect. Unfortunately, here is great uncertainty in model results and little guidance from observations.

Coupled climate models, for example the Hadley Center model HadCM3, show similar behaviour for the 21st century. We took a cursory look at the A1B scenario run where atmospheric CO_2 concentrations increase by 1% annually until doubling from pre-industrial values. For the first half of the 21st century the export rate of liquid fresh water from the Arctic remains rather constant although the fresh water content increases due to increasing precipitation and run-off. Only in the last decades of the 21st century do the exports gain in strength and counterbalance the increase in surface fluxes. The fresh water balance reaches a new equilibrium with larger fluxes of liquid fresh water through the Arctic Ocean and a thicker Arctic halocline.

Based on model results, long-term variability of liquid fresh water and sea ice export from the Arctic to the subpolar Atlantic are among the key variables for the large-scale ocean circulation. For a quantitative assessment, there are still many uncertainties. Some of the uncertainties have their origin in the numerical models, especially the treatment of surface boundary conditions and the quality of fresh water source data. This problem currently limits our ability to determine the strength of the feedbacks between the fresh water content in the interior Arctic (strongly related to its surface elevation) and the Fram Strait export or the Bering Strait inflow. How strongly will increasing future fresh water content lead to a decrease in Bering Strait inflow and an increase in fresh water export to the Atlantic?

The communication between the fresh boundary currents and the centers of deep water formation governs the large-scale impact of fresh water exported from the Arctic. What constrains these exchanges and what do we need to change in models to improve their representation? What are the most relevant processes for incorporating the fresh water into deep and intermediate water masses? What are the thresholds beyond which the downward transport of fresh water ceases?

Acknowledgements This work was partly funded by the NORDATLANTIK (BMBF contract 03 F 0443 A-E) project, the SFB 512 "Low pressure systems and the climate system of the North Atlantic" of the DFG, and the ASOF-N project of the EU. Further contributions came from INTAS "Nordic Seas in the global climate system" (INTAS ref. no. 03-51-4260) and the Arctic Ocean Model Intercomparison Project (AOMIP). Part of the work was done in the framework of Damocles. The Damocles project is financed by the European Union in the 6th Framework Programme for Research and Development.

References

Aagard K, Carmack EC (1989) The Role of Sea Ice and other fresh water in the Arctic Circulation. J. Geophys. Res., 94, 14,485–14,498

Belkin IM, Levitus S, Antonov JI, Malmberg S-A (1998) "Great Salinity Anomalies" in the North Atlantic. Prog. Oceanogr., 41, 1–68

Dickson RR, Meincke J, Malmberg S-A, Lee AJ (1988) The "Great Salinity Anomaly" in the northern North Atlantic, 1968–1982. Prog. Oceanogr., 20, 103–151

Dickson RR, Rudels B, Dye S, Karcher M, Meincke J, Yashayaev I (2007) Current Estimates of Freshwater Flux through Arctic and Subarctic seas. Progress in Oceanography, 73(3–4), (2007)

Gerdes R, Karcher MJ, Kauker F, Schauer U (2003) Causes and development of repeated Arctic Ocean warming events. Geophys. Res. Lett., 30(19), 1980, doi:10.1029/2003GL018080

Gerdes R, Hurka J, Karcher M, Kauker F, Köberle C (2005) Simulated history of convection in the Greenland and Labrador seas 1948–2001. In: Drange H, Dokken T, Furevik T, Gerdes R, Berger W (eds) The Nordic Seas: An Integrated Perspective. AGU, Geophysical monograph 158, 221–238

Haak H., Jungclaus J, Mikolajewicz U, Latif M (2003) Formation and propagation of great salinity anomalies. Geophys.Res.Lett., 30, 1473, doi:10.1029/2003GL017065

Häkkinen S (1993) An Arctic source for the Great Salinity Anomaly: A simulation of the Arctic ice ocean system for 1955–1975. J. Geophys. Res., 98, 16397–16410

Häkkinen S, Proshutinsky A (2004) Freshwater content variability in the Arctic Ocean. J. Geophys. Res., 109, C03051, doi:10.1029/2003JC001940

Hansen E, Holfort J, Karcher M (2006) Comparing observed and modelled freshwater fluxes: Preliminary results. ASOF-N WP4 report

Holfort J, Meincke J (2005) Time series of freshwater-transport on the East Greenland Shelf at 74° N. Meteorologische Zeitschrift, 14, 703–710

Holloway G, Dupont F, Golubeva E, Häkkinen S, Hunke E, Jin M, Karcher M, Kauker F, Maltrud M, Morales Maqueda MA, Maslowski W, Platov G, Stark D, Steele M, Suzuki T, Wang J, Zhang J (2007) Water properties and circulation in Arctic Ocean models. J. Geophys. Res, 112, C04S03, doi:10.1029/2006JC003642

Karcher MJ, Oberhuber JM (2002) Pathways and modification of the upper and intermediate waters of the Arctic Ocean. J. Geophys. Res., 107, 3049, doi:10.1029/2000JC000530

Karcher MJ, Gerdes R, Kauker F, Köberle C (2003) Arctic warming-evolution and spreading of the 1990s warm event in the Nordic seas and the Arctic Ocean. J. Geophys. Res., 108(C2), 3034, doi:10.1029/2001JC001265

Karcher MJ, Gerdes R, Kauker F, Köberle C, Yashayaev I (2005) Arctic Ocean change heralds North Atlantic freshening. Geophys. Res. Lett., 32, L21606, doi:10.1029/2005GL023861

Kauker F, Köberle C, Gerdes R, Karcher M (2007) Reconstructing atmospheric forcing data for an ocean-sea ice model of the North Atlantic for the period 1900–2003. J. Geophys. Res., (submitted)

Köberle C, Gerdes R (2003) Mechanisms determining the variability of Arctic sea ice conditions and export, J. Clim., 16, 2842– 2858

Köberle C, Gerdes R (2007) Simulated variability of the Arctic Ocean fresh water balance 1948–2001. J. Phys. Oceanogr., 37(6), 1628–1644, doi:10.1175/JPO3063.1

Komuro Y, Hasumi H (2005) Intensification of the Atlantic deep circulation by the Canadian Archipelago throughflow. J. Phys. Oceanogr., 35, 775–789

Large WG, Yeager SG (2004) Diurnal to decadal global forcing for ocean and sea-ice models: the data sets and flux climatologies, CGD Division of the National Center for Atmospheric Research, NCAR Technical Note: NCAR/TN-460+STR

Lohmann G, Gerdes R, Chen D (1996) Sensitivity of the thermohaline circulation in coupled oceanic GCM – atmospheric EBM experiments. Climate Dyn., 12, 403–416

McLaughlin F, Carmack E, MacDonald RW, Weaver AJ, Smith J (2002) The Canada Basin 1989–1995: Upstream events and farfield effects of the Barents Sea. J. Geophys. Res., 1077(C7), 3082, doi:10.1029/2001JC000904

Meredith MP, Heywood KJ, Dennis PF, Goldson LE, White RMP, Fahrbach E, Schauer U, Østerhus S (2001) Freshwater fluxes through the western Fram Strait. Geophys. Res. Lett., 28, 1615–1618

Myers, PG (2005) Impact of freshwater from the Canadian Arctic Archipelago on Labrador Sea Water formation. Geophys.Res.Lett., 32, L06605, doi:10.1029/2004GL022082

Peterson BJ, Holmes RM, McClelland JW, Vörösmarty CJ, Lammers RB, Shiklomanov AI, Shiklomanov IA, Rahmstorf S (2002) Increasing river discharge to the Arctic Ocean. Science, 298, 2171–2173

Prange M, Gerdes R (2006) The role of surface fresh water flux boundary conditions in prognostic Arctic ocean-sea ice models. Ocean Model, 13, 25–43

Prinsenberg SJ, Hamilton J (2004) The Oceanic fluxes through Lancaster Sound of the Canadian Archipelago. ASOF Newsletter No. 2, 8–11

Prinsenberg SJ, Hamilton J (2005) Monitoring the volume, freshwater and heat fluxes passing through Lancaster Sound of the Canadian Arctic Archipelago. Atmos-Ocean, 43, 1–22

Proshutinsky A, Bourke RH, McLaughlin FA (2002) The role of the Beaufort Gyre in Arctic climate variability: Seasonal to decadal climate scales Geophys. Res. Lett., 29(23), 2100, doi:10.1029/2002GL015847

Proshutinsky A, Ashik I, Häkkinen S, Hunke E, Krishfield R, Maltrud M, Maslowski W, Zhang J (2007) Sea level variability in the Arctic Ocean from AOMIP models. J. Geophys. Res., 112, C04S08, doi:10.1029/2006JC003916

Rahmstorf S, Willebrand J (1995) The role of temperature feedback in stabilizing the Thermohaline Circulation. J. Phys. Oceanogr., 25, 787–805

Röske F (2006) A global heat and freshwater forcing data set for ocean models. Ocean Model., 11, 235–297

Serreze M, Barrett AP, Slater AG, Woodgate RA, Aagaard K, Lammers RB, Steele M, Moriz R, Meredith M, Lee CM (2006) The large-scale freshwater cycle of the Arctic. J. Geophys. Res., 111, C11010, doi:10.1029/2005JC003424.

Shiklomanov IA, Shiklomanov AI, Lammers RB, Peterson BJ, Vorosmarty CJ (2000) The dynamics of river water inflow to the Arctic Ocean. In: Lewis EL (ed) Freshwater Budget of the Arctic Ocean. NATO/WCRP/AOSB, Kluwer, Boston, MA, 281–296

Steele M, Boyd T (1998) Retreat of the cold halocline layer in the Arctic Ocean. J. Geophys. Res., 103, 10,419– 10,435

Steele M, Ermold W, Häkkinen S, Holland D, Holloway G, Karcher M, Kauker F, Maslowski W, Steiner N, Zhang J (2001) Adrift in the Beaufort Gyre: A model intercomparison. Geophys. Res. Lett., 28, 2935–2838

Woodgate RA, Aagard K, Weingartner T (2005) Monthly temperature, salinity, and transport variability of the Bering Strait throughflow. Geophys. Res. Lett., 32, L04601, doi:10.1029/2004GL021880

Zhang S, Greatbatch RJ, Lin CA (1993) A reexamination of the polar halocline catastrophe and implications for coupled ocean-atmosphere modeling. J. Phys. Oceanogr. 23, 287–299

Chapter 18
The Overflow Transport East of Iceland

Svein Østerhus[1], Toby Sherwin[2], Detlef Quadfasel[3], and Bogi Hansen[4]

18.1 Introduction

East of Iceland, there are several areas in which overflow of dense water passes from the Norwegian Sea across the Iceland–Scotland Ridge into the Atlantic Ocean. Together, these overflows have been estimated (Hansen and Østerhus 2000) to yield a volume transport of similar magnitude to that through the Denmark Strait (Fig. 18.1). These eastern overflows thus contribute about 50% of the dense overflow from the Arctic Mediterranean into the North Atlantic.

The largest eastern overflow occurs in the deepest passage across the Greenland–Scotland Ridge, the Faroe Bank Channel, with a sill depth of 840 m. Some dense water also crosses the shallower ridges between Iceland and Scotland, the Iceland–Faroe Ridge with sill depth around 480 m, and the Wyville Thomson Ridge with a sill depth around 600 m.

During the ASOF period, observations of the eastern overflows were included in the ASOF-MOEN project. These observations included extension of the measurements of dense overflow through the Faroe Bank Channel and new instrumental records have been obtained for the Iceland–Faroe Ridge and Wyville Thomson Ridge.

Reviews on the history of overflow research and its accomplishments are given in Saunders (2001) and Hansen and Østerhus (2000), who focus on the eastern overflows. In this chapter, we offer a brief overview of the eastern overflows, including results from observations made during the last decade in a series of projects: Nordic WOCE, VEINS, ASOF-MOEN.

[1] Bjerknes Centre for Climate Research, University of Bergen, Bergen, Norway

[2] Scottish Association for Marine Science, Oban, UK

[3] Universität Hamburg, Zentrum für Meeres- und Klimaforschung, Hamburg, Germany

[4] Faroese Fisheries Laboratory, Tórshavn, Faroe Islands

R.R. Dickson et al. (eds.), *Arctic–Subarctic Ocean Fluxes*, 427–441

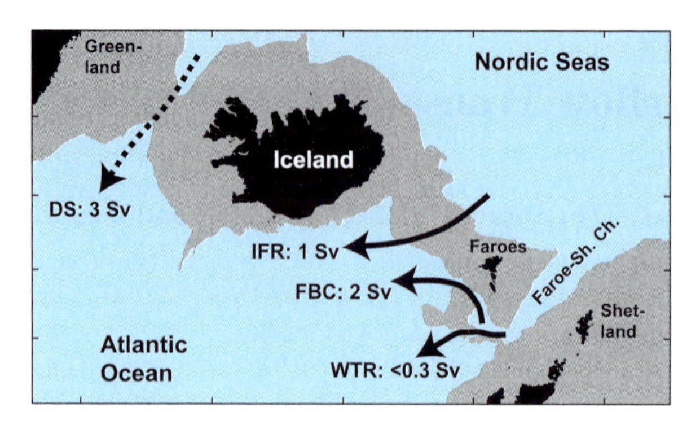

Fig. 18.1 The overflow of dense water between Greenland and Shetland consists of the Denmark Strait (DS) overflow (dashed arrow) and the three eastern overflows: the Iceland–Faroe Ridge (IFR), the Faroe Bank Channel (FBC), and Wyville Thomson Ridge (WTR) overflows. (Based on Hansen and Østerhus 2000)

18.2 The Iceland–Faroe Ridge Overflow

The Iceland–Faroe Ridge consists of a shallow plateau (Fig. 18.2) with several intersecting trenches. The interface between the warm Atlantic water and the cold, dense overflow water in the Nordic Seas usually lies below the sill depth of the plateau along most of its length (Fig. 18.3), but mesoscale eddies may lift the interface so that overflow events may occur intermittently in many locations along the Ridge (Fig. 18.2). Close to Iceland, the interface is often found to be higher and the trench closest to Iceland seems to carry a more persistent overflow (Perkins et al. 1998).

Using results from the Danish "Ingolf expedition" Knudsen (1898) observed for the first time the Iceland–Faroe Ridge overflow (IFR-overflow) and suggested that it was intermittent while Nielsen (1904) used observations from the 'Thor' expedition and argued that it must be continuous. Dietrich (1956) reported measurements from four sections across the ridge and Hermann (1959) and Steele (1959) published sections and maps of bottom temperatures. Between 1959 and 1971, a German research vessel occupied a standard section west of the Ridge on 14 occasions and Meincke (1972) has reported the average temperature distribution on this section (Fig. 18.4). The cold water that dominates the deep parts of this section must have crossed the Ridge and this supports the concept that overflow occurs widely along the Ridge, but variably.

In 1960 and 1973, the International Council for the Exploration of the Seas (ICES) coordinated two overflow expeditions contributing to the general topic of water mass exchange across the Greenland–Scotland Ridge (Tait 1967; Meincke 1974). From the results of the Overflow '60 expedition, Hermann (1967) estimated the IFR-overflow to have a volume transport of 1.1 Sv (1 Sv = 10^6 m^3 s^{-1}) and the results from the Overflow' 73 experiment were largely consistent with that (Meincke 1983). Based on observations close to the Icelandic continental rise,

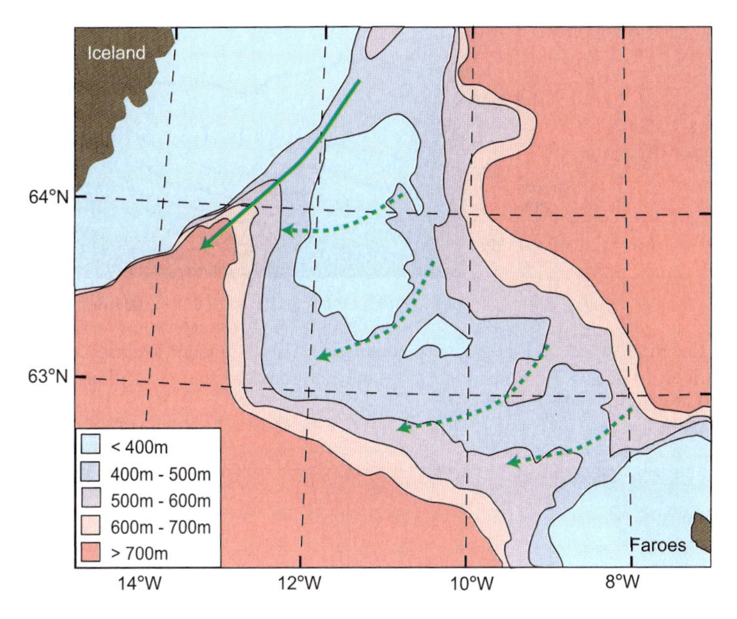

Fig. 18.2 Topography of the Iceland–Faroe Ridge. Green arrows indicate intermittent (dashed arrows) or more persistent (continuous arrow) overflow paths across the Ridge according to Hansen and Østerhus (2000)

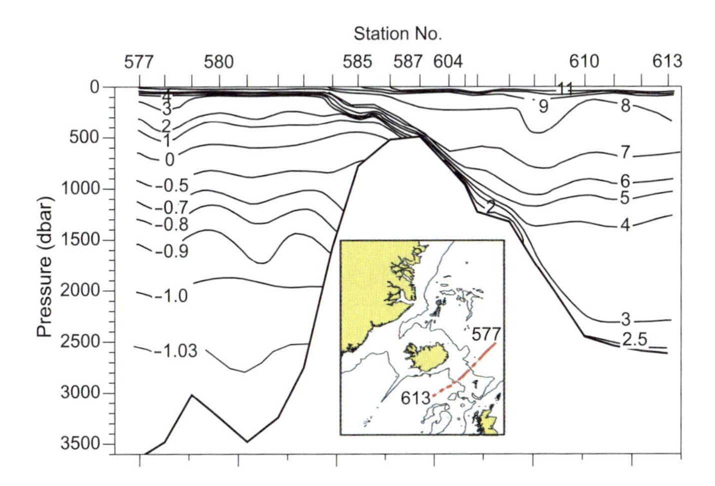

Fig. 18.3 Temperature section from the Norwegian Sea (Station no 577) over the Iceland–Faroe Ridge to the Iceland Basin (Station 613). Red dots on inset map indicate stations

Perkins et al. (1998) estimated a transport of undiluted overflow water of 0.7 Sv. They considered this an underestimate of the total IFR-overflow since they did not cover all the flow, and so 1 Sv is commonly cited as the IFR-overflow volume transport (Meincke 1983; Dickson and Brown 1994; Hansen and Østerhus 2000).

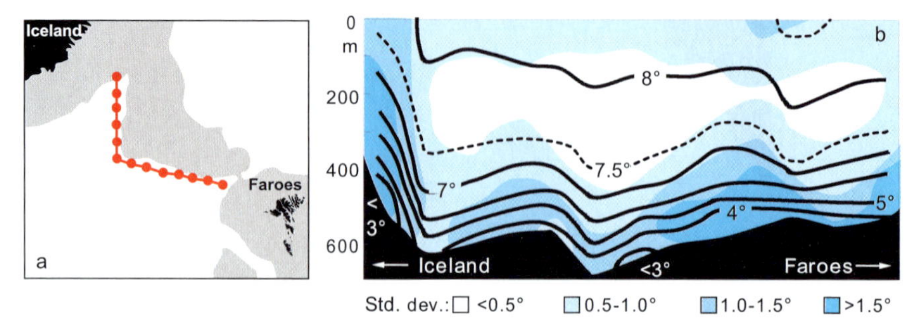

Fig. 18.4 Location of (a) and average temperature on (b) a standard section occupied by RV Anton Dohrn on 14 occasions. Red circles in (a) indicate standard stations. Colours in (b) indicate the standard deviation of the temperature

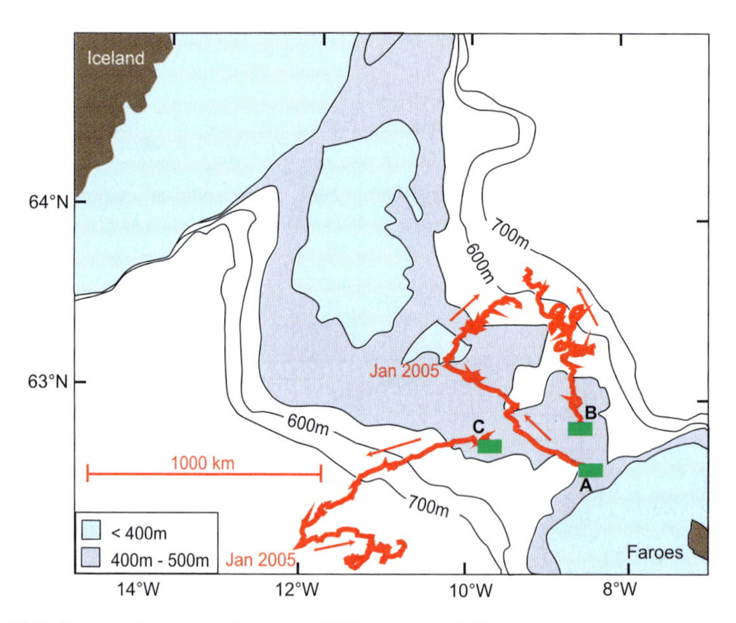

Fig. 18.5 Progressive vector diagrams (PVDs) from ADCP measurements on the Iceland–Faroe Ridge. ADCP locations are indicated by green rectangles, labelled A, B, and C. Red traces indicate PVDs 18–20 m above the bottom (bin 1) at sites A (period: 2004.09.07–2005.05.23), B (period: 2003.07.05–2004.06.10), and C (period: 2004.07.04–2005.05.23). Red length scale indicates scale for all the PVD traces

During the ASOF period, the IFR-overflow was observed as part of the ASOF-MOEN project by means of bottom mounted Acoustic Doppler Current Profilers (ADCPs) in addition to temperature and salinity sensors. Figure 18.5 shows near-bottom progressive vector diagrams (PVDs) from three deployments, which each lasted more than 9 months, and Fig. 18.6 shows simultaneous bottom temperatures measured at two of these.

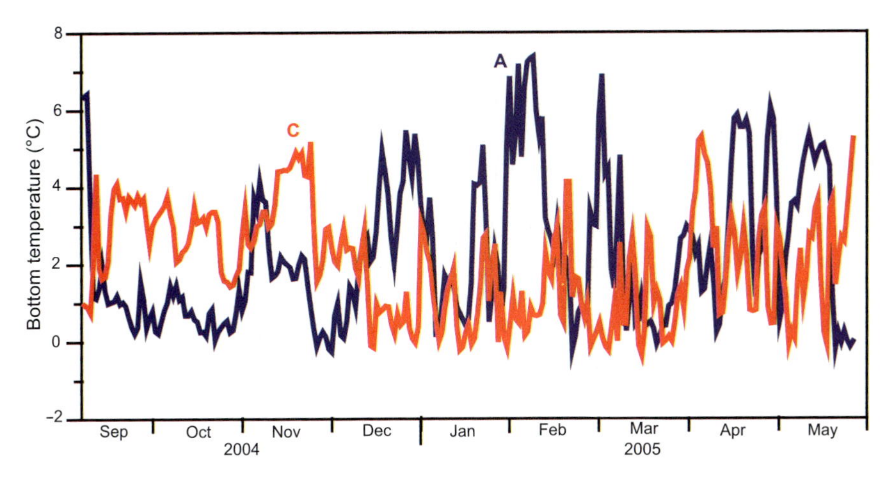

Fig. 18.6 Daily averaged bottom temperature at two ADCP sites (A and C in Fig. 18.5) on the Iceland–Faroe Ridge from 7 September 2004 to 23 May 2005

The bottom temperature at these two sites (and the one at site B, the previous year) show frequent occurrences of overflow water with temperatures below 2 °C, but there seems to be no clear relationship (either positive or negative) between the occurrences at the two sites. The PVDs from these two sites do, however, exhibit a certain similarity, since they both indicate a change in the general flow direction at about the same time (in January 2005, Fig. 18.5). This change seems to imply a transition from stronger to weaker overflow in this region, and it seems to involve time scales of several months, but the bottom temperatures (Fig. 18.6) do not show any obvious signs of a similar change.

The ASOF-MOEN ADCP deployments were intended to explore the overflow characteristics and dynamics in selected areas, rather than for transport estimates and, when the analysis has been completed, they will provide better statistics and better understanding of the IFR-overflow. At the same time they do, however, demonstrate the difficulties involved in measuring, not to say monitoring, the volume transport of IFR-overflow.

It may be argued that the IFR-overflow is the most difficult one of all the overflow branches to monitor because it occurs over such a wide area and intermittently in most locations. It is possible to deploy moorings farther downstream that may capture all the IFR-overflow, but then there may also be contributions from the FBC-overflow and entrained Atlantic water.

18.3 The Faroe Bank Channel Overflow

The occurrence of dense water overflow through the Faroe Bank Channel (FBC-overflow) is a long-established fact (Hermann 1959). There is also a comprehensive literature on this phenomenon, based both on observations and theory, as reviewed by Hansen and Østerhus (2000) and by Saunders (2001). A detailed study, based on recent current and hydrographic data has been submitted by Hansen et al.

The Faroe Bank Channel sill depth is 840 m (Fig. 18.7), which is more than 200 m deeper than any other passage across the Greenland–Scotland Ridge and it is the main outlet for the coldest deep water produced in the Arctic Mediterranean. The deep parts of this channel are always dominated by cold, dense water that flows with large speed towards the Atlantic (Fig. 18.8).

Since the late 1980s, the Faroese Fisheries Laboratory has monitored the water mass properties of the channel with regular CTD cruises and in 1995, monitoring of current velocities by moorings was initiated as part of the Nordic WOCE programme (Østerhus et al. 2001). This activity has been maintained since then and from these observations (Fig. 18.9), a comprehensive data set has been generated (Hansen and Østerhus 2007).

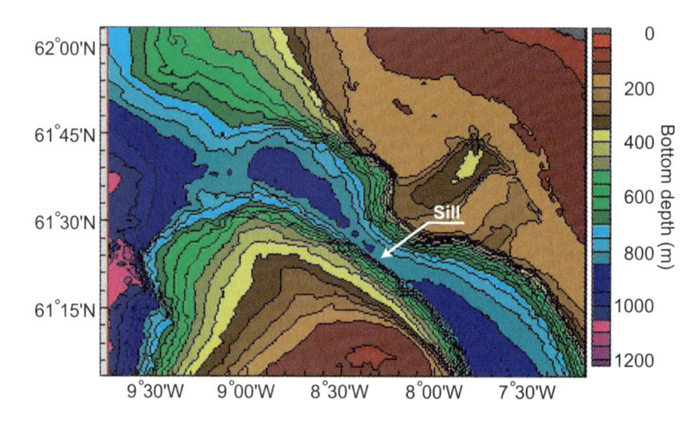

Fig. 18.7 Bottom topography of the Faroe Bank Channel (Courtesy of Knud Simonsen)

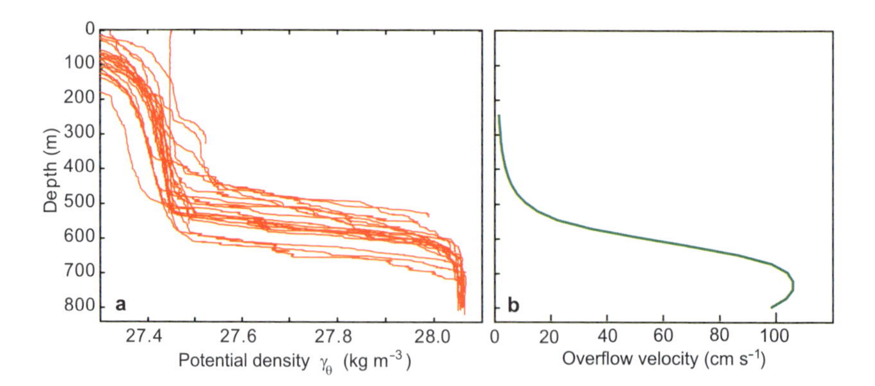

Fig. 18.8 Density (a) and velocity (b) profiles in the Faroe Bank Channel. Red traces in (a) indicate potential density from individual CTD profiles over the sill. Green curve in (b) shows the (vectorially) averaged velocity component in the along-channel direction (towards 304°) based on a typical ADCP deployment. (Adapted from Hansen and Østerhus 2007)

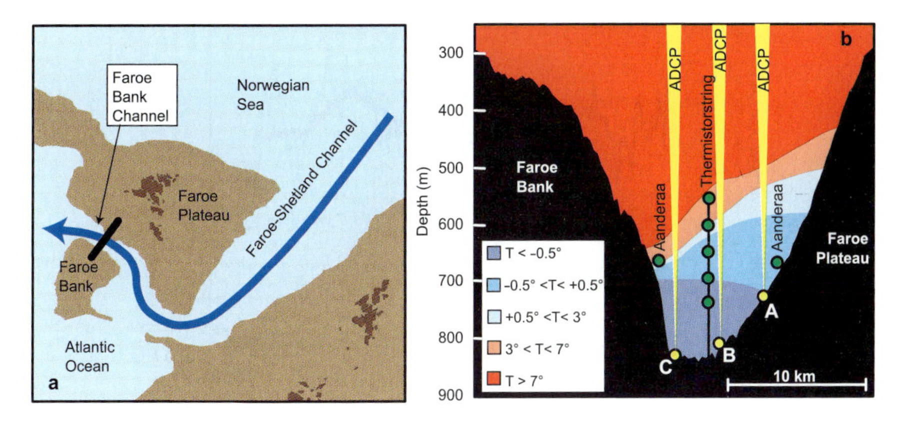

Fig. 18.9 A section crossing the Faroe Bank Channel at the sill (thick line in a) shows a typical temperature distribution and equipment that has been moored for longer or shorter periods (b). The ADCP at site B has been deployed continuously since November 1995, apart from short annual servicing periods. A CTD section across the channel has been occupied at least four times a year since the late 1980s

The narrowest part of the channel is at the sill, where the overflow-filled part is only about 10 km wide (Fig. 18.9). Since this width is comparable to the baroclinic Rossby radius, one might expect fluctuations in hydrography and current in the overflow-layer to be coherent across the channel. Observations have confirmed this, which makes it possible to monitor the volume transport through the channel by means of only one bottom mounted ADCP (Hansen and Østerhus 2007).

Based on this, a decade-long time series of the overflow volume transport has been generated with unprecedented accuracy (Hansen and Østerhus 2007). Figure 18.10 shows the monthly and annually averaged kinematic (defined from the velocity field only) volume transport of the FBC-overflow from 1995 to 2005. The kinematic volume transport exhibits a clear seasonal variation with maximum in August and minimum in February. The seasonal amplitude is about 10% of the average volume transport. The average kinematic volume transport for the whole 1995–2005 period was 2.1 Sv. Using the 27.8 kg m^{-3} isopycnal to define water sufficiently dense to be characterized as overflow, the volume transport for the period was estimated to 1.9 Sv with an average temperature of 0.25 °C and density (γ_θ) of 28.01 kg m^{-3} (Hansen and Østerhus 2007).

The kinematic volume transport exhibits some inter-annual variations but no persistent trends are seen over the 1995–2005 period (Fig. 18.10). Hansen et al. (2001) suggested a link between the strength of the overflow and the depth of isopycnal surfaces upstream in the Norwegian Sea at Ocean Weather Station M, but this suggestion is not supported by the more recent observations (Hansen and Østerhus 2007). The isopycnals 27.8 and 28.0 kg m^{-3} at Ocean Weather Station M are shown in Fig. 18.10. The depth of the 27.8 kg m^{-3} isopycnal has a trend over the 1995–2005 period, but the increased depth did not reduce the strength of the Faroe Bank Channel overflow. The bottom temperature at the sill also has clear seasonal

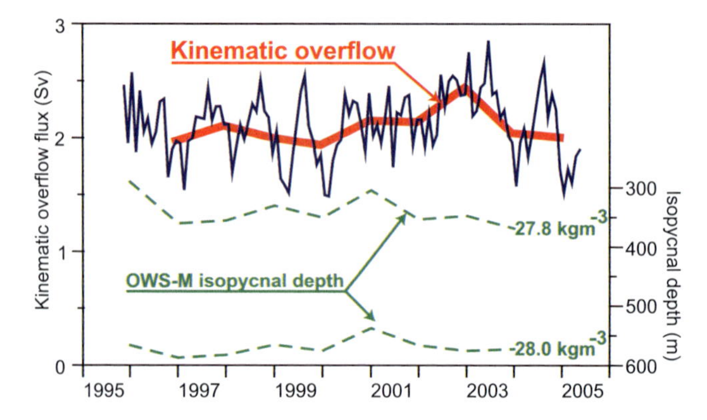

Fig. 18.10 Monthly (thin blue line) and annually (thick red line) averaged kinematic overflow transport through the Faroe Bank Channel and the depths of two isopycnal surfaces at Ocean Weather Ship M (dashed lines) in the Norwegian Sea. The kinematic overflow is defined as the volume transport below the level where the overflow velocity has been reduced to half the core velocity (Based on Hansen and Østerhus 2007)

and inter-annual variations, but no trend. The warmest parts of the FBC-overflow did, however, increase in salinity in this period, with a small density increase as a consequence (Hansen and Østerhus 2007).

18.4 The Wyville Thomson Ridge Overflow

The quantity of overflow water transported across the Wyville Thomson Ridge (WTR-overflow) has been a matter of controversy since the Ridge was first discovered in 1880. In recent times, observations by Ellett and Edwards (1978), Saunders (1990), and Ellett (1991) produced estimates that range from 1.2 Sv for the maximum combined transport of cold overflow water and entrained Atlantic water, through 0.35 Sv for water <3.5 °C, to 0.3 Sv for the mean combined transport, respectively. On the basis of these observations, Hansen and Østerhus (2000) concluded that the mean cold water transport across the ridge was in the range of 0.1–0.3 Sv.

Whilst this range is small compared with the total cold water transport through the Faroe Bank Channel, at its upper end it is comparable to that of the Mediterranean outflow (0.67 Sv, Bryden et al. 1994). Thus given the propensity for it to entrain over 2 times its own volume of Atlantic water, it is possible that the WTR-overflow is an important component of the circulation on the eastern side of the north-east Atlantic. For these reasons, and because its particular setting provides some interesting insights into the nature of overflows, sustained monitoring of the WTR-overflow was commenced in 2003.

Part of the difficulty of quantifying this transport is due to the complicated bathymetry that it flows across. In the first place, overflow can occur at two locations on the Wyville Thomson Ridge: (i) through the deepest point, a saddle of typically 600 m at 7° 45' W, and (ii) at the western end where bathymetric steering against the Faroe Bank can lift the cold water east of the Ridge to depths of 350 m. A detailed survey of a large overflow event by Sherwin and Turrell (2005) showed water crossing at both these points, before subsequently flowing westward and downward towards the Ellett Gully, a narrow gully between the Ymir Ridge and the Faroe Bank (Fig. 18.11).

The second problem is that the bathymetry of this gully is quite convoluted – its full detail was only realised following a swath bathymetry survey in October 2005. The existence of the Faroe Bank canyon on the northern side of the gully, coupled to theoretical considerations about the likely pathway of overflow water travelling along the flank of the Faroe Bank (e.g. Wåhlin 2002), and the ability of currents in the region to scour the topography (e.g. Kuijpers et al. 2002), leads one to conclude that the Ellett Gully collects water from several different directions. Thus, careful consideration needs to be given to the positioning of any recording instruments if they are to provide a true estimate of transport.

In the winters of 2003/04 and 2004/05 an ADCP was deployed at Station 1 (Fig. 18.11) and, following the swath bathymetry survey, a third deployment in the

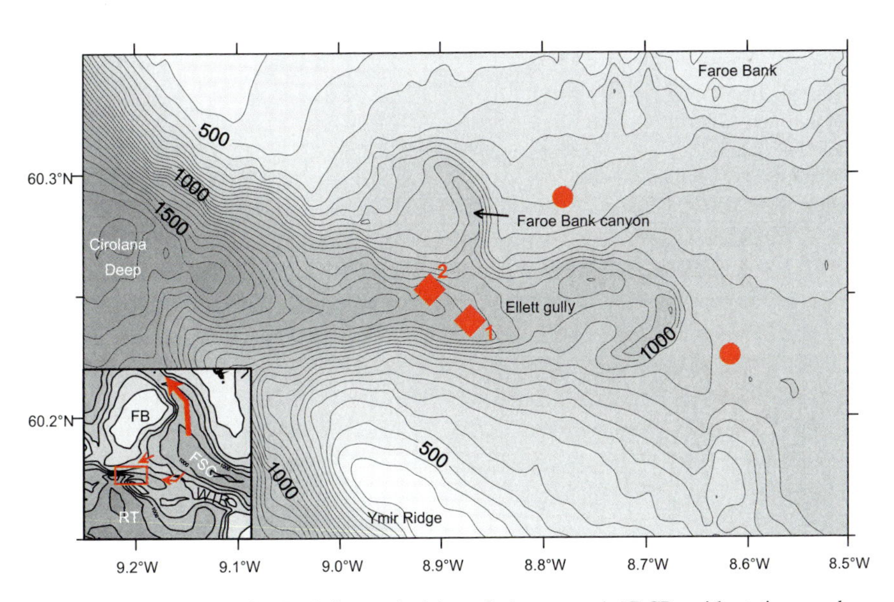

Fig. 18.11 The Ellett gully. Red diamonds (♦) mark the moored ADCPs with station numbers. Stations marked with red circles (•) were occupied by Ellett (1991). Inset: Red rectangle indicates position of this map relative to the Wyville Thomson Ridge (WTR), the Faroe–Shetland Channel (FSC), Faroe Bank (FB), and Rockall Trough (RT). Red arrows show the main pathways for overflow water, against the Faroe Bank and through the central gap on the WTR, and through the Faroe Bank Channel (to the north)

summer of 2005 was made at Station 2, downstream of the canyon. The results of these deployments are summarised in Table 18.1, and more detail can be found in Sherwin et al. (2007). Total transport calculations are based on the assumption that the overflow tended to fill the section of the gully at each station. By measuring the temperature at the ADCP and assuming a constant gradient to 8.5 °C at the top of the overflow, it is possible to estimate the transport of undiluted overflow water that has been mixed into it.

The record at Station 2 (Fig. 18.12) reveals a total transport that is remarkably steady compared with the observations at Station 1, see Table 18.1, and those reported by Ellett (1991) with a mean of 1.0 ± 0.34 Sv. The mean transport of undiluted overflow water (0.3 Sv) is at the top of the range mentioned in Hansen and Østerhus (2000). Using these data, Sherwin et al. (2007) suggested that a weighted average of the observations at Stations 1 and 2 should be used, and proposed a new estimate of between 0.8 and 0.9 Sv for the mean total transport, and between 0.2 and 0.3 Sv for undiluted overflow water.

Table 18.1 Summary of moored ADCP deployments downstream of the Wyville Thomson Ridge

	Station 1a	Station 1b	Station 2
Dates	26 Sept. 2003–5 April 2004	17 Oct. 2005–5 May 2006	7 May 2006–6 Oct. 2006
Location	60.24° N 8.87° W	60.24° N 8.87° W	60.25° N 8.91° W
Water depth (m)	1,140	1,140	1,200
Mean transport (Sv)	0.57	0.34	1.00
Standard deviation (Sv)	0.38	0.25	0.34
Mean transport of undiluted overflow (Sv)	0.11	0.056	0.30

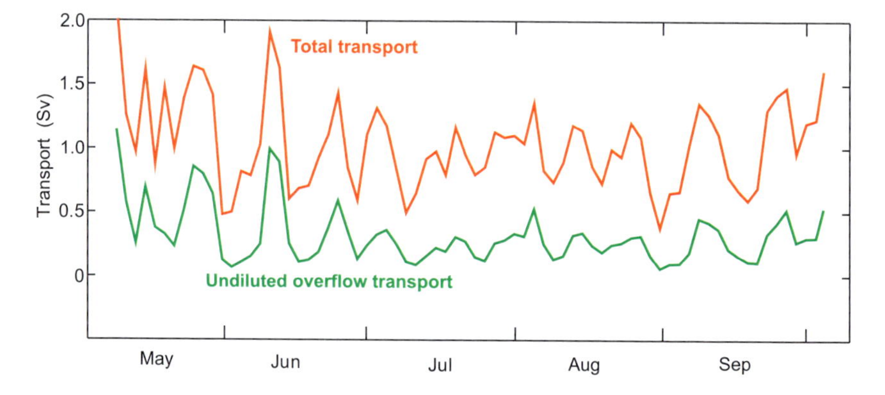

Fig. 18.12 Overflow westward transport at Station 2 west of the Wyville Thomson Ridge (Fig. 18.11) from May to October 2006. Red line: total transport; green line: transport of undiluted overflow water

However, it is still too early to call these values definitive. Even though it appears from inspection of Fig. 18.11 that the locations of the mooring sites used by Ellett (1991) were not particularly well suited for the task, his low values may still reflect a long term trend. More significantly, in the Faroe Bank Channel the strongest transport is in late summer so the difference between the summer deployment at Station 2 and the winter deployments at Station 1 may be seasonal rather than positional. Against this, Sherwin et al. (2007) argue that this explanation is too simplistic and that some of the increase must be due to transport in the Faroe Bank canyon. A more sustained period of deployment at Station 2 is required (and is taking place), and a better understanding of the transport paths of the overflow water is needed, before true confidence can be assigned to the significance of the WTR-overflow for the circulation of the north-east Atlantic.

18.5 Eastern Overflow Contribution to NADW

The flow that crosses the Iceland–Scotland Ridge towards the Atlantic includes water of different origins and with different properties. The coldest and densest water passes through the Faroe Bank Channel, where the deep parts of the channel are generally filled with water of temperatures well below 0 °C and potential densities exceeding 28 kg m^{-3} (Figs. 18.8 and 18.9). This water propagates westwards and southwards through the Atlantic Ocean, but only close to the Ridge do we find these low temperatures and high densities, because the overflow water entrains ambient waters from the Atlantic. The combined overflow and entrained water is commonly called Iceland–Scotland Overflow Water (ISOW) and this is the water mass that contributes to North Atlantic Deep Water (NADW).

The WTR-overflow seems to deliver most of its water to the Rockall Trough and on to the eastern part of the North Atlantic, but the water from the IFR-overflow and the FBC-overflow passes into the Iceland Basin (Fig. 18.13) and most of it is considered to continue into the western basin of the North Atlantic through the Charlie Gibbs Fracture Zone and other gaps in the Mid Atlantic Ridge (Hansen and Østerhus 2000). Using measurements from a series of current meter moorings (light blue circles on Fig. 18.13), Saunders (1996) identified a current core on the slope south of Iceland (light blue arrows on Fig. 18.13), but ISOW dominates the near-bottom layers of the Iceland Basin over much wider areas (Fig. 18.14b).

The ISOW in the Iceland Basin is, however, already strongly modified with temperatures typically above 2 °C (Fig. 18.14a). This indicates a large admixture of ambient water and Fogelqvist et al. (2003) used a multitracer survey by RV Johan Hjort in 1994 to establish that only 54% of the ISOW on a section extending south-eastwards from the southwestern tip of Iceland (Fig. 18.13) originated north of the Ridge. The remaining 46% was entrained water, including North East Atlantic Water (18%), Labrador Sea Water (25%), and North East Atlantic Deep Water (3%).

The strong component of Labrador Sea Water is to be expected when considering that this water mass is adjacent to the ISOW along much of its path (Fig. 18.14b)

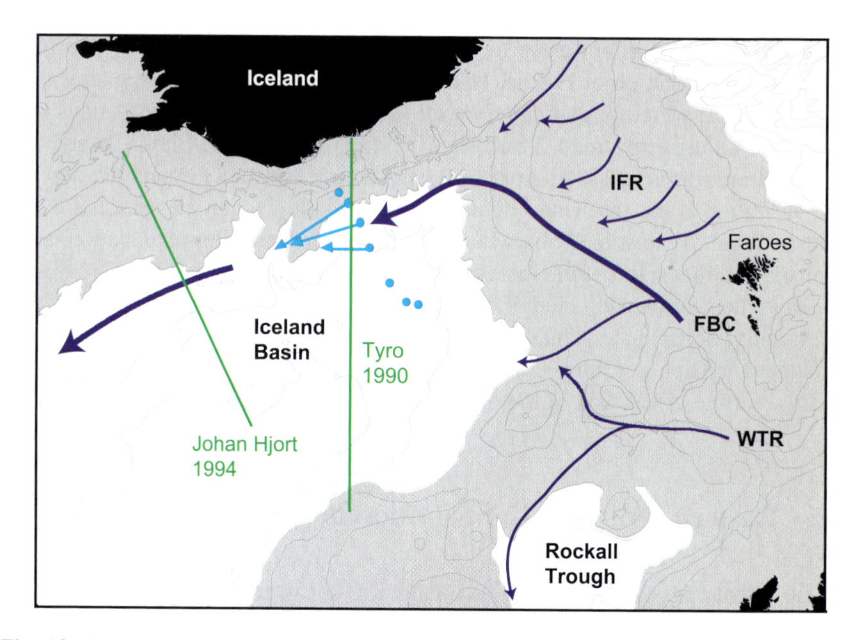

Fig. 18.13 The passage of overflow water into the eastern North Atlantic. Dark blue arrows indicate overflow paths. Light blue circles indicate moorings by Saunders (1996) and light blue arrows indicate the core of the overflow current through his section. Green lines indicate Research vessel sections referred to in the text. The eastern overflow areas are indicated: Iceland–Faroe Ridge (IFR), Faroe Bank Channel (FBC), and Wyville Thomson Ridge (WTR). Shaded areas are shallower than 1,500 m (Based on Hansen and Østerhus 2000)

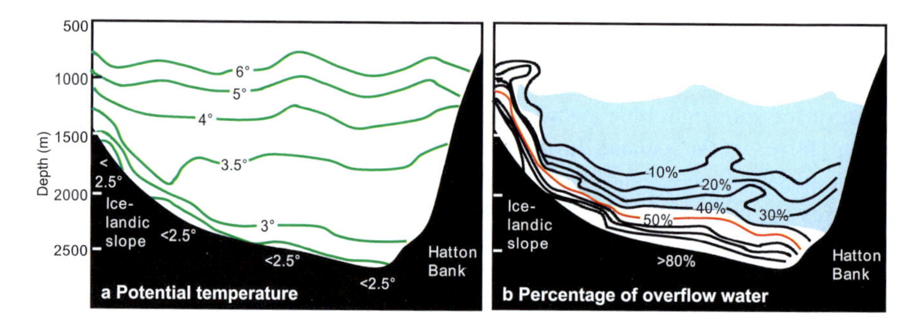

Fig. 18.14 Potential temperature (a) and percentage of overflow water (b) along a meridional section through the Iceland Basin, occupied by RV Tyro in 1990 (Fig. 18.13). Red line on (b) indicates 50% overflow content. Blue area on (b) indicates more than 50% Labrador Sea Water. (Based on de Boer et al. 1998 and de Boer 1998)

and ISOW can be expected to entrain more of it, and more deep water, on its further path to NADW. There is, however, also a considerable contribution from the shallower North East Atlantic Water and this must have been entrained before the overflow water has descended too deep. The IFR-overflow and the WTR-overflow can be expected to cross the Ridge in thin (vertically) plumes with low speeds, except possibly for the overflow passage closest to Iceland (Fig. 18.2). The water in these overflow branches will therefore be in close contact with Atlantic water.

The FBC-overflow, on the other hand, passes the sill in a layer that is more than 200 m thick and has speeds exceeding 1 m s^{-1}, but also this branch experiences intensive mixing shortly after passing the Ridge, as was demonstrated by an experiment in 1999–2001, funded by the Danish Research Council. In this experiment, a total of 25 moorings were deployed in three periods in the area immediately downstream of the Faroe Bank Channel exit (Fig. 18.15).

In this area, the bottom slopes steeply towards the depths of the Iceland Basin and the overflow will be strongly accelerated. Since the FBC-overflow is already hydraulically critical as it passes the sill (Hansen and Østerhus 2007), supercritical conditions and instabilities can be expected downstream of the exit and the experiment demonstrated very regular oscillations (period 88 h) in both temperature and current velocity close to the bottom (Høyer and Quadfasel 2001; Geyer et al. 2006).

The strong oscillatory motions are likely to induce intensive mixing and this may explain the rapid increase in bottom temperature as the overflow plume propagates away from the channel exit. At the sill, the bottom temperature was around −0.4 °C during this experiment. Some 250 km downstream, the coldest part of the mooring array close to 13° W had an average near-bottom temperature of 1.9 °C (Fig. 18.15).

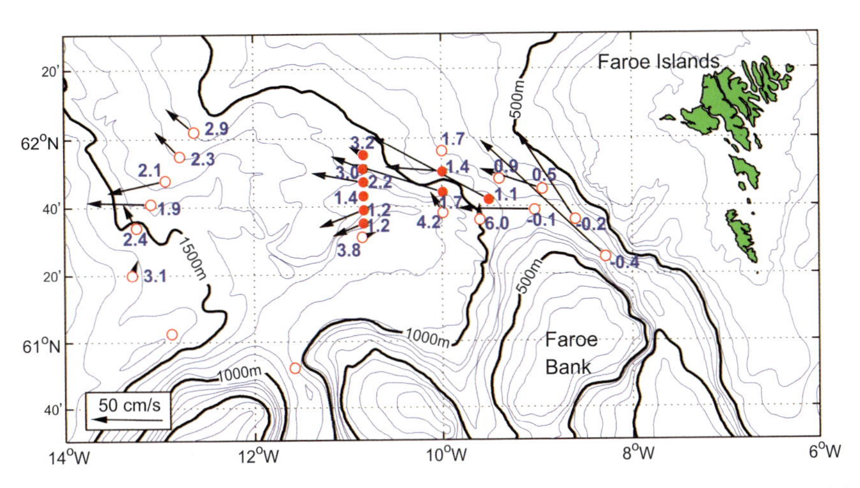

Fig. 18.15 Moorings with temperature and/or current measurements deployed downstream of the Faroe Bank Channel between July 1999 and February 2001 are indicated by red circles. Filled circles indicate sites where the regular (88 h) oscillations were observed. Arrows indicate average velocity and numbers indicate average temperature. All measurements were made within 30 m from the bottom

18.6 Conclusions and Outlook

During the last decade, observations of the Iceland–Scotland overflows have been obtained in a series of international and national projects, starting with Nordic WOCE, through VEINS to ASOF-MOEN. Systematic monitoring of the overflow through the Faroe Bank Channel was continued during the ASOF-MOEN project and new observations of overflows across the shallow ridges have been obtained.

The average volume transport of eastern overflow is estimated to about 3 Sv with the largest uncertainty deriving from the difficulty of measuring the overflow transport across the Iceland Faroe Ridge. For the 1995–2005 period, high-quality time series have been constructed for the overflow through the Faroe Bank Channel. These time series show seasonal and inter-annual variations, but no trends in volume transport, or in bottom temperature. The salinity of this overflow did, however, increase after the mid-1990s.

The total volume transport of the eastern overflows is of similar magnitude to the Denmark Strait overflow and they provide a considerable fraction of the total North Atlantic Deep Water when the entrained water is included. With future climate change, monitoring of these flows is therefore an important task. For the Faroe Bank Channel, a relatively simple system can provide accurate numbers and the overflow across the Wyville Thomson Ridge also seems manageable. The most demanding task will be monitoring of the overflow across the Iceland Faroe Ridge, but it is hard to imagine how that can be done without huge investments in instrumentation and ship-time. An alternative would be to monitor the total transport of eastern overflow and entrained water through the Iceland Basin, but even that will require large resources.

References

Bryden H, Candela J, Kinder TH (1994) Exchange through the Strait of Gibraltar. Progress in Oceanography 33:201–248

de Boer CJ (1998) Water mass distribution in the Iceland Basin calculated with an Optimal Parameter Analysis. ICES Cooperative Research Report No 225:228–246

de Boer CJ, van Aken HM, van Bennekom AJ (1998) Hydrographic variability of the overflow water in the Iceland Basin. ICES Cooperative Research Report No 225:136–149

Dietrich G. (1956) Überströmung des Island-Faröer-Rücken in Bodennahe nach Beobachtungen mit dem Forschungsschiff 'Anton Dohrn' 1955/56. Deutsche Hydrographische Zeitschrift 9:78–89

Dickson RR, Brown J (1994) The production of North Atlantic Deep Water: sources, rates, and pathways. Journal of Geophysical Research 99:12319–12341

Ellett DJ (1991) Norwegian Sea Deep Water overflow across the Wyville Thomson Ridge during 1987–88. ICES CM1991/C:41

Ellett DJ, Edwards A (1978) A volume transport estimate for Norwegian Sea overflow across the Wyville Thomson Ridge. Overflow' 73 expedition contribution nr 37. ICES CM1978/C:19, 12 pp

Fogelqvist E, Blindheim J, Tanhua T, Østerhus S, Buch E, Rey F (2003) Greenland-Scotland overflow studied by hydro-chemical multivariate analysis. Deep-Sea Research I 50:73–102

Geyer F, Østerhus S, Hansen B, Quadfasel D (2006) Observations of highly regular oscillations in the overflow plume downstream of the Faroe Bank Channel. Journal of Geophysical Research 111: C12020, doi:10.1029/2006JC003693

Hansen B, Østerhus S (2000) North Atlantic-Nordic Seas exchanges. Progress in Oceanography 45(2):109–208

Hansen B, Turrell WR, Østerhus S (2001) Decreasing overflow from the Nordic seas into the Atlantic Ocean through the Faroe Bank channel since 1950. Nature 411:927–930

Hansen B, Østerhus (2007) Faroe Bank Channel overflow 1995–2001. Progress in Oceanography. doi:10.1016/j.physletb.2003.10.071

Hermann F (1959) Hydrographic Observations in the Faroe Bank Channel and over the Faroe-Iceland Ridge June 1959. Journal du Conseil International pour l'Exploration de la Mer 118, 5 pp

Hermann F (1967) The T-S Diagram Analysis of the Water Masses over the Iceland-Faroe Ridge and in the Faroe Bank Channel (Overflow '60). Rapports et Procès-Verbaux des Réunions du Conseil International pour l'Exploration de la Mer 157:139–149

Høyer JL, Quadfasel D (2001) Detection of deep overflows with satellite altimetry. Geophysical Research Letters 28(8):1611–1614

Knudsen M (1898) Den Danske Ingolf-expedition. Bianco Lunos Kgl Hof-Bogtrykkeri (F. Dreyer). København, 1(2):21–154

Kuijpers A, Hansen B, Huhnerbach V, Larsen B, Nielsen T, Werner F (2002) Norwegian Sea overflow through the Faroe-Shetland gateway as documented by its bedforms. Marine Geology 188(1–2):147–164

Meincke J (1972) The Hydrographic Section along the Iceland-Faroe Ridge carried out by R.V. "Anton Dohrn" in 1959–1971. Berichte der Deutschen Wissenschaftlichen Kommission für Meeresforschung 22:372–384

Meincke J (1974) 'Overflow 73 – Large-scale Features of the Overflow across the Iceland-Faroe Ridge. ICES CM 1974/C:7, 10 pp

Meincke J (1983) The modern current regime across the Greenland-Scotland Ridge. In: Bott M, Saxov S, Talwani M, Thiede J (eds) Structure and Development of the Greenland-Scotland Ridge. Plenum, New York, pp. 637–650

Nielsen JN (1904) Hydrography of the waters by the Faroe Islands and Iceland during the cruises of the Danish research steamer 'Thor' in the summer 1903. Meddelelser fra Kommisionen for havundersøgelser. Serie Hydrografi I(4), 42 pp

Perkins H, Hopkins TS, Malmberg SA, Poulain PM, Warn-Varnas A (1998) Oceanographic conditions east of Iceland. Journal of Geophysical Research 103:21531–21542

Saunders P (1990) Cold outflow from the Faroe Bank Channel. Journal of Physical Oceanography 20:29–43

Saunders PM (1996) The flux of dense cold overflow water southeast of Iceland*. Journal of Physical Oceanography 26:85–95

Saunders PM (2001) The dense northern overflows. In: Siedler G et al. (eds) Ocean Circulation and Climate. Academic, New York, pp. 401–417

Steele JH (1959) Observations of deep water overflow across the Iceland-Faroe Ridge. Deep-Sea Research 6:70–72

Sherwin TJ, Turrell WR (2005) Mixing and advection of a cold water cascade over the Wyville Thomson Ridge. Deep-Sea Research I 52:1392–1413

Sherwin TJ, Griffiths CR, Inall ME, Turrell WR (2007) Quantifying the overflow across the Wyville Thomson Ridge into the Rockall Trough, In press Deep-Sea Research

Tait JB (1967) The Iceland-Faroe Ridge international (ICES) 'Overflow' expedition 1960. Rapports et Procès-Verbaux des Réunions du Conseil International pour l'Exploration de la Mer 157:38–149

Wåhlin AK (2002) Topographic steering of dense currents with application to submarine canyons. Deep-Sea Research I 49(2):305–320

Østerhus S, Turrell WR, Hansen B, Lundberg P, Buch E (2001) Observed transport estimates between the North Atlantic and the Arctic Mediterranean in the Iceland-Scotland region. Polar Research 20(4):169–175

Chapter 19
The Overflow Flux West of Iceland: Variability, Origins and Forcing

Bob Dickson[1], Stephen Dye[1], Steingrímur Jónsson[2,3], Armin Köhl[4], Andreas Macrander[5], Marika Marnela[6], Jens Meincke[4], Steffen Olsen[7], Bert Rudels[6], Héðinn Valdimarsson[3], and Gunnar Voet[4]

19.1 Introduction

The general introduction to this volume makes it clear that the overflow and descent of cold, dense water from the sills of the Denmark Strait and the Faroe–Shetland Channel into the North Atlantic represent key components of the global thermohaline circulation, ventilating and renewing the deep oceans and driving the abyssal limb of the Atlantic overturning cell. Though it is the whole full-latitude system of exchange between the Arctic and Atlantic Ocean that has to be addressed if we are to understand the full subtlety of the role of our Northern Seas in climate, it is their importance to climate that has justified the continued direct measurement of the overflow transport through Denmark Strait and its hydrographic characteristics until the longer-term variability of both can be understood.

The more specific argument for the study of overflow through Denmark Strait is the fact that it is the most potent site in the world-ocean for the transfer of ocean climate 'signals' between watermasses and to depth. The three watermasses which will eventually occupy much of the watercolumn of the NW Atlantic and which make the major contribution to North Atlantic Deep Water production flow together only in a limited zone along the Continental Slope off SE Greenland Slope; there,

[1]Centre for Environment, Fisheries and Aquaculture Science, Lowestoft, Suffolk NR33 0HT, UK

[2]University of Akureyri, Borgir v/ Norðurslóð, 600 Akureyri, Iceland

[3]Marine Research Institute, Skúlagata 4, 101 Reykjavík, Iceland

[4]University of Hamburg, Center of Marine and Climate Research, D-20146 Hamburg, Germany

[5]Alfred Wegener Institute for Polar and Marine Research, D-27515 Bremerhaven, Germany

[6]Finnish Institute of Marine Research, FI-00561 Helsinki, Finland

[7]Centre for Ocean and Ice, Danish Meteorological Institute, Lyngbyvej 100, 2100 København Ø, Denmark

[7]Marine Research Institute, Skúlagata 4, 101 Reykjavík, Iceland

R.R. Dickson et al. (eds.), *Arctic–Subarctic Ocean Fluxes*, 443–474

North East Atlantic Deep Water (NEADW) derived from the eastern overflow flows in a relatively narrow tongue between plumes of Labrador Sea Water spreading in at intermediate depths from the Labrador Basin and of Denmark Strait Overflow Water descending from its sill into the abyssal Atlantic. The factors promoting their mixing are also special to this zone. Through the descent of that plume, the stretching of the high potential vorticity watercolumn outflowing from Denmark Strait induces very strong cyclonic relative vorticity and '*a specific form of mesoscale variability that is unique to the Denmark Strait*', as Spall and Price (1998) have long pointed out; more recently (e.g., Pickart et al. 2003) we have become aware that the elevated heat-loss and intense wind stress curl associated with a 'tip jet' that forms episodically in the lee of Cape Farewell has the potential to promote small-scale but intense open-ocean deep convection that seems also to be unique to this zone.

For the types of reasons given above, the study of those aspects of the Denmark Strait Overflow that are of relevance to climate will include not only its transport, hydrographic character, forcing, and variability but also some consideration of the upstream influences and downstream impact of that variability, the comparison of properties between outflow(s) and the inflow to Nordic Seas, and the possible degree of covariance between the two principal deep outflows to the Deep Atlantic. These questions have formed the remit for the present chapter.

19.2 Sources

An appreciation of the importance of the Greenland–Scotland overflows for deep water formation in the North Atlantic and for the Atlantic thermohaline circulation was slow to develop. The presence of water from the Arctic Mediterranean south of the Greenland–Scotland Ridge was known and taken into account in Wüst's separation of the North Atlantic Deep Water (NADW) into three parts – an upper, saline part dominated by the Mediterranean outflow, a central part assumed to be supplied by open ocean deep convection in the Labrador and Irminger seas and a lower, colder contribution originating from north of the Greenland–Scotland Ridge (Wüst 1935).

That the overflow contribution was regarded as negligible is made clear in the table for the volume balance of the Arctic Mediterranean Sea given in *The Oceans* (Sverdrup et al. 1942). There, inflows of 3 Sv northwest of Scotland and 0.3 Sv through Bering Strait as well as the contribution from runoff were all assumed to exit as less dense Polar water through Denmark Strait. (As an aside we may note that the Arctic Mediterranean Sea was then considered more isolated than we now know it to be, the assessments of the inflows to that system were all less than half of the present-day estimates and the outflow through the Canadian Arctic Archipelago was regarded as too small to be taken into account).

All of the overflow passages east of Iceland were known in the early part of the 20th century and were discussed by Nansen (1912). However the Denmark Strait source was not considered in earnest until the 1950s (Dietrich 1957) when Cooper (1952, 1955) re-focused attention on their possible importance and when the

International Geophysical Year (1957–1958) intensified their study. By the end of the 1960s, estimates for the different outflow passages had been obtained: between the Faroes and Shetland, 1 Sv (Crease 1965), between Iceland and the Faroes 1 Sv (Hermann 1967) and through Denmark Strait 4 Sv (Dietrich 1957; Worthington 1970). These figures were largely confirmed by the subsequent overflow'73 experiment (Meincke 1983) and are remarkably close to current estimates.

In making his analogy with the Mediterranean outflow, Worthington (1970) assumed that the inflowing Atlantic water becomes cooler and denser primarily through heat loss in the Norwegian Sea, not in the stratified, ice-covered Arctic Ocean nor in the Greenland Sea, and the Norwegian Sea was considered the main source for deep and bottom waters in the Arctic Mediterranean (Helland-Hansen and Nansen 1909; Wüst 1942). However, the Norwegian Sea at that time was often taken to include the Greenland Sea and the Iceland Sea, just as Helland-Hansen and Nansen (1909) did in their classic work '*The Norwegian Sea*'. Worthington used the observations of overflow to estimate the inflow across the Greenland–Scotland Ridge and to formulate volume and heat budgets for the Arctic Mediterranean. This integrated approach, using heat and volume balances to estimate and characterise the overflows, was further developed by McCartney and Talley (1984).

From their closer study of the properties of Denmark Strait Overflow, Swift et al. (1980) found that the Norwegian Sea Deep Water characteristics present in the Faroe–Shetland outflow were not seen in the waters overflowing Denmark Strait. They recognised two intermediate watermasses in the Iceland Sea, one warmer and more saline, consisting mainly of Atlantic water recirculating from Fram Strait, and a second colder and less saline watermass identified as water created by winter convection in the Iceland Sea. Though Swift et al. (1980) called these two watermasses lower Arctic Intermediate Water (lAIW) and upper Arctic Intermediate Water (uAIW), respectively, we use instead the terms Recirculating Atlantic Water (RAW) for lAIW, first introduced by Bourke et al. (1988; see Ryder 1891 for its first attribution), and Iceland Sea Arctic Intermediate Water (IAIW) for uAIW.

By examining the seasonal changes in volume of the different waters in the Iceland Sea, Swift and Aagaard (1981) and Swift et al. (1980) concluded that the IAIW (uAIW) supplied most of the water for Denmark Strait Overflow. Observations of its tritium concentration indicated that the overflow water was recently ventilated, which again supported a nearby source such as the Iceland Sea. Later Smethie and Swift (1989) found that the overflow plume could be separated into two distinct layers, one less dense with a high tritium content, the other denser with lower tritium concentration. They assumed that the denser part originated farther to the north, presumably as Arctic Intermediate Water (AIW) formed in the Greenland Sea.

The Greenland Sea, considered the main area of deep convection in the Nordic Seas, was otherwise largely ignored as a source for overflow waters, especially for the Denmark Strait Overflow. The principal reason was that deep water formed in the Greenland Sea is too dense to cross the Greenland–Scotland Ridge and would therefore be constrained to circulate within the Arctic Mediterranean. During the 1980s, it was gradually realised that the deep waters of the Arctic Ocean had a character and spatial variability that are distinctly different from those observed in the Greenland Sea

(Aagaard 1980), implying that deep water is produced by brine rejection and shelf-slope convection in the Arctic Ocean; Aagaard et al. (1985) and Rudels (1986) suggested different plume concepts for this convection and constructed circulation schemes connecting the two deep water formation areas, the Arctic Ocean and the Greenland Sea, as well as the Norwegian Sea, where no deep water formation occurs.

However, this deep circulation was assumed to be confined to the deep basins of the Arctic Mediterranean, so that no explicit discussion of the connections to overflow was made. For continuity reasons, the deep water formed at the two sites has eventually to leave the deep circulation, either by diffusion into the upper layers or by flowing out of the domain. Rudels (1986) suggested a flow across the Jan Mayen Fracture Zone into the Iceland Sea and eventually to the North Atlantic. The more detailed mass balance models of the deep Nordic Seas–Arctic Ocean circulation formulated at that time (Smethie et al. 1988; Heinze et al. 1990) were also content to state that due to continuity, water has to leave the deep circulation into the layers above and eventually leave the Arctic Mediterranean Sea.

Aagaard et al. (1991) reported that after passing through Fram Strait, a part of the Arctic Ocean Deep Water brought south in the East Greenland Current did not enter the Greenland Sea but continued along the continental slope across the Jan Mayen Fracture Zone to enter the Iceland Sea. Buch et al. (1992, 1996) identified the presence of saline Arctic Ocean Deep Waters just north of the Denmark Strait sill, confirming that not only Recirculating Atlantic Water (RAW) but also the deep dense waters from the Arctic Ocean continued in the East Greenland Current into the Iceland Sea.

The importance of the East Greenland Current for the Denmark Strait Overflow was highlighted by Mauritzen (1996a, b). She concluded from her use of an inverse box model and from the reported constancy of the Denmark Strait Overflow (Dickson and Brown 1994) that the central gyres of the Greenland and Iceland seas contributed little to the overflow. Instead she proposed that its main sources were two Atlantic waters, a component of RAW recirculating direct from Fram Strait, and a component of Arctic Atlantic Water (AAW) that spreads on a longer circuit into and around the Arctic Ocean before returning through Fram Strait in the East Greenland Current. Since the AAW loses little heat in the Arctic Ocean, this scheme is essentially similar to the concept of Worthington (1970). Heat loss in the Norwegian Sea provides for most of the increase in density that is required to form Denmark Strait Overflow Water from Atlantic water; the pathways of this cooled water had now been specified.

Faroe–Shetland overflow water is initially denser than Denmark Strait Overflow Water and must have a different source. Mauritzen suggested that this was provided by the Atlantic water that enters the Arctic Ocean across the Barents Sea shelf where local cooling might increase its density sufficiently to create water for the Faroe–Shetland overflow. However, Rudels et al. (1994) and Schauer et al. (1997) have shown that the main Barents Sea contribution to the Arctic Ocean is a watermass (upper Polar Deep Water or uPDW) which occupies the same density range as the Arctic Intermediate waters formed in the Greenland Sea and the Iceland Sea and would therefore contribute to the Denmark Strait Overflow rather than that through Faroe–Shetland Channel. In fact, the Faroe–Shetland overflow mostly derives from Norwegian Sea Deep Water (NSDW) which is initially formed in the Greenland Sea as a mixture between locally produced Greenland Sea Deep Water and Arctic Ocean deep waters. The transformation of the watermasses in the Arctic Ocean and the Nordic Seas are summarised in Rudels (1995), Meincke et al. (1997) and Rudels et al. (1999a).

By examining the evolution of the East Greenland Current from Fram Strait to Denmark Strait, Rudels et al. (2002) described the different watermasses that make up the East Greenland Current and identified where they join and where they split (Fig. 19.1). The denser Arctic Ocean deep water (Eurasian Basin Deep water, EBDW) enters the Greenland Sea but does not cross the Jan Mayen Fracture Zone. The less dense Canadian Basin Deep Water (CBDW) is partly able to cross the fracture zone into the Iceland Sea but is too dense to cross the Denmark Strait sill. Only the uPDW, largely deriving from the Barents Sea branch inflow, and the RAW and AAW components are light enough to pass over the sill in Denmark Strait. The uPDW and the CBDW become less saline and colder as they flow south along the Greenland slope from Fram Strait to the Jan Mayen Fracture Zone, indicating that Arctic

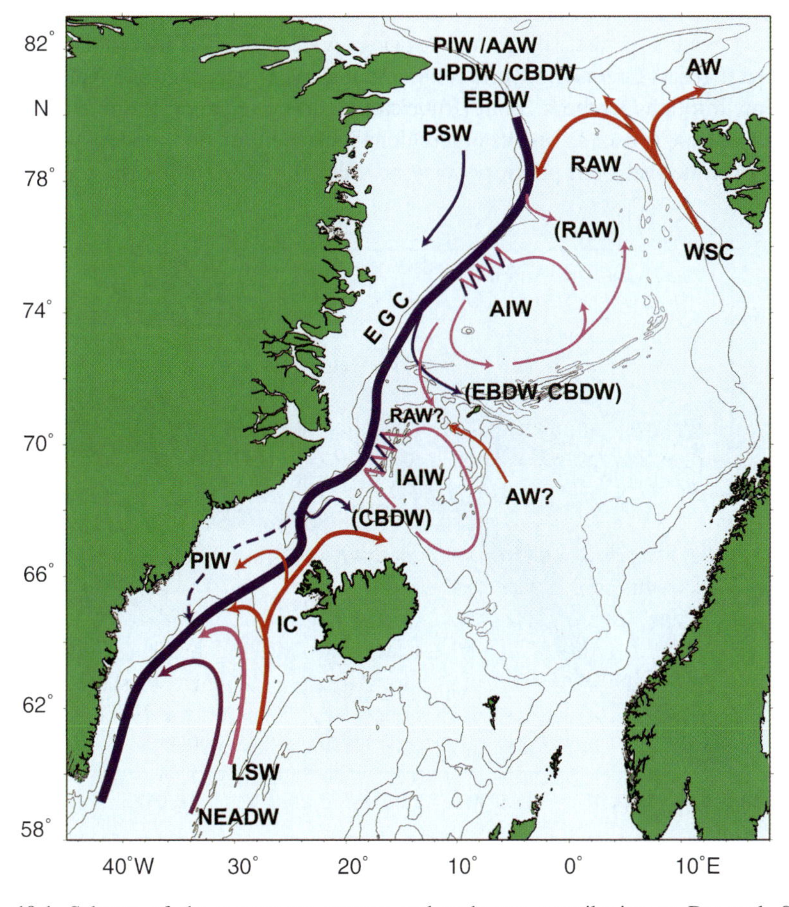

Fig. 19.1 Scheme of the watermass sources and pathways contributing to Denmark Strait Overflow according to Rudels et al. 2002, by permission Oxford University Press, ICES J. Mar. Sci.

Intermediate water from the Greenland Sea (AIW) is mixed isopycnally into the East Greenland Current; they are thus able to contribute to the Denmark Strait Overflow.

Though the East Greenland Current has the *potential* to provide the Denmark Strait Overflow Water, this does not automatically make it the prime overflow source. The waters below 100–200 m in the Iceland Sea are dense enough to supply the overflow water and Jónsson (1999) has shown that the densest overflow water and the most consistent southward flow, observed on a current meter array north of the strait, are found on the Iceland slope. Furthermore, a barotropic, southward flowing jet has been observed by ship-mounted ADCP over the 600 m isobath on the continental slope north of Iceland (Jónsson and Valdimarsson 2004). So far, it is not known if this jet originates in the central Iceland Sea, east of the Kolbeinsey Ridge, or if it originates from the deflection of part of the East Greenland Current toward Iceland as it approaches the sill; and of course if the main overflow source is the Iceland Sea as Jónsson (1999) and Jónsson and Valdimarsson (2004), and Swift et al. (1980) before them, advocate, there remains the problem of what happens to the denser part of the East Greenland Current.

In fact it is likely that both scenarios may occur. The changing characteristics of overflow waters at the sill provide strong evidence for a temporal switching between the East Greenland Current and the Iceland Sea as the dominant source of overflow through Denmark Strait (Rudels et al. 2003; see also Olsson et al. 2005). Köhl et al. (2007) use the results of a regional ocean circulation model (MIT GCM) to demonstrate that both pathways may be valid (Fig. 19.2), that the transports in

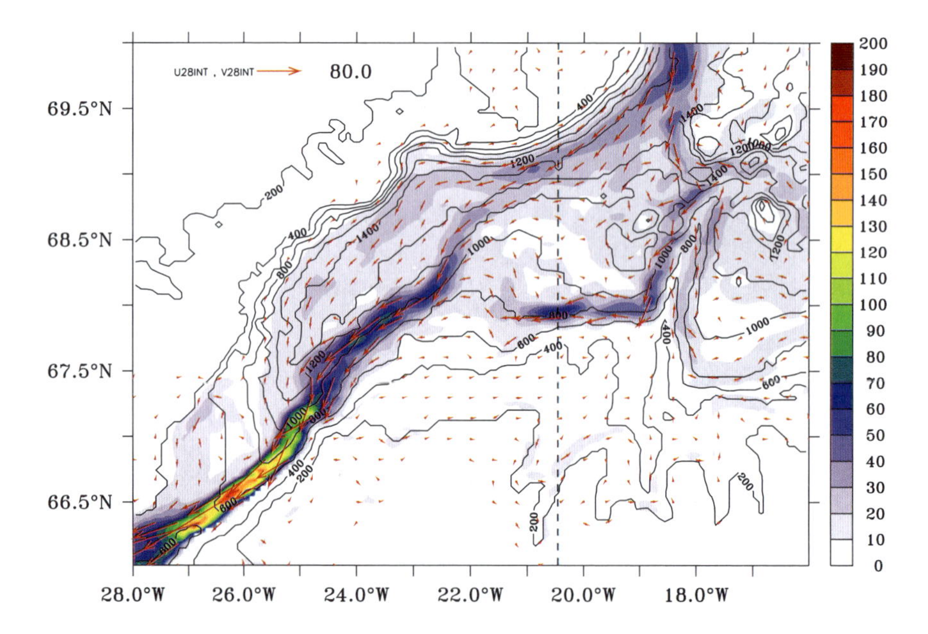

Fig. 19.2 Mean volume transports in $m^2 s^{-1}$ for the source region of the Denmark Strait Overflow north of the sill, from regional circulation modelling by Köhl et al. (2007). This transport scheme may resolve the issue of whether the East Greenland Current (Rudels et al. 2002) or the Iceland Sea (Jónsson and Valdimarsson 2004) acts as the source of overflow through Denmark Strait. Both pathways may be valid. From Köhl et al. 2007, © 2007 American Meteorological Society

the two branches appear to be inversely correlated so that their dense water supply to the Denmark Strait sill is nearly constant, and that both branches may thus have their upstream origins in the East Greenland Current.

19.3 The Nature of the Overflow Plume

The overflow plume is stratified. This is especially obvious during the first part of its descent, in water depths above 2,000 m. Its upper part consists of a low salinity lid which covers a temperature-stratified layer and an almost homogeneous, colder, bottom part of the plume (the salinity is almost the same for the two lower components). The lower part represents the largest fraction of the overflow plume and is probably homogenised by the mixing induced by the high velocities close to the bottom. It is most conspicuous in the first part of the descending plume, and as the plume approaches 3,000 m depth the bottom layer becomes thinner and more stratified.

In Θ–S diagrams the characteristics of the waters found at the sill form an envelope around those observed in the plume. The different waters are, however, distributed laterally over the sill. The coldest, densest water is found in the deep channel, just west of but deeper than the northward flowing Irminger Current, while the warmer and less dense RAW and AAW are found farther to the west and also over the shallower shelf areas. A cold, Polar watermass is usually observed at mid-depth in the central part of the strait, just west of the northward flowing Irminger Current. Farther to the west the waters of overflow densities are overlain by less dense and warmer waters that must derive from the Irminger Current.

The low salinity lid (Fig. 19.3b) derives from the Polar Intermediate Water (PIW) (Malmberg 1972). Arctic Intermediate Water from the Iceland Sea (IAIW) is generally too saline to contribute to the PIW, and noting that the Θ–S characteristics of the PIW were similar to those of the Arctic Ocean thermocline, Rudels et al. (2002) have recently suggested that PIW ultimately originates in the Arctic Ocean; it is usually well ventilated, implying recent contact with the atmosphere. Since the PIW is denser than the Irminger Current water it is able to join the AAW and the RAW contribution to the overflow and form the conspicuous low salinity lid of the overflow plume. The PIW layer is located between the warmer Atlantic overflow waters of the plume and the warm ambient waters above (Fig. 19.3). As the plume descends, its temperature will increase rapidly by mixing and only its salinity minimum will remain. The internal mixing between the different waters of the Denmark Strait Overflow plume would also increase the temperature of the densest component, making the homogeneous deep part of the plume warmer and less dense than at the sill. As the plume penetrates deeper, it eventually encounters the derivative of Iceland–Scotland Overflow Water. This is warmer and more saline and will mix isopycnally with the less dense part of the overflow plume, increasing its temperature and salinity and gradually removing the salinity minimum.

The location and rate of entrainment south of the sill has yet to be fully resolved. Comparing historic observations from the sill with all available observations downstream, Dickson and Brown (1994) suggested that '*much of the entrainment into the overflowing stream … takes place close to the point of overflow with a greatly reduced entrainment thereafter*'. Thus the 80% increase that they identify in densewater transport at $\sigma_\theta > 27.80$, from 2.9 Sv at the sill to ~5.2 Sv at Dohrn Bank, 160 km downstream (Fig. 19.4), was suggested to arise largely from entrainment, while the further increase to 10.7 Sv at the Angmagssalik Array further south

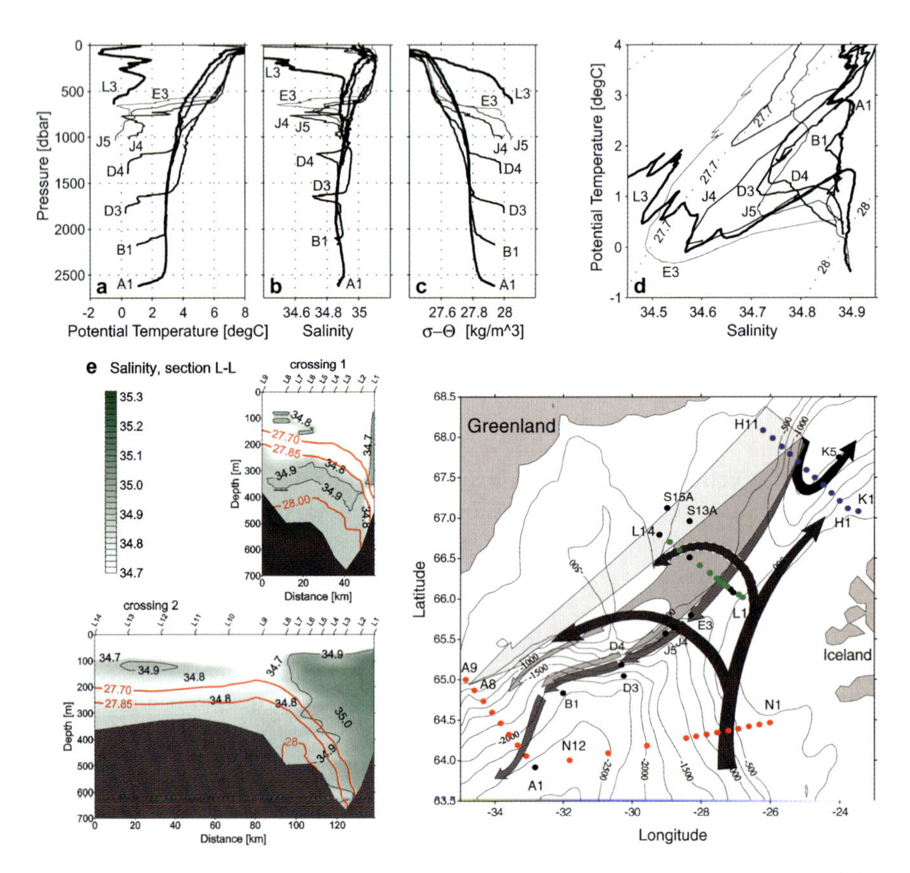

Fig. 19.3 The discovery of freshwater 'lids' on the overflow in 1997 by Rudels et al. (1999b) suggested that fresh but dense EGC water from the adjacent shelf (light grey tones on the circulation map) might contribute to the less-dense fractions of the Overflow south of the sill and implied that entrainment south of the sill might be less vigorous than supposed. The 'lids' are evident in the vertical profiles of θ, S and σθ on the different sections worked across the overflow (panels a, b, c; see map for section locations). The θ–S diagram in panel (d) confirms that the low salinity waters on section L are sufficiently dense to contribute to overflow, and shows that the low salinity lids further south are found at a similar density. The two salinity sections L–L (panel e) offer further confirmation that the relatively fresh lens of water at salinities <34.8 on the shelf lies within a density range (σθ = 27.70–27.85) that contributes to the descending overflow plume. Redrawn from Rudels et al. 1999b, © 1999 American Geophysical Union

seemed to result largely from confluence with the Iceland–Scotland Overflow-derivative arriving from the east. Using the same and later data, a new study by Voet (2007) lends general support to this idea in suggesting that horizontal eddy heat transport is insufficient to explain the initial warming-rate of the DSOW plume south of the sill (~0.40 °C per 100 km) as recorded by current meter thermistors and annual hydrography, but it *can* explain the lesser warming rate (~0.06 °C per 100 km) from the 'TTO array' southwards. Voet concludes from this that vertical turbulent processes dominate the entrainment into the overflow plume during the first phase of its descent into the deep Atlantic. Since it seems improbable that such delicate features as the 'freshwater lids' (Rudels et al. 1999) could survive such vigorous entrainment, we are left with two ideas to explain them – either they are imposed later on the descending plume, after the main entrainment process is complete; or entrainment is sufficiently patchy in space and time to allow these features to survive in places. The issue is currently unresolved.

Localised overspill and entrainment certainly occurs along the East Greenland Slope south of the sill. Some of the dense East Greenland Current water that was located too far west to join the main overflow plume descending into the deep Irminger Basin is known to spread south along the East Greenland Shelf, eventually crossing the shelf break farther to the south. Spilling from the shelf, it encounters and has to pass through the Irminger Current water at the Slope before reaching the overflow plume, resulting in enhanced local entrainment. Its density is thereby

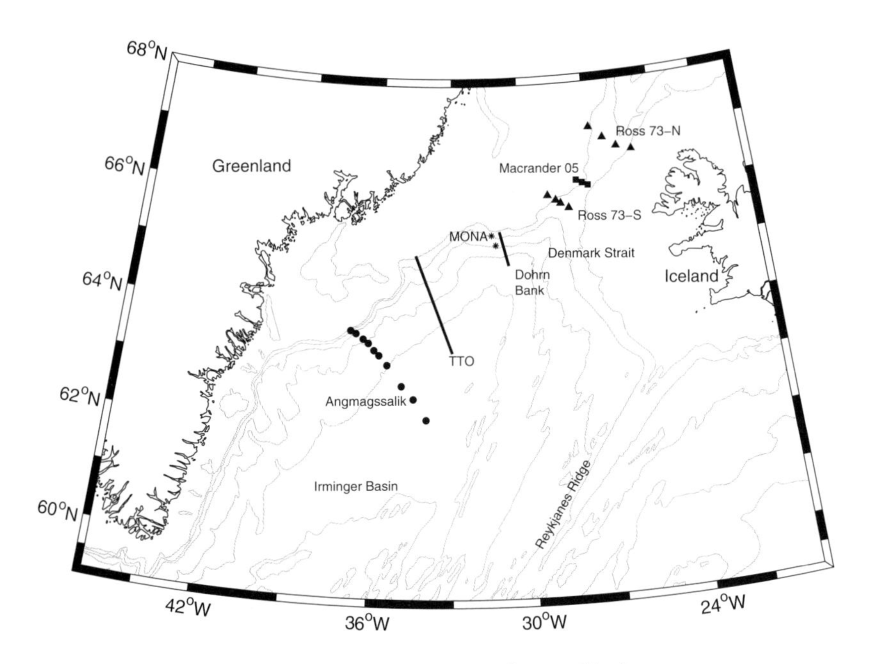

Fig. 19.4 Location of the moored current meter arrays discussed in the text

reduced and as a result, much of this overspill will not merge with the Denmark Strait Overflow plume (Rudels et al. 1999) but instead will form a separate, less dense stream higher up on the Slope (Fig. 19.3). This 'spill jet' (Pickart et al. 2005) would then, together with the Labrador Sea Water and the Subpolar Mode Water, contribute to the subpolar gyre.

19.4 Transport

Though we have long understood the importance of doing so, early attempts to measure the overflow transport were thwarted by the violence of the flow. In the first attempt by Worthington (1969), conducted over a period of 1 month in the winter of 1967, none of his moorings survived on the western slope of the Strait where the coldest, fastest flows were expected, and the one complete record, recovered from 760 m depth in the middle part of the Strait, showed a highly energetic and variable flow of up to 143 cm s^{-1} with a mean of 21.4 cm s^{-1} and a dominant timescale of a few days. On the basis of this, the transport of water colder than 4 °C was calculated to be 2.7 Sv. Later, during the ICES exercise "Overflow '73" a remarkable set of 25 5-week current records were recovered from three arrays located immediately to the north and south of the sill by Ross (1975, 1984; Fig. 19.4). The records from north of the sill show weak (mean <10 cm s^{-1}) and variable flows, in sharp contrast to those in the overflowing stream further south. The latter, about 55 km south of the sill (see Fig. 19.4), indicate a vigorous bottom-intensified flow following topography, with the core of the current (mean, >60 cm s^{-1}; maximum, 167 cm s^{-1}) lying midway up the slope on the Greenland side and with an energetic dominant fluctuation timescale of 1.8 days present in all records. Smith (1976) explained these observations not in terms of meteorological forcing but in terms of the steady movement of dense water south toward the strait from an upstream reservoir of constant pressure-head that intensifies as it is funnelled through the strait, becomes unstable hydrodynamically, and acquires its energetic fluctuating component through the development of baroclinic instability south of the sill. Ross (1984) estimated the time-averaged volume flux of water colder than 2 °C to be 2.9 Sv. Only much later in 1999–2003 were direct measurements to return to the Denmark Strait sill itself; then, the discontinuous use of bottom-mounted Acoustic Doppler Current Profilers (ADCPs) close to the sill provided values slightly higher than the short-term estimates of Ross, namely 3.7 Sv reducing to 3.1 Sv over the 5-year period of record (Macrander et al. 2005, 2007; see Fig. 19.4 for location, Fig. 19.8 for the model-optimised transport series based on these measurements).

In 1975, the first year-long records were obtained from the overflow with the recovery of moorings MONA 5 and 6 by Aagaard and Malmberg (1978; Fig. 19.4) from the exit of the Strait along 30°40′ W, slightly downstream from Ross's (1975) main array. Each mooring carried two instruments at heights of 25 and 100 m above the bed and all gave full 360-day records. These confirmed Ross's observations by showing a strong bottom-intensified flow directed along the isobaths with a mean speed of about 50 cm

s^{-1} and with various low-frequency variations superimposed, including a dominant and persistent fluctuation at 1.5–2.5 days with an amplitude comparable to the mean which Aagaard and Malmberg (1978) attributed (probably) to baroclinic instability. For the first time, these records were long enough to demonstrate the lack of a seasonal fluctuation in either the speed or the temperature of the overflow.

The first long-term measurements across the full width of the overflow were begun by the Lowestoft Laboratory in 1986 off Angmagssalik SE Greenland, were developed to their present form by a UK, German and Finnish team within the EC MAST3-VEINS Project in 1997–2000, and have continued through the EC-ASOF Project (2003–2005) to the present day. Typically, the data derive from annual deployments of a 'picket fence' array of 7 or 8 current meter moorings equipped in recent years with a variable number of SBE-37 salinity sensors, set normal to the SE Greenland Slope in a position to intercept the descending plume some 500 km south of the Denmark Strait sill (Figs. 19.4, 19.6). Although these moorings did extend through the Denmark Strait Overflow Water (DSOW) into the overlying layer of Iceland–Scotland Overflow Water (ISOW), the Angmagssalik array was primarily designed to cover the denser overflow from Denmark Strait. Thus although transports are calculated below *both* for densities >27.80 (which

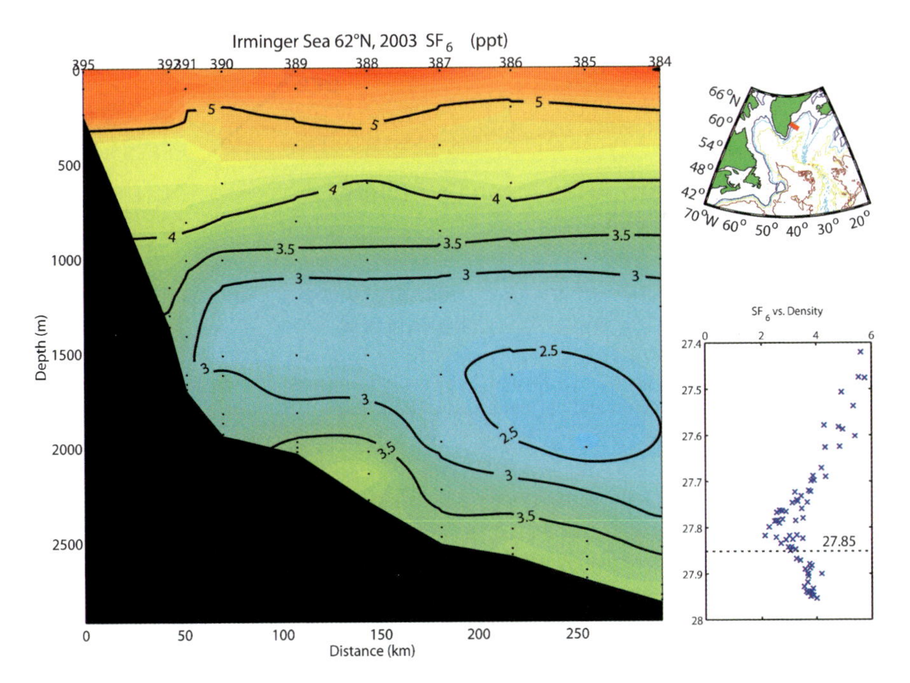

Fig. 19.5 Vertical distribution of SF6 along the line of the Angmagssalik Array (Section 19.3) during ASOF cruise M59/1 in 2003 (Tanhua, personal communication, 2006). The mean vertical concentration of SF6 versus density (inset panel) influenced the selection of the $\sigma_\theta = 27.85$ isopycnal as a mid-point interface between ISOW and DSOW, and as the upper bound of DSOW for the purpose of calculating overflow transports

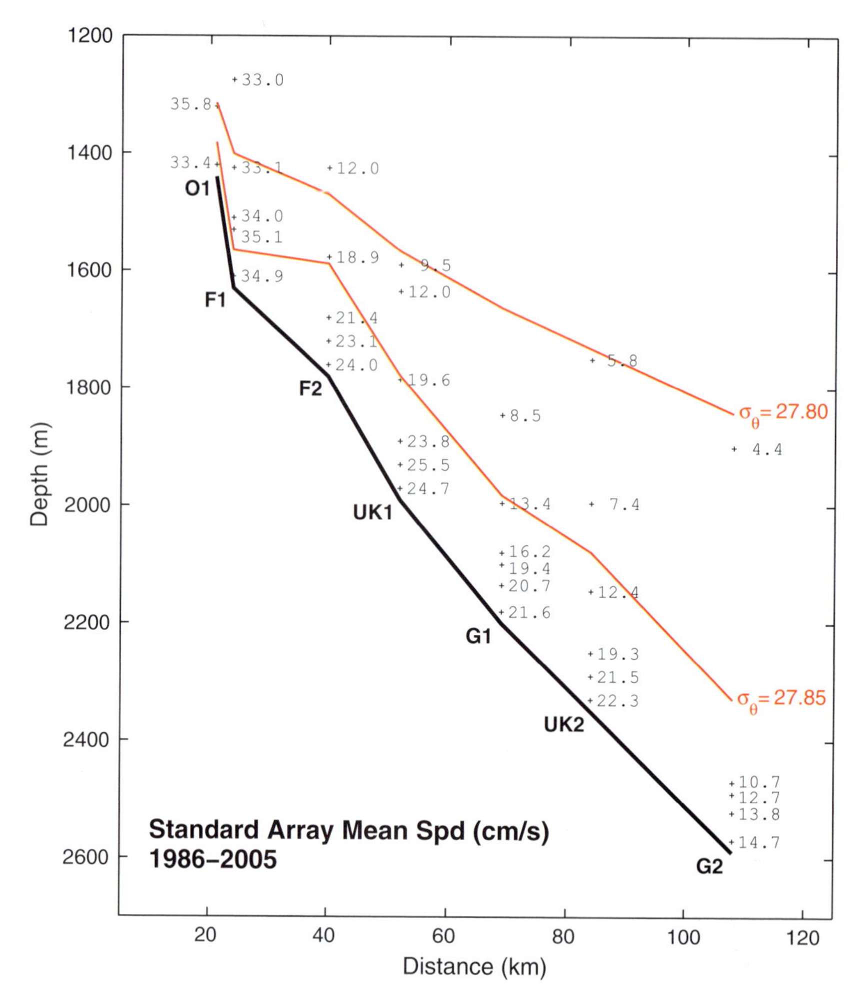

Fig. 19.6 Distribution of the σ_θ = 27.80 and 27.85 mean isopycnals on the Angmagssalik section in relationship to bottom depth and current speeds, 1986–2005

should contain the two overflow-derived layers passing south along the Greenland Slope), *and* for the near-bottom layer at densities >27.85 in which the Denmark Strait Overflow is concentrated, the basis for the former is less complete.

The density interval appropriate to each overflow layer was defined partly from hydrographic analysis, but chiefly from the distribution of the tracer sulphur hexafluoride (SF6) on the section as measured by F/S METEOR during ASOF cruise M59/1 in 2003 (Tanhua, personal communication, 2006). This tracer was released in 1996 on the $\sigma\theta$ = 28.0492 isopycnal in the central Greenland Sea under the EC-MAST-III ESOP-2 programme and has subsequently spread south of the sills in the Greenland–Scotland Ridge. Tanhua et al. (2005) calculate that by summer 2003, approximately 4 kg of excess SF6 had passed over the Denmark Strait sill, where – at the time of measurement – it provided excellent discrimination between the SF6-enriched DSOW plume and the overlying SF6-poor ISOW-derived layer. From the distribution of SF6 versus density on Repeat Section 3 (collinear with the mooring array and therefore of greatest relevance to our transport calculation; see Fig. 19.10), Fig. 19.5 provides a clear indication that the interface between the SF6-minimum ISOW layer and the SF6-maximum DSOW layer lies close to the $\sigma\theta = 27.85$ isopycnal. We therefore use this isopycnal as the upper boundary of DSOW transport. Figure 19.6 shows the distribution of our direct current meter observations and the mean current speeds encountered along the line of the Angmagssalik array for the entire period of measurement 1986–2005. From repeat hydrography, the mean position of the $\sigma_\theta = 27.85$ isopycnal (lower red line) is found to lie at a height of 200 m above bottom along much of the section and from Fig. 19.6, it is clear that the current meter effort is reasonably well distributed within this 200 m near-bottom layer and thus appropriate to cover the overflow plume from Denmark Strait.

19.5 Transport Variability

Various modes and scales of transport variability are claimed for the Denmark Strait Overflow in the literature. Bacon (1998; also 2002) made reference-level near-bottom flux calculations for occasional hydrographic sections in the vicinity of Cape Farewell since 1955 to claim that the overflow transport might vary by a factor of 2 or 3 on decadal timescales; some support for this view was recently provided by an analysis of Kieke and Rhein (2006). Both studies clearly show a period of high baroclinic transport in the late 1970s and 1980s and by including direct current measurements in two of these years (1978 and 1991), Bacon argues that this is a feature of the total overflow transport. However, the well-known problems of making geostrophic transport estimates from (relatively) widely spaced hydro-stations on steeply sloping topography, the difficulty of discriminating low-frequency change from high frequency variability on the basis of annual or semi-annual sections, and the unknown barotropic component of overflow transport, hidden to hydrography are grounds for doubting this assertion (Saunders 2001). In his complete reworking of the Angmagssalik data set, Saunders (2001) did identify brief episodes of weakened transport in late 1988–early 1989 that were not reported by Dickson and Brown (1994); and Dickson et al. (1999) have speculated that a later episode of slowing in the deeper part of their array in January–February 1997 might be due to transient extreme warmth at the sill, leading to a reduction in overflow density so that the descending plume lay higher than usual on the Slope – a further new mode of overflow variability. But Saunders' (2001) reworking of the Angmagssalik data set prompted essentially the same conclusion as Dickson and Brown (1994) – that in the longer-term, the DSO transport was essentially steady. As Saunders observes, "*Before these measurements were made, a conjecture that the overflow would show such overall steadiness would have been met with derision*".

Girton et al. (2001) supported this conclusion by comparing Overflow 73 with their modern, rapid, high-resolution XCP & XCTD survey of the sill to '*add more evidence to support the view of the DSO as an unchanging, hydraulically-controlled flow on timescales longer than a few days*'.

It is therefore of some interest whether and to what extent the complete modern data set agrees with such a statement. Figure 19.7 shows the full record of dense-water transports at $\sigma_\theta > 27.85$, based on measured current speed and mean isopycnal depth, which we take to be the transport of Denmark Strait Overflow Water through the Angmagssalik array. In the period with optimal coverage, the DSOW transport is 4.0 ± 0.4 Sv, a figure which is in close agreement with the geostrophic estimate of transport for this section and density interval (see below). (We note that Whitehead (1998) predicted an overflow flux of 3.8 Sv for the Denmark Strait, close to that observed, on the basis of hydraulic constraints and Käse and Oschlies (2000) have used hydrographic observations and modelling to demonstrate that the strength of the DSO is, to first order, in hydraulic balance).

Though the time series shows interannual changes, most notably the waxing then waning of the flow in the period 1998–2002, we have as yet no obvious or convincing evidence of a longer-term slowdown in DSOW transport over the 8–9-year period for which our records of flow are continuous (1997–2005). For this period the transport estimates show no significant trend. The question of whether the trend in overflow transports should have been calculated for the whole period of record since 1987, despite the data-gap in the early 1990s, was decided on the basis of modelling studies by Malcolm (2005) and by Olsen (personal communication, 2005). Both studies employ versions of the primitive equation Modular Ocean Model (MOM) of Pacanowski (1995), with the former used in the FLAME configuration (Family of Linked Atlantic Model Experiments; FLAME Group 1997) and the latter using the Max-Planck-Institut Ocean Model. Both appear to reproduce the observed maximum in overflow transport around the year 2000; neither shows any convincing sign of a longer-term trend in overflow transport over the entire period since the mid-1980s spanned by our records; and both show clearly enough that our earliest records in 1988–1990 are likely to have

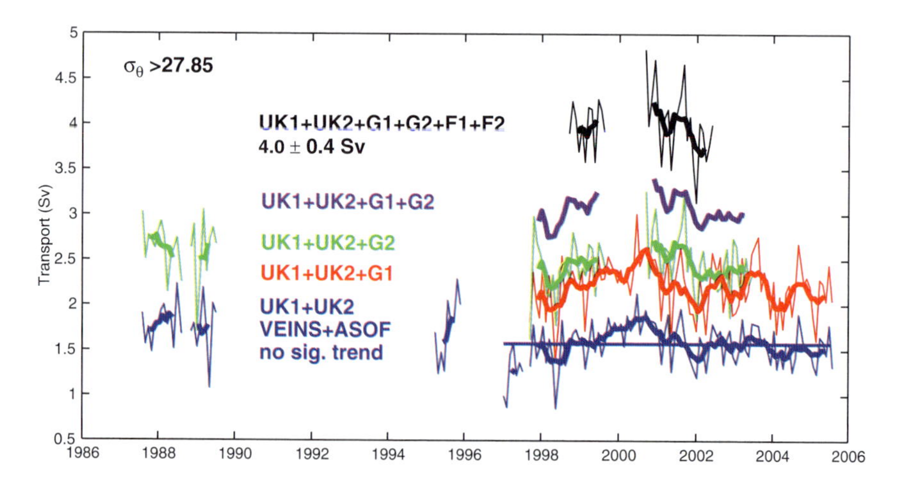

Fig. 19.7 The dense-water transports at $\sigma_\theta > 27.85$ through the Angmagssalik array

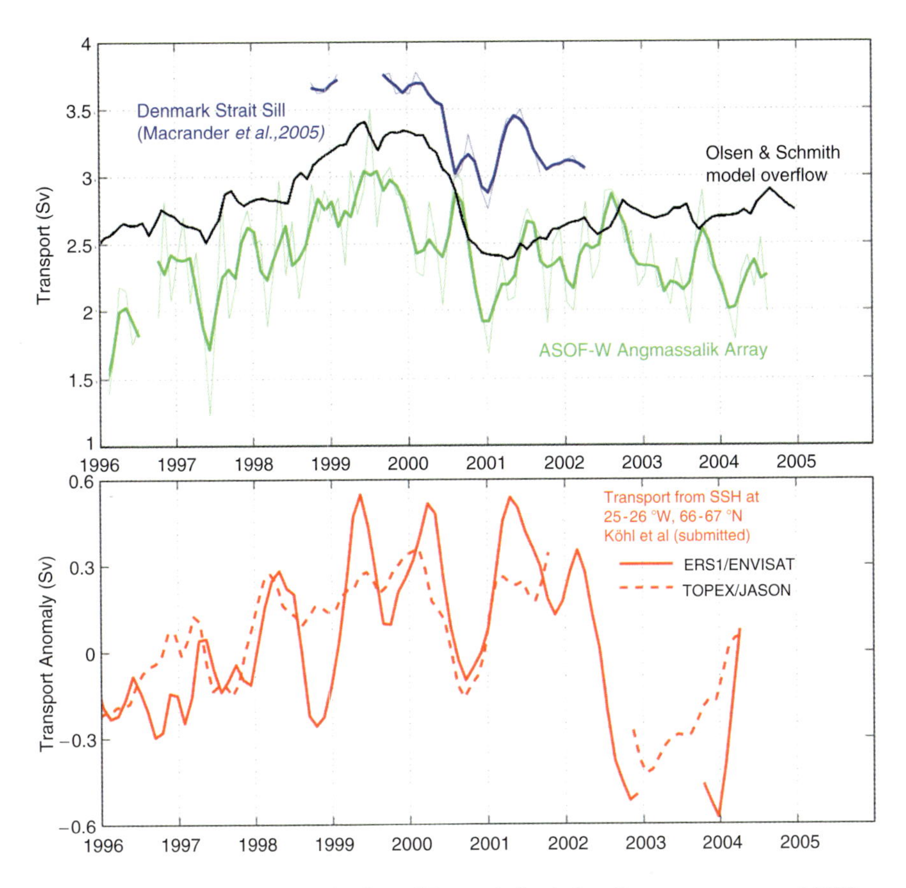

Fig. 19.8 Strengthening-then-weakening of Denmark Strait Overflow transport around 2000, as revealed by four independent data sets: (a) direct observations of flow through the long-term CEFAS-UHH-FIMR moored current meter array on the East Greenland Slope off Angmagssalik (green curve); (b) model-optimised estimates based on a discontinuous Acoustic Doppler Current Profiler (ADCP) array close to the sill (blue curve; see Macrander et al. 2005); (c) model estimates of DSOW transport calculated by Olsen, DMI using the MPI Ocean Model (black curve; see Olsen and Schmith 2007, for method); (d) transport estimates from sea surface height as measured by TOPEX/Jason & ERS2/ENVISAT satellite altimetry close to Denmark Strait by Köhl et al. (red curve; 2007). Note that in the case of the Angmagssalik array we are emphasising the change in transport rather than its magnitude, so using the moorings with the longest continuous time series from the core of the plume. When data from all moorings across the full width of the plume are considered, the transport of Denmark Strait Overflow Water (densities $\sigma_\theta > 27.85$) is around 4 Sv

been influenced by a real but short-term sub-maximum in transport to suggest that it would be unwise to use these records as our start point in estimating trends. We therefore base our trend estimate only on the continuous part of the record.

Since, even so, uncertainty is imposed on our transport calculations by the lack of direct data on overflow layer-thickness, a deficiency that still continues, it is fortunate that we have three separate methods of checking the general sense of overflow variability against independent data sets. In Fig. 19.8, that comparison is provided first by an array of between 1 and 3 bottom-mounted ADCPs that was discontinuously deployed on the Denmark Strait sill between 1999 and 2003 by the Leibniz Institute for Marine Sciences, Kiel and the Marine Research Institute, Reykjavik (Macrander et al. 2005). As shown, the same general slowdown in transport reported by Macrander et al. (2005) is evident passing through the Angmagssalik array ~500 km downstream after a short time lag (sill leading by 70 days), though the greater length of the Angmagssalik series identifies the apparent slowdown as part of the waxing then waning of overflow during a 9-year period centred around 2000. Second, Olsen's recent simulation of overflow transport using a MOM-based ocean circulation model with ECMWF/NCEP-forcing (see Olsen and Schmith 2007) shows a very similar pattern of interannual transport variability to that directly measured on the Angmagssalik array; for the low-passed series (compare black and green curves, Fig. 19.8), the correlation coefficient is +0.82. Third, overflow transport estimates based on sea surface height measurements close to Denmark Strait by Köhl et al. (2007) based on TOPEX/Jason & ERS2/ENVISAT satellite altimetry (red curve, Fig. 19.8), also appear to capture the same waxing then waning of overflow transport around 2000.

Thus, in summary, one of the principal results from this array over a decade of continuous observation is the finding that although the DSO transport time series may show interannual variability, there is no obvious or convincing evidence *yet* for any long-term trend in the DSOW transport. This appears to be confirmed in the full simulation, 1948–2005, by Olsen and Schmith (2007; see upper panel Fig. 19.9). We note moreover that in the case of the *eastern* deep overflow of ISOW through the Faroe Bank Channel, recent analysis of the full record from 1995–2005 likewise shows no convincing evidence of any long-term downturn in transport, in contrast with early preliminary reports (Hansen et al. 2001) though once again, a distinct interannual variability is shown (see Chapter 18, this volume).

The full simulation by Olsen and Schmith (2007) does however show an intriguing correspondence between the Denmark Strait Overflow and the Atlantic inflow across the Greenland–Scotland Ridge (Fig. 19.9). The apparent co-variance between these two key oceanic exchanges is under investigation. As one promising lead, the simulation by Olsen and Schmith suggests that a fast barotropic mode of variability in the cyclonic gyre of the GIN seas might link the eastern inflow and deep western overflow, while a baroclinic adjustment in the Nordic Seas on the scale of a few years might drive their variability on interannual to decadal scales.

19.6 Forcing

In his 1998 review, Whitehead summarized our basic understanding of the control of the Denmark Strait Overflow through hydraulic balance. His theoretical approach was based on the assumption of zero potential vorticity and rectangular sill geometry and gave a high upper bound (3.8 Sv) for the hydraulically controlled transport through the Strait. Stern (2000) discussed improvements to this approach in which

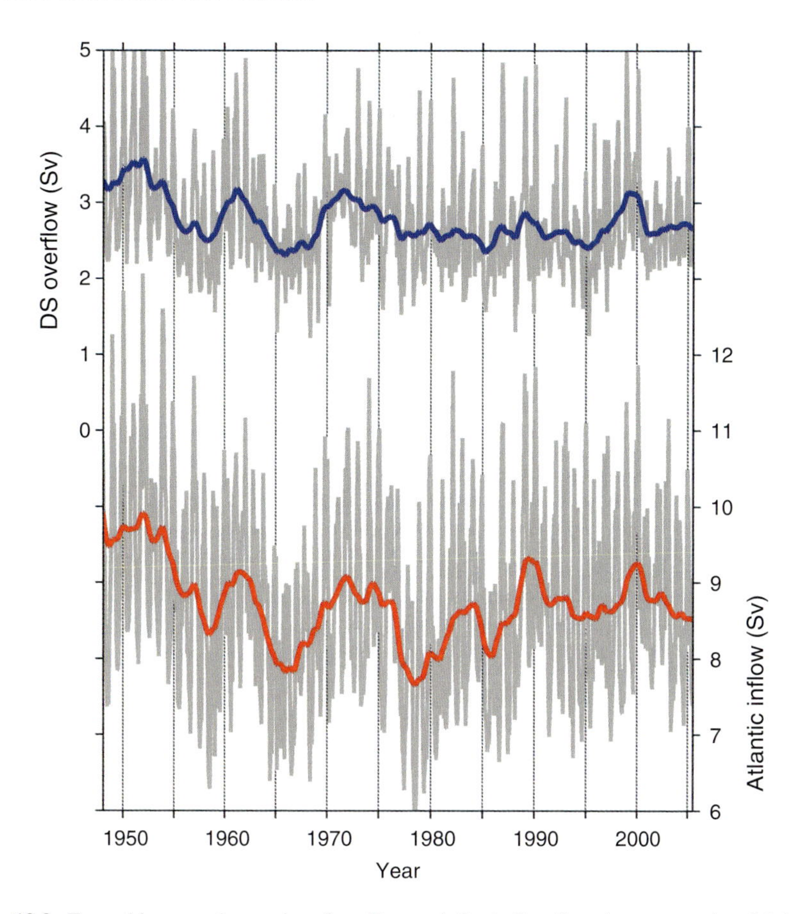

Fig. 19.9 Ensemble mean time series of net Denmark Strait Overflow (upper curve) and Atlantic inflow (lower) across the Greenland–Scotland Ridge, 1948–2005, from a hindcast simulation by Olsen and Schmith (2007) using a global coupled ocean/sea-ice model (MPI-OM, Marsland et al. 2003) constrained by atmospheric reanalysis data (Kistler et al. 2001) and observed Arctic river discharges (http://grdc.bafg.de). Curves show low-passed data using a first order Butterworth filter with a cut-off frequency of 1/24 months^{-1} and with the annual cycle removed prior to filtering

PV is unconstrained and sill geometry is improved, which led him to a substantially lower transport estimate, about half of Whitehead's figure. In recent years, Käse and his research groups at Kiel and Hamburg have tested and developed this theory further through the use of numerical models and by direct, modern observations of the dynamic properties of overflow.

Whilst high-resolution numerical process models had been used to investigate the dynamics of the overflow plume itself (Jungclaus and Backhaus 1994; Krauss and Käse 1998), a simulation by Käse and Oschlies (2000) included the broader region upstream of the sill and so had no need to prescribe a flux through the Strait. Their process model, with bottom following coordinates, was the first to replicate the flow rates predicted by Whitehead (1998); the hydraulically and topographically controlled flow was considered to be essentially constant with some small variation around the mean.

Observational studies made during the last 5 years have provided strong evidence that the Denmark Strait Overflow is indeed hydraulically controlled. The theoretical requirements now confirmed in our observations include: (1) the existence of a critical point (Froude No. >1) in the flow near the sill (but not at it due to friction). This critical point is observed approximately 100 km downstream from the sill (Girton 2001). (2) The deepening of isopycnals as the flow crosses the sill and descends the east Greenland Slope into the Irminger Basin. This too has been demonstrated by Girton (2001) who notes that the descent is strongly influenced by friction. (3) A geostrophically balanced flow at the sill. This was first demonstrated by Macrander et al. (2005, 2007).

Although the survey data of Girton et al. (2001) had led them to the '*view of the DSO as an unchanging, hydraulically-controlled flow on timescales longer than a few days*' (see above), a range of observational and modelling studies would now strongly assert (Fig. 19.8) that the overflow transport *is* subject to interannual variation. The change in the upstream 'reservoir' that permits such variability of a hydraulically controlled flow has been described by Macrander (2005). In essence, it was estimated that a decrease by 50 m in the height of the isopycnal $\sigma_\theta = 27.8$ at the Icelandic Kögur section, 200 km upstream of the sill over the Icelandic Slope, should lead to a decrease in the hydraulically controlled component of overflow from 2 to 1.7 Sv. Though the total overflow transport is actually a function both of hydraulics and wind-forcing, it is thought significant that as the isopycnal structure at Kögur deepened by 50 m, as it did between 1999 and 2003, the overflow transport observed at the Denmark Strait sill decreased by the predicted magnitude from 3.7 to 3.1 Sv.

In comparison with the view of Girton et al. (2001), we would now describe the Denmark Strait Overflow as the combination of an interannually varying transport of about 2 Sv due to hydraulic forcing, together with a barotropic component of around 1.4 Sv, driven by the slope of the sea surface height (SSH) across the Strait (Kösters et al. 2004). This is roughly consistent with the transport that we derive by multiplying the cross-sectional area of the DSO plume by the surface velocity of about 0.2 m s^{-1} (Macrander et al. 2007).

If we increase the geographic scale of the forcing to that of the Greenland–Scotland Ridge and Nordic Seas, two hypotheses for the control of overflow strength have been investigated in idealised models during the period of the ASOF study. First, Biastoch et al. (2003) have investigated the possibility of a large-scale wind-driven effect that would act inversely on the two main overflows. As the regional windstress curl increases, a strengthened overflow transport is forced though the Denmark Strait, with a balancing reduction in the eastern overflow through Faroe Bank Channel. Experimentally altering the cross-sill density contrast produced a strong response in the total cross-Ridge overflow but the DSO component was relatively insensitive to these changes in the model. Moreover, the model did not reproduce the critical flow conditions required for full hydraulic control across the Denmark Strait sill.

In an alternative approach, Wilkenskjeld and Quadfasel (2005) extend hydraulic control theory beyond the scale of the Greenland–Scotland Ridge to embrace the Nordic Seas, and more specifically investigate the effect on overflow of changes in the reservoir height of its source water. One purpose of this study was to investigate whether or not a sudden shut off of the thermohaline circulation over the Greenland–Scotland Ridge could be expected when the interface falls below a certain threshold level. According to their results this is not the case. Instead the non-linear dependence of the volume fluxes on the height of the interface upstream of the channel means that the overflow transport becomes slowly weaker as the interface is lowered. They also illustrate that this effect has a greater impact on the Denmark Strait component of the total overflow because as the interface deepens its northwest–southeast slope becomes smaller thus decreasing the local isopycnal height above the sill at the Denmark Strait end more quickly than at the Faroe Bank end.

The records from a long-term observing system over all parts of the ridge may help to separate the effects of buoyancy and wind forcing on the variability of the overflows. It is a final noteworthy point that at present, comparing the decade-long time series of directly observed transports, there is no evident co-variance between the two deep overflows.

19.7 Hydrographic Variability

While (see above) there seems no obvious or convincing evidence for any long term trend in the DSOW transport that might be linked or implicated in the recent reported slowdown of the Atlantic Meridional Overturning Circulation (MOC; Bryden et al. 2005), long-term changes in the thermohaline properties of overflow are a different matter and may well be involved in the circulation changes observed at lower latitudes. Annual repeat hydrography of high quality was begun in 1990 with the WOCE (World Ocean Circulation Experiment) annual coverage of Section A1E starting at Cape Farewell at the southern tip of Greenland and running eastward across the Reykjanes ridge to Porcupine Bank west of Ireland. Under the EC-VEINS project, a set of five sections normal to the SE Greenland slope was added between the WOCE A1E-section and Denmark Strait in 1997 and these have been continued throughout the ASOF-W study (see Fig. 19.10). These sections are supplemented by results from the Icelandic standard hydrographic sections Faxaflói (Fig. 19.10) and Látrabjarg, Kögur and Hornbanki.

The accumulated hydrographic record reveals the following:

(a) *Composition and transport*: despite the admitted difficulty of making valid geostrophic transport estimates of overflow due to the variability of the flow, the sloping bottom, the unknown level of no motion, the presence of eddies and other ageostrophic motions, etc., such computations are not without value in describing the changing constituents of the flow. Here the geostrophic velocities are computed using the method of Jacobsen and Jensen (1926) for stations of different depths. We assume no northward flow at the slope and the southward velocity is expected to

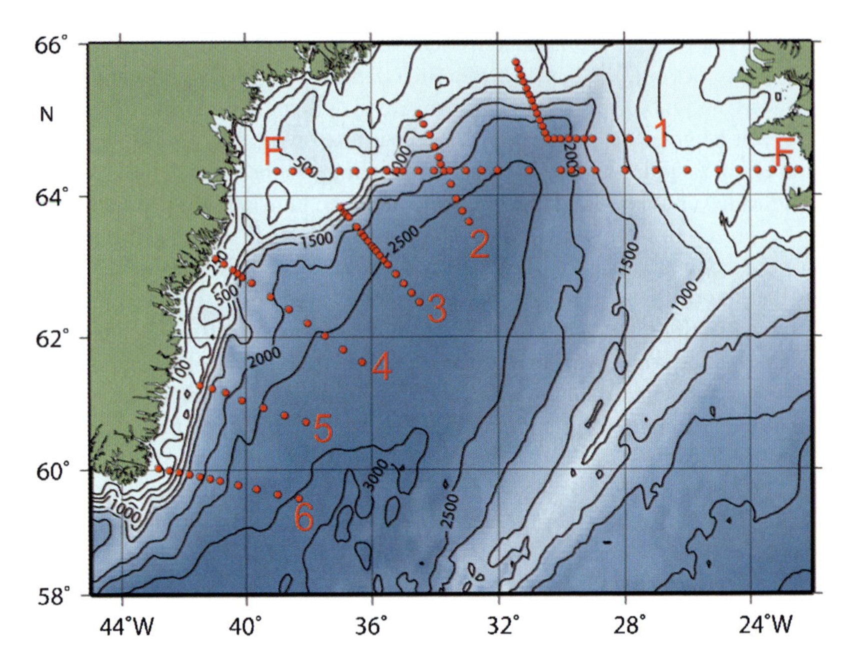

Fig. 19.10 Map of the ASOF-W annual hydrographic sections (F–F is the Icelandic Standard Section 'Faxaflói')

increase both towards the surface, because of the less dense water closer to the coast, and towards the bottom and the deep boundary current. The velocity profiles are adjusted to have no northward velocity and the minimum southward velocity is set to zero. This implies a varying level of no motion and the assumption that eddies present in the flow are not strong enough to reverse the flow but will drift with the main current.

Table 19.1 shows the average transports of watermasses in the density range $\sigma_\theta > 27.85$ for Sections 2–4 (see Fig. 19.10; note that we exclude Section 1 because it may have an ageostrophic down-slope component, and Section 6 since the transports appear to increase downstream between Sections 4 and 6, perhaps due to a recirculation loop within the Irminger Basin). As we have argued above, this density interval provides our best estimate of the Denmark Strait Overflow and the aggregate of sections taken within a week each year can be considered as a sparsely sampled time series spanning 1–2 months – the time needed for the overflow water to advect from Sections 1 to 6. As shown, the total transport varies between 2.5 Sv in 1997 and 5 Sv in 2005 with an average of 3.56 Sv, so that the mean and the variability of the transports obtained each year by geostrophic computations thus agree reasonably well with those measured by the current meter array. That the mean transports obtained from geostrophic calculations are smaller than those found from the direct current measurements is unsurprising since any barotropic velocity component present would not be accounted for by the geostrophy.

Table 19.1 Percentages and transports of different watermasses at $\sigma_\theta > 27.85$ for Sections 2–4 in the Deep Western Boundary Current off SE Greenland

	PIW Θ = 0°C; S = 34.6		DSOW Θ = 0.5 °C; S = 34.9		SPMW Θ = 5 °C; S = 34.95		ISOW Θ = 3.2 °C; S = 34.93		Tot transport
Year	Fraction	Sv	Fraction	Sv	Fraction	Sv	Fraction	Sv	Sv
1997	11.83	0.297	54.78	1.377	22.32	0.561	11.61	0.292	2.514
1998	14.19	0.370	45.19	1.177	1.04	0.027	39.57	1.030	2.604
1999	16.54	0.553	46.09	1.542	1.04	0.035	36.21	1.212	3.346
2000	12.16	0.445	47.19	1.727	5.85	0.214	34.04	1.246	3.659
2001	10.19	0.395	50.94	1.976	6.82	0.265	31.76	1.232	3.878
2002	16.74	0.476	37.84	1.077	0.45	0.013	44.96	1.280	2.846
2003	5.84	0.226	47.87	1.856	5.89	0.228	39.92	1.547	3.876
2004	16.84	0.726	52.01	2.243	11.00	0.474	18.68	0.806	4.313
2005	8.59	0.430	49.19	2.461	2.23	0.112	39.58	1.980	5.002

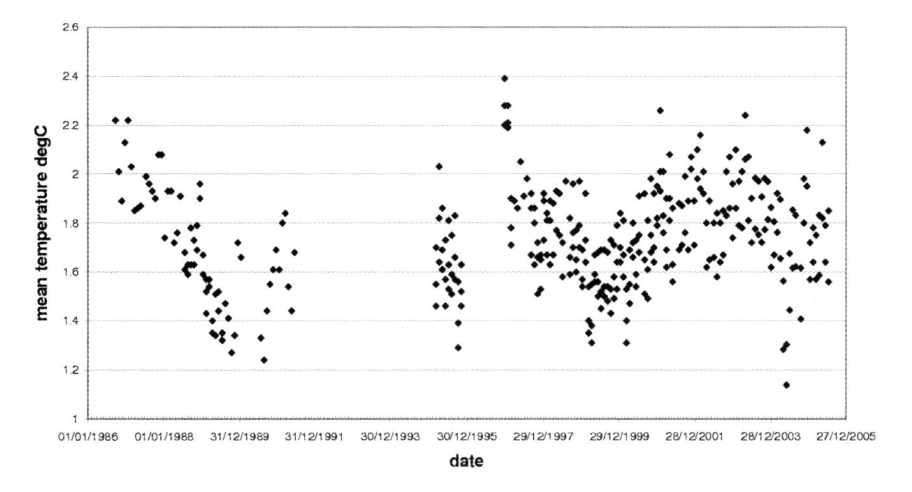

Fig. 19.11 Thirty-day mean hourly temperatures since 1986 in the bottom 100 m at mooring UK-1, set at 2,000 m water depth in the core of the Denmark Strait Overflow on the SE Greenland Slope off Angmagssalik

(b) *Interannual changes in temperature and salinity.* Near-continuous temperature records from current meter thermistors in the core of the overflow since 1986 and from SBE-37 salinity sensors deployed across the array since 1998 have provided clear evidence of extreme interannual change in both T and S on the Angmagssalik line and have provided clues as to the likely upstream sources and downstream impacts of these changes. Figure 19.11 illustrates the conspicuous interannual fluctuations in temperature over the last 2 decades, with record warmth in 1997 (see Section 19.5) and record cold in 2004.

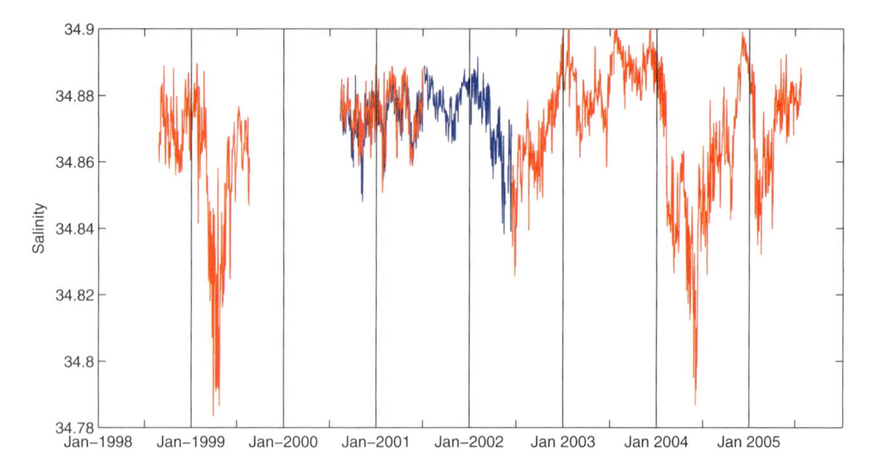

Fig. 19.12 The longest salinity records yet recovered from near-bottom depths (Z = 2,000 m; H = 20 m) in the core of the Denmark Strait Overflow, showing the extreme freshening by up to 0.1 psu that passed through the array in the first half of 1999 and 2004

No smoothing has been employed in compiling this figure. Each dot is the simple mean of 720 successive hourly values of thermistor temperature measured by current meter in the near-bottom layer. Following Dickson et al. 1999, this pattern of change in overflow temperature appears to be the lagged reflection of temperature variability in the upper 500 m of the eastern Fram Strait, some 2,500 km upstream and 3 years earlier.

Though our salinity records are shorter, our lengthening series from the near-bottom layer (+20 m) UK-1 and G-1 (Fig. 19.12) identifies a series of freshening events passing through the overflow core in the early part of most years but with extreme freshening by around 0.1 psu between January and July in 1999 and 2004. Comparison with Fig. 19.11 suggests that both extrema coincided with long-term minima in overflow temperature. However their spike-like nature and the fact that they begin each winter would seem to rule out a distant upstream source for these changes (as we suppose for temperature) but to indicate a much more local origin. In fact these episodes of extreme freshening are now convincingly attributed to the strengthening of the freshwater feed to the overflow from the East Greenland Current, arising locally from an anomalously strong north-wind component immediately to the north of Denmark Strait (Fig. 19.13; from Holfort and Albrecht 2007). These events are still recognizable features of the NEADW where it spreads through the eastern Labrador Sea (Yashayaev and Dickson 2007, Chapter 21, this volume), but are less obviously present by the Labrador Slope.

Such major and broadscale anomalies in salinity and temperature offer the potential to track the rates and pathways of overflow downstream. In Yashayaev and Dickson (2007, their Fig. 3), the 1-year-lagged correspondence between the

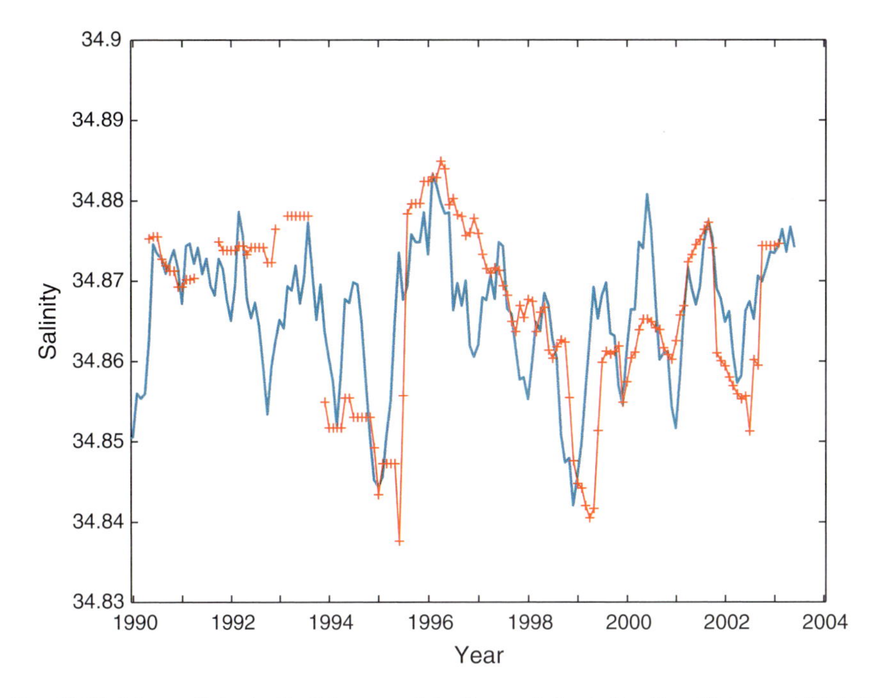

Fig. 19.13 Mean salinity (red) of the core of the Denmark Strait Overflow Water on ASOF-W Section 6, 1990–2004 (see Fig. 10) versus the mean meridional wind component through Denmark Strait (blue), represented here by the scaled difference of the mean sea level pressure (NCEP) between 67.5° N 30° W and Iceland. The overflow core was chosen to lie at the density level of $\sigma_2 = 37.12$ (From Holfort and Albrecht 2007)

temperature of the core of the Denmark Strait Overflow off Angmagssalik and the census by temperature classes across the abyssal Labrador Sea is shown with astonishing clarity, suggesting the real possibility of predicting the T, S and density fields of the abyssal Labrador Sea one year in advance from the hydrographic changes recorded off Angmagssalik. The passage of such identifiable 'events' opens the possibility of using the reduction in amplitude and increase in areal extent of these hydrographic 'signals' to derive a better figure than currently available for mixing and entrainment en route. Both of these studies are underway.

(c) *Long-term hydrographic trends and the THC.* Though the magnitude of the response differs between models, most computer simulations of the ocean system in a climate with increasing greenhouse gas concentrations predict a weakening of the thermohaline circulation through an increasing freshwater flux from high northern latitudes to the N Atlantic through subarctic seas. The large-scale, large amplitude freshening that has taken place in the upper 1–1.5 km of the Nordic Seas over the last 3–4 decades is therefore both remarkable in its own right and of potential climatic importance. Transferred south from the Nordic Seas by the two dense overflows through Denmark Strait and the Faroe–Shetland Channel, we find that the entire system of overflow and entrainment that ventilates the deep Atlantic has undergone a remarkably rapid and remarkably steady freshening by between 0.010 and 0.015 per decade over the last 4 decades (Fig. 14, from Dickson et al. 2002). And as deepening convection under the amplifying NAO transferred freshening to intermediate depths also, ultimately reaching to 2,300 m, the net result was a freshening of the entire watercolumn of the Labrador Sea over 4 decades which is believed to be the largest change in the modern oceanographic record, anywhere, equivalent by 1992 to adding and mixing-down an extra 6 m of freshwater at the sea surface (Lazier 1995). The figures for the freshwater loading of the entire Subpolar gyre and Nordic Seas are even more impressive. Curry and Mauritzen (2005) calculate that between the mid-1960s and the mid-1990s, an extra 19,000 km^3 of freshwater passed into this domain, with 4,000 km^3 remaining in the watercolumn of the Nordic Seas and the balance (15,000 km^3) passing into the watercolumn of the Subpolar Gyre.

A monumental set of changes in the hydrology of Arctic and subarctic seas had by these means been transferred to the deep and abyssal ocean at the headwaters of the "Great Conveyor".

Our changing view on the issue of whether and how such a major increase in the outflow of freshwater might have impacted the Atlantic MOC is discussed elsewhere in this volume (Vellinga et al., Chapter 12, this volume) and need not be repeated in detail here, beyond a brief statement of how the Denmark Strait Overflow might theoretically be involved. We recognize both a local and a regional mechanism for that involvement:

1. *Local*: recognising that it is the density contrast across the Denmark Strait sill that drives the overflow and noting the rapid freshening of both overflows over the past 4 decades (Dickson et al. 2002), Curry and Mauritzen (2005) use Whitehead's hydraulic equation to ask how much more freshwater would have to be added to the western parts of the Nordic seas to produce significant slowdown. They find that's not going to happen anytime soon: '*At the observed rate, it would take about a Century to accumulate enough freshwater (e.g. 9000 km³) to substantially affect the ocean exchanges across the Greenland–Scotland Ridge, and nearly two Centuries of continuous dilution to stop them. In this context, abrupt changes in ocean circulation do not appear imminent*'.

The fact that the freshening trend of both overflows at the sill has slowed to a stop over the last 10 years has merely reinforced that conclusion.

2. *Regional*: here the rationale follows the outcome of "hosing" experiments, in which the MOC in coupled climate models is deliberately shut down by spreading large quantities of freshwater across the surface of the N Atlantic south of the Greenland–Scotland Ridge. The reasoning behind such experiments is that capping the watercolumn with freshwater increases static stability, decreases vertical mixing and reduces production of the North Atlantic Deep Water that 'drives' the MOC. However, the involvement of the dense water overflow system in delivering that freshwater (Fig. 19.14) has meant that in practice, the increased accession of freshwater in recent decades has not simply been delivered to the surface of the NW Atlantic but has been spread unevenly over the entire watercolumn of the Labrador Sea. The importance of this fact is made evident in a new analysis of HadCM3 results

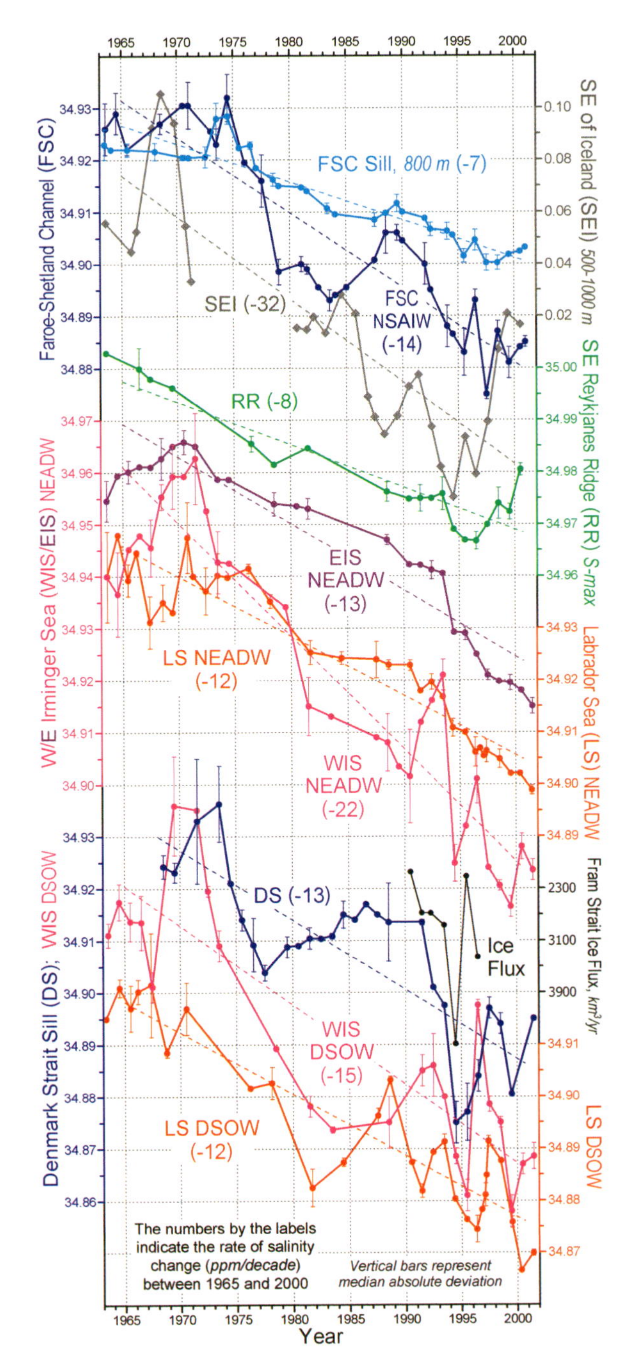

Fig. 19.14 Freshening of the overflow system of the northern N Atlantic since 1965 (From Dickson et al. 2002)

by Vellinga (2005; see also Vellinga et al. 2007) which clearly shows that the same freshwater anomaly (0.1 Sv*year or 3,000 km^3) when spread to depth, has much less effect on MOC weakening than if it were spread across the surface as in 'hosing experiments'.

19.8 The Denmark Strait Overflow: Summary, and Ideas for a Future Observing System

The overflow that forms the subject of this chapter remains of central importance to the Earth's climate system, acting as a principal driver of the abyssal limb of the Atlantic Meridional Overturning Circulation, occupying a uniquely important site for the transfer of ocean-climate signals between watermasses and to abyssal depths, and (if Fig. 19.9 is to be believed) providing a fairly direct balance for the warm water inflow to the Arctic Mediterranean. In summary (Table 19.2):

The strength of the overflow is essentially in hydraulic balance and its mean speed decreases from ~60 cm s^{-1} at the sill to ~25 cm s^{-1} off Angmagssalik. We have observed a complex of space–time variability in its transport out to interannual timescales, but with little evidence of seasonality (<0.1% of variability), no evidence (as yet) of any long-term trend and no convincing evidence of co-variance with the eastern dense overflow through Faroe Bank Channel. Observations over many decades have identified a complex of locally- and remotely-driven large-amplitude variations in the hydrographic character of overflow and its sources, including a long-sustained trend in salinity of 3–4 decades duration. Its conspicuous thermohaline anomalies are readily tracked downstream to the abyssal Labrador Basin with a transit time of about 1 year. Developing a more complete understanding of the longer-term variability in both the hydrography and transport of overflow and their forcing remains critical to the continued development of climate models.

To define that variability, there seems every likelihood that with support from satellite altimetry, the number and cost of our direct observations can be reduced to a mix of ADCP coverage at the sill and a reduced array of conventional moorings in the overflow core off Angmagssalik. At the sill, we strongly recommend that the optimum minimum mooring configuration should consist of two upward-looking ADCPs, one at the deepest point close to the recent Icelandic DS-1 deployment

Table 19.2 The mean characteristics of the overflow core on the Angmagssalik Line

	$\sigma_\theta > 27.8$	$\sigma_\theta > 27.85$
Volume flux	7.3 Sv	4 Sv
Average θ	2.1 °C	1.7 °C
Average S	34.89	34.88
Heat flux relative to input (8.5 °C; Østerhus et al. 2005)	−200 TW	−120 TW
Salt transport	260 kT/s	140 kT/s
Freshwater flux relative to input (35.25; Østerhus et al. 2005)	75 mSv	40 mSv

(ADCP 'A' in Macrander et al. 2005, 2007, at position 66° 04.96′ N, 27° 04.79′ W in 650 m depth), and the second 12 km to the NW of it (ADCP 'B' at 66° 07.60′ N, 27° 16.10′ W in 582 m depth), with one MicroCat T/S sensor to permit the monitoring of density; modelling by Macrander suggests that 2 ADCPs would capture 80% of the overflow passing the sill, while a single ADCP would capture just over 50% (too low a return for any reasonable transport monitoring), and three ADCPs would increase the return to only 84%. In order to maintain continuity with the transport time series downstream (in place discontinuously since 1986) we recommend the continued deployment of order 4 conventional current meter and MicroCat moorings centred on the UK-1 and G-1 sites off Angmagssalik which have provided the longest continuous series. This reduced picket fence array might be maintained to the point where decadal measurements of overflow transport and its hydrographic characteristics have been recovered. Though the array has already been largely proven and is close to being fully developed, there are two main exceptions. First the distribution of moored salinity sensors needs to be further extended in breadth and depth. Second, we have for too long lacked a direct measure of the thickness of the overflow plume, and though we believe, in our transport calculations, that we have overcome this deficiency using the historic hydrographic record, reliance on this assumption is both unwise and unnecessary; two bottom-mounted ADCP's and/or reliable profiling CTDs should be added in the core of the overflow, and the presence of a freshwater 'cap' merely strengthens that requirement. As a third element of the observing system, we concur with the conclusion of Köhl et al. (2007) "*that in principle, altimetric satellite observations of SSH should be appropriate to monitor changes in the Denmark Strait transport*". The composite map of (high-passed) sea-level for high Denmark Strait Overflow derived from Olsen's modeling study (left hand panel, Fig. 19.15)

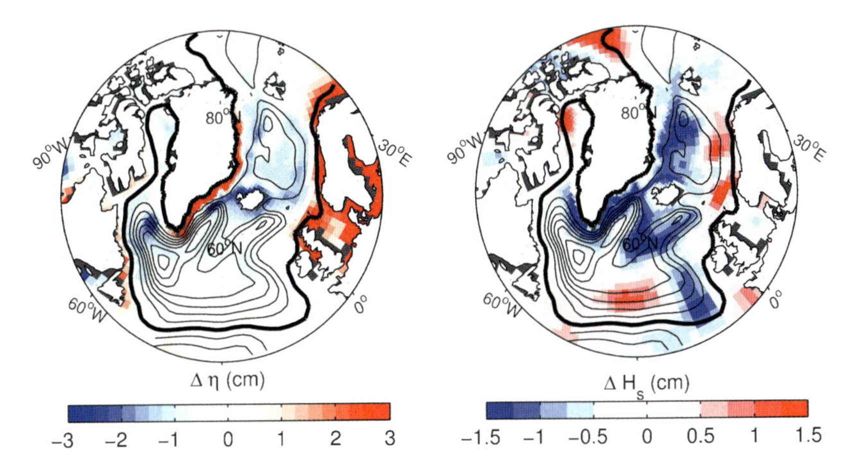

Fig. 19.15 Composite maps of high-passed sea-level (left) and low-passed steric height referenced to 1,000 m (right) relative to high Denmark Strait Overflow using a ±½ standard deviation criteria (of the overflow time series). The applied filter is a first order Butterworth filter with a cut-off frequency of 1/24 months^{-1}. Contours indicate the time-mean, ensemble mean, barotropic streamfunction (the interval between contours is 4 Sv)

reflects the fast barotropic time scale that underlies that link. Its pattern of correlation is essentially similar to the correlation map that Köhl et al. (2007) describe between SSH and overflow transport.

Olsen's study is of a sufficiently large geographic scale to reveal in addition that the pattern of (low-passed) steric height for high Denmark Strait Overflow is of much larger (gyre) scale (right hand panel, Fig. 19.15). This suggests that the slow baroclinic adjustments responsible for longer-term interannual to decadal changes in overflow may not be locally forced but may need to be discussed as an integral part of the large-scale ocean-atmosphere changes on both sides of the Ridge recently highlighted by Hátún et al. (2005). This would suggest that elements of the observing system for monitoring the longer-term changes in overflow transport would need to be of a similarly broad scale, perhaps through the continued development of Deep Glider systems (see Chapter 25) capable of routinely cruising the full watercolumn of the Northern Gyre. SeaGlider coverage, then, represents the fourth element of our recommended observing system for the Denmark Strait Overflow.

Acknowledgements This work has received support from the European Commission Framework Programme 5 project ASOF-W (contract EVK2-CT-2002-00149). SD & BD were also supported by Defra (SD0440), NOAA-CORC-ARCHES and WHOI-OCCI. SJ acknowledges the support of the Research fund of the University of Akureyri. The Deutsche Forschungsgemeinschaft is acknowledged by AK (SFB-512) and AM (SFB-460). The work of MM & BR also received support from the European Commission programme ASOF-N (contract EVK2-CT-2002-00139) and from the Academy of Finland (nr 210551).

References

Aagaard K (1980) On the deep circulation in the Arctic Ocean, Deep-Sea Res. 251–268

Aagaard K, Malmberg S-A (1978) Low frequency characteristics of the Denmark Strait overflow, ICES CM 1978/C: 47, International Council for the Exploration of the Sea, Copenhagen

Aagaard K, Swift JH, Carmack EC (1985) Thermohaline circulation in the Arctic Mediterranean Seas. J. Geophys. Res. 90, 4833–4846

Aagaard K, Fahrbach E, Meincke J, Swift JH (1991) Saline outflow from the Arctic Ocean: Its contribution to the deep waters of the Greenland, Norwegian and Icelandic seas. J. Geophys. Res. 96, 20433–20441

Bacon S (1998) Decadal variability in the outflow from the Nordic Seas to the deep Atlantic Ocean. Nature 394, 871–874

Bacon S (2002) The dense overflows from the Nordic Seas into the deep North Atlantic. ICES Mar. Sci. Symp. 215, 148–155

Biastoch A, Käse RH, Stammer DB (2003) The sensitivity of the Greenland-Scotland Ridge Overflow to Forcing Changes. J. Phys. Oceanogr. 33, 2307–2319

Bourke RH, Weigel AM, Paquette RG (1988) The westward turning branch of the West Spitsbergen Current. J. Geophys. Res. 93, 14065–14077

Buch E, Malmberg S-A, Kristmannsson SS (1992) Arctic Ocean deep water masses in the western Iceland Sea. ICES, C.M. 1992/C, Hydrography Committee, 18 pp

Buch E, Malmberg S-A, Kristmannsson SS (1996) Arctic Water masses in the western Iceland Sea. J. Geophys. Res. 101, 11965–11973

Cooper LHN (1952) Factors affecting the distribution of silicate in the North Atlantic Ocean and the formation of North Atlantic Deep Water. J. Mar. Biol. Assoc. UK 30, 511–526

Cooper LHN (1955) Deep water movements in the North Atlantic as a link between climatic changes around Iceland and the biological productivity of the English Channel and the Celtic Sea. J. Mar. Res. 14, 347–362
Crease J (1965) The flow of Norwegian Sea Deep water through the Faroe Bank Channel. Deep-Sea Res. 12, 143–150
Curry R, Mauritzen C (2005) Dilution of the Northern North Atlantic Ocean in Recent Decades. Science 308(5729): 1772–1774
Curry R, Dickson B, Yashayaev I (2003) A change in the freshwater balance of the Atlantic Ocean over the past four decades. Nature 426, 826–829
Dickson RR, Brown J (1994) The production of North Atlantic Deep Water: Sources, rates and pathways. J. Geophys. Res. 99, 12319–12341
Dickson RR, Lazier J, Meincke J, Rhines P, Swift J (1996) Long-term Coordinated Changes in the Convective Activity of the North Atlantic. Prog. Oceanogr. 38, 241–295
Dickson RR, Meincke J, Vassie IM, Jungclaus J, Østerhus S (1999) Possible predictability in overflow from the Denmark Strait. Nature Lond. 397, 243–246
Dickson RR, Yashayaev I, Meincke J, Turrell W, Dye S, Holfort J (2002) Rapid Freshening of the Deep North Atlantic over the past Four Decades. Nature Lond. 416, 832–837
Dietrich G (1957) Schichtung und Zirkulation der Irmingersee in Juni 1955. Berichte des Deutsche Wissenschaftlichen Kommission für Meeresforschung, Neue. Folge 14, 255–312.
FLAME Group (1998) FLAME – a Family of Linked Atlantic Model Experiments. J Dengg (ed), AWI Bremerhaven, unpublished manuscript
Girton JB (2001) Dynamics of Transport and Variability in the Denmark Strait Overflow. Ph.D. thesis, University of Washington, Seattle, WA
Girton JB, Sanford TB, Käse RH (2001) Synoptic sections of the Denmark Strait Overflow. Geophys. Res. Lett. 28, 1619–1622
Hansen B, Turrell WR, Østerhus S (2001) Decreasing overflow from the Nordic seas into the Atlantic Ocean through the Faroe-Shetland Channel since 1950. Nature 411, 927–930
Hátún H, Sandø AB, Drange H, Hansen B, Valdimarsson H (2005) Influence of the Atlantic Subpolar Gyre on the Thermohaline Circulation. Science 309, 1841–1844
Helland-Hansen B, Nansen F (1909) The Norwegian Sea. Its physical oceanography based upon the Norwegian researches 1900–1904. Report on Norwegian Fishery and Marine Investigations 2, 390 pp
Heinze Ch, Schlosser P, Koltermann K-P, Meincke J (1990) A tracer study of the deep water renewal in the European polar seas. Deep-Sea Res. 37, 1425–1453
Hermann F (1967) The T-S diagram analysis of the water masses over the Iceland – Faroe Ridge and in the Faroe Bank Channel. Rapports et Procès Verbaux des Réunions, Conseil Permanent International pour l'Exploration de la Mer 157 139–149
Holfort J, Albrecht T (2007) Atmospheric forcing of DSOW salinity. Ocean Science, 3, 411–416
Jacobsen JP, Jensen AJC (1926) Examination of hydrographic measurements from research vessels Explorer and Dana during the summer of 1924, Conseil Permanent International pour l'Exploration de la Mer, Rapports et Procès-Verbaux 39, 31–84
Jónsson S (1999) The circulation in the northern part of Denmark Strait and its variability. ICES CM 1999/L: 06.
Jónsson S, Valdimarsson H (2004) A new path for the Denmark Strait overflow water from the Iceland Sea to Denmark Strait. Geophys. Res. Lett. 31(3), L03305, doi: 10.1029/2003GL019214
Jungclaus JH, Backaus JO (1994) Application of a transient reduced gravity plume model to the Denmark Strait Overflow, J. Geophys. Res. 99, 12375–12396
Käse R, Oschlies A (2000) Flow through Denmark Strait. J. Geophys. Res. 10, 28527–28546
Kieke D, Rhein M (2006) Variability of the overflow water transport in the western North Atlantic, 1950–97. J. Phys. Oceanogr. 36, 435–456
Kistler R, Kalnay E, Collins W, Saha S, White G, Woollen J, Chelliah M, Ebisuzaki W, Kanamitsu M, Kousky V, van den Dool H, Jenne R, Fiorino M (2001) The NCEP-NCAR 50-year Reanalysis: Monthly means CD-ROM and documentation. Bull. Am. Met. Soc. 82, 247–268

Köhl A, Käse R, Stammer D, Serra N (2007) Causes of changes in the Denmark Strait Overflow, J. Phys. Oceanogr. 37(6), 1678–1696

Kösters F, Käse R, Fleming K, Wolf D (2004) Denmark Strait overflow for Last Glacial Maximum to Holocene conditions. Paleoceanography 19, PA2019, doi:10.1029/2003PA000972

Krauss W, Käse RH (1998) Eddy formation in the Denmark Strait overflow. J. Geophys. Res. 103, 15525–15538

McCartney MS, Talley LD (1984) Warm to Cold Water Conversions in the Northern North Atlantic Ocean. J. Phys. Oceanogr. 14(5), 922–935

Macrander A, Send U, Valdimarsson H, Jónsson S, Käse RH (2005) Interannual changes in the overflow from the Nordic Seas into the Atlantic Ocean through Denmark Strait. Geophys. Res. Lett. 32, L06606, doi:10.1029/2004GL021463

Macrander A, Send U, Valdimarsson H, Jónsson S, Käse RH (2007) Spatial and temporal structure of the Denmark Strait Overflow revealed by acoustic observations. Ocean Dynamics, doi:10.1007/s10236–007–0101-x

Malcolm G (2005) Variabilität des Dänemarkstraßen Overflow und ihre atmosphärische Anregung, analysiert aus Modellrechnungen und Beobachtungen. Diploma thesis, Universität Hamburg

Malmberg S-A (1972) Intermediate Polar Water in the Denmark Strait Overflow August 1971. ICES CM 1972/C: 6, 6pp + 11 figs (mimeo)

Marsland SJ, Haak H, Jungclaus JH, Latif M, Röske F (2003) The Max-Planck-Institute global ocean/sea-ice model with orthogonal curvilinear coordinates. Ocean Modelling 5, 91–127

Mauritzen C (1996a) Production of dense overflow waters feeding the North Atlantic across the Greenland – Scotland Ridge. Part I: Evidence for a revised circulation scheme. Deep-Sea Res. 43, 769–806

Mauritzen C (1996b) Production of dense overflow waters feeding the North Atlantic across the Greenland – Scotland Ridge. Part 2: An inverse model. Deep-Sea Res. 43, 807–835

Meincke J (1983) The modern current regime across the Greenland-Scotland Ridge. In Bott, Saxov, Talwani, Thiede (Eds.) Structure and development of the Greenland-Scotland Ridge. Plenum, pp 637–650

Meincke J, Rudels B, Friedrich HJ (1997) The Arctic Ocean-Nordic Seas thermohaline system. ICES J. Mar. Sci. 54(3), 283–299

Mikolajewicz U, Groger M, Maier-Reimer E, Schurgers G, Vizcaino M, Winguth AME (2007) Long-term effects of anthropogenic CO_2 emissions simulated with a complex earth system model. Clim. Dyn., doi:10.1007/s00382–006–0204-y

Nansen F (1912) Das Bodenwasser und die Abkühlung des Meeres. Internationale Revue des Gesamten Hydrobiologie und Hydrographie 5(1), 42

Olsen SM, Schmith T (2007) North Atlantic–Arctic Mediterranean exchanges in an ensemble hindcast experiment. JGR-Oceans, 112, C04010, doi:10.1029/2006JC003838

Olsson KA, Jeansson E, Tanhua T, Gascard J-C (2005) The East Greenland Current studied with CFCs and released sulphur hexafluoride. J. Mar. Syst. 55(1–2), 77–95

Østerhus S, Turrell WR, Jónsson S, Hansen B (2005) Measured volume, heat, and salt fluxes from the Atlantic to the Arctic Mediterranean. Geophys. Res. Lett. 32, L07603

Pacanowski RC (1995) MOM 2 Documentation, User's Guide and Reference Manual. Technischer Bericht 3 GFDL Oceans Group

Pickart RS, Spall MA, Ribergaard MH, Moore GWK, Millif RF (2003) Deep convection in the Irminger Sea forced by the Greenland tip jet. Nature 424, 152–156

Pickart RS, Torres DJ, Fratantoni PS (2005) The East Greenland spill jet. *J*. Phys. Oceanogr. 35, 1037–1053

Ross CK (1975) Overflow 73, current measurements in Denmark Strait. ICES 1975, CM 1975/C: 6, 7pp +6 Figs (mimeo).

Ross CK (1984) Temperature-salinity characteristics of the 'overflow' water in Denmark Strait during 'OVERFLOW '73', Rapp. P. V. Fisheries. Reun. Cons. Int. Explor. Mer. 185, 111–119

Rudels B (1986) The Θ–S relations in the northern seas: Implications for the deep circulation. Polar Res. 4 n.s., 133–159

Rudels B (1995) The thermohaline circulation of the Arctic Ocean and the Greenland Sea. Phil. Trans. R. Soc. Lond. A 352, 287–299
Rudels B, Jones EP, Anderson LG, Kattner G (1994) On the intermediate depth waters of the Arctic Ocean. In OM Johannessen, RD Muench, JE Overland (Eds.) The role of the Polar Oceans in Shaping the Global Climate. American Geophysical Union, Washington, DC, pp 33–46
Rudels B, Friedrich HJ, Quadfasel D (1999a) The Arctic circumpolar boundary current. Deep-Sea Res. II 46(6–7), 1023–1062
Rudels B, Eriksson P, Grönvall H, Hietala R, Launiainen J (1999b) Hydrographic observations in Denmark Strait in fall 1997, and their implications for the entrainment into the overflow plume. Geophys. Res. Lett. 26(9), 1325–1328, doi:10.1029/1999GL900212
Rudels B, Fahrbach E, Meincke J, Budéus G, Eriksson P (2002) The East Greenland Current and its contribution to the Denmark Strait overflow. ICES J. Mar. Sci. 59(6), 1133–1154
Rudels B, Eriksson P, Buch E, Budéus G, Fahrbach E, Malmberg S-A, Meincke J, Mälkki P (2003) Temporal switching between sources of the Denmark Strait overflow water. ICES Mar. Sci. Symp. 219, 319–325
Ryder C (1891–1892) Tidligere Ekspeditioner til Grønlands Østkyst nordfor 66° Nr. Br., Geogr. Tids. Bd. 11, 62–107, København
Saunders PM (2001) The dense northern overflows. Pp 401–417 In G Siedler, J Church, J Gould (Eds.) Ocean circulation and climate. Academic Press Int Geophys Ser 77, 715 pp.
Schauer U, Muench RD, Rudels B, Timokhov L (1997) Impact of eastern Arctic shelf water on the Nansen Basin intermediate layers. J. Geophys. Res. 102, 3371–3382
Smethie WM, Swift JH (1989) The tritium:krypton-85 age of the Denmark Strait overflow water and Gibbs Fracture zone water just south of Denmark Strait. J. Geophys. Res. 94, 8265–8275
Smethie WM, Chipman DW, Swift JH, Koltermann K-P (1988) Chlorofluoromethanes in the Arctic Mediterranean seas: evidence for formation of bottom water in the Eurasian Basin and the deep-water exchange through Fram Strait. Deep-Sea Res. 35, 347–369
Smith PC (1976) Baroclinic instability in the Denmark Strait overflow. J. Phys. Oceanogr. 6, 355–371
Spall MA, Price JF (1998) Mesoscale variability in Denmark Strait: the PV outflow hypothesis. J. Phys. Oceanogr. 28, 1598–1623
Stern ME (2004) Transport extremum through Denmark Strait. Geophys. Res. Lett. 31, L12303, doi:10.1029/2004GL020184
Sverdrup HU, Johnson MW, Fleming RH (1942) The Oceans: their Physics, Chemistry and General Biology. Prentice-Hall, New York, 1042 pp.
Swift JH, Aagaard K (1981) Seasonal transitions and water mass formation in the Iceland and Greenland seas. Deep-Sea Res. A 28A(10), 1107–1129
Swift JH, Aagaard K, Malmberg S-A (1980) Seasonal transitions and water mass formation in the Iceland and Greenland Seas. Deep-Sea Res. 28, 1107–1129
Tanhua T, Bulsiewicz K, Rhein M (2005) Spreading of overflow water from the Greenland to the Labrador Sea. Geophys. Res. Lett. 32, L10605, doi:10.1029/2002GLO227700
Vellinga M (2005) Sensitivity of MOC to Vertical Distribution of Freshwater Perturbations. ASOF Newsletter #3, pp 18–20
Voet G (2007) Entrainment in the Denmark Strait Overflow Plume by meso-scale eddies. Diploma thesis in Physics, University of Hamburg, 80 pp
Wilkenskjeld S, Quadfasel D (2005) Response of the Greenland-Scotland Overflow to changing deep water supply from the Arctic Mediterranean. Geophys. Res. Lett 32, L21607, doi:10.1029/2005GLO24140
Worthington LV (1969) An attempt to measure the volume transport of Norwegian Sea overflow water through the Denmark Strait. Deep-Sea Res. 16(Suppl.), 421–432
Worthington LV (1970) The Norwegian Sea as a Mediterranean basin. Deep-Sea Res. 17, 77–84
Whitehead JA (1998) Topographic control of oceanic flows in deep passages and straits. Rev. Geophys. 36, 423–440

Wüst G (1935) Schichtung und Zirkulation des Atlantischen Ozeans. Die Stratosphäre. In Wissenschaftliche Ergebnisse auf dem Forschungs und Vermessungsschiff "Meteor" 1925–1927, 6: 1st part, 2, 180 pp
Wüst G (1942) Die morphologischen und ozeanographischen Verhältnisses des Nordpolarbeckens. Veröff. Deutschen Wissensch. Inst. zu Kopenhagen. Reihe 1(6) Berlin

Chapter 20
Tracer Evidence of the Origin and Variability of Denmark Strait Overflow Water

Toste Tanhua[1], K. Anders Olsson[2], and Emil Jeansson[3]

20.1 Introduction

The overflow of dense water from the Nordic Seas to the North Atlantic through the Denmark Strait is an important part of the global thermohaline circulation. Denmark Strait Overflow Water (DSOW) has its sources in the Nordic Seas and the Arctic Ocean and is a complex mixture of several water masses.

The magnitude and variability of the overflow are significant not only for the local oceanography, but also for the global large-scale circulation. Just as the intensity of the overflow is temporally and geographically variable, so are the hydrographic and hydrochemical characteristics of the overflow shifting. Variations in these properties have two possible sources: (1) changes in the characteristics of water masses and, (2) changes in the water mass composition of the overflow. Changes in atmospheric forcing and convection within the source region for DSOW might change its water mass composition and characteristics, changes that in turn will propagate to the North Atlantic Deep Water.

The variability of the overflows has received significant attention the last couple of decades, not least through efforts by VEINS, ASOF and related projects. Although there has been significant progress during this time, as is evident from papers in this volume, many questions remain, at least partly, unresolved.

In this chapter, we have synthesised the knowledge of the characterisation and origin of DSOW from historical and recent studies, all using chemical tracers.

[1]Department of Marine Biogeochemistry, Leibniz Institute for Marine Sciences at Kiel University, Düsternbrooker Weg 20, DE-24105 Kiel, Germany. E-mail: ttanhua@ifm-geomar.de

[2]Bjerknes Centre for Climate Research, University of Bergen, Allégaten 55, NO-5007 Bergen, Norway

Present address: Department of Chemistry, Göteborg University, SE-412 96 Göteborg, Sweden. E-mail: andols@chem.gu.se

[3]Department of Chemistry, Göteborg University, SE-412 96 Göteborg, Sweden

Present address: Bjerknes Centre for Climate Research, University of Bergen, Allégaten 55, NO-5007 Bergen, Norway. E-mail: emil.jeansson@bjerknes.uib.no

R.R. Dickson et al. (eds.), *Arctic–Subarctic Ocean Fluxes*, 475–503

We are further focusing on the formation and variability of the Denmark Strait Overflow Water as found in the strait or in the nearby Irminger Basin. We are thus ignoring the extensive literature on tracers in the North Atlantic Deep Water further south, as well as those focusing solely on the Arctic Mediterranean. Similarly, results derived solely from "classical" hydrography are presented elsewhere in this volume (Dickson et al. 2008). The increased number of tracer observations and thus the increased spatial and temporal data coverage has enabled more sophisticated water-mass analysis. Although changes in the water mass composition and hydro-chemical characteristics of the DSOW is evident on annual basis, continued monitoring of tracers in the Denmark Strait will enable detection of changes in the source region for DSOW.

20.1.1 Water Masses

Since both the water mass composition and the properties of the water masses and their temporal variability are central for the produced DSOW, we give a summary of the significant water masses in the area (Table 20.1). Several studies have defined water masses in the region. These studies have often used different names for the water masses and have been differently detailed depending on the scope of the study and the available data. Since the nomenclature of water masses in the literature is somewhat variable we have tried to follow that of a recent, detailed investigation of all water masses in the East Greenland Current from north of Fram Strait to the Denmark Strait using a large set of parameters by Jeansson et al. (2008) and list the names and acronyms in Table 20.1. Tables of water mass properties based on hydro-chemical measurements relevant for DSOW are also presented by others (e.g., Fogelqvist et al. 2003, Olsson et al. 2005b; Tanhua et al. 2005b). It is obvious that water mass properties at times differ considerable between studies. There are several reasons for this discrepancy; the definitions are based on different data sets and varying sets of parameters; the data are from different regions and from different years; or the purposes of the studies differ. In general, it can be concluded that the farther away from the source region the water mass properties are defined, and the fewer parameters available, the less water masses can possibly be identified. It is not always of interest, or possible, to identify the "original" water mass, but rather the source region or water class. Sometimes, a water mass is defined to fill a certain depth or density interval while in other cases, for instance in quantitative water mass analysis, a water mass is defined from its original properties close to the formation area. In Table 20.1, we have tried to follow the latter of the two philosophies, the most natural when trying to solve the composition of a water mixture. The presented properties are hence those representing a water mass core in the source region and not always representative for a wider layer along its spreading path. Figure 20.1 shows the region of water mass formation of interest for DSOW.

Table 20.1 This table provides an overview of Characteristics of selected water masses and "clusters of water masses" reported to contribute to DSOW or being entrained into it in recent studies. Characteristics for the dowstream water masses entrained into DSOW (MIW, LSW, ISOW) are from Tanhua et al. (2005b) and represent conditions in 1997

Water mass	Parameter	Unit	Value
Polar Intermediate Water (PIW) (Stefansson 1962) is a recently ventilated, cold and relatively fresh water mass that is formed either in the Arctic Ocean or along the east coast of Greenland. **PIW** may also be characterized as a mixture of Polar Surface Water and Recirculated Atlantic Water (e.g., Tanhua et al. 2005b).	Pot. temp.	°C	−0.71
	Salinity		34.531
	Pot. density anomaly	kg m^{-3}	27.761
	O_2	μmol kg^{-1}	333
	PO_4	μmol kg^{-1}	0.79
	NO_3	μmol kg^{-1}	10.7
	SiO_2	μmol kg^{-1}	5.2
	CFC-12	pmol kg^{-1}	3.3
	CFC-11	pmol kg^{-1}	6.2
	F12 (TTD) age	Years	1
	F11 (TTD) age	Years	0
Iceland Sea Arctic Intermediate Water (ISAIW) (Rudels et al. 2002) is formed in the Iceland Sea and was originally called upper **AIW** (Swift et al. 1980, Swift and Aagaard 1981). **Arctic Intermediate Water (AIW)** is a class of water masses formed within the Nordic Sea (Stefansson 1962), divided into types formed in different regions. **ISAIW** is high in oxygen and tracers and contributes to the **AIW** layer of the Norwegian Sea (Blindheim 1990) where it is often included in what is called **NSAIW**.	Pot. temp.	°C	−1.06
	Salinity		34.779
	Pot. density anomaly	kg m^{-3}	27.977
	O_2	μmol kg^{-1}	351
	PO_4	μmol kg^{-1}	0.82
	NO_3	μmol kg^{-1}	11.5
	SiO_2	μmol kg^{-1}	4.2
	CFC-12	pmol kg^{-1}	3.4
	CFC-11	pmol kg^{-1}	6.8
	F12 (TTD) age	Years	0
	F11 (TTD) age	Years	0
Greenland Sea Arctic Intermediate Water (GSAIW) (Olsson et al. 2005b) is the saltiest and densest type of **AIW**, and was denoted uAIW2 by Blindheim (1990). GSAIW has in later years been identified by high levels of sulphur hexafluoride (SF_6), released in the Greenland Sea in 1996 (Watson et al. 1999). Together with **ISAIW** it constitutes the **AIW** layer of the Norwegian Sea (Blindheim 1990), i.e., **NSAIW**.	Pot. temp.	°C	−0.90
	Salinity		34.884
	Pot. density anomaly	kg m^{-3}	28.055
	O_2	μmol kg^{-1}	333
	PO_4	μmol kg^{-1}	0.88
	NO_3	μmol kg^{-1}	13.1
	SiO_2	μmol kg^{-1}	6.6
	CFC-12	pmol kg^{-1}	2.9
	CFC-11	pmol kg^{-1}	5.4
	F12 (TTD) age	Years	10
	F11 (TTD) age	Years	17

(continued)

Table 20.1 (continued)

Recirculating Atlantic Water (RAW) (Bourke et al. 1988) is transported by the West Spitsbergen Current to the Fram Strait where it re-circulates and joins the **East Greenland Current (EGC)**. **RAW** has also been denoted lower **AIW** (Swift et al. 1980, Swift and Aagaard 1981). **RAW** is the warmest and most saline water mass in the **EGC** with relatively high levels of CFCs. This reveals recent ventilation and shorter transport path of **RAW** compared to **AAW** that re-circulates within the Arctic Ocean (Jeansson et al. 2007).	Pot. temp.	°C	3.02
	Salinity		35.053
	Pot. density anomaly	kg m^{-3}	27.926
	O_2	µmol kg^{-1}	313
	PO_4	µmol kg^{-1}	0.79
	NO_3	µmol kg^{-1}	11.7
	SiO_2	µmol kg^{-1}	5.1
	CFC-12	pmol kg^{-1}	2.7
	CFC-11	pmol kg^{-1}	4.9
	F12 (TTD) age	Years	0
	F11 (TTD) age	Years	0
Arctic Atlantic Water (AAW) (Mauritzen 1996) is formed from Atlantic Water that enters the Arctic Ocean (Rudels et al. 1999b). **AAW** has also been called Modified Atlantic Water (Rudels et al. 2000). Fractions that have taken different circuits through the Arctic Ocean can be separated, and we present the properties of the Canadian Basin version, i.e., the longest route through the Arctic Ocean (Jeansson et al. 2007). Canadian Basin Intermediate Water (Jeansson et al. 2007) is similar to **AAW** and Olsson et al. (2005b) call the mixture of these two for **AAW**.	Pot. temp.	°C	0.70
	Salinity		34.832
	Pot. density anomaly	kg m^{-3}	27.930
	O_2	µmol kg^{-1}	293
	PO_4	µmol kg^{-1}	0.86
	NO_3	µmol kg^{-1}	12.9
	SiO_2	µmol kg^{-1}	7.0
	CFC-12	pmol kg^{-1}	1.7
	CFC-11	pmol kg^{-1}	3.2
	F12 (TTD) age	Years	37
	F11 (TTD) age	Years	45
upper Polar Deep Water (uPDW) (Rudels et al. 1999b) is formed in the Arctic Ocean and enters the **EGC** below the **AAW**. Different types of **uPDW** from the Canadian and Eurasian basins (Rudels et al. 2000) can be identified in the Arctic Ocean. **uPDW** has lower CFC content and higher salinity compared to intermediate waters from the Nordic Seas.	Pot. temp.	°C	–0.35
	Salinity		34.907
	Pot. density anomaly	kg m^{-3}	28.049
	O_2	µmol kg^{-1}	302
	PO_4	µmol kg^{-1}	0.93
	NO_3	µmol kg^{-1}	14.2
	SiO_2	µmol kg^{-1}	8.1
	CFC-12	pmol kg^{-1}	0.8
	CFC-11	pmol kg^{-1}	1.7
	F12 (TTD) age	Years	97
	F11 (TTD) age	Years	99

(continued)

Table 20.1 (continued)

Description	Parameter	Unit	Value
Nordic Seas Deep Water (NDW) (e.g., Rudels et al. 2005; Jeansson et al. 2007) includes most deepwater masses of the Nordic Seas and is hence more or less a mixture of Canadian Basin Deep Water (Aagaard et al. 1985), Eurasian Basin Deep Water (Aagaard et al. 1985) and Greenland Sea Bottom Water (Rudels et al. 2005). Different variations of local deepwater mixtures have also been called Norwegian Sea Deep Water (Swift and Koltermann 1988) and Arctic Ocean Deep Water.	Pot. temp.	°C	−0.92
	Salinity		34.912
	Pot. density anomaly	kg m^{-3}	28.079
	O_2	μmol kg^{-1}	299
	PO_4	μmol kg^{-1}	1.02
	NO_3	μmol kg^{-1}	14.9
	SiO_2	μmol kg^{-1}	11.4
	CFC-12	pmol kg^{-1}	0.4
	CFC-11	pmol kg^{-1}	0.9
	F12 (TTD) age	Years	162
	F11 (TTD) age	Years	154
Labrador Sea Water (LSW) is formed by wintertime convection in the Labrador and Irminger Seas (e.g., Rhein et al. 2002; Bacon et al. 2003; Pickart et al. 2003). **LSW** is characterized by low salinity and high CFC content.	Pot. temp.	°C	2.95
	Salinity		34.864
	Pot. density anomaly	kg m^{-3}	27.781
	O_2	μmol kg^{-1}	294
	PO_4	μmol kg^{-1}	1.07
	NO_3	μmol kg^{-1}	16.1
	SiO_2	μmol kg^{-1}	9.2
	CFC-12	pmol kg^{-1}	1.9
	CFC-11	pmol kg^{-1}	4.2
	F12 (TTD) age	Years	15
	F11 (TTD) age	Years	13
Middle Irminger Water (MIW) (Min 1999) is found at 1,000–1,200 m in Irminger Basin and is primarily recognized as a minimum in oxygen and CFCs. In the Iceland Basin this water mass was called Intermediate Water by van Aken and de Boer (1995). **MIW** is transported from the Iceland Basin to the Irminger Basin over the Reykjanes Ridge and modifies the overflow south of the sill.	Pot. temp.	°C	3.86
	Salinity		34.946
	Pot. density anomaly	kg m^{-3}	27.758
	O_2	μmol kg^{-1}	275
	PO_4	μmol kg^{-1}	1.07
	NO_3	μmol kg^{-1}	16.2
	SiO_2	μmol kg^{-1}	10.0
	CFC-12	pmol kg^{-1}	1.4
	CFC-11	pmol kg^{-1}	3.1
	F12 (TTD) age	Years	28
	F11 (TTD) age	Years	26

(continued)

Table 20.1 (continued)

Iceland–Scotland Overflow Water (ISOW) is also commonly named Northeast Atlantic Deep Water. This is the overflow water entering the North Atlantic through the Faroe Bank Channel and over the Iceland–Scotland Ridge. The ISOW flows through the Charlie-Gibbs Fracture Zone and over the southern Reykjanes Ridge into the Irminger Basin.	Pot. temp.	°C	2.76
	Salinity		34.912
	Pot. density anomaly	kg m^{-3}	27.837
	O_2	µmol kg^{-1}	281
	PO_4	µmol kg^{-1}	1.09
	NO_3	µmol kg^{-1}	16.4
	SiO_2	µmol kg^{-1}	12.3
	CFC-12	pmol kg^{-1}	1.1
	CFC-11	pmol kg^{-1}	2.4
	F12 (TTD) age	Years	45
	F11 (TTD) age	Years	44

The properties of the water masses from the Nordic Seas and the Arctic Ocean are from Jeansson et al. (2007), and represent the conditions in 2002. Characteristics for the dowstream water masses entrained into DSOW (MIW, LSW, ISOW) are from Tanhua et al. (2005b) and represent conditions in 1997.

The CFC ages are mean ages calculated with the TTD method (Waugh et al. 2003) assuming 90% saturation and $\Delta/\Upsilon = 1$. This method does take the effect of mixing into account for the age calculation. These ages are considerable higher than the "CFC-ages" that do not account for mixing, and represent a more realistic age distribution within a water mass. Note that the CFCs are poor proxies for water mass age for recently ventilated waters due to the decrease in the atmospheric trend in the 1990s.

The table of water masses contributing to the formation of DSOW is rather extensive. Often the water masses have similar characteristics in many respects, but they still differ in the formation area and circulation history, i.e., properties significant for understanding the dynamics and variability of the overflow. The table shows that the Denmark Strait, and the "pre-mixing" area upstream of the strait, have several contributing water masses that mix north of and within the strait to form DSOW. The Denmark Strait is a dynamically active area with documented short-term variability (e.g., Macrander et al. 2005) that makes it challenging to trace the sources and variability of DSOW. The formation is not completed when the overflow leaves the sill since active mixing within the overflow plume, as well as entrainment of surrounding water south of the sill is important for the final overflow product (Rudels et al. 1999a; Tanhua et al. 2005b). This product is commonly denoted North-West Atlantic Bottom Water (Lee and Ellett 1967), and makes up the dense portion of the North Atlantic Deep Water. In this work, we will focus on the origin of the overflow, i.e., the source regions within the Arctic Mediterranean, but will also briefly discuss the entrainment south of the sill, that will be covered in more detail elsewhere in this volume (Dickson et al. 2008).

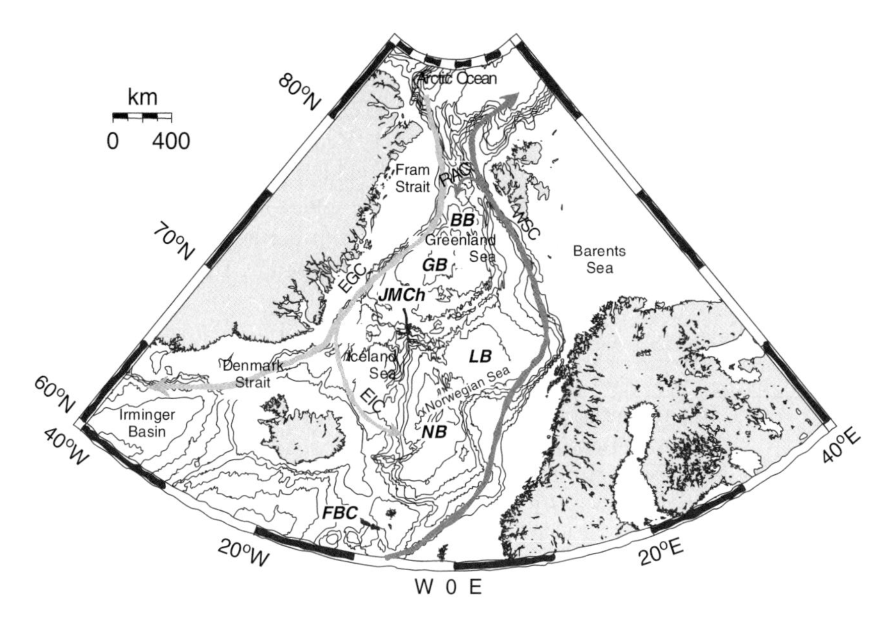

Fig. 20.1 Map over the Nordic Seas and the overflow areas to the North Atlantic. Abbreviations as follows: BB – Boreas Basin; RAC - Return Atlantic Current; WSC – West Spitsbergen Current; GB – Greenland Basin; LB – Lofoten Basin; NB – Norwegian Basin; EIC – East Icelandic Current; JMCh – Jan Mayen Channel; EGC – East Greenland Current; FBC – Faroe Bank Channel

20.2 State of the Art in 1990

In this section, we present some early tracer evidence for the origin of Denmark Strait Overflow Water (DSOW), published before 1990. Before chemical tracers were widely used and accepted tools in oceanography, it was generally accepted that Norwegian Sea Deep Water (NSDW) was the major source water mass for the Denmark Strait. As we shall see, this view has been abandoned since chemical tracer observations were pieced together (e.g., Swift et al. 1980).

An early attempt to characterise the DSOW with chemical parameters was published in 1968 by Stefánsson. Since salinity and temperature alone do not separate sufficiently between the source waters for a detailed analysis, the authors used chemical parameters, such as oxygen and silicate, to characterise the water masses in the complex area south of the Denmark Strait. However, suffering from a limited data set, particularly from lack of data north of Denmark Strait, quantification of the water mass composition of DSOW was limited in extent. However, the results from Stefánsson (1968) indicate considerable decadal changes in hydrochemical properties of the overflow when compared to the VEINS-era data (Tanhua et al. 2005b; Tanhua and Olsson 2006).

Setting the stage for transient tracer studies of the DSOW, Swift et al. (1980) suggested that the most important contributor to DSOW was Arctic Intermediate

Water (AIW) formed in the Iceland Sea (and possibly in the Greenland Sea) by winter cooling. Important for this conclusion were tritium (^3H) measurements from the GEOSECS (see Table 20.2) program in 1972. Tritium was mainly produced during atmospheric nuclear bomb tests in the 1950s and 1960s, and reached the ocean surface through precipitation. Consequently, high tritium concentration indicated a water parcel that had recently been in contact with the surface ocean. The tritium concentration within DSOW in the deep north-west Atlantic was as high as 4 T.U. (tritium units); see Fig. 20.2 (bottom at station 11). This was significantly

Table 20.2 List of projects mentioned in this chapter

Acronym	Project name	Approximate period
ARCICE	Sea Ice and Ocean Vertical Circulation	1999–2001
ASOF	Arctic/Subarctic Ocean Fluxes	1999–2006
ESOP-2	European Sub-polar Ocean Programme, phase 2	1996–1998
GEOSECS	Geochemical Ocean Sections	1972
Nordic WOCE	Nordic contribution to WOCE	1994–1997
TRACTOR	Tracer and Circulation in the Nordic Seas Region	2001–2004
TTO/TTO-NAS	Transient Tracers in the Ocean – North Atlantic Study	1981
VEINS	Variability of exchanges in the Northern Seas	1996–1999
WOCE	World Ocean Circulation Experiment	1990–1998

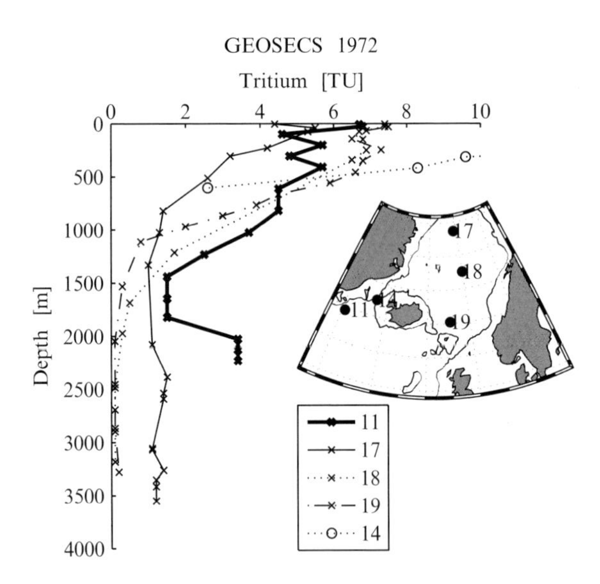

Fig. 20.2 Tritium profiles in the Nordic Seas and the Denmark Strait Overflow from four selected stations from the GEOSECS experiment in 1972 with the overflow station (#11) in bold. Data are downloaded from http://cdiac.esd.ornl.gov/, and are reported in tritium units (T.U. = tritium unit = $10^{18} \times [T]/[H]$) normalised to 1 January 1974. The high tritium concentration in the overflow water (the deepest part of station 11) proves that the overflow must have a component of relatively recently ventilated water, i.e., Arctic Intermediate Water (AIW)

higher than the typical levels in NSDW (<1 T.U.; deep layers at stations 18 and 19 in Fig. 20.2), the water mass previously believed to be the main source of DSOW. Therefore, another source water mass with higher tritium concentration was necessary to reproduce the measured DSOW tritium concentration. From the tritium profiles, as well as from hydrographic data, it was concluded that AIW must be a major component of DSOW (Fig. 20.2). However, even though no NSDW was found south of the ridge, low tritium NSDW was found at the Denmark Strait sill (profile 14 in Fig. 20.2), suggesting intermittent overflow of NSDW into the North Atlantic.

Whereas tritium enters the ocean mainly through precipitation from the atmosphere, ^{137}Cs (cesium-137) and ^{90}Sr (strontium-90) has an additional source from nuclear fuel reprocessing plants in Europe. The nuclide-enriched water is advected northwards to the Nordic Sea and the Arctic Ocean where it can be incorporated in the formation of overflow waters, thus providing a tracer with a different input history. The release from the nuclear fuel reprocessing plants can be reconstructed, and the tracer signal started to affect the Nordic Seas area in the late 1970s. Studies of multiple tracers with different input history are particularly helpful to interpret the data. Observations of tritium, ^{137}Cs and ^{90}Sr from GEOSECS and TTO-NAS in 1981 were reported for the overflow water by Livingston et al. (1985). In addition, the first complete tracer section (TTO-NAS in 1981) across the overflow in the Irminger Basin shows high tritium concentrations in DSOW, again suggesting that the bulk is recently formed water. Further, the authors reported drastically increased tracer concentrations in the overflow between the two surveys (GEOSECS and TTO-NAS). Similarly to Swift et al. (1980), they also noted the inverse correlation between tritium (and ^{137}Cs) and salinity for the overflow, and interpreted it as evidence for Iceland Sea Arctic Intermediate Water being the principal overflow source. Livingston et al. (1985) further noted that a large set of contributing water masses led to highly variable overflow characteristics. A limited data set on the signal progression of ^{137}Cs suggested a transport time of ~2 years from the surface of the Greenland Sea to the overflow (Livingston et al. 1985), an estimate that has not changed much since.

Additional tracer evidence for the source of DSOW was published by Smethie and Swift (1989), again using a multi-tracer approach: tritium and ^{85}Kr (krypton-85) measurements from 1981 (TTO-NAS), as well as the ratio between them. Krypton-85 is a radioactive gas with a half-life of 10.76 years that enters the ocean by rapid air–sea exchange and with its main source to the atmosphere from nuclear power plants. The authors concluded that two water types make up DSOW: one with low salinity and one with high salinity. Again, NSDW was rejected as the source of the high-salinity type, this time aided by the too high silicate concentrations of NSDW. The transient nature of the two tracers and their internal relationship led to the estimates that the residence time north of the Iceland–Scotland Ridge was 15 years for the high-saline overflow type, and less than 1 year for the low-saline type.

To summarise, we have seen how a number of tracers, most of them with strong transient signals, has invaded the surface ocean, including the Nordic Seas. Although there were relatively few tracer measurements in the Nordic Seas, and even fewer within the DSOW south of the sill, much information of the sources of DSOW was gathered even before the VEINS and ASOF programmes started. Around 1990, the overflow was assumed to be structured in two layers; the shallower with a tracer content corresponding to an origin at intermediate depth in more than one source area in the Nordic Seas; and the deeper with lower tracer content and therefore less recently ventilated. The limited geographical coverage of the data (mainly GEOSECS and TTO-NAS), however, did not allow a more precise identification of the source waters for the overflow in the Nordic Seas and the Arctic.

20.3 Changing Ideas and Capacity Since the Early 1990s

In this section, we present knowledge on the Denmark Strait Overflow Water (DSOW) resulting from tracer studies from the 1990s and onwards as well as from older samples or observations (e.g., GEOSECS and TTO-NAS) analysed during this period. Partly, we will describe results based on new tracers and methods that became available to the tracer community during the last approximately 15 years.

The first example is the evolution in ^{14}C (carbon-14) signal in the Nordic Seas over a time range of more than 20 years (from GEOSECS and onwards) used to evaluate the formation of DSOW (Nydal and Gislefoss 1996). Carbon-14 is a radioactive isotope with a half-life ($t_{1/2}$) of 5,730 years. Although this isotope is naturally present in the environment, the nuclear tests roughly doubled the atmospheric ^{14}C content. On the other hand, carbon dioxide (CO_2) from fossil fuel burning do not contain any ^{14}C, so there was a slight decrease in atmospheric ^{14}C before the nuclear bomb tests. This carbon isotope is transferred to the ocean by air–sea–gas exchange with small enrichment (fractionation), leading to higher ^{14}C signals in recently ventilated waters, with a strong signal from the bomb tests in the 1960s. Considering the ^{14}C data, Nydal and Gislefoss (1996) concluded that the overflow water mainly consisted of surface waters of the Nordic Seas and Arctic intermediate water. Profiles of ^{14}C are shown in Fig. 20.3, and the high concentration in DSOW (around 2,000 m) at Station 11 can not come from the deep layers in the Nordic Seas, since the concentration there is too low. Although it is slightly more complicated to interpret the ^{14}C signal than for instance tritium, the profiles largely confirm the conclusions made by Swift et al. (1980) from tritium data. However, Nydal and Gislefoss (1996) additionally make a strong case for the value of repeat measurements of ^{14}C (and other tracers) on the GEOSECS and/or the TTO stations.

20.3.1 *Isotopes from Nuclear Reprocessing Plants*

A relatively recently introduced set of tracers is the radioactive isotopes released from the nuclear reprocessing plants in Sellafield and La Hague (e.g., Dahlgaard 1995). These tracers are particularly valuable by having both a site-specific source

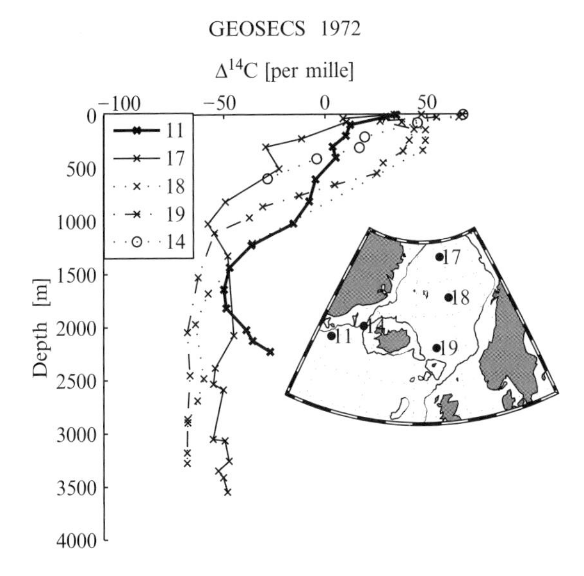

Fig. 20.3 ^{14}C profiles measured during the GEOSECS expedition in 1972 (Modified from Nydal and Gislefoss 1996. Data downloaded from http://cdiac.esd.ornl.gov/)

and a temporal trend. The most investigated of these tracers, ^{129}I (iodine-129), is a long-lived isotope ($t_{½}$ ~16 × 10^6 years), which has its overwhelmingly largest source from these facilities (Raisbeck et al. 1995). The tracer is transported with surface currents from coastal Europe to the Nordic Seas where it is mixed within the upper layers as well as downwards into the watercolumn. Subsequently, a fraction of the ^{129}I will eventually flow through the Denmark Strait (Raisbeck and Yiou 1999). During the early 1990s, the discharge of ^{129}I from the European reprocessing plants increased by 600% (Smith et al. 2005), thereby providing a pulse-like tracer input to the Nordic Seas. Thus, high levels of ^{129}I primarily indicate water from the North Sea, and secondarily water that is recently ventilated or has a short transport time. Among other tracers released from these reprocessing plants are ^{99}Tc (technetium-99), ^{90}Sr and ^{137}Cs, although for the latter two, other sources dominate (e.g., Dahlgaard 1995; Raisbeck et al. 1995).

The first study of ^{129}I, based on observations from 1993, suggested that Arctic intermediate water from the Greenland Sea was the source for the densest layer of DSOW (Zhou et al. 1995). Based on observations from the ESOP-2 project a few years later it was concluded that DSOW was a more complex mixture than being only water from the Greenland Sea (Raisbeck and Yiou 1999). Archived water samples collected during GEOSECS and TTO-NAS were analysed for ^{129}I in the 1990s when the analytical technique was refined enough to use small volume samples (Edmonds et al. 1998). This showed the potential for this isotope in tracing DSOW, and concluded that water passing the reprocessing facilities influenced the overflow. Further, samples collected in 1993 (Intergovernmental Oceanographic Commission's Baseline Survey of Contaminants) showed ^{129}I signals that were elevated seven times in DSOW compared to the overlaying water in the Irminger Basin (Edmonds et al. 2001). The high concentrations led to the conclusion that the DSOW originated from intermediate water in the Iceland Sea close to the Greenland Shelf, possibly influenced by Recirculating Atlantic Water (RAW) due to the high temperature, while intermediate water in the southern Norwegian Sea had too low ^{129}I concentration. The authors calculated that 99% intermediate water (AIW/RAW) observed in the Iceland Sea mixed with 1% polar surface water reproduced the signal of ^{129}I, salinity and temperature in DSOW in the Irminger Basin.

An extensive data set collected in the western Nordic Seas in 2002 (Alfimov et al. 2004) took advantage of the rapid discharge increase during the 1990s, and the authors concluded that the densest overflow in the Denmark Strait was neither from the Greenland Sea nor from the East Greenland Current since the ^{129}I was too low. However, the slightly lower observed ^{129}I could have been obtained by an admixture to Greenland Sea Arctic Intermediate Water by dense waters from the Arctic Ocean, which was situated too deep to be sampled by their shallow section within the Greenland Sea. This scenario also agrees with the mixing suggested by, for instance, Olsson et al. (2005b) and Jeansson et al. (2007). Alfimov et al. (2004) however linked the less dense part of the overflow to Arctic intermediate water from the Greenland Sea. A recent study (Smith et al. 2005) presents time series of ^{129}I and CFC-11 and tracks DSOW downstream of the ridge. These data alone put few constrains on the origin of DSOW but showed that ^{129}I is a useful tracer to observe the transport of DSOW and its variability in the North Atlantic.

20.3.2 Deliberately Released Tracers

Another "new" tracer in the region during this period is sulphur hexafluoride (SF_6). This non-reactive gas is introduced to the ocean via air–sea exchange and it has a linearly and rapidly increasing atmospheric history due to its anthropogenic sources and it has therefore been used as a transient tracer in a few studies (Law and Watson 2001; Watanabe et al. 2003; Tanhua et al. 2004; Bullister et al. 2006). In 1996 a total of 320 kg SF_6 was deliberately released into the intermediate layer of the central Greenland Sea gyre (Watson et al. 1999) within the project ESOP-2. Since the primary goal of the experiment was to study convection processes of the Greenland Sea, the tracer was injected at a larger potential density anomaly (28.049 kg m^{-3}) than normally found on the Denmark Strait sill. However, the spreading of tracer-tagged water did provide a unique possibility to follow and map the pathways of Greenland Sea Arctic Intermediate Water (GSAIW) to other basins (Messias et al. 2008). Also the relative magnitude of the contribution of GSAIW to the Denmark Strait Overflow (Olsson et al. 2005b; Tanhua et al. 2005a) and the Iceland–Scotland Overflow (Olsson et al. 2005a) could be estimated, as discussed below. Figure 20.4 shows the strong tracer signal within the East Greenland Current north of the sill (station 44) in 1999, a clear signal at the sill (station 56), but a lack of "excess" SF_6

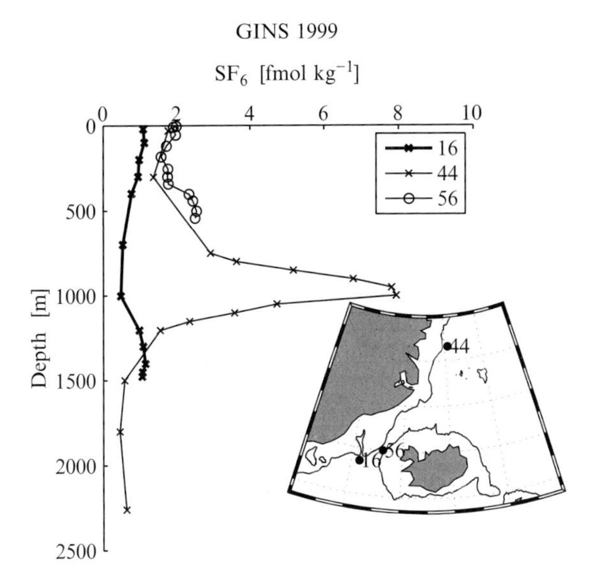

Fig. 20.4 SF_6 concentrations in the East Greenland Current and in DSOW sampled in 1999. The Greenland Sea Arctic Intermediate Water (GSAIW) is seen on station 44 as a mid-depth SF_6 maximum, and on station 56 at the sill. There is, however, no trace of GSAIW in the DSOW on station 16, just south of the sill

just south of the sill at station 16 (Olsson et al. 2005b). The tracer observations has also provided an upper limit for the transit time from the interior of the Greenland Sea to the Denmark Strait sill of 3 years (Olsson et al. 2005b), that can be compared with numerical modelling estimates of 2.5 years (Eldevik et al. 2005), and to the Labrador Sea of 7 years (Tanhua et al. 2005a).

Even though the addition of deliberately released SF_6 (excess SF_6) distorts the transient SF_6 signal (i.e., the anthropogenic SF_6 with its source in the atmosphere), the relation between the released and transient parts of the signal can be estimated with aid from other transient tracers such as chlorofluorocarbons (CFCs). The high SF_6 concentration in the DSOW plume on the Greenland slope is clear in a section sampled in 2003 during a cruise to the Irminger Sea, Fig. 20.5 (Tanhua et al. 2005a). The section is the northern most of the VEINS/ASOF standard sections and is very close to the TTO-NAS section. The excess SF_6 in the DSOW has been estimated to be roughly 0.12 fmol kg^{-1}, while the bulk of the tracer signal (about 1.2 fmol kg^{-1}) is of atmospheric origin (Tanhua et al. 2005a).

20.3.3 *Chlorofluorocarbons*

Chlorofluorocarbons (CFCs) are anthropogenic gases that enter the ocean through air–sea exchange, and whose rapidly increasing atmospheric concentration provides

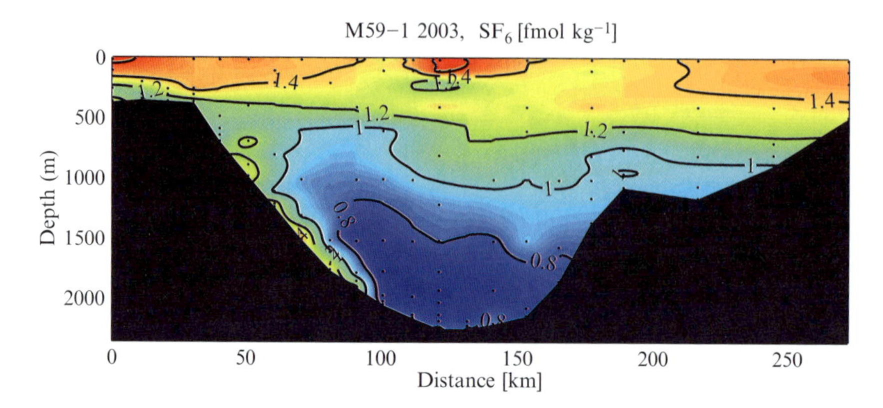

Fig. 20.5 A section of SF_6 at 65° N sampled in 2003 from the *RV Meteor* (Tanhua et al. 2005a) across the Irminger Basin just south of the Denmark Strait, viewed from the south. Approximately 0.12 fmol kg^{-1} of the SF_6 signal in the overflow along the Greenland sloop in the western part of the section originates from the Greenland Sea Trace Release Experiment (see Fig. 20.4); the bulk of tracer signal is thus the transient signal of SF_6

a powerful tool to trace the overflow water. The possibly most important advantage of measuring CFCs instead of the various isotopes discussed above is the relative ease and low cost of analysis, that can be done immediately onboard the ship. Hereby tracer data can be sampled and analysed onboard with high frequency, making a large spatial and temporal coverage possible. Just as the limited spatial coverage of GEOSECS and TTO-NAS made detailed assessment of the DSOW sources difficult, the extended coverage of CFC measurements provides a solid basis to resolve the source of DSOW.

At the time of TTO-NAS (1981), some of the first CFC measurements were made in the North Atlantic (Weiss et al. 1985) and the first profiles from the Nordic Seas were made the year after (Bullister and Weiss 1983). Although a number of studies have discussed the tracer signature in the North Atlantic Deep Water farther downstream (e.g., Rhein 1994; Pickart and Smethie 1998), few papers dealt with tracer measurements in the vicinity of Denmark Strait until the 1990s. During this period, substantially more tracer and hydrochemical data were collected in the Nordic Seas, the Denmark Strait and the Irminger Basin, and hereby more detailed, tracer-based analysis of the sources of the overflow became feasible. For comparison with other tracer measurements, we present a figure with two CFC-12 profiles (Fig. 20.6), one from north of the sill, and one from south of the sill. Just as for most tracers discussed so far, high CFC concentration indicates low ages, i.e., water that was recently in contact with the atmosphere, i.e., in the mixed layer of the ocean. The high concentration of the DSOW (close to the bottom) at station 440, supports the earlier conclusions of a source for the overflow in the upper, or intermediate, layers of the Nordic Seas. However, as we will discuss in more detail below, the combination of good spatial coverage together with measurements of several other hydrochemical parameters reveals a more complex picture.

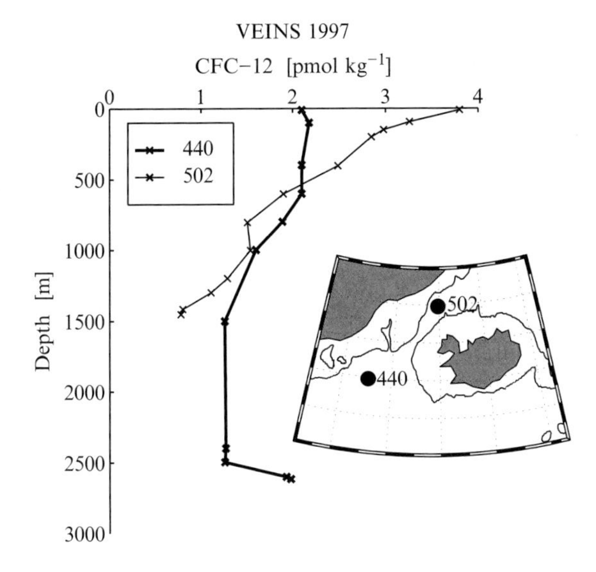

Fig. 20.6 CFC-12 in the Iceland Sea and in the Irminger Basin measured in 1997 during the VEINS cruise on *RV Aranda*. The DSOW is seen on station 440 as high tracer concentration close to the bottom

20.3.4 Other Tracers

Other innovative tracer methods also have provided information about aspects of the overflow. For instance, neodymium isotopic composition (Lacan and Jeandel 2004) suggests that the deepest layer of DSOW origins from intermediate depth in the EGC farther north. Neodymium observations south of the Greenland–Scotland Ridge led Lacan and Jeandel (2005) to question the mixing of DSOW and formation of North Atlantic Deep Water, as it was demonstrated how the various sources of the latter can be determined from paleorecords.

Similarly, oxygen isotope composition in the North Atlantic points out differences between the eastern and western overflow (Frew et al. 2000). The data also indicate entrainment of Labrador Sea Water into DSOW downstream the ridge and clearly show that oxygen isotopes can be valuable to distinguish water masses.

20.3.5 Water Mass Composition

Thanks to the large amount of data on different parameters, several attempts have recently been made to decompose the DSOW by multivariate techniques, i.e., mathematical methods working with a large number of variables. In multivariate analysis it is assumed that the water mass properties are equally affected by mixing and that the water mass distribution can be determined by a system of linear equations (Preisendorfer 1988; Tomczak and Large 1989). In the case of determining the composition of DSOW, the relatively short transport times involved allows the use of transient tracers as well as nutrients and oxygen. The assumption is that the chemical parameters and tracers change little, if at all, during the transport from the source regions to the overflow. A particular important method for water mass determination that has been applied a few times to the Denmark Strait Overflow is called optimum multiparameter (OMP) analysis (c.f. Tomczak 1999; Karstensen and Tomczak 2000).

We will now compare results from multi-parameter studies of DSOW evaluated by multivariate analysis. All these studies are based on data sets that include hydrography and chemistry (e.g., nutrients, oxygen, CFCs, and SF_6). We will discuss qualitative results such as sources and pathways together with quantitative results on the water mass composition. Finally, we will see if the available data allow us to make any statements about the variability of the water mass composition. It should be noted that even though there are several data sets in the Irminger Sea, as well as in the Nordic Seas, there are few that combine the two areas and include tracers. For instance, TTO-NAS is an excellent historical data set, which includes several tracers, but there are no samples from the East Greenland Current, and this will most likely bias a water mass analysis. We are therefore not attempting to do a water mass analysis on older data sets.

A Nordic WOCE cruise in 1993 benefited from a number of repeated CFC sections across the strait (Tanhua 1997). The observations manifested high variability in the water mass composition on weekly timescales. A striking feature was the intrusion of recently ventilated Polar Intermediate Water (PIW), which seems to contribute irregularly, and sometimes substantially, to the overflow. Since the coverage of the cruise was limited to the vicinity of the sill, no attempts were made to find the sources of DSOW farther away and Iceland Sea Arctic Intermediate Water (ISAIW) was identified as the densest component of the overflow.

Similarly, multivariate analysis was used on a hydrochemical data set, including four transient CFC tracers, from Nordic WOCE in 1994 (Fogelqvist et al. 2003). From a somewhat limited data set from the Denmark Strait, it was concluded that the DSOW immediately downstream the sill was composed to one third of Iceland Sea Deep Water, i.e., the water filling the deep layers of the Blosseville Basin north of the sill, and to the remaining 2/3 of less dense water that had properties similar to ISAIW, but that could as well has its source in the East Greenland Current.

Based on a much more comprehensive Nordic WOCE/VEINS data set from 1997 (Tanhua et al. 2005b), a more detailed analysis of the origin of DSOW was made. The cruise included nine sections near the sill, some of them occupied more than once, and a stepwise multivariate analysis was used to decompose the source water masses for the overflow. Although CFC measurements were crucial for this study, it was the combination with standard hydrochemical parameters such as temperature, salinity, oxygen and nutrients that allowed them to perform the multiple source analysis. This study also benefited from observations made during other cruises to the Nordic Seas, so that more distant source waters could be accounted for.

At the sill of the Denmark Strait, the authors found the following water-mass composition and ranges for the overflow (based on five repeats close to the sill): Arctic deepwater 18–31%, upper Polar Deep Water 5–12%, GSAIW 9–17%, Arctic Atlantic Water 7–15%, RAW 22–34%, ISAIW 5–6% and Polar Surface Water 4–12% (see Table 20.1 and Fig. 20.8). This reveals large variability, but also some degree of consistence between the repeats. This pattern can partly be explained by differences in sampling locations, but there were also actual changes in the water mass composition.

The composition on a section south of the sill, close to the position of the tracer section occupied during TTO-NAS, was also determined (Tanhua et al. 2005b). At this position, the overflow water is more homogeneous, and possibly less temporally variable, than at the sill. Here the overflow plume contained 18% dense Arctic Ocean water, 32% modified Atlantic water, 20% Arctic intermediate water from the Nordic Seas, and 30% of water entrained south of the sill.

The entrainment into the plume is visualized in Fig. 20.7 where the fractions of four water masses are shown as a function of distance from the sill. The water mass fractions are calculated with OMP analysis and the entrained water masses (reported by Tanhua et al. 2005b) are presented in Table 20.1. An intermediate water mass in the Irminger Sea, Middle Irminger Water (MIW) (see Table 20.1) was the most important water mass to entrain into the overflow close to the sill in 1997. Since this water mass is the least dense of the entrained water masses in the overflow, and thus the most affected by mixing from above, it can easily become excluded from the DSOW range. The steep decrease in MIW fraction ~350 km south of the sill is likely due to temporal variability as shown by (Dickson et al. 2008) (their Table 19.1).

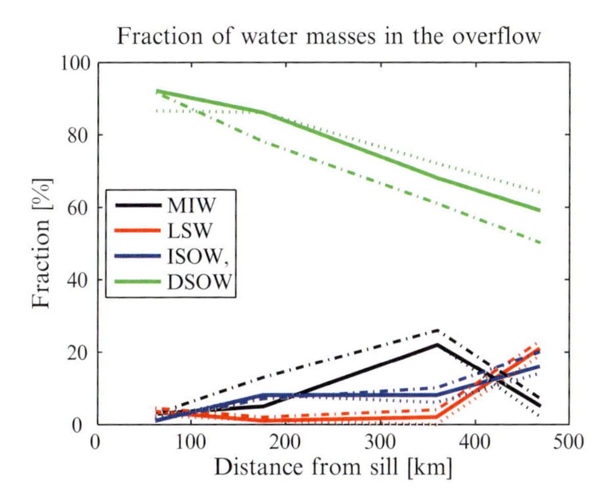

Fig. 20.7 Water mass composition as a function of distance south of the Denmark Strait sill. The different lines mark different potential temperature limits to define the DSOW: solid ≤2 °C; dotted ≤1.5 °C; dash/dotted ≤2.5 °C (for all the limit in potential density anomaly ≥27.8 kg m^{-3} is used) (Modified after Tanhua et al. 2005b)

Further south as the plume reached larger depth in the Irminger Basin, Iceland–Scotland Overflow Water and Labrador Sea Water (LSW) becomes the most important of the entraining water masses. The entrainment is clearly influenced by the transient properties of the entraining water masses as well as the DSOW. For instance, in 1997 larger volumes of relatively dense LSW penetrated deep into the Irminger Basin. Historic data suggest that the situation was very different in other years, which suggest that LSW was not always important for entrainment (see Fig. 6 in Tanhua et al. 2005b). The densification of LSW between the 1960s and the 1990s (Dickson et al. 2002) has thus likely a significant influence on the entrainment into the overflow, as well as on the deepest layers of North Atlantic Deep Water. In this volume, Dickson et al. (2008) discusses the processes and locations of entrainment downstream the sill and how this modifies the water and enhance the volume of the overflow. They also present transport time series for the different water masses southeast of Greenland that reveal large temporal variability.

Two years later, during a cruise in 1999, SF_6 as well as CFCs were measured at several sections across the East Greenland Current from northern Greenland Sea to the Irminger Basin (Olsson et al. 2005b). With this extensive data set and one intermediate water mass (GSAIW) tagged with released SF_6 (Fig. 20.4), the authors decomposed the source water masses for DSOW. Although the study focused on the deeper part of the overflow where the signal of SF_6 was present, they also attempted to quantify the less dense part of the overflow. The analysis revealed that the bulk of the overflow reached the Denmark Strait by the East Greenland Current, but due to few data in the strait and the large variability reported there, they avoided estimating the actual contributions to the overflow. However, at a short section across the Denmark Strait, the overflow was divided into three layers: (1) a fresh and recently ventilated layer; (2) a heterogeneous intermediate layer of many water masses; and (3) dense and saline water (Olsson et al. 2005b). The Greenland Sea and the Arctic Ocean contributed to about 90% of the denser layer while ISAIW, RAW, AAW and PIW became increasingly important water masses in the shallower layers (Fig. 7 in Olsson et al. 2005b). During this cruise no evidence of GSAIW, and enhanced SF_6 concentration, was found south of the sill. However, based on a cruise in 2003, with more extensive sampling for SF_6 and CFCs in the Irminger Basin and the Labrador Sea (Tanhua et al. 2005a), SF_6 from the tracer release was found in DSOW, confirming that water with a density anomaly above 28.045 kg m^{-3} is influencing the overflow.

A more recent and thorough analysis of the sources to the Denmark Strait Overflow has been made by Jeansson et al. (2008) using an extensive synoptic data set from north of Fram Strait to south of Denmark Strait in 2002 on the Swedish *IB Oden*. Benefiting from the large area covered and a comprehensive set of parameters, the authors revealed more details on the water mass composition than previously possible. At the sill of the Denmark Strait, it was found that many water masses contributed to the overflow. Based on the mean composition across the whole Denmark Strait, modified Atlantic water (RAW & AAW) dominated while the remaining part consisted of about equal parts of Polar water, Arctic intermediate water and Arctic deepwater (Jeansson et al. 2008).

An increase in sample coverage and number of tracers, together with the introduction of multivariate analysis has hence made it possible to do ever more detailed analyses of the water mass composition at the Denmark Strait sill. We can now say that we have a fair idea of what compose DSOW today. However, we have put only the first pieces of the puzzle of temporal changes in the water mass composition together, and have only an idea of the recent variability. So far, we have also difficulties in comparing various estimates of water mass composition, mainly due to variable refinements in water mass definitions and source water sampling. The Arctic Ocean is for instance not at all mentioned as a source to the overflow in pre-VEINS studies. This is likely due to lack of data from the Arctic Ocean that could support such a statement, but may also indicate a shift in the sources during the 1990s. It is well documented that Arctic Ocean deep water invaded the deep Greenland Sea during the 1990s (Blindheim and Rey 2004; Karstensen et al. 2005), so an increased influence of Arctic deep waters on the DSOW can certainly not be ruled out. On the other hand, the opposite conclusion is drawn from a comparison between the late 1980s and the late 1990s (Rudels et al. 2003) where it is suggested that the change in convection in the Greenland Sea has resulted in a larger portion of Arctic deepwater entering the Greenland Sea and that less continues south to the Denmark Strait.

A comparison of the water-mass composition is further complicated by the well-known, high short-term temporal variability at the sill and the strong mixing near the sill. A comparison of water mass analyses performed in the core of DSOW in northern Irminger Basin offers the best opportunity to reveal long-time trends, although such analyses are somewhat complicated by the intrusion of downstream water masses into and homogenisation of the overflow. However, due to scarcity of such data, we compare the water mass composition at the Denmark Strait sill calculated from data obtained in 1997 (Tanhua et al. 2005b), 1999 (Olsson et al. 2005b), and 2002 (Jeansson et al. 2007), Fig. 20.8. The 1997 estimates are averages based on seven sections, and the absolute standard errors of the water mass fractions for each of these sections were as follow: Polar water 2.8%; modified Atlantic water 8.8%; Arctic intermediate water 2.6%; and Arctic deepwater 6.3% (see text in Fig. 20.8 for details). These values are then indicative of the short-term temporal variability in water mass composition. Due to lack of repeat sections in during 1999 and 2002, we assume that the variability is of similar magnitude during these years as in 1997.

One striking difference in water mass composition between these 3 years is the high abundance of Arctic intermediate water in 1999 and accompanying low levels of modified Atlantic water. This is mainly due to a larger presence of Arctic intermediate water from the Iceland Sea in 1999 than most years during this period; something that is also observed by Rudels et al. (2003). Further, the DSOW seems to be less influenced of Arctic deepwater in 2002 compared to the late 1990s but it is, however, premature to conclude that this is a shift in sources to the Nordic Seas. More observations are needed to determine if this is short-time variability or a real shift.

Keeping in mind the high temporal variability at the sill, it might be useful to look at the composition in the overflow south of the sill, and assume that there is

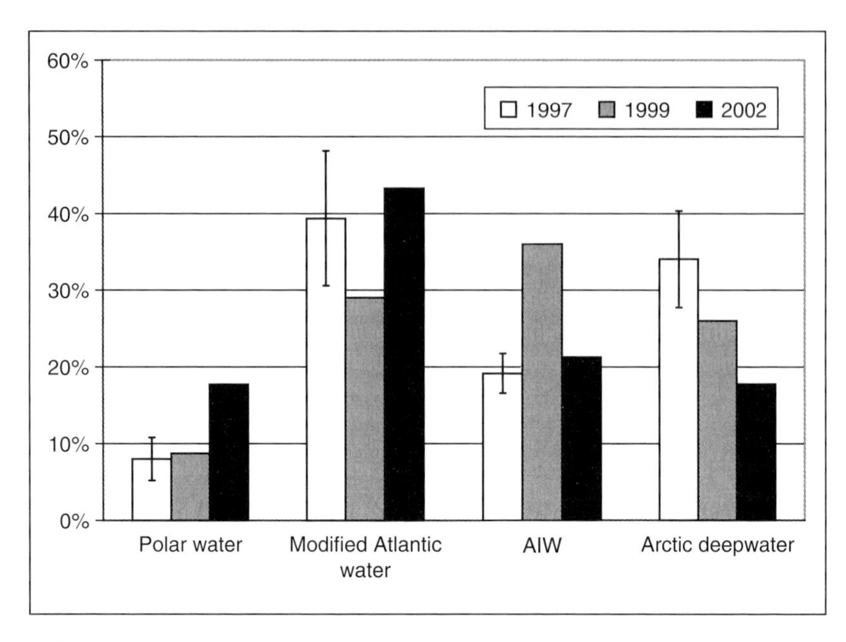

Fig. 20.8 Bar diagram showing the relative contribution of the main water mass classes at the Denmark Strait sill in 1997 (Tanhua et al. 2005b); 1999 (Olsson et al. 2005b); and 2002 (Jeansson et al. 2007). The following water masses (see Table 20.1) are included: Polar water: Polar Surface Water and Polar Intermediater Water; modified Atlantic water: Recirculating Atlantic Water, Arctic Atlantic Water and Canadian Basin Intermediate Water; Arctic intermediate water (AIW): Iceland Sea AIW and Greenland Sea AIW; Arctic deepwater: upper Polar Deep Water, Canadian Basin Deep Water, Eurasian Basin Deep Water and Greenland Sea Basin Water. The error bars for the 1997 results are assumed to be valid also for the years 1999 and 2001. For more details on some water masses see Table 20.1

less temporal variability. The DSOW (defined as a density anomaly >27.8 kg m^{-3} and $\theta < 2\,°C$) composition in 1997 in the Irminger Basin approximately 360 km south of the sill was estimated to be about 1/3 modified Atlantic water, 1/3 deepwater from the Arctic Ocean and Arctic intermediate water summed together. The remaining one third was water entrained into the plume south of the sill, complemented by a smaller fraction of Polar water (Tanhua et al. 2005b). By applying the transport estimates by (Ross 1976) the following water transports were calculated: modified Atlantic water 1.1 Sv (1 Sverdrup = 10^6 m^3 s^{-1}); Arctic intermediate water 0.6 Sv; and Arctic deepwater 0.7 Sv (Tanhua et al. 2005b). By comparing these data with two historical nearby sections (TTO-NAS in 1981 and an Icelandic cruise in 1965 (Stefánsson 1968)), significant changes in salinity as well as in chemical composition were revealed (Tanhua et al. 2005b). Similarly, decadal change in oxygen concentration in the northern Irminger Basin was discussed by Tanhua and Olsson (2006), and it was concluded that the properties of the water surrounding DSOW in the Irminger Basin is important for the final product as there is significant entrainment. The differences in water mass extension in the Irminger Basin seem to be of similar importance. The deepwater produced by entrainment to DSOW downstream the ridge is normally named North West Atlantic Bottom Water, that farther south becomes part of the North Atlantic Deep Water.

20.4 Tracer Age of DSOW

We have seen how more evidence on the composition of DSOW has been gathered gradually. A justified question to ask is; does it make sense to discuss the age, or ventilation time, of such a complex mixture as DSOW? The answer to the question might be twofold: (1) tracer ages can be determined for the components of the overflow individually, and this will give information on the response time of the overflow to changes in forcing and circulation in the Nordic Seas and the Arctic Ocean. (2) The tracer age of DSOW as observed in the Irminger Basin can be determined, and this will give an indication on the variability of DSOW itself, as a component to the North Atlantic Deep Water.

We start with the first answer, and draw your attention to Table 20.1, which presents some tracer properties of the DSOW source waters. We have included estimates of the mean age of the water mass calculated from the CFC content using the Transit Time Distribution (TTD) concept (Waugh et al. 2003). The age is determined by first calculating the equilibrium tracer mole fraction using the solubility functions by (Warner and Weiss 1985), assuming 90% saturation at the time of formation, and comparing those to the known atmospheric history of the CFCs (Walker et al. 2000). We then apply the TTD concept and assume that the TTD can be represented by an inverse Gaussian function. We further assume that the width of the TTD to be equal to the mean age (i.e., $\Delta/\Gamma=1$), as demonstrated by Waugh et al. (2004).

Due to the changing temporal trend in tracer concentration (that will shift the age estimate with time slightly) and difficulties in defining and sampling the source water in the source region, a comparison of temporal changes in the tracer age of the major water masses is not straightforward. We are therefore not attempting to elaborate on the first answer presented above, but rather look at the second alternative.

To graphically present the variability and mixing of water masses with various tracer content, we present CFC-12 data in the overflow from six cruises ranging from 1993 to 2002 (Fig. 20.9). This figure presents the CFC-12 concentration vs. latitude for all samples that fall in the definition of DSOW (>27.85 kg m^{-3}, see Dickson et al. 2008). The figure shows that north of the sill (located at 66.2° N), there is a wide range of tracer concentrations. Notably, there is an addition of young water (high CFC concentration) close to the sill, reflecting the contribution of recently ventilated PIW to the overflow. It further seems as the oldest, and most dense, component in the Iceland Sea does not influence the overflow significantly, a conclusion that is verified by multivariate analysis (e.g., Tanhua et al. 2005b). A wide range in CFC-12 concentration (~0.8–3 pmol kg^{-1}) is present even at the Denmark Strait sill. Due to vigorous mixing, the DSOW plume rapidly homogenises

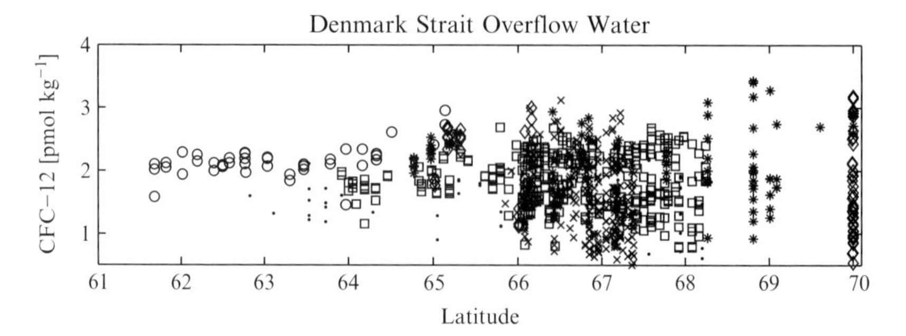

Fig. 20.9 CFC-12 concentrations (pmol kg^{-1}) in DSOW (>27.85 kg m^{-3}) from six cruises from 1993 to 2002 versus latitude. The data are from the following cruises, with project names in parenthesis; 1993 (crosses); *RV Aranda* (Nordic-WOCE), 1994 (dots); *RV Johan Hjort* (Nordic-WOCE), 1997 (squares); *RV Aranda* (VEINS), 1998 (circles); *RV Valdivia* (VEINS), 1999 (diamonds); *RV Marion Dufresne* (GINS), 2002 (stars); *IB Oden*. The Denmark Strait sill is located at approximately 66.2° N

downstream the sill, and at about 63° N the plume has a uniform tracer concentration of approximately 2.2 pmol kg^{-1} in 1997. Since most data presented here do not extend far enough south, it is difficult to draw any conclusions on the temporal variability in the age of DSOW. Additionally, a mixing model must be employed for such a calculation due to the nearly constant atmospheric CFC-12 concentrations since the early 1990s. When the plume descends into the Irminger Basin, the CFC concentration indicates a relatively low age for DSOW (the TTD mean age is ~30 years in 1997, but only about 20 years in 1998). The other overflow water mass, Iceland–Scotland Overflow Water, is considerably older.

20.5 Open Questions for the Denmark Strait Overflow

With this article we have tried to synthesise the current knowledge of the changing sources and characteristics of DSOW the last 30 years, as determined from relatively sparse and short time records of tracers (longer records are available for hydrographic and hydrochemical data, but such changes are beyond the scope of this work). We have shown that the water mass composition of the Denmark Strait Overflow seems to be changing over time, and that the characteristics of the source water masses might be changing as well. Both these processes will result in changing characteristics of the overflow, and it is not a trivial problem to distinguish between these two processes. Variations in the overflow are mainly and most likely due to variations in climatological factors such as freshwater input, wind forcing and temperature. Such variations are often found as a result of the North Atlantic Oscillation (NAO), but with the signal of climate change superimposed. Variations in the forcing are reflected in the DSOW characteristic and composition. Thus, monitoring of the DSOW characteristics has the potential to be an index of the conditions in the Nordic Seas. This is further discussed by Dickson et al. (2008).

The changing sources of the deep overflows over the Greenland–Scotland Ridge are, in our opinion, an important piece of information to understand the dynamics of the Nordic Seas region, as well as those of the North Atlantic. Even though these changes, to some extent, can be monitored by hydrography and current measurements, we have shown that the inclusion of a set of tracers provides additional information on the source water masses, and their pathways and transport times to the overflow, something that complements the physical measurements substantially.

Our suggestion is to monitor tracers in the overflow during hydrographic surveys on a regular basis. New approaches to regular sampling of water in the overflow are very interesting, and we would like to promote efforts in the direction of automated sampling arrays (e.g., moorings) as recently initiated, to complement hydrographic surveys by research vessels.

It is a bold endeavour to suggest sampling and measurement strategies for the future. Nonetheless, a few suggestions, based on the experience gained from a decade of tracer measurements in the Denmark Strait region, are presented in the following.

The first suggestion regards sampling strategies. The temporal variability at the sill of the Denmark Strait is large on short timescales, and one-time surveys on the sill will most likely never be able to representatively sample the overflow water, at least it is difficult to really know whether the conditions at the sill were representative or not. Rather, we suggest that the priority sampling is done at positions sufficiently far north and south of the sill to filter out most of the short time variability. We further suggest that sampling is concentrated on routinely repeated sections, which have a history of measurements to facilitate comparison (see Fig. 20.10).

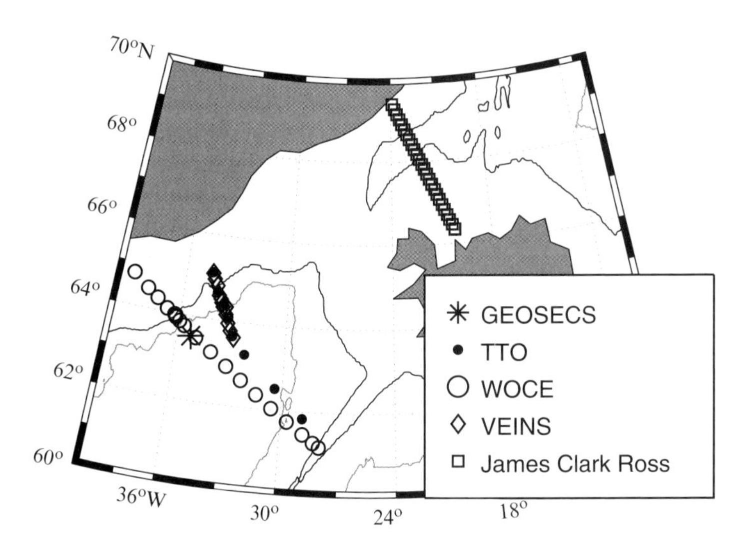

Fig. 20.10 Map of the Denmark Strait with the position of a few historical sampling locations; GEOSECS 1972, TTO-NAS 1981, WOCE A24N 1997, a VEINS standard section repeated several times and one section sampled from *RRS James Clark Ross* 1999 (ARCICE) north of the sill

South of the sill, it is reasonable to sample close to the TTO section, for three reasons: (1) There exists a significant historical record from this section, including TTO-NAS in 1981 and it is close to one GEOSECS station from 1972, as well as current moorings and more recent tracer measurements discussed above such as the WOCE section A24N and the standard VEINS section repeated several times, albeit not all the way across the basin. (2) The location is far enough downstream for the sill for DSOW to have homogenised sufficient to represent an "end-product". (3) The section extends not only over the Greenland shelf, but also over the Reykjanes Ridge, thus capturing the inflow of water important to the mixing south of the sill.

For the sampling upstream, there is not the same history of tracer measurements. Again, it is important that the section extends across the basin and up on both the Icelandic and Greenland shelf. Such a section was sampled by an ARCICE survey on the *RRS James Clark Ross* in 1999 (e.g., Messias et al. 2007). This section will be able to represent all the water masses transported with the East Greenland Current, as well as water masses formed locally in the Iceland Sea. The Icelandic standard section Kögur, located south of the ARCICE section, has the benefits of being a standard section, although it suffers from not reaching the Greenland shelf. However an extension of the Kögur section would be well-suited section for tracer measurements. Additionally, a section at the sill certainly has many benefits, and should also be sampled along with the two sections suggested above if possible.

We have here shown data and results obtained from a wide variety of tracers, and that the observed tracers have shifted over time. Also in the future it is likely that "new" tracers will be added and other disappear as when their transient signal decline, as in the case of cessation in CFC increase in the atmosphere. One example of an additional tracer with currently increasing source function is HCFC-22 that could prove to be complementary to the CFC measurements. Therefore, it is sensible to continue monitoring tracers such as CFCs, SF_6, ^{129}I, tritium, ^{137}C and ^{90}Sr in the overflow to connect to the historical records, and at the same time be alert to new tracers that might develop with time.

We also would like to stress the importance of including measurements of parameters such as oxygen and nutrients that are shown very valuable for water mass analysis (Tanhua et al. 2005b). In addition to water mass analysis, there is a scientific interest measuring the flux of nutrients and oxygen, as well as the carbonate system (i.e., anthropogenic carbon) across the sills. For these flux calculations hydrochemical measurements of high quality are very important, ideally calibrated against certified reference materials.

20.6 Conclusions

The understanding of the composition and variability of Denmark Strait Overflow Water (DSOW) has evolved considerably since the early 1990s, and part of this knowledge stems from tracer observations. Already in the 1980s, the general opinion on what was the main source of DSOW changed from the Norwegian Sea Deep Water to intermediate waters. Most pre-1990 studies, however, pointed out the Iceland Sea as the main source region of DSOW while it since 1990, has been realised that DSOW is a rather complex mixture of a large set of water masses formed by different processes and in different regions. This change of view is, at least partially, an effect of the introduction of new methods and parameters and of higher temporal and spatial data resolution, and might not reflect an actual change in water mass composition. However, inter-annual comparison indicates moderate variability in recent years although decadal variability might be considerably higher. The development and use of new methods and tracer compounds have been fundamental in understanding the water mass composition, and its variability. Examples of new tracers include the radioactive isotope ^{129}I and the SF_6 released in the Greenland Sea. The former has both a site-specific and temporal source implying its large potential while the later tagged one specific water mass, which has been followed into the overflow.

Since the 1990s, tracer data suggest that the bulk of the overflow has been supplied by the East Greenland Current with water from the Arctic Ocean, the Fram Strait and the Greenland Sea. The denser part of the overflow has two main sources: the Arctic Ocean and the Greenland Sea, of which the Arctic Ocean dominated during the last decade although it seemed to vary considerably (Fig. 20.8). For the shallower layers, modified Atlantic Water and Arctic intermediate water from the Nordic Seas were the dominating contributors, although the highly variable influence of Polar waters is important by making DSOW fresher. The contribution from the central Iceland Sea was minor, in general around 5%, except for in 1999 when it contributed to about one third of the less dense DSOW fraction. This extreme in fresh (i.e. low-saline) water is clear in the time series presented by Dickson et al. (2008)

It has been suggested by Rudels et al. (2003) that during periods of modest convection, the regional circulation affects the contribution to DSOW in a way that the denser layers will be more influenced by the Greenland Sea and less by the Arctic Ocean while the Atlantic layer will instead have a larger portion that has passed through the Arctic Ocean and a smaller that has been recirculated already in the Fram Strait. The supply of water from the Iceland Sea on the other hand seems to vary on shorter timescales. Throughout the short period of detailed, tracer-based, studies on the composition of DSOW, water formation processes in both the Arctic Ocean and the Greenland Sea, together with the transformation of Atlantic water in the Arctic Ocean and the Fram Strait, are of large importance for the DSOW. The properties of the water masses show temporal variability, in particular those of the more locally produced water masses, such as water from the Iceland Sea and Polar waters since they are more directly affected by changes in, for example, wind fields. The tracer-based water mass studies further suggest that a change in the production of one water mass, for example, caused by the shift in convection intensity in the Greenland Sea, may, at least initially, be compensated by a change in the supply of another water mass. As a result, the volume of the overflow can stay relatively constant, whereas, at the same time, the properties of the overflow may change significantly, which would affect the entrainment downstream the Denmark Strait and the further circulation. The notion of relatively constant strength of the Denmark Strait Overflow over decadal timescales is indeed supported by observations (e.g., Ross 1984; Dickson and Brown 1994; Dickson et al. 2008), as well as by models (e.g., Käse 2006), even though Macrander et al. (2005) found the variation in the transport to be about 30% over a 4-year period. Large-scale changes in forcing will likely affect more than one of the regions or processes of water mass formation. Thus, such changes may have long-term effects on the overflow, even if the overflow appears to be robust due to its origin in more than one process and region.

Acknowledgements We thank all investigators that under several decades with great effort and the uttermost care have collected tracer data relevant to the Denmark Strait Overflow. We particularly thank the authors of a as yet unpublished manuscript (Jeansson et al. 2008) for letting us use some of their results in this work. We gratefully acknowledge Peter Jones for letting us use some unpublished CFC data and for valuable comments on the manuscript. During the preparation of this manuscript, T.T. was supported by the Deutsche Forschungsgemeinschaft (DFG) through SFB460.

References

Aagaard, K., J. H. Swift and E. C. Carmack (1985) Thermohaline circulation in the Arctic Mediterranean Seas. Journal of Geophysical Research, 90 (C3):4833–4846.

Alfimov, V., A. Aldahan and G. Possnert (2004) Tracing water masses with ^{129}I in the western Nordic Seas in early spring 2002. Geophysical Research Letters, 31 (19):L19305.

Bacon, S., W. J. Gould and Y. Jia (2003) Open-ocean convection in the Irminger Sea. Geophysical Research Letters, 30 (5).

Blindheim, J. (1990) Arctic Intermediate Water in the Norwegian Sea. Deep-Sea Research A, 37 (9):1475–1489.

Blindheim, J. and F. Rey (2004) Water-mass formation and distribution in the Nordic Seas during the 1990s. ICES Journal of Marine Science, 61 (5):846–863.

Bourke, R. H., A. M. Weigel and R. G. Paquette (1988) The westward turning branch of the West Spitsbergen Current. Journal of Geophysical Research, 93 (C11):14065–14077.

Bullister, J. L. and R. F. Weiss (1983) Anthropogenic chlorofluoromethanes in the Greenland and Norwegian seas. Science, 221 (4607):265–268.

Bullister, J. L., D. P. Wisegarver and R. E. Sonnerup (2006) Sulfur hexafluoride as a transient tracer in the North Pacific Ocean. Geophysical Research Letters, 33 (18):L18603.

Dahlgaard, H. (1995) Transfer of European Coastal Pollution to the Arctic: Radioactive Tracers. Marine Pollution Bulletin, 31 (1–3):3–7.

Dickson, B., S. Dye, S. Jónsson, A. Köhl, A. Macrander, M. Marnela, J. Meincke, S. Olsen, B. Rudels, H. Valdimarsson and G. Voet (2008) The overflow flux west of Iceland: varaibility, origins and forcing. In: B. Dickson, J. Meincke and P. Rhines (eds). Arctic-Subarctic Ocean Fluxes: Defining the role of the Northern Seas in Climate. Springer.

Dickson, B., I. Yashayaev, J. Meincke, B. Turrell, S. Dye and J. Holfort (2002) Rapid freshening of the deep North Atlantic Ocean over the past four decades. Nature, 416 (6883):832–837.

Dickson, R. R. and J. Brown (1994) The production of North Atlantic Deep Water: Sources, rates, and pathways. Journal of Geophysical Research, 99 (C6):12319–12341.

Edmonds, H. N., Z. Q. Zhou, G. M. Raisbeck, F. Yiou, L. Kilius and J. M. Edmond (2001) Distribution and behavior of anthropogenic ^{129}I in water masses ventilating the North Atlantic Ocean. Journal of Geophysical Research, 106 (C4):6881–6894.

Eldevik, T., F. Straneo, A. B. Sandø and T. Furevik (2005) Pathways and export of Greenland Sea water. In: H. Drange, T. Dokken, T. Furevik, R. Gerdes and W. Berger (eds) The Nordic Seas: An integrated perspective, Geophysical Monograph, vol. 158. American Geophysical Union, Washington, DC, USA, pp. 89–103.

Fogelqvist, E., J. Blindheim, T. Tanhua, S. Østerhus, E. Buch and F. Rey (2003) Greenland-Scotland overflow studied by hydro-chemical multivariate analysis. Deep-Sea Research I, 50 (1):73–102.

Frew, R. D., P. F. Dennis, K. J. Heywood, M. P. Meredith and S. M. Boswell (2000) The oxygen isotope composition of water masses in the northern North Atlantic. Deep-Sea Research I, 47 (12):2265–2286.

Jeansson, E., S. Jutterström, B. Rudels, L. G. Anderson, K. A. Olsson, E. P. Jones, W. M. Smethie, Jr and J. H. Swift (2008) Sources to the East Greenland Current and its contribution to the Denmark Strait overflow. Progress in Oceanography accepted for publication, 2008.

Karstensen, J., P. Schlosser, D. W. R. Wallace, J. L. Bullister and J. Blindheim (2005) Water mass transformation in the Greenland Sea during the 1990s. Journal of Geophysical Research, 110 (C7):C07022.

Karstensen, J. and M. Tomczak (2000) OMP analysis package for MATLAB, Online available software, http://www.ldeo.columbia.edu/~jkarsten/omp_std/.

Käse, R. (2006) A Riccati model for the Denmark Strait Overflow Variability. Geophysical Research Letters, 33:L21S09.

Lacan, F. and C. Jeandel (2004) Denmark Strait water circulation traced by heterogeneity in neodymium isotopic compositions. Deep-Sea Research I, 51 (1):71–82.

Lacan, F. and C. Jeandel (2005) Aquisition of the neodymium isotopic composition of the North Atlantic Deep Water. Geochemistry, Geophysics, Geosystems, 6 (12):Q12008.

Law, C. S. and A. J. Watson (2001) Determination of Persian Gulf Water transport and oxygen utilisation rates using SF_6 as a novel transient tracer. Geophysical Research Letters, 28 (5):815–818.

Lee, A. and D. Ellett (1967) On the water masses of the northwest Atlantic Ocean. Deep-Sea Research, 14:183–190.

Livingston, H. D., J. H. Swift and H. G. Ostlund (1985) Artificial radionuclide tracer supply to the Denmark Strait Overflow between 1972 and 1981. Journal of Geophysical Research, 90 (C4):6971–6982.

Macrander, A., U. Send, H. Valdimarsson, S. Jónsson and R. H. Käse (2005) Interannual changes in the overflow from the Nordic Seas into the Atlantic Ocean through Denmark Strait. Geophysical Research Letters, 32 (6):L06606.

Mauritzen, C. (1996) Production of dense overflow waters feeding the North Atlantic across the Greenland-Scotland Ridge. Part 1: Evidence for a revised circulation scheme. Deep-Sea Research I, 43 (6):769–806.

Messias, M.-J., A. J. Watson, T. Johannessen, K. I. C. Oliver, K. A. Olsson, E. Fogelqvist, J. Olafsson, S. Bacon, J. Balle, N. Bergman, G. Budéus, M. Danielsen, J.-C. Gascard, E. Jeansson, S. R. Olafsdóttir, K. Simonsen, T. Tanhua, K. Van Scoy and J. R. Ledwell (2008) The Greenland Sea Tracer Experiment 1996–2002: horizontal mixing and transport of Greenland Sea Intermediate Water. Progress in Oceanography accepted for publication, 2008.

Min, D.-H. (1999) Studies of large-scale intermediate and deep water circulation and ventilation in the North Atlantic, South Indian and Northeast Pacific Oceans, and in the East Sea (Sea of Japan), using chlorofluorocarbons as tracers. Ph.D. thesis, Scripps Institution of Oceanography, University of California, San Diego, La Jolla, CA.

Nydal, R. and J. S. Gislefoss (1996) Further application of bomb C-14 as a tracer in the atmosphere and ocean. Radiocarbon, 38 (3):389–406.

Olsson, K. A., E. Jeansson, L. G. Anderson, B. Hansen, T. Eldevik, R. Kristiansen, M.-J. Messias, T. Johannessen and A. J. Watson (2005a) Intermediate water from the Greenland Sea in the Faroe Bank Channel: spreading of released sulphur hexafluoride. Deep-Sea Research I, 52 (2):279–294.

Olsson, K. A., E. Jeansson, T. Tanhua and J.-C. Gascard (2005b) The East Greenland Current studied with CFCs and released sulphur hexafluoride. Journal of Marine Systems, 55 (1–2):77–95.

Pickart, R. S. and W. M. Smethie, Jr. (1998) Temporal evolution of the deep western boundary current where it enters the sub-tropical domain. Deep-Sea Research I, 45 (7):1053–1083.

Pickart, R. S., F. Straneo and G. W. K. Moore (2003) Is Labrador Sea Water formed in the Irminger basin? Deep-Sea Research I, 50 (1):23–52.

Preisendorfer, R. W. (1988) Principal component analysis in meteorology and oceanography. Elsevier, Amsterdam, The Netherlands.

Raisbeck, G. M. and F. Yiou (1999) 129I in the oceans: origins and applications. Science of the Total Environment, 237–238:31–41.

Raisbeck, G. M., F. Yiou, Z. Q. Zhou and L. R. Kilius (1995) ^{129}I from nuclear fuel reprocessing facilities at Sellafield (UK) and La Hague (France); potential as an oceanographic tracer. Journal of Marine Systems, 6 (5–6):561–570.

Rhein, M. (1994) The Deep Western Boundary Current - tracers and velocities. Deep-Sea Research I, 41 (2):263–281.

Rhein, M., J. Fischer, W. M. Smethie, D. Smythe-Wright, R. F. Weiss, C. Mertens, D. H. Min, U. Fleischmann and A. Putzka, 2002: Labrador Sea Water: Pathways, CFC inventory, and formation rates. Journal of Physical Oceanography, 32: 648–665.

Ross, C. K. (1976) Transport of overflow water through Denmark Strait. ICES CM, 1976 (C:16).

Ross, C. K. (1984) Temperature – salinity characteristics of the "overflow" water in Denmark Strait during "OVERFLOW '73". Rapports et Procès-Verbaux des Réunions Conseil International pour l'Exploration de la Mer, 185:111–119.

Rudels, B., G. Björk, J. Nilsson, P. Winsor, I. Lake and C. Nohr (2005) The interaction between waters from the Arctic Ocean and the Nordic Seas north of Fram Strait and along the East Greenland Current: results from the Arctic Ocean-02 Oden expedition. Journal of Marine Systems, 55 (1–2):1–30.

Rudels, B., P. Eriksson, E. Buch, G. Budéus, E. Fahrbach, S.-A. Malmberg, J. Meincke and P. Mälkki (2003) Temporal switching between sources of the Denmark Strait overflow water. ICES Marine Science Symposia, 219:319–325.

Rudels, B., P. Eriksson, H. Grönvall, R. Hietala and J. Launiainen (1999a) Hydrographic observations in Denmark Strait in fall 1997, and their implications for the entrainment into the overflow plume. Geophysical Research Letters, 26 (9):1325–1328.

Rudels, B., E. Fahrbach, J. Meincke, G. Budéus and P. Eriksson (2002) The East Greenland Current and its contribution to the Denmark Strait overflow. ICES Journal of Marine Science, 59 (6):1133–1154.

Rudels, B., H. J. Friedrich and D. Quadfasel (1999b) The Arctic Circumpolar Boundary Current. Deep-Sea Research II, 46 (6–7):1023–1062.

Rudels, B., R. Meyer, E. Fahrbach, V. V. Ivanov, S. Østerhus, D. Quadfasel, U. Schauer, V. Tverberg and R. A. Woodgate (2000) Water mass distribution in Fram Strait and over the Yermak Plateau in summer 1997. Annales Geophysicae, 18 (6):687–705.

Smethie, W. M., Jr and J. H. Swift (1989) The tritium:krypton-85 age of Denmark Strait Overflow Water and Gibbs Fracture Zone Water just south of Denmark Strait. Journal of Geophysical Research, 94 (C6):8265–8275.

Smith, J. N., E. P. Jones, S. B. Moran, W. M. Smethie, Jr and W. E. Kieser (2005) Iodine 129/CFC 11 transit times for Denmark Strait Overflow Water in the Labrador and Irminger Seas. Journal of Geophysical Research, 110 (C5):C05006.

Stefansson, U. (1962) North Icelandic waters. Rit Fiskideildar, 3:269.

Stefánsson, U. (1968) Dissolved nutrients, oxygen and water masses in the Northern Irminger Sea. Deep-Sea Research, 15:541–575.

Swift, J. H. and K. Aagaard (1981) Seasonal transitions and water mass formation in the Iceland and Greenland seas. Deep-Sea Research A, 28A (10):1107–1129.

Swift, J. H., K. Aagaard and S.-A. Malmberg (1980) The contribution of the Denmark Strait overflow to the deep North Atlantic. Deep-Sea Research A, 27A (1):29–42.

Swift, J. H. and K. P. Koltermann (1988) The origin of Norwegian Sea Deep Water. Journal of Geophysical Research, 93 (C4):3563–3569.

Tanhua, T. (1997) Halogenated substances as marine tracers. Ph.D. thesis, Department of Analytical and Marine Chemistry, Göteborg University, Göteborg, Sweden.

Tanhua, T., K. Bulsiewicz and M. Rhein (2005a) Spreading of overflow water from the Greenland to the Labrador Sea. Geophysical Research Letters, 32 (10):L10605.

Tanhua, T. and K. A. Olsson (2006) A note on the oxygen flux in the deep northern overflows. ASOF Newsletter, 5:8–11.

Tanhua, T., K. A. Olsson and E. Fogelqvist (2004) A first study of SF_6 as a transient tracer in the Southern Ocean. Deep-Sea Research II, 51 (22–24):2683–2699.

Tanhua, T., K. A. Olsson and E. Jeansson (2005b) Formation of Denmark Strait overflow water and its hydro-chemical composition. Journal of Marine Systems, 57 (3–4):264–288.

Tomczak, M. (1999) Some historical, theoretical and applied aspects of quantitative water mass analysis. Journal of Marine Research, 57 (2):275–303.

Tomczak, M. and D. G. B. Large (1989) Optimum multiparameter analysis of mixing in the thermocline of the eastern Indian Ocean. Journal of Geophysical Research, 94 (C11):16141–16149.

Walker, S. J., R. F. Weiss and P. K. Salameh (2000) Reconstructed histories of the annual mean atmospheric mole fractions for the halocarbons CFC-11, CFC-12, CFC-113 and carbon tetrachloride. Journal of Geophysical Research 105 (C6):14285–14296.

van Aken, H. M. and C. J. de Boer (1995) On the synoptic hydrography of intermediate and deep water masses in the Iceland Basin. Deep-Sea Research I, 42 (2):165–189.

Warner, M. J. and R. F. Weiss (1985) Solubilities of chlorofluorocarbons 11 and 12 in water and sea water. Deep-Sea Research, 32 (12):1485–1497.

Watanabe, Y. W., A. Shimamoto and T. Ono (2003) Comparison of time-dependent tracer ages in the western North Pacific: Oceanic background levels of SF_6, CFC-11, CFC-12 and CFC-113. Journal of Oceanography, 59 (5):719–729.

Watson, A. J., M. J. Messias, E. Fogelqvist, K. A. Van Scoy, T. Johannessen, K. I. C. Oliver, D. P. Stevens, F. Rey, T. Tanhua, K. A. Olsson, F. Carse, K. Simonsen, J. R. Ledwell, E. Jansen, D. J. Cooper, J. A. Kruepke and E. Guilyardi (1999) Mixing and convection in the Greenland Sea from a tracer-release experiment. Nature, 401 (6756):902–904.

Waugh, D. W., T. W. N. Haine and T. M. Hall (2004) Transport times and anthropogenic carbon in the subpolar North Atlantic Ocean. Deep-Sea Research I, 51 (11):1475–1491.

Waugh, D. W., M. H. Hall and T. W. N. Haine (2003) Relationships among tracer ages. Journal of Geophysical Research, 108 (C5):3138.

Weiss, R. F., J. L. Bullister, R. H. Gammon and M. J. Warner (1985) Atmospheric chlorofluoromethanes in the deep equatorial Atlantic. Nature, 314 (6012):608–610.

Zhou, Z. Q., G. M. Raisbeck, F. Yiou, L. Kilius, H. N. Edmonds, J. M. Edmond, J. C. Gascard, C. I. Measures and J. Meincke (1995) ^{129}I as a tracer of North Atlantic deep water formation and transport. CSNSM Report 95–11, Centre de Spectrométrie Nucléaire et de Spectrométrie de Masse, Orsay, France.

Chapter 21
Transformation and Fate of Overflows in the Northern North Atlantic

Igor Yashayev[1] and Bob Dickson[2]

21.1 Introduction

The largest full-depth changes in the modern instrumented oceanographic record have taken place in the Labrador Basin of the northwest Atlantic over the last 4 decades. The extreme amplitude of anomalous conditions there and the importance of their claimed effects for the thermohaline circulation and for climate (e.g. Bryden et al. 2005) justify attempts to identify the origin of change throughout the watercolumn of the subpolar Atlantic. At depths in the Labrador Basin greater than the limits of open-ocean deep convection (2,300 m or so), change is necessarily imported to the Basin by the two main dense water overflows that cross the Greenland–Scotland Ridge via the Denmark Strait and Faroe–Shetland Channel. Each of the constituent watermasses that form these overflows (see, for example, Rudels et al. 2002) will carry with them the imprint of time-varying climatic forcing in their source regions and of modifications *en route*, and their properties will also be subject to alteration by the processes of horizontal and vertical exchange from their spillways to the Labrador Basin. The purpose of this chapter is to identify from the hydrographic record those locations that are of primary importance for the transfer of ocean climate 'signals' into and between the two spreading overflow plumes, and if possible to trace the influence of these changes downstream to the Newfoundland Basin and beyond in the Deep Western Boundary Current (DWBC; See Fig. 21.1).

[1]Ocean Circulation Section, Ocean Sciences Division, Bedford Institute of Oceanography, Fisheries and Oceans Canada, 1 Challenger Drive, P.O. Box 1006, Dartmouth, NS, B2Y 4A2, Canada

[2]Centre for Environment, Fisheries and Aquaculture Science (CEFAS), The Laboratory, Pakefield Road, Lowestoft Suffolk NR33 OHT, UK

R.R. Dickson et al. (eds.), *Arctic–Subarctic Ocean Fluxes*, 505–526

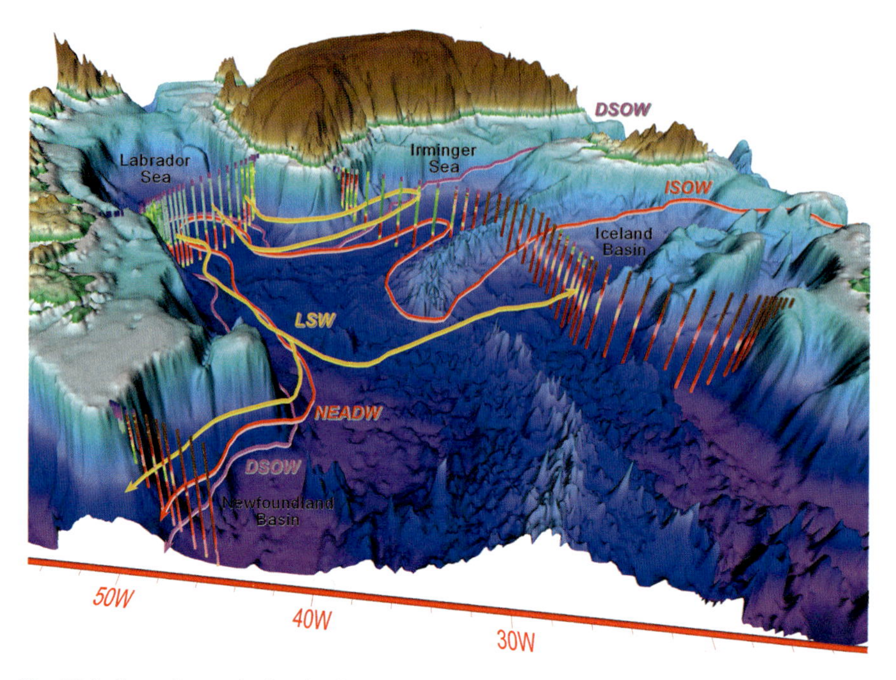

Fig. 21.1 Spreading paths for the three watermasses whose interactions form the basis for this chapter; the Denmark Strait Overflow Water (DSOW), Iceland–Scotland Overflow Water (ISOW) and Labrador Sea Water (LSW). Note that the ISOW substantially modifies *en route* to become Northeast Atlantic Deep Water (NEADW)

21.2 Denmark Strait Overflow Water

Variability in the cold dense overflow which descends from the Denmark Strait sill to ventilate the deep Atlantic is the main source of change in the deepest layers of the Labrador Basin. From our present partial understanding, that variability is more likely to reflect the relatively large changes in the hydrographic character of the overflow than the shorter-term changes observed in its transport.

Off the southeastern coast of Greenland, discontinuous direct flow measurements have been made in the core of this overflow since 1985, latterly under the EC-VEINS and ASOF programmes. Using tracers (T. Tanhua 2006, personal communication) to define the density range ($\sigma_\theta > 27.85$) occupied by the descending plume off Angmagssalik, the most recent study (see Dickson et al., Chapter 19, this volume) calculates a value of 4.0 Sv ± 0.4 Sv for the overflow transport passing through the main array at these densities, a figure that is in close agreement with their geostrophic estimates of transport for this section and density interval. Fluctuations in transport of a few year's duration have certainly been observed (Dickson et al., op. cit.), now corroborated by both satellite altimetry (Köhl et al. 2007) and by a model-optimized ADCP array close to the sill (Macrander et al. 2005), and other more-subtle changes in flow are suspected (for example, changes in the trajectory of the descending plume induced by temperature extrema at the sill; Dickson et al. 1999); but this most recent analysis agrees with earlier reports (e.g. Dickson and Brown 1994; Girton et al. 2001) in finding no significant or convincing long-term trend in overflow transport.

Changes in the hydrographic character of the overflow have certainly been observed however, reflecting the wide range of influences, both local and remote, that may drive change in its contributory watermasses. As recent examples, a tracer-based analysis by Tanhua et al. (2006) reveals that 400 km downstream from the sill, the DSOW is made up of ~30% of watermasses from the Arctic Ocean, ~20% from the Nordic seas and ~30% from the Irminger Basin, though this 'recipe' (and in fact each of its contributory watermasses) can be expected to vary. And even their pathways can vary; recent modelling by Köhl et al. (2007) suggests that a direct *and* an indirect feed from the East Greenland Current to the sill may both be valid and vary inversely in strength (c.f. Rudels et al. 2002, 2003, 2005; Jónsson and Valdimarsson 2004). Despite this broad spectrum of possible variability, three main types of change seem to dominate the DSOW off SE Greenland.

First, and common to both overflows, the long-term freshening of the DSOW since the mid-1960s (Dickson et al. 2002; see also Dickson et al. 2007, their Fig. 14) is assumed to reflect the broadscale freshening of the *Nordic Seas* reported by Curry and Mauritzen (2005), equivalent to an increased freshwater loading by ~4,000 km^3 since the mid-1960s. As with the eastern overflow, the updated series (Fig. 21.2) suggests that the freshening trend at the sill may have slowed or ceased over the past decade. Second, about this long-term trend, the core of the Denmark Strait Overflow at ~2,000 m off SE Greenland exhibits a well-defined variability in temperature on multiannual-to-decadal timescales which appears to be the lagged reflection of temperature variability in the upper 500 m of the *eastern Fram Strait*, some 2,500 km upstream and 3 years earlier, transferred south by the Recirculating Atlantic Water component from Fram Strait (Dickson et al. 1999, 2007; now successfully simulated in the A-W-I NAOSIM model by Karcher et al. 2003). Third, we have new and compelling evidence that short-term but intense freshening events in the deep layer off SE Greenland are attributable (Holfort and Albrecht 2007) to a strengthening of the freshwater feed to that Overflow from the East Greenland Current, arising from an anomalously strong north-wind component in *waters immediately to the north of Denmark Strait* (discussed and described in Dickson et al. 2007).

Tracking these changes downstream, we also find a clear correspondence between overflow hydrography off SE Greenland and the temperature, salinity, density and dissolved oxygen of the abyssal layer of the Labrador Sea a further 1 year later (Dickson et al. 2003 and Fig. 21.2). In a little more detail, we note in particular from Fig. 21.2 that: (a) the main hydrographic changes arrive first around the basin margins, then spread to the interior (the time-axes are lagged in Fig. 21.2 to reflect this); (b) the amplitude of hydrographic variability changes little from place to place along this circuit, suggesting that mixing and entrainment here are relatively weak; (c) the deep winter freshening episodes observed off SE Greenland, which could not be expected to reach the deep Labrador Basin until Autumn, are not well resolved – nor could they be – by the annual *spring* survey of the Basin.

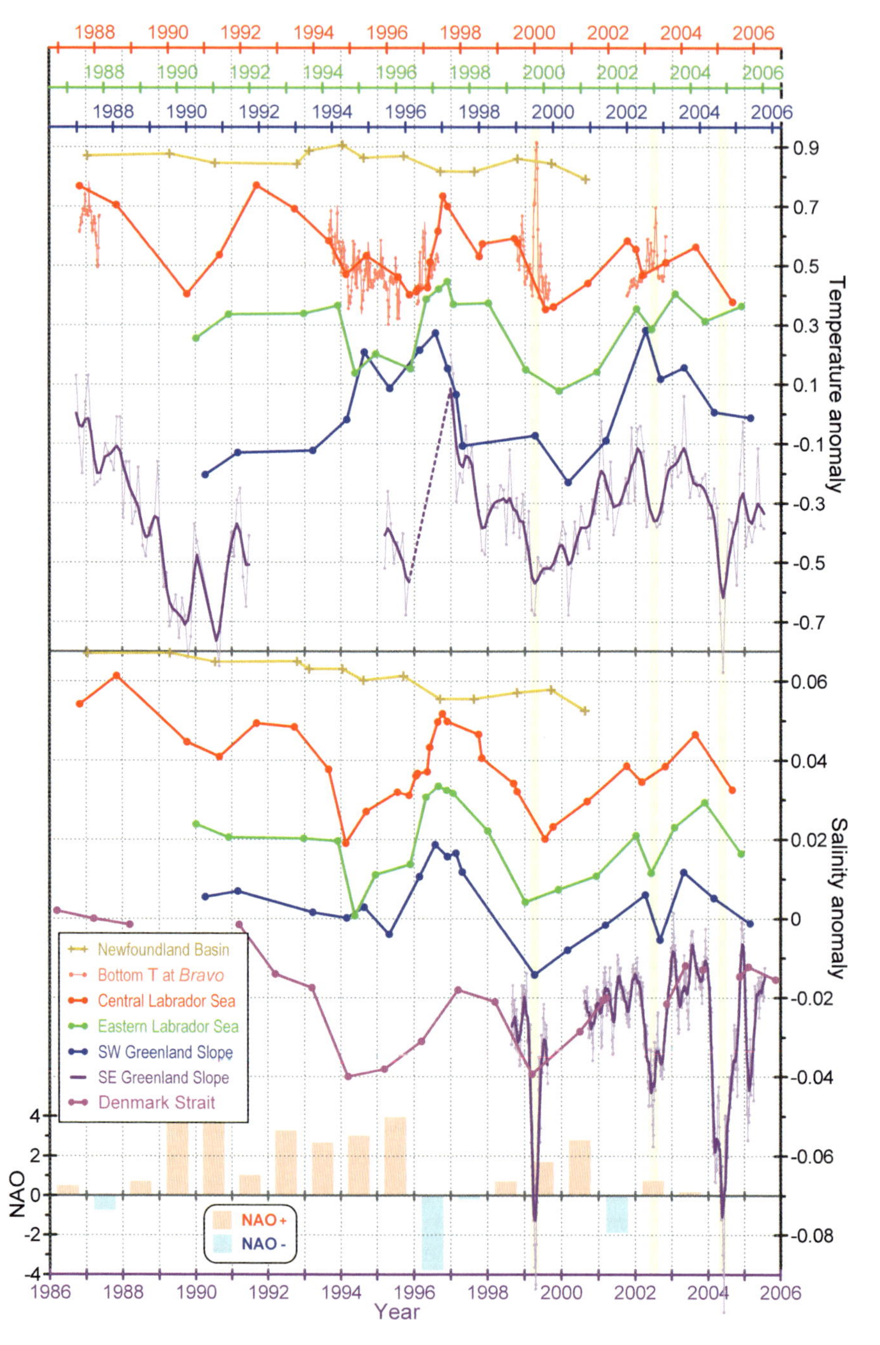

Fig. 21.2 Time series of the salinity and temperature of the Denmark Strait Overflow Water from the sill to the Labrador Sea and Newfoundland Basin since 1986. Note that the time-axes for series in the Labrador Basin are displaced in time to reflect the advective time-lag. Note that the Newfoundland Basin time series is given the same time-axis as the Central Labrador Sea. In general, there is a 1-year lag between changes at the sill and those in the abyssal Labrador Sea. A bar graph of the NAO index is also shown for comparison

In Fig. 21.3, we provide a further clear illustration of the approximate 1-year lag along this spreading path by comparing the mean temperature curve for the core of the Denmark Strait Overflow off Angmagssalik (lower panel) with the volumetric census of temperature by 0.1 °C classes for the DSOW-derived layer (3,000–3,700 m) in the Labrador Basin between 1986 and 2005. The fit, this time plotted against a common time-axis, is remarkable.

Thus of the three scales of variability that have dominated DSOW hydrography over the past several decades – trend, decadal and intra-annual change – all three have involved the transfer of ocean-climate ‘signals’ from a variety of ranges in the near-surface of the subarctic seas to the deep and abyssal ocean south of the Greenland–Scotland Ridge. While the advective time-lags we observe there largely confirm what we know of the mean circulation into and around the Labrador Basin (e.g. Clarke 1984, his Fig. 2), we are left with one conundrum: in Fig. 21.2, there is only a one-quarter-year delay between change at the boundary and the interior; in Fig. 21.3 we are able to treat the deepest part of the Basin as one uniform layer for the purpose of hydrographic census. It is still unclear how the signals of hydrographic change spread to the central Labrador Sea practically as fast as they spread around the boundary.

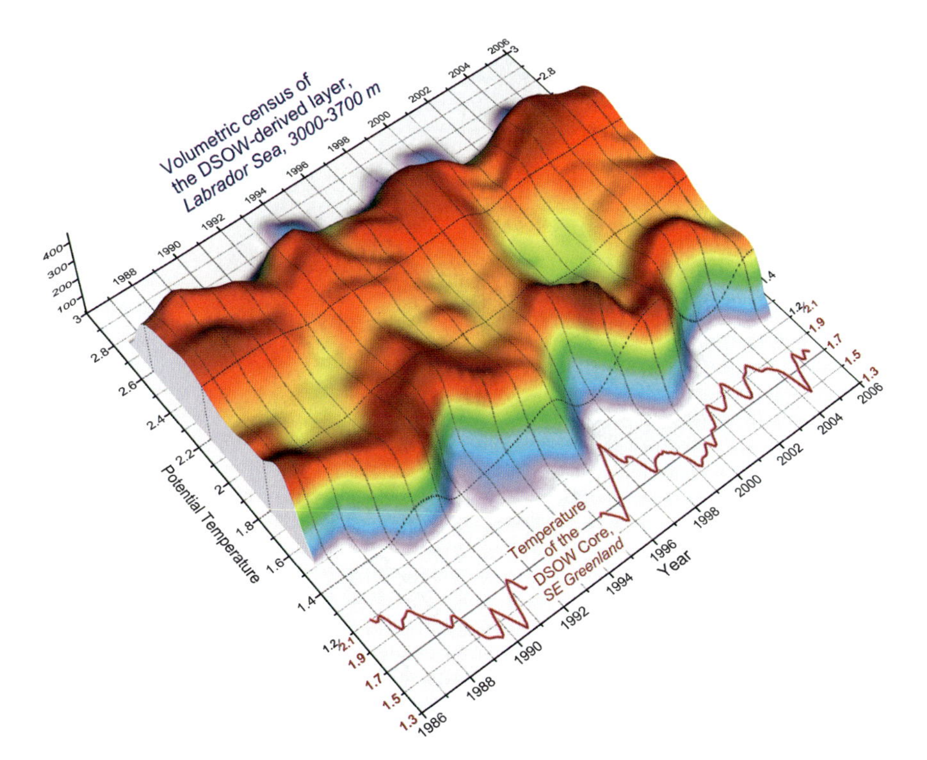

Fig. 21.3 Volumetric census of temperature by 0.1 °C classes for the DSOW-derived layer (3,000–3,700 m) of the Labrador Basin between 1986 and 2005. The mean temperature curve for the core of the Denmark Strait Overflow off Angmagssalik (based on Dickson et al. 2007, their Fig. 11) is shown below for comparison

21.3 The Northeast Atlantic Deep Water

Through direct measurement in the EC-VEINS, MAIA and ASOF-MOEN programmes, the transport of the cold, dense outflow from the Nordic Seas through Faroe Bank Channel has been soundly established (Hansen and Østerhus 2000; Østerhus et al. 2007, Chapter 18, this volume). Between 1995 and 2005, a ~200 m thick near-bottom layer (z = 600–800 m) of Iceland–Scotland Overflow Water (ISOW) with temperatures at or a little below 0 °C and with mean speeds at or a little above 1 m s^{-1} directed a transport of 1.5–2.5 Sv westward along the Iceland–Faroe Slope at the head of the Iceland Basin [1 Sv = 10^{6} m^{3} s^{-1}]. Aspects of the changing character of this eastern overflow have been described if not yet fully explained. Its sustained freshening tendency by between −7 and −14 ppm per decade between the mid-1960s and 2000, after decades of less-varying salinity

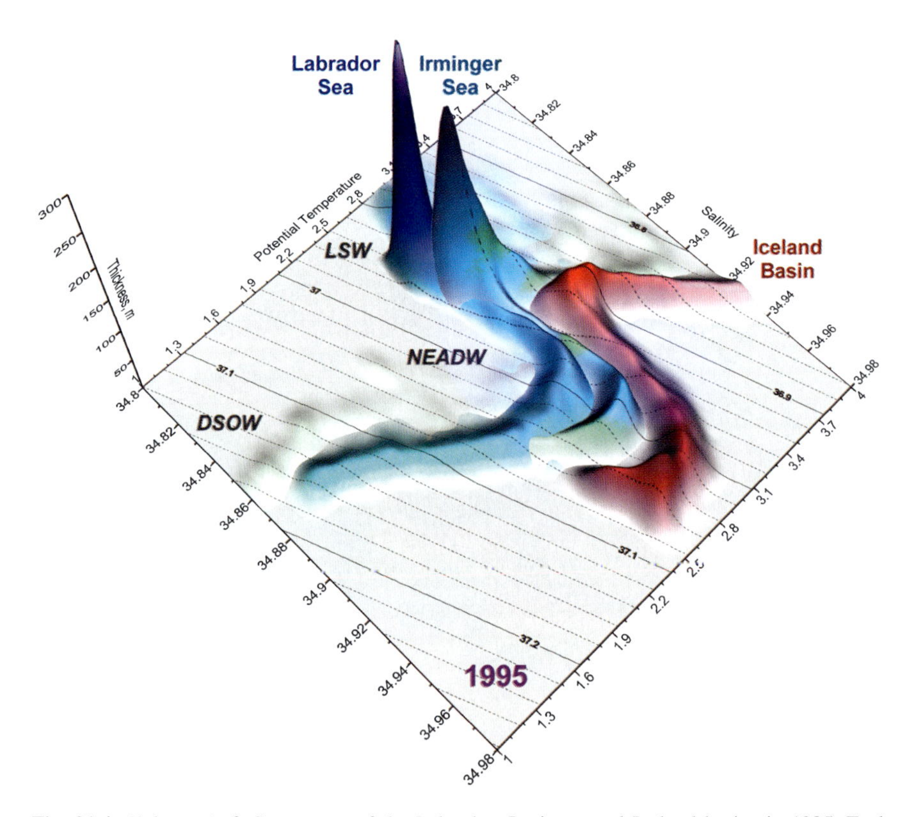

Fig. 21.4 *Volumetric* θ–S censuses of the Labrador, Irminger and Iceland basins in 1995. Each projection is based on the average vertical thicknesses (*m*) of θ–S layers defined by two-dimensional θ–S intervals with Δθ = 0.1 °C and ΔS = 0.01. The solid and dashed contours are isolines of σ_2 (kg m^{-3}) defined as a function of θ and S. (σ_2 is potential density anomaly referenced to 2,000 db.) Note that the North East Atlantic Deep Water core (the salinity maximum layer in each basin) occupies a slightly higher density-range in the Iceland Basin than west of the Mid Atlantic Ridge

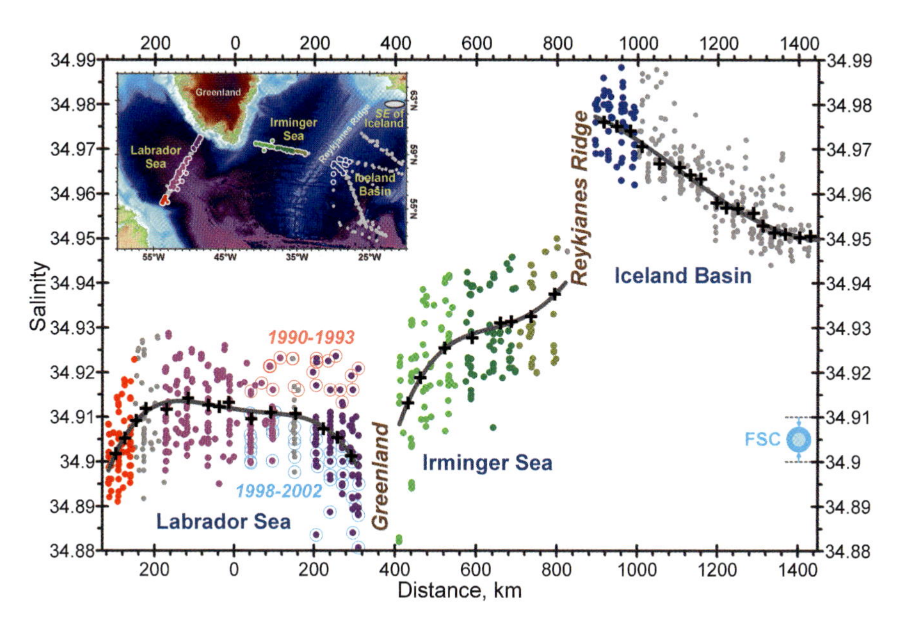

Fig. 21.5 Mean salinity of ISOW/NEADW at successive points along its spreading/transformation path from the western part of the Iceland Basin to the Labrador Sea (inset). Values are calculated for the density range $37.00 < \sigma_2 < 37.06$ in the Iceland Basin compared with $36.98 < \sigma_2 < 37.04$ west of the Mid Atlantic Ridge following the behaviour of the salinity maximum layer associated with the NEADW (Fig. 21.4). The populations of points are colour-coded both to show location (inset map) and to match the time series of salinity variation shown in Fig. 21.6. The black crosses indicate medians over stations grouped in 40 km spatial (distance) bins. The grey lines are polynomial fits of these salinity medians on distance, and provide the basis for the calculation of ISOW/NEADW anomalies elsewhere in this chapter. Since the AR7 Section followed here does not cover the head of the Iceland Basin, the mean salinity of ISOW as it overflows the Faroe–Shetland Channel is included for comparison (arrowed circle)

(Turrell et al. 1999), appears to reflect the broadscale freshening of the upper water-column of the Nordic Seas over this period and is correspondingly a feature of both overflows (Dickson et al. 2002).

Attempts to explain perceived changes in overflow *transport* as a response to slow changes in the density structure of the Norwegian Sea upstream (Hansen et al. 2001) or to changes in the strength of the regional windfield (Biastoch et al. 2003) have been less convincing. In fact there now appears to be no persistent or co-variant trend in the strength of either of the two main overflows (Dickson et al. 2007).

The downstream modification of ISOW as it spreads west to form the Northeast Atlantic Deep Water (NEADW) of the North Atlantic (Fig. 21.4) has not been fully described in either space or time. Figures 21.5 and 21.6 now address both by describing the mean trans-ocean salinity profile and the time-dependence of salinity over the last 2 decades along the spreading path of the (salinity maximum) NEADW layer from the Iceland Basin to the Labrador Sea. The sites and sense of the watermass interactions that modify the spreading plume *en route* are clear from these two

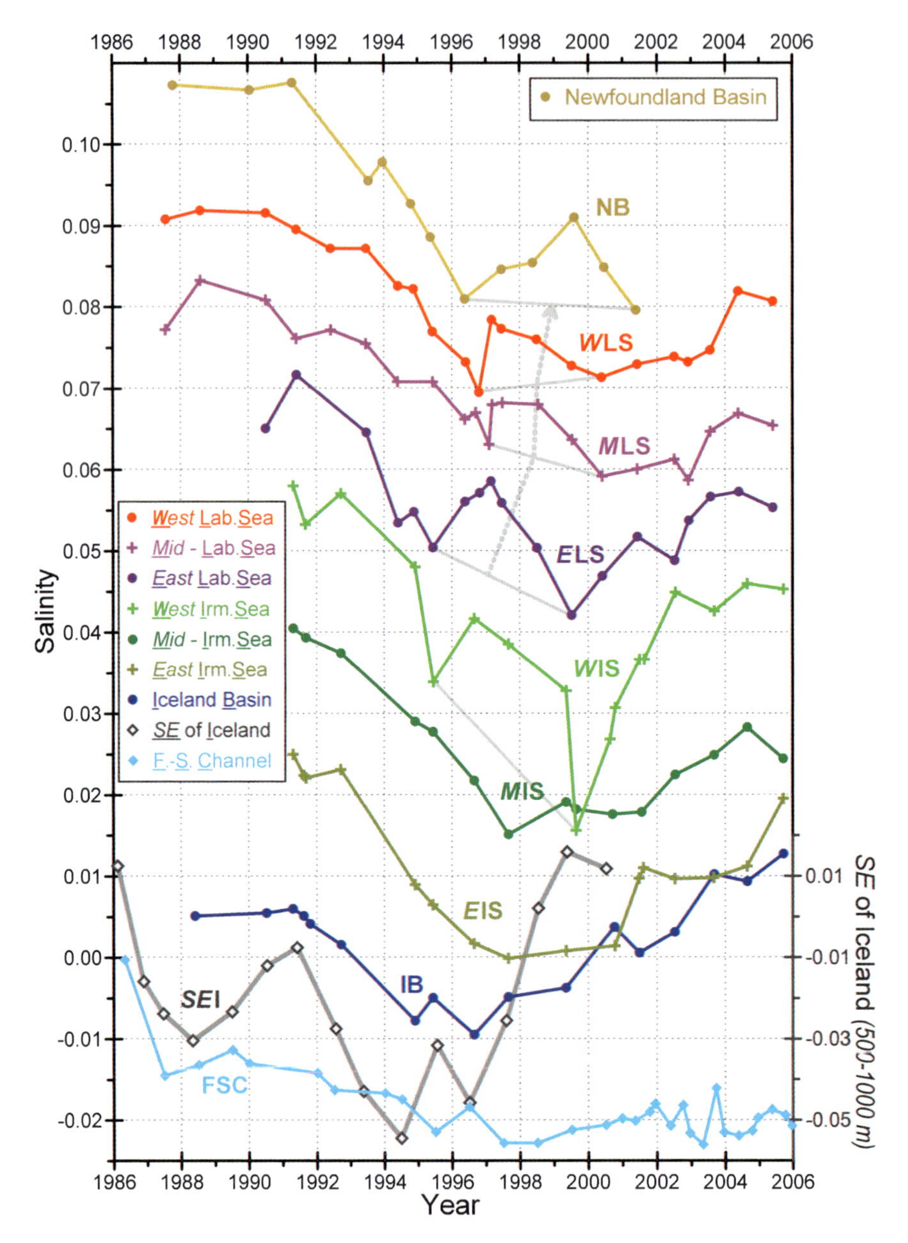

Fig. 21.6 Time series of salinity anomalies (relative to the 1987–2005 mean salinity profile composed from the grey lines shown in Fig. 21.5) for ISOW, becoming NEADW, at selected locations along its spreading pathway from the sill of the Faroe–Shetland Channel to the Labrador Sea. The density range occupied by this salinity-maximum layer is defined as $37.00 < \sigma_2 < 37.06$ in the Iceland Basin and $36.98 < \sigma_2 < 37.04$ west of the Mid Atlantic Ridge; the salinity of the Subpolar Mode Water that is entrained by the NEADW at the head of the Icelandic Basin (***S****outh* ***E****ast of Iceland*) is also shown. The time series are colour-coded to reflect locations along the mean salinity profile of Fig. 21.5

figures. Since the mean salinity profile of NEADW shown in Fig. 21.5 is largely restricted to the line of the AR7 section (inset map), it cannot strictly represent the initial changes in NEADW mean salinity from the overflow-sill to the Reykjanes Ridge. For this easternmost part of the transect, the grey points shown (Fig. 21.5) are merely the salinities at NEADW densities in the central Iceland Basin, and the salinity at the sill is indicated by the much lower value indicated by the arrowed circle. Comparing the latter with the point where AR7 intersects the path of the NEADW plume (blue dots east of the Ridge), it is clear that along the north wall of the Iceland Basin and in its initial descent along the Reykjanes Ridge, the ISOW/NEADW has increased rapidly in salinity from <34.91 to >34.97 (Fig. 21.5) and doubled in layer-thickness from 200 to 400 m through the rapid entrainment of the warm salty resident Subpolar Mode Water (SPMW).

Expressed as a function of time, Fig. 21.6 shows that as the overflow leaves the sill of the Faroe–Shetland Channel, its salinity has remained essentially unaltered for the past decade (lowest curve, Fig. 21.6); its long freshening trend (Turrell et al. 1999) has evidently halted. Passing around the Iceland Basin however, the ISOW/NEADW exhibits a steady salinification from the mid-1990s which we take to reflect its rapid entrainment of an SPMW watermass which was itself becoming rapidly more-saline over the same period (grey curve, Fig. 21.6). From that point, the westward spread of this freshening-then-salinifying turning-point in the NEADW is perhaps the major feature of this figure, passing to the Labrador Sea over the next 4–5 years and arriving along the Labrador Slope in 2000.

These figures show other features however. In particular (Fig. 21.5) the immediate and rapid drop in the mean salinity of NEADW from ~34.97 to ~34.94 as the spreading plume rounds the snout of the Reykjanes Ridge to enter the Irminger Sea provides a clear indication of a growing influence and interchange with LSW west of the Ridge (see below), and this decline in the mean salinity of the layer continues (to ~34.91) in the western Irminger Sea where a thinning NEADW plume passes south between both of the low-salinity watermasses of the Basin, the relatively fresh influence of Denmark Strait Overflow Water (DSOW) now adding to that of LSW. As with the mean, so with the variability. Mapping the change in salinity between 1964–1972 and 1995–1997 across the $\sigma_2 = 37$ surface associated with the NEADW (Fig. 21.7), we find that the freshening of the eastern overflow waters at source (Turrell et al. 1999; Dickson et al. 2002) becomes greatly amplified towards the head of the Irminger Basin, where the NEADW 'wedge' flows between layers of LSW and DSOW that were themselves becoming fresher at that time.

The specific influence of the Denmark Strait Overflow Water is perhaps most clearly seen by comparing the curves of NEADW salinity from the middle and western Irminger Sea (Fig. 21.6). In the latter, the NEADW has acquired deep salinity minima in 1995 and 1999 which are not present in the former; these brief but dramatic freshening episodes have already been described by Dickson et al. (2002, their Fig. 2) and further examined here (Fig. 21.2) as features of the Denmark Strait Overflow, and Holfort and Albrecht (2007) have recently attributed them to a strengthening of the freshwater feed to that overflow from the East

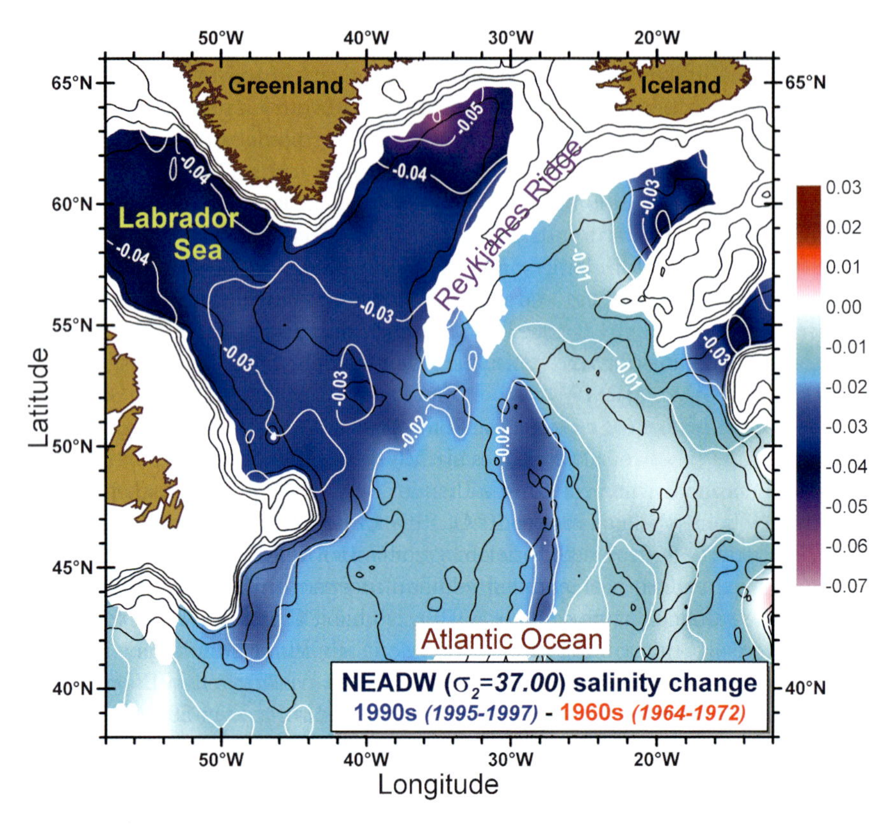

Fig. 21.7 Change in mean salinity from 1964–1972 to 1995–1997 on the $\sigma_2 = 37$ surface associated with the NEADW core

Greenland Current, arising from an anomalously strong north-wind component in these years immediately to the north of Denmark Strait. These two events are still recognizable features of the NEADW where it spreads through the eastern Labrador Sea (Fig. 21.6), but are less obviously present by the Labrador Slope.

Finally, Fig. 21.5 confirms that as it rounds Cape Farewell, the NEADW attains its minimum mean salinity of 34.90 or less, and this relative freshness is retained along the direct path of the inflowing watermass around the boundary of the Labrador Basin. As this watermass mixes with older saltier NEADW from the interior or spreads offshore to recirculate slowly in the basin interior at depths between 2,200 and 3,200 m, some renewed salinification evidently takes place (Fig. 21.5). The rates of this process are largely unknown but it may be instructive that the salinity contrast between the boundary and the interior was noticeably greater by a few hundredths in 1998–2002 than in 1990–1993 (these points are identified, and circled in Fig. 21.5). During the former period, LSW was at its freshest, coldest, deepest and densest of record, and the combined effect of this deep, dense 'plug' in the basin interior and a more-intense gyre circulation (Yashayaev and

Clarke 2005; Hakkinen and Rhines 2004) may well have acted to promote some degree of isolation between the interior and the boundary.

Though this chapter is focused on the overflow system and the locations where change is 'imported' to that system, it is already evident that the exchange of ocean-climate signals with the Labrador Sea Water (LSW) layer is one major influence, but one which has varied greatly in both space and time. A brief description of change in the LSW is necessary before summarising the net result of these exchanges on the principal deep watermasses of the northern North Atlantic.

21.4 The Transfer of Ocean Climate 'Signals' Between Labrador Sea Water and Northeast Atlantic Deep Water

Over the last 3–4 decades, repeat hydrographic transects of the Labrador Basin have described a remarkable suite of changes in its convectively formed mode water (LSW), so spanning the upper watercolumn to the limit of convection (~2,300 m). These dramatic changes (see Yashayaev et al., Chapter 24, this volume) are thought to reflect the sustained if non-steady evolution of the leading mode of wintertime atmospheric forcing in our sector, the North Atlantic Oscillation (NAO), from its most extreme negative state in the instrumental record during winters of the 1960s to its most extreme and prolonged positive state in the early 1990s (Hurrell 1995). Ultimately, in severe winters between 1987 and 1994, a 'vintage' of LSW was formed (referred to here as '$LSW_{1987-1994}$') that was fresher, colder, thicker and denser than at any other time in the history of deep measurements there (Lazier 1995; Dickson et al. 2002). Between 1960 and 1994, LSW became fresher by 0.08, colder by 0.9 °C and denser by 0.08 kg m^{-3} (Yashayaev et al. 2003; Yashayaev and Clarke 2005). Its layer thickness also increased. From the Atlantic-wide change in thickness

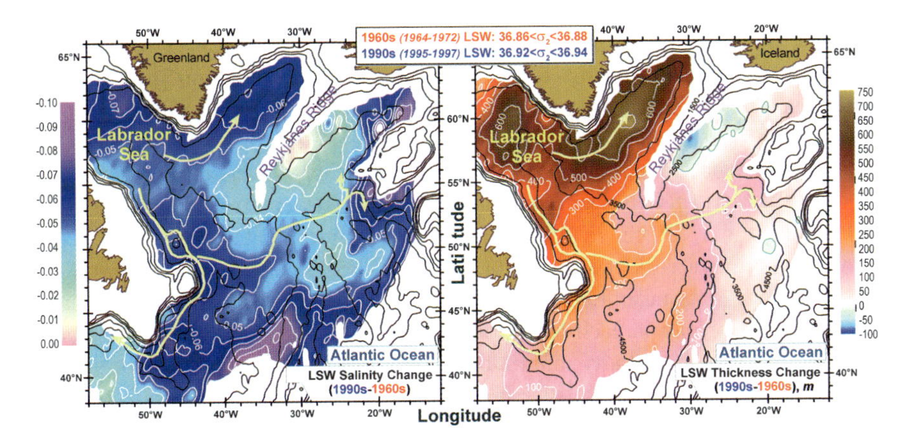

Fig. 21.8 Change in LSW salinity and thickness between the 1960s and 1990s in the northern North Atlantic

of the LSW layer between the late 1960s and mid-1990s, Yashayaev et al. (2004) estimate its minimum production rate to be 2 Sv ($2 \times 10^6\ m^3\ s^{-1}$) over this period.

Since the mid-1990s, with the return of the NAO index to near-normality, these trends have been reversed (Yashayaev and Clarke 2005; see also Avsic et al. 2006; Kieke et al. 2007); LSW has become steadily warmer by >0.6 °C and more saline by 0.04; the vast mass of LSW, ~1,800 m thick at its 1994 peak, has drained from the basin; from 2000, the production of that earlier vintage of LSW ('$LSW_{1987-1994}$'), is replaced by a thin layer of 'LSW_{2000}' of much lower density and at shallower depth; and as the net result of these changes, a long-sustained lowering of sea level in the basin has been reversed, associated by Hakkinen and Rhines (2004) with a slowdown and westward retraction of the Subpolar Gyre.

These monumental changes, the factors that forced them and their trans-ocean spreading are all fully described in Yashayaev et al. (2007, this volume) and need not be repeated here. In the present context, we aim instead to examine the areas of congruence between the eastward spreading LSW and westward spreading NEADW so as to identify the sites and instances where the transfer of ocean-climate signals may have taken place between the two.

Figure 21.8 provides important clues as to where such transfers are likely to take place and where they are not. Though LSW does spread to the eastern Atlantic, it does so along certain preferred pathways (Talley and McCartney 1982; Cunningham and Haine 1995) and these will determine, to a large extent, the scope for interaction with the ISOW/NEADW. In mapping the change in LSW salinity and thickness between the 1960s and 1990s (see Yashayaev 2007 for the actual distributions), Fig. 21.8 confirms that although strong freshening and thickening took place throughout the LSW layer in the Irminger Basin, these changes are only evident in the Iceland Basin along the main LSW spreading path from ~48° N 38° W to the tail of Rockall Bank (yellow line), too far south to have much effect on the plume of ISOW where it descends the east flank of the Reykjanes Ridge; in that location and over that time interval, Fig. 21.8 records no significant change in the LSW. Figure 21.5 has already confirmed this point in showing that the mean salinity of NEADW along its westward-spreading path does not decrease sharply until the plume rounds the snout of the Reykjanes Ridge to enter the Irminger basin, i.e. there is no *strong* contact with the freshening influence of LSW to the east of the Reykjanes Ridge.

Figure 21.9 now assembles these elements into time–distance plots of salinity anomaly for both LSW (upper panel) and NEADW (lower) over the period 1987–2006 along the 1,700 km length of the trans-Atlantic AR7 repeat-hydrography section, running across the Labrador, Irminger and Icelandic basins from the Labrador Slope to the tail of Rockall Bank via Cape Farewell (see inset map, Fig. 21.5). On each CTD profile in each basin, the core-depth of the LSW (NEADW) layer was identified by the salinity minimum (maximum) and the anomalies of salinity at these depths have been plotted and gridded.

Figure 21.9 exhibits the following features: (a) The dramatic freshening of the LSW to the early 1990s is observed to pass east to the Irminger Basin in 2 years or so, and to spread through the eastern Atlantic on timescales of 5–7 years (double

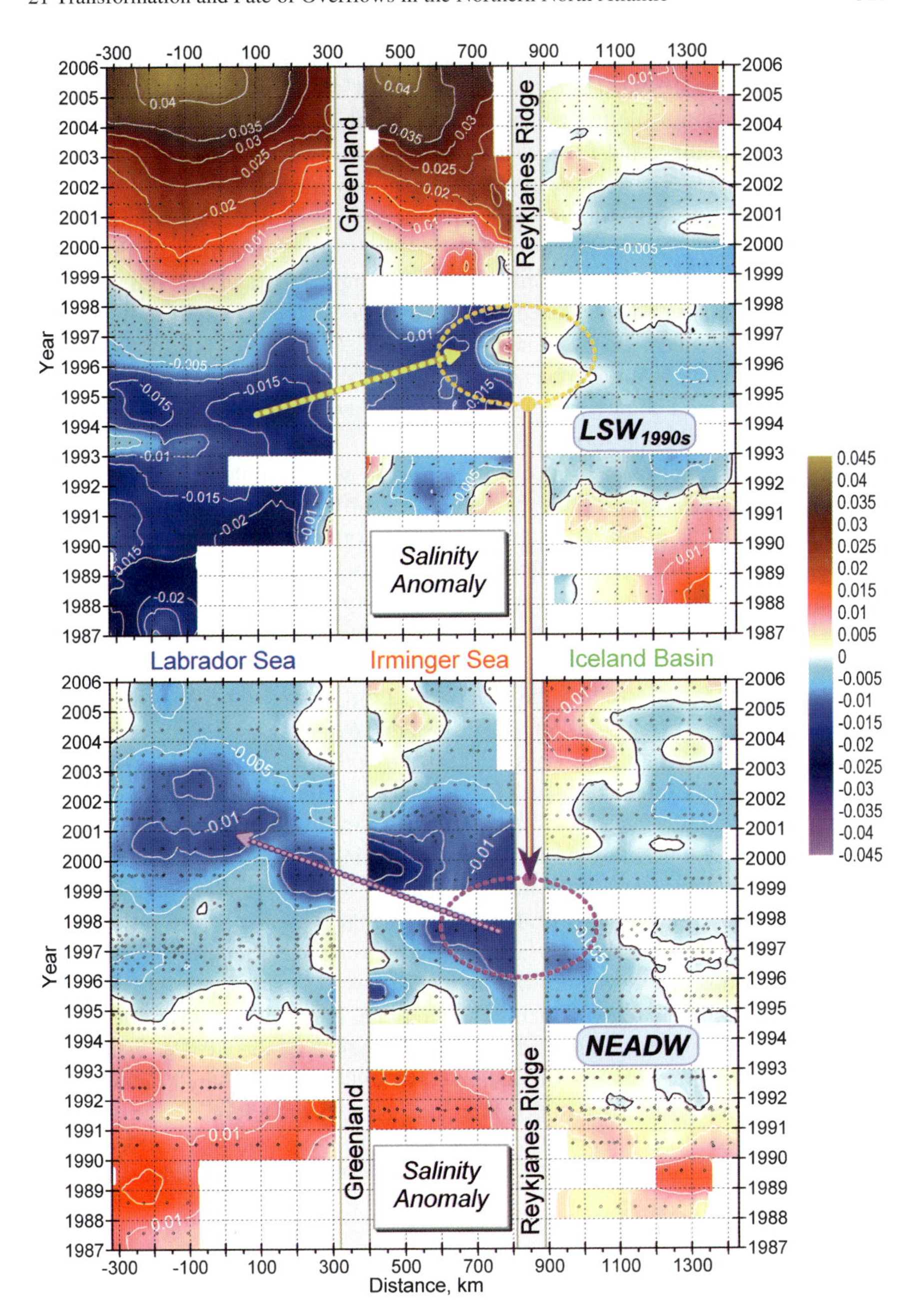

Fig. 21.9 Time: distance plots of salinity anomaly for both LSW (upper panel) and NEADW (lower) over the period 1987–2006 along the trans-Atlantic AR7 repeat-hydrography section, from the Labrador Slope to the tail of Rockall Bank via Cape Farewell. The major LSW salinity minimum of ~1994–1995 appears to be transferred to the NEADW in the vicinity of the Reykjanes Ridge around 1996–1997, and is then returned westward by 2000–2001

those suggested by Sy et al. 1997; see Yashayaev et al. 2004). (b) We note in particular that the LSW salinity minimum of ~1994–1995 at its source in the Labrador Sea reaches the Reykjanes Ridge around 1995–1997. (c) By entraining SPMW in its circuit of the Iceland basin, the NEADW has already picked up the freshening signal of SPMW as it dipped to its salinity minimum in 1994–1997 (see '$SEI_{500-1000}$' curve, Fig. 21.6). (d) But with the arrival of the LSW salinity minimum at the Reykjanes Ridge around 1995–1997, the freshening of the (now-underlying) NEADW core becomes strongly amplified, and (e) this strongly negative salinity anomaly is carried westward again by the NEADW to the deeper layers of the Labrador Sea by around 2000–2001. One of the greatest ocean-climate signals in the hydrographic record has apparently been 'exported', transferred, returned and thus carried to depths that were beyond the reach of the convective processes that formed it.

Figure 21.10 captures the effect of these events as if from the fixed 'viewpoint' of the Labrador Sea. It describes the long slow shifts in the volumetric census of most (200–3,000 m) of the watercolumn of the northwest Atlantic since the late 1950s, based on 0.005 salinity classes. The following interpretation is suggested to explain its principal features: a long, slow freshening of the LSW layer from the early 1970s, following the renewal of deep convection in 1972 and its subsequent intensification, leads to a general salinity minimum in the LSW during the late-1980s/early 1990s, followed by the steady re-salinification of LSW as convection in the Labrador Basin weakens and shallows once again.

This deep salinity minimum spreads eastward with the LSW to Mid-Atlantic, is transferred through exchange to the NEADW sublayer and returns westwards in that watermass, bringing a (at least 7-year delayed) salinity minimum to the NEADW core in the Labrador basin by around 2000–2001 (Figs. 21.9, Fig. 21.10).

In the most recent years, there are signs that the NEADW in the Labrador Basin is becoming more saline once again. This is not attributable to the influence of LSW. Though the LSW itself becomes rapidly more saline after 1994 (Fig. 21.10), it is always, in absolute terms, fresher than NEADW; so exchange with LSW west of the Ridge is unlikely to explain this change in the NEADW sublayer after 2000–2001. Rather, we attribute this to the remote influence of the SPMW watermass, entrained along the Reykjanes Ridge, that was itself becoming rapidly more saline after 1994–1995 (Fig. 21.6). By 2005 then, it is as the net result of change, drift and exchange between three watermasses over half a Century and half an Ocean that the modal salinity values of LSW and NEADW in the Labrador Sea have become almost merged (Fig. 21.10).

21.5 Discussion and Conclusions

It has been the concern of this chapter to 'dissect' the vertical stack of watermasses in the northern North Atlantic, layer by layer, south of the Greenland–Scotland Ridge, in order to describe the combination of local, regional and remote influences

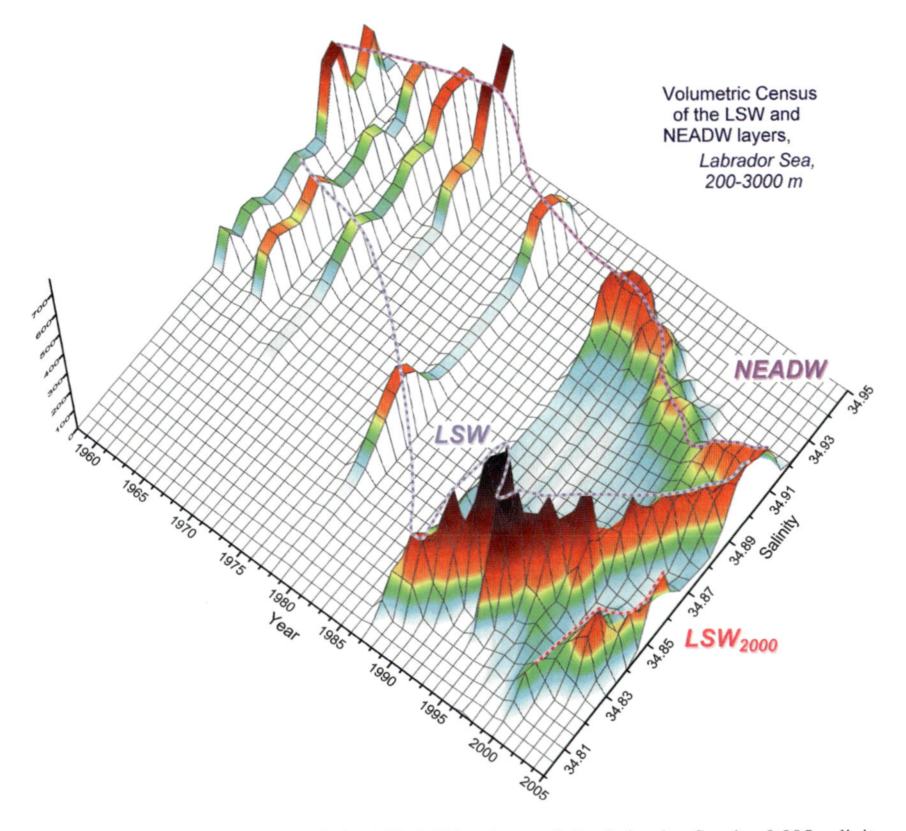

Fig. 21.10 Volumetric census of the 200–3,000 m layer of the Labrador Sea by 0.005 salinity classes over the period of the modern hydrographic record (1958–2005). The census is based on the thickness of salinity classes along the most frequently worked repeat hydrographic section, the annual AR7W section crossing the Labrador Sea between Labrador and Greenland. To convert from layer-thickness to volume, it is necessary to assume this optimal data set is representative of the surrounding basin, within limits set according to the station density and coverage of the entire AR7W section (>70%)

that have driven record hydrographic change through the watercolumn of the Northwest Atlantic in recent decades. The justification for the study lies in the fact that the great storage basins of the Labrador Sea, Irminger Sea and Iceland Basin not only receive the full range of inputs from Arctic and subarctic seas but together form the 'headwaters' of the global thermohaline circulation, where regional change may have global impact.

Our *first general conclusion concerns the apparently unique global importance of regional processes in the Irminger Sea for the transfer of ocean climate signals between watermasses and to great ocean depths.* The reasons have already been described in Chapter 19 of this volume but are worth reiterating here. The four main factors are these: (a) The three watermasses that will eventually occupy much of the watercolumn of the northwest Atlantic and which make the major contribution to

North Atlantic Deep Water production flow together in close contact *only* in a limited zone along the Continental Slope off SE Greenland; there, the North East Atlantic Deep Water (NEADW) derived from the eastern overflow passes in a relatively narrow tongue between plumes of Labrador Sea Water spreading in at intermediate depths from the Labrador Basin and of Denmark Strait Overflow Water descending from its sill into the abyssal Atlantic. (b) The factors promoting their mixing are also special to this zone. Through the descent of that plume, the stretching of the high potential vorticity watercolumn outflowing from Denmark Strait induces very strong cyclonic relative vorticity and '*a specific form of mesoscale variability that is unique to the Denmark Strait*', as Spall and Price (1998) have long pointed out; more recently (e.g. Pickart et al. 2003; Pickart et al. 2007, Chapter 26, this volume) we have become aware that the elevated heat-loss and intense wind stress curl associated with a 'tip jet' that forms episodically in the lee of Cape Farewell has the potential to promote small-scale but intense open-ocean deep convection that also seems to be unique to this zone. (c) Though LSW does spread to the eastern Atlantic, its spreading axis is normally too far south to affect the ISOW plume east of the Reykjanes Ridge. Exchange between the LSW and the ISOW/NEADW core and the rapid freshening that results (Figs. 21.5 and 21.7) is thus primarily a feature of the Irminger basin. (d) And as the conduit for the major oceanic inputs from Arctic and subarctic seas, large hydrographic changes do pass through the Irminger Sea. Quite apart from their impact downstream, the amplitude of the more-extreme of these features has enabled us to track certain of the exchanges between watermasses in this zone. Thus we are left in no great doubt that the two extreme freshening episodes that affected the DSOW in 1995 and 1999 can be traced upstream and earlier to a north-wind-induced strengthening of the freshwater feed from the East Greenland Current to the Overflow in the upper watercolumn; that these fresh events were passed on to the NEADW where it flows in close company with the DSOW through the western Irminger Sea; and that as a result, recognisable traces of these two features were to be found in both the deep and abyssal layers of the Labrador Sea up to 1 year later. Equally, we can hardly mistake the coherent eastward spreading of extreme freshening in the LSW from the mid-1990s, its arrival at the Mid Atlantic Ridge around 1995–1997, its freshening of the NEADW sublayer and its return in that watermass to greater depths in the Labrador Sea by 2000–2001.

For the present, we are not really equipped to use the passage of such identifiable 'events' and their changes in amplitude and areal extent to derive convincing figures for mixing and entrainment *en route*. Even in Fig. 21.5, where the mean salinity of the NEADW core drops over a short distance from ~34.97 to ~34.94 as the spreading plume passes from the salinifying influence of SPMW to the freshening influence of LSW around the snout of the Reykjanes Ridge, the large space-time shifts in the hydrographic character and thickness of all three watermasses prevent us from backing-out reliable mixing rates from a largely annual hydrographic record. For the present, the convincing evidence is mainly about *where* these transfers take place.

If the Irminger Sea remains the basin where signal transfer to depth is most efficiently carried out, *our second general conclusion is that the changing watercolumn of the Labrador Sea is of greater climatic importance*, acting as the receiving volume for time-varying inputs of watermasses from Northern Seas which are then stored, recirculated, transformed and discharged to modulate the Deep Western Boundary Current (DWBC) of the North Atlantic – the abyssal limb of the Atlantic Meridional Overturning Circulation (AMOC). It is not just through discharge to the DWBC that the Labrador Sea exerts its influence on climate. Wu and Wood (2006, submitted) suggest that the freshening recently observed in Subpolar seas may ultimately be triggered by Labrador Sea deep convection. And Hátún et al. (2005) are convincing in their view that the changes in steric height of the Labrador basin and the slowdown and retraction of the subpolar gyre that has resulted (see above) have been instrumental in directing an inflow of near-record warmth and salinity to the Nordic seas through the Faroe–Shetland Channel in recent years.

Annual research cruises centred on the AR7W line have provided an increasingly clear view of the long, slow shifts in Basin hydrography over half a Century, including the increase in production of LSW to the mid-1990s, its subsequent draining from the basin, and the lagged influence of this change on the NEADW (Fig. 21.10). The principal unknown, and perhaps the most important one from the viewpoint of the changing AMOC remains the question of whether and to what extent these extreme changes throughout the watercolumn of the Labrador Basin may have fed south to contribute to the changes encountered by Bryden et al. (2005) at 26° N.

Some transfer at depth between the deep and abyssal Labrador Sea and the Newfoundland basin already seemed evident in Fig. 21.6; there the NEADW layer of the Newfoundland basin appeared by 2000 to have undergone a similar freshening to that of the NEADW throughout its domain, although we lack the more recent hydrographic record that might have allowed us to set a timescale for its inter-basin transfer.

In fact, the evidence from south of the Greenland–Scotland Ridge over the past few decades is of a complex repartitioning, redistribution and restructuring of much of the watercolumn, not merely an alteration of the hydrographic characteristics of this or that watermass. Figure 21.11 shows something of the complexity of these changes by comparing vertical salinity sections, annotated with potential density contours, through the Newfoundland, Labrador and Irminger Basins for 1966 and 1994. For example, the NEADW core, which closely coincides with the $\sigma_2 = 37.00$ surface (see earlier, Fig. 21.4), "sank" by 300 m or more across most of its 'domain' between 1966 and 1994 (Yashayaev 2007). This deepening is largely an expression of the volume loss experienced by NEADW when LSW expanded to its record thickness partly at the expense of the less dense classes of NEADW. The remaining NEADW not only became fresher (due to increased entrainment of fresher LSW or to the change in its source waters), but also significantly decreased in volume, while its core deepened. Interestingly, this relative increase and decrease in the volumes of LSW and NEADW between 1966 and 1994 resulted in inverse changes in watercolumn density above and below 2,000 m. The water above 2,000 m became denser, while its lower counterpart became less dense (the $\sigma_2 = 37.00$ surface was also deeper in 1994).

In the abyssal layers too, the comparison of these sections clearly shows that a freshening and a decrease in densities has been transferred to the Newfoundland basin by the later date (1994; Fig. 21.11). There, the near-bottom salinities

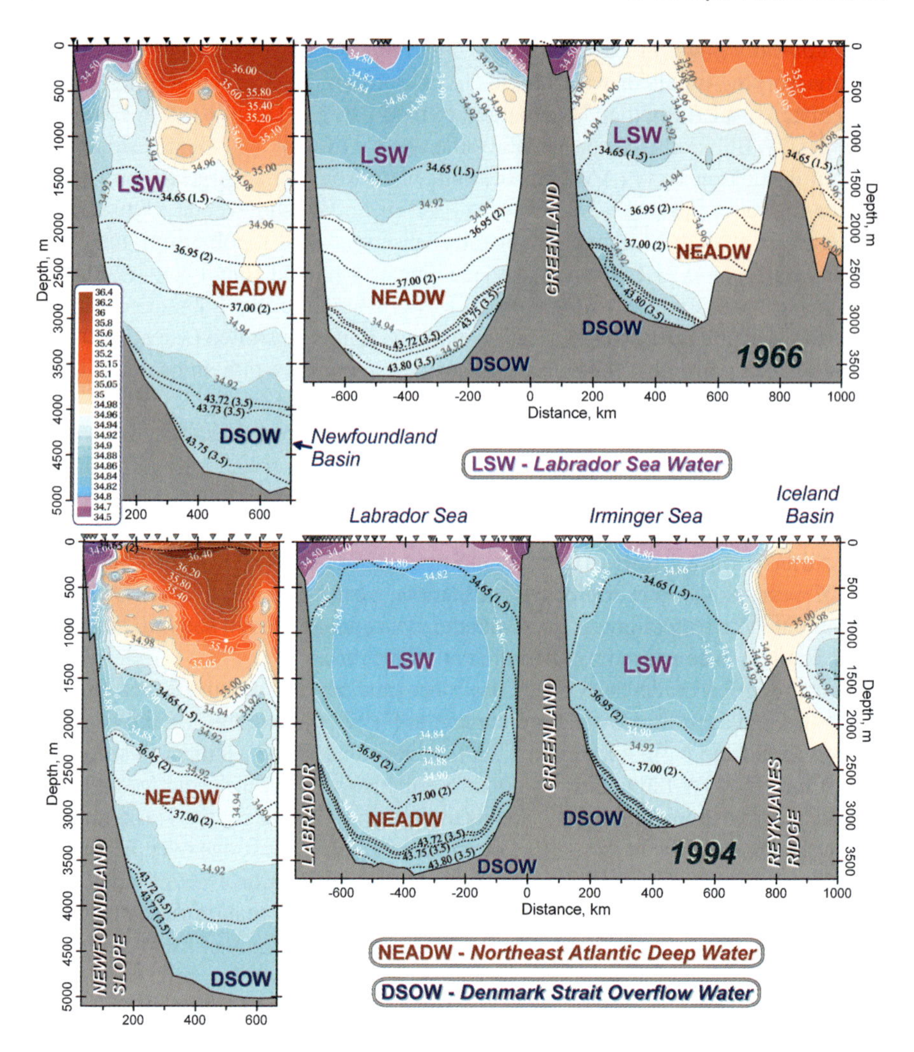

Fig. 21.11 Vertical salinity sections, with potential density contours superimposed, through the Newfoundland, Labrador and Irminger Basins in 1966 and 1994. The three main differences are the increase in density at intermediate depths, the decrease in density in the deep and abyssal layers and the lesser thickness of DSOW in 1994

decreased by 0.02–0.03 between the two occupations, reflecting either the freshening of its source waters in the Nordic seas (Dickson et al. 2002), or the influence of a freshening and expanding LSW production, or both; some dense DSOW classes ($\sigma_{3.5}$ > 43.75) seen in 1966 in the Newfoundland Basin were not present there 3 decades later; and as its densest classes disappeared, the volume of the DSOW-derived water ($\sigma_{3.5}$ > 43.73) decreases.

Figure 21.12 now sums up much of the content of this chapter by presenting, in a single Θ–S diagram, the change, transformation and mixing that has characterized the

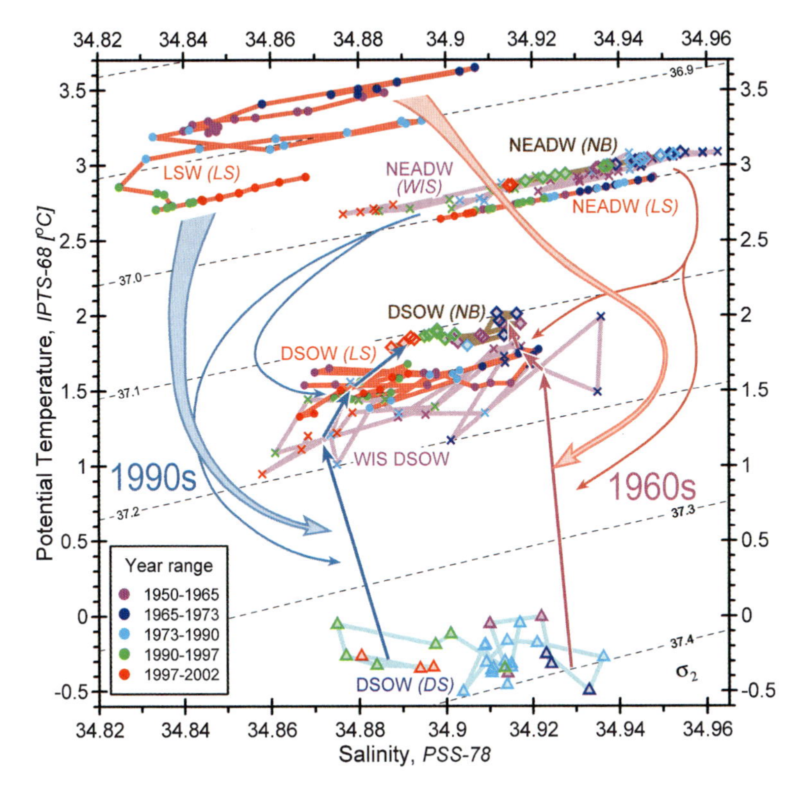

Fig. 21.12 Potential temperature-salinity diagram describing the transformation and mixing of watermasses of the northern North Atlantic between the Greenland–Scotland Ridge and the Newfoundland Basin since the 1960s. See text for explanation

intermediate, deep and abyssal watermasses of the northern North Atlantic from the Greenland–Scotland Ridge to the Newfoundland Basin since the 1960s. (Shapes denote areas: Δ = Denmark Strait; x = Irminger Sea; • = Labrador Sea; ◇ = Newfoundland Basin. Their colours denote years: see inset colour key. Annotations and line colours denote watermasses.) Starting with the coldest, we see towards the foot of the diagram, a cloud of DSOW points that vary little in temperature but freshen markedly from the 1960s to 1990s. The clouds of points representing NEADW (top right) and LSW (top left) undergo a similar freshening over this time-span. Between the Denmark Strait sill (DS) and the western Irminger Sea (WIS), the DSOW 'cloud' jumps in temperature by around 2 °C as its main entrainment takes place. Though a wider range of watermasses are involved in that initial rapid entrainment than are considered here (see Dickson et al., Chapter 19, this volume), the freshening LSW and NEADW layers contribute to that process and the DSOW points in the Western Irminger Sea show the same general distribution in salinity as at source, saltiest in the 1960s, freshest in the 1990s. Warming of the DSOW core does continue between the Labrador Sea and Newfoundland Basin, but at a much slower rate reflecting a much-weakened rate of entrainment and mixing (calculated from a simple mixing model at ~10%, largely from the overlying NEADW). In the Newfoundland Basin, the principal new feature is the damping down of interannual variability compared with the Labrador Sea and Irminger; though the same freshening *trend* is still evident, the span of salinity change in NB-DSOW and NB-NEADW is only about half the amplitude that these watermasses had exhibited upstream (this damping of interannual variability between the Labrador Sea and Newfoundland Basin is clearly evident in the T and S time series of Fig. 21.2).

For the present, our final discussion point must remain more of a conviction than a conclusion: *that as the largest changes in Oceanography were played out on a timescale of decades throughout most of the watercolumn and across all four of the great storage basins south of the Greenland–Scotland Ridge, it would seem probable that their influence contributed significantly to the climatically important changes in the MOC that Bryden et al. (2005) describe downstream.*

Resolving this conviction will require the resolution of some major problems: the changing rates of the great recirculating gyres of the Labrador Basin and off the eastern seaboard of North America that McCartney described as 'the greatest problem in Oceanography' in his Sverdrup Lecture to the AGU in 1996; the lateral and vertical mixing and entrainment of northern-source waters into these gyres as the Deep Western Boundary Current passes south; the effect on circulation of the apparent changes in the near-bottom density and stratification downstream from the Labrador Sea and Newfoundland Basin; even the effects of bottom topography and of the rapid change in the depth of the DSOW core as it passes south from the Labrador basin into the Deep Western Boundary Current (see Fig. 21.11). While these issues may be unresolved, what we already see of the complexity of these changes in the northern North Atlantic (Fig. 21.12) would suggest that the large-scale changes in ocean dynamics, mixing and inter-gyre exchange associated with substantial shifts in the MOC will not easily be diagnosed without them.

Acknowledgements The authors thank Allyn Clarke and Jens Meincke for valuable comments and suggestions during the work on the manuscript. Hendrik van Aken, Manfred Bersch, Detlef Quadfasel and John Mortensen generously provided hydrographic data from the recent AR7E occupation used to compute properties of ISOW/NEADW in the Irminger and Iceland basins.

References

Avsic, T., J. Karstensen, U. Send, and J. Fischer, 2006. Interannual variability of newly formed Labrador Sea Water from 1994 to 2005. GRL, 33, L21S02, doi:10.1029/2006gl026913,2006.

Biastoch, A., R.H. Käse, and D.B. Stammer, 2003. The sensitivity of the Greenland-Scotland Ridge Overflow to Forcing Changes. J Phys Oceanogr, 33, 2307–2319.

Bryden, H.L., H.R. Longworth, and S.A. Cunningham, 2005. Slowing of the Atlantic meridional overturning circulation at 25°N. Nature, 315, 21–26.

Clarke, R.A. 1984. Transport through the Cape Farewell-Flemish Cap section. Rapp P-v Reun Cons, Int Explor Mer, 185, 120–130.

Cunningham, S.A. and T.W.N. Haine, 1995. Labrador Sea Water in the Eastern North Atlantic. Part I: A Synoptic Circulation Inferred from a Minimum in Potential Vorticity. J Phys Oceanogr, 25(4) 649–665.
Curry, R. and C. Mauritzen, 2005. Dilution of the Northern North Atlantic Ocean in Recent Decades. Science, 308(5729), 1772–1774.
Dickson, R.R. and J. Brown, 1994. The production of North Atlantic Deep Water: Sources, Rates and Pathways. J Geophys Res, 99(C6), 12319–12341.
Dickson, R.R., J. Meincke, I.M Vassie, J Jungclaus, and S. Østerhus, 1999. Possible predictability in overflow from the Denmark Strait. Nature, 397, 243–246.
Dickson, R.R., I. Yashayaev, J. Meincke, W.R. Turrell, S.R. Dye, and J. Holfort, 2002. Rapid freshening of the deep North Atlantic Ocean over the past four decades. Nature, 416, 832–837.
Dickson, R.R., R. Curry, and I. Yashayaev, 2003. Recent changes in the North Atlantic. Phil Trans Roy Soc Lond A, 361, 1917–1934.
Girton, J.B., T.B. Sanford, and R.H. Käse, 2001. Synoptic sections of the Denmark Strait Overflow. Geophys Res Lett, 28, 1619–1622.
Hakkinen, S. and P. Rhines, 2004. Decline of the Subpolar North Atlantic Circulation During the 1990s. Science 304, 555
Hansen, B. and S. Østerhus, 2000. North Atlantic-Nordic Seas exchanges. Prog Oceanogr, 45, 109–208.
Hansen, B., W.R. Turrell, and S. Østerhus, 2001. Decreasing overflow from the Nordic seas into the Atlantic Ocean through the Faroe-Shetland Channel since 1950. Nature, 411, 927–930.
Hátún, H., A.B. Sandø, H. Drange, B. Hansen, and H. Valdimarsson, 2005. Influence of the Atlantic Subpolar Gyre on the Thermohaline Circulation. Science, 309, 1841–1844.
Holfort, J. and T. Albrecht, 2007. Atmospheric forcing of DSOW salinity. Ocean Sci, 3, 411–416.
Hurrell, J.W. 1995. Decadal trends in the North Atlantic Oscillation: regional temperatures and precipitation. Science, 269, 676–679.
Jónsson, S. and H. Valdimarsson, 2004. A new path for the Denmark Strait overflow water from the Iceland Sea to Denmark Strait. Geophys Res Lett 31(3), L03305, doi: 10.1029/2003GL019214.
Karcher, M.J., R. Gerdes, F. Kauker, and C. Köberle, 2003. Arctic warming: Evolution and spreading of the 1990s warm event in the Nordic seas and the Arctic Ocean. J Geophys Res, 108(C2), 3034
Kieke, D., M. Rhein, L. Stramma, W.M. Smethie, J.L. Bullister, and D. LeBel, 2007. Changes in the pool of Labrador Sea Water in the subpolar North Atlantic related to changes in the upper North Atlantic circulation. GRL, accepted.
Köhl, A., R. Käse, D. Stammer, and N. Sera, 2007. Causes of change in the Denmark Strait overflow. J Phys Oceanogr, 37(6), 1678–1696.
Lazier, J.R.N. 1995. The salinity decrease in the Labrador Sea over the past thirty years. In: D.G. Martinson, K. Bryan, M. Ghil, M.M. Hall, T.M. Karl, E.S. Sarachik, S. Sorooshian, and L.D. Talley (Eds.), *Natural Climate Variability on Decade-to-Century Time Scales* (pp. 295–304). Washington, DC: National Academy Press.
Macrander, A., U. Send, H. Valdimarsson, S. Jonsson, and R.H. Käse, 2005. Interannual changes in the overflow from the Nordic Seas into the Atlantic Ocean through Denmark Strait. Geophys Res Lett, 32, L06606, doi:10.1029/2004GL021463.
Pickart, R.S., M.A. Spall, M.H. Ribergaard, G.W.K. Moore, and R.F. Millif, 2003. Deep convection in the Irminger Sea forced by the Greenland tip jet. Nature, 424, 152–156.
Rudels, B., E. Fahrbach, J. Meincke, G. Budéus, and P. Eriksson, 2002. The East Greenland Current and its contribution to the Denmark Strait overflow. ICES J Mar Sci 59(6), 1133–1154.
Rudels, B., P. Eriksson, E. Buch, G. Budéus, E. Fahrbach, S.-A. Malmberg, J. Meincke, and P. Mälkki, 2003. Temporal switching between sources of the Denmark Strait overflow water. ICES Mar Sci Symp, 219, 319–325.
Rudels, B., G. Björk, J. Nilsson, P. Winsor, I. Lake, and C. Nohr, 2005. The interaction between waters from the Arctic Ocean and the Nordic Seas north of Fram Strait and along the East Greenland Current: results from the Arctic Ocean-02 Oden expedition. J Mar Sys, 55(1–2), 1–30.

Spall, M.A. and J.F. Price, 1998. Mesoscale variability in Denmark Strait: the PV outflow hypothesis. J Phys Oceanogr, 28, 1598–1623.

Sy, A., M. Rhein, J. Lazier, P. Koltermann, J. Meincke, A. Putzka, and M. Bersch, 1997. Surprisingly rapid spreading of newly formed intermediate waters across the North Atlantic Ocean. Nature, 386, 675–679.

Talley, L.D. and M.S. McCartney, 1982. Distribution and circulation of Labrador Sea Water. J Phys Oceanogr, 12, 1189–1205.

Turrell, W.R., G. Slesser, R.D. Adams, R. Payne, and P.A. Gillibrand, 1999. Decadal variability in the composition of Faroe-Shetland Channel bottom water. Deep-Sea Res I, 46, 1–25.

Wu, P. and R. Wood, 2006. Convection-induced long term freshening of the Subpolar North Atlantic Ocean. J Climate (submitted).

Yashayaev, I. and A. Clarke, 2005. Recent warming of the Labrador Sea. ASOF News Lett, 4, 17–18.

Yashayaev, I. 2007. Hydrographic changes in the Labrador Sea, 1960–2005. Prog Oceanogr, 73(3–4), 242–276.

Yashayaev, I., J.R.N. Lazier, and A. Clarke, 2003. Temperature and salinity in the Central Labrador Sea. ICES Mar Sci Symp Ser, 219, 32–39.

Yashayaev, I., M. Bersch, H. van Aken, and A. Clarke, 2004. A new study of the Production, Spreading and Fate of the Labrador Sea Water in the Subpolar North Atlantic. ASOF News Lett, 4, 20–22.

Chapter 22
Modelling the Overflows Across the Greenland–Scotland Ridge

Johann H. Jungclaus[1], Andreas Macrander[2], and Rolf H. Käse[3,4]

22.1 Introduction

The Atlantic Meridional Overturning Circulation (AMOC) is part of a global redistribution system in the ocean that carries vast amounts of mass, heat, and freshwater. Within the AMOC, water mass transformations in the Nordic Seas (NS) and the overflows across the Greenland–Scotland Ridge (GSR) contribute significantly to the overturning mass transport. The deep NS are separated by the GSR from direct exchange with the subpolar North Atlantic. Two deeper passages, Denmark Strait (DS, sill depth 630 m) and Faroe Bank Channel (FBC, sill depth 840 m), constrain the deep outflow. The outflow transports are assumed to be governed by hydraulic control (Whitehead 1989, 1998). According to the circulation scheme by Dickson and Brown (1994), there is an overflow of 2.9 Sv (1 Sv = 1 Sverdrup = 10^6 m^3 s^{-1}) through DS, 1.7 Sv through FBC and another 1 Sv from flow across the Iceland–Faroe Ridge (IFR). To the south of the GSR, the overflows sink to depth and then spread along the topography, eventually merging to form a deep boundary current in the western Irminger Sea. During the descent, the dense bottom water flow doubles its volume by entrainment of ambient waters (e.g. Price and Baringer 1994) so that there is a deep water transport of 13.3 Sv once the boundary current reaches Cape Farvel (Dickson and Brown 1994). Thus the overflows and the overflow-related part of the AMOC account for more than 70% of the maximum total overturning, which is estimated from observations to be about 18 Sv (e.g. Macdonald 1998).

Climate model studies (e.g. Schmittner et al. 2005) indicate a considerable decrease of the AMOC and a reduction of the heat transport under global warming

[1]Max Planck Institute for Meteorology, Bundesstrasse 53, 20146 Hamburg, Germany, e-mail: johann.jungclaus@zmaw.de

[2]Alfred Wegener Institute for Polar and Marine Research, Bussestrasse 24, 27570 Bremerhaven, Germany

[3]Institut für Meereskunde, Universität Hamburg, Bundesstrasse 53, 20146 Hamburg, Germany

[4]IfM-Geomar, Leibniz Institute for Marine Research, Düsternbrooker Weg 20, 24105 Kiel, Germany

R.R. Dickson et al. (eds.), *Arctic–Subarctic Ocean Fluxes*, 527–549

conditions, mainly due to increasing static stability of the upper ocean in high latitudes. Given the role of the overflow-related contribution to the present-day AMOC, it is of importance to elucidate the underlying processes in the present and future climate, and their proper representation in numerical models. This holds in particular for large-scale, coarse-resolution climate models. Summarizing the state-of-the-art at the time of the release of the International Panel of Climate Change (IPCC) third assessment report, Stocker et al. (2001) conclude that the "uncertainties in the representation of the flow across the GSR limit the ability of models to simulate situations that involve a rapid change in the THC". In relatively coarse resolution ocean general circulation models (OGCMs), the overflows across the GSR and the AMOC have been found to be sensitive to the geometry of the throughflow channels. At grid sizes of the order of 100 km, changes in the representation of the topography may result in considerable changes of the overflow transports and its water mass composition. For example, Roberts and Wood (1997) found a 50% change in heat flux at the GSR latitude when a single grid box was modified. Moreover, some OGCMs feature quite unrealistic depth and width of the channels (e.g. Schweckendiek and Willebrand 2005).

The throughflow transports are determined by hydraulic control and it is not clear which numerical resolution is required to properly simulate the partly non-linear processes involved. If the grid size is too large and the simulated velocities and Froude numbers are too small, the information exchange between the dense layers north and south of the GSR, which under appropriate physics is only in one direction, occurs in both directions.

The fate of the overflow waters once they have passed the sill depends on the amount of mixing they experience downstream. Observations indicate that there is an entrainment of the order of 100% of the original volume transport and that the mixing often occurs over steep slopes on very small horizontal distances (e.g. Baringer and Price 1997). In recent years, several attempts have been made (see Section 3.1) to parameterize the downslope flow and the entrainment in GCMs.

Here, we review first the development in theory and process modelling touching on hydraulics and dense gravity currents. Then, as a major part of the paper, we describe the representation of the overflow in high-resolution regional models and in a coarse-resolution state-of-the-art climate model. The sensitivity of the overflow to climate variations on interannual to decadal variations and to anthropogenic global warming is discussed.

22.2 Hydraulics of Sill and Strait Flow

Analytical theories of flows through passages and across sills (Gill 1977; Whitehead 1989; Killworth and McDonald 1993; Whitehead 1998) are important concepts to investigate transport limits of the outflow. Considerable progress has been made in the last decade to understand the nature of overflows in theory and by analysis of new observations. We mention here the special volume on "The Physical Oceanography of Sea Straits" (Pratt and Smeed 2004) that addresses topics, such as hydraulic control, mixing and friction as well as far-field effects and time dependence. In view of the importance of overflows for the stability of the AMOC, increasing interest was laid on the applicability of a 'weir' formula that would allow one to characterize the strength of the outflow by a relationship between the total outflow Q and a 'proxy' variable that could be measured without much instrumental efforts. The rationale for this approach is based on the fact that in hydraulically controlled flows, velocity and interface height are not independent, so that the determination of one would allow the calculation of the other. Unfortunately, analytical solutions are restricted to cases with simple bathymetry-like rectangular or parabolic cross-sections and certain assumptions on the upstream potential vorticity. An often-used prediction for a rotating rectangular channel (Whitehead 1989) is

$$Q = g'\frac{h_u^2}{2f}; \quad L > \sqrt{\left(2g'\frac{h_u}{f^2}\right)}; \tag{22.1}$$

and

$$Q = \left(\frac{2}{3}\right)^{\frac{3}{2}} L\sqrt{(g')}\left[h_u - \frac{f^2L^2}{8g'}\right]^{\left(\frac{3}{2}\right)}; \quad L \le \sqrt{\left(2g'\frac{h_u}{f^2}\right)} \tag{22.2}$$

with Coriolis parameter f, width of the channel L, the reduced gravity based on the density difference between an upstream and downstream reference profile g', and the height of the bifurcation depth h_u (Fig. 22.1). Whitehead (1998) investigated a large number of ocean gateways and found that his hydraulic predictions ranged from a factor 1 to 2.7 against volume flux estimates. For the passages of interest here, the Denmark Strait and the Faroe Bank Channel, the factor was 1.3 and 1.6, respectively.

Nikolopoulos et al. (2003) showed that the maximum volume flux for Denmark Strait was in much better agreement with observations if a more realistic geometry was included. A parabolic profile is also suggested to be a better approximation for the Faroe Bank Channel (Borenäs and Lundberg 1988), which results in an estimate for the volume flux

$$Q = \frac{h_u^2}{2+r}\sqrt{\left(\frac{3f^2}{2r}\right)}; \quad r = a\frac{f^2}{g'}$$

with the coefficient a of the parabolic depth profile.

Helfrich and Pratt (2003) revisited Gill's (1977) theory for rotating hydraulic sill flow, which assumes semigeostrophic flow of uniform potential vorticity through a

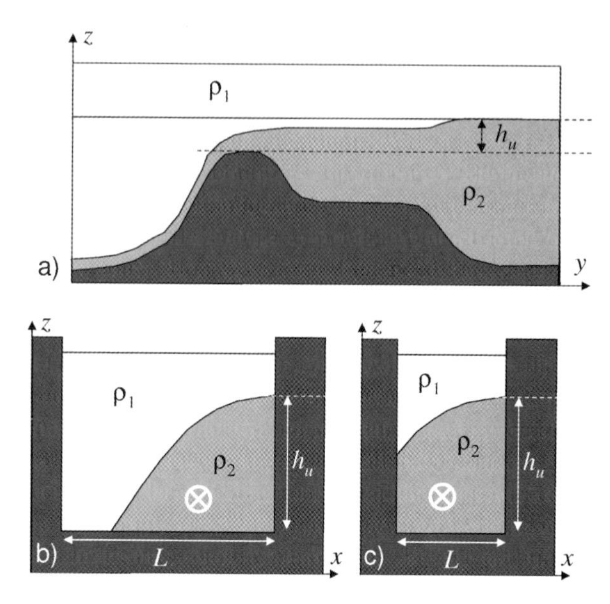

Fig. 22.1 Schematic of the hydraulic flow over a sill: along-channel view (a), and cross-channel view for (b) wide (Eq. 22.1) and (c) narrow (Eq. 22.2) upstream channel configuration, respectively

rectangular channel, and compared it with results from a reduced-gravity shallow water model for a combined basin and sill configuration. The coupled basin-strait system was shown to "select" an average overflow potential vorticity corresponding to Gill's solution. Another interesting finding is that robust estimates of the sill transports from upstream conditions can only be obtained if these conditions are taken from the strait entrance, not from the basin interior.

22.3 Modelling the Outflow Plume Downstream of the Sill

The numerical modelling of gravity currents was initiated by the stream-tube model of Smith (1975). This model, as well as refined versions, which take into account, for example, variable topography (Price and Baringer 1994), considers a stationary, laterally integrated stream-tube with variable cross-sectional area. Several applications showed that such a model is able to reproduce the pathway and the along-pathway evolution of water mass properties (e.g. Baringer and Price 1997). Girton and Sanford (2003) used observed quantities and were able to quantify the balance of terms that govern the descent of the Denmark Strait overflow plume. They found a balance between the terms describing loss of potential energy and bottom friction. It is interesting to note that no entrainment stress at the interface was taken into account. In view of the strong mixing due to the presence of eddies (Bruce 1995), the expected descent pathway agrees surprisingly well with the observations (see Fig. 2 of Girton and Sanford 2003). This rate differs markedly from other models such as that of Killworth (2001) who claims a constant descent rate of 1/400.

The stream-tube model does not give any information about the spatial structure of the plume. Therefore, Jungclaus and Backhaus (1994) developed a two-dimensional transient reduced-gravity model that resolves the horizontal structure of the plume and is able to simulate the splitting and merging of the plume. Substantial progress in the understanding of the dynamics of the overflow plume was achieved after including the vertical dimension in high-resolution process studies with three-dimensional general circulation models (GCM). Using a sigma-coordinate model, Jiang and Garwood (1996) were able to simulate the three-dimensional evolution of an overflow plume on an idealized continental slope. They showed that the plume breaks apart into a chain of eddies and that these eddies exhibit a pronounced surface expression, so that the overflow may be tracked by satellite imagery. The destabilization of the gravity current has been further investigated in the numerical studies of Jungclaus et al. (2001), Shi et al. (2001), and Ezer (2006). Using the MIT z-coordinate GCM Riemenschneider and Legg (2007) have investigated the dependence of the representation of the FBC overflow on grid resolution and mixing parameterisations. They show that the structure and properties of the overflow plume is comparable with observations at the highest resolution of 2 km where the Rossby radius is clearly resolved and the ratio between horizontal and vertical resolution allows for an adequate representation of the fluxes (see below). The numerical mixing in the model is found to be most sensitive to changes in horizontal resolution, and to a lesser extent on vertical resolution and vertical viscosity.

22.4 Parameterization of Dense Outflow and Bottom Boundary Layer Processes in Large-Scale Ocean Models

The ultimate goal of these idealized model exercises is to use the knowledge obtained from process studies to develop improved parameterizations for large-scale models. The model intercomparison study carried out in the Dynamics of North Atlantic Models (DYNAMO) project (Willebrand et al. 2001) showed that the proper representation of the mixing depends much on the choice of the vertical coordinate. Most climate OGCMs are formulated on depth coordinates and obtain far too much mixing for dense overflows at a sill. Due to the staircase-like representation of the bottom topography, downslope flow is simulated as a succession of horizontal advection and a (convective) overturning. Models with terrain-following vertical (sigma or s) coordinates perform better but often require a substantial smoothing of the topography in order to avoid spurious effects in the horizontal pressure gradient formulation. Isopycnic models have been shown to reproduce realistic dense outflow plumes once diapycnic mixing is properly parameterized (Hallberg 2000; Xu et al. 2006).

The correct simulation of the overflow pathways downstream of the sill and the entrainment of ambient waters into the plume is therefore challenging for coarse resolution *z*-coordinate models. Winton et al. (1998) estimated that a horizontal resolution of $\Delta x = \Delta z/\alpha$ is required, where α is the bottom slope, and that the bottom boundary layer should be resolved by several points.

In coarse-resolution models, more realistic down slope flow and water mass transformation have been achieved by parameterizing the turbulent bottom boundary layer (BBL). There are two different approaches to achieve this goal. One relatively simple 'plumbing' approach is to reduce the spurious mixing by connecting grid cells above and below a topographic step (Beckmann and Döscher 1997; Campin and Goosse 1999; Marsland et al. 2003). In the latter scheme for example (see Marsland et al. 2003, their Fig. 1), the dense water flow is redirected into a deeper level (but not necessarily the bottom level, as in Beckmann and Döscher 1997) of an adjacent grid. The target depth is determined by the stratification of the receiving grid cell, similar to the approach by Campin and Goosse (1999). On the other hand, Killworth and Edwards (1999) and Song and Chao (2000) couple a full two-dimensional bottom boundary layer model to a GCM. This allows for a more physically based approach to determine both detrainment and entrainment. For example, Killworth and Edwards (1999) employ a frictional bottom boundary layer to determine the mixing rates. Whereas the 'plumbing' schemes are used in a number of large-scale models, the BBL sub-modules have not found their way into climate GCMs, to our knowledge. Currently, there is a coordinated project founded by the US National Science Foundation to evaluate and improve overflow parameterisations. The Climate Process Team on Gravity Currents and Entrainment (http://cpt-gce.org/) aims to combine observational programmes and numerical modelling reaching from extremely high-resolution non-hydrostatic models (Özgökmem et al. 2006) to GCM analyses (Legg et al. 2006).

Another approach to parameterize the overflow in models with low resolution is to determine the transport directly from hydraulic relations (Kösters et al. 2005). Their application of a hydraulic transport parameterization to a coarse resolution ocean model gave a considerable increase in both the AMOC and the meridional heat transport that resulted in 1 K warmer air temperatures over Europe compared to the standard model set-up.

22.5 Regional Models of the Exchange Flow Across the GSR

High-resolution limited-area models with realistic bottom topography allow one to compare the results with theoretical estimates, such as upper bounds on hydraulic transports. In a numerical process model, Käse and Oschlies (2000) calculated the exchange between two basins that are connected by a strait with bathymetry mimicking the Denmark Strait. The application of a bottom-following vertical coordinate and a horizontal resolution of about 4 km allowed for a quite realistic representation of bottom boundary layer processes and eddies. Käse and Oschlies filled the upstream basin from the northern boundary, while permitting the lighter water to leave via the northern boundary to obtain a net zero mass flux. Käse and Oschlies (2000) found that the transport is topographically controlled and the predictions of Whitehead (1998) and Killworth and McDonald (1993) are consistent with the model results. The dense sill flow increased according to the increasing northern basin interface height, which became stationary when the prescribed inflow equaled about 60% of the Whitehead (Eq. 22.1) maximum flux. Stern (2004) offered an alternative explanation for this factor independent of hydraulic arguments. He investigated the maximum geostrophic flow through a parabolic bottom shape and found a theoretical factor of 9/16 of the Whitehead value. It has to be noted that the height scale h_u in Stern's formulation is taken at the right hand side wall of the parabolic channel.

A recent re-assessment of the Käse and Oschlies (2000) simulations revealed that the plume thickness and the sea surface elevation (SSH) are highly correlated. Figure 22.2 shows the correlation between these two quantities near the sill. The narrow flow path is related to a depression in SSH. The overflow path and its variability might therefore be observed by altimetry. This aspect was investigated in more detail by Köhl et al. (2007). They use a high-resolution (0.1°) regional model including the entire GSR that is embedded in the 1° global ECCO model (Stammer et al. 2002) and that is forced with NCEP reanalysis data for the 1990s. They found high correlations between interface height and SSH in the strait and around Iceland (their Fig. 10, and see below) and were able to reconstruct (simulated) overflow transports by regression to SSH. In general, the theories of Whitehead (1989) and Gill (1977) that describe the steady maximum transport over a sill were found to be

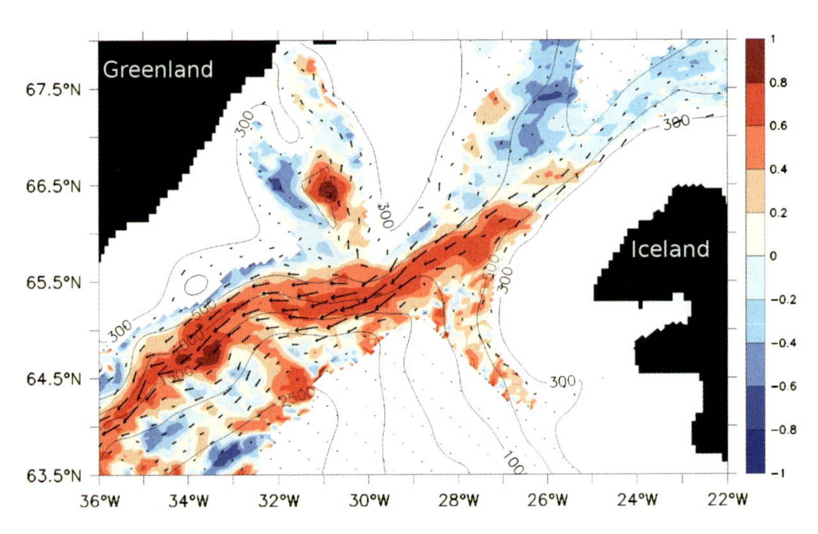

Fig. 22.2 Point-to-point correlations between the interface depth and the SSH in the Käse and Oschlies (2000) model. Note that positive correlations indicate negative SSH anomalies overlying positive thickness (negative interface depth) anomalies

consistent with the simulated mean transport and even with its variations as long as the timescale is much longer than the synoptic scale. Thus transport variations can be inferred from measuring the upstream interface height in the vicinity of the strait entrance, and possibly, altimetry. An attempt to apply the reconstruction of overflow transport by regression with SSH from observed data revealed unforeseen difficulties and Köhl et al. (2007) concluded that very high accuracy in the observation is required.

22.6 The Representation of the Overflows in Large-Scale Climate Models

22.6.1 The Mean State: Transports Across the GSR and Water Mass Properties

The simulation of the NS circulation with state-of-the-art ocean models driven by prescribed atmospheric fluxes obtained from reanalysis has recently been reviewed by Drange et al. (2005) and applications to the exchanges between the Nordic Seas and the North Atlantic have been published by Haak et al. (2003), Nilsen et al. (2003), and Zhang et al. (2004). Even though the simulation of the overflows was not the main focus of these papers, the studies showed that present-day OGCM can, given the correct forcing and its variability, quite realistically reproduce the observed exchange processes across the GSR.

Simulating these processes with a coupled atmosphere–ocean model is even more challenging. In climate modelling, one major breakthrough over the last few years has been that most models no longer need flux adjustments to maintain a stable climate. The reason for that improvement may be sought as well in the atmospheric or oceanic GCMs. A consequence is, however, that the (although stable) mean state of the surface climate may differ substantially from observations (e.g. Jungclaus et al. 2006a). As an example of a typical state-of-the-art model, we present results from the Max Planck Institute for Meteorology (MPI-M) climate model that participated in the IPCC Fourth Assessment Report (AR4). The simulations were conducted with the coupled atmosphere–ocean–sea ice general circulation model ECHAM5/MPIOM. The atmospheric part of the model, ECHAM5 (Roeckner et al. 2003), has a horizontal resolution of 1.875° by 1.875° (T63) and 31 vertical levels. The ocean/sea ice part of the model, MPIOM (Marsland et al. 2003), has a 1.5° by 1.5° horizontal resolution on a curvilinear grid with 40 vertical levels. MPIOM uses a curvilinear orthogonal grid and the two grid poles are placed upon Antarctica and Greenland thus avoiding the pole-singularity problem at the North Pole and providing relatively high resolution in the deep-water formation regions of the Labrador Sea and the Greenland Sea (see Jungclaus et al. 2006a, their Fig. 1). Denmark Strait (DS) is located close to the grid pole over Greenland and is resolved by several grid points at 15–25 km resolution (Fig. 22.3). DS sill depth is 600 m. Following the

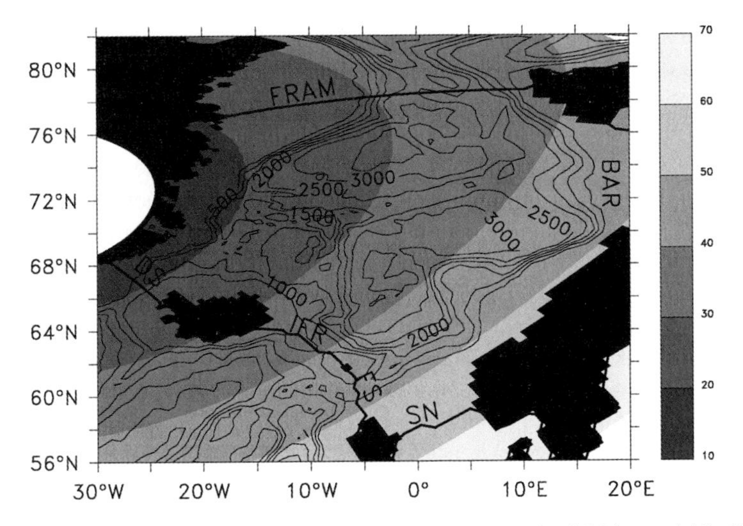

Fig. 22.3 Bottom topography (contours) of the Greenland–Scotland Ridge and Nordic Seas region in the global ocean model MPIOM. Gray shadings (units = km) show the numerical grid size and solid black lines indicate the sections across Denmark Strait (DS), along the Iceland–Faroe Ridge (IFR), the Faroe–Scotland section (FS), the Scotland–Norway section (SN), the Barents Sea opening (BAR), and the section across the Fram Strait (FRAM). Contour interval (ci) for topography = 500 m

Table 22.1 Overview over the exchanges across the GSR. Mean values and standard deviations of mass and heat transports

Section	Total (Sv)	Into NS (Sv)	Out NS (Sv)	$\sigma_\theta > 27.8$ (Sv)	$\sigma_\theta < 27.8$ (Sv)	HTR (TW, total)
DS	−4.7 ± 0.5	1.5 ± 0.16	−6.2 ± 0.6	−3.3 ± 0.5	−1.4 ± 0.3	−5 ± 8
IFR	4.5 ± 0.6	6.4 ± 0.6	−1.8 ± 0.2	0	4.5 ± 0.6	188 ± 25
FS	0.37 ± 0.6	3.9 ± 0.5	−3.5 ± 0.2	−3.0 ± 0.2	3.3 ± 0.6	123 ± 26
SN	0.2 ± 0.05	1.3 ± 0.1	−1.1 ± 0.1	0	0.2 ± 0.05	3 ± 2

Iceland–Scotland Ridge (ISR) to the east, the grid size widens and is about 50 km in the Faroe–Scotland section. The Faroe Bank Channel (FBC) is the deepest outlet with a depth of 890 m. The bottommost section is, however, just resolved by one velocity grid point. At the sill depths, the vertical resolution is 80 m at 600 m depth and 120 m at 900 m depth.

The results shown here stem from a 505-year long control integration under pre-industrial conditions for greenhouse gas concentration. It has been preceded by a multicentury spin-up run that was started from the PHC climatology (Steele et al. 2001). Various aspects of the mean state and internal variability in the ECHAM5/MPI-OM IPCC set-up have been discussed by Jungclaus et al. (2006a), Müller and Roeckner (2006), and Bengtson et al. (2006).

We begin the discussion of the exchanges across the GSR by an account of the mean fluxes in Table 22.1. The $\sigma_\theta = 27.8\,\mathrm{kg\ m^{-3}}$ interface is traditionally taken as the boundary between the warm inflow and the cold outflow. In the model, there seems to be some recirculation of AW across the IFR. In total, IFR and FS contribute 7.8 Sv (4.5 Sv and 3.3 Sv, respectively) to the warm inflow into the Nordic Seas. Direct measurements by Østerhus et al. (2005) gave 3.8 Sv for each section and an additional 0.8 Sv inflow through DS, where the model simulates 1.4 Sv. The deep outflow in DS of 3.3 Sv is somewhat higher than the often-quoted 2.9 Sv of Ross (1984) and of Dickson and Brown (1994) but compares well with the most recent measurements of 3–3.5 Sv by Macrander et al. (2005). These numbers depend, however, to some extend on the exact location of the section. There is an outflow of 3 Sv dense ($\sigma_\theta > 27.8$) water through FS but profiles of the flows averaged across the sections (c.f. Fig. 22.6c) indicate that the 27.8 criterion is too restrictive for the FS outflow. Hence, FS outflow is taken to be all outflow with density $\sigma_\theta > 27.6$, which increases the mean FS outflow transport to 3.6 Sv. This number is considerably higher than the observed transport estimates of 2 Sv (Saunders 1990; Borenäs and Lundberg 2004). However, since in the model there is no overflow across the IFR (assumed from observations to account for roughly 1 Sv (Dickson and Brown 1994)), the 3.6 Sv represent the total dense outflow through the Iceland–Scotland section. Most likely, the model IFR topography does not resolve small overflow channels and the dense water there is guided by the ridge towards the FS section and escapes through the FBC. In total, the model simulates a mass transport overturning of more than 6 Sv across the GSR. Since no deep water is flowing through the Canadian Archipelago this overturning reflects the water mass transformation in the NS and in the Arctic (not taking any contributions from the Bering Strait into account). From budget considerations the outflow of water $\sigma_\theta > 27.8$ must be balanced by production (flux through the $\sigma_\theta = 27.8$ interface) and the time derivative of the interface height (reservoir changes). Many previous coarse resolution climate models 'closed' the overturning cell with deep convection to the south of the GSR and did not include a proper contribution from the overflows and from water mass transformation to the north of the GSR. In the MPI-M model, 30% of the entire Atlantic overturning of about 22 Sv comes directly from the Nordic Seas. Together with the roughly 100% entrainment that occurs to the south of the sills, the northern branch forms the backbone of the overturning and plays therefore a more pronounced role compared to earlier climate models. Heat transports across the sections add up to 0.31 PW. Decomposition of the heat transport across the entire GSR into depth independent (gyre) and overturning contribution (e.g. Gulev et al. 2003) shows that the heat transport is dominated by the gyre (0.21 PW), but variations in gyre and overturning heat transports are of similar magnitude (standard deviations are about 0.015 PW)). The total heat transport is quite similar to those deduced from observations (e.g. 0.275 PW by Blindheim and Østerhus (2005), and 0.313 PW (Østerhus et al. 2005)).

The model simulated near-surface circulation (not shown) in the transition region between the Atlantic and the Nordic Seas (NS) shows the well-known features of the East Greenland Current (EGC), carrying cold and fresh Arctic water masses to the south, and the three branches of the Atlantic inflow, the Irminger Current to the west of Iceland, the Faroe Current that crosses the GSR near

Iceland, and the slope current along the northwestern European continental slope. To the north of the IFR, the currents move eastward and the North Irminger Current (NIC) and Faroe Current extensions join the slope current to form the Norwegian Atlantic Current. The simulated circulation broadly agrees with available observations (e.g. Jakobsen et al. 2003).

The potential density isosurface $\sigma_\theta = 27.8$ is often used as the boundary between the upper layer and the outflow. In the simulation, the maximum outflow densities exceed $\sigma_\theta = 27.9$ in DS and occasionally 28.0 in the FBC. The mean topography of the $\sigma_\theta = 27.8$ surface is depicted in Fig. 22.4. The interface is deep in the east where Atlantic water masses dominate and shallow in the Greenland Sea and Iceland Sea. Outcropping regions exhibit strong temporal variability over the simulation (not shown). The volume of waters denser than $\sigma_\theta = 27.8$ is maintained by inflow through Fram Strait, convection, diapycnal mixing in the basin, and the outflow through the DS and FS. In the simulation, there is no direct outflow of this water mass across the IFR and only a small amount of water with $\sigma_\theta > 27.7$ crosses the sill from north to south. In the interior NS, the interface height is several hundred meters higher than the sill depths and it is the potential energy stored here that drives the overflows. The transport integrated from the bottom to the $\sigma_\theta = 27.8$ surface gives some indications on the flow paths feeding the overflows (Fig. 22.4). There is inflow of dense waters from Fram Strait, and the cyclonic basin circulation features a pronounced boundary current below the EGC. Approaching DS, the flow splits up into two paths, one that continues on the East Greenland continental shelf and another branch that approaches DS from the northeast. As has been shown by Käse and Oschlies (2000) and Helfrich and Pratt (2003), the eastern branch is

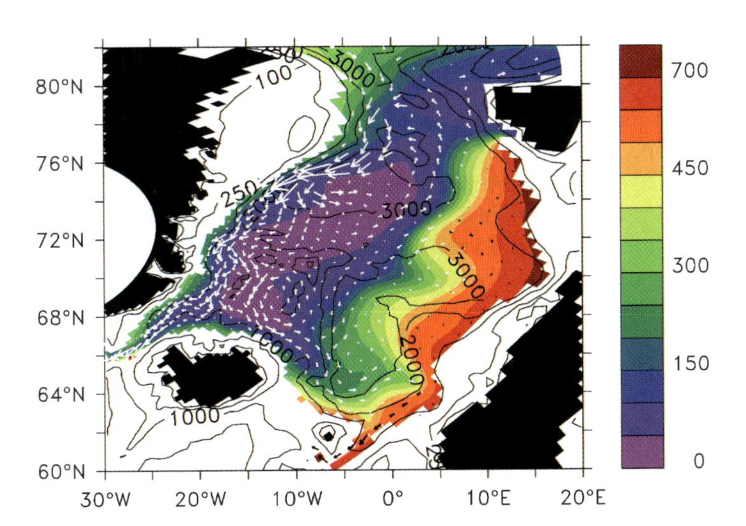

Fig. 22.4 Simulated mean depth (m) of the $\sigma = 27.8\,kg\,m^{-3}$ isosurface (shading) together with bottom topography (contours) and transports ($m^2\,s^{-1}$, every other vector shown) integrated from the bottom to the 27.8 surface. For interface depths below 500 m vectors are plotted in black

a consequence of potential vorticity conservation for a flow approaching a sill. This could be a dynamical explanation for the eastern flow path that has recently been described from observations (Jónsson and Valdimarsson 2004). There is eastward flow along the northern slope of the ISR and the FSC/FBC outflow is fed, in part, by these waters.

Sections through the outflow channels DS and FS (Fig. 22.5a and c, respectively; for locations see Fig. 22.3) are given for the simulated along-channel velocities and the potential densities. For DS, the respective fields are also available from a vessel-mounted acoustic doupler current profiler (ADCP) and hydrographic observations (Fig. 22.5b). At the DS sill (Fig. 22.5a), in the upper part of the strait, the western side is dominated by the light and fresh Arctic waters from the EGC separated from the warm Atlantic waters of the NIC. The dense overflow water can be found below 200 m and its density exceeds 27.9 below 400 m. Observations (Fig. 22.5b) indicate densities above 27.8 occupying a large part of the western slope. The thickness of the overflow layer (water denser than 27.8) is roughly 300 m in both observations and simulations. The isopycnal is steeply inclined directly above the deepest part of the channel but the slope changes sign above the western slope. The along-channel velocity has a maximum roughly 100 m above the bottom but the coarse vertical resolution does not allow for a proper representation of a well-mixed boundary layer. The along-channel velocity maximum near the surface is located about 60 km west of the channel axis but the velocity structure is quite barotropic. In the observations there is indication of an eddy but Macrander (2004) confirms that the general flow patterns have been robust for a number of other cruises. A remarkable feature is that the flow is quite barotropic despite of the large density contrasts with the steeply inclined isopycnals. In contrast to DS, the simulated current structure across the Faroe–Shetland channel (Fig. 22.5c) shows the features of a two-layer exchange flow with a zero crossing of the velocity profile at mid-depth. Again, the coarse horizontal and vertical resolution does not allow the detailed simulation of, for example, the velocity profile (Hansen et al. 2001) or the relatively thick, well-mixed boundary layers that are associated with cross-slope Ekman fluxes (Jungclaus and Vanicek 1999). It should be noted, however, that even though the model is able to simulate the density structure quite realistically, it fails to reproduce the water mass properties. The overflow waters are up to 2 K too warm and 0.3 psu too salty (not shown).

As has been discussed in Section 22.4, a proper representation of the sinking and mixing in the overflow plume is quite challenging for GCMs. In MPIOM, a slope convection parameterization was implemented and Marsland et al. (2003) showed that the scheme improved the near bottom water mass characteristics to the south of the GSR even though a vertical grid resolution of 100–300 m at 2,000 m depth does not allow for a 'realistic' vertical structure of the plume. Furthermore, the horizontal resolution is far too coarse to allow horizontal eddies with a radius of about 50–100 km to develop (Jiang and Garwood 1996; Jungclaus et al. 2001). In the coupled simulation presented here, the overflows mix with ambient waters downstream of the sills, but can be traced as a dense bottom current that follows the topographic contours (Fig. 22.6). The Faroese outflow flows along the southern

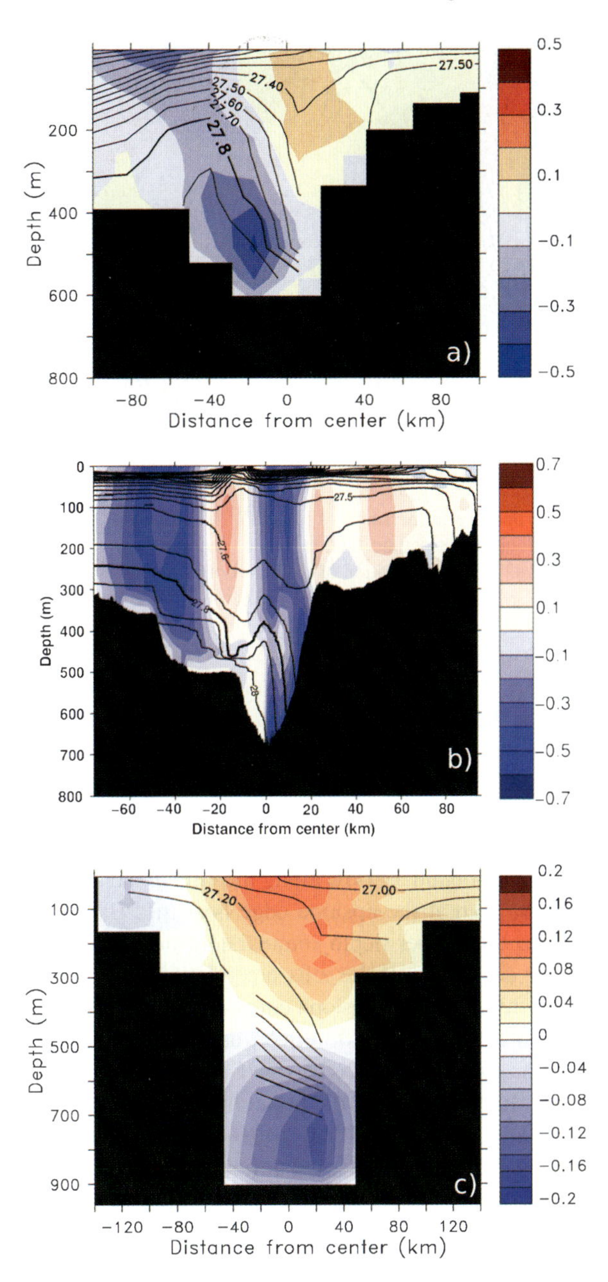

Fig. 22.5 Density (contours) and along-channel velocity (shading in m s^{-1}) sections through the DS and the FS: (**a**) simulated mean DS, (**b**) observed DS, and (**c**) simulated mean FS section. The observed along-channel velocities (**b**) were measured by vessel-mounted ADCP during Poseidon P262 cruise in July 2000 and potential density was derived from hydrographic CTD data. *X*-axis is in kilometers from west to east relative to the channel centre

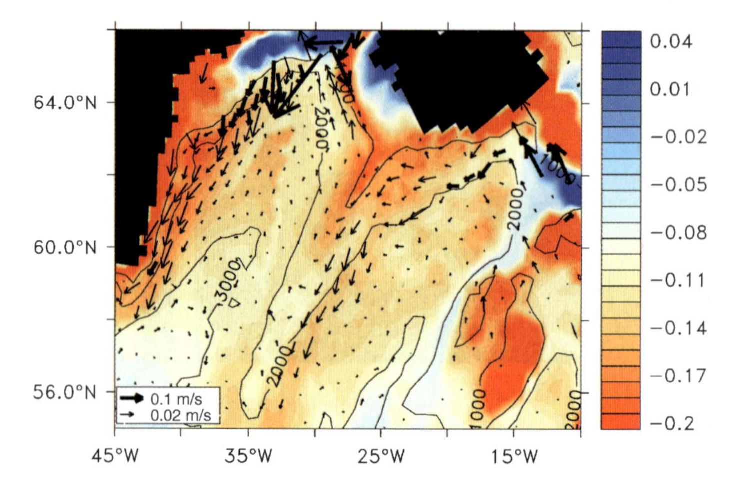

Fig. 22.6 Simulated properties to the south of the GSR: Mean near-bottom velocities (vectors). For clarity, current velocities exceeding 0.05 m s^{-1} are indicated by bold arrows with lengths downscaled by a factor of 3. Mean deviation of the near bottom density (kg m^{-3}) from the respective field from the PHC climatology shaded in the background

edge of the ISR and is then deviated southward by the Reykjanes Ridge. The bottom current occupies a depth range of 1,000–2,500 m. At various depths, the flow enters the Irminger Basin. However, the flow loses intensity and excess density by mixing and there is only weak northward flow along the western edge of the Reykjanes Ridge. The bottom density error is smallest near the exit of the FBC. The DS is characterized as a strong boundary flow along the Greenland continental slope. The bottom density is too high at the upslope edge near the strait, indicating that the flow doesn't sink to the proper depth immediately downstream of the strait. In general, the density deviations at the sills are relatively small but increase with distance, indicating that even with a slope convection parameterization the correct bottom density is hard to achieve. Moreover, as has been mentioned above, the simulated water mass properties in the Nordic Seas are off quite substantially due to too warm and salty Atlantic inflow but probably also due to insufficiencies in the parameterization of vertical mixing and convection. As a result, the overflow waters are also too warm and too salty, even though their density at the sill, owing to compensating effects of temperature and salt, compares relatively well with the observations. This will have effects on the density by ways of the nonlinear equation of state and the differential mixing of heat and salt. Errors in the water mass properties appear to be the major shortcomings in the present simulation. In comparison to earlier models using flux adjustment, the results are probably even worse because much of the (near surface) errors in SST, precipitation, etc. was masked by the corrections. A systematic review of this problem is beyond the scope of this paper and a thorough evaluation of several climate models using different parameterizations is necessary. Progress is also expected from detailed process studies, such as those currently carried out within the framework of the Climate Process Teams on gravity flows (see Section 22.4).

22.6.2 Variability of the GSR Exchange and Overflows in the Climate GCM

While the long-term overturning circulation across the GSR is determined by the production of dense water in the NS and in the Arctic, variations of the overflow transports are determined by changes in the wind-stress forcing and changes in water mass properties that determine the density contrast between the NS and the Atlantic. Annual mean time series (not shown) of the total (i.e. vertically integrated) transport anomalies from the east (IFR, FS, SN) and from the west (DS) of Iceland are almost perfectly anticorrelated ($r = -0.93$), indicating the importance of the wind-driven barotropic transports that has been described by Biastoch et al. (2003). Variations of the dense outflows through DS are very similar to the total DS transport changes and both time series are correlated with $r = 0.8$. In contrast, the dense FS outflow does not show that high correlation with the total flow, regardless of taking the boundary at 27.6 or 27.8. Therefore, the FS and DS overflow do not show the clear anticorrelated behaviour that was reported by Köhl et al. (2007).

For hydraulic control, Whitehead's formula (Eq. 22.1) predicts an upper limit of the overflow transports that is proportional to the density contrast and the square of the upstream reservoir height above the sill. In the presence of rotation and friction, smaller values of the transports are expected, but outflow variations will depend on reservoir height and density changes, where the former (due to the quadratic term) is probably more important, as discussed by Macrander et al. (2005). They were able to reconstruct the observed transports from variations of the 27.8 interface at the Kögur section. In the simulations, we find indeed high correlations ($r > 0.8$) between the transports and the interface variations near the entrance of the Denmark Strait (Fig. 22.7a). Correlations are also relatively high in the Iceland Sea and along the GSR. At lag 0 years, there are also regions with significantly high negative correlation (indicating relatively deep interface depth) in the Norwegian Sea, a finding that is further discussed below. Using (Eq. 22.1) and a density contrast of 0.39 kg m^{-3}, we are able to reconstruct (Fig. 22.8a) the transport changes from the variations of the $\sigma = 27.8$ interface depths taken from an average over 4 grid points entered near 67° N, 24° W (see Fig. 22.7a). For the FS outflow, we find significant correlations only for a few grid cells just upstream of the FS section (Fig. 22.7b). Using interface variations from that point, however, enables one to reconstruct also the FS outflow (Fig. 22.8b). Hansen et al. (2001) have tried to reproduce FBC outflow variation in the second half of the 20th century by relating them to the interface variations far upstream at weather ship Mike. Even though our results may be influenced by the relatively coarse resolution in FS, the findings support Helfrich and Pratt's (2003) conclusion that a strong relation between the outflow and the interface depth exists only close to the entrance of the outflow channel.

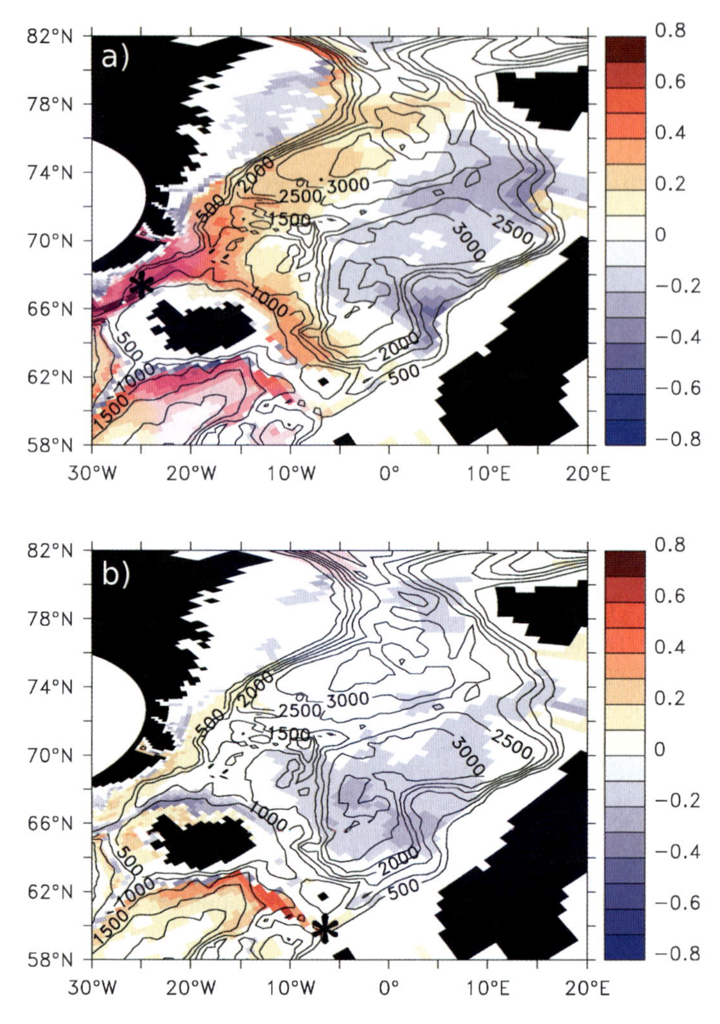

Fig. 22.7 Correlation coefficients between the depth of the 27.8 isopycnal and (**a**) the DS outflow, and (**b**) the FS outflow at zero lag. Positive correlations indicate negative depth (positive thickness) anomalies at enhanced (negative) outflow transports. A * symbol indicates the location where time series of the interface height were taken to reconstruct the respective outflow transports

Köhl et al. (2007) were able to reconstruct the modelled overflow transport from modelled SSH variations in DS. The correlation map from the 505-year annual mean time series from the coupled experiment (Fig. 22.9a) bears remarkable resemblance to the monthly data from Köhl et al.'s high-resolution regional model (Fig. 22.9b). Correlations are high around Iceland (indicating the enhancement of the circulation around the island). In the coupled climate model the correlations between the SSH and the DS overflow time series exceed 0.7 and allow for a reconstruction of the deep transports (Fig. 22.8a) from the regression, explaining more than 50% of the

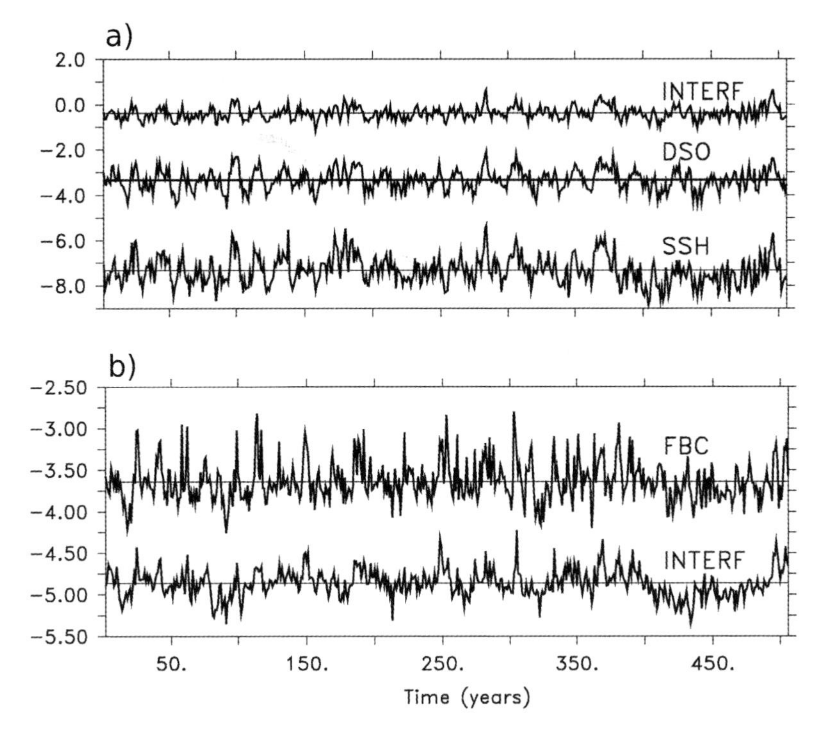

Fig. 22.8 Time series (annual means) of (**a**) simulated Denmark Strait overflow ($\sigma > 27.8$, DSO) and (**b**) Faroe Bank Channel overflow ($\sigma > 27.6$, FBC). Units are Sv, negative sign indicates flow out of the NS. Also included are reconstructed transports from the hydraulic relation (1) (INTERF) and, for DS also from the sea surface height (SSH) variations. Offsets for the DS reconstructions are +4 Sv and −4 Sv, respectively and for the FBC reconstruction −1.5 Sv. Correlations between the DSO and the SSH and the interface reconstructions are 0.73 and 0.8, respectively. Correlation between FBC and the interface reconstruction is 0.67

overflow's variance. For the FBC outflow, however, we do not find the expected relation of interface and surface elevation (i.e. a depression of SSH at times of strong outflow). Correlations are negative all over the northwest European Continental Slope (not shown). One might think that this is related to the coarse resolution of the throughflow channel in the coupled model. However, the respective data from the high-resolution model (kindly provided by A. Köhl) also gives only low correlation (albeit of various sign) over the FBC.

The total flow through DS is correlated (not shown) to the NAO with $r = -0.43$ and the total flow through the Iceland–Scotland section is similarly correlated with a reversed sign. A running correlation indicates that the relation between the NAO and the GSR exchange flows varies with time, indicating shifting atmospheric pressure patterns likely similar to the ones observed in the second half of the 20th century (Jung et al. 2003). The DS overflow is correlated to the NAO with $r = -0.32$ (correlations $r > 0.2$ are considered significant) at zero lag (Fig. 22.10). The lagged

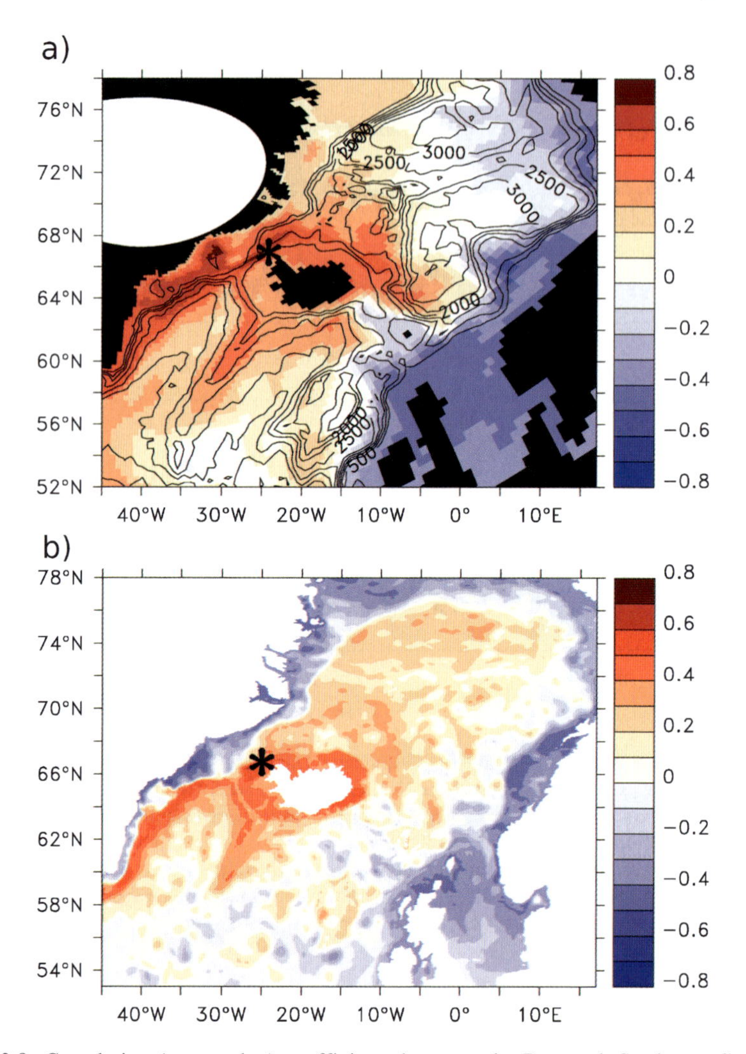

Fig. 22.9 Correlation (at zero lag) coefficients between the Denmark Strait overflow and the sea surface elevation (SSH) derived from (**a**) the coupled ECHAM5/MPIOM, and (**b**) from a high resolution regional model (redrawn after Köhl et al. 2007). The outflow has negative sign so that positive correlations mean depression of the sea surface. A * symbol indicates the location where SSH time series were taken to reconstruct the respective outflow transports

correlations show some asymmetry around zero, which might indicate a certain response to the (white noise) NAO forcing. The NAO may affect the overflows in several ways. First, there is the barotropic response to changes in the wind stress. Second, the heat flux pattern associated with a negative NAO index favours convection in the Greenland and Iceland Sea (Dickson et al. 2000) and the increase

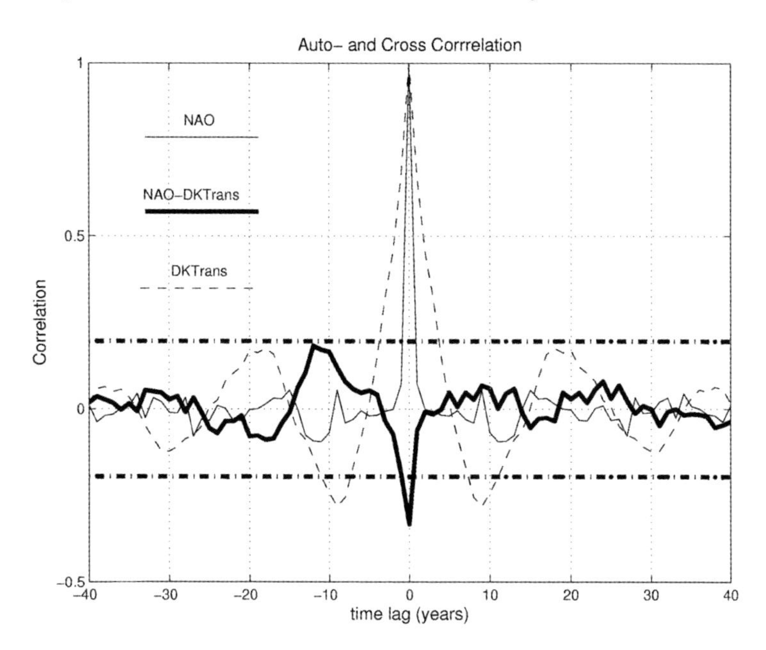

Fig. 22.10 (Lag) correlation coefficients between the annual mean NAO and (solid thick line) the dense overflow transport through DS. Note that overflows (outflows) are defined with a negative sign so that high positive NAO is associated with high overflow transports. Negative time lags indicate NAO leading. Thick dashed lines indicate the 99% confidence interval. Also included are the autocorrelation functions for the DS overflow (thin dashed line) and for the NAO (thin solid line)

in reservoir height would lead to more outflow. Third, the inflow of Atlantic waters also depends on the NAO. Given that the overflows consist of more or less of Atlantic Water, the changing conditions at the entrance may be traced, with a time lag of a few years, to the overflows (Dickson et al. 1999). While the first mechanism is fast and barotropic, the latter two will require some time lag for the anomalies to reach DS and may integrate the high-frequency forcing variability in time. Spectra of the overflow time series (not shown) indicate elevated energy in the interdecadal band with a peak at 20 years. A detailed investigation of the multidecadal variability is beyond the scope of this paper. One explanation, as to how a white-noise (NAO) forcing can generate a red-noise response has been given by Käse (2006): A controlled volume box model, where accumulation in volume is driven by net imbalances between prescribed inflow, outflow, and NAO-derived flux through the interface, results in a Riccati equation for filling and flushing. For small interface fluctuations with white-noise forcing, the overflow spectrum is red-noise with a timescale between 5 and 15 years. The proposed mechanism is an effective low-pass filter for higher frequency variations and the long-term changes in the DS transports are a reflection of coupled ocean–atmosphere interactions. There are several limitations in the simple model but comparisons to observations indicate that such a concept can be useful when a strongly limiting process such as hydraulic control dominates the outflow. However, further investigations using long integrations with models resolving the important processes are necessary to better understand the mechanisms behind the low-frequency variations in the overflows.

22.7 Summary and Conclusion

Various aspects of the representation of the overflows in state-of-the-art models have been reviewed. At high-enough resolution, numerical models are able to reproduce the structure and dynamics of the overflows. The representation of the overflows in coarse-resolution ocean models has been improved but is far from being perfect. In the ECHAM5/MIOM simulation analysed here, main shortcomings can be seen in the water mass properties of the overflows and the downslope evolution of the plume south of the GSR. On the other hand, the model reproduces the current structure and the hydraulic character of the overflows remarkably well. The model results confirm previous findings that there is a strong coupling between the dense overflow in the DS and the SSH variations above. This suggests that DS transport variations can be monitored by satellite altimetry. However, a first attempt by Köhl et al. (2007) revealed unforeseen difficulties. In order to be able to monitor small long-term transport changes from altimeter, very high accuracy is required.

The coupled ocean–atmosphere model is able to reproduce a realistic 6 Sv of overturning across the GSR, whereas many previous climate models closed their overturning cell by convection to the south of the GSR. This has consequences for the stability of the THC in climate change simulations, which, for the IPCC AR4, are presently being assessed. Results from the MPI-M model and from the NCAR model (Hu et al. 2004; Jungclaus et al. 2006b) show that there is a considerable decrease, but no breakdown of the THC in the greenhouse gas induced warmer climate. In contrast to the maximum of the overturning streamfunction, however, the overflow transports increase slightly in both models even though the water mass properties to the north and to the south of the GSR change dramatically. Hence, the reduction of the overturning takes place only to the south of the GSR, where open ocean convection in the Labrador Sea forms a more direct and apparently more vulnerable component of the THC. In contrast, the hydraulic system of the overflows keeps working and this stabilizes the THC. According to these studies, the overflows across the GSR provide the backbone of a substantial overturning circulation even in a warmer climate.

Acknowledgements Funding for this study was (in part) provided by the Deutsche Forschungsgemeinschaft Special Research Grants (SFB 460: IfM-Geomar Kiel and SFB 512: IfM Hamburg). The ECHAM5/MPIOM model simulations have been carried out at the German Climate Computing Center (DKRZ).

References

Baringer MO, Price, JF (1997) Mixing and spreading of the Mediterranean outflow. J Phys Oceanogr 27: 1654–1677.

Beckmann E, Döscher R (1997) A method for improved representation of dense water spreading over topography in geopotential coordinate models. J Phys Oceanogr 27: 581–591.

Bengtson L, Hodges KI, Roeckner E (2006) Storm tracks and climate change. J Climate, 19: 3518–3543.

Biastoch A, Käse RH, Stammer DB (2003) The sensitivity of the Greenland-Scotland overflow to forcing changes. J Phys Oceanogr 33: 2307–2319.

Blindheim J, Østerhus S (2005) The Nordic Seas, Main oceanographic features. In: Drange H et al. (eds): The Nordic Seas: An integrated perspective, oceanography, climatology, biogeochemistry, and modelling. American Geophysical Union, Geophysical Monograph 158, pp. 11–37.

Borenäs K, Lundberg P (1988) On the deep-water flow through the Faroe Bank Channel. J Geophys Res 93: 1281–1292.

Borenäs K, Lundberg P (2004) The Faroe-Bank Channel deep-water overflow. Deep-Sea Res II 51: 335–350.

Bruce JG (1995) Eddies southwest of Denmark Strait. Deep-Sea Res I 42: 13–29.

Campin J-M, Goosse H (1999) Parameterization of density-driven downslope flow for a coarse-resolution ocean model in z-coordinates. Tellus 51: 412–430.

Dickson RR, Brown J (1994) The production of North Atlantic Deep Water: Sources, rates, and pathways. J Geophys Res 99: 12319–12341.

Dickson RR, Meincke J, Vassie I, Jungclaus J, Østerhus S (1999) Possible predictability in overflow from the Denmark Strait. Nature 397: 243–246.

Dickson RR et al. (2000) The Arctic response to the North Atlantic Oscillation. J Climate 13: 2671–2696.

Drange H, Gerdes R, Gao Y, Karcher M, Kauker F, Bentsen M (2005) Ocean general circulation modelling of the Nordic Seas. In: Drange et al. (eds): The Nordic Seas: An integrated perspective, oceanography, climatology, biogeochemistry, and modelling. American Geophysical Union, Geophysical Monograph 158, pp. 199–219.

Ezer T (2006) Topographic influence on overflow dynamics: Idealized numerical simulations and the Faroe Bank Channel overflow. J Geophys Res 111: C02002, doi:10.1029/2005JC003195.

Gill AE (1977) The hydraulics of rotating channel flow. J Fluid Mech 80: 641–671.

Girton JB, Sanford TB (2003) Descent and modification of the overflow plume in the Denmark Strait. J Phys Oceanogr 33: 1351–1364.

Gulev SK, Barnier B, Knochel H, Molines J-M, Cottet M (2003) Water mass transformation in the North Atlantic and its impact on the meridional circulation: Insights from an ocean model forced by NCEP-NCAR reanalyses surface fluxes. J Climate 16: 3085–3110.

Haak H, Jungclaus JH, Mikolajewicz U, Latif M (2003) Formation and propagation of great salinity anomalies. Geophys Res Lett 30: 1473, doi:10.1029/2003GL017065.

Hallberg R (2000): Time integration of diapycnal diffusion and Richardson-number-dependent mixing in isopycnal coordinate ocean models. Mon Wea Rev 128: 1402–1419.

Hansen B, Turell WR, Østerhus S (2001) Decreasing overflow from the Nordic Seas into the Atlantic Ocean through the Faroe Bank Channel since 1950. Nature 411: 927–930.

Helfrich KR, Pratt L (2003) Rotating hydraulics and upstream basin circulation. J Phys Oceanogr 33: 1651–1663.

Hu A, Meehl GA, Washington WM, Dai A (2004) Response of the Atlantic Thermohaline Circulation to increased atmospheric CO_2 in a coupled model. J Climate 17: 4267–4279.

Jakobsen PK, Ribergaard MH, Quadfasel D, Schmith T, Hughes CW (2003) Near-surface circulation in the northern North Atlantic as inferred from Lagrangian drifters: Variability from the mesoscale to interannual. J Geophys Res 108: 3251, doi:10.1029/2002JC001554.

Jiang L, Garwood RW (1996) Three-dimensional simulations of overflows on the continental slopes. J Phys Oceanogr 26: 1214–1233.

Jónsson S, Valdimarsson H (2004) A new path for the Denmark Strait overflow water from the Iceland Sea to Denmark Strait. Geophys Res Lett 31: L03305, doi:10.1029/2003GL019214.
Jung T, Hillmer M, Ruprecht E, Kleppek S, Gulev SK, Zolina O (2003) Characteristics of the recent eastward shift of interannual NAO variability. J Climate 16: 3371–3382.
Jungclaus JH, Backhaus JO (1994) Application of a transient reduced gravity plume model to the Denmark Strait overflow. J Geophys Res 99: 12375–12396.
Jungclaus JH, Vanicek M (1999) Frictionally modified flow in a deep ocean channel: Application to the Vema Channel. J Geophys Res 104: 21123–21136.
Jungclaus JH, Hauser J, Käse RH (2001) Cyclogenesis in the Denmark Strait overflow plume. J Phys Oceanogr 31: 3214–3229.
Jungclaus JH, Botzet M, Haak H, Keenlyside N, Luo J-J, Latif M, Marotzke J, Mikolajewicz U, Roeckner E (2006a) Ocean circulation and tropical variability in the coupled model ECHAM5/MPI-OM. J Climate 19: 3952–3972.
Jungclaus JH, Haak H, Esch M, Roeckner E, Marotzke J (2006b) Will Greenland melting halt the thermohaline circulation? Geophys Res Lett 33: L17708, doi:10.1029/2006GL026815.
Käse RH, Oschlies A (2000) Flow through Denmark Strait. J Geophys Res 105: 28527–28546.
Käse RH (2006) A Riccati model for the Denmark Strait overflow variability. Geophys Res Lett 33: L21S09, doi:10.1029/2006GL026915.
Killworth PD, McDonald NR (1993) Maximal reduced gravity flux in rotating hydraulics. Geophys Astrophys Fluid Dyn 70: 31–40.
Killworth PD, Edwards NR (1999) A turbulent bottom boundary layer code for use in numerical ocean models. J Phys Oceanogr 29: 1221–1238.
Killworth PD (2001) On the rate of descent of overflows. J Geophys Res 106: 22267–22275.
Köhl A, Käse RH, Stammer D, Serra N (2007) Causes of changes in the Denmark Strait overflow. J Phys Oceanogr 37: 1678–1696.
Kösters F, Käse RH, Schmittner A, Herrmann P (2005) The effect of Denmark Srait overflow on the Atlantic Meridional Overturning Circulation. Geophys Res Lett 32: L04602, doi:10.1029/2004GL022112.
Legg S, Hallberg RW, Girton JB (2006) Comparison of entrainment in overflows simulated by z-coordinate, isopycnal and non-hydrostatic models. Ocean Modelling 11: 69–97.
Macdonald AM (1998) The global ocean circulation: a hydrographic estimate and regional analysis. Prog Oceanogr 41: 281–382.
Macrander A (2004) Variability and processes of the Denmark Strait overflow. Ph.D. thesis. IfM Geomar, Leibniz Institut für Meereswissenschaften an der Universität Kiel, 177 pp.
Macrander A, Send U, Valdimarsson H, Jónsson S, Käse RH (2005) Interannual changes in the overflow from the Nordic Seas into the Atlantic Ocean through Denmark Strait. Geophys Res Lett 32: L06606, doi:10.1029/2004GL021463.
Marsland SJ, Haak H, Jungclaus JH, Latif M, Röske F (2003) The Max- Planck- Institute global ocean/sea ice model with orthogonal curvilinear coordinates. Ocean Modelling 5: 91–127.
Müller WA, Roeckner E (2006) ENSO Impact on Mid-Latitude Circulation Patterns in Future Climate Change Projections. Geophys Res Lett 33: L05711, doi: 10.1029/2005GL025032.
Nikolopoulos AN, Borenäs K, Hietala R, Lundberg P (2003) Hydraulic estimates of the Denmark Strait overflow. J Geophys Res 108: 3095, doi:10.1029/2001JC001283.
Nilsen JEØ, Gao Y, Drange H, Furevik T, Bentsen M (2003) Simulated North Atlantic-Nordic Seas water mass exchange in an isopycnic coordinate OGCM. Geophys Res Lett 30: 1536, doi:10.1029/2002GL016597.
Østerhus S, Turrell WR, Jónsson S, Hansen B (2005) Measured volume, heat, and salt fluxes from the Atlantic to the Arctic Mediterranean. Geophys Res Lett 32: L07603, doi:10.1029/2004GL022188.
Özgökmem TM, Fischer PF, Johns WE (2006) Product water mass formation by turbulent density currents from a high-order nonhydrostatic spectral element model. Ocean Modelling 12: 237–267.
Pratt LJ, Smeed DA (2004) The physical oceanography of sea straits (Editorial). Deep-Sea Res II 51(4–5): 319.

Price JF, Baringer MO (1994) Outflow and deep water production by marginal seas. Progr Oceanogr 33: 161–200.
Riemenschneider U, Legg S (2007) Regional simulations of the Faroe Bank channel overflow in a level model. Ocean Modelling 17: 93–122.
Roberts MJ, Wood RA (1997) Topography sensitivity studies with a Bryan-Cox type ocean model. J Phys Oceanogr 27: 823–836.
Ross CK (1984) Temperature-salinity characteristics of the "overflow" water in Denmark Strait during "Overflow '73". Rapp P-v Reun Cons Int Explor Mer 185: 111–119.
Roeckner E, Bäuml G, Bonaventura L, Brokopf R, Esch M, Giorgetta M, Hagemann S, Kirchner I, Kornblueh L, Manzini E, Rhodin A, Schlese U, Schulzweida U, Tompkins A (2003) The atmospheric general circulation model ECHAM5, part I: Model description. Max- Planck-Institut für Meteorologie, Report No. 349, 127 pp.
Saunders P (1990) Cold outflow from the Faroe Bank Channel. J Phys Oceanogr 20: 29–43.
Schmittner A, Latif M, Schneider B (2005) Model projections of the North Atlantic thermohaline circulation for the 21st century assessed by observations. Geophys Res Lett 32: L23710, doi:10.1029/GL024368.
Schweckendiek U, Willebrand J (2005) Mechanisms affecting the overturning response in global warming simulations. J Climate 18: 4925–4936.
Shi XB, Roed LP, Hackett B (2001) Variability of the Denmark Strait overflow: A numerical study. J Geophys Res 106: 22277–22294.
Smith PC (1975) A stream-tube model for bottom boundary currents in the ocean. Deep-Sea Res 22: 853–873.
Song Y, Chao Y (2000) An embedded bottom boundary layer formulation for z-coordinate models. J Ocean Atmos Tech 17: 546–560.
Stammer D, Wunsch C, Fukumori I, Marshall J (2002) State estimation improves prospects for ocean research. EOS Transactions 83: 294–295.
Steele M, Morley R, Ermold W (2001) PHC: A global ocean hydrography with high-quality Arctic Ocean. J Climate 14: 2079–2087.
Stern ME (2004) Transport extremum through Denmark Strait. Geophys Res Lett 31: L12303, doi:10.1029/2004GL020184.
Stocker TF et al. (2001) Physical climate processes and feedbacks. In: Climate Change 2001: The scientific basis, Intergovernmental Panel of Climate Change (IPCC) Technical summary of the Working Group 1 Report, Cambridge University Press, New York, pp. 418–470.
Whitehead JA (1989) Internal hydraulic control in rotating fluids: Application to oceans. Geophys. Astrophys Fluid Dyn 36: 187–205.
Whitehead JA (1998) Topographic control of oceanic flows in deep passages and straits. Rev Geophys 36: 423–440.
Willebrand J et al. (2001) Circulation characteristics in three eddy-permitting models of the North Atlantic. Progr Oceanogr 48: 123–161.
Winton M, Hallberg RW, Gnanadesikan A (1998) Simulation of density-driven frictional downslope flow in z-coordinate ocean models. J Phys Oceanogr 28: 2163–2174
Xu X, Chang YS, Peters H, Özgökmem TM, Chassignet EP (2006) Parameterization of gravity current entrainment for ocean circulation models using a high-order 3d nonhydrostatic spectral element model. Ocean Modelling 14: 19–44.
Zhang J, Steele M, Rothrock DA, Lindsay RW (2004) Increasing exchanges at Greenland-Scotland Ridge and their links with the North Atlantic Oscillation and Arctic sea ice. Geophys Res Lett 31: L09307, doi:10.1029/2003GL019304.

Chapter 23
Satellite Evidence of Change in the Northern Gyre

Sirpa Häkkinen[1], Hjálmar Hátún[2], and Peter Rhines[3]

23.1 Introduction

The northern gyre is an economically important region because of its commercial fisheries which have waxed and waned over time, often in association with cooling and warming of the ocean. The northern, or subpolar, gyre is defined by strong boundary currents, East and West Greenland Current, Labrador Current and the North Atlantic Current, framing a cyclonic circulation between 50° N and 65° N. Particularly the Labrador Sea has been a subject of alternating periods of deep convection and periods when a freshwater cap prevents deep convection. The subpolar gyre belongs to the westerly wind domain where the North Atlantic Oscillation (NAO) is the dominant fluctuation of atmospheric forcing. The original definition of NAO is based on pressure difference between Iceland and Portugal (later Azores) and in effect NAO measures the strength of the westerlies (Hurrell 1995). NAO experienced both its lowest and highest recorded index values during the last 50 years of the century (Fig. 23.1). NAO forcing is considered to be the salient linkage to the Labrador Sea climate fluctuations whereby, for instance, strong westerly winds associated with the positive index phase of NAO bring in cold continental air masses. These cold air outbreaks are conducive to large heat flux from the ocean leading to deep convection. On the other hand, periods with freshwater cap in the Labrador Sea are manifestations of 'great' and less great salinity anomalies which have appeared nearly decadally in the subpolar gyre (Dickson et al. 1988, 1996; Reverdin et al. 1997; Belkin et al. 1998). The fresh event of 1968–1969 was called the 'Great Salinity Anomaly' by Dickson et al. (1988). Also there has been long-term freshening in the deep salinity in the Labrador Sea (Dickson et al. 2002), which was traced to the freshening of the Nordic Sills overflows. However, it is not clear at this time whether this change is about to reverse, because there are signs that the overflows have started to get more saline in the very recent years (see Osterhus et al., Chapter 18, this volume).

[1]NASA Goddard Space Flight Center, Greenbelt, MD, USA, e-mail: sirpa.hakkinen@nasa.gov

[2]Faroese Fisheries Laboratory, Torshavn, Faroe Islands, e-mail: hjalmarh@frs.fo

[3]University of Washington, Seattle, WA, USA, e-mail: rhines@ocean.Washington.edu

R.R. Dickson et al. (eds.), *Arctic–Subarctic Ocean Fluxes*, 551–567

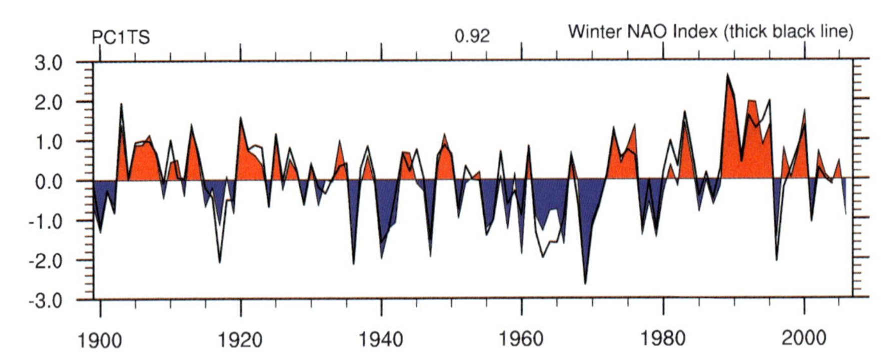

Fig. 23.1 Winter NAO from www.cgd.ucar.edu/cas/jhurrell/indices.info.html. The principal component (PC) (in color) is shown for the leading EOF of seasonal (December through March) SLP anomalies over the Atlantic sector (20–80° N, 90° W–40° E) (as in Hurrell 1995). The station-based index is given by the thick black line

While we know a great deal about the stratification changes in the subpolar gyre, we have much less direct information of variability in the wind and buoyancy driven currents there. Several research groups from Canada, USA and Europe have maintained current meter moorings, but most of them have duration of only a few years and measure currents only in a few locations along the Newfoundland Coast. An important step towards a larger scale, near-global view to variability in ocean circulation is provided by altimetry from satellites measuring the sea surface heights (SSH), from which we can derive the geostrophic circulation field. As a brief explanation why non-seasonal (average seasonal heating-cooling cycle is removed) SSH variations are interpreted as dynamic changes, we note that in the subtropics and tropics, SSH is mainly determined by adiabatic vertical movement of isopycnals (i.e. a change in heat storage) due to local and/or remotely forced dynamics (exceptions do exist even in tropics, e.g. from intense rainfall). Toward high latitudes, SSH is affected both by an increasing salinity contribution to sea-water density and by an increasingly barotropic flow structure due to weak stratification. In the subpolar gyre, both of these effects are active but we expect the stratification variability in the central dome to have a greater impact on the strength of the gyre. An independent source of data to validate findings from satellite altimetry is surface drifters, which have been available since 1989, however, their spatial and temporal concentration is highly variable.

This chapter is structured as follows: First we discuss altimeter missions to date and what has been established in early studies in describing the subpolar circulation. Then we proceed to more recent studies of the subpolar gyre variability, mainly related to the 1990s large changes in the stratification structure during which time we are fortunate to have had TOPEX/Poseidon mission. Finally, we present the latest developments as gathered from altimetry and drifters for the subpolar gyre and North Atlantic Current (NAC) variability.

23.2 Altimeter Missions to Date and Lessons Learned

23.2.1 Altimetry Before 1990

The first altimeter mission was launched on Seasat 1978 and operated for 3 months. In fact, Seasat was the first satellite mission devoted to measuring oceanographic parameters from space (in addition to ocean surface topography), such as sea surface winds and temperatures, ocean wave heights, internal waves, sea ice, and atmospheric water vapor. While being a demonstration of the altimeter capability of retrieving sea surface height with accuracy 8–10 cm, scientific accomplishments were remarkable. The study by Menard (1983) is of interest for the later discussion in which he showed the usefulness of eddy kinetic energy (EKE) based on geostrophic velocities from Seasat in locating the major current systems. (Scatterplots of EKE versus surface drifter velocity (not shown) suggests that high EKE is associated with high drifter velocity, but high drifter velocity is not necessarily associated with high EKE.) The altimetric SSH is referenced to the mean geoid, hence geostrophic velocities represent anomalies from the mean circulation, and may not reflect the actual path of major currents. By extracting the fluctuations of the current, i.e. EKE, one can trace the effect of eddies spawned due to baroclinic instability (or generated by winds) and thereby pinpoint the location of the current. This work was extended to the subpolar gyre by Heywood et al. (1994) and White and Heywood (1995) who used altimetric EKE to study subpolar North Atlantic circulation changes, changes in the NAC path and their relationship to wind stress (curl). Heywood et al. showed that altimetry can recognize the splitting of the NAC into the three known branches; the Rockall Trough and Iceland Basin branches and one on the western flank of the Reykjanes Ridge. Their studies used data from Geosat repeat orbit mission that lasted from 1986 to 1988. Geosat SSH accuracy is of the order of 8–10 cm. However, it is considered that EKE derived from only one satellite data underestimates the level of meso-scale activity (Ducet et al. 2000). If high resolution gridded SSH data is available, such as AVISO 1/3 degree data, the EKE can be computed from SSH, denoted as η, assuming geostrophy, as follows:

$$EKE = 1/2[<u_g^2> + <v_g^2>], \text{ where}$$
$$u_g = -(g/f)\,\Delta\eta/\Delta y, \text{ and } v_g = (g/f)\Delta\eta/\Delta x$$

are the zonal and meridional geostrophic velocity anomalies (relative to the mean over the given time period, e.g. 3 years, from January 2003 to December 2005), f is the Coriolis parameter, and g gravity. $< >$ is time average of the squared velocity anomalies. Lilly et al. (2003) show that eddies with diameters smaller than about 50 km are not seen even with the densely sampled trackline data of the altimeter, and even less so with the gridded SSH data, leaving satellite derived EKE as an underestimate in some regions. As an example of EKE maps, an EKE field is shown in Fig. 23.2 from years 2003 to 2005 based on the AVISO product which is compiled from retrievals from several altimeters. Figure 23.2 shows the

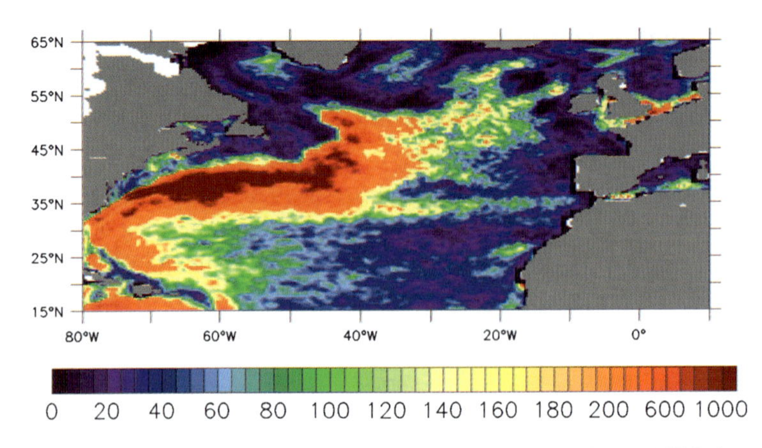

Fig. 23.2 EKE (in cm^2 s^{-2}) from years 2003 to 2005 (3 years of data) from AVISO data

familiar features of the Gulf Stream with the highest EKE, with an extension of high EKE including several branches towards the subpolar gyre as the signal of the North Atlantic Current, and the nearly zonal EKE feature at about 32° N as the Azores Current.

23.2.2 Altimetry After 1990

The 1990s brought several new altimeter missions, ERS-1/2 1991 onwards and TOPEX/Poseidon launched 1992 which lasted until 2005. Jason-1 launched 2001, ENVISAT 1999- and Geosat Follow-on 1998- continue to collect SSH data. With this suite of altimeters we have accumulated a nearly 14-year-long time series of ocean dynamic topography for studies of climatic changes. TOPEX and Jason-1 have an accuracy of 3–4 cm which is a significant improvement over the earlier generation of altimeters for detecting variability in the subpolar gyre.

As already noted, the Labrador Sea along with the northern North Atlantic Ocean has experienced major changes in stratification structure during the 1990s (see Chapter 24 by Yashayev). In the early 1990s deep convection reached to the overflow layers and the new formed Labrador Sea Water was colder and fresher than in the previous decades. However, by 1996 the deep convection had diminished considerably and a surface freshwater cap developed (Belkin 2004). The contemporaneous changes with the Labrador Sea events have been discovered in various parts of the subpolar gyre. Based on hydrographic data covering the 1990s, Bersch et al. (1999) and Bersch (2002) reported the movement of the subarctic front westward in the Iceland Basin indicating contraction of the subpolar gyre (and expansion of the subtropical gyre) in the late 1990s. These changes in hydrography are evident in altimetry also as in the study by Reverdin et al. (1999) who analyzed the altimeter data and dynamic height computed from a repeat WOCE XBT line (data extended to 700 m) between Iceland and Newfoundland. Both data sets show increasing surface- height and baroclinic-dynamic-height trends from 1992 to 1998 for the time series for the first EOF mode across the transect, with both data sets having nearly 50% of the variance in the first mode. Using hydrographic line AR7E line across the Irminger Sea and Iceland Basin (between Greenland and Ireland), Volkov and van Aken (2003) show that the close relationship between altimetric SSH and dynamic height changes applies also elsewhere in the subpolar basin. They used variations in the dynamic height computed over depths 20–2,000 dbar (1,200 dbar over Reykjanes Ridge), and they suggest that altimetric SSH changes reflect changes in the deep water column in the subpolar gyre.

The large stratification changes described above are bound to lead to circulation and transport changes. Han and Tang (2001) used TOPEX/Poseidon data and in situ hydrographic data to deduce that the total transport of the Labrador Current had decreased by a significant amount between 1992 and 1998 (the last year of their data analysis). Verbrugge and Reverdin (2003) note the slowing down of the NAC branches after 1996 based on altimetric currents and suggested weakening of the subarctic front and of the Irminger gyre which together would allow warm water to spread westward to the central subpolar gyre. Considering the work by Volkov and van Aken (2003), the near surface changes in the NAC, described by Verbrugge and Reverdin as determined by geostrophic velocity estimates from altimetry, are likely to reflect changes much deeper in the water column. The relationship between the NAO index phases and subpolar circulation was explored also by Flatau et al. (2003) using both altimetry and drifter data. Flatau et al. report that the north-eastward flow of NAC was stronger during a positive NAO phase and was associated with a farther eastward location of the subarctic front both in altimetric currents and in drifter currents corrected for Ekman drift. They also point out the stronger cyclonic circulation in the Irminger Sea associated with positive NAO. Examination of their figures on drifter data between NAO+ and NAO– years indicates that west of the mid-Atlantic Ridge (MAR), the Gulf Stream extension (with velocities >30 cm s^{-1}) was forced eastward south of 45° N in NAO+ years. These drifters of NAO+ years did not form north-eastward flow until in the far eastern Iceland Basin, but during the NAO– years, drifters with the highest velocity easily reached 50° N and formed a north-eastward flow from Flemish Cap to the Rockall Trough (in agreement with the westward movement of the subarctic front).

In summary, the above studies are consistent in suggesting a significant modulation of the subpolar gyre strength from the early 1990s to the late 1990s. Besides these observational signals, significant changes of altimetric SSH were discovered to have taken place from the 1980s, to the early 1990s and again in the late 1990s (Häkkinen 1999, 2001). The SSH changes can be typified by a dipole pattern where one center is over the Gulf Stream and the other center is over the subpolar gyre (Häkkinen 1999, 2001; Häkkinen and Rhines 2004). The updated first empirical orthogonal function (EOF) mode of altimetric SSH and its principal component (PC) are shown Fig. 23.3. The sea-level change has been particularly strong in the Irminger Sea where the sea-level rise from the early 1990s to 2005 has been nearly 15 cm.

This region of strong SSH variability is just beneath the lee cyclogenesis in the atmosphere associated with tip-jets round Cape Farewell (Pickart et al. 2003) and gap-jets over the top of Greenland (Jung and Rhines 2006). As was shown in Häkkinen and Rhines (2004) the SSH changes can be converted to geostrophic velocity field which can be subjected to EOF analysis. By normalizing the anomalous velocity vectors in each grid point by the magnitude of the velocity anomaly, one can suppress the large signal from Gulf Stream fluctuations and to emphasize more subtle changes around the gyres. The resulting first EOF mode and its PC are shown in Fig. 23.4. The updated analysis was done only for the T/P and Jason-1 period, but the time series from the Häkkinen and Rhines study is shown which include the earlier satellite data. This approach of normalizing the velocity fluctuations shows that the subpolar gyre has been continually gaining anticyclonic character, i.e. the subpolar gyre has been weakening. However, the first EOF mode depicted in Fig. 23.4 does not necessarily reflect all manifestations of the subpolar circulation which will weaken with the weakening phase of the EOF1.

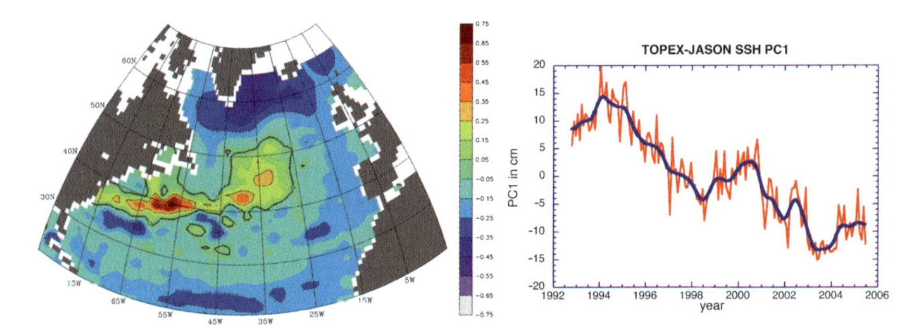

Fig. 23.3 The first EOF mode of the altimetric SSH: spatial pattern (left; non-dimensional) and time series (right, in cm)

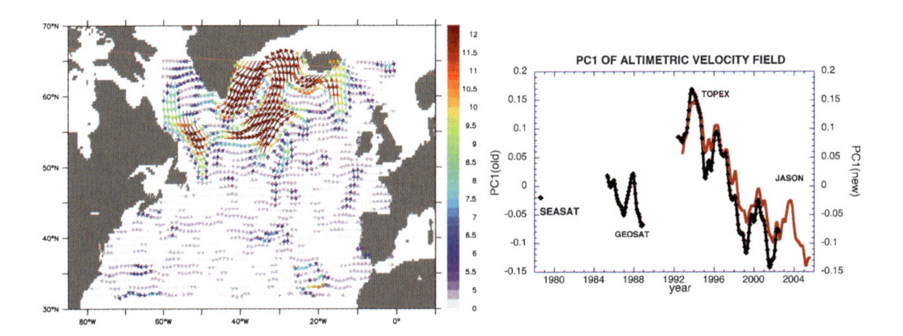

Fig. 23.4 The first EOF mode of the geostrophic velocity computed from altimetric SSH: spatial pattern (left) and time series (right). The red curve is update to the black curve from Häkkinen and Rhines (2004)

Recently Hátún et al. (2005) discovered that the gyre strength, or alternatively SSH PC1, can be linked to the salinity changes in the eastern North Atlantic in the NAC branches carrying the Atlantic waters to the Norwegian Sea (and also in the Irminger Current). This relationship in the model simulations can be traced back over several decades. The impact of the gyre strength on salinity is consistent with the observations that with a weak and contracted gyre, the subarctic front is located farther westward, hence more saline (eastern) Atlantic waters are able spread northward. This study emphasizes that decadal variability in the subpolar zone can involve changing configuration of the circulation, as well as changing strength.

In Hátún et al., the increased transport was linked to the highest salinity water masses flowing toward NE while the gyre index was weakening (their Figure 23.4a, b). Weakening gyre, more saline conditions and higher transport of the high salinity surface waters in the NE subpolar gyre do not at first sight seem to be consistent with each other, since the high salinity signal has to originate from the subtropics. However, the Hátún et al. analysis shows an increased transport of saline (southern branch) NAC water as it crosses the Reykjanes Ridge, related to the northward shift of the front.

Issues relating to the source waters of the NAC as well as the source waters of the Nordic Seas inflow have been hotly debated over the years (e.g. reviews by Rossby 1996; McCartney and Mauritzen 2001), and this issue is still open. Surface drifter and subsurface float fields compiled into an Eulerian average are showing that surface waters from the western Atlantic cross the mid-Atlantic Ridge as a wide and highly fluctuating current except for the branch crossing through the Charlie Gibbs Fracture Zone (Krauss 1986; Perez-Brunius et al. 2004) and proceeding towards north to the Iceland Basin and the Rockall Trough (Brügge 1995; Fratantoni 2001; Bower et al. 2002). On the other hand, Reverdin et al. (2003) and Brambilla and Talley (2006) show that very few surface drifters make it to the subpolar gyre from the subtropics. This issue brings a further complication of determining the source of the saline waters in the Rockall Trough which are more saline than the NAC branch in the Maury Channel (= Iceland Basin branch) and which flow northward to the Nordic Seas. Pollard et al. (2004) has called the saline waters masses in the Rockall Trough (Hatton Bank) as Eastern North Atlantic Water with origins in the eastern subtropics, yet in part cycling through a branch of the NAC before adding some salinity by mixing. Additionally Holliday et al. (2000) find significant long-term changes in the properties of the Rockall Trough water masses and attributes this variability arising from several processes; advection, ventilation and mixing with Mediterranean Waters upstream from the Rockall Trough. On the other hand, McCartney and Mauritzen (2001) attribute the source to be the NAC water masses from farther west which have gained some salinity by mixing with the Mediterranean Overflow Waters before entering the Rockall Trough. Based on subsurface float data, Bower et al. (2002) show that the NAC waters below 200 m eventually turn back to west, so they cannot be the source for the Nordic Seas inflow. They conclude that the source has to be waters in the top 200 m. The pathways of warm, saline subtropical water reaching the Greenland–Scotland Ridge and feeding the overflows are thus not clearly known, and may not involve pathways fixed in time. However the debate may be in part semantic: the NAC waters that become more saline by mixing as they reach the eastern basin and turn north involve, in a Lagrangian description, a branch of northward circulation in the eastern N. Atlantic, which is called mixing in an Eulerian description.

The differing views on the Nordic Seas inflow sources represent one of the many unresolved issues. The boundary currents in the Labrador Sea as measured by current meters at various levels appear to behave differently from the gyre index based on altimetry (F. Schott, private communication, 2005). From the following discussion it should be obvious that various regional and depth-stratified details of the subpolar system cannot be described solely by one index such as the gyre index based on the SSH PC1.

23.2.3 MOC, LSW and the Strength of the Subpolar Gyre

An obvious question related to these subpolar gyre and stratification changes is whether they reveal anything about the meridional overturning circulation (MOC). Recently Bryden et al. (2005) published analysis of hydrographic data estimating the Atlantic MOC at 25° N to be 30% weaker in 2004 than in 1957 with the steepest decline between 1992 and 2004, which overlaps with the altimetry data. Also their estimate of the northward heat transport at 25° N showed a decline from 1.3–1.4 PW in 1957, 1981 and 1992 to 1.1 PW in 2004. Bryden and his colleagues have recently updated the decrease to be about 10% during the last 25 years. Numerical hindcasts of the North Atlantic Ocean signal a major drop in the MOC and/or the subpolar gyre strength in the 1990s (Häkkinen 2001; Hátún et al. 2005; Treguier et al. 2005; Mauritzen et al. 2006; Boening et al. 2006). As a common feature of all these model experiments, they all show a steady increasing trend from 1960 (1980 onwards in four models reviewed in Treguier et al.) up to the peak in the early 1990s in either Labrador Current, MOC or SSH PC1. Hence they are in conflict with the result of Bryden et al. (2005) for 1957, which year should be reflective of conditions also in the early 1960s. Both Häkkinen (2001) and results from a higher resolution version of the same model using NCEP/NCAR Reanalysis, Fig. 23.5, show an upward trend since 1960 both in the SSH PC1 and in the MOC computed at 45° N.

Another question closely associated to our topic is whether the Labrador Sea Water (LSW) production or alternately deep convection or its lack has impact on the MOC. Using numerical models one can link the convective years to the strength of the deep western boundary current (WBC) (Treguier et al. 2005; Boening et al. 2006). Treguier et al. note that the linkage between convection and boundary current strength becomes apparent only if high resolution models are used for the hindcasts. Their simulated decline was of the same order as that found by Boening et al. during the 1990s, when the WBC declined 15%, 7–8 Sv in absolute terms.

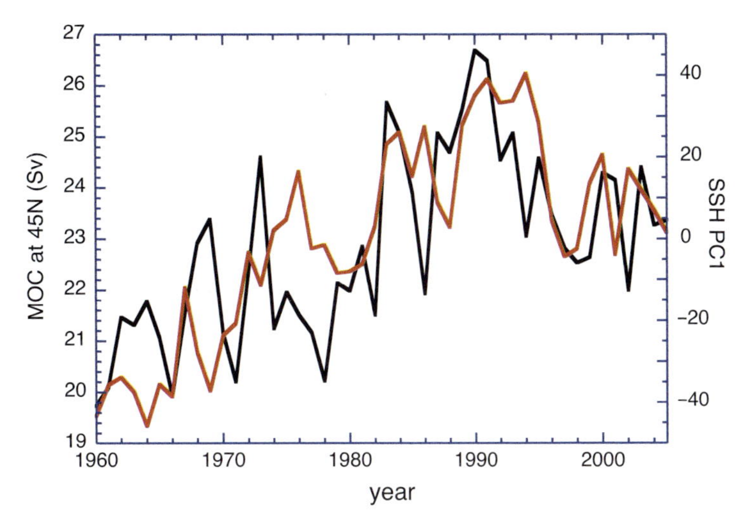

Fig. 23.5 MOC (in Sv) at 45° N (red) and time series of the 1st EOF mode of the North Atlantic SSH. Both are results from a GSFC model using NCEP/NCAR Reanalysis from 1948–2005

These changes manifest themselves in the mid-latitude MOC a couple of years later. It is another matter how to link the amount of LSW formed to overturning rates because of highly variable observational estimates of LSW (upper LSW and classical LSW) production: Smethie and Fine (2001) estimate 9.6 Sv, Pickart and Spall (2006) 2 Sv, Keike et al. (2006) estimates vary from 6.9 to 9.2 Sv for the 1990s, but only 3.3 to 4.78 Sv in 2000–2001. These widely varying water-mass transformation rates may reflect differing observational analyses, or in some case may relate to differing time periods. The definition of water-mass transformation through transport analysis on the T/S plane has been advocated by Bailey et al. (2005) as a more articulate expression of these processes.

23.3 Recent Developments: Variability of NAC and Its Branches in the Northern Gyre

The above analysis of altimetry data emphasizes the gyre circulation, and in many ways the resulting pattern reflects the dynamic height as predicted from combined GRACE gravity measurements and hydrography in Jayne (2006). The pressure field at 700 m (his Fig. 23.4) in particular has great similarity to the EOF1 of SSH. As noted earlier, increased transport of high salinity surface waters into the Rockall Tough as simulated by the Bergen model (Hátún et al. 2005) and the gyre index appear to behave differently. The focus in the following analysis is the surface currents during the last 14 years, to confirm the Hátún et al. model findings, although this analysis will not address the source of the surface waters in the Rockall Trough. We will use the term NAC to indicate those upper-ocean waters originating west of the Reykjanes Ridge and crossing it eastward, roughly in the latitude band 45–55° N. In fact the NAC is considered a deeper flow reaching 1,000 m in the eastern North Atlantic in Bower et al. (2002), who also show that most of the NAC turns westward to recross the Reykjanes Ridge, joining the subpolar gyre, and that only the top 200 m of NAC peels off to flow northward across the Nordic sills. Krauss (1986) used the term 'westwind drift' to describe near surface NAC when it crosses the MAR.

To gauge the intensity and location of the three branches of the North Atlantic Drift we use EKE. The differences from three different 3-year periods from T/P - Jason-1 data are displayed in Figs. 23.6a, b and Fig. 23.2. Compared to the early

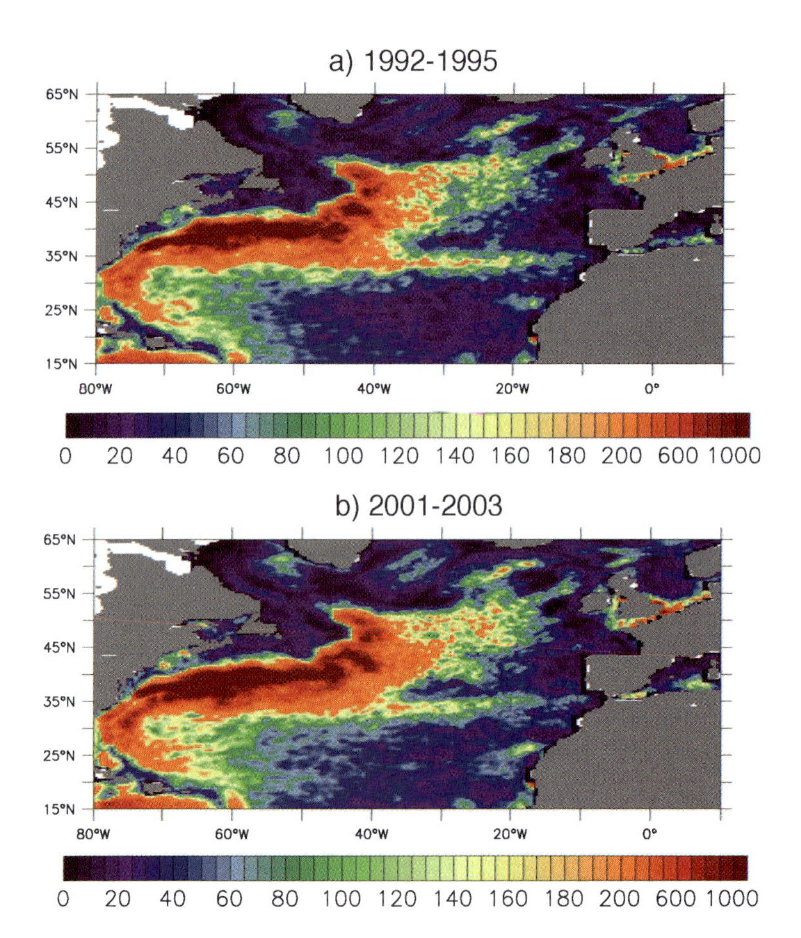

Fig. 23.6 EKE from AVISO data from periods 1992 to 1995 (a) and (2001–2003) (b). Units in (a–b) are $cm^2\ s^{-2}$

1990s period, the EKE maps after 2000 (Figs. 23.2, 23.6b) show high EKE values east of 35° W, north of 40° N, suggesting the stronger NAC crossing the MAR, although this point requires further analysis. Along its north-eastward path, the NAC EKE increased in the Iceland Basin, also EKE activity in the Irminger Basin is elevated in the later EKE maps (Figs. 23.2, 23.6b) compared to Fig. 23.6a. As NAC EKE has increased in strength north-eastward towards Nordic Seas, the mid-latitude Azores Current has lost some of its EKE and eastward penetration. These EKE changes between the early 1990s and 2000s are shown in Fig. 23.7a, b as difference maps which show increased EKE activity north of 45° N and east of the MAR while the western subpolar gyre has a decline in EKE. The difference maps show the changes in the Azores Current to be striking. Lacking of a clear dipole structure the EKE changes are more indicative of a weakening current.

Again we find the dissimilar behavior of the gyre index and increased EKE of the NAC branches. First, higher EKE values in the later period could be a reflection of the intensification of the surface fronts associated with the NAC branches, which does not necessarily mean higher transport in the NAC. In our view, the surface trapped currents and EKE can get stronger, even though the deeper integrated transport is slowing down as expressed by the first EOF mode of SSH and the geostrophic velocity. The altimetric EKE variability is supported by surface drifter

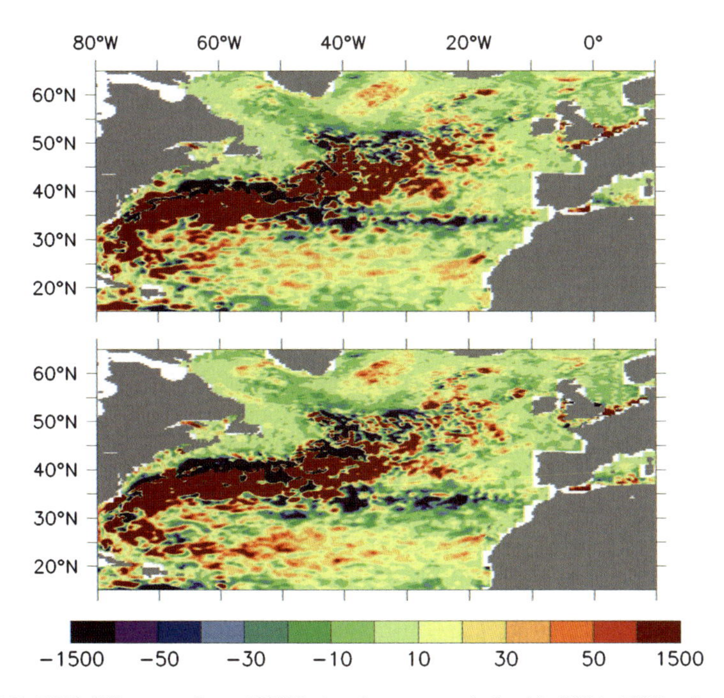

Fig. 23.7 EKE differences from AVISO data between periods: (a) (2001–2003) minus (1992–1995), and (b) (2003–2005) minus (1992–1995). Units in (a–b) are $cm^2 s^{-2}$

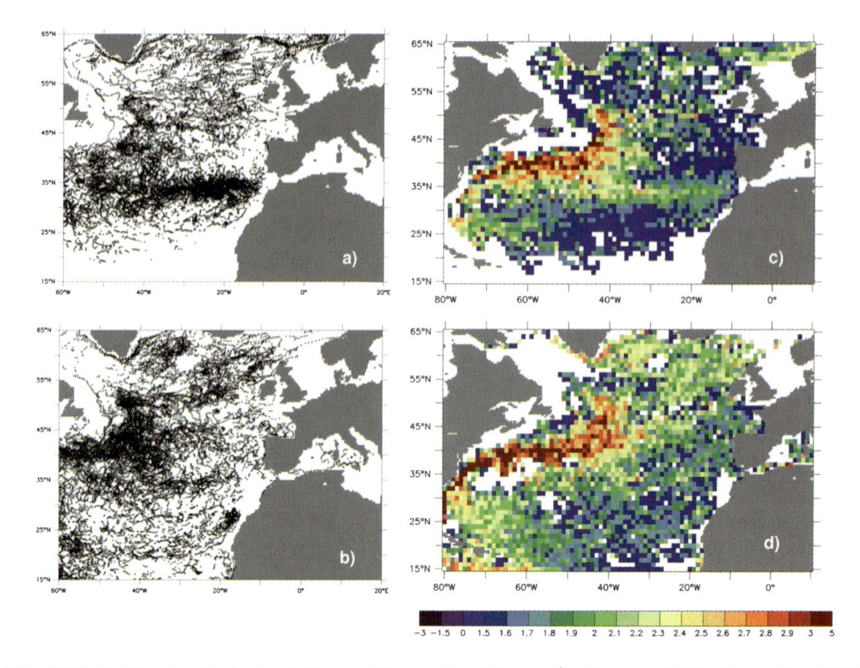

Fig. 23.8 Drifter tracks (only drifters with speeds >25 cm s^{-1}): October 1992–September 1995 (a; top-left), the corresponding drifter EKE (b; top-right); drifter tracks January 2001–December 2003 (c; bottom-left), the corresponding EKE (d; bottom-right). EKE (in $cm^2\ s^{-2}$) is expressed on log-10 scale

data. The surface drifters are not evenly distributed, nevertheless, the drifter EKE (Fig. 23.8a, b) shows the same tendencies as the altimetric EKE, highlighting the differences even more between the early 1990s and early 2000s: the Iceland Basin, Rockall Trough and Irminger branches show an order of magnitude higher EKE as well as the north-eastward flowing NAC EKE across MAR is larger than in the early period. Also the changes in the Azores Current are similar to the altimeter findings, although far less drifters are present in the region in the later period. These EKE changes are consistent with the modeling result of Hátún et al. (2005) that high salinity surface waters have increased in transport (their Fig. 4a), towards north-eastward, giving a high salinity signal in locations usually associated with the NAC branches.

The drifter velocity fields for the early 1990s and early 2000s period are displayed in Fig. 23.9 super-imposed on a background of the altimetry derived EKE. Figures show that indeed high EKE values are associated with stronger surface currents and give an impression of a much wider NAC in the later period. Particularly the flows in the Iceland Basin and Rockall Trough are better distinguishable in the later period than in the earlier one. Also in the later period, an increased EKE particularly distinguishes the branches of the NAC crossing over the MAR at 46–48° N (Maxwell Fracture Zone) and at 50–52° N (Faraday and Charlie

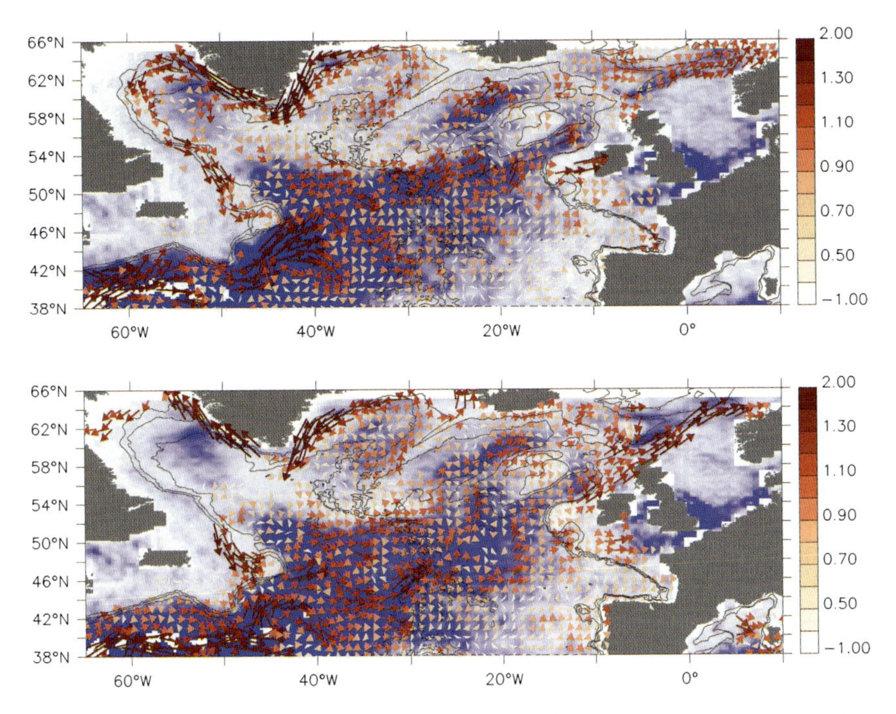

Fig. 23.9 Average surface drifter velocity (cm s^{-1}; magnitude on log-10 scale) from periods 1992 to 1995 (top), and 2001–2003 (bottom) when drifter data is interpolated to 1° × 1° grid. Velocities are superimposed on altimeter EKE from corresponding periods

Gibbs Fracture Zones). The branch into the Rockall Trough during 2001–2003 appears to originate from the path passing through in the Maxwell Fracture Zone and continuing E-SE. This branch turns sharply northward at about 18° W, 46–48° N, before turning into the Rockall Trough.

The order of magnitude increase of the drifter EKE in the eastern branches suggests stronger surface currents which can readily be computed from the surface drifter data. The average velocity (positive values are eastward) for the two periods at 19° W and between 42° N and 62° N are:

U (1992–1995) = 4.9 cm s^{-1}, with a range (along latitude) (−6–19 cm s^{-1}) and standard deviation of 7 cm s^{-1}

U (2001–2003) = 9 cm s^{-1}, with a range (−18–37.4 cm s^{-1}) and standard deviation of 12.4 cm s^{-1}.

The later period average velocity is nearly twice as large but statistically the two average velocities are not different (at 95%) due to the large standard deviation. As noted before, the apparent strengthening of surface currents does not mean that the deeper integrated transports will also increase. This reflects the difference between EKE and SSH variability: EKE defines the locations of the intense currents recording the eddy activity imprinted into SSH, but SSH (being an integrated quantity) illustrates the large-scale gyre changes. The increased strength of the NAC surface branches in the Iceland Basin is likely associated with the westward movement of the subarctic front allowing expansion of the subtropical gyre and increased baroclinicity in the upper ocean. The EKE changes suggest that more Gulf Stream waters proceeds across the MAR to the eastern basin to the area straddled between the subpolar and subtropical gyres. The supporting evidence for this scenario is the stunted eastward penetration of the Azores Current in years after 2000.

23.4 Summary

Recent variability of the North Atlantic subpolar gyre is reviewed based on altimetric results along with hydrographic and modeling findings. Our altimetry time series is still relatively short but its coverage has afforded us a glimpse of the remarkable changes in the northern North Atlantic Ocean that took place in the 1990s and which are still ongoing. These findings can be derived from the altimetric sea surface height and from geostrophic currents computed from sea surface height. As a result of weakened deep convection and changes in stratification, sea surface height has increased everywhere in the subpolar gyre reaching nearly 15 cm in the Irminger Sea. EOF analysis of the geostrophic currents reveals that the subpolar gyre has slowed down since beginning of the TOPEX/Poseidon launch and that the subpolar gyre remains in a weakened state as of late 2005. Based on numerical model hindcasts the strength of the gyre varies with the meridional overturning circulation. Models show that the MOC reached a maximum along with the NAO index in the early 1990s. After the NAO started to descend from its maximum, the MOC slowed down also. There is now some direct observational evidence from hydrography that there was a slowdown in the MOC in the 1990s. Unfortunately the models and the hydrographic analysis do not agree for the earlier decades.

Another interesting connection arises between the gyre strength and the size of the subpolar gyre which on interannual to interdecadal timescales appears to control the salinity and temperature of the Atlantic Inflow to the Nordic Seas. Reduction of the subpolar gyre allows expansion of the subtropical gyre and enhancement of the North Atlantic surface drift east of the mid-Atlantic Ridge allowing more saline products of Atlantic water to reach the Nordic Seas. Recent work suggests that pathways as well as current strength can change over decadal timescales; the NAC is generally thought to be a primary source of the Atlantic water entering the Nordic Seas, eventually to feed the dense overflows, yet its shifting branches and the high EKE-to-mean-KE ratio in the northeastern Atlantic suggest that mixing of saline eastern-Atlantic waters into the NAC can also vary decadally. The increased salinity of the Nordic Seas' inflow acts as a positive feedback to enhance or at least stabilize MOC through the overflows and to counterbalance the subpolar branch weakness. The drifter data confirms the altimetric EKE changes showing the enhancement of the NAC surface branches across the MAR, and increased surface drift in the eastern subpolar gyre. Of course, therein lies our contradiction to be resolved: Weakening gyre but stronger EKE in the NAC surface branches which should mean stronger surface currents. Understanding this difference could be improved by reconciliation of the various views on the sources of the NAC and Nordic Seas inflow which diverge significantly whether using hydrography, and Eulerian average of drifter velocities or Lagrangian drifter tracks. Classic descriptions of the inflow source have evolved based on climatology or at most decadal spans of data which both are unsatisfactory if the hydrographic conditions and circulation in the northern North Atlantic change within a decade, as we have seen with altimetry. Despite much uncertainty in detail, there is growing evidence that both the northward upper-ocean flows that feed the Nordic Seas and the deep-reaching subpolar gyre circulation involve pathways, as well as amplitudes, which shift in time.

Several questions remain to be answered and to settle interpretations of the forcing of gyre changes. Despite much work, understanding of the roles of buoyancy and wind stress in low-frequency variability of the NAC, its path and variability are lacking. Another important question is how the strengths of the subpolar gyre and NAC are associated with MOC changes. We expect that continued development of modeling and assimilation of the observations into models will illuminate the interaction between the subtropical and subpolar gyres and relationships between the atmosphere and high latitude ocean. In the effort to follow up the development in the subpolar gyre, we hope that altimeter missions from international space agencies continue to find support for years to come, to form a valuable climate data record.

Acknowledgments SH gratefully acknowledges the support from NASA Headquarters Physical Oceanography Program for this work. PBR and HH are supported by NASA through the OSTST Science Team. We thank Ms Denise Worthen for the invaluable technical assistance in data set analysis and graphics.

References

Bailey D, Rhines PB, and Hakkinen S (2005) Pathways and formation of North Atlantic Deep Water in a coupled ice-ocean model of the Arctic-North Atlantic Oceans. Climate Dynamics 24, 10.1007/s00382-005-0050-3.

Belkin IM, Levitus S, Antonov J, and Malmberg S-A (1998) "Great Salinity Anomalies" in the North Atlantic, Progress in Oceanography 41: 1–68.

Belkin IM (2004) Propagation of the 'Great Salinity Anomaly' of the 1990s around the northern North Atlantic, Geophysical Research Letters 31, Art. No. L08306.

Bersch M, Meincke J, and Sy A (1999), Interannual thermohaline changes in the northern North Atlantic 1991–1996, Deep-Sea Research Part II-Topical Studies In Oceanography 46(1–2): 55–75.

Bersch M (2002), North Atlantic Oscillation – induced changes of the upper layer circulation in the northern North Atlantic Ocean, Journal of Geophysical Research 107: 3156, doi:10.1029/2001JC000901.

Boening CW, Scheinert M, Dengg J, Biastoch A, and Funk A (2006), Decadal variability of subpolar gyre and its reverberation in the North Atlantic overturning, Geophysical Research Letters 33: Art. No. L21S01.

Bower AS, Le Cann B, Rossby T, Zenk W, Gould J, Speer K, Richardson PL, Prater MD, and Zhang H-M (2002) Directly measured mid-depth circulation in the north-eastern North Atlantic Ocean, Nature 419: 603–607.

Brambilla E and Talley LD (2006) Surface drifter exchange between the North Atlantic subtropical and subpolar gyres, Journal of Geophysical Research 111: C07026, doi:10.1029/2005JC003146.

Bryden HL, Longworth HR, and Cunningham SA (2005), Slowing of the Atlantic meridional overturning circulation at 25N, Nature 438: 655–667.

Brügge B (1995) Near-surface mean circulation and kinetic energy in the central North Atlantic from drifter data, Journal of Geophysical Research 100: 20543–20554.

Cuny J, Rhines PB, Niiler PP, and Bacon S (2002) Labrador Sea boundary currents and the fate of the Irminger Sea Water, Journal of Physical Oceanography 32: 627–647.

Dickson RR, Meincke J, Malmberg S-A, and Lee AJ (1988) The 'Great Salinity Anomaly' in the northern North Atlantic 1968–1982, Progress in Oceanography 20: 103–151.

Dickson RR, Lazier J, Meincke J, Rhines P, and Swift J (1996) Long-term coordinated changes in the convective activity of the North Atlantic, Progress in Oceanography 38: 241–295.

Dickson R, Yashayev I, Meincke J, Turrell B, Dye S, and Holfort J (2002) Rapid freshening of the deep North Atlantic Ocean over the past four decades, Nature 416: 832–837.

Ducet N, LeTraon PY, and Reverdin G (2000) Global high-resolution mapping of ocean circulation from TOPEX/Poseidon and ERS-1 and –2, Journal of Geophysical Research 105: 19477–19498.

Flatau MK, Talley L, and Niiler PP (2003) The North Atlantic Oscillation, Surface Current Velocities, and SST Changes in the Subpolar North Atlantic, Journal of Climate 16: 2355.

Fratantoni DM (2001) North Atlantic surface circulation during the 1990s observed with satellite-tracked drifters, Journal of Geophysical Research 106: 22067–22093.

Han G and Tang CL (2001) Interannual Variations of Volume Transport in the Western Labrador Sea Based on TOPEX/Poseidon and WOCE Data, Journal of Physical Oceanography 31: 199–211.

Häkkinen S (1999) Variability of the simulated meridional heat transport in the North Atlantic for the period 1951–1993, Journal of Geophysical Research 104: 10991–11007.

Häkkinen S (2001) Variability in sea surface height: A qualitative measure for the meridional overturning in the North Atlantic, Journal of Geophysical Research 106: 13837–13848.

Häkkinen S and Rhines PB (2004) Decline of subpolar North Atlantic gyre circulation during the 1990s, Science 304: 555–559.

Heywood KJ, McDonagh EL, and White MA (1994) Eddy kinetic energy of the North Atlantic subpolar gyre from satellite altimetry, Journal of Geophysical Research 99: 22525–22539.

Holliday NP, Pollard RT, Read JF, and Leach H (2000) Water mass properties and fluxes in the Rockall Trough, 1975–1998, Deep Sea Research, Part I 47: 1303–1332.

Hurrell JW (1995) Decadal Trends in the North Atlantic Oscillation regional Temperatures and precipitation, Science 269: 676–679.

Hátún H, Hansen B, Sandø AB, Drange H, and Valdimarsson H (2005) De-stabilization of the North Atlantic Thermohaline Circulation by a Gyre Mode, Science 309: 1841–1844.

Jayne SR (2006) Circulation of the North Atlantic Ocean from altimetry and the Gravity Recovery and Climate Experiment geoid, Journal of Geophysical Research 111, doi:10.1029/2005JC003128.

Keike D, Rhein M, Stramma L, Smethie WM, LeBel DA, and Zenk W (2006) Changes in the CFC inventories and formation rates of Upper Labrador Sea Water, 1997–2001, Journal of Physical Oceanography 36: 64–86.

Krauss W (1986) The North Atlantic Current, Journal of Geophysical Research 100: 5061–5074.

Lilly JM, Rhines PB, Schott F, Lavender K, Lazier J, Send U, and d'Asaro E (2003) Observations of the Labrador Sea eddy field, Progress in Oceanography 59: 75–176.

McCartney MS and Mauritzen C (2001) On the origin of the warm inflow to the Nordic Seas, Progress in Oceanography 51: 125–214.

Mauritzen C, Hjøllo SS, and Sandø AB (2006) Passive tracers and active dynamics: A model study of hydrography, and circulation in the northern North Atlantic, Journal of Geophysical Research 111: C08014, doi:10.1029/2005jc003252.

Menard Y (1983) Observations of eddy fields in the Northwest Atlantic and Northwest Pacific by SEASTA altimeter data, Journal of Geophysical Research 88: 1853–1866.

Perez-Brunius P, Rossby T, and Watts DR (2004) Absolute transports of mass and temperature for the North-Atlantic Current-Subpolar Front system, Journal of Physical Oceanography 34: 1870–1883.

Pickart RS, Spall MA, Ribergaard MH, Moore GWK, and Milliff RF (2003) Deep convection in the Irminger Sea forced by the Greenland tip jet, Nature 424: 152–156.

Pickart RS and Spall MA (2006) Impact of Labrador Sea convection on the North Atlantic meridional overturning circulation, Journal of Physical Oceanography, submitted.

Pollard RT, Read JF, and Holliday NP (2004) Water masses and circulation pathways thorugh the Iceland Basin during Vivaldi 1996, Journal of Geophysical Research 109: C04004, doi:10.1029/2003JC002067.

Reverdin G, Cayan D, and Kushnir Y (1997) Decadal variability of hydrography in the upper Northern Atlantic, 1948–1990, Journal of Geophysical Research 102: 8505–8532.

Reverdin G, Verbrugge N, and Valdimarsson H (1999) Upper ocean variability between Iceland and Newfoundland 1993–1998, Journal of Geophysical Research 104: 29599–29611.

Reverdin G, Niiler PP, and Valdimarsson H (2003) North Atlantic surface currents, Journal of Geophysical Research 108: 3002, doi:10.1029/2001JC001020.

Rossby T (1996) The North Atlantic Current and surrounding waters: At crossroads, Reviews of Geophysics 34: 463–481.

Treguier AM, Theetten S, Chassignet EP, Penduff T, Smith R, Talley L, Beismann JO, and Boening C (2005) The North Atlantic Subpolar Gyre in four high-resolution models, Journal of Physical Oceanography, 35, 757–774.

Smethie WM and Fine RA (2001) Rates of North Atlantic Deep Water formation calculated from chlorofluorocarbon inventories, Deep-Sea Research Part I 48: 189–215.

White MA and Heywood KJ (1995) Seasonal and interannual changes in the North Atlantic subpolar gyre from Geosat and TOPEX/Poseidon altimetry, Journal of Geophysical Research 100: 24931–24942.

Verbrugge N and Reverdin G (2003) Contribution of horizontal advection to the interannual variability of sea surface temperature in the North Atlantic, Journal of Physical Oceanography 33: 964–978.

Volkov DL and van Aken HM (2003) Annual and interannual variability of sea level in the northern North Atlantic Ocean, Journal of Geophysical Research 108: 3204, doi:10.1029/2002JC001459.

Chapter 24
The History of the Labrador Sea Water: Production, Spreading, Transformation and Loss

Igor Yashayaev[1], N. Penny Holliday[2], Manfred Bersch[3], and Hendrik M. van Aken[4]

24.1 Introduction

In winter, cold Arctic outbreaks from Labrador result in intense air–sea heat exchanges transferring large quantities of heat from the ocean to the atmosphere and intermittently exciting convective mixing known to form the most prominent water mass of the subpolar North Atlantic – the *Labrador Sea Water* (LSW).

Production, spreading and recirculation of LSW, and its transformation and loss through mixing and export are analyzed using the most extensive systematic collection of hydrographic observations across the Atlantic between 50° N and 65° N. We demonstrate striking changes in the subpolar North Atlantic caused by massive LSW production between the mid-1980s and mid-1990s and document recent salinification and warming that have already brought the whole subpolar gyre to a state that is as warm and saline as the reported extreme of the late 1960s.

Since the mid-1980s, two prominent LSW classes have been observed as they developed in the Labrador Sea and spread across the North Atlantic. The first, extremely dense, deep and voluminous class was progressively built by intense winter convection during the period of 1987–1994. Most of this LSW class has left the subpolar gyre, however its remnants can still be found there. The second, shallower class gained much of its strength in 2000, and over subsequent years it became thicker and deeper. The anomalous signals acquired by these LSW classes in their formation region arrive in the Irminger and Iceland basins with characteristic delays of 2 and 5 years for deeper LSW and 1 year and 4 years for shallower LSW.

[1] Ocean Circulation Section, Ocean Sciences Division, Bedford Institute of Oceanography, Fisheries and Oceans Canada, Challenger Drive, P.O. Box 1006, Dartmouth, NS, B2Y 4A2, Canada

[2] National Oceanography Centre, Southampton, UK

[3] Institute of Oceanography, University of Hamburg, Germany

[4] Royal Netherlands Institute for Sea Research

R.R. Dickson et al. (eds.), *Arctic–Subarctic Ocean Fluxes*, 569–612

It took between 9 and 11 years for the changes seen in the deep LSW class to reach the southern Rockall Trough and northern Iceland Basin, and at least 10 years to arrive at the northern Rockall Trough.

Once convection weakens, the dense and deep LSW loses "*communication*" with the atmosphere. Via transformation due to mixing, and uncompensated volume loss through export, it steadily loses its strength as well as its cold and fresh signature. This LSW volume loss resulted in the most striking restratification of the intermediate and upper layers ever documented, while the properties and shape of the deep LSW class are now being solely controlled by mixing and *advective–diffusive* exchange with the other intermediate waters of the North Atlantic.

The temperature–salinity–time projections constructed individually for each region of interest allow us to look at the whole development cycle of the most intriguing water mass of the Atlantic Ocean and to link each regional cycle with that in the LSW formation region. From these projections we are able to identify the points that characterize the transition of LSW from the development (formation) to decay (cessation of renewal) stage in each analyzed region. This "*tendency reversal*" point was found to "*propagate*" eastward across the subpolar North Atlantic, consistent with the stated LSW transit times, and then north towards the heads of the Iceland Basin and Rockall Trough.

The outline of this chapter is as follows. First we provide an overview of the Labrador Sea circulation and water masses, and introduce the *Labrador Sea Water* (LSW) and the important role it plays in the North Atlantic circulation and budgets of freshwater and heat. After describing the data set used in this study we summarize the variations in the LSW properties over time and across the subpolar North Atlantic. We introduce the technique of volumetric analysis to identify and study different classes of LSW, and use the results to describe the formation and decline of two recent classes. Finally we trace the spreading of those classes across the subpolar North Atlantic, showing how their distinguishing features can be consistently traced to the northern limits of the eastern basin despite very great modification of properties *en route*.

24.2 The Labrador Sea: Receiver, Transformer and Producer of the Intermediate and Deep North Atlantic Water Masses

The subpolar North Atlantic (Fig. 24.1) is a key arena for the low and high latitude regions to exchange their heat and freshwater signals and is an area where the major intermediate and deep waters of the North Atlantic develop, gaining their characteristic properties and signatures.

Analysis of oceanographic sections remains the most effective way of exploring both *water masses*, a basic element of our investigations, and spatio-temporal changes throughout the ocean's water columns. Loyal to this tradition, we will first look at composite temperature and salinity sections across the subpolar North Atlantic shown in Figs. 24.2 and 24.3. The hydrographic archive used to construct these vertical transoceanic sections and also property maps, time series, regional and

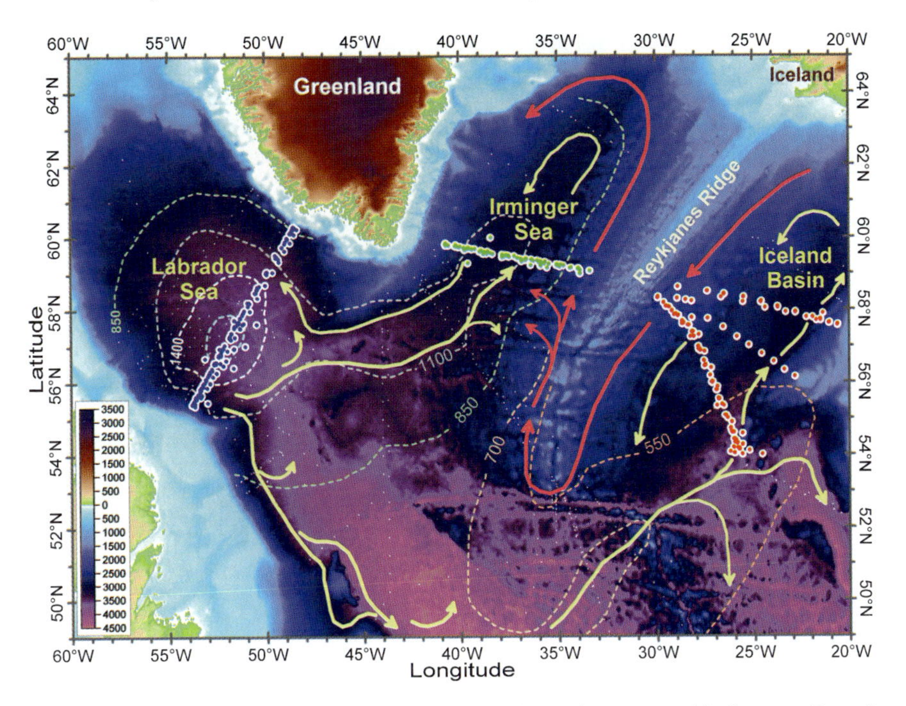

Fig. 24.1 Map of the subpolar North Atlantic showing major topographic features (the color legend indicates elevation/depth in meters). White-padded circles indicate hydrographic stations occupied between 1987 and 2005 along the trans-Atlantic section AR7 in the Labrador, Irminger and Iceland basins. Dashed lines represent thickness (meters) of the layer defined by the σ2 range best confining the core of deep LSW in 1995–1997 ($36.92 < \sigma_2 < 36.95\,\mathrm{kg\ m^{-3}}$, σ_2 is potential density anomaly referenced to 2000 db). This mapping was based on the 1995–1997 hydrographic profiles, whose positions are indicated in the figure by white dots. The yellow arrow headed lines follow the LSW spreading and recirculation pathways as inferred from the LSW thickness and vertical section plots, the red arrow headed lines indicate the spreading of ISW (Reproduced from Yashayaev et al. 2007a)

watermass specific syntheses and full-depth inventories is described in Section 24.3. The analyzed sections all cross the key subpolar basins in a similar way (Figs. 24.2 and 24.3, *insert maps*) and collectively demonstrate the most prominent hydrographic change in the deep ocean (discussed in Section 24.4). The acronyms ISOW and NEADW indicate the relatively salty *Iceland–Scotland Overflow Water*, subsequently developing into the *Northeast Atlantic Deep Water*; DSOW indicates the *Denmark Strait Overflow Water*, typically identifiable by the lowest temperatures over the deep and bottom layers of the Labrador and Irminger basins. These water masses arrive from the Arctic and are strongly regulated by and mixed with a water mass that is entirely formed in the subpolar North Atlantic, the *Labrador Sea Water* (LSW) (Boessenkool et al. 2006; Yashayaev and Dickson 2007, Chapter 21, this volume). LSW appears as a prominent feature in transoceanic sections such as those shown in Figs. 24.2, 24.3 and 24.14, a great thickness of homogenous mid-depth water (e.g., LSW, indicated with vertical double-headed arrows). LSW is largely formed in the Labrador Sea through deep convection during severe winters.

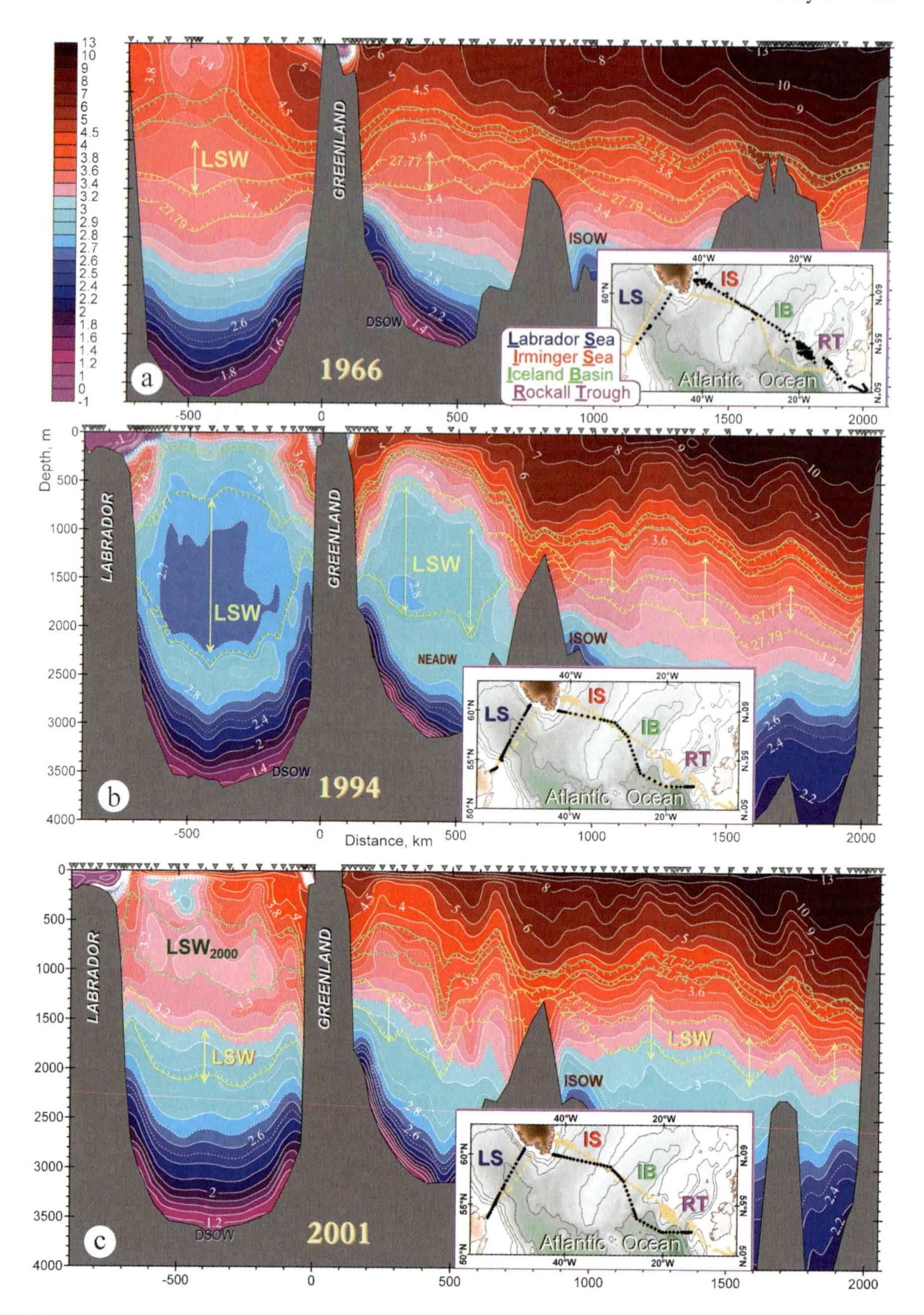

Fig. 24.2 Trans-Atlantic potential temperature in 1966 (a), 1994 (b), 2001 (c) and 2004 (d). Sampling/profiling sites are shown in inserts. The hatched contours indicate constant potential density levels of 24.72 and 24.74 (*green*) and 24.77 and 24.79 (*yellow*) kg m^{-3} (*referenced to the sea surface*)

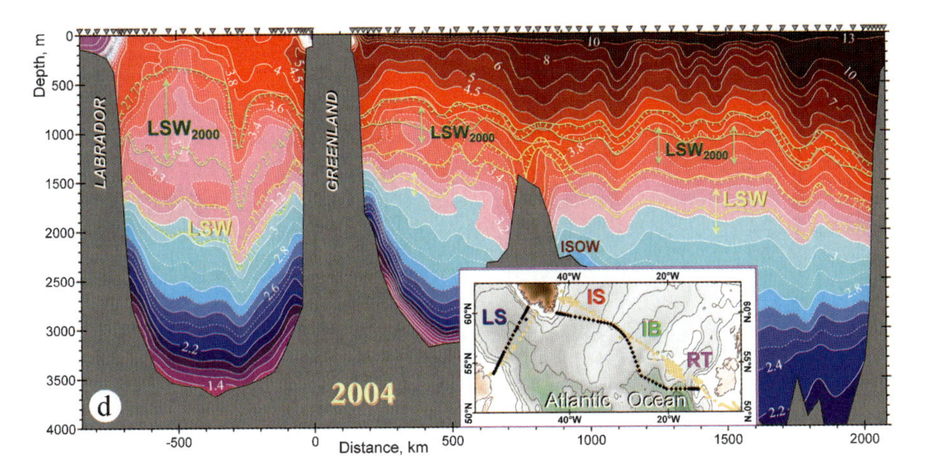

Fig. 24.2 (continued)

Consequently, the full-depth circulation, mixing and hydrographic changes in the formation region of this water mass are of high importance for the intermediate layers and inter-layer signal transfers in the subpolar North Atlantic and in the remote regions through which LSW is transported.

The Labrador Sea is commonly recognized as a crucial location for the development of regional to large-scale anomalies in ocean properties, and, consequently, as a region exerting a significant influence on the climate system. In order to better identify the role of this basin in the sub-Arctic hydrography and the ocean's long-term changes (e.g., Dickson et al. 2002) we first discuss the hydrographic composition of the Labrador Sea and mark the episodes of extreme atmospheric forcing that have caused substantial changes in the regional heat and freshwater contents, which in their turn regulate the inter-basin exchanges.

24.2.1 The Labrador Sea Composition

In temperature and salinity sections and maps, the Labrador Sea can be identified as the coldest and freshest region of the subpolar North Atlantic (Figs. 24.2 and 24.3). At the same time this basin serves as the final destination for the warm and salty waters originating from the North Atlantic Current. Cooled and freshened during their passage around the subpolar gyre, these waters still maintain significant temperature and salinity contrasts in the upper layer of the Labrador Sea. This leads to intense mixing, and vertical and horizontal transfer of heat and freshwater. Ultimately, these Atlantic waters become involved in the formation of the Labrador Sea characteristic water mass that we discuss below.

In the upper layer, the two main relatively fresh and cold inflows arrive in the North Atlantic from the Arctic Ocean by way of the Canadian Arctic Archipelago and the East Greenland shelf, and pass around the margins of the Labrador Sea forming its boundary currents. These currents, respectively known as the Labrador

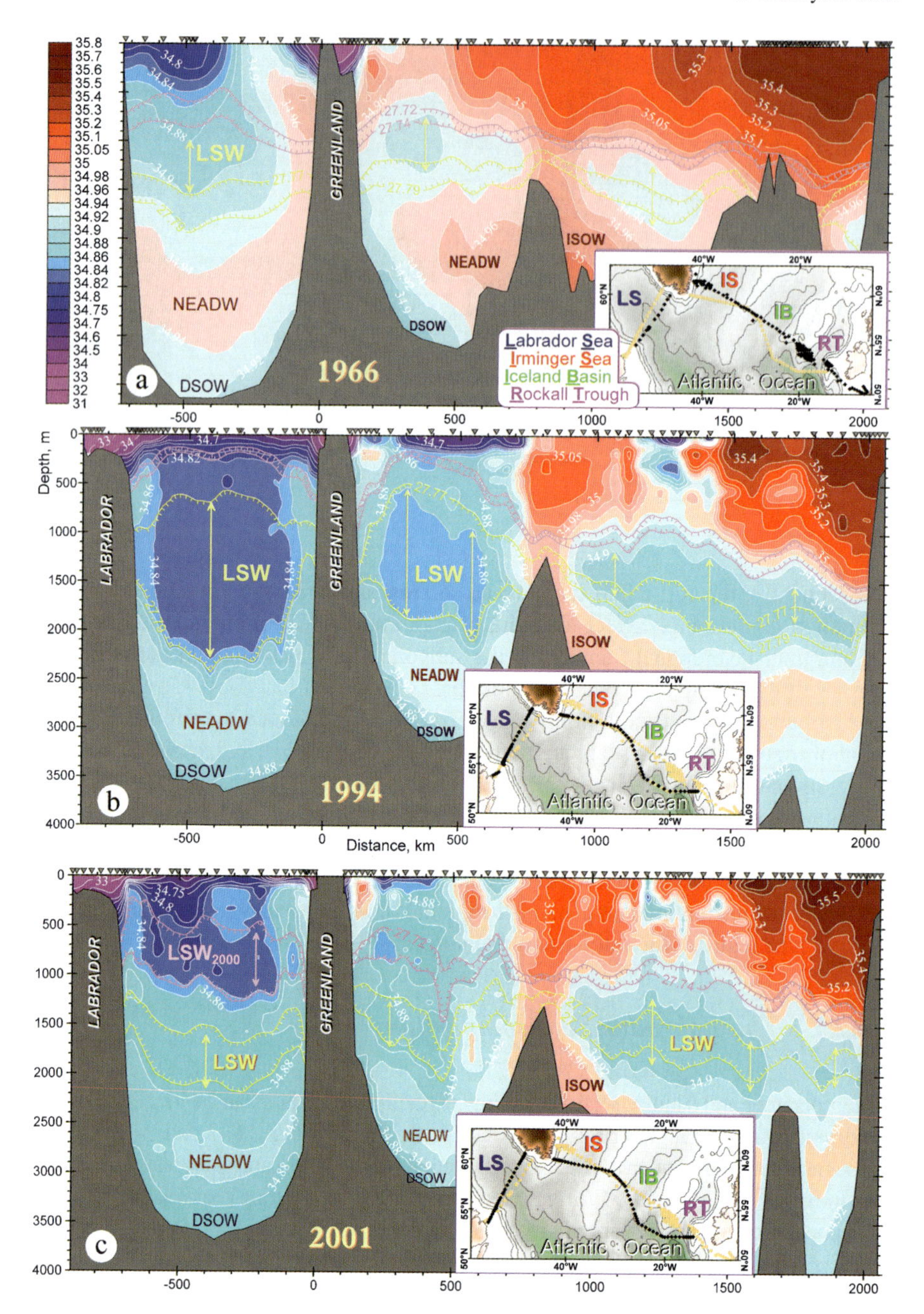

Fig. 24.3 Trans-Atlantic salinity in 1966 (a), 1994 (b), 2001 (c) and 2004 (d). Sampling/profiling sites are shown in inserts. The hatched contours indicate constant potential density levels of 24.72 and 24.74 (*violet*) and 24.77 and 24.79 (*yellow*) kg m^{-3} (*referenced to the sea surface*)

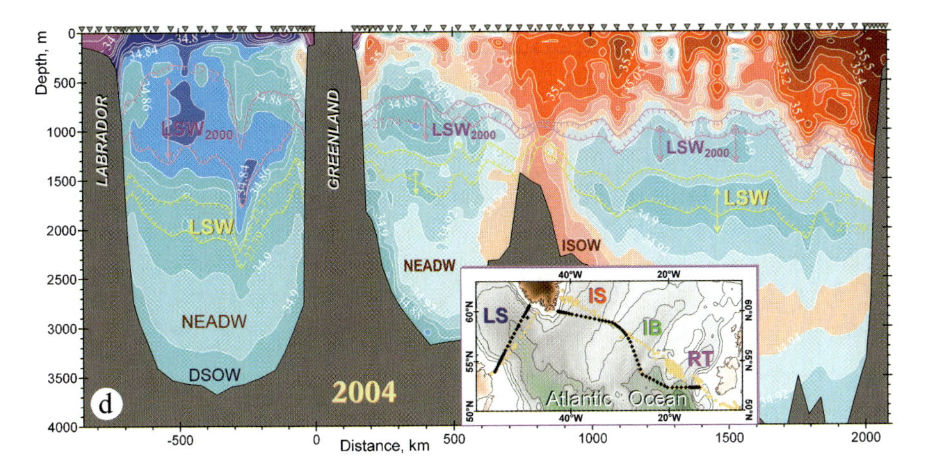

Fig. 24.3 (continued)

and West Greenland Currents, can be easily identified by their fresh and cold cores over the upper continental slopes (Figs. 24.1–24.3). The low-salinity water found over the Labrador continental shelf largely originates from southward flows out of Baffin Bay, two parts of which are consecutively referred to as the Baffin Island Current and the aforementioned Labrador Current. The Labrador Current in its turn forms the main pathway for the equatorward-flowing cold and fresh Arctic and subarctic waters which subsequently influence the hydrography and ecosystems of the shelf-slope regions downstream.

The zones of rapid transition between the low-salinity waters over the shelves and the high-salinity waters at the same depths further offshore reveal two strong baroclinic currents (Figs. 24.2 and 24.3). Patches of warmer and saltier water are found off the Greenland and Labrador continental slopes. These patches are associated with a flow, originating from a branch of the North Atlantic Current or its derived water mass, the *Subpolar Mode Water* (McCartney and Talley 1982). We define this flow as the Irminger Current. The deep Irminger Current is found offshore and deeper than the East and West Greenland Currents. Its warm and salty core is typically centered at about 500 m below the sea surface (Figs. 24.2 and 24.3). The Irminger Current and its associated eddies contribute to the overall heat, salt and freshwater budgets of the basin and, by maintaining the flux of heat and salt toward the center of the Labrador Sea, influence the development of winter convection.

In addition to holding the coldest and freshest water column of the whole North Atlantic, the Labrador Sea produces large temperature and salinity anomalies on annual to decadal timescales. These anomalies develop intermittently at the sea surface with some penetrating into deeper layers (Figs. 24.4 and 24.5). Following their formation, these anomalies move out of the Labrador Sea, join the subpolar circulation gyre and spread east-to-northeast across the North Atlantic and south-to-southwest along the western boundary of the North Atlantic. In some cases these anomalies cross the Subpolar Front and influence the Gulf Stream and North Atlantic Current (Yashayaev 2000).

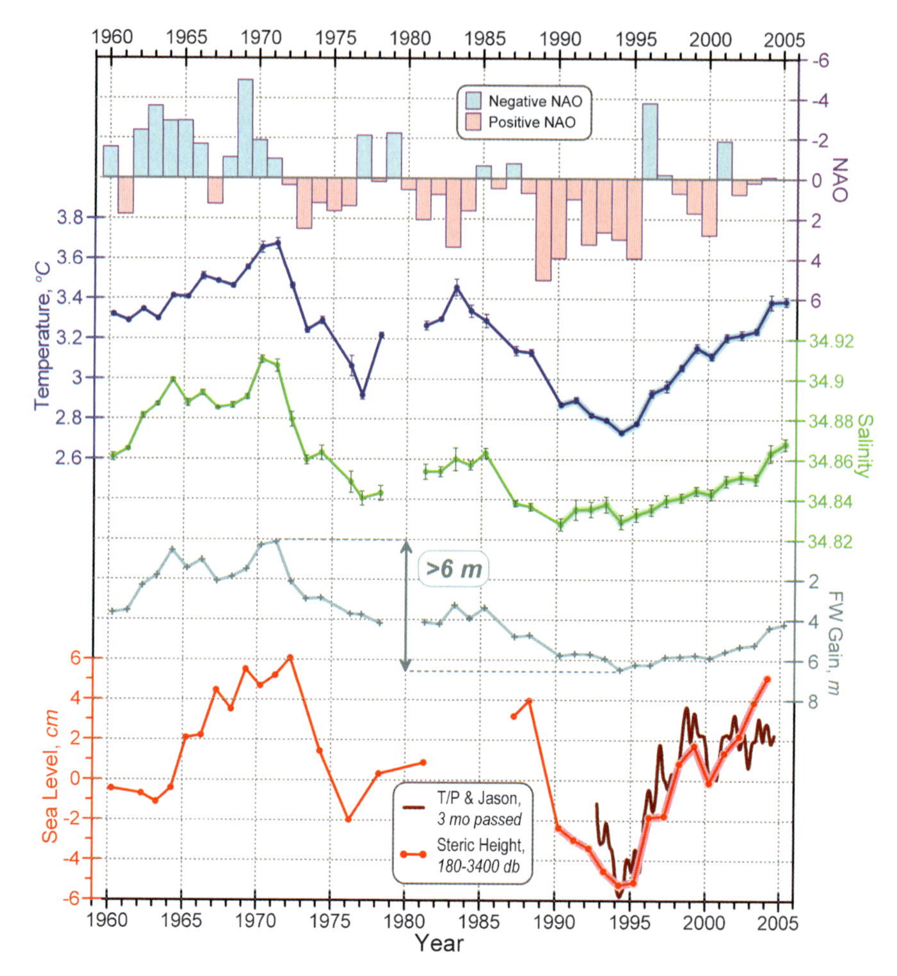

Fig. 24.4 The 150—2,000 m potential temperature and salinity means in the central Labrador Sea bounded by the 3,250 m isobath. The freshwater gain (FW, *note that the FW axis is inverted*) was calculated over the full water depth. The steric height (*lower plot, red line, 1960–2005*) represents the water column thickness; it was derived from all available temperature and salinity measurements in the central Labrador Sea. The observed sea-level anomalies (*brown line, 3-month low-pass filtered anomalies relative to the record mean, 1992–2004*) were calculated from satellite altimeter data acquired by the *Topex/Poseidon* and *Jason* missions. The period of AR7W occupation, *1990–2005*, is highlighted (Reproduced from Yashayaev 2007b)

24.2.2 Deep Convection in the Labrador Sea and Production of the Labrador Sea Water

Through production of its characteristic intermediate, deep and abyssal water masses, the northern North Atlantic contributes to the *meridional overturning circulation* (MOC) of the whole Atlantic Ocean. These waters form the lower limb of the great ocean conveyor and subsequently participate in the ventilation of the deep layers of the

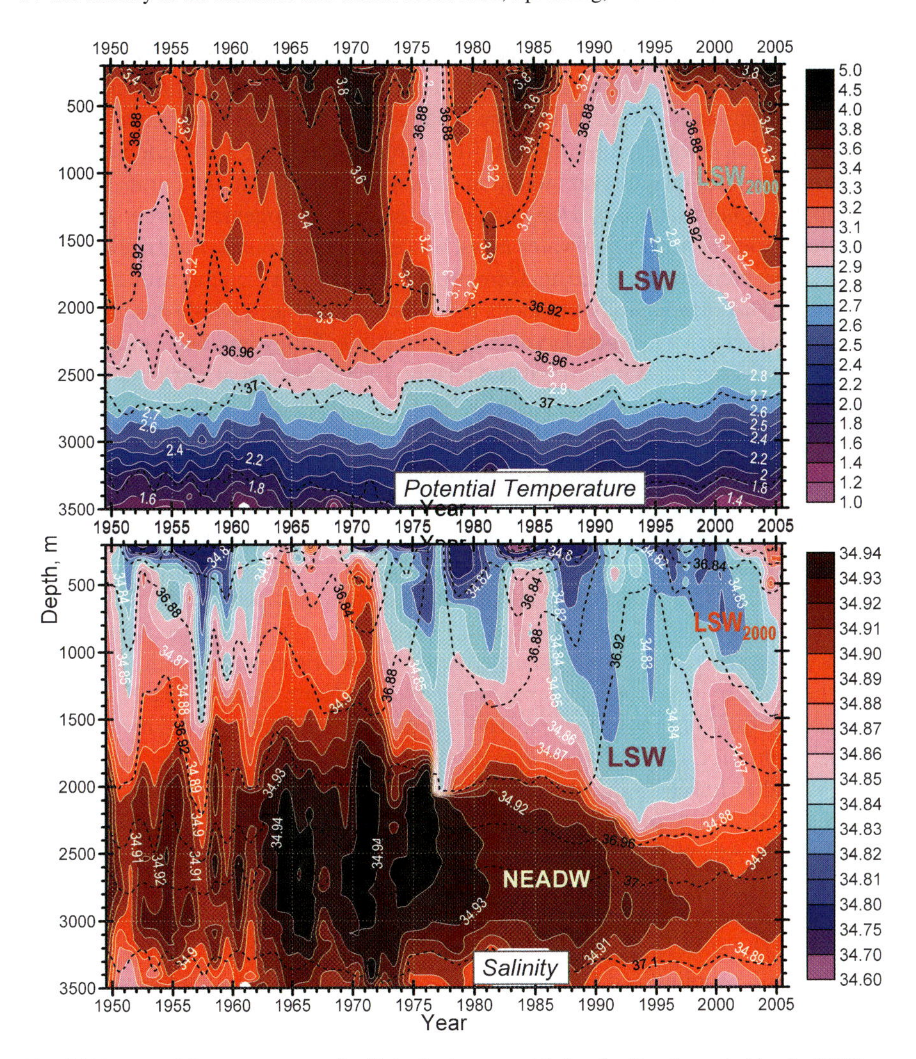

Fig. 24.5 Potential temperature and salinity in the central Labrador Sea (*bounded by the 3,250 m isobath and the 150 km distance range from the Labrador Sea section AR7W*), 1950–2005. The σ_2 (*potential density anomaly referenced to 2,000 db*) isolines (*dashed contours*) indicate that the Labrador Sea Water (LSW) produced between 1990 and 1994 was the record's densest

world ocean. This ocean-wide chain of water mass development and transformation is largely governed by specific processes taking place in some local *water mass formation* or *source regions*. One of these key regions intermittently produces a characteristic water mass that spreads out forming a remarkable intermediate layer in the North Atlantic. This layer can be unmistakably identified by the mid-depth salinity minimum (Figs. 24.3 and 24.14). The direction in which the thickness and properties of this water mass change between the basins implies that it originates in the Labrador Sea. The next two paragraphs link the development of the most renowned water mass of the North Atlantic with the strength of the atmospheric forcing in the Labrador Sea.

In winter, cold Arctic outbreaks result in intense atmosphere–ocean heat exchanges, transferring large amounts of heat from the ocean to the atmosphere. During exceptionally severe winters the ocean heat losses reach their extreme levels. These losses accumulated by the gyre-like circulation of the Labrador Sea create the densest winter mixed layers found in the subpolar North Atlantic or its southern neighbors. Episodically this deep convective mixing homogenizes large quantities of water, producing a unique water mass, named after its source, the *Labrador Sea Water*. The literature devoted to formation and variability of this water mass is extensive. The current study broadens the view on LSW as a whole based on both historic and recent hydrographic data (Lazier 1980; Talley and McCartney 1982; Clarke and Gascard 1983; Gascard and Clarke 1983; Lazier et al. 2002; Yashayaev et al. 2003, 2007a, b; Yashayaev 2007a, b).

Varying production, properties and thickness of convectively formed LSW largely determine the net supply of freshwater, nutrients (Clarke and Coote 1988) and gasses to the intermediate and deep layers of the ocean and affect the rate of the Atlantic overturning circulation both directly (Dickson et al., 2002) and indirectly (e.g., by affecting the *Iceland–Scotland Overflow Water*, Boessenkool et al. 2006). In its turn, the LSW production is believed to be strongly affected by the phase and persistence of the *North Atlantic Oscillation* (NAO) (Dickson et al. 1996). The North Atlantic Oscillation is the primary mode of climate variability that involves the winter atmospheric circulation over the northern hemisphere and principally over the North Atlantic. The NAO index is the normalized Azores-to-Iceland sea-level pressure difference (*since the meteorological record at Lisbon is longer than at Azores, Lisbon is often used in the place of Azores*) and is linked to the strength of the large-scale zonal atmospheric transport (Hurrell et al. 2001). A positive NAO means a stronger pressure difference over the central North Atlantic and hence stronger winter winds over this region. The fact that the correlation between the NAO index and the strength of Labrador Sea convection weakens on sub-decadal timescales emphasizes the role of the ocean dynamics and local atmospheric forcing in deep convection (Yashayaev 2007b).

Convective cooling and freshening of the Labrador Sea between the mid-1980s and mid-1990s have produced a characteristic LSW that by 1994 became the coldest, densest, deepest and most voluminous since the 1960s (Yashayaev 2007b) and indeed in the entire historical record going back to the 1930s (Yashayaev et al. 2003). The present article unfolds the most complete history of LSW production, spreading and transformation based on several decades of precise, routinely collected observations. It documents the changes in the deep basins of the subpolar North Atlantic caused by exceptionally large production of this water mass via extremely deep convection recurrent during the late 1980s – early 1990s.

24.2.3 LSW in the Atlantic Overturning Circulation

When produced in large quantities, LSW spreads across the subpolar North Atlantic (Fig. 24.1) filling its intermediate reservoir with water that is relatively fresh, cold (Figs. 24.2 and 24.3) and rich in dissolved gases. The varying production, volume and properties of this water mass determine the mid-depth circulation, mixing and signal propagation in the subpolar basins (Talley and McCartney 1982; Yashayaev et al. 2007a, b). In this way the Labrador Sea plays a role in the ocean's circulation, signal transfers and exchanges beyond the direct export of its newly formed intermediate-depth water mass loaded with freshwater, anthropogenic gases and other important substances. The Labrador Sea can regulate or influence the circulation and mixing in the Atlantic Ocean and, ultimately, the global ocean conveyor belt in several other ways, as follows.

The rate of LSW production is directly involved in determining the strength of exchange between the subtropical and subpolar gyres. The depth, thickness and density of all resident LSW formations contribute to the sea-level height of the Labrador Sea at the centre of the subpolar gyre. This leads to changes in the dynamic height gradient between the subpolar gyre and the subtropical gyre, influencing the volume transport of the North Atlantic Current (Curry and McCartney 2001). A weak gradient acts to contract the subpolar gyre reducing its transport, pulling the Subpolar Front westwards and drawing more subtropical water into the eastern subpolar gyre (Bersch 2002; Hakkinen and Rhines 2004; Hátún et al. 2005; Bersch et al. 2007). In addition, changes in the dynamic height pattern within the Labrador Sea regulate the geostrophic mass, heat and freshwater transports by the *Upper Labrador, Labrador Slope, Deep Western Boundary* and other key currents (Yashayaev 2007b).

The LSW directly contributes to the formation of the water masses that eventually fill the deep reservoirs of the North Atlantic, through mixing with the cold and dense overflows. The *Iceland–Scotland* and *Denmark Strait Overflow Waters* (ISOW and DSOW) enter the North Atlantic through the deepest trenches in the Greenland–Iceland–Faroe–Shetland–Scotland underwater ridges (referred to as the Greenland–Scotland Ridge). As they descend from the shallow ridge they mix briefly with the surface mode waters, and in a more prolonged way mix with the intermediate LSW (Dickson et al. 2002). As they do so they entrain the relatively fresh water, elevated gas and transient tracer signatures of LSW. The ISOW evolves along its spreading pathway, becoming markedly fresher as it crosses the Reykjanes Ridge and enters the Irminger Sea, where it transforms into the *Northeast Atlantic Deep Water* (NEADW).

The vigor of the deep-ocean flow that originates from ISOW and contributes to the lower limb of the Atlantic MOC is thought to be strongly controlled by the volume and properties of LSW (Boessenkool et al. 2006). Finally, all of the aforementioned components of the lower limb of the MOC pass through the Labrador basin where their evolving properties are being effectively monitored (Yashayaev 2007b).

24.3 The Subpolar Trans-Atlantic Watch: Hydrographic Data Used in Studies of the Intermediate and Deep Water Masses of the North Atlantic North of 50° N

The changes in production, properties and volumes of the intermediate (mostly LSW) and deep (ISOW/NEADW, DSOW) waters in the North Atlantic are key for diagnoses and prognoses of the state and vigor of the MOC and, thereby, the climate system. This fact underscores the observational programs led by research groups in the North Atlantic Ocean. These programs, essentially complementing one another, seek regular occupation of several principal oceanographic sections running across the subpolar basins. Such sections known as *repeat hydrography lines* have been regularly occupied under the aegis of the *World Ocean Circulation Experiment* (WOCE, 1990–1997) and *Climate Variability and Predictability* Program (CLIVAR, since 1997). Such a network of systematic observations in the northern North Atlantic is highly important to multidisciplinary studies of the oceanic climate and ecosystems. In particular, the repeat hydrography in the subpolar North Atlantic is invaluable for diagnosing and understanding the Arctic–Atlantic exchanges and their role in climate change.

The subpolar hydrographic survey is primarily aimed at systematic monitoring of the intermediate and deep water masses of the North Atlantic, and the rapidly changing *Labrador Sea Water* (LSW) has been the focus of attention since the beginning of WOCE (Lazier 1995; Sy et al. 1997). The maps in Figs. 24.2 and 24.3 (*inserts*) show typical locations of water column profiling and sampling for important seawater properties (e.g., *temperature, salinity, dissolved oxygen, nutrients, carbonates, CFCs*) on the trans-Atlantic repeat hydrography section, known as AR7 and its 1966 prototype. This key WOCE–CLIVAR section running across the *Labrador Sea, Irminger Sea, Iceland Basin* and *Rockall Trough* represents an exemplary case of international collaboration. The Labrador Sea part of the AR7 section (*Misery Point on the Labrador coast to Cape Desolation, Greenland*) is known as AR7W. This line has been annually occupied by the *Bedford Institute of Oceanography* (BIO, Canada) since 1990, while its counterpart, AR7E, has been occupied by the *University of Hamburg* (Germany), *German Hydrographic Office* and *Royal Netherlands Institute for Sea Research* every year since 1991, except 1993 and 1998. The Iceland Basin and Rockall Trough were also surveyed in 1990.

The observations on the AR7W section, when combined with the US Coast Guard's *Ocean Weather Ship (OWS) Bravo* time series (1950–1974) and other historic data (e.g., Lazier 1980), document large interannual and decadal changes through the entire depth in this key region (Figs. 24.4 and 24.5). Several oceanographic surveys complying with the data quality requirements of WOCE were conducted during the 1980s close to AR7. These surveys allow us to examine the early development of the LSW class that during the early-to-mid-1990s developed to its record voluminous state.

In addition to physical properties of seawater, recent studies of water mass renewal make use of transient anthropogenic tracers. Chlorofluorocarbons (CFCs), the most popular of these tracers, are taken up by the ocean via air–sea gas exchange. They enter intermittently deep layers via intense winter convection and are exported out of their entry region by major ocean currents. Since their atmospheric history is well documented, CFCs are effectively used to document production, evolution, relative age and spreading pathways of newly formed waters. Since 1991, CFCs have been measured annually at all water sampling depths as a part of the monitoring of the Labrador Sea (Azetsu-Scott et al. 2003). A complementary effort directed at producing CFC inventories has employed measurements from spatially extensive but less frequent, typically biennial, surveys conducted since 1997 (Kieke et al. 2006).

To portray LSW in a year typical of the low NAO phase of the 1960s, and also to highlight the extreme LSW development of the 1990s in the multi-decadal history of this water mass, we compare the AR7 transoceanic hydrographic compilations for 1994, 2001 and 2004 with their 1966s prototype/precursor (Figs. 24.2 and 24.3). The 1966 section was constructed from hydrographic stations found near AR7; the majority of the observations used to produce this composite section are from the most extensive pre-WOCE survey of the subpolar North Atlantic, led by John Lazier (the 1966 *CSS Hudson* expedition to the northern North Atlantic).

To compare the rapid restratification of the intermediate layers of the subpolar basin that we started to observe soon after the cessation of deep convection in the mid-1990s with a similar episode of the past, we "simulate" the AR7 section of 1962 by combining the closest observations from the "*Erica Dan*" sections (Fig. 24.14).

Finally we include some analysis of a repeated hydrographic section in the north-eastern corner of the subpolar gyre, the *Extended Ellett Line* (Fig. 24.12, *insert map*). The Ellett Line runs across the northern Rockall Trough from the west coast of Scotland to the tiny island of Rockall, and was begun in 1975 by David Ellett at the *Scottish Association for Marine Science* (SAMS) (Ellett et al. 1986; Holliday et al. 2000). Since 1996 the section has been extended by SAMS and the *National Oceanography Centre* (Southampton, UK). It now runs to Iceland across the Hatton Bank and northern Iceland Basin, adding to periodic meridional sections along 20°W made during the WOCE field program.

24.4 LSW Variations in Time

24.4.1 From Warm and Salty to Cool and Fresh

The Labrador Sea exhibits considerable variability in water properties and stratification (Figs. 24.2–24.4) and is characterized by extremely complex circulation and mixing. Time series of vertically averaged or integrated potential temperature, salinity and density can be used to provide an overview of major climatically significant changes in the LSW source observed during several decades (Fig. 24.4). When combined with the temporal evolution of the whole water column of the Labrador Sea (Fig. 24.5) and hydrographic sections across the subpolar North Atlantic (Figs. 24.2 and 24.3), the averaged properties of the 150–2,000 m layer reveal important changes in the LSW production and thickness, and the impacts of such on the heat and freshwater content, stratification and steric height throughout the whole subpolar domain.

Temporal evolution of a whole water column can be effectively visualized by producing an average or characteristic vertical profile of a given seawater property for each year or hydrographic survey and mapping a succession of such profiles in *time-depth* coordinates. Figure 24.5 shows such a progression of annually averaged vertical profiles constructed for the central region of the Labrador Sea for the period of 1949–2005, inclusive. Part of Fig. 24.5 is expanded in Fig. 24.8 which focuses on the recent full-depth hydrographic developments in the central Labrador and Irminger basins, discussed in Section 24.6. The method for generating the time series for analysis is as follows. The central regions of the Labrador and Irminger basins were defined by the bottom depths exceeding 3,250 and 2,830 m, respectively, and by the horizontal distance range from the AR7 line (Fig. 24.1) not exceeding 150 km. Each characteristic vertical profile was formed by robust averaging of temperature, salinity, pressure (depth) and $\Delta\sigma_2 = 0.01\,\mathrm{kg\,m^{-3}}$ layer thickness (σ_2 *is potential density anomaly referenced to 2,000 db*). This averaging was performed individually for each calendar year with available observations and over each σ_2 bin (layer), predefined by $\Delta\sigma_2 = 0.005\,\mathrm{kg\,m^{-3}}$. The techniques of robust averaging, vertical interpolation and other data analyses used in the present study are documented in Yashayaev (2007b).

The 55-year record clearly shows three periods of the Labrador Sea warming (1962–1971, 1977–1983, and 1994–1997, Figs. 24.4 and 24.5). The first warming period was preceded by a fairly significant renewal of LSW that occurred during the late 1950s to early 1960s. At the end of this warming period, in 1970–1971, the Labrador Sea reached its warmest and saltiest state ever observed. The most recent warming started in 1994 and has continued throughout the following years (1994–2007). The average temperature and salinity of the upper 2,000 m layer have already returned to the high levels observed in the late 1960s and are likely to surpass these record high levels in the near future. The tendencies of warming and salinification are maintained by continuous inflow of the warm and salty waters from outside of the Labrador Sea (see Section 24.6 for more details).

On two occasions since 1960, the warming and salinification of the Labrador Sea was interrupted and offset by significant cooling and freshening caused by strong winter convection. The most remarkable event of convective watermass renewal occurred between the mid-1980s and mid-1990s and led to the development of a characteristic LSW that turned out to be the coldest, freshest, densest, deepest and most voluminous since the 1930s (Lazier et al. 2002; Yashayaev et al. 2003; Yashayaev 2007b). At the same time, the recent change in oceanographic conditions can be recognized as part of a cycle in water mass development affecting the entire subpolar domain. Periodic changes within the Labrador Sea result from the interplay of LSW and intermediate waters of a similar density from outside the basin. The warmer and saltier LSW alternatives tend to reoccupy the mid-depths as LSW production loses its vigour and is unable to compensate the loss in the LSW volume resulting from its draining out of the Labrador Sea. In Section 24.6 we discuss in more detail the "*life cycle*" of LSW, comprising this water's production, development, transformation and loss (due to its mixing and export).

The long-term periodic changes in the LSW properties and the subpolar hydrography as a whole can be linked to the *North Atlantic Oscillation* (NAO; Fig. 24.4, *upper plot, note that the NAO axis is inverted*). A predominance of high positive values of the NAO index is reflected in periods of cooling and freshening of the Labrador Sea, associated with renewal of the intermediate waters to 2000 m and deeper (1972–1976 and 1988–1994). In contrast, a predominance of negative NAO years from 1962 through to 1971 coincides with the period of little convective renewal of LSW, when it was becoming warmer and more saline (e.g., the 1966 section in Figs. 24.2a and 24.3a). The relationship between the NAO index and LSW production and properties is not straightforward because there are significant local processes which force the ocean on the interannual timescales. For example the thermal inertia opposes short-term fluctuations in the ocean's heat losses to the atmosphere. In addition, the local wind field is not always directly related to the strength of the *westerlies* over the central North Atlantic (Yashayaev 2007b).

The cycles of LSW development, evident in Figs. 24.4 and 24.5, had similar signatures of their rise and decline rates, stratification losses and vertical redistributions of freshwater. On the other hand, they varied in strength (or intensity), persistence and interannual variability within a cycle. As a result, the mixed layers formed in different years ranged in their thickness, depth, density and other characteristics. The aforementioned LSW development of the 1990s has surpassed in its prominence and outreach any other known production cycle of this water mass, forming an extreme in the recorded history of water mass renewal in the subpolar North Atlantic.

Figures 24.2 and 24.3 highlight two extreme states, the warm saline 1960s with little convective renewal, and the cold fresh mid-1990s with prolonged deep convection. The changes observed in the Labrador Sea 150–2,000 m mean temperatures and salinities between these extremes are on the order of 1 °C and 0.08, respectively, while the sea level dropped by more than 10 cm. It is notable that the density decrease and therefore expansion of the water column due to this long-term freshening could compensate only about half of the density decrease and water column contraction that would have resulted from the cooling alone. The 1970–1994 accumulation of freshwater by the Labrador Sea inferred from the full column salinity change between these years is equivalent to mixing of at least 6 m of freshwater into the water column of 1970.

In addition to the noted differences, the two extreme LSW states are responsible for markedly different vertical stratification (Fig. 24.5) and overall water mass distribution throughout the whole subpolar domain (Figs. 24.2 and 24.3). The change from the extremely low to high LSW production may have also affected the vigour of the deeper flow of ISOW (Boessenkool et al. 2006).

24.4.2 A Panoramic View of Subpolar Changes

The compilation of composite trans-Atlantic hydrographic sections presented in Figs. 24.2 and 24.3 introduces all principal water masses of the subpolar North Atlantic evolving from their warmest and saltiest state of the mid-to-late 1960s to the extremely cold and fresh phase of 1994 and then to the generally warmer and saltier conditions of the recent years. The differences between the two extreme states have been thoroughly analyzed in several recent publications (Yashayaev 2007a, b; Yashayaev et al. 2007a; Boessenkool et al. 2006). Here we review and expand those analyses by considering two other trans-Atlantic sections (*2001* shown in Figs. 24.2c and 24.3c, and *1962* shown in Fig. 24.14) and by discussing temperature changes.

A distinct salinity minimum at the intermediate depths is associated with LSW and can be recognized in all sections crossing the subpolar basins (Figs. 24.3 and 24.14). At the same time, this water mass exhibits substantial changes in its characteristics, thickness and depth from basin to basin, within basins and between surveys. Here we make a basic assumption that the Labrador Sea largely leads in this process: the changes seen in its main reservoir (Figs. 24.2 and 24.3, 24.14) spread to the other subpolar basins, presumably via *advective–diffusive exchange* (Straneo et al. 2003). Next we describe the observations that provide evidence to support this assumption. The full-depth transoceanic sections demonstrate how well the changes in the characteristic LSW layers are coordinated across the subpolar region.

1966 is in the middle of a prolonged phase of negative NAO (Fig. 24.4, *1960–1971*) and period of record warm and salty Labrador Sea conditions (Figs. 24.4 and 24.5). The winter of 1965–1966 was particularly mild in the Labrador Sea region (Lazier 1980), so it is unlikely that there was significant convective renewal of LSW during the mid- and late 1960s. This caused the LSW lying at the intermediate depths to remain isolated from the upper layer, become warmer and saltier through its mixing with surrounding waters and also become advectively replaced by those waters. In 1966, deep LSW could be identified in the Labrador Sea as a nearly homogeneous layer with salinities between 34.88 and 34.90 (Fig. 24.3a, *the distance range – −700 to −240 km*). A retrospective analysis suggests that the last significant renewal of LSW occurred in the winter of 1962–1963 (Lazier 1980; Yashayaev et al. 2003).

A particularly large change occurred between the 1966 and 1994 surveys, with freshening occurring at virtually all depths across the subpolar gyre. In 1966 the deep LSW core was everywhere saltier, warmer and shallower than in any hydrographic survey of the subpolar North Atlantic during the 1990s. The change by 1994 was a result of production of large volumes of an exceptionally cold, fresh, dense, deep and vertically homogeneous LSW class by strong winter convection of the late 1980s–early 1990s (Lazier et al. 2002; Yashayaev 2007b). This LSW is, in fact, the most voluminous LSW observed in the historic record and is discussed in more detail later. In 1994 this water mass was the most prominent feature of the intermediate layers, filling the entire central part of the Labrador Sea basin within the depth range of 500–2,400 m (Figs. 24.2b and 24.3b). This means that within the Labrador and Irminger basins, and to some degree in the Iceland Basin, the well-mixed body of fresh LSW had penetrated to the depths previously occupied by more saline NEADW. As a result, this LSW exceeded both vertically and horizontally any other water mass seen in the subpolar North Atlantic since the beginning of the *International Ice Patrol* survey in the 1930s. As time progressed, temperature, salinity and density stratification re-established above the thinning patch of LSW (Figs. 24.2–24.5, 24.7 and 24.8). The isolation of LSW was a result of a substantial decrease in the net annual heat loss from the Labrador Sea to the atmosphere after 1994 (Lazier et al. 2002; Yashayaev 2007b).

By 2001, most of the excess volume of LSW had disappeared from the Labrador and Irminger Seas and the water column had restratified above its deep core. However, 7 years after the last convective renewal of LSW took place, remnants of this water mass could still be easily identified in these two basins inside the 1,500–2,200 m depth range. These remnants were warmer and saltier than at the time of formation in 1993–1994 (Figs. 24.2c and 24.3c). In the same year the Iceland Basin did not show any significant increase in salinity and decrease in the volume of LSW. Indeed, a fairly extensive patch of LSW with salinities similar to those observed in 1994 could still be seen in this basin; a newly formed vintage of dense LSW was still arriving there. However, during these years the low-salinity layer of the Iceland Basin became denser, consistent with the LSW source changes (details follow). In the same year the Iceland Basin did not show any significant increase in salinity and decrease in the volume of LSW relative to 1994 (as shown further in this chapter the LSW of the Iceland Basin was fresher between 1994 and 2001 and saltier after 2001).

By 2004 the Iceland Basin's reservoir of LSW had also begun to lose its volume while gaining heat and salt. This suggests that it took no longer than 10 years after the cessation of the very deep convection in the Labrador Sea, for this LSW to start disappearing from the eastern parts of AR7 (Figs. 24.2d and 24.3d). The other water columns continued to restratify above the deep LSW core causing stratification in the whole subpolar North Atlantic to change. At the same time, the remnants of LSW that had become warmer and saltier over the years could still be recognized in 2004 (and even in 2007, not shown) by their thinned and weakened salinity and potential vorticity minima.

The intermediate depth ranges of the 2001 and 2004 AR7 sections exhibit two recently developed hydrographic features not seen in 1994 (Figs. 24.2 and 24.3). The first feature is a new LSW class seen as a homogenous low-salinity layer in the 400–1,300 m depth range. This water was massively formed in the winter of 2000, was modified via lateral and possibly convective mixing during subsequent years, and can still be identified via a volumetric analysis (Section 24.5). Even though some mixed layers could be found in the Labrador and Irminger Seas during 1997–1999, it was only in the winter of 1999–2000 when a distinct and homogeneous LSW class, maintaining its integrity in space and time, was produced. The 2000's increase in winter convection coincides with a local high of the NAO index, meaning higher levels of winter-time heat losses and explaining why in the year 2000 the convectively formed water spread deeper and wider in the Labrador Sea than during the preceding pentad. What makes the 2004 AR7 section particularly interesting for our investigation of the 2000's LSW class is that it was the first time when this water mass was definitely observed in the Iceland Basin, identified by a distinct salinity minimum within the depth range of 1,000–1,300 m.

The second feature that appeared between 1994 and 2001 is a relatively warm and salty intermediate layer separating the two LSW layers in the Labrador and Irminger Seas. This is a modified core of the *Icelandic Slope Water* (ISW), which is usually found near the Reykjanes Ridge and is now spreading towards the basin's centers to replace the deep LSW class. ISW is known to be formed through a direct linear mixing process blending the original *Iceland–Scotland Overflow Water* (ISOW) with the overlying Atlantic thermocline water near the Faroes, not involving LSW (van Aken and de Boer 1995). ISW then follows the slopes of Iceland and the Reykjanes Ridge until it enters the Irminger Sea. From the western slope of the Reykjanes Ridge the ISW intrudes into the center of the Irminger gyre, forming a relatively thin, but noticeably salty and warm layer. This characteristic salinity maximum is typically 140–200 m deeper than its temperature companion (discussed in Section 24.6.4).

While the vertical average of properties is commonly used to provide an overview of the temporal evolution of a water mass, a more detailed analysis of the development of LSW is required to understand the processes that determine the variability. In Section 24.5 we discuss a new approach which gives greater insight into those processes by distinguishing between classes of LSW formed at different times, and allowing their temporal and spatial changes to be accurately tracked.

24.5 Identification of Labrador Sea Water Classes

24.5.1 Two Volumetric Approaches to Identification of Water Masses and Their Changes

It is clear from Section 24.5 that the term *LSW* covers more than one water type; the vintages of LSW convectively formed at different times take on different properties dependent on changes in various sources of freshwater and salt, and variations in the heat loss to the atmosphere. Only by distinguishing accurately between the water types can we truly understand their development. Simple averaging on depth or density levels across the subpolar gyre ignores the spatial and temporal evolution of LSW so presenting an inadequate picture. Here we describe a different approach that avoids those problems – the volumetric method.

The volumetric method is a particularly useful tool for identification and examination of specific water masses and studying their "*life cycles*". The "*life cycle*" of each water mass includes production, spreading and mixing. As soon as water mass production starts, a newly formed water mass begins to spread, advect and mix with other waters. The mixing process leads to both spatial and temporal transformation of a water mass, resulting in its structural and property changes. Both mixing and export are responsible for water mass dilution, dissipation and, ultimately, loss. The volumetric methods allow the identification of a water type and tracing of its development in time and space by automatically adjusting to those changes.

Several volumetric applications are discussed in Chapter 21 (Yashayaev and Dickson 2007). Here we summarise the LSW identification technique based on two complementary volumetric approaches (Yashayaev 2007b), the essence of which is reflected in Fig. 24.6. A convectively formed water mass can be reliably identified and monitored firstly by the *density layer volumetric* method. Each value in Fig. 24.6a represents the basin-mean thickness (in meters) of an individual σ_2 layer (σ_2 *is potential density anomaly referenced to 2,000 db*) defined by σ_2 and time ranges (0.01 kg m^{-3} year). This σ_2*-time* layer thickness plot was constructed by averaging $\Delta\sigma_2 = 0.01$ kg m^{-3} layer thicknesses from all hydrographic stations in a

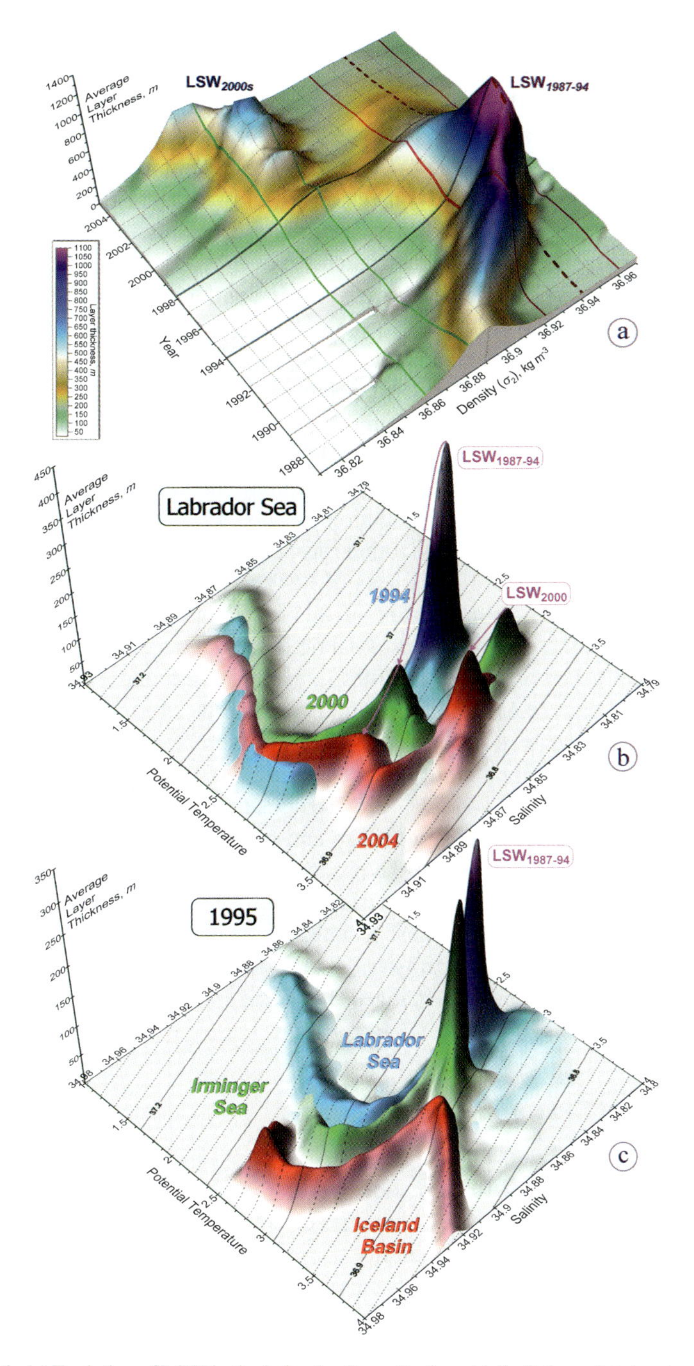

Fig. 24.6 (a) Evolution of LSW in the Labrador Sea: a "volumetric" σ2-time plot showing the average thickness (meters) of $\Delta\sigma_2 = 0.01\,\mathrm{kg\ m^{-3}}$ layers in the Labrador Sea (σ_2 *is potential density anomaly referenced to 2,000 db*) (Reproduced from Yashayaev et al. 2007). (b) Temporal volumetric changes: 1994, 2000 and 2004 "volumetric" potential temperature (θ)–salinity (S) censuses of the Labrador Sea. (c) Spatial volumetric changes: 1995 "volumetric" θ–S censuses of the Labrador, Irminger and Iceland basins (Fig. 24.1). Each value in (b) and (c) represents the average thickness (in meters) of a 0.1 °C × 0.01 Δθ × ΔS layer. The solid and dashed contours are isolines of σ2 (kg m^{-3}) defined by θ and S

given year weighted by the distance or area represented by these stations. Two examples of the second approach, *volumetric potential temperature* (θ) – *salinity (S) analysis*, are presented in Figs. 24.6b and 24.6c. Each analyzed layer in these θ–S diagrams was defined by two-dimensional θ–S intervals: $\Delta\theta \times \Delta S = 0.1\,°C \times 0.01$, set by ½Δθ and ½ΔS in the corresponding directions. This approach was applied individually to all available annual sets of hydrographic data from the Labrador, Irminger and Iceland basins, resulting in annual volumetric θ–S censuses for each of these three basins.

Figure 24.6b shows three annual hydrographic surveys of the Labrador Sea (1994, 2000, 2004), while Fig. 24.6c shows the three basins of interest (Labrador, Irminger, Iceland) "sampled" by AR7 in the same year, 1995. Figure 24.6b, therefore, reflects temporal transformation or evolution of LSW and other waters; while Fig. 24.6c illustrates spatial water mass transformation and change.

Strengthening and deepening convection creates, remixes and modifies LSW, causing its thickness, density and other properties to change. The processes responsible for transformation and losses of LSW, including mixing, entrainment and export, also change the properties of this water mass, altering its core and boundaries. Indeed, the subpolar trans-Atlantic section plots (e.g., Figs. 24.2, 24.3 and 24.14), time series of vertical profiles (e.g., Figs. 24.5 and 24.8), volumetric inventories (e.g., Fig. 24.6) and compilations (e.g., Fig. 24.9) imply that a fixed (static) range of any seawater property or a combination of such ranges can not be used as a universal criterion identifying a specific LSW core, vintage or class (the terms "*LSW core*" and "*LSW class*" are introduced below). On the contrary, a characteristic property (e.g., density) range and other LSW identification criteria need to comply with the changes in the properties used. The methods that we use in identification and analysis of characteristic water masses are primarily based on σ_2 and θ–S volumetric censuses. These techniques are capable of automatic *adjustment* to a specific LSW core, "*locking on*" its year-to-year transformation and thus revealing its spatial and temporal changes.

24.5.2 LSW Cores and Classes

Distinguishable isolated LSW formations can be identified in the time series of vertical profiles (Fig. 24.5), in the volumetric density census (σ_2, Fig. 24.6a), in the volumetric potential temperature–salinity censuses (θ–S, Figs. 24.6b and 24.6c), and in the compilation of annual θ–S curves (Fig. 24.9) constructed from corresponding annual volumetric θ–S projection. These methods applied to each *basin-survey* reveal the principal water masses residing in the studied basins. The σ_2 and θ–S volumetric approaches are the most appropriate for identification of a convectively formed water mass. A volumetric peak with its θ–S–σ_2 coordinates, area and also integrated and mean heights identifies a specific LSW formation, with the peak's maximum or its central point representing the "core" of the examined water mass.

The volumetric LSW cores were detected from all available full-column volumetric censuses of individual basin-surveys (separately for Labrador, Irminger and Iceland basins and for each survey). Such a census for the Labrador Sea is shown in Fig. 24.6a; examples of the LSW core σ_2 are $36.916|_{1990}$ and $36.940|_{1993}$ (*the subscript indicates the year of a survey*). Next, each LSW layer that was detected in a given basin-survey was individually identified in each vertical profile by selecting all measurements within a $\Delta\sigma_2 = \pm 0.017\,\text{kg m}^{-3}$ range centered at the σ_2 corresponding to the LSW volumetric core. (θ–S or combined θ–S and σ_2 ranges could also be used to define specific LSW bodies.)

All annual volumetric peaks like the examples shown in Fig. 24.6 form continuous progressions reflecting year-to-year development and transformation of a specific LSW core. This grouping allowed us to introduce a LSW "class" representing a sequence of LSW cores with a common development history.

Analysis of all annual hydrographic surveys of the subpolar North Atlantic allow us to build a complete history of formation, development, spatial and temporal evolutions, and decay of two LSW classes, $LSW_{1987-1994}$ and LSW_{2000}, characteristic for the last 2 decades (Figs. 24.5–24.9). These two classes were identified by their progressively "evolving" volumetric peaks (e.g., the ridges of high values in Fig. 24.6a or local maxima in Figs. 24.6b,c); the *subscript* names that were given to them reflect a time period (1987–1994) or a particularly extraordinary year (2000) during which they progressively developed to their extreme cold, dense and voluminous states.

The first *LSW class*, $LSW_{1987-1994}$, is associated with the most extraordinary production of LSW ever reported. A solitary volumetric θ–S peak (maximum) was first observed in 1987 at $\sigma_2 = 36.885\,\text{kg m}^{-3}$, and reached its all-time high in 1994 at $\sigma_2 = 36.940\,\text{kg m}^{-3}$ (Figs. 24.6a, 24.7 and 24.8), completing this unprecedented phase of LSW production. Having reached its record volume in 1994, this LSW class (the $LSW_{1987-1994}$ peak) has substantially diminished over the subsequent years, becoming barely identifiable in the volumetric diagrams constructed for the early-to-mid-2000s (Fig. 24.6a and the year 2004 in Fig. 24.6b). On the other hand, the remnants of the transformed $LSW_{1987-1994}$ still show a characteristic salinity minimum, helping with objective identification of this LSW class.

A thick weakly stratified layer (*high $\Delta\sigma_2$ layer thickness, weak density gradients and low vertical stability*) reappears in the Labrador Sea in 2000 as the second *LSW class*, LSW_{2000} (Figs. 24.5, 24.6a and 24.8). It was massively formed in 2000 and continued to develop and deepen over the subsequent years. It is found at shallower depths and lower densities than the $LSW_{1987-1994}$ class.

The two LSW classes defined here have much in common. Both of their developments were preceded by freshening of the upper layer (Fig. 24.5), followed by years of increased wintertime atmospheric forcing (Fig. 24.4). Both waters are surrounded by saltier and often warmer layers (Figs. 24.5 and 24.8) and water columns, and both became warmer and saltier over the years following their massive convective formation (Figs. 24.5, 24.7–24.9). When a certain LSW class loses its volumetric prominence (e.g., $LSW_{1987-1994}$ in recent years), additional criteria can be used to validate and refine its volumetric definitions (e.g., *salinity minimum*).

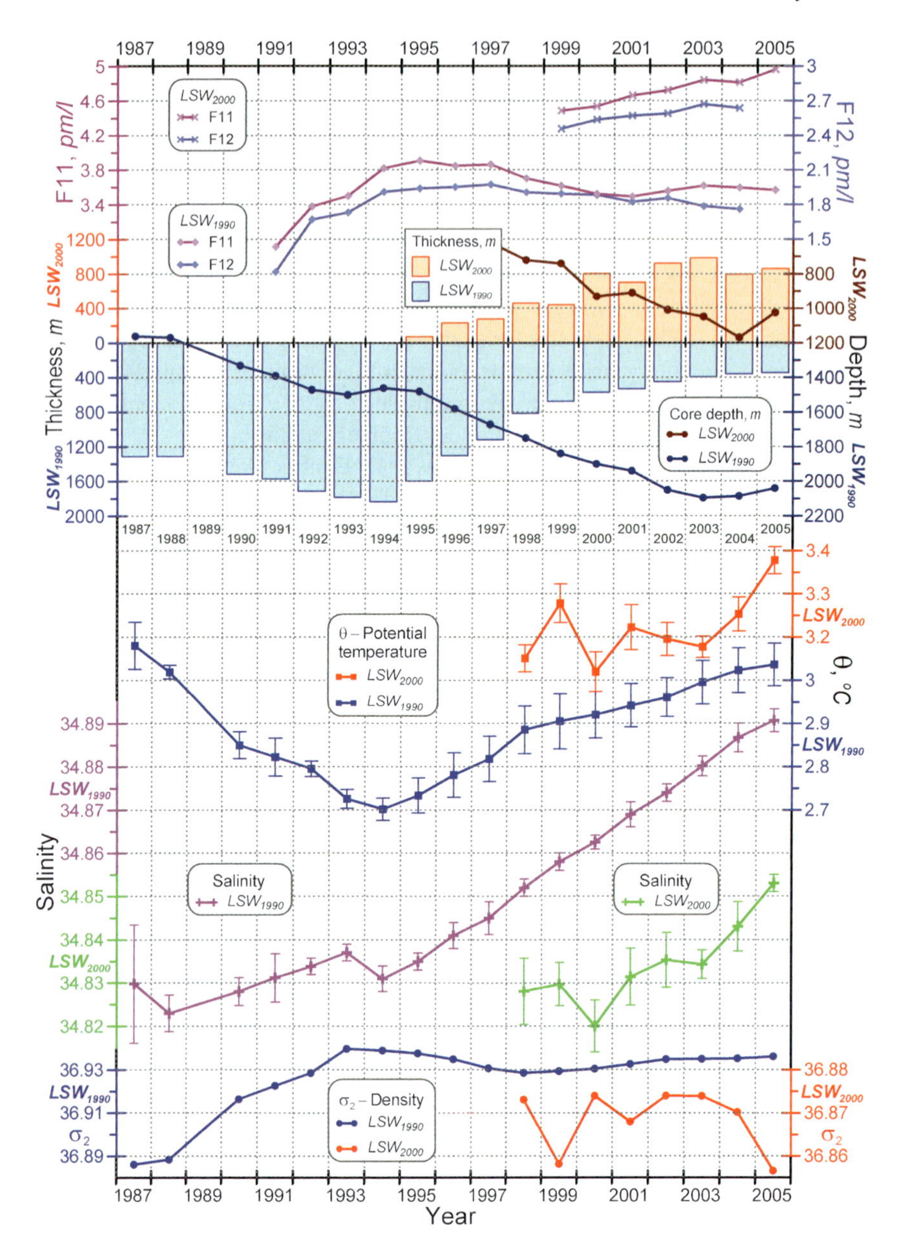

Fig. 24.7 Average potential density (σ_2), salinity, potential temperature (θ), thickness (*shown in vertical bars*), depth and CFCs (F11 and F12) of the $LSW_{1987-1994}$ (labeled as LSW_{1990}) and LSW_{2000} classes in the Labrador Sea (Reproduced from Yashayaev 2007b)

The development and transformation histories of the two LSW classes are well captured by temporal evolutions of their key properties; the values in Fig. 24.7 ('LSW_{1990}' indicates $LSW_{1987-1994}$) are based on the annual LSW

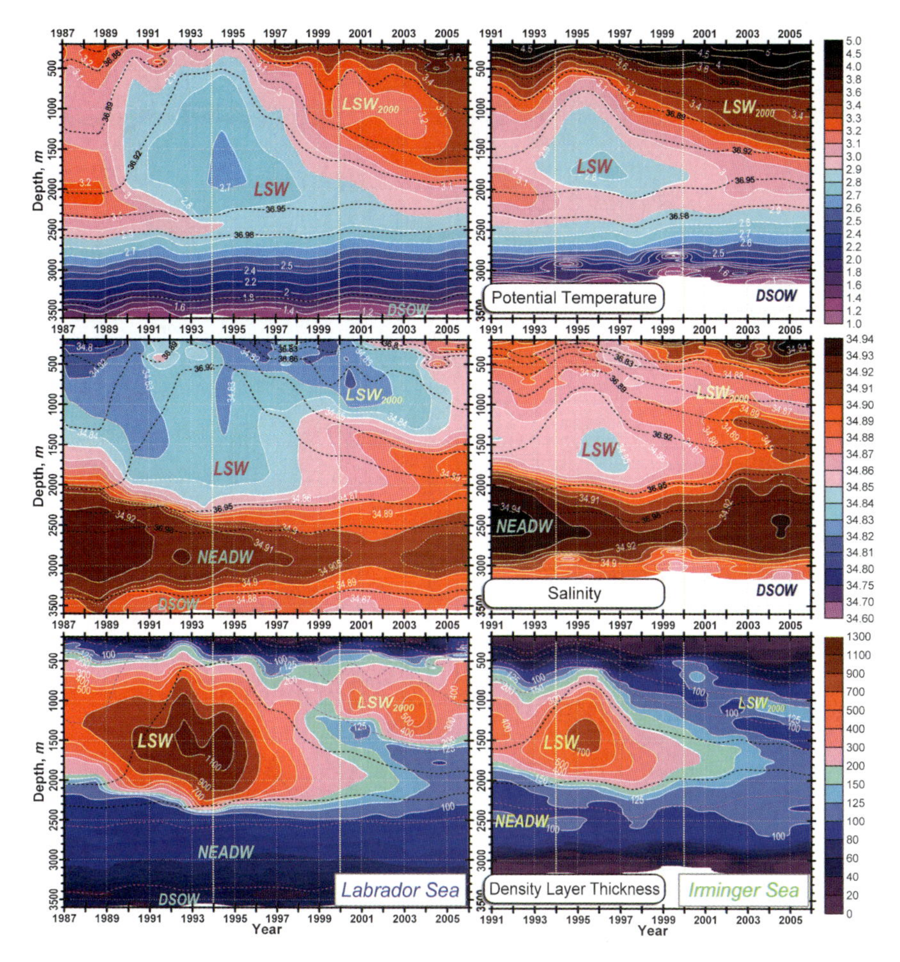

Fig. 24.8 Potential temperature (*upper*), salinity (*middle*) and thickness (in meters) of $\Delta\sigma_2 = 0.01\,\text{kg m}^{-3}$ layers (*lower*) in the central Labrador (*left column, 1987–2005*) and Irminger (*right column, 1991–2005*) basins. The $\Delta\sigma_2$ (*potential density anomaly referenced to 2,000 db*) isolines (*dashed contours*) indicate that the LSW produced between 1990 and 1994 was the densest on record (Reproduced from Yashayaev et al. 2007b)

inventories in the Labrador Sea (Yashayaev 2007b). These properties, with the exception of the LSW thickness, were computed for each LSW class through weighted-averaging over the LSW-specific layers. The σ_2 ranges defining such layers were confined by $\sigma_2|_{LSW}$–$0.017\,\text{kg m}^{-3}$ and $\sigma_2|_{LSW} + 0.017\,\text{kg m}^{-3}$ (here $\sigma_2|_{LSW}$ represents σ_2 of a given LSW core identified volumetrically for each basin and survey). The weights of individual measurements in each summation account for the depth and distance ranges represented by these measurements and for their θ–S "closeness" to the LSW core in the basin and year (survey) where and when they were taken. The

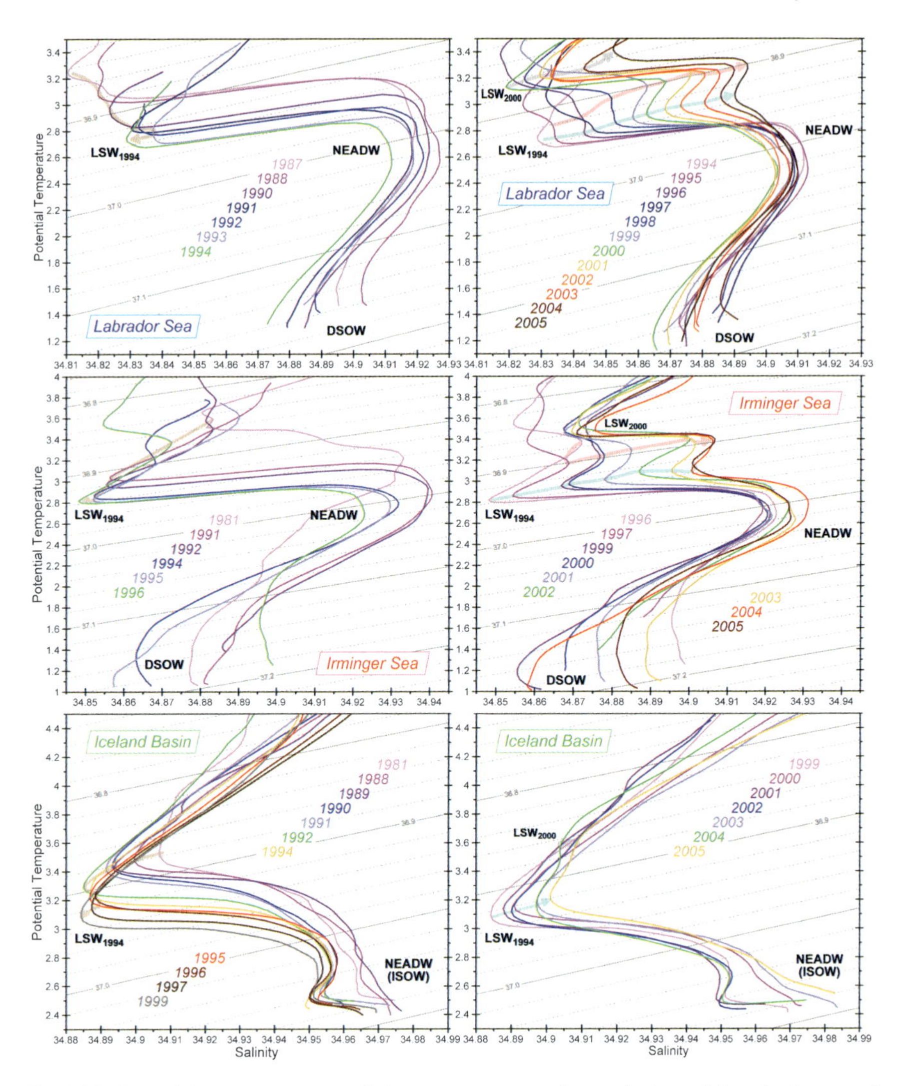

Fig. 24.9 Potential temperature vs salinity curves representing typical conditions in the central Labrador Sea (*bottom depth > 3,250 m*) for the years 1987–1994 (*upper left*) and 1994–2005 (*upper right*), central Irminger Sea (*bottom depth > 2,500 m*) for the years 1981–1996 (*middle left*) and 1996–2005 (*middle right*) and central Iceland Basin (*bottom depth > 2,300 m*) for the years 1981–1999 (*lower left*) and 1999–2005 (*lower right*). The light-colored trajectories indicate developments and subsequent transformations of $LSW_{1987-1994}$, LSW_{2000} and the warm and salty ISW-derived layer separating the two LSW classes. The contour lines show potential density anomaly referenced to 2,000 db ($\sigma 2$) as a function of salinity and potential temperature (Reproduced from Yashayaev et al. 2007b)

LSW series (Fig. 24.7) provide a convenient source reference for this water mass that can be used to interpret the LSW signals observed in other Atlantic basins, as well as changes in other water masses (see Chapter 21).

24.5.3 *Potential Temperature–Salinity Projections of Water Mass Developments*

A final tool for analysis is the construction and compilation of θ–S curves to represent yearly typical hydrographic conditions in the Labrador, Irminger and Iceland basins over the last 2 decades (Fig. 24.9). Two separate sets (panels) of θ–S curves are shown for each of these basins. The θ–S curves were grouped to allow us to illustrate the two principal phases in a full-record long history of the $LSW_{1987-1994}$ class. The left columns in Fig. 24.9 represent the *progressive development phase* of $LSW_{1987-1994}$ and the right columns represent the subsequent *transformation phase*, characterized by rapidly declining, decaying and ultimately vanishing of this LSW class. The procedure that we used for building an individual basin-survey θ–S is given in Yashayaev (2007b), and is briefly described here. First, all available measurements within a 50–150 km range of the AR7 line were used to construct annual volumetric θ–S censuses (0.1 °C × 0.01 θ–S *intervals*) for each basin. Then, taking these censuses one by one, all θ–S points from individual σ_2 ranges ($\Delta\sigma_2$ = 0.010 kg m^{-3}) were averaged with the weights based on the corresponding θ–S layer thicknesses.

In addition to documenting the progressive development (1987–1994) and the subsequent decay (1994–2005) of the $LSW_{1987-1994}$ class and the recent development of the LSW_{2000} class, Fig. 24.9 effectively demonstrates the temporal and basin-to-basin transformations and changes in the Northeast Atlantic Deep Water (NEADW) and Denmark Strait Overflow Water (DSOW) associated with large-scale mixing and signal transfer in these waters (see Chapter 21).

24.6 The Recent Production and Transformation History of the Labrador Sea Water in the Labrador, Irminger and Iceland Basins

In this section we will describe in detail the two LSW classes, discuss the stages of their development and transformation within the formation region, and examine the signatures of these stages in the other subpolar basins.

24.6.1 *Rise of $LSW_{1987-1994}$*

A strong freshwater anomaly had developed in the upper 500 m of the Labrador Sea over the late 1980s (Figs. 24.5 and 24.8). The excessive amount of freshwater initially stored in the upper layer was redistributed over a broader depth range in the subsequent years. This vertical redistribution of freshwater in the water column of the Labrador Sea was caused by the recurring strong winter convection events of the years 1987–1994 that collectively produced a large homogeneous volume of exceptionally cold, fresh and dense LSW reaching below 2,400 m (Figs. 24.2, 24.3 and 24.5–24.9). This water formed a distinct $LSW_{1987-1994}$ class. Between 1987 and 1994, this water mass became about 0.45 °C colder, 0.06 kg m^{-3} denser and almost doubled its volume. Although mid-depth layers were becoming fresher, when viewed as a series of properties in the deepening volumetric core of $LSW_{1987-1994}$ it can be seen that as convection was getting deeper between 1990 and 1993, the $LSW_{1987-1994}$ class was steadily becoming saltier. This salinity increase is best illustrated in the θ–S plot in Fig. 24.9 (see the *light-colored arrow-headed line* behind the 1987–1994 θ–S curves in *upper left panel*), where in 1993 the well-mixed layer of LSW can be identified by a narrow temperature and salinity minimum close to $\sigma_2 = 36.940$ kg m^{-3}. The increase in salinity within the core occurred as the body of freshwater mixed ever deeper into the warm saline NEADW below (Lazier et al., 2002; Yashayaev 2007b). It is possible that entrainment of warm and salty Irminger Sea waters also added salt to $LSW_{1987-1994}$. The warming due to mixing was more than compensated by heat loss to the atmosphere, but the additional salt mined by the convection prevailed in its effect on the LSW salinity over the source of freshwater from the inshore waters and the sea surface.

A reversal of the trend to higher salinities occurred in 1994 when the $LSW_{1987-1994}$ core became significantly fresher (by 0.01). We suggest the following explanation for the abrupt freshening. Despite further cooling of the whole mixed layer caused by winter convection in 1994, the mixed layer did not extend noticeably deeper in the spring of 1994 than the mixed layer of the previous year (Fig. 24.8). The convection of 1994 therefore did not bring up much new saline NEADW from below. On the other hand, an increased accumulation of less saline water in the upper 300 m layer between 1993 and 1994 is evident in Fig. 24.8. Thus, the freshening effect from convective entrainment of the less-saline upper-layer waters dominated over the salting effect of NEADW entrained from below. The amount of the warmer and saltier Irminger Sea waters that was brought into the Labrador Sea between 1993 and 1994 was also insufficient to compensate the freshening arriving from the less-saline entrainment. The result was the 1994 minimum in the $LSW_{1987-1994}$ salinity series (Fig. 24.7). This event was the second minimum in the deep LSW salinity record since the mid-1980s. The first minimum was observed in the late 1980s (Fig. 24.5). This indication of two consecutive $LSW_{1987-1994}$ salinity minima in the Labrador Sea will help to interpret the LSW changes in other regions.

$LSW_{1987-1994}$ showed two periods of increasing salinity, *the late 1980s to 1993* and *1994 to the present*. These similar trends, however, are different by nature. First, the sustained increase in the $LSW_{1987-1994}$ salinities until 1993 was maintained by entrainment of saltier NEADW into fresher LSW every time that convection deepened. It is also possible that convective entrainment of the warm and salty waters from the Irminger Sea had added extra salt to $LSW_{1987-1994}$. Second, after the cessation of convective renewal of $LSW_{1987-1994}$, isopycnal mixing with saltier intermediate waters from outside the Labrador Sea became the main agent for the salinification of this LSW class. This change in the dominant source of salt altered the rates of salinity increase (Fig. 24.7). Post-1994 changes are discussed in Section 24.6.2.

The history of the $LSW_{1987–1994}$ temperature changes is somewhat simpler than that of salinity. The cooling of $LSW_{1987–1994}$ that started in the mid- or late 1980s continued uninterrupted to form the all-time-coldest state observed in 1994 (Figs. 24.7–24.9), after which the $LSW_{1987–1994}$ temperature began to increase, resulting in a single temperature minimum. The matching tendencies in the temperature and salinity contributions between 1988 and 1993 (Fig. 24.7) explain why $LSW_{1987–1994}$ became notably denser in each year during this period (Figs. 24.6a, 24.7 and 24.9), which was also accompanied by exceptionally high buoyancy losses in the top 2,000 m.

The described progressive development of $LSW_{1987–1994}$ can be recapped by the θ–S–time trajectory shown with the *light-colored arrow-headed* line behind the 1987–1994 annual θ–S curves constructed for the Labrador Sea (Fig. 24.9, *upper left panel*, 'LSW_{1994}' indicates the coldest point in the history of the $LSW_{1987–1994}$). This trajectory indicates that the density increase observed during this LSW development was mostly due to the cooling caused by excessive high heat losses during the severe winters of the early 1990s, associated with the high-NAO phase and high heat losses from the sea surface to the atmosphere (Fig. 24.4).

The progressive developments of $LSW_{1987–1994}$ observed in the Irminger (Figs. 24.8 and 24.9, *middle left panel*) and Iceland (Fig. 24.9, *lower left panel*) basins had much in common with the reported build-up and development of the same water in its source or formation region in the Labrador Sea. Both in the Labrador and in the other two basins this remarkable water mass experienced substantial *cooling and freshening, accompanied by density, volume and depth increases*. Even though the $LSW_{1987–1994}$ property changes seen in the three basins were similar in their appearance (note the increase in the $LSW_{1987–1994}$ layer thickness in the Irminger Sea), there was one principal difference – the $LSW_{1987–1994}$ class reached its all-time densest/coldest state in 1993/1994 in the Labrador Sea, in 1995/1996 in the Irminger Sea and in 1999 in the Iceland Basin Figs. 24.9 and 24.13. This fact is highly important for understanding the overall spreading process of this water mass, as discussed later in this chapter.

24.6.2 Decline of $LSW_{1987–1994}$

After the body of LSW in the Labrador Sea achieved its greatest thickness and property anomalies in 1994, it began to thin out and weaken (Figs. 24.5, 24.6a, 24.7–24.9). Since 1994 the deep reservoir of $LSW_{1987–1994}$ has mostly remained isolated from the winter mixed layer. This was caused by a substantial decrease in the net annual heat loss from the Labrador Sea to the atmosphere after 1994, linked to changes in the NAO (Fig. 24.4). The weakening in the atmospheric forcing resulted in less intense convective mixing, mostly limited to the shallower reservoirs which were becoming filled with less-dense LSW. The mild winters allowed temperature and density stratification to re-established above the thinning patch of

$LSW_{1987-1994}$. Since its last convective renewal in 1994, the deep LSW core has evolved without much interaction with the layers above, becoming warmer and saltier and slightly changing its density (Figs. 24.5–24.9). The volume of the $LSW_{1987-1994}$ class declined as the water drained away from the Labrador Sea to other regions of the ocean (Figs. 24.6a, 24.7 and 24.8). After 1994 the thickness of $LSW_{1987-1994}$ in the density range of $36.92 < \sigma_2 < 36.95$ kg m^{-3} decreased from 1,900 to 250 m, a reduction by about 87%. The volumetric LSW identification technique and unsmoothed measurements suggests that the $LSW_{1987-1994}$ thickness decreased from 1835 m in 1994 to 350 m in 2005, suggesting 81% reduction (Fig. 24.7). Remarkably, even after more than a decade of isolation, the $LSW_{1987-1994}$ class could still be identified in the annual θ–S curves as a salinity minimum at $\sigma_2 = 36.935$ kg m^{-3} (Fig. 24.9, *upper right panel*). Between 1994 and 2005 the water at this deep salinity minimum in the Labrador Sea became warmer and more saline by 0.34 °C and 0.062, through isopycnal mixing (Yashayaev 2007b).

Even larger changes are seen in the layers above the deep LSW core. The combination of less heat loss to the atmosphere and continued horizontal, and possibly vertical, mixing of heat and salt into the LSW and shallower layers resulted in the entire upper 2000 m of the Labrador Sea becoming warmer, saltier and less dense (Fig. 24.4). This change was neither vertically uniform nor equally persistent at all depths (Figs. 24.5 and 24.8). The annual changes between 1,800 and 2,300 m are smaller but steadier than those that we see at the shallower intermediate depths. Although we do not dismiss the possibility that there might have been occasional short convective events penetrating deeper than 1,500 m and reaching into the deep and dense LSW classes after 1994, there is no evidence that such events made any significant change in the overall properties and volume of the deep LSW created in 1994 and earlier.

The fact that the deepest LSW showed consistent annual increases in temperature and salinity while it was gradually thinning out after 1994, adds to our understanding of the LSW export and transformation processes as follows. The whole body of $LSW_{1987-1994}$ loses mass through draining or export from its main subpolar reservoir to the other basins of the Atlantic Ocean. The volume lost is replaced by warmer, saltier and less dense waters that are imported from outside the Labrador Sea, and which accumulate above and adjacent to the deep LSW core. Despite the volume loss, $LSW_{1987-1994}$ remains in the Labrador and Irminger seas, recirculating within these basins and steadily becoming warmer and saltier as it mixes along the reservoir margins. The consistent annual temperature and salinity increases observed in the deep LSW core (Fig. 24.8, ~2,000 m) imply a steadiness of the horizontal (and, when stratification established, vertical) heat and salt fluxes into the LSW core.

The post-development transformation of $LSW_{1987-1994}$ is illustrated in Fig. 24.9 (*upper right panel*). The *light-blue arrow-headed* line shown behind the 1994–2005 annual θ–S curves constructed for the Labrador Sea indicates that this water is changing systematically over these years, *converging* with the body of NEADW also seen in Fig. 24.8. We expect that $LSW_{1987-1994}$ will eventually become indistinguishable from NEADW.

While in the Labrador Sea, the $LSW_{1987-1994}$ class began its *volumetric decline* in 1994 (Figs. 24.6–24.9), the same water mass found in the Irminger Basin started to decline after 1996 and in the Iceland basin after 1999 (Fig. 24.9). Therefore, after 1999, in addition to becoming warmer and saltier, the $LSW_{1987-1994}$ class was withdrawing from the three named subpolar basins.

24.6.3 Development of LSW_{2000}

The next most significant winter convection occurred in 2000. A freshwater anomaly that appeared in the upper 500 m in the previous years was carried down through the convection event of 2000 and those of some following years (salinities <34.82 in Figs. 24.5 and 24.8). This convectively driven vertical redistribution of freshwater was similar to the development of deep freshening that we saw in the late 1980s–early 1990s.

The 2000 convection reached 1600 m and was extensive enough to produce a distinct cold and fresh LSW class that is still seen in the Labrador and Irminger seas (LSW_{2000} in Figs. 24.2, 24.3, 24.5–24.9). While the volume of $LSW_{1987-1994}$ has substantially declined and its signature has diminished over the past decade, the volume of the less-dense LSW classes, dominated by the LSW_{2000}, has increased. Since 2000, the LSW_{2000} class has always been thicker than its deeper, gradually decaying companion.

Even though LSW_{2000} is warmer, less dense and shallower than $LSW_{1987-1994}$, these two LSW classes have similar rates of annual change (Figs. 24.5–24.9) and transit times to the Irminger and Iceland basins (Yashayaev et al. 2007a, b). The post-production evolution of LSW_{2000} is similar to that of $LSW_{1987-1994}$ – both LSW classes are steadily becoming warmer and saltier. The light-gray trajectory shown in Fig. 24.9 (*upper right panel*) follows the LSW_{2000} transformation history. Yashayaev et al. (2007a, b) reconstructed the LSW_{2000} signals between Labrador and Iceland basins, and suggested that this change may result from mixing with warmer and saltier waters from outside the Labrador Sea. However, there are indications that some of the post-2000 events of winter convection had also brought warmer and saltier waters into the LSW_{2000} layer. In 2001–2003 LSW_{2000} became thicker and showed irregular year-to-year changes in salinity, CFCs (Fig. 24.7) and oxygen, suggesting that this water was renewed, at least partially, after 2000.

On the other hand, the salinity increases observed in the upper 500 m after 2003 are apparently insufficient to compensate the buoyancy gain in this layer caused by its warming (Figs. 24.2d, 24.3d and 24.8). This increases the potential for the LSW_{2000} class to be replaced by a new LSW class that could be less dense than any previous vintage. In addition, the necessary conservation of mass within the Labrador Sea requires that the draining LSW is replaced by inflowing water. The layer immediately above the deep $LSW_{1987-1994}$ is becoming filled by the Icelandic Slope Water, a narrow temperature and salinity maximum (Figs. 24.2d, 24.3d, 24.8 and 24.9), while the upper water column is filled with LSW_{2000} (Figs. 24.2, 24.3, 24.7 and 24.8). However the LSW_{2000} core had evidently deepened between 2000 and 2004 possibly as a response to the draining deeper waters. As a result, the "sinking"

LSW_{2000} is now progressively withdrawing from the upper layers, requiring shallower, warmer and less dense waters to be imported and to form a new LSW class. Indeed, becoming regularly remixed by weak convection these new warm low-density waters are presently building up, expanding both vertically and horizontally. Over the past several years we have observed an increase in the density stratification in the top layers of the Labrador Sea (>3.4 °C in Fig. 24.8; Yashayaev 2007b) caused by this recent water mass development. Meanwhile it is becoming more and more difficult for a winter renewal of the LSW_{2000} to take place. Will we see the LSW_{2000} class becoming fully isolated from the upper layers in the near future and thus close another page in the history of LSW developments? Will deep convection renew the waters of the Labrador Sea to 1,500 or even 2,000 m as it did in the 1990s, but operating at much lower densities? Several years will pass before we will be able to answer these questions with high confidence, but now we are at least able to monitor the arrival and transformation of LSW_{2000} in other Atlantic basins.

The LSW_{2000} arrived in the Irminger (Figs. 24.8 and 24.9, *middle right panel*) and Iceland (Fig. 24.3d, *secondary salinity minimum*, and Fig. 24.9, *lower right panel*) basins after transit times about a year shorter than those determined for the $LSW_{1987–1994}$. The faster spreading can be explained by the increasing circulation strength towards the upper layer of the subpolar gyre. In the Irminger Basin LSW_{2000} became deeper between 2001 and 2005 (compare with the Labrador Sea, Fig. 24.8). The LSW_{2000} class entered the Iceland basin in 2004, with the same densities ($36.86 < \sigma_2 < 36.87$ kg m^{-3}) as those it had 4 years earlier in the Labrador Sea.

24.6.4 Sub-LSW Temperature Maxima

Two other characteristic features of the Labrador Sea water column and their formation were discussed in Yashayaev (2007b). Here we show that the same features can be seen in the Irminger Sea and their development is similar to that observed in the Labrador Sea.

There are thin but distinct temperature maxima beneath each of the LSW classes. In the Labrador Sea, the sub-$LSW_{1987–1994}$ temperature maximum was located at about 2,500 m in 1994 (Fig. 24.2b), where it first appeared between 1,900 and 2,400 m in the mid-1980s as a result of the increased strength of winter convection (the deep LSW produced between 1887 and 1994 turned out to be colder than the underlying NEADW). This temperature maximum can be nicely seen at about 2,400 m inside the 2.9 °C isotherm between 1990 and 1994 (Figs. 24.5 and 24.8). This layer narrowed, deepened and became colder as convection developed to 2,400 m (Figs. 24.2, 24.5 and 24.8). Between the mid-1980s and the early 2000s, this feature was typically 0.02–0.03 kg m^{-3} (σ_2) denser than the deep LSW core (Figs. 24.8 and 24.9).

As the contrast between the deep LSW and NEADW decreases and the sharp interface separating these water masses diminishes, this maximum also fades away. The disappearance of the temperature maximum can be attributed to the warming of the deep LSW inverting the thermal contrast at the upper boundary of NEADW and mixing between these water masses. On the other hand, the multidecadal freshening of NEADW was also accompanied by its cooling, including the upper part of this water mass where the temperature maximum was formerly seen; this cooling tendency made the last pentad the coldest period for NEADW yet observed (Yashayaev 2007b).

Figures 24.8 (*upper right panel*) and 24.9 (*middle row*) show a surprisingly similar development of the sub-$LSW_{1987–1994}$ temperature maximum in the Irminger Sea. However it was slightly warmer and lasted about 1 year longer than in the Labrador Sea, consistent with the differences between the $LSW_{1987–1994}$ temperatures in these two basins.

Although there is no distinct sub-$LSW_{1987–1994}$ temperature maximum in the Iceland Basin, there was a noticeable broadening in the separation between the isotherms at and below the deep LSW core (e.g., 3.1 and 3.2 °C isotherms in 1994, Fig. 24.2b).

The sub-LSW_{2000} salinity and temperature maxima are associated with the relatively saline and warm intermediate layer separating the $LSW_{1987–1994}$ and LSW_{2000} in the Labrador and Irminger basins (Figs. 24.2c, d, 24.3c, d, 24.5, 24.8 and 24.8). This layer is formed by the core of saltier and warmer water that can be traced back to the *Icelandic Slope Water* (ISW) seen near the Reykjanes Ridge. At the end of its passage around the subpolar gyre, ISW arrives in the Labrador Sea where it partially replaces the underlying LSW. Remarkably, the ISW-characteristic temperature maximum is typically 140–200 m shallower in the Labrador Sea than its salinity companion (Figs. 24.2c, d, 24.3c, d, and 24.8) – this signature was inherited by ISW from the Atlantic thermocline waters at its formation stage and preserved during transit.

While the mid-depth temperature maximum appeared in our annual surveys only in 2000, the saltier water started to form a weak characteristic feature between 500 and 1,000 m in 1995, identifiable in both Labrador and Irminger basins. In the Labrador Sea, this happened almost immediately after winter convection had lost its strength and was not able to compensate the losses in the discharging $LSW_{1987–1994}$. As time progressed, the ISW-derived water developed into prominent θ–S maxima in both basins, also characterized by a volumetric peak, and forming a unique water mass derived from the warmer and saltier ISW. The corresponding *light-pink arrow-headed* lines in Fig. 24.9 (*right upper and right middle panels*) follow its evolution in the Labrador and Irminger basins. As for the $LSW_{1987–1994}$ and LSW_{2000} cores, the core of ISW (σ_2 = 36.90 kg m^{-3}) is steadily becoming warmer, saltier and deeper, tending to replace the $LSW_{1987–1994}$, which in its turn has substantially drained away, strongly diminished in size and lost much of its contrast with the surrounding waters. While the sinking of the warm and salty ISW-derived layer stopped in 2004, this water still continues to become warmer and saltier.

Despite their visible similarity, the two sub-LSW maxima are different in origin. The sub-$LSW_{1987–1994}$ temperature maximum resulted from the $LSW_{1987–1994}$ temperature decrease creating a temperature minimum layer above NEADW, while the sub-LSW_{2000} salinity and temperature maxima are associated with a new water mass arriving in the region and occupying a niche between LSW_{2000} and $LSW_{1987–1994}$.

24.6.5 "Coordinated Changes" in the Deeper Layers

The subpolar hydrographic summaries presented in Figs. 24.8 and 24.9 support and complement the statements made in Chapter 21 about the NEADW and DSOW transfer from the Irminger to Labrador basins. While the deep LSW signals arrive in the Irminger Sea from the Labrador Sea with about a 2-year delay, NEADW first reversed its freshening trend in the Irminger Sea in 2000 (Fig. 24.8, *right middle panel*) and about 2 years later in the Labrador Sea (Fig. 24.8, *left middle panel*). Both Figs. 24.8 and 24.9 also suggest that in the Irminger Sea, NEADW is presently more advanced in regaining its salt than it is in the Labrador Sea. This means that the NEADW salinity minimum did not just occur after the $LSW_{1987-1994}$ salinity minimum, the time interval separating these two minima changes with region: for the Irminger Sea it is about 5 years, while for the Labrador Sea it is on the order of 10 years (Fig. 24.8). These results fit well with the concept of the deep water mass formation introduced in Section 24.2; we suggest that LSW is entrained into ISOW in the Iceland Basin and flows back to the north-western North Atlantic as NEADW. Indeed, the LSW transit time to the Iceland Basin is about 5 years; it takes between 5 and 8 years for the NEADW signals to travel back to the Labrador Sea (Chapter 21). Since LSW is a principal contributor to NEADW (Dickson et al. 2002), a signal detected in the former is expected to affect the latter, and the timings of the LSW and NEADW salinity minima in the Labrador and Irminger basins (Fig. 24.8) agree with this idea.

Finally, DSOW, whose interannual variations notably exceed those in NEADW, experiences anomalous signals that first appear in the Irminger basin and then, about 1 year later, arrive in the Labrador basin (Figs. 24.8 and 24.9).

24.7 Propagation of the Labrador Sea Water Anomalies to the Irminger and Iceland Basins

The views presented in Figs. 24.4–24.9 deal with the basin-mean properties and thereby reflect temporal changes in certain (*water mass, depth or density constrained*) layers. Here we complement these views by introducing *horizontal distance along the AR7 line* as an independent parameter that will be used in our analysis of the LSW signals spreading across the North Atlantic.

24.7.1 Temperature and Salinity of LSW Across the Subpolar North Atlantic

In this section, we discuss results from an analysis of observations in which we have identified the $LSW_{1987-1994}$ and LSW_{2000} cores in each available basin-survey and constructed the multiyear development, spreading and decline histories for each LSW class. Here, each LSW property is regarded as a function of two variables, *distance and time*, rather than just a function of time independently representing each basin. This approach allows us to better map arrivals of the LSW signals along their pathways and to analyze spatial signatures of LSW in the subpolar basins.

All density (σ_2) values related to the same LSW class (e.g., $LSW_{1987-1994}$, LSW_{2000}) were grouped by individual basin-surveys and then used to construct distance–time distributions of σ_2 for the two LSW cores. The corresponding LSW properties (layer thickness, depth, temperature, salinity, dissolved oxygen, etc.) were then calculated separately for each hydrographic profile in all available basin-surveys. Distributions of the $LSW_{1987-1994}$ temperature, salinity and thickness along the AR7 hydrographic section are shown in Fig. 24.10. The values compiled in this figure were computed as follows: at each hydrographic station we selected all measurements captured by the ±0.017 kg m^{-3} $\Delta\sigma_2$ range centered at the LSW core's σ_2 (the σ_2 value having been defined for the time and location of this station); the selected points were then integrated or averaged, providing the LSW properties at the examined station.

The continuity that can be seen in the $LSW_{1987-1994}$ thickness, temperature and salinity distributions over the Labrador and Irminger Seas (Fig. 24.10) implies that these two basins actively "*communicate*" with each other and exchange their waters (Yashayaev 2007b sees these two regions as jointly forming the *main LSW reservoir*). We recall that this deeper and denser LSW class ($LSW_{1987-1994}$) can typically be found within the 700–2,500 m depth range along the trans-Atlantic section AR7 (Figs. 24.2 and 24.3).

The smaller volume and higher level of transformation of this water entering the Iceland Basin is a result of the longer (in both space and time) LSW transit from its source region to the Iceland Basin. Recall that LSW enters the Iceland Basin much further south, at the latitude of the *Charlie–Gibbs Fracture Zone*, near 52° N (Fig. 24.1). This explains why the LSW anomalies observed in the Iceland Basin are usually "blurred" in time, having longer duration and smaller amplitude (Figs. 24.9 and 24.11).

24.7.2 Calculation of LSW Anomalies

In order to separate the temporal changes of LSW from the changes set by *en route* mixing and resulting in spatial transformation of this water, we have calculated LSW anomalies along AR7. The anomalies will better reflect the spreading nature of the LSW events than the original property values. To compute the LSW anomalies presented in the *distance–time* coordinates in Fig. 24.11, a *special function of distance* reflecting the multiyear (20 years for LSW_{1994} and 7 years for LSW_{2000}) average state of LSW was subtracted from all LSW properties, which in their turn were based on individual hydrographic profiles. Such a "*special function of distance*" used as the reference in our anomaly calculations was constructed piece by piece for each deep basin as follows: (1) *all individual property values from the*

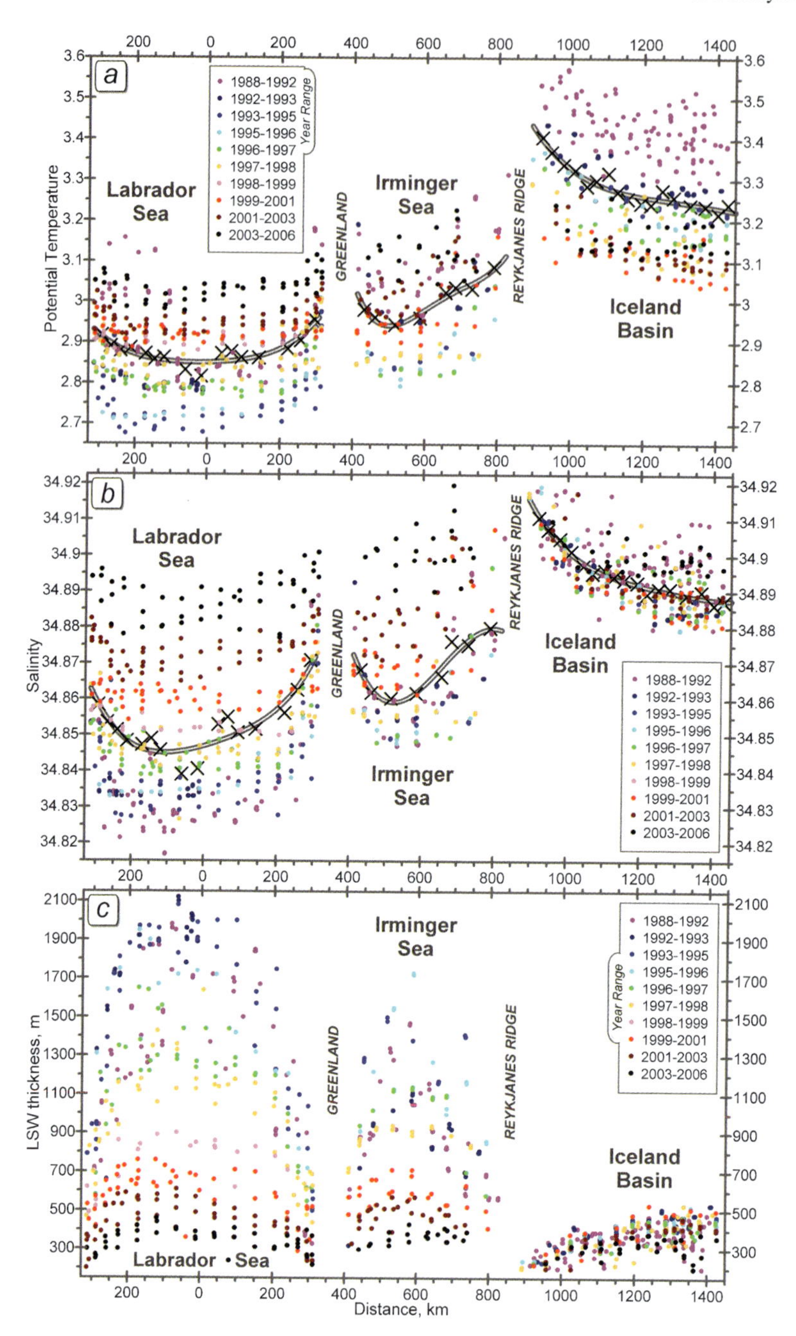

Fig. 24.10 Scatter plots of (a) potential temperature, (b) salinity, and (c) thickness (in meters) of $LSW_{1987-1994}$ "extracted" from individual hydrographic profiles along AR7 (Fig. 24.1). The circles indicate medians over stations grouped in 40 km spatial (distance) bins. The gray lines are polynomial fits of the temperature and salinity medians as functions of distance; these curves provide a continuous norm used to calculate the LSW anomalies shown in Fig. 24.11

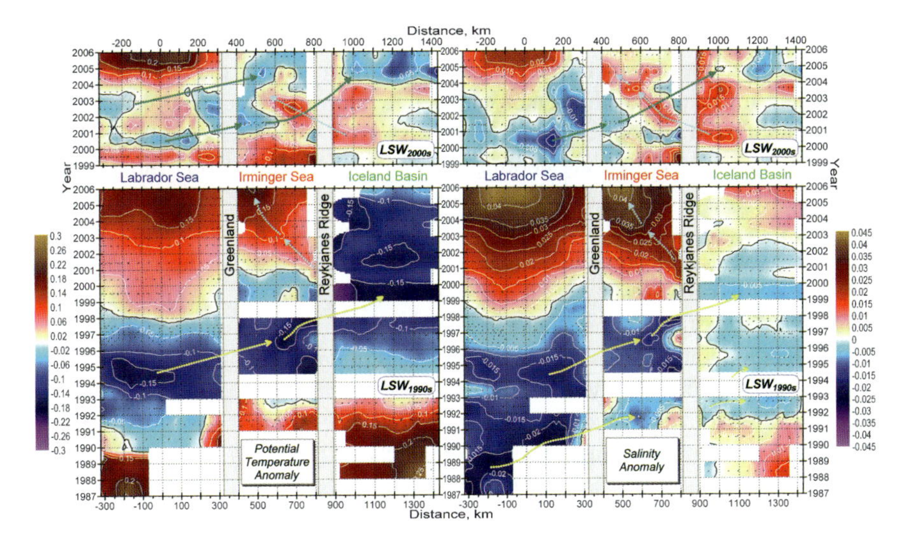

Fig. 24.11 Anomalies of potential temperature (*left column*) and salinity (*right column*) averaged over time- and distance-dependent vertical density ranges identifying $LSW_{1987–1994}$ (*lower row*) and LSW_{2000} (*upper row*), 1987–2005 (*inclusive*). The distance indicates the position along the composite AR7 section, with its origin in the central Labrador Sea. The eastward (*right*) pointing arrows indicate how record cold and fresh classes of LSW spread across the ocean, reaching first the Irminger Sea and then the Iceland Basin. The westward (*left*) pointing arrows indicate how a bulk of anomalously warm and saline water appeared near the western flank of the Reykjanes Ridge in 2000 and over subsequent years spread west along AR7, to the central and then western parts of the Irminger and Labrador basins. The anomalies were computed for hydrographic profiles indicated by dark dots. A distance-dependent long-term mean reference state used to compute these anomalies was comprised of three segments, derived for each basin separately to prevent any influence of adjacent basins (Fig. 24.10). Each basin-wide norm was constructed by grouping all individual property values in 40 km distance bins; calculating medians in each group; and then fitting a polynomial function of distance to these medians to achieve continuity of the reference state in each basin (Reproduced from Yashayaev et al. 2007a)

same basin were used to populate spatial bins (size: 40 km, overlap: 50%); (2) *all values from each bin were reduced to their median value*; (3) *the medians were approximated by a polynomial function of distance*; and finally; (4) *the polynomial functions from all basins of study were lined up along the same distance axis to construct the sought reference or long-term norm*. The latter was used in the calculations of LSW anomalies. (*Notes: (1) the reason for splitting the basins was to prevent the influence of neighboring seas; (2) polynomial fitting was employed because it provides continuity of a norm within each basin*).

24.7.3 Transit of LSW Anomalies to the Irminger Sea and Iceland Basin

The four panels of Fig. 24.11 (Yashayaev et al. 2007a) show potential temperature (*left*) and salinity (*right*) anomalies in the two LSW classes – $LSW_{1987–1994}$ (*lower*) and LSW_{2000} (*upper*) and also warmer and saltier waters surrounding each LSW class in each subpolar basin. The distance is measured along the composite AR7 section and referenced to the center of the Labrador Sea; fractional year reflects actual date of each station.

Here we again see a steady cooling in the Labrador Sea during the period of 1987–1994 (~0.45 °C, Fig. 24.11, *lower left panel*). During all these years except 1994, $LSW_{1987-1994}$ was steadily becoming saltier (Fig. 24.11, *lower right, Labrador Sea, 1988–1993*). As previously noted, these changes explain the rapid annual increases in the LSW density 1988 and 1993 (Fig. 24.6a).

The characteristic points in the development and transformation of $LSW_{1987-1994}$ identified in the Labrador Sea provide effective references for timing the arrivals of the same events spreading to the other Atlantic basins. For example, the buildup and following rapid decline of $LSW_{1987-1994}$ class expressed in the increase and then decrease in the corresponding density layer thickness (Figs. 24.6a and 24.8, *lower left panel*) had their 2-year delayed imprint in an analogous volumetric compilation for the Irminger Sea (Fig. 24.8, *lower right panel*). In addition, the single temperature and dual salinity minima seen during the $LSW_{1987-1994}$ development in the Labrador Sea are identifiable in the other two basins.

By linking the $LSW_{1987-1994}$ signals observed in the LSW formation region with the signals arriving in the Irminger Sea and Iceland Basin (Fig. 24.11) we draw two conclusions. Firstly the $LSW_{1987-1994}$ temperature anomalies were at a record low in the Labrador Sea in 1994, while the coldest LSW invaded the entire Irminger Basin 2 years later, in 1996. Secondly the sustained cooling of $LSW_{1987-1994}$ in the Iceland Basin ended only in 1999, suggesting a 5-year delay. Since 1999, this deep LSW of the Iceland Basin shows a slight warming.

The double salinity minima of $LSW_{1987-1994}$ seen in the Labrador Sea can also be recognized in the two other basins. Focusing only on the second low in the $LSW_{1987-1994}$ salinities, it can be seen in the Labrador, Irminger and Iceland basins in 1994, 1996 and 1999. This supports the 2- and 5-year transits of $LSW_{1987-1994}$ to the two latter basins.

Between 2000 and 2003 $LSW_{1987-1994}$ withdrew from the eastern part of the Irminger Sea leaving it to warmer and saltier waters, which include ISW, Subpolar Mode Water (McCartney and Talley 1982) and also strongly modified LSW. For more than a decade now, the $LSW_{1987-1994}$ layers of the Irminger and Labrador basins have been persistently becoming warmer and saltier. These signals can be traced back to the western flank of the Reykjanes Ridge, where in 2000 a strong positive anomaly was first seen replacing a strong cold and fresh anomaly associated with $LSW_{1987-1994}$. This shift in the water masses is also reflected in the hydrographic sections shown in Figs. 24.2 and 24.3 (*west of the Reykjanes Ridge*).

LSW_{2000} has been sporadically renewed during its short history, resulting in patchier temperature and salinity anomaly fields than those for the deeper water (Fig. 24.11). Even if our sections and volumetric estimates suggest that LSW_{2000} and its anomalies arrive from the Labrador Sea, in some years this water could be remixed and its signals altered outside of the formation region. Nevertheless, the tendency in its development is clear in the Labrador Sea and traceable to the

Irminger and Iceland basins, suggesting that it took 1 year and 4 years for LSW_{2000} to reach each of these basins in order. These transit times are roughly 1 year shorter than the $LSW_{1987–1994}$ transit times.

24.8 Spreading of the Labrador Sea Water to the Rockall Trough and Northern Iceland Basin

The furthest corners of the subpolar North Atlantic reachable by LSW are the northern parts of the Iceland Basin and Rockall Trough. Using hydrographic profiles collected at the deepest station of the northern Iceland Basin monitoring line (60° N; 20° W) and all profiles from *Station M* (57.3° N, 10.383° W; 2,340 m), the deepest station of the northern Rockall Trough section, we have constructed *time-depth* distributions of temperature, salinity and density for these water columns (Fig. 24.12). Even though the versions of LSW found in these two regions are more strongly modified and diluted compared to those seen along AR7 (Figs. 24.2 and 24.3), this water mass can still be unmistakably identified there by its characteristic salinity and/or potential vorticity minima. Also of interest is the fact that the σ_2 levels associated with these salinity/potential vorticity minima (*yellow or red dotted contours* in Fig. 24.12) belong to the range of historic LSW densities in the formation region of this water.

A close look reveals the same tendency as the "LSW upstream" regions, namely, as time progressed the deep core of what is thought to be the $LSW_{1987–1994}$ class was steadily becoming denser through to the present, implying that the densest modification of $LSW_{1987–1994}$ arrived at its final destinations just recently. In addition to the $LSW_{1987–1994}$ density increases matching those seen in the western basins much earlier, each of the two northern corners shows consecutive arrival of two salinity minima separated by 6–8 years. We suggest that these minima are substantially modified but yet identifiable replicas of the two minima recorded in the Labrador Sea between the mid-1980s and mid-1990s. The arrival of the second minima in the northern Rockall Trough in 2004 gives a transit time to that basin of 10 years. It is not fully understood why it takes 4–5 years for LSW to travel through the Iceland Basin. At present we can only link the delays in arrival of LSW in the northern Iceland Basin and Rockall Trough with its crossing the pathway of the *North Atlantic Current* within these two basins.

In summary, the appearance and passage of these low-salinity LSW events was revealed by our various analysis methods, in sequence, first in the Labrador Sea, then in the Irminger Sea (Figs. 24.5, 24.7, 24.8 and 24.11), then in the Iceland Basin (Fig. 24.11), and now in the northern Iceland Basin and northern Rockall Trough (Figs. 24.12 and 24.13) at the two points closing the LSW passage along its main LSW subpolar trajectory.

In contrast to the deep LSW layer in the eastern basins, the upper layers of the northern Rockall Trough and Iceland Basin regions show extraordinarily high warming and salinification since the mid-1990s (Fig. 24.12). A similar increase in

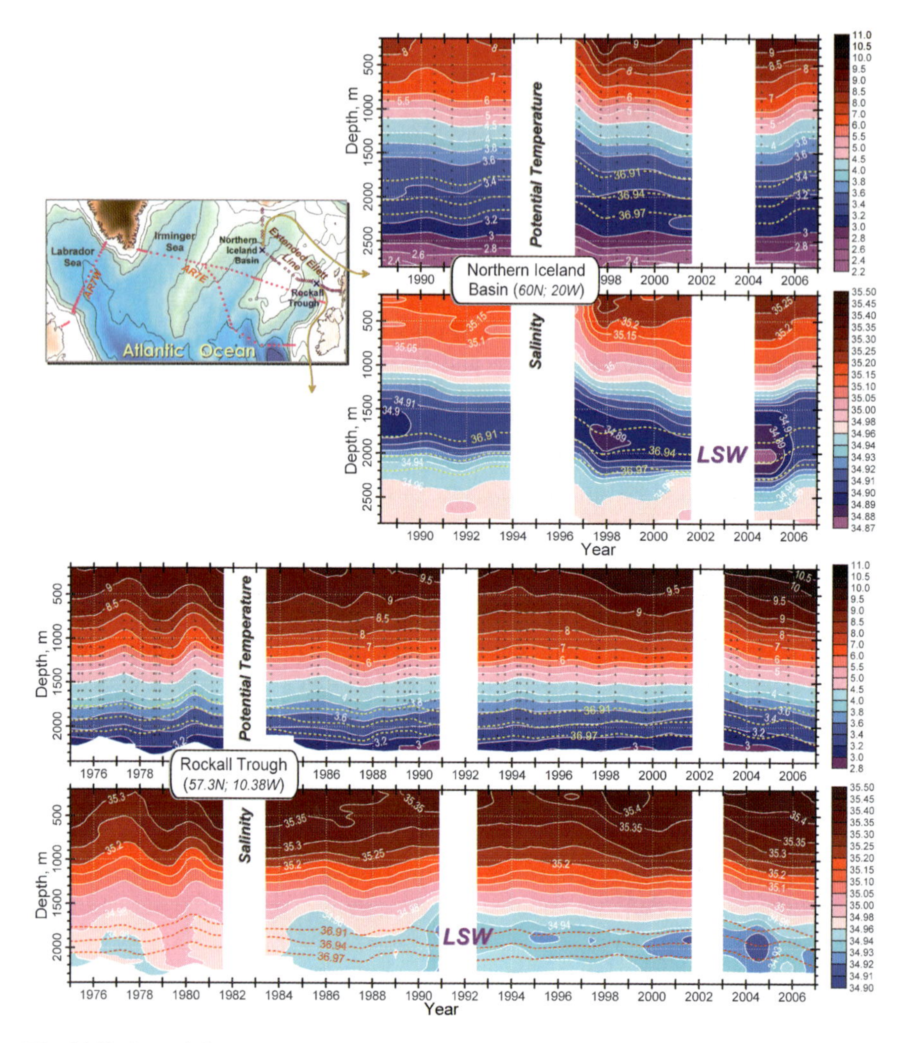

Fig. 24.12 Potential temperature (*first and third from the top*) and salinity (*second and fourth*) in the northern Iceland Basin (*first and second from the top*) and the Rockall Trough (*third and fourth*) in the coordinates of *time* and *depth* (Reproduced from Yashayaev et al. 2007b)

salinity and temperature is being observed in the upper layers of the Labrador Sea. Our knowledge of the cyclonic nature of the subpolar gyre suggests that these temperature and salinity increases in the upper 1,000 m layer of the eastern basins are an upstream source of additional heat and salt for the Labrador Sea. However it is far from clear as to which other large-scale or local processes are contributing to the upper layer changes in the Labrador Sea. But whatever the source of additional heat and salt, it is clear that they are affecting and shaping upcoming LSW developments.

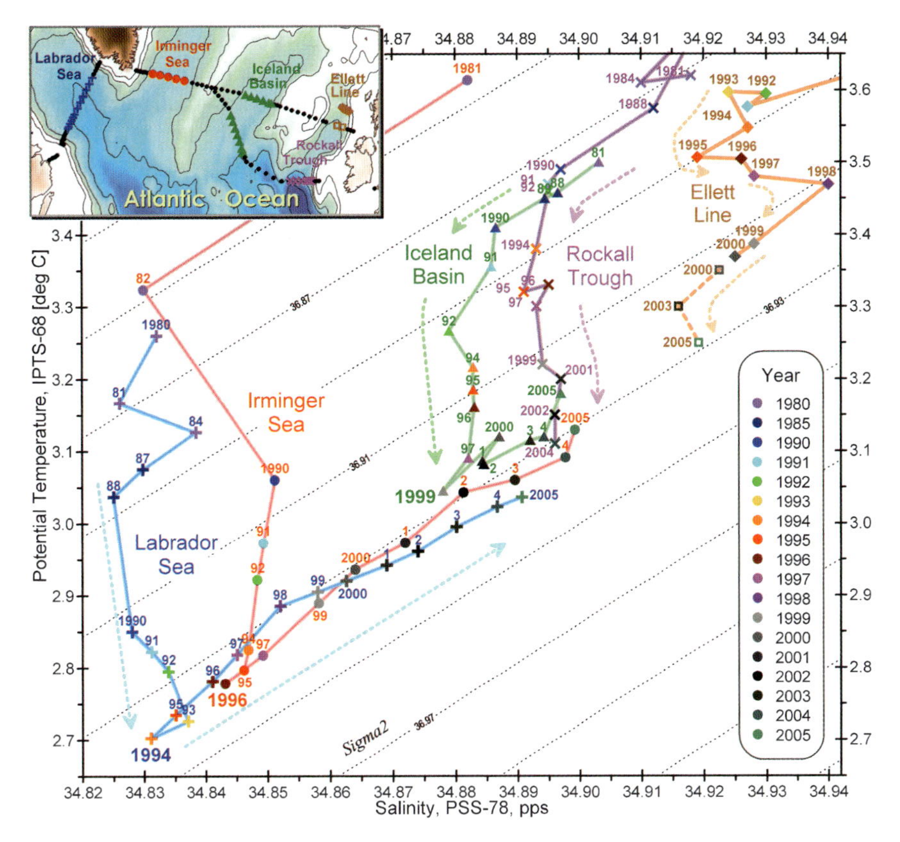

Fig. 24.13 *Potential temperature – salinity – time θ-S-time "trajectories"* of the $LSW_{1987-1994}$ class in the Labrador, Irminger, Iceland, Rockall Trough, and "Ellett Line" basins (*left to right*). These "*trajectories*" follow development, spreading and transformation of the most voluminous LSW class recorded by oceanographic measurements. The *arrow-headed* lines indicate the direction of evolution of the $LSW_{1987-1994}$ class created from the late 1980s through the mid-1990s by winter convection in the Labrador Sea. The dotted contours are σ_2 (*kg m*$^{-3}$) isolines

24.9 Discussion and Conclusions

The Labrador Sea Water (LSW) deserves to be ranked as the most prominent water mass of the subpolar North Atlantic Ocean. Formed by deep convective mixing in the Labrador Sea it takes in large quantities of freshwater, nutrients and gases, subsequently exporting these substances to the mid-depth layers of the ocean and, through mixing with the deeper waters, to the Atlantic abyss (Lazier 1980; Talley and McCartney 1982; Lazier et al. 2002; Yashayaev et al. 2003, 2007a, 2007a, b; Azetsu-Scott et al. 2003). This, in fact, is the most effective pathway for freshwater and a variety of other substances (particularly, carbon dioxide and other dissolved anthropogenic gases) to transit from the surface of the subpolar gyre to the deep Atlantic basin. However, since its functionality is directly controlled by annual-to-decadal variations in the production, character and thickness of LSW, this mid-depth conveyor is essentially unsteady. Its state can range from an intense and massive water mass renewal triggering ventilation of the oceanic mid-depths to a long-lasting nearly complete disruption in exchange between the upper and deeper layers resulting in isolation of deeper LSW and re-stratification over the entire sub-polar gyre. Since the varying LSW production, volume and properties ultimately influence the Atlantic overturning circulation, the present extensive synthesis on various aspects of LSW suggests many new research directions involving the ocean hydrography, circulation and climate science.

24.9.1 From Extreme LSWs to Extreme Subpolar Hydrographies

Summarizing our discussion of the long-term changes in LSW and expanding the conclusions of our previous time series (Dickson et al. 2002, Yashayaev et al. 2003; Yashayaev 2007b), we report a strong salinity and temperature contrast between the 1966 and 1994 sections, which are representative of the historic LSW property ranges. The warmer and saltier conditions recorded between the mid-1960s and early 1970s and the fresher and colder conditions recorded between the late 1980s and late 1990s define two extreme states in the history of reliable hydrographic observations in the Labrador Sea (*since the 1930s*). The intermediate layers of the subpolar North Atlantic have already approached the salinity and temperature levels of 1966 (typical of the salty and warm state of the mid-1960s–early 1970s), consistent with the recent decline in NAO and winter convection.

The convective cooling and freshening of the mid-depth Labrador Sea between the late 1980s and early 1990s have produced a characteristic LSW that by 1994 became the coldest, densest, deepest and most voluminous in the entire historical record going back to the 1930s. The LSW production averaged over the years 1987–1994 is equivalent to the volume flux of about 4.5 Sv; the corresponding export from the subpolar gyre is on the order of 3 Sv (Yashayaev 2007b). However, in the *extreme-convection years*, the individual *annual LSW production* and *export rates* were likely to exceed 7 and 5 Sv while the rate of the *LSW accumulation* in the subpolar reservoir could well reach 2 Sv. These estimates were based on the annual volumetric $LSW_{1987-1994}$ series from the whole subpolar gyre (Yashayaev 2007b); in order to establish a relationship between the average LSW thickness on the AR7 line and the volume of the entire LSW reservoir, the AR7 hydrography was used in conjunction with several large-scale surveys.

24.9.2 Transit of LSW Anomalies to the Irminger Sea and Iceland Basin

Temperature and salinity anomalies associated with the cores of $LSW_{1987-1994}$ and LSW_{2000} (Fig. 24.11) advect eastward, reaching the Irminger Sea in 2 years and

1 year, respectively, and arriving in the Iceland Basin in 5 and 4 years from the times of their formation in the Labrador Sea. It is not surprising to us that these transit times are significantly longer than those suggested by Sy et al. (1997) since their conclusions were derived from a rather short observational record, insufficient to register the true arrival of LSW into the two eastern basins.

We conclude that the true major source of LSW is the Labrador Sea and provide strong evidence that the cold, dense and deep LSW class ($LSW_{1987-1994}$) arrives in the Irminger and Iceland basins after being predominantly formed in the Labrador Sea.

In addition to documenting the spreading of new vintages of LSW across the ocean, our analysis reveals the source and spreading of the recent warming and salinification at the intermediate depths of the subpolar regions (Fig. 24.11). The westward spreading of warm and salty anomalies underlines the essence of the *three-dimensional exchange* between the Labrador and Irminger basins maintained by the cyclonic circulation within each basin – after most of $LSW_{1987-1994}$ has drained from the Labrador Sea, the anomalously warm and salty waters entering the Labrador-Irminger gyre from the east and southeast (e.g., the *Icelandic Slope Water* arriving from the Reykjanes Ridge) become noticeable and their pathway can be mapped (Fig. 24.11). This confirms the expectation that the Irminger-Labrador gyre receives waters from multiple sources and passes their anomalous features in both eastward (LSW) and westward (e.g., ISW) directions.

24.9.3 A Composite View of the Recent LSW Spreading History

A compilation of the $LSW_{1987-1994}$ *θ–S–time trajectories* (Fig. 24.13) for each of the basins discussed recaps the entire history of build-up, development, spreading and transformation of this most intriguing and revealing water mass class ever documented and reported. Together with the *distance–time* distributions of the LSW anomalies (Fig. 24.11), the θ–*S* projection of the LSW history (Fig. 24.13) presents our main observational evidence that LSW spreads across the *North Atlantic Ocean*, affecting regional hydrography. Even though this figure contains a rich message, we let it talk for itself and express its primary content in a single statement: *it took at least a decade for the coldest densest* $LSW_{1987-1994}$ *to travel from the Labrador Sea to the easternmost subpolar basin, the Rockall Trough.*

24.9.4 Where the Times Meet: Back to 1962 with "Erica Dan"

The recent hydrographic surveys of the subpolar basins (Figs. 24.2 and 24.3) have revealed complex variability in the top 2,000 m of water. In particular, there are several layers different in age or origin coexisting in narrow depth ranges. One specific layer is a remnant of cold, deep and dense LSW, formed before 1995. Can

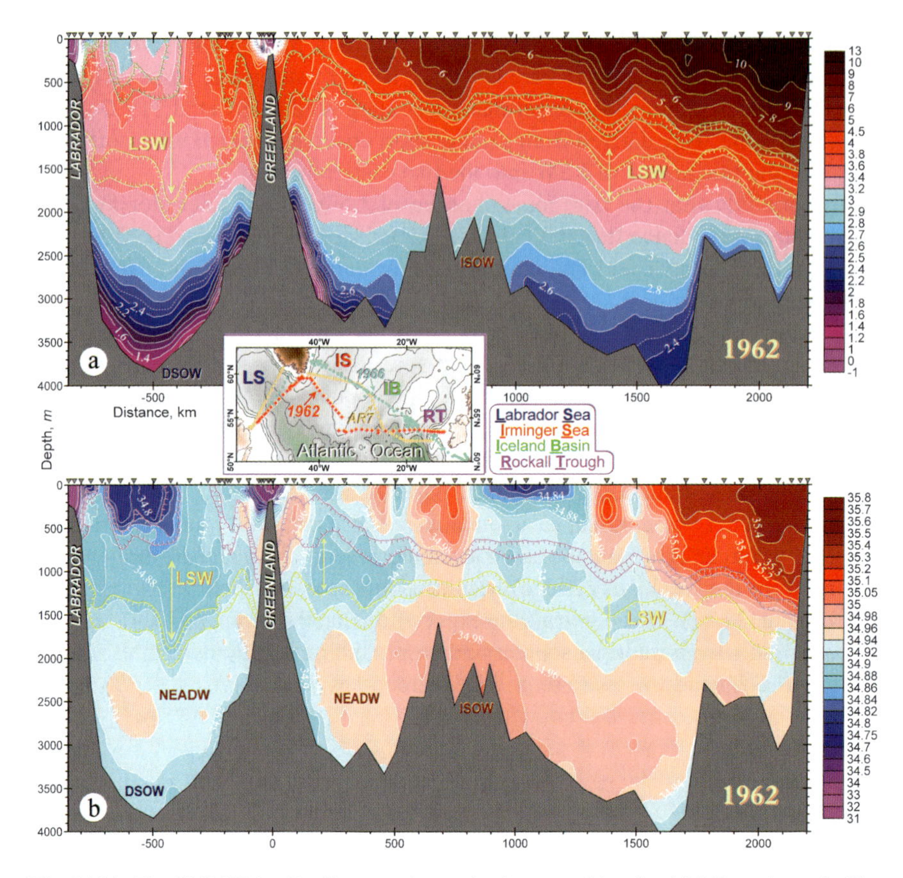

Fig. 24.14 The 1962 "*Erica Dan*" composite section best matching the AR7 lines shown in Figs. 24.2 and 24.3: (a) potential temperature and (b) salinity. The same notations as used in Figs. 24.2 and 24.3

we extrapolate our accumulated knowledge of the recent LSW development and transformation histories over occasional or *one-time* hydrographic surveys of the past and reconstruct a preceding LSW history? Even though, we can not give a complete answer, there is a case when such reconstruction is possible.

Selecting the fragments of the 1962 "*Erica Dan*" sections, we were able to construct one composite section (Fig. 24.14) providing a reasonable match to the AR7 lines (Figs. 24.2 and 24.3). This section (Fig. 24.14) shows generally fresher and colder conditions than the survey of 1966 (Figs. 24.2a and 24.3a). Similar to 2001 and 2004, several LSW classes can be identified in the subpolar North Atlantic in 1962. Some of these waters are likely to represent remnants of the colder and fresher LSW produced in the 1950s (Yashayaev et al. 2003). We believe that 1962 had the same relation to the 1950s as the recent years to the early-to-mid-1990s. Similar to the recent observations, in 1962 we could observe warm and saline layers separating fresher and colder LSW classes, probably originating from ISW. A closer look at spatial gradients and layer thickness distribution reveals more similarities in the deep LSW classes of the past and present ... but for now we are leaving a comparative LSW history out, thus leaving it for future investigation.

24.9.5 *Curtains*

The most remarkable LSW development of the late 1980s – mid-1990s left notable footprints in all subpolar basins of the North Atlantic: it took these basins one by one, influencing their water mass composition, stratification and circulation. Studies of the evolving hydrography have led to significant progress in our understanding of subpolar processes, but the past and present changes in LSW are yet to be fully studied and comprehended. We look forward to anticipated future developments.

Acknowledgements The authors thank Dan Wright for reviewing the manuscript and providing valuable comments and suggestions. We are grateful to Allyn Clarke, John Lazier, Jens Meincke, Bob Dickson and many others who over several decades surveyed, explored and monitored the subpolar basins of the North Atlantic.

Fisheries and Oceans Canada (DFO) at the Bedford Institute of Oceanography is gratefully acknowledged for its ongoing support of oceanographic surveys of the Labrador Sea under its Ocean Climate Monitorig program. The 2004 A1E hydrographic data are courtesy of Detlef Quadfasel and John Mortensen, and were collected with funding by the *Bundesminister für Bildung und Wissenschaft* (*German CLIVAR*) and the *EU Commission* (*ASOF-E*). Finally, many our friends and colleagues on land and, especially, at sea supported our study by generously sharing their skill, spirit and a kind word of encouragement.

References

Azetsu-Scott, K., E. P. Jones, I. Yashayaev, and R. M. Gershey (2003), Time series study of CFC concentrations in the Labrador Sea during deep and shallow convection regimes (1991–2000), *J. Geophys. Res.*, 108(C11), 3354, 10.

Bersch, M (2002), North Atlantic Oscillation-induced changes of the upper layer circulation in the northern North Atlantic Ocean, *J. Geophys. Res.*, 107 (C10), doi:10.1029/2001JC000901.

Bersch, M., I. Yashayaev, and K. P. Koltermann (2007), Recent changes of the thermohaline circulation in the subpolar North Atlantic, *Ocean Dynamics*, 57, 223–235, doi:10.1007/s10236-007-0104-7.

Boessenkool, K. P., I. R. Hall, H. Elderfield, and I. Yashayaev (2006), North Atlantic climate and deep-ocean flow speed changes during the last 230 years, *Geophys. Res. Lett.*, 34, L13614, doi:10.1029/2007GL030285.

Clarke, R. A. and J.-C. Gascard (1983), The formation of Labrador Sea Water. Part I: Large-scale processes, *J. Phys. Oceanogr.*, 13(10), 1764–1778.

Clarke, R. A. and A. R. Coote (1988), The formation of Labrador Sea Water. Part III: The evolution of oxygen and nutrient concentration, *J. Phys. Oceanogr.*, 18, 469–480.

Curry, R. G. and M. S. McCartney (2001), Ocean gyre circulation changes associated with the North Atlantic Oscillation, *J. Phys. Oceanogr.*, 31, 3374–3400.

Dickson, R. R., J. R. N. Lazier, J. Meincke, P. Rhines, and J. Swift (1996), Long-term coordinated changes in convective activity of the North Atlantic, *Prog. Oceanogr.*, 38, 241–295.

Dickson, R. R., I. Yashayaev, J. Meincke, B. Turrell, S. Dye, and J. Holfort (2002), Rapid freshening of the deep North Atlantic Ocean over the past four decades, *Nature*, 416, 832–837.

Ellett, D. J., A. Edwards, and R. Bowers (1986), The hydrography of the Rockall Channel – an overview. *Proc. R. Soc. Edinb*, 88(B): 61–81.

Gascard, J.-C. and R. A. Clarke (1983), The formation of Labrador Sea water. Part II: Mesoscale and smaller-scale processes, *J. Phys. Oceanogr.*, 13(10), 1780–1797.

Hakkinen, S. and P. B. Rhines (2004). Decline of subpolar North Atlantic circulation during the 1990s, *Science*, 304, 555–559.

Hátún, H., A. B. Sando, H. Drange, B. Hansen, and H. Valdimarsson (2005), Influence of the Atlantic subpolar gyre on the thermohaline circulation, *Science*, 309, 1841–1844.

Holliday, N. P., R. T. Pollard, J. F. Read, and H. Leach (2000), Water mass properties and fluxes in the Rockall Trough; 1975 to 1998, *Deep-Sea Res. I*, 47(7), 1303–1332.

Hurrell, J. W., Y. Kushnir, and M. Visbeck (2001), The North Atlantic Oscillation, *Science*, 291(5504), 603–605.

Kieke, D., M. Rhein, L. Stramma, W. M. Smethie, D. LeBel, and W. Zenk (2006), CFC inventories and formation rates of Upper Labrador Sea Water, 1997–2001, *J. Phys. Oceanogr.*, 36, 64–86.

Lazier, J. R. N. (1980), Oceanographic Conditions at OWS Bravo 1964–1974. *Atmos. Ocean*, 18(3), 227–238.

Lazier, J. R. N. (1995), The salinity decrease in the Labrador Sea over the past thirty years. In: D.G. Martinson, K. Bryan, M. Ghil, M.M. Hall, T.M. Karl, E.S. Sarachik, S. Sorooshian, and L.D. Talley (Eds.), *Natural Climate Variability on Decade-to-Century Time Scales* (pp. 295–304). Washington, DC: National Academy Press.

Lazier, J. R. N., R. M. Hendry, R. A. Clarke, I. Yashayaev, and P. Rhines (2002), Convection and restratification in the Labrador Sea, 1990–2000, *Deep-Sea Res.*, 49A(10), 1819–1835.

McCartney, M. S. and L. D. Talley (1982), The Subpolar Mode Water of the North Atlantic Ocean, *J. Phys. Oceanogr.*, 12(11), 1169–1188.

Straneo, F., R. S. Pickart, and K. Lavender (2003), Spreading of Labrador sea water: an advective-diffusive study based on Lagrangian data, *Deep-Sea Res. I*, 50(2003), 701–719.

Sy, A., M. Rhein, J. R. N. Lazier, K. P. Koltermann, J. Meincke, A. Putzka, and M. Bersch (1997), Suprisingly rapid spreading of newly formed intermediate waters across the North Atlantic Ocean, *Nature*, 386, 675–679.

Talley, L. D. and M. S. McCartney (1982), Distribution and circulation of Labrador Sea Water, *J. Phys. Oceanogr.*, 12(11), 1189–1205.

van Aken, H. M. and C. J. de Boer (1995), On the synoptic hydrography of intermediate and deep water masses in the Iceland Basin, *Deep-Sea Res.*, 42A, 165–189.

Yashayaev, I. (2000), 12-Year Hydrographic Survey of the Newfoundland Basin: Seasonal Cycle and Interannual Variability of Water Masses, *ICES (International Council for the Exploration of the Sea)*, CM 2000, L:17, 19 p.

Yashayaev, I. (2007a), Changing Freshwater Content: Insights from the Subpolar North Atlantic and New Oceanographic Challenges, *Prog. Oceanogr.*, 73(3–4), 203–209, 10.1016/j.pocean.2007.04.014.

Yashayaev, I. (2007b), Hydrographic changes in the Labrador Sea, 1960–2005, *Prog. Oceanogr.*, 73(3–4), 242–276, 10.1016/j.pocean.2007.04.015.

Yashayaev, I., M. Bersch, and H. M. van Aken (2007a), Spreading of the Labrador Sea Water to the Irminger and Iceland basins, *Geophys. Res. Lett.*, 34(10), L10602, doi:10.1029/2006GL028999.

Yashayaev, I. H. M. van Aken, N. P. Holliday, and M. Bersch (2007b), Transformation of the Labrador Sea Water in the Subpolar North Atlantic, *Geophys. Res. Lett.*, 34(22), L22605, doi:10.1029/2007GL031812.

Yashayaev, I., J. R. N. Lazier, and R. A. Clarke (2003), Temperature and salinity in the central Labrador Sea. ICES Mar. Symp. Ser., 219, 32–39.

Chapter 25
Convective to Gyre-Scale Dynamics: Seaglider Campaigns in the Labrador Sea 2003–2005

Charles C. Eriksen and Peter B. Rhines

25.1 Introduction

Observations of the ocean interior are, in general, sparse relative to the continuum of eddy noise that masks circulation on seasonal and interannual scales. The paucity of observations is particularly acute in remote heavy-weather regions, of which the Labrador Sea is an example. Although the Labrador Sea is a relatively small regional basin in the North Atlantic, few ships traverse it at any time of year, particularly in winter, because its harsh climate limits human habitation around its edges and inhibits exploration of its interior. The lack of intensive observations is in spite of the interesting and important climate signals heavy weather imparts to the region. We report here the application of long-range autonomous underwater glider vehicles in a pilot demonstration of this technology in addressing the need for year-round in situ observations in the Labrador Sea.

The Labrador Sea is well known as a region of intense air–sea interaction in winter. Surface mixed layers deepen to a thousand meters and more in winter, forming a distinct water mass that populates much of the North Atlantic subpolar gyre at shallow and intermediate depths. A portion of this water mass is also exported to the subtropical North Atlantic as well, through boundary currents and interaction with the extension of the Gulf Stream eastward from the Grand Banks across the ocean. In summer, the sea is observed to re-stratify more quickly than can be accounted for by local air–sea fluxes. This puzzle, and others in climate, such as the contribution of boundary currents around the basin to seasonal and interannual changes in heat, freshwater, and volume transport in the global ocean remain unanswered largely because in situ observations are limited by cost and difficulty.

Historically and until very recently, ship-based surveys have been the dominant source of in situ observations in the Labrador Sea. Because of the resources they demand, these observations have been restricted to a few transects and largely to fair-weather (or, more accurately, less foul weather) parts of the year. One transect, the World Ocean Circulation Experiment (WOCE) AR7W line between Hamilton

School of Oceanography, University of Washington, Seattle WA 98195-5351

R.R. Dickson et al. (eds.), *Arctic–Subarctic Ocean Fluxes*, 613–628

Bank off Labrador and Cape Desolation, Greenland, has been occupied annually in summer for almost 20 years. The interior structure of the Labrador Sea in other seasons and at other locations has not benefited from regular observation. Over the course of the past several years, Argo profiling floats have reported temperature and salinity structure at 10-day intervals at random locations within the basin, which has begun to reduce the fair-weather bias of in situ observations. With the advent of gliders, the location of conductivity-temperature-depth (CTD) profiles is controllable, at least to some degree, so observations can be extended into boundary currents and across mesoscale eddy features, yet still with autonomy from ships and through the severe weather and seas of non-summer seasons.

Trans-basin hydrographic sections are the traditional basis for estimates of volume, heat, and freshwater transports in the ocean. Because ship-based surveys are sparse in the Labrador Sea, the only interannual estimates of these quantities we have are from the AR7W line, and they are subject to the usual uncertainties in geostrophic reference level. The reference problem aside, the annual summer survey does not resolve seasonal variations which inevitably will alias variability on interannual timescales.

Important processes occur with time and space scales that are short compared to typical trans-basin hydrographic and Argo float resolution. These include convective mixing and eddy fluxes from boundary currents. Finer resolution observations are needed to detect these processes, if for no other reason than to avoid aliased estimates of large-scale impact.

Seaglider long-range autonomous underwater gliders were used to remotely survey the Labrador Sea over two winter periods and through one summer period in the 2 years starting October 2003. These vehicles were controlled to make basin-wide sections and to survey Davis Strait. As a demonstration project, the campaigns illustrated both strengths and limitations of autonomous gliders.

25.2 Seaglider Technology

Autonomous underwater gliders can be thought of as profiling floats with wings and also as buoyancy-driven relatives of traditional propeller-driven autonomous underwater vehicles (AUVs). They provide a compromise between the attributes of high endurance with which profiling floats have come to be associated (deployments of multiple years) and the spatial-temporal sampling control (minutes and tens to hundreds of meters horizontally) of AUVs. Glider endurance and spatial resolution falls between that of profiling floats and AUVs, yet their sampling is fully remotely controllable and their range far exceeds that of propeller-driven AUVs.

Seaglider is one of three glider vehicles with roughly similar size and capability (see Rudnick et al. 2004). It was designed to collect open-ocean observations at relatively low cost and high resolution over missions of several-month duration and several thousand kilometer range with enough control to navigate effectively through typical upper ocean current fields. Details of its original design can be found in Eriksen et al. 2001. While Seaglider missions in the Labrador Sea were challenging and not without shortcomings, the vehicles proved themselves effective at remotely surveying the basin through the most severe wind and sea conditions. Like any tool, they have advantages and disadvantages, which are briefly discussed below.

Seagliders are small (1.8 m long, 52 kg) reusable long-range AUVs that can be launched and recovered manually by two persons. They dive to 1 km depth along a glide slope as shallow as 1:5 or as steep as ~1:1, turn, and climb similarly to the sea surface. In order to maximize range and endurance while maintaining the ability to navigate through ocean currents, they typically maintain a speed of about ½ kt while consuming about ½W on average from their supply of primary lithium batteries. About 85% of the power budget is devoted to propulsion, while the remainder operates the vehicle microprocessor, sensors, GPS receiver, modem, and satellite transceiver. Propulsion is effected through vehicle volume displacement adjustment and attitude control. The resulting buoyancy forces balance wing lift and drag to produce forward motion. Internal trim adjustments (battery pack displacement along and across the vehicle axis) are used to control pitch and roll attitude, the latter causing the glider to turn. Seagliders dead-reckon underwater to maintain roughly constant magnetic heading. When they reach the sea surface, they pitch their nose down, exposing a trailing antenna stalk. A pair of GPS fixes brackets a communications session using the Iridium network. Seagliders both store all mission data on board and send it ashore to a computer base station between dive/climb cycles. They can receive control commands after each dive cycle, but operate autonomously while submerged. Seagliders routinely collect profiles of temperature, salinity, dissolved oxygen, chlorophyll fluorescence, and optical backscatter along saw-tooth paths. Through knowledge of their flight characteristics, profiles of the vertical current component and average horizontal current over the depth range of each dive cycle can be estimated.

The principal advantage of Seagliders is their modest cost. The cost of a 6-month Seaglider mission is comparable to the cost of 2 days of research vessel operation. Moreover, the acquisition cost is roughly double the cost of each mission, so that loss of a vehicle, while undesirable, is not devastating to a research program. Small size is a significant contributor to low cost, both from manufacturing and handling. A second advantage is that data is reported in near real time, so that sampling can be adjusted adaptively and vehicle operation can be monitored. In many cases, operational degradation has been detected and the vehicle rescued or software repaired in response. The use of the Iridium satellite communication system enables global remote control. Finally, Seagliders operate robustly in severe sea conditions and are impervious to the icing problem encountered by ships operating in sub-freezing air temperatures.

Limitations of Seagliders stem largely from the source of their economy: their size. Small size implies a limited payload to area ratio, hence a lower possible range and endurance compared to a larger vehicle. The extensive range and endurance of gliders is obtained at the expense of speed (quadratic drag implies range is inversely proportional to speed through the water), so synopticity is traded against survey extent. Small size limits the instrumentation carried in both size and power consumption. Not only must sensors be low power, but they must be hydrodynamically unobtrusive to avoid contributing unduly to vehicle drag, hence performance. Small size also contributes to fragility: collision with ships or icebergs and entanglement in fish nets are likely to be fatal encounters. Seagliders are limited to operating in the upper kilometer of the water column by the strength of their aluminum pressure hull. Finally, if communications are lost, so are gliders. They are susceptible to antenna breakage, electronic malfunction, and the ability of the vehicle to surface and raise its antenna. The ultimate limitation is reliability of the buoyancy engine.

25.3 Campaign in the Labrador Sea: A Tale of Persistence and Luck

Seagliders were deployed in the Labrador Sea in pairs in October 2003, September 2004, and April 2005. This choice was driven by the need to avoid field deployment and recovery operations during the most severe weather periods of winter. We based field operations in Nuuk, Greenland to take advantage of the logistic support of the Greenland Institute of Natural Resources (GINR), the availability of vessels for charter, isolation from ice floes, and the relative narrowness of the continental shelf. GINR furnished shelter and sustenance for a small field team of three as well as modest laboratory facilities in which Seagliders could be assembled and tested, assistance with transportation, and advice and contacts with the community as well as extensive local knowledge of the maritime environment. Basing at GINR combined with charter of a small vessel provided an alternative to the normal model of research vessel-based oceanography, although we gladly used the services of research vessels when available on an ancillary basis.

The first deployment was of two Seagliders at the edge of Fyllas Bank, offshore Nuuk after waiting nearly 2 weeks for a suitable weather window in October, 2003. The deployment was made on a day trip from shore aboard M/V Hans Egede, a small vessel normally used for tourism charters capable of cruising at ~20 kt in sufficiently smooth seas. The intent was to occupy trans-basin sections in the western Labrador Sea, cross the basin along the AR7W WOCE line to the Greenland side, then survey the West Greenland Current on the return to offshore Nuuk in 6-month missions. One of the gliders, designated sg004, was sent west to 58° W, then south to the Labrador continental shelf edge, east along ~55.5°W, then south again to the Labrador continental shelf edge off Hamilton Bank before heading to the Greenland side. The other glider, sg008, was sent southwest from offshore Nuuk to 55° W, then south to the continental shelf off edge off Hamilton Bank, then back north along the same parallel. The second deployment of a pair of gliders was carried out in Davis Strait by our University of Washington colleagues Dr. Craig Lee and Dr. Jason Gobat from the R/V Knorr in September, 2004. After several transects across Davis Strait, these two, sg014 and sg015, repeated the 58° W and 55° W sections. Then sg014 crossed the basin roughly along the AR7W line, while sg015 returned north along 58° W. The final deployment of a pair was made off Fyllas Bank in April, 2005, with the intent of repeating the sampling plan of the previous two winter deployments. The tracks of all the glider missions are shown in Fig. 25.1.

That Seagliders are able to execute trans-basin sections in the Labrador Sea is evident from the tracks in Fig. 25.1. The transects are neither as straight as typical

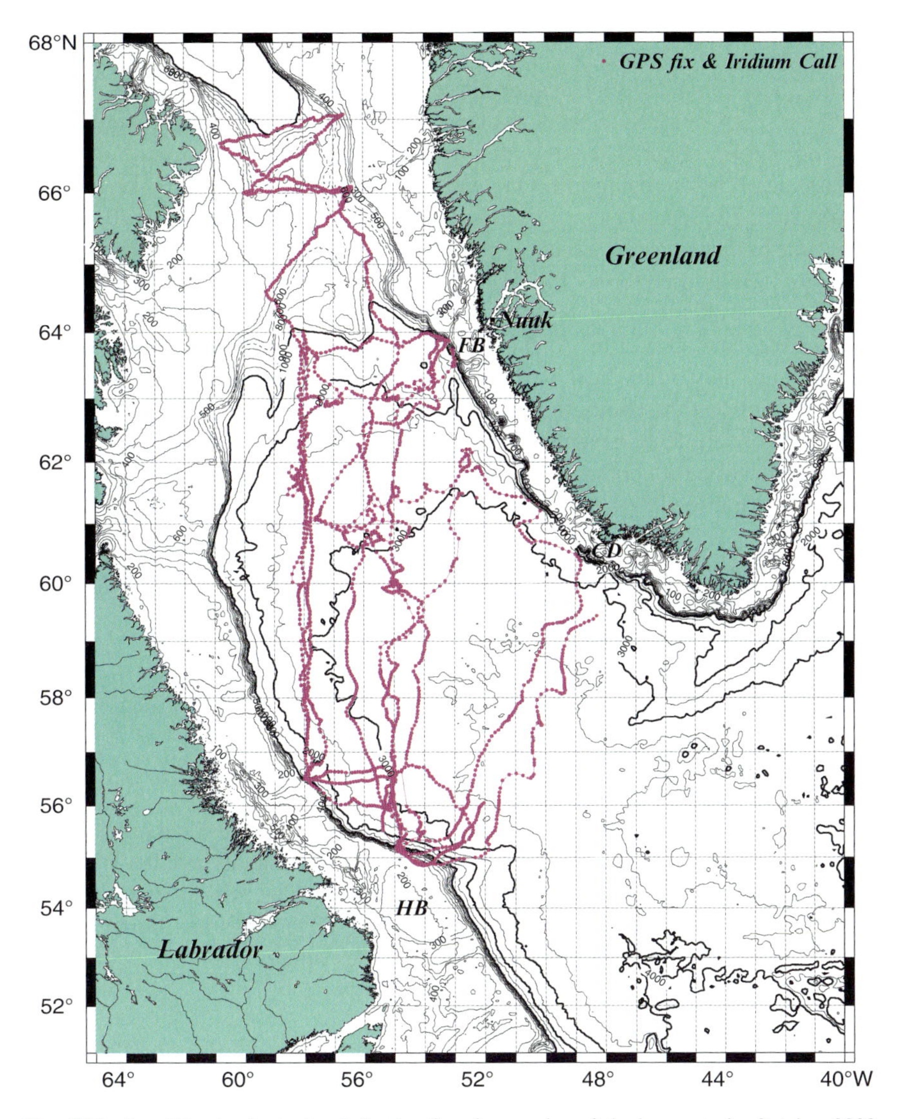

Fig. 25.1 Seaglider tracks in the Labrador Sea from pairs of deployments in October 2003, September 2004, and April 2005. Symbols are drawn at the location of each communication session that returned a GPS fix. Dive/climb cycles in water depths greater than 1 km typically lasted 8–9 h. Bathymetry contour interval is 100 m to 1,000 m depth, 500 m thereafter

shipboard hydrographic sections, nor as synoptic on basin scales. Irregularities in the sections are due to mesoscale eddy currents comparable to, and often exceeding, glider speed. Occasionally, the vehicles had to be redirected from their desired path to avoid being stalled by eddy currents. Instead of taking place over several weeks, as would be the case with a research vessel, the survey mission of each glider lasted several months. The horizontal resolution of glider sections, however, is considerably higher than for typical hydrographic surveys. Gliders reached the surface and 1 km depth every few km, in contrast to typical AR7W station spacing of ~30 km.

The glider missions were neither all complete nor free of difficulties. Neither sg004 nor sg008 returned to Nuuk as planned. Both vehicles developed communications difficulties about 3 months after launch and managed to operate only for another month before being lost. One was recovered on a beach on Disko Island 6 months later. The second pair of deployments was more successful, with sg014 completing the basin navigation after 7 months, 1 week at sea, traveling 3,750 km through the water, setting the records for AUV endurance and range on a single mission. Sg015 was lost after draining its battery trying to cross the eddy rich region off south-central Greenland. One of the final pair deployed (the one rescued on Disko Island) was lost after only a few days, but its companion managed to complete the circuit, despite having lost the ability to acquire GPS fixes shortly after starting the AR7W transect. We were able, nevertheless, to direct it to Nuuk by recognizing when it grounded itself on the continental slope of southern Greenland and using navigation information furnished to us by operators of the Iridium system. It was recovered by our colleagues C. Lee and J. Gobat from CCGS Hudson on Fyllas Bank. Despite having lost three gliders in six attempted missions, Seagliders collected over 5,000 CTD profiles over the course of two winters and one summer, demonstrating the feasibility of using long-range autonomous underwater vehicles to survey a remote, heavy weather region of the ocean.

25.4 Preliminary Results

The usefulness of Seaglider data to understanding aspects of physical processes in the Labrador Sea can be gauged by examining some preliminary results. We show here examples of trans-basin sections, individual profiles, estimates of volume transport, and sections through an eddy field. Both advantages and limitations of glider sampling are readily apparent in the examples offered.

25.4.1 Basin-Wide Sections

Meridional sections in the central Labrador Sea made in successive autumn seasons roughly along 55° W are shown in Fig. 25.2. All individual data points are plotted in the panels of this figure arranged by depth and range from a reference location. In situ horizontal locations were found by interpolating between GPS fixes at the

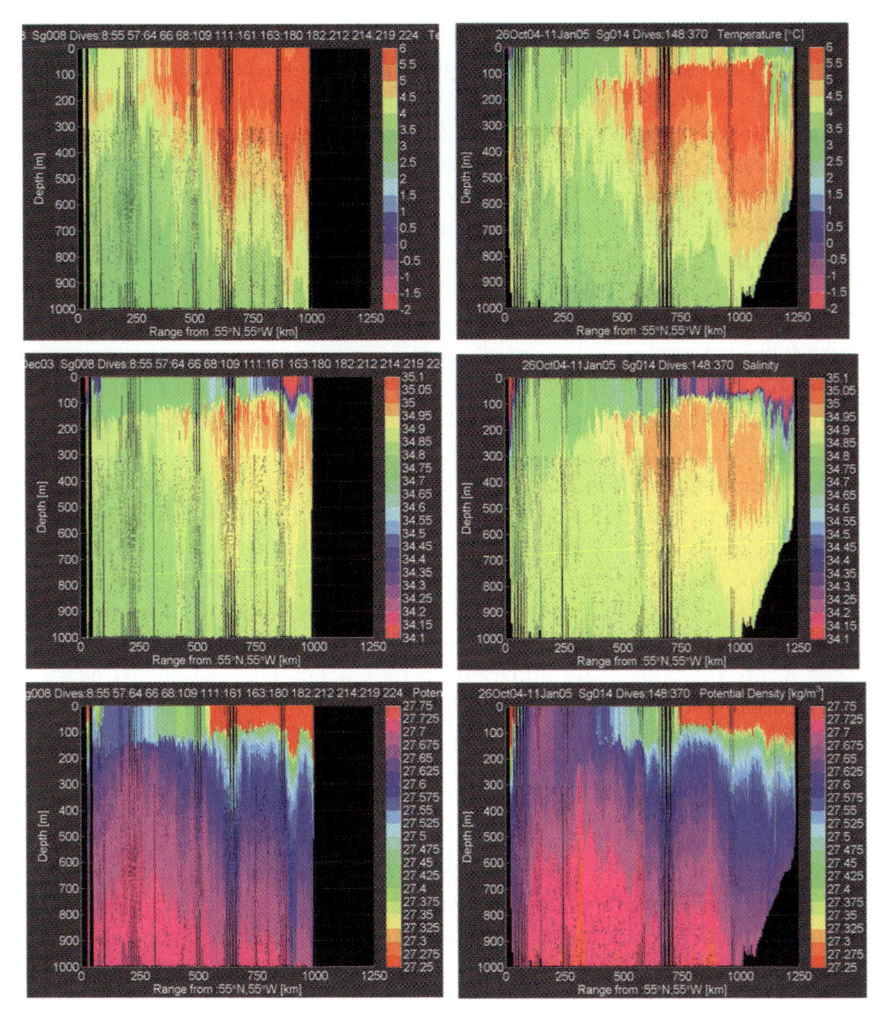

Fig. 25.2 Sections along 55° W in the Labrador Sea from 3 October–3 December 2003 (left column) and 26 October 2004–11 January 2005 (right column) of temperature (°C) (top panels), salinity (middle panels), and σ_t (kg m^{-3}) (bottom panels). Sections are plotted as depth (m) vs. range (km) from 55° N to identical scale. Values beyond color bar range are plotted as the appropriate endpoint color. The Labrador continental shelf near Hamilton Bank is plotted to the left and the entrance to Davis Strait on the right of each section

ends of each dive/climb cycle and applying the estimated depth-averaged current as a uniform drift with time, independent of depth. Depth-averaged current was computed from the difference between displacement between fixes and the time-integrated dead-reckoned vehicle velocity. Model vehicle speed is derived using observed vehicle buoyancy and pitch attitude assuming unaccelerated flight based on lift and drag coefficients regressed from minimizing the difference between predicted vertical speed and that inferred from pressure depth. Details of the model can be found in Eriksen et al. (2001). The 1,000+ km sections shown are sufficiently long to hide the saw-tooth sampling path of gliders except where strong along-transect currents reduce horizontal resolution markedly.

The sections in Fig. 25.2 required a season or more to complete, hence some apparent spatial changes are due to temporal evolution. In particular, the onset of winter brings the erosion of upper ocean stratification with the deepening, cooling and transition to higher salinity of the surface mixed layer. Because the sections were both carried out in a southward sense, mixed layer deepening appears as a tilt downward to the south (left) in the panels. Despite the temporal aliasing, the sections both indicate fresher, cooler near surface waters than at intermediate depths (~100–800 m). The contrast is particularly large at the northern end of the sections, presumably due to the influence of relatively warm saline water of Irminger Sea origin overlain by cooler fresher water of marginal origin, presumably largely from Greenlandic ice and runoff.

Clear interannual differences are evident in these two autumn sections. Near surface waters are warmer and saltier in autumn 2003 (left panels, Fig. 25.2) than in autumn 2004, while intermediate waters are somewhat fresher in the north (right side of panels) and cooler and fresher in the south in 2003 than in 2004. The attendant pycnocline height is elevated in the later year in the northern part of the section and extends to the sea surface in the southern portion. The deepest mixed layers found by these gliders were only a few hundred meters depth in winter 2004, but exceeded 1 km depth in winter 2005.

25.4.2 Davis Strait

Seaglider sections across Davis Strait are short enough to give a more synoptic view of internal structure, since the crossing between shelf edges takes about a fortnight or less. Three of the five sections collected from late September to early November 2004 are shown in Fig. 25.3. Depth-averaged currents for each dive/climb cycle are aligned roughly parallel to the channel walls and are of tidal frequency. Subtidal depth-averaged currents are relatively small, with the exception of narrow poleward flow on the Greenland (right) side of the sections. The northernmost section shows a thick layer of cool freshwater across the western and central portions of the Strait in the upper 200–300 m, capped by a thin layer of slightly warmer but even fresher water at the surface. Warm, salty water

Fig. 25.3 Three sections across Davis Strait, as indicated by the track segments highlighted in green in the right panels. Depth-averaged current vectors for each dive/climb cycle (indicated by blue arrows) are dominated by tidal flow through the Strait. Sections of temperature, salinity, and σ_t are drawn as in Fig. 25.2 with Baffin Island on the left and Greenland on the right in each section (note different scales than in Fig. 25.2 and arrangement of fields by column and sections by row). Dates of the northern, middle, and southern sections are 27 September–10 October, 9–22 October, and 22 October–5 November, 2004. Depths are contoured at 100 m interval to 1 km and the range reference for each section is indicated by a red cross

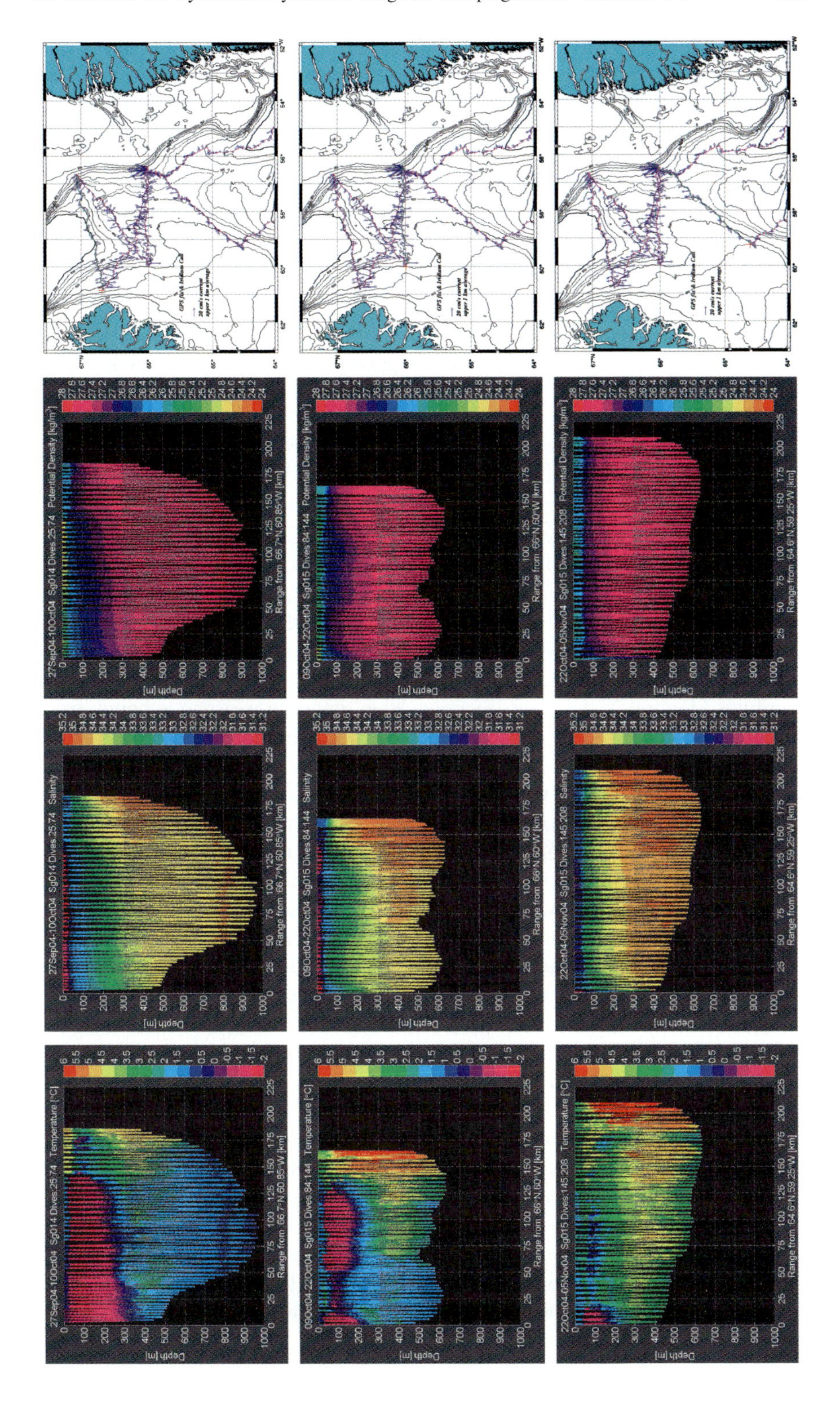
27Sep04-10Oct04 Sg014 Dives:25:74 Potential Density [kg/m^3]
27Sep04-10Oct04 Sg014 Dives:25:74 Salinity
27Sep04-10Oct04 Sg014 Dives:25:74 Temperature [°C]
Range from 66.7°N,60.85°W [km]
09Oct04-22Oct04 Sg015 Dives:84:144 Potential Density [kg/m^3]
09Oct04-22Oct04 Sg015 Dives:84:144 Salinity
09Oct04-22Oct04 Sg015 Dives:84:144 Temperature [°C]
Range from 66°N,60°W [km]
22Oct04-05Nov04 Sg015 Dives:145:208 Potential Density [kg/m^3]
22Oct04-05Nov04 Sg015 Dives:145:208 Salinity
22Oct04-05Nov04 Sg015 Dives:145:208 Temperature [°C]
Range from 64.6°N,59.25°W [km]
Depth [m]

is found at depth, particularly against the eastern (Greenland) side of the Strait. Density surfaces tilt down to the west, indicative of equatorward geostrophic shear, consistent with inflow to Baffin Bay to the north at depth and surface outflow to the Labrador Sea to the south.

Farther south, the Davis Strait sections show a progressive absence of the cool fresh intermediate layer (colored magenta in the temperature sections), and a progressively more prominent warm, salty water mass hugging the Greenland (right) side at depth (colored yellow and orange in the temperature and salinity sections). The overall picture is one of exchange flow as in an estuary, where salty water enters (Baffin Bay) at depth and freshwater exits nearer the surface. Interestingly, there is little evidence for baroclinic shear in the southernmost (bottom) section, but depth-averaged currents are generally northward across the deeper part of the passage (on the Greenland side).

25.4.3 Labrador Current

Gliders crossed the offshore branch of the Labrador Current several times in the southwest Labrador Sea (Fig. 25.4). The strategy was to cross obliquely so that current speed would not overwhelm glider speed. Not surprisingly, eddy activity aliased observations of the current, so although volume transports in the top 1 km were observed in individual sections to range from 2 to 35 Sverdrups (1 Sverdrup = 10^6 m^3 s^{-1}), most commonly the transport of the offshore branch was ~20 Sverdrups. Figure 25.4 shows sections from opposite seasons running from Hamilton Bank and heading roughly northeastwards in an attempt to follow the AR7W line. While the sections have similarly little structure offshore from the continental shelf and differ only in the appearance of a warm, slightly fresh layer in summer (bottom row, Fig. 25.4, in contrast to the top row), the biggest difference is the appearance of very cold (<0 °C) freshwater over the continental shelf.

The upper layers of the offshore branch of the Labrador Current showed a tendency to transport water onto the continental shelf near Hamilton Bank from offshore, particularly in the summertime crossings. Currents over Hamilton Bank were found to exceed 1 kt, more than double the glider speed, hence navigating the gliders off the continental shelf was challenging. Operating Seagliders over the continental shelf is several times less efficient energetically than in deep water, since although the buoyancy engine consumes about half as much battery energy pumping at 100 m depth as at 1,000 m depth, the distance the vehicle glides is proportional to its dive depth. Additionally, fresh surface layers commonly found on subpolar continental shelves increase the buoyancy barrier a glider must cross.

Our brief experience with gliders in the Labrador Current suggests that to be effective, gliders would need to regularly cross the current system as often as bimonthly on the same section to obtain a stable estimate of structure in a given season. A time series of sections across the Labrador Current in the southwest Labrador Sea would become feasible if gliders were launched locally and could endure the heavy ice flows that commonly reach the shelf edge.

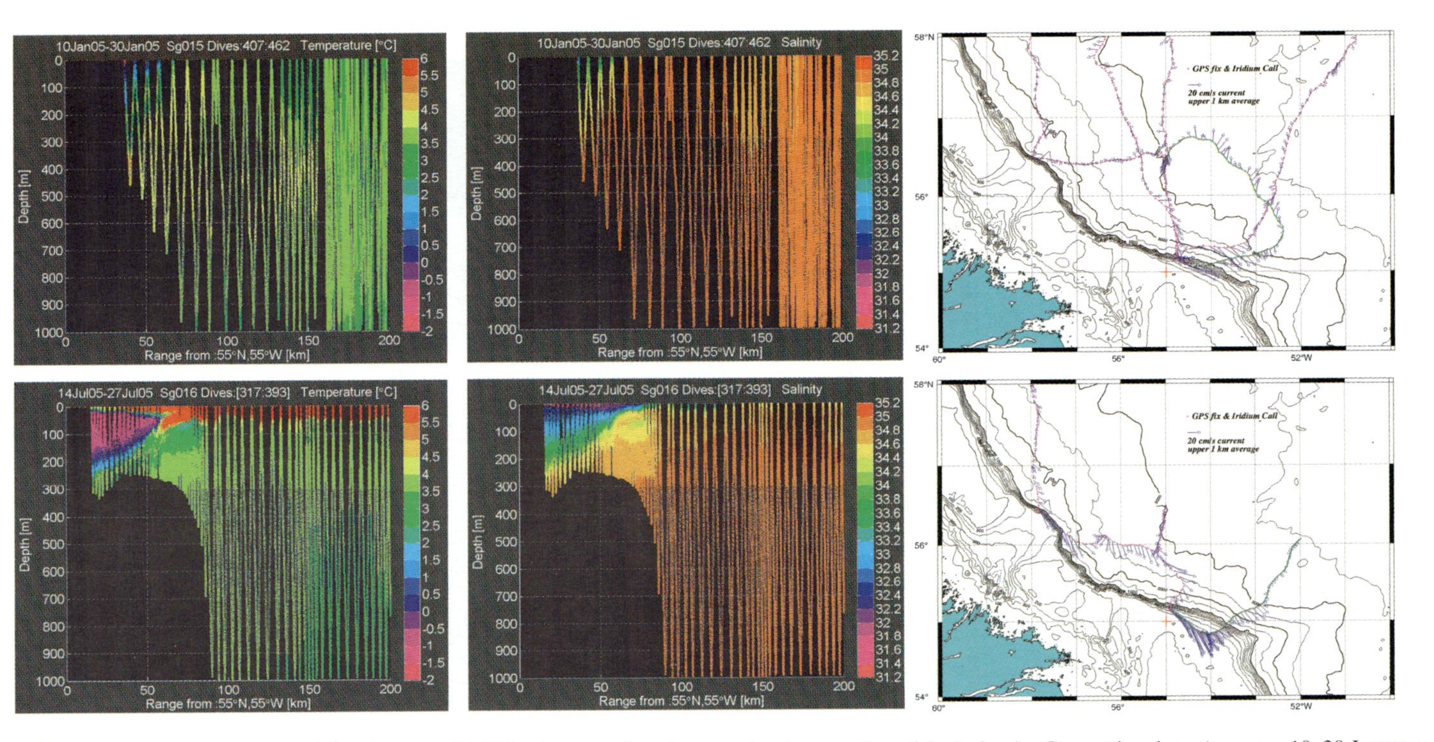

Fig. 25.4 Sections of temperature (left column) and salinity (center column) across the slope portion of the Labrador Current in winter (top row, 10–30 January 2005) and summer (bottom row, 14–27 July 2005) plotted as in Fig. 25.3. Sections are plotted against range from 55° W, 55° N (marked by a red cross) with the tracks highlighted in green. Depth-averaged currents (right column) are shown by blue arrows for these and nearly contemporaneous glider tracks in the region

25.4.4 Eddies in the Northeastern Labrador Sea

Mesoscale eddy activity is particularly rich in the northeastern Labrador Sea, so poses particular challenges to glider observations. As the tracks in Fig. 25.1 suggest, along the long transects in the central and western Labrador Sea (e.g. 58° W and 55° W), encounters with eddies were relatively rare. In the eastern half of the basin between about 60° N and 64° N, eddies are more numerous and energetic. Glider tracks and accompanying estimates of depth-averaged currents in this region are shown in Fig. 25.5. While it is common for a glider to encounter currents weak compared to its 0.2 m s^{-1} speed through the water for distances of 100 km or more, it is also common to encounter eddy currents that are much more intense, with depth-averaged speeds in the 0.4–0.6 m s^{-1} range. The tracks in Fig. 25.5 demonstrate how challenging it is to navigate with a vehicle whose speed is often exceeded by transient currents.

A particularly severe example of the loss of effective navigational control of a glider by currents can be seen in the track of a glider heading generally poleward along the West Greenland continental slope (lower right corner of Fig. 25.5). We were attempting to execute a sequence of short sections across the West Greenland Current on the way northwest toward Nuuk. Depth-averaged currents drew sg014 toward the shelf edge

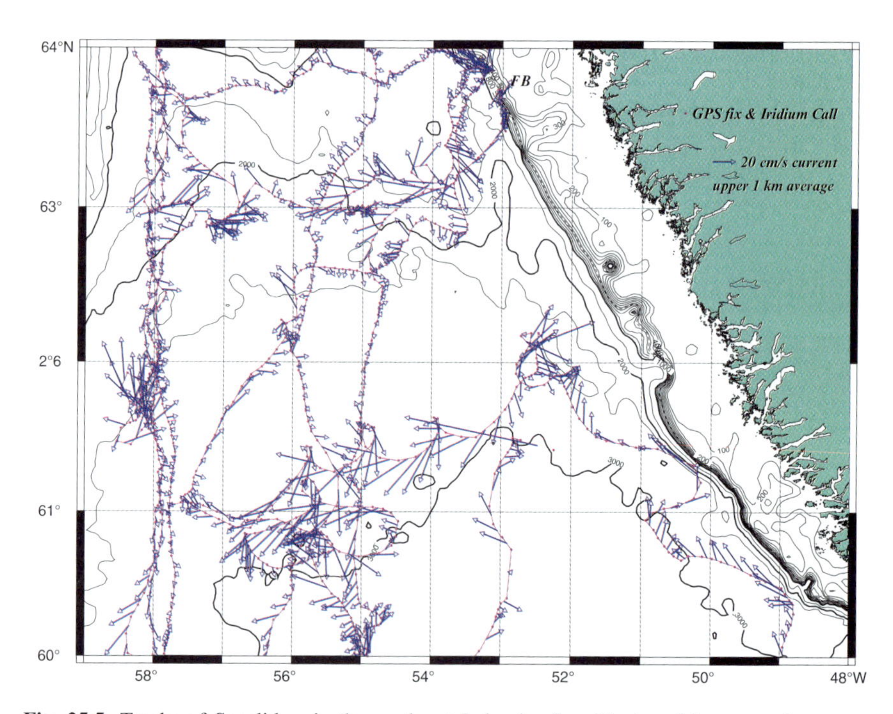

Fig. 25.5 Tracks of Seagliders in the northeast Labrador Sea. Vectors (blue arrows) represent depth-averaged current for each dive cycle. Dots indicate positions of GPS fixes at the start and end of each dive/climb cycle. Bathymetry contour interval is 100 m to 1,000 m depth, 500 m thereafter

near 61° N, 50° W, and equally abruptly ejected it to the west just ~50 km farther along the basin boundary. Once drawn seaward near 62° N, 52° W, sg014 was entrained in a powerful eddy that proceeded to take it offshore. The cusp-like trajectory of the glider and the anticyclonic turning of depth-averaged currents were analyzed by Hátún et al. (2008) to demonstrate the role of eddies in injecting cool freshwaters at shallow depths and warm salty waters at intermediate depths from the West Greenland Current and West Greenland Coastal Current into the Labrador Sea interior. After four trips from the eddy margin to near its center, we managed finally to extricate the glider by directing it radially outward across the anticyclonic circulation. Finally, near 61° N, 57° W, the glider left the eddy and was directed northeastward toward Fyllas Bank for recovery. A second glider, sg015, was not so lucky, and ran out of battery power before it could pass around this same eddy. It was left to drift at the surface.

25.5 New Developments

As with any technology, improvements and enhancements to Seaglider technology are underway to address some of its performance limitations. As the discussion above implies, the Labrador Sea environment tested Seaglider performance to its limits. Greater mission endurance, ability to navigate against swift upper ocean currents, and the ability to work near and under ice are all desirable extensions of Seaglider technology. Three derivative designs have been developed and implemented to address these extensions.

The Seagliders used in the missions discussed above are capable of missions of 6 months or more in the Labrador Sea, limited by their battery energy. In such regions of harsh weather and seas much of the year, missions as long as 1 year are desirable. Not only would 1-year missions simplify logistics, they hold the promise of nearly halving the operational cost of gliders, since the fixed costs of service, deployment, and recovery dominate the per mission cost. Fortunately, the original Seaglider end caps were slightly over-designed and safely could be made roughly 1.5 kg lighter, allowing additional batteries. In addition, higher energy density batteries were identified. Together the increase in battery capacity implies Seaglider mission endurance in the Labrador Sea of ~13 months is feasible.

A faster glider would navigate more effectively across swift boundary currents and through intense eddies, but, as mentioned above, glider range and endurance scale inversely with speed with quadratic drag. An alternative strategy is to reduce the depth-averaged flow a glider encounters by increasing the depth of its dives, since ocean currents are typically surface-intensified. Increasing the glider depth range also makes possible the study of deep circulation. Prominent deep circulation signals are a feature of the North Atlantic and the Labrador Sea in particular. To this end, a new vehicle called Deepglider has been developed and tested.

Deepglider uses a carbon fiber hull with a low drag shape in place of the aluminum hull – fiberglass fairing combination used in Seaglider. The composite hull not only is strong enough to withstand bottom pressure, but light enough to allow a larger battery complement than even the extended range version of Seaglider, mentioned above, enabling deep missions of 1 year or more. Deepglider is designed for 6,000 m capability, though because of manufacturing issues, to date we have laboratory-tested a hull to 4,000 dbar. We flew the first Deepglider with this hull to 2,750 m depth with a 150-dive mission across the continental slope offshore Washington in November–December 2006. An additional hull made with thermoplastic rather than thermoset resin is being prepared for testing to 6,000 dbar.

The need to operate long-range autonomous vehicles under ice has been addressed with an under-ice capable Seaglider developed by C. Lee and J. Gobat at the Applied Physics Laboratory, University of Washington. This variant uses RAFOS acoustic navigation and an altimeter to find its way when under ice. One of these gliders successfully reached open water after making an under-ice transect in Davis Strait in late 2006. Unlike standard Seagliders, which obtain navigational fixes and communicate upon visiting the sea surface, this variant is designed to detect and avoid ice while profiling beneath it. It stores data on board until it reaches a region free enough of ice to communicate via satellite. This version was the first to use the long-range battery pack.

25.6 New Observations from the Iceland–Faroe Ridge

Seagliders have most recently been deployed to observe inflow to and overflow from the Nordic Seas across the Iceland–Faroe Ridge. The intent of our study is to resolve the small-scale flows along the Ridge that exchange heat and freshwater with the North Atlantic by making full-depth transects with Seagliders. Originally, we envisioned repeating a sequence of fixed sections both along and across the Ridge (magenta line segments, Fig. 25.6, upper panel), but depth-averaged currents proved similar in magnitude to the upper limit on glider speed necessary for the required endurance of 3–4 months between the regular cruises of the Faroese Fisheries Laboratory R/V Magnus Heinason, used for deployment and recovery.

The first deployment, November 2006–February 2007, saw a pair of Seagliders survey the eastern portion of the Ridge between its crest and the 1 km Atlantic-side isobath. One of the two also crossed the Ridge before being recovered north of the Faroes. In the subsequent deployment, February–June 2007, a section nominally along the 900 m isobath on the Atlantic side of the Ridge was collected (Fig. 25.6, lower panel). The glider was unable to follow the 900 m isobath more closely than to within 100 m or so because of vigorous across-isobath currents, often swifter than glider speed through the water. Despite low stratification and the constraint of potential vorticity conservation, depth-averaged current components in the cross-isobath direction exceeding 0.1 m s^{-1} were commonly found along the section. Also found sporadically were cold, freshwater parcels in the bottom 50–100 m or so of the water column. The salinity section shown in Fig. 25.6 shows these distributed widely along the ridge.

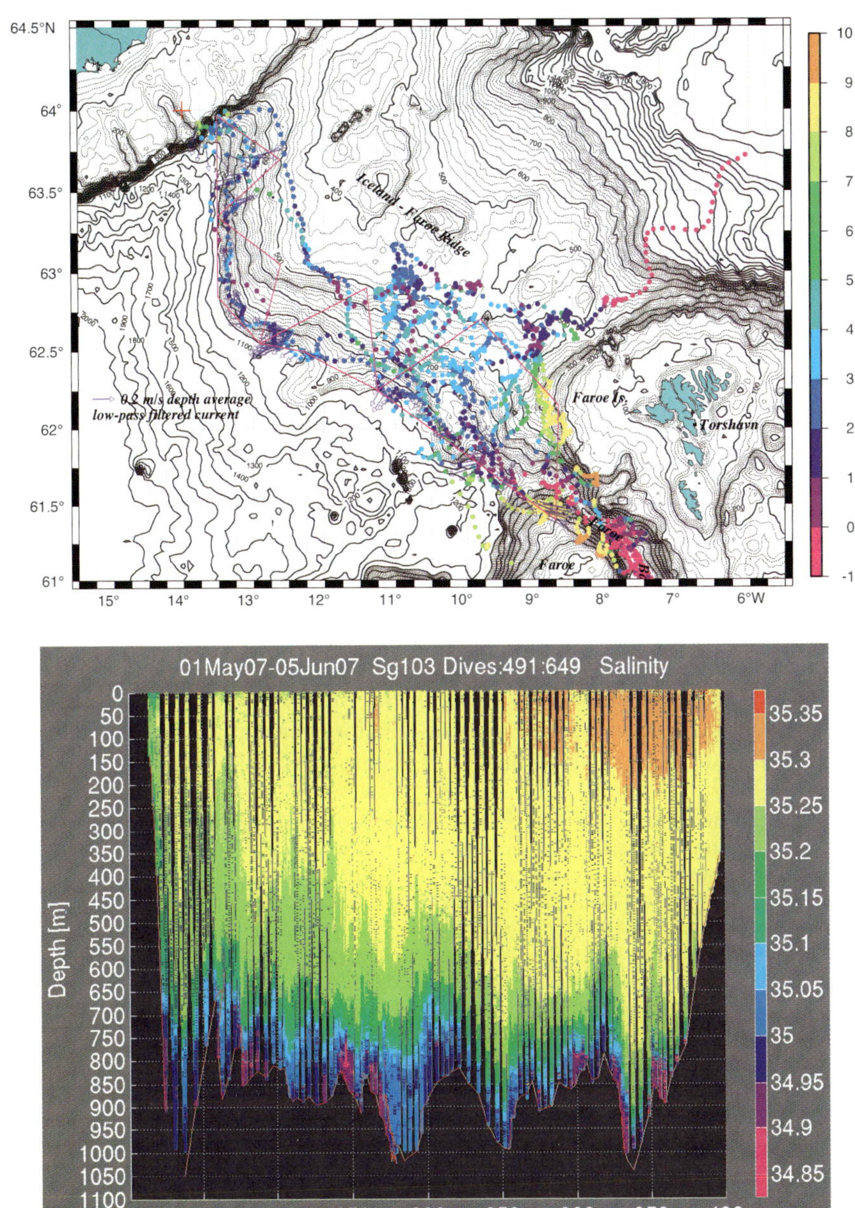

Fig. 25.6 Upper panel: Bottom temperature samples obtained from two glider missions in winter 2006–2007 and one in spring 2007 over the Iceland–Faroe Ridge. Magenta line segments indicate the planned sampling tracks, found to be unrealistic due to strong currents. Depth is contoured at 100 m intervals to 2 km depth, and additionally at 20 m intervals shallower than 1 km. Depth-averaged currents are indicated by blue arrows for dive/climb cycles used in drawing the salinity section in the lower panel. Lower panel: Salinity section along the Atlantic side of the Iceland–Faroe Ridge, nominally following the 900 m isobath. The section ends on the right at Faroe Bank. Horizontal distances are plotted as range from the reference location given by the red cross in the upper panel near the Iceland continental shelf edge

25.7 Discussion

Seagliders have proved an effective low-cost means to survey upper ocean internal structure in a severe environment. The ASOF Labrador Sea missions described here were the first experience for this technology in an environment where the ability to respond to vehicle malfunctions was severely limited by its remoteness and severe weather and sea conditions. The experience of losing the first two gliders as their communications became more and more intermittent motivated considerable changes in Seaglider software, including the ability to download entirely new versions of the glider operating code and reboot to it remotely. Despite difficulties and hardware losses, the data collection could not have been obtained by traditional shipboard means without extraordinary expense. The Labrador Sea experience validated the model of sending a small team to a remote location, basing operations on land, while using small boats on day trips to launch and recover gliders. We also departed from that model to use large research vessels when they were made available on an ancillary basis. The launches from R/V Knorr were the first Seaglider operations from a conventional oceanographic vessel.

We learned that to sustain effective glider measurements in the Labrador Sea, vehicles capable of year-long missions and ones less hampered by strong currents would be helpful. We look to the extended range version of Seaglider and to Deepglider to provide these advantages in the future. The problem of making synoptic measurements with a slowly moving platform remains and appears addressable by the use of more vehicles (than 2), repeating transects more often.

Acknowledgements Glider operations in the Labrador Sea benefited greatly from the support of the Greenland Institute of Natural Resources, Nuuk, our U.W. colleagues C. Lee and J. Gobat and their research group, the officers and crew of R/V Knorr and CCGS Hudson, and N. Bogue, T. Swanson, J. Bennett, and W. Fredericks. The work was supported by the US Office of Naval Research under grant N00014-02-1-0791 and the US National Oceanic and Atmospheric Administration Arctic Research Office. Work on the Iceland–Faroe Ridge is supported by the National Science Foundation through grant OCE-0550584. We thank the Faroese Fisheries Laboratory, Drs. B. Hansen and H. Hátún, and the Faroese Coast Guard for their considerable assistance.

References

Eriksen, C. C., T. J. Osse, R. D. Light, T. Wen, T. W. Lehman, P. L. Sabin, J. W. Ballard, and A. M. Chiodi, 2001, Seaglider: a long range autonomous underwater vehicle for oceanographic research. IEEE J. Ocean. Eng., 26, 424–436.

Rudnick, D. L., R. E. Davis, C. C. Eriksen, D. M. Fratantoni, and M. J. Perry, 2004, Underwater gliders for Ocean Research. J. Mar. Tech. Soc., 38, 73–84.

Hátún, H., C. C. Eriksen, P. B. Rhines, and J. M. Lilly, 2008, Buoyant eddies entering the Labrador Sea observed with gliders and altimetry. J. Phys. Oceanogr., in press.

Chapter 26
Convection in the Western North Atlantic Sub-Polar Gyre: Do Small-Scale Wind Events Matter?

Robert S. Pickart[1], Kjetil Våge[1,5], G.W.K. Moore[2], Ian A. Renfrew[3], Mads Hvid Ribergaard[4], and Huw C. Davies[5]

26.1 Introduction

In 1912, the polar explorer and scientist Fridtjof Nansen published an article entitled "Bottom Water and the Cooling of the Ocean", in which he discussed the origin of the deep water in the North Atlantic south of the Greenland–Scotland Ridge (Nansen 1912). It was known at the time that dense water spills over the ridge system, both through Denmark Strait and between Iceland and the Faroes. Nansen argued, however, that these sources were insufficient to ventilate the vast body of deep water in the North Atlantic basin. He postulated, therefore, that open-ocean convection must be occurring south of the ridge. Furthermore, he suspected that this process was taking place in the Irminger Sea, east of Greenland. He noted that the cyclonic circulation in the Irminger basin (originally documented by Knudsen 1899) would help keep restratifying waters at the fringes, and that the center of the gyre, where the circulation was weak, would be conducive for deep convection.

To test this hypothesis, Nansen organized a research cruise on the Norwegian Gunboat *Frithjof* in the summer of 1910. His proposed cruise track included a section across the Irminger gyre to the East Greenland shelf near 62° N. Unfortunately, the *Frithjof* ran short on coal and never made it to the area. Nansen instead turned to previously collected data from late-winter/early-spring in order to look for evidence of convection. His criterion for overturning was by necessity crude: convection was thought to occur where near-surface properties matched those of the deep water (keep in mind that the measurement uncertainties were quite large in those days). Furthermore, he had very few data to work with, and some of the data

[1]Woods Hole Oceanographic Institution, Woods Hole, MA 02543, USA, e-mail: rpickart@whoi.edu

[2]University of Toronto, Toronto, Canada M5S 1A1, e-mail: gwk.moore@utoronto.ca

[3]University of East Anglia, Norwich, United Kingdom NR4 7TJ, e-mail: i.renfrew@uea.ac.uk

[4]Danish Meteorological Institute, Copenhagen, Denmark DK-2100,
e-mail: kjetil@whoi.edu, mhri@dmi.dk

[5]Swiss Federal Institute of Technology, 8092 Zürich, Switzerland, e-mail: huw.davies@env.ethz.ch

R.R. Dickson et al. (eds.), *Arctic–Subarctic Ocean Fluxes*, 629–652

were collected outside the immediate area of interest. To his credit Nansen noted this and admitted that his conclusions were tentative. Nonetheless, he stated in the paper that "it is safe to assume that a significant part of the bottom water of the Northern Atlantic Ocean is created in this area." Figure 26.1 shows Nansen's schematic circulation of the western North Atlantic, and the region where he believed wintertime convection took place. It is worth noting that the April 1906 data that Nansen used were collected during an extended positive phase of the North Atlantic Oscillation (NAO), which means that the conditions were likely favorable for convective overturning in the western sub-polar gyre (e.g. Dickson et al. 1996). In addition, one of the 1906 stations was located very close to the A1E line of the World Ocean Circulation Experiment (WOCE) within the Irminger gyre. Data from this same area collected in April 1991 – during another high-NAO period – were used to argue that deep convection occurs in this vicinity (Pickart et al. 2003a).

Some 21 years after Nansen's article was published, a late-winter cruise to the Irminger Sea was carried out on the German survey vessel *Meteor*. This was the first wintertime survey of the area, and one of the expressed goals was to evaluate Nansen's ideas regarding convection in the Irminger Sea (Defant 1936). A follow-on winter cruise was conducted 2 years later. Based on the vertically uniform distributions of temperature and oxygen from stations occupied during these cruises, to the south and southeast of Cape Farewell (southern tip of Greenland), Wattenberg (1938) concluded that "there can be no doubt that Nansen's opinion voiced in 1912 is accurate." Wüst (1943) later analyzed the density data from the 1933 and 1935 winter cruises and highlighted evidence of overturning to 2,000 m due to the near-zero stratification. He noted that this provided "evidence for the correctness of Nansen's hypothesis."

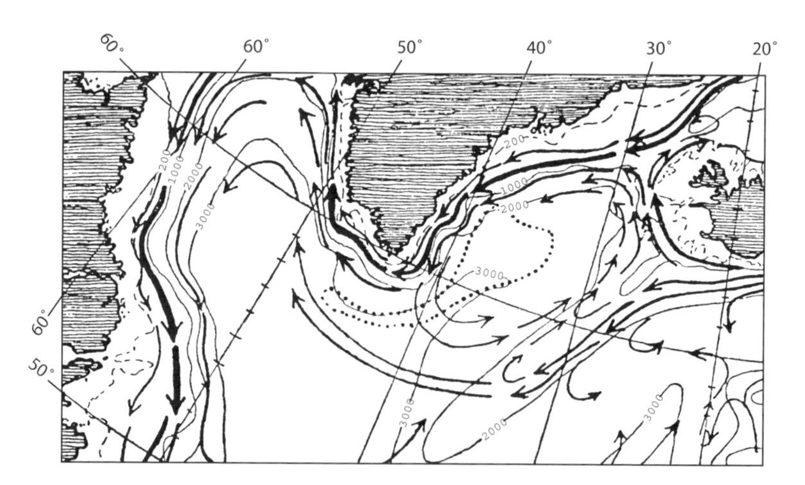

Fig. 26.1 Nansen's (1912) schematic of the circulation and region of deep convection (his Fig. 8). The area of overturning is delimited by the dotted line, most of which was thought to occur adjacent to southeast Greenland within the recirculating gyre of the Irminger Sea

It seems clear then that in the early part of the last century the oceanographic community believed that deep overturning and ventilation occurred southeast of Greenland (additional studies using summertime data supported this notion as well, for instance, Baggesgaard-Rasmussen and Jacobsen 1930; Smith et al. 1937). However, a few years before the German *Meteor* expeditions to the Irminger Sea, evidence was published that deep convection also occurred in the Labrador Sea (Nielsen 1928). In fact, Nielsen stated that the Labrador Sea was where "the greater part of the bottom water of the North Atlantic is then evidently formed", which seemed to contradict Nansen's (1912) earlier conclusions. Following Nielsen's (1928) study, the US Coast Guard carried out a series of summertime cruises to the Labrador Sea, from 1928 to 1935. Results from these expeditions were reported by Smith et al. (1937), who were the first to distinguish between intermediate water (which today is known as Labrador Sea Water) and deep and bottom waters. Smith et al. (1937) noted that intermediate water seemed to be formed in the Irminger basin; however, since the German *Meteor* results were not yet published, Smith et al. (1937) did not comment much further about the situation east of Greenland. They did argue that bottom water was formed via vertical convection only in the Labrador Sea, and their schematic highlighted an isolated area in the central Labrador Sea where this supposedly happened. Since the community was emphasizing bottom water formation at the time, the results of Smith et al. (1937), which were based on extensive profile data, likely had a big impact.

Despite the fact that published papers showed evidence of open-ocean convection on both sides of Greenland, subsequent studies of deep overturning in the western North Atlantic over the following decades were focused primarily on the Labrador Sea (e.g. Lazier 1973; Worthington 1976). It is known today that the Labrador basin is indeed the primary source of sub-polar mode water in the North Atlantic (e.g. Talley and McCartney 1982; Rhein et al. 2002). However, it is both interesting and curious that the notion of convection in the Irminger Sea fell completely out of favor. For example, this idea was discounted during planning stages of the North Atlantic WOCE experiment. In recent years, however, the notion has been re-kindled in a series of studies (Pickart et al. 2003a, b; Straneo et al. 2003; Bacon et al. 2003; Falina et al. 2007; Våge et al. 2008). In fact it has been argued that during strong positive phases of the NAO, the Irminger Sea may be a significant source of sub-polar mode water (Pickart et al. 2003b). If this is the case, then it means that there is a second location, outside of the Labrador Sea, where the atmosphere communicates directly with the deep ocean. This in turn might influence the meridional overturning circulation. It would also require us to revisit Labrador Sea Water formation rates and ventilation times, and compel us to interpret both observations and modeling results in a new perspective.

The purpose of this chapter is to summarize the latest thinking about deep convection in the western North Atlantic, emphasizing the region adjacent to southern Greenland studied almost a century ago by Nansen. Much has been published in the intervening years about convection in the western Labrador Sea, including results from the joint Canadian/French field program in the late 1970s (e.g. Clarke and Gascard 1983) and the recent Labrador Sea Deep Convection Experiment (Labsea

Group 1998). These and many other smaller programs have greatly enhanced our understanding of the convective process in the North Atlantic sub-polar gyre. However, new revelations about the meteorology around southern Greenland, together with recent oceanographic station data and mooring time series, have called into question the notion of a Labrador Sea-only source of sub-polar mode water. The evolution and dynamics of the atmospheric phenomena associated with the high orography of Greenland are different than those for the western Labrador Sea, and the spatial scales are significantly smaller and hence not sufficiently captured in low-resolution meteorological fields. The chapter begins with a brief consideration of convection in the western Labrador Sea. This is followed by a review of the atmospheric patterns associated with southern Greenland, various aspects of which are still emerging. In the remaining part of the chapter we address the ability of the intense winds in this region to force convective overturning in the eastern Labrador Sea and the western Irminger Sea. The ultimate question to be answered is, do these small-scale wind events have a large-scale climatic impact on the ventilation of the North Atlantic?

26.2 Convection in the Western Labrador Sea

Although the occurrence of deep convection in the ocean is fairly easy to discern after the fact, it is a difficult process to observe in real time. This is partly because of the harsh wintertime conditions surrounding this phenomenon (high winds, cold temperatures, rough sea state, and often times ice), and also due to the fact that the lateral scales of the convective plumes are very small (Marshall and Schott 1999). It was not until March 1976 that deep convection was directly observed in the Labrador Sea using a shipboard conductivity/temperature/depth (CTD) profiler (Clarke and Gascard 1983). A typical storm that drives overturning in the western Labrador Sea is shown in Fig. 26.2. The storms generally follow the North Atlantic storm track past Newfoundland toward Iceland (Hoskins and Hodges 2002), and the cyclonic circulation draws bitterly cold air off of the Labrador landmass. Total ocean-to-atmosphere heat fluxes from the storms often exceed 500 W m^{-2}, with the largest fluxes occurring near the marginal ice zone (Renfrew and Moore 1999; Pagowski and Moore 2001; Renfrew et al. 2002).

During positive phases of the NAO, the storm track tends to shift to the northeast and the frequency of cyclones increases (Rogers 1990). While this makes it more conducive for overturning to occur in the western Labrador Sea (Dickson et al. 1996), numerous other factors come into play. These include: (1) advection of freshwater from the Arctic, for instance the great salinity anomaly (Dickson et al. 1988) that shut down convection in the early 1970s (Talley and McCartney 1982), as well as other smaller events in the 1990s (Belkin et al. 1998); (2) interannually varying input of warm and salty subtropical water (Curry et al. 1998); and (3) the "memory" of the system (e.g. Straneo and Pickart 2001). Since it takes roughly 5–6

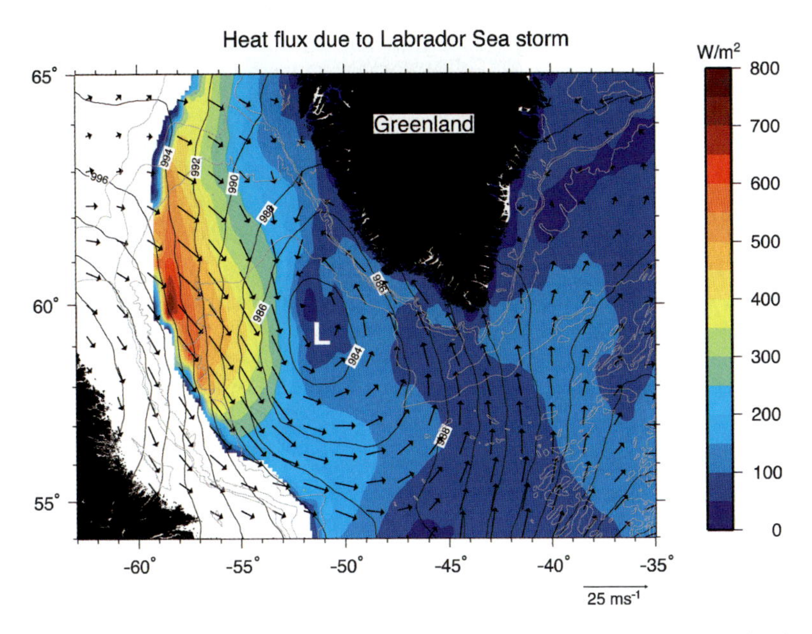

Fig. 26.2 Typical Labrador Sea winter storm (16 February 1997) from NCEP. The sea level pressure is contoured, and the vectors are the 10 m winds. The total heat flux (sensible + latent) is in color, where positive flux corresponds to heat loss from the ocean. The flux corrected product of Moore and Renfrew (2002) has been used. The center of the storm is denoted by the L, and the marginal ice zone along the Labrador shelf is colored white. The isobaths are 1,000, 2,000, and 3,000 m (gray lines)

years for all of the Labrador Sea Water to be flushed from the basin after formation (Yashayaev 2007), this means that repeated winters of deep convection will prime the system for continued overturning even if a subsequent winter is not very cold or windy. This was the case for the winter of 1996–1997 which produced deep convection with only moderate atmospheric forcing over much of the season (Pickart et al. 2002).

Although the region of strong heat flux from storms such as that in Fig. 26.2 is fairly broad, convection does not readily occur over the entire Labrador basin. This is partly due to the circulation of the sea. As explained in Marshall and Schott (1999), one of the factors, in addition to the atmospheric forcing, that promotes convection is the presence of cyclonic circulation. This both weakens the upper-layer stratification due to the doming of the isopycnals, and traps the water thereby allowing numerous storms to influence the same water parcels. The circulation of the western sub-polar gyre consists of a strong boundary current over the continental slope, and a series of closed cyclonic recirculations adjacent to this (Lavender et al. 2000; Fig. 26.3). It has been argued that these recirculations are driven by the enhanced wintertime windstress curl to the east of Greenland (see below), governed by the dynamics of topographic beta plumes (Spall and Pickart 2003). As seen in Fig. 26.3, it is clear that the deepest mixed-layers in the Labrador Sea occur within

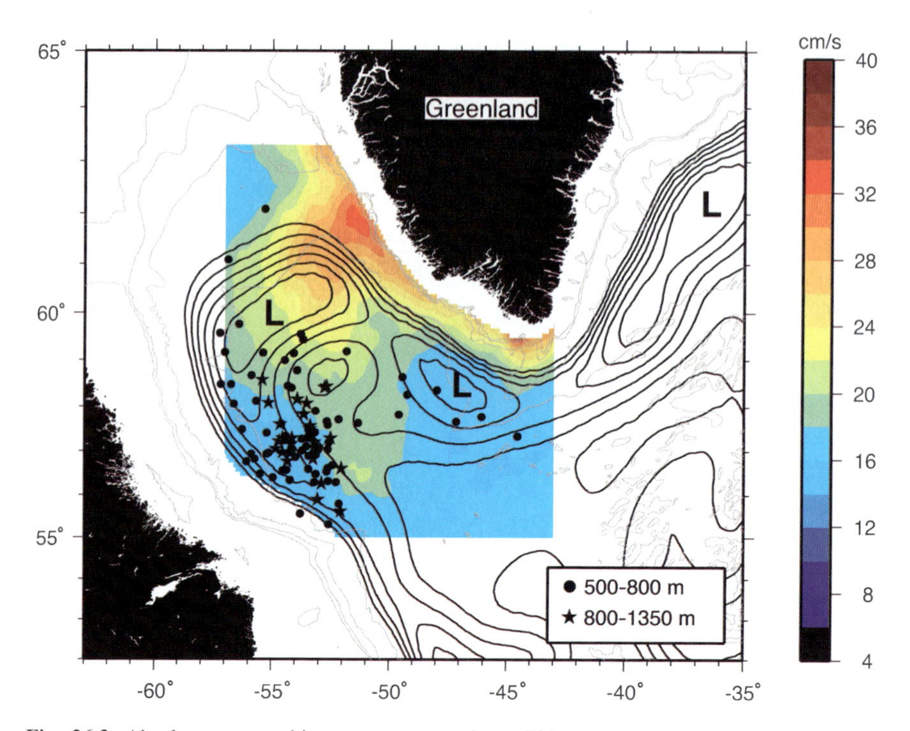

Fig. 26.3 Absolute geostrophic pressure anomaly at 700 m (contours) from Lavender et al. (2000), overlaid on the distribution of surface eddy speed (color) from Lilly et al. (2003). The locations of convection measured by profiling floats in winter 1997 are denoted by the symbols (see legend). The contour interval for the pressure anomaly is 1 cm, and the isobaths are 1,000, 2,000, and 3,000 m. Regions of low geostrophic pressure are indicated by an L

the recirculations. This notion is supported as well by the observations of Clarke and Gascard (1983), Pickart et al. (2002), and Lavender et al. (2002).

A second factor controlling the spatial extent of convection in the Labrador Sea is the eddy field. It is now known that the eastern boundary of the sea, near 61–62° N, is a site of eddy formation, apparently due to both barotropic and baroclinic instability of the boundary current (Eden and Böning 2002; Katsman et al. 2004; Bracco and Pedlosky 2003). One of the main factors is the local variation in topographic slope, which is conducive for instability (Wolfe and Cenedese 2006). The anti-cyclones formed from this region are long-lived and generally translate to the southwest (Prater 2002; Lilly et al. 2003). These features contain warm and salty boundary current water in their cores (hence the name Irminger Rings, Lilly et al. 2003), and they are an effective means of transporting buoyant water into the interior. This is believed to play an important role in the restratification after convection (Katsman et al. 2004), and also seems to influence the location where the deepest convection occurs in the basin. Note in Fig. 26.2 that the heat loss due to the storms is strong in the northern Labrador Sea, within one of the regions of cyclonic circulation (Fig. 26.3). This implies that deep convection should occur there, but observations show that the spreading of buoyant water from the boundary by the Irminger rings inhibits deep overturning (Pickart et al. 2002). This is consistent with the distribution of surface eddy speed (Fig. 26.3) which shows that the northeast part of the basin is strongly influenced by the eddies. The low occurrence of deep mixed-layers in this region (Fig. 26.3) implies that the eddy field helps to confine the deepest overturning to the western part of the basin.

The newly convected Labrador Sea Water leaves the basin by one of three general pathways (Talley and McCartney 1982). The first pathway is in the Deep Western Boundary Current, which is an effective means of transporting the water to the subtropics (e.g. Molinari et al. 1998; Pickart et al. 1997). The second pathway is with the North Atlantic Current, which advects the water to the eastern Atlantic (e.g. Read and Gould 1992). The third pathway is into the Irminger basin. It is this pathway that lies at the heart of the issue of whether or not Labrador Sea Water is formed entirely within the Labrador basin. Based on hydrographic data and models, it has recently been established that the travel time for Labrador Sea Water to reach the Irminger Sea via this pathway is approximately 2 years (Pickart et al. 2003a; Yashayaev et al. 2007; Falina et al. 2007; Kvaleberg et al. 2008). This raises the possibility that past observations of relatively newly convected Labrador Sea Water in the Irminger basin might have been incorrectly interpreted as local formation south and east of Greenland. Judging by the station map of Defant (1936), it is possible that some of the data from the 1935 *Meteor* cruise that validated Nansen's hypothesis were in fact taken within this pathway from the Labrador Sea. However, this is unlikely the case for all of the stations. Furthermore, based on recently collected hydrographic data, the distribution of Labrador Sea Water within the Irminger basin is inconsistent with a Labrador Sea-only source.

Sy et al. (1997) argued that, in the mid-1990s, Labrador Sea Water took only 6 months to reach the Irminger basin, which is inconsistent with the above estimates and the model results. Furthermore, using early springtime measurements, Pickart et al. (2003a) showed that if this pathway were the sole means by which Labrador Sea Water entered the Irminger basin, then the advective timescale would at times have to be less than 3 months. This is clearly unrealistic, especially since the observations of Pickart et al. (2003a) were taken inside the Irminger gyre. Advective–diffusive models (Straneo et al. 2003; Kvaleberg et al. 2008) show that it takes on the order of 3 years for Labrador Sea Water to penetrate the gyre from the outside. Additionally, observations collected during the most recent positive phase of the NAO during the early 1990s show two separate extrema of weak mid-depth stratification: one in the western Labrador Sea, and one in the southwest Irminger Sea (Fig. 26.4). It is impossible to reproduce such a lateral tracer pattern from a single source of convection in the Labrador Sea (Straneo et al. 2003), and the double-source of Labrador Sea Water considered by Kvaleberg et al. (2008) fits the observations better. Next, we discuss new insights regarding the atmospheric patterns associated with the high orography of Greenland – patterns that are conducive for convective overturning in the vicinity of Cape Farewell.

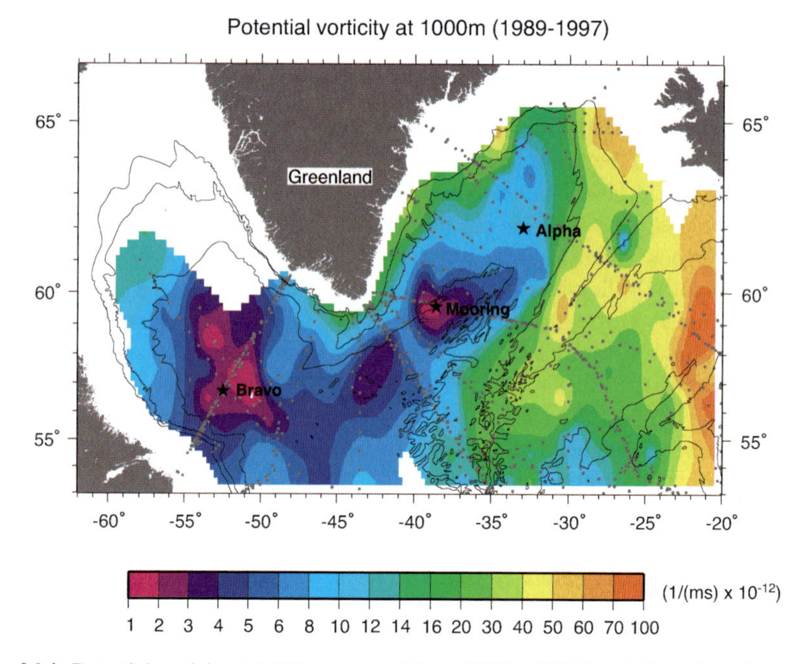

Fig. 26.4 Potential vorticity at 1,000 m averaged from 1989 to 1997 (excluding wintertime measurements) from Pickart et al. (2003a). This shows the distinct regions of weak stratification in the Labrador and Irminger Seas, respectively. Ocean weather stations Bravo and Alpha are marked, as well as the location of the profiling mooring discussed in Section 26.4. The isobaths are 1,000, 2,000, and 3,000 m, and the CTD station locations are marked by the small grey dots

26.3 Wintertime Atmospheric Circulation near Southern Greenland

The high topography of Greenland (>3,000 m) plays a critical role in the atmospheric patterns affecting the western North Atlantic Ocean. In a broad-scale sense the orography of Greenland influences the Icelandic low (e.g. Kristjansson and McInnes 1999; Petersen et al. 2004) and the associated North Atlantic Oscillation, which in turn impacts the state of the sub-polar gyre and its interannual variability (e.g. Dickson et al. 1996; Häkkinen and Rhines 2004). On smaller scales, the topography of Greenland exerts an enormous influence on local weather systems and on the passage of individual storms (e.g. Cappelen et al. 2001; Doyle and Shapiro 1999). Recent advances from improved atmospheric mesoscale models (e.g. MM5, HIRLAM) and high-resolution observations (QuikSCAT) have offered new glimpses into the structure and dynamics of these features. In wintertime, several distinct atmospheric patterns dominate, each of them associated with strong low level winds (>20 m s^{-1}) over different parts of the western Irminger Sea and eastern Labrador Sea. Here we limit the discussion of these patterns to the region near Cape Farewell. One should keep in mind that the length scales of the features in question (often as small as tens of kilometers) make them difficult to detect in the relatively low-resolution global meteorological analyses (e.g. from NCEP or ECMWF). At the same time, proper representation of these features and the associated air–sea heat transfer can be important for short- and medium-term weather prediction. Sensitivity analyses often indicate that the specification of the initial conditions in this region can exert a significant impact on the subsequent forecast.

26.3.1 Forward Greenland Tip Jet

The North Atlantic storm track generally steers low pressure systems northeastward past Newfoundland towards the southern Labrador Sea, and into the vicinity of Greenland and Iceland (Hoskins and Hodges 2002). Most of the storms pass east of Greenland into the Irminger Sea (Serreze et al. 1997), although sometimes they enter the Labrador Sea (Våge et al. 2008). Occasionally a storm splits due to the presence of Greenland, and the secondary low enters the Labrador Sea (Petersen et al. 2003). Along the storm track, intense cyclogenesis occurs in the region between Newfoundland and Greenland (Serreze et al. 1997; Tsukernik et al. 2007). There is, however, a second region of storm enhancement in the southern Irminger Sea, which is largely fueled by the strong surface gradient between the cold waters of the east Greenland shelf (and sometimes ice) and the warmer interior waters of the Irminger basin (Tsukernik et al. 2007). While the large-scale circulation associated with these intense cyclones is strong, the magnitude of the winds around southern Greenland is often significantly greater than would be expected from the storms alone. This is because of interaction of the cyclones with the high orography of Greenland.

In the vicinity of Cape Farewell, the wintertime winds tend to be bi-modal – either predominantly westerly or northeasterly. This is seen in the long-term weather station data from Prins Christian Sund (Fig. 26.5). While there may be fjord effects associated with this record, it is consistent with the shorter-term QuikSCAT wind measurements from this area collected since 1999 (Moore and Renfrew 2005). The first of these modes is associated with a phenomenon known as the forward Greenland tip jet, a name given by Doyle and Shapiro (1999) who were the first to study this phenomenon (although weather forecasters have known about this condition for decades, L. Rasmussen, 2002, personal communication). A forward tip jet is an intense, episodic westerly wind that often develops when the center of a low pressure system passes to the northeast of Cape Farewell (Fig. 26.6a). Based on nearly 40 years of Prins Christian Sund weather data, the average duration of a tip jet event is 3 days, and, climatologically, they occur every 10 days during the months of December through March (Pickart et al. 2003b). However, the QuikSCAT data suggest that they occur even more frequently (because their path sometimes misses the weather station), while the strongest

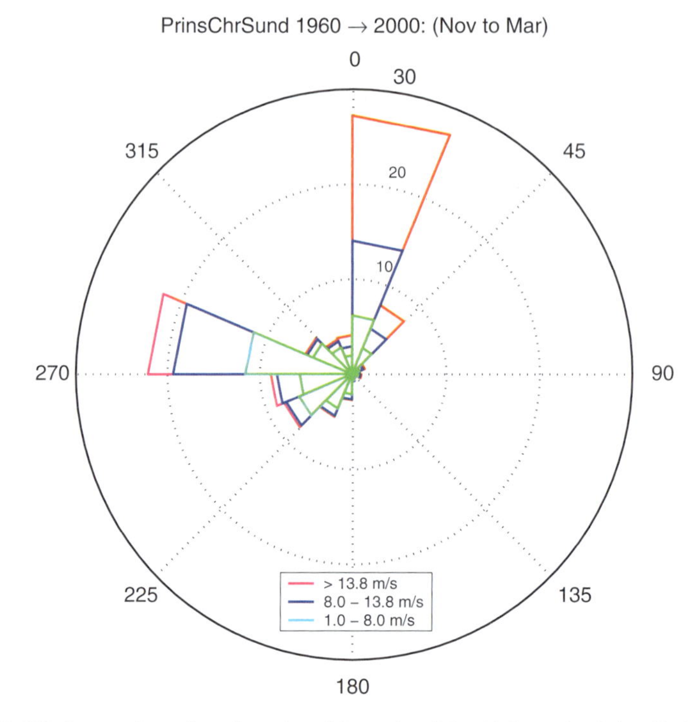

Fig. 26.5 Windrose using wintertime data (November–March) from the Prins Christian Sund meteorological station near Cape Farewell. The time period is 1960–2000

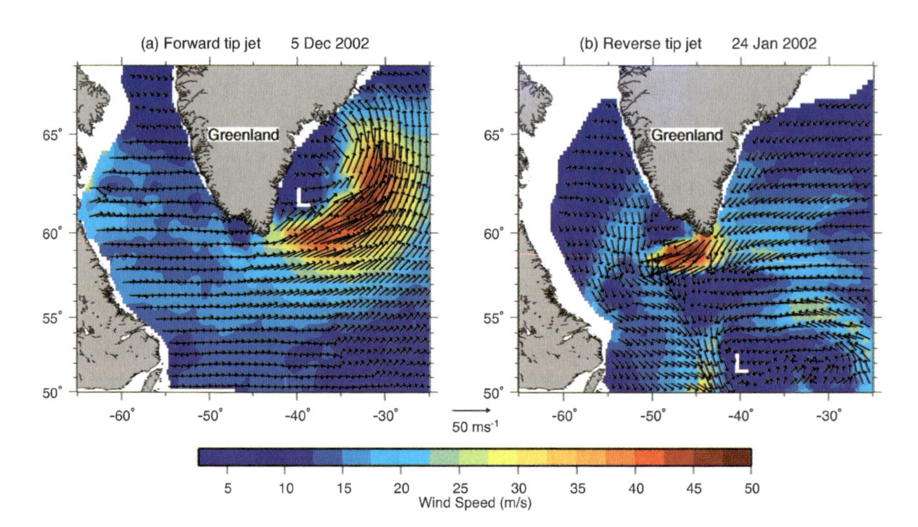

Fig. 26.6 Example of the two types of Greenland tip jet: (a) Forward tip jet; (b) Reverse tip jet. The surface wind speed (color) and vectors from QuikSCAT are plotted, along with the center of the parent low pressure system (denoted by the L)

winds generally persist for less than 24 h (Våge et al. 2008). The number of tip jets in a given winter varies interannually and is significantly correlated with the NAO index. More events tend to occur during a higher NAO winter (Pickart et al. 2003b), and the latitude of the center of action of the NAO also impacts their frequency (Bakalian et al. 2007).

What specific factors cause the tip jets to develop? There are two contributing circumstances. Doyle and Shapiro (1999) argue that the westerly winds intensify via acceleration during orographic descent from the Greenland plateau, due to the air parcels conserving Bernoulli function. Doyle and Shapiro (1999), Petersen et al. (2004) and Moore and Renfrew (2005) also discuss a second mechanism of acceleration associated with flow splitting or flow distortion. This occurs as the air parcels intercept the topographic barrier of Greenland and are forced around and/or over it, depending on the atmospheric stability and barrier dimensions (i.e. the Froude number). Neither Doyle and Shapiro (1999) nor Moore and Renfrew (2005) were able to sort out the precise relationship between, or relative contributions of, these two mechanisms.

In an effort to shed light on this, Våge and Davies (2008) studied tip jet events using the ERA-40 reanalysis data set. Using an empirical orthogonal function approach, more than 500 events were identified over the time period 1957–2002. The 3D trajectory model Lagranto (Wernli and Davies 1997) was applied to a subset of these events to compute backward air parcel trajectories terminating over the southern Irminger Sea. Nearly 3,000 back-trajectories were considered, emanating from approximately 100 tip jet events. The results demonstrate that the vast majority of the air parcels curve around the southern tip of Greenland and accelerate (Fig. 26.7a). There is, however, some vertical descent involved, associated with the edge of the Greenland landmass (Fig. 26.7b). This suggests that both flow splitting and orographic descent play a role in the development of the tip jet, but that most air parcels remain over the ocean. Figure 26.7a shows that maximum tip jet velocities exceeding 20 m s^{-1} occur east of Cape Farewell. It should be noted, however, that the wind speeds from ERA-40 are significantly less than those observed concurrently from QuikSCAT (Våge and Davies 2008), and individual tip jet events can be as strong as 50 m s^{-1} (Fig. 26.6a).

26.3.2 Reverse Greenland Tip Jet

The second dominant wintertime air pattern near Cape Farewell is an intermittent northeasterly wind known as a reverse tip jet (Fig. 26.6b), a name given by Moore (2003) and Moore and Renfrew (2005) who studied this phenomenon using QuikSCAT data. These events are also associated with the passage of cyclones, although the storm centers are located south of Greenland (Fig. 26.6b), in contrast to the forward tip jet case where the storms are located northeast of Cape Farewell. As mentioned above, the occurrence of strong northeasterly winds in this region has long been established, and is related to the barrier winds along the southeast

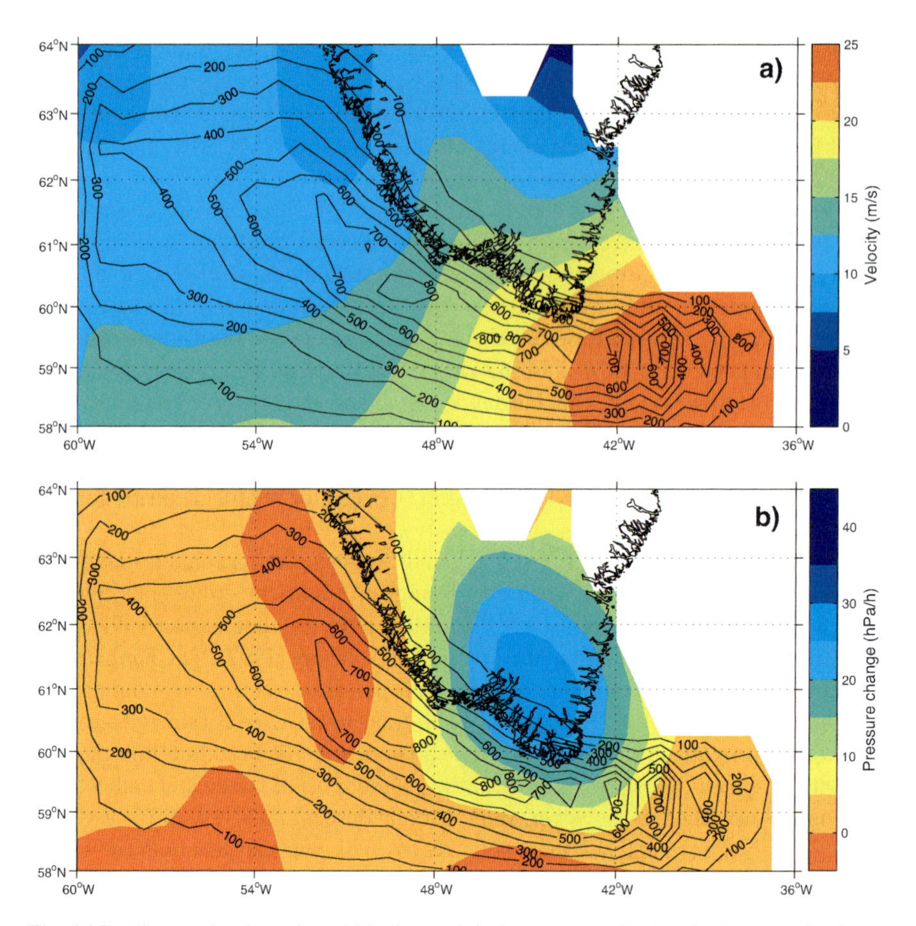

Fig. 26.7 Air parcel trajectories within forward tip jets computed using the Lagranto back trajectory model, from Våge and Davies (2008). (a) Top panel: Velocity of the air parcels (color) showing the acceleration near the tip of Greenland. The contours denote the number of realizations, indicating that most of the trajectories pass to the south of the land mass. (b) Bottom panel: Change in pressure along the trajectories, showing the descent near Greenland

coast of Greenland (Cappelen et al. 2001). Essentially, the cyclonic winds associated with the storm impinge upon the topographic barrier of Greenland, which causes a damming of the cold air and the establishment of a cross-shelf pressure gradient. This leads to strong northeasterly winds that accelerate further near the tip of Greenland where the barrier disappears. As explained in Moore and Renfrew (2005), the increase in wind speed is related to the anticyclonic curvature of the trajectories in the absence of the barrier, consistent with an inertial force balance in this region. Wind speeds immediately south and west of Cape Farewell can be very strong during these events (Fig. 26.6b). While less is known about the climatology of the reverse tip jets – for instance their frequency, typical duration, and relationship to the NAO – both the meteorological station measurements near Cape

Farewell (Fig. 26.5) and the QuikSCAT satellite data (Moore and Renfrew 2005) indicate that they are a common wintertime phenomenon.

The kinematics, dynamics and air–sea fluxes associated with tip jets, reverse tip jets and barrier winds around southern Greenland are the subject of considerable interest at the moment. An international aircraft-based field program took place during February–March 2007 called the Greenland Flow Distortion experiment, with one aim being to collect the first comprehensive in situ observations of such features. The experiment obtained an array of observations via dropsondes at high- and low-levels (down to 100 ft) of a reverse tip jet event, several barrier flow events, and also a lee cyclogenesis event that took place just east of Cape Farewell. It is too early to highlight any significant results, but a preliminary look at the observations reveals, for example, reverse tip jet core wind speeds near 50 m s^{-1} and significant off-ice barrier flow heat fluxes.

26.4 Impact of Small-Scale Wind Patterns on Oceanic Convection

There is increasing evidence that the small-scale atmospheric flow patterns described above have a significant impact on the western North Atlantic sub-polar gyre – both in terms of convective overturning and in regards to certain aspects of the circulation. Since more research has been done on the consequences of the forward tip jet, we discuss this first.

26.4.1 Convection in the Western Irminger Sea

Two aspects of the forward Greenland tip jet are of particular importance to the ocean. The first is the meridional length scale of the jet, and in particular the sharp gradient in wind speed to the north of the jet axis. Often times the wind decreases significantly over a very short distance. For example, in Fig. 26.6a, to the east of Cape Farewell, the westerly wind speed diminishes to the north by 15 m s^{-1} in just 50 km. Such sharp gradients result in very large synoptic values of cyclonic wind-stress curl, nearly three orders of magnitude larger than the broad-scale curl of the North Atlantic (Pickart et al. 2003b). Figure 26.8 shows the composite wind stress curl from 7 years of QuikSCAT data, where the year has been divided into winter (November–April) and summer (May–October). The frequent storms in winter result in a band of strong cyclonic curl along southeast Greenland (Fig. 26.8a), due largely to the barrier winds. The strongest positive curl occurs east of Cape Farewell and is clearly the result of the forward tip jet (as is the enhanced negative curl west of Cape Farewell). Note that this curl signature largely vanishes during the summer (Fig. 26.8b). The forward tip jet is thus a major contributor to the enhanced seasonal cyclonic curl pattern near southern Greenland.

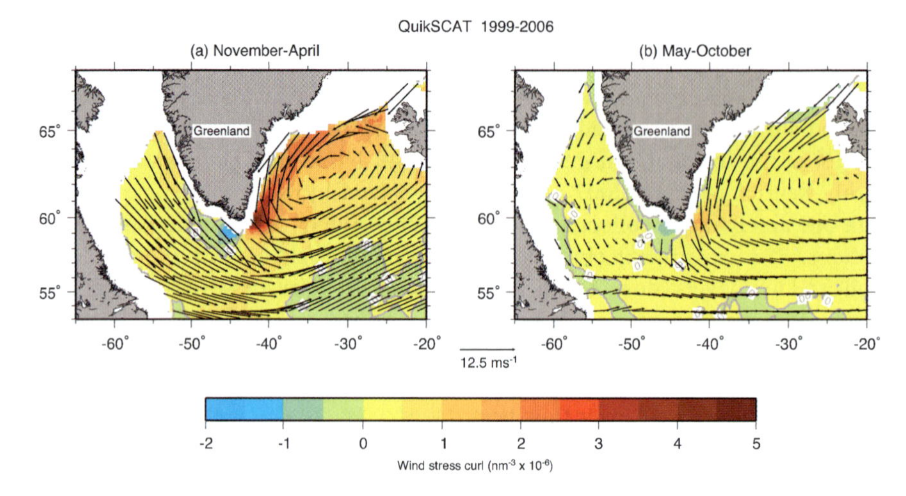

Fig. 26.8 Climatological average surface wind vectors and wind stress curl (color) over the period 1999–2006 from QuikSCAT. The year has been split into two 6-month averages. (a) November–April; (b) May–October

This vorticity distribution in turn has a significant impact on the circulation of the sub-polar gyre. According to the numerical model study of Spall and Pickart (2003), the enhanced positive curl drives the cyclonic recirculation in the western Irminger and Labrador Seas (Fig. 26.3). Even though the wind forcing is seasonal, a steady circulation develops because of the slow baroclinic wave response at this latitude (wave speeds roughly 1 cm s^{-1}), together with the effect of the bottom topography which causes the deep circulation to dampen the seasonal response. The bottom topography also helps to form the multiple areas of closed streamlines along the lower continental slope (Kvaleberg and Haine 2008). Pickart et al. (2003b) showed that frequent tip jets alone (i.e. without any barrier winds) can drive the Irminger gyre. Hence, the tip jet is largely responsible for the trapping of water near the region of southern Greenland, as well as for the doming of the isopycnals in this area. Both of these factors help facilitate deep convection (Marshall and Schott 1999).

The second crucial aspect of forward tip jets is the large heat flux that results from the cold air being advected over the warm ocean. This was recognized by Doyle and Shapiro (1999), and subsequently studied by Pickart et al. (2003b), Centurioni and Gould (2004), and Våge et al. (2008). Using a numerical model forced by a sequence of tip jets associated with a strong winter, Pickart et al. (2003b) showed that deep convection can occur in the southwest Irminger Sea. The area of overturning in the model corresponded with the observed extremum in mid-depth potential vorticity to the east of Cape Farewell (Fig. 26.9). This provided compelling evidence that the Labrador Sea is not the sole source of the sub-polar

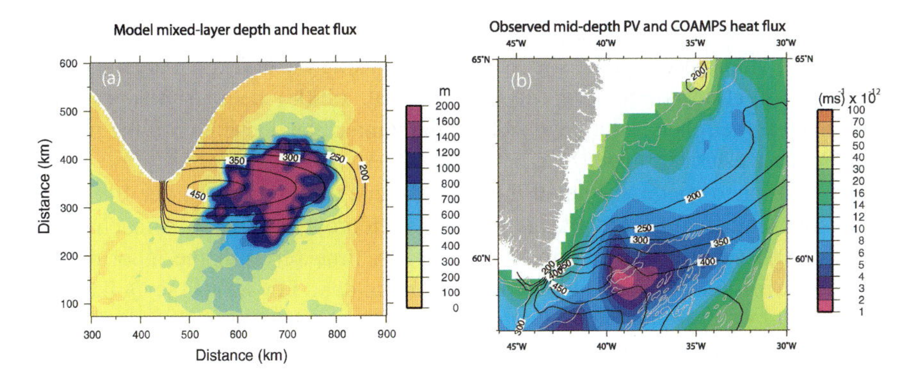

Fig. 26.9 Results from the study of Pickart et al. (2003b). (a) Final depth of the winter mixed-layer (color) in a regional ocean model forced by repeated occurrences of the forward tip jet. The heat flux of the tip jet is shown by the contours (W m^{-2}). (b) Observed potential vorticity (color) showing newly ventilated water east of Cape Farewell. The heat flux of a forward tip jet event from the COAMPS model is shown by the contours

mode water of the western North Atlantic, and solved the puzzle regarding the unrealistically fast travel times into the Irminger Sea deduced from measurements during the 1990s high phase of the NAO. However, the model configuration as well as the forcing used by Pickart et al. (2003b) were idealized, and direct wintertime measurements of deep convection in the Irminger Sea are still lacking today.

Unfortunately, during the period of active mode water formation in the early 1990s there were no wintertime cruises to the Irminger Sea. Furthermore, the PALACE/ARGO profiling float programs (e.g. Lavender et al. 2000, 2002; Centurioni and Gould 2004) had not yet begun. It was not until 1997 onward that the floats were able to measure the seasonal development of the mixed-layer in the western sub-polar gyre, but by this time the winters had become more moderate and convection had diminished considerably in the Labrador Sea (Lazier et al. 2002). Nonetheless, the float data were used by Bacon et al. (2003) and Centurioni and Gould (2004) to demonstrate that overturning to depths of 400–700 m did occur during this period in the western Irminger Sea. Centurioni and Gould (2004) also used several 1D mixed-layer models, forced by an idealized representation of the forward tip jet, to show that tip jets were likely responsible for the observed convection.

In an effort to elucidate the role of the Greenland tip jet on convection in the Irminger Sea, a subsurface mooring was deployed east of Cape Farewell in August 2001, in the region of the low potential vorticity (see Fig. 26.4). The mooring contained a McLane CTD profiler with an acoustic current meter, and was programmed to return two vertical traces per day between 55 and 1,800 m. Unfortunately the profiler failed the first year, but did successfully profile through the winter during the next two deployments (2002–2003 and 2003–2004). Since these winters were characterized by a low value of the NAO index, it was not expected that deep convection would occur at the site. Nonetheless, using the CTD data together with a variety of atmospheric data sets, Våge et al. (2008) demonstrated that the Greenland tip jet plays a dominant role in the wintertime deepening of the mixed-layer in the western Irminger Sea.

As mentioned above, the small meridional scale of the tip jet (order 100 km or less) makes it difficult to resolve in the global meteorological fields. As such, Våge et al. (2008) constructed an improved heat flux time series at the mooring site using bulk formulae together with various surface data. For winds the QuikSCAT data were used, for sea surface temperature the (extrapolated) mooring time series was used, for air temperature the weather station data from Cape Farewell were used, and for relative humidity the NCEP data were used. The latter two time series were adjusted for the mooring site using meteorological buoy data collected at the site during the fall of 2004 (see Våge 2006). The resulting total heat flux, averaged over the winter of 2002–2003, was more than 30% larger than that from NCEP alone. The biggest discrepancy occurred during the tip jet events. For the 12 robust events between December and April, the average heat flux from NCEP was 267 W m^{-2}, compared to a value of 413 W m^{-2} from the improved estimate – an increase of 55%.

Not surprisingly, this extra heat flux has a significant impact on the evolution of the mixed-layer. To demonstrate this Våge et al. (2008) ran the Price et al. (1986) 1-D mixed-layer model on a CTD profile from November 2002, forced with both the NCEP heat flux time series and the improved heat flux product. As seen in Fig. 26.10, the mixed-layer depth predicted from NCEP alone (red curve) is too shallow compared to the observations from the mooring, whereas the depth from the improved heat flux time series (blue curve) does a much better job tracking the envelope of deepest observed mixed-layer depth (black curve). (The high frequency signal in the observations is likely due to the effects of lateral advection, which is not captured by the 1D model.) To quantify the effect of the intermittent wind events, a third model run was performed in which the tip jets were removed from the improved heat flux time series (green curve in Fig. 26.10). It is clear that the heat loss due to the succession of tip jet events over the course of the winter had a sizable impact on the final depth of convection.

One of the remaining questions is, can the forward tip jet cause deep convection during high NAO winters? It will be impossible to address this with observations until the return of cold and stormy winters to the western North Atlantic. However, Våge et al. (2008) have shown that the answer is likely yes. During the early 1990s there were on average more robust tip jet events per winter and overall stormier conditions (Pickart et al. 2003b), plus the water column in the Irminger Sea was better preconditioned for overturning. Våge et al. (2008) initialized the mixed-layer model with a CTD profile from fall 1994, and forced the model with a similarly computed improved heat flux time series for the winter of 1994–1995. The predicted final depth of convection for this calculation was nearly 1,800 m (Fig. 26.11), consistent with hydrographic data collected in this

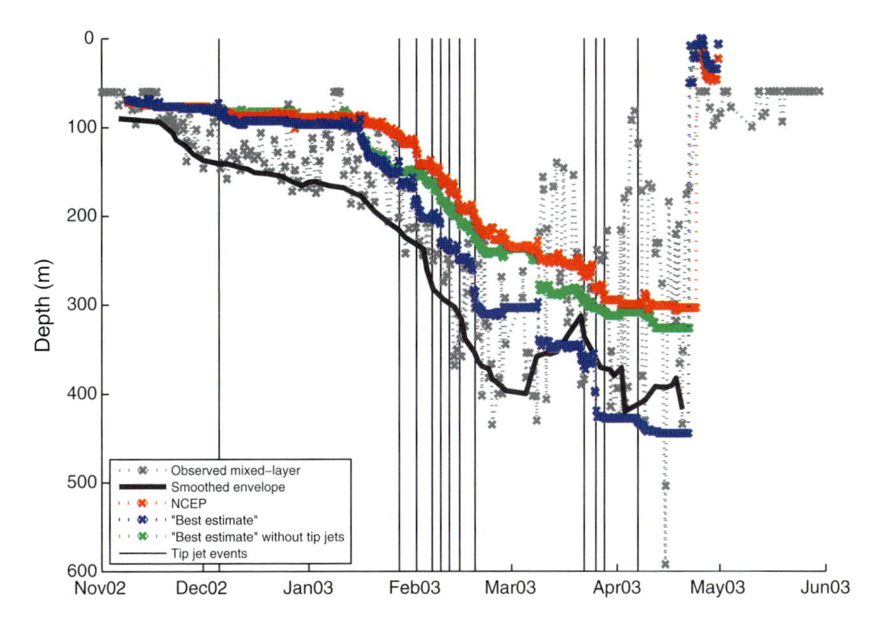

Fig. 26.10 Comparison of observed and modeled mixed-layers for winter 2002–2003 in the southwestern Irminger Sea, from Våge et al. (2008). Figure 26.4 shows the location of the mooring. See the key and the discussion in the text for an explanation of the different curves. The vertical lines denote the tip jet events

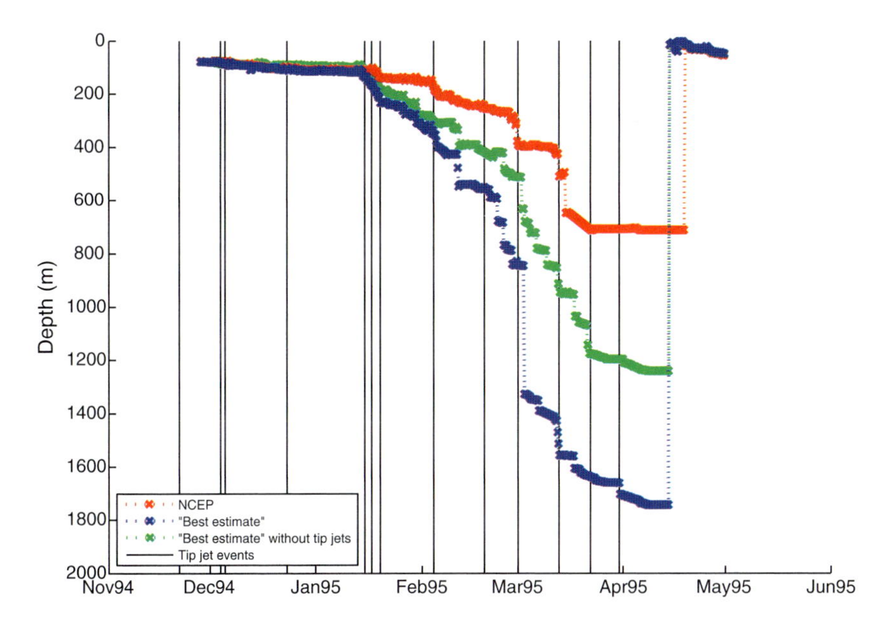

Fig. 26.11 Modeled mixed-layer for the winter of 1994–1995, from Våge et al. (2008). The different curves are the same as in Fig. 26.10 (see key)

region the following summer (Pickart et al. 2003a). This is roughly 1,000 m deeper than predicted using NCEP alone, and, as was true for the winter of 2002–2003, the presence of the tip jets had a significant impact.

26.4.2 Convection in the Eastern Labrador Sea

One of the interesting results from Lavender et al.'s (2002) study was the occurrence of relatively deep convection in the eastern Labrador Sea (east of 50° W, see Fig. 26.3). While the observations are few (the NAO index was low during the time period of their measurements), the deep mixed-layers occurred within the recirculation gyre to the southwest of Cape Farewell. This sub-basin scale gyre seems to be a robust feature of the Labrador Sea circulation, since it was also measured by surface drifters in the area (Jakobsen et al. 2003). Does the reverse tip jet play a role in the convection here? As discussed by Moore and Renfrew (2005), reverse tip jets are intimately tied to the barrier winds along the southeast Greenland coast. The winds tend to veer toward the Labrador Sea after the barrier ends, bringing the air carried by the jet directly over the region of closed oceanic circulation (Fig. 26.12). If the air is cold and dry enough, this would result in enhanced heat loss within the gyre, satisfying the set of conditions for convection: trapped water, domed isopycnals, and strong ocean-to-atmosphere heat loss.

While the reverse tip jet is a dominant feature in the wintertime wind climatology of the region (Moore and Renfrew 2005), there has been no in-depth study yet carried out of the characteristics and impact of the individual events. Hence, unlike the case of the forward tip jet, no quantitative conclusions can be drawn regarding the ability of the reverse tip jets to force convection. However, circumstantial evidence suggests that, during high NAO winters, this may be the case. During winter, pack-ice forms along the East Greenland shelf all the way to Cape Farewell, and, in strong winters, the ice cover can extend the full width of the shelf. As an example consider Fig. 26.13, which shows the ice concentration in mid-March 2007. At this time a band of 70–90% concentration extended to the tip of Greenland. Hence, during any reverse tip jet events, the barrier winds would flow along the ice before veering into the eastern Labrador Sea (compare Figs. 26.12 and 26.13). The strongest heat loss would thus occur beyond the ice edge, directly over the closed gyre. This is akin to the western Labrador Sea where the strongest heat loss is immediately east of the ice edge (Fig. 26.2).

This scenario of course needs verification, and there are other factors that require consideration. For example, the southeast Greenland shelf is a dynamic area with strong surface currents (e.g. Bacon et al. 2002; Sutherland and Pickart 2007). The pack-ice can vary synoptically depending on these currents as well as on the local wind field (K. Hansen, 2006, personal communication), and it is not obvious how important such variability is over an entire winter. Another factor is the

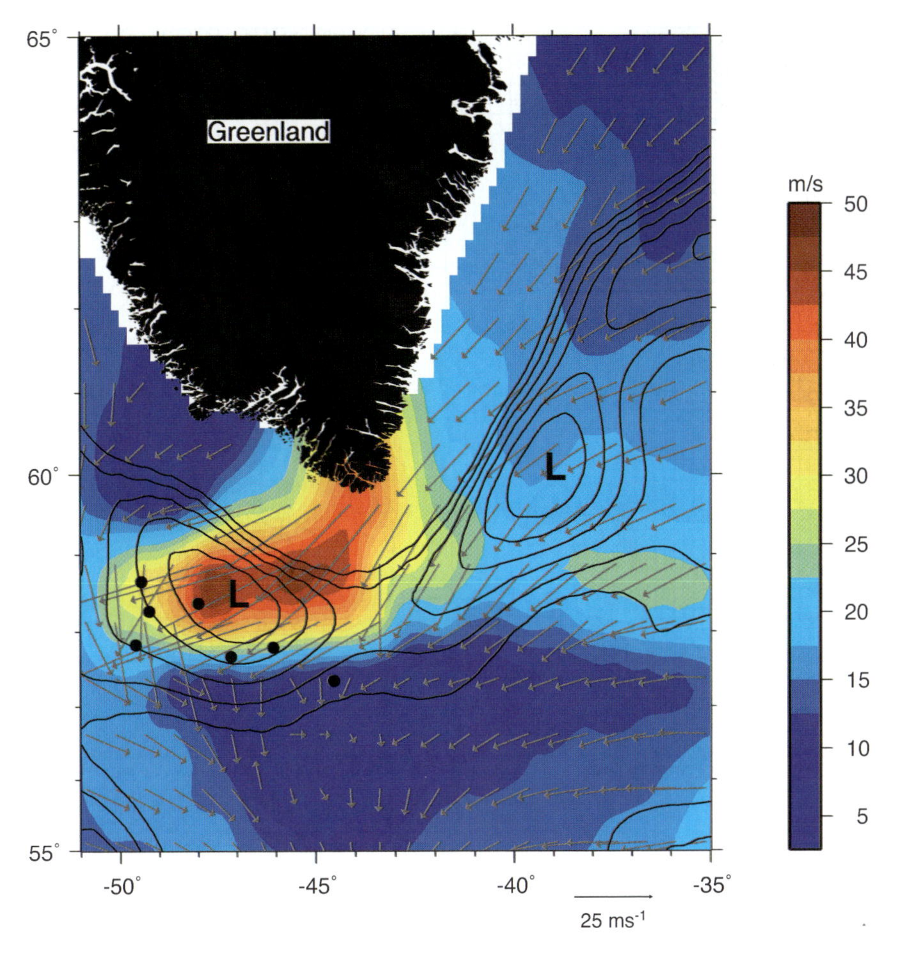

Fig. 26.12 Reverse tip jet of Fig. 26.6b overlaid on the geostrophic streamfunction of Fig. 26.3 (contours). The wind speed (color) and vectors are shown, as well as the locations of deep convection in winter 1997 from Fig. 26.3 (black dots)

upstream history of the air parcels that form the barrier winds, and whether or not the air emanating from the tip of Greenland is indeed cold and dry enough to cause significant heat loss. A third factor to consider is the impact of the westerly winds that blow over the sea during typical Labrador Sea storms. In the example of Fig. 26.2 the heat loss in the eastern part of the basin is moderate, but one could envision stronger storms with enhanced heat flux in the vicinity of Cape Farewell. Hence, the relative roles of the reverse tip jets, the ice field, and the general storminess of the Labrador Sea in driving convection in the eastern part of the basin remains to be determined. It is worth noting that convection here might also help precondition the water that makes it to the western side of the basin later in the winter via the boundary current.

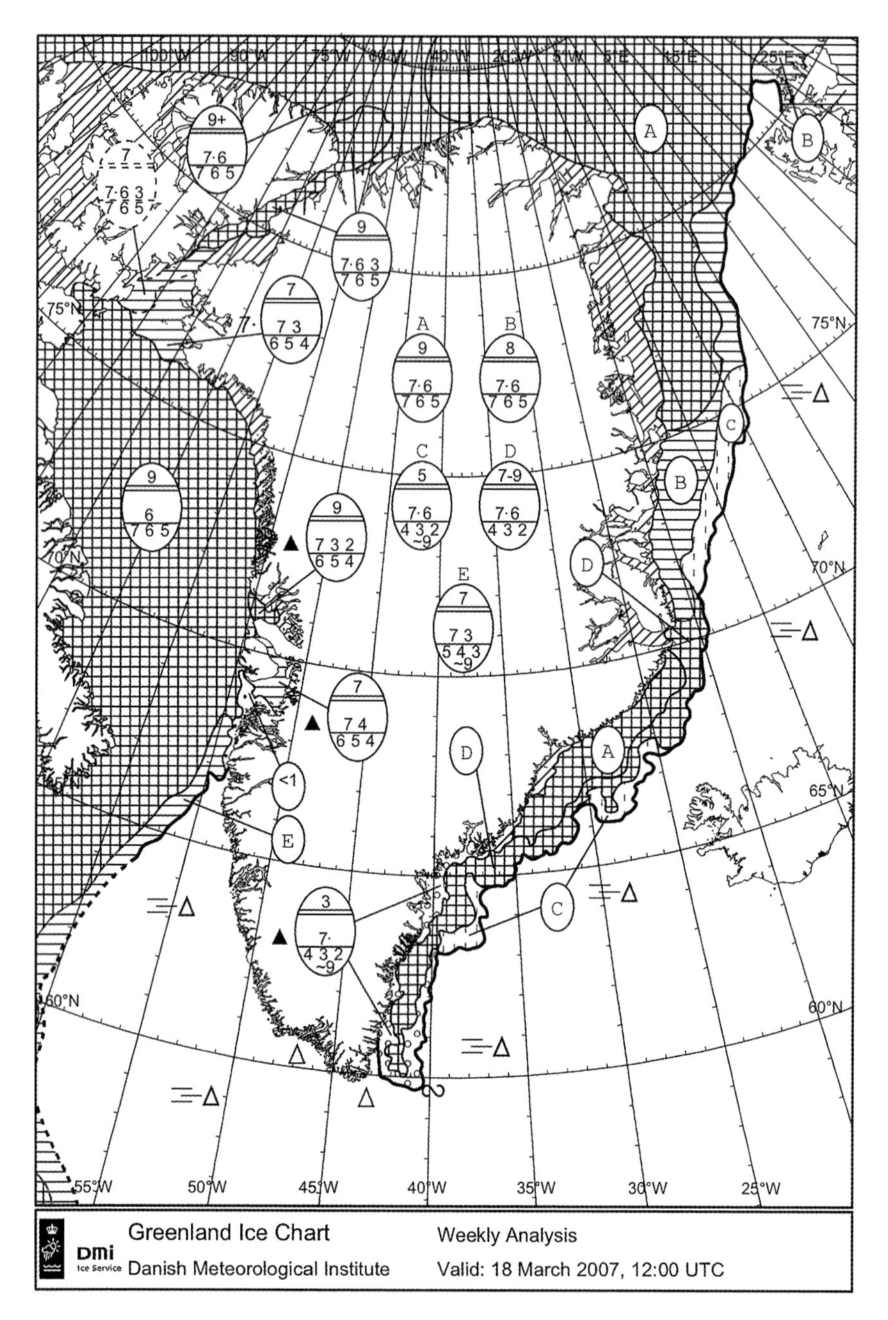

Fig. 26.13 Ice chart for 18 March 2007, from the Danish Meteorological Institute

26.5 Conclusions

It has been shown in this chapter that measurements taken nearly a century ago near Cape Farewell, which hinted of deep convection, may in fact have revealed the ocean's response to small-scale wind patterns caused by the high topography of Greenland. Recent satellite-based observations from QuikSCAT, high-resolution aircraft measurements, and mesoscale atmospheric models have begun to shed light on the complex nature of the atmospheric circulation near the tip of Greenland. Both forward tip jets and reverse tip jets are associated with intense winds that geographically correspond to regions of deep oceanic mixed-layers. The impact of the forward tip jets on the southwest Irminger Sea is more established at this point, including the importance of the strong cyclonic windstress curl and enhanced heat flux. While the reverse tip jets are an obvious candidate for the observed deep mixing in the eastern Labrador Sea, the heat flux resulting from these events still needs to be quantified.

Further research on a variety of fronts is necessary to establish more concretely the connection between these atmospheric phenomena and the ocean circulation and convection. For example, the dynamics of the air patterns – particularly the impact of the orography of Greenland – needs to be elucidated to understand why these patterns arise in the first place. The role of the large-scale storm climate in dictating their frequency and strength needs to be better established. The ability of the air patterns to drive the ocean, both through windstress curl input and buoyancy forcing, requires more detailed consideration. This is especially true for the reverse tip jets, including the role of the pack-ice. Finally, concurrent observations of the atmosphere and ocean, especially during robust winters, are crucial if we are to understand better how such small-scale atmospheric patterns impact the ocean on climatically relevant time and space scales.

Acknowledgments The authors acknowledge the following funding agencies for this work. National Science Foundation grant OCE-0450658 (RP and KV); Canadian Foundation for Climate and Atmospheric Sciences (GWKM); Nordic Council of Ministers–West-Nordic Ocean Climate (MHR).

References

Bacon S et al. (2002) A freshwater jet on the east Greenland shelf. Journal of Geophysical Research 107, doi:10.1029/2001JC000935

Bacon S et al. (2003) Open-ocean convection in the Irminger Sea. Geophysical Research Letters 30, doi:10.1029/2002GL016271

Baggesgaard-Rasmussen, Jacobsen JP (1930) Contribution to the hydrography of the waters round Greenland in the year 1925. Medd. fra Kommissionen for Havundersogelser, Serie I, Hydrografi, Bd I, 2:10, p 24

Bakalian F et al. (2007) Influence of the Icelandic Low latitude on the frequency of Greenland tip jet events: Implications for Irminger Sea convection. Journal of Geophysical Research 112: C04020, doi:10.1029/2006JC003807

Belkin IM et al. (1998) "Great Salinity Anomalies" in the North Atlantic. Progress in Oceanography 41:1–68
Bracco A, Pedlosky J (2003) Vortex generation by topography in locally unstable baroclinic flows. Journal of Physical Oceanography 33:207–219.
Cappelen J et al. (2001) The observed climate of Greenland, 1958–99, with climatological standard normals, 1961–90. Technical Report 00–18, Danish Meteorological Institute, Copenhagen, Denmark
Centurioni LR, Gould WJ (2004) Winter conditions in the Irminger Sea observed with profiling floats. Journal of Marine Research 62:313–336
Clarke RA, Gascard JC (1983) The formation of Labrador Sea water. Part I: Large-scale processes. Journal of Physical Oceanography 33:1764–1778
Curry RG et al. (1998) Oceanic transport of sub-polar climate signals to mid-depth subtropical waters. Nature 391:575–577
Defant A (1936) Bericht über die ozeanographischen Untersuchungen des Vermessungsschiffes "Meteor" in der Dänemarkstrasse und in der Imingersee. Preuss. Akad.d. Wiss., Sitz., Phys.-Math. Klasse XIX:232–242
Dickson RR et al. (1988) The "Great Salinity Anomaly" in the Northern North Atlantic 1968–1982. Progress in Oceanography 20:103–151
Dickson RR et al. (1996) Long-term coordinated changes in the convective activity of the North Atlantic. Progress in Oceanography 38:241–295
Doyle JD, Shapiro MA (1999) Flow response to large-scale topography: The Greenland tip jet. Tellus 51MA:728–748
Eden C and Böning C (2002) Sources of eddy kinetic energy in the Labrador Sea. Journal of Physical Oceanography 32:3346–3363
Falina A et al. (2007) Variability and renewal of Labrador Sea Water in the Irminger Basin in 1991–2004. Journal of Geophysical Research 112:C01006, doi:10.1029/2005JC003348
Häkkinen S, Rhines PB (2004) Decline of the sub-polar North Atlantic circulation during the 1990s. Science 304:555–559
Hoskins B, Hodges K (2002) New perspectives on the northern hemisphere winter storm tracks. Journal of the Atmospheric Sciences 59:1041–1061
Jakobsen PK et al. (2003) Near-surface circulation in the northern North Atlantic as inferred from Lagrangian drifters: Variability from the mesoscale of interannual. Journal of Geophysical Research 108, doi:10.1029/2002JC001554
Katsman CA et al. (2004) Boundary current eddies and their role in the restratification of the Labrador Sea Journal of Physical Oceanography 34:967–1983
Kvaleberg E et al. (2008) Spreading of CFC-11 in the sub-polar North Atlantic Ocean. Journal of Geophysical Research, submitted
Kvaleberg E and Haine TWN (2008) Recirculating flow in the Labrador and Irminger Seas: Impact of Bathymetry. Journal of Physical Oceanography, submitted
Knudsen M (1899) Hydrography. Danish Ingolf Expedition. Bd. I, No. 2. Kopenhagen
Kristjansson JE, McInnes H (1999) The impact of Greenland on cyclone evolution in the North Atlantic. W J R Meteorological Society 125:2819–2834
Labsea Group (1998) The Labrador Sea deep convection experiment. Bulletin of the American Meteorological Society 79:2033–2058
Lavender KL et al. (2000) Mid-depth recirculation observed in the interior Labrador and Irminger Seas by direct velocity measurements. Nature 407:66–69
Lavender KL et al. (2002) Observations of open-ocean deep convection in the Labrador Sea from subsurface floats. Journal of Physical Oceanography 32:511–526
Lazier JRN (1973) The renewal of Labrador Sea water. Deep-Sea Research 20:341–353
Lazier JRN et al. (2002) Convection and restratification in the Labrador Sea, 1990–2000. Deep Sea Research 49:1819–1835
Lilly JM et al. (2003) Observations of the Labrador Sea eddy field. Progress in Oceanography 59:75–176

Marshall J, Schott F (1999) Open-ocean convection: Observations, theory, and models. Reviews of Geophysics 37:1–64
Molinari RL et al. (1998) The arrival of recently formed Labrador Sea Water in the deep western boundary current at 26.5N. Geophysical Research Letters 25:2249–2252
Moore GWK (2003) Gale force winds over the Irminger Sea to the east of Cape Farewell Greenland. Geophysical Research Letters 30:184–187
Moore GWK, Renfrew IA (2005) Tip jets and barrier winds: A QuickSCAT climatology of high wind speed events around Greenland. Journal of Climate 18:3713–3725
Nansen F (1912) Das bodenwasser und die abkuhlung des meeres. Internationale Revue der gesamten Hydrobiologie und Hydrographie Band V, Nr. 1:1–42
Nielsen J (1928) The waters around Greenland. In: Greenland. The discovery of Greenland, exploration, and the nature of the country. I:185–230
Pagowski M, Moore GWK (2001) A numerical study of an extreme cold-air outbreak over the Labrador Sea: Sea-ice air-sea interaction and the development of polar lows. Monthly Weather Review 129:47–72
Petersen GN et al. (2003) Flow in the lee of idealized mountains and Greenland. Journal of the Atmospheric Sciences 60:2183–2195
Petersen GN et al. (2004) Numerical simulations of Greenland's impact on the northern hemisphere winter circulation. Tellus 56A:102–111
Pickart RS et al. (1997) Mid-depth ventilation in the western boundary current system of the sub-polar gyre. Deep-Sea Research I 44:1025–1054
Pickart RS et al. (2002) Hydrography of the Labrador Sea during active convection. Journal of Physical Oceanography 32:428–457
Pickart RS et al. (2003a) Is Labrador Sea Water formed in the Irminger Basin? Deep-Sea Research I 50:23–52
Pickart RS et al. (2003b) Deep convection in the Irminger Sea forced by the Greenland tip jet. Nature 424:152–156
Prater MD (2002) Eddies in the Labrador Sea as observed by profiling RAFOS floats and remote sensing. Journal of Physical Oceanography 32:411–427
Price J et al. (1986) Diurnal cycling: Observations and models of the upper ocean response to diurnal heating, cooling, and wind mixing. Journal of Geophysical Research 91:8411–8427
Read J F, Gould WJ (1992) Cooling and freshening of the sub-polar North Atlantic ocean since the 1960's. Nature 360:55–57
Renfrew IA, Moore GWK (1999) An extreme cold-air outbreak over the Labrador Sea: Roll vortices and air-sea interaction. Monthly Weather Review 127:2379–2394
Renfrew IA et al. (2002) A comparison of surface-layer heat flux and surface momentum flux observations over the Labrador Sea with ECMWF analyses and NCEP reanalyses. Journal of Physical Oceanography 32:383–400
Rhein M et al. (2002) Labrador Sea water: Pathways, CFC-inventory, and formation rates. Journal of Physical Oceanography 32:648–665
Rogers JC (1990) Patterns of low-frequency monthly sea level pressure variability (1899–1986) and associated wave cyclone frequencies. Journal of Climate 3:1364–1379
Serreze MC et al. (1997) Icelandic low cyclone activity: Climatological features, linkages with the NAO, and relationships with recent changes in the northern hemisphere circulation. Journal of Climate 10:453–464
Smith EH et al. (1937) The Marion and General Greene expeditions to Davis Strait and Labrador Sea. Scientific results, Part 2, physical oceanography. Bulletin US Coast Guard 19:1–259
Spall MA, Pickart RS (2003) Wind-driven recirculations and exchange in the Labrador and Irminger Seas. Journal of Physical Oceanography 33:1829–1845
Straneo F, Pickart RS (2001) Interannual variability in Labrador Sea Water formation and export: How does it correlate to the atmospheric forcing? US CLIVAR Meeting, June 2001, extended abstract volume

Straneo F et al. (2003) Spreading of Labrador Sea water: An advective-diffusive study based on Lagrangian data. Deep-Sea Research I 50:701–719

Sutherland DA, Pickart RS (2007) The East Greenland coastal current: Structure, variability, and forcing. Progress in Oceanography, submitted

Sy A et al. (1997) Surprisingly rapid spreading of newly formed intermediate waters across the North Atlantic Ocean. Nature 386:675–679

Talley LD, McCartney MS (1982) Distribution and circulation of Labrador Sea Water. Journal of Physical Oceanography 12:1189–1205

Tsukernik M et al. (2007) Characteristics of winter cyclone activity in the northern North Atlantic: Insights from observations and regional modeling. Journal of Geophysical Research, 112, doi:10.1029/2006JD007184

Våge K (2006) Winter mixed-layer development in the central Irminger Sea: The effect of strong, intermittent wind events. Master's Thesis, Massachusetts Institute of Technology and Woods Hole Oceanographic Institution Joint Program in Oceanography and Oceanographic Engineering, Woods Hole Oceanographic Institution, p 79

Våge K et al. (2008) Winter mixed-layer development in the central Irminger Sea: The effect of strong, intermittent wind events. Journal of Physical Oceanography, submitted

Våge K, Davies HC (2008) The Greenland tip jet from the ERA-40 reanalysis. Manuscript in preparation

Wattenberg H (1938) Die Verteilung des sauerstoffs im Atlantishcen Ozean. Von Walter de Gruyter & Co., Berlin, p 132

Wermli H and Davies HC (1997) A Lagrangian-based analysis of extratropical cyclones 1: The methods and some applications. Quarterly Journal of the Royal Meteorological Society, 123:467–489

Wolfe CL, Cenedese C (2006) Laboratory experiments on eddy generation by a buoyant coastal current flowing over variable topography. Journal of Physical Oceanography 36:395–411

Worthington LV (1976) On the North Atlantic circulation. The Johns Hopkins Oceanographic Studies, Vol 6, The Johns Hopkins University Press, Baltimore, MD, 110 pp

Wüst G (1943) Der subarktische Bodenstrom in der westatlantischen Mulde. Annalen der Hydrographie und Maritimen Meteorologie Heft IV/VI:249–256

Yashayaev I et al. (2007) Spreading of the Labrador Sea Water to the Irminger and Iceland basins. Geophysical Research Letters 34:L10602, doi:10.1029/2006GL028999

Yashayaev I (2007) Hydrographic changes in the Labrador Sea, 1960–2005. Progress in Oceanography, 73:242–276

Chapter 27
North Atlantic Deep Water Formation in the Labrador Sea, Recirculation Through the Subpolar Gyre, and Discharge to the Subtropics

Thomas Haine[1], Claus Böning[2], Peter Brandt[2], Jürgen Fischer[2], Andreas Funk[2], Dagmar Kieke[3], Erik Kvaleberg[1], Monika Rhein[3], and Martin Visbeck[2]

27.1 Introduction

North Atlantic Deep Water (NADW) is a water mass that is central to the oceanography of the deep Atlantic, the global meridional overturning circulation (MOC), and the climate of the Earth itself. The subpolar Atlantic is an especially important place for these phenomena because of the large changes wrought on NADW in these basins.[1] Indeed, once it is discharged past 45°N, NADW temperature and salinity are altered at substantially slower rates before encountering Circumpolar Deep Waters in the subpolar ocean of the southern hemisphere (McCartney and Talley 1984; Reid et al. 1977). Formation of NADW, recirculation through the subpolar gyre, and injection into the subtropical ocean past Newfoundland are therefore central issues to ASOF science and are discussed here.

The subpolar North Atlantic Ocean is arguably the best understood and most intensively studied of all the ocean basins. Indeed, some gross features of the surface circulation, such as the boundary current system, were discovered by sea-farers when the first Norse colonists reached the New World over a millennium ago (Haine 2007). The characteristics of the primary water masses contributing to NADW have been known since the mid-20th century (see, e.g., Warren 1981 or Worthington 1976): NADW consists mainly of Greenland–Scotland Overflow Water, originating from mid-depths in the Nordic Seas, and Labrador Sea Water formed by intense winter-time air/sea interaction in the western subpolar Atlantic. There are also lesser contributions from Antarctic

[1] Department of Earth and Planetary Sciences, 329 Olin Hall, The Johns Hopkins University, Baltimore, MD 21218-2682, USA, e-mail: Thomas.Haine@jhu.edu

[2] IfM-GEOMAR, Kiel, Germany

[3] Institut für Umweltphysik, Universität Bremen, Germany

[1] We define the subpolar North Atlantic Ocean as the region North of 45°N, south of the Greenland-Scotland Ridge and bordered to the east by the European continental shelf and to the west by Davis Strait, Hudson Strait and the Canadian continental shelf.

R.R. Dickson et al. (eds.), *Arctic–Subarctic Ocean Fluxes*, 653–701

Bottom Water, Sub-Polar Mode Water and Mediterranean Water (explicit definitions of these water types are given below). Since the mid-20th century, the most sustained effort has been to describe and then quantify the gross time-averaged circulation patterns of the full water column. In the last 10–15 years, advances in observing technology and refinements in ocean circulation models have allowed improved estimates of the time mean state, a better understanding of the mechanisms involved, and a first estimate of the variability. These issues are the topic of this chapter, but a comprehensive review of all recent advances in subpolar physical oceanography is beyond our scope. Instead, we concentrate attention on the following questions:

1. What are typical formation rates of Labrador Sea Water and how have they varied over time?
2. What are the advective/diffusive transport[2] pathways and transport timescales of Labrador Sea Water in the subpolar Atlantic?
3. What are the characteristics of the NADW passing Newfoundland and the Grand Banks into the subtropical Atlantic?
4. What do numerical circulation models tell us about the processes involved?

In keeping with the overall theme of this volume the two cross-cutting questions that run through these issues are:

1. What have we learned in the last 15–20 years?
2. Where are the areas of greatest uncertainty and where is future progress most likely?

The overarching purpose is to understand how subpolar NADW formation and subsequent transformation occurs and why. Ultimately, we want to understand how water mass anomalies entering the subpolar North Atlantic from multiple possible sources are retained, stored, modified, and discharged to the subtropics. We eventually seek predictable information on these processes that may be exploited to forecast the future state of the ocean. Neither of these long-term goals are yet clearly in sight, but we attempt to describe the recent steps towards them and the best way ahead.

To address these questions the chapter is laid out as follows: The historical starting point around 1990 is described in Section 27.2. In Section 27.3, we discuss water mass formation (especially in the Labrador Sea) through air/sea interaction and interior mixing. We review estimates of formation and discuss the processes involved. In Section 27.4, Labrador Sea Water circulation pathways and rates are considered. The emphasis is on how deep water climate signals are carried through the subpolar gyre and modified through retention there before being discharged to the subtropics. In Section 27.5 the western boundary current is discussed and the associated NADW export to the subtropics. In particular, we emphasize the properties of deep water rounding Newfoundland. Section 27.6 explores the evidence from numerical models on the processes involved in these issues (a brief discussion of model results also

[2] Note that "transport" refers to advective/diffusive movement. Purely advective movements – traditionally called "transports" – are here called "volume fluxes"; the distinction and reasoning is made clear in sections 3 and 4.

appears in a few other sections where this makes most sense) and Section 27.7 contains concluding remarks. Some commentary on technical issues is in the footnotes, and other relevant chapters are 12, 18–22 (Section C) on overflows, and the companion chapters in Section D on the subpolar "receiving volume."

27.2 State of Knowledge Circa 1990

We have chosen the year 1990 as the nominal starting point for this discussion. By this we mean that the relevant fieldwork was completed by 1990, although some of the papers we cite here appeared a few years later. Clearly, there is no adequate single point in time to use as a baseline. Yet 1990 seems a reasonable choice for a number of reasons: First, it immediately precedes the substantial field campaigns in the subpolar North Atlantic coordinated by programmes such as WOCE, the Labrador Sea Project, the German national SFB460 project, and CLIVAR. The data collected in the last 15 years have easily doubled the number of deep CTD hydrographic measurements in the subpolar basin, for example.[3] Moreover, new measurement methods have proliferated too, such as chlorofluorocarbon (CFC) tracers and basin-scale deployment of autonomous float squadrons. Second, 1990 immediately predates the era of routine satellite altimetry, of mesoscale eddy resolving numerical general circulation models (GCMs), and of global coupled ocean/atmosphere climate models. Finally, and for these reasons, an intellectual shift was underway at that time. Studies on subpolar circulation prior to 1990 were distinguished by the desire to provide a basic description of the large-scale (O(100) km), low-frequency (O(10) year) general circulation. Often information was provided in semi-quantitative, schematic flow diagrams. Interannual and shorter-period variability was not treated (with some exceptions for seasonal changes); mesoscale and smaller-scale variability was also neglected. Since 1990, variability in the ocean state has been of foremost importance. Physical oceanographers are also now moving towards more detailed, quantitative accounts of the circulation as a continuous flow field – such as a geostrophic streamfunction – rather than as a cartoon of connected pipes.

We present this discussion in two stages: first, water masses and circulation pathways are described, then volume flux estimates are discussed. In what follows we draw on several papers by Woods Hole descriptive physical oceanographers (Worthington 1976; Warren 1981; Talley and McCartney 1982; McCartney and Talley 1984; McCartney 1992; Schmitz and McCartney 1993; Schmitz 1996); Table 27.1 defines the water mass acronyms and their properties.[4]

---

[3] There are about 1600 CTD T and S data deeper than 2000 m available at the National Oceanographic Data Center database prior to 1990, and 4100 prior to 2005.

[4] It is difficult to assign unambiguous definitions to water mass types. This is partly because water masses display different characteristics in different places and at different times. Convenient definitions for one author with one dataset are therefore sometimes awkward in other circumstances. Water mass names are also used to variously emphasize the geographic location of the water (for example, North Atlantic Central Water), its geographic source (for example, Mediterranean

Table 27.1 Water mass acronyms and properties

Water mass	Acronym	Temperature (°C)	Salinity (psu)	Density (σ_θ) (kg m^{-3})
North Atlantic Central Water[a]	NACW	8–19	35.1–36.7	
Sub-Polar Mode Water[b]	SPMW	4–14.7	34.95–36.08	
Mediterranean Water[c]	MW	>3	>35	
Labrador Sea Water[d]	LSW	3–4	<34.94	
Upper Labrador Sea Water[e]	ULSW			27.68–27.74
Classical Labrador Sea Water[e]	CLSW			27.74–27.8
Iceland–Scotland Overflow Water[f]	ISOW		34.98–35.03	
Denmark Strait Overflow Water[f]	DSOW	0–2	34.88–34.93	
North Atlantic Deep Water[d]	NADW	1.8–4	34.88–35	
Antarctic Bottom Water[g]	AABW	<1.8	<34.88	

[a] Main thermocline waters of the North Atlantic Ocean (Sverdrup et al. 1942, p. 667).

[b] See Talley and McCartney (1982); Mode waters are also characterised by weak stratification.

[c] See Worthington (1976); pure MW overflowing at Gibraltar has a temperature and salinity of around 11.9 °C, 36.50 psu, respectively (Wüst 1935).

[d] See Worthington (1976).

[e] See Kieke et al. (2006b).

[f] See Warren (1981).

[g] See Worthington (1976) and McCartney (1992). AABW here refers to abyssal water of southern origin in the mid-latitude North Atlantic.

27.2.1 Water Masses and Circulation Pathways

By 1990 the basic elements of the subpolar circulation and the identities of the main water masses were established. First, consider the near-surface ocean, nominally the upper 1,000 m. At these depths, relatively warm (roughly, 8–15 °C), saline (35–36 psu) thermocline waters enter the Newfoundland and then the West European Basins in the North Atlantic Current (NAC). On the southern (warmer, T > 10 °C) side of the NAC some fluid separates to circulate south and leave the

Water), or another property such as stratification (for example, Sub-Polar Mode Water). Finally, the water mass definitions are not intended to be exclusive of each other. Table 1 should be read with these caveats in mind. In ambiguous cases, the reader should refer to the primary papers involved (see also McCartney (1992)).

subpolar region, for example, in the Portugal Current. On the cooler northwestern side of the NAC some fluid detrains and circulates along the Reykjanes Ridge to form the Irminger Current. The remaining NAC water flows past Ireland, the Faroes, and into the Norwegian Sea. Cold, fresh (roughly, $T < 4\,^{\circ}C$, $S < 34.60$ psu) water enters the area from the north mainly through Denmark Strait (the East Greenland Current) and Davis Strait (the Baffin Island Current). The East Greenland Current partly merges with the Irminger Current by Cape Farewell forming a relatively strong cyclonic boundary current system around the Irminger and Labrador Basins. Past Cape Farewell the outer part of the jet is called the Irminger Current, the inner part is called the West Greenland Current, and off Canada the system is collectively called the Labrador Current. As water moves around this circuit it is progressively cooled and freshened by air/sea exchange – especially in winter – and mixing with the northern source waters. Starting from cooler varieties of North Atlantic Central Water (NACW; $T < 10\,^{\circ}C$, $S < 35.5$ psu) the transformation in the T/S plane is to progressively denser types of Sub-Polar Mode Water (SPMW) and finally, in the southeastern Labrador Sea, Labrador Sea Water (LSW) at, or colder than, 4 °C (McCartney and Talley 1982). The mode waters (including LSW) are associated with 200–2,000 m deep well-mixed convective layers in late winter (Clarke and Gascard 1983) and so are weakly stratified and form a voluminous "mode" in the T/S plane. Near-surface flow out of the Labrador Sea in the Labrador Current follows the shelf-break and upper continental slope. Some of the Labrador Current detaches from the bathymetry and joins the NAC northwest wall near Flemish Cap, some flows as far as the Grand Banks before recirculating, and some passes south of Newfoundland into the Gulf of St. Lawrence through Cabot Strait.

At mid-depths (nominally 1,000–3,000 m) the subpolar gyre is dominated by the circulation of LSW. LSW is the densest of the subpolar mode waters formed by deep convection in the subpolar basins and exhibits local minima in both salinity and stratification. It is often taken to have a temperature between 3 °C and 4 °C and a salinity less than 34.94 psu (for example, see Worthington 1976). LSW is considered to be the lightest constituent of NADW and penetrates as far as the equatorial Atlantic along the western boundary (Talley and McCartney 1982; Weiss et al. 1985). The other influential mid-depth subpolar water mass is Mediterranean Water (MW). Pure MW overflows into the eastern North Atlantic through the Strait of Gibraltar at a temperature near 11.9 °C and salinity 36.50 psu (Wüst 1935). It is diluted with ambient water during descent into the deep sea then spreads out as a high salinity water mass with temperature above about 3 °C.

Talley and McCartney (1982) identify the main mid-depth transport pathways for LSW. Some LSW flows in the lower part of the Labrador Current described above. It splits near Flemish Cap where some is entrained into the deep NAC, while the remainder flows west round Grand Banks into the subtropics. The eastward flowing LSW in the deep NAC subsequently splits into a part that passes into the Irminger Sea (some LSW also seems to flow directly into the Irminger Sea from the convection area) and a part that crosses the mid-Atlantic Ridge into the West European Basin. There, some of the LSW is lost south to the subtropics while progressively mixing with MW. The rest circulates past Rockall and then back across the mid-Atlantic Ridge into the Irminger Basin. There is no evidence for LSW crossing the Iceland–Scotland Ridge into the Norwegian Sea (McCartney 1992; Dickson and Brown 1994). Finally, the LSW subpolar circuit is completed by recirculation into the Labrador Sea following the Irminger Current.

Three main water masses enter the subpolar region to occupy the deep and abyssal basins (nominally, deeper than 2,500 m). They are Denmark Strait Overflow Water (DSOW), Iceland–Scotland Overflow Water (ISOW) – both derived from the upper 1,000 m in the Nordic Seas – and Antarctic Bottom Water (AABW) from the south. (The Greenland–Scotland Overflows are also discussed in detail in Chapters 18–22 of this volume.) DSOW passes across the 600 m deep saddle between Greenland and Iceland with temperature and salinity in the range 0–2 °C and 34.88–34.93 psu, respectively (Warren 1981). ISOW crosses the 850 m deep saddle in the Faroe Bank Channel at a slightly higher temperature and salinity, around 1.8–3 °C and 34.98–35.03 psu. Some water also crosses the Iceland–Faroe Ridge (at 450 m) and the Wyville–Thomson Ridge (at 500–600 m), but is less important than ISOW for the formation of NADW (see also Chapter 18). Both DSOW and ISOW descend into the abyssal ocean as bathymetry-following turbulent boundary currents. They both entrain substantial amounts of warmer, mid-depth, ambient water – mainly SPMW at temperatures ≥8 °C for ISOW and ≥6 °C for DSOW and LSW near 4 °C (McCartney 1992) – and approximately double their volume flux by the time they reach the abyssal floor. ISOW flows along the eastern flank of the Reykjanes Ridge and then passes through the 3,600 m deep Charlie-Gibbs Fracture Zone (near 52°N) into the Irminger Basin. It then circulates along the bathymetry to join the slightly denser DSOW although some water from the Charlie-Gibbs Fracture Zone may also enter the Labrador Sea directly (Clarke 1984).

AABW that has penetrated the North Atlantic is also an important component of the abyssal water mass structure. McCartney (1992) has shown how AABW enters the subpolar domain along the eastern slopes of both the Newfoundland and West European Basins. He describes how AABW is the precursor to the deep northern boundary current and is entrained into DSOW and ISOW to form the boundary jet in the Iceland, Labrador, Newfoundland, and, possibly, the Irminger Basins. Together, these water masses circulate around Cape Farewell, the Labrador Basin, the Flemish Cap, and finally the Grand Banks of Newfoundland. In this view, the abyssal flow recirculates in gyres that actually carry heat (weakly) equatorward. Consequently, the West European and Iceland Basins exhibit increasing bottom potential temperatures to the North while the Irminger, Labrador, and Newfoundland Basins have the opposite gradient. West of the mid-Atlantic Ridge, bottom potential temperatures and salinities are lower than on the eastern side, and densities are greater (Dietrich 1969). At the Grand Banks, Swift (1984) estimates that the NADW consists of, roughly, 37% ISOW (comprising 15% entrained SPMW and 22% eastern overflow), 32% LSW, and 31% western overflow – a more or less even split between overflow waters and waters formed in the subpolar region – although he was not aware of the AABW contribution highlighted by McCartney (1992) and so included no AABW component. Once past Grand Banks the southward flow of NADW is conventionally called the Deep Western Boundary Current and is lost from the subpolar system.

27.2.2 Volume Flux Estimates

By 1990 NADW volume flux estimates also existed at a few locations, with various certainties. A prime example is Schmitz (1996)'s Fig. 1–86 (based on Schmitz and McCartney 1993's Fig. 12b) which is reproduced here as Fig. 27.1a. This schematic shows the flow of NADW in the temperature range 1.6–4 °C and is based mainly on hydrographic data from the 1950s and 1960s, plus review of the literature.

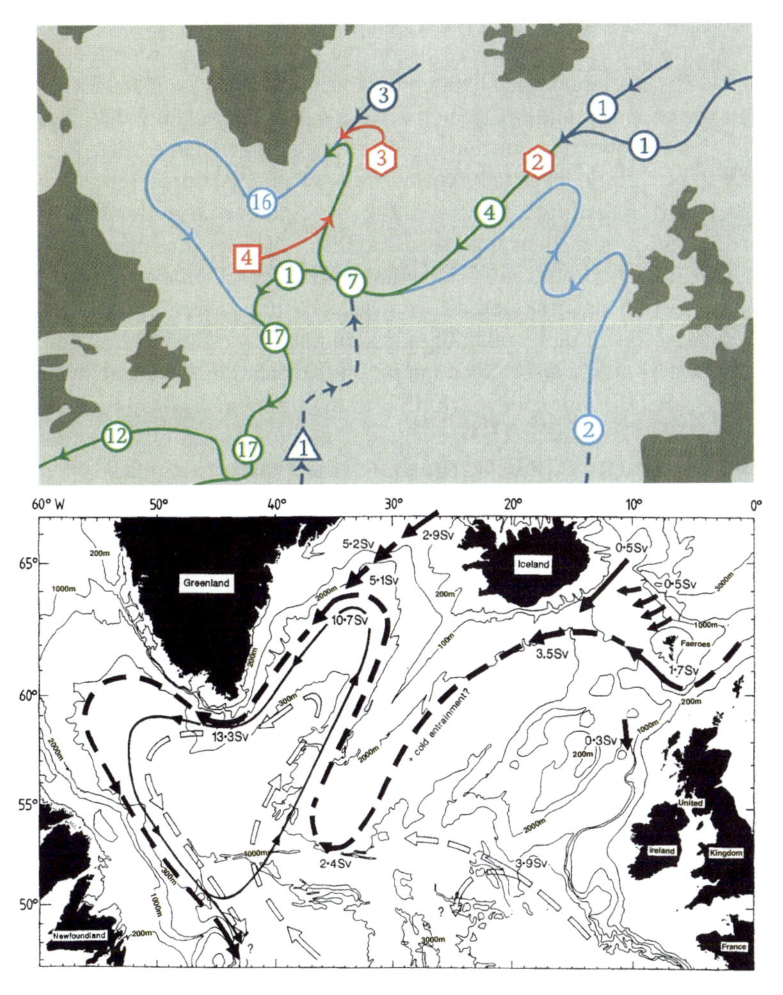

Fig. 27.1 North Atlantic Deep Water volume flux schemes circa 1990. (a) Circulation schematic for NADW(1.6–4 °C) adapted from Schmitz (1996), Fig. 1–86 (reproduced with permission of Woods Hole Oceanographic Institution; Jack Cook drew the original diagram). Green denotes NADW, dark blue is bottom water, and light blue indicates NADW at the bottom. Volume fluxes are shown in Sverdrups (10^6 m^3 s^{-1}); hexagons indicate entrainment, squares represent sinking, and the triangle represents upwelling. (b) Circulation schematic for waters denser than $\sigma_\theta = 27.80$ taken from Dickson and Brown (1994), Fig. 13 (reproduced with permission of the American Geophysical Union).

Another good example is due to Dickson and Brown (1994) (their Fig. 13, reproduced here as Fig. 27.1b). It shows flow at densities greater than $\sigma_\theta = 27.80$ (around 3 °C at these salinities) and so explicitly excludes LSW. This scheme was based on all available measurements at that time with strong emphasis on current meter arrays off east Greenland to measure DSOW. The deep and abyssal circulation described above is clearly picked out in both these cartoons. There is agreement on DSOW and ISOW volume flux at the sills (3 Sv and 2–2.7 Sv, respectively; 1 Sv is 10^6 m^3 s^{-1}). Both schemes show doubling of these fluxes during descent to the deep ocean through entrainment of SPMW. At Cape Farewell the volume flux estimate denser than $\sigma_\theta = 27.80$ is 13.3 Sv (due to the mooring array reported by Clarke (1984), and R. A. Clarke's personal communication to Dickson and Brown 1994) and 16 Sv between 1.6° and 4 °C which is more or less consistent with the combined NADW and LSW boundary current transport reported by Clarke (1984) of 19.5 Sv. There are some significant disagreements, however. For example, the flux of ISOW through Charlie-Gibbs Fracture Zone differs by a factor of three, and there are different pathways proposed in the Labrador Basin. These differences presumably reflect the real uncertainties in such semi-quantitative schematics; while the basic NADW circulation patterns and water mass structures were known circa 1990, there were very few reliable, unambiguous volume flux estimates to work with.

Schmitz (1996) includes 4 Sv of sinking in the Labrador Sea to represent LSW formation[5] which then merges with ISOW in the Irminger Basin. It is unclear why he did not depict the subpolar LSW circulation proposed by Mc-Cartney and Talley (1982) in his picture of 1.6–4 °C water flow, although his depth-averaged view clearly obscures depth-dependent details.[6] In any case, 4 Sv of LSW formation was a widely held estimate at that time. Table 27.2 shows values for LSW formation rates between 2 and 8.6 Sv published by the early 1990s. We present these numbers in detail to compare with more recent rates discussed in Section 27.3. All these baseline estimates presume a steady circulation with no interannual changes in LSW formation except that from the detailed survey of Clarke and Gascard (1983) in 1976. Most of them were based on thermal wind calculations using hydrography from just a few cruise sections in the 1950s and 1960s and crude arguments about

[5] "Formation rate" of a water mass is a concept which requires care to be precisely defined. The main problem is to directly relate it to the fluid velocity field and the fluid property fields (temperature, salinity, etc.). Here, "formation rate" means the net volume flux across the material surface bounding the water mass in the geographical region of formation. As material surfaces move with the flow, this flux is entirely diabatic (diffusive, or radiative in the euphotic zone). For instance, if we crudely define LSW as having temperature less than 4oC (ignoring salinity and the lower bound on temperature), "formation rate" means the flux of water mixing across the 4oC isotherm in a specific area such as the southeastern Labrador Sea. This definition basically follows the ideas in Nurser and Marshall (1991) on subduction rates, but other diagnostics are also possible; see Hall et al. (2006) on "ventilation rates", for example, and also section 3.

[6] Schmitz and McCartney (1993) Fig. 11 splits out LSW circulation around 1000 m from NADW flow around 2500 m and could have been used instead of the examples in Fig. 1. Their estimate for LSW formation is 7 Sv, but the number has an obscure basis. It seems most likely to derive from McCartney and Talley (1984) although they estimate LSW formation at 8.6 Sv (see McCartney (1992).

Table 27.2 Annual mean volumetric formation rates of Labrador Sea Water from the literature. See also Fig. 27.5 and Sections 27.2 and 27.3

Reference	Rate[a] (Sv)	Period	Method	LSW definition	Notes
Classic estimates prior to circa 1990 (Section 27.2)					
Sverdrup et al. (1942)	4	<1940	Mass budget	3–3.5 °C 34.86–34.94 psu	
Wright (1972)	3.5	<1970	Heat budget	3.45 °C, 34.92 psu	
Worthington (1976)	2	1964	Salt budget	<(4 °C, 34.96 psu)	
Clarke and Gascard (1983)	3.9	1976	Detailed survey	2.9 °C, 34.84 psu	
McCartney and Talley (1984)	8.6	1957–1967	Mass budget	3.5 °C	
Speer and Tziperman (1992)[b]	6.5	1941–1972	Surface fluxes	$\sigma_\theta = 27.60–27.80$	Isemer and Hasse (1987) fluxes
Schmitz and McCartney (1993)	7	1957–1967?	Mass budget?	3–4 °C	
Schmitz (1996)	4	<1960	Mass budget	?	
Modern estimates (Section 27.3)					
Böning et al. (1996)	3.5	1941–1972	1/3° GCM	2.6–3.0 °C	
Mauritzen and Häkkinen (1999)	5.9	1980–1986	90 km GCM	$\sigma_0 > 27.4$, $\sigma_1 < 32.3$	
Marshall and Schott (1999)	12.7	early 1990s	Hydrography	?	
Marsh (2000)	(3.4)	1980–1997	Surface fluxes	$\sigma_\theta = 27.65–27.775$	SOC fluxes
Khatiwala and Visbeck (2000)	1.3	1964–1974	Eddy overturning	?	
Smethie and Fine (2001)	2.2	1970–1990	CFC inventory	(see reference)	ULSW
	7.4			(see reference)	CLSW
Khatiwala et al. (2002)	(2.7)	1960–1998	Surface fluxes	$\sigma_\theta = 27.70–27.90$	NCEP fluxes
Rhein et al. (2002)	8.1–10.8	1988–1994	CFC inventory	$\sigma_\theta = 27.74–27.80$	CLSW
	1.8–2.4	1995–1997			
	4.4–5.6	1970–1997			
Haine et al. (2003)	7	1986–1988	GCM CFC inversion	$\sigma_\theta = 27.68–27.78$	
Böning et al. (2003)	(4.3)	1970–1997	1/3° GCM	$\sigma_\theta = 27.74–27.80$	CLSW
Yashayaev et al. (2004)	2	1970–1995	Hydrographic changes	(see reference)	
Gerdes et al. (2005)	(3.1)	1948–2001	1/4° GCM	(see reference)	
Marsh et al. (2005)	(7.4)	1985–2002	1/4° GCM	$\sigma_\theta = 27.7–27.8$	

(continued)

Table 27.2 (continued)

Reference	Rate[a] (Sv)	Period	Method	LSW definition	Notes
Kieke et al. (2006b)	6.9–9.2	1997–1999	CFC inventory	$\sigma_\theta = 27.68–27.74$	ULSW
	3.3–4.7	1999–2001			
Kieke et al. (2006a)	2.5	2001–2003	CFC inventory	$\sigma_\theta = 27.68–27.74$	ULSW
Brandt et al. (2007)	7.9	1986–1988	1/12° GCM	$\sigma_\theta = 27.74–27.80$	CLSW; resolves seasonal cycle
Yashayaev and Clarke (2006)	4.5	1987–1992	Hydrographic changes	?	
Pickart and Spall (2006)	2	1990–1997	Hydrography	$\sigma_\theta = 27.6–27.80$	
Böning et al. (2006)[c]	(3.0)	1959–2000	1/3° GCM	$\sigma_\theta = 27.74–27.80$	CLSW
Böning et al. (2006)[d]	(3.8)	1987–2003	1/12° GCM	$\sigma_\theta = 27.74–27.80$	CLSW
Myers and Donnelly (2007)	(4.4)	1949–1999	Surface fluxes	$\sigma_\theta > 27.65$	CLSW + ULSW; NCEP fluxes

[a] Parentheses mean the average formation rate is given over the period shown when a time series was also reported; the time series are shown in Fig. 27.5.
[b] See also Speer et al. (1995).
[c] These estimates supercede the time series reported by Böning et al. (2003) using an earlier version of the 1/3° FLAME model. In Fig. 27.5 we show the Böning et al. (2003) average formation rate and the 1/3° time series from the experiments reported by Böning et al. (2006).
[d] The Böning et al. (2006) 1/12° results closely follow the 1/3° results for the period of overlap and are thus not shown in Fig. 27.5.

mass or heat budgets. Notably, Speer and Tziperman (1992) pioneered the use of climatological air/sea flux and sea-surface hydrographic data to infer North Atlantic formation rates, and the modern view in Section 27.3 is adapted from their notions. Several of the classic estimates in Table 27.2 have specific problems to the modern eye: Sverdrup et al. (1942) believed that all NADW was formed convectively in the Labrador and Irminger Seas, for example, with no significant contribution from the overflows (whose existence was known at that time; see Warren 1981 or Worthington 1976 for historical notes). The heat budget of Wright (1972) has been criticised as unphysical by McCartney and Talley (1984). The geostrophic balance fundamental to large-scale ocean currents was partly abandoned by Worthington (1976). Speer et al. (1995) show significant differences from the results of Speer and Tziperman (1992) using a different climatology. Schmitz (1996) does not justify his value of 4 Sv, but it appears to be taken from Dietrich et al. (1980) whose basis is also obscure. Finally, the problem of an adequate reference level for thermal wind calculations is a constant worry in these early studies. So, despite several published values, no truly adequate LSW formation rate estimate existed in 1990 and formation rate variability was almost always ignored (the few important exceptions are cited below). We now discuss what is known today and where the grand challenges remain.

27.3 LSW Formation Rates and Their Variability

In this section we focus on LSW formation rates and variability, because LSW formation is central to NADW formation in the subpolar Atlantic. Indeed, some climate forecasts predict that LSW formation is the most vulnerable element of the dense limb of the North Atlantic overturning circulation in the coming decades (Wood et al. 1999). Much is now known about LSW formation processes and the history of LSW properties (for example, see The Lab Sea Group (1998) and Marshall and Schott (1999) for the former and Chapter 24 for the latter). Estimating formation rates is harder, however, and – as we see below – a quantitative time series of LSW formation rate has only recently begun to emerge. The existence of interannual changes in formation rate has been known for more than a quarter of a century. Lazier (1980) first reported changes in Labrador Sea convection depths from Ocean Weather Ship Bravo (56° 30' N, 51° 00' W) data in the period 1964–1974. He linked these variations to the synoptic atmospheric situation over Greenland. Talley and McCartney (1982) also identified decade long changes in LSW formation rate in the period 1948–1974, but could not quantify these variations. More recently, Dickson et al. (1996) have proposed that these changes in LSW convection depths (and hence formation rates) are coordinated by the North Atlantic Oscillation (NAO). Deep convection was suppressed in the late 1960s because of surface freshwater anomalies inherited from the East Greenland Current and feeble winter heat loss in the Labrador Sea due to an extreme negative NAO index. Again, quantifying the changes in formation rate was not possible. In the last 10 years important new data sets have become available to address this question, however.

In particular, repeat basin-wide CFC surveys have provided valuable insight into LSW production variations. These are discussed in Section 27.3.1. Eddy-resolving GCMs also offer intriguing evidence on the formation processes themselves and highlight the need for a consistent definition of "formation rate" (Section 27.3.2). Finally, we summarise the existing LSW formation rate estimates (Table 27.2 and Fig. 27.5) and attempt to reconcile them in Section 27.3.3.

27.3.1 Evidence from CFC Inventories

In the past decade, the use of CFC inventories has proven to be a valuable tool to estimate water mass formation rates. The method is based on the well-known atmospheric increase of CFCs since 1930. CFCs enter the surface of the ocean by air–sea gas exchange.[7] The components CFC-11 and CFC-12 behave in seawater like a noble gas with very long lifetimes and have no known natural sources. CFC-tagged surface water is transferred into the interior of the ocean by deep convection, for instance in the Labrador Sea. The deeper the convection, and the longer it lasts, the more CFCs are sequestered causing an increase in the CFC inventory of LSW.

Orsi et al. (1999) were the first to use observed CFC inventories to infer (Southern Ocean) water mass formation rates. In their work, and in the papers that followed their lead, the water mass formation rate is considered to be the transport of newly formed source water that sinks across the upper boundary of the respective water mass. Some CFC may be lost by diapycnal mixing, but this flux is usually assumed to be negligible (Rhein et al. 2002; Walter et al. 2005). The CFC inventory within a water mass volume is thus directly related to the formation rate according to Orsi et al. (1999).

Smethie and Fine (2001) applied a similar approach to infer formation rates of water masses being formed in the North Atlantic. A major challenge they faced was to assemble a large-scale data set containing enough hydrographic and tracer profiles to calculate Atlantic-wide CFC inventories of the respective water masses. Smethie et al. (2000) analyzed data collected between 1986 and 1992 and made corrections to infer a consistent data set for 1990.[8] They provided maps of

---

[7] Air-sea gas exchange rates are quantified with piston velocities, k. A typical value for CFC gases is k ? 3–4 mday?1. This rate yields a time scale of H/k to bring a one-dimensional column of seawater of depth H into saturation equilibrium with the atmosphere. The time scale is thus around one month for H = 100 m. Deep convection in the Labrador Sea, where H ranges from a few 100 m to 2000 m, lasts only a few weeks each year and air/sea gas exchange is not fast enough to bring the CFC concentration close to saturation equilibrium with the atmosphere. In practice, CFC measurements show that newly-homogenised LSW has saturations of 60–85%, however, depending on the depth of convection (Azetsu-Scott et al. 2003, Rhein et al. 2002). This observation makes it clear that CFC sequestration in LSWmust occur by re-exposing the same waters to the atmosphere over multiple years; for example, by CFC uptake into progressively denser SPMW varieties that eventually become LSW. See also Haine (2006) and references therein.

[8] Adjusting seawater CFC concentrations to a time different to the measurement time is a delicate issue. Mathematically, the problem is ill-posed without unrealistic assumptions about the ocean circulation. In practice, the error incurred is probably not overwhelming for corrections of a few years or less, although Figure 2 provides a salutary example of how hard estimating this correction can be.

CFC-11 concentration for Upper and Classical LSW (ULSW and CLSW, respectively)[9] in the North Atlantic which were defined using complex criteria based on local CFC minima and isopycnals that varied from basin to basin. Smethie and Fine (2001) provided first estimates of the CFC-11 inventories for these two types of NADW. They inferred water mass formation rates of 2.2 Sv and 7.4 Sv for ULSW and CLSW, respectively, both estimates representing the period 1970–1990. For ULSW, the inventory estimate covered latitudes 20°S–46°N so the source region in the western subpolar North Atlantic was excluded. For CLSW, an inventory estimate representative of the formation region was subtracted from the total North Atlantic CFC-11 inventory of this water mass. It is unclear how these issues affect the accuracy of the formation rates Smethie and Fine (2001) quote. The uncertainties in their analysis are probably overwhelmed by substantial problems with data gaps and uneven station coverage, however.

Rhein et al. (2002) chose a somewhat different approach by focusing on the subpolar gyre. They highlighted the formation and spreading of CLSW in the subpolar North Atlantic with data from, or adjusted to represent, the year 1997. During the summer months of 1997 various field programs overlapped, resulting in the best spatially resolved CFC data set measured yet in this region. Using this data, Rhein et al. (2002) estimated the CFC-11 inventory as a function of layer thickness and CFC-11 mean concentration and made detailed analyses of the uncertainties in the inventory estimate. In this study, CLSW was defined in the density range $\sigma_\theta = 27.74–27.80$. For the region 40–65°N Rhein et al. (2002) found 16.6 million moles of CFC-11 in the CLSW layer. The corresponding mean annual formation rate, representative of the period 1970–1997, was 4.4–5.6 Sv using the Orsi et al. (1999) approach. The NAO during the 1990s changed from a high positive phase in the early 1990s to a high negative phase after 1995. To take this source of variability into account, Rhein et al. (2002) estimated formation rates of 8.1–10.8 Sv for the high NAO index phase (1988–1994), and 1.8–2.4 Sv for the low NAO index phase (1995–1997). Böning et al. (2003) applied the same method to an eddy-permitting ocean model of the North Atlantic (the 1/3° resolution FLAME model, forced by heat flux anomalies derived from the NCEP/NCAR data set that were superimposed on the climatological forcing fields; see Sections 27.3.2 and 27.6 for model details). The synthetic CFC field gave good agreement with the observed CFC distribution (see Section 27.6.3). The formation rates of CLSW calculated from the synthetic data according to the method of Rhein et al. (2002) were 3.4–4.4 Sv, and

[9] ULSW is distinguished from CLSW by its relative freshness, warmth, and shallow depth. ULSW was first named by Pickart et al. (1996) although it was known for several years before then (for example, see Molinari and Fine (1988)). Pickart et al. (1996) and Pickart et al. (1997) identified the Labrador Current in the southeast Labrador Sea as the source region for ULSW. In the last 10 years, ULSW has replaced CLSW in the central Labrador Sea, being formed in the place of early 1990s CLSW formation. From this perspective, ULSW and CLSW are simply different versions of the continuum of LSW formed at different places and times. The ULSW/CLSW nomenclature is therefore not adopted by all authors. See Chapter 24 and Kieke et al. (2006b) for more details.

agreed well with the model's 1970–1995 average volumetric formation rate of 4.3 Sv diagnosed from the volume of newly homogenised water during a winter season. The year-to-year variability in the model's volumetric formation rate was high, however (Fig. 27.5), and the good agreement is probably sensitive to methodological details (see Hall et al. in press who explore the definitions and meanings of various tracer "ventilation rates").

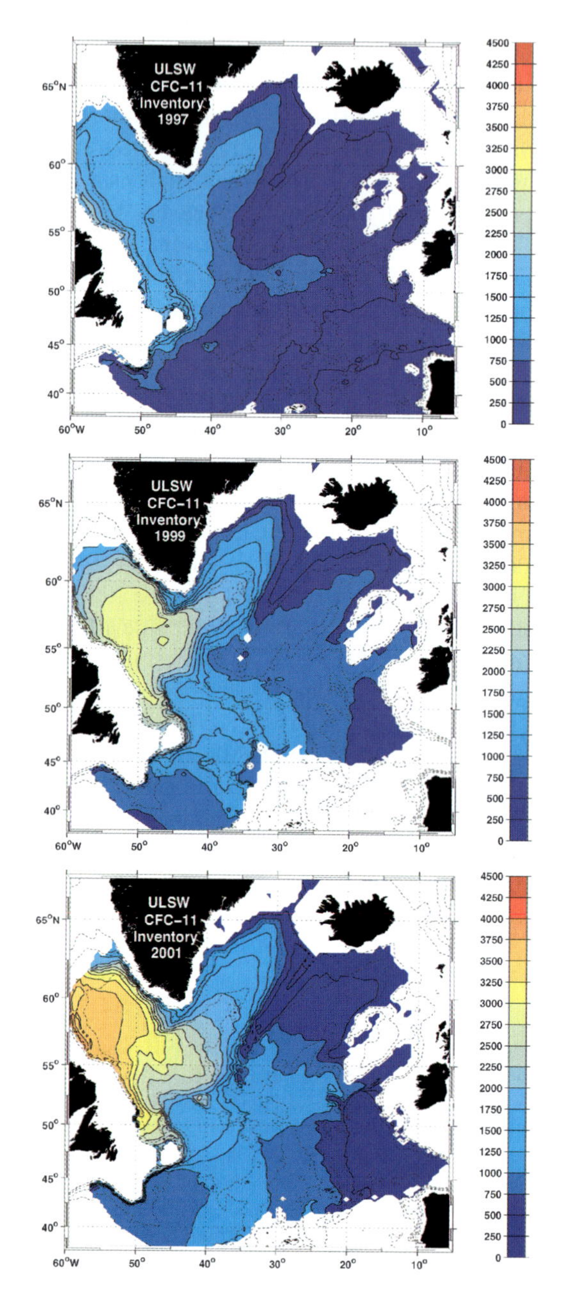

Fig. 27.2 CFC-11 inventory (Moles per grid cell) for ULSW in 1997, 1999, and 2001 (Adapted from Kieke et al. 2006b).

Kieke et al. (2006b) further extended the Rhein et al. (2002) analysis by exploring changes in the CFC-11 inventories of ULSW between the years 1997, 1999, and 2001. Here, ULSW is defined in the density range $\sigma_\theta = 27.68–27.74$. So, for the first time, changes in the water mass formation rates between two different 2-year periods could be estimated from tracer observations. The CFC-11 inventory in ULSW is shown in Fig. 27.2 for 1997, 1999, and 2001. The CFC-11 inventory of ULSW increased considerably from 1997 to 1999 and increased somewhat from 1999 to 2001. It is interesting to see the pattern of CFC-11 burden changing; some of this is due to the changing CFC input through different convection depths in successive winters, but some may also be due to changes in the LSW circulation (see Section 27.4 and Chapter 23). The formation rate estimates corresponding to these inventories are 6.9–9.2 Sv for the first and 3.3–4.7 Sv for the second period. In a recent paper, Kieke et al. (2006a) added observations from 2003. ULSW formation rates derived from CFC-12 inventories for the years 1997, 1999, and 2001 yielded very similar results, and the inventory difference between 2001 and 2003 indicated a continuing decrease in the ULSW formation to 2.5 Sv.

27.3.2 Evidence from Circulation Models

Several studies using numerical models of various kinds have analysed deep water formation in the Labrador Sea. They have focused on active physical processes, quantitative formation rate estimates, and ventilation (gas uptake) of the ocean. The classical picture as described by Marshall and Schott (1999) is that convection occurs in the central Labrador Sea and that a complex set of plumes and rim current eddies evolve during convection and restratification. It is also possible that a significant part of the water mass transformation takes places in, or very close to, the boundary current, which clearly plays a key role in deep water formation (Spall and Pickart (2001); see also Section 27.6.2). Spall (2004) analyzed an idealized model of a marginal sea with a cyclonic boundary current and uniform buoyancy loss to the atmosphere. He found downwelling located within a narrow boundary layer over the sloping bottom and along the offshore edge of the boundary current. The density of waters formed in the interior of the marginal sea is not directly related to the amount of downwelling in the boundary current, however. Cuny et al. (2005) made a detailed analysis of heat fluxes and mixed layer depths in the Labrador Sea based on observed CTD-profiles and moored stations. They applied air–sea buoyancy fluxes to observed early-winter density profiles and modeled mixed layer depths in late winter/early spring. Starting with a profile at the position of their mooring B1244 (see Fig. 27.3 for the location) they could not simulate any significant mixed layer deepening. Starting with a profile that during the winter is advected cyclonically around the Labrador Sea from the northern Labrador Sea to the position of mooring B1244, they obtained a good agreement between the simulated and observed profile. This finding shows that the Lagrangian heat flux history cannot be substituted for the Eulerian history over timescales longer then a few days.

To explore these processes in more detail, a thorough analysis of LSW formation in a state-of-the-art GCM has recently been performed by the Family of Linked

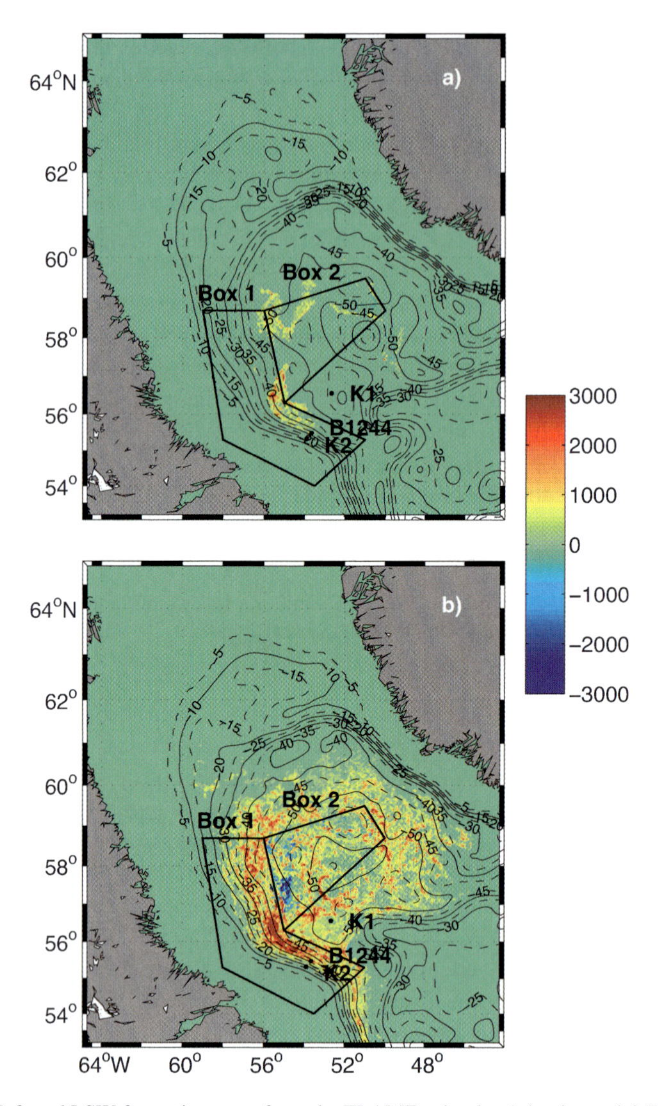

Fig. 27.3 Inferred LSW formation rates from the FLAME subpolar Atlantic model (Brandt et al. 2007). The increase in LSW (σ_θ = 27.77 – 27.80) layer thickness (m; colorbar) due to negative surface buoyancy flux is shown for (a) January and (b) January–March. The barotropic streamfunction (Sv) is overlaid in black contours and the positions of the moorings K2, B1244 and K1 are also shown. Boxes 1 and 2 represent the near boundary current regime and central Labrador Sea regime, respectively. See Section 27.3.2 for details.

Atlantic Ocean Model Experiments (FLAME) group (Brandt et al. 2007). They developed a regional, eddy-resolving (45 level, 1/12° longitude, 1/12° × cos ϕ latitude, where ϕ is latitude) model for the subpolar North Atlantic covering the domain from 43–70°N and 71°W–16°E. The model setup is essentially the same as that of Eden and Böning (2002), but to control the Labrador Sea salinity the following changes were applied (see Section 27.6): First, the northern boundary in the Greenland Sea was opened allowing for a more realistic inflow of freshwater in the East Greenland Current and outflow of salty water with the Norwegian Current. Second, the model used isopycnal diffusion (using a diffusivity of $50\,m^2\,s^{-1}$), biharmonic friction (using a viscosity coefficient of $2 \times 10^{10}\,m^4\,s^{-1}$), and the bottom boundary layer parameterization of Beckmann and Döscher (1997). A climatological run was analysed using 1986–1988 6-hourly ECMWF forcing which yielded winter-time convection down to 1,600 m.

Brandt et al. (2007) adopt a meaning of water mass formation as a time and space-varying field that follows ideas of Speer and Tziperman (1992) and Walin (1982) on water mass transformation.[10] Essentially, they calculate the volumetric flux, T, into the LSW density range caused by diabatic air/sea buoyancy forcing. This computation is performed using the GCM output every day, and for every grid point, using a one-dimensional convective adjustment scheme to vertically mix unstable profiles (that is, it assumes 100% efficient homogenisation of static instabilities in each vertical column separately). The results of these calculations are presented in terms of the LSW layer thickness change at each grid cell over January (Fig. 27.3a) and January–March (Fig. 27.3b). (To get formation rates in Sverdrups multiply by grid cell area and divide by the relevant period; at 56°N 2,000 m deep mixed layers correspond to about 20 mSv per grid cell in Fig. 27.3a). In January the formation of LSW starts just offshore the deep Labrador Current in a region around 55.5°W, 56.5°N, somewhat upstream of the AR7W section (Fig. 27.3a). At the end of March this is also the region where the largest amount of water has been transformed into LSW (Fig. 27.3b). The region of transformation reaches over large parts of the Labrador Sea limited mostly by the 45 Sv isoline of the barotropic streamfunction and reaching farther onshore at the southwestern boundary. The densest water is formed in the more central region around 55°W, 57.5°N, however.

The LSW formation rate in Fig. 27.3 can be used to infer changes in the model volume budget of LSW. The volume of LSW, V, over a horizontal area, A, evolves according to:

$$\frac{\partial V}{\partial t} + \int_A \int_z \nabla_h \cdot \boldsymbol{u}_h \, dz \, dA = \int_A T + R \, dA \qquad (27.1)$$

where R is a residual term due to diabatic fluxes not accounted for in T (for example, internal wave breaking) plus errors. (Recall, T is defined using the one-dimensional convective adjustment algorithm described above. Subscript h means horizontal and (Eq. 27.1) follows simply from conservation of volume in the Boussinesq GCM) In Fig. 27.4 the two left hand terms and the first right hand term in this balance are shown over time for A defined by the two boxes shown in Fig. 27.3a. For both boxes in January and February, inflation of LSW volume and formation by air/sea forcing

[10] Speer and Tziperman (1992) use "transformation" and "formation" in slightly different senses to what we call here "formation." In particular, they define water mass boundaries with density surfaces rather than material surfaces. See also Nurser et al. (1999).

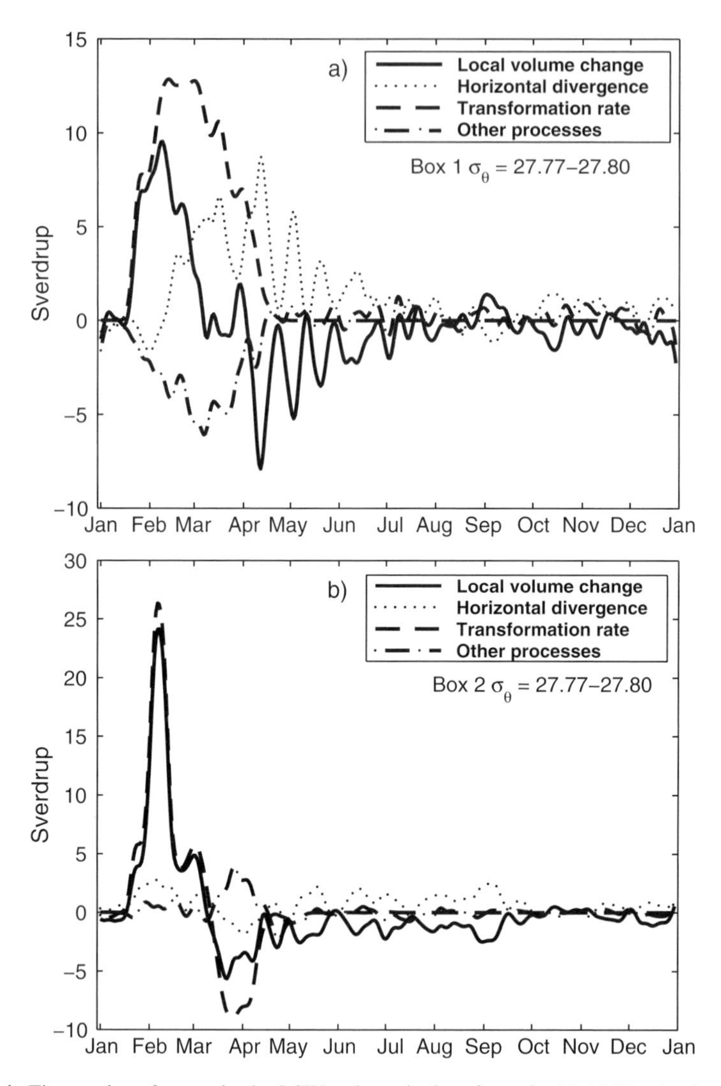

Fig. 27.4 Time series of terms in the LSW volume budget from the FLAME subpolar Atlantic model (Eq. 27.1); Brandt et al. 2007). The terms shown are the local volume change (solid line), horizontal divergence (dotted line), formation rate due to negative surface buoyancy flux *T* (dashed line), and other processes *R* (dashed-dotted line) within (a) the boundary current box, and (b) the central Labrador Sea box – marked in Fig. 27.3 – for LSW in the range $\sigma_\theta = 27.77–27.80$.

are balanced. In the central box at the end of March the formation rate becomes negative as LSW ($\sigma_\theta = 27.77 - 27.80$, the dense varieties of CLSW) is transformed into denser water with $\sigma_\theta > 27.80$.[11] In the boundary current box the horizontal divergence term, which is a measure of LSW export from the box, has its maximum of

[11] Note that convective formation of waters at densities greater than 27.80 is not very realistic, and is due, at least in part, to excessive model salinity in the Labrador Sea (see section 6).

about 9 Sv in April and shows positive values from February to June. In the central box the horizontal divergence shows much smaller values, up to 2 Sv with positive values from February to September, suggesting a more steady and continuous export of LSW. The residual *R* (labelled "other processes" on Fig. 27.4) is the smallest term in both boxes, although it is comparable to the horizontal divergence. This gives moderate confidence that the main processes forming LSW are captured by the air/sea diabatic term *T* in this analysis.

Overall, Fig. 27.4 shows that the GCM formation rate varies between February maxima of 10 and 24 Sv and March/April minima of –8 and –5 Sv for the boundary current and the interior boxes, respectively. Averaged over the whole year the formation rates are 2.1 and 0.7 Sv. For CLSW (σ_θ = 27.74 – 27.80) over the whole Labrador Sea (northwest of 52°N, 43°W) the annual average formation rate is 7.9 Sv, for a year of moderately deep convection to 1,600 m.

This discussion also draws attention to the difference between "ventilation" of LSW and "formation" of LSW. We usually speak of a water mass being ventilated when it is in contact with the atmosphere and gas exchange is occurring (Hall et al. 2007), but we speak of volumetric formation when there is a flux of water into the water mass which, other things being equal, leads to an increase of total LSW volume. Diagnosing ventilation rates is essential to understand and monitor ocean gas uptake and the global CO_2 cycle, while knowledge of formation rates is essential for the Atlantic MOC and meridional heat fluxes. Measures of convection depth, ventilation rate, and formation rate all contain sensible and valuable information, but they are clearly distinct diagnostics.

27.3.3 Summary of LSW Formation Rate Estimates

Since 1990 the number of LSW formation rate estimates has grown substantially. These estimates are summarised in Table 27.2 and Fig. 27.5. In addition to the CFC-based and model-based estimates discussed above some other notable studies have explored this issue. They include: Böning et al. (1996), Mauritzen and Häkkinen (1999), Gerdes et al. (2005), and Marsh et al. (2005), who analysed coarse resolution or eddy-permitting GCM solutions driven with climatological winds or reanalysed winds; Marshall and Schott (1999), who made a rough estimate of peak formation in the early 1990s based on hydrographic measurements; Marsh (2000), who derived a time series for 1980–1997 using the Speer and Tziperman (1992) method and the SOC air/sea flux data; Khatiwala and Visbeck (2000), who diagnosed an eddy-driven overturning cell in the Labrador Sea; Khatiwala et al. (2002) and Myers and Donnelly (2007), who applied the approach of Speer and Tziperman (1992) to NCEP reanalysis fluxes yielding 40-year-long time series; Haine et al. (2003), who diagnosed formation rates from a 4/3° resolution North Atlantic GCM constrained by CFC data (expanding an earlier study by Gray and Haine 2001); Yashayaev and Clarke (2006), Yashayaev et al. (2004), and Pickart and Spall (2006) who analysed AR7W repeat transects; and Böning et al. (2006) who analysed 1/3° and 1/12° resolution FLAME hindcasts. For convenience,

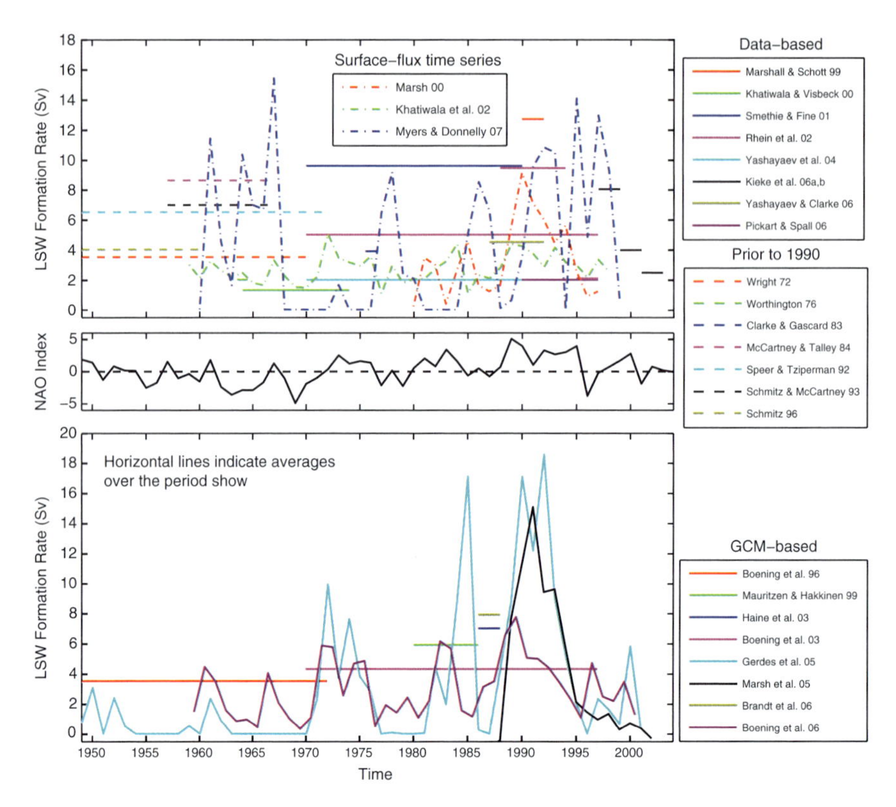

Fig. 27.5 LSW formation rate estimates from the literature (see also Table 27.2). The various rates have been grouped according to whether they were made prior to 1990 (dashes, upper panel; see Section 27.2), based directly on *in situ* data (continuous lines, upper panel), using the Speer and Tziperman (1992) surface-flux method (dash-dotted, upper panel), or from GCMs (continuous lines, lower panel). Long horizontal lines indicate averages over the periods shown. The December–March North Atlantic Oscillation Index is also shown (From J. Hurrell's website); see Section 27.3.3 for details.

we have omitted the indirect estimates of LSW formation from analyses of LSW volume flux across coast-to-coast transects using hydrographic inverses (for example, see Lumpkin and Speer 2003; Talley 2003 and references therein).

These estimates in Table 27.2 and Fig. 27.5 reveal several interesting issues. First, prior to 1990 the LSW formation rate estimates spanned a factor of about four. This range has not diminished in recent years: for example, the Smethie and Fine (2001) and Yashayaev et al. (2004) estimates for 1970s–1990s are 9.6 (for ULSW and CLSW together) and 2 Sv, respectively. For the period of deep convection in the early 1990s, the estimates based on field data are generally higher than in previous years, but they still span a factor of six. There is more consistency in the estimates from GCMs: the Böning et al. (1996) 1941–1972 average rate agrees well with the Böning et al. (2003) 1970–1997 average, and Brandt et al. (2007) agree well with Haine et al. (2003) for 1986–1988, while the 1980–1986 estimate of Mauritzen and Häkkinen (1999) is only a little lower. The six time series estimates from GCMs and the surface-flux method show relatively high formation during the early 1970s and 1990s, but exhibit qualitative disagreements too. For example, the Khatiwala et al. (2002) and Myers and Donnelly (2007) time series differ significantly during 1960–1980, although they are based on the same flux data (Myers and Donnelly 2007 allow the sea-surface properties to vary from climatology, which is more realistic, but adds uncertainty). Also, Khatiwala et al. (2002) shows muted interannual variations whereas the Gerdes et al. (2005) time series consists of years of strong formation punctuated with years of zero formation. Although the estimates for the last 10 years are fewer in number, they show qualitative agreement with declining formation that is occurring at lighter density.

Some of these differences are unsurprising. In particular, there is no consistent definition of LSW among the entries in Table 27.2. Narrow criteria on LSW will naturally lead to smaller formation estimates than more inclusive definitions, for instance. Moreover, there exist about ten distinct methods to infer LSW formation rate, and each of them has random and systematic errors which are essentially unknown. Indeed, there are few robust attempts to quantify formation uncertainties among the works in Table 27.2 – the exceptions being Rhein et al. (2002) and Kieke et al. (2006, 2007) – and this is a topic ripe for future work. As discussed in Sections 27.3.1 and 27.3.2, there are also multiple notions about the precise definition of the formation rate diagnostic itself. A study that combines the various published estimates in Table 27.2 to produce a coherent synthesis of LSW formation rate over the last 50 years would therefore be very worthwhile. Nevertheless, the semi-quantitative agreement found here is encouraging, and the salient feature of Fig. 27.5 is that we have in hand the beginnings of a reliable formation rate time series. There are intriguing connections to the NAO index, as one might expect (the estimates of Khatiwala et al. 2002; Marsh 2000; Böning et al. 2006 are in phase with the index), but no universal relation (for example, the Myers and Donnelly 2007 time series). Looking ahead, we can anticipate convergence of these estimates if, first, the data density is maintained (annual occupations of AR7W are particularly important, as are periodic CFC surveys), and, second, there is convergence in diagnostic methodology and a consistent definition of "formation rate." Arguably, the best strategy is to synthesize the field data and air/sea flux data with the GCMs through data assimilation. Such approaches are conspicuous by their absence in Table 27.2, although Haine et al. (2003) make a preliminary attempt and several North Atlantic assimilation products are now routinely available in near real time and could be used for this purpose.

27.4 Pathways and Timescales of LSW Transport in the Subpolar Atlantic

Since 1990 three important advances concerning LSW circulation and transport in the subpolar North Atlantic have occurred. These are: the detailed description of variability in LSW formation rates and properties (Section 27.3, Chapter 24); the mapping of anthropogenic CFC gases in LSW throughout the North Atlantic (Section 27.3); and the greatly improved view of the mid-depth geostrophic circulation from subpolar floats (discussed in this section). We now have a much better picture of the time-averaged LSW circulation, at least during the late 1990s. In turn, this improved view raises deeper questions about LSW transport pathways, and the associated timescales. Dispersion of LSW anomalies through the subpolar Atlantic can be now addressed directly. These can be hydrographic anomalies or anomalies in dissolved chemical concentration, such as anthropogenic carbon or oxygen, and they are the focus of this section. Variability in subpolar currents themselves is also known to exist (see Chapter 23). There is not yet enough data to cleanly identify the sources of mid-depth property changes, however. They could arise from LSW property variations in the Labrador Sea and/or from changes in circulation patterns. Understanding this issue in detail lies in the future and we have only preliminary results to present.

Before discussing pathways and timescales in the time-averaged LSW flow a couple of remarks on fundamental matters are needed. First, in this context, "transport" does not mean "volume transport" of traditional physical oceanography. By "transport" we mean both the advective and the diffusive motion of a dynamically inactive tracer. Clearly, advection by ocean currents is a critical part of this motion (sometimes called "stirring"; Eckart 1948), but diffusive dispersal of tracer is also important (sometimes called "mixing"). Irreversible mixing by diffusion ultimately occurs at the Batchelor scale (typically O(1–1,000) μm for temperature, salinity, and dissolved chemical species, Thorpe 2005), but stirring through all turbulent scales of oceanic motion steepens tracer gradients and thereby accelerates the diffusive process. At scales larger than the Batchelor scale the processes causing the forward cascade of tracer variance are traditionally lumped into an enhanced "eddy" diffusivity, which is the approach taken here. Second, this distinction between advective and advective/diffusive movements is not merely academic. In fact, it leads directly to some conclusions that affect the basic way in which oceanic transport of anomalies is described and understood. Most important here is the notion that no single timescale exists for movement of an anomaly from its source (for example, the southeast Labrador Sea) to a remote location (such as the eastern basin). Instead, a continuous range of timescales exists between these two points (see Haine and Hall 2002 for the theory and Waugh et al. 2004 for simple applications to the subpolar Atlantic). This range of timescales is quantified using a transit-time distribution (TTD).[12] As we see below, this deep shift in perspective has some important repercussions. Of course, the traditional idea of advective volume transport is not replaced by the advective/diffusive transports discussed here. Advective volume transport is still a well-defined quantity that captures essential information about the flow. It is not fundamentally suitable to describe movement of temperature, salinity, or dissolved chemical anomalies, however. Instead, advective/diffusive transport is the appropriate framework for these cases.

[12] The TTD is a type of Green's function to the tracer equation and is sometimes also called an age spectrum or boundary propagator. See Haine and Hall (2002) and references therein.

As explained in Section 27.2.1, LSW is the densest of the subpolar mode waters, characterized by a mid-depth local minimum in both salinity and stratification (hence planetary potential vorticity). In addition, it carries high levels of atmospheric gases, such as dissolved oxygen, anthropogenic carbon, and CFCs, and low nutrient concentrations. Talley and McCartney (1982) used the potential vorticity minimum to trace LSW spreading through the subpolar region, and several studies have followed their lead to exploit these tracer signals and infer LSW transport pathways and timescales. None of these works have found significant departure from the three LSW transport pathways identified by Talley and McCartney (1982) (Section 27.2.1), but the associated spreading timescale estimates vary widely. For instance, Sy et al. (1997) used hydrographic anomalies and CFC data to trace LSW from its formation region to the eastern basin. Their timescale estimate is 4–5.5 years along this path. Cunningham and Haine (1995) inferred the circulation of LSW by using potential vorticity and salinity distributions, and estimated a transit time of 12 ± 7 years to the eastern basin. Finally, Read and Gould (1992) estimated a timescale of 18–19 years for this trajectory by tracing LSW hydrographic anomalies from the source region. Estimates of transit times from the Labrador Sea into the Irminger Sea also vary widely. Sy et al. (1997) found that it takes 0.5 years, while Lavender et al. (2005), using direct velocity measurements from subsurface profiling floats, estimated the timescale to be 1–1.5 years. Straneo et al. (2003) simulated spreading of an ideal tracer with an advective–diffusive numerical model, and found a transit time of 2 years to the Irminger Sea. The main goal of this section is to reconcile these different estimates of subpolar LSW spreading rates, and to present new results that provide insight into these discrepancies and their origin. Knowledge of the mid-depth circulation in the subpolar North Atlantic has been greatly improved by the deployment of several dozen autonomous floats in recent years (nominally at 1,500 m; vertical current shear appears to be quite weak). Faure and Speer (2005) combined data from profiling PALACE, and acoustically tracked isobaric and isopycnal floats to derive a mean circulation field for the subpolar gyre at LSW depths for the period 1996–2001.[13] Figure 27.6 shows streaklines of simulated trajectories in the geostrophic streamfunction from this work (corresponding to their Fig. 7a). The major pathways identified by Talley and McCartney (1982) are visible, although export around Newfoundland and Grand Banks is not obvious (see Section 27.5). The main advance from Talley and McCartney (1982) is that the level of detail is substantially improved so that O(100) km scale features in the time-averaged flow now emerge. For example, recirculation cells are prevalent offshore of the western boundary currents (see the recirculating trajectories in the Labrador Basin and in the Irminger Sea). These cells have attracted attention since Lavender et al. (2000) first announced them (Käse et al. 2001; Spall and Pickart 2003; Kvaleberg and Haine in revision in 2007), but a decisive dynamical explanation is still pending.

[13] Faure and Speer (2005) use 57 PALACE floats from October 1996–October 2001 at 1500 m, 21 MARVOR floats from September 1996–June 2003 at 1750 m, and 34 RAFOS floats from May 1997–August 2000 at 1500 m. Their work is a synthesis of previous studies by Lavender et al. (2000); Bower and Hunt (2000); Lavender et al. (2005); Fischer and Schott (2002) and Bower et al. (2002).

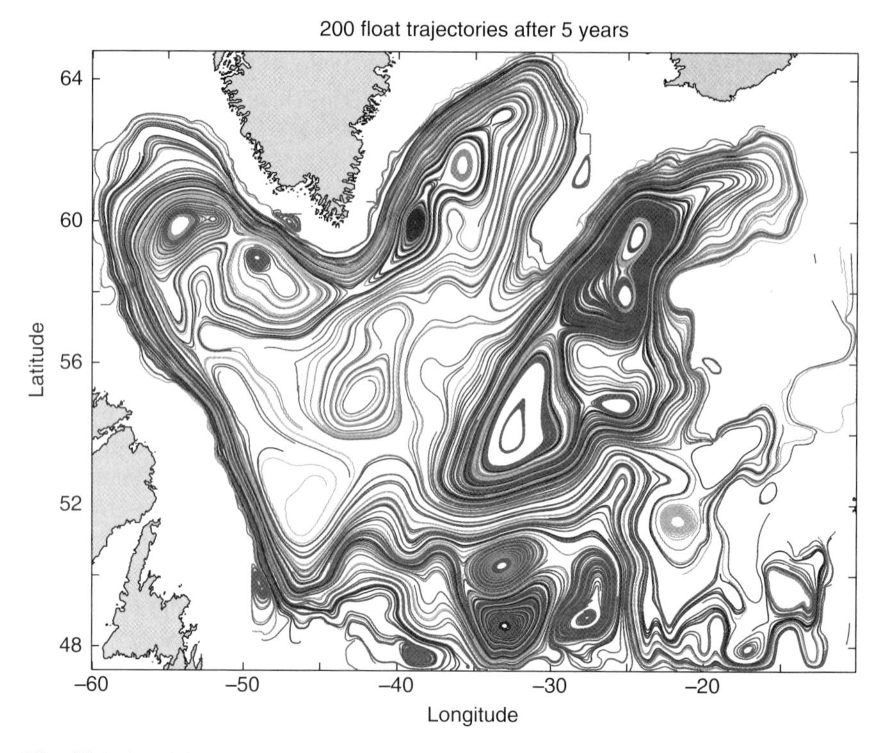

Fig. 27.6 Streakline trajectories in the subpolar North Atlantic Ocean at the depth of LSW (approximately 1,500 m) based on a synthesis of float data during 1996–2001 (Faure and Speer 2005). The gridded geostrophic streamfunction data used to create the trajectories were provided by V. Faure. The 200 trajectories last 5 years and start at random initial conditions in the area.

In order to diagnose LSW transport pathways and rates in detail an advective–diffusive numerical model of subpolar tracer dispersion is driven by the geostrophic flow field used in Fig. 27.6. The model is based on the MITgcm, but only solves the kinematic equations (Marshall et al. 1997). An ideal tracer is released continuously in the deep convection region of the Labrador Sea for 1 month, after which the region has its tracer concentration reset to zero. The eddy diffusivity is 500 $m^2\ s^{-1}$ (experiments with different constant diffusivities, and with variable diffusivity fields based on the float data themselves, show broadly similar results to what follows). As explained above, LSW is carried away from the Labrador Sea by both advection and diffusion. A distribution of transit times (the TTD) therefore exists between the LSW source region and any remote point in the subpolar gyre. The ideal tracer simulation in our advective–diffusive model provides the TTD for the float-derived circulation.

Results from the numerical model show that LSW is spread through the subpolar gyre along the principal pathways identified above. Figure 27.7 shows the TTD as a function of transit time at four places in the domain; the northern Labrador Sea, the Irminger Sea, within the deep western boundary current (DWBC), and in the

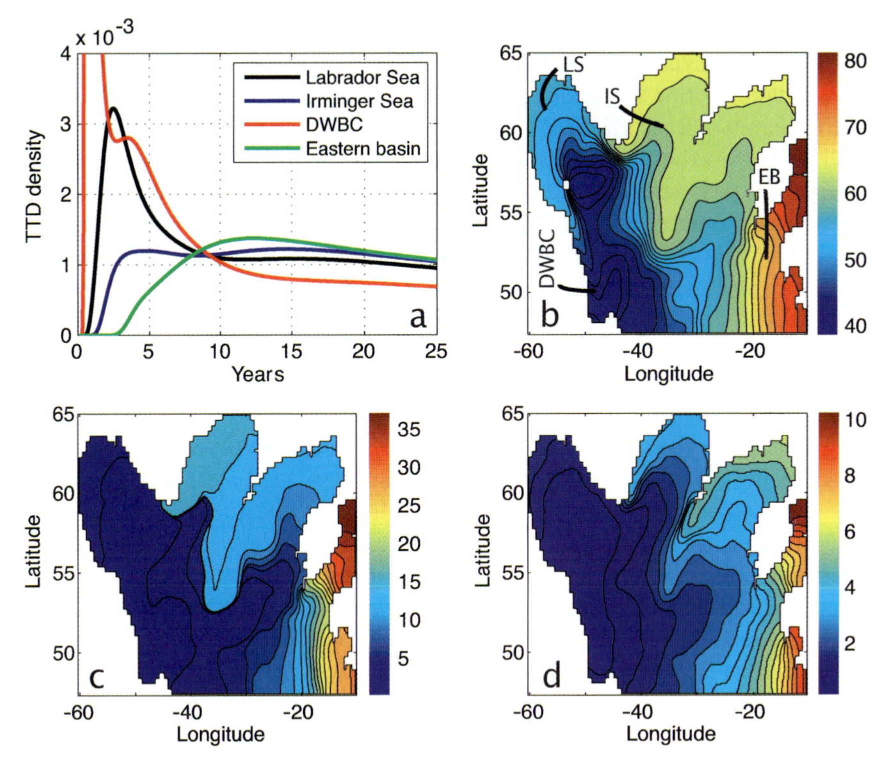

Fig. 27.7 Timescale diagnostics of LSW spreading based on the trajectories in Fig. 27.6. (a) Transit-time distributions (TTDs) at four different locations. (b) Mean transit-time (years). The locations of the TTDs in (a) are marked: LS = Labrador Sea, IS = Irminger Sea, DWBC = Deep Western Boundary Current, and EB = Eastern Basin. The white rectangle shows the release point. (c) Modal transit-time (years), that is the arrival times of the TTD peaks. (d) The time when the TTD density first reaches 10% of its modal value (years).

eastern basin. The TTD time series shows the range of transit times from the deep convection source region to the remote point in question (called the field point). Note that the TTD is typically a continuous smooth curve with non-zero density over a wide range of transit times. This property reflects the fact that the LSW at the field point consists of a continuous blend of different LSW from different years (meaning a range of transit times from the LSW source in the southeast Labrador Sea). Peaks in the TTDs indicate the most likely transit time (that is, the time it takes for most LSW to arrive at that location). Multiple peaks in a single time series indicate that there is more than one important timescale and therefore more than one main pathway. Multiple peaks from distinct pathways occur largely because of the sub-basin recirculations in the Labrador and Irminger Seas (Lavender et al. 2000): a significant fraction of newly formed LSW is peeled away from the Labrador Current and then delayed by passage round a recirculation cell. The delayed fluid then merges back into the Labrador Current and proceeds to the field point in question. Animations of the TTD show this process clearly.

The black and red curves show time series taken from the Labrador Sea and the DWBC where advection is relatively strong, hence they show rapid arrival of new LSW within the first few years, with well-defined peaks. The most likely transit time to Hamilton Bank via the DWBC is less than 1 year. The TTD time series from the Irminger Sea (blue curve) also shows that LSW is carried into this region within the first 2–3 years, but the broad maximum after this time is maintained by diffusion and an internal recirculation off the east Greenland coast. The time series from the eastern basin (green curve) indicates a diffusively dominated pathway; the TTD rises to a broad maximum at about 12 years then declines slowly.

A TTD time series exists for every point in the subpolar gyre, which is unwieldy for some purposes (for convenience, we show only four locations in Fig. 27.7a). One way to summarise the TTD information is to calculate its first moment over transit time at every point. This gives a map of the mean LSW transit time (not the most likely transit time) and is shown in Fig. 27.7b. The deep convection source region is shown by the white square in the Labrador Sea. The smallest mean transit times are near this site and are a few decades. These values are much longer than the most likely transit times in Fig. 27.7a because of the strongly skewed TTD tail to long transit times. This long tail means that some of the water takes a very long time to reach the field point following diffusive pathways. The primary LSW spreading pathways are visible in Fig. 27.7b as tongues of low mean transit time. Physically, this means that LSW is transported along these pathways at a relatively fast rate, and that advection is more important than diffusion. The highest mean transit times are seen near the domain edges and in the eastern basin. Transport of LSW to these regions is diffusive and may take more than 80 years on average.

To exclude the contribution to the mean transit times from the long tail in the TTD time series, we can instead map the peak arrival times, or the mode of the TTD, at each field point (Fig. 27.7c). The modal transit time is the most likely transit time from the source region to any point in the domain. Transport timescales now range from less than 10 years over much of the domain to 35 years in the eastern basin. The TTD peaks are broad in the east, however, and from the green curve in Fig. 27.7a it is clear that significant amounts of LSW arrive well before the modal transit time. A third method for inferring transit times from the TTDs is therefore to map the "first arrival times", or the times that the TTD density at each point first reaches a certain value. For illustration we have chosen this value to be 10% of the modal value (Fig. 27.7d). The resulting transit times are now less than 4 years in the DWBC and the Labrador and Irminger Seas, with a maximum of 10 years in the eastern part. Clearly, these values depend sensitively on the arbitrary threshold used to define the first arrival time.

Given the wide range of transit times present at any particular field point, the apparently severe contradiction in the prior timescale estimates cited above fades away. Once we accept that transport of material properties includes a diffusive component, the logical deduction is that a continuous distribution of transit times exists. There is no single timescale for LSW propagation into the interior. Different diagnostics of the TTD (mean transit time, modal time, first arrival time, etc.) yield widely different timescale estimates. Different "tracer age" diagnostics, and tracers with different source histories, also yield different timescales (Waugh et al. 2003). These differences almost certainly explain the factor 2–4 discrepancy between the published timescale estimates. Recall that the float-based flow is, nominally, an average for the period 1996–2001 and does not overlap the periods studied in most of the papers cited above. Therefore, no appeal to variability in the subpolar geostrophic streamfunction appears necessary to reconcile these timescale differences (Kvaleberg et al. under revision in 2007).

Before turning to the boundary current circulation in Section 27.5 a few final remarks are in order. First, using the float-based circulation we can easily compute how LSW anomalies are recirculated and mixed in the subpolar gyre. This information is vital for accurate estimates of anthropogenic carbon uptake by LSW and is a substantial advance over the state of knowledge circa 1990. Second, an obvious next question is to ask if the TTD results from the float circulation are consistent with the observed CFC field in the subpolar region (Section 27.3). Preliminary results on this issue suggest that indeed the float circulation is consistent with the CFC field. That is, simulation of CFC tracer in the flow of Fig. 27.6 gives reasonably good agreement with the available CFC database shown in Fig. 27.2 for 2001 (Kvaleberg et al. in revision in 2007). Again, this means that variability in the mid-depth subpolar circulation, while clearly present (Chapter 21), is not strong enough to be clearly seen in the current LSW CFC data. The problem is that there are three sources of variability in the CFC data (LSW source changes, circulation changes, and the variable CFC atmospheric history), and we cannot yet clearly distinguish which one of them is causing far-field LSW CFC anomalies. Third, the related question about consistency between the float circulation and the observed hydrographic anomalies needs to be tackled. Given the excellent hydrographic data available over the last 15 years (Chapter 21) this is a ripe issue and a priority for early attention. Finally, the entire discussion here has neglected errors in the float-based circulation fields. These errors are known to exist and are probably large, especially in the boundary currents (Section 27.5). Nevertheless, they are not large enough to severely corrupt the pathway and timescale estimates here. Although the order of magnitude improvement in subpolar LSW circulation estimates achieved since 1990 is a great step forward, there are clearly further refinements to hope for. In this regard, ocean model and data synthesis systems are a promising development (see, e.g., Menemenlis et al. 2005).

27.5 Boundary Currents and Export to the Subtropics

The most prominent feature of the deep and abyssal subpolar circulation is the cyclonic boundary current following the continental slopes of Greenland, Labrador, and Newfoundland (Section 27.2). Strong thermohaline variability is observed in all three constituents of NADW since 1990 (that is, in LSW, ISOW, and DSOW; Lazier et al. 2002; Stramma et al. 2004) with most dramatic changes occurring in the Labrador Sea Water through the last decade (Chapter 21). Claims have been made that changes can be traced far downstream along the pathways of the NADW(see Section 27.4 for subpolar spreading; Koltermann et al. (1999), Curry et al. (1998), and Molinari et al. (1998) for spreading to the subtropics; and Stramma and Rhein (2002) for spreading to the Equator). Whether these changes in water mass characteristics are accompanied by equivalent multi-year changes in the circulation of the deep subpolar gyre is presently under discussion (see, e.g., Chapter 23, Section 27.6.4, Bersch et al. 1999; Häkkinen and Rhines 2004). A long-term decay of the Atlantic MOC has been postulated by Bryden et al. (2005) from five subtropical hydrographic transects over the period 1957–2004. There is observational and modelling evidence that these variations indicate interannual to decadal variability and not long-term trends, however (Hirschi et al. 2006; Baehr et al. 2006). Furthermore, open questions remain about the relationship between the subtropical Atlantic MOC strength and the subpolar circulation (see Section 27.6.4 for a modelling perspective on this issue). The circulation along the western boundary of the subpolar North Atlantic and large-scale internal recirculations of NADW clearly play an important role, however, as they directly feed the deep limb of the subtropical MOC.

Due to the large barotropic component in these weakly stratified waters, and the corresponding absence of a level of no motion, geostrophic shear from hydrographic sections contains little information about the total volume flux of NADW along the boundaries. Correspondingly, there was an urgent need for direct velocity measurements at the boundaries in the early 1990s. The success of the DSOW mooring arrays off southeast Greenland clearly pointed to the value and potential of other direct measurements of boundary current velocities in the subpolar Atlantic (Dickson et al. 1990; Dickson and Brown 1994).[14] This velocity information has come from various sources in the last 15 years. The most comprehensive basin-scale description of the mid-depth velocity field has been provided by analyses of subsurface float data (Section 27.4; Lavender et al. 2000; Bower et al. 2002; Faure and Speer 2005). Although these observations provided important new insight into the O(100) km scale LSW subpolar circulation they are clearly not ideal. In particular, recent assessments in high-resolution GCMs conclude that "gridded float data cannot be used to infer a boundary current transport (*volume flux*) or to assess the boundary current strength in models" (our italics are inserted; Treguier et al. 2005). The error seems to be at least a factor of 2–3 and arises from insufficient coverage by floats of the narrow rapid boundary pathways.[15] More quantitative information on the structure, variability, and volume flux of the boundary current system has also been obtained from repeated ship-based sections of direct current observations and multi-year current meter mooring arrays at key locations along the continental slope off Labrador and Newfoundland. Much of this effort is part of a long-term

[14] The significant difficulty of extracting interannual volume flux variability was also apparent, however. Instrument failures and the need to re-deploy moorings annually caused changes in the array coverage which was particularly challenging.

[15] A qualitative feeling for the possible error can be gleaned from Fig. 6. This diagram shows 1000 simulated float-years in a smooth steady flow and still exhibits uneven coverage. In the real system only about 270 float-years are available (Faure and Speer, 2005) and the circulation is considerably more complicated.

German research effort (SFB460) and is now briefly described (see also Fischer et al. 2004; Schott et al. 2004; Schott et al. 2006).

27.5.1 Labrador Sea Boundary Current Volume Fluxes

The boundary current structure, flux, and variability has been observed at two important new transects off Labrador in the last decade. First, attention has focused on the Labrador Current at the AR7W section near Hamilton Bank (around 56°N). Pickart et al. (2002) reports a mean volume flux of 44 Sv from three Lowered Acoustic Doppler Current Profiler (LADCP) sections in winter 1997 at this line. This value is somewhat larger than the Dengler et al. (2006) estimate of 35 Sv at AR7W based on annual summer-time repeats of the section from 1996–2003. The agreement is reasonable given the various sources of variability and error, however. Dengler et al. (2006) also estimate that the ISOW and DSOW flow ($\sigma_\theta > 27.80$) across this section was 11.5 Sv while it was 17.2 Sv for CLSW and ULSW ($27.68 < \sigma_\theta < 27.80$). The second important new transect is at 53°N off northern Newfoundland where a moored current meter array was maintained for the period 1996–2005. The 1996–2003 average ISOW and DSOW flux here was 12.7 Sv, and for ULSW and CLSW it was 18.3 Sv (Dengler et al. 2006, based on LADCP data – see below). Intraseasonal variability at periods less than 60 days dominates the flow in the Labrador Current. Longer-term variability is much weaker, but annual mean currents from both Hamilton Bank and 53°N current meters show variations of the order of 10% (Häkkinen and Rhines 2004; Fischer et al. 2004). It appears that there was somewhat stronger deep flow after 1999 compared to the years 1996–1999 (Dengler et al. 2006), perhaps associated with low-frequency variations in the wind (see Section 27.6.1; Böning et al. 2006).

Figure 27.8a shows the average LADCP velocity normal to the 53°N section for 1996–2003.[16] The Labrador Current is seen as a succession of denser water masses draped over the continental slope; a shallow surface jet near the shelf-break gives way to LSW at depths greater than 500 m, then to eastern overflow water around 2,000 m, and finally to western overflow water deeper than about 3,000 m. For comparison, Fig. 27.8b shows mean results at the same section from the 1/12° resolution FLAME model (Böning et al. 2006; see Sections 27.3.2 and 27.6). The GCM field shows a surface jet that is too intensified compared to the data and hence a density structure that is too baroclinic (excessive downward sloping isopycnals near the Labrador coast, probably due to excessive model viscosity). The deep water is also too dense and the corresponding ISOW and DSOW volume fluxes exceed those from the average LADCP observations. Nevertheless, the basic structure of the model Labrador Current is quite accurate and the volume fluxes agree with data reasonably well overall. Offshore of the Labrador Current, direct recirculations appear as robust features

[16] Dengler et al. (2006) omit the data from 2005 because a vigorous eddy was present in the deep Labrador Current at that time. Including the 2005 data increases the fluxes by 10–20%, which is within the range of uncertainty.

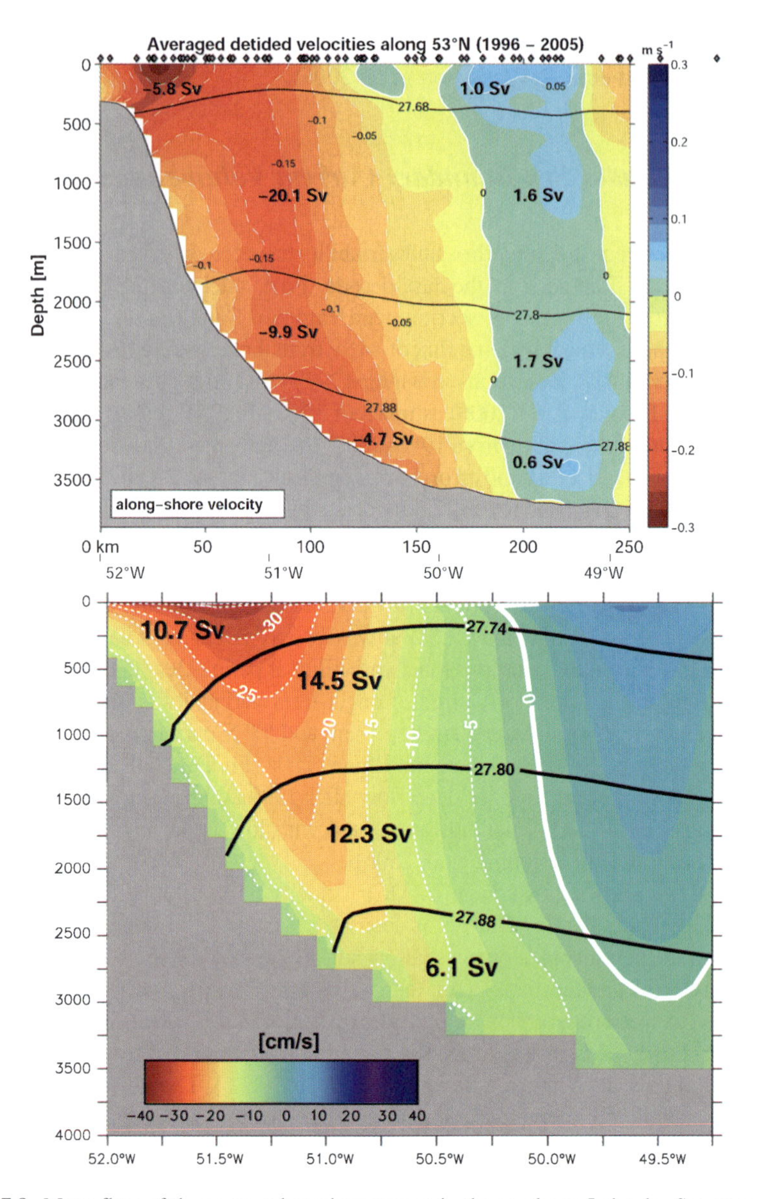

Fig. 27.8 Mean flow of the western boundary current in the southeast Labrador Sea near 53°N (a) Average LADCP section from seven cruises during 1996–2005. Color indicates speed normal to the section (m s^{-1}; positive to the northwest), large numbers indicate volume fluxes (Sv), and black lines indicate isopycnals (σ_θ). See Dengler et al. (2006) for details; this version of their Fig. 3 includes updated observations to 2005. (b) Equivalent cross-section from the 1/12° FLAME GCM (positive fluxes are to the southeast; from Böning et al. (2006) with permission of the American Geophysical Union).

in both the data and the GCM results (Section 27.4). The recirculating NADW volume flux is relatively weak, however; approximately 8 Sv at AR7W and 4.5 Sv at 53°N (for $\sigma_\theta > 27.68$; Dengler et al. 2006). The net NADW flux at 53°N leaving the Labrador Sea is around 26 Sv. As described in Section 27.2.1, this flux appears to feed three routes: eastward along the northern flank of the NAC then through the Charlie-Gibbs Fracture Zone into the eastern basin (Sy et al. 1997; Schott et al. 2004); from the interior Labrador Sea northward to the Irminger Sea, confirmed by individual float trajectories (Schott et al. 2004); and around Flemish Cap to the Grand Banks and the subtropical Atlantic. We now address this third NADW pathway (see Section 27.4 for LSW transport along the first two pathways).

27.5.2 Deep Water Export to the Subtropics

The passage of NADW round Grand Banks has long been thought as the primary export route to the subtropical circulation. The CFC field in Fig. 27.2 hints at this pathway, for example. During the WOCE period a section from the tail of the Grand banks towards Europe was established and a Canadian mooring array was present from 1993–1995 (the hydrographic line is called A2 and the mooring line is ACM-6; Clarke et al. 1998; Meinen and Watts 2000). This effort was later continued by German SFB activities from 1999 to 2005 (Schott et al. 2004, 2006). Figure 27.9 shows the average absolute velocity section from the mooring measurements. The southward NADW water flow along the western boundary (namely, the DWBC) is remarkably stable at 12 Sv over the 12-year period spanned by the data. From these measurements, the NADW denser than $\sigma_\theta = 27.74$ comprised 27% CLSW, 35% ISOW (called Gibbs Fracture Zone Water, GFZW, on Fig. 27.9; $27.80 \le \sigma_\theta \le 27.88$), and 38% DSOW. This composition is roughly consistent with Swift (1984) (Section 27.2.1), although with (relatively) less LSW and more DSOW. Offshore of the southward flow there is a much stronger, and variable, northward flow in the deep extension of the NAC and Mann Eddy as illustrated by individual current meter records and sections with direct LADCP observations of velocity (Schott et al. 2004). The volume flux in this branch reaches 51 Sv giving a net northward NADW flux of about 39 Sv at the mooring line. Published estimates of MOC strength at A2 are of 14–17 Sv (Ganachaud and Wunsch 2000; Schmitz and McCartney 1993), so one may anticipate substantial recirculation of the deep NAC between the eastern end of the mooring line and the mid-Atlantic Ridge.

The NADW export pathways into the subtropics have also been investigated by profiling floats seeded directly in the deep Labrador Current in the Labrador Sea (in LSW at 1,500 m during the period 1997–2000; Fischer and Schott 2002). Unexpectedly, the floats did not follow the DWBC into the subtropics, and none of the floats travelled farther south than the Grand Banks. This finding has stimulated studies with acoustically tracked isobaric RAFOS floats (A. Bower and S. Lozier, 2006, personal communication). RAFOS floats have the advantage that they do not need to surface to telemeter data, avoiding a potential cause of bias in the displacements of profiling floats. Preliminary results show that some floats follow

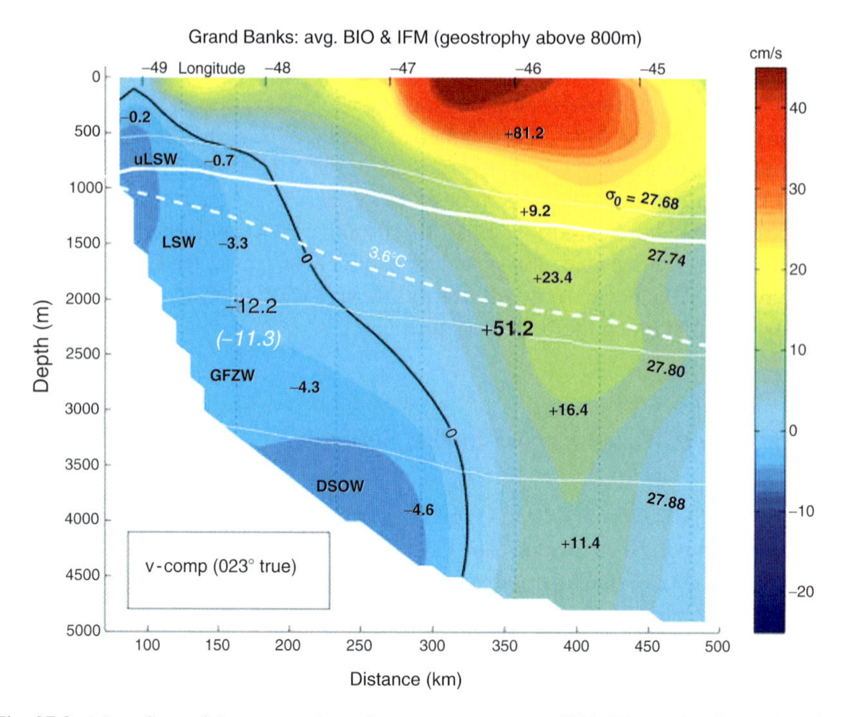

Fig. 27.9 Mean flow of the western boundary current system at WOCE mooring line ACM-6 off Grand Banks, Newfoundland (42–43°N). The measurements were made by the Bedford Institute of Oceanography (1993–1995) and the Institut für Meereskunde (1999–2005). The color indicates speed normal to the mooring line (cm s^{-1}; positive northwards), large numbers indicate volume fluxes (Sv), and white lines indicate isopycnals (σ_θ). See Schott et al. (2004) for details; this version of their Fig. 27.8 includes updated observations to 2005.

the classical DWBC path into the subtropics past Grand Banks but several recirculate. There are also some floats that enter the subtropics by interior paths avoiding the DWBC at Grand Banks. Recent results from simulated floats in the 1/12° resolution FLAME model also support these findings. Getzlaff et al. (2006) find that most floats (80%) seeded in the DWBC at 53°N are detrained from the boundary and stay in the subpolar gyre. Of those floats that reach the subtropics, only 60% do so in the DWBC; the other floats pass through the interior Newfoundland Basin and along the western flank of the mid-Atlantic Ridge. The numerical experiments show that eddy variability is very influential in dispersing floats from the boundary off Newfoundland, and that even RAFOS floats are biased compared to truly Lagrangian floats that descend by a few 100 m in the DWBC past Grand Banks. These intricate pathways are fascinating and a synthesis of results from real and synthetic floats is anticipated soon. This synthesis will, hopefully, resolve several important questions concerning the export routes of NADW from the subpolar North Atlantic.

27.5.3 Updated View of Deep Water Circulation

With these results in mind, an updated view of deep ($\sigma_\theta > 27.80$) circulation in the subpolar region is presented in Fig. 27.10. The figure focuses on DSOW and ISOW volume flux estimates from all available sources (expanding Fig. 27.1b with the recent findings described above and in Chapters 18 and 19). The picture of Greenland–Scotland overflow and entrainment has not changed radically since Dickson and Brown (1994) published their study (Section 27.2.2), although some details are different. Important new results are available downstream, however: At Cape Farewell recent hydrographic inverse studies by Lumpkin and Speer (2003) estimated a deep water volume flux of 13.1 Sv. This value agrees, perhaps luckily, with the early study of Clarke (1984) (Section 27.2.2). Mooring arrays are now in place at Cape Farewell to determine this flux, and its variability, with greater accuracy and precision (P. Lherminier and S. Bacon, 2006, personal communication). At Hamilton Bank offshore Labrador (near 56°N) Dengler et al. (2006) estimate the mean 1996–2005 volume flux to be 11.5 Sv and at 53°N their 1996–2003 estimate is 12.7 Sv (Section 27.5.1). The closeness of these three values suggests there are negligible changes in deep volume flux during passage through the Labrador Sea. At Grand Banks, the 1993–2005 estimate of Schott et al. (2006) is 8.9 Sv. As discussed in Section 27.5.2 above, this implies detrainment of 3–4 Sv from the DWBC at $\sigma_\theta > 27.80$ between 53°N and Grand Banks. This flux must either recirculate in the western subpolar gyre or enter the subtropics between the mid-Atlantic Ridge and the eastern end of the ACM-6 mooring line.[17] The Grand Banks array also shows strong, variable northward flux in the deep NAC with an average of about 27.8 Sv (Fig. 27.8). Lumpkin and Speer (2003) estimate a net flux of 9.4 Sv flowing south across A2 which is in reasonable agreement with the net MOC strength of 14–17 Sv cited above (recall that the flux with $\sigma_\theta < 27.80$ – about 4 Sv from Fig. 27.8 – plus the eastern basin part – perhaps 1 Sv – must also be added; see Section 27.5.2). The deep NAC must therefore also recirculate somewhere in the eastern Newfoundland Basin as shown on Fig. 27.10. This recirculation has not yet been definitively identified.

These recent measurements have substantially improved our knowledge of deep water circulation. Several obvious deficiencies persist, however. These include inadequate measurements of flow through Charlie-Gibbs Fracture Zone (and other deep gaps) and along the eastern side of the mid-Atlantic Ridge. The role of AABW is still obscure. There is also very little information available about the interior circulation away from boundaries. It seems inevitable that recirculation cells exist, but almost nothing is known about them. Given the advances in understanding LSW circulation since 1990 (Section 27.4), the schematic Fig. 27.10 is still clearly provisional.

[17] We reasonably assume here that there is no eastward flow of water back through Charlie-Gibbs Fracture Zone (Schott et al., 2004). Upwelling across $\sigma_\theta = 27.80$ is also assumed to be weak. This process must occur to some extent, perhaps over the western flank of the mid-Atlantic Ridge, but it seems doubtful that it would dominate the volume budget

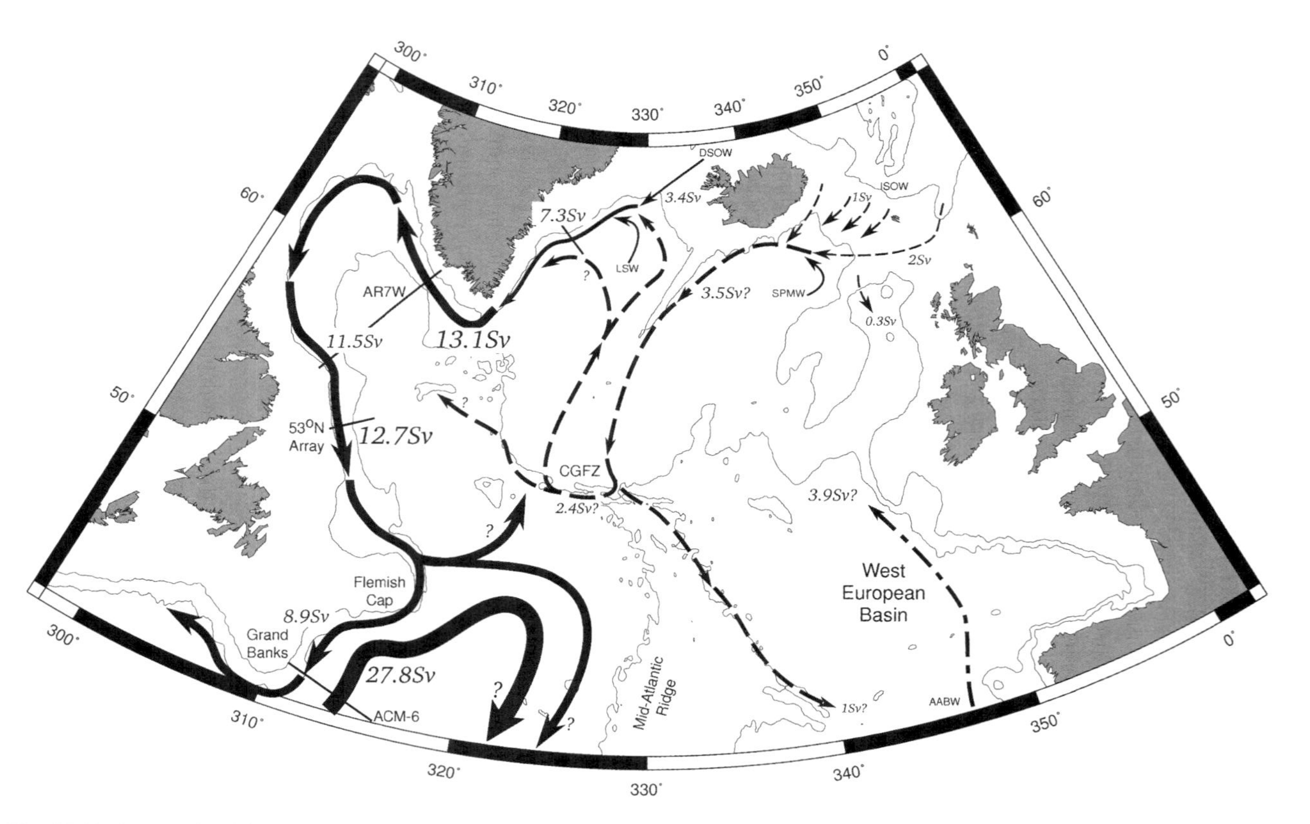

Fig. 27.10 Schematic of the deep circulation in the subpolar North Atlantic Ocean for, nominally, 1995–2005 ($\sigma_\theta > 27.80$, namely, ISOW and DSOW). The volume flux numbers of Dickson and Brown (1994) (Fig. 27.1b) have been updated with recent results from mooring arrays and repeat ship-based sections at the Greenland–Scotland ridge, in the Irminger and Labrador Seas, and at Grand Banks (see Sections 27.5.1–27.5.3). Volume flux (Sv) numbers are given where estimates exist; uncertain and/or old values carry question marks.

27.6 Evidence on Dynamical Mechanisms from Circulation Models

In this section on GCM results we consider how current North Atlantic models contribute evidence on the dynamical mechanisms controlling NADW in the subpolar gyre. We begin with a short discussion of how recent models compare with measurements of mean circulation and variability. Then we consider what lessons have been learned from GCMs about the three main questions of interest.

27.6.1 How Accurate are Modern Subpolar Circulation Models?

General circulation model studies of the subpolar gyre circulation are challenging because of the prime role of buoyancy forcing in the dynamics of the system and the difficulty of realistically simulating the small, but dynamically important, temperature and salinity contrasts in the water masses of the region. The important role of sea ice processes, the small deformation radius (typically 10–50 km), and uncertain air/sea fluxes, all exacerbate these challenges. A particular problem is the maintenance of the salinity structure in the central Labrador Sea, with implications for the conditioning of deep winter convection and the properties of LSW. Present state-of-the-art high resolution North Atlantic basin models (with around 1/10° horizontal grid spacing) typically have a positive bias in Labrador Sea salinity. They also have trouble accurately representing deep winter-time convection patterns in the Labrador Sea. It seems that the problem partly stems from the non-linear response of convection depth to cumulative buoyancy loss and partly from difficulties representing horizontal eddy fluxes which are important for both the preconditioning and re-stratification phases (see Section 27.6.2; Czeschel 2004).

Nevertheless, there is generally good agreement with the salient features of the observed low-frequency velocity field (Treguier et al. 2005). Specifically, 10–20 km horizontal resolution appears adequate to capture the main boundary current system, but 100–200 km resolution – used by current state-of-the-art climate GCMs – gives a gyre circulation that is significantly too weak. For example, three of the four high-resolution GCMs analysed by Treguier et al. (2005) (the 1/10° POP, the 1/6° CLIPPER, and the 1/12° FLAME models) simulate gyre volume fluxes just higher than 40 Sv in the Labrador Sea, while the streamfunction reaches 60 Sv in the 1/12° MICOM case. For the 1° CLIPPER model the value is just 26 Sv, well below the recent estimate from measurements of 35–44 Sv (see Section 27.5.1). One difference between these solutions is in the strength of the recirculation cell offshore of the Labrador Current which may explain the variations. Other work has showed a strong sensitivity of this feature to the representation of bathymetry (Käse et al. 2001; Kvaleberg and Haine 2006). Changes in the deep density field along the continental slope – formally represented by the coupling of baroclinicity and topography in the barotropic vorticity equation (the so-called JEBAR effect; Salmon 1998) –also lead to a response of the barotropic volume flux over a few years (see, e.g., Döscher et al. 1994; Eden and Willebrand 2001). Another likely reason that model gyre volume fluxes are too weak in GCMs that do not resolve mesoscale eddies is poor overflow simulation (e.g., Böning et al. 1996; Treguier et al. 2005). Accurate representation of the Nordic Seas outflows includes the water mass transformation processes north of the sills as well as the overflow and entrainment processes. Adequate resolution of the overflow processes probably requires grid spacing near 1 km and 10 m in the horizontal and vertical directions, at least in the vicinity of overflow descent (see Chapter 22; Legg et al. 2004). This resolution is still well beyond current resources for uniformly resolved basin-scale GCMs, but is likely possible in the coming decade. Interestingly, the role of the thermohaline processes that determine the Nordic Seas overflows and the winter deep convection in the Labrador Sea appears relatively large compared to the wind stress in setting the mean shape and volume flux of the gyre (Böning et al. 2006). Indeed, analyses of GCM vorticity budgets suggest that the subpolar gyre is not governed by the wind through simple Sverdrup dynamics (Bryan et al. 1995).

Modern eddy-resolving GCMs seem to capture important aspects of the observed variability in subpolar circulation, as well as the time-averaged flow. Different model hindcasts forced with air/sea flux products from atmospheric reanalyses show striking similarities in the simulated changes in subpolar gyre barotropic streamfunction over the last few decades (Eden and Willebrand 2001; Hátún et al. 2005; Treguier et al. 2005; Böning et al. 2006). In particular, they typically show a decrease in the gyre strength during the latter half of the 1990s of about 8 Sv (or 15% of the long-term mean), then an increase again by 4–5 Sv to 2003. These changes are consistent with recent in situ field results of Dengler et al. (2006) (see also Sections 27.5 and 27.6.4). Although grid spacing near 10 km is needed to capture the basic circulation field, accurate variability can be simulated with a 1/3° resolution model (Böning et al. 2006). This robustness in GCM behaviour points to a common dynamical origin that is well represented by the models. In particular, Eden and Willebrand (2001) noted that an increase of deep convection was followed, 2–3 years later, by an increase in subpolar gyre volume flux in their 4/3° resolution model. Higher-resolution hindcasts show similar responses (for example, in the 1/6° CLIPPER model presented by Treguier et al. 2005 and in both the 1/3° and 1/12° North Atlantic FLAME models discussed by Böning et al. 2006).

These results confirm the following views: First, changes in GCM gyre volume flux are mainly linked to thermodynamic changes rather than variations in mechanical forcing. Second, GCMs with around 1/10° resolution seem to capture both the basic mean and time-varying components of the subpolar North Atlantic circulation. With these findings in mind, we now consider how GCM results bring evidence to bear on the key questions of this chapter.

27.6.2 What Controls Labrador Sea Water Formation Rates?

Model intercomparison studies assessing the depth and area of deep winter mixing in the Labrador Sea have repeatedly shown sensitivity to the details of air/sea fluxes, grid configuration, and parameterization of subgrid-scale mixing processes (Willebrand et al. 2002; Treguier et al. 2005; see also Section 27.3.2). Different model hindcasts (for example, Häkkinen 1999; Eden and Willebrand 2001; Böning et al. 2003; Mizoguchi et al. 2003; Bentsen et al. 2004; Gerdes et al. 2005) generally concur on the main features of decadal changes, however, with enhanced convection during the early 1970s, mid-1980s, and, most strongly, during the first half of the 1990s (see Section 27.3.3 and Fig. 27.5). This conspicuous agreement across a wide range of model configurations reinforces the notion that decadal variability in LSW formation is controlled by the atmospheric conditions associated with the state of the NAO (and is in sharp contrast to the situation in the Greenland Sea, as demonstrated in the model analysis of Gerdes et al. 2005). Both the increased surface buoyancy fluxes and increased doming of isopycnals during positive NAO phases contribute to the robust correlation between NAO index and deep Labrador Sea convection. The main factor capable of disturbing this atmospheric control is the import of major freshening pulses from the Arctic (for example, Houghton and Visbeck 2002). A clear example of this process is the 1983/1984 event which appeared to have suppressed or, at least, delayed deep convective mixing, even in the presence of positive NAO forcing at that time (Curry et al. 1998; Gerdes et al. 2005). This source of variability is excluded in the basin-scale GCMs that resort to specifying climatological hydrography in the Nordic Seas, however.

A related process influencing LSW convection depths, and hence formation rates, is lateral buoyancy mixing in the Labrador Sea. In particular, "Irminger eddies" carry a large heat flux from the buoyant boundary current system into the deep convection region (Lilly et al. 2003; Katsman et al. 2004).[18] The narrow continental shelf and steep continental slope off southwest Greenland is a well-known source of these eddies (for example, Cuny et al. 2002; see also Chapter 23). The eddy buoyancy flux can vary according to changes in eddy intensity and also to the presence of fresh pulses in the East and West Greenland Currents, as mentioned above. Figure 27.11 reveals the importance of this flux. It shows a snapshot of the surface currents and the mixed layer depth during a winter convection phase (March) in the 1/12° FLAME model. Clearly visible is the strong Irminger/East

[18] The boundary current structure off southwest Greenland, and its interaction with the interior Labrador Sea, is not entirely understood. Relatively warm and salty water is imported from the Irminger Sea in a jet following the continental slope and within about 100 km of the shelf-break. This jet is often called the Irminger Current and probably forms the major source of the Irminger eddies. Inshore of this feature is a strong surface front associated with the shelf-break that transitions to cold, fresh, and even more buoyant, water of Arctic origin. The jet associated with this front is often called the West Greenland Current. The shelf circulation remains obscure, but of key importance because it is the primary meltwater conduit (see Chapter 28).

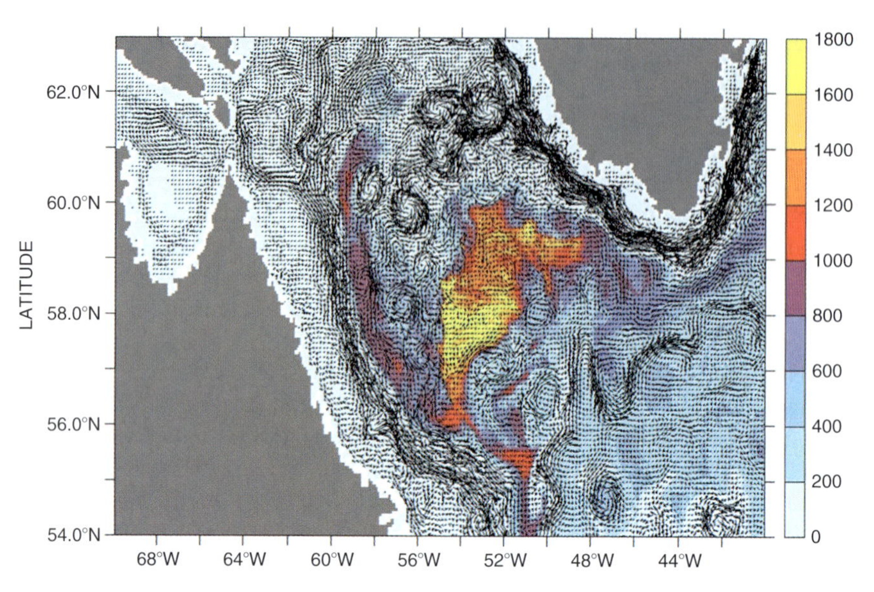

Fig. 27.11 Snapshot (March) of simulated near-surface currents and mixed layer depth in the Labrador Sea, illustrating the stabilizing effect of the Irminger eddies, effectively setting the northern extent of the deep convection region. From the 1/12° FLAME simulation of Czeschel (2004).

Greenland/West Greenland Current system plus a train of four or five coherent anticyclonic vortices originating off southwest Greenland. Scrutiny of the figure suggests that this wedge of eddy activity may be the main factor that determines the northern extent of the deep convection region. In a sequence of numerical experiments, Czeschel (2004) examined the effect of a varying intensity of these eddies by perturbing the strength of the boundary current in the model (by artificially changing the wind stress forcing over the subpolar gyre). He found that an increased (decreased) boundary current instability and eddy formation led to a weaker (stronger) convection depth and LSW formation. This result is consistent with the idealized model of Katsman et al. (2004), who point out the important role of Irminger eddies in restratifying the deep convection region in spring.

These are promising results, although it is unclear if 1/10° resolution is enough to adequately capture the eddy buoyancy flux or the details of the boundary current system. Indeed, transfer of fresh Arctic waters off the shelfbreak into the deep Irminger and Labrador Seas appears to involve O(1) km scale cascades (Pickart et al. 2005) that presumably require grid spacing near 1/100° to be properly resolved. Also, air/sea flux products to force ocean models are unavailable at the scales of the deep convection plumes themselves. Some evidence suggests that the ocean is sensitive to high frequencies and wavenumbers in forcing, however (for example, the Greenland Tip Jet intermittently drives very deep mixed layers southeast of Cape Farewell; Pickart et al. 2003, Chapter 26). Clearly, LSW formation rates are governed in complex ways by local and remote buoyancy sources, and processes on scales between 1 and 100 km are competing for control. Progress on understanding these processes from models and observations remains an important challenge.

27.6.3 *What are the Transport Pathways and Timescales in Modern GCMs?*

Diagnosing transport pathways and timescales in circulation models is a challenging task. The main reason is that (advective and diffusive) transport is a complicated and subtle diagnostic of a model's flow and mixing parameterisations. Moreover, consistent observational estimates of transport pathways and timescales are not readily available for comparison (see Section 27.4). Nevertheless, model simulations exhibit an LSW export from the convection region along the same basic routes as inferred from observations; along the western boundary, into the Irminger Basin, and through the Charlie-Gibbs Fracture Zone into the eastern basin. For example, Fig. 27.12

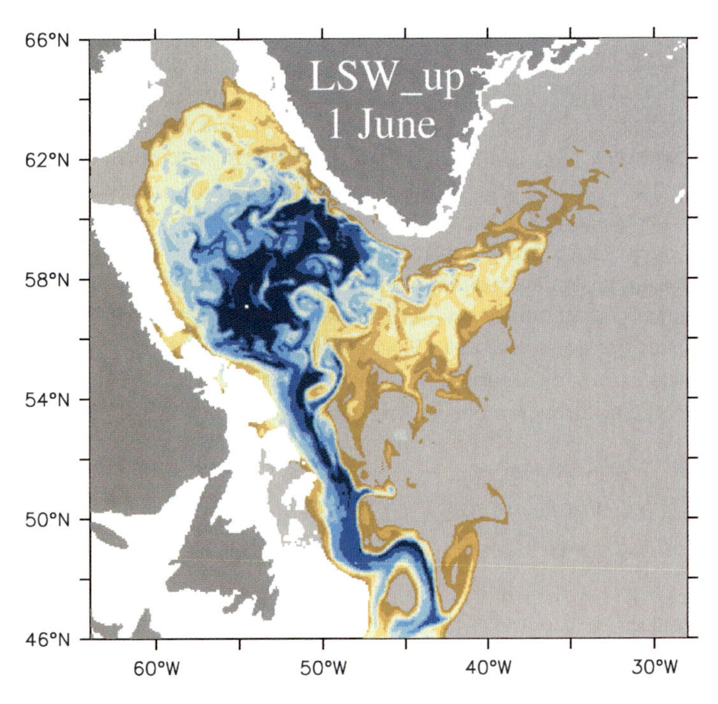

Fig. 27.12 Initial spreading of LSW illustrated by the evolution of an idealized "convection tracer" during spring and summer in a 1/12° FLAME simulation. The tracer was initialized within the deep winter mixed layer in the central Labrador Sea and is shown here at the start of summer. Blue indicates high concentrations and brown indicates low concentrations (From Czeschel 2004).

illustrates the initial LSW spreading in the 1/12° FLAME model by depicting the distribution of an idealized "convection tracer" a few months after deep mixing. The tracer distribution emphasizes the dominant export route along the western boundary, and a second broader path apparently governed by eddying flows into the Irminger Basin. A third pathway, along the (lower) NAC into the eastern basin, develops near Flemish Cap where a considerable fraction of the boundary flow is entrained into the NAC by the strong eddy activity in that region.

The convection tracer is revealing in qualitative ways, but such ideal tracers have not yet been consistently applied in state-of-the-art Atlantic GCMs.[19] There has been more attention on simulating anthropogenic transient tracer fields (most commonly CFCs) at non-eddy-resolving or eddy-permitting resolution. Successful simulation of these tracers provides reasonable confidence in the model transport, although only a subset of the model transport pathways and timescales are being probed (Zhang et al. 2005). A few papers have reported careful comparisons between model transient tracer fields and data (see Zhao et al. 2006 and references therein). For example, Böning et al. (2003) found a general agreement with the observed CFC patterns, but noted a much too weak spreading into the eastern basin in their 1/3° FLAME solution (see also comments in Section 27.3.1). This result highlighted a potential effect of the drift towards higher salinities described in Section 27.6.1; model LSW that is too dense, and therefore too deep, may no longer be able to negotiate the passages of the mid-Atlantic Ridge. Haine et al. (2003) and Gray and Haine (2001) performed detailed comparisons between a 4/3° resolution model prediction and CFC data. They performed an inverse calculation to determine the air/sea CFC flux that gave the best fit to data in each of a sequence of GCM calculations. They found reasonably consistent agreement with data in all their experiments, indicating that the primary transport pathways are robustly captured even at 4/3° resolution. Both the flow smoothness and the inferred LSW formation rates varied significantly between experiments, however (see Section 27.3.3). This suggests that matching the CFC data reasonably well is not a stringent test of a model's circulation, even though valuable constraints on transport pathways and timescales are being exerted. Clearly, the next steps are to: (i) study transient tracer dispersal in eddy-resolving GCMs and compare to data, and (ii) diagnose transport explicitly in such models using ideal tracers such as the TTD.

27.6.4 What Controls the Volume Flux of NADW into the Subtropics and How Is this Related to the MOC?

Finally, we consider how GCMs inform us about control of NADW volume flux into the subtropics and the relation to the MOC. We highlight a recent paper by Böning et al. (2006) who address this question directly and make links between

[19] The transit-time distribution from a 1/10o class of Atlantic model has not yet been published, for example, although the ideal convection tracer in Fig. 12 is similar to the Green's function tracers used to diagnose the TTD.

several of the topics considered above (see their bibliography for other relevant studies). Böning et al. (2006) show intriguing GCM evidence that connects decadal variations in subpolar sea level with integrated horizontal circulation of the subpolar gyre, the anomalies in LSW formation, the 53°N boundary volume flux, and the subtropical MOC strength at 26.5°N.

First, they show how the 1992–2002 variations in central subpolar sea level identified by Häkkinen and Rhines (2004) from altimetry are reproduced by the 1/12° FLAME model (see Section 27.6.1). Next, they show that this subpolar sea level co-varies with the barotropic streamfunction in the central Labrador Sea. They show that 1/3° resolution is adequate to capture the decadal variations in barotropic streamfunction seen at 1/12° resolution through the 1990s (Section 27.6.1). Both air/sea heat fluxes and air/sea stresses have comparable importance for driving these changes in the 1/3° model between 1960 and 2000. Böning et al. (2006) then claim that the western boundary current volume flux at 53°N co-varies with the barotropic streamfunction fluctuations. The deep part of this western boundary current – that is, the NADW – is controlled mainly by air/sea fluxes and less by wind stress. Therefore, the NADW boundary volume flux follows anomalies in central Labrador Sea winter mixed layer depth (and presumably LSW formation rate) with a lag of 1–2 years. At 26.5°N in the subtropical Atlantic the imprint of these 53°N boundary flux changes are seen in the MOC after about 1 year (via a wave mechanism, not advective/diffusive transport of anomalies). Subtropical wind forcing drives 26.5°N MOC changes that are about twice as strong, but the subpolar signals remain coherent.

These interesting links hint at dynamical mechanisms that require more detailed study. If robust, they promise a valuable way in which subpolar gyre strength and NADW export could be monitored via altimetry and the 53°N section. Moreover, the causal chain suggests that the 26.5°N MOC strength may be partly predictable from such measurements.

27.7 Summary and Outlook

The main findings on NADW formation since 1990 in the context of the questions in Section 27.1 are:

- There is now a greatly expanded set of estimates of LSW formation rate and its variability (Section 27.3). These estimates are shown in Table 27.2 and Fig. 27.5. There is now recognition that defining "formation rate" is a subtle issue that requires care. Although there has not yet been consistent use of this diagnostic, or robust uncertainty analyses, a more reliable formation rate time series is starting to emerge. The well-known variations in Labrador Sea deep convection are clearly related to formation rate changes in Fig. 27.5 and the NAO, for example.
- Much has been learned about the mechanisms of LSW formation rate variability from numerical circulation models. Figs. 27.3 and 27.4 show in detail how the 1/12° FLAME model forms LSW, for instance (Section 27.3.2). Although accurate simulation of LSW is challenging (Section 27.6.1), GCMs point to the important role of

the long-period air/sea buoyancy forcing and the import of buoyant water to the Labrador Sea from the Irminger Sea and Greenland shelf (Section 27.6.2).

- A much more detailed understanding of the pathways of LSW circulation through the subpolar North Atlantic is now available (Section 27.4). Two new instrumental methods have permitted this substantial advance: profiling floats and CFC tracers (Section 27.3.1). They have provided a modern picture of mid-depth circulation (at least during 1996–2001; Fig. 27.6) and LSW spreading (Fig. 27.2) that has swept away the old view circa 1990 (Fig. 27.1a). Details in the interior flow at O(100) km scales have been revealed and new questions about the variability of the LSW recirculation through the subpolar gyre are now being raised. We also have a much better understanding of the rates of LSW circulation (Section 27.4). As with LSW formation rates, a fundamentally robust notion of spreading timescale was missing until recently. This idea – the transit-time distribution – is now in place and formerly disparate timescale estimates are being reconciled (Section 27.4). Full exploitation of the TTD idea has not been realised yet and GCM studies to understand LSW transport dynamics are still at an early stage.
- Important progress has been achieved in observing and understanding the deep Labrador Current in the southeast Labrador Sea at the AR7W line (56°N) and 53°N (Section 27.5.1). Seven-year-long current records have been acquired yielding unprecedented insight into the mean structure of the Labrador Current, its variability, and volume flux. Recent modelling work suggests that the NADW volume flux at these sections is a good proxy for overall subpolar gyre strength, and may predictably lead subtropical Atlantic MOC changes by about 1 year (Section 27.6.4).
- Important progress has also been achieved by observing the boundary current system off the Grand Banks (Section 27.5.2). Direct current measurements have been made there for most years in the period 1993–2005. Accurate estimates of NADW composition, volume flux and variability are available, but the data clearly show the dominant part played by the northward NAC at this section. Large recirculation of the deep NAC seems inevitable in the eastern Newfoundland Basin. The classical view of NADW export to the subtropics exclusively via Grand Banks in the DWBC (Fig. 27.1a) is being revised; floats and model experiments show multiple pathways for NADW export and questions about NADW export to the subtropics persist.
- A new schematic circulation diagram for ISOW and DSOW ($\sigma_\theta > 27.80$) has been drawn (Fig. 27.10, Section 27.5.3). Compared to the view circa 1990 (Fig. 27.1b), the number of reliable boundary flux estimates has more or less doubled with important new measurements in the Labrador Sea and at Grand Banks. Nevertheless, basic questions about the mean circulation, especially away from boundaries, remain.

Sustained observation from space and by *in situ* instruments will continue to reveal the nature of NADW formation, recirculation, and export to the subtropics. Particularly important is accurate coverage of sea-surface height, surface temperature, surface wind speed, and sea ice properties. Continued annual repeats of the

AR7W section (including transient tracer data), sampling by O(10) ARGO floats, and moored arrays at, for example, 53°N and the central Labrador Sea are equally important. Another priority is to repeat sections outside the Labrador Sea periodically (for example, WOCE lines AR7E, A02, A25, A16N). These measurements will provide insight into the response to anticipated changes in air/sea forcing in the coming years. Together with the rapidly improving ocean synthesis systems we expect a significant decrease in uncertainties in the key NADW diagnostics discussed here.

Scientifically, the main challenges for future work are as follows: For LSW, the challenges revolve around synthesising variability in formation rate, variability in properties of newly formed LSW, and variability in mid-depth subpolar circulation. We now have an excellent picture of the evolving state of LSW since 1990 in the deep convection regions where it is formed, mainly from the annually repeated AR7W transect and the long-term mooring time series at Ocean Weather Station Bravo (Chapter 24). We also have a much improved view of LSW circulation and spreading pathways. Progress has been made fitting these pieces of evidence together, but so far the LSW circulation has been treated as steady and anomalies in LSW formation and properties have been treated as dynamically passive. So, the next challenge is to observe and understand how the mid-depth subpolar circulation is changing at the scale of the Labrador Sea recirculation cells (O(100) km). These changes in flow must be related to changes in deep convective activity, LSW formation rate, properties of newly formed LSW, and propagation of LSW into the interior subpolar gyre. The dynamical relationships must be elucidated and the predictable mechanisms identified. The relative importance of the Irminger Sea versus the Labrador Sea as a source of LSW should also be determined. Resolving these issues will allow us to say how LSW anomalies are caused, how they propagate through the subpolar gyre, and how they are altered through mixing with ambient waters in the mid-depth subpolar North Atlantic.

For overflow-derived NADW (ISOW and DSOW), the challenges are more basic. Although significant progress has been made in understanding the boundary current system off Labrador and Newfoundland since 1990, the level of detail is still uneven and little is known about the O(100) km resolution deep and abyssal flow patterns, especially in the interior. There is no geostrophic streamfunction for the near-bottom currents unlike the mid-depth circulation, for example. The interaction between the boundary currents and the interior flow is unknown. We also have only an immature understanding of the exchange of eastern and western overflow waters through the mid-Atlantic Ridge or the roles of AABW and the deep NAC. Therefore, an improved picture of the basic time-averaged deep NADW circulation at O(100) km scales is needed before we can confidently address the question of how subpolar NADW anomalies are generated and evolve. The issue of variability in the basin-wide deep circulation can then also be tackled.

Finally, we need to better understand the processes controlling NADW (including LSW) transit around Newfoundland. Mid-depth floats are surprisingly reluctant to enter the subtropics in the DWBC past Grand Banks. This pathway is clear in surveys of NADW property distributions, however, and has been central to understanding of the deep North Atlantic circulation for decades. The origin of this seeming contradiction – perhaps unrecognized pathways, low-frequency variability in the flow, or inadequate sampling by the floats – must be identified in the near future. Only then can we talk with confidence about how deep-western boundary current NADW is drawn from the reservoirs of the subpolar North Atlantic and how its anomalies are inherited.

Acknowledgments We are supported by the Physical Oceanography program at NSF (grants 0136327 and 0326670), Sondersforschungbereich SFB460 of the German Science Foundation, and the 'NORDATLANTIK' BMBF program. Samar Khatiwala, Robert Marsh, Paul Myers, and Rüdiger Gerdes kindly made their LSW formation rate data available for Fig. 27.5. Vincent Faure supplied the geostrophic streamfunction data for the trajectory calculation in Fig. 27.6. Lars Czeschel provided Figs. 27.11 and 27.12 from his Ph.D. thesis.

References

Azetsu-Scott, K., E. P. Jones, I. Yashayaev, and R. M. Gershey (2003), Time series study of CFC concentrations in the Labrador Sea during deep and shallow convection regimes (1991–2000), J. Geophys. Res., 108, doi:10.1029/2002JC001,317.

Baehr, J., K. Keller, and J. Marotzke (2007), Detecting potential changes in the meridional overturning circulation at 26°N in the Atlantic, Clim. Change, doi:10.1007/S10584-006-9153-7.

Beckmann, A., and R. Döscher (1997), A method for improved representation of dense water spreading over topography in geopotential-coordinate models, J. Phys. Oceanogr., 27, 581–591.

Bentsen, M., H. Drange, T. Furevik, and T. Zhou (2004), Simulated variability of the Atlantic meridional overturning circulation, Clim. Dyn., 22, 701–720.

Bersch, M., J. Meincke, and A. Sy (1999), Interannual thermohaline changes in the northern North Atlantic 1991–1996, Deep Sea Res., Part II, 46, 55–75.

Böning, C. W., F. O. Bryan, W. R. Holland, and R. Döscher (1996), Deep water formation and meridional overturning in a high resolution model of the North Atlantic, J. Phys. Oceanogr., 26, 1141–1164.

Böning, C. W., M. Rhein, J. Dengg, and C. Dorow (2003), Modeling CFC inventories and formation rates of Labrador Sea Water, Geophys. Res. Lett., 30, 1050, doi:10.1029/2002GL014,855.

Böning, C. W., M. Scheinert, J. Dengg, A. Biastoch, and A. Funk (2006), Decadal variability of subpolar gyre transport and its reverberation in the North Atlantic overturning, Geophys. Res. Lett., 33, L21S01, doi:10.1029/2006GL026,906.

Bower, A. S. and H. D. Hunt (2000), Lagrangian observations of the Deep Western Boundary Current in the North Atlantic Ocean. Part I: Large-scale pathways and spreading rates, J. Phys. Oceanogr., 30, 764–783.

Bower, A. S., B. le Cann, T. Rossby, W. Zenk, J. Gould, K. Speer, P. L. Richardson, M. D. Prater, and H.-M. Zhang (2002), Directly measured middepth circulation in the northeastern North Atlantic Ocean, Nature, 419, 603–607.

Brandt, P., A. Funk, L. Czeschel, C. Eden, and C. Böning (2007), Ventilation and transformation of Labrador Sea Water and its rapid export in the deep Labrador Current, J. Phys. Oceanogr., 73, 946–961.

Bryan, F. O., C. W. Böning, and W. R. Holland (1995), On the midlatitude circulation in a high resolution model of the North Atlantic, J. Phys. Oceanogr., 25, 289–305.

Bryden, H. L., H. R. Longworth, and S. A. Cunningham (2005), Slowing of the Atlantic meridional overturning circulation at 25°N, Nature, 438, doi:10.1038/nature04,385.

Clarke, R. A. (1984), Transport through the Cape Farewell-Flemish Cap section, Rapp. P.-v. Reun. Cons. Int. Explor. Mer., 185, 120–130.

Clarke, R. A. and J. C. Gascard (1983), The formation of Labrador Sea Water: Part 1, large scale processes, J. Phys. Oceanogr., 13, 1779–1797.

Clarke, R. A., R. M. Hendry, and I. Yashayaev (1998), A western boundary current meter array in the North Atlantic near 42°N, Int. WOCE Newsl., 33, 33–34.

Cunningham, S. A. and T. W. N. Haine (1995), On Labrador Sea Water in the Eastern North Atlantic. Part I: A synoptic circulation inferred from a minimum in potential vorticity, J. Phys. Oceanogr., 25, 649–665.

Cuny, J., P. B. Rhines, P. P. Niiler, and S. Bacon (2002), Labrador Sea boundary currents and the fate of the Irminger Sea Water, J. Phys. Oceanogr., 32, 627–647.

Cuny, J., P. B. Rhines, F. Schott, and J. Lazier (2005), Convection above the Labrador Continental Slope, J. Phys. Oceanogr., 35, 489–511.

Curry, R. G., M. S. McCartney, and T. M. Joyce (1998), Oceanic transport of subpolar climate signals to mid-depth subtropical waters, Nature, 391, 575–577.

Czeschel, L. (2004), The role of eddies for the deep water formation in the Labrador Sea, Ph.D. thesis, Kiel University, Leibniz-Institut für Meereswissenschaften, 101 pp.

Dengler, M., J. Fischer, F. A. Schott, and R. Zantopp (2006), Variability of the Deep Western Boundary Current east of the Grand Banks, Geophys. Res. Lett., 33, doi:10.1029/2006GL026,702.

Dickson, R. R. and J. Brown (1994), The production of North Atlantic Deep Water: Sources, rates and pathways, J. Geophys. Res., 99, 12319–12341.

Dickson, R. R., E. M. Gmitrovic, and A. J. Watson (1990), Deep-water renewal in the northern North Atlantic, Nature, 344, 848–850.

Dickson, R. R., J. Lazier, J. Meincke, P. Rhines, and J. Swift (1996), Long-term coordinated changes in the convective activity of the North Atlantic, Prog. Oceanogr., 38, 241–295.

Dietrich, G. (1969), A new atlas of the northern North Atlantic Ocean, Deep Sea Res., supplement to 16, 31–34.

Dietrich, G., K. Kalle, W. Kraus, and G. Siedler (1980), General oceanography, an introduction, 2nd ed., Wiley, New York, 626 pp.

Döscher, R., C. W. Böning, and P. Herrmann (1994), Response of circulation and heat transport in the North Atlantic to changes in thermohaline forcing in northern latitudes: A model study, J. Phys. Oceanogr., 24, 2306–2320.

Eckart, C. (1948), An analysis of the stirring and mixing processes in incompressible fluids, J. Mar. Res., 7, 265–275.

Eden, C. and C. Böning (2002), Sources of eddy kinetic energy in the Labrador Sea, J. Phys. Oceanogr., 32, 3346–3363.

Eden, C. and J. Willebrand (2001), Mechanism of interannual to decadal variability of the North Atlantic circulation, J. Climate, 14, 29–70.

Faure, V. and K. Speer (2005), Labrador Sea Water Circulation in the Northern North Atlantic Ocean, Deep Sea Res., Part II, 52, 565–581.

Fischer, J. and F. A. Schott (2002), Labrador Sea Water tracked by profiling floats – From the boundary current into the open North Atlantic, J. Phys. Oceanogr., 32, 573–584.

Fischer, J., F. A. Schott, and M. Dengler (2004), Boundary circulation at the exit of the Labrador Sea, J. Phys. Oceanogr., 34, 1548–1570.

Ganachaud, A. and C. Wunsch (2000), Improved estimated of global ocean circulation, heat transport and mixing from hydrographic data, Nature, 408, 453–457.

Gerdes, R., J. Hurka, M. Karcher, F. Kauker, and C. Kööberle (2005), Simulated history of convection in the Greenland and Labrador Seas 1948–2001, AGU, Geophysical Monograph Series 158, 370 pp.

Getzlaff, K., C. Böning, and J. Dengg (2006), Lagrangian perspectives of deep water export from the subpolar North Atlantic, Geophys. Res. Lett., 38, doi:10.1029/2006GLO26470.

Gray, S. L. and T. W. N. Haine (2001), Constraining a North Atlantic ocean general circulation model with chlorofluorocarbon observations, J. Phys. Oceanogr., 31, 1157–1181.

Haine, T. W. N. (2006), On tracer boundary conditions for geophysical reservoirs: How to find the boundary concentration from a mixed condition, J. Geophys. Res., 111, C05003, doi:10.1029/2005JC003,215.

Haine, T. W. N. (2007), What did the Viking discoverers of America know of the North Atlantic environment? Weather, in press.

Haine, T. W. N. and T. M. Hall (2002), A generalized transport theory: Water-mass composition and age, J. Phys. Oceanogr., 32, 1932–1946.

Haine, T. W. N., K. J. Richards, and Y. Jia (2003), Chlorofluorocarbon constraints on North Atlantic ocean ventilation, J. Phys. Oceanogr., 33, 1798– 1814.

Häkkinen, S. (1999), Variability of the simulated meridional heat transport in the North Atlantic for the period 1951–1993, J. Geophys. Res., 104, 10991–11007.

Häkkinen, S. and P. B. Rhines (2004), Decline of subpolar North Atlantic circulation during the 1990s, Science, 304, 555–559.

Hall, T. M., T. W. N. Haine, M. Holzer, D. A. LeBel, F. Terenzi, and D. W. Waugh (2007), Ventilation rates estimated from tracers in the presence of mixing, J. Phys. Oceanogr., in press.

Hátún, H., A. B. Sandø, H. Drange, B. Hansen, and H. Valdimarsson (2005), Influence of the Atlantic subpolar gyre on the thermohaline circulation, Science, 309, 1841–1844.

Hirschi, J., P. D. Killworth, and J. R. Blundell (2007), Subannual, seasonal and interannual variability of the North Atlantic meridional overturning circulation, J. Phys. Oceanogr., 37, 1246–1265.

Houghton, R. W. and M. Visbeck (2002), Quasi-decadal salinity fluctuations in the Labrador Sea, J. Phys. Oceanogr., 32, 687–701.

Isemer, H.-J. and L. Hasse (1987), The Bunker climate atlas of the North Atlantic ocean, Vol. 2: Air-sea interactions, Springer-Verlag New York Inc., New York, NY 218 pp.

Käse, R. H., A. Biastoch, and D. B. Stammer (2001), On the mid-depth circulation in the Labrador and Irminger seas, Geophys. Res. Lett., 28, 3433–3436.

Katsman, C. A., M. A. Spall, and R. S. Pickart (2004), Boundary current eddies and their role in the restratification of the Labrador Sea, J. Phys. Oceanogr., 34, 1967–1983.

Khatiwala, S. and M. Visbeck (2000), An estimate of the eddy-induced circulation in the Labrador Sea, Geophys. Res. Lett., 27, 2277–2280.

Khatiwala, S., P. Schlosser, and M. Visbeck (2002), Rates and mechanisms of water mass transformation in the Labrador Sea as inferred from tracer observations, J. Phys. Oceanogr., 32, 666–686.

Kieke, D., M. Rhein, L. Stramma, W. M. Smethie, J. L. Bullister, and D. A. LeBel (2007), Changes in the pool of Labrador Sea Water in the subpolar North Atlantic, Geophys. Res. Lett., 34, L06605, doi:10.1029/2006GL028959.

Kieke, D., M. Rhein, L. Stramma, W. M. Smethie, D. A. LeBel, and W. Zenk (2006b), Changes in the CFC inventories and formation rates of Upper Labrador Sea Water, 1997–2001, J. Phys. Oceanogr., 36, 64–86.

Koltermann, K. P., A. V. Sokov, V. P. Tereschenko, S. A. Bobroliubov, K. Lorbacher, and A. Sy (1999), Decadal changes in the thermohaline circulation of the North Atlantic, Deep Sea Res., Part II, 46, 109–138.

Kvaleberg, E. and T. W. N. Haine (2007), Recirculating flow in the Labrador and Irminger Seas: Impact of bathymetry, J. Phys. Oceanogr., under revision.

Kvaleberg, E., T. W. N. Haine, and D. W. Waugh (2007), Labrador Sea Water transport rates and pathways in the subpolar North Atlantic ocean, J. Geophys. Res., under revision.

Lavender, K. L., R. E. Davis, and W. B. Owens (2000), Mid-depth recirculation observed in the interior Labrador and Irminger Seas by direct velocity measurements, Nature, 407, 66–69.

Lavender, K. L., W. B. Owens, and R. E. Davis (2005), The mid-depth circulation of the subpolar North Atlantic Ocean as measured by subsurface floats, Deep Sea Res., Part I, 52, 767–785.

Lazier, J. R. N. (1980), Oceanographic conditions at O.W.S. Bravo, 1964–1974, Atmosphere-Ocean, 18, 227–238.

Lazier, J. R. N., R. Hendry, A. Clarke, I. Yashayaev, and P. Rhines (2002), Convection and restratification in the Labrador Sea, Deep Sea Res., Part I, 49, 1819–1835.

Legg, S., R. W. Hallberg, and J. B. Girton (2006), Comparison of entrainment in overflows simulated by z-coordinate, isopycnal and nonhydrostatic models, Ocean Modelling, 11, 69–97.

Lilly, J. M., P. B. Rhines, F. Schott, K. Lavender, J. Lazier, U. Send, and E. D'Asaro (2003), Observations of the Labrador Sea eddy field, Prog. Oceanogr., 59, 75–176.

Lumpkin, R. and K. Speer (2003), Large-scale vertical and horizontal circulation in the North Atlantic ocean, J. Phys. Oceanogr., 33, 1902–1920.

Marsh, R. (2000), Recent variability of the North Atlantic thermohaline circulation inferred from surface heat and freshwater fluxes, J. Climate, 13, 3239–3260.

Marsh, R., S. A. Josey, A. J. G. Nurser, B. A. de Cuevas, and A. C. Coward (2005), Water mass transformation in the North Atlantic over 1985–2002 simulated in an eddy-permitting model, Ocean Sci., 1, 127–144.

Marshall, J. and F. Schott (1999), Open-ocean convection: Observations, theory and models, Rev. Geophys., 37, 1–64.

Marshall, J., A. Adcroft, C. Hill, L. Perelman, and C. Heisey (1997), A finite volume, incompressible Navier Stokes model for studies of the ocean on parallel computers, J. Geophys. Res., 102, 5753–5766.

Mauritzen, C. and S. Häkkinen (1999), On the relationship between dense water formation and the "meridional overturning cell" in the North Atlantic Ocean, Deep Sea Res., Part I, 46, 877–894.

McCartney, M. S. (1992), Recirculating components to the deep boundary current of the northern North Atlantic, Prog. Oceanogr., 29, 283–383.

McCartney, M. S. and L. D. Talley (1982), The subpolar mode water of the North Atlantic, J. Phys. Oceanogr., 12, 1169–1188.

McCartney, M. S., and L. D. Talley (1984), Warm-to-cold water conversion in the northern North Atlantic Ocean, J. Phys. Oceanogr., 14, 922–935.

Meinen, C. S. and D. R. Watts (2000), Vertical structure and transport on a transect across the North Atlantic Current near 42°N: Time series and mean, J. Geophys. Res., 105, 21869–21891.

Menemenlis, D., et al. (2005), NASA supercomputer improves prospects for ocean climate research, EOS, 86, 89, 96.

Mizoguchi, K., S. L. Morey, J. Zavala-Hidalgo, N. Suginohara, S. Häkkinen, and J. J. O'Brien (2003), Convective activity in the Labrador Sea: Preconditioning associated with decadal variability in subsurface ocean stratification, J. Geophys. Res., 108, doi:10.1029/2002JC001,735.

Molinari, R. L. and R. A. Fine (1988), A continuous deep western boundary current between Abaco (26.5°N) and Barbados (13°N), Deep Sea Res., 35, 1441–1450.

Molinari, R. L., R. A. Fine, W. D. Wilson, R. Curry, J. Abell, and M. McCartney (1998), The arrival of recently formed Labrador Sea Water in the Deep Western Boundary Current at 26.5N, Geophys. Res. Lett., 25, 2249–2252.

Myers, P. G. and C. Donnelly (2007), Water mass transformation and formation in the Labrador Sea, J. Climate, submitted.

Nurser, A. J. G. and J. C. Marshall (1991), On the relationship between subduction rates and diabatic forcing of the mixed layer, J. Phys. Oceanogr., 21, 1793–1802.

Nurser, A. J. G., R. Marsh, and R. G. Williams (1999), Diagnosing water mass formation from air-sea fluxes and surface mixing, J. Phys. Oceanogr., 29, 1468–1487.

Orsi, A. H., G. C. Johnson, and J. L. Bullister (1999), Circulation, mixing, and production of Antarctic bottom water, Prog. Oceanogr., 43, 55–109.

Pickart, R. S. and M. A. Spall (2007), Impact of Labrador Sea convection on the North Atlantic meridional overturning circulation, J. Phys. Oceanogr., 37, 2207–2227.

Pickart, R. S., W. M. Smethie, J. R. N. Lazier, E. P. Jones, and W. J. Jenkins (1996), Eddies of newly formed upper Labrador Sea water, J. Geophys. Res., 101, 20711–20726.

Pickart, R. S., M. A. Spall, and J. R. N. Lazier (1997), Mid-depth ventilation in the western boundary current system of the sub-polar gyre, Deep Sea Res., Part I, 44, 1025–1054.

Pickart, R. S., D. J. Torres, and R. A. Clarke (2002), Hydrography of the Labrador Sea during active convection, J. Phys. Oceanogr., 32, 428–457.

Pickart, R. S., M. A. Spall, M. H. Ribergaard, G. W. K. Moore, and R. F. Milliff (2003), Deep convection in the Irminger Sea forced by the Greenland tip jet, Nature, 424, 152–156.
Pickart, R. S., D. J. Torres, and P. S. Fratantoni (2005), The East Greenland Spill Jet, J. Phys. Oceanogr., 35, 1037–1053.
Read, J. F. and W. J. Gould (1992), Cooling and freshening of the subpolar North Atlantic Ocean since the 1960s, Nature, 360, 55–57.
Reid, J. L., W. D. Nowlin, and W. C. Patzert (1977), On the characteristics and circulation of the southwestern Atlantic Ocean, J. Phys. Oceanogr., 7, 62–91.
Rhein, M., J. Fischer, W. M. Smethie, D. Smythe-Wright, R. F. Weiss, C. Mertens, D. H. Min, U. Fleischmann, and A. Putzka (2002), Labrador Sea Water: Pathways, CFC-inventory and formation rates, J. Phys. Oceanogr., 32, 648–665.
Salmon, R. (1998), Lectures on geophysical fluid dynamics, Oxford University Press, Oxford.
Schmitz, W. J. (1996), On the world ocean circulation, some global features/North Atlantic circulation, Vol. 1, Technical Reports, Woods Hole Oceanographic Institution.
Schmitz, W. J. and M. S. McCartney (1993), On the North Atlantic circulation, Rev. Geophys., 31, 29–49.
Schott, F. A., R. Zantopp, L. Stramma, M. Dengler, J. Fischer, and M. Wibaux (2004), Circulation and deep-water export at the western exit of the subpolar North Atlantic, J. Phys. Oceanogr., 34, 817–843.
Schott, F. A., J. Fischer, M. Dengler, and R. Zantopp (2006), The deep Labrador Current and its variability 1996–2005, Geophys. Res. Lett., 33, doi:10.1029/2006GL026,563.
Smethie, W. M. and R. A. Fine (2001), Rates of North Atlantic Deep Water formation calculated from chlorofluorocarbon inventories, Deep Sea Res., Part I, 48, 189–215.
Smethie, W. M., R. A. Fine, A. Putzka, and E. P. Jones (2000), Tracing the flow of North Atlantic Deep Water using chlorofluorocarbons, J. Geophys. Res., 105, 14297–14323.
Spall, M. A. (2004), Boundary currents and watermass transformation in marginal seas, J. Phys. Oceanogr., 34, 1197–1213.
Spall, M. A. and R. S. Pickart (2001), Where does dense water sink? A subpolar gyre example, J. Phys. Oceanogr., 31, 810–826.
Spall, M. A. and R. S. Pickart (2003), Wind-driven recirculations and exchange in the Labrador and Irminger Seas, J. Phys. Oceanogr., 33, 1829– 1845.
Speer, K., and E. Tziperman (1992), Rates of water mass formation in the North Atlantic ocean, J. Phys. Oceanogr., 22, 93–104.
Speer, K., H.-J. Isemer, and A. Biastoch (1995), Water mass formation from revised COADS data, J. Phys. Oceanogr., 25, 2444–2457.
Stramma, L. and M. Rhein (2002), Variability in the Deep Western Boundary Current in the equatorial Atlantic at 43°W, Geophys. Res. Lett., 28, 1623– 1626.
Stramma, L., D. Kieke, M. Rhein, F. Schott, I. Yashayaev, and K. P. Koltermann (2004), Deep water changes at the western boundary of the subpolar North Atlantic during 1996 to 2001, Deep Sea Res., Part I, 51, 1033–1056.
Straneo, F., R. S. Pickart, and K. Lavender (2003), Spreading of Labrador Sea Water: An advective-diffusive study based on Lagrangian data, Deep Sea Res., Part I, 50, 701–719.
Sverdrup, H. U., M. W. Johnson, and R. H. Fleming (1942), The oceans: Their physics, chemistry and general biology, Prentice-Hall, Englewood Cliffs, NJ, 1087 pp.
Swift, J. H. (1984), The circulation of the Denmark Strait and Iceland- Scotland overflow waters in the North Atlantic, Deep Sea Res., 31, 1339– 1355.
Sy, A., M. Rhein, J. R. N. Lazier, K. P. Koltermann, J. Meincke, A. Putzka, and M. Bersch (1997), Surprisingly rapid spreading of newly formed intermediate waters across the North Atlantic Ocean, Nature, 386, 675–679.
Talley, L. D. (2003), Shallow, intermediate, and deep overturning components of the global heat budget, J. Phys. Oceanogr., 33, 530–560.
Talley, L. D. and M. S. McCartney (1982), Distribution and circulation of Labrador Sea Water, J. Phys. Oceanogr., 12, 1189–1205.

The Lab Sea Group (1998), The Labrador Sea deep convection experiment, Bull. Am. Meteor. Soc., 79, 2033–2058.

Thorpe, S. A. (2005), The turbulent ocean, Cambridge University Press, Cambridge/New York.

Treguier, A. M., S. Theetten, E. Chassignet, T. Penduff, R. Smith, L. Talley, J. O. Beismann, and C. Böning (2005), The North Atlantic subpolar gyre in four high resolution models, J. Phys. Oceanogr., 35, 757–774.

Walin, G. (1982), On the relation between sea-surface heat flow and thermal circulation in the ocean, Tellus, 34, 187–195.

Walter, M., C. Mertens, and M. Rhein (2005), Mixing estimates from a large-scale hydrographic survey in the North Atlantic, Geophys. Res. Lett., 32(13), L13, 605, doi:10.1029/2005GL022,471.

Warren, B. A. (1981), Deep circulation of the world ocean, in B. A. Warren and C. Wunsch (eds.), Evolution of physical oceanography, MIT, Cambridge, MA, pp. 6–41.

Waugh, D. W., T. M. Hall, and T. W. N. Haine (2003), Relationship among tracer ages, J. Geophys. Res., 108, doi:10.1029/2002JC001,325.

Waugh, D. W., T. W. N. Haine, and T. M. Hall (2004), Transport times and anthropogenic carbon in the subpolar North Atlantic, Deep Sea Res., Part I, 51, 1475–1491.

Weiss, R. F., J. L. Bullister, R. H. Gammon, and M. J. Warner (1985), Atmospheric chlorofluoromethanes in the deep equatorial Atlantic, Nature, 314, 608–610.

Willebrand, J., et al. (2002), Circulation characteristics in three eddy permitting models of the North Atlantic, Prog. Oceanogr., 48, 123–161.

Wood, R. A., A. B. Keen, J. F. B. Mitchell, and J. M. Gregory (1999), Changing spatial structure of the thermohaline circulation in response to atmospheric CO_2 forcing in a climate model, Nature, 399, 572–575.

Worthington, L. V. (1976), On the North Atlantic circulation, 6, The Johns Hopkins Oceanographic Studies, The Johns Hopkins University Press, Baltimore, MD, 110 pp.

Wright, D. G. (1972), Northern sources of energy for the deep Atlantic, Deep Sea Res., 19, 865–877.

Wüst, G. (1935), Aschichtung und Zirkulation des Atlantischen Ozeans, Vol. 6:1st Part, Engl. Transl., in W. J. Emery (ed.), The Stratosphere of the Atlantic Ocean, Amerind, New Delhi, 1978, 112 pp.

Yashayaev, I. and A. Clarke (2006), Recent warming of the Labrador Sea, AZMP Bull. PMZA, 5, 12–20.

Yashayaev, I., M. Bersch, H. van Aken, and A. Clarke (2004), A new study of the production, spreading and fate of the Labrador Sea Water in the subpolar North Atlantic, ASOF Newsl., 2, 20–23.

Zhang, H., T. W. N. Haine, and D. W. Waugh (2005), Relationships between tracer age and dynamical fields in double gyre circulation, J. Phys. Oceanogr., 35, 2250–2267.

Zhao, J., J. Sheng, R. J. Greatbatch, K. Azetsu-Scott, and E. P. Jones (2006), Simulation of CFCs in the North Atlantic Ocean using an adiabatically corrected ocean circulation model, J. Geophys. Res., 111, doi:10.1029/2004JC002,814.

Chapter 28
Accessing the Inaccessible: Buoyancy-Driven Coastal Currents on the Shelves of Greenland and Eastern Canada

Sheldon Bacon[1], Paul G. Myers[2], Bert Rudels[3], and David A. Sutherland[4]

28.1 Introduction

One reason why the Polar and Sub-polar shelf seas are an important component of the global climate system is that they support the fluxes of large volumes of both solid and liquid freshwater supplied from the cryospheres, the hydrosphere and the atmosphere.

This chapter is about sub-Arctic shelf waters in the western Atlantic sector, the extent of which is illustrated in Fig. 28.1. We will discuss the relevant coasts of Greenland and eastern Canada: specifically, east Greenland from Belgica Bank through Denmark Strait to Cape Farewell; then west Greenland from the Labrador Sea through Davis Strait to Baffin Bay; then Baffin Island and the coast of Labrador. Finally, we will summarise what we think we know, and also what is important that we do not know.

The shelves under consideration display highly variable bottom topography. Their widths vary widely, from a few tens to over a hundred km. Their depths are typically 200–500 m, but offlying banks can be as shallow as a few tens of metres (e.g. Belgica Bank, north-east Greenland – see below), and, in the many troughs – evidence of past glaciation – as deep as 1,000 m (Melville Bugt, north-west Greenland). While topographic steering is likely to be important in determining the path of any shelf currents, the most obviously important dynamical feature of high-latitude shelf seas is horizontal salinity contrast. The shelf seas in which we are interested here are typically adjacent to open oceanic waters with salinities >34, while the shelf seas themselves can have salinities below 30. Now a horizontal salinity difference of 1 approximates to a density difference of 1 kg m^{-3}, while a temperature difference of 1 °C

[1] National Oceanography Centre, Southampton, UK, e-mail: S.Bacon@noc.soton.ac.uk

[2] Department of Earth and Atmospheric Sciences, University of Alberta, Edmonton, Alberta, Canada

[3] Finnish Institute of Marine Research, Helsinki, Finland

[4] Woods Hole Oceanographic Institution, Woods Hole, MA, USA

R.R. Dickson et al. (eds.), *Arctic–Subarctic Ocean Fluxes*, 703–722

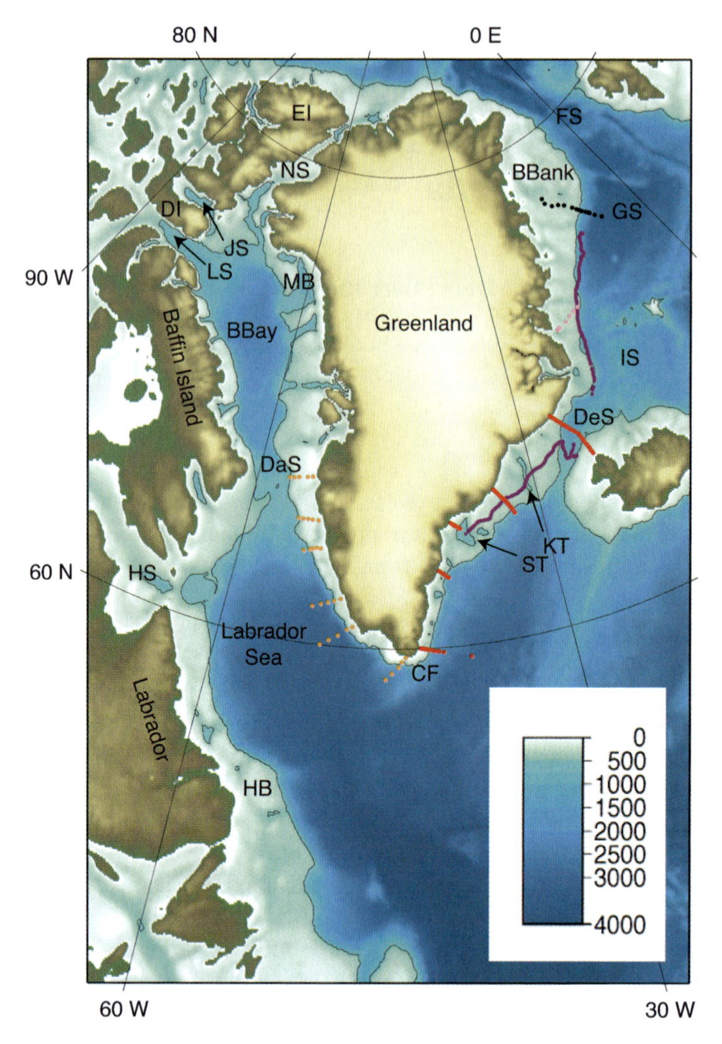

Fig. 28.1 The Greenland/eastern Canada region; feature abbreviations are, reading clockwise from top right: Fram Strait (FS); Belgica Bank (BBank); Greenland Sea (GS); Iceland Sea (IS); Denmark Strait (DeS); Kangerdlussuaq Trough (KT); Sermilik Trough (ST); Cape Farewell (CF); Hamilton Bank (HB); Hudson Strait (HS); Davis Strait (DaS); Baffin Bay (BBay); Melville Bugt (MB); Lancaster Sound (LS); Jones Sound (JS); Devon Island (DI); Nares Strait (NS); Ellesmere Island (EI). The locations of data discussed in the text are: cruise JR44 (black dots); PIMMs drifter (purple dots); *Oden* cruise (pink dots); cruise JR105 (red dots; sections numbered 1–5 from south to north); west Greenland repeat hydrography (orange dots; sections named, from south to north, *Cape Farewell, Cape Desolation, Paamiut, Fylla Bank, Maniitsoq, Sisimiut*). The scale bar shows depths in metres

approximates to 0.1 kg m^{-3}, and the relative importance of salinity to density increases as temperature decreases. In high latitude seas where temperature contrasts are seldom more than a few degree Celsius, salinity differences are the dominant cause of density differences and so of geostrophic currents.

28.2 North-East Greenland

We begin with Belgica Bank, the shelf at the north-eastern corner of Greenland. Belgica Bank was recognised by Mikkelsen (1922) as an "area usually covered with unbroken ice". Kiilerich (1945), in reviewing the state of knowledge of the Nordic Seas at the time, said that "the position and extent of the Belgica Bank is still almost unknown". Present knowledge of the hydrography and bathymetry of Belgica Bank is based largely on work carried out in the last 30 years. Figure 28.2 shows the International Bathymetric Chart of the Arctic Ocean (IBCAO) data for the area at 2.5 km nominal resolution (Jakobsson et al. 2000).

Belgica Bank has an average depth of about 300 m and an area of about 150,000 km^2. Confusingly the name is applied in the literature both to the whole north-east shelf region and to the specific bathymetric feature indicated on Fig. 28.2. We use the former meaning unless specified otherwise. Greatest depths exceed 500 m in the ring of troughs encircling the central banks. The least charted depth is less than 20 m, in the shoal patch to the west of the central bank which is also called

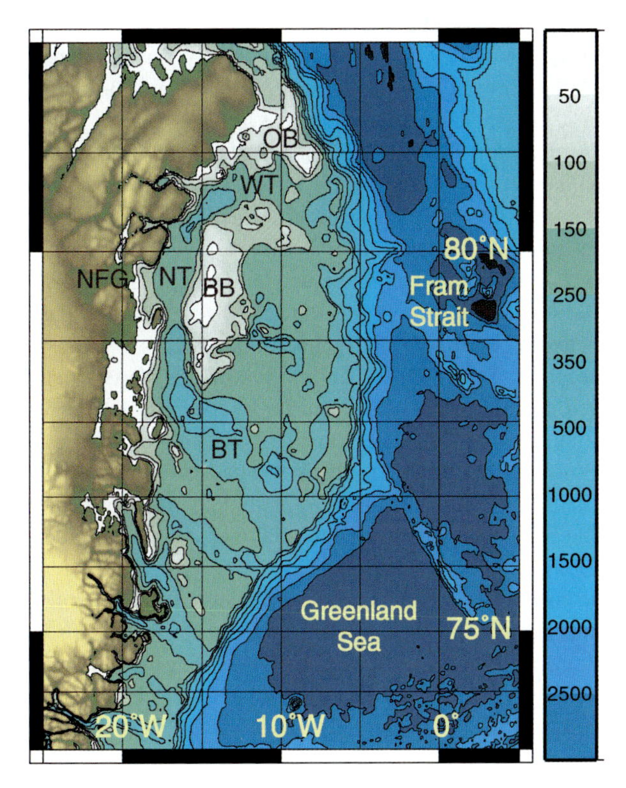

Fig. 28.2 Belgica Bank bathymetry. Indicated features are: Ob Bank (OB); Belgica Bank (BB); Westwind Trough (WT); Norske Trough (NT); Belgica Trough (BT). The scale bar shows depths in metres

"Belgica Bank" ("BB" on Fig. 28.2). The northern extremity of the region is about 5 nm wide; southwards, it bulges to over 120 nm wide at about 78° N, then narrows to about 60 nm south of 74°N.

The major oceanographic feature of the region is the East Greenland Current (EGC), which carries ice and freshwater out of the Arctic through Fram Strait. It runs along the shelf break roughly over the 2,000 m contour but the shelf region is largely protected from its direct influence, particularly by Ob Bank in the north. The EGC comprises cold, fresh polar waters near the surface, overlying a layer of warmer, more saline Atlantic waters which have recirculated in Fram Strait.

The Nioghalvfjerdsfjorden Glacier (NFG) is important. It is also known as the 79° N Glacier (Mayer et al. 2000). The NFG is a major outlet glacier in Greenland which drains over 8% of its ice sheet area. It forms a floating ice tongue about 60 km long and 20 km wide, situated at 79.5° N, 19–22° W: see Fig. 28.2.

The bathymetry and hydrography of Belgica Bank have been described largely thanks to two groups of cruises. The first group of three cruises were by the icebreakers *Westwind*, in 1979, and *Northwind*, in 1981 and 1984 (Bourke et al. 1987). The second group comprised cruises by the USCGC *Polar Sea* and the FS *Polarstern* in 1992/93: there is a broadly anticyclonic circulation determined from geostrophic calculations referenced to zero at 200 m (Budeus and Schneider 1995). The upper layer waters (0–100 m) are largely locally formed. The deeper waters (below 150 m), including those which fill the troughs, are thought to be modified waters of Atlantic origin, although some may have a source north of Ob Bank. There is an intermediate layer, called Knee Water (Paquette et al. 1985), which spreads diagonally across the shelf with increasing westward penetration to the south. It is presumed to be Arctic water of Atlantic origin (Budeus et al. 1997). There is a north-going coastal current which is responsible for the opening of the North-East Water (NEW) polynya (Schneider and Budeus 1995), with the assistance of fast ice at 79° N (the Norske Ø Ice Barrier). A sketch of the circulation over Belgica Bank is given by Budeus and Schneider (1995).

The NEW polynya (Wadhams 1981; Smith et al. 1990) is located south of Ob Bank and its area is about 40,000 km^2. It usually opens in the spring and closes in the autumn. It is supposed to be a mechanical phenomenon: the north-going current near the coast is swept clear of ice, in the melting season, by the Norske Ø ice barrier. Norske Ø is the island on Fig. 28.2 immediately south of the letter 'N' of the legend 'NT'. The barrier bulges out around the island to the edge of Belgica Bank ('BB' on Fig. 28.2).

A level-of-no-motion calculation in shallow waters is not an entirely convincing way to determine circulation. Geostrophic currents are O(10 cm s^{-1}); the few current meter measurements made in the area suggest that those geostrophic currents could be enhanced by barotropic flows of the same order (Topp and Johnson 1997). The closure of the anticyclonic flow to the south of 77° N is not understood. Input from north of Ob Bank is not understood.

Vertical circulation has been addressed very little. There seems to be a *prima facie* case for exchange between the shelf waters and the open waters of the Greenland Sea (Fig. 28.3; Bacon and Yelland 2000; Hawker 2005). The EGC flows

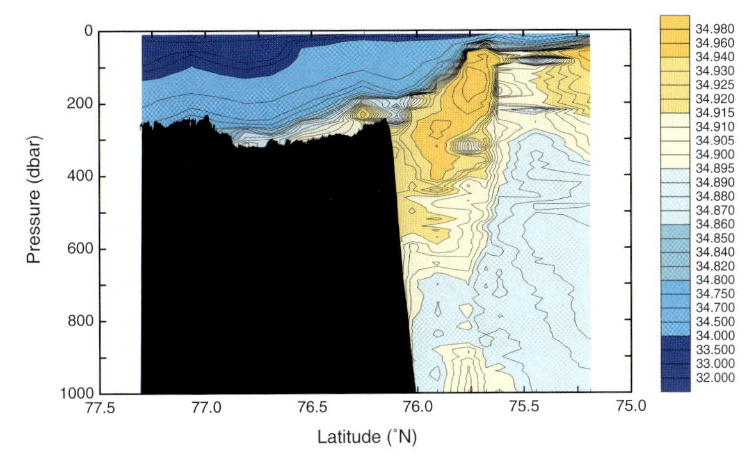

Fig. 28.3 Upper-ocean salinity section off north-east Greenland, from cruise JR44 in 1999. See Fig. 28.1 for station locations. Belgica Bank is on the left

southwards with a cap of freshwater, some of which appears to have originated from the shelf to the west; and towards the bottom is a layer of saltier water which appears to have moved westwards from the EGC onto the shelf. These are summer measurements. It is possible that in winter, the vertical circulation reverses, or is at least modified, by ice formation at the surface causing brine rejection and sinking, so that surface waters are drawn westwards to replace the sinking water, with the now less-buoyant deeper waters able to escape off-shelf (e.g. Chen et al. 2003).

The remainder of the north-east Greenland shelf between the southern end of Belgica Bank and Denmark Strait is difficult to access, with ice expected to be present in some quantity for most of the year (e.g. Parkinson et al. 1999; Kvingedal 2005). Kiilerich (1945) summarises the results of all pre-Second-World-War scientific expeditions to the Nordic Seas, including a view of the flows on the north-east Greenland shelf. He interprets the measurements to show generally southward currents directionally influenced by local bathymetry; the small number of calculated on-shelf surface currents are of order 10 cm s^{-1}. Subsequent reviews of Nordic Seas physical oceanography have tended to concentrate on the deep waters beyond the shelf break and to ignore the shelf waters. The most recent work to examine directly the north-east Greenland shelf are Rudels et al. (2002) and Rudels et al. (2005), who clearly show salinity decreasing on the shelf towards the coast.

We examine more closely a section from the 2002 expedition on the Swedish icebreaker *Oden* (Rudels et al. 2005). The expedition ran sections from Fram Strait to south of Denmark Strait off the east coast of Greenland. The most northerly section with the closest approach to the coast is section 5, at ~72° N, roughly halfway between Belgica Bank and Denmark Strait; locations are shown in Fig. 28.1. Salinity from stations 68–73 is shown in Fig. 28.4, which also shows geostrophic velocity calculated with a level of no motion at 250 db (or the sea bed, whichever is shallower). We impose this level of no motion because there is a deep trough (>500 m) near the shore.

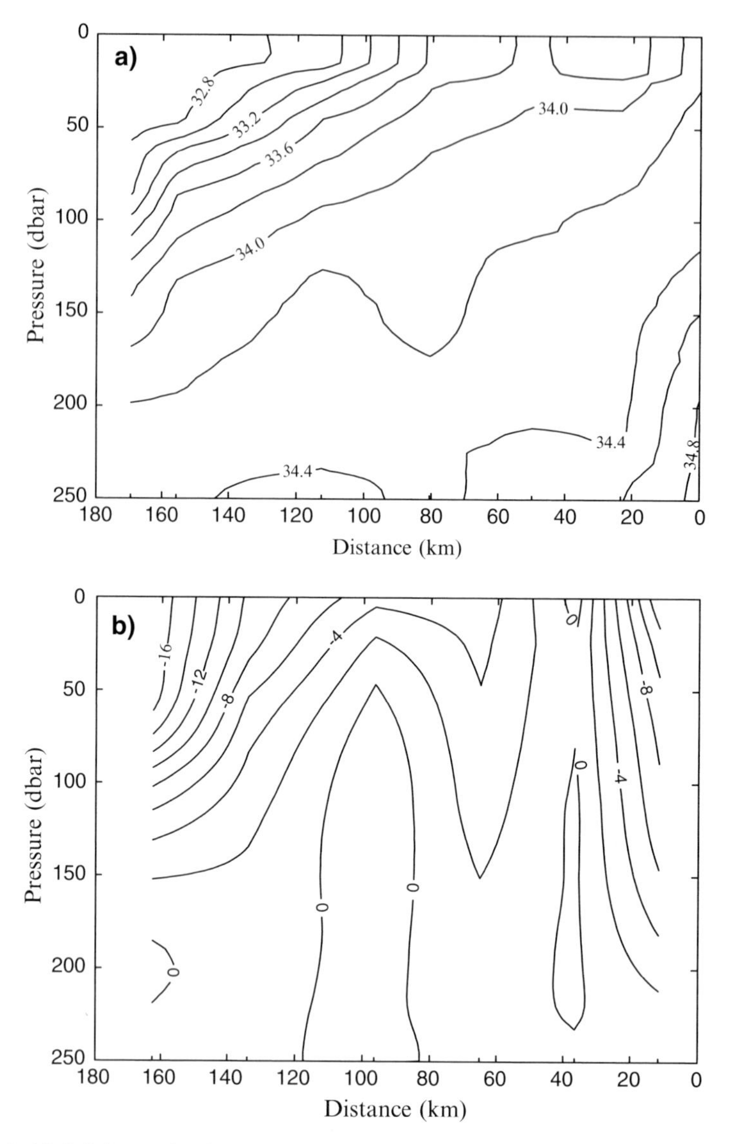

Fig. 28.4 (a) Salinity section from R/V *Oden* in 2002. See Fig. 28.1 for station locations. The Greenland coast is on the left. (b) Geostrophic velocity section from R/V *Oden* in 2002. See Fig. 28.1 for station locations. The Greenland coast is on the left

Potential temperature (not shown) is nearly isothermal: $-1.75\,^{\circ}\mathrm{C} < \theta < -1.85\,^{\circ}\mathrm{C}$ between the surface and 200 db. However, salinity shows the nearshore fresh wedge associated with the East Greenland Coastal Current (EGCC: Bacon et al. 2002). Pending further study, we can suggest that this is evidence for the EGCC existing on the shelf along the north-east coast of Greenland as well as the south-east. We calculate the flux of seawater of this current to be 0.77 Sv south, above 250 m (or the seabed). Using a reference salinity of 34.4, which is the salinity at 250 m, the freshwater flux is 24 mSv (760 km^3 year^{-1}). Using a high reference salinity of 34.8, a typical offshore ambient salinity, the freshwater flux is proportionately higher: 34 mSv (1,090 km^3 year^{-1}). Interestingly, Sutherland and Pickart (2007), in their section north of Denmark Strait (see below) find transports very similar to the 2002 *Oden* results: *ca.* 0.8 Sv southwards seawater flux and 30 mSv associated freshwater flux.

Rudels et al. (2005) observe that the prevailing winds during their expedition were northerly and would therefore push the sea ice towards the coast; subsequent melting of this sea ice could then give rise to the observed salinity distribution. Prevailing wind directions will be considered below.

28.3 South-East Greenland

This section takes us from Denmark Strait to Cape Farewell. We consider the broad shelf at Denmark Strait first, which was the one of the subjects of joint Norwegian–Icelandic expeditions in 1963 and 1965 (Malmberg et al. 1972). They showed fast (>20 cm s^{-1}), southwards flows inshore (the EGCC) and off the shelf break (the EGC), and a slower current (10–20 cm s^{-1}) trending south-westwards from the shelf edge to join the EGCC, a pattern broadly confirmed by a recent (2004) survey of the EGCC on RRS *James Clark Ross*: see Fig. 28.5, and Sutherland and Pickart (2007). It is likely that topographic steering affects the course of the EGCC, as there is a deep trench aligned north-south in the centre of the broad shelf here.

The broad shelf at Denmark Strait has a "vertical" circulation analogous to that inferred for Belgica Bank. Pickart et al. (2005) observed the East Greenland Spill Jet in hydrographic sections south of Denmark Strait. The spill jet is an intense, narrow current banked against the upper continental slope, and is believed to be the result of dense water cascading over the shelf edge and entraining ambient water. It remains uncertain, however, whether the dense water is a locally formed winter water mass, or a remotely (northern-) sourced water mass. It is logical to assume that a deep outflow of dense water from the shelf will be balanced by inflowing lighter, upper waters. It may be that the offshore-to-onshore current branch described above, and supported by the drifter tracks in Bacon et al. (2002), performs this role.

For the rest of the shelf to Cape Farewell, Bacon et al. (2002) returned attention to the existence of a buoyant shelf current (the EGCC) separate from the EGC using hydrographic measurements near Cape Farewell. Surface drifter data (Bacon et al. 2002) showed very high continuous speeds >1 m s^{-1} centrally over the south-east Greenland shelf between 65–60° N. Wilkinson and Bacon (2005) demonstrated the multi-decadal persistence (using historical data 1932–1997), variability (typically 0.5–2.0 Sv) and extent (between Cape Farewell and Denmark Strait) of the EGCC, and derived freshwater fluxes up to 0.1 Sv (3,200 km^3 year^{-1}). The best measurements to date of the EGCC are described by Sutherland and Pickart (2007), comprising a

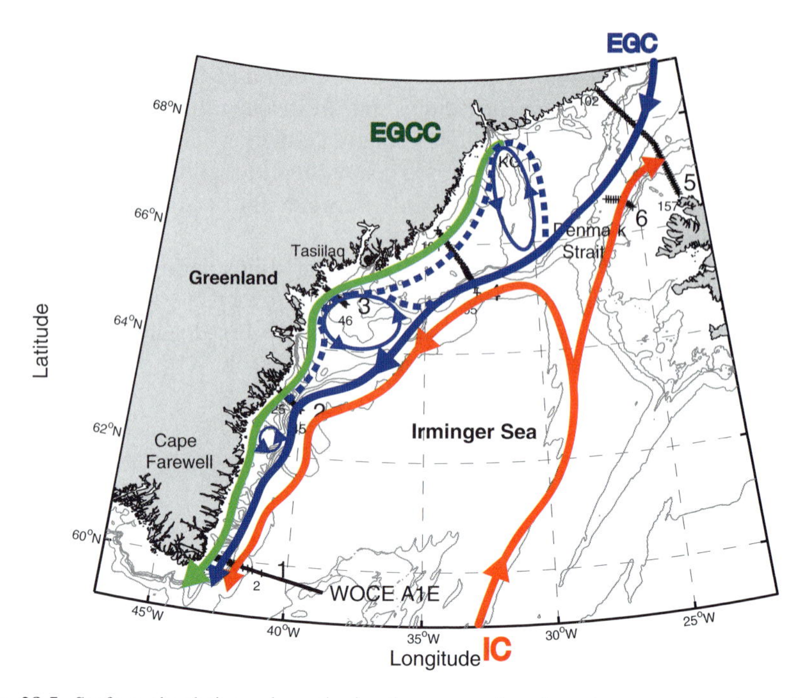

Fig. 28.5 Surface circulation schematic for the summertime boundary current system of the Irminger Sea. Numbered hydrographic sections are from JR105, also indicated on Fig. 28.1. Solid lines show the paths of the East Greenland Current (EGC, blue), the East Greenland Coastal Current (EGCC, green) and the Irminger Current (IC, red; not discussed here – see Sutherland and Pickart 2007); dashed lines indicate possible flow paths induced by bathymetric or wind effects; thin blue ellipses show inferred recirculations around major troughs on the shelf. North of the Kangerdlussuaq Trough (KT) the EGCC's presence is uncertain, though it is likely weaker than what is observed farther south

series of high-resolution hydrographic and velocity sections. They suggest that the EGCC is an inner branch of the EGC that forms south of Denmark Strait, and they argue that bathymetric steering and strong along-shelf wind forcing cause the EGC to split at Denmark Strait. By combining the EGCC and EGC transports, they find a roughly constant seawater flux over their study area (~2 Sv), and a southwards increase of freshwater flux, by ~60%, from 59 to 96 mSv, that is explained by a budget accounting for meltwater runoff, melting sea-ice, melting icebergs, and precipitation minus evaporation (P–E).

Thus far, all observations of the EGCC were made in summer months when the region was accessible to research vessels. We offer next the first observation of the EGCC's likely existence in the winter. In February–March 2000, a group of sea-ice-capable surface drifting buoys (PIMMs: Polar Ice Motion Monitors; Hawker 2005) were deployed from RV *Jan Mayen* onto ice floes in the sea ice in the Greenland Sea between 72–75° N. The PIMMs were equipped with air temperature sensors on their top plate and sea surface temperature sensors on the underside.

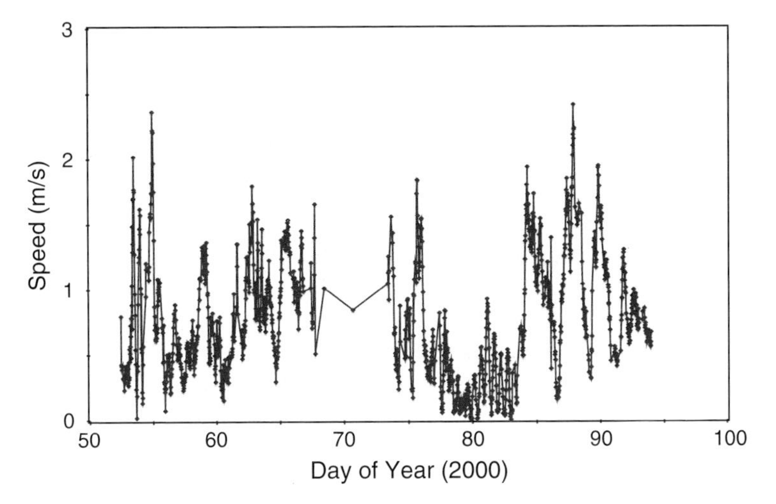

Fig. 28.6 Speed of PIMMs drifter in winter 2000. See text (Section 28.3) for concordance between day number and track (Fig. 28.1)

One of these (the track is shown in Fig. 28.1) followed the EGC as far as Denmark Strait, then crossed onto the shelf and remained roughly in the centre of the shelf, taking 20 days to travel between 67–65° N. Its temperature records confirmed that it remained on its ice floe (and not in the water) until it expired. Its speed is shown in Fig. 28.6.

The first half of the record is when the drifter is in the EGC in the Greenland and Iceland Seas. Between days 76 and 84 it is moving slowly, in northern Denmark Strait, and then on the eastern edge of the shelf. From day 84 to the end of the record, it is moving south-westwards in the centre of the shelf at about 1 m s^{-1}, with high variability. The speed is too high to be due directly to wind forcing so we attribute this (mainly) to advection by the EGCC, but we suggest that the variability in speed may be a response to synoptic meteorological variability, probably by forcing the drifter (and its ice floe) back and forth across the EGCC and so, occasionally, it is outside the current core.

28.4 EGCC Forcing

Bacon et al. (2002) suggests that the EGCC's freshwater wedge is derived from seasonal terrestrial runoff: i.e. summer land ice melt. While this is undoubtedly present in summer, it is probably not the main cause of the EGCC. Sutherland and Pickart (2007) suggest that feeding by the EGC combined with topographic steering is important; and both Sutherland and Pickart (2007) and Rudels et al. (2005) mention the likely importance of wind forcing. We offer here a further perspective on the EGCC's generation.

The EGCC is usually seen as a freshwater wedge, with surface salinity decreasing towards the coast. Next, the sea ice band down the whole of east Greenland is always adjacent to the coast, narrowing, presumably also thinning, and certainly melting as it goes. Finally, consider the mean atmospheric sea level pressure for the region (Fig. 28.7).

Isobars are near-parallel to the east Greenland coast everywhere, the most relevant consequence of which is that the sea ice will be forced to the right – towards the coast – by the geostrophic wind. In the Nordic Seas, the under-ice shelf waters are cold (surface-to-bottom temperature difference *ca.* 0.3 °C) and the surface air temperature is low; in the North Atlantic, under-ice waters are rather warmer (surface-to-bottom temperature difference *ca.* 3 °C; e.g. Bacon et al. 2002) and surface air temperatures, given the lower latitude, correspondingly higher. Lower sea and air temperatures towards the north imply weaker sea ice melting, and *vice versa* towards the south. We also note that the variability study of Wilkinson and Bacon (2005) estimate freshwater fluxes in the EGCC south of Denmark Strait to be up to 0.1 Sv, or 3,200 km^3 $year^{-1}$ (more typically half that), which is similar to the Fram Strait sea ice flux (e.g. Vinje 2001). We speculate that the foregoing, taken together, imply that the wind (particularly the Iceland Low) is the organising principle for the EGCC, which is formed from sea ice melting near the coast, with more melting, lower salinities and a stronger current towards the south. Other effects, as described

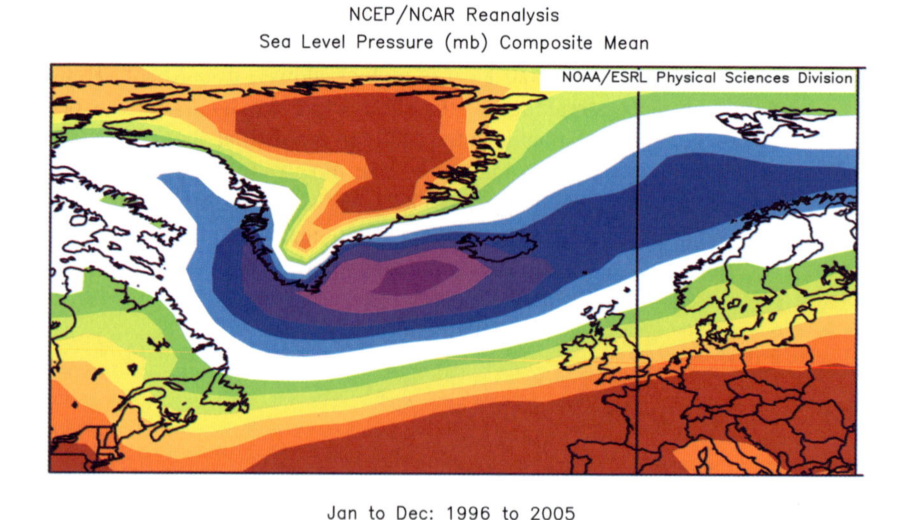

Fig. 28.7 Decade-mean (1996–2005, January–December) sea level pressure in the North Atlantic region from the US National Centers for Environmental Prediction (NCEP)/National Center for Atmospheric Research (NCAR) reanalysis project

by Sutherland and Pickart (2007) – iceberg melt, P–E, terrestrial runoff – are contributory factors, but, we suggest, not the dominant ones. Lastly, we do not mean to imply that all the Fram Strait sea ice (whether in solid or liquid form) necessarily reaches Cape Farewell: some will recirculate within the Nordic Seas.

28.5 Cape Farewell and West Greenland

As the currents "turn the corner" at Cape Farewell, the EGC becomes the West Greenland Current (WGC). However, Holliday et al. (2007), using data from a cruise on RRS *Discovery* in 2005 (Bacon 2006), show that part of the EGC (about 30%, or 5 Sv) retroflects back into the Irminger Basin at Cape Farewell, and also that the EGCC actually leaves the shelf and moves out into the deeper waters, on top of the WGC. This latter point is supported by the drifter tracks in Bacon et al. (2002) and Cuny et al. (2002), which do the same.

The west Greenland coastal waters have been surveyed annually for many years, and in the case of some hydrographic sections, decades (Buch 2000; Stein 2005), and the temperature and salinity variability there is well established. However, there is a lack of geostrophic transport estimates, and no measurements of currents. Kulan and Myers (2007) have addressed this deficiency in the following way.

The west Greenland hydrographic data (positions shown on Fig. 28.1) were obtained from the International Council for the Exploration of the Sea (ICES) database. Normally five stations were taken each year on each section but occasionally fewer were performed. As long as at least three stations were present, the section was used in this analysis. As a first step, the section data were used to calculate geostrophic, baroclinic velocities for the flow between the inner and outer station on each section, relative to 700 db (the deepest level common to all years). To obtain estimates of the barotropic component of the velocity, a mean summer (April–June) climatology of the Labrador Sea, produced by objectively analyzing (in an isopyncal framework) all available measurements (Kulan and Myers 2007), was used as input to an ocean general circulation model run in diagnostic mode. The resulting model barotropic velocities were then interpolated to each station.

An issue with this approach was that none of the sections went to the coast and thus missed parts of the coastal circulation. The distance from last station to coast was *ca.* 10 km, similar to the radius of deformation (e.g. Bacon et al. 2002). This is important since that is where much of the freshwater transport occurs. To address this problem, a simple analytic frontal model of density (Webb 1995) was fitted to each section each year (Webb 1995; Bacon et al. 2002). High resolution salinity values were reversed out of the equation of state, as discussed in Wilkinson and Bacon (2005). The revised properties based on the frontal model were used to re-estimate the geostrophic velocities and transports and then merged with the original transports estimates. Figure 28.8 shows the resulting transports of seawater (Fig. 28.8a) and freshwater (Fig. 28.8b; reference salinity 34.8) inshore of the 300 m isobath for all section except Cape Farewell, for which 500 m was used.

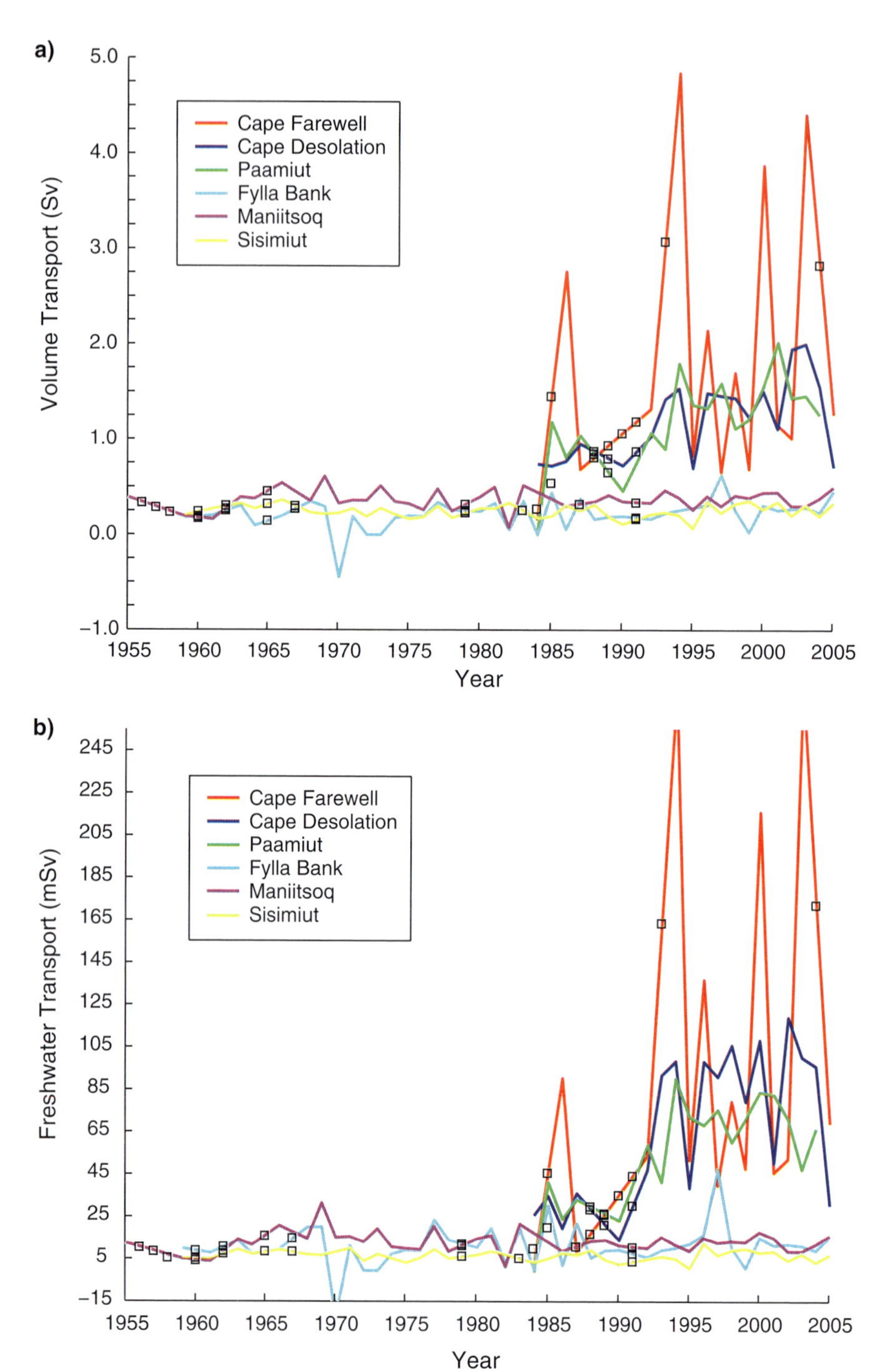

Fig. 28.8 Time series of (a) seawater, and (b) freshwater transports in six repeat hydrographic sections off the south-west coast of Greenland; see text (Section 28.5) for description

We consider first the southernmost of the west Greenland sections (*Cape Farewell* on Fig. 28.1), because it exhibits odd behaviour. We chose a slightly deeper isobath with which to illustrate its seawater and freshwater transport because moving the small additional distance shorewards to the 300 m isobath removes most of the transport signal. The small remaining seawater and freshwater transports are, in this case, close to those for the northernmost three sections, discussed below. Using the 500 m isobath, we see high transport variability, with a base level around 1.5 Sv punctuated by occasional values up to 5 Sv. We interpret this to be caused by frontal movement whereby the inshore edge of the EGC pushes up close to, but not quite onto, the shelf, while the EGCC, in rounding Cape Farewell, has mostly moved off-shelf to join the EGC (cf. Holliday et al. 2007). The differences between the 300 and 500 m transport calculations, and the differences within the 500 m transport calculations, show that the *Cape Farewell* section has much the highest variability of the six West Greenland sections discussed here. We attribute this variability to the effects of the dynamics of two fast boundary currents rounding Cape Farewell, with its complex bathymetry.

Next, the southernmost three sections all show very similar transports: the *Cape Farewell* base level, *Cape Desolation* and *Paamiut* are all about 1.5 Sv. Finally, the northernmost three, *Fylla Bank, Maniitsoq* and *Sisimiut* are all similar at the much lower level of ~0.25 Sv. Interannual variability decreases as transport magnitude decreases, and there are no obvious trends over the 20–50 years of data. More work is needed to understand this behaviour, but it is likely that topography and/or wind stress are important. The northernmost three sections are at or to the north of the location where the Labrador Sea bathymetry curves westwards (Fig. 28.1); they are also in the region where isobars of mean sea level pressure are near-normal to the coast. In either case, we suppose that the data show that of the ~1.5 Sv of seawater heading northwards in the southern sections, most continue round within the Labrador Sea, with only ~0.25 Sv continuing northwards through Davis Strait into Baffin Bay.

Considering freshwater transports (Fig. 28.8b), the seawater transport pattern is repeated, with a separation between the northern and southern three sections, and the high variability in the *Cape Farewell* section. The northern three sections' freshwater transport is ~10 mSv. Interestingly, there is some indication of recent enhancement in freshwater flux in the southern three sections. Ignoring the anomalous high values at the *Cape Farewell* section for the same reason as above, we see that the southern three freshwater transports prior to 1993 are all ~30 mSv; after then, they are elevated to higher values, ~85 mSv. This elevation is coincident with increases in sea surface temperature and surface air temperature in the region of south-west Greenland (Stein 2005), so we infer this to be the consequence of enhanced ice melt, but we do not know in what proportions it may be due to sea or land ice melt.

Finally, there is a contradiction between the observations of seawater and freshwater fluxes using *in situ* current measurements off south-east Greenland and Cape Farewell, and those off Cape Farewell and south-west Greenland using modelled barotropic velocity components, given (i) that hydrographic and drifter observations indicate that the EGCC leaves the shelf after Cape Farewell, and (ii) that the south-west Greenland transports described above are of similar magnitude to those in the EGCC. Does the model overestimate the barotropic contribution to the west Greenland fluxes? Might the observations of the EGCC leaving the shelf be, in some way, unrepresentative? Is the addition of "new" freshwater to the south-west Greenland shelf, whether from sea ice, land ice or iceberg melt, sufficient to account either for all or for part of the difference?

28.6 Baffin Bay and Labrador

There is a fine recent survey of Baffin Bay hydrography (Tang et al. 2004; T04 hereafter), so we will only briefly summarise their findings. Baffin Bay is partially covered by sea ice all year except for July and August. The ice penetrates further south on the west side than on the east, because the northward-flowing west Greenland waters are relatively warm, and the south-flowing waters off Baffin Island are very cold. The bay itself is 2,500 m deep in the centre. Its connections to adjacent seas and basins are to the Labrador Sea through Davis Strait, which is over 600 m deep, and to the Arctic Ocean and the waters of the Canadian Archipelago through Nares Strait (Robeson Channel, Kennedy Channel, Kane Basin, Smith South) between Greenland and Ellesmere Island, Jones Sound between Ellesmere and Devon Islands, and Lancaster Sound between Devon and Baffin Islands. Circulation in Baffin Bay is cyclonic but asymmetrical; while the north-going waters sourced from south-west Greenland recirculate and turn south at the north end of the bay, these waters are enhanced by ice and seawater fluxes through Nares Strait and Jones and Lancaster Sounds. The south-going current on the west side of the bay is called the Baffin Current, and as it flows offshore of the 500 m isobath, we will not consider it further here. See T04 for all relevant seawater and freshwater flux estimates.

As an interesting supplement to T04, Zweng and Münchow (2006) identify a multi-decadal freshening trend of the surface waters (50–200 m) over the Baffin Island shelf of *ca.* –0.09 ± 0.04 per decade. This is likely to have (as yet undetermined) consequences for freshwater fluxes. While they present several hypotheses to explain this freshening, they are unable to determine the cause. Their analysis is complicated by the (inevitable) seasonal bias in their hydrographic data. They also find a similar freshening trend over the west Greenland shelf, of *ca.* –0.04 ± 0.02 per decade; no manifestation of this is evident in our freshwater flux calculations reported above.

While much of Baffin Bay is adequately covered by current meter and hydrographic measurements, T04 have greatest difficulty estimating the northwards fluxes of seawater and freshwater on the broad north-west Greenland shelf. They estimate 0.7 Sv northwards with a range of 0.2 Sv southward (March, minimum) to 2 Sv northward (November, maximum), where reference currents for geostrophic calculations are obtained from a mooring on the shelf edge. Their associated freshwater flux estimates are 35 mSv northwards (mean) and a range of 12 mSv southwards to 78 mSv northwards. We note the consistency of our estimates of seawater and freshwater fluxes, described in the preceding section and Fig. 28.8, on the west Greenland shelf at the eastern side of Davis Strait (*Maniitsoq* and *Sisimiut* sections) of ~0.4 Sv seawater, ~10 mSv freshwater, with standard deviations ~30% of the mean. There is an obvious inconsistency between our estimates and T04's. It may be that their shelf-edge reference current is too strong, being located in the West Greenland Current. It is unlikely that we are aliasing aspects of the seasonal cycle in this particular respect, because the *Maniitsoq* and *Sisimiut* section data span 5 decades and show very low variability, while T04's seasonal (fall) maximum freshwater transport, roughly temporally coincident with our sections, is over seven times higher. More information is needed.

The Baffin Current passes southwards though Davis Strait; it receives a net seawater and freshwater flux enhancement in passing Hudson Strait, which separates southern Baffin Island from the northern Labrador coast of mainland Canada; it then becomes the Labrador Current. Loder et al. (1998) is the best review of this region, supplemented by references in Colbourne (2004), but it exposes the deficiency in measurements here. We believe that (surprisingly) the most comprehensive study of the Labrador Sea to include the Canadian shelf waters is still the description of the several *Marion* and *General Greene* expeditions between 1928 and 1935, supplemented by other expedition data, in Smith et al. (1937). Lazier and Wright (1993) is the most useful modern work, describing the wide shelf area of the Hamilton Bank section. The Labrador Current is offshore of the shelf break and will not be considered here. The several sections of Smith et al. (1937) to cross both the Labrador Current and the shelf waters demonstrate clear separation of the Labrador Current from a freshwater current over the shallow waters of the shelf. The measurement program of Lazier and Wright (1993) included one long-term current meter mooring on the shelf: their M1, 1978–1987, with occasional interruptions. Surface salinities from hydrographic section data decline from ~33 over the shelf break to <28 at the most nearshore station, and there is the appearance of a freshwater wedge structure. Their surface velocity plots, with barotropic corrections derived from current meter measurements, show a modest jet with mean core speed 12 cm s^{-1} southwards, to flow over the centre of the shelf. This is consistent with Smith et al. (1937), whose shelf current is 10–20 cm s^{-1}. Lazier and Wright (1993) estimate a seawater flux of 0.8 Sv. There is no corresponding estimate of freshwater flux. This current is important to net freshwater flux balance estimation. It should be given its own name, the Labrador Coastal Current, as it is geographically and most likely dynamically distinct from the Labrador Current, and not merely an "inshore branch". Indeed, Myers et al. (1990) demonstrate that salinities on the Newfoundland shelf, off St. Johns ("Station 27"), are correlated (with a lag of a few months) both with Hudson Bay runoff anomalies and with ice-melt anomalies in Hudson Bay and on the Labrador shelf, further prompting the hypothesis that this current, similar to the others discussed in this chapter, is largely the result of ice melt of one sort or another. The model study of Greenberg and Petrie (1988) has a clear representation of the shelf current, but their northern boundary is in the vicinity of Hamilton Bank, so it does not represent most of the Labrador shelf. Nevertheless, it does illuminate the importance of topographic steering.

28.7 Discussion

Having reviewed fluxes of seawater and freshwater on the Greenland and Canada shelves, we find that the shelves generally support geostrophic currents flowing with the coast on their right, and that these currents are generally manifested along the front between a nearshore wedge of freshwater and offshore, more saline "ambient" water. We note that almost everywhere, the mean wind is parallel to the coast and oriented to push any sea ice towards the coast. A notable exception to this generalisation is found in the vicinity Cape Farewell, where synoptic-scale cyclones can interact with the topography of Greenland to cause mesoscale wind events known as tip jets and barrier winds (e.g. Doyle and Shapiro 1999; Pickart et al. 2003; Moore and Renfrew 2005). Tip jets are typically very strong westerly winds off Cape Farewell, which will tend to push water and ice offshore.

Direct liquid freshwater inputs from the Arctic and the Canadian Archipelago contribute to off-shelf boundary currents, such as the EGC and the Baffin and Labrador Currents, but the extent to which they contribute to the shelf currents seems unclear. We propose that the shelf currents are the geostrophic consequence of the freshwater wedge structure resulting in part from terrestrial runoff, P–E and iceberg melt, but mainly, through the organising principle of the wind, from sea ice melt, the main sources of sea ice being Fram Strait, Nares Strait, Baffin Bay, Hudson Bay and the Canadian Archipelago. If the generation of these currents requires sea ice melt, then that implies that the currents should be stronger as surface air temperature and sea surface temperature increase: i.e. generally they should be stronger to the south, and we have indications that this is so, for the EGCC. Also, there is evidence of dense water formation on some of the broader shelves: Belgica Bank and Denmark Strait, for example; but we have no confirmation or quantification of any such processes. We also recognise that Chapman and Beardsley (1989) made a similar inference about a continuous coastal current, 5,000 km long, buoyancy-driven, originating (in their case) along the southern coast of Greenland. We both support and extend their hypothesis, to include most of the east coast of Greenland.

This chapter has made as much as possible of rather little information, so what needs to be done?

- We need to resolve the spatial continuity of the currents in many areas; only the EGCC off south-east Greenland is now well-described, and then only for summer months, apart from one small drifter data set.
- Temporal continuity is largely unknown: how many of these currents exist in winter, and what are the magnitudes of their mean and annual cycles of seawater and freshwater fluxes?
- There is a general lack (with a few exceptions) of direct current measurements.

- In spite of the plausible hypothesis described above, it is necessary to elucidate the shelf current forcings and dynamics: what is the relative importance of sea ice, land ice, P–E, winds, topography, air–sea heat flux?

Some help will be gained from improved knowledge of sea ice mass fluxes and divergences from remote sensing, using new platforms such as ICESat (Zwally et al. 2002) and CryoSat (to be launched in 2009), and new analysis techniques (e.g. Laxon et al. 2003). However these shelf waters remain difficult to access even in ice-strengthened research vessels, the environment is hostile to tall moorings which can be destroyed by passing icebergs, surface drifters can only enter and survive in the shelf waters when clear of ice, and subsurface floats would spend too little time in the fairly strong currents encountered there to be useful. What remains, therefore, is the prospect of suitable seabed installations with some profiling capability, whether mechanical or acoustic.

This chapter has concentrated on examining the small number of available measurements in the western Atlantic sector of the sub-Arctic. Space does not permit a review of the extensive theoretical literature which may be employed to assist the interpretation of future measurements. However, we note a few key references in this field; all of the following are for the rotational case. Griffiths and Linden (1981) is an important study of the stability of buoyant boundary currents, a subject pursued more recently by Cenedese and Linden (2002), who incorporate varying bottom topography, and by Dahl (2005), who examines the dynamics of small perturbations on a buoyant coastal current. Yankovsky and Chapman (1997) formulate a theory to predict the vertical structure and offshore spreading of a localised buoyant inflow onto a continental shelf. Williams et al. (2001) consider the influence of a channel normal and adjacent to the shelf (like a river mouth or fjord). Scaling theories for buoyant currents are developed by Lentz and Helfrich (2002) along a sloping bottom, and by Avicola and Huq (2002) for the interaction between the current and the continental shelf. Chapman (2003) studies the separation of a buoyancy current at a bathymetric bend; and eddy generation by buoyant boundary currents are considered by Cenedese and Whitehead (2000) and Wolfe and Cenedese (2006) in the cases of flow around a cape, and flow over variable bathymetry, respectively.

Finally, we note that it should be recognised that these buoyant coastal currents are distinct dynamically and geographically from their (generally) adjacent deep-water boundary currents, and not just a curiosity, or "inshore branch". And in any event, it is clear that in the western Atlantic sector, the sub-Arctic shelf waters are a rectified pipeline sending large volumes of freshwater eventually southwards, and they require further study.

Acknowledgments Our thanks to: the database of the International Council for the Exploration of the Sea (ICES: http://www.ices.dk/); the US NOAA–CIRES Climate Diagnostics Center, Boulder, Colorado (http://www.cdc.noaa.gov/); author support from: UK Natural Environment Research Council (SB); US National Science Foundation and WHOI Academic Programs (DS); Canadian NSERC and CFCAS grants (the latter through the Canadian CLIVAR network: PGM). We also like to thank Mads Ribergaard and Chris Donnelly for assistance with some of the processing of the ICES data set.

References

Avicola G, Huq P (2002) Scaling analysis for the interaction between a buoyant coastal current and the continental shelf: experiments and observations. J. Phys. Oceanogr. 32:3233–3248.

Bacon S (2006) RRS *Discovery* Cruise 298, 23 Aug–25 Sept 2005. Cape Farewell and Eirik Ridge (CFER-1). National Oceanography Centre, Southampton, Cruise Report No. 10, 113 pp.

Bacon S, Yelland MJ (2000) RRS *James Clark Ross* Cruise 44, 23 July–31 Aug 1999. Circulation and Thermocline Structure – Mixing, Ice and Ocean Weather: CATS-MIAOW. Southampton Oceanography Centre Cruise Report No. 33, 140 pp.

Bacon S, Reverdin G, Rigor IG, Snaith HM (2002) A freshwater jet on the East Greenland Shelf. J. Geophys. Res. 107, doi:10.1029/2001JC000935.

Bourke RH, Newton JL, Paquette RG, Tunnicliffe MD (1987) Circulation and water masses of the East Greenland Shelf. J. Geophys. Res. 92:6729–6740.

Buch E (2000) Air-sea-ice conditions off Southwest Greenland, 1981–97. J. Northw. Atl. Fish. Sci. 26:123–136.

Budeus G, Schneider W (1995) On the hydrography of the Northeast Water Polynya. J. Geophys. Res. 100:4287–4299.

Budeus G, Schneider W, Kattner G (1997) Distribution and exchange of water masses in the Northeast Water Polynya (Greenland Sea). J. Mar. Syst. 10:123–138.

Cenedese C, Linden PF (2002) Stability of a buoyancy-driven coastal current at the shelf break. J. Fluid Mech. 452:97–121.

Cenedese C, Whitehead JA (2000) Eddy shedding from a boundary current around a cape over a sloping bottom. J. Phys. Oceanogr. 30:1514–1531.

Chapman DC (2003) Separation of an advectively trapped buoyancy current at a bathymetric bend. J. Phys. Oceanogr. 33:1108–1121.

Chapman DC, Beardsley RC (1989) On the origin of shelf water in the Middle Atlantic Bight. J. Phys. Oceanogr. 19:384–391.

Chen CTA, Liu KK, Macdonald R (2003) Continental margin exchanges. In Fasham MJR (ed.), Ocean biogeochemistry: the role of the Ocean Carbon Cycle in Global Change, Springer, pp 53–97.

Colbourne EB (2004) Decadal changes in the ocean climate in Newfoundland and Labrador waters from the 1950s to the 1990s. J. Northw. Atl. Fish. Sci. 34:41–59.

Cuny J, Rhines PB, Niiler PP, Bacon S (2002) Labrador Sea boundary currents and the fate of the Irminger Sea water. J. Phys. Oceanogr. 32:627–647.

Dahl OH (2005) Development of perturbations on a buoyant coastal current. J. Fluid Mech. 527:337–351.

Doyle JD, Shapiro MA (1999) Flow response to large-scale topography: the Greenland tip jet. Tellus 51A:728–748.

Greenberg DA, Petrie BD (1988) The mean barotropic circulation on the Newfoundland shelf and slope. J. Geophys. Res. 93:15541–15550.

Griffiths RW, Linden PF (1981) The stability of buoyancy-driven coastal currents. Dyn. Atmos. Oceans 5:281–306.

Hawker EJ (2005) Nordic Seas Circulation and Exchanges. Ph.D. thesis, University of Southampton, School of Ocean and Earth Science, 239 pp.

Holliday NP, Meyer A, Bacon S, Alderson SG, de Cuevas B (2007) Retroflection of part of the East Greenland Current at Cape Farewell. Geophys. Res. Lett. 34:L07609, doi:10.1029/2006GL029085.

Jakobsson M, Cherkis N, Woodward J, Macnab R, Coakley B (2000) New grid of Arctic bathymetry aids scientists and mapmakers. Eos 81:89, 93, 96.

Kiilerich A (1945) On the hydrography of the Greenland Sea. Meddelelser om Gronland, vol 144, Appendix 2, pp 1–63.

Kulan N, Myers PG (2007) Comparing two climatologies of the Labrador Sea: geopotential vs. isopyncal. Atmos. Ocean, in press.

Kvingedal B (2005) Sea ice extent and variability in the Nordic Seas, 1967–2002. In H. Drange, T. Dokken, T. Furevik, R. Gerdes, W. Berger (eds.), The Nordic Seas: an integrated perspective, pp 51–64. Geophysical Monograph 158, American Geophysical Union, Washington, DC.

Laxon S, Peacock N, Smith D (2003) High interannual variability of sea ice thickness in the Arctic region. Nature 425:947–950.

Lazier JRN, Wright DG (1993) Annual velocity variations in the Labrador Current. J. Phys. Oceanogr. 23:659–678.

Lentz SJ, Helfrich KR (2002) Buoyant gravity currents along a sloping bottom in a rotating fluid. J. Fluid Mech. 464:251–278.

Loder JW, Petrie B, Gawarkiewicz G (1998) The coastal ocean off northeastern North America: a large-scale view. In Robinson AR, Brink KH (eds.), The sea, vol 11, Wiley, New York, pp 105–133.

Malmberg, S-A, Gade HG, Sweers HE (1972) Current velocities and volume transports in the East Greenland Current off Cape Nordenskjöld in August-September 1965. Sea Ice Conference Proceedings, Reykjavik, pp 130–139.

Mayer C, Reeh N, Jung-Rothenhaüsler F, Huybrechts P, Oerter H (2000) The subglacial cavity and implied dynamics under Nioghalvfjerdsfjorden Glacier, NE-Greenland. Geophys. Res. Lett. 27:2289–2292.

Mikkelsen E (1922) Alabama-expeditionen til Grønlands Nordøstkyst 1909–1912. Meddelelser om Grønland, vol 52, 295 pp.

Myers RA, Akenhead SA, Drinkwater K (1990) The influence of Hudson Bay runoff and ice-melt on the salinity of the inner Newfoundland shelf. Atmos. Ocean 28:241–256.

Moore GWK, Renfrew IA (2005) Tip jets and barrier winds: a QuikSCAT climatology of high wind speed events around Greenland. J. Climate 18:3713–3725.

Paquette RG, Bourke RH, Newton JF, Perdue WF (1985) The East Greenland Polar Front in Autumn. J. Geophys. Res. 90:4866–4882.

Parkinson CL, Cavalieri DJ, Gloersen P, Zwally HJ, Comiso JC (1999) Arctic sea ice extents, areas and trends, 1978–1996. J. Geophys. Res. 104:20837–20856.

Pickart RS, Spall MA, Ribergaard MH, Moore GWK, Milliff RF (2003) Deep convection in the Irminger Sea forced by the Greenland Tip Jet. Nature 424:152–156.

Pickart RS, Torres DJ, Fratantoni PS (2005) The East Greenland Spill Jet. J. Phys. Oceanogr. 35:1037–1053.

Rudels B, Fahrbach E, Meincke J, Budéus G, Eriksson P (2002) The East Greenland Current and its contribution to the Denmark Strait Overflow. ICES J. Mar. Sci. 59:1133–1154.

Rudels B, Björk G, Nilsson J, Winsor P, Lake I, Nohr C (2005) The interaction between waters from the Arctic Ocean and the Nordic Seas north of Fram Strait and along the East Greenland Current: results from the Arctic Ocean–02 Oden Expedition. J. Mar. Syst. 55:1–30.

Schneider W, Budeus G (1995) On the generation of the Northeast Water Polynya. J. Geophys. Res. 100:4269–4286.

Smith SD, Muench RD, Pease CH (1990) Polynyas and leads: an overview of physical processes and environment. J. Geophys. Res. 95:9461–9479.

Smith EH, Soule FM, Mosby O (1937) The *Marion* and *General Greene* Expeditions to Davis Strait and Labrador Sea under direction of the United States Coast Guard 1928–1931–1933–1934–1935. Scientific Results. Part 2. Physical Oceanography. US Treasury Department, Coast Guard Bulletin No. 19. US Government Printing Office, Washington, DC, 258 pp. (Also available as Woods Hole Oceanographic Institution Collected Reprints, 1937 Part II, dated March 1938; Contribution No. 107).

Stein M (2005) North Atlantic Subpolar Gyre Warming – impacts on Greenland Offshore Waters. J. Northw. Atl. Fish. Sci. 36:43–54.

Sutherland DA, Pickart RS (2007) The East Greenland Coastal Current: structure, variability and forcing. Progress in Oceanography, in press.

Tang CCL, Ross CK, Yao T, Petrie B, DeTracey BM, Dunlap, E (2004) The circulation, water masses and sea ice of Baffin Bay. Prog. Oceanogr. 63:183–228.

Topp R, Johnson, M (1997) Winter intensification and water mass evolution from yearlong current meters in the Northeast Water Polynya. J. Mar. Syst. 10:157–173.
Vinje T (2001) Fram Strait ice fluxes and atmospheric circulation: 1950–2000. J. Climate 14:3508–3517.
Wadhams P (1981) The ice cover in the Greenland and Norwegian Seas. Rev. Geophys. 19:345–393.
Webb DJ (1995) The vertical advection of momentum in Bryan-Cox-Semtner ocean general circulation models. J. Phys. Oceanogr. 25:3186–3195.
Wilkinson D, Bacon S (2005) The spatial and temporal variability of the East Greenland Coastal Current from historic data. Geophys. Res. Lett. 32:L24618, doi:10.1029/2005GL024232.
Williams WJ, Gawarkiewicz GG, Beardsley, RC (2001) The adjustment of a shelfbreak jet to cross-shelf topography. Deep-Sea Res. I 48:373–393.
Wolfe C, Cenedese, C (2006) Laboratory experiments on eddy generation by a buoyant coastal current flowing over variable bathymetry. J. Phys. Oceanogr. 36:395–411.
Yankovsky AE, Chapman, DC (1997) A simple theory for buoyant coastal discharges. J. Phys. Oceanogr. 27:1386–1401.
Zwally HJ, Schutz B, Abdalati W, Abshire J, Bentley C, Brenner A, Bufton J, Dezio J, Hancock D, Harding D, Herring T, Minster B, Quinn K, Palm S, Spinhirne J, Thomas R (2002) ICESat's laser measurements of polar ice, atmosphere, ocean and land. J. Geodyn. 34:405–445.
Zweng MM, Münchow, A (2006) Warming and freshening of Baffin Bay, 1916–2003. J. Geophys. Res. 111:C07016, doi:10.1029/2005JC003093.

List of Contributors

Tom A. Agnew
Environment Canada, Meteorological Service of Canada, 4905 Dufferin St., Downsview ON, Canada M3H 5T4, tom.agnew@ec.gc.ca

Leif G. Anderson
Department of Chemistry, Göteborg University, SE-412 96 Göteborg, Sweden, leifand@cham.gu.se

Sheldon Bacon
National Oceanography Centre, Southampton, UK, shb@noc.soton.ac.uk

Manfred Bersch
Institute of Oceanography, University of Hamburg, Germany, bersch@ifm.uni-hamburg.de

Andrew P. Barrett
National Snow and Ice Data Center, Cooperative Institute for Research in Environmental Sciences, Campus Box 449, University of Colorado, Boulder CO, 80309–0449, USA, apbarret@kryos.colorado.edu

Agnieszka Beszczynska-Möller
Alfred Wegener Institute for Polar and Marine Research, Bremerhaven, Germany, abeszczynska@awi-bremerhaven.de

Claus Böning
IfM-GEOMAR, Kiel, Germany, cboening@ifm-geomar.de

Peter Brandt
IfM-GEOMAR, Kiel, Germany, pbrandt@ifm-geomar.de

Eddy Carmack
Fisheries and Oceans Canada, Institute of Ocean Science, 9860 W. Saanich Road, Sidney, B.C., V8L 4B2, CarmackE@pac.dfo-mpo.gc.ca

Ruth Curry
Woods Hole Oceanographic Institution, Woods Hole, MA 02543, USA, rcurry@whoi.edu

Huw C. Davies
Swiss Federal Institute of Technology, 8092 Zürich, Switzerland,
huw.davies@env.ethz.ch

Bob Dickson
Centre for Environment, Fisheries and Aquaculture Science, Lowestoft, Suffolk
NR33 0HT UK, r.r.dickson@cefas.co.uk

Stephen Dye
Centre for Environment, Fisheries and Aquaculture Sciences, Lowestoft, UK,
s.r.dye@cefas.co.uk

Charlie C. Eriksen
School of Oceanography, University of Washington, Seattle WA 98195-5351,
charlie@ocean.washington.edu

Patrick Eriksson
Finnish Institute of Marine Research, Erik Palménin aukio 1, P.O. Box 2,
FI-00561 Helsinki, Finland, Patrick.Eriksson@fimr.fi

Eberhard Fahrbach
Alfred Wegener Institute for Polar and Marine Research, Bremerhaven, Germany,
efahrbach@awi-bremerhaven.de

Kelly K. Falkner
College of Ocean and Atmospheric Science, Oregon State University, Corvallis,
OR 97331-5503, USA, kfalkner@coas.oregonstate.edu

Kerstin Fieg
Alfred-Wegener-Institut für Polar- und Meeresforschung, Bremerhaven, Germany,
Kerstin.Fieg@awi.de

Jürgen Fischer
IfM-GEOMAR, Kiel, Germany, jfischer@ifm-geomar.de

Andreas Funk
IfM-GEOMAR, Kiel, Germany, Afunk@ifm-geomar.de

Tore Furevik
Geophysical Institute, University of Bergen, Norway and Bjerknes Centre
for Climate Research, Norway, tore.furevik@gfi.uib.no

Jean-Claude Gascard
LOCEAN, Université Pierre et Marie Curie, Paris, France,
gascard@locean-ipsl.upmc.fr

Ruediger Gerdes
Alfred Wegener Institute for Polar and Marine Research, Bremerhaven, Germany,
rgerdes@awi-bremerhaven.de

David A. Greenberg
Fisheries and Oceans Canada, Bedford Institute of Oceanography, Box 1006 Dartmouth NS, Canada B2Y 4A2, GreenbergD@mar.dfo-mpo.gc.ca

Helmuth Haak
Max-Planck-Institut für Meteorologie, Hamburg, Germany, haak@dkrz.de

Thomas Haine
Department of Earth and Planetary Sciences, 329 Olin Hall, The Johns Hopkins University, Baltimore, MD 21218-2682, USA, Thomas.Haine@jhu.edu

Sirpa Hakkinen
NASA Goddard Space Flight Center, Greenbelt, MD, USA, sirpa@ltpmail.gsfc.nasa.gov

Bogi Hansen
Faroese Fisheries Laboratory, Tórshavn, Faroe Islands, Bogihan@frs.fo

Edmond Hansen
Norwegian Polar Institute, Tromsoe, Norway, edmond.hansen@npolar.no

Hjálmar Hátún
Faroese Fisheries Laboratory, Tórshavn, Faroe Islands, hv@hafro.is

Jürgen Holfort
Norwegian Polar Institute, Tromsø, Norway and now at BSH, Germany, juergen.holfort@bsh.de

Penny Holliday, National Oceanography Centre, Southampton, UK, nph@noc.soton.ac.uk

Randi Ingvaldsen
Institute of Marine Research, Norway and Bjerknes Centre for Climate Research, Norway, randi.ingvaldsen@imr.no

Motoyo Itoh
Japan Agency for Marine-Earth Science and Technology, Yokohama, Japan, motoyo@jamstec.go.jp

Emil Jeansson
Department of Chemistry, Göteborg University, SE-412 96 Göteborg, Sweden, emilj@chem.gu.se

Peter Jones
Bedford Institute of Oceanography, Dartmouth, NS, B2Y 4A2, Canada, JonesP@mar.dfo-mpo.gc.ca

Steingrímur Jónsson
Marine Research Institute, Reykjavík, Iceland and University of Akureyri, Akureyri, Iceland, steing@unak.is

Simon Josey
National Oceanography Centre, Southampton, England,
Simon.A.Josey@noc.soton.ac.uk

Johann Jungclaus
Max Planck Institute for Meteorology, Hamburg,
Germany, johann.jungclaus@zmaw.de

Rolf H. Käse
Institut für Meereskunde, Universität Hamburg, Bundesstrasse 53, 20146
Hamburg, Germany and IfM-Geomar, Leibniz Institute for Marine Research,
Düsternbrooker Weg 20, 24105 Kiel, Germany, kaese@ifm.uni-hamburg.de

Michael Karcher
Alfred Wegener Institute for Polar and Marine Research, Bremerhaven, Germany,
mkarcher@awi-bremerhaven.de

Frank Kauker
Alfred Wegener Institute for Polar and Marine Research, Bremerhaven, Germany,
Frank.Kauker@awi.de

Dagmar Kieke
Institut für Umweltphysik, Universität Bremen, Germany,
dkieke@physik.uni-bremen.de

Cornelia Köberle
Alfred-Wegener-Institut für Polar- und Meeresforschung, Bremerhaven, Germany,
Cornelia.Koeberle@awi.de

Armin Köhl
University of Hamburg, Center of Marine and Climate Research, D-20146
Hamburg, Germany, armin.koehl@zmaw.de

Torben Koenigk
Max-Planck-Institut für Meteorologie, Bundesstraβe 53, 20146 Hamburg,
Germany, torben.koenigk@zmaw.de

Richard Krishfield
Woods Hole Oceanographic Institution, Woods Hole, MA, USA,
rkrishfield@whoi.edu

Erik Kvaleberg
Department of Earth and Planetary Sciences, 329 Olin Hall, The Johns Hopkins
University, Baltimore, MD 21218-2682, USA, kvaleberg@jhu.edu

Craig M. Lee
Applied Physics Laboratory, University of Washington, 1013 NE 40th St., Seattle
WA 98105, USA, craig@apl.washington.edu

Harald Loeng
Institute of Marine Research, Norway and Bjerknes Centre for Climate Research,
Norway, harald.loeng@imr.no

Steffen Malskær Olsen
Danish Meteorological Institute, Copenhagen, Denmark, smo@dmi.dk

Andreas Macrander
Alfred Wegener Institute for Polar and Marine Research, Bremerhaven, Germany, amacrander@ifm-geomar.de

Marika Marnela
Finnish Institute of Marine Research, Erik Palménin aukio 1, P.O. Box 2, FI-00561 Helsinki, Finland, Marika.Marnela@fimr.fi

Fiona McLaughlin
Fisheries and Oceans Canada, Institute of Ocean Science, 9860 W. Saanich Road, Sidney, B.C., V8L 4B2, McLaughlinF@pac.dfo-mpo.gc.ca

Jens Meincke
University of Hamburg, Centre of Marine and Climate Research, Hamburg, Germany, jens.meincke@zmaw.de

Humfrey Melling
Fisheries and Oceans Canada, Institute of Ocean Sciences, Box 6000 Sidney BC Canada V8S 3J2, MellingH@pac.dfo-mpo.gc.ca

Mike Meredith
British Antarctic Survey, Cambridge, UK, mmm@bas.ac.uk

Uwe Mikolajewicz
Max-Planck-Institut für Meteorologie, Bundesstraβe 53, 20146 Hamburg, Germany, uwe.mikolajewicz@zmaw.de

GWK Moore
University of Toronto, Toronto, Canada M5S 1A1, moore@atmosp.physics.utoronto.ca

Kjell Arne Mork
Geophysical Institute, University of Bergen, Norway and Institute of Marine Research and Bjerknes Centre for Climate Research, Bergen, Norway, kjell.arne.mork@imr.no

John Mortensen
University of Hamburg, Hamburg, Germany, jomo@natur.gl

Andreas Münchow
College of Marine Studies, University of Delaware, 44112 Robinson Hall, Newark DE 19716, USA, muenchow@udel.edu

Paul G. Myers
Department of Earth and Atmospheric Sciences, University of Alberta, Edmonton, Alberta, Canada, myers@sumeria.eas.ualberta.ca

Anders Olsson
Bjerknes Centre for Climate Research, University of Bergen, Allégaten 55, NO-5007 Bergen, Norway, anders.olsson@bjerknes.uib.no

Kjell Arild Orvik
Geophysical Institute, University of Bergen, Norway, kjell.orvik@gfi.uib.no

Svein Østerhus
Bjerknes Centre for Climate Research, University of Bergen, Bergen, Norway, Svein.Osterhus@gfi.uib.no

Vladimir Ozhigin
Knipovich Polar Research Institute of Marine Fisheries and Oceanography (PINRO), Murmansk, Russia, ozhigin@pinro.ru

Brian Petrie
Fisheries and Oceans Canada, Bedford Institute of Oceanography, Box 1006 Dartmouth NS. Canada B2Y 4A2, PetrieB@mar.dfo-mpo.gc.ca

Robert S. Pickart
Woods Hole Oceanographic Institution, Woods Hole, MA 02543, USA, rpickart@whoi.edu

Jan Piechura
Institute of Oceanology, Polish Academy of Sciences, Sopot, Poland, piechura@iopan.gda.pl

Simon J. Prinsenberg
Fisheries and Oceans Canada, Bedford Institute of Oceanography, Box 1006 Dartmouth NS, Canada B2Y 4A2, PrinsenbergS@mar.dfo-mpo.gc.ca

Andrey Proshutinsky
Woods Hole Oceanographic Institution, Woods Hole, MA, USA, aproshutinsky@whoi.edu

Detlef Quadfasel
Universität Hamburg, Zentrum für Meeres- und Klimaforschung, Hamburg, Germany, detlef.quadfasel@zmaw.de

Monika Rhein
Institut für Umweltphysik, Universität Bremen, Germany, mrhein@physik.uni-bremen.de

Ian A. Renfrew
University of East Anglia, Norwich, United Kingdom NR4 7TJ, i.renfrew@uea.ac.uk

Mads Hvid Ribergaard
Danish Meteorological Institute, Copenhagen, Denmark DK-2100, mhri@dmi.dk

Peter Rhines
Department of Oceanography and Atmospheric Sciences, University of Washington, Seattle, WA, USA, rhines@ocean.washington.edu

Roger M. Samelson
College of Ocean and Atmospheric Science, Oregon State University, Corvallis OR 97331-5503, USA, rsamelson@coas.oregonstate.edu

Bert Rudels
Finnish Institute of Marine Research, Erik Palménin aukio 1, FI-00561 Helsinki, Finland, Bert.Rudels@fimr.fi

François J. Saucier
Université du Québec à Rimouski, Canada, francois_saucier@uqar.qc.ca

Ursula Schauer
Alfred Wegener Institute for Polar and Marine Research, Bremerhaven, Germany, uschauer@awi-bremerhaven.de

Mark C. Serreze
National Snow and Ice Data Center, Cooperative Institute for Research in Environmental Sciences, Campus Box 449, University of Colorado, Boulder CO, 80309–0449, USA, serreze@nsidc.org

Toby Sherwin
Scottish Association for Marine Science, Oban, UK, Toby.Sherwin@sams.ac.uk

Koji Shimada
Japan Agency for Marine-Earth Science and Technology, Yokohama, Japan, shimadak@jamstec.go.jp

Øystein Skagseth
Institute of Marine Research, Norway and Bjerknes Centre for Climate Research, Norway, skagseth@gfi.uib.no

Andrew G. Slater
National Snow and Ice Data Center, Cooperative Institute for Research in Environmental Sciences, Campus Box 449, University of Colorado, Boulder CO, 80309–0449, USA, aslater@cires.colorado.edu

Fiammetta Straneo
Woods Hole Oceanographic Institution, Woods Hole, MA, USA, fstraneo@whoi.edu

David A. Sutherland
Woods Hole Oceanographic Institution, Woods Hole, MA, USA, dsutherland@whoi.edu

Toste Tanhua
Department of Marine Biogeochemistry, Leibniz Institute for Marine Sciences at Kiel University, Düsternbrooker Weg 20, DE-24105 Kiel, Germany, ttanhua@ifm-geomar.de

William R. Turrell
Marine Laboratory, Fisheries Research Services, Aberdeen, UK, turrellb@marlab.ac.uk

Héðinn Valdimarsson
Marine Research Institute, Reykjavík, Iceland, hv@hafro.is

Hendrik van Aken
Royal Netherlands Institute for Sea Research, aken@nioz.nl

Kjetil Våge
Woods Hole Oceanographic Institution, Woods Hole, MA 02543, USA and Swiss Federal Institute of Technology, 8092 Zürich, Switzerland, kjetil@whoi.edu

Michael Vellinga
Met Office Hadley Centre, Fitzroy Road, Exeter EX1 3PB UK, michael.vellinga@metoffice.gov.uk

Martin Visbeck
IfM-GEOMAR, Kiel, Germany, mvisbeck@ifm-geomar.de

Gunnar Voet
University of Hamburg, Center of Marine and Climate Research, D-20146 Hamburg, Germany, voet@ifm.uni-hamburg.de

Waldemar Walczowski
Institute of Oceanology, Polish Academy of Sciences, Sopot, Poland, walczows@iopan.gda.pl

Richard Wood
Met Office Hadley Centre, Exeter, UK, richard.wood@metoffice.gov.uk

Rebecca A. Woodgate
Applied Physics Laboratory, University of Washington, 1013 NE 40th St., Seattle WA 98105, USA, woodgate@apl.washington.edu

Peili Wu
Met Office Hadley Centre, Exeter, UK, peili.wu@metoffice.gov.uk

Michiyo Yamamoto-Kawai
Fisheries and Oceans Canada, Institute of Ocean Science, 9860 W. Saanich Road, Sidney, B.C., V8L 4B2, KawaiM@pac.dfo-mpo.gc.ca

Igor Yashayaev
Ocean Circulation Section, Ocean Sciences Division, Bedford Institute of Oceanography, Fisheries and Oceans Canada, 1 Challenger Drive, P.O. Box 1006, Dartmouth, NS, B2Y 4A2, Canada, YashayaevI@mar.dfo-mpo.gc.ca

Index

A

ADCP (Acoustic Doppler Current Profiler) 20, 48, 67, 132, 198, 269, 407, 430, 448, 506, 538, 681
Advection 66, 93, 118, 272, 291, 531, 557, 632, 674, 711
Advective-diffusive 570, 635, 675
Aerological 344, 365
AIW (Arctic Intermediate Water) 134, 329, 445, 477
Alaskan Coastal Current 99, 162, 199, 346
Alkalinity 156, 227, 281, 386
Altimetry 10, 38, 89, 141, 243, 374, 457, 506, 533, 552, 628, 655, 693
Ammonium 157, 229
AMSR (Advanced Microwave Scanning Radiometer) 90, 221
Amundsen Gulf 221, 239
Antarctic Bottom Water (AABW) 658, 695
Anthropogenic tracer(s) 134, 580
AOMIP (Arctic Ocean Model Intercomparison Project) 422
AR7 (W & E) (repeat hydrography, section, line) 511, 571, 604, 613, 669
Arctic Atlantic Water (AAW) 326, 446, 478, 491
Arctic Climate System Study (ACSYS) 3
Arctic Front 46, 69, 131, 147
Arctic Intermediate Water (AIW) 47, 65, 125, 134, 329, 445, 477
Arctic Mediterranean 15, 147, 265, 323, 427, 444, 476, 480
Arctic Ocean Oscillation (AOO) 206, 231
Arctic Oscillation (AO) 231, 354
Arctic sea ice 66, 179, 371
Arctic water 23, 47, 99, 132, 194, 227, 259, 536, 690, 706
Argo float(s) 89, 136, 614, 695
Atlantic inflow 6, 15, 23, 34, 47, 78, 138, 317, 343, 364, 410, 458, 536, 564
Atlantic Meridional Overturning Circulation (AMOC) 7, 289, 376, 461, 521, 527
Atlantic Water inflow 6, 111, 397
Atlantic Water 1, 15, 40, 45, 65, 111, 131, 147, 197, 265, 272, 289, 315, 378, 386, 413, 428, 445, 478, 507, 537, 557, 570, 706
Atmosphere-ocean general circulation model(s) 7, 61, 289
Atmospheric planetary waves 179
Atmospheric Reanalysis 29, 35, 376, 459
Autonomous Underwater Vehicle (AUV) 614, 628
Auxiliary constraints 320

B

Baffin Bay 150, 178, 197, 225, 250, 315, 348, 398, 409, 420, 575, 622, 703, 718
Baffin Current 232, 716
Barents Sea Opening 46, 61, 73, 115, 122, 131, 315, 535
Barents Sea 6, 19, 45, 57, 65, 82, 96, 111, 122, 131, 156, 172, 289, 315, 348, 405, 446, 535
Barents shelf 186, 364, 377
Baroclinic instability 452, 553, 634
Baroclinicity 209, 240, 564, 688
Barotropic Currents 48, 66, 273, 336
Barotropic eddies 315, 339
Barotropic transport(s) 336, 415, 541
Barrier winds 639, 647
Barrow Strait 196, 229
Beaufort Sea 124, 176, 222, 401, 417
Belgica Bank 703, 709
Bellot Strait 197
Bering Sea 77, 149, 200

Bering Strait 31, 77, 90, 148, 178, 194, 242, 315, 346, 364, 378, 388, 405, 444, 536
boundary current(s) 6, 65, 91, 111, 159, 202, 307, 332, 419, 462, 505, 524, 551, 573, 613, 633, 653, 675, 693, 710, 719
buffering capacity 128
buoyancy frequency 150

C

Canada Basin 145, 196, 320, 398, 417
Canadian Arctic Archipelago (CAA) 2, 149, 243, 305, 315, 353, 386, 400, 444, 573
Canadian Basin Deep Water (CBDW) 326, 447, 479, 494
Canadian Basin 113, 127, 148, 165, 326, 388
Cape Farewell 1, 268, 444, 514, 556, 630, 649, 657, 685, 703, 715
Cardigan Strait 196, 241
Chlorofluorocarbon(s) 227, 360, 487, 580, 655
Chukchi Sea 157, 179, 195, 322
Classical Labrador Sea Water (CLSW) 656, 683
Climate change 2, 34, 62, 165, 177, 195, 260, 296, 367, 385, 402, 440, 496, 580
Constraints 7, 17, 77, 103, 145, 194, 290, 315, 365, 456, 692
convective (mixing, process) 4, 97, 149, 197, 232, 291, 304, 348, 515, 531, 558, 569, 613, 630, 657
current meter moorings 16, 280, 437, 453, 552
cyclonic circulation 136, 182, 211, 551, 609, 625, 629

D

Davis Strait 7, 149, 195, 232, 249, 306, 353, 389, 420, 614, 626, 657, 703
decadal (changes, variability) 88, 113, 152, 172, 223, 268, 292, 419, 470, 499, 545, 557, 565, 680, 689
deep convection 6, 97, 172, 282, 291, 380, 402, 419, 444, 505, 536, 551, 571, 630, 657
Deep water exchange(s) 320
deep water formation 8, 89, 232, 297, 320, 376, 419, 444, 534, 653
Deep water production 320,420, 443, 520
Deep western boundary current (DWBC) 5, 463, 505, 558, 635, 658
de-nitrification 157
Denmark Strait Overflow Water (DSOW) 264, 444, 475, 506, 571, 658
Denmark Strait 2, 16, 105, 120, 151, 219, 264, 304, 399, 416, 427, 443, 461, 475, 497, 505, 527, 629, 657, 703
drift speed 136
Driving force 32, 56

E

East Greenland Current (EGC) 17, 70, 121, 171, 267, 317, 388, 476, 507, 536, 657, 706
East Icelandic Current (EIC) 23, 133, 267, 481
Eastern North Atlantic Water 19, 118, 557
ECHAM5/MPI-OM 10, 172, 305, 363, 534
Eddy activity 133, 563, 622, 690
Eddy heat transport 451
Eddy kinetic energy (EKE) 553
Entrainment 82, 197, 291, 450, 480, 507, 527, 588, 659
Eurasian Basin Deep Water (EBDW) 326, 447, 479,
Eurasian Basin 66, 122, 149, 334, 389, 416, 478
Evaporation 61, 90, 145, 250, 343, 363, 385, 410, 710
Export event 171, 406

F

Faroe Bank Channel 8, 16, 427, 434, 458, 480, 510, 527, 658
Faroe Current 15, 536
Faroe-Shetland Channel 1, 15, 75, 131, 435, 443, 505, 538
Fast ice 195, 706
FLAME model 665, 692
Floats 91, 132, 142, 614, 634, 674, 695, 719
Flux adjustment 172, 411, 534
Flux calculation 265, 455, 498, 716
Forcing mechanisms 34, 56, 210
Fram Strait 3, 62, 65, 113, 121, 131, 171, 188, 223, 259, 263, 285, 304, 315, 338, 343, 364, 386, 405, 445, 476, 507, 535, 706
Fram Strait liquid fresh water transport 410, 416
Fram Strait sea ice export 173, 184, 376
Fresh water anomalies 10, 145, 194, 231, 250, 298, 302, 468, 593, 663
Fresh water balance 11, 33, 105, 254, 265, 316, 405, 423

Fresh water budget(s) 145, 201, 252, 264, 316, 343, 363, 381, 385, 575
Fresh water component(s) 146, 243, 282, 389
Fresh water content 147, 164, 265, 281, 333, 363, 416, 423, 573
Fresh water flux(es) 7, 33, 83, 100, 146, 193, 227, 252, 263, 283, 296, 323, 366, 385, 405, 422, 465, 709
Fresh water inventory 157, 203, 389
Fresh water outflow 7, 152, 231, 257, 305, 333, 385
Fresh water storage 7, 125, 145, 267, 296, 346, 376
Fresh water transport(s) 88, 103, 149, 201, 254, 264, 285, 304, 315, 378, 407, 579, 614, 713
Freshening event 263, 419, 464, 507
Freshening 33, 89, 157, 231, 263, 290, 327, 344, 367, 402, 417, 464, 507, 608, 689, 716
Frontal current 75

G

Geochemical 155, 202
Geomagnetic pole 195
Geostrophic balance 211, 663
Geostrophic calculation(s) 194, 276, 319, 407, 462, 706
Geostrophic transport(s) 318, 455, 713
geostrophic velocities 255, 282, 338, 461, 553, 556, 707
Geostrophy 201, 318, 462, 553, 684
(Sea) Glider(s) 2, 91, 132, 234, 470, 613, 628
Global Conveyor 385, 402
Global hydrological cycle 145, 367
Global warming 89, 165, 294, 365, 380, 420, 527
Gravity Recovery and Climate Experiment (GRACE) 10, 375, 559
Great Salinity Anomaly (GSA) 24, 52, 152, 171, 263, 308, 358, 551, 632
Greenland Scotland Ridge 2, 15, 45, 106, 289, 427, 444, 489, 505, 527, 557, 629, 686
Greenland Sea 45, 70, 105, 133, 148, 172, 282, 304, 320, 402, 420, 445, 482, 534, 669, 706
Greenland shelf 68, 232, 268, 326, 400, 405, 451, 486, 573, 629, 694, 707
Greenland Tip Jet 102, 637, 690
Greenland-Iceland-Norwegian Seas 359
Greenland-Scotland Overflow Water 444, 653,
Gyre index 24, 557

H

Halocline water formation 338
Halocline 124, 145, 228, 330, 378, 401, 405
Heat balance 76, 93, 126
Heat budget(s) 32, 62, 66, 95, 125, 152, 252, 315, 445, 661
Heat flux(es) 3, 55, 67, 89, 106, 121, 193, 253, 293, 335, 359, 365, 468, 528, 551, 632, 649, 665, 719
Heat loss(es) 33, 47, 91, 113, 218, 253, 296, 331, 415, 444, 520, 633, 663
Heat transport 6, 15, 45, 66, 83, 87, 118, 252, 263, 291, 318, 451, 537, 558
Hell Gate 196, 241
Hindcast simulation 28, 412, 459
Historic observation(s) 450
Hornbanki section 20, 26
Hosing experiments 7, 297, 466
Hudson Bay 149, 229, 249, 353, 717
Hudson Strait 7, 229, 249, 260, 717
Hydraulic balance 456
Hydraulic control 460, 527, 546
Hydrography 2, 19, 47, 89, 118, 131, 164, 228, 267, 305, 411, 433, 451, 476, 507, 554, 573, 660, 704
Hydrological budget 7, 363
Hydrological cycle 3, 145, 363

I

Ice arch 195
Ice export 162, 171, 189, 195, 264, 317, 376, 402, 412
Ice Profiling Ssonar (IPS) 203, 239
Ice sheet 228, 267, 302, 350, 374, 380, 706
Ice thickness 173, 188, 220, 242, 257, 317, 355, 372
Iceberg(s) 2, 194, 255, 264, 616, 710
Iceland Basin 118, 429, 479, 510, 553, 569, 609, 658
Iceland Sea Arctic Intermediate Water (IAIW) 445, 483
Iceland Sea 19, 45, 282, 385, 445, 482, 537, 578, 711
Iceland-Faroe Front 23
Iceland-Faroe Ridge 10, 17, 113, 427, 527, 535, 626, 658

Iceland–Scotland Overflow Water (ISOW) 437, 449, 492, 506, 571, 585, 658
ICYCLER 202, 239
interannual changes 11, 155, 456, 613, 660
interannual variability 75, 155, 173, 252, 266, 293, 347, 372, 405, 414, 458, 524, 583, 636, 715
Internal feedback mechanism 59
International Geophysical Year (1957-1958) 445
Intrusion of signals 118
Iodine 133, 141, 485
Irminger Basin 451, 476, 507, 540, 561, 571, 609, 629, 658, 713
Irminger Current 15, 449, 536, 557, 575, 657, 710
Irminger Sea 8, 25, 149, 389, 444, 487, 513, 527, 555, 579, 629, 657, 710
isopycnal mixing 594
isotope 227, 279, 484, 499

J

Jan Mayen Fracture Zone 282, 446
Jet(s) 46, 66, 92, 102, 131, 211, 230, 268, 444, 556, 637, 649, 657, 709
Jones Sound 219, 315, 716

K

Keel(s) 198, 237
Kennedy Channel 197, 389, 716
Knowledge gaps 165

L

Labrador Basin 305, 444, 505, 524, 579, 631, 657
Labrador Current 249, 420, 551, 575, 622, 657, 694, 717
Labrador Sea Water (LSW) 9, 260, 306, 420, 437, 444, 489, 506, 554, 569, 631, 653
Labrador Sea Water formation rate 631, 689
Labrador Sea Water transport 673, 694
Labrador Sea 8, 89, 118, 149, 171, 197, 249, 292, 315, 385, 419, 437, 444, 479, 507, 524, 534, 551, 569, 609, 613, 628, 631, 649, 653, 695, 703
Labrador shelf 229, 252, 633, 717
Lagrangian 131, 195, 558, 667
Lancaster Sound 2, 196, 315, 389, 409, 704
Land Surface Models (LSMs) 346, 353
Liquid freshwater export 321, 335, 405, 422
Lower Arctic Intermediate Water (lAIW) 445

M

Magnetic 2, 194, 315, 615
Mass balance 10, 121, 231, 280, 302, 317, 355, 374, 446
Mass budget 31, 661
Mass transport 90, 142, 227, 326, 378, 527
M'Clure Strait 221, 239
Mean state 47, 178, 378, 534, 654
Mediterranean Water 18, 557, 654
Meridional Overturning Circulation (MOC) 87, 263, 558, 576, 631, 653
Mesoscale eddy 81, 139, 428, 614, 624, 655, 688
Mesoscale turbulence 137, 142
Mid-seventies Anomaly 23
Mixing 6, 25, 46, 70, 88, 118, 134, 157, 195, 259, 264, 304, 327, 439, 444, 480, 507, 528, 557, 569, 649, 654
Model Experiment(s) 10, 29, 97, 111, 183, 291, 558, 668
Model intercomparison 128, 294, 422, 531, 689
Model Projections 357, 380
Model(ing) uncertainty 40, 290, 308
Modified East Icelandic Water (MEIW) 23
Modified North Atlantic Water (MNAW) 18
MOEN (Meridional Overturning Exchange with the Nordic seas) 16, 40, 289, 427, 510
Mooring 2, 16, 51, 67, 91, 117, 165, 198, 253, 268, 285, 317, 407, 431, 452, 497, 552, 632, 660, 716
Multidecadal variability 173, 419, 422, 545

N

NAOSIM (North Atlantic/Arctic Ocean Sea Ice Model) 112, 266, 406, 422, 507
Nares Strait 149, 195, 241, 315, 392, 409, 716
NCC (Norwegian Coastal Current) 47, 132, 316
Nitrate 155, 228, 386
Nordic Seas 1, 15, 37, 45, 66, 105, 111, 131, 149, 264, 289, 315, 355, 368, 411, 428, 444, 475, 507, 527, 557, 626, 653, 705
Nordic WOCE 16, 427, 482
North Atlantic Current (NAC) 19, 46, 113, 551, 573, 635
North Atlantic Deep Water (NADW) 7, 100, 290, 437, 443, 475, 520, 653

North Atlantic Oscillation (NAO) index 31, 47, 125, 178, 205, 354, 368, 419, 544, 555, 578, 639, 644, 663
North Atlantic Oscillation (NAO) 29, 47, 97, 105, 112, 172, 177, 205, 266, 292, 354, 368, 406, 411, 466, 496, 507, 515, 543, 551, 578, 630, 663
North Atlantic Water (NAW) 1, 18, 65, 118, 557, 570
North East Atlantic Deep Water (NEADW) 437, 444, 510, 520
North Icelandic Irminger Current 15
North Pacific Oscillation (NPO) 356
North-Atlantic central water 657
Northeast Atlantic Deep Water (NEADW) 480, 510, 571, 593
Northern Irminger Sea 396, 400
Norwegian Atlantic Current (NwAC) 6, 28, 46, 113, 131, 537
Norwegian Sea Arctic Intermediate Water (NSAIW) 23
Norwegian Sea Deep Water (NSDW) 73, 136, 440, 445, 479
Norwegian Sea 9, 19, 45, 131, 148, 289, 415, 427, 445, 477, 511, 541, 557, 657
Numerical model(s) 16, 111, 189, 206, 284, 291, 405, 459, 487, 528, 558, 642, 654
Nutrient relationships 387
Nutrient(s) 91, 155, 227, 281, 386, 490, 578, 675

O

Observational constraints 290, 307
Ocean general circulation model (OGCM) 7, 28, 61, 289, 528, 713
Open-ocean deep convection 444, 505, 520
Orography 224, 632, 649
Overflow density 127, 455
Overflow plume 439, 445, 480, 505, 530
Overturning streamfunction 9, 88, 546
Oxygen 91, 136, 227, 279, 388, 477, 507, 615, 630, 674

P

Pacific Decadal Oscillation (PDO) 356
Pacific fresh water 387, 401
Pacific water 66, 152, 165, 229, 281, 322, 385, 401
Pack ice 194, 240, 646
Phosphate 155, 228, 386
Potential predictability 112
Precipitation Minus Evaporation (P-E) 33, 99, 343, 363, 379, 410, 710
Precipitation 46, 89, 145, 231, 253, 263, 302, 316, 343, 358, 365, 380, 385, 405, 482, 540
Predictive potential 116, 422
Process model(s) 459, 528, 532
Production of LSW 521, 589
Propagation speed 127

R

Radarsat 220
RAFOS 138, 142, 626, 683
Recirculating Atlantic Water (RAW) 272, 327, 445, 478, 507
Recirculation 8, 28, 70, 120, 141, 165, 232, 252, 326, 462, 536, 569, 633, 653, 695, 710
Reference salinity 145, 193, 264, 318, 345, 363, 406, 709
Reference temperature 77, 100, 121, 318, 338
Renewal 39, 406, 518, 570, 608
Repeated hydrographic section(s) 48, 581
Residence time 133, 142, 162, 232, 483
Restoring 411, 418
River Discharge 29, 152, 251, 343, 355, 363, 380, 405, 459
River water(s) 47, 156, 252, 385, 403, 411,
Robeson Channel 212, 716
Rockall Trough (RT) 16, 118, 435, 553, 563, 570, 609
Rossby radius 83, 211, 433, 531
Runoff 33, 67, 90, 146, 172, 202, 263, 316, 344, 365, 444, 620, 710, 718

S

Salinity anomalies 52, 112, 171, 294, 333, 412, 512, 551, 575, 608
Salinity distribution(s) 415, 601, 675, 709
Salinity stratification 98
Salt balances 318, 410
Salt budget 26, 33, 661
Satellite altimetry 10, 89, 141, 243, 457, 506, 546, 552, 635
Scenario calculation 420, 421
Scoresby Sound 400
Sea ice export 162, 171, 267, 376, 414, 423
Sea ice meltwater 157, 385, 403
Sea Ice Transport 177, 189, 416
Sea level anomalies 60, 576
Sea level forcing 37, 38

Sea level 34, 57, 105, 140, 173, 194, 289, 339, 354, 374, 465, 516, 555, 576, 633, 693, 712
Sea surface height (SSH) 10, 90, 118, 231, 344, 415, 457, 533, 543, 552, 563, 694
Seaglider(s) 2, 233, 470, 613, 628
Shear 203, 282, 622, 675
Shelf Edge Current 17, 46, 118
Silicate 156, 230, 481
Slope current 17, 46, 66, 131, 537
Smith Sound 211, 230, 398, 400
Snowpack Water Equivalent (SWE) 347, 352
Southern Faroe Current (SFC) 16, 18
Steric height 81, 469, 521, 576, 581
Storm track(s) 62, 87, 172, 182, 348, 365, 632
Stratospheric polar vortex 178
Subpolar Gyre 8, 19, 54, 75, 99, 118, 263, 293, 452, 516, 551, 565, 569, 613, 653, 695
Subpolar Mode Water 452, 513, 575, 631, 657
Subpolar North Atlantic 112, 367, 410, 527, 553, 569, 653
Sulphur hexafluoride (SF6) 454, 477, 486
Surface boundary condition 410, 412, 423
Surface drifters 131, 552, 646, 719
Sverdrup Basin 218, 239
Synoptic 183, 209, 243, 389, 492, 534, 615, 641, 663, 711

T

Temperature anomalies 73, 112, 188, 277, 302, 604
Temporal variability 6, 111, 152, 228, 265, 315, 422, 476, 499, 537,
Tendency reversal 570
Thermohaline Circulation (THC) 3, 15, 97, 127, 131, 173, 282, 290, 380, 385, 443, 475, 505
Thermohaline slowdown 6, 290
Tide(s) 88, 207, 218, 255, 375
Tip jet 102, 444, 520, 556, 637, 649, 690, 718
Topographically trapped 56
Total fresh water (TFW) 386, 401, 413
Tracer(s) 8, 29, 91, 132, 155, 202, 231, 266, 282, 306, 335, 385, 410, 475, 500, 506, 579, 635, 655, 695
Trajectories 135, 142, 308, 592, 609, 625, 639, 675
Transit-time distribution (TTD) 495, 674, 694
Turbulence 137, 240

U

upper Arctic Intermediate Water (uAIW) 445
Upper Labrador Sea Water (uLSW) 656, 664
upper Polar Deep Water (uPDW) 329, 446, 478, 494

V

Variational approach 320, 339
Ventilation 15, 47, 228, 320, 478, 557, 576, 608, 631, 666
Vertical distribution of freshening 302
Virtual salt flux 410
Volume budget 32, 322, 669
Volume transports 9, 47, 66, 113, 133, 179, 238, 254, 264, 300, 321, 339, 376, 409, 427, 440, 448, 528, 579, 613, 674, 714
Volumetric analysis (census) 509, 518, 570, 585, 589

W

Warming 1, 51, 74, 88, 111, 165, 202, 257, 294, 356, 365, 420, 451, 523, 527, 551, 569, 609
Water mass formation rate 664, 667
Water mass properties 29, 99, 293, 326, 432, 476, 530
Water mass transformation (evolution) 7, 99, 113, 218, 259, 304, 326, 527, 532, 559, 588, 667
Water Vapor Pathways 350
West Greenland Current 202, 237, 249, 402, 419, 551, 575, 616, 625, 657, 713
Western North Atlantic Water 19
Wind forcing 34, 152, 176, 197, 240, 460, 496, 642, 693, 710
Winter convection 445, 569, 608, 687
WOCE AR7W 613
World Ocean Circulation Experiment (WOCE) 2, 16,40, 91, 427, 461, 482, 555, 580, 613, 630, 655, 710
Wyville Thomson Ridge (WTR) 16, 427, 440, 658

Y

Yukon River 199, 344